# North American Tunneling
## 2010 Proceedings

Edited by Lawrence R. Eckert, Matthew E. Fowler
Michael F. Smithson, Jr., Bradford F. Townsend

Published by:

Society for
Mining, Metallurgy
& Exploration

Society for Mining, Metallurgy, and Exploration, Inc. (SME)
8307 Shaffer Parkway
Littleton, Colorado, USA 80127
(303) 948-4200 / (800) 763-3132
www.smenet.org

SME advances the worldwide mining and minerals community through information exchange and professional development. With members in more than 50 countries, SME is the world's largest association of mining and minerals professionals.

Copyright © 2010 Society for Mining, Metallurgy, and Exploration, Inc.
Electronic edition published 2010.

All Rights Reserved. Printed in the United States of America.

Information contained in this work has been obtained by SME, Inc. from sources believed to be reliable. However, neither SME nor its authors guarantee the accuracy or completeness of any information published herein, and neither SME nor its authors shall be responsible for any errors, omissions, or damages arising out of use of this information. This work is published with the understanding that SME and its authors are supplying information but are not attempting to render engineering or other professional services. Any statement or views presented here are those of individual authors and are not necessarily those of the SME. The mention of trade names for commercial products does not imply the approval or endorsement of SME.

No part of this publication may be reproduced, stored in a retrieval system, or transmitted in any form or by any means, electronic, mechanical, photocopying, recording, or otherwise, without the prior written permission of the publisher.

ISBN: 978-0-87335-331-1

**On the cover:** Cross-section of the Transbay Transit Center Program Downtown Rail Extension Project. **Architect:** Pelli Clarke Pelli. Rendering courtesy of Transbay Joint Powers Authority.

# Contents

Preface .................................................................................................................. ix

## TRACK 1: TECHNOLOGY

### Session 1: Applied Technologies ................................................................................ 1

Deep Inclined Water Intake Shafts ................................................................................ 3
*Albert Ruiz, David Jurich, Guy Leary*

Construction of the Railway Bosphorus Tube Crossing, Tunnels, and Stations ..................... 10
*Yosuke Taguchi, Fumio Koyama, Atsushi Imaishi*

Ground Freezing Challenges for Horizontal Connection Between Shafts Under Difficult
Geologic and Hydrostatic Conditions ............................................................................ 26
*David K. Mueller, Joseph M. McCann, Kenneth E. Wigg, Paul C. Schmall, Matthew L. Bartlett*

Final Lining at Devil's Slide Tunnel .............................................................................. 35
*Stephen Liu*

The History of Tunneling in Portland: Rail, Highways, and the Environment ..................... 43
*Susan L. Bednarz, Paul T. Gribbon, Joseph P. Gildner*

San Vicente Pipeline Tunnel: A Sophisticated Ventilation System ...................................... 55
*Jean-Marc Wehrli, Richard McLane, Brett Zernich*

### Session 2: Innovation ................................................................................................ 63

Onsite, First Time Assembly of TBMs: Merging 3D Digital Modeling, Quality Control,
and Logistical Planning ............................................................................................... 65
*Joe Roby, Desiree Willis*

Large Diameter Segmentally Lined Shafts ..................................................................... 75
*Darin R. Kruse, Rodney Meadth*

Large Diameter TBM Development .............................................................................. 89
*Martin Herrenknecht, Karin Bäppler*

ADECO as an Alternative to NATM: 22 m Wide, 14 m High, Full Face Tunnel Excavation
in Clays .................................................................................................................... 96
*Fulvio Tonon*

Cutter Instrumentation System for Tunnel Boring Machines ........................................... 110
*Aaron Shanahan*

Towards Precise Under Ground Mapping System in Canada .......................................... 116
*Mike Ghassemi, D. Zoldy, H. Javady*

### Session 3: Pressurized Face Tunneling .................................................................... 123

Lake Mead Intake No 3 Tunnel: Geotechnical Aspects of TBM Operation ....................... 125
*Georg Anagnostou, Linard Cantieni, Marco Ramoni, Antonio Nicola*

Continuous Conveyor Design in EPB TBM Applications ............................................... 136
*Dean Workman, Desiree Willis*

International Practices for Connecting One Pass Precast Segmental Tunnel Linings .......... 140
*Christophe Delus, Bruno Jeanroy, David R. Klug*

Geotechnical and Design Challenges for TBM Selection on the ICE Tunnel .................... 147
*Steve Dubnewych, Stephen Klein, Paul Guptill*

Soft Ground Tunneling on a Mexico City Wastewater Project ................................. 158
*Doug Harding, Desiree Willis*

Small Diameter Tunneling: Ems-Dollard Crossing ......................................... 164
*Klaus Rieker*

## Session 4: Sustainability ................................................................... 171

Sustainable Tunnel Linings: Asset Protection That Will Not Cost the Earth ...................... 173
*Colin Eddie, Mike Harper, Sotiris Psomas*

Design for Sustainable and Economical Tunnels ........................................... 183
*Derek J. Penrice, Bradford F. Townsend*

Sacramento UNWI Sections 1 and 2 Project: Special Tunnel Construction with
Plastic-Lined PCC Segments ............................................................. 191
*Jeremy Theys, Michael Lewis, Greg Watson*

Sustainable Underground Structure Design ................................................ 200
*Lei Fu*

Use of Underground Space in a Pristine Watershed: Chester Morse Lake Pump Plant and
Intake, North Bend, WA ................................................................. 206
*Joe Clare*

## Session 5: Tunnel Lining and Remediation ................................................... 213

High-Pressure Concrete Plug Leakage Remediation .......................................... 215
*Carlos A. Jaramillo, Camilo Quinones-Rozo, Robert A. McManus, Andrew Yu*

Structural Inspections of Colorado's Eisenhower Johnson Memorial Tunnel,
Hanging Lake Tunnel, and Reverse Curve Tunnel ........................................... 224
*Ralph Trapani, Jon Kaneshiro, David Jurich, Michael Salamon, Steve Quick*

Corrosion Protected Systems for Tunnels and Underground Structures ......................... 233
*James B. Carroll, Heather M. Ivory*

Lined Concrete Segments: An Alternative Construction Method for a Large Diameter
Sewer Tunnel .......................................................................... 242
*Galen Samuelson Klein, Michaele Monaghan, Samuel Gambino, Dwayne Deutscher,
Rigoberto Guizar*

Portal Slope Stability and Tunnel Leakage Remediation .................................... 249
*Carlos A. Jaramillo, Camilo Quinones-Rozo, Robert A. McManus, Andrew Yu*

Inspection and Rehabilitation of Heroes Highway Tunnel in Connecticut ....................... 257
*Mohammad R. Jafari, Larry Murphy, Michael Gilbert*

## TRACK 2: DESIGN

## Session 1: Design Validation by Instrumentation, Monitoring, and Mapping ..................... 267

Instrumentation of Freight Tunnel in Chicago ............................................. 269
*Alireza Ayoubian, Richard J. Finno*

Field Mapping and Photo Documentation of the Southern Nevada Water Authority's
Lake Mead Intake No. 3 Project, Saddle Island, Lake Mead, Nevada ......................... 283
*Scott Ball, Richard Coon, Erika Moonin*

Tunnel-Induced Surface Settlement on the Brightwater Conveyance East Contract ............... 298
*Michael A. Lach, Farid Sariosseiri*

Strategic Tunnel Enhancement Programme, Abu Dhabi, UAE: An Overview of Geology
and Anticipated Geotechnical Conditions Along the Deep Sewer Tunnel Alignment .............. 304
*Mubarak Obaid Al Dhaheri, Mark Marshall, Panagiota Asprouda, Suresh Parashar*

Geotechnical Variability and Uncertainty in Long Tunnels . . . . . . . . . . . . . . . . . . . . . . . . . . . . . . . . . . 316
*Jack Raymer*

## Session 2: Challenging Conditions and Site Constraints . . . . . . . . . . . . . . . . . . . . . . . . . . . . . . . . . . . . . . . . . 323

New Irvington Tunnel Design Challenges . . . . . . . . . . . . . . . . . . . . . . . . . . . . . . . . . . . . . . . . . . . . . . . 325
*Glenn Boyce, Steve Klein, Yiming Sun, Theodore Feldsher, David Tsztoo*

The Sunnydale CSO Tunnel: Dealing with Urban Infrastructure . . . . . . . . . . . . . . . . . . . . . . . . . . . 337
*Renée L. Fippin, Heather E. Stewart, Richard M. Nolting III, Manfred M. Wong*

New York Harbor Siphon Project . . . . . . . . . . . . . . . . . . . . . . . . . . . . . . . . . . . . . . . . . . . . . . . . . . . . . 346
*Colin Lawrence, David Watson, Michael Schultz*

Rehabilitation of Rail Tunnels with Widening the Cross-Section While Maintaining the
Regular Rail Traffic . . . . . . . . . . . . . . . . . . . . . . . . . . . . . . . . . . . . . . . . . . . . . . . . . . . . . . . . . . . . . . . . . 355
*Christian Wawrzyniak, Heiner Fromm*

Development of Seismic Design Criteria for the Coronado Highway Tunnel . . . . . . . . . . . . . . . . . 365
*David Young, Anil Dean, James R. Gingery, Tomas Gregor*

## Session 3: Managing Risk, Safety, and Security Through Design . . . . . . . . . . . . . . . . . . . . . . . . . . . . . . . 373

Performance-Based Design Using Tunnel Fire Suppression . . . . . . . . . . . . . . . . . . . . . . . . . . . . . . . 375
*Fathi Tarada, James Bertwistle*

First Comprehensive Tunnel Design Manual in the United States . . . . . . . . . . . . . . . . . . . . . . . . . . 387
*James E. Monsees, Nasri A. Munfah, C. Jeremy Hung*

Geotechnical Investigations for the Anacostia River Projects . . . . . . . . . . . . . . . . . . . . . . . . . . . . . . 397
*Amanda Morgan, Kevin Fu, Ronald E. Bizzarri, Carlton M. Ray*

## Session 4: Strength, Stresses, and Stability Assessment Selection . . . . . . . . . . . . . . . . . . . . . . . . . . . . . . . 407

Lining Design Issues Associated with the Storage of Cryogenic Fluids in Rock Caverns . . . . . . . . . . . 409
*Christopher Laughton*

Design Guidelines for Sequential Excavations Method (SEM) Practices for Road Tunnels
in the United States . . . . . . . . . . . . . . . . . . . . . . . . . . . . . . . . . . . . . . . . . . . . . . . . . . . . . . . . . . . . . . . . . 415
*Vojtech Gall, Nasri Munfah*

Continuum and Discontinuum Modeling of Second Avenue Subway Caverns . . . . . . . . . . . . . . . . 423
*Verya Nasri, William Bergeson, Nils Pettersson*

Shaft, Cavern, and Starter Tunnel Construction for Lake Mead Intake No. 3: Temporary
Support and Permanent Lining Solutions . . . . . . . . . . . . . . . . . . . . . . . . . . . . . . . . . . . . . . . . . . . . . . 433
*Luis Piek, Rob Harding, Jeff Hammer*

Methodology for Structural Analysis of Large-Span Caverns in Rock . . . . . . . . . . . . . . . . . . . . . . 441
*Sanja Zlatanic, Patrick H.C. Chan, Ruben Manuelyan*

## Session 5: Design Optimization and Alignment Selection . . . . . . . . . . . . . . . . . . . . . . . . . . . . . . . . . . . . . . 459

Factors Influencing Tunnel Design in Mass Transit Applications . . . . . . . . . . . . . . . . . . . . . . . . . . . 461
*Irwan Halim, Marci Benson*

Transbay Transit Center Program Downtown Rail Extension Project . . . . . . . . . . . . . . . . . . . . . . . 475
*Meghan M. Murphy, Derek J. Penrice, Bradford F. Townsend*

A Tale of Two Capitals: Modeling Helps Designers Manage Strong Surge and Pneumatic
Forces in Deep Combined Sewer Storage Tunnels . . . . . . . . . . . . . . . . . . . . . . . . . . . . . . . . . . . . . . . 483
*Daniel J. Lautenbach, John K. Marr, Peter Klaver, John F. Cassidy, David Crawford*

McCook Reservoir Main Tunnel Connection Marks Another Significant Milestone in
Chicago's TARP . . . . . . . . . . . . . . . . . . . . . . . . . . . . . . . . . . . . . . . . . . . . . . . . . . . . . . . . . . . . . . . . . . . 502
*Faruk Oksuz, Megan Puncke, Jeffrey Bair, Paul Headland*

Integration of Operations and Underground Construction: Sound Transit University Link............ 517
*John Sleavin, Peter Raleigh, Samuel Swartz, Phaidra Campbell*

## TRACK 3: PLANNING

*Session 1: Project Cost Estimating/Finance* .................................................... **525**

    Size Matters If You're a Tunnel.................................................................. 527
      *Lee W. Abramson*

    Setting the Owner's Budget: A Guideline................................................... 536
      *Paul T. Gribbon, Julius Strid*

    Show Me the Money: The Real Savings in Tunnel Contract Payment Provisions.............. 541
      *John M. Stolz*

    Planning Level Tunnel Cost Estimation...................................................... 548
      *Hamid Javady, Jamal Rostami*

*Session 2: Project Delivery* ..................................................................... **557**

    Lake Mead Intake No. 3, Las Vegas, NV: A Transparent Risk Management Approach Adopted by the Owner and the Design-Build Contractor and Accepted by the Insurer............... 559
      *Michael Feroz, Erika Moonin, James Grayson*

    Tunneling MegaProjects: They Are Different............................................... 566
      *David J. Hatem, David H. Corkum*

    Alternative Contracting and Delivery Methods............................................ 570
      *Janette Keiser, John J. Riley*

    Design of the Waller Creek Tunnel, Austin, Texas........................................ 576
      *Tom Saczynski, Nieves Alfaro, Prakash Donde*

    Tunneling as a PPP-Project: Risks from the Viewpoint of the Insurer on a Case Study of a Tunnel Collapse............................................................................. 588
      *Christian Wawrzyniak, Winfried Luig, Achim Dohmen*

    DCWASA's Project Delivery Approach for the Washington DC CSO Program.................. 593
      *Ronald E. Bizzarri, Carlton Ray, William W. Edgerton*

    How to Deliver Your Project On Time: An Owners Procurement Strategy .................. 601
      *Wayne Green*

*Session 3: Project Planning and Implementation I* .......................................... **607**

    Sustainability Drives Jollyville Transmission Main Tunnel Design........................ 609
      *Clay Haynes, Tom Knox*

    Sustainable Underground Solutions for an Above Ground Problem.......................... 613
      *Laura S. Cabiness, Steven A. Kirk, Stephen A. O'Connell, Jason T. Swartz*

    Large Diameter TBM Solution for Subway Systems......................................... 621
      *Verya Nasri*

    Digging Deep to Save Green While Being Green and Sustainable........................... 629
      *S. Nielsen, J. Morgan, J. McKelvey, J. Trypus*

    Design and Construction Considerations for Shafts at Grand Central Terminal for the MTACC's East Side Access Project............................................................ 635
      *A.J. Thompson*

    The Urban Ring Project: Planning a New Bus Rapid Transit Tunnel for Boston............ 641
      *D.M. Watson, J.P. Davies, J.A. Doyle*

*Session 4: Project Risk, Budget, and Schedule* .................................................. 653

    Blast and Post Blast Behavior of Tunnels .................................................. 655
    *Sunghoon Choi, Ian Chaney, Taehyun Moon*

    Decision-Making Case History for Municipal Infrastructure Improvements ...................... 664
    *Lizan N. Gilbert, Jennifer Stark, Gopal Guthikonda, Joe Hoepken*

    Risk Management to Make Informed, Contingency-Based CIP Decisions ......................... 676
    *Paul Gribbon, Gregory Colzani, Christa Overby, Julius Strid*

    Linear Schedules for Tunnel Projects ....................................................... 683
    *Mun Wei Leong, Daniel E. Kass*

    Building Mined Underground Stations in Soft Ground with NATM Construction Practices ......... 695
    *Gerhard Sauer, Juergen Laubbichler, Sebastian Kumpfmueller, Thomas Schwind*

    Cost and Schedule Contingency for Large Underground Projects: What the Owner Needs to Know ... 710
    *Christopher Laughton*

    Management of Cost and Risk to Meet Budget and Schedule ................................... 718
    *John Reilly, Dwight Sangrey, Steve Warhoe*

    Overhead and Uncertainty in Cost Estimates: A Guide to Their Review ........................ 725
    *John M. Stolz*

*Session 5: Project Planning and Implementation II* .............................................. 731

    Sedimentary Rock Tunnel for CSO Storage and Conveyance in Cincinnati, Ohio ................ 733
    *Michael Deutscher, Samer Sadek, Roger Ward*

    Tunneling to Preserve Tollgate Creek ....................................................... 740
    *Michael Gilbert, Jacqueline Wesley, Swirvine Nyirenda*

    Tunneling Under Downtown Los Angeles ...................................................... 750
    *Mohammad Jafari, Jeffrey Woon, Osman Pekin, Amanda Elioff, Girish Roy*

    Selecting an Alignment for the Blacklick Creek Sanitary Interceptor Sewer Tunnel—
    Columbus, Ohio............................................................................ 758
    *Valerie R. Rebar, Heather M. Ivory*

## TRACK 4: CASE HISTORIES

*Session 1: Small Diameter* ...................................................................... 769

    Microtunneling Challenges: Crossing Under Major Railroad and Highways in Very Soft
    Glacial Soils—The Evolution of a Ground Treatment Assessment Process ...................... 771
    *Philip W. Lloyd, Glenn Duyvestyn, Zhenqi Cai*

    Marysville Trunk Interceptor Project: A Case History ....................................... 788
    *Paul de Verteiul*

    Case History: Innovative CSO Pipe Installation in a Congested Urban Setting ................ 796
    *Emad Farouz*

    Pipe Jacking Through Hardpan: A Case History—North Gratiot Interceptor Drain Phase I ....... 804
    *Joseph B. Alberts, Jason R. Edberg, Keith Graboske, Gordon Wilson, Steven Mancini*

    Construction Challenges for Small Diameter Soft Ground Tunnels ............................. 813
    *William Bergeson, Verya Nasri, Alan Pelletier, Leo Martin, Richard Palmer*

*Session 2: NATM/SEM* ............................................................................ 823

    Past and Present Soft Ground NATM for Tunnel and Shaft Construction for the
    Washington, D.C. Metro .................................................................... 825
    *John Rudolf, Vojtech Gall, Timothy O'Brien*

The Lincoln Square Tunnel: Tunneling Between Two Parking Garages Using Sequential
Excavation Mining .................................................................... 836
   *Chris D. Breeds, Larry Leone, Don Gonzales*

Case History: Complex Design and Construction of Tunnel and SOE to Accommodate
Challenging Site Conditions ............................................................ 843
   *Emad Farouz, John I. Yao*

## Session 3: Challenging Conditions ........................................................ 853

Tunneling on Brightwater West........................................................... 855
   *Glen Frank, Mina M. Shinouda, Greg Hauser*

New York City Transit No. 7 Subway Extension Underpinning and Construction Under the
8th Ave. Subway ...................................................................... 863
   *Aram Grigoryan, Raymond J. Castelli, Fuat Topcubasi, Sankar Chakraborty*

Consolidation Grouting of the Riverbank Filtration Tunnel ................................. 876
   *Adam L. Bedell, Steve Holtermann, Kay Ball*

Gotthard Base Tunnel: Micro Tremors and Rock Bursts Encountered During Construction ......... 880
   *Michael Rehbock-Sander, Rolf Stadelmann*

Optimization in Blasting Production and Vibration Mitigation for Shaft and Tunnel
Construction at Lake Mead ............................................................. 889
   *Jordan Hoover, Dan Brown, Caroline Boerner, Shimi Tzobery, Erika Moonin*

## Session 4: Conventional Tunneling ....................................................... 895

A Conventionally Tunneled River Undercrossing........................................... 897
   *Adrian A.J. Holmes, Sangyoon Min, Klaus G. Winkler, Jim Brunkhorst*

Tunneling Ground Reinforcement by TAM Grouting: A Case History.......................... 906
   *Ahmad Samadi, Gary Seifert*

The Construction of the Tunnels and Shafts for the Project XFEL (X-Ray Free Electron Laser) ...... 916
   *Paul Erdmann*

Canadian Fast-Track Drill and Blast: Excavating the Rupert Transfer Tunnel at James Bay,
Québec, Canada....................................................................... 923
   *C.H. Murdock, R.W. Glowe*

Re-Design of Water Tunnels for Croton Water Treatment Plant, New York City................ 926
   *Jozef F. Zurawski, Paul J. Scagnelli*

Keys to Success in Managing a Complicated Tunnel Project: City of Columbus—
Big Walnut Sanitary Trunk Sewer Extension .............................................. 938
   *Michael J. Hall, John G. Newsome*

Drop Structures and Diversion Structures for the East Side Combined Sewer Overflow
Project, Portland, Oregon ............................................................... 944
   *Roy F. Cook, Tammy R. Cleys, Tony O'Donnell, Tom Corry*

## Special Session: Operational Criteria and Functionality for Highway Tunnels ..................... 955

National Tunnel Inspection Standards (NTIS) ............................................. 957
   *Jesus M. Rohena y Correa*

U.S. Domestic Scan Program—Best Practices for Roadway Tunnel Design, Construction,
Operation, and Maintenance ............................................................ 960
   *Jesus Rohena*

Index................................................................................. 965

# Preface

The chair, executive committee, and program committee welcome all authors, attendees, and exhibitors to Portland for the 10th North American Tunneling Conference (NAT 2010).

The conference theme, "Tunneling: Sustainable Infrastructure," underscores the important role that the tunneling industry plays worldwide in the development of underground space, transportation systems, conveyance systems, and other forms of sustainable infrastructure. This describes the evolving nature of our underground work, methods, and technology and serves to document the challenges we face and the lessons learned while advancing our projects in support of a sustainable future for our society. This conference reflects our ability to adapt and excel in the environment of continual evolution that characterizes the tunneling industry today. The selection of program committee chairs, each representing a segment of our industry (i.e., manufacturers, designers, owners, and contractors), demonstrates this ability to adapt and excel.

This conference covers a wide range of subjects dealing with nearly all aspects of underground construction. The papers accepted for NAT 2010 describe projects and experiences in North America as well as projects from around the world. NAT 2010 strives to keep pace with the ever-evolving practice of underground construction in an effort to disseminate knowledge of both success and failure within our industry. Accordingly, four tracks, each with five sessions, run concurrently to address a broad range of issues important to team members in the tunneling arena.

The program committee chairs extend an especially warm welcome to students attending this conference. We wish to encourage a new generation of engineers to explore the endless opportunities and innovations that are offered within the underground construction industry. As with previous conferences, we hope the professional attendees will take this opportunity to introduce themselves and their companies to the students and younger engineers, and share some of the challenges and successes that they have encountered.

We would like to express our sincere appreciation to the NAT 2010 organizing committee, technical program chairs, session chairs, and authors for their contribution to the conference. Members of the special programs committee, speaker committee, student outreach committee, and field trip committee are also gratefully acknowledged for their respective efforts. In addition, we offer our sincere gratitude for the support and contributions of the SME staff. Their patience and dedicated professionalism, as well as their experience and efficiency, have been critical to this conference and the publication of these proceedings.

Lawrence R. Eckert—Lachel, Felice and Associates
Matthew E. Fowler—Parsons Brinckerhoff
Michael F. Smithson, Jr.—Kenny Construction Company
Bradford F. Townsend—Hatch Mott MacDonald

## UCA of SME Executive Committee

Brenda Bohlke
Myers Bohlke Enterprise LLC

Lester Bradshaw
Bradshaw Construction Corp

Thomas Clemens
DSI Underground

William Edgerton
Jacobs Associates

Refik Elibay
Jordan, Jones, & Goulding

Robert Goodfellow
Black & Veatch Corp

LokHome
The Robbins Co.

Marcus Jensen
Black & Veatch Corp.

David Klug
David R Klug & Associates

Marc Kritzer
Northeast Ohio Reg Sewer District

Rick Lovat
Lovat Inc

Robert Palenno
GZA GeoEnvironmental Inc

Jeffery Petersen
Kiewit Construction Co

Gregory Raines
MWH Americas

Paul Scagnelli
Schivone Construction Co. Inc

Paul Schmall
Moretrench American Corp

Arthur Silber
New Jersey Transit

Rob Tumbleson
Akkerman

Donald Zeni

## NAT 2010 Executive Committee

Gregg Davidson
Jacobs Associates

Larry Eckert
Lachel, Felice & Associates

Matthew Fowler
Parsons Brinckerhoff

Robert Freeman
Cellular Concrete LLC

Paul Gribbon
City of Portland

Alan Howard
Brierley Associates LLC

Heather Ivory
URS Corp

Marc Kritzer
Northeast Ohio Reg Sewer District

Colin Lawrence
Hatch Mott MacDonald

Dennis Ofiara
The Robbins Co

Robert Palermo
GZA GeoEnvironmental Inc

Robert Pintabona
MWH

Greg Raines
MWH

Mark Ramsey
Kiewit Construction Co

Mike Smithson
Kenny Construction

Bradford Townsend
Hatch Mott MacDonald

Michael Vitale
Hatch Mott MacDonald

Brett Zernich
Traylor Brothers Inc.

# NAT 2010 Session Chairs

Robert Beck
TJPA

Eric Bier
Mining Equipment

James Brady
Lachel Felice & Associates

Geoff Clemens
DSI Underground

Gregg Davidson
Jacobs Associates

Daniel Dobbels
Jacobs Associates

Lisa Dwyer
Parsons Brinckerhoff

Christian Heinz
Kenny Construction

Galen Klein
URS Corp

Colin Lawrence
Hatch Mott MacDonald

Daniel Louis
MTACCIURS Corp

James McKelvey
Black & Veatch

Chris Mueller
Black & Veatch

Dennis Ofiara
The Robbins Company

David Pease
Traylor Brothers

Derek Penrice
Hatch Mott MacDonald

Samuel Swartz
Jacobs Associates

Michael Vitale
Hatch Mott MacDonald

Brett Zernich
Traylor Brothers

Sanja Zlantanic
Parsons Brinckerhoff

# TRACK 1: TECHNOLOGY

Mike Smithson, Chair

*Session 1: Applied Technologies*

# Deep Inclined Water Intake Shafts

## Albert Ruiz, David Jurich
Hatch Mott MacDonald, Phoenix, Arizona

## Guy Leary
Salt River Project, Phoenix, Arizona

**ABSTRACT:** The 2,250 MW coal fired Navajo Generating Station located in Page, AZ draws water from Lake Powell using submersible pumps installed in five inclined shafts. Drought threatens to lower the reservoir to below the existing intakes. A system of new steel lined intake shafts located on the same site was required to ensure uninterrupted plant operation. The small size of the site and the need to keep the existing system in operation required detailed designs and placed significant limitations on construction equipment and activities. The Navajo Sandstone forms the near vertical shoreline of the reservoir, is 98% very fine grained quartz, highly abrasive, and contains highly fractured intervals that presented numerous challenges to drilling the new 500-foot deep 43-inch diameter intake shafts inclined 23 to 26 degrees from vertical. Spatial constraints, submersible pump design, and operational criteria required each shaft to have a unique inclination and orientation and hit a small breakout target located 250 feet underwater. Sophisticated drilling equipment and techniques and state of the practice downhole survey technology and methods were used to make adjustments and maintain drilling accuracy with less than 1% deviation. Environmental requirements mandated drilling fluids introduced to the lake were kept to a minimum and grout was not allowed in to the lake under any circumstances. The shafts were video taped and the breakout locations were examined using an underwater ROV to verify rock conditions and location and to guide placement of pneumatic packers to ensure grout did not transmit to the lake during the steel liner installation. In anticipation of the invasion of Quagga mussels, the shaft design included an allowance for a chemical dosing system and a copper rich alloy at the exposed portions of the steel liner.

## BACKGROUND

The Navajo Generating Station (NGS) is a coal fired power plant located on the Navajo Indian Reservation near Page, AZ. Built in 1974, the 2,250 Megawatt power plant supplies electricity to the states of Arizona, Nevada, and California. The plant also provides the energy necessary to operate the pumps for the vital Central Arizona Project which supplies 1.5 million acre-feet of water to arid Arizona. Scrubbers were installed to the plant in the 1990s as part of a $420 million environmental upgrade. Due to this the NGS is now one of the cleanest operating coal fired power plants in the country. The plant also is essential to the economy of northern Arizona, providing many jobs for residents of Page and the Navajo Indian Tribe.

The NGS currently draws cooling water necessary for operation from nearby Lake Powell. An ongoing multiyear drought has been steadily lowering the water levels in the Lake Powell reservoir and began to approach the level of the existing cooling water intakes. In 2003 the managing owner of the plant, Salt River Project (SRP) developed a plan to add new lower water intakes. After completing the initial Environmental Impact Report, the owners contracted consultant Hatch Mott MacDonald in 2004 to develop and assess different alternatives and to fully design the selected concept.

The five existing 42" diameter cooling intakes are the only source of cooling water available to the plant. Four of the five intakes must always be in service to ensure continued operation of the plant, while the fifth serves as a backup. The intakes are composed of low-carbon (mild) steel liners with cement grout in the annular space. Each intake houses a single pump. When all five pumps are operational the combined flow of the system is approximately 30,000 gpm. The tops of the shafts are evenly spaced 20 feet on centers, positioned inside a pump house located at the top of a cliff overlooking Lake Powell at an elevation of 3733 feet. The site of the pump station is approximately 35,000 square feet and recessed into the ground as much as 16 feet so that it is not visible from the lake below. The original shafts were drilled at an incline and penetrate the cliff at elevations ranging from 3473 feet for Intake #1 to

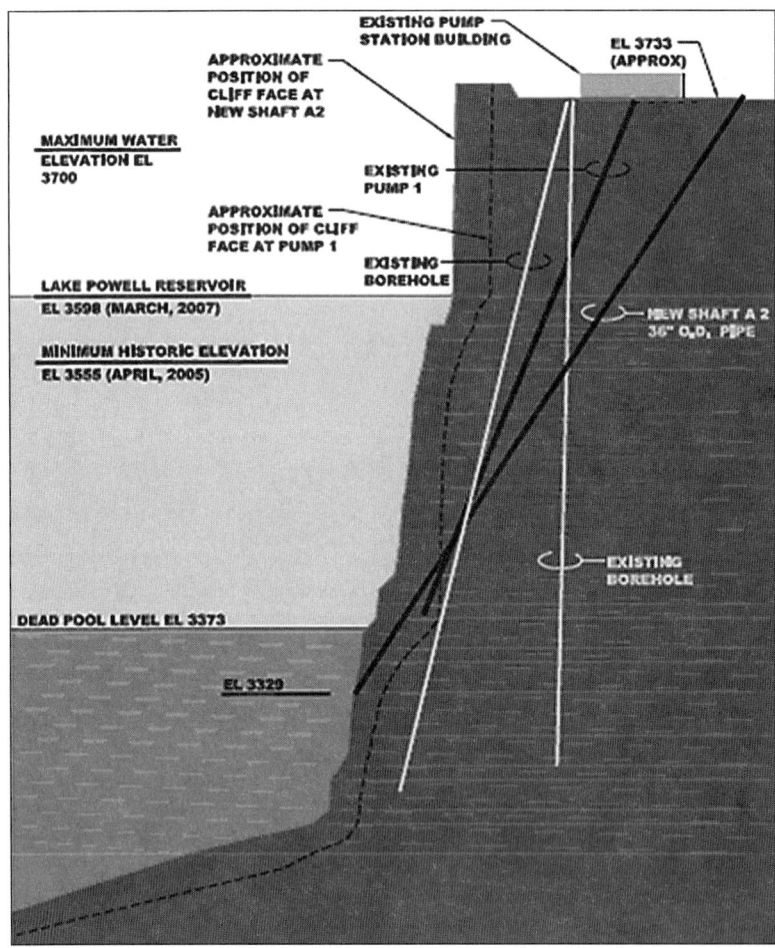

Figure 1. Shaft A2 typical elevation showing existing shafts and boreholes

3383 feet for Intake #5. The water surface elevation for Lake Powell can range from a maximum Full Pool elevation of 3700 to the absolute minimum Dead Pool Level of 3373 feet. In April 2005 the water surface was at an elevation of 3555 feet, bringing it within approximately 57 feet of the minimum operational level for Intake #5. Figure 1 shows an elevation of Shaft A2 and existing shafts and boreholes which needed to be avoided, which was typical of all the new shafts.

The options available for the new intake system were limited by a number of constraints. All land based components of the new system had to be located within the pump station site to remain within the existing Navajo Lease Boundary (NLB) land easement. Approximately one quarter of the existing square shaped pump station site was available for use, as the existing system that had to remain in service at all times during construction occupied the remainder of the site. The easement in Lake Powell available from the Department of National Park Services for the water intakes to penetrate into the reservoir was located immediately adjacent to the pump station site and was 400 feet wide and centered with respect to the existing pump station building. Within this easement the new water intake system had to penetrate the cliff in an area with a near vertical slope and at least 25 feet lower than the dead pool elevation of the reservoir at 3374 feet. This penetration elevation is necessary to provide sufficient pump inlet pressure for the five new pumps that would be drawing water at up to 30,000 gpm from the lake. The pumps station site with easements and shaft arrangement is shown in Figure 2. Requirements also dictated that no part of the intake system could be visible from Lake Powell. Furthermore the client wished that the new intake system not be readily accessible to the public due to security concerns.

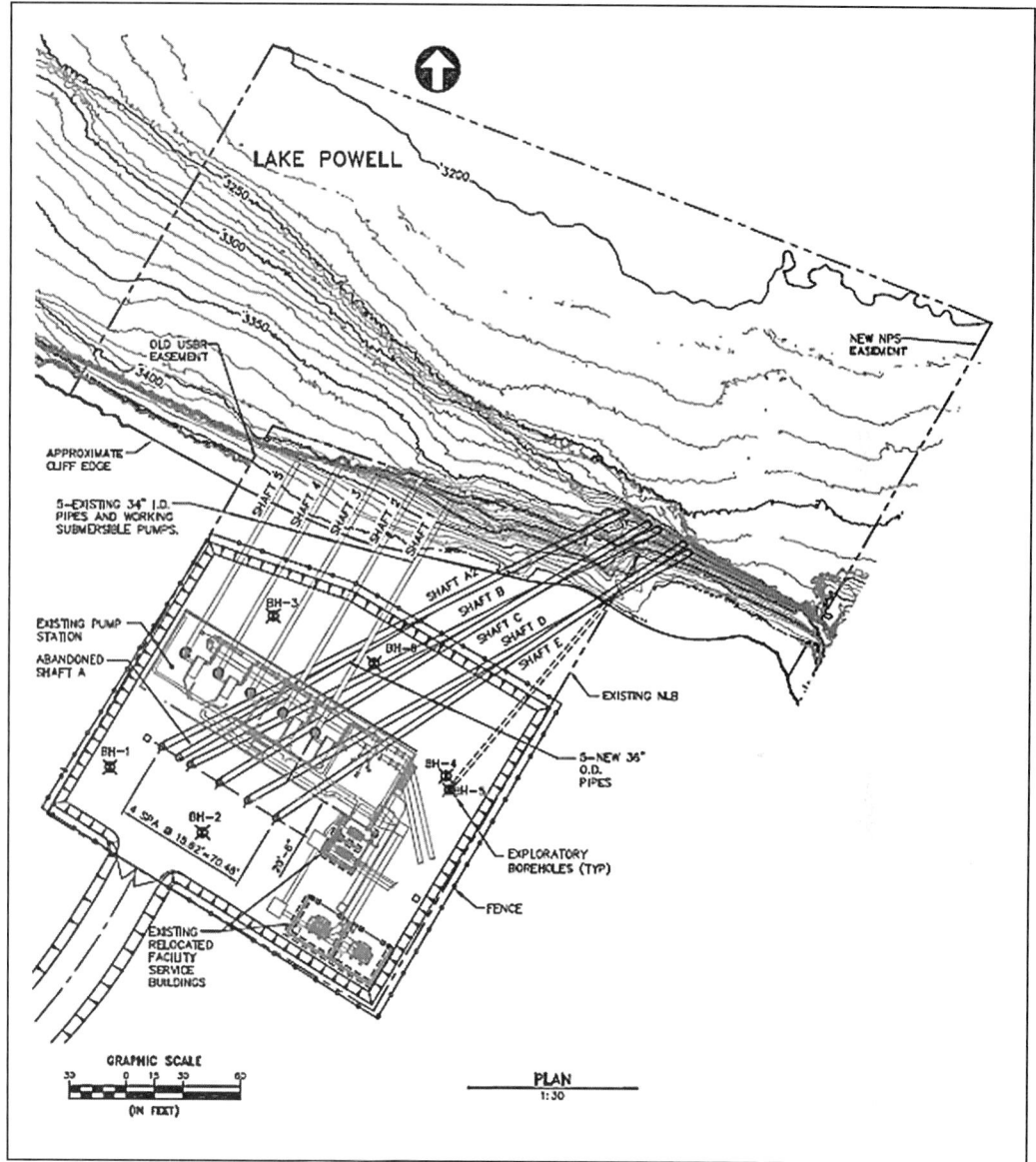

Figure 2. Pump station site with easements and shafts arrangement

Hatch Mott MacDonald and SRP developed 15 different alternative solutions for a new cooling water intake system for the plant. These concepts ranged from floating intake pump barges, to pipes installed to the side of the cliff, to a single vertical large diameter shaft with multiple lateral tunnels to tap into the lake. After comparing the concepts based upon the constraints provided by the owners, it was decided that the best solution would be five new inclined drilled shafts in a similar configuration to the existing system.

## GEOLOGY

Numerous exploratory methods were undertaken to investigate the geologic conditions at the site. The geotechnical investigation program was comprised of field inspections, exploratory core holes, packer permeability tests, underwater remote operated vehicle (ROV) and sonar surveys. The field observations taken from the top of the cliff and of the cliff face from a boat on Lake Powell revealed that there were large potentially unstable blocks on the cliff face adjacent to the pump house building. A risk assessment was completed and it was determined that it was necessary to remove them prior to construction activities.

Six exploratory boreholes were completed as part of the geotechnical program on the pump station site. Four of the six holes were inclined from vertical, and all of the boreholes were advanced to approximately 400 feet below ground surface (bgs). During drilling operations for two of the boreholes, drilling rod became stuck in the hole. Part of the rod was recovered from both holes, however 120 feet and 35 feet of drill rod were abandoned in the respective holes and new core holes were re-drilled within 10 feet of the abandoned holes. Caliper surveys including guard resistivity were performed on all of the holes as well as optical televiewer and video surveys. All core holes were backfill grouted.

Laboratory testing was performed on select segments of the recovered core for its unconfined compressive strength, split tensile strength, density, porosity, permeability and determination of Young's modulus and Poisson's ratio. The corrosive potential of the rock was tested by measuring the pH, resistivity, sulfate content, and chloride content to determine if there would be any effects on the shaft steel casing.

The Navajo Sandstone Formation that composes the cliffs that form the shoreline of Lake Powell is a massive, medium grained, quartz-rich sandstone deposit. The Navajo sandstone recovered from the boreholes at the pump station generally consisted of moderately indurated fine to medium grained quartz sand. The rock is moderately soft to moderately hard with an unconfined compressive strength ranging from 451 to 4,672 psi and an average value of 2,378 psi. The average dry density is 122.9 pounds per cubic foot and the average tensile strength is 554 psi. Core recovery during drilling operations was close to 100 percent, and the sandstone had a high rock quality designation (RQD) with values ranging from 80 to 100 percent, with less than 10 percent of the cores having values lower than 80.

Drilling fluid recovery was generally 90 to 100 percent with the exception of a zone where circulation was completely lost ranging from 330 to 362 feet bgs that was encountered in at least 3 of the boreholes. This was attributed to a zone of ½ to 5 inch diameter rock fragments at 337 feet bgs in boring BH-1. In boring BH-2 a 5 inch wide fracture filled with dark brown silty fine grained sand located at 365 feet bgs caused the loss of circulation. In boring BH-3 the loss of circulation was due to an approximately 2 foot thick broken limestone lens at a depth of 362 feet bgs. It was determined that these features were likely not related to each other based upon reviews of the core and downhole survey data.

Underwater ROV surveys were used as part of the preliminary geotechnical investigation to inspect the condition of the existing intakes and to locate features of the cliff that could possibly create a hazard to construction of the new intakes. It was found that the exterior of the existing intakes had strong encrustation, while the interior had no encrustation at all. Side scan underwater sonar was used to map the underwater profile of the sandstone. Matching the underwater contours with the land based surveys produced a 3D model of the terrain in the area of the pump station.

## DESIGN CHALLENGES

The new cooling water intakes system consists of five new 43-inch diameter inclined shafts designated A through E. Each was drilled at a unique inclination and bearing (about 32 degrees from vertical) to intersect a small target zone approximately 250 feet underwater. The shafts are lined with a 36-inch diameter, ½-inch thick steel pipe with the annular space between the pipe and shaft wall grouted. The shafts are equally spaced roughly 15.5 feet center to center and are 20.5 feet from the edge of the existing pump station building. The shafts range from approximately 485 to 517 feet long. They penetrate the cliff face elevations between 3329 and 3300 feet.

The lake pump station site is located on the edge of a cliff overlooking the Lake Powell Reservoir. Only 9,200 square feet of the site was available to be used as a work area for drilling and grouting operations. An adjacent area approximately 116,000 square feet in size was available as a staging yard during construction. The small size of the drilling area was a challenge in placing the drill rig, and the drill rod carrier of the rig was within inches of the building in some cases. Access to the pump building and surrounding equipment had to be maintained during the entire length of construction for inspection and maintenance of the existing intake system, placing further restrictions on the usable space during construction.

Investigation and analysis of the underwater cliff profile revealed that the cliff face at the minimum breakout elevation in front of and west of the pumping site was sloped at too great of an angle to be used for breakout. The only portion of the cliff face that was suitable for breakout started at the east edge of the property line of the existing pump station site and extended to the east edge of the NPS easement. Therefore, the shafts had to angle across the property towards the northeast corner. Two of the shafts also had to pass right underneath the corner of the existing Navajo land easement.

The target area where the new inclined shafts had to breakout in the cliff was 45 feet high by 70 feet wide. All shafts had to breakout within this

**Figure 3. The RD-20 drill rig during drilling of the new 43-inch diameter inclined shafts**

zone and remain at least 10 feet apart from each other for optimum pump performance. The drilling contractor was allowed a 2% tolerance in deviation from the planned alignment and could not have any kinks or sharp turns within the hole that would prevent the 40 foot long—24 inch diameter pump assembly from being lowered down the shaft.

During drilling, no drilling fluids were allowed to be transmitted into the lake. When the hole was close to breaking out of the cliff face, drilling had to be stopped and all drilling fluids pumped out of the hole and replaced with water. The hole then could be drilled through to complete the lake breakout. Similarly, during grouting, absolutely no grout could be deposited into the lake. An inflatable pneumatic packer was used during grouting of the annular space between the drilled shaft and the steel liner to prevent grout from being pumped into the lake.

In recent years the invasive Quagga Mussel species has been appearing in reservoirs around the United States. This species' tendency to rapidly colonize surfaces that it attaches to has caused numerous problems and reductions in capability for other intake and underwater structures. Due to this, the client wished to have in place a system to address the Quagga Mussel issue. Each shaft was designed to be installed with a ¾-inch diameter PVC pipe grouted in the annular space for a chemical dosing delivery system to prevent quagga mussels from attaching to the intakes.

## CONSTRUCTION—DRILLING

The contractor's initial approach to drilling the new 43-inch diameter inclined shafts was to use a single pass method starting on Shaft A. The RD-20 drill rig shown in Figure 3 utilized a dual rotary reverse circulation system to drill the shafts with an 18,000 lb, 9 foot long downhole hammer. The procedure was to drill a 44-inch diameter shaft to approximately 25 feet and grout in a steel surface casing. Once this was complete the drilling of the 43-inch diameter shaft would be advanced out of the bottom of the surface casing. As is the tendency for inclined drilled shafts, significant downward gravitational forces act on the drill string and cause the hole to droop off alignment, the problem worsening the further the hole is drilled. In an effort to mitigate the effects of this the contractor stiffened the drill string by adding 41-inch diameter stabilizers in place of drill rod. Initially, the drill string consisted of the 9-foot long hammer, a 5-foot long stabilizer and a 20-foot long stabilizer, then alternating segments of 20 foot drill rod and 5 foot long stabilizers for 120 feet, and then followed by drill rods for the rest of the hole.

The hole was surveyed every 20 feet to monitor its position relative to the designed alignment. If the hole started to deviate out of tolerance, a steering bit attachment could be added to the hammer to try to correct the alignment. In order to determine if using the steering unit was needed, it was necessary to monitor the progress of the drilling in real time using a Gyro Smart tool to obtain data that was then forwarded to Hatch Mott MacDonald. HMM created real time 3 dimensional models of the shaft position compared to designed alignment as well as the other possible obstructions such as the existing drilled shafts and abandoned core hole with drill rod. HMM also created exhibits of the area of the cliff face that the shaft was projected to breakout at. These models and exhibits were used to determine whether corrections were necessary and what actions should be taken.

During the drilling of Shaft A the hole began to dip exponentially and dropped almost 2 feet from 120 to 140 feet bgs. The contractor attempted unsuccessfully to solve the problem by adding more stabilizers to stiffen the drill string. At 281 feet bgs the hole was 7 feet low and was projected to breakout as much as 28 feet below the target area. At this point the contractor grouted the hole up to 67 feet bgs and attempted to re-drill the hole after making the drill string stiffer and increasing the diameter of the

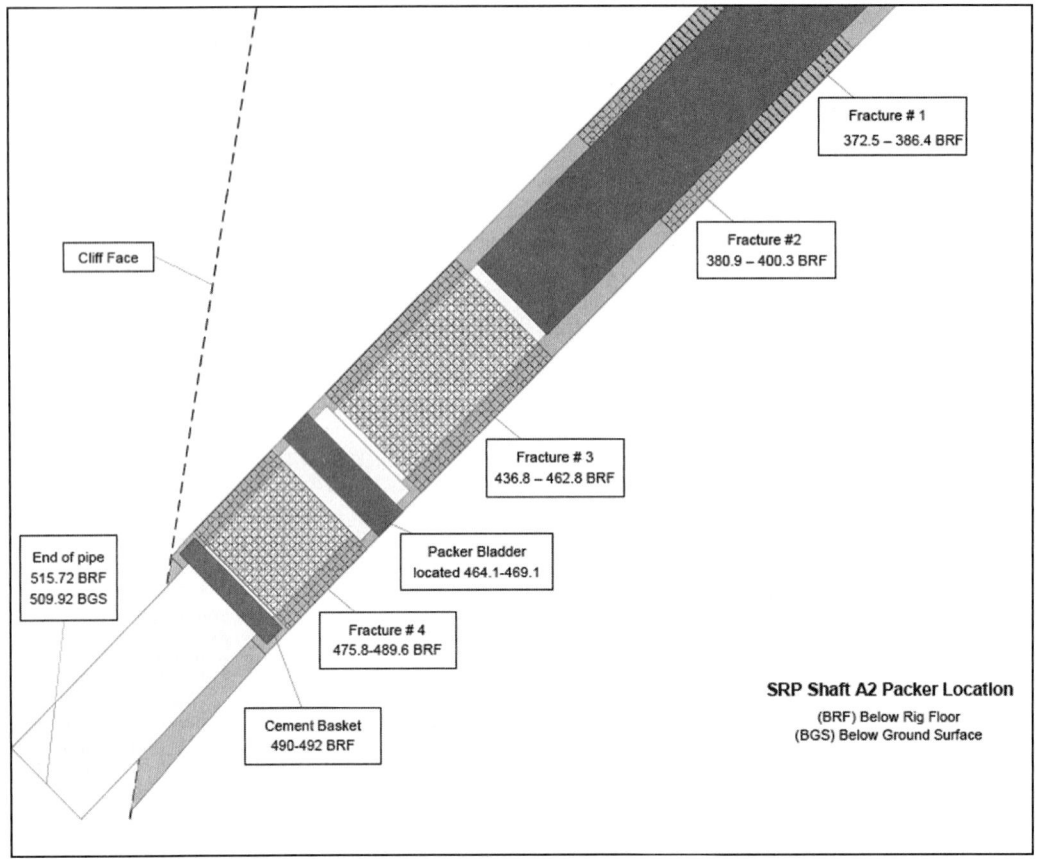

**Figure 4. Diagram showing fractured intervals at the breakout of Shaft A2**

hammer to 42.75 inches and the stabilizer diameter to 42 inches. After drilling to 126 feet bgs it became apparent that the drill was following the previously drilled hole. The contractor re-grouted the hole to 35 feet bgs and allowed the grout more time to cure before attempting to re-drill the hole. Once again, however, the drill followed the previous hole. It was decided at this time that Shaft A would be backfilled with cement grout and abandoned and a replacement shaft labeled A2 would be attempted in its place.

The contractor proposed drilling Shaft A2 using a two pass method by first drilling a 22.25-inch diameter pilot hole to breakout into the lake before reaming it to the full 43-inch diameter. The pilot hole utilized 80 feet of stabilizers directly behind the hammer with an additional stabilizer every 60 feet to reduce rod droop. The contractor aimed 13 feet above the target breakout to account for droop in the hole. This combined with constant monitoring of the pilot bore during drilling as was done with Shaft A ensured that Shaft A2 hit the target breakout elevation in over 250 feet of water.

The reamer consisted of the original hammer with a custom bit attached to the front to make certain that the reamer followed the pilot hole. The attachment had a polymer ring seal at the base to ensure that no cuttings or fluids traveled down the pilot bore during the reaming process to keep within requirements of the NPS construction permit.

The remainder of the shafts were completed successfully on the first attempt using the same two pass method. All of the shafts broke out within the target area and within the allowed tolerance. Upon drilling completion, each shaft was tested to make sure that they met the required design parameters. Prior to installation of the casing, a downhole caliper test was performed to make sure that the shafts were roughly circular and not overly out of shape. Downhole videos were taken the entire length of each shaft and each fracture zone was logged. Underwater ROV videos were taken of each shaft breakout to check the status of each and verify that there were no underwater features that would pose possible hazards or impairments to the future intake structures.

## CONSTRUCTION—GROUTING

Once drilling of each shaft was complete and tests verified that it was within design parameters, the bore had to be lined and grouted before drilling could commence on the next shaft. Using a combination of the videos from the underwater ROV survey and the downhole survey, the pneumatic packer for each shaft had to be precisely located with respect to the rock fractures to guarantee that no grout was deposited in the lake as required per the NPS permit. To accomplish this, each fracture location was recorded and then mapped as seen in Figure 4 to create a diagram for acceptable packer locations.

The 36-inch diameter, ½-inch wall thickness steel pipe was lowered into place and centered in the hole. The grout was poured in less than 100-foot lifts so that the hydrostatic pressure of the wet cement did not crush the steel liner. Lastly, a 40-foot long, 24-inch diameter mandrel modeled to be the same size as the intake pumps that the shafts would house was lowered and raised within each of the shafts to ensure that the pump would fit and there were no kinks that would impede installation or removal.

Each shaft was grouted successfully and no grout was deposited into Lake Powell.

## CONCLUSION

Construction of the 5 inclined shafts was completed successfully in March 2009 and SRP began construction of an interface to the existing pump station conveyance system. The new intakes were put into service in December 2009. The chosen alternative provided the owner with a new water intake system that met all of the numerous design constraints as well as costing significantly less than all of the other options.

## REFERENCES

Bansberg, R. and DePonty, J. (2004). Lake Pump Station Project Final Geotechnical Investigation Report, AMEC Earth and Environmental.

Bureau of Reclamation. (2004). Lake Powell Vertical Wall 2004 Survey, U.S. Department of the Interior.

# Construction of the Railway Bosphorus Tube Crossing, Tunnels, and Stations

Yosuke Taguchi, Fumio Koyama, Atsushi Imaishi
Taisei Corporation, Tokyo, Japan

**ABSTRACT:** The Marmary project is currently under construction in Istanbul, Turkey to provide a rail link between Europe and Asia beneath the Bosphorus Strait. The $1.03 billion project includes the design and construction of 4 railway stations and 13.6km of double track tunnel. The design build construction contract has been undertaken by the Taisei-Gama-Nurol Joint Venture and is financed by JICA. Tunneling elements include a 1.4km immersed tube segment, 18.5km of TBM tunnel, 1 station constructed by SEM/NATM and two stations constructed by open cut. The paper will provide an overview and update of the various tunneling technologies employed on the project.

## INTRODUCTION

The Bosphorous Strait lies between Asia and Europe in Istanbul, Turkey, and extends 30km from the Black Sea to the Sea of Marmara (Figure 1).

Because the railway network in Istanbul is inadequate, passengers and freight must rely on road transport, which has resulted in chronic traffic congestion and atmospheric pollution. The Railway Bosphorus Tube Crossing is being constructed with the objective of relieving traffic congestion by the use of rail, and therefore relieving the pollution.

## OUTLINE OF THE PROJECT

The overall project is referred to as the "Marmaray" project, which is a contraction of "Marmara + Ray (the Turkish word for rail)," and consists of modernizing the railways along the Sea of Marmara, to provide a total of 76km of railway connected by a tunnel under the strait. Of this, the Taisei Gama Nurol joint venture is carrying out the design and construction of a total length of 13.6km from Kazlicesme to Ayrilikcesme, including the part under the strait (Figures 2 and 3 and Table 1). The contract is an EPC (engineering, procurement, construction) contract, and as a rule the design methods, construction methods, and procurement methods were incorporated into the "Client's Requirements" in the contract document.

Tunnel will be constructed over 11km of the work section, by the shield, immersed, and NATM tunneling methods.

The use of the immersed tunneling method in the strait was a contractual condition that was decided in the client's basic plan. The installation

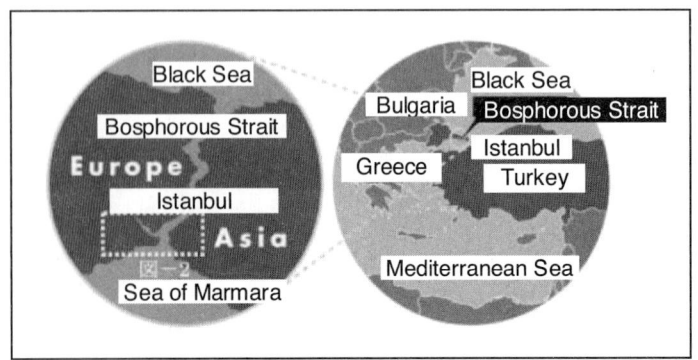

Figure 1. Bosphorous Strait location diagram

**Figure 2. Plan of the route**

water depth reaches 60m, so this immersed tunnel is the deepest in the world for water depth.

The shield tunnel is being constructed in the area with mainly sandy soil (west of Yenikapi Station) using one shield, and the area with mainly rock (east of Yenikapi Station) with four shields and a total of 19km will be constructed including the lines in both directions. Between the lines in both directions, connecting tunnels will be constructed at 200m intervals by the NATM method for passenger evacuation in an emergency. In addition, at two locations on the tunnel, crossovers are being provided (near Uskudar Station and Sirkeci Station), in which the lines in the two directions cross over. To incorporate branch lines in these areas the cross-section was made larger than the shield tunnel, so these areas were constructed by the NATM method.

In Istanbul, which includes areas designated as a World Heritage site, an archeological survey is mandatory, and the start of construction must wait until the Historical Conservation Committee of the Ministry of Culture and Tourism arrives at a conclusion. Therefore archeological surveys are being carried out at all the open excavation areas, which is causing a long time delay prior to start of construction.

Besides the tunnel, the contract includes stations at four locations, three of which are underground (Yenikapi, Sirkeci, Uskudar). Yenikapi and Uskudar Stations will be excavated by the cut and cover method, and Sirkeci Station will be constructed by a combination of the cut and cover method and the NATM method. At Yenikapi Station, two shafts will be used to start the shield tunneling towards the immersed tunnel, but excavation must await the completion of the archeological survey, so it has not started. As a result, the start of shield tunneling from Yenikapi has been delayed by more than three years.

At Uskudar Station, the three-year long archeological survey has been completed, so excavation is scheduled to start soon.

At Sirkeci Station, the archeological survey is ongoing at the location of the ventilation shaft, so excavation has not started yet.

## GEOLOGICAL OUTLINE

The geology over the total length of 13.6km can be broadly divided into a rock portion (bedrock), an onshore sandy soil portion, and a sandy soil portion in the strait. Figure 4 shows the vertical geological profile.

### Bedrock Portion

Istanbul is covered by carboniferous period bedrock known as the Trakya Formation, whose thickness reaches at least 2km. This consists mainly of sandstone and mudstone, but in the upper parts mudstone is predominant, with stratiform or block form limestone dispersed locally, interspersed with stratiform limestone shale. As a result of Hercynian mountain-forming activity and alpine mountain-forming activity, it was subjected to folding and faulting, so that overall it is a finely jointed fractured structure, the top portion of which has formed a weathered zone due to surface water. It is anticipated that the shield tunnels

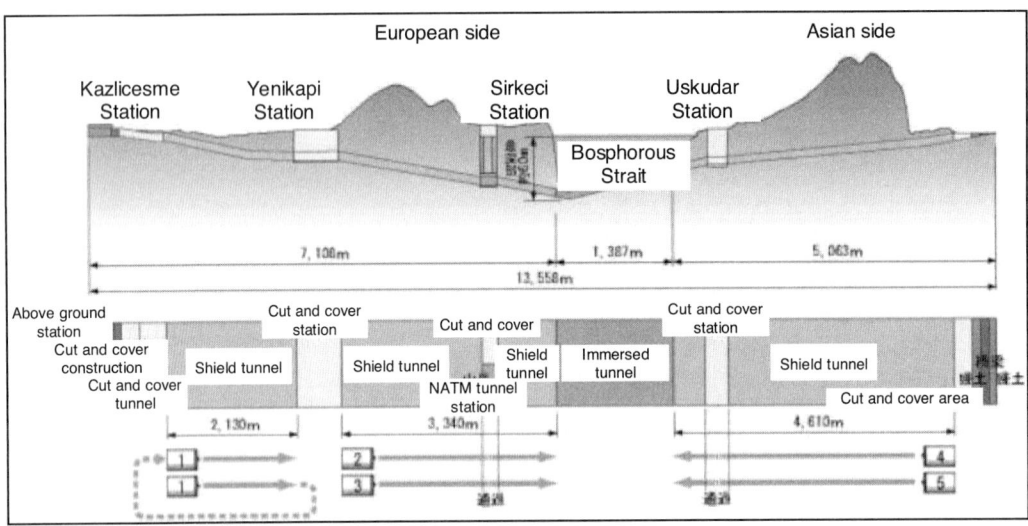

Figure 3. Schematic section through the route

Table 1. Project outline

| | |
|---|---|
| Client | DG Construction of Railways, Ports and Airports, Ministry of Transport and Communication, Republic of Turkey |
| Contractor | Taisei, Gama, Nurol Joint Venture (Gama and Nurol are Turkish companies) |
| Client's Representative | Avrasya Consult JV |
| Contract amount | JPY102.3 billion |
| Financing | Japan Bank for International Cooperation (JBIC), environmental yen loan |
| Contract period | 56 months (construction period is due to be extended as a result of lengthening of the archeological survey) |
| Details of contract | • Immersed tunnel: 1,387m (under the strait)<br>• Shield tunnel: 9,360m multiple lines<br>• NATM tunnel: Line crossovers, connecting passages in both directions<br>• Underground stations: Yenikapi Station, Sirkeci Station, Uskudar Station<br>• Above ground station: Kazlicesme Station<br>• Tracks, bridges, ventilation buildings, electrical equipment<br>• Archeological survey (the construction delays caused by this will be subject to compensation by the client) |
| Design conditions | • Earthquake: Moment magnitude 7.5, measures against liquefaction<br>• Train fires: Heat source energy 100MW (freight train carrying fuel)<br>In addition, water depths, current velocities, and other environmental conditions to be taken into consideration |

and the NATM tunnels will frequently encounter this fractured structure and weathered zone when excavating through the bedrock.

**Onshore Sandy Portion**

The surface strata of the city and along the banks of the Bosphorous Strait has been covered with a fill stratum from ancient times, which is distributed along the whole line apart from in the strait. The thickness of the stratum is about 2–10m, and it is in this layer that the archeological relics referred to previously are found. In addition to backfill sand, gravel, and silt, this stratum contains much man-made waste, such as bricks, mortar fragments, timber fragments, seashells, etc. To the west of Yenikapi Station the shield tunnel will pass through a sandy

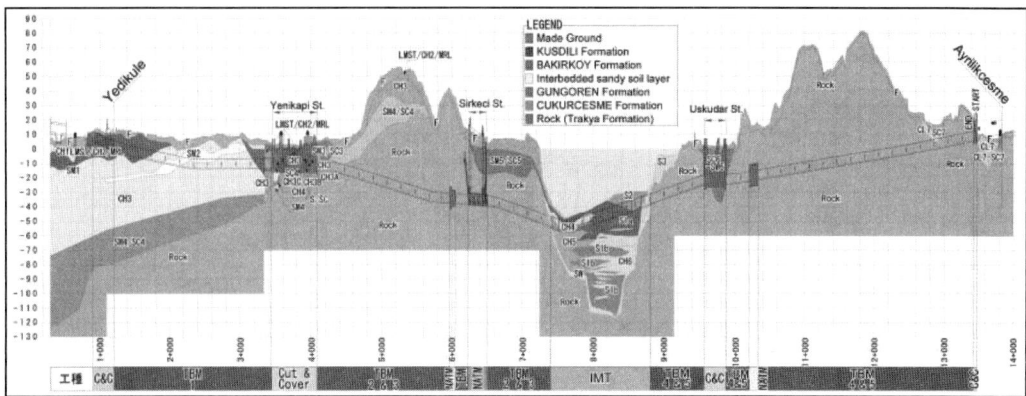

**Figure 4. Overall geological section**

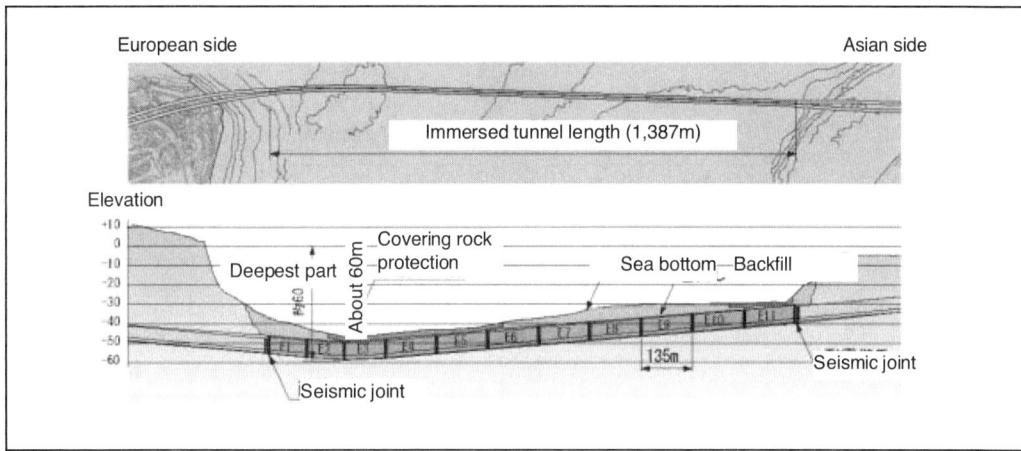

**Figure 5. Immersed tunnel, plan, and section**

soil stratum called the Bakirkoy Formation, and the Gungoren Formation.

The Bakirkoy Formation consists mainly of limestone and is a stratum distributed below the fill stratum, but depending on the location is also exposed on the surface. The thickness of the stratum is about 10–20m on the line of the tunnel, and in places there are stratified inclusions of clay or marl. The Gungoren Formation is a very stiff clay layer distributed below the Bakirkoy Formation. The stratum thickness reaches 40–60m, but a 5–10m thick very dense sand stratum is interspersed in this stratum, and the tunnel will also pass through this sand stratum.

**Sandy Soil in the Strait**

The alluvial strata in the strait area consists of a group of three geological strata (upper stratum, middle stratum, lower stratum). The upper stratum is distributed along the surface of the shore, and is a gravelly sand (the main component of the gravel is calcareous seashell fragments), and also includes charcoal and bricks, etc. The middle stratum is mainly an alluvial stratum of the strait, and consists of shelly sand, fine sand, silty fine sand, and sandy mud. It is thought that the sandy soil can easily liquefy. As described later, measures against liquefaction in an earthquake were carried out in this stratum, which forms part of the bearing stratum for the immersed tunnel elements. The lower stratum is a muddy sandy gravel covering the bedrock, with a thickness of 5–10m.

**CONSTRUCTION OF THE IMMERSED TUNNEL**

The immersed tunnel is formed from 11 RC elements with a maximum length of 135m, as shown in Figure 5, with a 2-cell rectangular cross-section (Figure 6). The internal space includes the rail

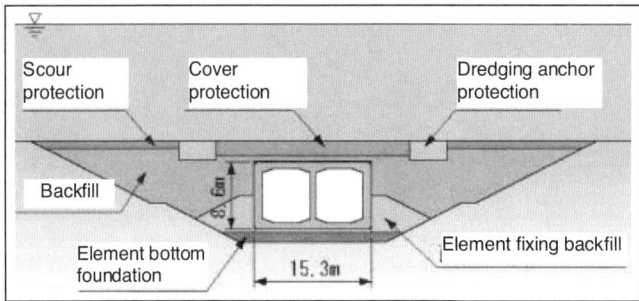

**Figure 6. Immersed tunnel cross section**

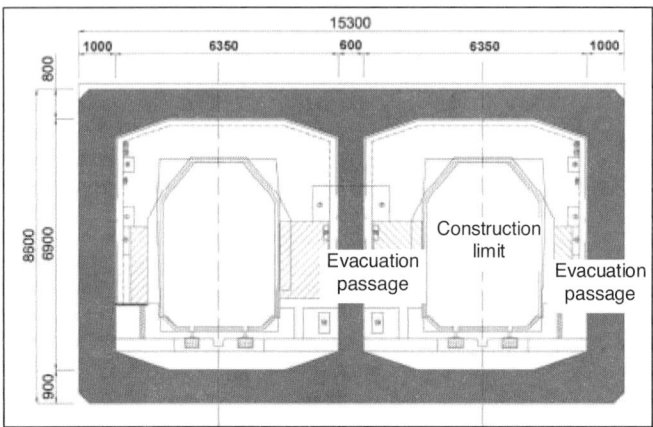

**Figure 7. Standard diagram of tunnel internal space**

construction limits, inspection passage, evacuation passage, and fireproof coating, as well as a 10–15cm allowance for installation tolerances of the elements (Figure 7).

The elements were constructed at a newly constructed temporary dry dock about 40km from the strait, and at a marine mooring facility. After completion of construction of the elements, they are floated and towed to the strait by tugboat, and immersed. After immersion, the gap between the bottom surface of the elements and the foundation mound is filled (foundation filling). Then the outsides of the elements are backfilled to a certain height with crushed stone. Figure 8 shows the flow from construction of the elements, immersion, and backfilling.

**Construction of the Immersed Tunnel Elements**

Figure 9 shows an outline of the elements. The side and bottom surfaces of the elements are covered with steel plate, to which galvanic anode electrical corrosion protection is applied. Waterproof sheeting is applied to the top surface, which is then covered with protective concrete. The structural concrete (design standard strength 40MPa) uses a mixture of moderate heat Portland cement and fly ash to prevent initial cracking, and silica fume is added to improve the durability.

In addition, a comparatively high density of reinforcement is required for the high water pressures and seismic loads, so head bar (Photo 1) was adopted for shear reinforcement. Head bar is structural shear reinforcement steel in which instead of double hooks in high density reinforcement which is difficult for placement, anchorage of the bar is obtained with frictional pressure welded plate[1]). In this project a total of about 1.3 million head bars will be used.

The element fabrication yard consisted of a temporary dry dock (Photo 2) and a marine mooring facility (Photo 3) for floating construction. Temporary dry docks were newly constructed at two locations near to each other, and each dry dock could simultaneously construct two elements, so a

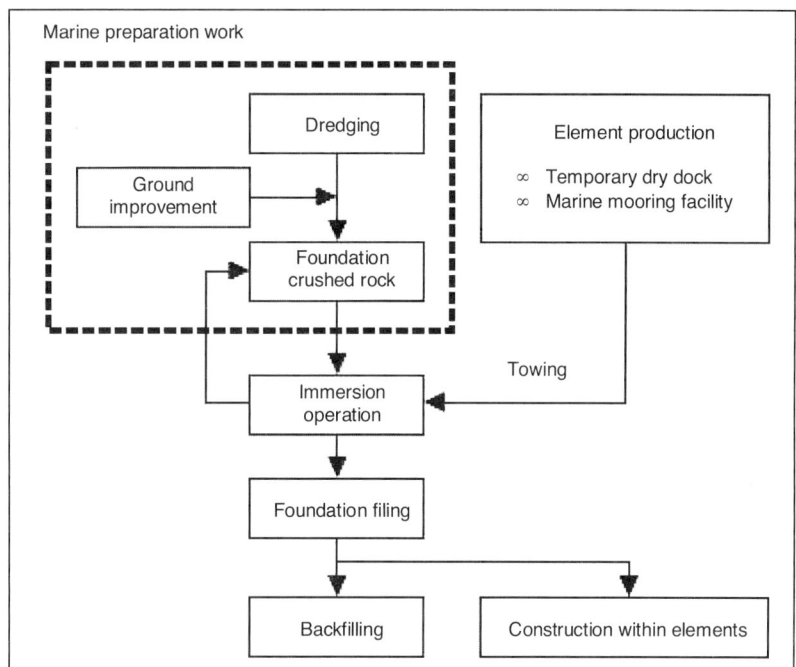

**Figure 8. Immersed tunnel construction flow chart**

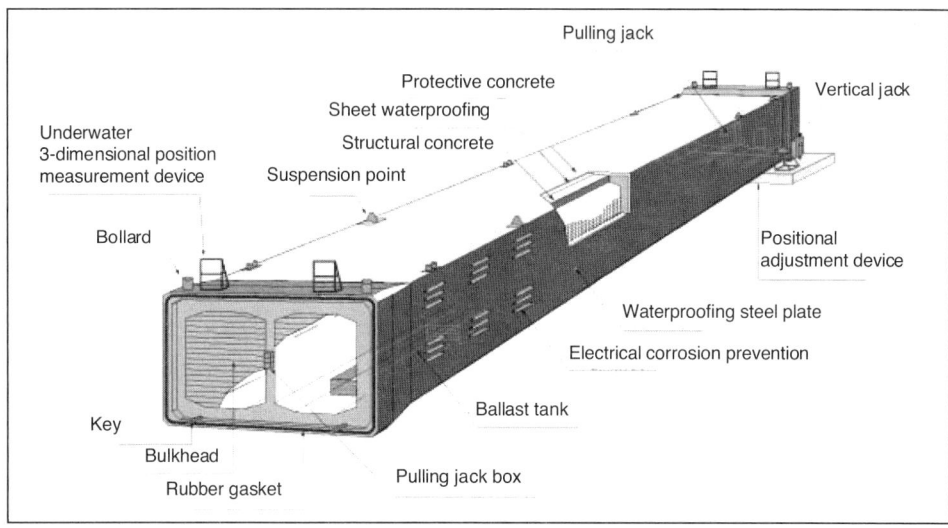

**Figure 9. Aerial view of element**

four-element system was established. The first step in constructing the elements is to construct the end steel shells, waterproofing steel plates, the bottom plates, and the bottom half of the walls in the dry dock. The second step is to flood the dry dock with seawater in this condition to float the elements, and after moving to the marine mooring facility the remaining top half of the walls and the top plate is constructed. This 2-step construction method was adopted to improve the efficiency of use of the dry dock, and to promote efficient progress of the construction process.

**Photo 1. Head bar and reinforcement layout**

**Photo 3. Element construction at marine mooring facility**

**Photo 2. Element construction in temporary dry dock**

**Photo 4. Work at the installation location on the sea (front: ground improvement, distance: dredging)**

## Marine Preparation Work

In parallel with the construction of the elements, preparatory work such as dredging, ground improvement, and emplacement of crushed rock, etc., is carried out at the location of immersion of the elements in the Bosphorous Strait (Photo 4).

Turkey is a seismic country, and the 1999 the Kocaeli earthquake caused damage in Istanbul. The immersed tunnel is installed in sandy soil sediments, so ground improvement was carried out as a measure against liquefaction in earthquakes, to increase the density of the soils (compaction pile grout) from elements E8 to E11, where there was concern over liquefaction.

Also, in the foundation crushed rock operations, because of the great water depth and the fast tidal currents, leveling the crushed rock by divers is both dangerous and inefficient. Therefore the work is being carried out using an underwater robot (Photo 5). In this method the underwater robot is being placed on the dredged floor surface, and the surface of the crushed rock leveled by remote

**Photo 5. Underwater crushed rock leveling robot**

control from the surface of the sea. The actual accuracy of the finished surface is within +10cm from the design value.

## Towing and Immersion

After completion of construction of a element, the element is carried by a twin-hulled type immersion operation vessel, towed by tugboats to a sea area that

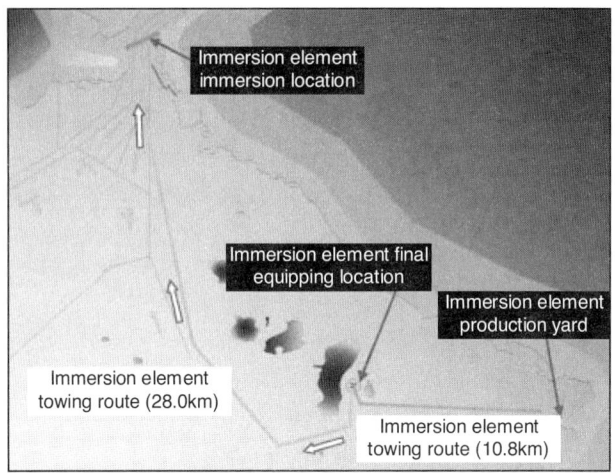

**Figure 10. Towing route for immersion element**

is the element's final immersion rigging location, and moored, as shown in Figure 10. This rigging sea area is a calm sea area behind an island where it is possible to maintain calm conditions while carrying out the equipping operations within and outside the element necessary for immersion.

After completion of equipping, the element is towed about 28km to the element immersion area by two tugboats to the front of the element and one tugboat to the rear. From start of towing to completion of immersion and water pressure connecting requires two days. The condition for enabling implementation of immersion was set to an average tidal current of 3 knots or less from the sea surface to a depth of 15m. Therefore accurately predicting the tidal current speeds and their variations during the two days of towing and immersion is an important factor to enable immersion. For this purpose tidal current measurement data in the Bosphorous Strait and meteorological data over a wide area from the Black Sea to the Sea of Marmara was collected over two years, and correlation analysis was carried out. To this, simulation analysis technology has been incorporated, so that at the present time a system for predicting from the meteorological and sea data the changes in the tidal currents in the 48 hours required for immersion has been constructed and is being operated[2,3]. The probability of a correct prediction is about 90%.

Immersion is being carried out by accurately determining the position of the immersion operation vessel using GPS installed on the operation vessel, and proceeding with the immersion operation while monitoring the element position and the shape of the sea bottom using multibeam (Photo 6, Figure 11). At the stage of making the final connection between

**Photo 6. Immersion operation vessel**

elements, the element is lowered to the bottom while measuring the distance between the two elements, the axial deviation, and the orientation deviation using ultrasonic distance measurement sensors installed on the opposing end surfaces of the two elements, and correcting the position.

After the element is installed in the specific location, the connecting is carried out using the procedure shown in Figure 12. First, a jacking rod is inserted into the element being immersed from the existing immersed element side, and the element is pulled towards the existing element using a tensioning jack. Then, the seawater between the bulkheads at the two end surfaces is evacuated, and the element being immersed is connected to the existing immersed element by the compression and waterproofing action of a rubber gasket by the imbalance of the seawater pressure.

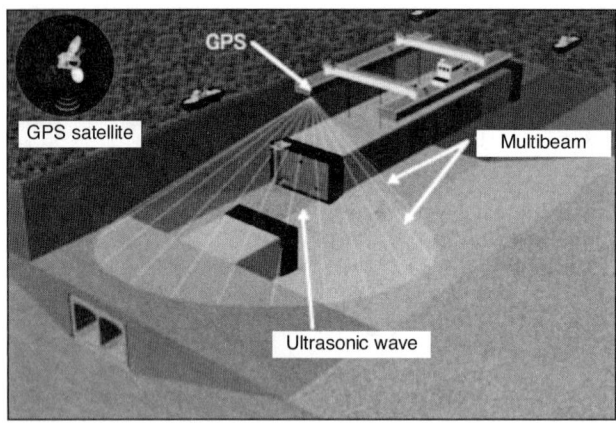

Figure 11. Outline diagram of monitoring during immersion

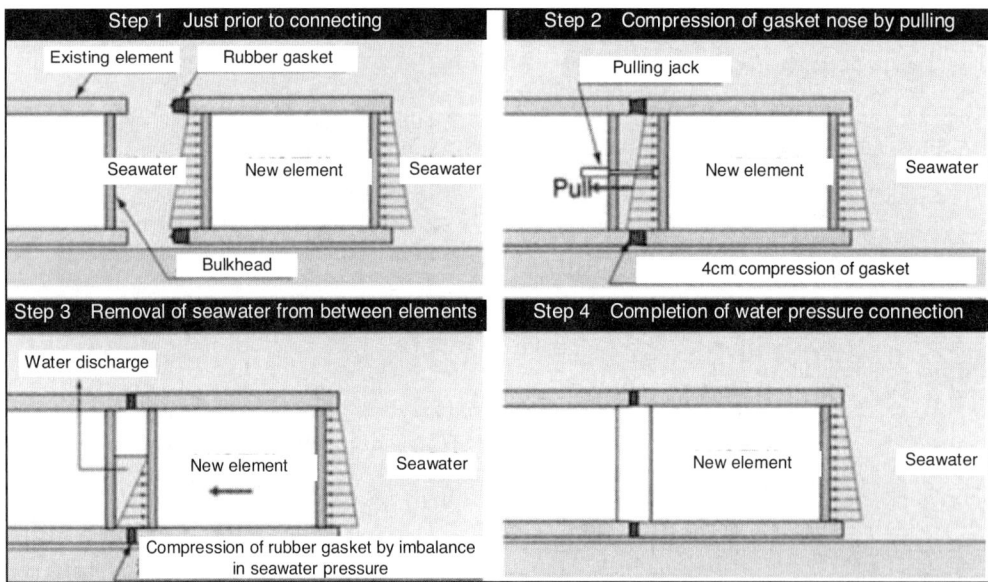

Figure 12. Element connecting procedure

**Foundation Filling, Construction Within Elements, Backfilling Work**

After completion of immersion of the elements, the gap (about 45cm) between the bottom surface of each element and the foundation, crushed rock is filled with 2-liquid mixture mortar from within the elements.

Construction work within the elements started after completion of backfilling around the elements so that foundation displacements are minimized, and consisted firstly of constructing the connections between elements, and then the other construction work is successively carried out.

Backfilling is carried out using tremie pipes in order to prevent impact or movement of the elements due to dropped backfill material. Materials, equipment, ventilation, and personnel access the tunnel from the temporary shaft at element E11 on the Asian side during the period before connecting with and penetration of the shield tunnel. Figure 13 shows a schematic diagram of the area around the connection between the immersed tunnel and the shield tunnel on the Asian side.

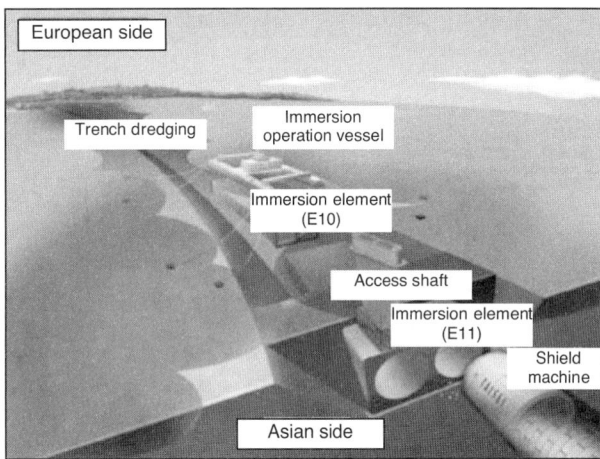

**Figure 13. Schematic diagram of the area around the connection between the immersed tunnel and the shield tunnel on the Asian side**

## CONSTRUCTION OF THE SHIELD TUNNEL
### Shield

In this project, five shields are being used, numbered as shown in the bottom of Figure 3. The No. 1 shield on the European side is for sandy soil with low water pressure and little earth cover, so an earth pressure balanced type shield (dia. 7,890) was adopted. However the ground for shield Nos. 2 and 3 on the European side and Nos. 4 and 5 on the Asian side is rock and is under high water pressure, including the portion under the strait, so a slurry type shield (dia. 7,850) was adopted. Also, in all cases the construction is below urban high density housing areas, so the closed shield method was adopted to minimize the effect on the surrounding environment.

The slurry type shield (Photo 7) for use in rock was an articulated type provided with fifty-five 17-inch disk cutters, and eight 250kW inverter motors capable of varying the speed up to four revolutions per minute, provided with twenty-five 3,000kN jacks for forward propulsion, and twenty-two 3,000kN jacks for articulation. An erector is provided to which an excavator for ground improvement can be fitted when changing bits, and injection holes are provided within the shield for waterproofing the connection with the immersed tunnel element.

### RC Segments

The tunnel has a finished internal diameter of 7,040mm, and RC segments are used for all shield tunnel lines. The segments thickness is 320mm in sandy soil areas, 300mm in rock areas, and the segment width is 1,500mm. One ring is formed from seven pieces, connections between rings are tenon groove joints, connections between segments are

**Photo 7. Slurry type shield**

butt joints, and all-taper segments are used on all lines. Photo 8 shows a view where the installation between RC segments is completed.

The elements for the evacuation passages provided every 200m between the tunnels in both directions are a bolted connection type, and steel segments were adopted for the pieces that are scheduled to be excavated by the NATM method. Photo 9 shows a view of steel segments installed.

Flexible segments were provided near the connection between the shield tunnel and the immersed tunnel, in order to absorb seismic displacements (Figure 14).

A temporary factory building has been constructed in Gebze, which is located about 60km from the outskirts of Istanbul, to produce the RC segments. This has a maximum daily production capacity of 22 rings, as a result of the use of steam curing. Photo

Photo 8. View of RC segments at completion of construction

Photo 10. Segment factory

Photo 9. View of steel segments at completion of construction

10 shows an overall view of the segment factory. A total of 12,900 rings are scheduled to be produced at this factory.

**Method of Connecting with the Immersed Tunnel**

An F-PAS device was adopted as waterproofing device for the connection between the immersed elements and the shield for the connecting operation underneath the strait under the pressure of the artificial backfill. Figures 15 and 16 show diagrams of the F-PAS procedure and the structure, respectively. A waterproofing seal device is fitted within a steel shell sleeve tube shaped like glasses fitted to the end of the immersed element, and the interior of the sleeve tube is filled with LW material (2-liquid ground improvement material). The LW material has the property that after freezing and melting, it changes from a solid to a liquid that can flow, so before the arrival of the shield the area around the F-PAS device is frozen into a donut like shape. After arrival, the area around the device is liquefied by melting using hot

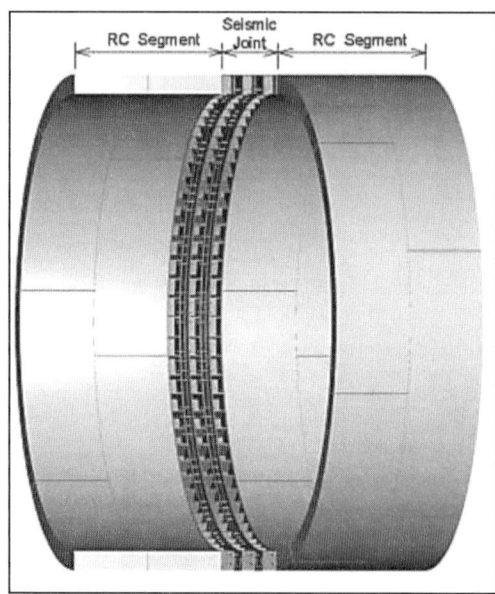

Figure 14. Structure of flexible segment

water, the waterproof seal is activated by pumping pressurized water into a pressurizing tube, to achieve close contact with the skin plate of the shield. The LW material prevents foreign matter from penetrating, so a secure waterproof seal can be formed with the skin plate.

**CONSTRUCTION OF THE NATM TUNNEL**

As described above, in the tunnel crossing the Bosphorous Strait, the immersed tunnel for the part under the strait, and the shield tunnel for the connecting on land parts was adopted as the standard construction for the tracked parts, but the NATM tunneling method was adopted in some parts. The

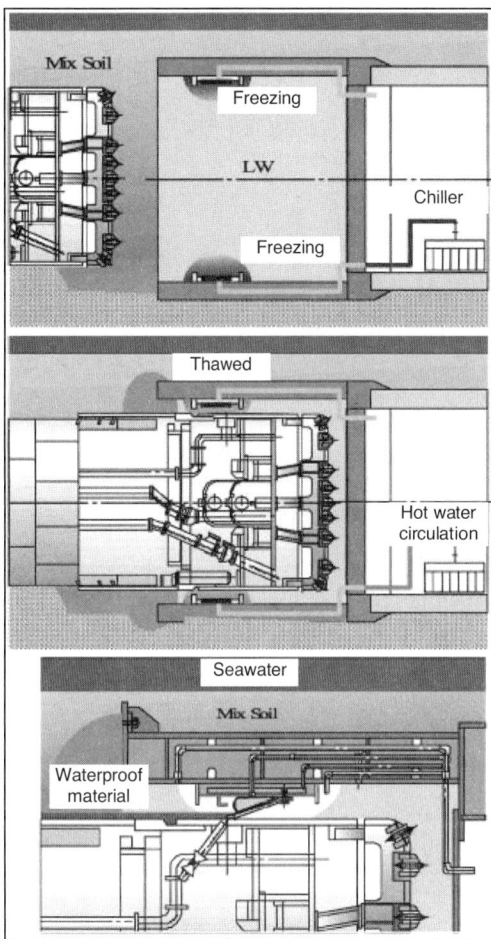

**Figure 15. F-PAS procedure**

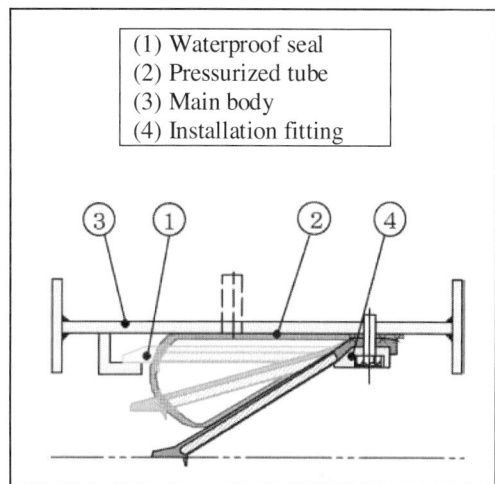

**Figure 16. F-PAS structure**

parts where the NATM method was adopted can be broadly divided into two areas. The first was at Sirkeci Station, where the cut and cover method was difficult due to restrictions from the surrounding environment, so a shaft will be used, and the underground station will be constructed using the NATM method. The other area is where a cross-section that is different from that of the shield tunnel is necessary in an enclosed underground space, namely at the two crossovers (places where the lines in the two directions cross over), the 43 passenger evacuation passages provided every 200m, and pump pits in three locations. There are a total of 17 types of cross-section (maximum excavated cross-sectional area 236m$^2$), with a total length of shaft plus tunnel of 2,130m, and excavated volume of 240,000m$^3$.

Another major feature of this construction is that all underground structures, including the tunnels, were required to be waterproof. The NATM structures, including the connections with the shield tunnels, are also watertight structures, so water pressure resistant lining concrete using RC structures is essential. In particular, at Sirkeci Station near the strait, and in the evacuation passages and pumping pits, it is anticipated that the water level is high, so structures to deal with the high water pressure were required. A maximum design water pressure of about 800kPa was assumed. In particular stress concentrations occur at gable walls and intersection points so the thickness and amount of reinforcement is increased, and it is difficult to appropriately lay waterproof sheeting, so careful construction is required.

At present the archeological survey is in progress, so in this paper Sirkeci underground station which is due to start construction in the future and Uskudar crossover tunnel, for which excavation is already complete are introduced.

**Construction of Sirkeci Underground Station and Present Status**

Sirkeci Station is located in a historic tourist and commercial area close to the Topkapi Palace, which was the residence of the Sultans of the Ottoman Empire. Figure 17 shows the land use at the planned location of the station. From the figure, the difficulty of excavating directly below the dense concentration of commercial buildings such as hotels, etc., from a shaft on a limited site can be appreciated.

About 60% of the existing buildings on the ground are of RC construction, but of these about 70% are more than 30 years old. The remaining 40% that are not RC construction are brick buildings which are mostly more than 50 years old. This

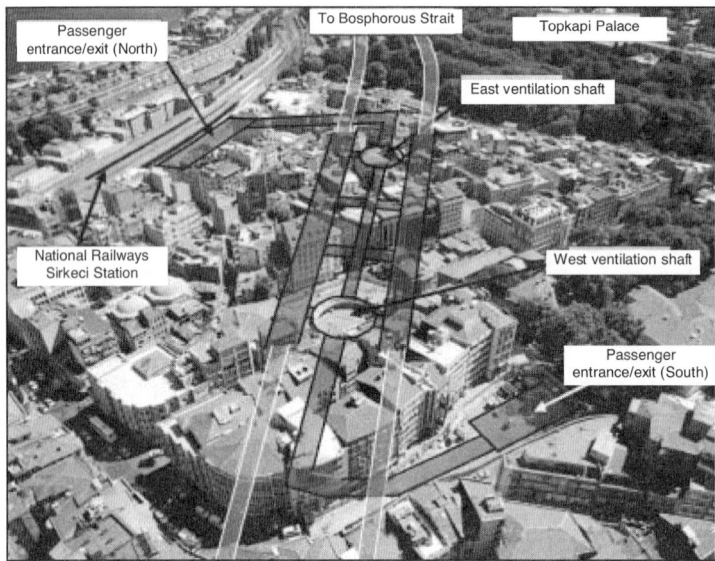

**Figure 17. Aerial view at the planned location of Sirkeci station**

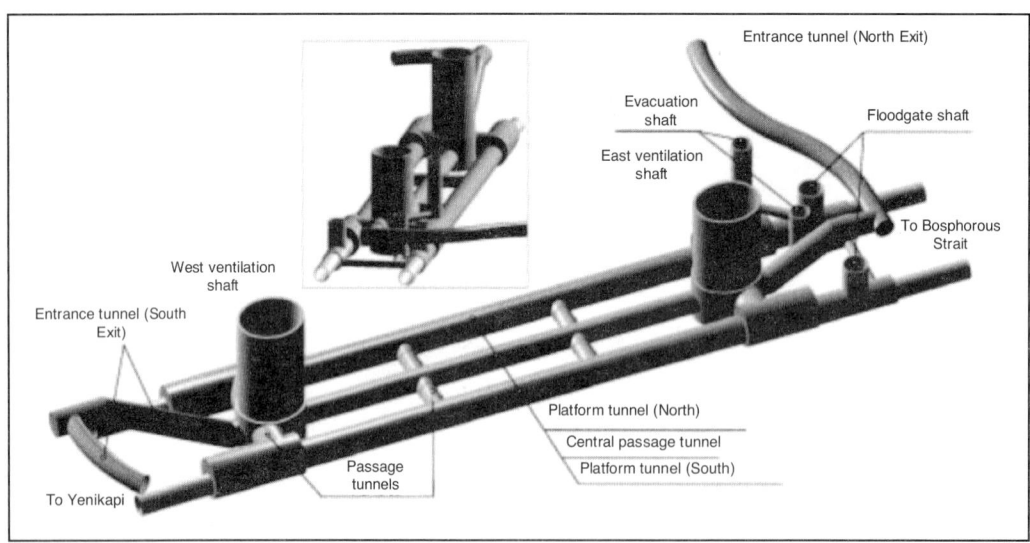

**Figure 18. Three-dimensional view of Sirkeci station**

is difficult construction requiring careful attention to the traffic situation and the nearby environment.

Figure 18 shows an aerial view of the underground station which will be excavated using the NATM method. A central passage tunnel parallel to the two platform tunnels and four intersecting passage tunnels will be constructed about 50m underground. Two entrance tunnels (North exit, South exit) will connect to the platforms from the ground. In addition ancillary services include the east and west shafts which provide the ventilation function and a duct tunnel. The reason the passage tunnel connected to the east and west shafts and the platform tunnels have large cross-section is because ventilation ducts are provided in the top portions of these tunnels. Other ancillary equipment includes a floodgate, which is characteristic of tunnels under the sea, and an evacuation shaft. In terms of function, as can be seen from the figure, because of the problem with the site, it is necessary to carry out

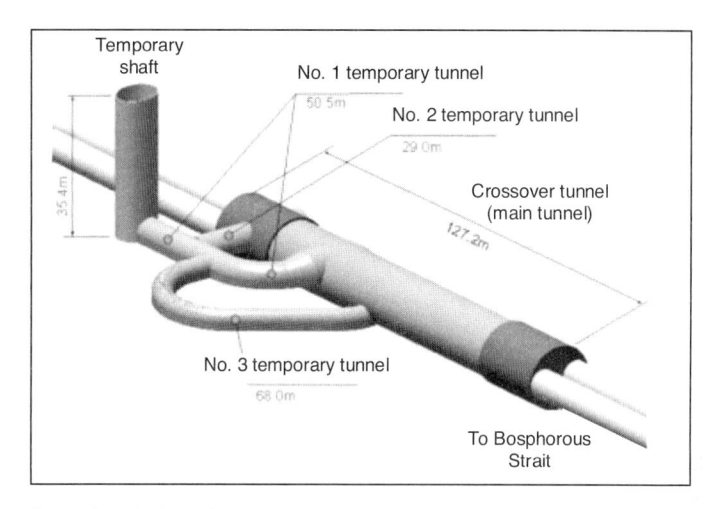

Figure 19. Three-dimensional view of Uskudar crossover

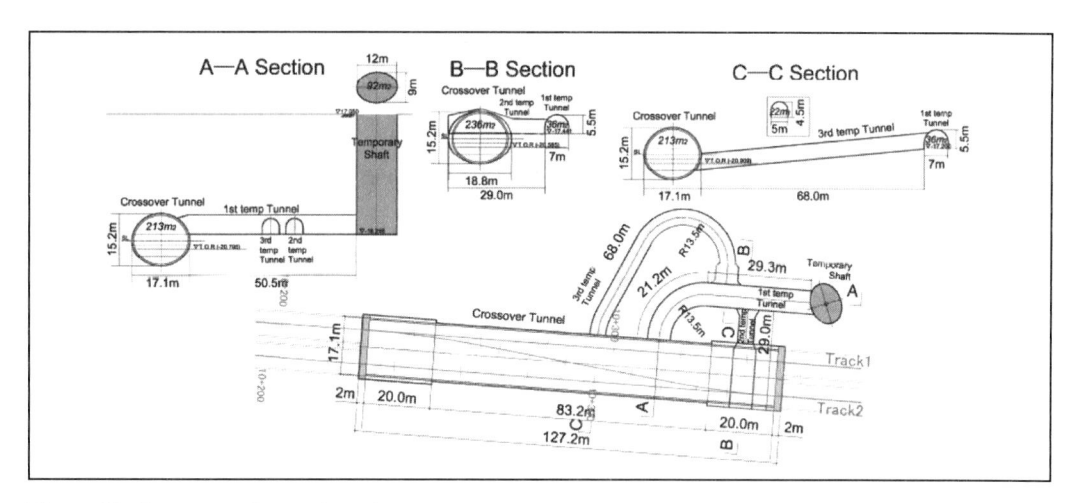

Figure 20. Crossover plan and section

construction in locations very close to other tunnels. Due to the necessity of also providing the services described above, more than 30 points of intersection between tunnels and shafts must be processed at Sirkeci Station, as well as the necessity of constructing points with major changes in cross-section and a shaft inclined at 30 degrees, the construction sequence and construction management has difficulties not seen in the other stations.

**Construction of the Uskudar Crossover Tunnel**

The Uskudar crossover tunnel is about 300m inland from Uskudar, the station on the Asian side of the strait. At this location the two cross-sections of the independent single line shield tunnels are merged to form a large cross-section, the location for the tracks to cross over. At the location there is an urban residential area at ground level, and the depth to the track level is about 40m, so the cut and cover construction method would be difficult, so the NATM method was adopted. Figure 19 shows a 3-dimensional overall drawing, Figure 20 shows a plan and section through the work section, and Figure 21 shows the cross-section of the crossover tunnel (main tunnel).

The construction was carried out in advance of the shield tunneling. A temporary shaft for access from ground level (depth 35m) was excavated, and a temporary approach tunnel was constructed from the bottom of the shaft. The temporary approach tunnel branches into three, and all materials and equipment

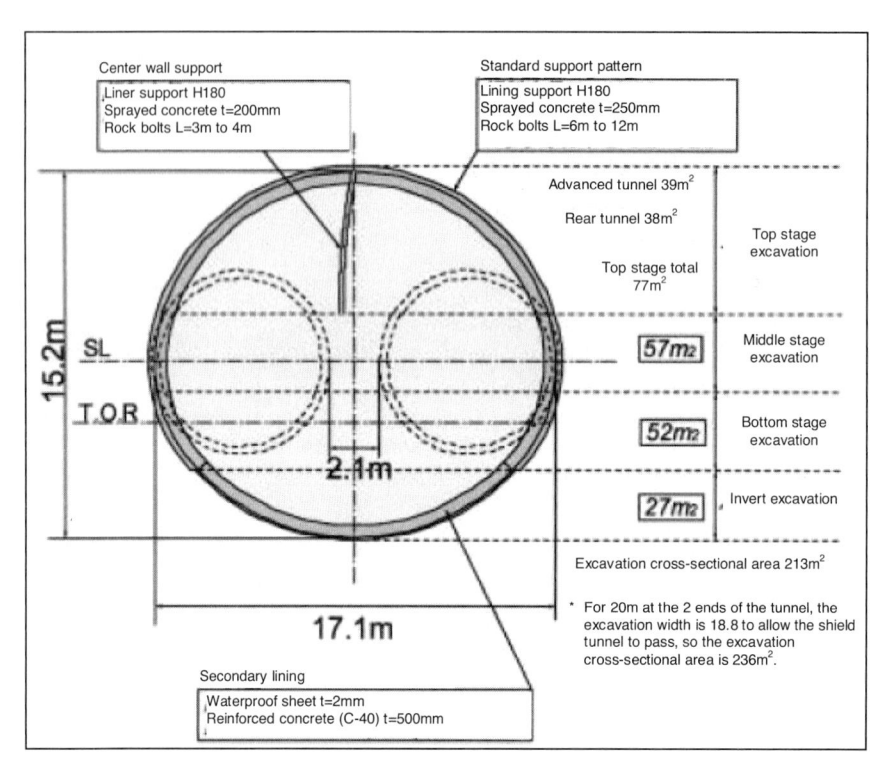

**Figure 21. Cross section of crossover main tunnel**

were brought in and out using these temporary shaft and tunnels. The crossover tunnel is a large cross-section tunnel with a maximum excavated cross-section of 236m². Therefore the bench cut method was used, dividing the excavation into top, middle, bottom, and invert stages, and the top stage was excavated using the central wall method, with the excavation divided into an advance tunnel and a rear tunnel. The middle stage excavation was divided into a left and right excavation with the top stage central wall left in position as a rule, and at the stage that the arch was completed to the middle, the central wall was removed. Photo 11 shows a view of the removal of the central wall in progress. Further, at the two end portions the shield tunnel will arrive and start. For this reason the temporary approach tunnel is connected to the side of the crossover tunnel; the 1st and 2nd tunnels access the top half, and the 3rd tunnel accesses the bottom half of the crossover tunnel. As a result of this arrangement the supports and rock bolts of the temporary tunnels will not interfere with the shield, and it was possible to excavate safely and by a good procedure with section headings divided into six areas. The detailed excavation procedure, the measurement control, and the supplementary measures to reduce the tunnel displacements are described in Reference 4. However, in addition to the conditions of excavating a large cross-section of width 17–19m and height about 15m, with a soil cover of 25m in an urban area, the work was impeded by the crushed and weathered mudstone. Therefore displacement within the space and settlement at the ground surface continued, so it was necessary to use an excessive number of additional supports. Specifically, in the areas with significant displacement, the following measures were taken.

1. The number of rock bolts was increased and lengthened (6m → 8–12m), and placed earlier (at an incline)
2. At the top (advance, rear tunnels), middle, and bottom excavation stages, the temporary inverts were sprayed closed
3. Steel supports were introduced to the invert sprayed parts
4. Construction of foot piles (3.5B, L 3.4 to 5.1m)
5. Fore poling, including resin injection, and face bolts

Although a maximum local settlement of the ground surface of 100mm was measured, all the excavation has been completed, and the lining concrete has been

**Photo 11. View of excavation of the crossover tunnel**

**Photo 12. View of crossover tunnel lining**

completed except in part. Photo 12 shows a view of a place where the lining is completed.

It is necessary that the experience and data from this construction be analyzed and used for the construction of Sirkeci Station and the other remaining construction.

**REFERENCES**

Taisei Corporation: Plate anchorage type shear reinforcement "head-bar," civil engineering materials technology and technical approval verification report in accordance with the Private Development Construction Technology Technical Approval and Evidence Approval Regulations, No. 1104, Public Works Research Center, 1999.9.

Oda and Ito: Development of Flow Prediction method Taking Dynamic Fluctuations in a Two Layer Flow Field into Consideration, Journal of the Japanese Association for Coastal Zone Studies, Vol. 19, No. 4, pp. 13–24, 2007.

Oda, Ito, Ueno, Koyama, Sakae: System and Accuracy Verification of Current Predictions for the Construction of the Railway Bosphorus Tube Crossing, Civil Engineering in the Ocean, Vol. 23, pp. 345–350, 2007.

G. Akay, Y. Taguchi, Y. Shimizu: Excavation and its behavior of extra large tunnel in urban area, The 2nd Symposium on Underground Excavations for Transportation, Istanbul, 2007. 11.

# Ground Freezing Challenges for Horizontal Connection Between Shafts Under Difficult Geologic and Hydrostatic Conditions

**David K. Mueller, Joseph M. McCann, Kenneth E. Wigg, Paul C. Schmall**
Moretrench, Rockaway, New Jersey

**Matthew L. Bartlett**
Kiewit Construction Company, Omaha, Nebraska

**ABSTRACT:** Ground freezing to assist in horizontal tunneling presents unusual challenges to the ground freezing design-build contractor. Stresses on the frozen ground vary with depth and the strength and water tightness of the interface between the frozen ground and existing structures is critical. For the East Side CSO Tunnel project, in Portland, Oregon a hand-mined tunnel was completed between two existing concrete slurry wall shafts at 42.7 m (140 feet) below ground surface. This paper discusses the design, installation and operation of the ground freezing system, together with QA/QC measures employed to ensure full closure and maintain structural competency of the frozen ground during the tunneling activities.

## BACKGROUND

Scheduled for completion in 2011, the Willamette River Combined Sewer Overflow Control Project, currently under construction in Portland, Oregon, will significantly reduce the frequency and volume of overflow into the Willamette River by providing additional storage and conveyance. The East Side CSO tunnel is the final major component of the overall project. The 6.7 m (22 ft) ID tunnel will be approximately 9,600 m (31,680 ft) long on completion. Seven new shafts along the tunnel alignment will transfer flows from existing outfalls to the tunnel. The tunnel terminates at the Port Center Shaft which connects to the existing Confluent Shaft between 34.4 m (113 ft) and 38.7 m (127 ft) below the surface by means of a 4.3-m (14-ft) high, 3.0 m (10 ft) wide, 3.7-m (12-ft) long tie-in. Ground freezing was utilized to provide groundwater cut-off and excavation support during hand-mining and construction of the tie-in.

The two shafts requiring the interconnection are in close proximity to the Willamette River. To complicate matters, the connection lies directly within the Troutdale formation, a permeable geologic layer consisting of sands, gravels and cobbles. The ground freezing system was therefore designed to take into account the following potential conditions:

- Persistent groundwater flows, caused by tidal effects, funneled along the path between the two shafts. Groundwater movement introduces heat into the soils being cooled and will delay, and sometimes prevent, full freezing of the ground.
- The development of trapped frozen water within the unfrozen soil volume that could ultimately generate undesirable pressures on the adjacent shaft structures.

## GEOTECHNICAL PROFILE

The Port Center Shaft has a 13.7 m (45 ft) finished inside diameter and a 17.1 m (56 ft) outside diameter and extends to a depth of approximately 42.4 m (139 ft) below the ground surface. No suitable impermeable horizon was identified beneath the shaft therefore a tremie concrete plug was installed below the tunnel invert inside the slurry wall. Subsurface conditions, based on borings in the area and excavation of the adjacent Confluent Shaft, are as follows:

- Artificial fill consisting of medium dense to dense poorly graded sand and silty sand interbedded with layers of wood debris and gravel to a depth of 6.7 m (22 ft).
- Sand/silt alluvium consisting of interbedded soft to stiff silt, organic silt, and silty clay with organics between 6.7 m (22 ft) and 36.9 m (121 ft).
- Gravel alluvium consisting of very dense poorly graded gravel with silt, sand, cobbles and boulders from a depth of approximately 36.9 to 41.8 m (121 to 137 ft).

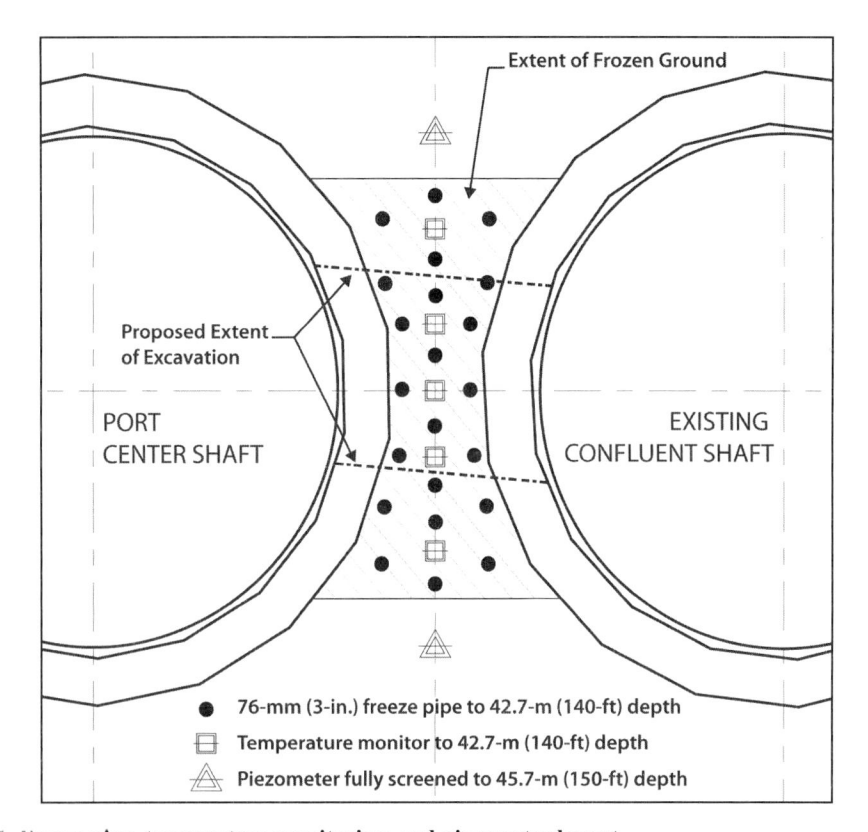

**Figure 1. Freeze pipe, temperature monitoring, and piezometer layout**

- Troutdale Formation consisting of very dense poorly graded gravel with sand, cobbles, and boulders from a depth of approximately 41.8 to 94.8 m (137 to 311 ft).

## THE GROUND FREEZING CONCEPT

When in situ pore water is frozen, it acts as a bonding agent, fusing together particles of soil or rock to create a frozen soil mass with significantly improved compressive strength and impermeability. Small-diameter, closed-end freeze pipes are inserted into vertical drilled holes in a pattern consistent with the shape of the area to be improved and the required thickness of the wall or mass. As the cooling agent, typically chilled brine, is circulated through the pipes, heat is extracted from the soil, causing the ground to freeze around the pipes. The brine is returned to the refrigeration plant where it is again cooled.

The frozen earth forms around the freeze pipes in the shape of vertical, elliptical cylinders. As the cylinders gradually enlarge, they intersect to form a continuous wall. With heat extraction continuing at a rate greater than heat replenishment, the thickness of the frozen wall will expand with time. Once the frozen wall has achieved its design thickness, the freeze plant may be operated at a reduced rate to maintain the condition during shaft or tunnel excavation and liner placement. Following excavation and completion of construction, refrigeration is discontinued, allowing the ground to return to its normal state (Powers et al., 2007)

## FREEZE SYSTEM INSTALLATION PROCEDURE

The design called for 22 freeze pipes, field located, to be installed in the area between the Port Center Shaft and the existing Confluent Shaft (Figure 1). In addition, five pipes designed to obtain ground temperatures and piezometric data were installed in proximity. These temperature monitors were field located based on the actual freeze pipe alignment. Exterior piezometer arrays were installed to monitor groundwater levels. A rotary drill rig utilizing a bentonite-based drilling fluid was mobilized for the installation.

Freeze pipes were installed within PVC insulating sleeves designed to focus the freezing effort between 30.5 and 42.7 m (100 and 140 ft) below ground surface. The finished hole dimensions were sufficient to allow installation of steel pipes previously capped at their lower ends. Pipe connections were made by welded joints and pressure tested to approximately three times their working pressure. Directional surveys of all holes were made using an inclinometer-type sensor mounted on lightweight, single-axis alignment rods. A high-density, flexible polyethylene tube was inserted in every freeze pipe so that its open end was suspended within a short distance from the steel pipe base cap. A freeze head assembly enabled the chilled brine to enter via the polyethylene tube and return up the annular space between the inner tube and the outer casing.

Temperature geotechnical monitoring holes were specially equipped to provide the temperature data required for monitoring the progress of the ground freezing operation. Temperatures were measured at several pre-selected depths by stationing a multi-wire thermocouple harness in a screened PVC pipe similar to the piezometers. The temperatures were recorded daily and inputted into a spreadsheet for computer graphing for home-office analysis. Groundwater elevations were recorded from data logging probes installed inside the temperature monitoring holes for computer graphing and analysis. The monitor holes also acted as vertical pressure relief for any trapped water within the freeze zone.

**Coolant Circulation System**

The coolant was a calcium chloride solution with a freezing point safely below the coldest anticipated temperature in the refrigeration plant. Chilled brine was pumped to the freeze location and returned to the refrigeration plant using custom built, pre-insulated manifolds arranged between the two shafts. From these manifolds, 38-mm (1.5-in.) diameter, valved supply and return hoses connected individually to the freeze head assemblies. Monitoring of the brine circulation system consisted of supply and return temperature measurement, pressure measurement, brine flow measurement and air purging devices. The brine networks formed a closed system with the addition of a brine holding tank installed to displace brine into air pockets purged from the system. The brine holding tank also served as a leakage detection device.

The refrigeration plant was fitted with two centrifugal brine pumps to provide efficient heat transfer both at the freeze pipes and in the chiller within the refrigeration plants. Flow-actuated switches acted to automatically protect the system in the event of pump failure.

**Refrigeration Plant**

The main components of the refrigeration unit used were:

- A single-stage refrigerant compressor.
- An evaporative condenser and potable water circulating pump.
- Refrigerant expansion control valves and oil recovery systems.
- A multi-pass refrigerant-to-brine heat exchanger or chiller.
- Centrifugal brine pumps.
- 110V electrical control system with automatically actuated fail-safe monitors, and
- Auto dialer in the event of power outage or plant shut down.

The freeze plant mobilized for this project was a custom-built, fully automated unit requiring field connections to external brine pipelines, cooling water and electric power lines. The maximum power requirement was approximately 500 HP. The freeze plant set-up is shown in Figure 2.

The average cooling water requirement for the refrigeration plant was approximately 94.6 L/min (25 gpm) and was furnished from city water mains. The cooling water was non-contact and was discharged to the city sewer system.

## CRITICAL DESIGN AND CONSTRUCTION ELEMENTS

### Geotechnical Design Considerations

The ground freezing system was designed to take account of, and mitigate, possible persistent ground water flows and the potential for development of trapped water within the unfrozen soil structure. There are several methods of handling excessive groundwater flows and the choice is best facilitated by an early detection monitoring program. Groundwater moves in response to hydraulic gradients and these can be monitored by comparing piezometric levels along a suspected flow path. Accordingly, piezometers (screened over the vertical extent of the future tunneled connection) were established in the throat of the funnel and on each side of the soil freeze area. The piezometers were active for at least a week before freezing began and their reaction pattern throughout each tidal cycle were monitored and compared. Hypothetically, when the direct hydraulic connection between the two outer piezometer screens is sealed by the frozen structure a change will be evident based on the reduced attenuation of the tidal fluctuation.

To reduce the possibility of unfrozen, saturated soils becoming isolated within the frozen mass, the geometric layout of the freeze pipes and the order in which they were activated was carefully considered. The first stage of freezing occurred within freeze pipes grouped within the center of the frozen structure. The central, closely-spaced pipes generated a core of frozen ground and as the freeze zones

**Figure 2. Freeze plant set-up**

continued to grow, unfrozen water found relief between the wider adjacent spaced pipes. After seven days, the second stage of the freeze pipes was activated, allowing the formation process to advance. After another five days, the third and final stage of freeze pipes was activated. The fully activated system is shown in Figure 3. During the formation period, interior piezometers and temperature monitors were monitored closely for signs of trapped water and both shafts were closely monitored for any movement.

**Structural Design Considerations**

*Frozen Soil Strength*

At a target soil core temperature of –15°C (5°F) and after nine weeks of freezing, the strength of the frozen soil was estimated to be approximately 4136.8 kPa (600 psi). This was based upon recent frozen soil testing results from a project in Seattle, WA in a very similar geologic setting with silt, sand and gravel with cobbles. The testing yielded a frozen compressive soil strength of approximately 4474.7 kPa (649 psi) after nine weeks (Brightwater Conveyance System: Geotechnical Baseline Report, 2006).

*Frozen Soil Creep*

Frozen soil has a tendency to creep over time. Creep is usually a consideration if the frozen soil is under stress for greater than eight weeks. Since the duration of the tunneling operations was approximately eight weeks, there was little concern with significant creep of the frozen soil, and in fact, no creep was observed during the tunneling.

**Figure 3. Activated freeze system**

*Lateral Forces on Slurry Walls*

Prior to construction, it was decided that potential lateral forces exerted on the adjacent slurry walls as a result of the ground freezing process were not quantifiable since no practical methodology exists to calculate such loadings. Forces due to frost expansion are the result of the formation of ice lenses and the corresponding volumetric increase associated with the phase change from water to ice. This condition can occur in a geologic setting with silty gravels, silty sands or silty clay soil. The expansion pressures developed are a complex relationship of permeability, geology and segregation potential.

It was determined that the following conditions present at the site served to mitigate the potential for excessive loading on the adjacent slurry walls:

- The top 30.5 m (100 ft) of soils was largely unfrozen, as freeze pipes were installed with an insulating casing.
- Frost susceptible soils represented only approximately half of the proposed frozen zone. Furthermore, all of the silt layers in the Troutdale Formation contained sand.
- The zones above and below the frost susceptible soils had a higher thermal conductivity and would tend to freeze more quickly, thus significantly reducing the source of groundwater recharge to the soils with volumetric expansion potential.
- The tremie slab and its close proximity to the frozen zone will act as a strut, providing additional structural support to the slurry wall.
- The bottom 9.1 m (30 ft) of the final concrete lining would be poured prior to the start up of the ground freezing. This, too, would provide additional structural support for the Port Center Shaft during the ground freeze and the tie-in tunnel excavation.
- The short duration of the potential loading (8 weeks to reach formation and approximately 10 weeks to complete the adit) and the progressive nature of the freeze.

An observational approach was recommended to be taken regarding the potential lateral loading issue. It was decided that both the adjacent slurry wall shafts be monitored for concrete strain during the formation of the freeze. Monitoring locations included approximately half of each of the shaft perimeters. Monitoring consisted of surveying fixed points set into the interior surfaces of the Port Center Shaft slurry wall and the top surface of the Confluent Shaft.

*Insulation of the Frozen Excavated Face*

A spray-on, cellular plastic foam insulation was recommended for insulation of the frozen excavated faces. Installation of the insulation consisted of a base of reinforcing wire attached to the frozen face which acted to rigidly affix the insulation to the frozen wall.

*Structural Analysis*

The structural design considered a frozen block of soil with the outer dimensions of 12.2 m (40 ft) tall by 11.0 m (36 ft) wide. In order to reduce the complexity of the calculations, the design analyzed the stresses of a frozen circle inscribed within the 12.2-m by 11.0-m (40-ft by 36-ft) block with an effective circular opening of 7.8 m (25.6 ft) to account for the rectangular excavated tunnel dimensions of 6.1 m (20 ft) high by 4.9 m (16 ft) wide and a uniform pressure as defined by the highest pressure on the circle. Factors of safety of 2.3 and 1.8 respectively were calculated for hoop stresses generated within the structure and 1.8 for stresses generated due to the structure deformation cause by eccentricity generated by the structure deforming. Once excavation occurred, the frozen structure had grown at least two times thicker than was required, thereby doubling the factor of safety.

**Thermal Design**

*Groundwater Movement*

Because of the tidal influence and close proximity to the river, groundwater movement was considered in the thermal design. To obtain estimates of the times required for freezing, the ground freezing contractor adapted the empirical formulae of Sanger and Sayles which are conservative, and have been used successfully over the last 30 years (Sanger and Sayles, 1979). The thermal design also considered closer than typical freeze pipe spacing and higher than typical refrigeration capacity in an effort to overcome the tidal fluctuation in a reasonable time. As a result, the tidal fluctuation was quickly overcome. Surprisingly there were no significant groundwater gradients observed across the site as was expected and therefore formation was unhindered in that respect.

*Final Concrete Liner*

The method of construction for the Port Center Shaft included the installation of a final concrete lining installed just prior to activating the freezing operation. The thermal design examined the heat gain from the concrete lining and estimated that an additional 15 days of freezing time would be required. Because of the relatively small area and the amount of refrigeration capability available, the heat gain from the lining was overcome easily. The ground freezing contractor also recommended against using "high early" strength concrete for the lining as this would generate a higher temperature spike in the concrete mass. It was recommended that a slow release of heat-of-hydration concrete be utilized.

## INSTRUMENTATION AND MONITORING

Freeze pipes were surveyed for location at ground surface and also surveyed with an inclinometer for subsurface location. Subsurface plots were generated to depict the location of each freeze pipe with depth.

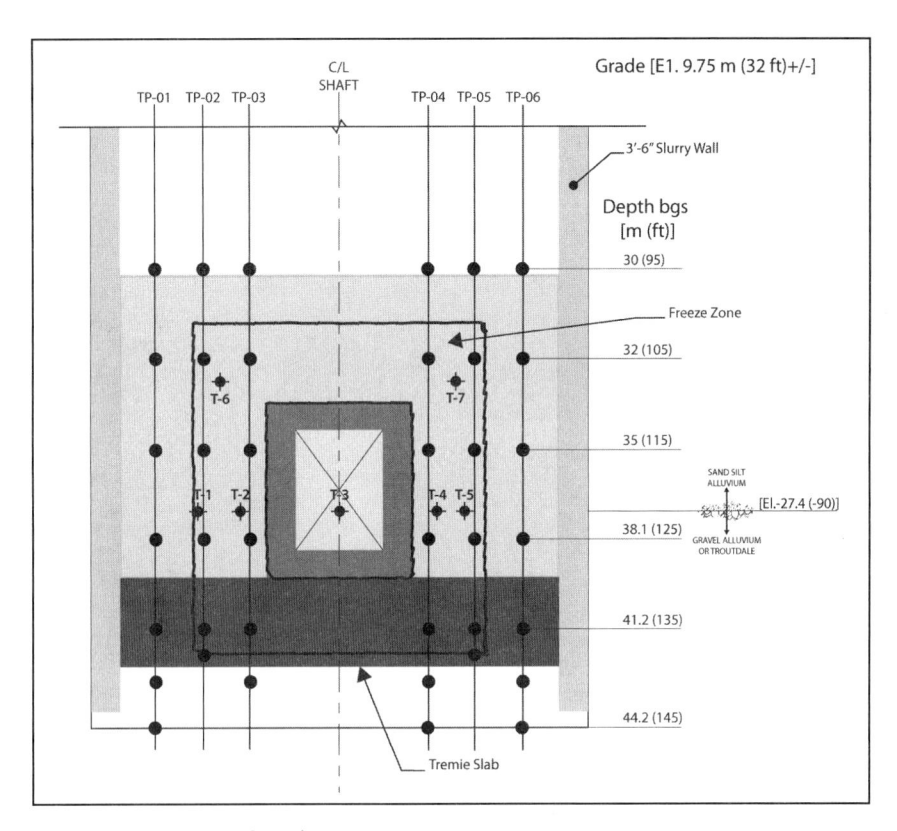

**Figure 4. Temperature monitoring pipe and sensor locations**

Temperature monitoring consisted of pipes drilled in or adjacent to the frozen barrier at selected locations and fitted with temperature sensors at approximately 4.6 m (15 ft) intervals within and adjacent to the frozen zone (Figure 4). The exact location of each temperature monitor was selected once the subsurface plots indicating freeze pipes locations were generated. Temperatures were recorded manually. Piezometers were installed outside the freeze zone to monitor the groundwater regime at the site and ultimately the formation of the frozen wall.

Growth of the frozen wall was determined by a combination of the data generated from the instrumentation. Formation of the frozen wall was most clearly determined through the analysis of the temperature monitor data. Graphical plots of temperature versus time and temperature versus distance were developed to analyze frozen wall growth. Monitoring of the groundwater data also indicated growth of the frozen structure. As the growth of the frozen mass increased, the tidal fluctuation slowly dampened because of the cut-off created.

## EXCAVATION THROUGH FROZEN GROUND

The tie-in adit between the Port Center Shaft and the Confluent Shaft was approximately 3.7 m (12 ft) long. This length included the two 1.1-m (3.5-ft) thick concrete slurry walls and approximately 1.5 m (5 ft) of frozen soil in between the two slurry walls. The finished concrete dimensions of the tie-in adit were designed as 4.3 m (14 ft) high by 3.0 m (10 ft) wide. The designed thickness of the concrete lining was 0.9 m (3 ft). The excavation cross-section was therefore 6.1 m (20 ft) high by 4.89 m (16 ft) wide.

To complete the tie-in tunnel excavation, the general contractor performed the following major activities:

- Removal of the Port Center Shaft slurry wall.
- Excavation and insulation of the frozen soil.
- Application of shotcrete as required for additional temporary ground support.
- Re-plumbing of the freeze pipes inside the excavation cross-section in order to maintain the freeze.
- Removal of the Confluent Shaft slurry wall.

**Figure 5. Coring in preparation for wire saw cutting**

### Removal of the Port Center Shaft Slurry Wall

When the temperature monitors showed that the soil mass within the tie-in tunnel excavation had reached the target temperature of –15°C (5°F), the contractor began removal of the concrete slurry wall at the Port Center Shaft. Initially, it was planned to use an excavator with a hoe-ram hammer attachment to demolish the slurry wall. However, given concerns about damaging the freeze pipes and the final concrete lining the decision was made to wire saw the slurry wall.

The opening in the slurry wall would be about 6.1 m (20 ft) high by 4.9 m (16 ft) wide. This reinforced concrete slurry wall panel would weigh nearly 68.0 tonnes (150,000 lbs) and could not be cut and removed in one piece. The Contractor therefore divided the opening into four quadrants to be removed one at a time, starting at the top. But first, the wire saw required 254-mm (10-in.) diameter core holes at each corner of each cut, extending through the slurry wall and approximately 1.2 m (4 ft) into the frozen ground (Figure 5). A total of nine holes were required.

The first core hole confirmed that the ground was indeed frozen. After all nine of the 254-mm (10-in.) diameter holes were cored, approximately 3.0 m (10 ft) of fill material was placed at the bottom of the shaft for safer access to the two upper slurry wall panels [still 3 m (10-ft) high]. The wire saw unit was then anchored to the first panel and completed a "plunge cut" through the reinforced concrete slurry wall. After all four cuts were made in the upper left panel, steel wedges and greased teflon sheets were inserted in each of the cuts. Four hoist rings were then anchored to the panel [3.0 m (10 ft) tall by 2.4 m (8 ft) wide] and rigged to a 12.2 m (40-ft) sling reaved through a 54.4-tonne (60-ton) load block attached to a 40.8 tonne (45-ton) sling that was cast into the

**Figure 6. Cut slurry wall panel prepared for removal**

concrete shaft lining opposite the tie-in adit. With the hoist block reaved and secured, the main fall block of a 158.8-tonne (175-ton) crane at the top of the shaft was hooked to the loose end of the sling. The crane then pulled the slurry wall panel horizontally about 6.1 m (20 ft) until it could be re-rigged and hoisted out of the shaft for demolition and load-out at the surface (Figure 6). When the first panel did not come loose immediately, an inflatable jack pack bladder was inserted in the bottom cut to provide additional force to break the panel loose. The upper right slurry wall panel was removed in precisely the same manner. The frozen soil behind the slurry wall stood up well. Nonetheless, the contractor applied insulation foam and blankets onto the face and proceeded with removal of the two bottom panels. The 3.0 m (10 ft) of backfill had to be removed first but the wire saw was anchored to the bottom panels and completed the "plunge cuts" similar to the work on the upper panels. After the two bottom panels were removed, the 6.1 m by 4.9 m (20-ft by 16-ft) excavation face was fully exposed.

Figure 7. Excavation of frozen soil

**Excavation of the Frozen Soil**

An excavator with a hoe-ram hammer attachment was used to excavate the frozen soil (Figure 7). Before excavation began, however, the freeze pipes inside the excavation limits were turned off and flushed. These freeze pipes would be carefully exposed, cut and re-plumbed in order to maintain the freeze until the concrete liner could be completed. Excavation proceeded from the top down. As excavation approached a freeze pipe, the contractor hand-mined around the freeze pipe with a chipping gun and used a 68,947.6-kPa (10,000-psi) pressure washer to remove any remaining soil. After all of the freeze pipes were exposed (Figure 8), they were cut and re-plumbed as described in more detail below.

**Application of Shotcrete as Required for Temporary Ground Support**

While the ground freeze design concluded that temporary ground support measures would not be required during excavation of the tie-in tunnel, the contractor planned to apply a 127-mm (5-in.) thick layer of shotcrete to the exposed excavation face at the end of each shift. During excavation, however, the frozen ground stood up well so this contingency measure was not employed. When the excavation was completed, prior to cutting into the Confluent Shaft slurry wall, the contractor did apply a 127-mm (5-in.) thick layer of shotcrete to support and insulate the exposed frozen soil surface.

**Re-plumbing of the Freeze Pipes to Maintain the Freeze**

The freeze pipes inside the excavation limits were carefully exposed with a chipping gun and 68,947.6-kPa (10,000-psi) pressure washer. The steel freeze pipes were cut about 152.4 mm (6 in.) below the excavation crown and about 152.4 mm

Figure 8. Exposed pipes showing insulation above and shotcrete on walls

(6 in.) above the invert. Freeze heads were installed at the top and bottom of each pipe and connected with rubber hoses. After the hoses were secured to the excavation walls, the freeze pipes were pressure tested and refilled with coolant. The system was then brought on line slowly to avoid thermal shock to the equipment at the freeze plant. Once on line, the flow was adjusted from the manifold at the surface to equalize flow to all of the re-plumbed freeze pipes. The re-established freeze was then maintained until excavation and concrete was completed.

**Removal of the Confluent Shaft Slurry Wall**

The contractor planned to remove the Confluent Shaft slurry wall with a wire saw in the same manner as the Port Center Shaft slurry wall. However, the Confluent Shaft was part of a live CSO system. As such, a 50.8-mm (2-in.) diameter probe hole was cored through the slurry wall to check for water on the other side. With flow inside the Confluent Shaft determined to be minimal, coring of the 254-mm (10-in.) diameter holes required for the wire saw proceeded. Since there was free air on the other side of the Confluent Shaft slurry wall (as opposed to frozen soil on the other side of the Port Center Shaft slurry wall), it was possible to "pull cut" with the wire saw instead of "plunge cutting." Pull cutting was a significantly faster way of cutting through the concrete; but access was required to the Confluent Shaft side in order to rig the wire saw properly. The contractor therefore "plunge cut" an access panel about 1.2 m (4 ft) tall by 1.2 m (4 ft) wide in the middle of the Confluent Shaft slurry wall face (Figure 9). This opening allowed the wire saw to be rigged to complete the remaining "pull cuts." The Confluent Shaft slurry wall was removed in a five panels instead of just four so additional core holes were required but the panel removal process was the same.

With the tie-in tunnel excavation completed, the contractor then proceeded with construction of the reinforced concrete lining. The excavation and concrete placement took a total of 10 weeks to accomplish. After the concrete lining was completed, the freeze system was turned off, decommissioned and demobilized.

**Figure 9. Access panel to the active Confluent Shaft to check for water presence**

## CONCLUSION

Ground freezing for the horizontal connection between the Port Center and Confluent shafts presented a number of challenges, both in the freeze design and implementation, and in the excavation and completion of the shaft. These challenges were overcome with careful planning by all parties and by comprehensive instrumentation and monitoring during the freeze formation and connecting adit excavation and completion. The experience of the ground freezing contractor and the general contractor, and close cooperation throughout the work, were instrumental in the success of this project.

## ACKNOWLEDGMENTS

The authors extend their thanks and appreciation to the staff and management of Moretrench American Corporation and Kiewit-Bilfinger Berger Joint Venture for their cooperation and support in the production of this paper, and to Derek Maishman of Moretrench and Steve Spencer of GZA for their contribution to the successful completion of this project.

## REFERENCES

Powers et al. (2007). *Construction Dewatering and Groundwater Control: New Methods and Applications*, John Wiley & Sons, New York, NY. pp 508.

Sanger, F.J and Sayles, F.H. (1979). "Thermal and Rheological Computations for Artificially Frozen Ground Construction." Proceedings of the 1st International Symposium on Ground Freezing, Bochum, Germany.

Brightwater Conveyance System; Geotechnical Baseline Report (2006). "Table 6-4, Engineering Properties of Frozen Soil."

# Final Lining at Devil's Slide Tunnel

Stephen Liu
Kiewit Construction, Pacifica, California

## INTRODUCTION

The purpose of this presentation is to explain and illustrate the complexity and challenges of constructing the final arch lining operation while performing tunnel excavation at Devil's Slide Tunnel.

This presentation will cover the following topics:

1. Surface preparation and surface application operations initiates the lining process and sets the pace for follow-on work.
2. Subcontracted work involving the coordination and installation of waterproofing; rebar and electrical and mechanical embeds.
3. Erection of 500 tns of steel gantry and formwork to be utilized behind the smoothing and subcontract operations.
4. Arch Concrete Placement Cycles
5. Interfacing and combining formwork involving niches, cross passages, and final arched inverts.

Tunnel excavation while mining is not often performed in the North America primarily due to access constraints in both operations. Referencing European approaches for long distance tunneling, the Devil's Slide Tunnel Project has applied innovation and technology to provide the ability to overcome this unique challenge. The Devil's Slide tunnel requires the construction of twin bore tunnels excavated based SEM (Sequential Excavation Method) construction methods using NATM principles. The twin tunnels are approximately 30' in diameter and 4,100 lf in length. The purpose of this project is to re-route Route 1 from around the Devil's Slide through the San Pedro Mountains crossing from Montara, CA along the south and into Pacifica, California in the north.

The Caltrans project was awarded in December 2006 for a total of $272,366,000. The job is an A+B job thus the estimate scheme of final lining while tunnel excavation is important to save time. A+B is a method of rewarding a contractor for completing a project as quickly as possible. By providing a cost each working day, the contract combines the cost to perform the work (A) with the cost of impact to the public (B) to provide the lowest cost to the public.

Concurrent mining while lining is achieved through a series of gantries and a set of tunnel forms allowing the pass-through of the excavation equipment. The tunnel lining operation passes through several phases of work that is dependent upon the progress of tunnel excavation process. This involves unique milestones such as the evolution of the ventilation system, development of an enlarged cross passage to allow equipment to access each tunnel from within the tunnel and finally the installation of an invert prior to arch lining that limits the continuity of the final lining operation.

## TUNNEL FINAL LINING

The tunnel liner is a 15" thick liner using a 4000 psi concrete pea-stone mix. Rebar reinforcement calls for double matted rebar. Both inner and outer mat is detailed at #4 rebar at 6" spacing. The electrical embed installation calls out for an average of 475 LF of conduit per 40' pour along with an average of 4 junction boxes per block. Each concrete "block" pour is 40' (12m) in length. These block lengths were chosen based on the location of a series of 42 repeating service niches, formation drain cleanout (FDC) niches as well as the 10 cross passages located throughout the alignment of the tunnel. With every block pour, all of the niches and cross passages are projected to fall within the limits of each of block pour. There are a total of 200 block pours, 99 on the Southbound Tunnel (SBT) and 101 along the Northbound Tunnel (NBT).

### Sequence of Installation

The entire final liner goes through many installation steps before the final product is achieved. The following is a brief description of the construction sequence for an independent final lining, block pour:

1. The shotcrete initial liner is prepared via smoothing shotcrete. This involves both the surface preparation and the application of smoothing shotcrete.

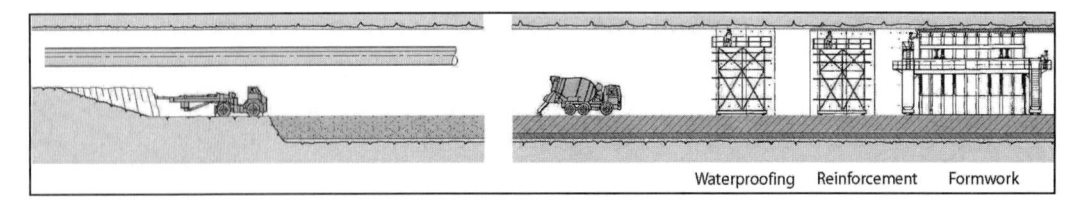

**Figure 1. Service gantry**

**Figure 2. Excavation equipment drive through**

2. This is followed by the installation of 24" abutments that span the length of the tunnels.
3. A 10" continuous formation drain is cast outside the abutments which ties in the PVC waterproofing membrane.
4. Once the smoothing and surface preparation steps are completed, a PVC waterproofing membrane is installed along the arc-length of the initial liner. This waterproofing membrane ties into the abutment and over the formation drain system to allow water seepage into the drain system.
5. Once the waterproofing membrane is installed, a contact grout system is attached to the membrane along the arches.
6. This is followed up by the installation of a double mat of rebar over two distinct steps. Once before the electrical sub installs the conduit and once after the conduit is installed.
7. The concrete is then placed into the arch liner forms with the mechanical and electrical conduits, anchors and junction boxes embedded.
8. Prior to the arch forms arriving, on every 5th pour a set of FDC and service niche forms are erected using separate formwork.
9. On every 10th arch pour, a cross passage adapter is mounted on the Cross Passage (CP) inverts in order to mate it with each passing arch pour.
10. Stretched across various locations throughout each tunnels are areas where a full arched invert must be placed in order for the continuity of the arch concrete to pass through these areas. Thirty-five percent (35%) of the length of the tunnels are estimated to require a full arched invert.

Waterproofing the initial shotcrete liner involves a detailed process of both surface preparation and surface application of smoothing shotcrete. The contract requires the application of a universal 1" layer of NFRS (non-fiber reinforced shotcrete) to cover any protrusions or sharp edges. Another criterion is the depth to wavelength ratio of the initial liner which helps ensures pronounced dips and undulations are smoothed out, thus preventing the waterproofing from stretching too far. Upon final surface acceptance, a 1" thick PVC membrane is pinned to the initial liner in a series of 2×75' transverse strips. These strips are then heat welded together. Upon its completion, a series of template anchors are installed through the PVC membrane to serve as the template bar for the rebar mats that are installed immediately after this.

Following the lining placement the Mechanical and Electrical packages represent a large portion of work post tunnel concrete. The mechanical package requires the installation of a 12" DIA water main and a 10" fire main down the length of each tunnel. The electrical package consists of 6 sets of electrical plans (utility and grounding plan; lighting plan; power plan; traffic system plan; fire life safety plan; instrumentation plan) . These 6 systems consist of approximately 130,000' of conduit and

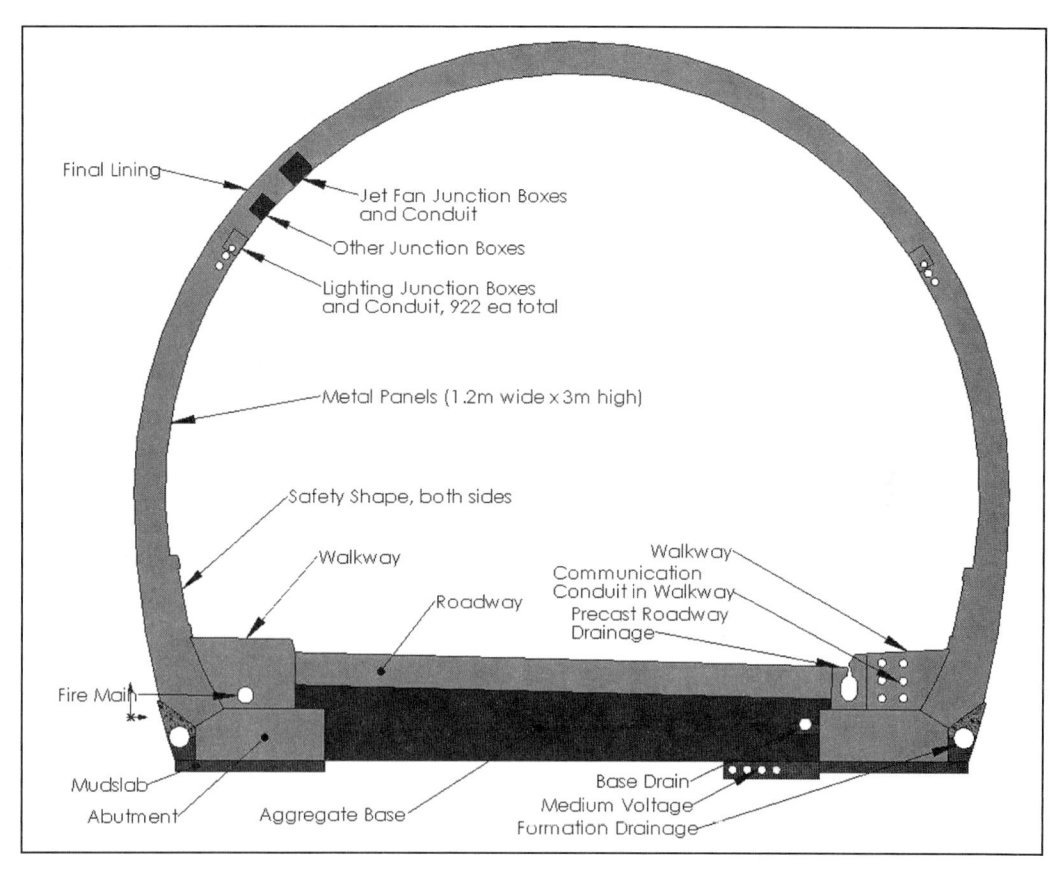

Figure 3. Devil's slide tunnel liner schematic

Figure 4. Waterproofing membrane installation

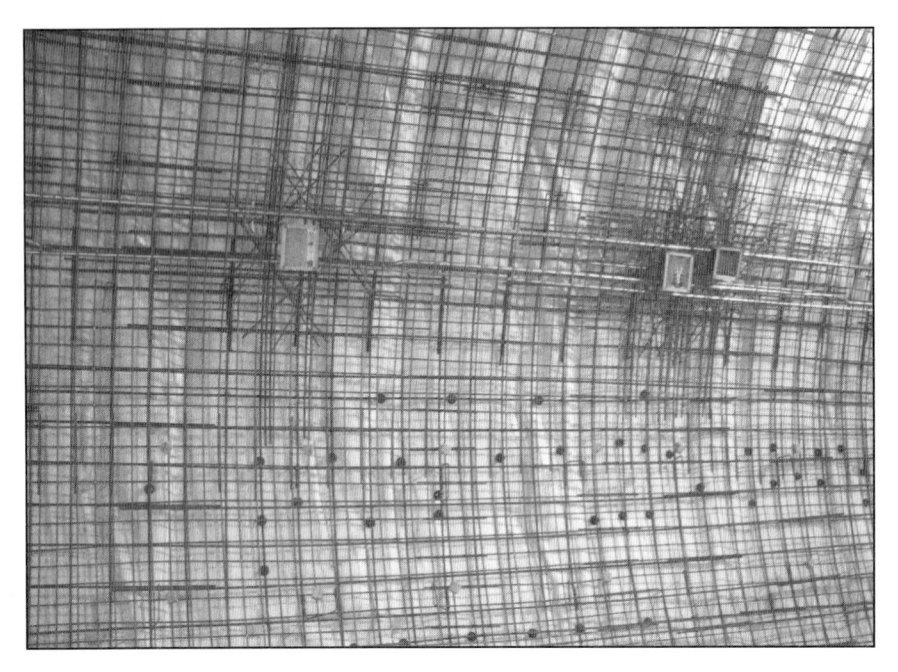

**Figure 5. Electrical embeds showing continuous conduit and junction boxes**

approximately 800 junction boxes of various sizes. A thorough QC program for verifying embed location, quantity and its secure placement is critical to the success of the factory testing and commissioning stage of this project.

**Gantries and Tunnel Forms**

The Devil's Slide Tunnel purchased a series of 6 identical working gantries and 2 tunnel forms from Ceresola tls. All of the respective gantries and forms are fabricated out of steel.

Each working gantry is 50' in length, 26' wide and 30 tns in weight. The advantage to having large gantries of this size allows for increased material stocking of up to 20,000 # and the ability to have up to 10 persons working simultaneously. Up to 6 working gantries are planned to be in use in both tunnels. This allows for concurrent work to occur for waterproofing installation, rebar installation (inner vs. outer mat); mechanical and electrical embeds. The 4 post gantries are run hydraulic over electric using a 25kw diesel generator which is incorporated in the body of each gantry. They are designed to travel up the 2% slope of the tunnel at a maximum speed of 20'/minute. There are several unique features to the gantries. Access can be made of several different levels along the arch working at different levels concurrently; second, a 24' long trolley is mounted onto an I-beam that can stock both the rebar bundles as well as electrical conduits for a capacity of 20,000#.

The tunnel arch forms are 130 tns in weight and 40' long. The forms also are mounted on a four post carrier frame that rides on 85# rail that is gauged at 20'. The tunnel form operates hydraulic over electric and is capable of 1700 psf of concrete loading. The steel form is hinged at 2 locations that allow for stripping of the tunnel forms. The forms are secured during a concrete pour using 8 jacked legs and two sets of float pins—four on each end of the form. The unique aspect of these float pins are that they are placed entirely outside each pour. One set on the trailing end on pre-existing concrete. The other set on the initial liner. Once the arch concrete is placed, the forms are stripped and moved forward in its entirety with all of the form accessories and bulkheads and re-established in the next pour.

The procurement process for both the gantries and the tunnel forms required many months of pre-engineering and pre-planning. Each gantry needs to be flexible enough to perform different tasks such as storage of rebar as well as the installation of it. Meanwhile, the electrical and mechanical subs need the gantries to install embeds. Each gantry was analyzed and determined to have specific requirements that matched its intended operational goals based on the original estimate. Meanwhile specific height and width clearances need to be addressed to not only allow traffic to pass underneath but also be capable of sustained utility connections throughout the gantry operations.

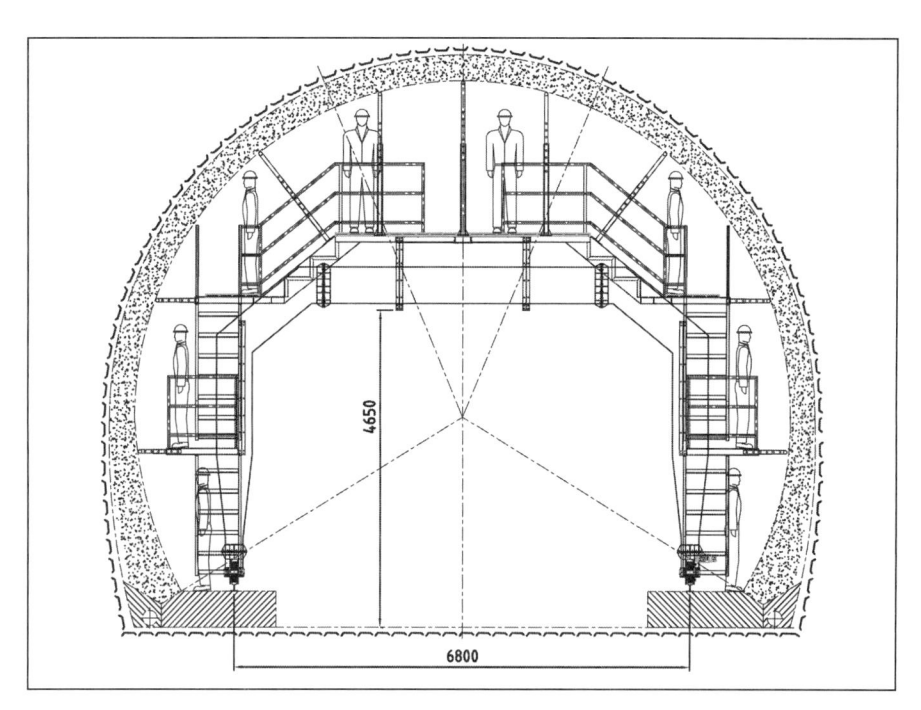

Figure 6. Service gantry schematic

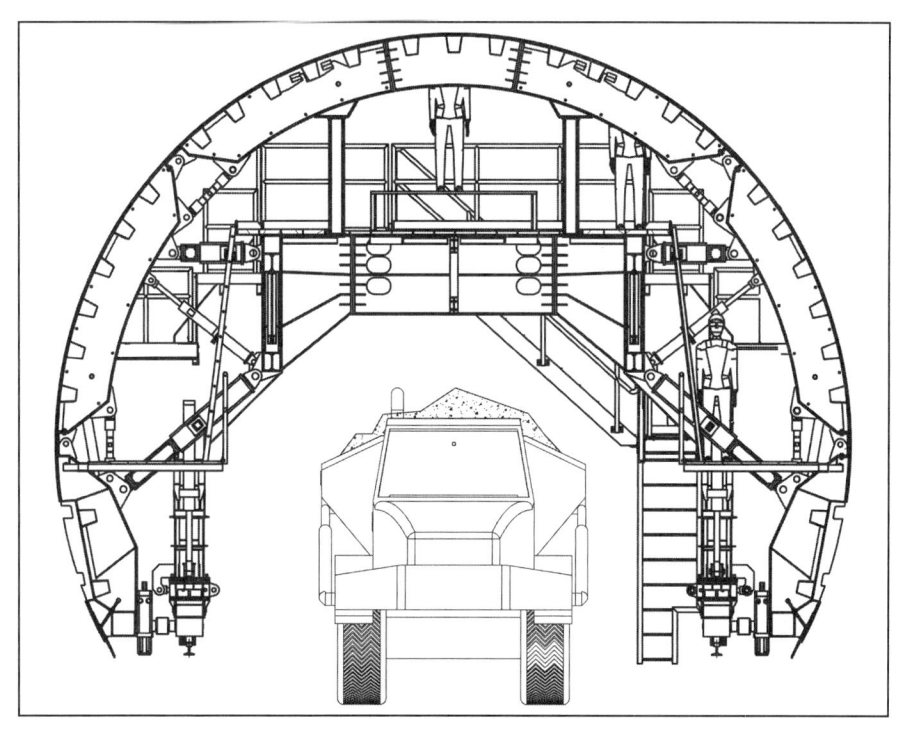

Figure 7. Tunnel arch form with equipment driving through

**Figure 8. Textured panels attached to arch tunnel forms**

**Figure 10. Fantail bulkhead braced by C-channel and coil rod**

**Figure 9. Fantail bulkhead**

The arch forms required even more engineering and attention to detail as each form is intended to perform 100 concrete placements. It also had to sustain a liquid head around the entire perimeter of the forms while providing for a bulkhead design that allow both the rebar and the electrical/mechanical embeds to pass through. Along the lower quarter arch, a textured elastomeric panel was mounted onto the side elements to produce a textured look above the safety walkway. In addition to this requirement was the need to be able to remove these panels on a regular basis (every 5th pour) to allow for the installation and concrete placement of our niche formwork that repeated 42 times in the tunnel.

The forms were also designed to accommodate the differential settlement post the pour. Based on past experiences with this type of lining, the forms were oversized by 18 mm to allow for this settlement. Other unique features was the inclusion of a steel chamfer system that produces clean defined CJ's around the perimeter of each joint.

A key takeaway in the procurement stage is ensuring that all intended users of either the gantry system or in the formwork system have thoroughly scrubbed the original design for their scope of work or involvement in the system. This includes everything from the supplier of the elastomeric panels and the arch formwork designer to the rebar installer and the gantry capabilities. Specific conflicts like fit-up, access, code regulation or integration of the intended installation process versus physical obstacles become difficult to manage once the process has started.

**Arch Concrete Placement**

The final liner concrete is placed via concrete mixer trucks and pumped using an electric Schwing 750. The concrete pump is capable of pumping up through a 5 inch DIA concrete slickline. The concrete is distributed through the formwork using a unique concrete distribution unit by ACME. The concrete distributor has eight outlets that is fed by a main turret that can swivel around and be connected to a series of ports that are plumbed to different guillotines. Concrete is deposited in the formwork through a series of guillotine ports that allow differential settlement along each side of the form. The forms are designed to accommodate 3' lifts loaded uniformly until the arch form is full. The leading end

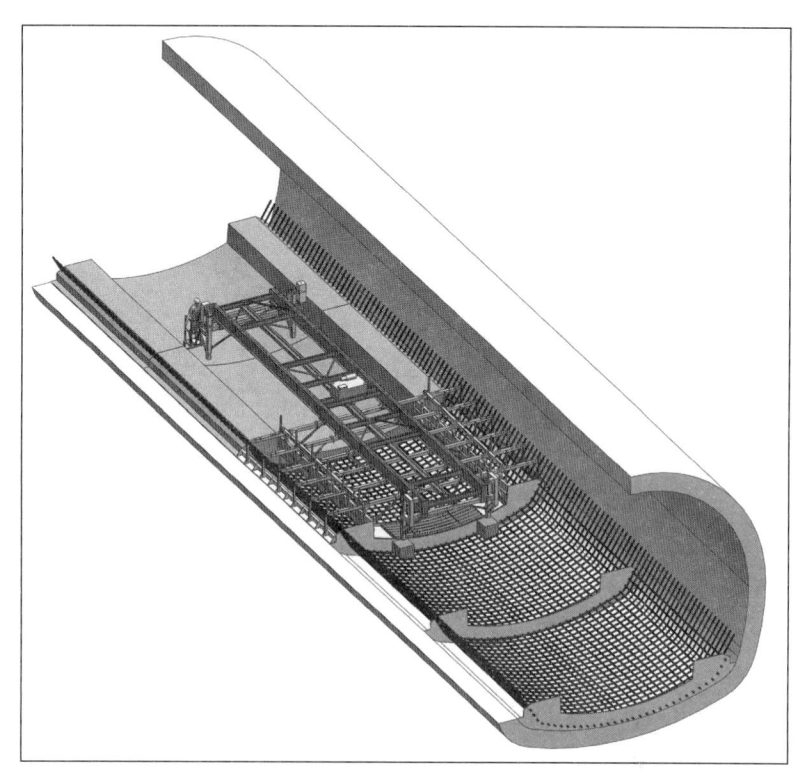

**Figure 11. Customized invert arch liner form shows the need to install the invert concrete prior to the arch form tunnel liner**

of each new block is contained using a 2×4 "fantail" bulkhead stacked radially along the arc-length of the tunnel form. These "fantails" are braced using C-channels and coil rod along its entire radius. The bulkhead must also allow for the double matted rebar to penetrate through along with all the conduit lines for the electrical and mechanical systems. Out of the entire erect and strip process, the majority of complexity and man-hours are directed to this specific piece of work.

## KEY CHALLENGES

### Access

One of the primary challenges for the final lining operation is access. Each tunnel drive contains a top heading excavation as well as a bench excavation spread. Each of these heading operations requires support equipment on a regular basis. The following are examples of challenges that are addressed daily:

- The construction of concrete abutments concurrent to the excavation process and the impacts to access during each respective cycle.
- A minimum of 500' is required to adequately park and stage all excavation equipment in each tunnel drive.
- Haul truck traffic is impeded when placing arch concrete
- Ventilation system orientation does not allow lining operations to proceed past ⅓ way through the tunnel (Cross Passage 3)
- The installation of a non-continuous final arch invert in each of the tunnels.

### Material Handling

The quantity of material to be installed and coordinated concurrently within each tunnel varies between subcontractor and between different operations. In order to balance the production advancement between subcontractors, each day is carefully coordinated between operations.

- Tunnel waterproofing requires the daily handling of either fleece or waterproofing membrane. These deliveries consist of stocking 2×75' rolls of waterproofing membrane and 10–12 rolls of fleece onto a tube scaffold.

**Figure 12. Staging waterproofing membrane on scaffolds**

**Figure 13. Final arch liner form set in the northbound tunnel**

- The stocking of #4 rebar × 30' long directly from the delivery truck onto the steel 15m long gantries.
- Requires the use of a intermodal trailer towing a flatbed of 30,000# twice-weekly during the day shift.
- Rebar is bundled into 2000# loads 30' long. 10 bundles are hoisted onto the gantries via a trolley hoist that lifts the bundles up onto the top deck of the gantries.
- Electrical conduit, junction boxes and other electrical embedded items are also required to be stocked weekly for the electrical subcontractor.
- The unique aspect of coordination between electrical and rebar subcontractor is keeping them well paced apart so that they do not run into each other.

## CONCLUSION

The complexity of final lining reflects the many challenges for building a final product "right the first time." The key lesson has been to plan the final lining operations with accurate detail to reflect the complexity of the work. This requires the need for detailed brainstorm sessions early in the pre-engineering phase.

The number of concurrent operations between different subcontractors and Kiewit operations all sharing the same space but working at different paces require constant surveillance and adaptation. Success involves early coordinated planning with the subcontractors and getting their buy-in on the proposed means and methods for access and material handling.

When choosing the right form design and engineering issues, it is important to ensure that the completed engineering design performs as intended. Taking advantage of key opportunities such as a *preliminary mockup* and having our key operations personnel stay with both the design and the fabrication phase ensures any fit-up conflicts are minimized before it departs the supplier.

Finally, the benefits to concurrent mining while lining provide a key advantage to future tunnel projects. Understanding the sequence and flow of work on what is achievable versus what is impractical will be a valuable lesson.

## GENERAL FACTS

- 23,670 m$^3$ of final lining concrete ; 2,200 m$^3$ of portal concrete
  - 135 NCY per 12 m placement (8.6 m$^3$/TM)
- Concrete mix design
  - 7 sack mix (4000 psi)
  - ½" and ⅜" aggregate
  - Chemical admixtures (VMA and Glenium) for dealing with high slump
  - High early strength testing and monitoring (strip at 8 MPa)
- Placement rate (30 CY per lift) @ 1 M displacement per side
- 2,331,000 kg of rebar (1604 tns) OR 400 #/LF
  - Outer mat is #19 @ 325 mm spacing (variance from Specification)
  - Inner mat is #13 @ 150 mm spacing
- 65,900 m$^2$ of waterproofing
- Cross passage concrete (10 each @ 17 meters long)
- 3 chambers (SEC, CEC, NEC)
- Curing requirements
  - 7 day form in place method
  - Curing compound

# The History of Tunneling in Portland: Rail, Highways, and the Environment

**Susan L. Bednarz**
Jacobs Associates, Portland, Oregon

**Paul T. Gribbon**
City of Portland Bureau of Environmental Services, Portland, Oregon

**Joseph P. Gildner**
Sound Transit, Seattle, Washington

**ABSTRACT:** Tunneling in Portland mirrors the industrialization and urbanization of America. Beginning with rail tunnels in the early 1900s, tunneling has evolved as a tool to protect the environment by reducing combined sewer overflows into the Willamette River. At least 14 tunnel projects exist in the Portland area, ranging from a 1909 rail tunnel to the East Side CSO Tunnel, currently under construction. The variety of tunneling methods used to construct these tunnels reflects the diverse local geology, ranging from basalt bedrock to open gravel and boulders to soft silt. Challenging ground conditions have led to tunneling innovation, including the first use of a slurry mixshield tunnel boring machine (TBM) in North America and the longest microtunneling drive in the United States.

## INTRODUCTION

Portland tunnels have been built for heavy rail, highways, and light rail; however, some of the largest of the tunnels have been built to convey wastewater for treatment and to prevent combined sanitary/storm sewer overflows into the Willamette River and Columbia Slough. Diverse tunneling methods have been used to tunnel through Portland's complex geology, including excavating by drill and shoot, slurry TBMs and microtunneling tunnel boring machines (MTBMs), hard rock TBMs, and earth pressure balance (EPB) TBMs and MTBMs, TBMs run in open mode, shield, and hand excavation with and without ground freezing. Solving Portland's variable ground conditions has led to innovations that have benefited the tunneling industry.

## PORTLAND'S OLDEST TUNNELS

Beginning in the 1860s and ending around WWI, Portland folklore tells of "Shanghai Tunnels" in Portland's Old Town, which conveyed men out through connected basement passageways to sailing ships moored along the Portland waterfront. The men, who had been allegedly drugged in illicit boarding houses, were forced to work onboard as sailors (Fraizer 2001). This creative use of subterranean passageways illustrates Portland's early recognition of the usefulness of tunnels.

Portland's two earliest excavated tunnels were constructed by the railroads to facilitate passenger and freight traffic in and out of the city. The Oregon Railroad and Navigation Company built the 1,654-m (5,425-ft) long Peninsula Railroad Tunnel in north Portland from 1909 to 1911 (Anderson 2005). This concrete-lined tunnel, which cuts through the north Portland highlands, shortens freight movement north over the Columbia River to Washington State. The Peninsula Tunnel was excavated through catastrophic glacial flood deposits of gravel and sand above groundwater. The tunnel and approaches cost $800,000 to construct at the time ($74 million in 2009 dollars); however, no record of the tunneling method can be found. The tunnel was subsequently acquired by Union Pacific Railroad (UPRR), and it is still in use 100 years later.

In 1921, the 425-m (1,395-ft) long Elk Rock Tunnel was constructed in south Portland by Southern Pacific for the Red Electric East Side local passenger train. Originally, the Red Electric crossed a trestle along the base of a steep rock cliff along the west shoreline of the Willamette River. Rockfall onto the trestle became unacceptable when Mrs. Ella Newlans, the wife of the president of the Oswego Cement Company, was hit in the forehead

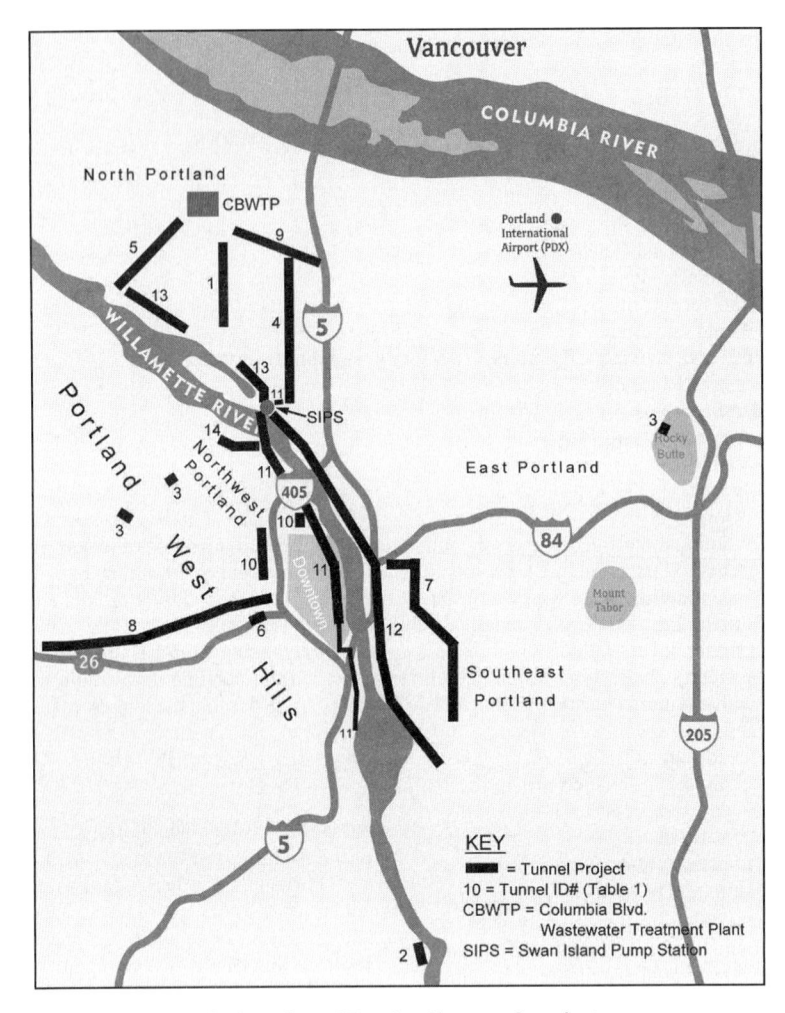

**Figure 1. Vicinity map showing the location of Portland's tunnel projects**

by a falling rock that crashed into her coach (*The Webfooter* 2008). Mrs. Newlans received several stitches, and the Elk Rock tunnel was excavated within the offending cliff to move passenger traffic out of harm's way. The Lake Oswego Trolley has used the Elk Rock Tunnel for many years, and it is currently being upgraded to expand Portland's extensive streetcar system to Lake Oswego.

These two early rail tunnels are precursors to the later wastewater, highway, and light rail tunnel projects that have been built along the east and west banks of the Willamette River, along the Columbia Slough, through the West Hills, and in east Portland.

## PORTLAND'S UNIQUE GEOLOGY AND GROUND CONDITIONS

Portland's unique and diverse geology has complicated the construction of both hard rock and soft ground tunnels over the last 100 years. Between 15 and 17 million years ago, Portland was inundated by thick Columbia River Basalt flows, which underlie the down-warped Portland Basin and the uplifted Portland West Hills (Tualatin Mountains). Approximately 13,000 years ago, the final catastrophic glacial flood scoured the Columbia River channel and the Portland Basin, leaving behind thick deposits of both coarse and fine flood deposits over most of the City's lower elevations. These glacial flood deposits form steep bluffs that parallel the east bank of the Willamette River north and south of Portland's downtown area. Portland is the only major metropolitan area in the continental United States that is populated by volcanoes. Interbedded volcanic cinders and basalt flows from these Boring Lava vents underlie potions of the West Hills and east Portland highlands. Thick deposits of windblown glacial loess, known as the West Hills Silt, blanket the Portland West Hills.

Since the end of the last glacial epoch, Portland's ancient river channels and lowlands have been infilled by sand and silt alluvium that has been deposited as the glaciers melted and sea level rose. Thick deposits of Quaternary Alluvium extend down to El. –42.6 m (–140 ft) beneath Portland's central eastside riverfront and Swan Island in North Portland. Between the Columbia River Basalt flows and the glacial deposits and recent alluvium, the Troutdale Formation gravel and the Sandy River Mudstone document channel and overbank sediments of a much larger, ancestral Columbia River. These old alluvial units have been weathered and eroded and were later scoured by the catastrophic glacial floods.

Tunneling ground conditions in Portland are as unique and diverse as the local geology. The Willamette and Columbia rivers' shorelines are underlain by very soft silt and sand that flows into excavations is prone to excessive settlement and complicates the maintaining of tunneling grade. Open matrix catastrophic glacial flood deposits below the groundwater level in lowlands preclude the use of EPB tunneling methods and cause excessive slurry loss with the use of slurry TBMs. Hard gravel, cobbles, and boulders in these deposits cause significant cutterhead and crusher wear. Cemented gravel horizons within the Troutdale Formation slow shaft excavation also cause excessive cutterhead wear. Weathered and faulted Columbia River Basalt in the West Hills contains highly fractured zones and closely spaced cooling, clay-filled joints that have a very short stand-up time.

## SUMMARY OF PORTLAND TUNNEL PROJECTS

At least 14 tunnel projects have been built in Portland over the last 100 years. These projects include highway, heavy rail, light rail, and water/wastewater tunnels. Table 1 provides a summary of available information on Portland's tunnel projects, including tunnel length, diameter, ground type, tunneling method, owner, contractor, and cost. Figure 1 shows the location of each of these tunnels. The following summaries provide highlights of each of the projects from the 1930s to present. The summaries highlight adaptations and innovations that were developed to mine through Portland's complex ground conditions. The completion date or expected completion date for each tunnel or tunnel project is shown in parentheses next to its name in Table 1.

### Early Portland Highway Tunnels (1939 to 1941)

The Federal Works Progress Administration (WPA) built three tunnels in Portland between 1939 and 1941 (Hadlow 2008). The West Burnside Tunnel and NW Cornell Road tunnels improved traffic through the Portland West Hills by straightening the roadways and removing steep grades. Rocky Butte Tunnel was built to improve residential and scenic access on Rocky Butte volcano in east Portland.

The Rocky Butte Tunnel, which took 16 months to complete, was excavated by WPA workers through Boring Lava cinders and basalt flows, using hand mining and drill and shoot methods, respectively. Tunnel muck was removed in one-cubic yard "Swede" cars on light-gauge rails, pulled by cable on a stationary winch. The portals of the curved 114.3-m (375-ft) long Rocky Butte Tunnel almost overlie each other due to the steep (5 percent) grade on the volcano. Special traveling steel forms were used to apply a 50.8-cm (20-in.) thick reinforced CIP concrete lining over the timber initial support.

The 74.4-m (244-ft) long West Burnside Tunnel was excavated through clay (decomposed Columbia River Basalt and/or West Hills Silt). Two 1.8 to 2.4 m (6 to 8 ft) high sill drifts and one crown drift were driven before the remainder of the tunnel was excavated by hand using "Swede" cars. The two 76.2-m (250 ft) and 152.4-m (500-ft) long NW Cornell Road tunnels were excavated through highly weathered Columbia River Basalt using drill and shoot methods. "Swede" cars on rail were used to remove tunnel muck. Both the West Burnside and NW Cornell Road tunnels were lined with CIP reinforced concrete.

Mason Ralph Curcio and his crew designed and built beautiful masonry portals for all three tunnels using Rocky Butte basalt (Boring Lava). Mr. Curcio, an Italian immigrant who learned masonry in Europe, is responsible for the spectacular masonry work along the historic Columbia River Gorge Highway, the Crown Point Vista House, Multnomah Falls Lodge, and Timberline Lodge (Hadlow 2008). Figure 2 shows the masonry portal at the West Burnside Tunnel.

### Peninsular Tunnel (1950)

An interconnected tunneled and open-cut sewer pipeline was built in the 1950s along the east side of the Willamette River to convey flows from southeast Portland to the Columbia Boulevard Wastewater Treatment Plant in north Portland. The pipeline consists of three elements (from south to north): the Southeast Interceptor, the Central Eastside Interceptor, and the Peninsular Tunnel. The 5,059-m (16,600-ft) long and up to 50.3-m (165-ft) deep Peninsular Tunnel, which cuts east west through the north Portland highlands, was both the longest and deepest Portland tunnel until the 1990s. The tunnel was excavated through catastrophic glacial flood deposits consisting of sand, gravel, and boulders above groundwater. Breastboards were required for

Table 1. Summary of 100 years of Portland tunnel projects

| Tunnel Name (Purpose) | ID # | Year Completed[1] | Length | Diameter | Ground Type | Method | Current Owner | Contractor | Construction Cost (Millions) |
|---|---|---|---|---|---|---|---|---|---|
| Peninsula Tunnel (Freight rail) | 1 | 1911 (tunneling started in 1909) | 1,658 m (5,438 ft) | 6.9 m (22.5 ft) high (ID) 4.9 m (16 ft) wide (ID) | Soft ground (no groundwater) | Steel ribs and timber spiling were placed over ribs and breast boards. Finished as a concrete-lined horseshoe. | UPRR | | $0.8 |
| Elk Rock Tunnel (Passenger rail) | 2 | 1921 | 425 m (1,395 ft) | 7 m (23 ft) high (ID) 5.5 m (18 ft) wide (ID) horseshoe tunnel | Hard rock | Drill and shoot | City of Portland | Southern Pacific | Unknown |
| Multnomah County Highway Tunnels Rocky Butte, West Burnside, and NW Cornell Road Tunnels | 3 | 1939, 1940, 1941 | 70 to 152 m (230 to 500 ft) | 8.2 m to 9.6 (27 to 31.5 ft) wide (ID) | Hard rock, soft rock, and soft ground. | Mixture of hand dug and drill and shoot. Reinforced CIP lining constructed with special traveling slip forms. WPA masonry portals. | City of Portland | WPA laborers and Ralph Curcio (masonry) | $0.5 (Rocky Butte Tunnel) |
| Peninsular Tunnel (Sewer) | 4 | 1950 | 5,060 m (16,600 ft) | 2,438 mm (96 in.) (ID) horseshoe tunnel | Soft ground (no significant groundwater) | Excavated using breastboards, rib and board support, CIP concrete lining. | BES | NA | NA |
| Portsmouth Tunnel (Sewer) | 5 | 1967 | 2,179 m (7,149 ft) | 2.4 m (8 ft) (OD), 1.8 m (6 ft) (ID) | Soft ground (no groundwater) | Horseshoe open-face pneumatic shield, with steel sets and wood lagging, CIP concrete liner. | BES | NA | NA |
| Vista Ridge Tunnel (Highway) | 6 | 1969–1970 | EB: 305 m (1,000 ft) WB: 151 m (494 ft) (curved) | 4.8 m (15.6 ft) high (ID) 12.5 (41 ft) wide (ID) | Hard rock | Drill and shoot (assumed) | ODOT | Coat Contractors (pilot tunnel), Donald M. Drake and Winston Brothers (main bores) | $4.2 (EB) $3.7 (WB) |
| Southeast Relieving Interceptor—Phases 2 through 4 (Sewer) | 7 | 1980s | 5,547 m (18,200 ft) (total length of tunneled section in Phases 2 through 4 | 1,981–2,591 mm (78–102 in.) (ID). | Soft ground (with and without groundwater) | 3.7 m (12 ft) diameter open face shield, circular rib & board support, CIP concrete lining (Phases 2 and 4). 2,896 mm (114 in.) diameter close-faced shield, circular rib & board support, RCP final lining (Phase 3). | BES | Dick Schumann (Phase 2 and 4) Frank Coluccio Construction Co. (Phase 3) | $25 (includes open cut work) |

(continues)

**Table 1. (continued)**

| Tunnel Name (Purpose) | ID # | Year Completed[1] | Length | Diameter | Ground Type | Method | Current Owner | Contractor | Construction Cost (Millions) |
|---|---|---|---|---|---|---|---|---|---|
| Westside Light Rail Transit Tunnel (Light rail) | 8 | 1994 and 1996 | 4,542 m (14,900 ft) (twin tunnels) | 6.4 m (21 ft) | Hard rock and soft ground (groundwater) | Hard Rock TBM (modified), drill and shoot | TRIMET | Frontier, Kemper/ Traylor Brothers | $190 |
| Columbia Slough Consolidation Conduit—Segment 2 (CSO) | 9 | 2000 | 2,548 m (8,360 ft) | 4.6 m (15 ft) (OD) 3.7 m (12 ft) (ID) | Soft ground (no groundwater) | 4,597 mm (181 in.) wheel EPB TBM (mixed face—modified for erecting ribs and lagging) CIP reinforced concrete final liner | BES | Frank Coluccio Construction Co. (Phase 3) | $25.5 |
| Tanner Creek Stream Diversion Project, Phases 2 and 5 (Stormwater) | 10 | 2002 | 1,158 m (3,800 ft) (Phase II) 457 m (1,500 ft) (Phase 5) | 2,261 mm (89 in.) (OD) 1,829 mm (72 in.) (ID) | Soft ground (groundwater) | Slurry MTBM with RCP | BES | Robison Construction Inc. | $13 |
| WSCSO Program WSCSO Tunnel, SWPI Segment 3, Peninsular Force Main, and Tanner Extension (CSO[2]) | 11 | 2006 | 6,706 m (22,000 ft) (tunnel) 2,246 m (7,370 ft) (SWPI) 997 m (3,270 ft) (PFM) 421 (1,380 ft) (TE) | 4.3 m (14 ft) (tunnel) 2,134 & 1,829 mm (84 in. & 72 in.) (SWPI) 2,438 mm (96 in.) (PFM) 1,829 mm (72 in.) (TE) | Soft ground (groundwater) | Slurry TBM with one pass segmental liner MTBM with RCP (SWPI & TE) Steel casing with carrier pipe (PFM) | BES | Impregilo/Healy Joint Venture | $80 (main tunnel) $50 (micro-tunnels) $70 (shafts) $105 pump station (shaft) |
| ESCSO Program ESCSO main tunnel with connected Outfall 28, 37 and 38, 40, 41 and 46 diversion pipelines. (CSO) | 12 | 2011[1] | 8,918 m (29,260 ft) (tunnel) 76 to 914 m (250 to 3,000 ft) (pipelines) | 7.6 m (25 ft) (OD), 6.7 m (22 ft) (ID) main tunnel Five 2,134 mm (84-in.) (ID) microtunnels | Soft ground (groundwater) | Slurry TBM with one pass segmental liner, MTBM with RCP, conventional mined connection through frozen ground | BES | Kiewit/ Bilfinger Berger | $220 (main tunnel) $35 (micro-tunnels) $105 (shafts) |
| Portsmouth FM Project (Segments 1 and 2) (CSO[2]) | 13 | 2011[1] | 1,782 m (5,848 ft) (Segment 2 tunnel). 899 m (2,950 ft) (Segment 1 microtunnel) | Segment 2: 2,642 mm (104 in.) (OD) circular tunnel Segment 1: 2,134 mm (84 in.) (OD) microtunnel. Both segments: 1,676 mm (66 in.) (ID). | Soft ground (groundwater) | Segment 2: TBM run in open mode, ribs and boards. Segment 1: slurry MTBM with RCP. Both segments: 1,676 mm (66 in.) (ID) Hobas carrier pipe. | BES | Michels Tunneling (Segment 2). Mountain Cascade, Inc. (Segment 1). | $19.4 (Segment 2) $28 (Segment 1) |
| Balch Consolidation Conduit (CSO) | 14 | 2011[1] | 1,951 m (6,400 ft) | 2,134 mm (84 in.) (ID) | Soft ground (groundwater) | Slurry MTBM with RCP | BES | James W. Fowler Co. | $57 |

[1] Project is currently under construction. Estimated completion date is listed.
[2] Combined Sewer Overflow.

**Figure 2. West Burnside Tunnel showing historic WPA stone masonry portal (photo used with permission of Robert Hadlow)**

tunneling to prevent running ground. The tunnel was excavated by placing permanently located steel ribs with timber spiling over the ribs and breast boards on the heading. Tunnel excavation was initially supported using ribs and boards. Collapsible steel forms were used to construct the CIP concrete horseshoe-shaped lining for the 244-cm (96-in.) finished diameter tunnel. Initially, a double pumping system was employed to deliver concrete to the lining forms. This method was abandoned when the line plugged; concrete was then delivered to the form via a rail-mounted car pulled by a small locomotive. The entire project was completed in 554 days, mining simultaneously from both portals.

### Portsmouth Tunnel (1967)

The Portsmouth Tunnel was constructed in north Portland to convey sewer flows through the highland along the east side of the Willamette River to the Columbia Boulevard Wastewater Treatment Facility. The tunnel was excavated using an open-face pneumatic shield with an overhanging canopy and conveyor belt. Figure 3 shows a photograph of the shield outside the tunnel.

The 2,179-m (7,149-ft) long tunnel was excavated through fine-grained catastrophic glacial flood deposits above the groundwater table in sand that was "so clean that it was sold as aggregate" (*The Oregonian* 1967). The tunnel was supported during construction by steel sets and wood lagging (Figure 4) and was completed with a circular CIP concrete lining. Tunnel excavation took eight months to complete, and five men worked at the heading during each of the three shifts per day.

### Vista Ridge Tunnel (1969–1970)

The Vista Ridge Tunnel was built through the eastern edge of the Portland West Hills to significantly increase traffic flow between Portland to the Tualatin Valley. The tunnel connects I-405 (Stadium Freeway) with US 26 (Sunset Highway). The curved Vista Ridge Tunnel consists of a three-lane, 305-m (1001-ft) long eastbound tunnel, and a three-lane, 289-m (949-ft) long westbound tunnel. The Oregon State Highway Commission contracted the construction of a pilot tunnel near the crown of the proposed eastbound three-lane tunnel prior to awarding the contract for final tunnel construction (Hadlow 2004). Contractors were allowed the opportunity to study the Columbia River Basalt rock structure within the pilot tunnel as they prepared their bids. Since the tunnels are described as two "semi-circular bores" (Hadlow 2004), they were most likely heading and bench excavations using drill and shoot methods through basalt bedrock. The tunnels were finished with CIP concrete linings, ceramic tile interiors, and daytime lighting systems (Hadlow 2004).

Figure 3. Front view of Portsmouth Tunnel open-face pneumatic shield with overhanging canopy (photo used with permission of the City of Portland Archives and Records)

Figure 4. Workers inside the Portsmouth Tunnel shield during tunnel excavation (photo used with permission of the City of Portland Archives and Records)

## Southeast Relieving Interceptor (mid- to late-1980s)

The Southeast Relieving Interceptor was built in four phases to convey overflows from the original Southeast Interceptor (Singleterry 2009). The tunnel portion of the pipeline alignment extends approximately 5,550 m (18,200 ft) through southeast Portland northward to the Sullivan Pump Station, located in Sullivan Gulch beneath the I-84/I-5 elevated interchange.

The Southeast Relieving Interceptor was constructed in Phases 1, 2, 3, and 4, from north to south. Part or all of Phases 2, 3, and 4 include tunnel sections with tunnel cover ranging from 12.2 to 24.4 m (30 to 80 ft). A 3.7-m (12-ft) diameter open face shield was used to excavate Phases 2 and 4. Phase 3, the only section located below groundwater, was excavated with a close-faced shield in conjunction with dewatering wells. Rib and board initial support was used for all three phases. Concrete CIP final linings were constructed in Phases 2 and 4, while the Phase 3 lining consists of pulled-in reinforced concrete pipe (RCP).

Phase 3 crosses through the historic Hawthorne Slough, which was filled during the urbanization of east Portland. Geology along the entire alignment includes artificial fill, Quaternary Alluvium, Catastrophic Glacial Flood Deposits, and Troutdale Formation. Sand lenses in the Troutdale Formation ran into the tunnel excavation, while the cemented gravel had excellent stand-up time.

## Westside Light Rail Transit Tunnels (1996)

The twin 4,542-m (14,900-ft) long, 6.4-m (21-ft) diameter Westside Light Rail Transit Tunnels were built to convey commuter trains through the Portland West Hills between downtown and the Tualatin Basin suburbs (Gildner et al. 1997). The project also included the construction of the deepest commuter train station in North America, to date. The tunnels and station shaft were excavated through the Grande Ronde Basalt Member (GRB) of the Columbia River Basalt Group, Boring Lava basalt, and Sandy River Mudstone. Numerous fault zones and associated highly fractured rock were encountered during tunneling.

The tunnels were excavated in three reaches: A, B, and C (from west to east). Reach A, approximately 1,585 m (5,200 ft), was mined using drill and blast techniques for the rock and earth excavation equipment for the soft ground due to extremely variable ground conditions. Contract Documents precluded TBM use in this reach of the tunnels (Gildner et al. 1997). Reaches B and C, approximately 2,970 m (9,750 ft) combined, were mined primarily through the GRB using a full-face hard-rock TBM. The geologic conditions in these reaches included units of the GRB that were considered less variable than those encountered in western portions of the alignment.

During mining of the first tunnel with the TBM, stand-up time in the first unit of the GRB (Sentinel Bluffs) proved inadequate for TBM operations. Raveling ground conditions created up to 6-m (20-ft) high voids above the TBM. Additionally, the hard basalt caused excessive wear to the cutterhead and cutters. To correct these problems, the contractors modified the TBM by redesigning the grippers, adding reverse rotation, increasing torque, adding protective wear plating to the cutterhead, and installing a "poor man's EPB" system to stabilize the heading (Gildner and others 1997).

During the excavation of the Washington Park Station, the soil-nail excavation for the headhouses encountered excessive ground movements in the soil (Portland Hills Silt). The excessive movements were tied to the station excavation being located within a massive landslide. The contractor had to install a series of long horizontal drains to lower the groundwater conditions in order to control soil movements prior to completing the excavation.

## Columbia Slough Consolidation Conduit (2000)

The Columbia Slough Consolidation Conduit (CSCC) was constructed in north Portland to reduce combined sewer overflows into the Columbia Slough as part of Portland's state-mandated CSO abatement program. The project included a 2,560-m (8,400-ft) long, 3.7-m (12-ft) diameter tunneled section that was excavated by a mixed-face wheel EPBM that was modified to erect ribs and lagging as initial support (Feroz et al. 2000). The contractor installed a hood on the TBM, and after modification to prevent diving, the hood mitigated raveling ground and gauge cutter overcut (Feroz et al. 2000). The final lining consisted of CIP reinforced concrete.

The tunnel was excavated through Quaternary Alluvium and Catastrophic Glacial Flood Deposits that consisted of slow raveling sand and fast raveling gravel above groundwater. Tunneling was complicated by the presence of oversized boulders within the coarse-grained Catastrophic Glacial Flood Deposits.

## Tanner Creek Stream Diversion Project, Phases 2 and 5 (2002)

The Tanner Creek Stream Diversion Project was built to convey creek flows and storm water from Portland's West Hills directly to the Willamette River (Klein et al. 2001). The project, also part of Portland's mandated CSO abatement program, was constructed in five phases through the north side of downtown.

Phases 2 and 5 included microtunneled sections through the historic Tanner Creek channel, which was filled with debris during the urbanization of Portland. The 1,158-m (3,800-ft) long Phase 2 alignment passes close to over 100 existing buildings, including brick and masonry structures. Phase 5 extends 457 m (1,500 ft) through an industrial area beneath several mainline railroad tracks. The 183-cm (72-in.) (ID) microtunnels were excavated below groundwater through extremely variable ground conditions including artificial fill, soft silt marsh deposits (Quaternary Alluvium), and Fine-grained Catastrophic Glacial Flood Deposits (Klein et al. 2001).

Artificial fill, which had been dumped to fill in the historic Tanner Creek channel, consisted of a heterogeneous mixture of soil and debris, including bricks, boulders, concrete, wood, and other manmade materials (Klein et al. 2001). The designer conducted ground penetrating radar (GPR) investigations and large-diameter borings to locate and characterize potential buried obstructions. Historical research identified buried timber planked roads supported by timber piles crossing the microtunnel alignment. Portland's first geotechnical baseline report for microtunneling was prepared, which set baselines for project ground conditions and obstructions.

To reduce construction risks, the contractor was required meet the following requirements:

- Provide an MTBM that was equipped to handle large, hard obstructions and fibrous wood debris
- Use a slurry MTBM to reduce ground settlement associated with tunneling
- Complete microtunneling through fill deposits containing obstructions that could damage the MTBM after all other microtunneling has been completed
- Install watertight support for Phase 5 shafts through fill deposits that contain buried obstructions

Even with these required precautions, buried obstructions were encountered during both tunneling and shaft excavation that delayed the project and increased the cost. The MTBM hit a buried tree stump in the artificial fill during Phase 2 tunneling that was too large to ingest. The MTBM continued "plowing" the stump off line and grade until the ground fractured to the surface and the construction of a recovery shaft was required to rescue the MTBM. Similarly, a buried tree is thought to have caused deflection of Phase 5 sheet pile shaft support that allowed flowing sand to enter the shaft excavation and caused excessive settlement of NW Front Avenue, adjacent rail lines, and buried utilities beneath Front Avenue.

## West Side CSO Project (2006)

The West Side CSO Project consists of a 5,541-m (18,180-ft) long, 4.3-m (14-ft) diameter tunnel, three microtunneled pipelines, and six major shafts, including the 42.7-m (140-ft) diameter, 50.3-m (165-ft) deep Swan Island Pump Station shaft located at the tunnel's northern terminus (McDonald 2007). The West Side CSO Project is one of two large CSO abatement projects included in the Willamette River CSO Program. The approximately 30.5-m (100-ft) deep tunnel parallels the Willamette River's west shoreline through downtown and northwest Portland before crossing beneath the river channel to connect to the Swan Island Pump Station in north Portland. The 2,246-m (7,370-ft) long Southwest Parallel Interceptor (Segment 3) pipeline conveys CSO flows from southwest Portland into the tunnel, while the 420.6-m (1,380-ft) long Tanner Extension collects CSO flows from Tanner basin directly north of downtown. The 997-m (3,270-ft) long Peninsular Force Main pipeline conveys flows from the Swan Island Pump Station to the existing Peninsular Tunnel.

Geologic units encountered during project excavations included artificial fill, Quaternary Alluvium, Fine and Coarse-grained Catastrophic Flood Deposits, and the Troutdale Formation (Fong et al. 2002). The Sandy River Mudstone, which is located 45.7-m (150 ft) below the base of the Swan Island Pump Station shaft, was used as an impervious layer for groundwater cut-off for a jet grout curtain. Extensive historical research was conducted during project design to identify and baseline potential buried obstructions along the tunnel and pipeline alignments and within shaft excavations. The tunnel and shaft locations were realigned because of the discovery of abandoned bridge foundations, abandoned steel dolphins, and buried foundations and contamination associated with a demolished flourmill.

The West Side CSO Project was the first large tunnel in North America to use slurry mixshield TBMs, which were utilized to tunnel through open gravel, cobbles, and boulders in the coarse-grained Catastrophic Flood Deposits and the cemented gravels of the Troutdale Formation. Similarly, slurry MTBMs were required for microtunneling through the same material. The tunnel has a reinforced concrete, gasketed, segmented liner to withstand up to three bars of groundwater pressure. To meet the tight project deadline, two 4.9-m (16-ft) (OD) TBMs were launched in either direction from the tunneling shaft to provide schedule flexibility for the construction of the Swan Island Pump Station. The TBMs were required to handle mixed-face ground conditions that often included soft silt and sand over open gravel with up to 36.6 m (120 ft) of groundwater head. The contractor experienced significant difficulty separating the bentonite slurry from the silt and fine sand alluvium during tunneling, causing increased bentonite usage. Tunnel muck was barged to Ross Island in south Portland, where it was used to partially fill an abandoned gravel pit in the Willamette River.

Slurry walls and secant pile walls were constructed as watertight support for tunnel and microtunnel shafts, respectively. Break-in and breakout areas outside the shaft walls were jet grouted to seal the shafts during tunneling into and out of shafts with significant groundwater head. Even with these precautions and the use of a seal at the tunnel eye, groundwater inflows and limited ground loss due to flowing silt and sand alluvium occurred at the tunneling shaft, endangering the TBM. The excavation of the Swan Island Pump Station shaft was delayed by six months because of the difficulty in achieving groundwater cutoff. Remedial jet grouting was performed, and limited dewatering was conducted before shaft excavation could be completed. To get back on schedule, the pump station was redesigned to move the Operations and Maintenance Building off the top of the structure. This move permitted concurrent construction of both structures. That, coupled with accelerated shaft excavation, brought the project back on schedule.

As expected, timber piles and Catastrophic Glacial Flood boulders were encountered during shaft excavation and microtunneling. A boulder, similar to those shown in Figure 5, fell into a slurry wall panel excavation at the Swan Island Pump Station shaft before concrete was placed to fill the panel. When this boulder was removed from the wall during shaft excavation, it created a hole that allowed groundwater to enter the shaft. The shaft was then flooded and repaired using jet grout.

West Side CSO microtunneling innovations included the use of a slurry MTBM equipped with an airlock to permit face interventions during tunneling. The airlock was used to access and remove an unidentified steel pile obstruction that was supporting an electrical duct bank along the Southwest Parallel Interceptor alignment, eliminating the need to construct a costly rescue shaft. Microtunneling costs were also reduced by using a single MTBM that was retrofitted on site for 183-cm (72-in.), 213-cm (84-in.), and 243-cm (96-in.) (ID) microtunneling. Timber piles and logs encountered by the MTBM were handled, and microtunneling through very soft silt did not significantly impact line and grade. Controlled density fill placed inside steel "top hat" enclosures was used at the tunnel eye during microtunnel shaft break-ins to reduce the potential for groundwater inflow and flowing ground.

**East Side CSO Project (2011)**

The East Side CSO Project, which is the largest project in the Willamette River CSO Program, is currently under construction along the east side of the Willamette River. The 8,918-m (29,260-ft) long, 6.7-m (22-ft) diameter tunnel extends from southeast Portland to the Swan Island Pump Station. Five microtunneled outfall diversion pipelines, ranging from 76.2 to 914.4 m (250 to 3,000 ft) long, convey CSO flows into the main tunnel. Seven shafts have been built along the tunnel alignment, which extends up to 51.8 m (170 ft) below the ground surface. The shafts permit TBM access at atmospheric conditions and provide drop structures for the diversion pipelines. Tunneling is being conducted in two directions from the Opera Shaft using a single 7.6-m (25-ft) diameter slurry mixshield TBM. The tunnel is supported by a gasketed, segmented liner of reinforced concrete. To reduce costs, steel fibers are being used to reinforce segments installed within the dense Troutdale Formation, while rebar cages are being used for segment reinforcement through the soft sands and silts and at shaft connections.

Geologic units and ground conditions are similar to those encountered during the construction of the West Side CSO Tunnel. The majority of the main tunnel is located in dense, sometimes cemented Troutdale Formation gravel with sand lenses. Soft silt and sand Quaternary Alluvium was encountered in ancient river and stream channels that cut across the alignment. Open gravel, cobbles, and boulders in the coarse-grained Catastrophic Flood Deposits were encountered in the Opera Shaft excavation and along a segment at the north end of the tunnel alignment. The tunnel and shaft flooded, and ground loss occurred during the breakout of the Opera Shaft at the start of tunneling. The TBM was damaged, and construction was delayed. To mitigate against future groundwater inflows and ground loss, intermediate tunneling shafts are flooded prior to TBM break-in.

Buried obstructions, including timber piles, abandoned building and bridge foundations, rip rap, and logs were identified and baselined during project design. Ground penetrating radar was used during design to locate deep steel piles supporting the Interstate 5/Interstate 84 interchange within an ancient channel. The main tunnel was realigned when a deviating steel pile was detected within the tunnel horizon.

Microtunneling costs were reduced when the owner selected a single pipeline diameter (213 cm [84 in.]) for all outfall diversion pipelines, preventing the need for an additional MTBM. One of the microtunneling drives on the East Side CSO Project won an award as the longest microtunnel drive in the United States (916-m [3,005 ft]). The owner chose to take the risk associated with lengthening the drive to avoid the cost of constructing an intermediate jacking shaft. The drive was successful, even though the MTBM advanced through wood debris, timber piles,

**Figure 5. Catastrophic glacial flood boulders encountered within the Swan Island Pump Station shaft excavation (photograph by Sue Bednarz, Jacobs Associates)**

and large steel spikes beneath a railroad yard and historic dock area.

**Portsmouth Force Main (2011)**

The Portsmouth Force Main, which is currently under construction, is part of the City of Portland's Willamette River CSO Program. The force main will convey flows through north Portland from the Swan Island Pump Station north to the Portsmouth Tunnel. The project is divided into two segments. Segment 1 is primarily an open-cut pipeline through the Swan Island lowlands, but also includes a microtunneled section. Segment 2 is a tunneled section through the highland bluff that borders the east bank of the Willamette River.

An 899-m (2,950-ft) long, 213-cm (84-in.) diameter microtunnel is under construction through sand fill and soft silt Quaternary Alluvium along the south end of Segment 1. The remainder of the Segment 1 force main will be installed in an open-cut excavation. A 168-cm (66-in.) diameter steel carrier pipe will be installed within the Segment 1 excavations. The Segment 1 microtunnel and shafts were located to avoid large buried obstructions, including a demolished flour mill and three 122-cm (48-in.) diameter, 173.7-m (570-ft) long large drainage pipes. Smaller obstructions, including timber piles, logs, abandoned dredge pipes, and rip rap, have been quantified based on a detailed review of historic maps and photographs.

The 1,783-m (5,850-ft) long Segment 2 tunnel extends up to 42.7-m (140 ft) deep through sandy Catastrophic Glacial Flood Deposits and Troutdale Formation gravel between the south portal shaft and the north connection shaft. A 264-cm (104-in.) diameter TBM run in open mode above groundwater is being used to excavate Segment 2 through potentially raveling ground. Temporary support consists of steel ribs and timber lagging. A 168-cm (66-in.) Hobas carrier pipe will be installed following tunnel excavation. Although extensive research was conducted to identify Segment 2 buried obstructions, unidentified cobbles and boulders were encountered beneath the bluff slope at the start of tunneling. At the time of this paper, this has resulted in project time loss due to boulder removal complicated by soft sands flooding the machine face.

**Balch Consolidation Conduit (2011)**

The Balch Consolidation Conduit is currently under construction in the industrial Guilds Lake area of northwest Portland. The conduit will convey sewer and stormwater flows from the Guilds Lake area into the West Side CSO Tunnel. The conduit crosses beneath historic Guilds Lake and the site of the 1905 World's Fair. After the fair, the lake was filled and an incinerator was constructed to burn Portland's trash.

The project consists of a 1,951-m (6,400-ft) long, 213-cm (84-in.) diameter microtunnel through fill, soft silt lake deposits (Quaternary Alluvium), open gravel, cobbles, and boulders (Catastrophic Glacial Flood Deposits), and Troutdale Formation gravel. Shafts are located to avoid buried obstructions and contamination associated with the incinerator site.

The contractor is installing a soil mix wall for shaft support using a cutter soil mixing (CSM) machine. A slurry MTBM was selected to facilitate tunneling through open gravel, cobbles, and boulders. Based on experience with the West Side CSO Project, an airlock has been installed in the project's MTBM to permit access to the face for obstruction removal. A secondary steering joint has also been installed for better steering control in the soft silt lake deposits.

## CONCLUSIONS

Portland's long history of tunnel projects has produced innovations in tunneling, microtunneling, and shaft construction that have benefited the tunneling industry. These innovations have been developed to handle the unique and diverse geology and ground conditions encountered in the Portland area. Although Portland's large Willamette River CSO Program tunneling projects will be completed by 2011, future projects will continue Portland's tunneling tradition.

## REFERENCES

Anderson, T. 2005. Breaking historical news, *The Columbia Gorge Gazette*, Vol. XXV(1):3–4.

Feroz, M.S., Gribbon, P.T., and Buehler, F.S. 2000. Columbia Slough Consolidation Conduit, a soft ground tunnel in Portland, Oregon, USA. *Proceedings of the GeoEng 2000 Conference, November 2000, Melbourne Australia.*

Fong, M.L., Bednarz, S.L., Boyce, G.M., and Irwin, G.L. 2002 History and exploration redefine Portland's West Side CSO Tunnel alignment. *North American Tunneling 2002*, Ozdemir (ed), A.A. Balkema (pub), pp. 79–87.

Fraizer, J.B. 2001. Portland's tunnels aided 'crimps' Shanghaiing sailors. Associated Press article in *The Oakland Tribune*, Sunday, May 6, 2001, Travel Section, p. 5.

Gildner, J.P., Nowak, D, Painter, D.Z., Revey, G.F., Wilmoth, P., and Yanagisawa, S. 1997. Punk rock in Portland, *1997 RETC Proceedings*, Ed. J.E. Carlson and T.H. Budd, SME, Littleton, CO, Chapter 9, pp. 151–182.

Hadlow, R.W. 2008. *Oregon Highway Tunnels, Determination of Eligibility for the National Register of Historic Places*, Section 106 documentation form—multiple property listing, Prepared by the Oregon Department of Transportation, July 8, 2004.

Klein, S., Havekost, M., Hutchinson, M, and Jossis, B. 2001. Microtunneling in Downtown Portland—Tanner Creek Stream Diversion Project, Phases 2 and 5, Unpublished article.

McDonald, J.A., Kabat, J.J. 2007. Slurry shield tunneling in Portland—West Side CSO Tunnel Project. *2007 RETC Proceedings*, Ed. M.T. Traylor and J.W. Townsend, SME, Littleton, CO, pp. 830–842.

Singleterry, D. 2009. Personal communication.

*The Oregonian*. 1967. New Portland sewer tunnel to punch through on Tuesday, February 28, 1967 edition, p. 2G.

*The Webfooter Newsletter*. August 2008. The story of Southern Pacific's Red Electrics, 42(8):3–8.

# San Vicente Pipeline Tunnel: A Sophisticated Ventilation System

**Jean-Marc Wehrli**
Traylor Bros., Inc., New York, New York

**Richard McLane**
Traylor Bros., Inc., Seattle, Washington

**Brett Zernich**
Traylor Bros., Inc., San Diego, California

**ABSTRACT:** Ventilation of tunnels during construction has always been a challenge, in particular where the internal tunnel diameter is relatively small, where the tunnel drives are long, where intermediate ventilation shafts are impractical, and where muck haulage equipment is powered with diesel engines. The ventilation system designed for Reach 4 East of the San Vicente to Second Aqueduct Pipeline Project in San Diego, CA was designed to supply 35,000 CFM of fresh air to the heading over a total length of 23,500 ft. The initial tunnel liner had an internal diameter of 10.5 ft with a final liner diameter of 8.5 ft, intended for use as a regional raw water supply pipeline. A Digger Shield Machine excavated this section of tunnel. This paper will examine the engineering challenges of a ventilation system in the areas of fan selection, fan arrangement, fan-line design, as well as a sophisticated fan control system implemented on this particular project.

## INTRODUCTION

Due to the length of this drive and the use of diesel locomotives, a large volume of air was the first requirement of ventilation design. It was necessary to maximize fan line size, limited by the height of the rolling stock procured for this project. It was clear that booster fans would be needed along the alignment. This meant that a sophisticated control system would have to be designed to prevent fan line collapse, startup overload, or catastrophic fan failure. That control system allowed project engineers to monitor the electric load on fans, monitor airflow and system pressure, and reduce power consumption during low demand periods. This paper will highlight the ventilation design for Reach 4E, discuss lessons learned, and potential energy savings for future projects to provide more sustainable tunneling projects.

## PROJECT DESCRIPTION

The San Vicente Pipeline Tunnel (SVPT) is a 17.45km (57,230ft) tunnel that is part of a major undertaking by the San Diego County Water Authority (SDCWA) to create emergency storage for San Diego's regional raw water supply. Once complete, this Emergency Storage Project (ESP) will create a system of reservoirs, pipelines, and pump stations to comprise a storage and distribution system for an approximate six-month water supply for the San Diego region should a natural disaster disrupt water deliveries from Northern California or the Colorado River. A regional map showing the project's locations is shown in Figure 1.

The SVPT project was awarded to a Joint Venture between Traylor Bros., Inc., of Evansville, IN and J.F. Shea Construction, of Walnut, CA. The Traylor Shea Joint Venture (TSJV) was given notice to proceed on July 14, 2005. The initial contract value was $198,266,900 and the original contract duration was 1,250 calendar days. The designer of record was Jacobs Associates, of San Francisco, CA and Construction Management was contracted to Parsons, of Pasadena, CA.

## REQUIREMENTS AND CONSTRAINTS

### Legal Requirements

Since the project was undertaken in the State of California, it fell under the authority of the Division of Occupational Safety and Health (DOSH or better known as Cal/OSHA) and therefore compliance with the Cal/OSHA Tunnel Safety Orders [2] was mandated. This meant that the ventilation system had to be reversible from the surface and exhausting using rigid duct. In terms of air quantity, 200 cfm of air

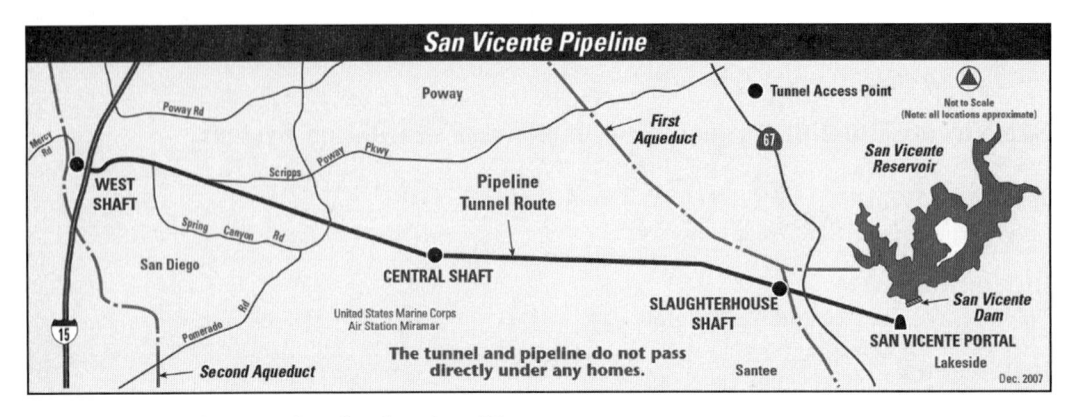

**Figure 1. Regional map and project locations [1]**

had to be provided for each person underground and 100 cfm per brake horsepower installed on each of the diesel-powered equipment utilized below surface. In addition, the velocity of fresh air in the tunnel had to be at least 60 feet per minute.

Furthermore, the tunnel had been classified Potentially Gassy by Cal/OSHA, which required all electrical installations to be Class I Division 2 certified. Therefore, all fans and control devices installed underground had to be explosion proof.

## Constraints

Because the tunnel alignment crisscrosses a nature preserve and residential areas, the sinking of intermediate shafts for the purpose of ventilation during construction was prohibited per the Contract. This implicated that relatively long stretches of tunnel had to be ventilated from a single access point. This constraint posed a great challenge to the design of the ventilation system because the losses of ventilation energy due to friction within the duct increase with the length of the duct linearly.

Another constraint was the limited headroom available for running a fan-line in the 10.5 ft diameter tunnel. Because the friction loss is proportional to the wetted perimeter of the duct, maximizing the duct size is the best way to reduce operational cost. Designers must balance the operational costs of the ventilation system with the capital costs of the fan duct. On this project, the maximum possible duct diameter was limited to 38 inches by the clearances required in the tunnel for the rolling stock used for mucking.

With a grade of up to 2%, the haulage locomotives had to be of a particular weight and rated to a specific horsepower to haul muck out of the tunnel. In addition, the relatively long tunnel drives required the provision of multiple trains that had to be able to pass each other at switches installed within the tunnel. Thus, the high demand in brake horsepower necessitated a high air-flow. Results for cycle time and traction force calculations for the muck trains led to the decision to use four muck trains pulled by 8-ton Balco locomotives. These locomotives were equipped with 82 horsepower Deutz diesel engines from a previous project. Real conditions proved these locomotives too light and underpowered to master the 2% grade. Seepage water running down the invert caused the rails to be wet for most of the line, which reduced the available traction force significantly compared to dry conditions. Consequently, the Balco locomotives were re-powered on site with 110 horsepower Tier III Cummins engines and their weight increased to 12 tons by adding steel plates. Due to the modifications, more air had to be provided in the tunnel to comply with the Tunnel Safety Orders [2] if all four muck trains were to be used. Unfortunately, all the fans had already been procured at the time this issue came up. To accommodate for increased horsepower and ventilation designed for smaller engines, we decided to run only three re-powered Balco locomotives in the tunnel on the 2% grade and an additional electric driven locomotive towards the end of the drive where the grade was shallower. This battery operated Brookville locomotive was used to shuttle cars in between the heading and the last switch.

## VENTILATION DESIGN

A general tunnel ventilation schematic is shown in Figure 2 depicting the east half of the San Vicente Tunnel project, including Reach 4E, 5W, 5E, and 6. The schematic shows the general layout of the shafts and tunnel, along with the ventilation features. A bulkhead sealed the Slaughterhouse Shaft so that fresh air would draft from the Portal. The main surface fan was located at Slaughterhouse Shaft—the closest surface access while the tunnel was driven. The surface fan, along with the other booster fans, was used to ventilate the completed Reaches while

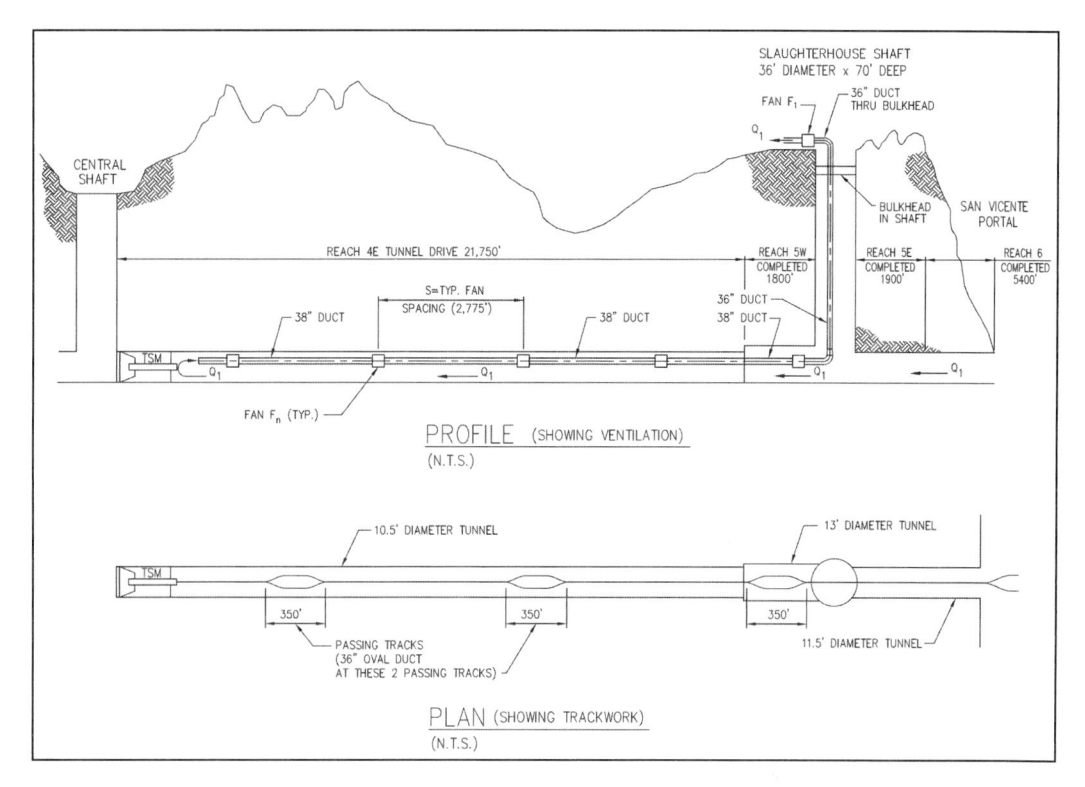

**Figure 2. Ventilation schematic**

tunnel excavation operations continued in Reach 4 East with the Digger Shield.

After reviewing all of the constraints with regard to tunnel size and airflow requirements, the project team decided a 38 inch nominal duct size would be the chosen duct size for the running tunnel. A 36 inch fiberglass oval line would be installed over the permanent tunnel switches in two separate locations. The minimum airflow was calculated to be approximately 35,000 cfm as described above. Table 1 shows a total system-pressure-loss calculation. Fans were chosen to provide the 35,000 cfm; the pressure each fan could overcome for that given volume would influence the spacing and quantity.

## EQUIPMENT SELECTION

### Vent Duct

Because it was mandatory by law for a negative pressure system with exhaust to the outside atmosphere, the duct had to be able to withstand pressures below atmospheric. Therefore, it was decided to use non-galvanized steel ducts tested to fail at 56 inches of water pressure with a wall thickness of 20 gauge manufactured by Mining Equipment. To minimize freight costs, the 30 feet long duct sections were rolled on the job-site using a truck-mounted rolling machine. Each end of the duct was fitted with a 1 inch wide flange. The ducts were joined with each other using gasketed draw-bolt steel couplings.

Where passing tracks for haulage trains were installed, the clearance between top of rail and tunnel crown was reduced further due to the elevated double-track. In order to maintain a steady duct cross-sectional area and therefore to minimize the static and dynamic losses in the duct across these areas it was decided to install oval shaped fiber-glass duct sections manufactured by Schauenburg Flexadux Corporation. The transition pieces from round to oval and vice-versa were fabricated out of fiberglass as well. The oval shaped duct sections were joined together using rubber-sealing bands.

### Fans

For generating the minimum required amount of airflow in the tunnel, high performance variable pitch vane axial fans rated at 125 HP were selected. The maximum spacing between the fans was derived from fan curves provided by the manufacturer. The 33 inch diameter units manufactured by Spendrup Fan Co. were direct driven—the motor shaft was directly connected to the aluminum fan rotor. To comply with the Class I Division 2 requirements,

Table 1. Ventilation duct pressure loss analysis

**VENTILATION ANALYSIS FOR REACH 4 EAST**

**Required Air Qty**

|  | Qty | HP | CFM per unit | Total CFM |
|---|---|---|---|---|
| Men in tunnel | 10 | N/A | 200 | 2,000 |
| 12-ton Locies | 3 | 110 | 100 | 33,000 |
|  |  |  | Total | 35,000 |

**Duct Pressure Loss**

| Duct Element | Length [ft] | Dia. [in] | No. [-] | Area [sft] | Friction Factor [-] | Velocity [fpm] | Velocity Pressure [in w.g.] | Loss Coefficient [-] | Pressure Loss [in w.g.] |
|---|---|---|---|---|---|---|---|---|---|
| Inlet | N/A | 36 | 1 | 7.07 | N/A | 4,951 | 1.53 | 0.85 | 1.30 |
| Elbow 90° | N/A | 36 | 2 | 7.07 | N/A | 4,951 | 1.53 | 0.33 | 1.01 |
| Duct (Shaft) | 60 | 36 | 1 | 7.07 | 12.0 | 4,951 | N/A | N/A | 0.45 |
| Expansion (36 to 38) | N/A | 36 | 1 | 7.07 | N/A | 4,951 | 1.53 | 0.59 | 0.90 |
| Duct (Tunnel) | 23,200 | 38 | 1 | 7.88 | 12.0 | 4,444 | N/A | N/A | 133.56 |
| Transition 38 / 36 (Oval) | N/A | N/A | 2 | 7.88 | N/A | 4,444 | 1.23 | 0.15 | 0.37 |
| Transition 36 (Oval) / 38 | N/A | N/A | 2 | 7.88 | N/A | 4,444 | 1.23 | 0.15 | 0.37 |
| Outlet | N/A | 38 | 1 | 7.88 | N/A | 4,444 | 1.23 | 1.00 | 1.23 |
| Velocity Head Loss | N/A | 36 | 1 | 7.07 | N/A | 4,951 | 1.53 | N/A | 1.53 |
|  |  |  |  |  |  |  | **Total Pressure Loss** |  | **140.72** |

**Fan Selection**

| Fan Model | Blade-Pitch [-] | Fan Pres. [in w.g.] |
|---|---|---|
| 125 Hp Axial-Fan | 4 | 15 |

**Required Number of Fans**

| Total | 9 |
|---|---|
| On top | 1 |
| In the tunnel | 8 |

**Required Fan Spacing**

| S [ft] | 2,775 |
|---|---|

the electric driven fan motors were built and rated explosion-proof. In addition, the fans were bi-directional to comply with the reversibility requirement imposed by Cal/OSHA.

The fan installed on the surface was equipped with a center-pod silencer on the inlet as well as outlet side to reduce noise. It was not necessary to install silencers on the in-line booster fans in the tunnel because they were not located near working areas nor considered to be a nuisance to the public. However, each fan was fitted with a screen on both ends to prevent objects accidentally sucked into the ducting system and permanently damaging the fan.

The power for the booster fans in the tunnel was drawn from the 13.2 kV main power supply cable for the Digger Shield and stepped down to 480 V using transformers supplied by Tunnel Electric—installed on the tunnel wall at each booster fan location. The transformer cabinet also housed the VFD for each fan as well as communication equipment.

**FAN CONTROL**

Any time industrial machinery needs to be precisely controlled, it is useful to use a Programmable Logic Controller or PLC. The PLC allows software to be programmed and read inputs and send outputs. In the most basic sense, the PLC is just a series of switches, and how the switches are turned on and off is done by the software, which is programmed by the user. The ability to make software changes in a short time reduces downtime and hardware costs. Traylor Shea JV chose to use an Allen Bradley CompactLogix L32 PLC. The smaller PLC includes a variety of I/O, while still allowing Ethernet communication.

Because the tunnel ventilation required a ramped startup, and reversible directions, the PLC needed to communicate to each fan individually but also simultaneously and quickly. Traylor Shea JV chose to use G.SHDSL communication network. G.SHDSL is a subset of the DSL technology, which is similar to a modem connection, but transmission is at a much higher frequency. G.SHDSL has a maximum bandwidth of 4.6 Megabits/sec, which was sufficient for the transfer of data to and from the VFD. After trying several different brands, Patton Electronics, proved to be the most reliable.

To use the Patton G.SHDSL modems, there needed to be a modem at each end of the line to convert the Ethernet connection to the G.SHDSL protocol. On the surface, a "Central" modem was installed for each fan. Each "Central" modem was connected to a "Remote" modem over a multi-conductor cable down to each transformer where the "Remote" modem would transform the G.SHDSL protocol back to Ethernet. Each modem had its own IP address and integrated web server, which aided in troubleshooting the network.

A Yaskawa VFD drove the fans and each had an optional Ethernet card installed. The option card allowed the VFD to be controlled by the PLC via Ethernet protocol instead of the analog or digital inputs on the drive. The option card had its own IP address and web server. By entering the IP address of a VFD into a web browser, such as Microsoft Internet Explorer or Mozilla Firefox, the webpage displays the status of each fan.

During instances when communication to a fan was lost, typically a broken wire, the PLC could no longer communicate with the drive—the drive would retain the last speed commands it received from the PLC and the PLC would retain the last information it received from the VFD. Once the communication link was restored, the protocol between the modems needed to be restored and this required that the power to the drive be recycled. Consequently, when a fan lost communication, the entire system would have to be shutoff and restarted to protect the ventilation duct from collapse. An updated option card from Yaskawa solved the problem of re-establishing communications and protocols so when a link went down and came back up, communication between the drive and PLC were established automatically.

To communicate with the Yaskawa VFD, word bytes were programmed into the PLC. These words are like little packets of information that include such commands as Start/Stop, Forward/Reverse direction; data from the drive include status information like Running State—Running/Stopped, Running Mode—Auto/Manual, Alarm or Fault Conditions, and current draw.

On fan startup, each fan was given a run command to ramp to 30Hz. This was done to get air moving in a slower fashion. If not programmed properly, a VFD has the function of jumping to the given speed command. After running at 30Hz for 30 seconds, the PLC issued the command to ramp to the second stage; the second stage would take the fans to the maximum current draw without tripping the circuit breaker. The circuit breakers were set at 135 amps, and the fan set points were at 130 amps. The VFD can only be given a speed command, and will return the current draw. The PLC was programmed to analyze the current draw each fan was pulling and compare against the programmed 130-amp set point in the PLC. If the current was too high, the PLC re-issued a speed command that was 0.02Hz less than the previous value, if the current was lower than 130 amps, it would increase speed by 0.01Hz. By analyzing the current fluctuation and adjusting the speed accordingly, the fans operated at maximum power without tripping the circuit. In the event that the current dropped below 120 amps on any single fan (such as a fan burning up), the system would automatically send a speed command to the other fans to reduce speed to 30 Hz; this was protection against fan-line collapse.

The performance and status of each fan was shown on a computer screen running RSView32 on both the Diggershield's Operator Screen, and in the Superintendant's trailer. A screenshot can be seen in Figure 3. The screen showed the status of each fan, including: Location, Status, Speed, Direction, Alarm State, Current Draw, Control—Auto/Manual, Fault Status, and Communication. The Yaskawa VFDs do not have a heartbeat with the PLC so in order to establish whether or not the communication link was up, the PLC analyzed the current fluctuation. If communication was lost between a fan and the PLC, the fan would run using the last commands it received from the PLC. The PLC would hold the last state of the Fan that it received. By looking at the fluctuating current, a quasi heartbeat could be established and could be used for communication status.

The tunneling operations were done in three shifts, six days per week. Sunday was an off day, so to save on power a weekend mode for the fans was programmed. The Superintendent would click the "Toggle Fan Mode" button on the fan control screen and a prompt for password would show. He would enter the password and click OK. As soon as the OK button was clicked, the fans would all ramp down to 30Hz, thus reducing the amount of energy used when no one was in the tunnel, but still maintaining airflow to prevent accidental gas build up. The password protections helped keep changing of the fan speed inadvertently. The weekend mode was also used if a single fan had to be stopped and then restarted.

Fan reversal was done at the fan control station. Outside of where the PLC was mounted, a fan control box showed the status via illuminated bulbs. It also had a stop button and direction control buttons. To reverse the fans in case of an Emergency, at the direction of the Superintendent, the top man could stop the fans. A light on the panel would indicate that all the fans have come to a complete stop. He could push the Run REV button. The PLC would not allow the reversal of fans until all the fans came to

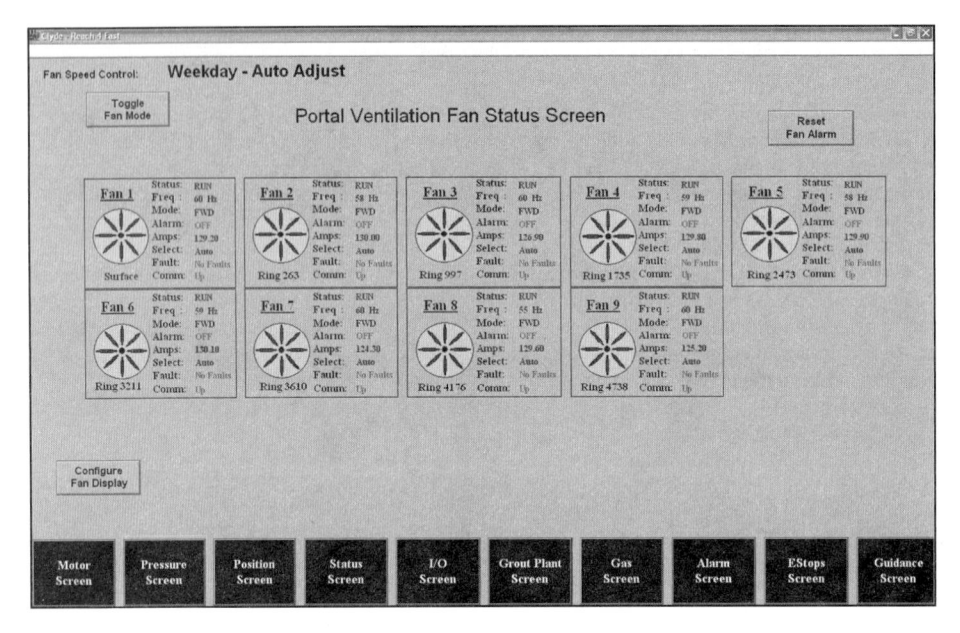

**Figure 3. Fan control screen**

a complete stop. A bit within the message from the VFDs gave the "Zero Speed" indication to the PLC. The fan reversal was tested at intervals to ensure that the system would work properly in the event of an emergency.

## ECONOMIC ANALYSIS/POTENTIAL

By utilizing a PLC, VFDs and the network between them, a real economical savings can be realized. It has become common in the mining industry to control ventilation where people and machines are working and to shutdown areas where no work is being performed. This is typically not the case of tunneling projects because of their linear nature. However, the opportunities to maximize airflow and minimize energy usage and costs when needed do arrive. In the cases in which tunneling operations change, such as off times, during cross passage work, or even after hole through, ventilation systems can adapt to benefit a contractor.

Some contractors may avoid PLC controlled systems for various reasons, such as initial startup costs of hardware, programming costs, and maintenance. These same arguments were the same for PLC controlled Tunnel Boring machines, now viewed as standard technology. The initial investment in the PLC and VFD drives can be substantial, but the realization of electrical costs soon outweighs the capital purchases. Furthermore, the system can be reused on any job with minor modifications. Additionally, clients informed about what technologies are available, are specifying PLC controlled ventilation systems.

With ever-rising energy costs and the need to keep "tunneling-sustainable" at a construction level, contractors have to continue to explore energy savings opportunities. A sophisticated ventilation system is one avenue that should be pursued in the future without sacrificing the safety of the tunnel workers.

Many mines around the world have already adapted demand based ventilation systems [3]. These systems control ventilation based on where the mining equipment and personnel are located in the mine at any given time. Much of this is controlled by PLCs, which monitor equipment and personnel's location through transponder monitoring systems, such as leaky feeder. If equipment and personnel are not working or traveling through a section of the mine, then the fans are shut down or speeds lowered to save energy. The cost savings can be very substantial, especially on a project with long headings.

If the project decides to operate the fans at half speed during non-work hours, such as over a weekend, the fans could be reduced to 50% of their load demand from control on the surface. For instance, if the ventilation system has six 125 HP fans and they are drawing about 130 amps each with a power rate of $0.147/kW-hr, that system costs about $81 dollars an hour during full load and $40 per hour during reduced mode. This equates to a savings of $984 per day for just those six fans. The SVT project had as many as twenty—125 HP fans operating at one time over two separate headings, the saving in the reduced mode was substantial.

The SVP shows that this same concept from the mining industry can be applied to the tunneling industry; given the right project and effort. Of course, resistance by the workers is the first obstacle to overcome, but by designing a sound and well thought out system, the opportunity is there to design a system based on demand in the tunnel without sacrificing worker's safety. Airflow requirements, gas monitoring, temperature, and pressure can all be monitored by the PLC and provide a system sophisticated enough to create an energy efficient ventilation system without risking the tunnel worker's safety.

**FAN-LINE COLLAPSE INCIDENT**

Near the end of the tunnel drive, half way between the planned interval distance between fan 6 and fan 7, the fan line suffered a catastrophic collapse at fans 4 and 5. There was already obvious concern for this problem, but was compounded due to the wear on the duct after driving this tunnel over a longer than anticipated schedule duration. In this event, several pieces of duct had collapsed directly in front of fan 5 and many more in front of fan 4. The mystery of this collapse was that all of the fans seemed to continue to operate and so there was no smoking gun as to the cause.

An inspection of the fans' screens for a blockage that could have caused the problem was made and none was found. The duct was also inspected for significant dents and deterioration and nothing out of the ordinary was discovered. Not knowing what the problem was immediately, TSJV was forced to install an additional fan at a less than anticipated fan spacing. This would increase the system horsepower and reduce the pressure in front of fan #6. In addition, interlocks were established in the PLC to cause a system slow down in the event that any of the fans lost communication or were faulted for any reason. Prior to the collapse, the VFD's were programmed to run in their last known state after a loss in communication. Furthermore, a pressure transducer was installed in front of the forward most fan—the suction side; this transducer tracked the pressure increase as the duct length increased. Another pressure transducer was installed in front of fan #5 to monitor the negative pressure in the duct.

After replacing the duct, inserting another fan, and programming the additional PLC interlocks, the tunnel was successfully finished, however several other interruptions to incoming power did occur. By monitoring and logging the pressure inside the duct and with the safety interlocks established, it was clear that the interlocks did protect the system from another catastrophic duct loss during these power loss events.

In hindsight, the likely cause of the collapse was a malfunction at fan #6, which in turn caused a collapse at the highest negative point in front of fan #5, which in turn caused the collapse in front of fan #4. This could have progressed all the way to the shaft, but luckily, it stopped at fan #4. With a duct system this long, and had the failure caused more damage, the duct collapse could have caused a huge schedule and cost impact to the project. Several other control measures were discussed over the course of the project, such as negative pressure blast doors in front of the fans that could relieve a high negative pressure. Contractors should consider the use of heavier gauge fan line in front of the fans to prevent fan line collapse at the highest negative pressure zone.

**CONCLUSION**

As tunneling projects become ever more complex, the ability to move large quantities of air becomes a larger constraint to the constructability of the project. This paper highlighted the design, implementation, operation, and failures of a multifaceted ventilation system in a small diameter tunnel. Furthermore, to demonstrate how these engineering challenges can be met and how a real economical savings can be achieved through industrial process control. As the implementation of PLCs on Tunnel Boring and Digger Shield Machines has changed the way today's contractor excavates tunnels, the PLC will take ventilation systems further in the 21st century.

**REFERENCES**

[1] San Diego County Water Authority. *San Vicente Pipeline Projects Fact Sheet*. July 2009.
[2] Tunnel Safety Orders (Rev. April 1, 1990). Division of Industrial Safety. California Code of Regulations.
[3] O'Connor D.F. *Ventilation on demand (VOD) auxiliary fan project—Vale Inco Limited, Creighton Mine*. 12th U.S./North American Mine Ventilation Symposium 2008. Wallace (ed).

**SELECTED READING**

Hartman, Mutmansky, Ramani, and Wang. *Mine Ventilation and Air Conditioning*. 1997. John Wiley & Sons, New York, NY.

Schauenburg Flexadux Corp. *Designing a Mine Auxiliary Ventilation System*. Grand Junction, CO.

C.J. Bise. *Mining Engineering Analysis*. Society of Mining, Metallurgy, and Exploration. 2003. Littleton, CO.

Rob Pope Jr. *Making the right ventilation choice*. January 2003. Tunnels & Tunneling International. UK.

R. Jorgenson. *Fan Engineering Handbook*. 9th Edition. Howden Buffalo. Buffalo, NY.

# TRACK 1: TECHNOLOGY

Mike Smithson, Chair

*Session 2: Innovation*

# Onsite, First Time Assembly of TBMs: Merging 3D Digital Modeling, Quality Control, and Logistical Planning

**Joe Roby, Desiree Willis**
The Robbins Company, Seattle, Washington

**ABSTRACT:** Traditionally, the delivery of Tunnel Boring Machines (TBMs) has been preceded by full assembly and testing of the TBM at the manufacturer's facility before dismantling and shipping to site. Recent years have seen a rapid development of 3D CAD tools, modular TBM and back-up designs, and more advanced Quality Assurance procedures. These advances can now prevent clashes of components and incompatibility of equipment without the need for full workshop assembly. To realize the benefits, a new method of TBM delivery has been developed called Onsite First Time Assembly (OFTA). This paper discusses the challenges and benefits of OFTA with specific examples given from several recent projects that employed this method of delivery.

## WHY IS OUR INDUSTRY LAGGING?

Traditionally, hard rock and earth pressure balance (EPB) tunnel boring machines have been fully assembled in a factory, tested, disassembled, transported to the job site and reassembled onsite prior to starting operation. Conversely, other large scale industrial equipment is rarely fully preassembled and tested prior to being installed in the final intended location. For examples, think of small to midsized gas-fired power plants, bucket wheel excavators and specialized small manufacturing factories. Like tunnel boring machines, for all of these examples, time is of the essence. In the period from ordering equipment to having operational equipment, nothing is being produced and cash flow is negative. While elimination of the first three steps of the process (factory assembly, no-load testing and disassembly) does not result in a 100% equivalent savings in time and labor, it does result in a substantial savings of both. In addition, there is savings for reduced transport cost.

Given the similarities in complexity and delivery times for industrial equipment and tunneling equipment, it begs the question: Why have tunneling machine manufacturers, civil constructors and project owners resisted for so long the direct shipment of components to site for first time assembly? It is easy to understand why machine manufacturers with a large investment in fixed manufacturing facilities would be resistant to on site, first time assembly (OFTA), preferring the status quo which keeps prices higher and their facilities full of equipment. It is more difficult to understand the motivations of owners and consultants. Even today, in the face of mounting evidence of the benefits of OFTA, many consultants continue to stipulate in tender documents that the tunneling machine must be fully factory assembled and tested. This resistance to OFTA might be caused by the generally conservative nature of project owners and their consultants; however, it is more likely to be a simple habit from the past which it is time to break.

## A CHANGING WORLD

### The Evolution of Design and Manufacturing Tools

Thirty years ago tunneling machines were designed manually, on drawing boards with pencil and paper, with design calculations performed on what are, by today's standards, antique calculators. Project management software was only a dream. Twenty years ago, things improved somewhat with 2D CAD becoming the norm and the first wave of project management software becoming available. Over the past 10 years, the expansion of 3D CAD and improvements in project management software, including links to many enterprise/manufacturing resource planning programs (ERP/MRP), have given manufactures invaluable tools with which to insure the quality of design, the fit up of complex parts and the delivery of complex systems.

### A Mature Industry

Thirty years ago, nearly every tunnel boring machine was unique in its design, custom built and manufactured specifically for the project. Today, that is very rarely the case. Thirty years ago perhaps scores of tunneling machines were made annually. Today hundreds of tunneling machines are made annually and, like automobiles, many of them are of the same make and model. Due in no small part to the sharing of knowledge through professional organizations such as those supporting this conference, today there is much common agreement regarding the "type" of tunneling machine best suited to certain geological conditions. Standard types are open hard rock, hard rock single shield or double shield, EPB or slurry shield. As a result, the design of each of these machine types have moved from their uniquely designed, custom origins to a vastly superior product today which is mature in design and subject to continuous improvement through incremental changes.

As a result of this maturation and evolution of the tunneling machine, when a civil contractor receives a machine today the probability is very high that at minimum the core of the machine has been produced many times previously. This provides both a more reliable product and the potential to eliminate the in-factory assembly phase with minimal risk for all involved.

**A Total Program for OFTA**

The use of 3D CAD software today makes it possible to accurately check the fit up of the component parts of complex machinery in the design phase. A thorough and time-tested quality assurance program insures that components are made per print, continuing the fit up guarantee through the next step of production. The availability of project management (PM) software with the capability to plan and monitor resources throughout the design and production process, and the linking of PM software to the ERP software for an entire company, provides a powerful tool for insuring that every component of these complex systems is delivered to the job site when it is needed during the assembly process.

But, it takes humans, experienced humans, as well as software to make OFTA work. While this is true throughout the design and manufacturing process, it is especially true on the job site, where the complex tunneling machine must be assembled safely, quickly and correctly to achieve the targeted schedule and cost savings. Fortunately, the widespread global growth in the use of all types of tunneling machines over the past twenty plus years has resulted in a worldwide pool of highly experienced tunneling machine professionals. Wherever in the world a project might be, it is possible today to put onsite a team of professionals who can direct the assembly and operation of every type of machine.

**Why OFTA?**

Depending on the size and complexity of the tunneling machine being produced, and whether the machine is new or refurbished, the savings in both schedule and cost can be substantial. On a small, 3.0 meter simple machine the savings in schedule can be as little as a month or so and perhaps 5,000 man-hours and 100,000 US dollars in transport cost. On a complex 10 meter or larger machine the savings in schedule can be as much as several months and possibly 15,000 man-hours can be saved as well. Eliminating the transport to factory for preassembly of such large machines can reduce transport costs by more than a million US dollars. However, the reduced costs noted here are generally dwarfed by the large commercial gain inherent in delivering a major tunneling project on a shorter schedule.

The remainder of this paper discusses some recent projects on which OFTA delivery was employed, noting problems encountered and their resolution. In closing, the paper lists the requirements necessary for a manufacturer to provide a successful OFTA program while minimizing risks associated with the program.

**RECENT PROJECTS EMPLOYING OFTA DELIVERY**

**Niagara Project—Canada**

In 2005, Austrian civil contractor STRABAG was awarded a 600 million Canadian dollar design–construct contract for the Niagara tunnel project. The TBM bored tunnel is concrete lined at 12.7 m (41.7 ft) internal diameter and 10.4 km (6.5 mile) long. The project funnels water past the famous falls to the Sir Adam Beck Power Station, providing power to the province of Quebec. Figure 1 shows the tunnel route. STRABAG purchased a 14.4 m (47.4 ft) open, hard rock, high-performance TBM (HP-TBM) from The Robbins Company for the project, and specified OFTA delivery. The large, custom designed machine was contractually specified to be delivered, ready to bore at the job site, 12 months from the signing of the TBM supply contract.

**The Niagara TBM—Design and Manufacture**

The 14.4 m Robbins HP-TBM, nicknamed "Big Becky," is the largest hard rock TBM ever produced. The TBM is fitted with 254 mm (20 inch) diameter, back-loading cutters and the cutterhead is powered by $15 \times 315$ kW (4725 kW, 6330 HP) motors with variable frequency speed control. The TBM, without backup, weighs over 1100 t (1210 st).

The TBM was designed in Robbins USA offices with major components being manufactured in the

**Figure 1. Niagara tunnel route**

USA, Canada and Europe. The 400 t (1210 st) cutterhead was manufactured in the UK. Subcomponents were, where possible within the tight schedule, pre-assembled in workshops as they were manufactured. The timing of exworks delivery of the components was tightly controlled in order to assure arrival of components at the job site in the order required for assembly, without undue handling and storage in the limited space available at site.

**The Niagara TBM—Onsite Assembly**

For onsite assembly, the TBM manufacturer provided a team of experienced supervisors and specialist technicians. The contractor provided local labor and tools. Assembly was carried out in the open cut leading to the bored tunnel. Figure 2 shows the key components in place with the gripper cylinders and carrier being lowered into position for installation. Figure 3 shows a special tool used for installation of the main drive pinions.

The core design of the Niagara machine is similar to previous machines manufactured by Robbins. Many of the components had been fit up on previous jobs and it was only necessary to provide a high level of inspections following component manufacturing to ensure proper fit up at the Niagara job site. In spite of a very aggressive in-factory quality control program, a few problems were encountered during the assembly. Fortunately, the errors were rapidly corrected. Two of the issues involved interferences found on fit up of parts. In one case, during

**Figure 2. Installation of gripper cylinders and carrier onsite**

the manufacture of a large and complex weldment/machining, the cutterhead support, a single rough machining step had been overlooked. While hundreds of dimensions had been properly checked during the factory inspection, this single dimension had not been checked. In both cases of interference, a local specialist in onsite machining was employed to make the necessary small corrections. It must be stressed that had OFTA not been employed, these same errors would not have been discovered until in-factory assembly and the resultant repair time would have been the same as it was in the field. The extremely large size of the parts makes it easier to

**Figure 3. Special tool for installation of final drive pinions**

bring the machining tools and personal to the part than to take the part to a machine shop. In any case, in spite of the few problems encountered, the onsite assembly proceeded on schedule. Figure 4 shows the TBM cutterhead being installed onsite.

**The Niagara Result**

The TBM supply contract stipulated a 12 month, "ready to bore" OFTA delivery which was in fact achieved. This saved approximately 4 to 5 months when compared to a traditional 11 to 12 month factory assembly schedule, followed by disassembly, transport and reassembly onsite. In addition, it is estimated that 1.3 to 1.8 million dollars were saved in labor and transport costs by eliminating the factory pre-assembly for this very large machine. Figure 5 shows the machine fully assembled at site, ready to bore.

**Alimineti Madhava Reddy (AMR) Project—India**

The Indian civil contractor Jaiprakash Associates Ltd. was awarded the construction contract for the Srisailam Left Bank Canal (SLBC) Tunnel, which is part of the Alimineti Madhava Reddy (AMR) Project in Andhra Pradesh in southern India. During monsoon season, water will be transferred from the Srisailam Reservoir to 300,000 acres of farmland as well as providing drinking water to many villages. The tunnel is approximately 43.9 km (27.3 miles) long with no possibility for intermediate access. Above the tunnel route are a tiger reserve, wild life sanctuary and areas of protected forestry. In order to minimize disturbances to these sensitive environments, Jaiprakash elected to employ tunnel boring machines to excavate the tunnels. In 2005, Jaiprakash ordered two 10 m diameter, hard rock, double-shield

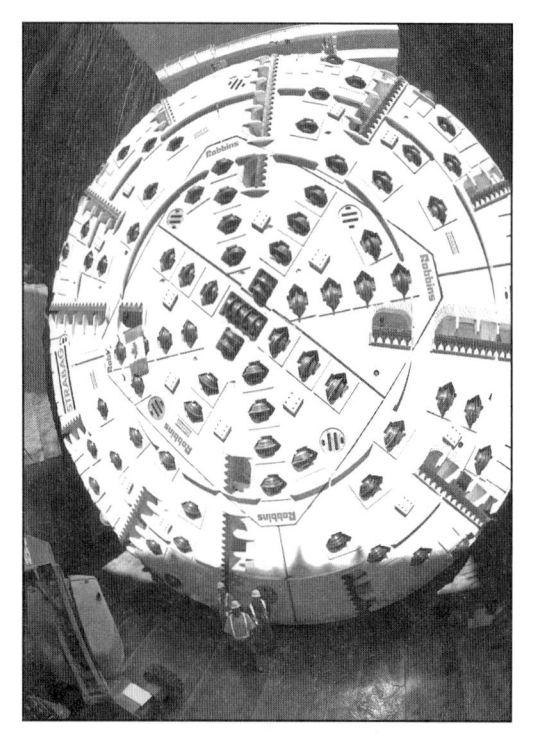

**Figure 4. TBM cutterhead installed onsite**

TBMs from The Robbins Company. The tunnel is to be completely lined with concrete, and 60% of the tunnel is in very blocky, layered shale and quartzite. Because of these factors, double shield TBMs were selected in order to get the highest advance rates with simultaneous tunnel lining, ensuring the quickest delivery of the final operational tunnel. Figure 6 shows an elevation view of one of the tunnel boring machines.

*The AMR TBM—Design and Manufacture*

The TBM supply contract required the first components to arrive onsite not later than eight months after order, with all components to be delivered not later than thirteen months after order. Contractor and machine supplier agreed to an OFTA delivery in order to have the machines ready to bore in the shortest possible time. Figure 7 is a photograph of the open cut assembly and startup area at the job site.

Unusual for modern TBMs, the 10 m double shield machines were in large part a completely new design for Robbins and so particular care was given in the design stage to ensure proper fit up of all parts at the job site. The major TBM structural components were manufactured in China, while the backup structure was manufactured in India. As is typical for modern TBMs, the other components

**Figure 5. TBM and back-up assembled in launch chamber**

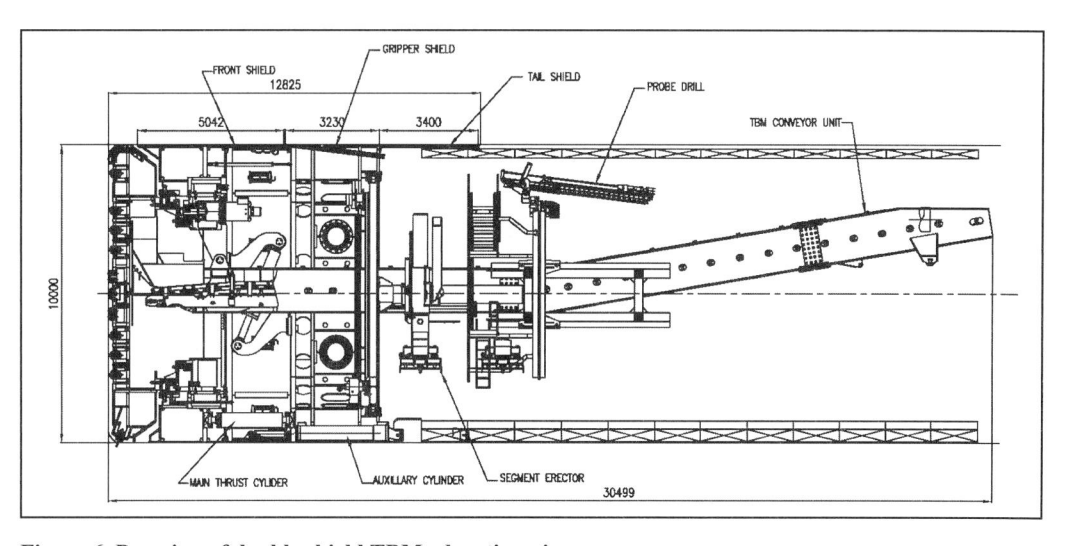

**Figure 6. Drawing of double shield TBM, elevation view**

(e.g., electric motors, gear reducers, main bearing, seals, hydraulic cylinders, etc.) were sourced from locations in the USA, Japan and Europe. Every component of the tunneling systems was tracked through design—from design drawing release through every step of manufacturing, transport and delivery to the job site. Figure 8 shows installation of the segment erector rotation ring at the job site, while Figure 9 shows cutterhead assembly.

**The AMR OFTA Result**

The shipping dates required per contract were met. It has been estimated that a schedule savings of 4 to 5 months was achieved when compared to a traditional

Figure 7. Assembly proceeding in the granite open cut

Figure 8. Installing the erector rotation ring—AMR, India

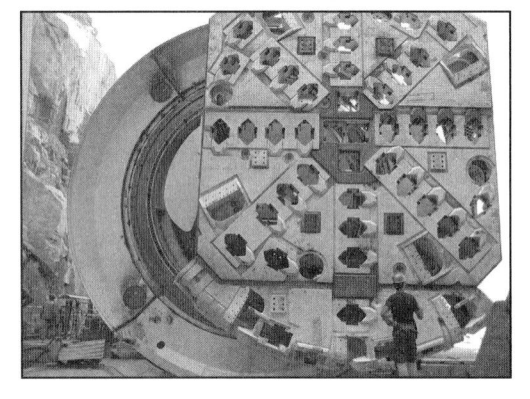

Figure 9. Installing the center cutterhead section—AMR, India

"factory assembled" delivery. Cost savings in reduced labor and transport costs were estimated to be greater than 3.5 million dollars for the two machines.

Again, the TBM manufacturer provided a large number of supervisors and technicians to direct and aid in the onsite assembly and testing of each machine. In spite of the machines being a new design, fit up problems at site were minimal and corrected rapidly without impacting the assembly schedule.

**Jin Ping II Hydroelectric Project—China**

At the Jin Ping II Hydroelectric project, four (4) parallel headrace tunnels are being driven with an average length of 16.6 km. Two of the tunnels are being excavated by 12.4 m open, hard rock TBMs and two by drill and blast. From intake structures near Jingfeng Bridge water will flow through the four Jinping headrace tunnels downgrade at 3.65% to the underground Dashuigou powerhouse. Eight 600 MW turbine generators will be installed in the power house for a total generating capacity of 4800 MW.

**Figure 10. Pre-assembly of the cutterhead support and shields—Jin Ping II, China**

**Figure 11. Onsite boring repair of gripper carrier way bushing**

China Railway 18th Bureau (Group) Co Ltd. (CR18) won the construction contract for headrace tunnel Nos. 1 and 2, which includes one drill and blast and one TBM bored tunnel. CR18 awarded the TBM supply contract to The Robbins Company.

*The Jin Ping TBM—Design and Manufacture*

Limited road access to the job site meant that the largest TBM components needed to be delivered by river. However, seasonal low and high water flows on the river made for a short seasonal window within which the components had to be delivered to site. If the parts could not be delivered to site within this window, it would be several months before there would be another opportunity. As a result, OFTA delivery was specified in order to reduce delivery time and reduce the risk of missing the transport window.

In this case, many of the core TBM component designs were the same as those employed on Niagara and previous projects, which reduced the risk of fit up problems onsite. The backup system was, however, a completely new design. Robbins designed the machine at their facilities in the USA and China. The main TBM structural elements were manufactured in the city of Dalian in northeast China. Where the schedule allowed, some factory preassembly was done to check critical component fits and reduce onsite assembly time. For example, the main bearing, gear, and pinions were installed in the cutterhead support and the ring gear-pinion mesh was checked. Also, the muck chute, side supports, roof support, and front support were temporarily installed in factory to check fits. Figure 10 shows factory preassembly of components on the cutterhead support. All remaining components were assembled for the first time onsite.

**The Jin Ping OFTA Result**

The onsite assembly was not 100% error free but once again it was proved that correction can be made onsite nearly as quickly as in a factory, even on a job site as remote as the Jin Ping, China site. Early in the assembly it was discovered that the bushings in the gripper carrier ways had not been finished machined in the factory. Shipping the part to a Chengdu factory, the nearest large city, was not possible due to severe damage to local roads and machine tools, which occurred in the severe Sichuan province earthquake in 2008. A machining contractor in Shanghai was employed to bring a portable boring unit to the job site and make corrections to the part. The part was line bored onsite in only 3 days. Figure 11 shows the line boring machine in use at site.

At the peak of effort, Robbins provided 42 people to support the assembly: 16 supervisory personnel from the USA and Europe and 26 engineers, mechanics and electricians from China. Despite record breaking snowstorms and a magnitude 8 earthquake, the TBM and backup was fully assembled and ready to bore in only three months. People experienced with field assembly of large diameter hard rock TBMs opine that the three month assembly period could not be improved upon by pre-assembly of the machine in a factory prior to delivery to site. Again, the savings in time with OFTA is estimated to be in the 4 to 5 month range and cost savings are estimated to be approximately 2.3 million dollars in labor and transport costs. Figure 12 shows the TBM in the onsite assembly hall, ready to walk to the face to begin boring operations.

**Mexico City Metro Line 12—Mexico**

Mexico City's metro has the 5th highest ridership in the world, carrying 1.46 billion passengers in 2008. In 2008 a joint-venture of Alstom and Mexican partners

**Figure 12. OFTA—ready to bore**

ICA and CICSA was awarded a 1.74 billion US dollar contract to build metro Line 12 in the southeast sector of the city. The new line 12 will be 24 km long from Mixcoac to Mexicaltzingo, have 22 stations and connections to lines 2, 3, 7 and 8. The contractors will utilize a Robbins 10.2 m (33.5 ft) diameter EPB on a 6.2 km (3.9 mi) long section of tunnel for the new line. The tunnel passes under the water table through high water content clays as well as sands, silt and gravel with the potential for boulders up to 800 mm (30 inches). Figure 13 shows a 3D CAD drawing, section view through the EPB machine.

### The Mexico Line 12 EPB—Design and Manufacture

The 10.2 m EPB was designed in Robbins' USA and China offices. Primary structural components were manufactured in Japan, China, Korea and Mexico. The main bearing, drive components and hydraulic and electrical components were sourced in the USA, Europe and Japan. This machine contained a combination of previously produced designs (cutterhead support/main bearing and seal assembly/main drives) and new designs. The main bearing and seal were installed in the cutterhead support at the factory in Korea prior to being transported to the job site. Again, an extensive quality control program was in place and rigorous dimensional checks were performed on major components prior to their delivery from the factories.

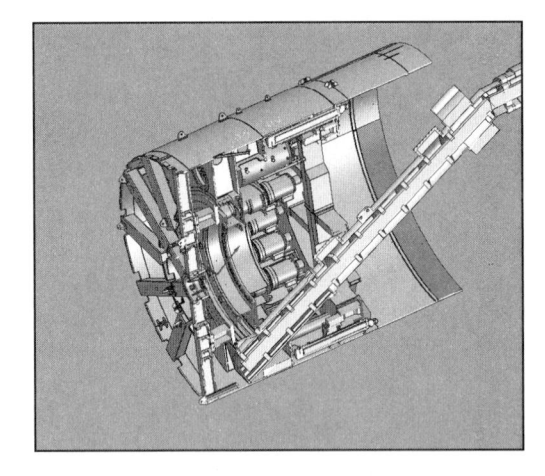

**Figure 13. 3D CAD section view of MX 12 EPB—Mexico MX12**

### The Mexico Line 12 OFTA Result

At the time of writing, the MX 12 on site assembly had just started with the primary major components arriving at site. Figure 14 shows the lowering into place of the bottom half of the front shield –Ring A. Figure 15 shows the cutterhead sections being joined with the pedestal at the top of the assembly shaft on site. Figure 16 includes two views of the cutterhead support: the upper photo is of the forward pinion support bearing and a bit of the ring gear, while the lower photo shows the same view after installation of the drive pinion.

Assembly is proceeding on schedule and as of this writing no major fit up problems have occurred.

## SUMMMARY

Specialist TBM tunneling contractors frequently own several TBMs that are refurbished and moved from job to job. Having been fully assembled once or more previously, it is extremely rare for these *used* TBMs to ever be fully assembled in a factory. They go straight from the contractor's storage and repair facility to the job site. Robbins has now successfully demonstrated the potential for Onsite, First Time Assembly of *new* TBMs. Several TBMs of different types (hard rock—open, double shield and EPB machines) have been delivered using the OFTA method. In every case there has been a substantial reduction in the time required to start boring—as much as 5 months. There has also been great savings in cost—more than two million dollars for a large diameter rock machine. There may be a further, though hard to quantify, advantage in the in-depth training the contractor's personnel receive during the onsite assembly, when working closely with the larger supervision staff provided by the machine manufacturer with OFTA delivery.

It cannot be argued that there is no increase in risk with OFTA, but experience has shown that the risk is definable, largely controllable and most important recoverable. The primary risk is, of course,

Figure 14. Installing lower section of A Ring—Mexico MX12

Figure 15. Assembling cutterhead sections with pedestal—Mexico MX12

Figure 16. Installation of drive pinions—Mexico MX12

errors in design or manufacturing which result in a misfit of components during assembly at site. This risk is mitigated through the use of previous designs, 3D CAD, the implementation of proper design procedures and checks, and the implementation of an aggressive quality control program. Finally, as has been shown on the OFTA deliveries to date, when problems are encountered during onsite assembly, the schedule can generally be recovered through onsite correction carried out with the assistance of specialist fabrication and machining companies.

Key components of a successful OFTA program include:

- Use of prior, proven designs where possible
- 3D design and computer aided test fitting of critical components
- 100% dimensional inspection of critical components at the fabricators
- Pre-assembly of subcomponents/modules when schedule allows or is not impacted by pre-assembly
- Aggressive quality control of all components manufactured to ensure proper fit up at site
- Absolute control of the total tunneling machine system bill-of-materials, to ensure that every part, large and small, which is required for the system is sent to the job site
- Logistical planning and control, to ensure that every part arrives at the job site, when it is required, in the order that it is required for efficient assembly and use of storage space
- Resource planning, to ensure that all tools and personnel of every type, qualification and quantity required for assembly are onsite when needed
- Advance alternative recovery planning, in order to be ready to react quickly to possible failures in any of the above steps

- A larger than usual team of highly experienced personnel must be provided by the machine manufacturer to supervise and assist with the onsite assembly

Project owners have an obligation to the public to deliver underground infrastructure at competitive prices, on the quickest practical schedule and without undue risk. Contractors need every tool available to meet these demands. It is time for tunnel owners and contractors to move into the new millennium. OFTA offers an opportunity to save both time and cost. Experience to date prove that it can be done with little risk with the use of modern design, quality control and project management tools. It is time for all involved to eliminate from future tunneling construction contracts the requirement for TBMs to be fully shop assembled and tested prior to delivery to site. Onsite, first time assembly is a practical, cost and schedule saving alternative.

**REFERENCES**

Gschnitzer, E. and Harding, D. (2008). Niagara Tunnel Project. Proceedings of the Tunnelling Association of Canada conference.

Willis, D. (2008). Beating the Clock: On-site First Time Assembly. *Tunnels & Tunnelling International,* September, 29–31.

Brundan, W. (2009). Robbins 10m Double Shield Tunnel Boring Machines on Srisailam Left Bank Canal Tunnel Scheme, Alimineti Madhava Reddy Project, Andhra Pradesh, India. Proceedings of the Rapid Excavation and Tunneling Conference.

Smading, S., Roby, J. and Willis, D. (2009). Onsite Assembly and Hard Rock Tunneling at the Jinping-II Hydropower Station Power Tunnel Project. Proceedings of the Rapid Excavation and Tunneling Conference.

# Large Diameter Segmentally Lined Shafts

## Darin R. Kruse, Rodney Meadth
Cobalt Construction Company, Simi Valley, California

**ABSTRACT:** This paper explores some of the unique engineering challenges contemplated in the building of shallow large diameter (up to 90 m [300 ft]) segmentally lined shafts for non-traditional heavy civil or commercial uses, such as parking, storage, transportation, or even housing facilities. The proposed design and construction approach addresses a number of inefficiencies currently present in commercial practice. The studied design considers excavation depths up to 15 m (50 ft), and includes internal structural bracing (floors, ring beams, etc.) and post-tensioning elements in its final form. The results of economic modeling, field testing, prototype grouting methods and 2D and 3D finite element models are discussed.

## INTRODUCTION

Typical building industry practice utilized in engineering and construction of relatively shallow underground structures constructed in soft ground (that require some form of temporary earth shoring) generally incorporates a two-step process: installation of temporary shoring to support earth loads during excavation, followed by construction of the permanent structure. These structures are typically rectilinear in shape.

From both a time and cost of materials standpoint, it would be advantageous to utilize a single-step process if available. The authors have undertaken a study examining the use of a wide circular shaft, consisting of an assemblage of precast concrete segments, which, when fitted together, provide both temporary earth support during excavation and permanent structural support for below grade facilities or structures. The structural system consists of a stacked series of circular rings. Each ring in turn is made up of curved precast concrete segments.

In simplified terms, the process of constructing this type of structure will usually begin with some form of ground improvement followed by structural excavation of the soil up to 1.8 m (6 ft) deep, and installation of dampproofing material against the soil face. Precast segments are then installed end-to-end forming a complete circle or ring. Following ring completion, grout is applied under prescribed pressure to fill the space (annulus) between the ring and the dampproofing/soil behind the ring, thereby engaging the ring in resisting lateral soil pressure.

During and following pressure grouting, the lateral soil pressure bearing on the ring applies compressive forces that are carried by hoop stress throughout the ring. The friction between the segments and the soil (resulting from the soil pressure), together with construction means and methods, resists the gravitational weight of the segments, enabling the next phase of excavation below each completed ring (underpinning), as shown in Figure 1. Any number of additional rings can be constructed below a completed ring by repeating these steps one ring at a time, until the design depth is achieved. This construction sequencing is defined as a "top down" construction method.

The purpose of this paper is to document the work accomplished to date exploring the unique engineering challenges faced in the design and anticipated in the construction of a large diameter segmentally lined shaft structure. 2D and 3D finite element models have been used, along with comparative economic analyses, field test results, and prototype grouting methods.

## PRELIMINARY INVESTIGATION OF CONCEPT

### Sample Study Projects

For comparative purposes, three proposed projects at different stages of design were selected for analysis. These project examples incorporated conventional engineering and construction systems, were sited in urban areas, and included underground parking structures designed to support four stories of above grade wood framed housing units. These designs were then compared to segmentally lined structures designed to provide an equal amount of parking spaces and building support. This analysis was intended to compare parking efficiency, construction time, and cost of the two different designs (conventional and proposed circular segmentally lined structure).

## Current Methods of Underground Parking Construction

Firstly, traditional underground parking designs are engineered specifically for particular sites; the plan is a "one off" that is tailored to soil type, above grade building structural requirement, lot dimensions, and other factors. The completed design is optimal for the intended site, but generally may not be used again elsewhere. Accurate budgeting is difficult until design work is practically complete.

The work required to complete a conventional urban underground facility can be separated into two distinct efforts. The first phase, temporary shoring, is intended to retain and support perimeter soil that is exposed during excavation. In most cases, it involves installation of vertical support members (beams or caissons) and then during excavation, installation of interim (between vertical member) soil retaining members (timber lagging, steel plate, etc.). Typically, all of this work is abandoned following installation of the permanent structure.

The second phase encompasses the actual construction of the underground facility, including foundations, exterior or perimeter retaining walls, interior walls or columns, and possibly structural interim height decks. These structural decks are intended to both support the vertical dead and live loads acting on them, and also become the permanent horizontal bracing for the exterior walls, thereby resisting the soil pressure.

## Advantages of Circular Shaft Design

Aside from the benefit of using the ring for both temporary shoring and the permanent structure's perimeter walls, a circular lay-out was found to be a more effective use of area, with no space wasted by corners. To highlight these efficiencies, comparisons of proposed underground parking structures located in greater Los Angeles were conducted. A summary of one of the comparison studies is shown in Table 1. for a 3 level (32 foot deep) 357 stall parking structure designed to support 4 stories of above grade wood framed construction. The markers of efficiency here are cost per stall, area per stall, and days to completion.

These advantages are firstly due to the fact that the corners of a rectilinear structure are inefficient from a vehicular circulation standpoint and wasted for parking due to access issues. Secondly, the interior core of the circular design provides an efficient area for vehicle ramps, pedestrian movement (stairs and elevators), and utility and ventilation services. These circulation and design efficiencies translate into a reduction in required building area resulting in decreased soil excavation for a circular excavation compared to the square excavation with the same width/diameter. Lastly, as mentioned earlier, the effort and cost required to install the temporary shoring is eliminated.

To explore further the circulation and parking efficiencies of the circular design, using the City of Los Angeles parking dimensions as a constraint, the circular design becomes more efficient with

**Figure 1. Underpinning with second ring of segments**

**Table 1. Efficiency analysis comparing Cobalt Construction's Encino project to an alternative ShorWall design on the same site**

|  | Number of Stalls | Cost Per Stall | Area Per Stall, m$^2$ (ft$^2$) | Days to Completion |
| --- | --- | --- | --- | --- |
| Conventional Design | 357 | $35,019 | 42.5 (457.5) | 197 |
| Circular Shaft Design | 361 | $30,153 | 30.5 (327.9) | 130 |
| Improvement | 1.1% | 13.9% | 28.3% | 34.0% |

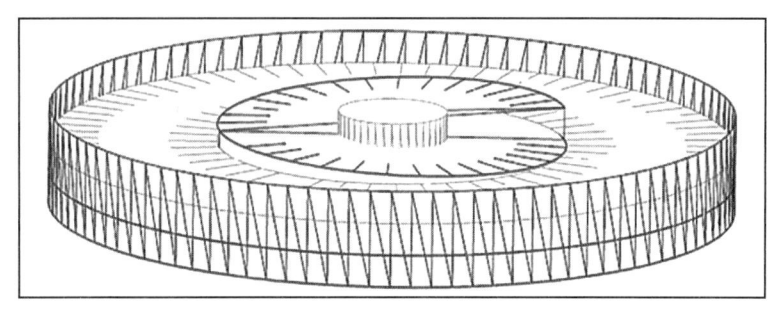

Figure 2. Optimized design for the circular parking structure, with helical access ramp around central core

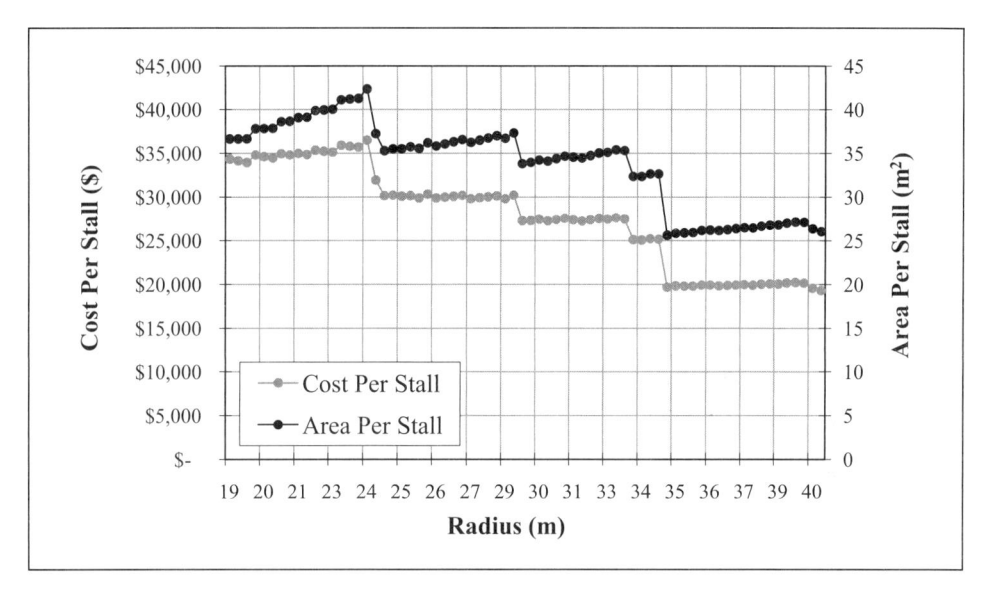

Figure 3. Cost and efficiency variation for different radii; note optimum size at 34.7 m (114 ft) radius

increasing radius. For a radius from 19.5 to 23.8 m (64 to 78 ft), there can only be a single drive aisle, with one set of stalls along the perimeter of the structure. The parking deck in this case is a continuous spiral.

As the radius increases, parking stalls may be placed on both sides of the drive aisle, and the continuous spiral deck is replaced with horizontal decks accessed by a spiral ramp around the central core (Figure 2). The maximum efficiency for parking stalls is found at a 34.7 m (114 ft) radius or greater, where the cost per stall is around $20,000 and the area per stall is around 26 m² (280 ft²). This variation in efficiency is shown in Figure 3.

As a further benefit, the circular shaft approach is essentially modular and scalable; a given design may be successfully used for a wide range of soil types and lot sizes, with only minimal redesign work. Certain parameters of the shaft may be changed (radius, wall thickness, depth, parking arrangement, etc.), but the overall concept and construction method remains the same. This systemized approach to the design of underground structures can provide significant benefits in estimating both budgets and construction durations for subterranean construction.

## DESIGN REQUIREMENTS AND CONSTRAINTS

Having established that a circular structure is more efficient in terms of construction duration, site planning, and in the elimination of temporary shoring, the options for the design of the liner itself were considered. Various design requirements and constraints were taken into account, as listed below.

- Minimize construction time
- Cost-efficiency

Figure 4. Construction of small-diameter shaft (July 2008)

Figure 5. Load testing of two-ring small-diameter shaft (July 2008); filled water tanker resting on a timber mat on top of the shaft, weight of 54,000 kg (60 ton)

- Sustainable design (minimize material, energy, social impacts)
- Simplified design using systemized construction means and methods
- Compatible with urban setting
- Radius and depth scalable to any size
- Modular design requiring little re-engineering for any given application
- Construction means and methods suitable to variable soft ground conditions
- Aesthetic design for public use
- Water impermeability
- Ability to accommodate a variety of internal uses

Investigations of similarly sized shafts revealed unique characteristics that disqualified their design from the intended industrial, commercial, and civil applications (underground storage, transportation, parking, or even housing). For example, some large diameter shafts make use of shotcrete walls, auger cast piles or secant piles, but these were deemed unsightly and generally are not 100% impermeable to water infiltration (a prerequisite for public use structures) (Celestino 2005). Other shafts are built using thick reinforced concrete segments, with diameter-to-thickness ratios in the range of 25:1 to 50:1. Concrete segments designed accordingly would be over 1.5 m (4.9 ft) thick, meaning that 1) they would be excessively expensive, neutralizing the economic benefits already stated, and 2) the extreme weight of each segment (in the lengths proposed) would create transportation and handling issues.

With these requirements in mind, the designers opted for a reinforced precast concrete segmentally lined shaft, using very high diameter-to-thickness ratios. Such a design ensures that the benefits of a circular design are not negated by the cost of producing the structure. Several complications to this design present themselves which are addressed later in this paper.

## INITIAL FIELD TRIALS

Having established economical and schedule advantages, and having considered the constraints listed above, the team desired to gain some practical experience with the proposed design.

### Small-Diameter Shaft

Small-scale testing was carried out in July 2008, to gauge the viability of certain assumptions and methods. A 4.6 m (15 ft) diameter segmentally lined shaft was constructed, with two rings going to a total depth of 3.7 m (12 ft) (see Figure 4). The small shaft was successfully installed, grouted, and load tested to 54,000 kg (60 ton), as shown in Figure 5. The learning centered on:

- excavation and soil trimming techniques;
- ring assembly and fit up;
- dampproofing installation techniques;
- grout mix design engineering;
- grout delivery, permeation, and containment;
- packer materials;
- segment waterproofing; and
- soil-to-segment shear capacity.

The lessons learned provided much of the impetus for further research and refinement of shaft design and construction methods.

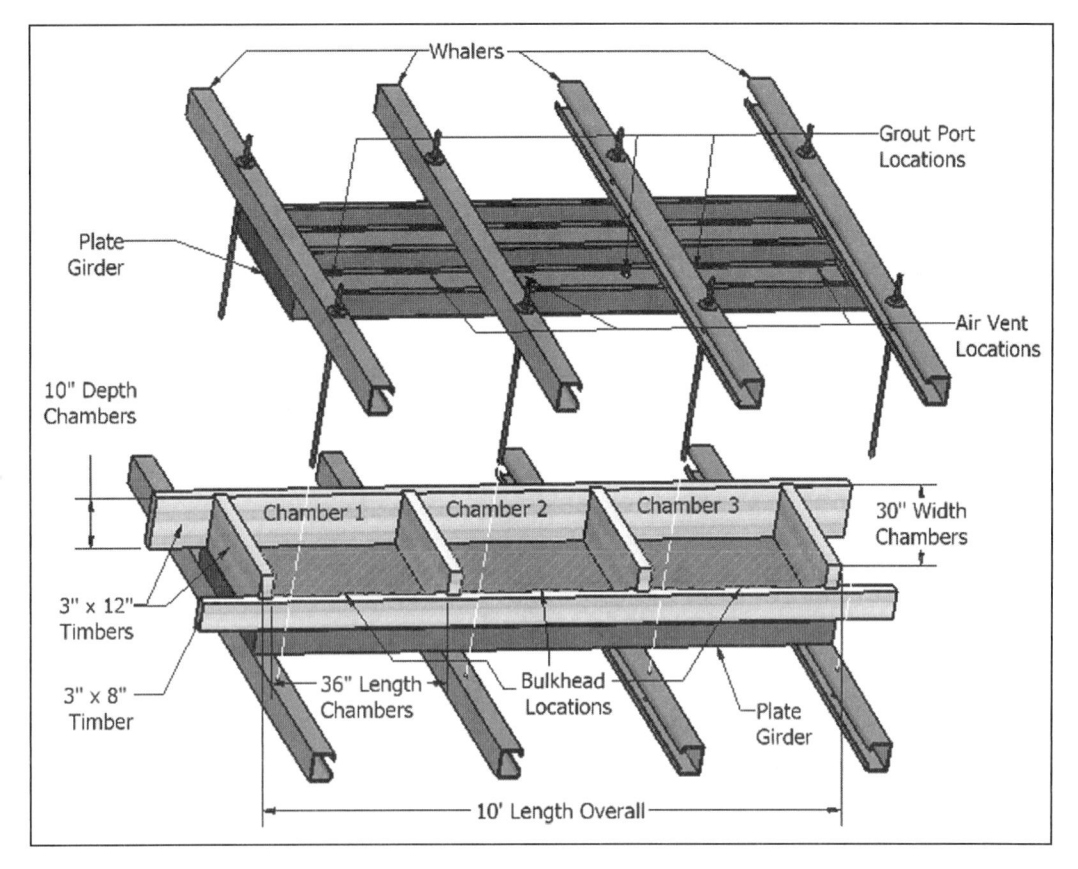

**Figure 6. Pressure vessel schematic**

### Pressure Vessel Tests

One such topic of follow-up research centered on the ability of the system to contain the pressurized grout in the annulus. The methods used in the small-diameter test were only partially successful at keeping the grout contained, and so work was undertaken to find materials and combinations of materials that could keep the annulus sealed. The pressure vessel apparatus is shown in Figures 6 and 7. A layer of soil filled the bottom of each test chamber.

A short list of materials was chosen for segment-to-segment packers, as well as for bulkhead material for the upper and lower segment-to-soil boundaries. The selected polyethylene foam gaskets were successful at resisting pressures up to 20 psi, with complete grout containment.

### Direct Shear Tests

An effort was also made to understand the shear capacity of the segment-to-soil interface, as this is the primary means of ring support against self-weight. As shown in Figure 5, the shear capacity of the small-diameter shaft was quite high, and more than enough for the weight of the segments themselves. The team set out to test a variety of specially prepared samples, depicted in Figure 8 using different

**Figure 7. One chamber in the pressure vessel (before grouting), with the successful polyethylene foam gasket**

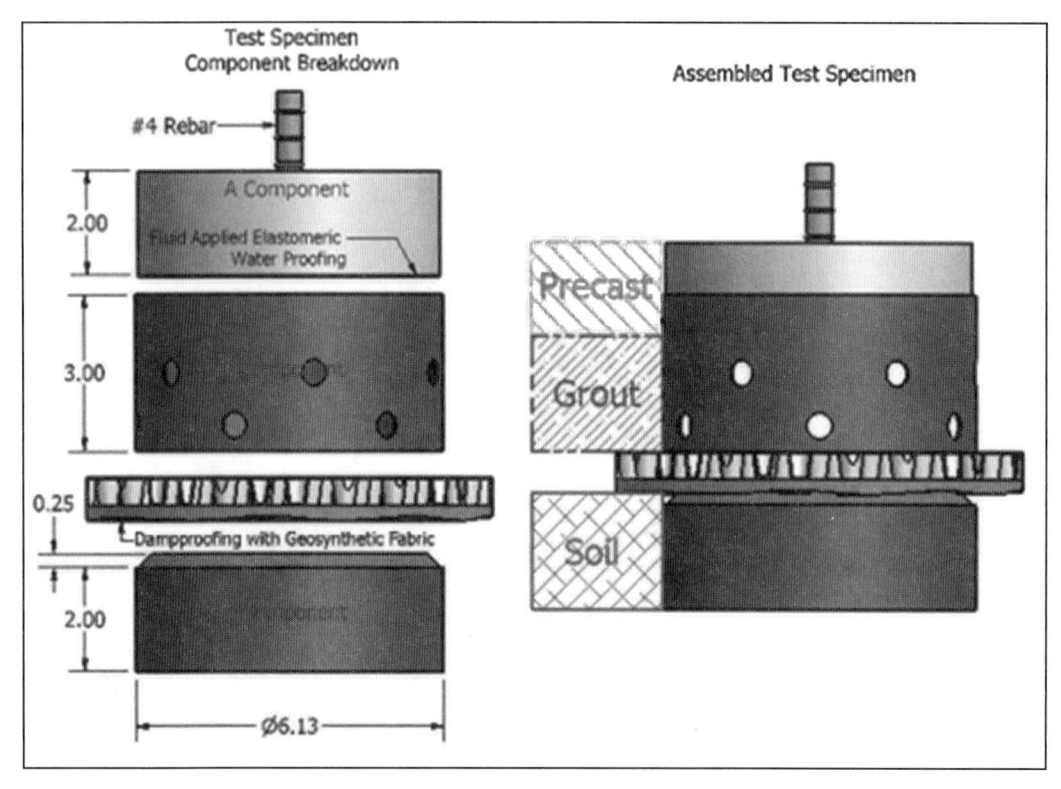

Figure 8. Specimen schematic diagram showing the concrete-to-grout-to-dampproofing-to-soil assembly

soil types, grout mixes, dampproofing materials, and at different ages (3 day and 7 day). The testing apparatus is shown in Figure 9.

Several conclusions were made from these tests:

- Increased normal pressure (simulating lateral earth pressure) on the sample results in increased shear strength.
- The soil-to-dampproofing interface tended to be the critical plane of shear failure.
- The preferred dampproofing material was a chain-link style mesh with geosynthetic fabric on both sides.
- The shear strength capacity is more than enough for the self-weight of the segments; for example, at the minimum tested 34 kPa (5.0 psi) normal stress, the ultimate shear strength was an average of 67 kPa (9.7 psi).

## INNOVATIVE DESIGN SOLUTIONS

### High Diameter-to-Thickness Ratio

The proposed structure departs from traditional segmentally lined shaft design primarily in that it depends on a much higher diameter-to-thickness ratio. Consider several typical larger shafts and tunnels in use today:

- The Lake Mead intake No. 3 shaft at Lake Mead, Las Vegas has a 9.1 m (30 ft) internal diameter, with a lining diameter of no less than 450 mm (1.5 ft), giving a diameter-to-thickness ratio of 6.1:1 (Hurt 2009).
- The Shanghai Yangtze River Tunnel has an inner diameter of 13.7 m (45 ft) and a wall thickness of 1.3 m (4.3 ft), giving it a ratio of 10.5:1 (Di et al 2008).
- The Seymour Shaft in Western Canada has a diameter of 11 m (36 ft) and a 0.25 m (0.82 ft) thick shotcrete layer, giving it a ratio of 44:1 (Prucker et al 2008).

The approximate range of dimensions considered so far is diameters of 49.0–90.8 m (160.8–298.0 ft), with segment thicknesses of 305–356 mm (12–14 in). This puts the diameter-to-thickness ratios anywhere from 140:1 to 300:1. As already mentioned, the crucial factor of parking efficiency means the diameter must be maximized, while the

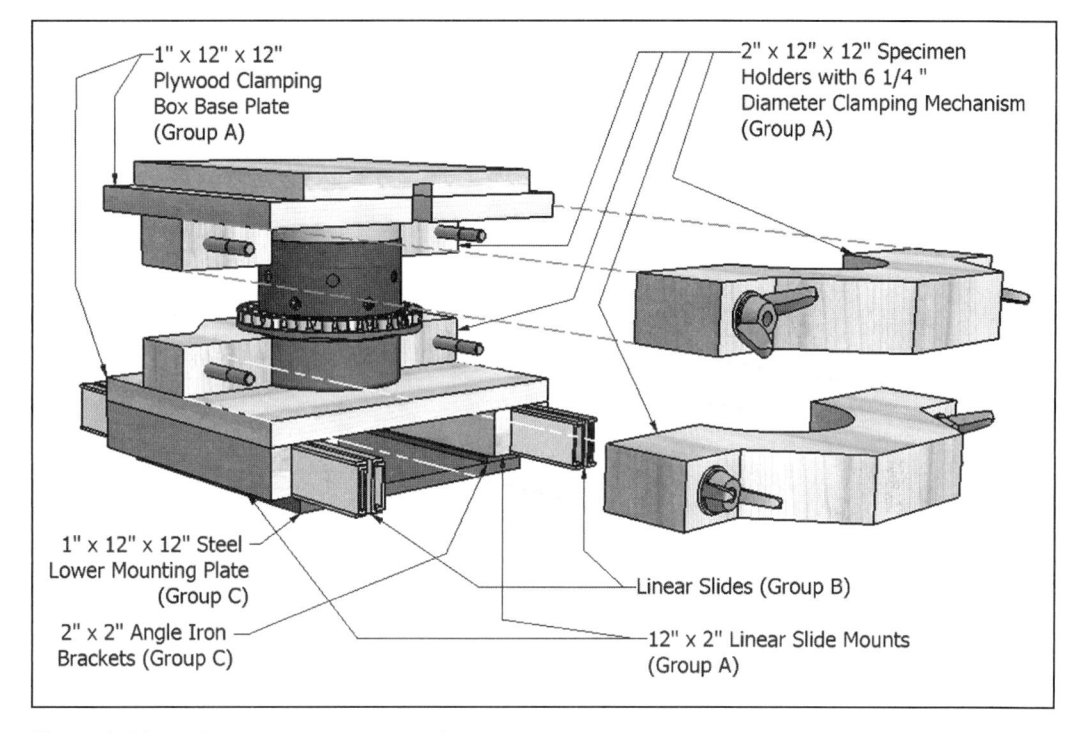

**Figure 9. Direct shear test apparatus used**

cost per segment is decreased with thinner segments. Currently, a ratio of 230:1 is being used as a baseline for engineering models.

Such geometric ratios are well into the range of thin-walled structures, and so the design and construction methods must account for the possibility of ring buckling, that is, where the entire ring becomes catastrophically and suddenly unstable. As a comparison, if a 10 m (30.5 ft) diameter shaft was similarly designed, it would have walls only 43 mm (1.7 in) thick!

Several options have been considered as possible solutions to the inherent instability of such a structure. The methods that appear to be the most cost-effective and risk-mitigating are as follows, as illustrated in Figure 10:

- The use of temporary post shores around the intrados of the ring, with soil anchors positioned around the excavated floor.
- Ground improvement (remove and re-compact with addition of 3–5% cement binder) around the immediate vicinity of the ring.
- Post-tensioning strands running horizontally and vertically through all segments.
- Reinforced concrete ring beams around the entire ring at every second circumferential joint (parking deck elevation).
- A starter ring or collar at ground level.

The stability issues were studied through several rounds of analytical and numerical analysis, using parametric studies, and the contributing effect of the various extra components were determined. A summary of those results are given in the *Finite Element Modeling* section.

**Grouting Challenges**

Another point of departure from current industry practice is in the method used to grout the annulus. The usual method of grouting segmentally lined shafts involves grout ports precast into the segments themselves, or in the case of some shafts, the installation of drop pipes into the annulus.

A circular shaft of the discussed diameters has a grout volume requirement of 19 to 23 m$^3$ (25 to 30 yd$^3$) per ring, which exceeds typical small diameter shaft designs by as much as nine times. In addition, the physical grout delivery system could require as many as 40 to 50 separate grout ports and hoses, with each port requiring exact pressure and flow control to maintain ring stability. It was determined that such a system was infeasible.

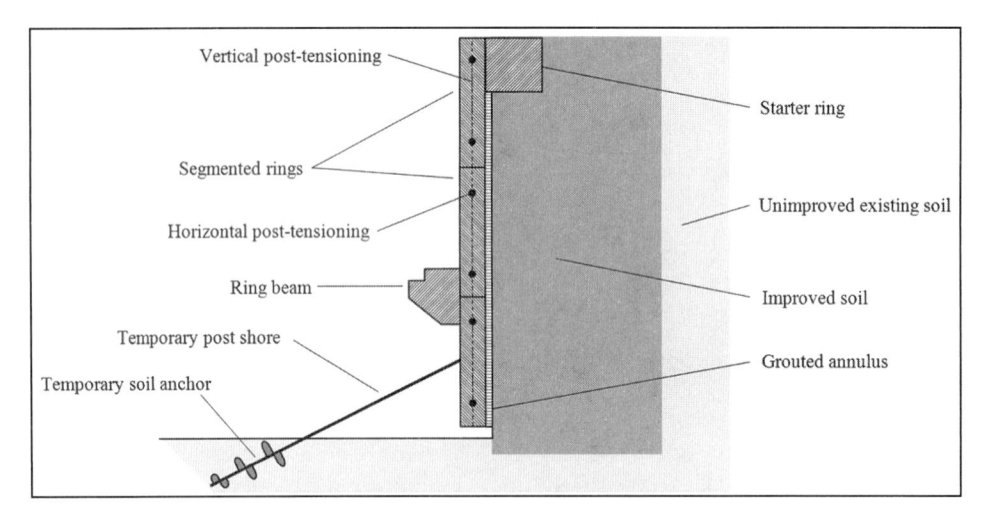

**Figure 10. Stabilization options to guard against buckling failure**

### Intra-Annulus Transmission Conduit (IATC)

A solution was proposed whereby the requirements of constant pressure and flow might be met, while still supplying adequate volumes of fill to the annulus. The designers conceived a circulating grout system that pumps the grout into an intra-annulus transmission conduit (IATC) that passes around the entire circumference and then returns back to its original starting point of the pump hopper. The IATC is perforated such that the grout escapes into the annulus, turning the entire void into a pressure vessel with dimensions in the range of 0.05 × 1.5 × 275 m (0.17 × 5 × 900 ft).

An IATC runs circumferentially around every ring, positioned in the center of the segment height. As such, the grout must travel less than 0.75 m (2.5 ft) up or down before it encounters the prior ring's grout (upper barrier) or the current ring's temporary grout barrier bulkhead (lower barrier). This very short distance and multitude of perforations will help assure full and complete grout penetration of all areas of the annulus. When the annulus is filled, and then pressurized, the supply and exit valves are closed, and the IATC is cast permanently into the grout. Figure 11 shows a schematic diagram of the IATC.

The designers have also begun a series of tests in order to verify the potential benefits of the IATC system, the first of which is described later in this paper.

### CONSTRUCTION SEQUENCE

Due to the inherently unstable nature of thin-walled structures, careful planning and execution of the construction sequence must be made. The goal is to always provide at least one method of support or strengthening throughout every stage.

1. Pre-excavation improvements: dewatering, pile/caisson drilling & construction, and/or surface performed ground improvement/consolidation grouting. Depending on project requirements and ground conditions, the following work may be required prior to shaft construction, dependent on site conditions.
   1.1 Installation of dewatering system
   1.2 Ground improvement
      1.2.1 *Option 1:* Perimeter soil removal and re-compaction—soil cement ground conditioning outside perimeter of shaft in cases where soil will not "stand up"
      1.2.2 *Option 2:* Consolidation grouting—for cases where excavation will occur in poorly graded or loose non-cohesive sands and gravels where no soil binder is present
   1.3 Podium "outside corner" foundation supports
   1.4 Temporary construction equipment staging platform foundation
2. Starter ring construction: excavate, form and pour
3. Excavation, panel erection, and ring completion sequence
   3.1 Rough excavation of soil 1.7 m (5.5 ft) below prior ring or starter ring, export soil
   3.2 Trim vertical face and remove spoils

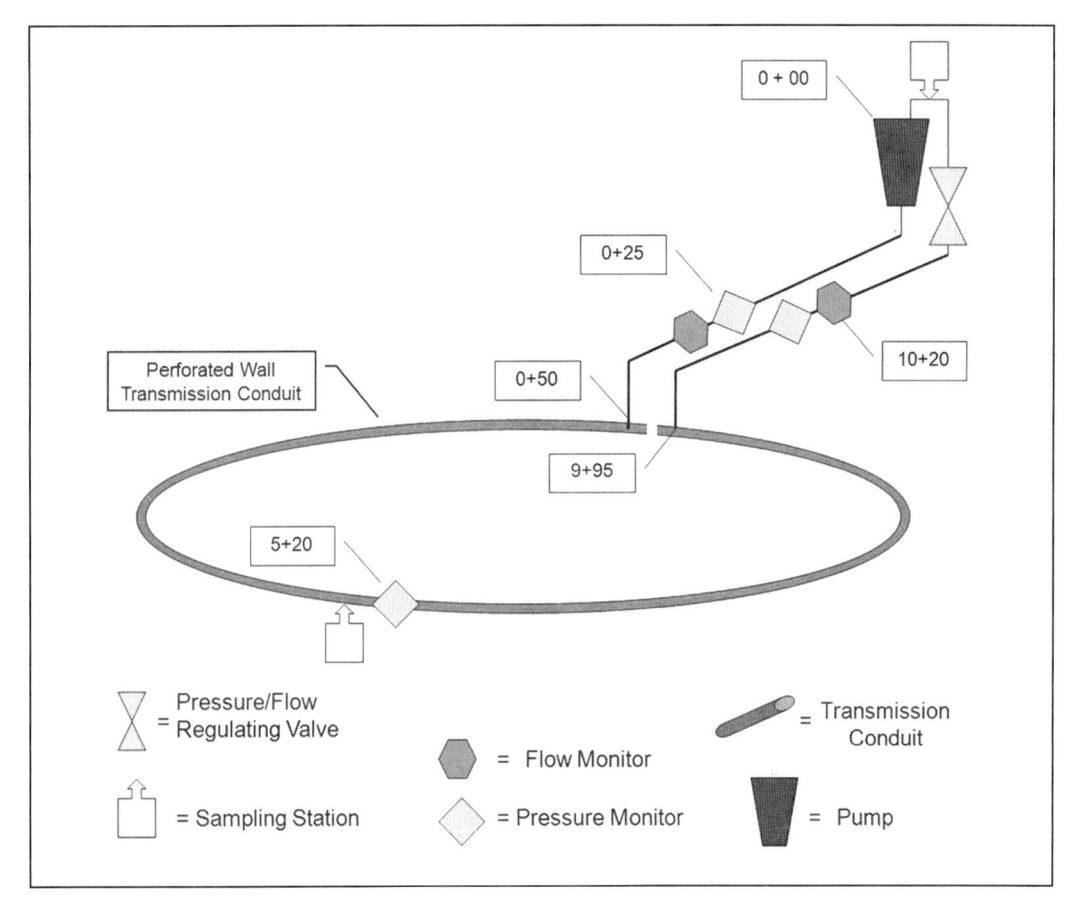

**Figure 11. Schematic diagram of the IATC**

3.3 Install vertical grout supply and return bypass piping
3.4 Install drainage composite (dampproofing) on vertical soil face
3.5 Install IATC (intra-annulus transmission conduit)
3.6 Set and connect panels (use of post tension cables for lifting and radial joint compression)
3.7 Install soil anchors as needed and fit post shores to segments, to assure accurate ring geometry and provide ring stabilization
3.8 Install "keystone" panel, completing ring
3.9 Set temporary grout barrier on underside of completed ring
3.10 Grout annulus using IATC method
3.11 Strip anchors and post shores upon reaching engineered grout unconfined compressive strength
3.12 Install ring beams every two rings (if required by design)
3.13 Repeat sequence until design depths are achieved

## GROUT EXPERIMENTS

As mentioned earlier, a range of tests were planned to verify the proposed IATC system. The goal of the first test was to obtain pressure and flow data around a full-length conduit using proper electronic instrumentation. This test is covered in this section, along with the criteria for the grout mix itself.

### Mix Design Targets

During the design process, it became increasingly apparent that the ability to properly grout the annulus behind the segmented ring was not only dependent upon the physical delivery method, but also upon the properties of the grout itself. Much research effort was made to optimize a grout mix design that would meet or exceed the following characteristics:

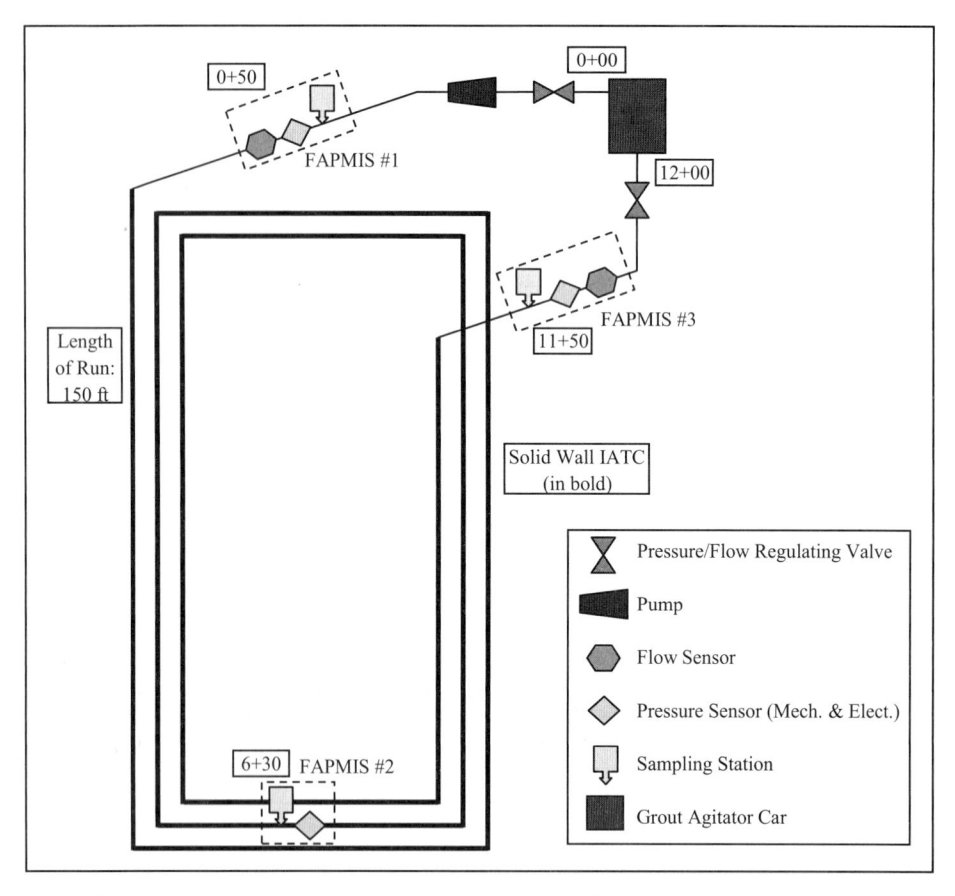

**Figure 12. Schematic of one circuit within the racetrack (FAPMIS: flow and pressure measuring instrument system; measurements in feet)**

- Low viscosity, to minimize head losses within the conduits. This was quantified using the ASTM D6103 flow test, with 300 mm (12 inches) being the minimum accepted flow diameter.
- Low density, to minimize the work required to move the fluid. Grouts tested had densities as low as 320 kg/m$^3$ (20 lb/ft$^3$), achieved using cellular concrete.
- A twenty-four-hour unconfined compressive strength of at least 140 kPa (20 psi), and seven-day strength of at least 690 kPa (100 psi), for the worst case of a 15.2 m (50 ft) deep excavation in clay or sand.
- An eighteen-hour shear strength of at least 69 kPa (10 psi), to enable the segment weight to be transferred properly to the surrounding soil.
- Volumetric strain (shrinkage) of less than 2%.
- Water impermeability once cured.
- Cost effectiveness.

## Grout Delivery Experiment

Once a suitable grout mix design was obtained, a large-scale experiment was carried out to gain an understanding of the behavior of the grout inside a full-sized intra-annulus transmission conduit (IATC). This was done using pressure sensors and electromagnetic flowmeters positioned around a variety of 366 m (1200 ft) long circuits (near the beginning, middle and end). The purpose was to obtain a pressure profile around the circuit, as well as to sample the grout at various stations to determine if the nature of the grout had changed due to pumping effects (change in density or viscosity, segregation, etc.).

A schematic diagram of one circuit is shown here in Figure 12. There were five circuits in total, and they nested into each other to form a "racetrack" layout, as shown in Figure 13. The five circuits had five different profiles, in order to gauge the effect of conduit diameter and shape on flow characteristics:

- 75 mm (3 in) uncompressed PVC
- 100 mm (4 in) uncompressed PVC
- 75 mm (3 in) compressed PVC
- 100 mm (4 in) compressed PVC
- 50 mm × 150 mm (2 in × 6 in) PVC rectangular fencing conduit

The conduits were compressed down to 50 mm (2 inches), being intended to simulate the IATC as it would be installed behind the concrete segments.

Approximately 6.5 m$^3$ (8.5 yd$^3$) of the optimum grout mix was batched, and sent into one of the five conduits. Once the line was charged, the grout was pumped at a range of flow rates, from between 1.9 m$^3$/hr (2.5 yd$^3$/hr) up to 11.5 m$^3$/hr (15 yd$^3$/hr), allowing time at each flow rate level for the system to stabilize so that constant flow and pressure readings could be taken.

**Pressure Drop Results**

A minimal pressure drop around the IATC is desirable, because a non-uniform pressure around the circumference of the ring is more likely to promote ring instability.

Out of two months' work, four days of testing were selected as being representative of the entire regime. The key identifying parameters of the grout were taken to be the density and flow test results (ASTM D6103). Pressure loss is linked to Reynolds number, which depends on flow speed, density, viscosity and pipe size.

The four grouts in Table 2 and Figure 14 all have different mix designs. Mixes 1028, 1029, 1030 and the water baseline were pumped inside a 75 mm (3 in) pipe, and mix 1217 was in a 100 mm (4 in) pipe. The pressure drop recorded is the difference between Station 1 and Station 3 (see Figure 12), which is a distance of approximately 335 m (1,100 ft).

**Pressure Equalization Results**

After the dynamic grout testing was completed for a given conduit, additional tests were carried out to observe the tendency of the system toward equilibrium. If the grouted annulus has a large variation in pressure around the circumference as the grout is curing, it could prove detrimental to the stability of the ring.

To conduct this test, the pump was shut off, and the terminal valve leading back to the agitator car was closed, thereby sealing the system. At the time of powering off, the pressures at the three stations were recorded. The results of one test were recorded as being 208 kPa (30.2 psi), 134 kPa (19.5 psi), and 34 kPa (5.0 psi) at power off, giving a total pressure difference of 174 kPa (25.2 psi). After ten minutes, the pressure at the three stations were 135 kPa (19.6 psi), 132 kPa (19.1 psi), and 120 kPa (17.4 psi), giving a total pressure difference of only 15 kPa (2.2 psi), which is practically equal. The graph over time as recorded is shown in Figure 15.

**Final Comments**

The grout mix design used in these experiments performs satisfactorily in all categories, and the pressure results are within an acceptable range. In particular, the rapid approach of the closed system toward equilibrium gives confidence that the ring can be grouted

**Figure 13. Prototype IATC test circuit—the "racetrack"**

**Table 2. Pressure drops across circuit for four representative grouts and water as a baseline (in order of density)**

| Mix Code | Density, kg/m$^3$ (lb/ft$^3$) | Flow, mm (in) | 3.8 m$^3$/hr (5 yd$^3$/hr) | 7.6 m$^3$/hr (10 yd$^3$/hr) | 9.6 m$^3$/hr (12.5 yd$^3$/hr) |
|---|---|---|---|---|---|
| 1217 | 328 (20.5) | 305 (12.0) | 120 (17) | 150 (22) | 160 (23) |
| 1029 | 345 (21.5) | 275 (10.8) | 255 (37) | 352 (51) | 386 (56) |
| 1028 | 500 (31.0) | 290 (11.5) | 303 (44) | 407 (59) | 434 (63) |
| 1030 | 630 (39.5) | 330 (13.0) | 448 (65) | 531 (77) | 558 (81) |
| 1015 (Water) | 1000 (62.4) | — | 18 (2.6) | 25 (3.6) | 30 (4.4) |

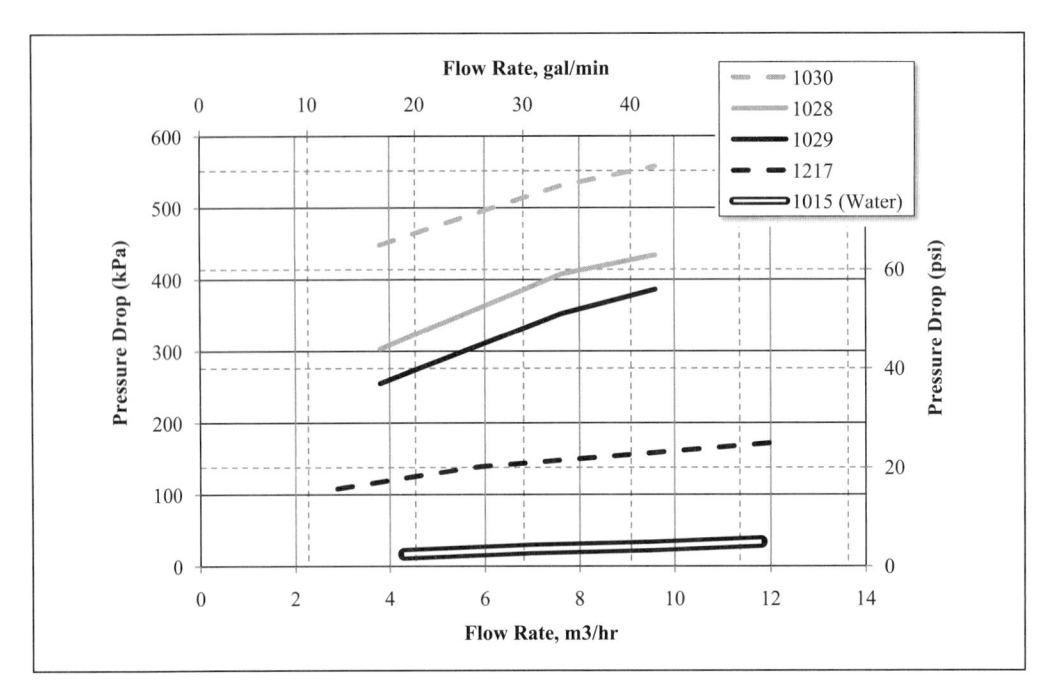

Figure 14. Pressure drops across circuit for four representative grouts and water as a baseline

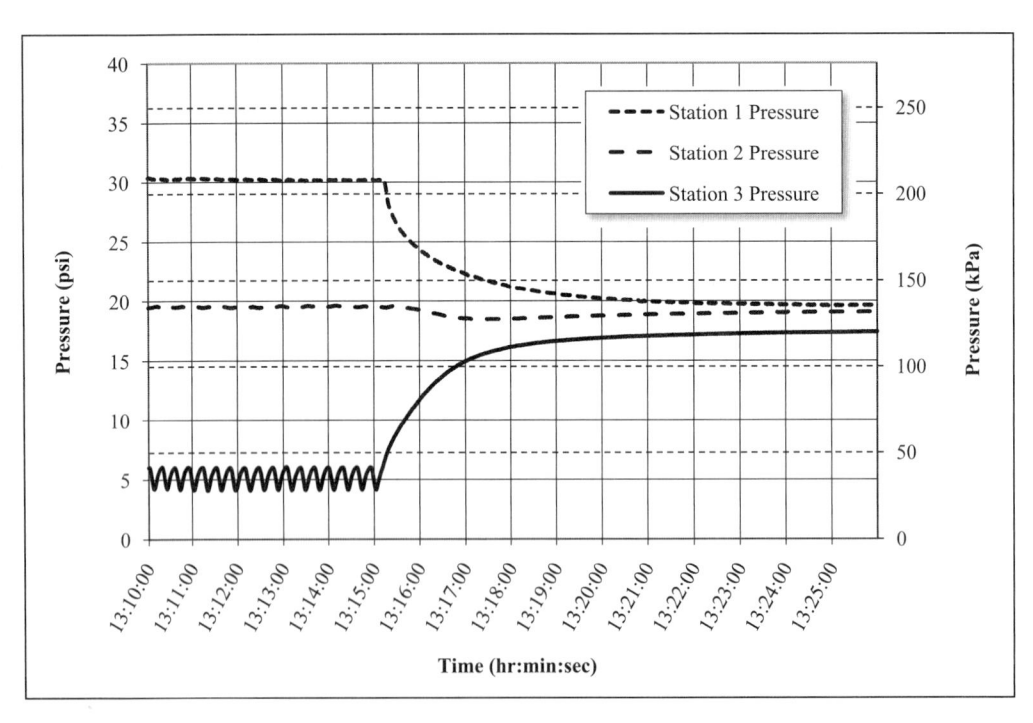

Figure 15. Pressure equalization over time with closed 366 m (1200 ft) long system

at even pressures once installed maintaining ring stability.

**FINITE ELEMENT MODELING**

Finite element models were used as an aid to determine the fundamental relationships governing the design of the proposed structure. Models were aimed not so much at determining safety factors and limit loads as probing the nature of the contribution of each type of structural element and parameter. It became immediately apparent that the construction means and methods as well as the construction sequence also had a significant impact on ring stability; as such, both construction means and methods and structural members were considered in the analysis.

The finite element work has been done in two main stages, the first in 2006 by Professors Aschheim, Hight, and Brandenberg (Santa Clara/UCLA), and the more recently in 2009 by Halcrow. All modeling was done with *SAP-2000*.

**Summary of 2006 Finite Element Results (Santa Clara/UCLA)**

The numerical and analytical work done in 2006 focused on the "sensitivity" of a single 90.8 m (298 ft) diameter ring to non-uniform loads, firstly as an inward point load on one of the liner joints, and then as a "mode four" displacement (the same shape as a ring in the fourth mode of vibration) (Aschheim et al 2006).

In another analysis, the ring was made "out of round" by displacing one joint node inward by 100 mm (4 inches). All three analyses were examined using two soil types corresponding to a medium-dense sand and a clay (CH). The results may be summarized as follows:

- The ring might beneficially deform to adjust to uneven earth pressures.
- The flexural loads imposed on the panels are much lower than the flexural capacity of the reinforced concrete (anywhere from 7–20% of capacity, depending on soil type).
- The joints should remain sufficiently flexible, in order to mitigate the bending moments in the panels.
- Grout pressures must exceed an upper estimate of the active soil pressure, and should be at least in the at-rest pressure range.
- Construction tolerances are very important, as the ring is sensitive to being "out of round."
- The soil itself helps to stabilize the system, reducing the bending moments that might otherwise develop.

**Summary of 2009 Finite Element Results (Halcrow)**

The numerical and analytical work done in 2009 focused on the threat of buckling, and the effect of various structural parameters on that failure mode. A baseline for buckling failure was obtained using Timoshenko's classic formula (Timoshenko 1956),

$$q_{crit} = \frac{3EI}{R^3}$$

where $q_{crit}$ is the critical buckling pressure for the first mode of failure, $E$ is the Young's Modulus of the ring, $I$ is the second moment of area in the buckling direction, and $R$ is the radius of the ring. A value of less than 6 kPa (0.8 psi) was found for a diameter of 70 m (230 ft) and a ring thickness of 305 mm (12 inches), those dimensions being somewhere in the optimum range according to the economic analysis already presented in this paper.

Timoshenko's formula is very conservative, allowing no contribution of strength from any source (soil resistance, starter rings, bracing, etc.), but served as a starting point. Various temporary and permanent structural elements were added into the analysis one at a time over different models, and the following conclusions were made.

- This diameter-to-thickness ratio is highly unstable for a simple theoretical ring, as evidenced by Timoshenko's formula.
- For the ideal case of a perfectly round ring, there is no possibility of buckling (in the finite element program). From that starting point, even small variations in geometry in the order of 25 mm (1 inch) cause a tendency toward catastrophic failure at low pressures, as predicted by Timoshenko.
- Section stiffness ($EI$) is a fundamental property and should be maximized using high-stiffness materials and adequate segment dimensions within the economic constraints.
- The soil-structure interaction helps to stabilize the system, guarding against catastrophic failure in a way that is practically independent of soil type/strength for the range of soil types commonly encountered, even for imperfect ring geometry. The soil stiffness, simplified to a modulus of subgrade reaction, was varied from 10,000–1,000,000 kPa/m (36.8–3,680 psi/in) with almost no change in stability.
- The use of temporary ring bracing, such as post shores fixed to soil anchors, effectively forces the ring into higher modes of buckling failure, thereby raising the critical buckling pressure, $q_{crit}$.

- The presence of interior ring beams around the circumference raise the critical buckling pressure by effectively increasing the section properties of the segment, and may also be used later to attach the horizontal decks to the walls.
- The presence of joints does not seem to weaken the ring, even when no tensile strength is assigned to the joints.
- If a joint were to open between two segments, it would cause elongation of the ring, with a corresponding increase in hoop stress which provides a restoring moment to the joint.
- The combination of these elements cumulatively increase ring stability, well above assumed soil pressures.

## Conclusions and Limitations of Finite Element Results

The finite element work carried out thus far should be taken in caution, as the results are highly dependent on the input values and assumptions. However, the qualitative results listed above are the results of parametric studies that have sought to determine the *nature* of the proposed structure. By and large, the results to date indicate that the design is both structurally feasible while remaining economically viable.

## CONCLUSIONS

- Economics: the proposed circular design for underground structures has a clear economic benefit, costing 13.9% less than a conventional rectilinear system, and being completed in two-thirds of the time.
- Initial Trials: small-diameter testing has given confidence as to construction methods; pressure vessel testing has enabled selection of appropriate gasket materials; shear test results suggest that failure in shear is not likely to be a primary failure mode, especially for improved soil.
- Diameter-to-Thickness Ratio: although problematic from a simplified theoretical perspective, the use of appropriate construction means and methods can allow structural dimensions that are both safe and cost-effective.
- Grouting Methods: the innovative IATC grouting method shows promise as a solution for rapid delivery of large volumes of grout with minimal variation in pressure around the ring.

## FUTURE WORK

Considerable work remains to be done on the project. To list some representative points:

- Further optimization of grout mix designs, aimed at reducing pressure losses and rapid strength gain.
- Further testing of the IATC "racetrack" system, and progressing to perforated pipe testing.
- A second phase of small-diameter shaft field trials, utilizing best practices and the results of research.
- Refined finite element models aimed at non-uniform grout pressure, construction sequence, surcharge loads, and seismic behavior.

## ACKNOWLEDGMENTS

The authors wish to thank the following people for their contributions:

Mike King, Mark Johnson, Ho Jung Lee (Halcrow); Gary Kramer (Hatch Mott MacDonald); James Warner; Scott Brandenberg (UCLA); Mark Aschheim (SCU).

## REFERENCES

Aschheim, M., Hight, T., and Brandenberg, S. 2006. Feasibility and Preliminary Design Analyses for Silo ShorWall System Development. UCLA/SCU.

Celestino, T.B. 2005. Shotcrete and Waterproofing for Operational Tunnels. International Tunnelling Association, Working Group on Shotcrete Use. http://www.ita-aites.org/cms/fileadmin/filemounts/general/pdf/ItaAssociation/ProductAndPublication/Training/TrainingCourses/SS1_2005.pdf. Accessed January 2010.

Di, Y.M., Yang, Z.H., Xu, Y. 2008. Integrated design and study of internal structure of Shanghai Yangzte River Tunnel. In The Shanghai Yangtze River Tunnel: Theory, Design and Construction. Edited by Huang, R. CRC Press.

Hurt, J., McDonald, J., Sherry, G., McGinn, A.J., and Piek, L. 2009. Design and construction of Lake Mead intake No. 3 shafts and tunnel. Rapid Excavation and Tunneling Conference Proceedings. Littleton, CO: SME.

Prucker, A., Rotzien, J., Saltis, A. 2008. Shaft sinking at the Seymour-Capilano Filtration Plant. North American Tunneling Proceedings. Littleton, CO: SME.

Timoshenko, S. 1947. Strength of Materials: Part II: Advanced Theory and Problems. 2nd Ed. New York City, NY: D. Van Nostrand Company, Inc.

# Large Diameter TBM Development

## Martin Herrenknecht, Karin Bäppler
Herrenknecht AG, Schwanau, Germany

**ABSTRACT:** Various municipalities throughout the world have realized the need to improve transportation infrastructure requires moving larger and larger traffic volumes; whether they be rail, truck or automobile. This has led to the need of larger diameter tunnels now approaching up to 16m diameter. This paper discusses how these MegaTBMs have evolved, which projects are currently underway, and what the future holds for large TBM development. Project histories will be reviewed and new projects in the planning stages will be discussed with specific recommendations for the type of TBM technology envisioned.

## INTRODUCTION

The world's growing population and rapidly proceeding urbanization are boosting an enormous demand for new and high-capacity infrastructures to secure the mobility of goods and people.

First and foremost, the densely populated areas face the challenge to provide efficient traffic infrastructure, e.g., a modern public transport system. At the same time, the maintenance and modernization of supply and disposal structures for water, sewage, energy and communication are essential. Last but not least, supra-regional transportation routes or, for example, long distance water diversion schemes are the challenges for the future. All these projects are as demanding as they are characterized by a tight time schedule.

In this context the demand for efficient tunnels for traffic and utility lines is increasing as the way of new infrastructures leads in most cases underground because of limited space above ground. Also, tunnels are the obvious choice to cross natural barriers like mountain ranges.

The tendency of the upcoming traffic tunnel projects shows the demand for large to very large profiles. The increasing need for high-performance infrastructure in the sector of transport and utility tunnelling favours tunnel solutions and thus the mechanized tunnelling. With the manufacturing of one of the first largest tunnel boring machine (Mixshield, Ø14.2m) for the 4th tube of the Elbe road tunnel in Hamburg in Germany and the currently largest tunnel boring machines for the inner-city tunnelling for the M30 highway project in Madrid (EPB-Shield, Ø15.2m) and the two machines for the river crossing near Shanghai (2 Mixshields, Ø15.43m), the feasibility of large diameter tunnels and the outstanding examples for applied technical engineering are given.

The SMART tunnel project in Kuala Lumpur, where two Mixshields with a diameter of 13.21m have been used, is one of the first pioneering examples to show that the tunnels can take over more complex service functions. With its dual-usage, as preventing flooding and alleviating traffic congestion, the project presents the tendency of extending the utilization ratio of the future tunnels.

This paper is to present the large diameter TBM development for traffic tunnelling projects.

## TREND OF VERY LARGE DIAMETER TUNNEL PROFILES

Up to now more than 70 machines with diameters larger 10 m have been delivered by the end of 2009. More often tunnel projects are planned with diameters exceeding the 10 m diameter limit. The large diameter TBMs are not restricted to special ground types. They are applied for both soft and hard rock or mixed face conditions. Mechanized tunnelling with diameters even larger 15m are today state of the art and can be coped with safely. Compared to conventional construction methods the mechanized shield tunnelling with larger diameters is considerably faster and its limits are set rather by logistical issues (e.g. removal of excavated material) than by construction safety or financial questions. Large tunnel profiles allow contractors and planners the possible installation of additional service and safety facilities for the operation of the tunnel. Herrenknecht's large diameter TBMs operating worldwide show that extremely large tunnel diameters can be safely and efficiently produced with the chosen tunnelling

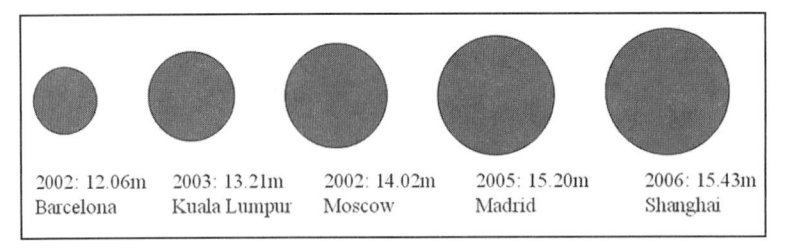

| 2002: 12.06m | 2003: 13.21m | 2002: 14.02m | 2005: 15.20m | 2006: 15.43m |
|---|---|---|---|---|
| Barcelona | Kuala Lumpur | Moscow | Madrid | Shanghai |

**Figure 1. Largest TBM by year (Herrenknecht AG)**

technology such as the machines applied for the 4th Elbe Tunnel in Hamburg, the Lefortovo and Silberwaldtunnel in Russia, the SMART tunnel in Kuala Lumpur, the M30 highway project in Spain and the currently two largest machines with a diameter of 15.43m which excavated two parallel tunnels near Shanghai.

A challenge for example was the M30 highway North tunnel in Madrid. An EPB-Shield Ø15.20m excavated and lined a three-lane, 3.65 kilometer long highway tunnel in the center of Madrid with an extremely tight time schedule. The target construction time of 12 months could clearly be reduced, and the 8-month tunnel construction time equals an excellent TBM performance of more than 450m per month. The completely unique TBM with two concentrically arranged cutting wheels and three screw conveyors for material discharge out of the working chamber achieved top daily performances of up to 36 meters of excavated and lined tunnel. However, not the size of the tunnel profile was a challenge but rather the logistics. During the construction of the 3.65 kilometer long highway tunnel, an average of 60 trucks a day went to the inner-city construction site for the delivery of the segments used to line the tunnel. At peak times, 720 trucks passed the construction site on one day to remove the excavated material.

Currently the largest machines, the two Mixshields Ø15.43m for the Changjiang Under River Tunnel Project in Shanghai excavated two parallel three-lane highway tunnels, each having a length of 7.47 kilometers. They excavated at a depth of up to 65 meters. The tunnels connect the Changxing River Island with the mainland of Pudong/Shanghai. The innovative features of the Shields are the cutting wheels which are accessible in free air for the replacement of the cutting tools.

Two further very large diameter Mixshields with a diameter of 14.93 meters excavated and lined 2.9 kilometers of road tunnel each. Both machines crossed parallel beneath the Yangtze River in Nanjing (Jiangsu Province, eastern China). The two machines were also equipped with the feature of cutting wheel arms which are accessible under atmospheric conditions.

## CURRENTLY WORLD'S LARGEST TBMS Ø15.43M APPLIED IN SHANGHAI

In November 2006 and January 2007 respectively, the two largest tunnel boring machines in the world with a diameter of 15.43m started the construction of the gigantic project "Shanghai Changxing Under River Tunnel" in China.

The river banks, Pudong mainland and the island of Chongming were connected. Two parallel motorway tunnels, each tunnel having a length of 7,170 meters, were built between the mainland of Pudong and the island of Changxing because the waterway in between is a very busy main shipping route. The connection between Changxing and the island of Chongming are achieved by a bridge construction.

The parallel motorway tunnels have two levels; the upper level contains three lanes for road traffic and the lower level is planned to integrate a rescue lane in the centre and a safety passage.

The main challenges of this project were the large shield diameter of 15.43 meters and the predicted geological and hydrological conditions with high groundwater pressures of up to 6.5bar.

The tunnels were built in clayey formations below the groundwater table. At the deepest point the tunnels run about 65 meters below the surface. Therefore both Mixshields were designed for a maximum working pressure of 6.5bar. To avoid adhesion of sticky clay at the cutting wheel, its center area was equipped with its own slurry circuit. Large openings in the cutting wheel optimize the material flow and reduce the risk of blockage of material in the centre.

A special feature of the soft ground cutting wheel are six accessible main spokes, sealed against the water pressure. The design of the cutting wheel was conceived in order to allow man access to its interior space in free air, sealed from the ground water pressure outside.

To handle the clayey soil conditions the cutting wheel was equipped with soft ground tools and buckets. Tool change devices integrated in the cutting wheel, allow the personnel to replace tools under atmospheric conditions from the interior of the cutting wheel.

**Figure 2.** World's largest Mixshield Ø15.43 m for the Shanghai Changxing Under River Tunnel

The tunnel is lined with reinforced concrete segments. The heavy segments which weigh up to 16.7 tons each were delivered by two special trucks from the segment fabrication yard which was about 1.5km away from the jobsite. The tunnel lining has an inside diameter of 13.7 meters. Each tunnel ring consists of 9+1 segments and has a length of 2m.

The breakthrough of each 7,170m long tunnel tunnels was in May 2008 and September 2008 and thus 12 and 10 month earlier than scheduled. The commissioning of the BOT-tunnels is planned in 2010.

The structural steelwork for the two new 15.43 meter diameter Mixshields was manufactured in China and an assembly time of only four month displayed a high technical standard in the field of tunnelling technology and this for projects with huge demands like the excavation of the two parallel motorway tunnels below the Yangtze with high ground water pressures.

The following cited TBM project is also characteristic of the demanding conditions of tunnelling with large diameter TBMs in extremely heterogeneous geological formations.

## TWO MIXSHIELDS FOR THE RAIL TUNNEL ACCESS ROUTE TO THE BRENNER BASE TUNNEL

In Austria two Mixshields with diameters of 13m were used for the construction of the northern rail access to the future Brenner base tunnel which will form a key link between Germany, Austria and Italy.

The sections concerned are situated in the Lower Inn Valley where the existing 40km double-track railway had to handle not only north-south traffic but also the east-west traffic between Vienna and western Austria. It is thus an important junction especially taking an increase in traffic both for freight and passenger capacity into consideration which cannot be handled with the existing infrastructure of today.

One Mixshield excavated a section of 5,835 meter for a double-track railway tunnel on Lot 3–4 (Münster-Wiesing) and the second Mixshield, which was used before in the SMART tunnel project in Kuala Lumpur, was used for the 3,470 meter long Lot 8 (Jenbach). Along this section the 13m diameter shield passed under the Jenbach station, a power station channel and the motorway. The machine which was used before in one section of the SMART tunnel project in Kuala Lumpur had to be adapted from 13.21m to a shield diameter of 13m to fit the demands for the railway tunnel project Lot 8. The shield was equipped with a 4,400kW hydraulic cutterhead drive system.

The Mixshield used for the lot H3-4 excavated the 5.8 km long main tunnel which is the longest tunnel section of the new Lower Inn Valley rail. This shield in contrary was equipped with 20 electric motors generating a power of 3,200kW. The shield started from a 30m deep shaft. Over a length of approx. 250 meters it passed beneath the river Inn with minimal distance between the tunnel crown and the river bottom. Also the motorway A 12 and the existing railway line was undercut. The shield drive ended in a cavern. The shield skin remained in the tunnel and the rest of the TBM was dismantled through the already built tunnel.

The extremely heterogeneous geological formations in the bottom of the valley of the Lower Inn, comprising alluvial sands, clays, gravels and boulders with the groundwater level just below the surface, was a particular structural challenge.

Both Mixshields with shield diameters of 13 meters are among the largest tunnel boring machines used to date in Europe.

Concerning the design of the tunnel profile a system with two independent sealing levels was demanded for reasons of operational-technical requirements, whereby one seal level must maintain the pressure. As a standard profile a double shell lining in form of a circular cross-section with segmental lining and an additional fire protection shell of in-situ concrete was preferred.

The tunnel boring machine was designed and manufactured according to the predicted geological conditions. The Mixshield technology presented the best solution for the handling of the prevailing changing geological conditions with permeability of $10^{-5}$ m/s in the gravel formations.

The tunnel face was stabilized with a bentonite suspension which functions not only as a support medium but also as a transport medium. In a conventional Mixshield as used for Lot H3-4 (Münster-Wiesing) a submerged wall separates the working chamber from the bulkhead and enables to regulate the quantity and pressure of the supporting medium separately from each other. The substantial advantage of the divided working chamber with air cushion in the rear chamber for the regulation of the support pressure at the tunnel face is the decoupling of the support pressure regulation from the total circulating quantity of the suspension in the slurry circuit.

The Mixshield for Lot 8 (Jenbach) in contrary was designed with an isolated invert segment. This innovative patent protected version of the Mixshield is predominantly for application in cohesive soils. This technology was used for the first time in the Mixshield Ø11.67m for the Weser Road Tunnel Project near Bremen in Germany.

With an isolated invert segment the function of support pressure control is separated from the soil conveying. Due to the isolation of the invert area, the prepared bentonite suspension is injected directly into the working chamber. The slurry circulates towards the suction nozzle via the isolated invert through the working chamber. The pressure control at the tunnel face is no longer exercised by means of the submerged wall opening as is usual, but through two pressure compensation pipes (see 9 in Figure 3) situated between the working chamber and the excavation chamber. The connecting pipes ensure that the support pressure control is still guaranteed by the air cushion and secondary compressed air equipment.

The isolation of the invert area of a TBM ensures the safe and controlled transportation of the excavated material even in cohesive and sticky ground. This makes a continuously high excavation speed possible, no matter the quality of the ground.

The excavated soil which is mixed with the suspension is pumped via a slurry line to a separation plant outside the tunnel. There the excavated soil is separated from the transport medium. It is planned to recycle the material as far as possible and to dump the non-usable material.

Except for the planned downtime for maintenance the Mixshield for the double-track railway tunnel Münster-Wiesing advanced 24 hours, including the weekends and holidays. The 13m Mixshield for the Lot 3-4 (Münster-Wiesing) of the Lower Inn Valley finished its 5,835m long drive after approximately 19 month of excavation and six month faster than schedule.

## DOUBLE SHIELDED HARD ROCK TBMS FOR THE BRISBANE NORTH SOUTH BYPASS TUNNEL (NSBT)

The NSBT project is a Public Private Partnership (PPP) project. The main benefit of the PPP is that the RiverCity Motorway company is responsible for delivering the project on time and on budget, reducing the overall cost and construction risk to Council. The RiverCity Motorway has contracted the design and construction of the North-South Bypass Tunnel to the Leighton Contractors and Baulderstone Hornibrook Bilfinger Berger Joint Venture (LBB JV).

The project includes two parallel bored twin-lane tunnels that were excavated and lined in rock below the City of Brisbane and under the Brisbane River.

The major benefits of the tunnel are the link between the Inner City Bypass and Lutwyche Road in the North with Ipswich Road and the South-East Freeway in the South and the additional Brisbane River crossing. The northbound and southbound tunnels bypass 18 existing sets of traffic lights. Moreover they take a significant number of vehicles underneath the city each day reducing surface congestion and thus enabling a series of urban enhancements to be completed in adjacent suburbs.

The geological conditions at tunnel level comprise Brisbane Tuff and Neranleigh Fervale (NF) beds. The NF beds are characterized by arenites and phyllites with quartz veins. Both the tuff and the rocks of the NF beds are generally of high to very high strength.

Due to the predicted local geological conditions along the excavation of the tunnels, a combination of tunnel excavation methods were used. They included

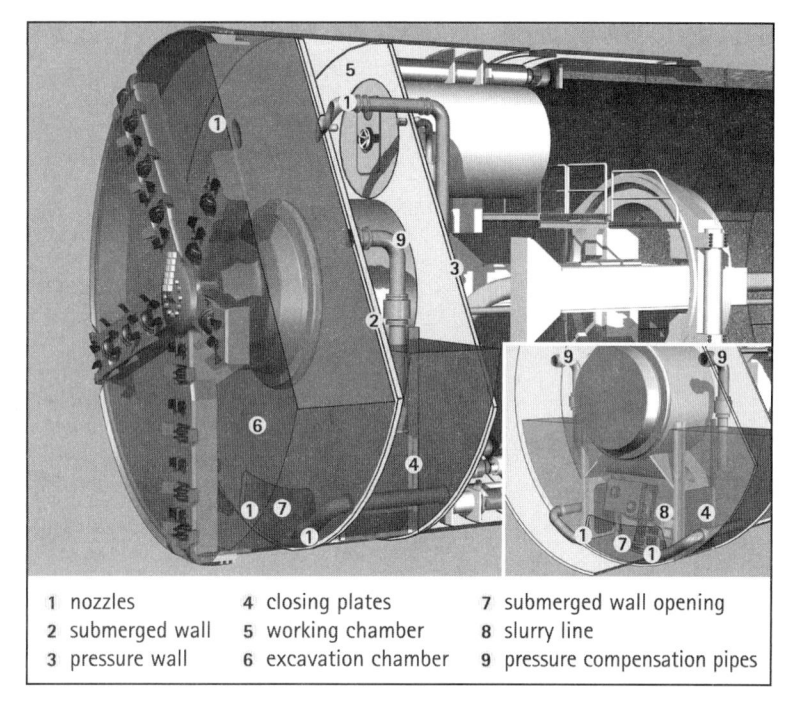

| | | |
|---|---|---|
| 1 nozzles | 4 closing plates | 7 submerged wall opening |
| 2 submerged wall | 5 working chamber | 8 slurry line |
| 3 pressure wall | 6 excavation chamber | 9 pressure compensation pipes |

**Figure 3. Isolated invert. Increasing tunnelling performance through controlled flow.**

cut-and-cover sections and sections driven by two tunnel boring machines and six roadheaders.

The overall excavated tunnel length by means of tunnel boring machines comprised 8.4km. For the two parallel bored tunnel sections of 4,067m and 4,348m respectively, two Double Shielded Hard Rock TBMs Ø12.34m of identical design were used. They excavated about 70% of the tunnel sections.

Double Shielded TBMs are one of the sophisticated TBM types in tunnelling because two applications—shield TBM and Gripper TBM—are combined in one and the same machine. Pronouncedly changing ground conditions can be handled with this type of machine because the shield can relatively easily be adapted to the geological conditions without any major setbacks affecting its progress even if poorer rock zones are encountered. This excavation method is characterized by safe working conditions. Moreover high and continuous production rates can be achieved in good rock conditions because the tunnel support can be placed whilst excavating.

The two Double Shields were ordered in July 2006. The cutterhead with a diameter of 12.4m was fitted with 74 19-inch back-loading disc cutters and 12 buckets. The Double Shields were fitted with an electrical 6m-diameter main drive and installed with a power of 4,200kW.

The geological condition, the inner-city location of the tunnel and the undercut of the Brisbane River required to equip the TBMs with drillings for probing ahead and taking core samples from the TBM. A full pre excavation grouting pattern as well as 2 probing ports were arranged in an angle of 8° through the gripper shield skin between approximately 11 and 1 o'clock in the crown. The machine was equipped with a percussive drill rig mounted on top of the first trailer. This drilling unit was installed on a moveable (180°) ring carrier to drill on a length of 22 m ahead of the TBM.

Mucking out of the tunnel was realized via a conveyor belt which was equipped with a weight measuring system and a volume measuring system (scanner device). The excavated material was taken to a purpose-built load-out facility and transported mainly via the arterial road network.

The tunnel is lined with a sealed segmental lining which consists of 8+1 reinforced concrete elements each having a length of 2 meters. Each segment is provided with an all-round seal which prevents ground water entering the tunnel. The tunnel lining has an outer diameter of 12.0m and an inner diameter of 11.20m.

The machine was equipped with a segment storage magazine which holds one complete tunnel ring (8+1) to avoid downtimes due to a delay in segment delivery. The segments were manufactured in a segment factory installed in a distance of 10 kilometers from the jobsite. The equipment was supplied by Herrenknecht Formwork Technology GmbH, a 100% subsidiary of Herrenknecht AG, who delivered a turn-key lining segment production facility for the NSBT project. In addition Herrenknecht Formwork provided and installed all the associated facilities and equipment. This includes handling equipment to turn, orientate, remove, deliver and store the segments, as well as equipment to install seals and produce the surface finish of the segments.

The segment factory included a carrousel system capable of carrying 5 sets of moulds (45 moulds), a curing tunnel, a concreting station, a reinforcement assembly area and a mould preparation line. The segment plant produced ten complete rings in 2 × 10 hours shifts using the 5 sets of moulds in production.

The design of the Herrenknecht Tunnelling Systems for the NSBT in Brisbane was based on the contract specifications of the client which define the technical basis and requirements for the Double Shielded Hard Rock TBM, the back-up systems and peripheral equipment such as tunnel belt conveyor and segment plant.

**Figure 4. Cutterhead of double shielded hard rock TBM Ø12.34 m for the north bound tunnel in Brisbane**

## HARD ROCK TUNNELLING FOR THE LONGEST TRAFFIC TUNNEL— THE GOTTHARD BASE TUNNEL IN SWITZERLAND

The new Gotthard base rail tunnel is currently under construction. The project is a future-oriented flat railway through the Alps and will be then the longest rail tunnel in the world with its two tunnels of each 57km. The tunnel will be put into service at the end of 2017. This pioneer work of the 21st century will lead to a prominent improvement of travel and transport possibilities in the heart of Europe.

The concept for the Gotthard Base Tunnel provides a simultaneous advance in five parts of different lengths comprising tunnel boring machines and drill and blast.

The mechanized tunnel sections excavated by means of Gripper TBMs comprise in total following four subsections:

- Erstfeld (2 × 7,178m)
- Amsteg (2 × 11,350m)
- Faido (1 × 12.4km, 1 × 11.9km)
- Bodio (2 × 14km)

The first mechanized tunnel of the subsection Amsteg was completed in June 2006, and the parallel section was excavated by the beginning of October 2006 about half a year ahead of schedule. The two approximately 14km long parallel tunnels of the subsection Bodio were completed at the beginning of September 2006 and the end of October 2006 respectively.

The four Gripper TBMs which excavated the often quite demanding rock massif and fault zones nevertheless finished the total of about 50km on time.

For the subsection Faido to Sedrun the two Gripper TBMs used in Bodio were completely refurbished. The geology along this section comprises two tectonic units, the Penninic Gneiss zone (approx. 5km) and the Gotthard Massif (approx. 10km). The Piora zone was predicted to comprise solid, compact and partially metamorphic dolomite anhydrite rocks at tunnel level. The TBMs applied for the subsection Faido have been modified. Apart from an increase in excavation diameter to 9.40m, to be prepared for the greater overburden from 1,200m up to 2,470m and thus greater rock pressures along this section, 12 instead of 8 buckets were applied and the 17-inch disc cutters have been replaced by 18-inch cutters. To support the diameter of 9.50m, the gripper and the walking legs were adapted. Modifications were also done on the cutterhead dust control system with an increase from 600m$^3$ per minute to 1,100m$^3$ per minute.

For the subsection Erstfeld the geology is characterized by mainly solid and geotechnically favorable highly metamorphic gneisses (Erstfelder gneiss). The AlpTransit Gotthard AG administrative council awarded the subsection Erstfeld to the Joint Venture Gotthard Base Tunnel North. The JV consists of the companies Murer-Strabag AG, of Erstfeld, Switzerland, and Strabag AG, of Spittal/ Drau, Austria. The TBM section comprises the excavation of two single-track tunnels of 7.2km from Erstfeld to Amsteg. The tunnels are excavated and secured by the two TBMs that have driven the 2 × 11.35km long subsection Amsteg. This section includes an underground junction to permit a future extension of the tunnel towards the north without interrupting the operation.

In connection with the TBM drives of a long tunnel with a large overburden (>2,500 m) and in a tectonically active rock mass (folding of the Alps), one can draw the conclusions that despite extensive clarifications in the run-up to the project, there can be a great difference between geological prediction and geological finding. The rock behaviour and the hazard scenarios can prove to be less favourable than expected, which could make an optimal use of the drive systems designed according to the hazard scenarios dominating the service contract impossible.

The constructional relevance can change very quickly on site, correspondingly the mountain only forgives faults in exceptional cases and sometimes requires quick decisions of all persons involved in the project and the prompt realization of immediate measures.

The applied TBM technology proved, however, that it is in a position to master technically essentially more critical situations than were provided in the service contract. The construction of these TBMs and trailers were subject to extensive adjustments (among others due to the extraordinary conditions) and optimizations during the drive for more than altogether 8 years and nearly 30 km each.

Extraordinary conditions can additionally aggravate the already very demanding technical and logistical challenges. A close and constructive cooperation between client, author of the project, supervisor of works and enterprise is of essential importance for the success of the project of the structure of the century.

## OUTLOOK

The cited projects show the multitude of pioneering references in large diameter mechanized tunnelling development such as Shanghai (Mixshield Ø15.43m) and M30 Madrid (EPB-Shield Ø15.20m). They support the feasibility of the construction of very large tunnels. The performances that have been achieved by the current largest tunnel boring machines include also an excellent logistical concept which presents a good basis for administrative authorities, project owners and contractors regarding the feasibility,

**Figure 5. Hard rock gripper TBM at Amsteg**

reliability, safety and speed of upcoming large diameter projects.

The tendency of future large diameter tunnel projects are in direct relation to the progressing urbanization and the possible impending total gridlock especially in metropolitan areas or larger cities and also at junctions such as the access to the Brenner Base tunnel.

To summarize the current state of the art in TBM technology, the TBMs range from Ø100mm to Ø16m. They are today reliably used for the realization of complex projects. In the future, tunnels with diameters of more than 16 meters are envisaged not only in densely populated areas but also through natural barriers like mountain ranges or under rivers and estuaries. The market requires practical engineering skills under toughest conditions.

Innovations such as seismic probing ahead, cutting wheels accessible under free air, muck control, drill units for ground stabilization measures from the TBM and cutter wear detection systems were designed and further developed. Information technology and extremely sophisticated measuring techniques in tunnelling are dramatically increasing safety as well as economic profitability.

# ADECO as an Alternative to NATM: 22 m Wide, 14 m High, Full Face Tunnel Excavation in Clays

Fulvio Tonon
University of Texas, Austin, Texas

ABSTRACT: With a cover of 4 to 30 m, the Cassia twin tunnels underpass Cassia road (main road to Rome), the remnants of an ancient Roman villa, and two existing tunnels. Stiff silty clay dominates the alignment. According to the ADECO (Analysis of Controlled DEformations) principles, Cassia tunnels were excavated full-face to 260 m$^2$ (2800 ft$^2$). With respect to SEM this highly simplifies construction and allows for full control of the ground ahead of the face, which is used as a stabilization measure. In order to preconfine the tunnel core, Trevi roto-injection technique was used to create sub-horizontal jet-grouting columns and to install reinforcing steel pipes, while avoiding up-heave or emptying of the columns. The tunnels were finished ahead of schedule and within budget.

## INTRODUCTION

The GRA or Grande Raccordo Anulare (literally, "Big Ring Junction") is a toll-free ring road, 68.2 km (42.6 mi) in circumference that encircles Rome, Italy. Currently, the GRA carries 160,000 vehicles daily; as shown in Figure 1, the two lanes in each direction built after World War II were substandard, over capacity and traffic jams were continuous throughout the day (probably the most famous is the one depicted in 1972 Federico Fellini's "Rome"). In 2002, a massive project called for the enlargement of 18.5 km of GRA from two to three lanes in each direction between Aurelia and Castel Giubileo exits; 3 km included completely new horizontal and vertical alignments meant to preserve parks and improve safety. The new 2-bore Cassia tunnel is part of this new alignment. In order to accommodate three lanes and a full emergency lane (Figure 2), an exceptionally large cross-section was needed: 22 m wide and 14 m high. The net pillar width between the two bores is about 4 m. The parallel bores of 232 m and 125 m length undercross Cassia Road (one of the main roads to and from Rome); the outer bore undercrosses an ancient Roman villa (with a net 5 m cover), and the inner bore is directly underneath the existing tunnels for the Ring Road (Figure 1 and Figure 3).

## DESIGN AND CONSTRUCTION APPROACH

Cassia tunnels were excavated mainly in clays, with pockets of sand in the invert and crown at some locations. Figure 4 shows how one should proceed to construct a road tunnel in these conditions according to the FHWA Technical Manual for Design and Construction of Road Tunnels (2009), which embraces Sequential Excavation Method (SEM) and/or New Austrian Tunneling Method (NATM).

The author has already pointed out that SEM dates back to the early 1800s (when there was no electricity, no compressed air, and people went around by horse and buggy wearing crinoline and top hats), and that Rabcewicz (1964–5) since his first NATM paper insisted that one should proceed full face (Tonon 2009). However, NATM can not proceed full face under all circumstances (and mainly in difficult stress-strain conditions where full-face excavation is key to success) because of technical limitations in the 1960s and because it does not recognize the

Figure 1. Driving along the old inner bore of Cassia tunnels

Figure 2. Driving along the finished external bore of Cassia tunnel

Figure 3. Existing tunnels for Rome Ring Road, and new Cassia tunnels

Elements of Commonly Used Soft Ground Excavation and Support Classes (ESC) in Soft Ground

| Ground Mass Quality – Soil | Excavation Sequence | Initial Shotcrete Lining | Installation Location | Pre-Support | Support Installation | Remarks |
|---|---|---|---|---|---|---|
| Stiff/hard cohesive soil - above groundwater table | Top heading, bench & invert; dependent on tunnel size, further sub-divisions into drifts may be required | Systematic reinforced (welded wire fabric or fibers) shell with full ring closure in invert; dependent on tunnel size 6 in (150 mm) to 16 in (400 mm) typical; for initial stabilization and to prevent desiccation, a layer of flashcrete may be required | Installation of shotcrete support immediately after excavation in each round. Early support ring closure required. Either temporary ring closure (e.g. temporary top heading invert) or final ring closure to be installed within one tunnel diameter behind excavation face. | Typically none; local spiling to limit over-break | Support installation dictates progress | Overall sufficient stand-up time to install support without pre-support or ground modification |
| Stiff/hard cohesive soil - below groundwater table | Top heading, bench and invert; dependent on ground strength, smaller drifts required than above | Systematic reinforced (welded wire fabric or fibers) shell with full ring closure in invert; dependent on tunnel size 6 in (150 mm) to 16 in (400 mm) typical; for initial stabilization and to prevent desiccation, a layer of flashcrete may be required; frequently more invert curvature than above | Installation of shotcrete support immediately after excavation in each round. Early support ring closure required. Either temporary ring closure (e.g. temporary top heading invert) or final ring closure to be installed within less than one tunnel diameter behind excavation face; typically earlier ring closure required than above | Typically none; locally pre-spiling to limit over-break | Support installation dictates progress | Sufficient stand-up time to install support without pre-support or ground improvement; dependent on water saturation, swelling or squeezing can occur |
| Well consolidated non-cohesive soil - above groundwater table | Top heading, bench & invert; dependent on tunnel size, further sub-divisions into drifts may be required | Systematic reinforced (welded wire fabric or fibers) shell with full ring closure in invert; dependent on tunnel size 6 in (150 mm) to 16 in (400 mm) typical; for initial stabilization and to prevent desiccation, a layer of flashcrete is required | Installation of shotcrete support immediately after excavation in each round. Early support ring closure required. Either temporary ring closure (e.g. temporary top heading invert) or final ring closure to be installed within less than one tunnel diameter behind excavation face | Frequently systematic pre-support required by grouted pipe spiling or grouted pipe arch canopy; alternatively ground improvement | Support installation dictates progress | Stand-up time insufficient to safely install support without pre-support or ground improvement |
| Well consolidated | Top heading, bench & | Systematic reinforced | Installation of shotcrete support | Frequently | Support | Stand-up time |

**Figure 4. Recommended support for tunnels in cohesive ground. FHWA (2009)**

importance of the ground ahead of the tunnel face as a stabilization measure.

An alternative to SEM/NATM was used for the Mt. Baker Ridge Tunnel, illustrated in Figure 6, along I-90 in Seattle, WA. The tunnel carries two lanes on the bottom level, three lanes in the middle level, and pedestrians on the top level. Built between January 1983 and May 1986 20 m to the side of two existing tunnels (Figure 7), the Mt. Baker Ridge Tunnel has an interior diameter of 19.35 m (63.5 ft), an effective outside diameter of 24.5 m (80 ft), and a length of 457 m (4,500 ft). The geology of the site is shown in Figure 8 and comprises sand in the invert and stiff clays elsewhere. Clays were overconsolidated by a

**Example SEM Excavation and Support Classes in Soft Ground**

| Description | Cross Section | Longitudinal Section | Photo |
|---|---|---|---|
| **Soft Ground – shallow cover:**<br>• Systematic pre-support<br>• Systematic shotcrete initial lining support with early ring closure<br>• Top heading excavation (with temporary invert), bench and invert excavation<br>• Round Length<br>  Top Heading: I – 3'-3" (1 m)<br>  Top Heading: II – 6'-6" (2 m)<br>  Bench III/Invert IV – 6'-6" (2 m)<br>• Dimensions<br>  Height: 38'-0" (11.6 m)<br>  Width: 48'-0" (14.7 m)<br>**Example:** Fort Canning Tunnel, Singapore | 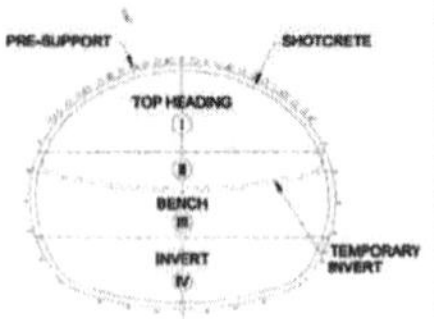 | 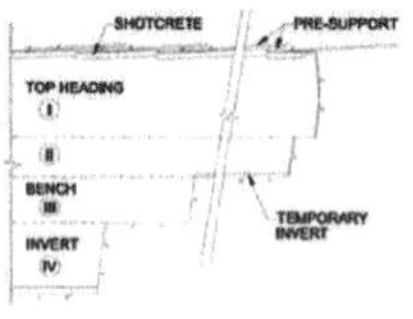 |  |
| **Soft Ground – deep level:**<br>• Systematic shotcrete support with early ring closure<br>• Top heading excavation closely followed by bench/invert excavation<br>• Round Length<br>  Top Heading: 3'-3" (1 m)<br>  Bench: 6'-6" (2 m)<br>• Dimensions<br>  Height: 20'-3" (6.3 m)<br>  Width: 20'-3" (6.3 m)<br>**Example:** London Bridge Station, London, UK | 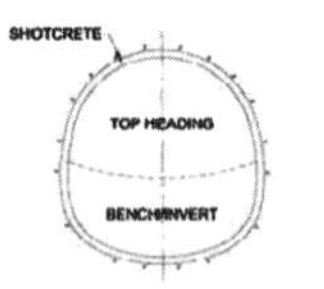 | 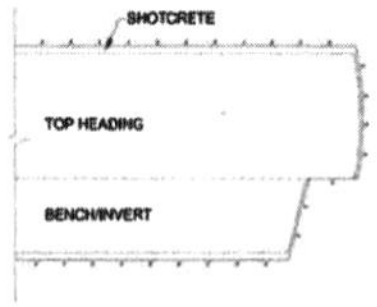 | 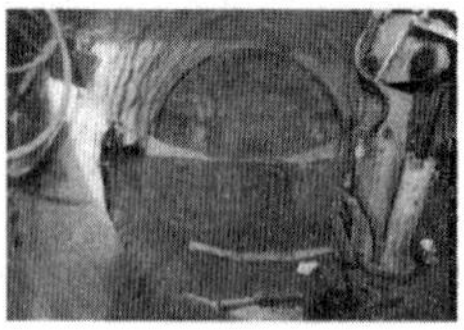 |

| Description | Cross Section | Longitudinal Section | Photo |
|---|---|---|---|
| **Soft Ground – deep level:**<br>• Systematic shotcrete support with early ring closure<br>• Sub-division into sidewall drifts<br>• Top heading excavation closely followed by bench and invert excavation<br>• Round Length<br>  Top Heading: 3'-3" (1 m)<br>  Bench: 6'-6" (2 m)<br>  Invert: 6'-6" (2 m)<br>• Dimensions<br>  Height: 30'-2" (9.2 m)<br>  Width: 37'-0" (11.3 m)<br>**Example:** London Bridge Station, London, UK | 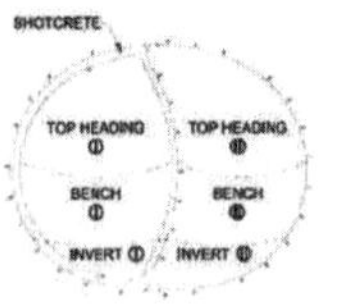 | 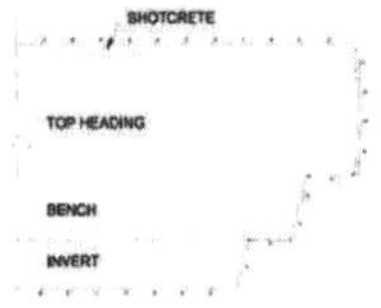 | 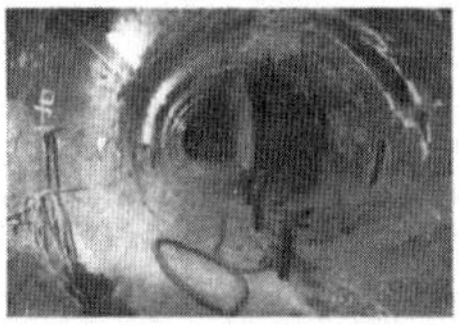 |

Figure 5. Typical examples for tunnels in cohesive ground. FHWA (2009).

1200 m thick ice sheet in the last Ice Age to produce $K_0 = 2$ (Holloway and Kjerbol 1988). A variant of the multiple-drift method was used for constructing the old tunnels (Figure 7) with 30 cm × 40 cm timbers at 1 m spacing and concrete placed in between. Given the difficult advance even with such a heavy support and considering the size of the new tunnel, Holloway and Kjerbol (1988, page 3) noticed that "a construction method that would permit minimum face exposure at any one time was necessary." Construction entailed shield excavation of 24 3-m diameter drifts, each supported with a 5-piece segmental lining (1.2 m long, 12.5 cm thick), sequentially excavated as depicted in Figure 9, and immediately backfilled with concrete and grouted before resuming excavation of the subsequent drift. The core of soil inside the stacked drifts was subsequently excavated in five lifts. In order to minimize risk, the owner, Washington Department of Transportation, purchased the land on top of the tunnel, including all properties. Compared to 30 cm predicted settlements, observed settlements were a maximum of 5 cm along the eastern two thirds of the tunnel and 20 cm in the west end of the tunnel. Most of the settlement occurred during construction of the drifts.

The two tubes of the Cassia tunnel were built *full face* (260 m$^2$) according to the ADECO (Analysis of Controlled DEformations). This design approach coupled with state-of-the-art preconfinement, allowed to achieve the following results: construction of a stiff preliminary lining right at the face, which rested on sub-horizontal jet-grouting columns, final invert construction within 6 m behind the face, minimization of ground disturbance and settlements, simplification of all construction operations and greatly improved construction safety, high advance rates, and risk and cost minimization. Figure 10 compares the cross-section of Mt. Baker Ridge Tunnel and the cross section of one of the Cassia tunnel bores: each of the two Cassia bores was wider than the Mt. Baker Ridge Tunnel!

## ADECO APPROACH

The ADECO workflow is illustrated in Figure 12. In the Diagnosis Phase, the unlined/unreinforced tunnel is modeled in its *in situ* state of stress with the aim of subdividing the entire alignment into the three face/core behavior categories: A, B, and C: these depend on the stress-strain behavior of the core (ground strength, deformability and permeability + *in situ* stress), not only on the ground class. The site investigation must be detailed and informative enough to carry out such quantitative analyses: this clearly defines what the investigation should produce.

In the Therapy phase, the ground is engineered to control the deformations found in the Diagnosis Phase. For tunnel category A, the ground remains in an elastic condition, and one needs to worry about rock block stability (face and cavity) and rock bursts; typically, rock bolts, shotcrete, steel sets and forepoling are used to this effect. In categories B and C yielding occurs in the ground; an arch effect must be artificially created *ahead* of the tunnel face (preconfinement) when a large yielded zone forms in category B, and in all cases in category C. By looking at the Mohr plane (Figure 13) two courses of action clearly arise:

- Protecting the core by reducing the size of the Mohr circle: this can be achieved either by providing confinement (increasing $\sigma_3$) or by reducing the maximum principal stress (reducing $\sigma_1$).

**Figure 6. Mt. Baker Ridge Tunnel, after Holloway and Kjerbol (1988)**

**Figure 7. Existing road tunnels at Mt. Baker Ridge, after Holloway and Kjerbol (1988)**

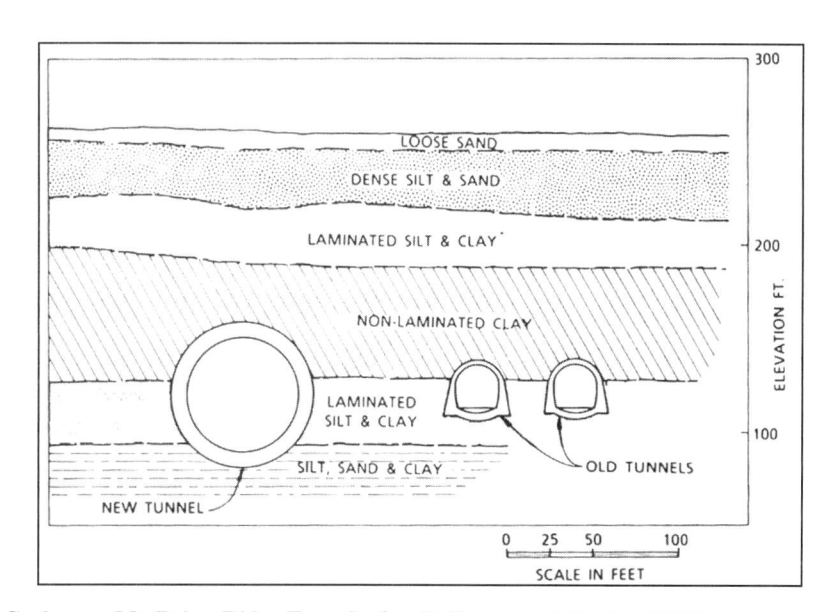

**Figure 8. Geology at Mt. Baker Ridge Tunnel, after Holloway and Kjerbol (1988)**

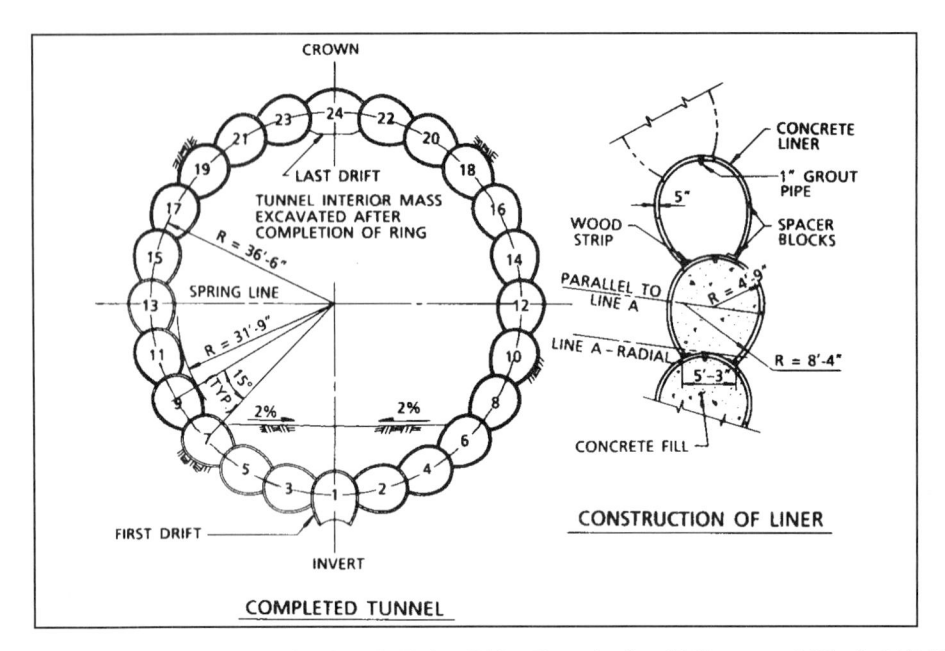

Figure 9. Construction sequence for the Mt. Baker Ridge Tunnel, after Holloway and Kjerbol (1988)

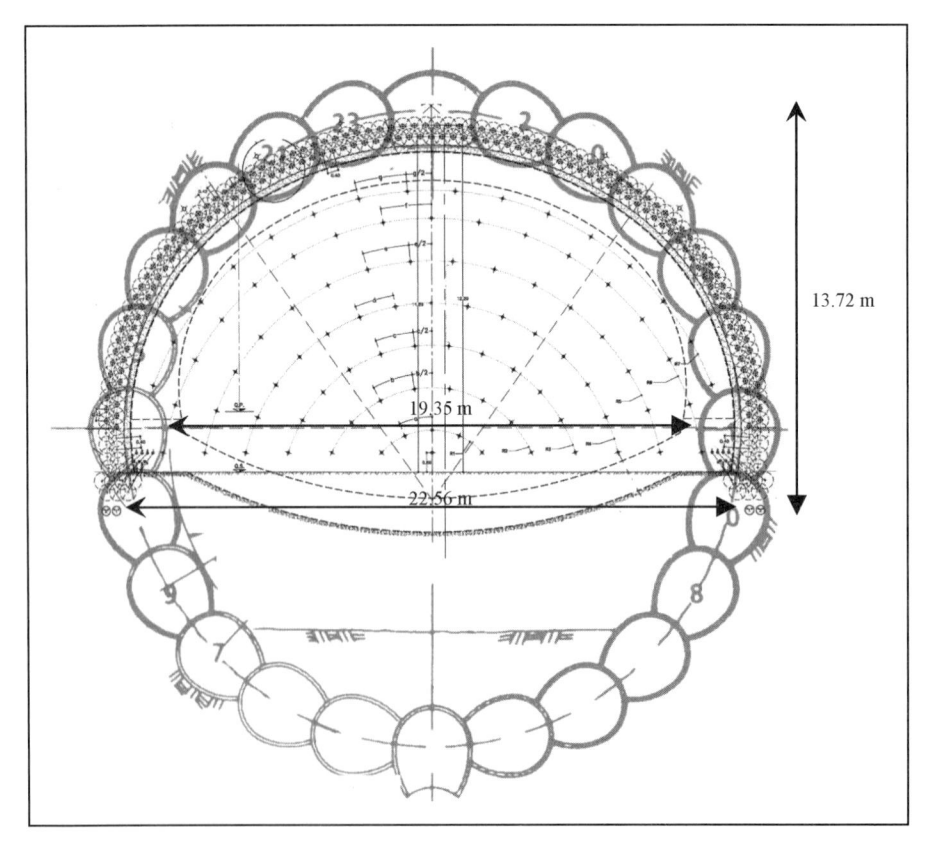

Figure 10. Comparison between the cross-sections of Mt. Baker Ridge Tunnel and of the Cassia tunnel bores

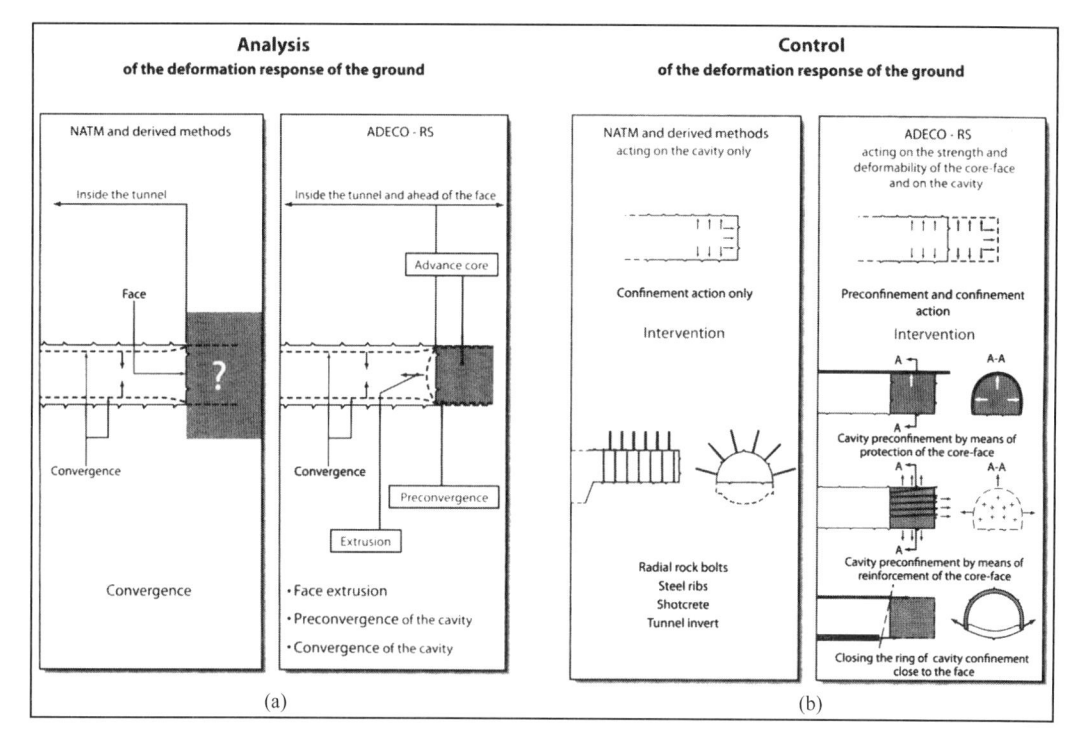

**Figure 11. NATM vs. ADECO, after Lunardi (2008)**

- Reinforcing the core, thereby pushing up and tilting upwards the failure envelope.

The rightmost column in Figure 11 depicts the actual implementation of these two ideas as pre-confinement actions. The third line of action consists of controlling the convergence at the face by using the stiffness of the lining (preliminary or even final, if needed), which may also longitudinally confine the core. It is only in this context that the different technologies currently available and listed in Figure 14 take their appropriate role. Notice that, at difference with the NATM, the ADECO embraces tunnels excavated with and without a tunnel boring machine.

Once the confinement and pre-confinement measures have been chosen, the cross-section is composed both in the transverse and longitudinal directions, and then analyzed. In all cases, full face advance is specified in all stress-strain conditions, thus fulfilling Rabcevicz's dream.

For each cross-section, displacement ranges are predicted in terms of convergence and extrusion (Figure 15). Besides plans and specs, construction guidelines are also produced during the design stage. The construction guidelines are used at the construction site to make prompt decisions based on the displacement readings. If the readings are in the middle of the predicted ranges, then the nominal cross-section in the plans and specs is adopted; if reading values fall to the lower end of the predicted displacement ranges, then the minimum quantities specified in the guidelines are adopted for the stabilization measures (Figure 14). Likewise, if reading values are on the upper end of the predicted displacement ranges, then the maximum quantities specified in the guidelines are adopted. Finally, if the readings are outside the predicted displacement ranges, the guidelines specify the new section to be adopted. In this way, ADECO clearly distinguishes between design and construction stages because no improvisation (design-as-you-go) is adopted during construction.

Monitoring plays a major role in the ADECO, but with two main differences with respect to the NATM:

- In categories B and C, not only convergence but also extrusion is measured because the cause of instability is the deformation of the core, and because stability of the core by pre-confinement actions is a necessary condition for the stability of the cavity.
- Monitoring is used to fine tune the design, not to improvise cavity stabilization measures, so that construction time and cost can be reliably predicted.

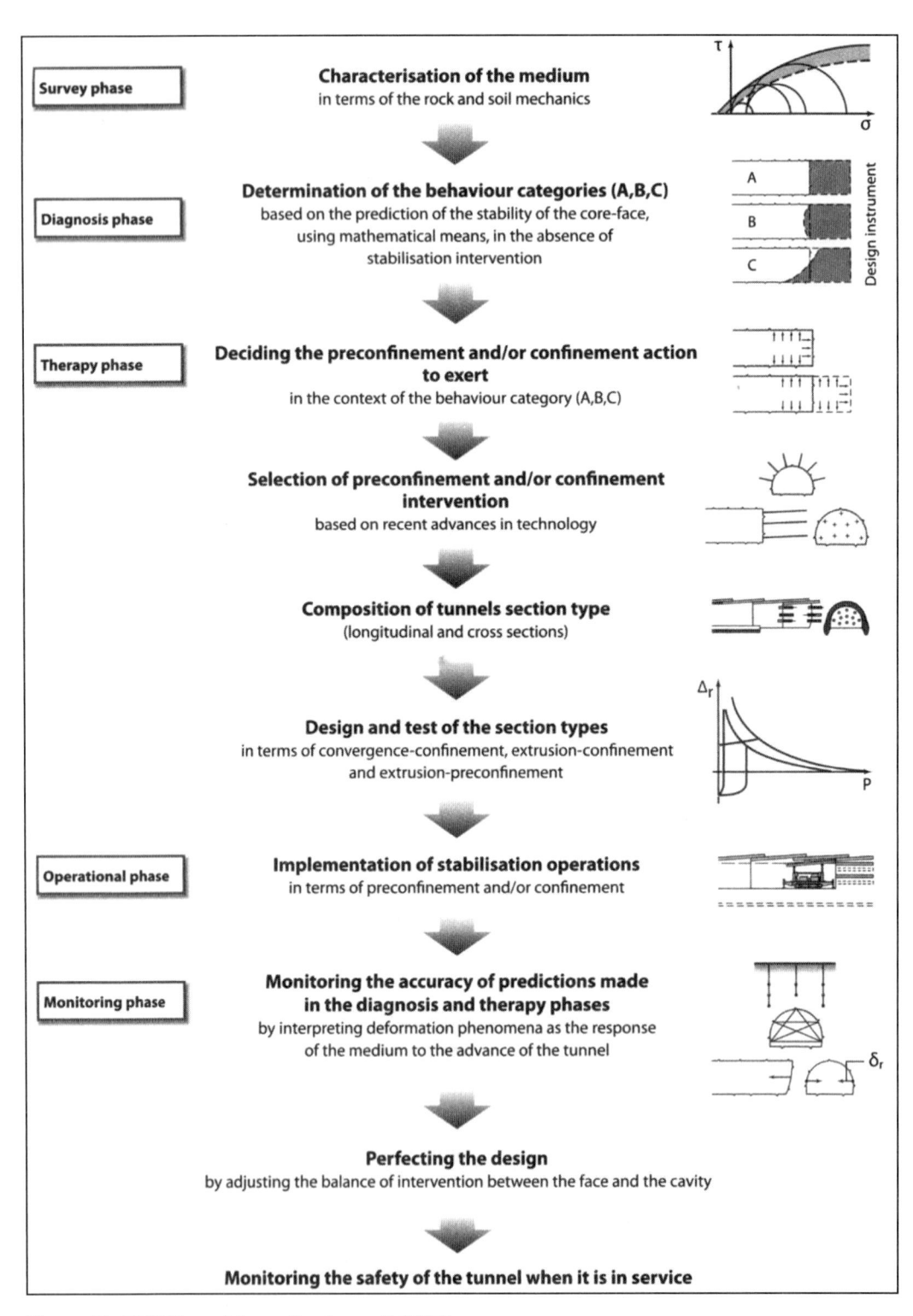

|  |  | |
|---|---|---|
| **Survey phase** | **Characterisation of the medium** in terms of the rock and soil mechanics | |
| **Diagnosis phase** | **Determination of the behaviour categories (A,B,C)** based on the prediction of the stability of the core-face, using mathematical means, in the absence of stabilisation intervention | Design instrument |
| **Therapy phase** | **Deciding the preconfinement and/or confinement action to exert** in the context of the behaviour category (A,B,C) | |
| | **Selection of preconfinement and/or confinement intervention** based on recent advances in technology | |
| | **Composition of tunnels section type** (longitudinal and cross sections) | |
| | **Design and test of the section types** in terms of convergence-confinement, extrusion-confinement and extrusion-preconfinement | |
| **Operational phase** | **Implementation of stabilisation operations** in terms of preconfinement and/or confinement | |
| **Monitoring phase** | **Monitoring the accuracy of predictions made in the diagnosis and therapy phases** by interpreting deformation phenomena as the response of the medium to the advance of the tunnel | |
| | **Perfecting the design** by adjusting the balance of intervention between the face and the cavity | |
| | **Monitoring the safety of the tunnel when it is in service** | |

Figure 12. ADECO workflow, after Lunardi (2008)

Tunnels are thus paid for how much they deform, which, unlike rock mass classifications carried out at the face, is an objective measure void of any interpretation. In addition, rock mass classifications are inapplicable to soils and complex rock mass conditions not included in classifications' databases. Experience in over 500 km of tunnels indicates that, when the ADECO has been adopted and tunnels were paid for how much they deformed, claims have decreased to a minimum.

## TUNNEL DESIGN

### Survey Phase

Cassia tunnel falls within the south-western outskirts of the Sabatino Volcanic Apparatus, which, mainly composed of pyroclastic deposits, underlies the entire region of Rome and its surroundings. The soils encountered may be described as follows (from the ground surface):

- Fill
- Detrital layer: remolded sandy clays, locally mixed with highly weathered pyroclastic sediments.
- Yellow to yellow-red medium-fine sand, very silty at places. Symbol "$s_1$."
- Gray-blue silty clay; yellow oxidized band at top of layer or close to the outcrops. Symbol "$la_2$."
- Gray-yellow silty medium-fine sand, with lenses of organic matter at places. Symbol "$s_2$."

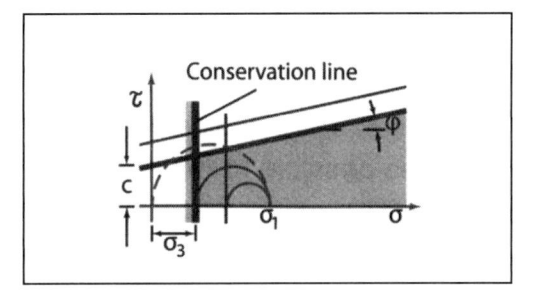

Figure 13. Mohr-plane explanation of approaches to stabilize/stiffen the core, after Lunardi (2008)

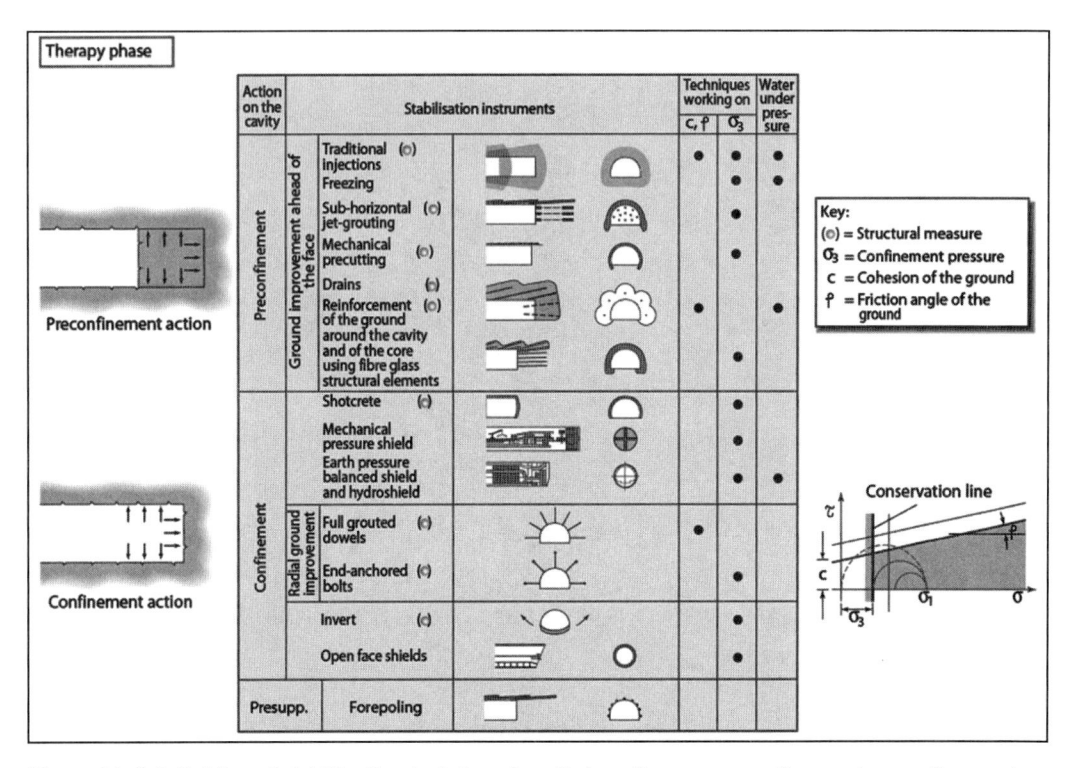

Figure 14. Subdivision of stabilization tools based on their action as pre-confinement or confinement, after Lunardi (2008)

| Section Types | Geology | Convergence (cm) | Extrusion (cm) |
|---|---|---|---|
| A | Monte Modino Sandstones | 2-3 | Negligible |
| B0 | | 3-5 | Negligible |
| B0V | | 5-10 | < 3 |
| B2 | Scaly Clays | 8-12 | < 6 |
| B2V | | 6-10 | < 5 |
| C2 | | 10-14 | < 10 |
| C6 | | 8-12 | < 8 |

| Section Types | Intervention | Variabilities | | |
|---|---|---|---|---|
| | | Minimum | Nominal | Maximum |
| C2 | Steel rib step | 1.2 m | 1.0 m | 0.8 m |
| | N°VTR face | 50 | 70 | 90 |
| | VTR face overl. | 10.0 m | 12.0 m | 14.0 m |
| | Excavation | 14.0 m | 12.0 m | 10.0 m |
| | Invert-face (°) | < 2.0∅ | < 1.5∅ | < 0.5∅ |
| | Crown-face | < 3.0∅ | < 5.0∅ | < 7.0∅ |

Figure 15. Displacement predictions and design guidelines, after Lunardi et al. (2008)

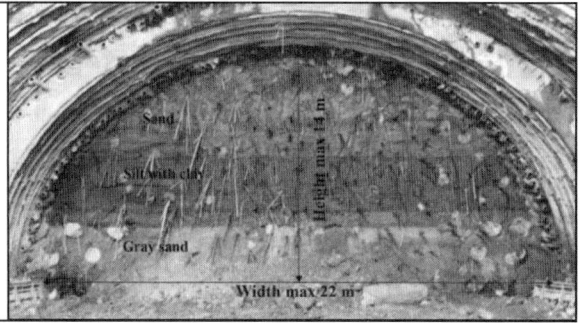

Figure 16. Soil at the tunnel face: (a) Typical "la$_2$" encountered along most of tunnel alignment (b) At east portal of outer bore

- Very dense to extremely dense, fine silty to slighlty silty sands with lenses of sandy silt. Symbol "s$_4$."

The groundwater table is located below the invert. With a cover depth between 5 to 25 m, Cassia tunnels were mainly excavated in "la$_2$" (Figure 16a), with "s$_1$" and "s$_2$" appearing at crown and invert only for a few meters starting from the East portal of the outer tube (Figure 16b).

Two substantial geotechnical investigations were carried out in 2004 (final design) and 2007 (shop drawings). *In situ* tests included: Standard Penetration Tests, Cone Penetration Tests, Menard Pressuremeter, pocket penetrometer, torvane, and down-the-hole shear wave velocity tests. Laboratory tests included: direct shear tests, resonant column, and CD, CU, and UU triaxial tests. Shear strength parameters of cohesionless soils were mainly determined based on *in situ* test results, whereas laboratory test results were mainly used for cohesive soils.

As for deformability parameters, highest weight was given to results of resonant column tests and shear wave velocity profiles. The final parameter ranges used in the design are given in Table 1.

**Diagnosis Phase**

When tunneling at shallow depths under important roads, buildings or archaeological sites, i.e. when deformations must be minimized, tunnels are always classified as Category C because the tunnel core must be preserved and stiffened in order to minimize the settlements (Lunardi, 2008). However, since the maximum cover is smaller than 3 equivalent diameters, the stability of the tunnel face was also studied by resorting to limiting equilibrium methods. Under maximum cover, the factor of safety was equal to 0.65 even when the maximum values for the strength parameters in Table 1 were adopted. This highlights the difficulty of the stress-strain conditions. As a consequence, the tunnel was classified

Table 1. Design parameter ranges for the soils within the mined cross-section of Cassia tunnels

| Formation | Soil Type | γ (KN/m³) | Dr (%) | $s_u$ (Kpa) | c' (Kpa) | φ' (°) | E' (MPa) |
|---|---|---|---|---|---|---|---|
| s1 | Loose silty sand | 16–19 | 34–60 | — | 0–5 | 34–37 | 91–134 |
| s4 | Silty fine sand to slightly silty sand | 19–20 | 44–58 | — | 0–5 | 37–39 | 175–259 |
| la2 | Silty clay | 17–19 | — | 55–170 | 3–57 | 20–29 | 130–219 |

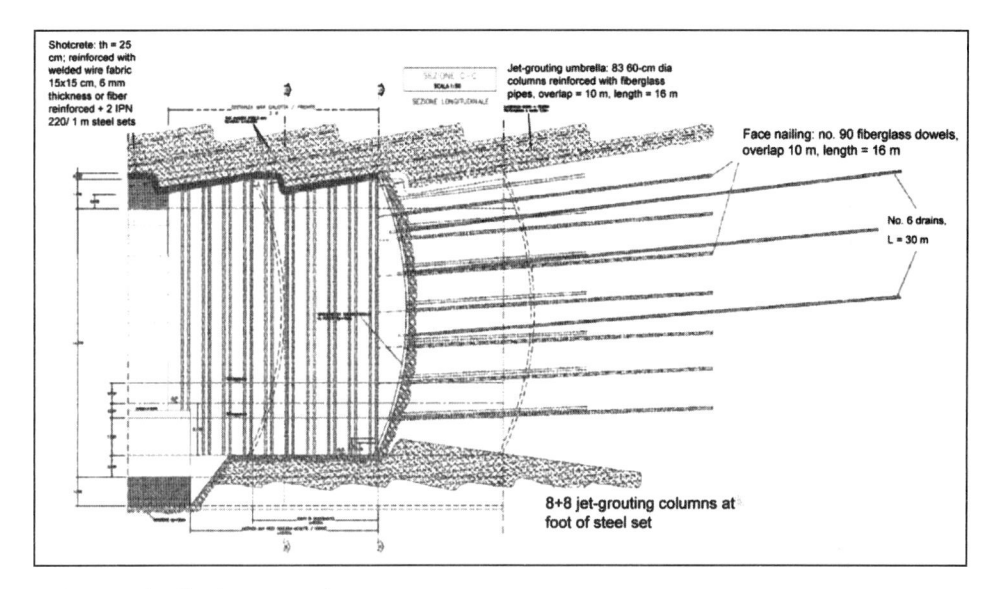

Figure 17. Longitudinal cross-section

as Category C, i.e. unstable even in the short term because an arch effect can develop neither ahead of the tunnel face nor around the excavation because the ground does not have enough residual strength. The deformation is unacceptable because it develops immediately into failure range leading to face and cavity collapses without time to install radial confinement. Preconfinement actions must be launched ahead of the tunnel face to create an arch effect ahead of the tunnel face itself.

**Therapy Phase**

In order to minimize settlements, the tunnel core was stiffened and protected so as to minimize extrusion (Figure 17 and Figure 18):

- Stiffening of the tunnel core was accomplished by 90 fiberglass dowels, 16 m in length, and overlapped 10 m. Between 80 and 90 dowels could be used at any given time based on the monitoring results.
- Sub-horizontal jet-grouting umbrella composed of 81 columns, 16 m in length and 60 cm in diameter, reinforced with steel pipes (168.3 mm diameter, 8 mm thickness) and realized with Trevi patented "roto-injection" technique described later. On either side of the cross-section, four additional (unreinforced) columns were realized as foundations for the steel sets.

Besides the preconfinement above, the following was specified:

- Sub-horizontal drains to be installed ahead of the tunnel face (3 + 3 pipes, L = 30 m).
- 1 m advance, with 5 cm of sealing shotcrete applied immediately and additional 20 cm of shotcrete applied after welded wire fabric (wwf) installation. The designer left it to the contractor to decide whether to use wwf or fiber-reinforced shotcrete.
- Primary lining composed of 25 cm of shotcrete (reinforced with wwf) and steel sets composed of 2 IPN 220 at 1 m spacing.
- Excavation and pouring of the invert (1.2 m thickness) and kicker walls (both of them

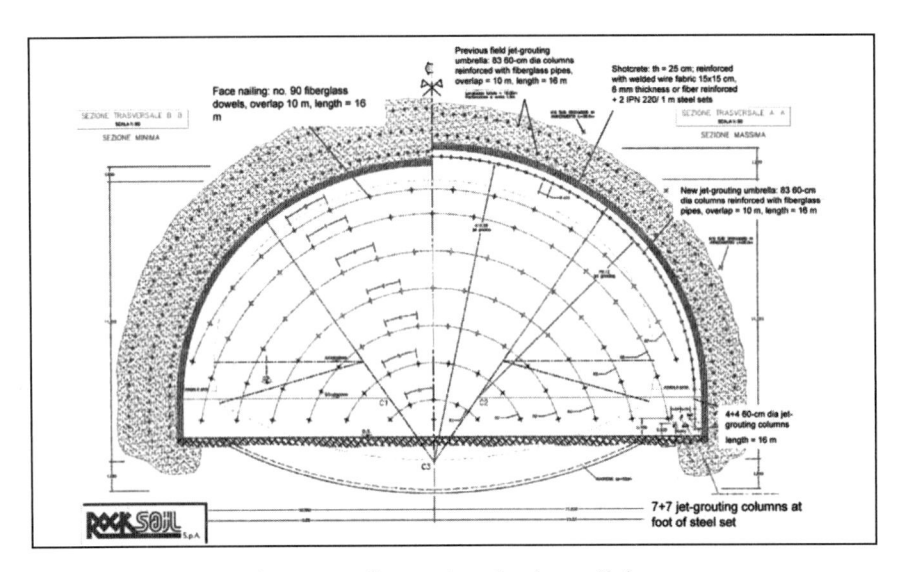

**Figure 18. Transverse cross-section, preconfinement, and primary lining**

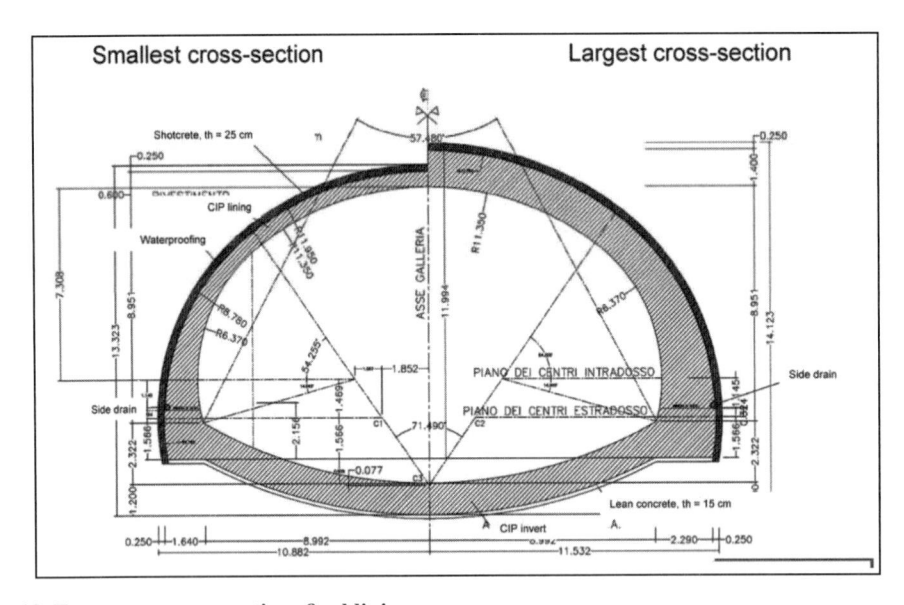

**Figure 19. Transverse cross-section: final lining**

reinforced) at a maximum distance of 9 m. The final distance was decided in the field based on the deformations measured.
- Waterproofing composed of a geosynthetic mat and a PVC membrane hot-welded *in situ*.
- Reinforced concrete final lining in crown and sidewalls to be poured within 3 diameters from the face (Figure 19). The final distance was decided in the field based on the deformations measured.

**Construction Sequence and Technological Aspects**

The construction sequence repeated itself every 6 m and consisted of the following operations (3 8-hour shifts) that lasted from 8 to 10 days:

- Core reinforcement with fiberglass dowels: 1.5 days
- Jet-grouting umbrella: 3.5 days
- Invert: 1.5 days (40 m$^3$/h pour): 6 hours for the excavation, 8 hours for reinforcement

**Figure 20. Soilmec SM 605 DT mobilized at tunnel face**

installation, 8 hours for pouring about 190 m$^3$. The first advance started 6 hours after the last concrete truck had left
- Excavation (1 m/advance), erection of 6 steel sets (1/m), shotcrete on walls and face (each advance): 3 days (2 advances/day)

Rig Soilmec SM 605 DT (Figure 20) was used both for the core reinforcement and the jet grouting umbrella, thus minimizing the downtime for equipment mobilization. It took about 2 hours to mobilize and demobilize the rig. The single-fluid jet-grouting technique patented by Trevi prevents overpressure development and emptying of the columns, the two typical problems occurring in sub-horizontal jet-grouting in cohesive soils. As shown in Figure 19a, the drilling rod is equipped with a 1.5 mm nozzle located just behind the 200 mm drill bit and with two jet-grouting nozzles (monitor) located further behind. The drilling rod is first used to make a hole in the center of the column; low pressure is used, and therefore only the small nozzle works by removing the spoil. The drill rod is then withdrawn at 1 m\min while, at the same time, 40 MPa water pressure is injected through the monitor: this phase, called water pre-cutting, is fundamental in creating a 35–40 cm diameter cavity that then minimizes overpressures and in ensuring the proper column diameter. The final step consists of re-inserting the drilling rods with the reinforcement following at a constant minimum distance of 30–50 cm necessary to leave the monitor unobstructed. This time, the grouting mix is injected at 40 MPa and 205 l/min and exits mainly through the monitor while the system moves downhole at a speed of 1.5 min/m to realize the final size jet grouting. The drill rod the reinforcement pipe rotate in opposite directions and the jet-grouting spoil is simply evacuated through the annulus between the drilling rod and the reinforcement pipe, which, together with the water pre-cut, reduces the overpressures in the jet-grouting column. The reinforcement pipe avoids that the fresh jet-grouting mix flows out of the drillhole (column emptying). When the drill bit reaches the end of the borehole, the pipe is inserted into the ground face at the end of the borehole to prevent backflow and emptying of the column.

The system also includes a special monitor, called ETJ, equipped with 5 mm diameter tubes that deliver the grout directly to the monitor nozzle, thus reducing turbulence and the aperture angle of the jet-grouting. The reduced angle improves the cutting-ability of the system.

## MONITORING

The entire monitoring system is based on the installation of monitoring stations placed perpendicular to the tunnel axis, and equipped with instruments outside, on the surface, and inside the tunnel. It consists of:

- Systematic convergence stations for the entire length of the tunnel.
- Extrusion measurements, by means of extrusion-meter up to 40 metres long.

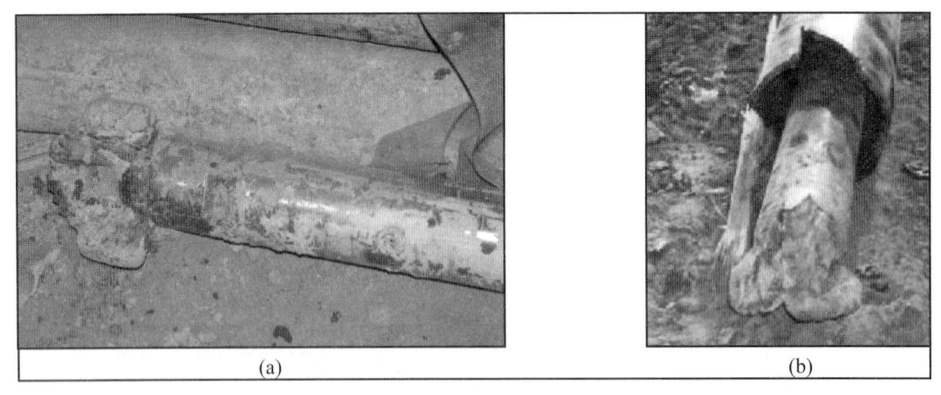

**Figure 21.** (a) Drill bit, water-pressure nozzle and monitor, (b) Drill bit with reinforcement pipe following at short distance

**Figure 22.** (a) Large equipment works concurrently at tunnel face, (b) Powerful equipment carries out several ground improvement operations at the same time

- Topographic measures composed of survey on monuments placed on the surface.

Moreover, for each tunnel there were special monitoring stations. Each station was generally equipped as follows:

- Measures of convergence, extrusion and topographic leveling.
- Extensimetric gauges placed on the steel ribs and on the final lining reinforcement (where prescribed), pressure cells (where prescribed) under the steel ribs footings, extenso-inclinometers, and piezometres.

Settlements were minimized to 1 cm, and no up heave was created by the jet grouting operations.

## CONCLUSIONS

The widest tunnel in the world in clay has been successfully completed under budget and ahead of schedule. With two tubes separated by a mere 4 m pillar, each tube has a width of over 22 m, a height of 14 m, and a cross-section of 260 m$^2$. The tubes under passed Cassia Road while it was open to traffic, a Roman villa under 5 m of cover, and two existing tunnels causing only 1 cm of settlement. This is no miracle, but the result of the ADECO in conjunction with state-of-the-art ground improvement for preconfining the ground in the core so that an arch effect could develop well ahead of the tunnel face, a concept not included in SEM/NATM. Complete industrialization was achieved in this 260 m$^2$ tunnel construction by adopting extremely powerful equipment, experienced construction personnel, and concurrent construction operations at the tunnel face (Figure 20). The achieved industrialization is confirmed by the advance rate for the Cassia tunnel that was constant and comprised between 0.75 m/day and 0.6 m/day for the entire tunnel while advancing full face and the final lining following suit at 50 m from the face. Construction proceeded without accident, which confirms the ability of the ADECO to minimize the risk.

As a comparison, the 457-m long Mount Baker Tunnel was built in 40 months at an advance rate of 0.37 m/day, it forced the Owner to purchase and to evacuate the entire land over the tunnel, and caused 5–20 cm of settlements.

**REFERENCES**

FHWA. 2009. FHWA Technical Manual for Design and Construction of Road Tunnels.

Holloway, L.J., and Kyerbol, G. 1988. Completion of the world's largest soft-ground tunnel bore. Transportation Research Record 1150: 1–10.

Lunardi, P. (2008). Design and Construction of Tunnels. Springer.

Lunardi, P., Cassani, G. and Gatti, M.C. (2008) Design aspects of the construction of the new Apennines crossing on the A1 Milan-Naples motorway: the base tunnel. Proc. AFTES International Congress "Le souterrain, espace d'avenir"; Monaco October 6–8, 2008.

Rabcewicz L. (1964). The New Austrian Tunnelling Method, Part one, Water Power, November 1964, 453–457, Part two, Water Power, December 1964, 511–515

Rabcewicz L. (1965). The New Austrian Tunnelling Method, Part Three, Water Power, January 1965, 19–24.

Tonon, F. 2009. ADECO as an alternative to NATM: how it works, why it works. In Proc. Rapid Excavation and Tunneling Conference (RETC), Las Vegas, NV, June 14–17, 2009.

# Cutter Instrumentation System for Tunnel Boring Machines

## Aaron Shanahan
The Robbins Company, Kent, Washington

**ABSTRACT:** Installing instruments on TBM cutters increases the efficiency of boring and improves cutter life. The benefits of an instrumentation system and findings from field tests will be presented. Instrumentation systems allow the operator to view in real-time how adjusting operating parameters dynamically impacts the cutting environment. By analyzing vibration data, cutter rpm, and cutter temperature, it is possible to infer the rock face condition and how it is affecting cutter operation. Knowing these data provides the operator an indication of cutter wear without entering the cutterhead to inspect the cutters and also alerts the operator to any abnormal cutter conditions.

## OVERVIEW

Since the advent of tunnel boring machines, operators and manufacturers of these machines have desired to know how the cutting device interacts with the material being bored. With this information, an operator can achieve maximum efficiency of the tunnel boring operation by varying the operating conditions of the machine. Costly delays due to cutter failures can be avoided by monitoring boring conditions in real-time.

Additionally, cutter manufacturers can adjust the design of components based on this data. Previous attempts at measuring cutter conditions have included theoretical mathematical models, simple force measurement devices, and inference of the interactions through ancillary evidence provided by operating conditions of the machine itself.

## HISTORY

Attempts have been made in the past to obtain some data about the loads and characteristics of a cutter operating on mechanical excavation machines. Strain gages have been placed inside the shafts of cutters, with the deflections measured and equated to a force acting on the cutter. Wireless transmitters have been installed on cutters, which would only be powered while the cutter is turning. With this system, when no signal is detected from the device, it can be inferred that the cutter is not rotating.

Most of the cutter information becomes available only after the machine has finished its boring stroke and the cutter inspection crew enters the cutterhead to check the cutters. Therefore, no immediate remediation of undesirable cutter conditions is possible while the machine is boring because there is no real-time information available. When damaged cutters are found, the inspection crew can only speculate on when the damage occurred—not only in reference to the time it occurred, but also at what point in the cutterhead rotation the damage happened.

## NEW SYSTEM

Previous efforts to instrument cutters have offered the operator limited data, but to date, no comprehensive approach to placing instruments on cutters has been attempted. It is not feasible to run electrical cables to each cutter to relay the signals, which has precluded the operation of electrical instruments on cutters.

The Robbins Company has developed a patent-pending wireless instrumentation system designed to provide the TBM operator with the most complete idea of cutter performance ever available. With the advances made in wireless technology within the last several years, it has become possible to create a network of wireless devices, with one installed on each cutter. Packages consisting of measurement devices, a power source, and a wireless transmitter housed in a protective case can be installed in each cutter housing to detect the specific operating characteristics of each cutter (see Figure 1).

### Cutter System Components

The instrumentation system consists of the components described below.

#### *Cutter Sensor*

Each cutter is fitted with an electronic sensor and a bracket meant to fix the sensor in close proximity to the cutter. The cutter sensor consists of a power

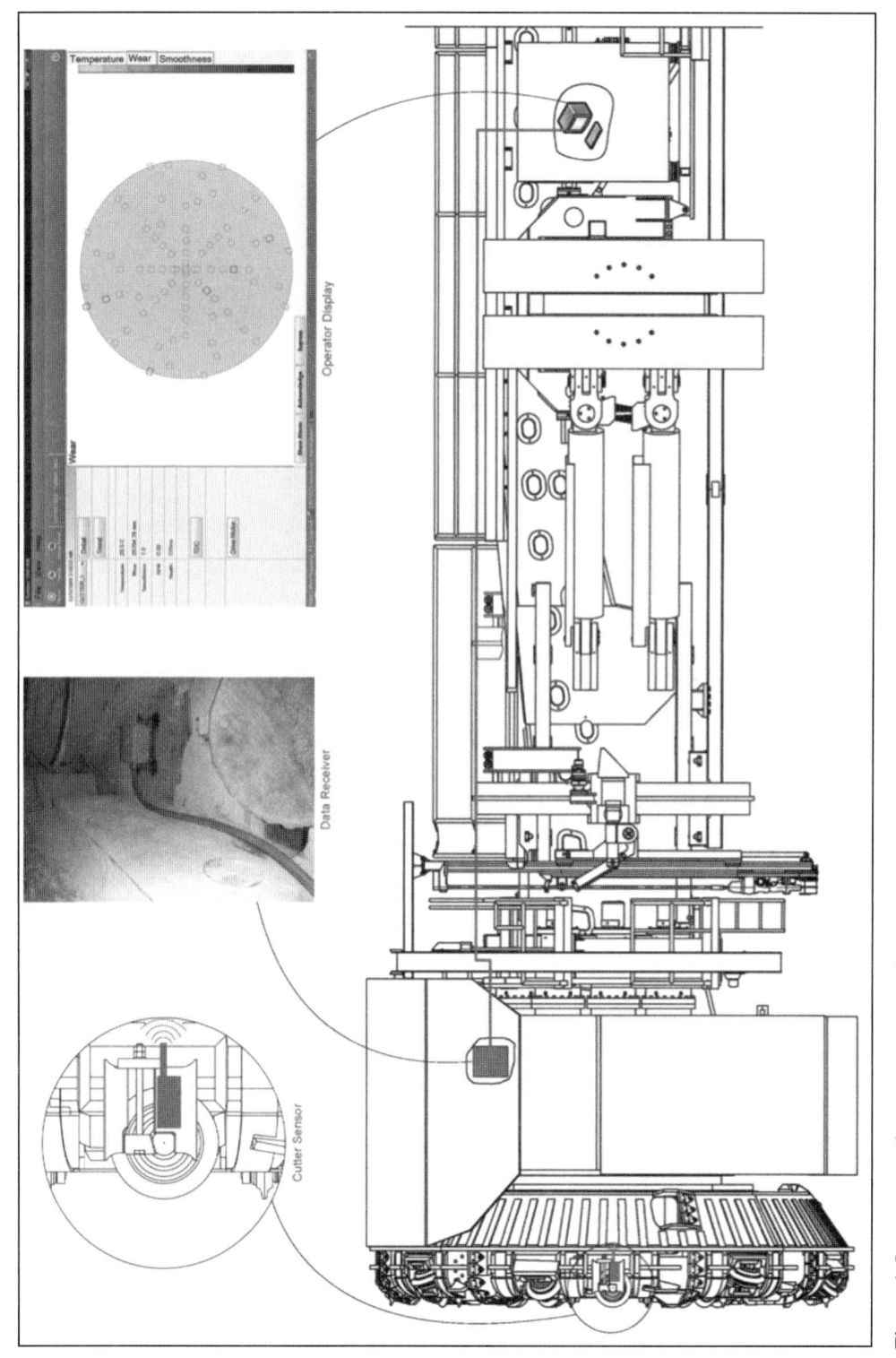

Figure 1. Instrumentation system schematic

source, measurement devices, a processor, and a wireless transmitter. The power source could be batteries or even a small motion-activated electrical generator. The measurement devices on the current generation of sensors include accelerometers, rotational sensors, and temperature sensors. Expanding the suite of sensors in the devices is an option, with inclusion of a camera and microphone as possible enhancements.

The processor is downloaded with operating parameters which can be adjusted for optimal data collection. The processor polls the measurement devices for data at a pre-defined sample rate and then prepares the data for transmission. The wireless transmitter takes the encoded data from the processor and emits an electro-magnetic wave containing the encoded data.

*Data Receiver*

The data receiver is mounted in a protective housing and installed behind the cutterhead. The receiver detects and processes the transmissions emitted from the cutter sensor. The data receiver contains a processor that is programmed with the same information as the cutter sensors and so is able to detect and process the signal from each cutter sensor. The receiver then takes the received data and transmits it to the operators display computer either over wire or wirelessly.

*Operator Display*

The operators display takes the data relayed from the data receiver and presents it in a manner which allows the operator to identify the cutter operational parameters currently of interest. The data is available in several formats. A chart with a column for each cutter can display the data of interest for all cutters at once. Additionally, a graphical image of the cutterhead, with cutter locations noted, can flag cutters which are operating outside a pre-established safety range. A touch screen is used for the display, allowing the operator to touch the cutter that is flagging for an anomalous condition and read the data values causing the alarms directly.

*Database*

The computer running the operator's display keeps an archive of all information received on a local database. The database is searchable for historical data, and data can be displayed in several formats, including graphically or in a table. The database can also be accessed remotely if the computer has an internet connection. This setup makes it possible to monitor the cutter performance without traveling to the jobsite.

**Machine System Components**

In addition to the information provided by the sensors installed on the cutters, some machine operating data needs to be collected to correlate the data from the cutter sensors. In particular, the head rpm and some type of reference point on the cutterhead must be known in order to identify where a particular cutter is located on the cutter face at any given time. For a machine with variable frequency drives (VFDs), the rpm of the cutterhead can be calculated by getting the motor speed from the VFD and the gear ratios of the motor to cutterhead interface. The reference point is still needed even if tying in to the VFDs, because the head rpm alone cannot identify how far along in one rotation of the cutterhead the cutter has traveled.

**DATA**

Various sensors can be included in a cutter sensor to detect physical characteristics helpful to cutter operations. Data applicable to cutters operating on a tunnel boring machine includes rotational data, vibration data, and temperature.

**Rotational Data**

Knowing the cutter rpm provides an indication of how smoothly the cutters are performing. Rotational data can be used to detect cutter conditions but can also be used to infer the conditions of the rock face. A constant cutter rpm indicates that the cutter bearings and seals are functioning properly, and also indicates that the rock being bored is competent and without major variations. A variable rpm provides a warning to the operator of many possible undesirable situations, including a locked-up cutter, improper machine thrust, or blocky ground.

By measuring the rotational speed of a cutter and knowing at what distance away from the axis of the cutterhead a cutter is located, along with knowing the cutterhead rpm, the diameter of the disc ring can be calculated:

$$D = \frac{2rd}{R}$$

where
$D$ = disc ring diameter
$r$ = cutterhead rpm
$d$ = distance to the cutter from the center of the cutterhead
$R$ = cutter rpm

**Vibration Data**

By measuring the amount of vibration a cutter is experiencing, some idea of the loading that a cutter is under can be determined. Changes in how much vibration the cutter is under could indicate a change in the geology, a blocky face, or a possible problem with the cutter, among many other things. With proper filtering, detailed analysis of cutter vibration can reveal such conditions as a bearing being overloaded, mounting system problems, or damaged disc rings.

**Temperature Data**

Measuring the temperature of a cutter can provide indications of anomalous operating conditions. If a cutter is blocked and is not rotating, it will show a rapid increase in temperature. The effectiveness of water spray or chemical spray systems intended to cool the cutters can be tested by measuring cutter temperature in real-time.

**BENEFITS**

There are many benefits to installing sensors on cutters. In general, with an instrumentation system, it is now possible for machine operators to obtain a true picture of how the cutters are operating in real-time and how changes in machine operating parameters affect the cutters.

**Identify Problems Before Failure**

If an anomalous condition is detected on a cutter, the operator will be notified of the condition immediately, and proper action can be taken before the anomaly can lead to a failure. Some of the anomalies which could arise and would be detected by the instrumentation system include a non-turning or intermittently turning cutter, high vibration shocks, and high temperatures.

A non-rolling cutter could indicate a problem with the cutter bearings, which would be a very serious problem needing to be addressed immediately. A cutter turning only intermittently could be due to a bearing problem as well, but it could also indicate that there are voids in the face of the rock where the cutter is not contacting any rock. When a void is present, there are fewer cutters to take the machine thrust load, which loads the bearings of the cutters in contact with the face greater than the nominal load. TBMs specify a maximum allowable thrust force, which is often calculated simply as the cutter maximum bearing load multiplied by the number of cutters. If the machine operator is operating at the maximum specified thrust rating and voids are present, the cutters in contact with the face are exceeding their design load limits (see Figure 2).

High vibration shocks could be indicative of a mixed face/blocky face condition and could also indicate that there is debris in the invert of the tunnel. A shock would occur every time the cutter came back into contact with the rock face after traveling

**Figure 2. Failed cutter bearing**

in a void. If a cutter is seeing a series of vibration shocks and the operator can identify where in the cutterhead rotation this is occurring, the options for what may be causing the shocks can be narrowed and fewer options for remediation would need to be considered.

High temperatures on one cutter would indicate a problem with that specific cutter and, if severe enough, the machine could be stopped and the cutter inspected before failure occurs. If all cutters show a temperature approximately equal and not elevated, then no problems with the cutters are indicated. Also, the effectiveness of any water or chemicals sprayed in front of the head, for instance to reduce wear or cool the cutters, while monitoring the cutters in real-time, can be evaluated as it is being sprayed on the face through real-time monitoring.

**Improved Cutter Life**

When a TBM is operating, it is often difficult to tell how efficiently the excavation is being performed. TBMs are very solidly supported in the tunnel while boring, so it can be difficult to tell from the machine reactions how well the rock fracturing process is being accomplished. Looking at the rock chips being formed can help indicate how well the cutters are performing, but it does not provide the complete a picture that is possible with an instrumentation system.

Tunnel boring machines usually have enough power installed that the cutterhead will turn no matter how efficiently the rock chipping is performed—non-rotating cutters will not cause enough drag force on the cutterhead rotation to stop the cutterhead. By receiving real-time information about the cutting environment, the machine operator can be alerted to any anomalous situations and also adjust the head rpm and machine thrust to achieve the ideal operational envelope.

**Fewer Cutter Inspections**

Disc ring wear is the most common reason for replacing cutters. Maintenance crews spend a portion of their shift in the cutterhead or in front of the machine measuring the wear on each cutter. Cutter inspections are usually a daily occurrence while boring, with each cutter's disc wear measured and recorded.

As described previously, when real-time cutter rotational speed data is available, the diameter of the cutter's disc ring can be calculated. Knowing the disc ring wear in real-time while boring allows the operator to eliminate some cutter inspections. Fewer cutter inspections means the machine has more available time to bore. Additionally, on machines employing cutters that are loaded on the head from the front of the machine, fewer cutter inspections means that the maintenance crews spend less time in front of the machine, where they are exposed to an unsupported section of tunnel. This reduces the risks to workers when the geology of a tunnel is unstable and rock cave-ins or collapses are possible.

**Stop Wipe Outs**

When one cutter gets blocked and stops rotating, it leads to a higher load on adjacent cutters, with a possibility of a cascading failure (wipe out) of all the cutters in the worst cases. Wipe outs are costly both because of the damaged cutter components and because of the time needed to replace all of the failed cutters. Severe wipe outs can also cause damage on the cutterhead, which could possibly require extensive head repairs. With the real-time rotational data, an operator can be notified immediately when a cutter stops rotating (see Figure 3).

## CHALLENGES

As with any new technology, challenges to fully realizing the potential of the invention will arise during the development and testing phases. Installing electronic instruments in a cutter housing, ensuring that the instruments can survive the rugged tunneling environment, and maintaining signal transmission and receipt are some of the specific challenges experienced when developing the cutter instrumentation system.

**Installation of Instruments**

There is a certain minimum size of protective enclosure which must be designed to house the electronics of the instruments. Finding a location on the cutter housing to install the sensor while keeping the device

**Figure 3. Non-rotating cutter damage**

protected and maintaining a clear wireless transmission path are all key factors which need to be adequately addressed to ensure a reliable system.

**Survivability**

The cutterhead of an operating tunnel boring machine is not a hospitable place for sensitive electronic devices. Dust, water, chemicals, and chipped rock fragments are all found on most machines. Electronic devices are regularly subjected to shock, wear, moisture, and heat, and therefore must be designed to withstand this extremely harsh environment. The material chosen for constructing the protective enclosure must be selected with all of these concerns in mind. Steel is inexpensive and provides protection against impact and abrasion but can interfere with wireless communications. Highly-engineered plastics with good wear resistance are available which do not interfere with wireless transmissions, but they cannot provide the same level of wear protection as steel and are often prohibitively expensive. Using a mix of steel and plastic components provides the best solution to addressing the short-comings of each kind of material.

**Signal Detection**

The cutterhead on a TBM contains a great deal of steel, which can interfere with detection of the wireless signals emitted from the cutter instruments. Additionally, wireless instruments can usually be expected to be reliable as long as there is a line-of-sight pathway between the transmitter and the receiver. This is not always possible for all cutter positions on a rotating cutterhead. On many machines, there is a hopper for channeling the muck from the buckets onto the conveyor, and this area

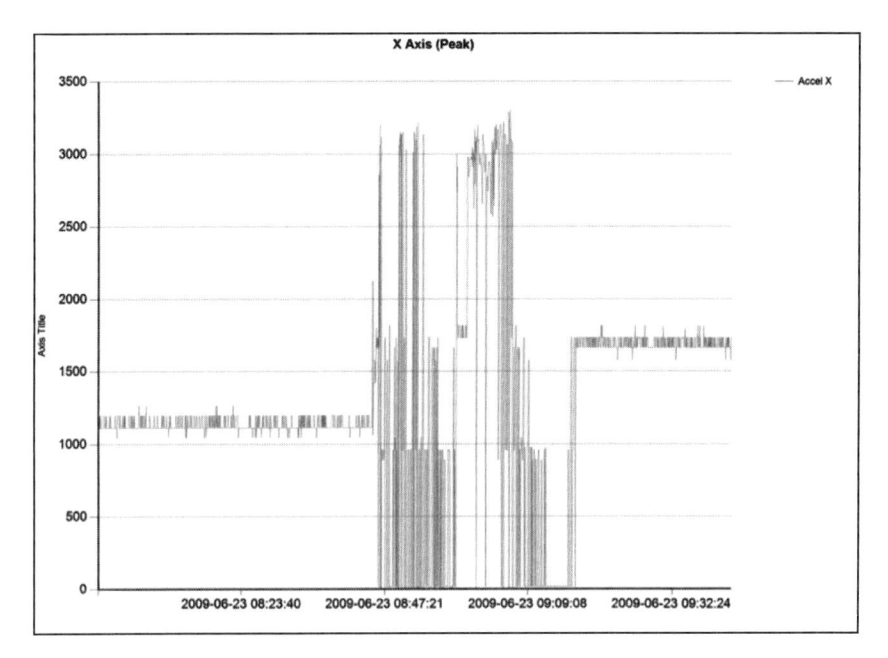

**Figure 4. Raw vibration data**

could possibly isolate instruments for a period of time when they are near the top of the cutterhead. Also, on very large machines, the outer cutter positions may be greater than 6 meters from the axis of the machine and the signal may have to travel between a series of steel plates to reach a receiver.

**Vibration Filtering**

Tunnel boring machines operate in a very dynamic environment which causes a background level of vibration which will always be present. Additionally, the centrifugal and tangential accelerations associated with the cutters rotating around on the cutterhead are detected by the vibration sensors.

Some of the issues associated with these extra forces acting on a cutter can be mitigated in part by orienting the sensor so that only the cutter forces attributed to actual rock fracturing cause data to register on the sensor. Sensor orientation alone, though, won't eliminate enough of the environmental interference to allow direct analysis of the vibration data. It is also necessary to filter out some of the noise by computational algorithms designed to identify and remove repetitive noise. Noise filtering is a well-developed field with many models available, but each application of noise filtering methods is unique. Placing vibration sensors on cutters is a new concept and, therefore, a large amount of data must be collected and analyzed to determine the best way to filter out noise and get the most useful data for analyzing cutter performance (see Figure 4).

**LOOKING FORWARD**

Much of the data obtained from the cutter instrumentation system can help TBM operators right now, from calculating the amount of cutter wear to indicating difficult geological conditions. Rotational data indicates how well the cutter is turning, which in turn clues the operator into the amount of wear on a cutter. Temperature data can indicate when a potentially harmful operating condition has arisen. Vibration data provides an indication of how well the cutters are fracturing the rock and the condition of the rock face.

Ultimately, these data will make it possible to generate a real-time map of the geological profile of the tunnel. Operators will be able to combine all of the possible data and note how certain phenomenon relate to the machine operational parameters as well as the condition of the material being excavated. The instrumentation data can also be tied into the TBM control system, which allows for conditions to be set in the machine's operation. The conditions will modify operational parameters if certain data are received. For example, thresholds might be set that would automatically change the thrust force or cutterhead rpm if a high vibration condition is detected.

What remains for the instrumentation system is to develop corollaries between filtered data and what physical phenomena this data represents and then to fully integrate the data with the operation of the machines, putting tunnel boring machines one step closer to an automated method of excavation.

# Towards Precise Under Ground Mapping System in Canada

Mike Ghassemi, D. Zoldy, H. Javady
AECOM, Toronto, Ontario

## INTRODUCTION

It is clear that precise over ground and underground mapping is an important factor for all designers. Due to rapid changes in underground construction, tunnel designers are increasingly in need of updating project information. This paper explains the latest techniques for detecting underground utilities as well as underground mapping.

GPS (Global Positioning System) is a very accurate and popular navigation and mapping system that has been used extensively aboveground, but is not used for underground mapping as signals are not strong enough to pass through ground or water.

This paper explains the advantages and limitation of each technique and finally will focus on the latest underground mapping techniques in Canada and North America.

Most underground techniques were developed for military applications during the First and Second World Wars, later the techniques became public. For example, Ground Penetrating Radar was developed for the military in 1970 to be used for locating tunnels under the demilitarized zone between North and South Korea. Later the same technique was used for detecting unexploded ordinance such as plastic land mines.

## POSITIONING SYSTEM

Generally, positioning systems are categorized into three main areas:

- Above ground positioning systems
- Underwater positioning systems
- Underground positioning systems

### Above Ground Positioning System

*Old Technology*

- Stars and astronomical tables
- Surveying equipment such as measuring tapes, theodolites, distomats, levels, compasses, etc.
- Sextants

*New Technology*

- Laser total station
- Aerial photography by unidentified airplanes
- LIDAR surveying
- Satellite positioning and mapping (GPS, GLONASS, GALILEO)

### GPS

Global Positioning System (GPS) is a U.S. space-based global navigation satellite system. It provides reliable positioning, navigation, and timing services to worldwide users on a continuous basis in all weather, day and night, anywhere on or near the earth.

Developed by the Department of Defense in 1973, GPS was originally designed to assist soldiers and military vehicles, planes, and ships in accurately determining their locations world-wide.

GPS is made up of three parts: between 24 and 32 satellites orbiting the Earth, four control and monitoring stations on Earth, and the GPS receivers owned by users. GPS satellites broadcast signals from space that are used by GPS receivers to provide three-dimensional location (latitude, longitude, and altitude) as well as the time.

### GLONASS

Global Navigation Satellite System: A system of satellites operated by the Russian government, enabling someone with an appropriate receiver to determine their position—some of the time. Development on the GLONASS began in 1976, with a goal of global coverage by 1991. The complete nominal constellation consists of 24 satellites, 21 operating and three on-orbit "spares," in three orbital planes, at a mean orbital height of 19100 km.

### GALILEO

Galileo is the informal name for the European Global Navigation Satellite System (GNSS), a system that will offer users anywhere in the world "near pin-point" geographic positioning when it becomes fully operational by 2009. Designed to be interoperable with the other two such systems, the United States' Global Positioning System (GPS) and Russia's Global Orbiting Navigation Satellite System (GLONASS), Galileo will enable a user to take a position from any combination of satellites with a single receiver. Both GLONASS and GPS are run by the defense departments of their respective countries. Galileo will be civilian-operated.

The Galileo system, which consists of 30 satellites orbiting the earth at a height of 15,000 miles, is expected to pinpoint a geographical position to within a single meter.

High accuracy (as low as ±1cm), positioning anywhere on the earth and working in any weather conditions are the advantages of satellite positioning.

## Limitations

- Minimum 5 satellite are required for precise positioning
- Accuracy can be lowered to 1 to 10 meters near tall buildings
- Low accuracy in forests and woods
- Signal can be jammed by other interfering signal users
- Does not work under water
- Does not work underground or in tunnels
- Satellite signals can be switched off in specific areas or accuracy can be changed by satellite owners
- Operated by military forces

## Underwater Positioning System

### Old Technology

- Diver searching
- Compasses
- Gyro stations combined with mechanical inertial systems

### New Technology

- Remotely Operated Underwater Vehicles (ROV)
- New Gyro stations combined with digital inertial systems
- Balloon GPS
- Side Scan sonar
- Magneto metering
- Sub bottom profiler
- Sonar positioning combined with GPS

Sound signal (Sonar) is the best way for positioning the underwater objects. If the utility pipelines are located above the sea floor then ROV and Side Scan sonar is recommended. If the utility pipelines

**Figure 1. Underground utilities**

are located under the sea floor then using magnetometer and sub bottom profiler is recommended.

## Underground Positioning System

Currently, mapping the utility lines including water, sewer, telecom, fiber optics, electrical, oil and gas is a challenge for surveyors and engineers. See Figure 1.

The oldest way for detecting the objects and mapping them is test pit but most of the times are not possible to dig the area due to the environmental and political issue.

Since GPS signals cannot travel through the ground it is necessary to combine other techniques to detect and map the underground objects. Underground objects can be categorized as:

- Ferrous objects
- Non-ferrous objects

### Ferrous Objects

Ferrous objects can be detected by metal detectors (Magnetometer) if they are located near the surface Current metal locator can detect the objects up to maximum depth of 3 meters.

Most of the metal locators are not able to determine the depth of objects but the depth of ferrous objects can be determined by using the Vector Magnetometer technique which is very expensive.

Another technique for detecting underground ferrous objects is using a pulse transmitter and receiver. In this technique a generator emits a pulse which can be detected by a receiver on the ground surface. This technique is used for horizontal directional drilling (HDD) and detecting the metal utilities. Detecting and mapping metal objects is much easier than non-metal objects.

### Non-Ferrous Objects

Non-ferrous objects, such as concrete pipes, brick pipes, PVC and HDPE pipes cannot be detected by

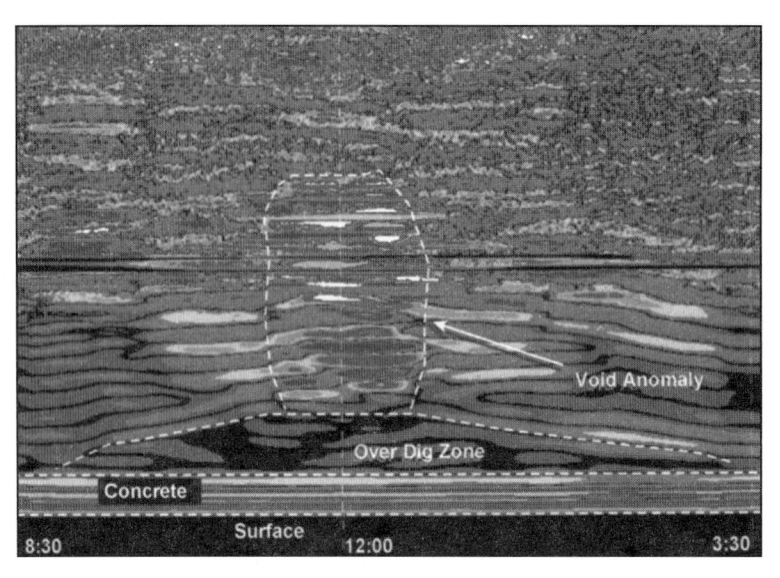

**Figure 2. Void detected by GPR in Hamilton Tunnel**

magnetic detector techniques. Depending on the size and depth of the non-ferrous pipe there are several techniques for locating the non ferrous pipes as follows:

### GPR (Ground Penetrating Radar)

In the early 1970s several different teams of scientists began to develop radars for viewing into the earth. GPR use in locating and mapping utility lines has been the subject of much on-going research conducted by both military and commercial organizations. GPR's are also known as "impulse radars" because the transmitted pulse is very short and is ordinarily generated by the transient voltage pulse generated from an overloaded avalanche transistor.

Ground-penetrating radars in principal are capable of locating plastic pipes as easily as metallic pipes since the radar signal reflection from the pipe depends on contrasting dielectric properties of the soil and pipe, not just a high electrical conductivity for the pipe. GPR normally has accuracy of several feet or less when measuring the depth of a buried object.

The performance capability of this type of radar is strongly dependent on the soil electrical conductivity at the site. If the soil conductivity is high, attenuation of the radar signal in the soil can severely restrict the maximum penetration depth of the radar signal. In California, where soils in many areas have a high clay content the soil absorptive losses can be quite high. Soil moisture, especially in soils with high clay content, only increases the radar attenuation rates, further limiting the radar performance.

To detect the sinkhole or void above the tunnel and also the voids beyond the segments, GPR has been used in a tunneling project located in Hamilton, Ontario, Canada a few years ago and the results were acceptable.

In this method, GPR was used inside the tunnel as well as outside of tunnel. See Figure 2.

As mentioned the GPR signals are not able to pass the metal objects, therefore if the tunnel is covered by liner plates then it is not recommended to use the GPR inside the tunnel. It is not recommend to use GPR from the ground surface for tunnels deeper than 10 meter.

Briefly, before using the GPR technology for tunneling projects the designer must consider many factors otherwise GPR cannot give the desired results.

GPR has been used by some mining companies like Kawasaki heavy industries for detecting objects in front of tunnel boring machines (TBM) so that ground conditions can be anticipated.

Kawasaki has performed some experimental work on TBM cutter head mounted Radar and the effective forward looking distance is only a couple of meters which doesn't provide an adequate warning period.

*Advantages*

- Locating buried objects
- Recommended for structural investigation
- Good results for shallow investigation

*Limitations*

- Material of objects not determinable
- Limitation in penetration (shallow depth of observation)
- Interpretation is not simple
- Depth of the object is not very accurate
- Useless in the tunnels with metal liner plate or metal casing

**Gyroscope Probe**

Gyroscope probes were using in the oil companies for the last two decades and recently some of the horizontal drilling companies are using this techniques for their drilling machines. Gyroscope probes are used for directional drilling rigs in inshore and offshore as well as resurveying the oil wells. See Figure 3.

Recently a few companies are used this technique for mapping the underground pipe lines and it can be used for pipes with diameters of 2 inches or more. This technique is being used more frequently in North America in recent years.

In this technique, the probe is sent from one end of the pipe and is received at the opposite end of the pipe. The gyroscope probe technique can be used for pipes up to 2 kilometers in length and acquires around 500 data points per second. Once the probe reaches to the end of the pipe the collected data can be downloaded to a computer. By applying the launching position and retrieval position to the probe to the software, the exact pipe alignment and elevations will be shown on the drawings.

New gyro probes do not need the launching and retrieval position and can calculate the position of the drilling head via gyro data and progress distance. In subsea operations, ROV can be used to place the survey probe in conductors.

Recently gyroscope probes have been combined with closed circuit televisions cameras (CCTV) to provide a visual inspection of the pipe that matches the tunnel alignment. If damage to the existing pipe is seen on the CCTV, then the exact position of the damage is known.

*Advantages*

- Positioning of pipe with any material is possible (PVC, iron, concrete, brick, steel, HDPE, etc)
- Magnetic field can not affect the Gyro data
- X,Y, Z can be extracted along the pipe alignment in centimeter intervals
- Applicable in small size pipe (2 inches and above)
- Can be used in live pipes as well as abandoned pipes

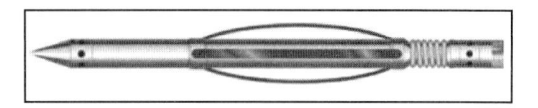

**Figure 3. Gyro probe**

- Applicable for micro tunnels and horizontal directional drilling

*Limitations*

- Exclusive market (Gyro probes are not for sale to the public, only leasing is possible)
- Calibration is very costly
- Leasing is very costly
- The precision of the data will be decreased with increasing the distance from the source

**3D Laser Scanner-LIDAR**

This is a new technique for mapping the new and existing tunnels. In this technique the size of the tunnel diameter must be more than five feet. This technique is usually used in new tunnels. High precision mapping (±5mm) and full coverage of the tunnel are the main advantages of this technique over traditional surveying techniques.

This technique was used in the 19th Avenue sewer tunnel in Toronto, Ontario. See Figure 4.

*Advantages*

- Very fast
- 3D modeling of as build
- Accurate volume calculations, especially for rock tunnels
- High accuracy
- Fly through is possible
- Excellent for presentation to the client and public
- User friendly with GIS software

*Limitations*

- Costly
- Not recommended for tunnel with very smooth surface
- Limitation in distance
- Cannot be used in the live tunnel and live pipes
- Activities in the tunnel must be shut down during the data acquisition

**Gyroscope Stations**

Gyroscope stations (North finder) were used widely in submarines, ships, radar sites and came to tunneling projects in 1940.

Figure 4. 3D laser scanner image from 19th Avenue tunnel project in Toronto

At that time the weight of a gyroscope station was around 150 kg and warming up time was around 4 hours. Modern gyroscopes stations are very light (around 15kg) and need only 15 minutes to warm up.

This technique was used in many tunnels around the world. Euro Tunnel is one of the examples.

It was used in the 19th Avenue tunnel project in Toronto by McNally Construction. The break through accuracy was less than one inch for 1.5 kilometer mining by a 3.20 meter EPB machine.

*Advantages*

- Precise north finder
- Works under any conditions
- Magnetic field doesn't affect the data
- Works in any kind of tunnel and pipe (concrete lining, steel lining, rock tunnel, etc)
- Saves the cost and time of the underground project
- The alignment holes are not required
- Recommended for long tunnels (more than 2 kilometers)
- Recommended for underwater tunnels

*Limitations*

- Expensive
- Difficult for transport
- Very sensitive to temperature, wind, and motion
- Transportation needs lots of care
- Calibration is costly
- Must be operated by experts
- Needs to enter the latitude of the area
- Must be setup on the stable station and a little bit of vibration can gives wrong results

## CONCLUSION

As populations increase, the quantity and complexity of the underground utilities rises to meet the demand. Various non destructive detection techniques including the magnetometers, GPR, Electromagnetic line locator, acoustic system etc are available to address these issues.

Briefly before choosing the detecting tools and using mapping software, the following items must be reviewed carefully:

- Geology of the area
- Depth of the object
- Object material
- Access to the site
- Political issues
- Traffic issues
- Environmental issues

And totally time saving, cost saving and achieving the best results needs experienced designers and contractors.

**REFERENCES**

Handbook of Offshore Surveying: Vol 1 (Preparation and Positioning)
*Ground Penetrating Radar* by David Daniels
Guide to GPS Positioning (Spiral-bound) by David Wells
GMT Gmbh, Gyromat DMT3000 manual
*The least squares adjustment of gyro-theodolite observations*/G.G. Bennett
New methods of observation with the Wild GAK 1 gyro-theodolite
Geophysic GPR international Inc.
Sensor and Software Inc.
Gyro data Ltd
Geospatial Ltd
FARO
GPR and unexploded ordinance
Fugro Aperio
Geomodel Surveys

# TRACK 1: TECHNOLOGY

Mike Smithson, Chair

*Session 3: Pressurized Face Tunneling*

# Lake Mead Intake No 3 Tunnel: Geotechnical Aspects of TBM Operation

**Georg Anagnostou, Linard Cantieni, Marco Ramoni**
ETH Zurich, Switzerland

**Antonio Nicola**
Impregilo SpA, Italy

ABSTRACT: The new Lake Mead No 3 Intake Tunnel will be constructed using a hybrid TBM (both slurry shield and open mode operation are possible) mostly through tertiary sedimentary rocks. Due to the very poor quality of the ground and the high pore pressures prevailing in the 4 km long subaqueous section of the tunnel (up to 14 bar, the highest pressures seen to date in closed shield tunneling worldwide), particular attention must be given to the risk of shield jamming or face collapse during boring or during the performance of maintenance activities in the working chamber. The paper outlines the expected geological-geotechnical conditions and discusses their potential impact on the operation of the hybrid TBM (e.g., mode of operation, face support pressures), as well as proposed auxiliary measures (e.g., advance drainage, grouting) and decision-making during construction.

## PROJECT OVERVIEW

Lake Mead is located approximately 30 km east of Las Vegas behind the Hoover Dam (Figure 1). It supplies about 90% of Las Vegas valley's water. Over the last nine years, drought has caused the lake level to decline by more than 30 m. A further drop of the lake level may render the existing intakes unusable. In order to maintain the water supply, a third intake will be constructed about 40–60 m deeper than the existing two intakes, i.e., deep enough to function at the lowest lake levels (Feroz et al. 2007). The main structures of the new intake are a 170 m deep access shaft, an approximately 4,700 m long intake tunnel with an internal diameter of 6.10 m and an intake structure in the middle of the lake (Figure 1).

## GEOLOGICAL CONDITIONS

The geology along the tunnel alignment has been explored by drilling 55 borings, 38 of them offshore. As shown in Figure 2a, the major part of the tunnel (including the subaqueous section) is located in tertiary sedimentary rocks of the so-called "Muddy Creek Formation" (conglomerates, breccias, sandstones, siltstones and gypsiferous mudstones of very variable quality). The tunnel alignment also crosses an older tertiary conglomerate of the Red Sandstone Unit, metamorphic rocks (amphibolites, schist and gneiss) and, close to the intake structure, basalts of the Callville Mesa Unit.

Furthermore, there are several faults in the project area. These are particularly critical, as they create the potential for water recharge directly from Lake Mead. Considerable water ingress must therefore be expected during construction. One well-known fault in the project area is the Detachment Fault, which has already been encountered in the access shaft (Hurt et al. 2009). This fault is located at the beginning of the tunnel alignment and consists of strongly foliated Phyllonite with zones of crushed and brecciated rock. The tunnel will cross this fault over a length of about 50 m. The exploratory boreholes showed that the centre of the fault consists of a gravel-like cohesionless material (for about 10 m). Special attention has also to be given to the submerged continuation of the Las Vegas Wash, which will be crossed by the tunnel drive at a small depth beneath the lake bed.

The maximum depth beneath the current lake level is around 140 m. The rock cover decreases from its maximum of 170 m at the beginning to just 20–30 m in the last portion of the alignment (in the Las Vegas wash as well as close to the intake structure, see Figure 2a). Table 1 summarizes the most important parameters of the prevailing geological units.

## CONSTRUCTION METHOD

Due to the high hydrostatic pressures and the very variable quality of the sedimentary rocks prevailing

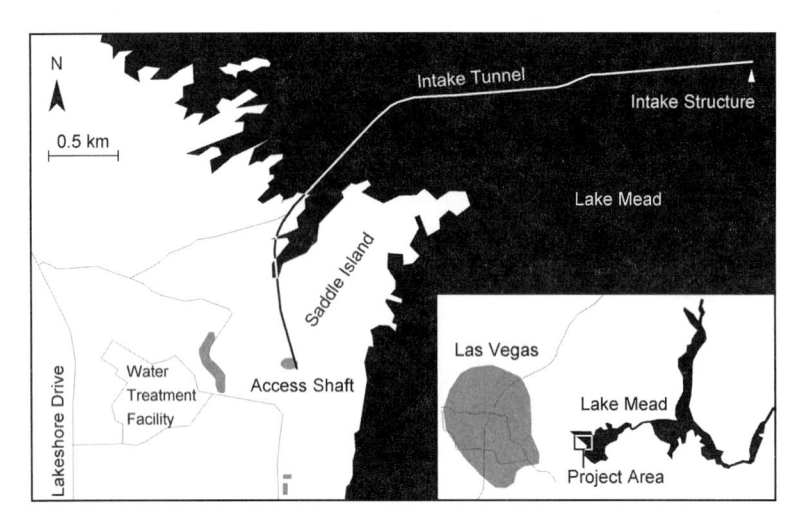

Figure 1. Project situation after Hurt et al. (2009)

Table 1. Ground parameters (from Vegas Tunnel Constructors, 2009)

| Geological unit | Young's modulus E [GPa] | Cohesion c [kPa] | Friction angle φ [°] | Permeability k [m/s] |
|---|---|---|---|---|
| Saddle island lower plate (Pcl) | 34–68 | 500–1500 | 35–40 | $2 \times 10^{-9}$–$10^{-6}$ |
| Saddle island detachment fault | 7–14 | 0–40 | 25–30 | $2 \times 10^{-6}$–$10^{-5}$ |
| Saddle island upper plate (Pcu) | 14–48 | 300–1000 | 35–40 | $2 \times 10^{-8}$–$10^{-6}$ |
| Muddy creek formation (Tmc 3) | 1.4–2.8 | 50–300 | 26–35 | $10^{-10}$–$3 \times 10^{-7}$ |
| Muddy creek formation (Tmc 2) | 0.3–1.4 | 50–300 | 26–35 | $2 \times 10^{-9}$–$10^{-7}$ |
| Muddy creek formation (Tmc 1/Tmc 2) | 0.7–2.8 | 50–300 | 26–35 | $10^{-9}$–$10^{-7}$ |
| Tmc 1 to Tmc 2, fault zones | 1.4–4.1 | 30–200 | 25–30 | $10^{-5}$ |
| Muddy creek formation (Tmc 4) | 1.4–4.1 | 100–500 | 28–35 | $10^{-11}$–$10^{-6}$ |
| Tmc 4 to Trs, fault zones | 0.7–1.4 | 30–200 | 25–30 | $10^{-5}$ |
| Red sandstone unit (Trs) | 1.4–3.4 | 30–150 | 25–28 | $3 \times 10^{-9}$–$2 \times 10^{-5}$ |
| Pcu to Tmc 4, fault zone | 1.4–4.1 | 30–200 | 25–30 | $10^{-8}$–$10^{-6}$ |
| Tmc 4 beneath the Las Vegas Wash | 1.4–4.1 | 30–200 | 25–30 | $10^{-8}$–$10^{-6}$ |
| Calville mesa formation (basalts) | 12–43 | 50–200 | 28–35 | $10^{-5}$–$10^{-4}$ |

over long portions of the alignment, attention was paid right from the start to the potential hazards of a cave-in at the working face or a flooding of the tunnel. The decision was therefore taken to construct the intake tunnel using a closed shield (Feroz et al. 2007).

The tunnel will be constructed using a convertible hybrid single shield TBM manufactured by Herrenknecht with a maximum installed thrust force of 100 MN (McDonald and Burger 2009). The TBM has a boring diameter of 7.22 m and can be operated either in open or in closed mode. In open mode, the face is not supported and a screw conveyor extracts the excavated rock from the working chamber (Figure 3a). In closed mode, the screw conveyor is retracted from the cutter head, mucking-out is done via the hydraulic circuit and the TBM supports the face with a pressurized bentonite slurry (Figure 3b). The TBM can be operated with partial, full or overcompensation of the water pressure and is designed to cope with hydrostatic pressures up to 17 bar—the highest ever pressures to date in closed shield tunneling worldwide. Due to the importance of advance probing and the possible need for pre-excavation

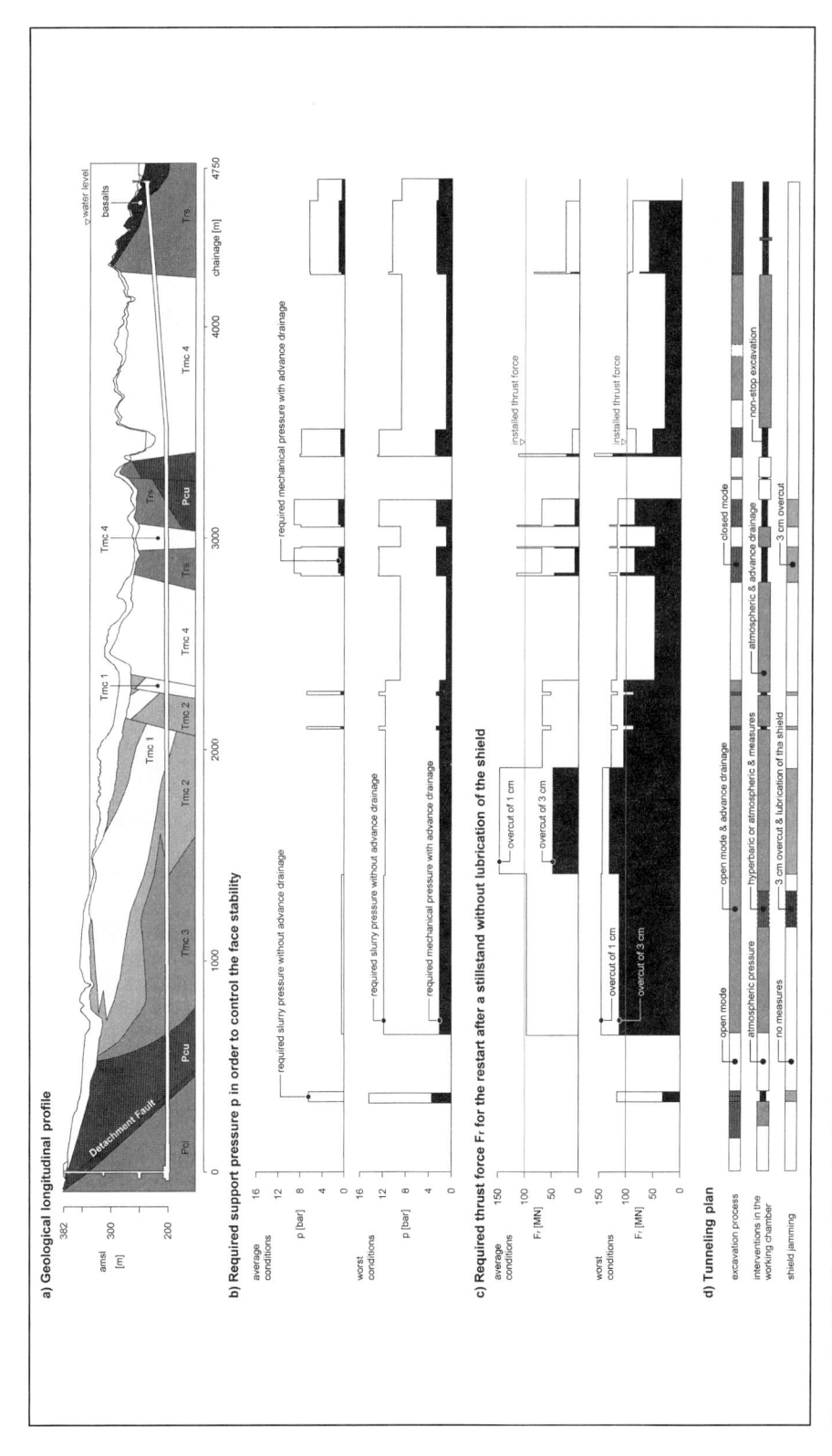

Figure 2. a) Geological longitudinal profile after Vegas Tunnel Constructors (2009); b) Required support pressure without (white columns) and with (black columns) drainage ahead of the face; c) Required thrust force in order to avoid shield jamming for overcuts of 1 cm (white columns) and 3 cm (black columns); d) Tunneling plan of the TBM including measures during the excavation process (top row), measures during interventions in the working chamber (middle row), and measures against shield jamming (bottom row). The dashed bars denote portions of a section where worse or better conditions may prevail than assumed for the rest of the section.

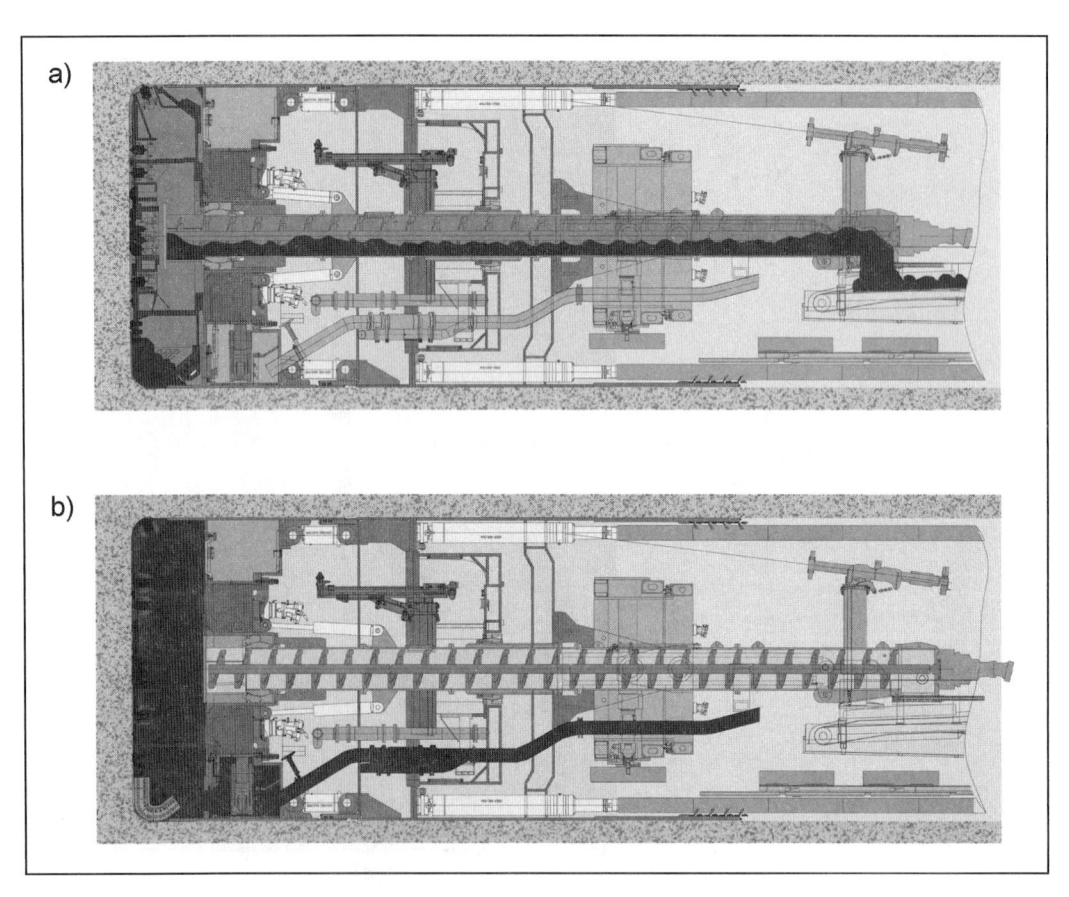

**Figure 3. TBM configuration for open (a) and closed (b) mode operation after McDonald and Burger (2009)**

ground improvement at least locally, the TBM is equipped with three permanent drill rigs and one mountable drill rig. Probing and drilling can also be carried out in closed mode using a blow-out preventer unit (McDonald and Burger 2009). Nevertheless, even if the TBM allows for boring in closed mode, the high hydrostatic pressures will make it extremely difficult to perform inspections and maintenance activities in the working chamber. In order to ensure stability during interventions, the face will have to be supported by applying compressed air. At the high pressures that are expected, however, professional divers will be required to perform the hyperbaric interventions and this will be very time-consuming. In addition, the stretches with closed-mode operation must be kept short because closed-mode operation generally results in lower TBM performances.

These considerations, in combination with the lack of experience with closed-mode TBM operation at such high hydrostatic pressures, made it necessary to conduct an investigation into the limits of open mode operation, i.e., working under atmospheric pressure in the chamber, possibly in combination with auxiliary measures such as grouting or drainage.

## POTENTIAL HAZARDS

A high damage potential, relatively high pore pressures and limited accessibility in the pre-construction phase are main features of subaqueous tunnels (Anagnostou 2009).

The high damage potential results from the possibility of a complete flooding of the tunnel in the case of a hydraulic connection to the lake. The risks associated with large water inflows can be mitigated to a large degree by installing extensive pumping capabilities and through a TBM design that will allow rapid conversion from open mode to closed mode by installing a screw conveyor for the removal of the excavated ground in open mode.

The high hydrostatic head leads, in combination with the small depth of cover in places, to the development of high seepage forces that increase the risk of face instability in a low strength ground. A collapse of the working face represents the most serious hazard scenario in the present case. Furthermore, given the sedimentary character of the prevailing rocks, jamming of the shield due to squeezing (Ramoni and Anagnostou 2009a) represents an additional hazard scenario.

The next two sections of this paper concern the geomechanical calculations and the assessment of these two potential hazards. For the purpose of assessing tunneling conditions, the tunnel has been subdivided in sections with practically uniform conditions. In order to check the sensitivity of the results, all of the calculations were performed for three sets of parameters, representing the so called "best," "average" and "worst" conditions for each section.

## COLLAPSE OF THE WORKING FACE

Details of the face stability assessment for the Lake Mead Intake No 3 Tunnel can be found in Anagnostou et al. (2010). Here only the most important assumptions, geotechnical considerations and investigation results will be presented.

With the exception of tunneling through cohesionless, granular soil, the stability of the tunnel face is in general time-dependent, i.e., a face that is stable in the short-term may collapse in the long-term. The time-dependency can be traced back to the rheological behavior of the ground (tertiary creep) or to the generation and subsequent dissipation of excess pore pressures (consolidation, cf., e.g., Anagnostou 2007b). The latter is particularly relevant in the case of low-permeability sedimentary rocks. The short-term (so-called "undrained") conditions are more favorable than the so-called "drained" conditions which affect the long-term behavior of the ground and are characterized by the development of destabilizing seepage forces.

In general, the less permeable the ground, the more rapid the excavation and the shorter the standstills, the more reasonable it is to assume favorable short-term conditions. In the case of high ground permeability, no favorable short-term behavior can be observed and unfavorable drained conditions will prevail in the face area already during excavation (Ramoni and Anagnostou 2007). The influence of ground permeability $k$ on the distinction between "undrained" and "drained" conditions during TBM excavation has been studied with numerical calculations that simultaneously take in account both the stress re-distribution and the consolidation process around the advancing tunnel heading (Anagnostou 2007a). Assuming an average TBM advance rate of 10 m/day, the computational results indicate that favorable undrained conditions apply only where there is low ground permeability ($k \leq 10^{-8}$ m/s) and only during the excavation process, including short standstills of up to 0.5–1 day (Anagnostou et al. 2009). For higher permeabilities or for longer standstills, unfavorable drained conditions must be expected. Over long portions of the alignment, the expected range of ground permeabilities (Table 1) is in the geotechnically demanding transition zone between drained and the undrained conditions. Face stability analyses have been carried out for both conditions, and the prediction uncertainties that exist with regard to the time-dependency of the ground behavior were taken into account in the tunneling plan.

Short-term face stability was investigated with the computational model of Anagnostou and Kovári (1994), while the calculations concerning long-term face stability were made by applying the nomograms of Anagnostou and Kovári (1996). In both cases, the assumed three-dimensional collapse mechanism consists of a wedge ahead of the tunnel face and an overlying prism (both in a state of limit equilibrium). In short-term the stability of the face is governed by the undrained shear strength of the ground, in long-term by the effective strength parameters ($c'$ and $\varphi'$). The calculations showed that for the given range of ground parameters the face would be stable in the short-term over the entire tunnel alignment. The white columns in Figure 2b apply to long-term stability conditions and show the minimum slurry pressure (or compressed air pressure) required in the working chamber in order to avoid face instability in the absence of a mechanical support. In long-term (which, as mentioned above, concern, e.g., a standstill longer than 1 day or a ground permeability higher than about $10^{-8}$ m/s), for both the "average" and "worst" ground strength parameters, closed mode operation with a stabilizing slurry pressure would be necessary for an extended portion of the alignment.

Operation and maintenance in closed mode at high pressures are very demanding and result in low advance rates. In order to operate the TBM in open mode over long portions of the tunnel, additional measures are necessary. With this in mind, we investigated whether advance drainage of the ground ahead of the tunnel face would result in significantly greater stability of an unsupported face.

Advance drainage—which, in the present case, can be carried out by means of boreholes drilled in the tunnel face through the cutter head—reduces pore pressures and their gradients in the core ahead of the face and thus also reduces the destabilizing seepage forces acting within the ground towards the opening. Once again, the computations were based upon the limit equilibrium mechanism proposed in Anagnostou and Kovári (1994). Seepage flow was taken into account by introducing the seepage forces into the equilibrium equations. In order to estimate

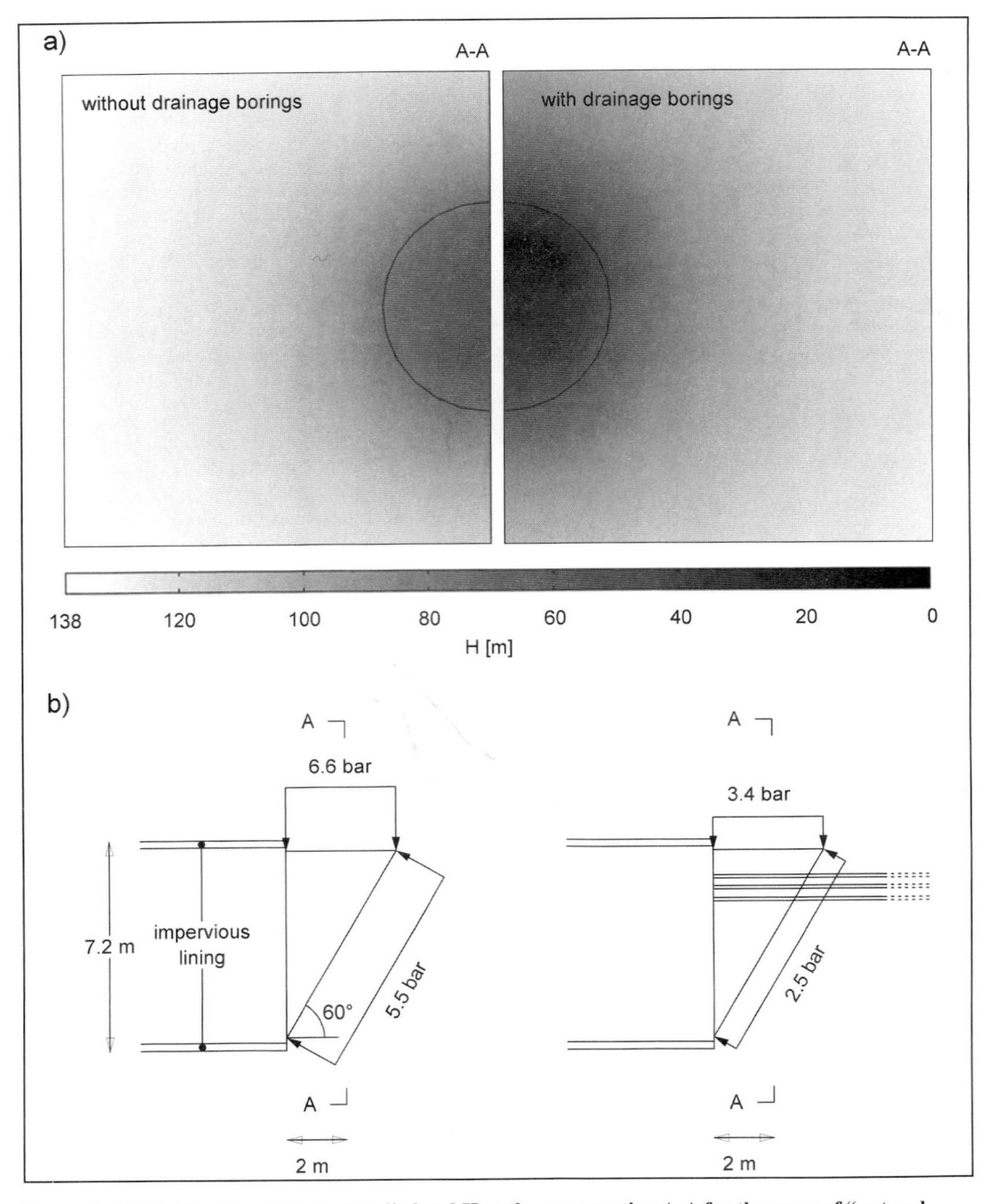

**Figure 4.** (a) Distribution of the hydraulic head H at the cross section A-A for the cases of "natural drainage" (left) and drainage using six boreholes ahead of the face (right). Both cases apply for a drainage time of t = 4 h assuming a permeability of the ground of k = $10^{-7}$ m/s and a storage coefficient of s = $1.3*10^{-5}$ m$^{-1}$; (b) Average pore pressure acting on a potentially unstable wedge (sliding plane inclined by 60°) for the two cases.

the seepage forces, three-dimensional, transient seepage flow calculations were carried out with the Finite Element Code COMSOL Multiphysics (formerly FEMLAB; COMSOL 2009), taking into account the incomplete drainage of the ground due to time-effects (drainage takes more or less time depending on the permeability of the ground) and due to the spacing of the drainage boreholes. Figure 4a illustrates the effect of drainage of the core on the hydraulic head field in a cross-section 2 m ahead of the tunnel face (Section A-A in Figure 4b). The figures on the left apply to the case of "natural" drainage through the open face, while the figures on the right apply to the case of drainage via six horizontal boreholes drilled in the upper part of the tunnel face. The drill pattern was selected according to McDonald and Burger (2009). As shown in Figure 4a (where darker tones apply for a lower hydraulic head $H$), the pore water pressures within the ground ahead of the face can be reduced significantly by advance drainage. Such a reduction also leads to lower seepage forces acting on the potentially unstable wedge in front of the atmospherical tunnel face. Figure 4b shows the average water pressure acting upon a potentially unstable wedge. For the present case, drainage over four hours by six boreholes halves the pore pressure acting on the wedge.

The reduction in pore pressures observed in the ground ahead of the face is very helpful in terms of face stability. The black columns in Figure 2b show the necessary mechanical support pressure in the case of advance drainage by six boreholes (under atmospheric conditions in the working chamber). The needed support pressure is significantly lower than the slurry pressure which would be needed in the absence of advance drainage (white columns). According to Figure 2b, advance drainage represents a very powerful improvement method and extends the feasibility range of open mode operation.

## JAMMING OF THE SHIELD DUE TO SQUEEZING

When using a TBM, relatively small convergences (in the order of one or two decimeters) may lead to considerable difficulties, due to the geometrical constraints of the equipment. On account of the poor ground conditions that are expected along some stretches of the alignment, jamming of the shield due to squeezing ground could therefore not be excluded a priori and it was accordingly investigated computationally.

The hazard scenario was assessed by computing the thrust force required for each tunnel section under uniform conditions. The calculations were carried out systematically for different operational modes, stages and measures. More specifically they were performed: (i) both for open and closed mode operations (the latter assuming full compensation of water pressure); (ii) both for restart after a standstill (static skin friction) and for ongoing excavation (lower sliding skin friction, but additional cutter head force for boring taken into account); (iii) with and without lubrication of the shield extrados (lubrication reduces the friction by about 50% and is automatically applied in the case of closed mode operations with bentonite suspension supporting the face); (iv) three values (3, 2 and 1 cm) for the radial gap size between shield and ground in order to study the effects of a reduction in the overcut (caused by the wear of the gauge cutters or by the packing of fines between the shield and the ground). The positive effects of a possible delayed ground response (i.e., time-dependent behavior due to consolidation or creep) were not taken into account in the calculations. This is a reasonable simplification in view of the difficulty of making a reliable forecast of the time-dependent development of ground deformations.

Concerning open mode operation, the required thrust force $F_r$ was taken as equal to:

$$F_r = F_b + F_f \quad (1)$$

where $F_b$ is the boring thrust force and $F_f$ the thrust force required for overcoming shield skin friction considering the friction coefficient $\mu = 0.30$ for sliding friction and $\mu = 0.45$ for static friction, respectively (Gehring 1996). The boring thrust force $F_b$ was considered only for the operational stage of "ongoing excavation" and assumed as:

$$F_b = F_c n_c = 13 \text{ MN} \quad (2)$$

where $F_c = 267$ kN is the bearing capacity of the cutters after Wehrmeyer et al. (2001) and $n_c = 48$ is the number of cutters.

The calculation of the thrust force required during closed mode operation has to consider additionally the thrust force required due to the face support pressure. Therefore, Equation 1 has to be enhanced with the additional term $F_p$:

$$F_r = F_b + F_f + F_p \quad (3)$$

where $F_p$ is equal to the integration of the support pressure over the face. Assuming full compensation of the water pressure,

$$F_p = H_w \gamma_w \pi D^2/4 \quad (4)$$

where $H_w$ is the depth of the tunnel beneath the lake level or groundwater table, $\gamma_w$ the unit weight of the water and $D$ the boring diameter.

Another difference between open and closed mode operation concerns initial stress, which has been considered in the calculations. For open mode operation, total stress shall be considered (cf. Anagnostou and Kovári 2003), i.e.

$$\sigma_0 = H\gamma' + H_w\gamma_w$$
if $H_w > H$ (subaqueous portion), and

$$\sigma_0 = H_w\gamma' + (H - H_w)\gamma_d + H_w\gamma_w$$
if $H_w < H$ (land portion), (5)

where $H$, $\gamma'$ and $\gamma_d$ denote the depth of cover, the submerged unit weight and the dry unit weight of the ground, respectively. For closed mode operation, the effective rather than the total initial stress must be taken into account:

$$\sigma_0 = H\gamma' \text{ if } H_w > H \text{ (subaqueous portion), and}$$

$$\sigma_0 = H_w\gamma' + (H - H_w)\gamma_d$$
if $H_w < H$ (land portion). (6)

The effective initial stress is lower than the total initial stress (which is favorable and leads to a lower frictional resistance $F_f$) but on the other hand the face support pressure must also be taken into account ($F_p$ in Equation 3).

A total of 1512 input parameter sets has been considered in the calculations for the required thrust force (Anagnostou et al. 2009). It was possible to make such a comprehensive investigation only on the basis of the design nomograms presented by Ramoni and Anagnostou (2009b). These nomograms assume a constant overcut along the shield, while the actual shield becomes smaller stepwise. This simplifying assumption tends to be unsafe concerning the loading of the front portion of the shield but is generally safe for the rear shield, which is the most critical part of the machine with respect to jamming.

Figure 5b shows the convergence $\Delta u$ of the bored profile according to a comparative numerical calculation with a more realistic modeling of the actual shield geometry (Figure 5c). Due to the conicity of the shield, the gap between ground and shield (dashed lines in Figure 5b) becomes closed three times: in the front part of the shield at a distance of about 2 m from the working face and, later, also in the middle and in the rear part of the shield. When the ground establishes contact with the shield by closing the gap, a pressure $p$ develops upon the shield. The thrust force required to overcome shield skin friction can be calculated by integrating the ground pressure over the shield surface and taking into account the skin friction coefficient. The simplified computational model of the nomograms is generally safe concerning the required thrust force.

Figure 2c shows the required thrust force over the entire tunnel alignment for the case of restarting after a standstill during open mode operation without lubrication of the shield. The results are presented both for the average and for the worst parameter combinations. The effect of the amount of overcut is illustrated by the black and the white columns (for 3 cm and 1 cm radial gap sizes, respectively). The positive effects of lubricating the shield can easily be understood if we bear in mind that the thrust force required to overcome shield skin friction depends linearly on the assumed skin friction coefficient $\mu$ between shield and ground.

The results of the computational investigations described in this section indicate that the potential problem of shield jamming is far less critical than that of face instability but must be taken into account in isolated portions of the alignment. As discussed later, the application of standard counter-measures is anticipated, such as lubrication of the shield mantle and the installation of new gauge cutters (in order to assure enough clearance between shield and ground) before entering the critical stretches.

## TUNNELING PLAN

The tunneling plan defines the TBM operational modes (open or closed), the operating pressures and whatever auxiliary measures are required. It is based upon a qualitative evaluation of the geological profile, the geomechanical calculations mentioned above, engineering judgment and risk considerations. As is the case for any tunneling project, there are uncertainties with respect to, (i), the structure of the formations (e.g., the sequence of the lithological units and the extent and location of fault zones) and, (ii), the response of the ground to tunneling operations (e.g., the stand-up time of the ground or the intensity of excavation-induced convergences).

The consequences of type (i) uncertainties can be reduced by systematic advance probing during the TBM drive. On the whole, advance probing is recommended for the entire tunnel. At each drilling station, two boreholes without core recovery shall be drilled. The timely and reliable identification of critical zones will also necessitate, however, core drilling on some occasions. A reliable geological pre-exploration of the conditions prevailing ahead of the face will reduce the need for precautions (such as closed mode operation). If the advance probing is less reliable, a greater number of protective measures will be required in order to handle risks which will possibly never materialize. The data from percussive drillings (water quantities and drilling data such as penetration rate, penetration force and torque during the drilling process) is relevant for determining water circulation or the presence of sharp transitions between hard rock and soft ground and it is therefore reliable only with respect to specific geological features (for example, highly fractured water bearing zones).

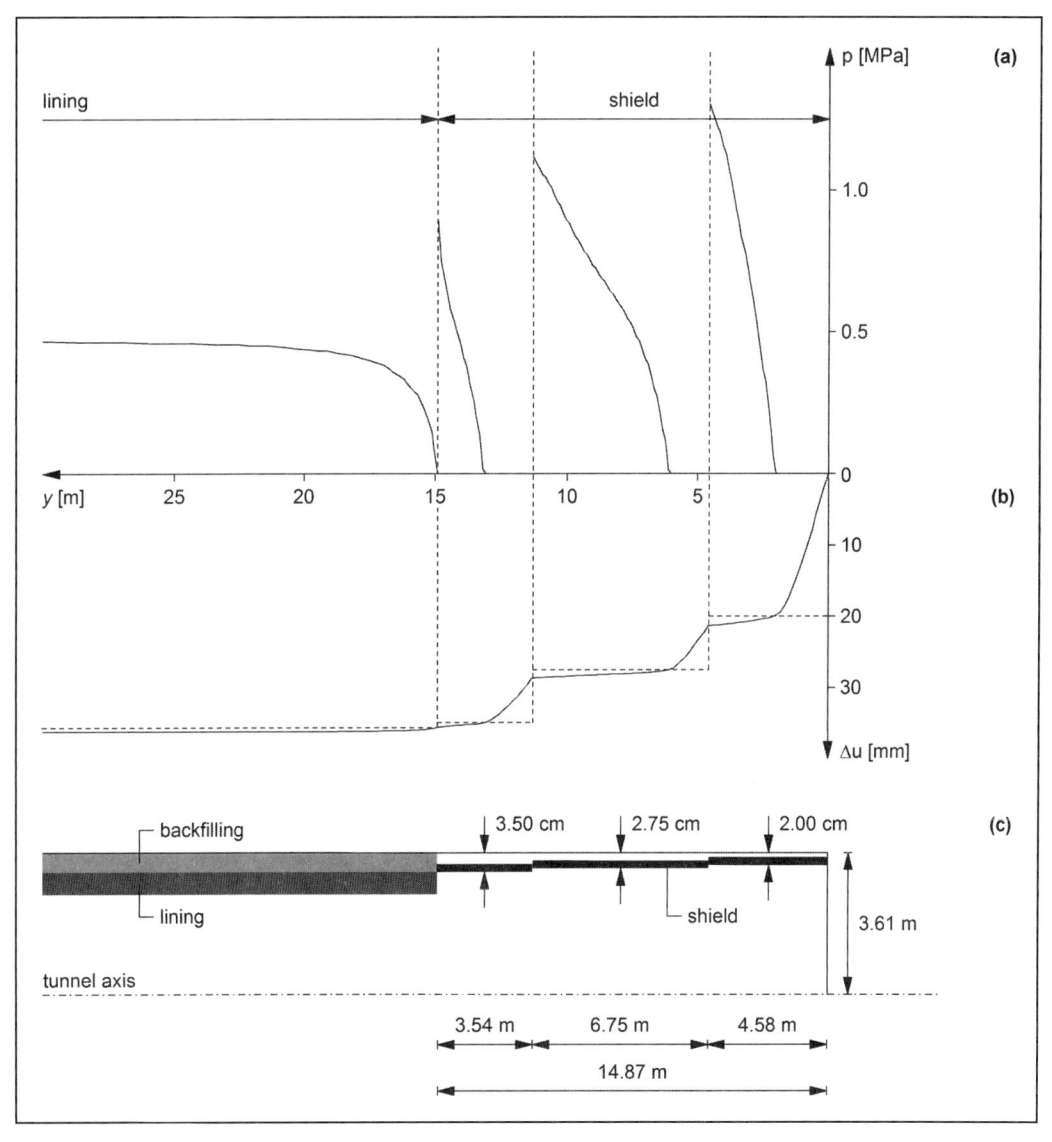

Figure 5. Ground pressure $p$ acting upon the shield and the lining (a) and convergences $\Delta u$ (b) for the actual geometry of the shield (c); the computations apply for the Conglomerate of the Muddy Creek Formation ("worst" conditions).

Even in the case of a well-known sequence of geological formations, there can be uncertainties with respect to ground behavior, i.e., the above-mentioned type (ii) uncertainties. In the present project, such uncertainties are relatively large due to the character of the ground and, more specifically, the difficulties of assessing the effects of the "time" factor. As already mentioned, for the expected range of ground permeabilities long portions of the alignment fall into the geotechnically-demanding intermediate stage between so-called "drained conditions" and so-called "undrained conditions." In this intermediate stage it cannot be said with certainty whether favorable short-term conditions or unfavorable long-term conditions will apply. This introduces an element of uncertainty concerning the stand-up time of the tunnel face and has therefore a direct consequence for the operating mode of the machine. The geomechanical calculations indicate that the effect of this uncertainty can be reduced significantly (but not entirely) by advance drainage.

According to Figure 2b, assuming that the "average" conditions prevail over the entire alignment, a pressurized face would be necessary over about 35% of the tunnel length. In the case of the "worst" conditions (a highly improbable hypothesis, of course) this figure increases to about 85%. The advance drainage of the ground reduces the amount of support required considerably. For the "average" conditions, only the faults, the Red Sandstones and the basalts would require an additional mechanical face support of only 0.5–1.1 bars. Assuming "worst" conditions a mechanical face support of 0.9–3.4 bars is required (instead of the 9–14 bars of slurry pressure). This emphasizes the huge importance of careful ground evaluation and decision-making during construction.

The top row of Figure 2d gives an overview of TBM operational modes for the excavation process which can reasonably be assumed taking into account the information available at present. In order to mitigate the risk of a face collapse, the faults, the Red Sandstones and the basalts have to be excavated in closed mode (red bars). Considerable slurry loss and a subsequent loss of the support pressure in the gravel like core of the Detachment Fault must be avoided by operating the TBM in closed mode in combination with advance ground improvement by grouting. The remaining portion of the tunnel alignment can either be excavated in open mode (white bars) or in a combination of open mode with advance drainage of the ground ahead of the face (orange bars). Figure 2b indicates that the face would not be stable in the long-term for the "worst-case" strength parameters, but one should consider that the adverse combination of high permeability ($k > 10^{-8}$ m/s) and low strength is rather improbable as low strength values apply to the more clayey units which exhibit rather low permeabilities. One may consider, furthermore, that the risk of face instability during ongoing excavation may be acceptable (no people in the working chamber) as long there is no connection to the lake (sufficient depth of cover, low-permeability ground).

Nevertheless, during the performance of inspections and maintenance activities in the working chamber the risk of a face collapse is clearly unacceptable. Due to the uncertainties concerning high water pressures, hyperbaric interventions should be avoided and continuous (non-stop) TBM operation in closed mode is recommended in the most critical stretches (black bars in the middle row of Figure 2d). Continuous operation should be possible at least for the shorter critical portions, provided that careful maintenance is carried out just before entering these stretches. In the relatively long portion through the Red Sandstone Unit and the basalts at the end of the tunnel alignment, however, one or more maintenance stops will probably be necessary. The work will then have to be carried out either under hyperbaric pressure or after grouting the ground (red bar). The possibility of non-stop excavation in this particularly adverse tunnel portion will be re-evaluated later, taking account of the experience gained during the TBM drive. In the remaining portions of the tunnel alignment, the interventions in the working chamber can be performed in open mode. In the metamorphic rock sections the working chamber can be accessed without any additional measures (white bars), whereas in the tertiary sedimentary rocks the working chamber can be accessed only after finding an appropriate location and after draining the ground ahead of the face (orange bars). For these stretches, it is recommended first of all that the face be inspected and its stability evaluated. If the ground conditions are good enough, drainage shall be carried out before entering the chamber. In the case of adverse face conditions, the TBM drive should be continued and then stopped again after few meters for a new inspection and assessment of the face. Based upon the frequency and extent of poor rock intervals found in the exploratory boreholes, it is reasonable to expect that one will find an appropriate location after one or two restarts. However, the possibility of longer stretches with poor ground conditions cannot be excluded entirely. If a safe location cannot be found after a number of stops and restarts, core drilling is recommended in order to find a safe spot for maintenance work. Where it is not possible to identify such a place with sufficient reliability, measures such as face bolts or grouting will be necessary in addition to drainage, in order to ensure the stability of the face (the alternative is for divers to perform the work under hyperbaric conditions).

Regarding the potential hazard of shield jamming due to squeezing, the major portion of the tunnel alignment can be excavated without taking any measures (see white bars in the lower row of Figure 2d). An overcut of at least 3 cm has to be provided in the so-called "Tmc 2" unit of the Muddy Creek Formation, in the Red Sandstones Unit and in most of the fault zones (light blue bars). Regarding the overcut of 3 cm, it should be noted that, due to the packing of fines or gauge cutter wear, the actual gap may be lower than the theoretical one. A lubrication of the shield in open mode will be necessary only if the TBM has to be stopped and restarted in the very low quality ground of the Muddy Creek Formation (dark blue bars). As already mentioned, this can be avoided through the use of core-drilling to identify appropriate locations for maintenance work.

## FINAL REMARKS

According to our investigations, a considerable portion of the tunnel can be constructed by open-mode

TBM operation in combination with advance drainage of the ground ahead of the face and systematic advance probing, including core-drilling on some occasions. The risk of shield-jamming is less critical than face instability but must be taken into account, particularly in the sedimentary rocks before the Las Vegas Wash.

The recommended operational modes are reasonable from the perspective of a qualitative risk analysis (the risk of an undesired event occurring is considered acceptable if its impact or probability of occurrence are small). A systematic evaluation of the experience gained during the TBM drive will enable better management of the uncertainties concerning ground behavior.

## REFERENCES

Anagnostou, G., and Kovári, K. 1994. The face stability of slurry-shield-driven tunnels. *Tunnelling and Underground Space Technology* 9 (1994) No. 2, 165–174.

Anagnostou, G., and Kovári, K. 1996. Face stability conditions with earth-pressure balanced shields. *Tunnelling and Underground Space Technology* 11 (1996) No. 2, 165–173.

Anagnostou, G., and Kovari, K. 2003. The stability of tunnels in grouted fault zones. Mitteilungen des Instituts für Geotechnik der ETH Zürich, Vol. 220.

Anagnostou, G. 2007a. Continuous tunnel excavation in a poroelastoplastic medium. Tenth international symposium on numerical models in geomechanics, NUMOG X, Rhodes, 183–188.

Anagnostou, G. 2007b. Practical consequences of the time-dependency of ground behavior for tunneling. RETC, Toronto, 255–265.

Anagnostou, G. 2009. Some rock mechanics aspects of subaqueous tunnels. EUROCK 09, Dubrovnik, in press.

Anagnostou, G., Cantieni, L., and Ramoni, M. 2009. Lake Mead Intake No 3 Tunnel—Assessment of the tunnelling conditions. Report to Vegas Tunnel Constructors, Boulder City, July 2009, unpublished.

Anagnostou, G., Cantieni, L., Nicola, A., and Ramoni, M. 2010. Face Stability Assessment for the Lake Mead Intake No 3 Tunnel. Tunnel Vision Towards 2020, ITA-AITES World Tunnel Congress 2010, Vancouver, in press.

COMSOL 2009. COMSOL Multiphysics 3.4. http://www.comsol.com/. Accessed October 2009.

Feroz, M., Jensen, M., and Lindell, J.E. 2007. The Lake Mead Intake 3 Water Tunnel and Pumping Station. RETC, Toronto, 647–662.

Gehring, K.H. 1996. Design criteria for TBM's with respect to real rock pressure. Tunnel boring machines—Trends in design & construction of mechanized tunneling. International lecture series TBM tunnelling trends, Hagenberg, 43–53.

Hurt, J., McDonald, J., Sherry, G., McGinn, A.J., and Piek, L. 2009. Design and construction of Lake Mead intake no. 3 shafts and tunnel. RETC, Las Vegas, 488–502.

McDonald, J., and Burger, W. 2009. Lake Mead Intake Tunnel No. 3. *Tunnel* No. 4, 43–48.

Ramoni, M., and Anagnostou, G. 2007. The effect of advance rate on shield loading in squeezing ground. Underground space—The 4th dimension of metropolises, ITA World Tunnel Congress 2007, Prague, 673–677.

Ramoni, M., and Anagnostou. G. 2009a. Tunnel boring machines under squeezing conditions. *Tunnelling and Underground Space Technology*, accepted for publication.

Ramoni, M., and Anagnostou, G. 2009b. Design nomograms for the assessment of the required thrust force for TBMs in squeezing ground. Research Project FGU 2007/005, Report 090202 of the Institute for Geotechnical Engineering (IGT) of the ETH Zurich to the Swiss Federal Roads Office (FEDRO).

Vegas Tunnel Constructors. 2009. Lake Mead Intake No. 3. Intake Tunnel—Geologic Profile. Drawings VTC-41 to 47. November 25, 2007, unpublished.

Wehrmeyer, G., Burger, W., and Knabe, M. 2001. Herausforderungen im Hartgestein: TBM-Entwicklungen, Projekte und trends. *Tunnel* No. 2, 12–29.

# Continuous Conveyor Design in EPB TBM Applications

**Dean Workman**
The Robbins Company, Oak Hill, West Virginia

**Desiree Willis**
The Robbins Company, Kent, Washington

**ABSTRACT:** While continuous conveyors have become the muck removal system of choice in long, hard rock TBM tunnels, they have gained acceptance in soft ground tunnels only recently. Soft ground, EPB TBM-driven tunnels provide a challenging environment for continuous conveyors due to the variety of materials present. The design of both horizontal and vertical soft ground conveyors will vary depending on the types of excavated material, amount of water present, and other factors. This paper will address the challenges of effective conveyance in different ground materials by analyzing conveyor performance in several recent EPB TBM projects.

## INTRODUCTION

Continuous conveyor systems were first utilized regularly in mining applications, and have since been adopted as one of the primary means of muck removal on hard rock tunneling projects. In recent years, conveyor system design has advanced considerably, from computerized monitoring systems to self-adjusting curve idlers to km long steel cable belt systems. Continuous conveyors for soft ground TBM projects are a fairly recent edition in the tunneling industry, with considerable advantages over muck cars including increased safety and efficiency.

## CASE STUDY #1: LOWER NORTHWEST INTERCEPTOR SEWER, SACRAMENTO, CA

One of the first EPB TBM projects to utilize continuous conveyor design was the Lower Northwest Interceptor Sewer (LNWI) Tunnels, excavated by Affholder, Inc. in 2005. The project is part of the Sacramento County Regional Sanitation District's (SRCSD) interceptor expansion project, which will ultimately extend approximately 320 km (200 mi) and provide service throughout the region (Togan et al, 2007).

Two 610 m (2000 ft) long river crossings were tunneled by a 4.59 m (15.1 ft) EPB TBM at steep grades through stiff clay, silt and sand. The first tunnel was excavated a 6% downgrade below the river, while the following section was completed at a 6% upgrade from below the riverbed, making muck removal via muck car a difficult proposition. Affholder settled on a continuous conveyor design (see Figure 1).

During machine operation, an extensible fabric-belt conveyor (30" BW) was constructed behind the TBM and back-up system at the tailpiece and back-up car assembly at the last deck. The muck was discharged from the screw conveyor into the loading hopper of the tailpiece at the front of the tailpiece back-up car, which elevated the conveyor into the crown of the tunnel. The belt structure was assembled in the tailpiece assembly, which was mounted on top of a tailpiece back-up car. The assembled belt structure then came out of the rear of the tailpiece as the TBM advanced forward. The extensible conveyor was equipped with a 500ft capacity belt cassette with a 100 HP hydraulic power unit and a splicing stand to allow conveyor belting to be added to the system in the open cut. The conveyor system was powered by a 250HP Main Drive, which brought the muck out of the tunnel and through the open cut and discharged it into the muck pit. The conveyor was designed for 2,150 ft and for 2,175 ft lengths at 685 USTPH while running at 600 FPM.

The continuous conveyor equipment was used on the first crossing, then removed and set up for the second tunnel section as well. Both sections of tunnel were mined within a one year time frame. Downtime for the duration of the project was minimal, with no conveyor problems and all downtime occurring due to TBM or segment issues (see Table 1).

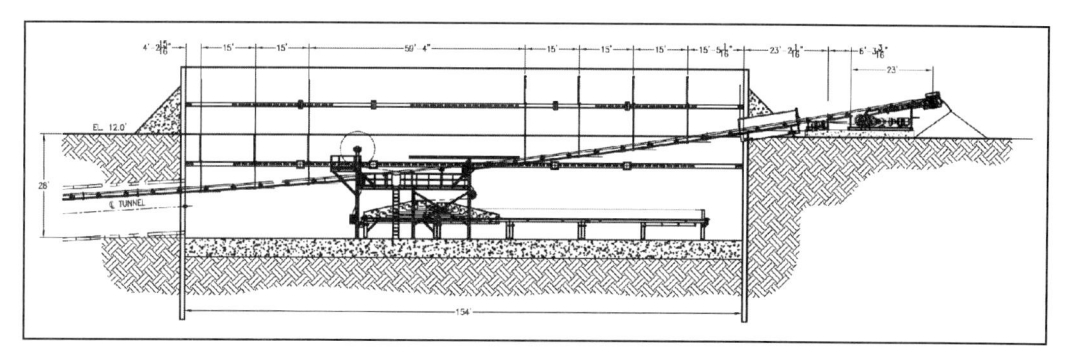

Figure 1. LNWI conveyor system setup

Table 1. LNWI conveyor system specifications

| | |
|---|---|
| Conveyor Length (south crossing) | 2,150 ft |
| Conveyor Length (north crossing) | 2,175 ft |
| Conveyor Capacity | 685 USTPH |
| Belt Width | 30 in |
| Belt Speed | 600 FPM |
| Main Drive | 250 HP |
| Cassette Capacity | 500 ft |
| Cassette Power Pack | 100 HP |

## DESIGN CONSIDERATIONS FOR SOFT GROUND CONVEYOR SYSTEMS

### Minimization of Start-Up Time and Maximization of Efficiency

The layout of conveyor systems is designed with swift setup in mind. Unlike many hard rock projects, EPB tunnels are relatively shallow and begin from an open cut. The use of a conveyor system set up at the surface can allow for initial use of the system at startup without having to mine a long starter tunnel.

With all components pre-assembled in place at the surface, switching from an initial muck box setup to continuous conveyor often takes a day or less. Crews simply pull the belt onto the system to start mining. By comparison, installation of a rail muck car system can take much longer. Once mining begins, reliability and system availability of a conveyor system are typically much greater. Even in tunnels using up to five muck trains and multiple California switches, the time required to remove muck from the tunnel cannot compare to conveyor muck removal. In addition, muck cars generally require a much higher level of routine maintenance.

### Variable Ground and the Role of Additives

Variable ground is very often a given in soft ground tunnels. One project can range from weathered rock to sand to clays with changing permeability and ground water. Injection of additives through the cutterhead, such as bentonite, foam, or polymer, can aid in consolidation of muck and eliminate many of the problems associated with conveying fluidized muck. Maintenance of a smooth flow through the cutterhead and screw conveyor onto the belt conveyor system minimizes belt stoppage and material spillage. Additives also have the ability to control the fluidity of very wet ground and help solidify loose, watery material.

Depending on the ground conditions, different additives are used to maximize efficacy. The type of additive used is based on a standardized curve comparing particle size and distribution based on filtering samples of material through differently sized screens. Ground with less than 30% fines, or particles less than 0.2 mm in diameter, is difficult to fluidize. In this type of non-cohesive ground, bentonite is used for consolidation. For other types of ground with fewer fines, foam consisting of water, surfactant, and additive is used. If water pressure is high and small particles are present, a polymer can be injected in addition to the foam to increase cohesiveness of the material (see Figure 2).

### Water-bearing Ground and Resulting Design Modifications

If a high amount of water-bearing ground is expected, continuous conveyor systems can be designed to minimize associated risks. Incline is kept relatively low for EPB conveyors—a maximum of about 10 degrees, compared to 18 degrees in hard rock tunnels. materials in EPB applications when the material is very fluid, keep the incline to a minimum. In addition, transfer points are entirely enclosed to keep material from spilling out. The enclosed points are equipped with additional belt skirting, a urethane material that seals the edges.

**Conveyor Cleaning**

Further design modifications minimize the wear of conveyor belt and prevent stoppage due to sticky material. Primary and secondary bore scrapers clean off very heavy material, while a belt wash box installed on the surface near the main drive effectively removes fine material from the conveyor before it cycles through the belt storage cassette. The wash box consists of water spray in combination with 'air knives'—pressurized jets of air that remove material from the belt without direct contact. The use of air knives eliminates the need for consumable components that come in direct contact with the belt and must be replaced.

**CASE STUDY #2: UPPER NORTHWEST INTERCEPTOR SEWER, SECTIONS 1 & 2, SACRAMENTO, CA**

Northern Sacramento's Upper Northwest Interceptor Sewer (UNWI) Project is unique in several respects—the 5.8 km long EPB TBM driven tunnel is fairly long for soft ground projects, passing through a number of manholes, and the tunnel liner includes pre-cast concrete segments with an imbedded PVC inner liner never before used in North America. The contractor, the Traylor/Shea JV opted for a continuous conveyor system rather than muck cars, because of the tunnel length and the increase in efficiency possible when compared to muck cars.

The new UNWI system will convey up to 560 million liters of wastewater per day from various areas of Sacramento. The entire project, for the Sacramento Regional County Sanitation District (SRCSD), includes nine sections totaling over 30 km. Tunneling is on schedule for completion in early 2010, and the pipeline is planned to begin operation that November.

The 5.8 km long conveyor system was specially designed for varying ground conditions and water inflows. Design features include sealed transfer points and receiving hoppers. Urethane rubber is used to seal the points and minimize spillage. Additives mixed with the wet ground, such as foam and bentonite, maintain a smooth consistency of muck that will flow on the conveyor even when significant ground water is present. Cutterhead design aids in injection of the additives, using four independent foam injection points that mitigate the risk of clogged lines. Each line continues operation if one is down, keeping cutterhead wear and ground consolidation even at the tunnel face and ensuring a smooth flow of muck (see Figure 3).

During TBM operation, an extensible fabric-belt conveyor (24" belt width) is constructed behind the TBM and over the top of the back-up system in

**Figure 2. Conveyor system with consolidated muck and additive: UNWI Project, Sacramento, CA**

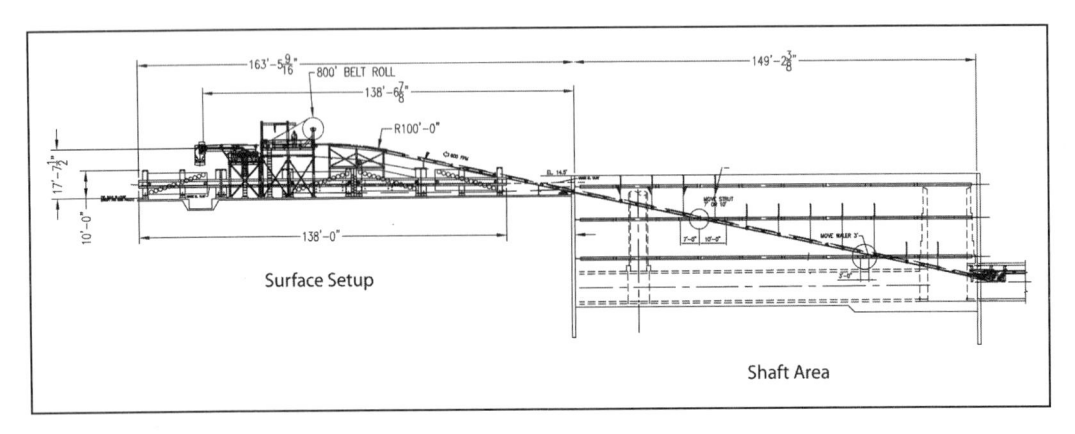

**Figure 3. UNWI conveyor system setup**

**Table 2. UNWI conveyor system specifications**

| | |
|---|---|
| Conveyor Length | 19,498 ft |
| Length of Straight Conveyor | 17,260 ft |
| Length of Curved Conveyor | 2,238 ft |
| Curve Radius | 1,200 ft |
| Conveyor Capacity | 250 USTPH |
| Belt Width | 24 in |
| Belt Speed | 600 FPM |
| Main Drive | 200 HP |
| Carrying Booster Drive (×2) | 200 HP |
| Belt Cassette Capacity | 1,150 ft |
| Cassette Hydraulic Power Unit | 150 HP |

the crown of the tunnel. The muck is discharged from the screw conveyor onto the conveyor at the front of the back-up system. From there, the conveyor system is elevated over top of the back-up system and into the crown of the tunnel. The belt structure is assembled in the installation window assembly which is mounted on top of a back-up deck, while the assembled belt structure comes out of the installation window as the TBM advances forward. A tripper assembly is located at the open cut in the tunnel and redirects the conveyor up thru the open cut at a 12 degree incline. The extensible conveyor is equipped with a 1,150ft capacity belt cassette with a 150HP hydraulic power unit and a splicing stand, which allows conveyor belting to be added to the system. The conveyor system is powered by a 200HP Main Drive, which discharges onto the stacking conveyor. Two 200HP carrying booster drives are added into the system as the conveyor advances, and are mounted in the crown of the tunnel by using an integrated drive support frame. The frame is designed with adjustable length legs that mount to the tunnel wall at designated distances in the tunnel. The conveyor is designed for a total length of 19,498 ft at 250 USTPH while running the conveyor at 600 FPM.

As the conveyor travels through radii down to 400 m, patented self-adjusting curve idlers transfer the load and enable the system to run through curves. Tunnel muck is discharged from the conveyor into a muck holding bin at the surface adjacent to the launching shaft (see Table 2).

**Conveyor System Performance**

As of November 2009, the conveyor system had performed at high availability and enabled very good advance rates. The best mining day (24 hours in three 8-hour shifts) was 160 linear feet, which occurred multiple times. The best week (five 24-hour days) was 690 feet, also occurring multiple times (see Figure 4).

**Figure 4. Surface setup conveying muck**

## SUMMARY

Continuous conveyor systems offer a distinct advantage over muck cars in soft ground conditions. Despite variable geology and ground water, the use of additives can effectively consolidate muck flow and prevent spillage on conveyor belts. Further modifications including water tight transfer points have mitigated the risk of muck loss in all but the most water-logged conditions. Though many equipment suppliers and contractors only recognize a difference in system availability between conveyors and muck cars over a 6,000 ft tunnel length, the advantages are being seen at shorter tunnel lengths as well. Increased safety, reliability and short start-up times are making soft ground conveyor systems a competitive option that may well replace muck car systems in the coming years.

## REFERENCES

Togan, et al. (2007). Construction of the Sacramento River Tunnels on the Lower Northwest Interceptor Sewer, Sacramento, California. Proceedings of the Rapid Excavation and Tunneling Conference, San Francisco, CA. 741–756.

# International Practices for Connecting One Pass Precast Segmental Tunnel Linings

**Christophe Delus, Bruno Jeanroy**
Anixter-Sofrasar, Sarreguemines, France

**David R. Klug**
David R. Klug and Associates, Pittsburgh, Pennsylvania

**ABSTRACT:** use of one pass precast segmental tunnel linings has advanced throughout the world to the point where they are used in not only soft ground applications, but in mixed geology and hard rock tunnels. The paper will give a short history of the use of precast linings and the associated connectors used to connect the segments in the ring and ring to ring, beginning with curved steel "banana" bolts to the current industry trend of using high performance polymer plastic dowels and straight bolts with polymer plastic socket embeds.

The paper will review the tunnel lining practices being used throughout the world as various countries have developed practices and quality control requirements to meet specific geotechnical and national requirements.

## HISTORICAL REMINDER

Up to 1930, TBM-driven tunnels were mainly lined using cast iron segments (Figure 1). Thereafter, precast concrete segments tunnels lining started to appear, mainly in Great Britain, for small diameter tunnels (1.5 to 3.0m) driven in London clay for use as sewers.

Since that period, several hundred kilometers of generally small diameter tunnels driven in the London area have been lined with concrete segments of various shapes and types. Oftentimes they were ribbed. In other words, their shape stemmed from that of cast iron segments. It should be noted that, most of the time, these underground structures were built in very low permeability ground in which the excavated periphery offered short-term stability (London clay).

In time, British manufacturers offered a whole range of standard off-the-shelf tunnel lining segments covering a wide range of diameters (1.5 to 6m internal diameters). One of the significant features of these segments was their small size and reduced weight (100 to 400 kg per segment) which resulted in a large number of ring elements for the largest diameter tunnels (12 segments per ring for a diameter of the order of 6.0m). The main reason for this large number of elements is that at that time the construction process was not mechanized and only made by hand labor.

Since 1965, major developments in the use of concrete segments linings in Europe (Germany, Austria, France, UK and Belgium) and Japan is noteworthy, in parallel with the development of TBMs for excavating large diameters tunnels (approximately 5.0 to 10.0m) in soft and water-bearing ground conditions. Specifically, mechanized erectors, larger size

Figure 1. Construction of the Tower Subway, London, 1869, using cast iron segments

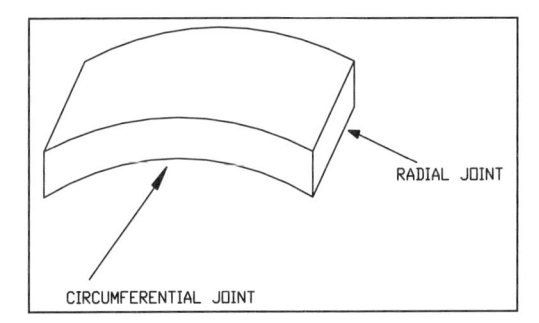

**Figure 2. Segment joints**

segments with very low precasting tolerances, elastomeric gaskets capable of guaranteeing lining water tightness even in heavily water-bearing ground and new connectors systems, have advanced this type of tunnel lining.

## PURPOSES AND EVOLUTION OF THE CONNECTORS

First it is important to notice that there are two kinds of connections (Figure 2): the connections in the circumferential joint, to connect one ring to the next. The other connection is in the radial joint, from one segment to another segment. Their purpose is not the same and the systems used in both joints can be different one from each other. We have listed 5 main purposes for the assembly systems.

### Purposes of Assembly Systems

Indeed, in the circumferential joint (connection ring to ring), their first purpose is to prevent any ring opening due to the gasket reaction load or to the pressure of the bentonite face slurry (in the case of a slurry TBM) when the TBM thrust cylinders are removed to install the next ring. Over time, this operation evolved as in the past, segments were almost pre-stressed by the connectors, when nowadays they only keep the force applied by the erector. The progress in the design of the erectors changed the philosophy and the parameters in how to design the assembly systems. Of course this feature is very important as it is linked to the sealing of the tunnel. In the radial joint, as a perfect ring is stable, their purpose is first to prevent any ovalization and to keep the ring in its original geometry when it comes out of the TBM tail shield and before the annulus back-grouting process.

One of the other main purposes of the assembly systems is to provide a good erection accuracy to prevent any offset between the segments and the rings. This is very important as it has significant influence on the water tightness, the more the offsets are reduced, the better the sealing gasket will work. We will see that with time, some specific systems were developed to ensure a high accuracy in the installation for both joints ensuring stability at the ring building stage even when no load is exerted by the TBM thrust cylinders.

The linking systems purpose is not limited to the construction stage. For example, in water conveyance projects, when the tunnel is put under internal hydrostatic pressure by the water, they have to prevent any deformations or openings of the ring. In this specific case, their purpose is still to keep the gasket compressed, as in the case of a primary lining, the gasket must have double action capabilities to prevent the ground water from coming inside the tunnel and preventing any polluted water to migrate in the ground outside of the lining. In some seismic areas, the connectors can also have specific features to allow the lining to deform with the earthquake and avoid any breaking of the lining.

In general, circumferential assembly systems are regularly spaced around the ring. Their number varies from one project to another depending on:

- The force to be balanced (reaction load of the gasket)
- The desired possibilities for relative rotation of a ring with respect to the last one installed (universal ring)

The number of connectors in the radial joint may vary from one to three elements (combination of different type of connectors) depending on the length of the ring and the type of connector used. In standard international ring design, the connectors are only designed for the purpose in the construction stage (except in specific cases, like described previously for the water conveyance or seismic areas or in North America), therefore it is now standard in continental Europe that the bolts are removed when the TBM is about 150 meters away and all the grouting operations have been completed. If the connectors are designed for a permanent use, it is important that they are designed with material able to provide the same durability as that of the structure itself.

### Evolution of Assembly Systems

In 1869, the first subway was built in London, using cast iron segments. These segments were linked using standard straight bolts and nuts. The first concrete segments were produced in the '30s using the same design (with hollow and ribbed). The connection systems and design of the segments evolved from this start.

#### *Straight Bolts*

This system was the first to be used with concrete segments, there was no significant innovation from the cast iron segments. The main issue is safety as the result of the significant number of bolts. Indeed numerous human operations are needed below the erected segment to connect the lining. Due to the narrow concrete thickness next to the bearing surface a specific reinforcement must also be considered.

*Straight Bolts with Steel Plates (Figure 3)*

This system, which is still used in Japan or Korea, is nearly the same as the previous described. In this case, the load bearing surface is decomposed between the concrete and the steel plates. It solves the problem of the reinforcement but does not improve the safety during the installation process. Furthermore using steel plates may induce some durability issues (corrosion).

*Curved Bolts (Figure 4)*

The curved bolts were used from the beginning of the '50s and it is still a connection method which is mainly used in Asia. In using this technology and compared to the two previous, the number of pockets is not reduced but their size and volume can be reduced. The installation of these bolts is not easy as the threaded bolt end sections must remain straight and therefore the tolerances of assembly are very large. This may induce some steps and lips. Furthermore during tightening operation the curved bolts tends to straighten and it may induce local stresses below the bolt towards the intrados of the segment, for this reason the reinforcement needs to be strengthened at this location.

*Straight Bolts with Sockets (Figure 5)*

This was a major evolution in the design of segments, for the first time, sockets were embedded in the segments. This change provided a reduction of the number of pockets by two in each segment. The installation is also safer for the worker as it can be made below a segment already installed. The main advantage is that the force exerted by the bolting system can be defined thanks to the relation between the torque and the tensile strength (according to the Norm NF E25-030, Figure 6). This calculation helps to define the adequate linking systems and therefore saves money in not over-sizing the connection. This has also some influence on the reinforcement design and on the global behavior of the segment as the smaller the connection will be the more concrete

**Figure 3. Detail on steel plate pocket**

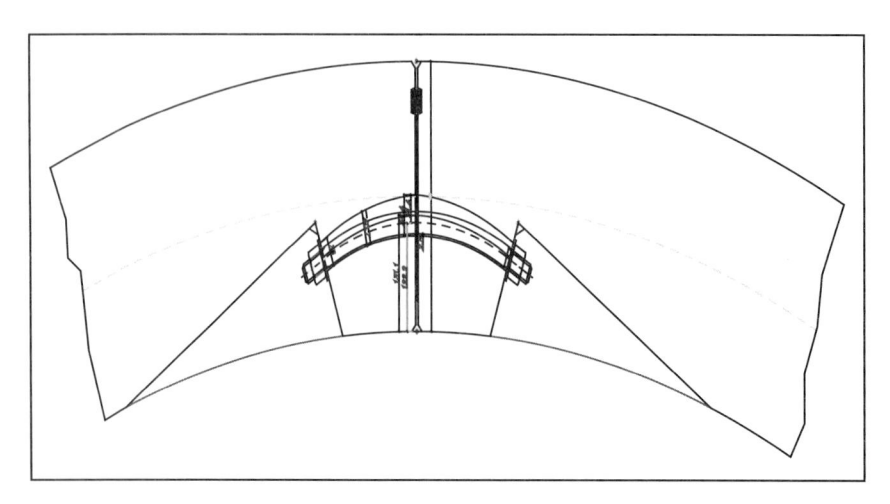

**Figure 4. Detail on curved bolt assembly**

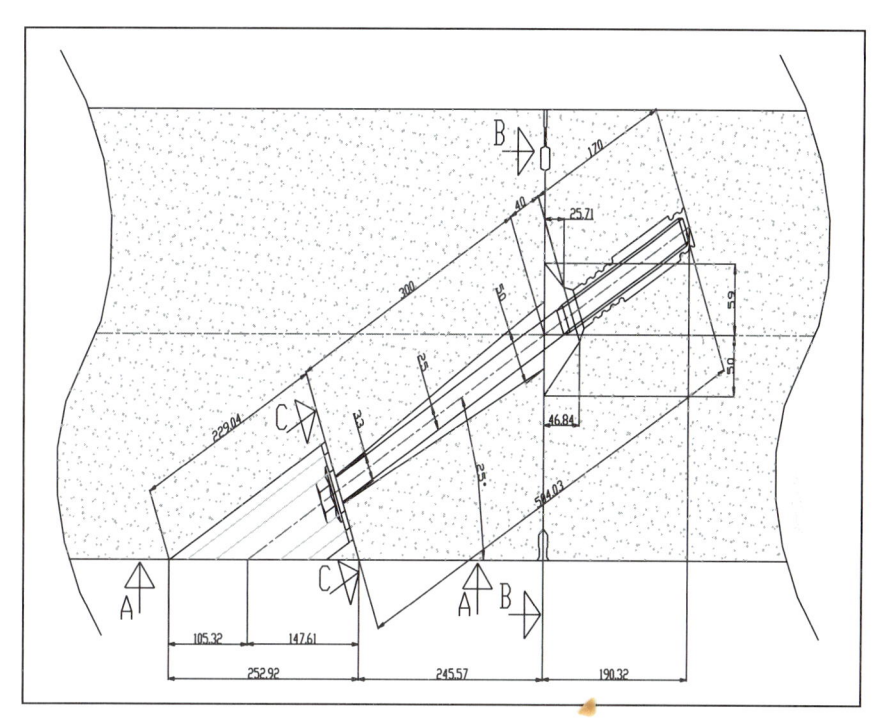

**Figure 5. Detail on straight bolt with socket**

there will be to provide a better resistance and a larger cover for the reinforcement.

The next step in this evolution is the change of the material to produce the sockets. In the early '80s, plastic sockets were used for the first time. These sockets were not designed for the tunnel construction. In France or Germany plastic sockets from the railway industry, designed for the concrete sleepers, were used. The sockets were produced out of High Density Polyethylene (HDPE) and the bolts used were coach screw bolts with a sharp thread. This socket from the railway industry offered many advantages, such as a good flexibility and a sufficient pull-out resistance for small transportation projects (metro). But on the other hand, the flexibility of the socket or the ability for the bolt to be easily installed was also a disadvantage from the engineer's point of view. In fact, if the bolt was not properly aligned, the bolt could have cut its own thread in the socket and in this case the maximum pull-out resistance could not be reached. Properties of the plastic were also studied and it was showed that the HDPE is not the proper material for a tunnel bolting system. In fact, the HDPE socket under a load stage creeps and therefore does not offer any safety to keep the gasket compressed. This creep leads to a release of the pressure on the gasket once

---

Norm NF E25-030 (French Standard)
$T = F(0.16P+\mu(0.583D2+rm))$
T: Torque applied
F: Tensile force exerted by the bolting system
P: pitch of the bolt thread
$\mu$: mean friction coefficient under the bolt head and in the bolt thread
D2: diameter on the flank thread
rm: mean radius of the bearing surface under the bolt head

**Figure 6. Norm E25-030, defining the torque to be applied depending on the required tensile strength**

the TBM rams are removed and before the installation of the next segment.

In a later stage, at the beginning of the 90s, new bolting systems were designed for the tunnel industry. The thread design was different and developed with non cutting threads. The sockets were produced out of polyamide that is stronger than HDPE and is not subject to creep. This new design provided the ability to install a bolt with slight misalignments (up to ±10°) and also to make sure that the theoretical designed pull-out resistance was reached and maintained.

*Dowels*

This kind of connection can only be installed in the circumferential joint, for the ring to ring connection. It is a great evolution in the connectors, as there is no human intervention below the segment and furthermore there is no pocket in the concrete, which has many advantages among them the durability of the connection and a smooth concrete surface which is a key criteria for water conveyance one-pass lining project. The use of the dowel in the tunnel construction started in the first half of the 20th century where some mined tunnels in Switzerland used dowels made of wood. The same kind of dowel was still used 50 years later on a non gasketed project in Munich, Germany (Hofoldinger Stollen). The development of dowels mainly started in the beginning of the 90s. Plastic dowels were developed and used for the first time in Italy on the "Passante Ferroviaro" project in Milano, Italy.

The first type of dowels were friction dowels, there was no embed receiving socket, only a recessed hole in the concrete segments; the dowel was pushed in the hole by the TBM rams and thanks to its geometry or material, it provided a sufficient pull-out resistance to keep the gasket compressed. Because there is no embedment, and the dowel during its installation induces radial stresses in the concrete, it is important to carefully design the reinforcement around the reservation. To achieve higher pull-out resistance, several types of dowels were developed by using different combinations of material (steel and plastic) and/or geometry. Nowadays, most of the dowels are made of polymer plastics and are made of two different components for the dowel and sockets. With this new design of locked in dowels, higher pull-out resistances are reached and less stresses are transmitted to the concrete during the installation because of the improved socket design.

**Figure 7. Segment installation with dowel system on circumferential joint**

*Alignment Dowels (Figure 8)*

For large diameter tunnels, where the technical requirements (pull-out and shear resistance) of the dowels are not sufficient for dowels but where the

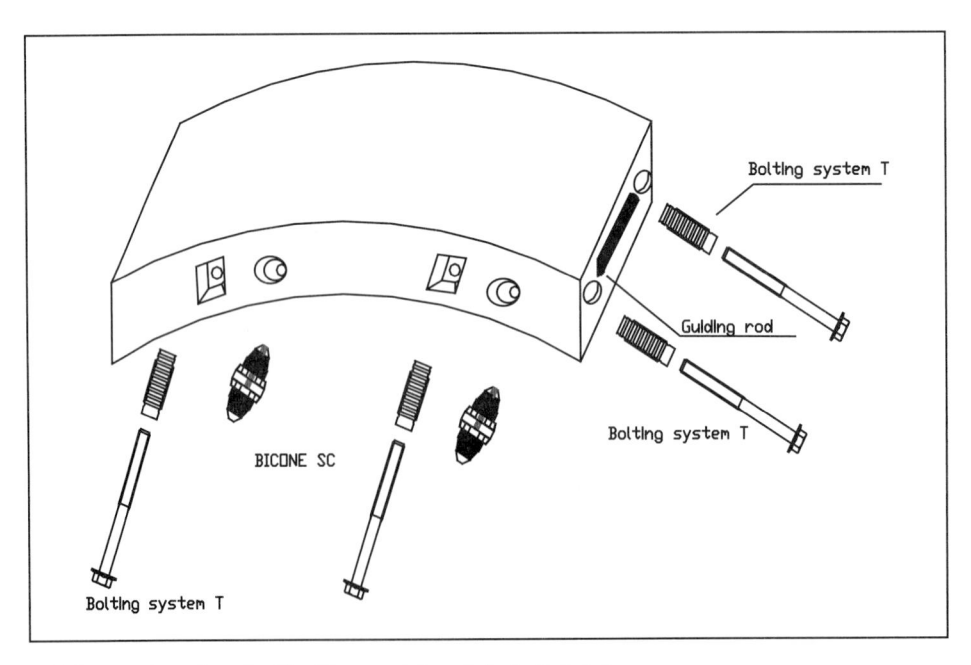

**Figure 8. Segment equipped with alignment dowels in both joints**

steps and lips could be a main issue because for example of the ground water pressure: a combination of bolts and alignment dowels can be used. This combined solution offers a higher pull-out resistance thanks to the steel bolting system and a good alignment with higher shear resistance thanks to the plastic alignment dowel. This solution also presents specific interests in Europe where bolts are mainly removed. In this case, the alignment dowels, which always remain in place provides a final shear resistance between the rings but also between the segments, as the guiding rod can be considered as an alignment dowel to be installed in the radial joint. The alignment dowels do not comprise any sockets as the pull-out resistance is not an issue. In this case, their only function is for guidance and shear resistance. The dowels are installed in a recess in the concrete segment in the circumferential joint and the guiding rods are glued in a groove in the radial joint.

## SELECTION OF THE LINK SYSTEM

First, it is essential to state that there is no unique design for a segmental lining. On most tunnel projects the design is based on the experience and skills acquired by Consulting Engineers and Contractors on previous projects, applicable design codes and accepted practices. For this reason the purpose of this section is only to review the key factors entering into the selection of the connector system. The main technical features of the connector (pull-out and shear resistance) must be evaluated on a project specific basis. We make the following comparisons considering that all the systems match the project requirements. The three main types of connectors are compared: curved bolts, straight bolts (with or w/o alignment dowel) and dowels.

### Guiding Function

How the connector will facilitate the erection of the segment rings by holding the different elements in place? The curved bolt does not offer any guidance, as the bolts are only inserted when the segment is fully positioned and in place. It is the same for straight bolt with socket; however the plastic socket offers much more flexibility than the curved bolt. The dowels are self-adjusting to the proper position of the segment in the ring while being pushed into position.

### Time of Assembly

Curved bolts are typically installed in the pocket with a hammer which can lead to a damage bolt thread and therefore the installation can be very time consuming. The straight bolting systems offers more flexibility and therefore less time for the installation, if they are combined with alignment dowel the assembling is even faster but still requires human intervention. The dowels systems need less time for the assembling as there is no operation required to secure the circumferential joint connection.

### Flexibility

The connection system must allow enough flexibility for tolerances in segment design and to the segment during the erection process. The curved bolt does not offer any flexibility, the segments have to be properly aligned otherwise the installation is not possible. Furthermore this kind of connection is 100% metallic and if it comes in contact with concrete in the guide hole, cracks may appear. The straight bolting system offers a little bit more flexibility, as the strong plastic socket is manufactured of polyamide and can absorb some misalignment ($\pm 10°$) thus the stresses are transmitted to the concrete. Plastic dowels cater for maximum flexibility during erection and are self-adjusting.

### Durability

The curved bolt is twice more likely to corrode than a straight bolting system, as the bolt head and nut are exposed, when only the head is exposed for the straight bolting system. Dowel systems are protected and not subject to corrosion as they are fully embedded in the middle of the concrete segments. Currently most of the dowels are 100% plastic made and are therefore not subject to corrosion and offer a long design life.

### Safety

The safety of the worker can also be one key criterion for the choice of the connector. Bolting systems require a worker to go down and insert the steel screw during the erection of the ring. With curved bolts, the worker needs to be beneath the moving and already installed segment. In using dowels, workers can assemble the rings with the help of the erector and of the TBM rams.

### Cost Benefit Analysis

The decision on what connector to be used should not be based solely on the base cost of the connector delivered to the precaster. Many different criteria must be analyzed prior to specifying or purchasing a connector such as service life required, is the connector going to be in a corrosive environment, will a dowel connector system provide multiple benefits (i.e., alignment mechanism during installation, faster ring installation time, connector of segments, elimination of post installation filling of bolt pockets), will the connector provide adequate tensile strength with a proper safety factor to keep the gaskets compressed during the ring installation process, does the connector system provide adequate shear and safety factor to keep the segments aligned in the tunnel geology to be encountered. Dowel connector systems may not be applicable in meeting all project requirements.

**FUTURE DEVELOPMENTS**

In the past 10 years, the use of dowels expanded in all international tunnel markets. It has been used on various projects, covering a range of diameter from 3.0 m up to 7.0 m ID, mainly for water conveyance projects due to strong market requirements and also on transportation projects (metro). At this stage the larger diameter projects, are still built more traditionally, using straight bolts with plastic sockets, but the use of alignment dowels (in combination with bolts) becomes more accepted. In the future, in order to be able to automate the ring installation, the use of dowel connections on large diameter projects will be needed. Manufacturers will have to look into new dowel designs and materials in order to achieve higher pull-out and shear resistance corresponding to these projects requirements

**BIBLIOGRAPHY**

AFTES Recommendations 1999

http://www.subbrit.org.uk/sb-sites/sites/t/tower_subway/index.shtml

Development of Dowelled Connectors for Segmental Linings, Davorin Kolic, Harald Wagner and Alfred Schulter, Felsbau 6/2000

# Geotechnical and Design Challenges for TBM Selection on the ICE Tunnel

**Steve Dubnewych, Stephen Klein**
Jacobs Associates, San Francisco, California

**Paul Guptill**
Kleinfelder, Irvine, California

**ABSTRACT:** The Irvine-Corona Expressway (ICE) tunnels consist of two 16.0-m-diameter (52.5-ft) road tunnels and, potentially, one rail tunnel extending 17.4 km (11.5 miles) between Riverside and Orange counties in Southern California. The purpose of the tunnels is to relieve traffic congestion along the SR-91 corridor.

Among the geotechnical challenges are variable and poor quality ground conditions, including weak, highly fractured rock, numerous fault and shear zones, high groundwater pressures, potential for gassy ground conditions, and a corrosive groundwater environment. Other significant challenges include protection of sensitive groundwater resources in the Cleveland National Forest. This paper discusses some of the results of the feasibility study recently completed for the project with a focus on tunnel boring machine (TBM) selection.

## INTRODUCTION

The Irvine-Corona Expressway (ICE) is a transportation corridor including tunnels and surface roads proposed between Interstate-15 near Cajalco Road in Corona and the interchange of the SR-133 and SR-241 toll roads in Irvine, California (see Figure 1). The tunnels evaluated in this study include highway and rail tunnels approximately 17.4 km (11.5 miles) long through the metamorphic and sedimentary rock formations of the Santa Ana Mountains separating Riverside and Orange counties. The ICE Tunnel Study considered highway and rail configurations relieving traffic congestion on the SR-91 through Santa Ana Canyon. According to California transportation authorities, traffic is projected to grow so much between now and 2030 that the SR-91 highway would have to expand from 12 lanes to 22 lanes in order to handle the increased demand. The highway tunnels, if constructed, are expected to remove roughly 60,000 to 70,000 average daily trips (ADT) from SR-91.

Funding for the ICE Tunnel feasibility evaluation was secured through the Safe, Accountable, Flexible, Efficient Transportation Equity Act—Legacy for Users (SAFETEA-LU). This paper summarizes both the geotechnical conditions to be encountered by the tunnels and the challenges posed to tunnel construction and tunnel boring machine (TBM) selection.

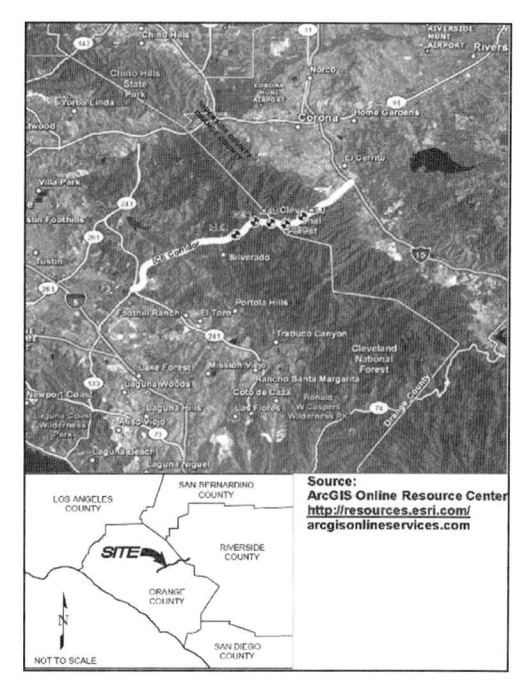

Figure 1. ICE tunnels and coreholes, location map

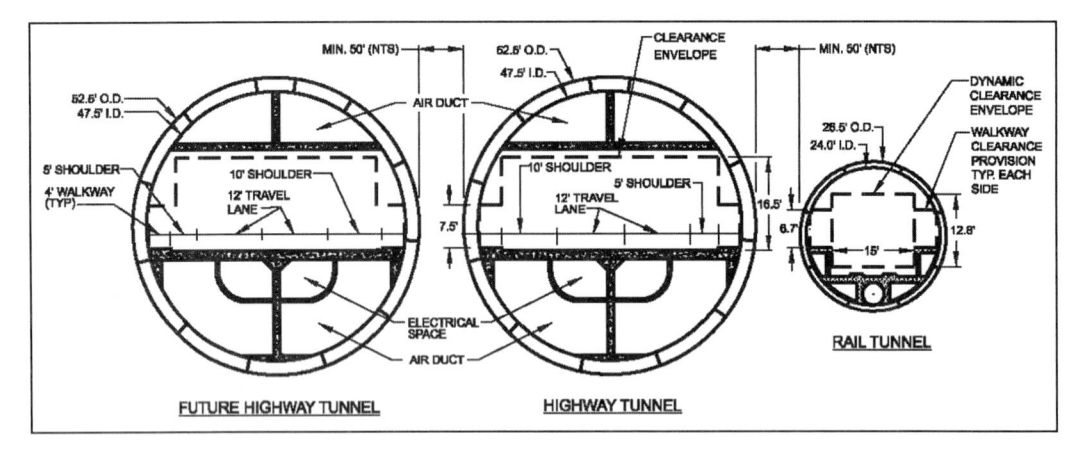

**Figure 2. Tunnel configuration**

## TUNNEL CONCEPTS CONSIDERED

Several tunnel concepts were evaluated, including a deep tunnel concept and a second concept that consists of a combination of surface roads and tunnels. The combined surface road/ tunnel concept is likely to present significant environmental challenges since the surface roads and associated construction activities would take place in the Cleveland National Forest and on nearby Irvine Ranch Conservancy land. Although this concept might be technically possible, it does not seem to be a viable approach at this time.

The deep tunnel concept considered four different tunnel configurations:

1. Twin-bore highway tunnels connected by emergency cross passages
2. A single two-lane reversible direction highway tunnel paired with a single track rail tunnel connected by emergency cross passages
3. Staged construction of twin-bore, two-lane highway tunnels paired with a single track rail tunnel connected by emergency cross passages, with the second highway tunnel being constructed at a later date
4. Three single-lane highway tunnels, two dedicated to one-way traffic and one reversible, all connected by emergency cross passages

This paper will focus on the third configuration, as shown in Figure 2, which includes twin-bore, two-lane highway tunnels (the second highway tunnel to be constructed at later date) and one rail tunnel, each with a total length of approximately 18.5 km (60,000 ft, or 11.5 miles). These tunnels would be connected to I-15 to the east and the SR-241/SR-133 interchange to the west by relatively short sections of surface highway. Each of the two or three tunnels would have two portals. The tunnel plan and profile are shown in Figures 1, 3a, and 3b.

The twin-bore tunnels start at an approximate elevation of 204 m (670 ft) above mean sea level (msl) at the West Portal and reach a maximum elevation of 649 m (2,130 ft) msl at approximately Station 510+00; the tunnels end at an approximate elevation of 515 m (1,690 ft) msl at the East Portal. The tunnel grade varies along the alignment and ranges from 0.1% to 5.0% (although inclusion of a rail tunnel in the project will likely limit the maximum grade to 3%). The ground cover above the tunnels ranges from a minimum of 6.1 m (20 ft) at Station 25+00 to a maximum of 408 m (1,340 ft) at Station 440+00. The minimum tunnel cover under the major canyons and creeks is approximately 15.2 m (50 ft).

Tunnel size exceeds 15.2 m (50 ft) in diameter, based on Caltrans clearance requirements for a highway tunnel of this length. Assuming two 3.7-m-wide (12-ft) traffic lanes in each bore and 1.5 and 3 m (5 and 10 ft) wide shoulders plus two 1.2 m (4-ft) walkways, a finished tunnel diameter (ID) of approximately 14.5 m (47.5 ft) is required for the project (Kleinfelder 2009b). Clearances for the rail tunnel indicate that a finished tunnel diameter of 7.3 m (24 ft) is required for a single track tunnel. These preliminary clearances have been adopted for the feasibility study, and they will be revisited in more detailed design studies for the project.

## REGIONAL GEOLOGY

The Santa Ana Mountains are the northern portion of the crystalline bedrock Peninsular Ranges that extend south into Mexico. The northeast side of the Santa Ana Mountains forms a steep scarp that rises from the Elsinore and Temescal valleys along the active Elsinore fault zone. The western side of the Santa Ana Mountains is less abrupt and slopes down

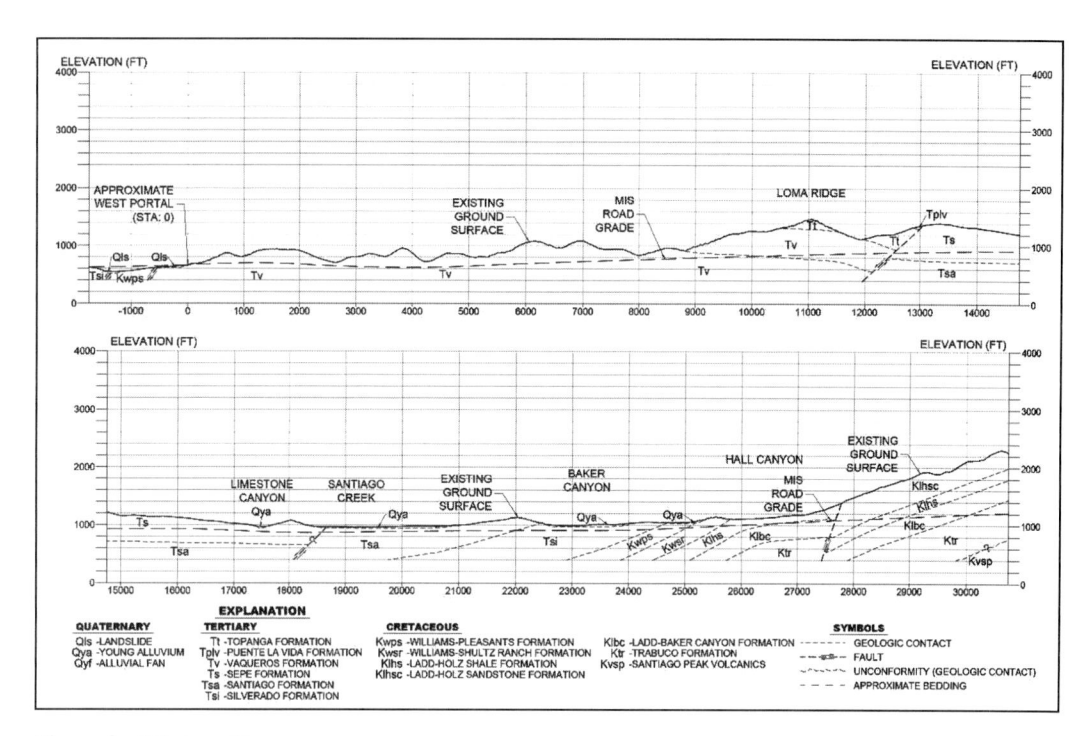

Figure 3a. West profile

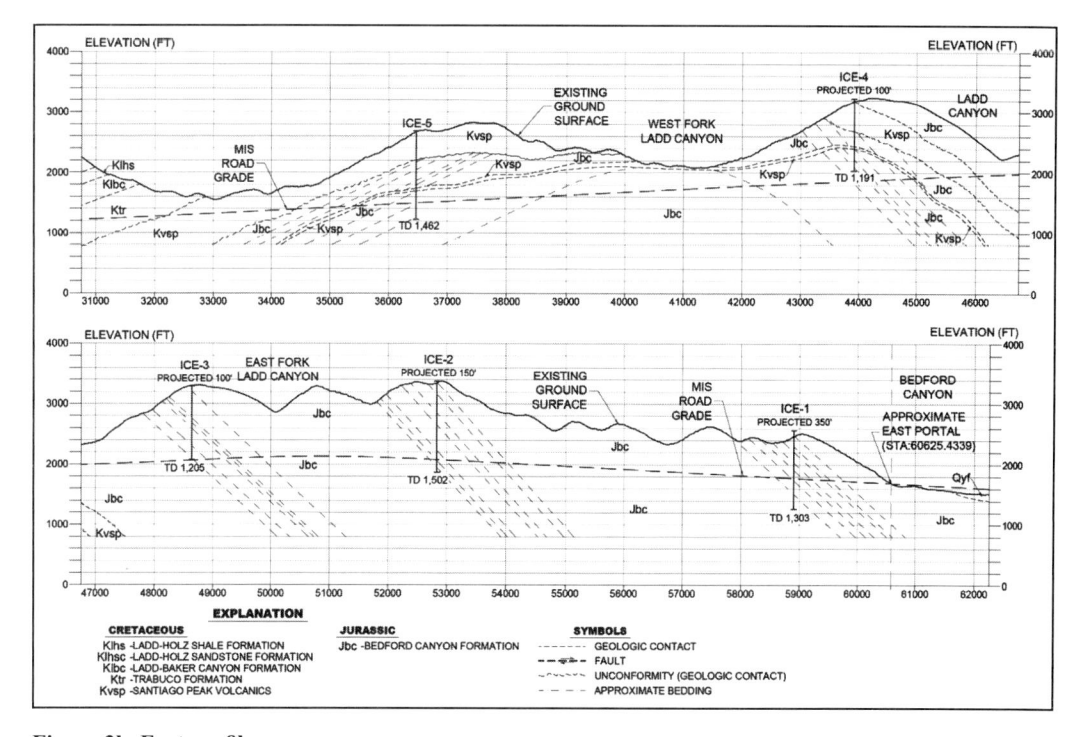

Figure 3b. East profile

to the Santa Ana Plain (see Figure 3a). The core of the Santa Ana Mountains consists of Mesozoic metasedimentary and igneous rocks that are flanked on the west by younger Late Cretaceous and Tertiary-aged clastic sedimentary rocks (see Figure 3b). The dip of strata in the eastern part of the Santa Ana Mountains is generally steep and to the east along the tunnel corridor (50 to 70 degrees) but is sometimes near vertical and locally overturned. The bedding dips of the western sedimentary strata are gentler, about 15 to 30 degrees, generally dipping to the west, although several anticlines and synclines have been mapped within the western strata (Schoellhamer et al. 1981).

The Santa Ana Mountains contain numerous faults and folds that generally trend northwest-southeast, parallel to the strike of the Tertiary sedimentary strata. The majority of the mapped faults demonstrate a down-to-the-west displacement (Schoellhamer et al. 1981), although the Elsinore fault, which is the dominant structural fault in the area, demonstrates secondary down-to-the-east displacement (i.e., thousands of feet). The predominant structural displacement along the Elsinore fault is right-lateral strike-slip displacement (i.e., tens of miles). The eastern tunnel portals have been strategically placed west of the Elsinore fault to avoid potential fault displacement across the tunnel.

## SEISMICITY

The Elsinore fault forms the eastern boundary of the Santa Ana Mountains. At its northern end, the Elsinore fault splays into two branches, the Chino fault and the Whittier fault. The maximum magnitude of an earthquake on the Elsinore fault is estimated to be M7.1 (Cao et al. 2003). There has only been one large earthquake on the Elsinore fault during historical times: the earthquake of 1910, an M6 near Temescal Valley, which produced no known surface rupture (SCEDC 2008). During the field investigations of this study, the M5.4 Chino Hills earthquake occurred on July 29, 2008, on a suspected "blind thrust fault" beneath the Puente Hills 25.7 km (16 miles) north of the site.

## FEASIBILITY-LEVEL FIELD INVESTIGATION RESULTS

The purpose of the feasibility-level field investigations was to evaluate geotechnical and hydrogeological conditions in the interior of the Santa Ana Mountains, where rock and groundwater conditions are least known. Therefore, the investigations focused on the eastern half of the ICE tunnel corridor, where high groundwater pressures and high overburden pressures are expected to define the most difficult design and construction challenges. The geologic setting and geotechnical condition of the sedimentary formations at the western end of the tunnel has been interpreted from the literature and other available geotechnical data.

The field investigations involved five deep coreholes (ICE-1, ICE-2, ICE-3, ICE-4 and ICE-5) completed at select sites along the ICE corridor (see Figure 1). The geotechnical data collected include continuous rock core (2,057 m [6,750 ft]); in situ geophysical logs; in situ hydraulic testing; and laboratory test data on rock samples.

In ICE-1, ICE-2, and ICE-3, the rock mass is composed of the Bedford Canyon Formation (see Figure 3b), which is a sedimentary flysch deposit consisting of alternating sandstone, argillite, pebbly mudstone, pebble conglomerate, mudstone, and shales that have undergone low-grade metamorphism followed by extensive shearing. In ICE-4 and ICE-5, the Bedford Canyon Formation has been locally intruded by the Santiago Peak Volcanics, a suite of volcanic and shallow plutonic igneous rocks that consist of basalt, andesite, diorite, and volcaniclastics that have also undergone low-grade metamorphism (see Figure 3b).

Data from vibrating-wire piezometers installed in the coreholes indicate that groundwater pressures at the tunnel invert range from 0.7 to 2.2 MPa (6.7 to 21.3 bar) after a year of equilibration. These pressures are less than expected, as a constant hydrostatic pressure gradient from the shallowest groundwater elevation to tunnel depth would result in pressures of 3.4 MPa (33.3 bar). Lower pressures are advantageous for tunneling and tunnel lining design; however, peizometer readings may vary seasonally, and long-term monitoring is required to confirm these initial findings.

The RQD values for 2,057 m (6,750 ft) of core do not exhibit a strong dependency upon lithology or depth (see Figure 4). Observed trends in the RQD do change considerably with corehole location, however. For example, at ICE-1, ICE-2 and ICE-3, approximately 90% of the RQD values are less than Fair (RQD <50), compared with 44% and 37% of the RQD values from ICE-4 and ICE-5, respectively, being less than Fair. Over 42% of the RQD values are Poor (RQD <25), irrespective of lithology or location.

Data from laboratory tests (Unconfined Compression, Brazilian Tensile tests) and field tests (point load index) of rock core indicate a wide range of intact rock strengths for both the Bedford Canyon Formation (Jbc) and the Santiago Peak Volcanics (Kvsp). The Bedford Canyon metasandstone ranges from moderately strong to extremely strong (25 to >250 MPa [3,500 to >35,000 psi]). The interbedded metasandstone and argillite ranges from weak to very strong (5 to >100 MPa; 750 to >15,000 psi). Strengths of the pebbly mudstones of the Bedford

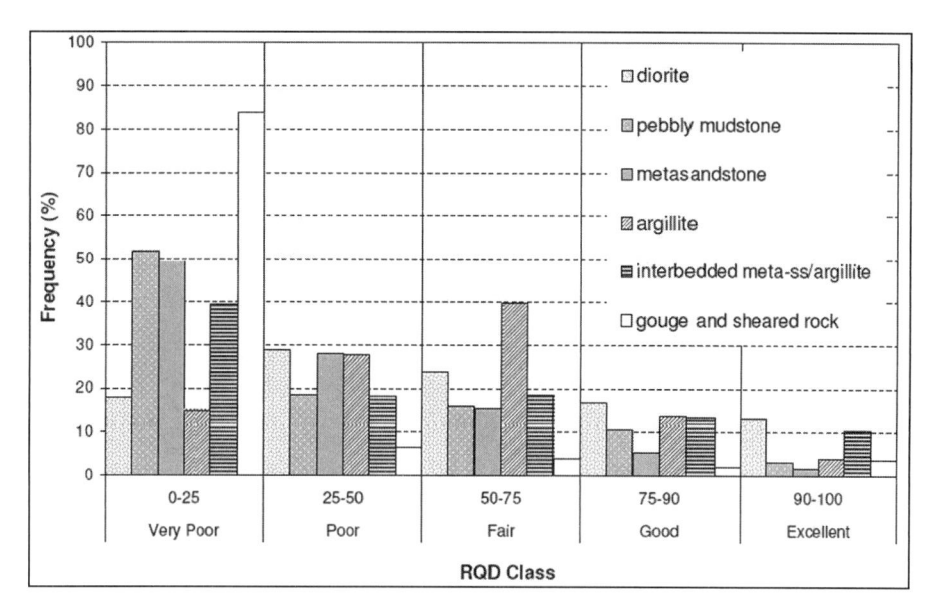

**Figure 4. RQD vs. rock lithology**

Canyon Formation ranges from very weak to moderately strong (1 to 50 MPa [150 to 7,500 psi]). The intact strength of the Santiago Peak Volcanics (diorite) also ranges widely from moderately strong to very strong (25 to 250 MPa [3,500 to 35,000 psi]). No testing of the sedimentary formations in the West Tunnel Segment was conducted under this study, but formations are estimated to range from extremely weak (e.g., shales) to moderately strong (shales, sandstone, and conglomerate) based on general lithology and strength-test results on rock cores from nearby projects (i.e., Bowerman Landfill and SR-241 Toll Road).

Rock mass classification systems indicate generally poor rock conditions for tunneling in the Bedford Canyon Formation and the Santiago Peak Volcanics, as suggested by RMR, Q, and GSI indicators. From 9,315 calculated RMR values, the rock mass character can be described as Poor to Fair rock, with more than 85% of the RMR values within the ranges defined by these two categories (21 < RMR < 60) (see Figure 5). From 3,056 calculated Q values, nearly 84% of the Q values occur in the Extremely Poor to Very Poor (0.004 < Q < 1) rock mass classes (see Figure 5). Figure 6 illustrates RMR versus Q values for the rock within the tunnel envelope only (15.2 m [50 ft] envelope). Nearly 83% of the GSI values for the entire rock core are less than Fair (GSI < 41) (see Figure 5).

The in situ hydraulic conductivity testing (i.e., packer testing) indicates that effective hydraulic conductivities at the ICE Tunnel envelope depths are on the order of 2.5E-05 cm/sec (Corehole ICE-1 between 198.7 and 228.8 m [652.1 and 750.6 ft] beneath ground surface [bgs]) to 2.9E-08 cm/sec (Corehole ICE-5 at 328.5 to 352.9 m [1,077.9 to 1,157.9 ft] bgs). The data suggest low groundwater inflows during tunneling in the Bedford Canyon Formation and the Santiago Peak Volcanics, although localized higher inflows should be expected.

**Groundwater Conditions**

Potentially adverse geochemistry of the Bedford Canyon Formation includes an abundance of sulfides, including pyrite, marcasite, and chalcopyrite yielding hydrogen sulfide gas noticeable during field exploration. Additionally, field testing of water samples from two mountain springs yielded pH readings as low 2.8 and 3.5; however, the majority of readings are in the neutral pH range.

**Geologic Profile**

The ICE Tunnels have been subdivided into a West and East Tunnel Segments based upon the anticipated geologic and groundwater conditions (see Figures 3a and 3b).

*West Tunnel Segment (Sta 000+00 to 322+00)*

The West Segment of the ICE Tunnels is anticipated to be located in sedimentary rocks that consist of shale, sandstone, and conglomerate that are estimated to be extremely weak to moderately strong and under moderate hydrostatic pressure 0 to 0.5 MPa (0 to 5 bar), with most below 0.3 MPa (3 bar). The geologic and hydrogeologic conditions along the West

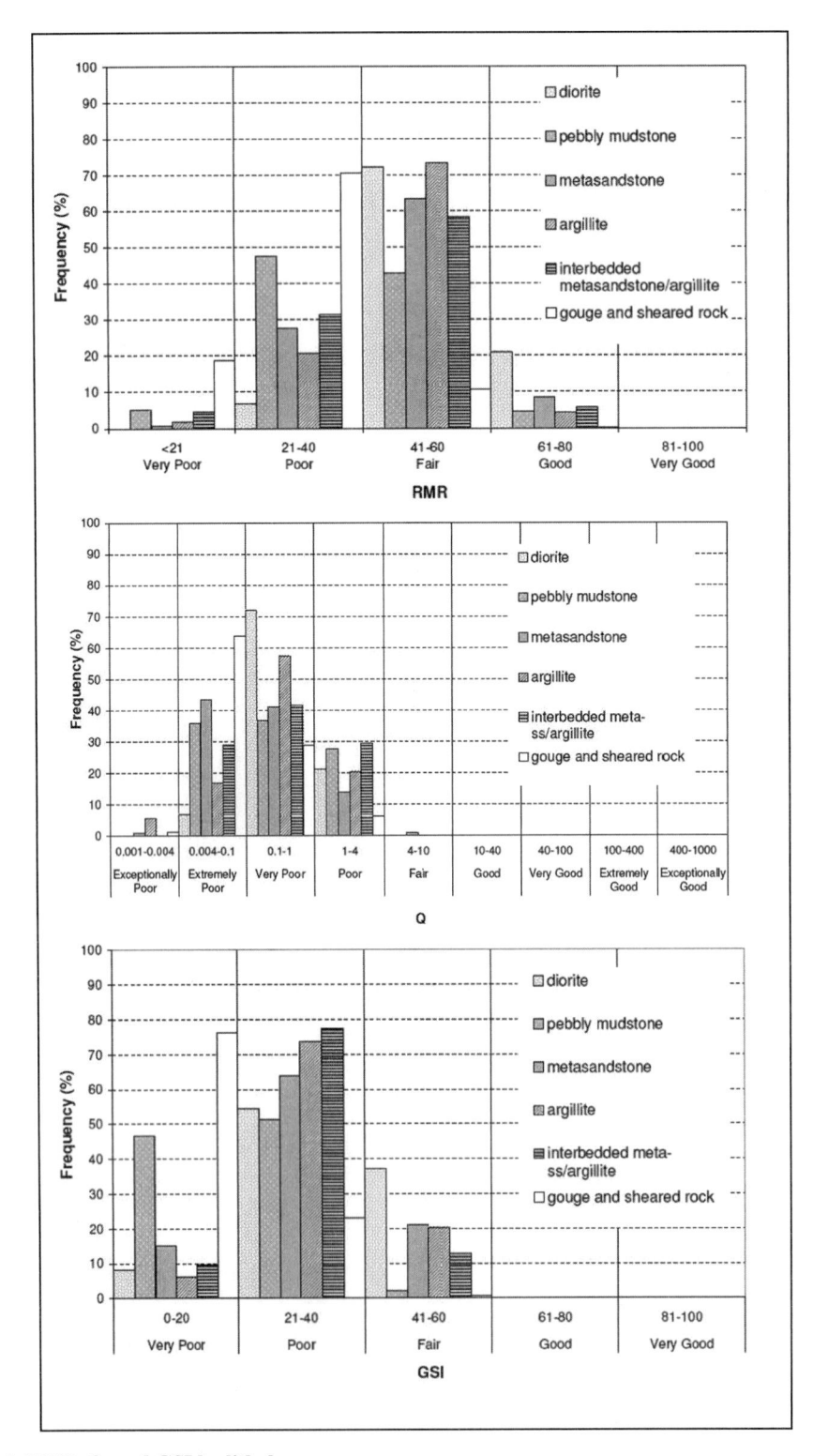

Figure 5. RMR, Q, and GSI by lithology

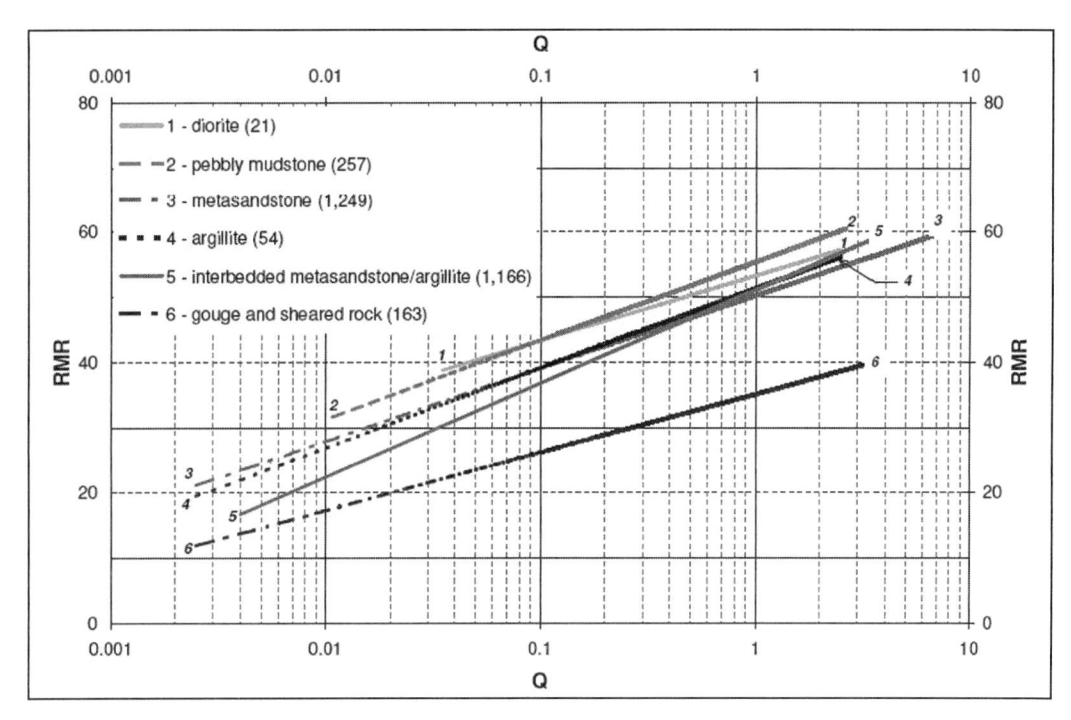

**Figure 6. RMR vs. Q**

Segment corridor are expected to be fairly uniform but with local shearing along bedding and at a few mapped fault zones. When tunneling through the West Segment, the ground is expected to be slow to fast raveling because many of these geologic formations are anticipated to be soft or weakly cemented. Some of the formations may exhibit soil-like behavior during tunneling, and flowing conditions could be encountered in isolated areas where the sedimentary formations are uncemented and the tunnel is below groundwater. The potential for groundwater inflows generally ranges from low to moderately low on the basis of the anticipated rock types.

According to published geologic maps (Schoellhamer et al. 1981) three fault traces have been identified. Squeezing ground conditions could be associated with these faults because the rock mass is weakened significantly. Also, groundwater inflows can be high in fault zones because of the increase in fracturing typically associated with fault activity.

*East Tunnel Segment (Sta 322+00 to 602+25)*

The East Segment runs through the core of the Santa Ana Mountains, and at tunnel depth is expected to encounter igneous and sedimentary to metasedimentary rocks under potential hydrostatic pressures up to 2.1 MPa (21 bar). Ground conditions are inherently variable in terms of lithology and composition. Some lithologies are extremely weak, while others have intact rock strengths that are extremely strong. Ground conditions are expected to range from massive to blocky and seamy to raveling. Potential squeezing conditions are expected in sheared and fault zones where the overburden is thick, and interbeds where the rock mass is predominantly argillite or pebbly mudstone.

**PROJECT CONSTRUCTION CHALLENGES**

The entire study area crosses a complex geologic zone with variable ground conditions ranging from sedimentary rock under relatively low groundwater pressures in the west to volcanic and metasedimentary rock under high groundwater pressures to the east. Potential design and construction challenges that are related to the geotechnical conditions include:

- Variable and difficult ground conditions
- High external water pressures
- Gassy ground
- Corrosive groundwater

Other significant design challenges include lining design and protection of groundwater resources.

**Variable and Difficult Ground Conditions**

Because of the potentially long tunnel lengths, a broad range of ground conditions may be encountered along the tunnel alignments. A particularly undesirable condition is a mixed face condition where the face is in both rock and soft ground or several materials of widely differing density and hardness. However, a given TBM will generally perform optimally in a relatively narrow range of ground conditions. If the rock has very high strength, the TBM may be designed for efficient mining of the strong rock, but will be less effective in mining poor quality rock. The opposite can also be true. For the ICE tunnels, the overall best performance may be achieved by tailoring the TBM to address the intensely fractured rock conditions that are currently estimated to comprise at least 53% of the alignment in the metamorphic terrain (approximately 9.7 km [6 miles]).

To overcome these challenging ground conditions and behaviors, the TBM should be designed with these considerations:

- The muck handling should be compatible with high water inflows and weak ground, and be able to efficiently collect the material under all conditions.
- The TBM should have exceptional thrust capacity to overcome high ground loads or muck-packing conditions.
- The cutterhead should be able to limit or control the flow of material through the head (both from the outside through the head or out of the head) and aid in maintaining face support under weak ground conditions.
- The TBM should be able to maintain line and grade in variable ground, including weak ground, and in curves.
- In squeezing ground conditions, special design provisions should be included, such as increasing the overcut, lubricating the TBM shield skin, reducing the TBM shield length, using a tapered shield, limiting TBM stops at critical stations, monitoring tunnel deformation and earth pressure, and having the ability to flush out material from the annulus back towards the cutterhead.

Technological advancements and additional practical experience with hybrid-style TBMs may eventually improve the performance of the TBM for the anticipated conditions of the ICE project.

**High External Water Pressures**

The maximum groundwater head is expected to be in excess of 2 MPa (20 bar). Excavating a tunnel under pressures of this magnitude presents health and safety hazards as well as challenges in designing a machine and initial lining to withstand the pressure. While tunneling under pressures of 0.4 MPa (4 bar) is routinely performed, pressures in excess of 0.5 MPa (5 bar) for this size of excavation will require state-of-the-art techniques. It should be noted that it would not be possible to operate a TBM with a closed, pressurized face under such high water pressure, as the machine could not be pushed forward against such pressure. The current concept is that the tunnels would be mined using a slurry TBM. Under this concept, in areas of lower groundwater pressure the heading area would be pressurized to control the potential water inflows and the primary lining would be erected and grouted in place within the rear of the TBM.

Recognizing that water inflow through some fractures, faults, and shear zones could potentially exceed the TBM's capacity to control water inflows, the TBMs will have to incorporate provisions to perform systematic probing (i.e., drilling ahead of the advancing TBM), and pre-excavation grouting ahead of the TBM. Systematic probing ahead of the tunnel face with probe holes will be required along the tunnel alignment where significant inflows may occur in order to mitigate the risk of encountering high flush flows that might exceed the water handling capacity of the TBM. Probing may also be used to detect areas of weak, unstable ground. When these conditions are found, TBM operation procedures may be modified or pretreatment may be warranted. Various methods can be employed to alter the operation of the TBM, such as closed or pressurized mode, to enhance its compatibility with unfavorable water or ground conditions. Pretreatment may include reducing the driving head through drainage of the groundwater to reduce impacts on the tunneling operations, or performing pre-excavation grouting ahead of the TBM. To sufficiently treat the problem areas ahead of the tunnel, the TBM will need to have a sufficient number of ports (openings) around the circumference of the machine to facilitate drilling grout holes ahead of the face. Alternative access ports through the TBM shield or concrete segments further back from the face facilitate treatment of the rock mass surrounding the TBM or immediately at or ahead of the tunnel face. Having an enhanced level of accessibility adds flexibility and options to the treatment of groundwater and ground behavior problems.

Interventions will need to be performed both routinely (planned interventions) and when the progress of the TBM is slower than expected (due to worn cutters). To access the cutterhead for maintenance while tunneling in closed mode, the interventions will need to be performed under free air, compressed air, or a mixed-gas environment.

Table 1. Results of lining analyses for roadway tunnels

| Tunnel Reach | Groundwater Head | Concrete Strength 762 mm (30 in.) Thick Segment | 914 mm (36 in.) Thick Segment |
|---|---|---|---|
| 0+00 to 323+60 | 0 to 49 m (0 to 160 ft) | Class I Lining 41.3 to 55.2 MPa (6,000 to 8,000 psi) | Class I Lining 41.3 to 55.2 MPa (6,000 to 8,000 psi) |
| 323+60 to 344+50 | 0 to 195 m (0 to 640 ft) | Class II Lining 55.2 to 96.5 MPa (8,000 to 14,000 psi) | |
| 344+50 to 540+00 | 110 to 244 m (360 to 800 ft) | Class III Lining 96.5 to 117.2 MPa (14,000 to 17,000 psi) | Class II Lining 55.2 to 96.5 MPa (8,000 to 14,000 psi) |
| 540+00 to 596+00 | 104 to 460 m (340 to 460 ft) | Class II Lining 55.2 to 96.5 MPa (8,000 to 14,000 psi) | Class I Lining 41.3 to 55.2 MPa (6,000 to 8,000 psi) |
| 596+00 to 606+25 | 0 m (0 ft) | Class I Lining 41.3 to 55.2 MPa (6,000 to 8,000 psi) | |

In locations where interventions need to occur under high pressures (e.g., blocky and seamy and crushed rock, sheared or faulted ground with high permeability), compressed air and/or mixed gas may be required; however, for interventions under high head in rock, the rock should be stable enough for the performance of interventions in free-air. Additionally, ground improvement methods could be employed to reduce or eliminate the need for compressed air by making the surrounding ground more stable. Depending on the pressures expected, the TBM may need to be fitted with a decompression chamber.

**Corrosive Groundwater**

During the groundwater monitoring program water samples were chemically tested, and two spring/stream monitoring sites in the middle fork of Ladd Canyon exhibited pH levels of less than 5.5. The results of an acid generation potential test performed during the geotechnical investigation (Kleinfelder 2009a) indicated intrinsic buffering capacity in a composited, sulfide-rich core sample. The buffering capacity is attributable to the neutralizing action of calcite in the rock mass. Therefore, it is anticipated that water collected during the tunnel excavation will have acidic properties along certain portions of the alignment. The corrosion potential of such water should be considered with regards to the design, operation, and maintenance of the TBM; health and safety of the crew; and the design of the ground support systems within the tunnel.

**Gassy Ground Conditions**

Within the ICE tunnel corridor, the lignitic shales of the Silverado Formation (Tsi) are a potential source of methane gas (see Figure 3a). Methane ($CH_4$) is the most common gas that occurs within gassy ground, and it is both highly flammable and an asphyxiant. The lower explosive limit (LEL) of methane is 5% by volume, while the upper explosive limit (UEL) is 15% by volume (Kissell 2006). According to the California Code of Regulations (CCR), a methane concentration greater than 0.25% by volume near any surface within the tunnel would warrant a gassy ground classification.

The potential for hydrogen sulfide gas is inferred from several intervals of Santiago Peak Volcanics and Bedford Canyon Formation that contained abundant sulfide metals (i.e., pyrite), and from the strong sulfurous odor noted in the exploration boreholes (Kleinfelder 2009a). Hydrogen sulfide ($H_2S$) has a strong odor similar to rotten eggs and is both corrosive and toxic. Hydrogen sulfide gas is also combustible, but at concentrations that are much higher than the 0.1% concentration that is toxic (Doyle 2001).

On the basis of the findings from the ICE Geotechnical Report (Kleinfelder 2008), the ICE tunnel corridor will likely be classified as "gassy" or "potentially gassy" ground according to California Occupational Safety and Health Administration (Cal/OSHA) criteria because of the presence of methane and hydrogen sulfide. In these conditions, a slurry TBM would be advantageous since it operates in a "closed circuit," minimizing workers' exposure to gas underground. Also, a slurry TBM provides more safety for the expected high pressures, especially in cohesionless ground.

## OTHER SIGNIFICANT DESIGN CONSIDERATIONS

**Lining Design**

The permanent tunnel lining will need to be watertight in order to avoid any long-term adverse impacts on groundwater levels in the Santa Ana Mountains. The proposed deep tunnels linking Riverside County to Orange County will be constructed using a TBM and will require the installation of a permanent watertight lining to control groundwater inflows into the tunnel. To achieve a watertight lining, a segmental precast concrete lining with gasketed joints will need to be used. Such linings were designed to withstand groundwater pressures up to almost 2.7 MPa (392 psi) for the Arrowhead Tunnels project near San Bernardino (Swartz et al. 2002). A two-pass lining would require an excavated diameter that exceeds the current state of the art design for TBMs.

Soil structure interaction methods were used to develop design concepts for the tunnel linings. The tunnel lining was analyzed at a section along the alignment corresponding to the highest water pressure (approximately 244 m [800 ft]), as well as squeezing ground loads. Lining analyses were performed for 762 mm (30 in.) and 914 mm (36 in.) segment thicknesses. Results of the tunnel lining analyses indicating anticipated lining thickness and concrete strength are summarized in Table 1.

Local concrete suppliers indicate that precast concrete with 56- to 90-day unconfined compressive strengths ranging between 89.6 and 96.5 MPa (13,000 and 14,000 psi) are readily achievable and that 117.2 MPa (17,000 psi) is feasible with materials available in southern California. Strengths up to about 137.9 MPa (20,000 psi) are a possibility, but higher quality aggregates from sources outside California may be required. In addition to high-quality aggregates, other key factors in producing high strength concrete include a low water-cement ratio, high-quality cement, additives such as silica fume and super plasticizers to improve workability, and a very high degree of quality control. Premium costs are associated with these high strength concretes. Factoring in the cost of materials, additives, and an increased level of quality control, the cost of 117.2 to 137.9 MPa (17,000 to 20,000 psi) concrete is approximately three times the cost of 41.4 MPa (6,000 psi) concrete, and 96.5 MPa (14,000 psi) concrete is approximately twice as expensive as 41.4 MPa (6,000 psi) concrete.

**Protection of Groundwater Resources**

The majority of the ICE tunnel alignments will be located beneath the Cleveland National Forest and private lands adjacent to the western forest boundary. In addition, for the most part the proposed tunnels will be constructed beneath the groundwater table. As a result, the tunnels have the potential to drain groundwater from the rock mass, and perhaps adversely affect water resources available to the overlying land. The extent to which drainage may occur during construction will be dependent on the hydraulic conductivity of the rock mass, and also the construction methods used.

At tunnel depth, the rock mass generally is of low to very low hydraulic conductivity, and groundwater flow through the rock mass is generally expected to occur at a very slow rate. This condition is favorable in terms of limiting the potential effects that tunnel construction could have on water resources in the vicinity of the project. Fault, shear, or fracture zones that are present in the rock mass could introduce relatively high water flows into the tunnels, causing significant hazards and/or difficulty during construction. Considering the water head present at tunnel depths, uncontrolled inflows could potentially be in the range of thousands of gallons per minute (gpm).

Pre-excavation grouting is expected to be necessary for tunnel excavations along significant portions of the ICE corridor, in particular along the east portion of the corridor within the Bedford Canyon Formation, and possibly the Santiago Peak Volcanics, where water head in excess of 2 MPa (20 bar) is anticipated at some locations. For the West Segment of the ICE Tunnels, the maximum groundwater head is expected to be substantially less than for the East Segment. Therefore, pre-excavation grouting may not be necessary along this portion of the tunnel alignments if pressurized face TBMs are used that are compatible with the ground and groundwater conditions. The objectives of pre-excavation grouting include minimizing the effects of tunnel excavation on the groundwater resources in the project area to satisfy any special permit requirements, and reducing groundwater inflows into the tunnel to improve ground conditions and facilitate tunnel excavation.

Under the assumption that a TBM will be used to excavate the tunnels, inflows may come from the heading area and through the completed tunnel lining. Therefore, some short-term water ingress will inevitably occur during construction, although this is unlikely to have significant effect on surface groundwater levels, and it is expected that recharge would occur relatively quickly after the tunnel face has passed any given location.

**PROJECT STATUS**

Constructing the proposed ICE tunnels appears to be geotechnically feasible on the basis of the information collected to date for the project. Many engineering and construction challenges would be encountered, such as variable and difficult ground conditions, high groundwater pressures, and gassy ground conditions. In addition, lining design and protection of groundwater resources are significant concerns. Special design considerations and state-of-the-art practice would be required to overcome these challenges with respect to TBM selection and tunnel construction. Upon presentation of the feasibility evaluation findings, the Orange and Riverside transportation agencies will decide if the project should move forward with additional engineering and environmental investigations. Funding mechanisms for future work are not available at the present; however, public and private partnerships may be explored in 2010.

**ACKNOWLEDGMENTS**

The authors would like to thank their respective employers for permission to publish this paper. Appreciation is also due to the design team: M. McKenna, J. Waggoner, and M. Torsiello of Jacobs Associates; A. Williams of Kleinfelder; and T. Rahimian of RMC, Inc.

**REFERENCES**

Cao, T., Bryant, W.A., Rowshandel, B., Branum, D., and Wills, C.J. 2003. *The Revised 2002 California Probabilistic Seismic Hazard Maps*. California Geological Survey.

Doyle, B.R. 2001. *Hazardous Gases Underground: Applications to Tunnel Engineering*. CRC Press, New York.

Kissell, F.N. 2006. *Handbook for Methane Control in Mining*. Department of Health and Human Services, Centers for Disease Control and Prevention, National Institute for Occupational Safety and Health, Pittsburgh Research Laboratory, Pittsburgh, PA.

Kleinfelder. 2008. *Geotechnical Data Report. Geotechnical Field Exploration and Testing Services in Support of Tunnel Evaluation Studies for Metropolitan Water District Central Pool Augmentation Project, Orange and Riverside Counties, California*.

Kleinfelder. 2009a. *Geotechnical Investigation in Support of a Feasibility Assessment for the Irvine Corona Expressway Tunnels*. Prepared for Riverside Orange Corridor Authority through Riverside County Transportation Commission, Document Control Number 2310-00041.

Kleinfelder, 2009b, *Feasibility Evaluation Report, for Irvine-Corona Expressway Tunnels*. Prepared for Riverside County Transportation Commission.

Schoellhamer, J.E. Vedder, J.G. Yerkes, R.F. and Kinney, D.M. 1981. Geology of the northern Santa Ana Mountains, California. United States Geological Survey, Professional Paper 420-D.

Southern California Earthquake Data Center (SCEDC). 2008. Alphabetical Fault Index, http://www.data.scec.org/fault_index/alphadex.html.

Swartz, S., Lum, H., McRae, M.,Curtis, D.J., and Shamma, J., 2002. *Structural Design and Testing of a Bolted and Gasketed Pre-Cast Concrete Segmental Lining for High External Hydrostatic Pressure*. North American Tunneling Conference.

# Soft Ground Tunneling on a Mexico City Wastewater Project

**Doug Harding**
The Robbins Company, Solon, Ohio

**Desiree Willis**
The Robbins Company, Kent, Washington

**ABSTRACT:** Ground settlement in Mexico City has caused the existing gravity feed wastewater system, built in 1975, to lose its slope. In addition to infiltration and corrosion, the system is severely undersized. To remedy the problems, the Mexico National Water Commission released a contract for a 7.8m ID × 62 km long pipeline known as the "Emisor Oriente" Wastewater Tunnel Project. To meet the demanding schedule, six Earth Pressure Balance (EPB) TBMs will be required. This paper will address the overall importance of the project to Mexico City as well as the unique design of the EPB TBMs needed for excavation of varying geology in pressures up to 10 bar. The current status of the project and any problems encountered to date will also be covered.

## PROJECT BACKGROUND

Mexico City, a metropolis of over 22 million people, is sinking at the rate of 10 cm per year. The world's second largest city was founded by the Aztecs in the Valley of Mexico, on what was once an island in the middle of Lake Texcoco. Spanish conquistadors later drained the lake bed using a system of canals, but the soft lake clays remained underneath the city's infrastructure. A combination of booming population and compression on the city's main sewer lines has necessitated the construction of one of Mexico's largest infrastructure projects—a 62 km long pipeline known as the Emisor Oriente, or Eastern Wastewater, tunnel.

Mexico City's wastewater system is almost exclusively served by the Emisor Central, a 68 km long line built in 1975. Over the past three and a half decades, ground settlement has caused a decrease in slope in the gravity sewer line and a reduction in capacity. Severe corrosion and nearly continuous groundwater infiltration have also made it impossible for the Emisor Central to be inspected and maintained between 1995 and 2008.

Once inspection was made possible, it was found that the overall system capacity had been reduced by 40% since 1975—from 280 m³/sec to just 165 m³/sec in 2008. Over the same time period the city's population more than doubled from 10 million to over 20 million inhabitants, increasing demand on the system.

## PROJECT DESCRIPTION AND LOCATION

Mexico's National Water Commission (CONAGUA) recommended immediate construction of a new line to help supplement the struggling system. The Emisor Oriente, or Eastern Wastewater Tunnel, will increase the city's current sewer capacity by 150 m³/sec once complete in September 2012. The line will carry wastewater from Mexico City to several water treatment plants currently under construction in the state of Hidalgo (see Figure 1).

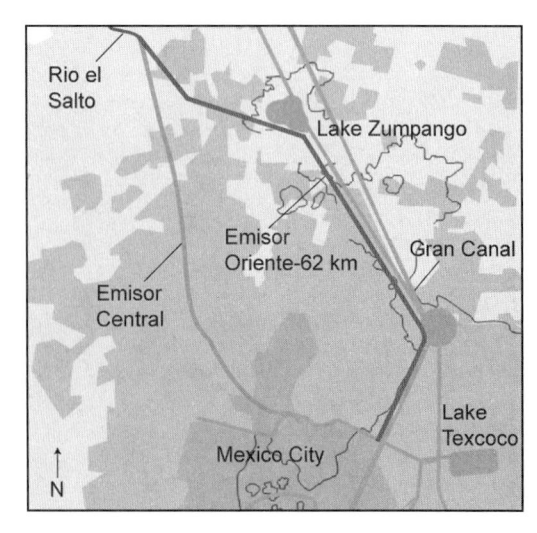

**Figure 1. Emisor Oriente pipeline layout**

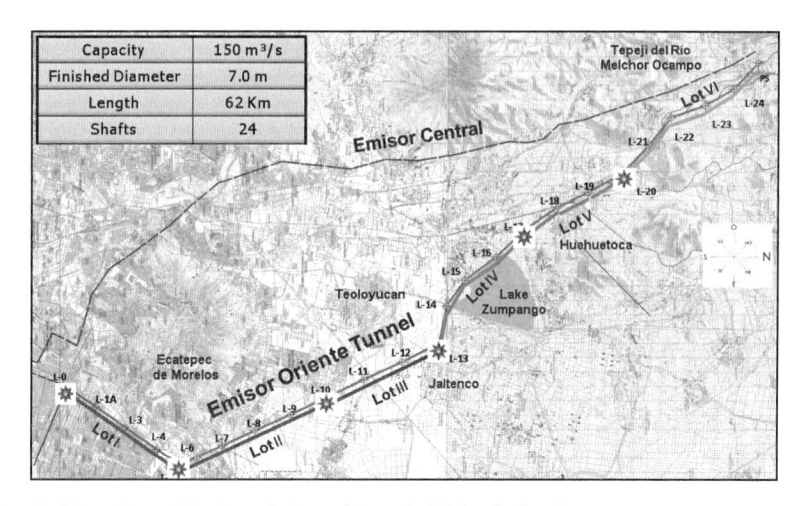

Figure 2. Detailed location of Emisor Oriente Tunnel. Main shafts shown as stars.

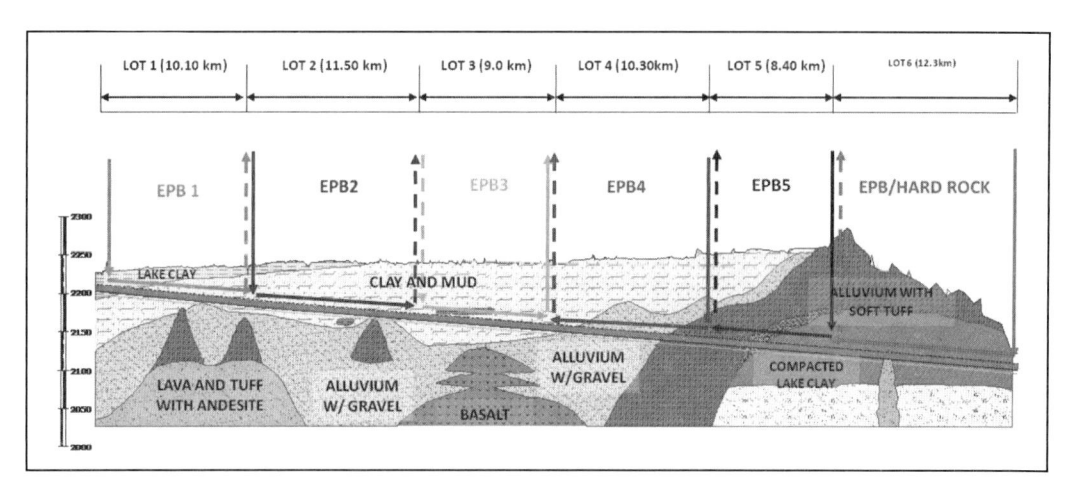

Figure 3. Geological profile of the Emisor Oriente Tunnel

The construction of the line was divided into six lots—lots 1, 2, and 6 under Mexican contractor Ingenieros Civiles Asociados (ICA) S.A. de C.V, and lots 3, 4, and 5 under Carso Infraestructura y Construcción, S.A. de C.V. All six lots will be bored with EPB machines— Robbins was awarded lots 3,4 and 5 which will be bored using three 8.93m diameter EPB TBMs (see Figure 2).

## GEOLOGICAL CONDITIONS

The Robbins machines are set to bore in alternating sections of compacted sand, gravel and clay with basalt rock, and are designed accordingly. The geology of the Valley of Mexico is also unique in that large boulders up to 600 mm in diameter are predicted throughout the drives. The particular set of geologic characteristics is found only in Mexico and in certain areas of Japan. The varied conditions consist of sections of lake clays, alluvium, and lava with tuff and andesite (see Figures 3–4).

## PROJECT APPROACH

The three Earth Pressure Balance Machines supplied by Robbins will be 8.93 m in diameter. All of the machines were optimally designed for mixed ground conditions (see Figure 5 and Table 1).

### Cutterhead Design

The machines are utilizing mixed ground, back-loading cutterheads for the variable geology. The design allows for a change in cutting tools between sections of soft ground and rock.

The seven-piece spoke-type cutterheads will utilize six outer segments plus a hexagonal-shaped center section to maximize the opening ratio of the face. The machines were designed with the largest possible opening ratios to ensure a smooth flow of muck into the cutterhead chamber.

Crews will switch out between carbide knife-edge bits and 17-inch, carbide disc cutters depending on the ground conditions. A number of small shafts, spaced every 3 km between the larger launch shafts, will be used to perform cutter inspection and changes. Specialized wear detection bits will lose pressure at specified wear points to notify crews a cutting tool change is needed. The knife edge bits are arranged at several different heights to allow for effective excavation at various levels of wear.

The design also allows for bearing and seal removal from either the front or back of the cutterhead. Twenty-five injection ports spaced around the periphery of the machine will be used for injection of various additives depending on ground conditions, and for probe drilling (see Figure 6).

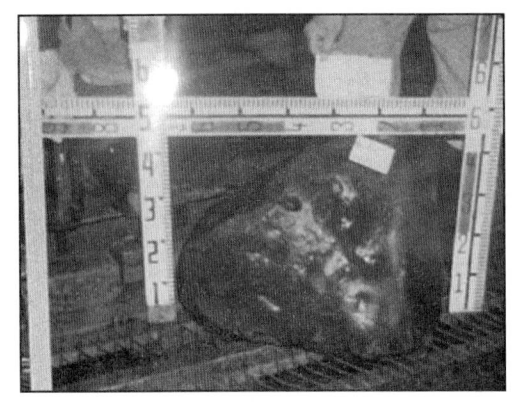

**Figure 4. Boulders taken from similar ground at the Sapporo Metro Project, Japan**

## Screw Conveyor and Muck Removal

Each machine will be fitted with a ribbon-type screw conveyor 900 mm in diameter. The screw conveyors allow boulders up to two-thirds the screw diameter (up to 600 mm) to travel up the shaft, where they are disposed of through a boulder collecting gate. Each of the three machines may encounter pressures of up to 10 bar, necessitating a two-screw setup with a ribbon screw and shaft-type screw in order to smoothly regulate pressure (see Figure 7).

Muck will be deposited from the screw to a rubber belt conveyor mounted on the trailing gear, which transfers to a side-mounted continuous conveyor. The continuous conveyor carries the muck to a 150 m long vertical belt conveyor located at the

**Table 1. Specifications for Emisor Oriente EPB TBMs**

| 8.93 m Diameter EPB TBMs | |
|---|---|
| Excavation Diameter Cutterhead for Soil Cutterhead for Rock | 8,910 mm 8,930 mm |
| Main Cutting Tools | Special knife-edge bit (soil) Single and Double-row Disc Cutters (rock) |
| Cutterhead Drive | Electric, variable speed |
| Cutterhead Power | 1,900 kW |
| Machine Thrust | 84,000 kN |
| Stroke | 2,300 mm |
| Max Torque | 17,900 kNm |
| Screw Conveyor #1 | Ribbon type, 900 mm diameter |
| Screw Conveyor #2 | Shaft type, 900 mm diameter |
| Articulation | Active |
| Segments | Reinforced concrete, 400 mm thick |
| Back-filling System | Two-Liquid Type |

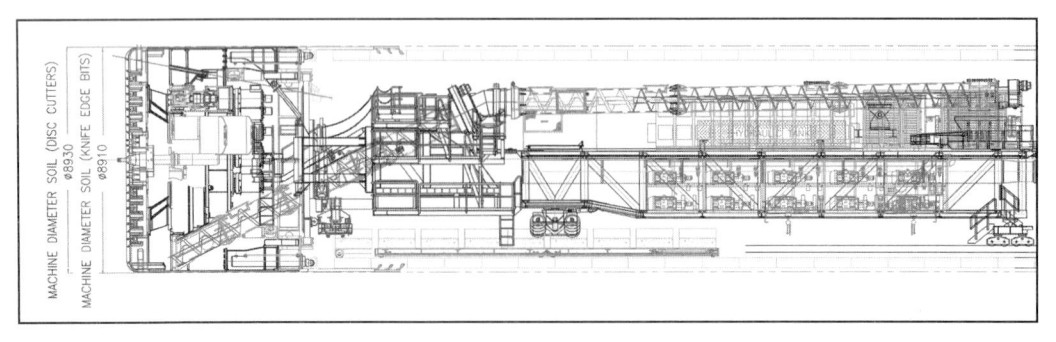

**Figure 5. EPB TBM general assembly**

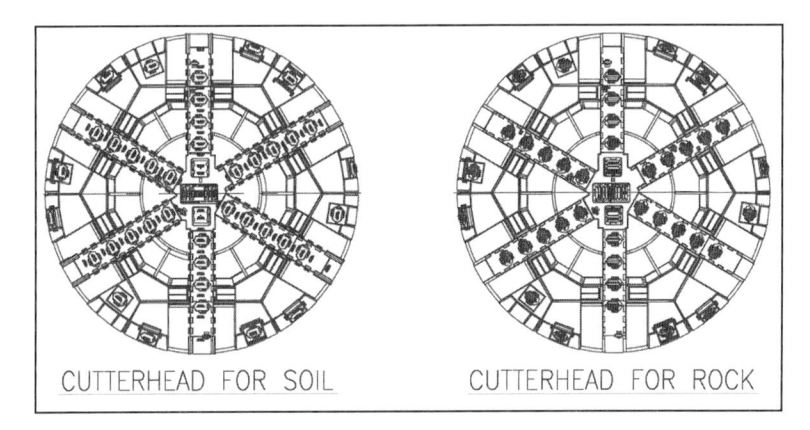

**Figure 6. Cutterhead design with interchangeable cutting tools for soil and hard rock**

**Figure 7. Example ribbon-type screw conveyor**

**Figure 8. Partial cutterhead assembly**

launch shaft. Once at the surface, a radial stacker will deposit muck in a kidney-shaped pile for temporary storage. This system will be used on all three lots.

The three continuous conveyor systems, also provided by The Robbins Company, consist of 762 mm wide fabric belt at a 3,200 m length and 900 MTPH capacity. Approximately 22% of the conveyor systems will be traveling through curves, with a minimum 700 m curve radius. To better handle curves, the systems will utilize patented self-adjusting curve idlers. The idlers help by pivoting to accommodate changing load tensions around curves. The pivoting action is also favorable because it does not unnecessarily alter the carrying capacity of the conveyor or the belt tension.

## Articulation

For accurate tunneling through curves, each machine will feature active articulation. Active articulation engages articulation cylinders between the front and rear shields to steer the machine independently of the thrust cylinders. The process allows the thrust cylinders to react evenly against all sides of the segment ring during a TBM stroke in a curve. Typical configurations, which use flat joints to articulate the shield, are capable of making 2 to 3 degree curve adjustments over the length of the segment or stroke.

Another reason active articulation was chose for this project was the risk of segment deformation, or racking. A common cause of project delays, deformation toccurs most commonly when the passive articulation system is used in curves. Passive articulation does not utilize articulation cylinders independent of the machine's thrust cylinders, so the TBM reacts against sides of the segments unevenly in curves.

## Segments and Back-filling System

The machines will line the tunnels with reinforced concrete segments 400 mm thick, in a 7+1

**Figure 9. Forward shield assembly**

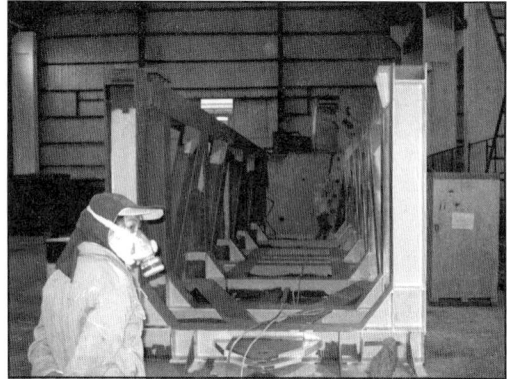

**Figure 10. Back-up system assembly**

arrangement. Each segment is 1,500 mm in length and weighs approximately 60 kN. The finished tunnel diameter will be 7,800 mm.

To back-fill any voids behind the segments and minimize ground settlement, the machines are utilizing two-liquid back-filling. Two-component backfill, made up of cement plus an accelerant, is used to harden ground rapidly. Grout is injected and the two separate components are mixed where the completed rings exit the tail shield. The mixture fills the annulus between the completed segment rings and surrounding soil. Volume and pressure of the backfill grout injection are constantly monitored and controlled to minimize surface subsidence, a concern in tunnels with low cover and in urban areas. After each injection, water is forced through the pipes to prevent clogging.

## PROJECT SPECIFICATIONS

### Current Status of Machines

The machines were designed by Robbins with manufacturing done in various Robbins manufacturing plants worldwide. Various components were sub-assembled and shipped for full workshop assembly in a Robbins workshop located in Corpus Christi, Texas, USA. The Corpus Christi workshop in Texas was selected due to its close proximity to the jobsite. Large assemblies will be loaded onto a barge and shipped directly to a Mexican port for ease of transportation to the jobsite. As of November 2009, partial cutterhead assembly for the Lot 4 machine had been completed, as well as assembly of the Lot 3 forward shield and back-up system (see Figures 8–10).

### Shaft Construction

The three 16 m diameter launch shafts are 80 m, 100 m, and 150 m deep. Machine launch and breakthrough will be as follows:

**Figure 11. Slurry wall construction, Shaft 20, November 2009**

- Lot 3 EPB TBM starting from shaft 13 and boring upwards to shaft 10.
- Lot 4 EPB TBM starting from shaft 17 and boring upwards to shaft 13.
- Lot 5 EPB TBM starting from shaft 20 and boring upwards to shaft 17.

Shaft construction is currently underway. At the surface, shaft construction begins by building slurry walls. Several of the slurry walls for the excavation support, including shaft 20, are being excavated using a hydromill (see Figures 11–12). Once the slurry walls are constructed, most of the shafts are being excavated conventionally using a backhoe. Stable ground below the slurry walls is supported using wire mesh and shotcrete. Material is removed from the bottom of a shaft using a crawler crane (see Figure 13). To excavate each shaft dewatering is also needed. Outside each shaft are four installed pumping stations that operate during the course of excavation. Volumes of water being pumped are as follows:

- Shaft 13—5 liters per second
- Shaft 17—2 to 3 liters per second
- Shaft 20—2 to 3 liters per second

**Figure 12. Hydro excavator machine, shaft 20**

## CURRENT SCHEDULE

After assembly is completed, the machines will be shipped to the jobsites in Spring 2010 where they will be lowered into separate deep shafts using mobile boom cranes. Each machine will start from a different 16 m diameter shaft at either 80 m, 100 m, or 150 m deep. Current schedule milestones are documented in Table 2.

Partially assembled machine components weighing as much as 120 metric tons will be lowered down the shafts to reduce assembly time underground. The 14.5 m long ribbon screw must be altered since its length exceeds available space in the shaft. The screw will be lowered in two halves and welded together through an inspection hatch on the casing. Once the ribbon screw has been installed the machine will be pushed a minimum of 30 m into a pre-excavated starting chamber. This arrangement will make space for installation of the machine's rear shield, bridge section, and the shaft-type screw conveyor. A shortened back-up system, including electrical cabinets, transformers, and the hydraulic system will also be assembled. Upon completion the TBM will bore forward approximately 50 m with a temporary mucking system until the remaining 65 m of back-up equipment can be installed.

**Figure 13. Shaft 13, November 2009**

**Table 2. Current scheduling milestones for Emisor Oriente project**

|  | Lot 3 | Lot 4 | Lot 5 |
|---|---|---|---|
| Forward shield assembly | Nov-09 | Dec-09 | Jan-10 |
| Rear shield assembly | Dec-09 | Jan-10 | Feb-10 |
| Main drive assembly | Dec-09 | Dec-09 | Jan-10 |
| Back-up system assembly | Dec-09 | Dec-09 | Jan-10 |
| Conveyor system | Dec-09 | Jan-10 | Feb-10 |
| Cutterhead assembly | Jan-10 | Dec-09 | Feb-10 |
| Final testing | Jan-10 | Feb-10 | Mar-10 |
| Delivery to jobsite | Feb-10 | Mar-10 | Mar-10 |

## REFERENCES

Willis, D. (2009). Robbins Trio dig deep in Mexico. *World Tunnelling,* April, 12.

# Small Diameter Tunneling: Ems-Dollard Crossing

## Klaus Rieker
Wayss and Freytag Ingenieurbau AG, Frankfurt, Germany

ABSTRACT: To adapt gas transportation in Europe to meet future requirements and circumstances, a tunnel is currently being built under the River Ems between Germany and the Netherlands. On its completion, the 4,016 m long tunnel with an outside diameter of 3.60 m and an inside diameter of 3.00 m will accommodate a 48-inch gas pipeline.

Due to the geological conditions a slurry shield has been chosen to excavate the tunnel. Concrete segments that are reinforced with steel fibers instead of conventional steel bars are being used to line the tunnel.

An approx. 4 kilometer long tunnel is currently being built to the west of the city of Emden, in an area called "Rysumer Nacken." This tunnel will cross the Ems-Dollard estuary in a southerly direction and surface again in the Netherlands. A 48-inch gas pipeline will be installed in the tunnel, which will subsequently be backfilled. The following article describes the constructional and logistical measures necessary for the implementation of the project.

In summer 2008, the joint venture "BAM Combinatie Eemstunnel," consisting of Wayss & Freytag Ingenieurbau AG, Frankfurt am Main, Tunnelling Division and its Dutch affiliated company BAM, was awarded the contract for the construction of the four kilometer long section of the gas pipeline, which will have a total length of 500 km and will run across the Netherlands. The gas pipeline is needed to adapt European gas supply to circumstances and future requirements. The contract comprises the mechanized excavation of the segment-lined tunnel, the installation of the 48-inch gas pipe and the backfilling of the tunnel.

The length of the tunnel is exactly 4,016 m. It starts in Germany in the "Rysumer Nacken," an area in the city of Emden, crosses under the Ems-Dollard estuary and ends at the Dutch town of Borgsweer (see Figure 1). In the area of the tunnel crossing the shipping channel of the Ems is 12 m deep. On account of this constraint and the geological circumstances, the tunnel axis lies at a depth of 23.5 m below sea level in the area of the shipping channel. The tunnel has its maximum gradient of 5.0% right at the start of the drive.

## GEOLOGY

The tunnel alignment lies in the Ems estuary zone in various friable, cohesive and organic soft ground deposits originating in quaternary, glacial and postglacial periods. The soil cover above the tunnel roof varies between 4 m and 6 m below the mainland and between 6 m and 17 m below the River Ems. Elongated erosion channels, filled with sand, silt and clay, were formed during the Elster ice age. During the subsequent Saale ice age, the deposits from the Elster ice age partially eroded, with first and foremost sands and occasional cohesive boulder soils being deposited. Subsequently, part of the older Pleistocene deposits, in turn, eroded during the Weichsel ice age and deep meltwater drainage channels were formed in them. In the course of the Weichsel ice age and as a result of the general rise in sea and groundwater levels these meltwater drainage channels were filled with meltwater sands of various compositions and local layers of gravel.

During the post-ice age the coastline moved landwards as a result of the ongoing rise in sea level, and watt sands with, in some cases, clay layers settled on top of the older Pleistocene deposits. Organic soft strata, mainly derived from bog and marsh formations, with occasional sand layers were deposited on top. Finally, artificial elevations, shaped as areas or lines, were created in the course of human settlement of the Ems estuary region, e.g., land reclamation areas and dykes, which extend to above the high-water mark of the Ems River.

Mainly confined groundwater is found in the Holocene and Weichselian sands of the Ems marsh due to the overburden of soils with low water conductivity. It is hydraulically connected to the water table of the Ems, which is subject to tidal fluctuations. Furthermore, confined groundwater and/or stratum water is found in the Elster ice age sands beneath the Lauenburg Clay that retains the groundwater. The maximum water pressure in the tunnel invert is approx. 3.0 bar at high tide.

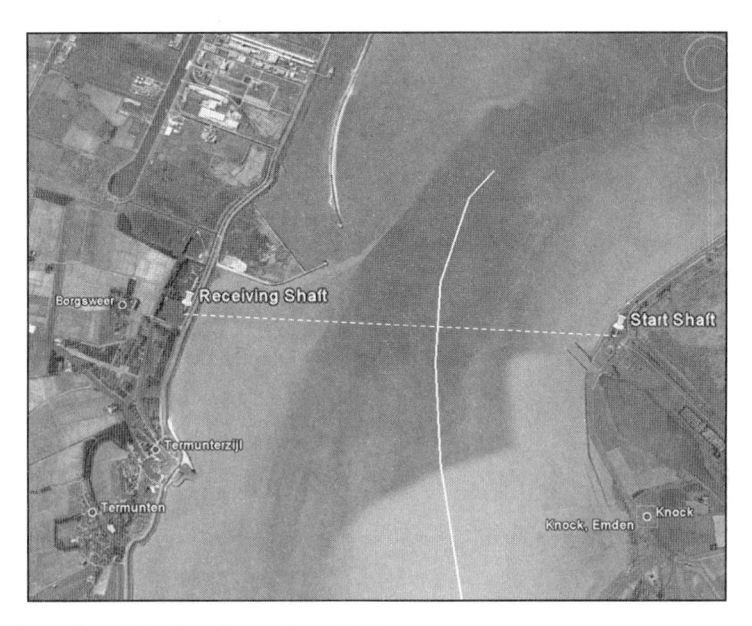

**Figure 1. Overview of start shaft and receiving shaft**

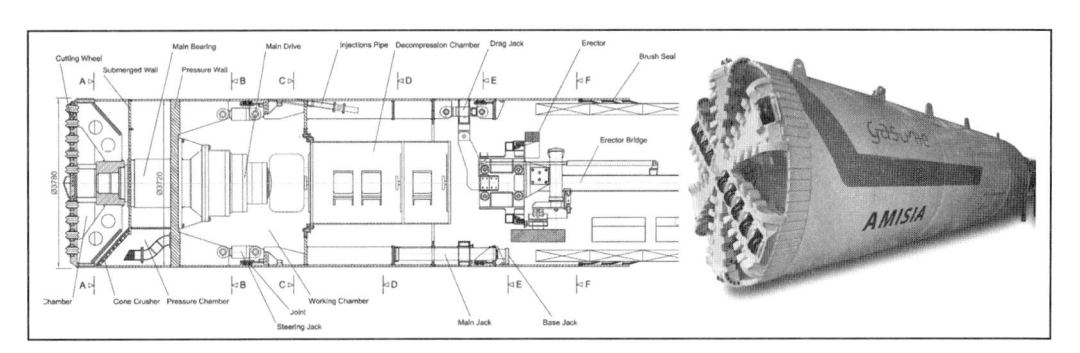

**Figure 2. Tunnel boring machine**

## TUNNEL BORING MACHINE

The BCE joint venture decided to use a slurry shield to suit the geological conditions mentioned above. In co-operation with Smet Boring, an existing tunnel boring machine, which had already be used on several projects, was refurbished and modified to match the geological conditions. Above all, additional medium and high pressure flushing nozzles were installed in the excavation chamber and the cutting wheel to be able to remove the Lauenburg Clay where it clogs the machine.

The TBM has a length of 12.68 m; the outer diameter of the cutting wheel is 3.78 m (see Figure 2). The back-up trailers, most of which were newly manufactured, have an overall length of approx. 80 m. The segments are supplied by train to trailer 5 of the backup system where they are lifted from the wagon by means of rapid unloading equipment. From there the individual segments are transported to trailer 3 by means of a monorail conveyor and deposited on the segment shuttle. The segment shuttle passes beneath trailers 1 to 3, carrying the segments to the segment feeder, which turns them by 90 degrees and passes them to the erector. The erector picks the individual segments up and assembles them to form a ring under the protection of the shield skin. Subsequently, a total of 11 driving jacks support themselves on the segment ring. On the one hand, the driving jacks serve to force the tunnel boring machine into the soil, on the other hand, they are used to hold the individual segments in place during the ring building operations.

Due to the geological conditions described above, the TBM is equipped for slurry mode

**Figure 3. Site installations**

operation. For effective working face support a pressurized bentonite suspension is used in the excavation chamber. The supporting pressure is controlled by a compressed air bubble in the working chamber. Excavation chamber and working chamber are linked to each other according to the principle of communicating vessels.

Together with the excavated material, the suspension is pumped to the start shaft at the German end of the tunnel, where the separation plant is situated. The excavated material is separated from the suspension in this separation plant that was specifically developed by Wayss & Freytag for the expected soil. The suspension is treated and used again, whereas the excavated material is temporarily stored at a dump on site before it is transported by truck to its final storage place in the vicinity of the construction site.

## CONSTRUCTION SITE INSTALLATIONS

Work on setting up the site installations at the German end of the tunnel started in September 2008 (see Figure 3). The area directly at the River Ems, provided by the Client for this purpose, is large enough to accommodate all the components required for the tunneling work.

The start shaft with the dimensions 8 m × 14 m and a maximum depth of 8 m consists of sheet pile walls and an underwater concrete bottom slab. It was already completed in December 2008. Then the shield cradle was concreted on the bottom slab of the shaft and early in 2009 the tunnel boring machine was assembled on the shield cradle. To the north the start shaft is connected to the start trench, which is approx. 65 m long and approx. 5 m wide. The start trench, which has a maximum depth of 6 m, also consists of sheet pile walls and an underwater concrete bottom slab. The backup trailers were assembled and attached to the TBM in this trench, which also accommodates the station for the tunnel railway, which is required for the tunneling work. Here the trains are loaded with segments, mortar, pipes and rails. The bottom slabs of the start shaft and the start trench have a gradient of 5%. With the same gradient, the tunnel goes down under the Ems River.

The mechanical installations required for the tunneling work, such as bentonite mixing plant, separation plant, centrifuge, etc. were assembled on site, to the west of the start trench, and tested in accordance with the time schedule. The segment store, the site office and the crew accommodations are located to the east of the start excavation.

A sealing block, consisting of a mixture of bentonite and cement, was installed at the southern end of the start shaft. Site erection work at the Dutch end of the tunnel started in October 2008. Here, too, a sealing block made of a cement-bentonite mixture was constructed, followed by a receiving shaft with sheet pile walls and an underwater concrete bottom slab, onto which a lean concrete block was built. On its arrival, the TBM will cut its shield cradle in this block on its own. After completion of the concrete work and as soon as the sealing block had hardened, the sheet pile wall was cut open for the passage of the TBM. Subsequently, the shaft was filled with water.

One special feature should be mentioned: The sheet pile walls of the receiving shaft are approx. 6 m higher than the ground surface to prevent the

**Figure 4. Double tracks in the start shaft**

surrounding terrain from being flooded in the case of water inflow into the tunnel or shaft, as it lies at 0.0 m or 2.0 m below sea level.

## LOGISTICS

In view of the small tunnel diameter the supply of the TBM with segments and mortar and the transport of the excavated material from the tunnel is a special challenge.

Mühlhäuser wagons, specially manufactured for this tunnel, and Schöma locomotives were purchased. Material supply to the TBM was done with two trains until approx. 2000 m were driven (see Figure 4). One train was loaded in the shaft while the second train was on its way to the TBM where it was unloaded. As soon as an empty train left the tunnel, the loaded train pulled out of the shaft into the tunnel. After 2000 m, a passing point for the trains was installed in the tunnel. This shortened the transport times again, as it was then possible for two trains to pass each other in the tunnel. With the installation of this passing point, the third train was put into operation. Now one train is unloaded at the TBM, another fully loaded one waits at the passing point and the third train is loaded in the shaft.

At the start of the tunneling work the power output of each locomotive was 75 kW. After approx. 2.5 km of the drive were completed, the output of the locomotives had to be reduced to 60 kW. The small diameter of the ventilation pipes caused such pressure and friction losses over this length that it was not possible to blow enough fresh air into the tunnel for a higher diesel engine power output.

## SEGMENTS

The tunnel is lined with segments. After a thorough search for the most efficient solution for the cross-section and the reinforcement a decision was made in favor of an inner tunnel diameter of 3.0 m.

A segment ring consists of 6 segments. The segments are 1.2 m wide and 0.25 m thick (see Figure 5). Universal rings, each with a taper of 30 mm, are produced to enable space curves to be driven and compensate for driving tolerances. As regards reinforcement, it was decided to deviate from the conventional type of reinforcement and to use steel fibers instead of steel bars. The target set for the accuracy of the segments was that prescribed in the relevant German regulations "ZTV-ING, Teil 5 Tunnelbau, Abschnitt 3 Maschinelle Schildvortriebsverfahren" [Additional Technical Terms of Contract and Guidelines for Civil Engineering, Part 5 Tunneling, Section 3 Mechanized Shield Tunneling].

The segments are produced by Rekers. Segment production at their factory in Spelle, Germany, started in September 2008. Equipped with 6 sets of moulds, their circulation plant can satisfy the site's demand for segments. The segments are transported by rail from the factory in Spelle to the intermediate store in Emden, which has a storage capacity of up to 150 rings. From there the rings are taken to the site by truck. The site has a maximum storage capacity of 120 rings. At the beginning of the tunneling work approx. 800 rings had been pre-produced at the Spelle factory. Due to the high tunnel advance rate segment production had to be increased from the planned 90 rings per week to 108 rings per week from mid-2009 on.

## SAFETY AT WORK

From the outset, the companies building the tunnel and the Client have placed especially high demands on safety at work so that not only the recommendations of the relevant code of practice for the planning and implementation of a health and safety concept on underground construction sites were fulfilled, but also further measures were taken in coordination with the local emergency services and the fire brigades. When the equipment of the TBM was chosen, special attention was paid to noncombustible or flame-resistant equipment and plastics. All locomotives, transformers, the TBM and all back-up trailers were provided with a Fogmaker 80 bar high-pressure fire-extinguishing system. The air lock of the TBM is so equipped that it can be used as a rescue chamber in case of fire.

A mobile firefighting support unit, specially designed and built according to the joint venture's specifications, was made available to the fire brigade. With this mobile firefighting support unit, which is mounted on a flat bed wagon of the tunnel train, it will be possible for the firemen to get as close to the seat of fire as possible and extinguish it under the protection of a firefighting water curtain.

At the same time, all employees were trained to be fire prevention assistants and first-aiders at the beginning of the site operations, as all parties involved are aware that in the case of such a narrow tunnel there is not much time to wait for help from the outside, but people on site have to start taking their own emergency response measures at once. For this reason the construction site has its own emergency plan, which not only covers the preventive measures to be taken, but also describes every possible emergency and provides a procedure for everyone on the site. Then the procedures were briefly summarized on emergency cards. Every person was given a card outlining the procedure to be followed by him/her in accordance with his/her function. The emergency cards are only meant to be a small memory aid for an emergency. The procedure in an emergency was therefore rehearsed in a number of emergency drills, in some of which the emergency services took part as well.

It was repeatedly pointed out to all crew members—and continues to be—among others during weekly toolbox meetings—that apart from entrepreneurial success safety at work has top priority and that the site management staff is to be notified of every "unsafe situation" so that a lesson can be learned from it and measures taken to ensure that the situation does not result in an accident.

## TUNNEL DRIVING

After initial problems and delays in the site set-up and the assembly of the TBM, the driving work started on 1 April 2009 (see Figure 6). From the beginning, it was possible to maintain or slightly exceed the planned advance rate of 12 m or 10 rings per day on average.

**Figure 5. Precast concrete segments**

The first inspection of the cutting tools was scheduled to take place after approx. 300 m, but due to the loose, sandy soil, which was partially interstratified with peat lenses, it was not possible to develop a compressed air bubble in the excavation chamber. When the compressed air supply line was opened, the air escaped into the soil and a blowout occurred. Further on, in the Lauenburg Clay, it was then possible to develop an air bubble in the excavation chamber, but despite the installed flushing nozzles the cutting wheel was clogged to such an extent that is was impossible to inspect the cutting tools.

Since the driving forces had varied only slightly on the first 1,000 m and no wear was visible on the pumps, valves and pipes of the discharge pipeline, it was assumed that the wear on the cutting tools would not be too bad.

Only after approx. one third of the drive was completed, at 1,300 m, was it possible to inspect the cutting tools. The assumption that there would be hardly any wear on the tools was fully confirmed; large areas of the cutting wheel spokes were still covered with paint.

Due to the good geology—fine-grained sands with little or no fines—it was possible to increase

**Figure 6. Inside view of tunnel**

the advance rate continuously after the successful compressed air intervention. Peak advance rates of 33 rings or 40 m per day were achieved. An average advance rate of 20 rings per day was maintained over a period of several months.

As the second inspection of the cutting wheel at 2,700 m revealed a similar situation as the first intervention, it was decided that the next cutting tool inspection would not be carried out until the sealing block in front of the receiving shaft was reached.

Owing to the crews' and the tunnel boring machine's good performance it was possible to make up for the initial delay and even get ahead of the contractual time schedule. Breakthrough into the receiving shaft on the Dutch side was achieved on 23 November 2009, i.e., 4 weeks ahead of time. Still within 2009, the machine was completely dismantled and sent off site back to the manufacturer. Clearing of the tunnel started and by now all utility pipes, the ventilation ducting and the HV cable have been removed. Installation of wooden deflection beams required for the gas pipe pull-in has begun. Within January 2010, installation of a gravel layer in the invert and at the same time the dismantling of the lights and the removal of the rails will take place.

## GAS PIPELINE WORKS

In mid 2009 assembly of the gas pipeline began on site. All in all, 4 strings of 1,000 m each were welded together on a piece of land next to the start shaft. Meanwhile the welds are being epoxy-coated to get the strings ready for insertion into the tunnel. Pipe pulling is currently scheduled for end of February 2010. Later on the tunnel will be filled with grout. Connection of the gas pipeline to the land-laid sections at the boundaries of our contract is scheduled for July 2010.

## CONCLUSION

Although tunneling started with a 3 months' delay, tunneling was finished 4 weeks ahead of schedule. This is owed to detailed work planning, equipment selection and a determined and well trained working crew. About 70% of the staff and workers on site are/were Wayss & Freytag permanent employees. It was proven that a 3 m internal diameter tunnel lined with precast steel-fiber-reinforced segments can be driven over 4 km without any incident and with higher than expected advance rates. If gas pipe installation and backfilling of the tunnel go on as planned, the works can be handed over to the client ahead of schedule.

# TRACK 1: TECHNOLOGY

Mike Smithson, Chair

*Session 4: Sustainability*

# Sustainable Tunnel Linings: Asset Protection That Will Not Cost the Earth

**Colin Eddie**
Morgan Est PLC, Rugby, UK

**Mike Harper**
Stirling Lloyd Polychem Ltd., Knutsford, UK

**Sotiris Psomas**
Underground Professional Services, Rugby, UK

## INTRODUCTION

The current and future societal and economic needs for advanced transportation and services networks mean that increasing demands will be placed on the development of the underground space. The UK Tunnelling market is set to expand significantly over the next few years. A major development of the subway system is planned in London along with plans for numerous utility tunnels. The new build nuclear power plants will also require the construction of many long-sea outfall tunnels for the cooling water. In addition to the construction of new tunnels, the ageing infrastructure of existing road, rail and utility tunnels requires continual maintenance and repair.

In parallel with this increasing demand, the environmental impact of new construction is on the political agenda and this has been the driver for specific legislation which focuses on the reduction of waste and 'life-cycle' cost of construction methods and materials. Since the early 1990s, Morgan Est Plc and Beton und Monierbau GmbH and more recently with Stirling Lloyd Ltd, have been introducing a number of innovations in tunnelling construction in an effort to improve safety, durability and at the same time optimise tunnel lining performance. These innovations are associated primarily with Permanent Tunnel Linings and include:

- A new sprayed concrete tunnelling process LaserShellTM suitable for permanent lining
- Integritank® HF seamless sprayable waterproofing membranes for tunnel SCL (Sprayed Concrete Linings)
- The development of High Performance Fibre Reinforced Cement Composites (HPFRCC) as permanent linings, sprayed or cast-in-place (CIP)

These innovations offer direct cost savings over the short-term and long-term life of the tunnel, as well as several environmental benefits as follows:

- Development of thinner linings with potentially less energy intensive cementitious mixes
- Design and construction of waterproof membranes with improved effectiveness, providing extensive durability of the structure
- Substitution of materials, such as steel rebars, that are energy intensive with synthetic reinforcement

Traditionally the construction of Sprayed Concrete Tunnels (SCL/SEM/NATM) incorporate a Primary (structurally considered as "Temporary") and a Secondary (structurally considered as "Permanent") linings, separated by a PVC sheet waterproof membrane. This inherently conservative philosophy discounts any contribution from the Primary lining in the long-term and relies only on the Secondary lining in the long-term. Whilst the use of permanent sprayed concrete for both the primary and secondary linings has been used in recent years (CombiShell™ method) these generally had limited ability to prevent water ingress as no waterproofing system was included. The systems presented in this paper constitute an innovative tunnel lining system which combines an UltraShell™ lining with a sprayable waterproof membrane. This system can significantly reduce the overall thickness and improve the durability of the tunnel linings. A comparison of this methodology against the traditional approach is given in Table 1.

**Table 1. Brief comparison of the traditional SCL and the UltraShell™ tunnel lining**

| Lining component | Traditional SCL/SEM/NATM | UltraShell™ lining system with sprayable waterproof membrane |
|---|---|---|
| Primary lining | Temporary lining<br>Medium strength sprayed concrete<br>+mesh+lattice girders | Permanent lining<br>High strength steel fibre reinforced sprayed concrete—*LaserShell*™ |
| Waterproofing membrane | PVC sheets with welded seams | Seamless sprayable membrane<br>*IntegritankHF*™ |
| Secondary lining | Permanent lining<br>CIP concrete with steel rebar | Permanent lining<br>CIP or SC *HPFRCC* |

**Photograph 1.** LaserShell™ application with Multi Shot TunnelBeamer™

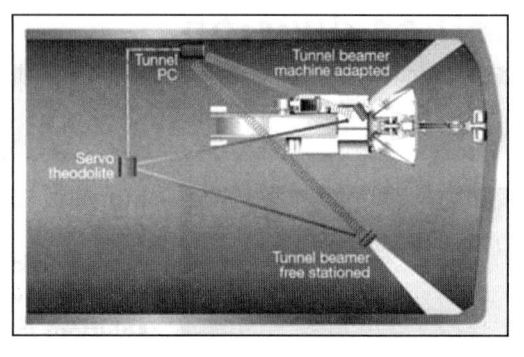

**Figure 1.** TunnelBeamer™ operation principle—system integration

## PERMANENT TUNNEL LININGS USING SPRAYED CONCRETE

### LaserShell™ Tunnel Process

This system has been developed by Morgan Est (UK) and Beton-und Monierbau (Austria) to improve both the safety and quality of underground works using sprayed concrete. The profile of both the excavation and spraying operations is controlled using an innovative real time survey system known as TunnelBeamer™.

Traditionally, sprayed concrete lined tunnels in soils, unstable ground or shallow tunnels require the use of lattice girders to provide profile control of the lining and secure the mesh during the application of the primary lining. However, it is the installation of the girders and mesh together with profile checks, which place the tunnel workers in the exposed vault to an unacceptable risk. To eliminate this risk, a method of controlling the lining shape, thickness and position remotely and in real time LaserShell™ has been developed. By removing the lattice girders, the excavation and spraying operations have no existing orientation line or physical profile control mechanism. The TunnelBeamer™ system is designed to operate in an underground environment and consists of either a single laser or a number of lasers grouped together to act as a distometer, which are directed at the excavation or sprayed concrete lining faces as required. The 3D-tunnel geometry information from these lasers is linked continuously to a computer (situated in the tunnel), which produces a comparison to the theoretical position, displayed on a monitor in the operators cab. The TunnelBeamer instrument can be mounted on a fixed point or mounted on the moving excavation or spraying equipment. When mounted on a moving object a servo-theodolite is used to locate the TunnelBeamer™ and allow the tunnel computer to relate the TunnelBeamer™ information to the theoretical tunnel alignment/profile.

The LaserShell™ methodology employs an inclined face excavation for increased stability and improved safety for tunnel workers. To minimise the number of construction joints and improve productivity, for tunnels up to 6.5m diameter it is proposed that LaserShell™ will be constructed full face. For the larger tunnels, only the crown or pilot excavation would be undertaken using full face LaserShell™ techniques.

Apart from the safety improvement for the personnel there are many other benefits (Eddie and Neumann 2003) such as speed of advance and improved ring closure times, in conjunction with an inclined face significantly reduce surface settlement. Removal of lattice girders and the replacement of mesh with high carbon steel fibre substantially improve the quality and durability of sprayed concrete linings by eliminating shadowing. Systematic capture of profile data relating to both the excavation and the sprayed concrete lining gives absolute confidence with respect to lining shape, thickness and position. Compared to traditional SCL/SEM construction methods, cost savings of up to 50% has been achieved in certain applications where often multiple-stage excavation processes can be rationalised.

**Quality Control and Performance**

The LaserShell™ system requires the use of a high quality permanent sprayed concrete with excellent bond strength between layers and at joints. Although the use of permanent SCL has been well documented for some time (Franzen, Garshol and Tomisawa 2001), a rigorous testing programme was performed in UK and Austria that met all the requirements of a permanent sprayed concrete lining system and prove the structural integrity from application (15 minutes strength) up to 120 years. Retention of sufficient workability to enable efficient application in a tunnel environment was also essential. Detailed and onerous performance criteria in respect of strength gain, flexural toughness, permeability, bond characteristics between layers and durability were set and benchmarked against comparable high quality cast-in-place structural concrete. The tests showed (Eddie and Neumann 2003) that high integrity joints and layers can be formed with high levels of structural integrity and low permeability. Attention was paid to the early-age strength development relative to workability retention times and into the effects of high early-age loading on immature sprayed concrete. Furthermore, durability testing was performed on several samples (with and without accelerator) at 1 month, 6 months and 1 year to determine compressive strength, modulus of elasticity, porosity and for signs of deleterious behaviour. On completion of the trials, no adverse strength or stiffness values or trends have been recorded on any sample and the SEM/Petrographic analyses have shown no deleterious processes.

In summary, it was demonstrated that the sprayed concrete mix developed was capable of achieving the permanent lining criteria (that is "comparable to cast in place"). This mix can be used in conjunction with the sprayable waterproof membranes, which are presented in the ensuing section.

**SPRAYABLE WATERPROOF MEMBRANES**

**General Requirements**

Stirling Lloyd with Morgan Est have developed a system to waterproof SCL tunnels. Traditionally,

**Photograph 2. Finished LaserShell™ SCL in Heathrow T5 underground works**

SCL tunnels have been waterproofed using PVC (or HDPE) sheets, joined together through extensive seaming, formed by heat welding. Although this methodology is cost effective in long tunnels of constant cross section, it has some significant drawbacks, which can be summarised in the followings:

- Difficult to overspray
- Complex geometries require single seam welding which cannot be tested
- Complex grouting systems are required
- Membrane not bonded to concrete leading to water-paths around the structure
- Leak point identification and treatment is difficult

Problems have been associated with leaks through the seams and it is generally accepted that leaks are commonplace using this method.

The use of spray applied waterproofing addresses the above issues, because seams are eliminated and detailing is simplified. However, attempts by a number of companies to produce a spray applied system have had mixed results. One of the fundamental requirements of a waterproofing membrane in a tunnel environment is its crack bridging capability. Concrete structures are prone to some form of cracking and the membrane must be capable of bridging these cracks as they form, and not cracking with the concrete. Waterproofing systems without a crack bridging capability are unlikely to achieve waterproofing integrity.

The Integritank® HF membrane is an adaptation for the SCL tunnelling environment of proven waterproofing technology that can be proven to be waterproof in-situ after application and prior to encapsulation by secondary concrete. The system has been tested for waterproofing integrity under pressure (up to 20 bar). Integritank® HF is also robust enough to receive sprayed or cast in-situ concrete as a secondary lining.

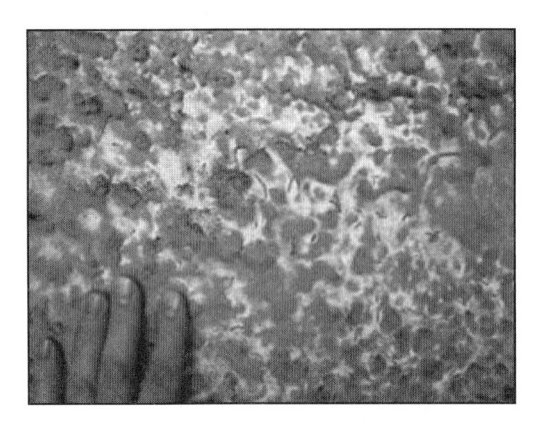

Photograph 3. "Mud cracking," pinholes, and uneven coverage of a one coat water based sprayed membrane, applied over "as shot" shotcrete

Photograph 4. For comparison, the first (yellow) coat of Integritank® HF membrane, applied to the rendered concrete surface

**Waterproofing Design Considerations**

There are a number of factors to be considered when evaluating waterproofing systems for a tunnelling project:

- How difficult will it be to go back to repair the waterproofing if necessary?
- What might it cost, both in terms of money and disruption, to undertake the repairs, if the waterproofing is leaking when in service?
- Can the chosen system be proven to provide the desired degree of waterproofing?
- Since any concrete is prone to some cracking, is the membrane capable of bridging cracks?
- Can the membrane be checked and proven to be waterproof before it is encapsulated by the secondary concrete lining?
- Can the membrane be installed safely?
- Can the membrane resist water pressure build up in the ground?
- Can a sprayed membrane be applied in such a way that the thickness is controlled and there is no risk of any area being missed, being under thickness, suffering from pinholes or shadowing?

There are also the following parameters to consider.

***Substrate Preparation—"As Shot" vs "Finished" Primary Concrete***

Variations in the "as shot" surface of sprayed concrete can make it an unsuitable substrate to receive a sprayed waterproofing membrane. There have been suggestions in the industry that membranes can be sprayed effectively directly onto 'as shot' sprayed concrete, but fibres protruding from the concrete and voids in the concrete surface will both lead to the waterproofing being compromised. The performance of the waterproofing membrane is highly dependent on the quality of the substrate preparation.

Some membranes are prone to shrinkage and cracking after application—particularly when membranes are water/cement based and where thickness varies because of the "as shot" concrete surface below.

An increase in the concrete surface roughness increases the chances of waterproofing leaking, but also voids in the surface and even "shadow" areas which are difficult to spray over. A waterproofing membrane is not designed to regulate or infill a surface—this should be done first with a concrete render. Despite exhaustive testing by tunnelling contractors to waterproof "as shot" concrete, Stirling Lloyd concluded that effective waterproofing was only achieved on a finished concrete surface. It is recommended that a sprayed concrete surface should be treated with a fibre free render of up to 30mm (1.2in) thick, to even out the surface, and encapsulate any fibres protruding from the primary concrete. The surface does not have to be completely flat, undulations are quite acceptable, but peaks and troughs and voids in the surface should all be evened out and closed up. Photograph 6 shows a concrete render being applied in panels to the primary lining. The application of alternate panels, followed by infill afterwards, makes levelling the system easier and allows any shrinkage in the render to take place. The render is smoothed off using a wood float to produce a closed surface finish.

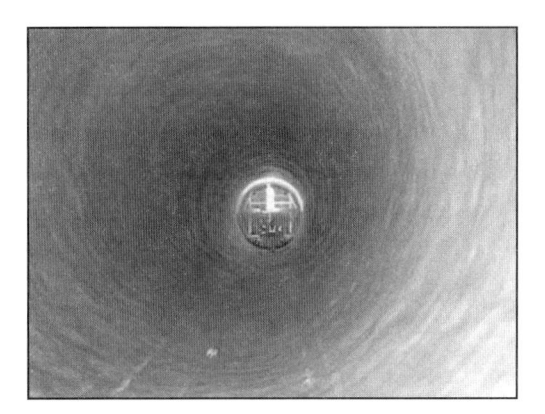

**Photograph 5.** Honor Oak (London, UK) water tunnel. Successful application on an uneven but closed surface.

**Photograph 6.** Application of concrete levelling render

### Active Water Ingress

Active water ingress, where water is flowing over the surface, will damage any sprayed waterproofing membrane during application. It is a common phenomenon for water ingress through the primary lining, which may exhibit itself in varying degrees, from discreet spots and general dampness, to significant running water. For damp substrates, Stirling Lloyd's Integritank® HF system utilises a damp resistant primer, developed specifically for SCL tunnels, to seal the surface prior to membrane application. This promotes a high bond strength of the membrane to even a damp concrete surface. However, active water ingress, where water is freely flowing over the concrete surface, needs to be stemmed prior to primer and membrane application. If this is discreet points, this is achieved using remedial waterproofing materials, such as rapid setting cement plugging mortars and resin injection techniques. In cases where active water ingress is widespread, the render smoothing layer referred to previously, is replaced by a high density water resistant render, designed not only to regulate the surface, but also to stem the flow of water over the short term, whilst the membrane system is installed and the secondary concrete placed over to encapsulate the whole.

### Surface Preparation and Testing

The concrete surface should be free from dust and laitance. Dust can mostly be removed using de-dusting equipment that is standard in tunnelling operations. The waterproofing contractor, using an air lance, will remove any remaining dust and surface water immediately before primer application. With this preparation done, tensile adhesion checks are carried out on the concrete surface. One of the advantages of using a sprayed membrane is that it is fully bonded to the concrete substrate. A good bond means that sprayed concrete can be used for the secondary lining also, which usually speeds up construction compared to cast in-place concrete. The bond is checked on site, during application, by the waterproofing contractor. Tensile adhesion tests are carried out on site to check the bond that can be achieved between the waterproofing and the substrate concrete. The mode of failure is also checked. A minimum value of 0.3MPa (43.5psi) should be achieved between the water proofing membrane and concrete interface. Failure should be in the concrete substrate.

### Priming

Provided that acceptable adhesion values have been achieved and the substrate prepared, the concrete will be primed with a primer that is suitable for use on damp substrates. The primer is sprayed to the whole concrete surface and allowed to dry. The use of the Primer has two functions: it improves the adhesion of the membrane to the concrete surface and it seals the concrete surface against its tendency for "out-gassing." A liquid product applied to the concrete surface is at risk of "pin-holing" as gasses escape the concrete surface and bubble through the membrane, leaving a hole and a potential water path. Therefore, in order to achieve the benefits of a fully bonded seamless membrane, a primer is applied first to the concrete that seals the surface against "out gassing" and minimises "pin-holing" in subsequent membrane application.

### Membrane Application

The membrane is delivered to site in a liquid state in purpose made stainless steel demountable tanks, ready for offload directly into the tunnel. The material should be factory blended under ISO 9001:2000

**Photograph 7. Application of membrane—yellow first coat**

**Photograph 8. Application of membrane—white second coat**

procedures and delivered to site ready to use. The system is a two-component resin system, sprayed on site through bespoke computer controlled spray equipment. The two-liquid resin components are mixed in the pump system and react together to form the fully cured membrane in less than 1 hour. The system is 100% solids and does not contain any solvents or water, which ensures the tunnel atmosphere remains safe, and the membrane does not suffer cracking during drying. The system should be sprayed by experienced operators using hand held spray equipment. This is more effective than robotic spray equipment as a spray operative can make allowances for undulations in the surface and ensure continuous visual inspection during spraying to ensure no area is missed. In addition, the operative should be within 1.5m (4.9ft) of the surface at all times and can achieve a good standard of visual inspection. Thickness is monitored in 3 ways:

- Measuring continuously during spraying by using a wet film dip method. This means coverage can be monitored in real time and adjusted as necessary to ensure the specified thickness is achieved.
- Utilising light coloured materials of contrasting colours. The pigmentation of these materials is such that opacity is only reached when sufficient thickness is applied. Essentially if the membrane is too thin, the substrate can still be seen through it.
- Recording the material quantity against the area that has been sprayed to ensure sufficient coverage is used as a secondary check of the system.

The first coat of membrane is yellow, a colour that best shows up defects under artificial lighting conditions, so visual inspection ensures no areas are

**Photograph 9. Integrity testing in a SCL tunnel**

missed. The completed membrane is then visually checked for any defects, which are made good at this stage. The second coat is white, which contrasts well with the yellow as a form of coverage control. The second coat will also cover even very small defects in the first coat, to ensure the complete system is waterproof. The completed system is then electrically tested for waterproofing integrity.

**Integrity Testing**

Integrity testing after the waterproofing membrane is complete, but prior to the application of the secondary concrete lining is essential. This is achieved by using a holiday detection equipment, which is capable of detecting even a pinhole sized defect that may not be visible to the eye. The whole surface should be checked in this way and any defects found should be made good by hand applied patch repair material at that point. The patch repair system is the same resin and bonds seamlessly to the membrane at the molecular level to form a seamless finished membrane. The

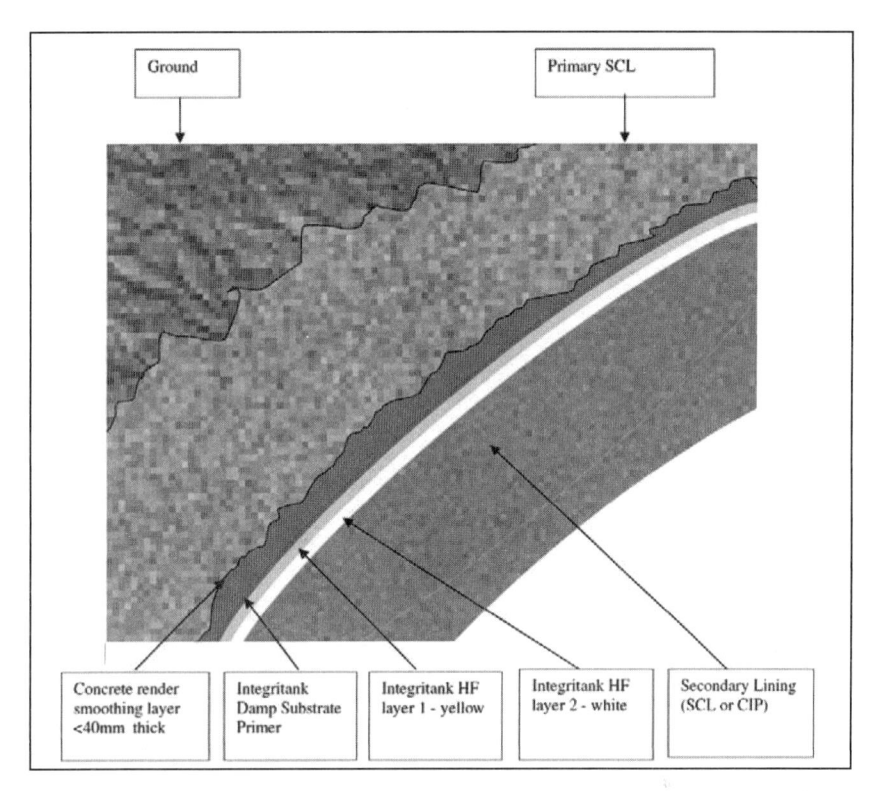

Figure 2. Cross section sketch to illustrate the build-up of a seamless sprayed waterproof membrane

application of the system by specialised workforce is necessary to ensure the highest possible level of confidence in the waterproofing integrity and reduce the risk of leaks.

*Concrete Application*

The membrane is fully cured and ready to receive further concrete within one hour of the application being complete. If SFR is to be used for the secondary lining, a fibre free first pass is applied, to protect the membrane from the effects of the fibres. Fibre free sprayed concrete and cast in-situ concrete can be applied directly to the membrane surface. Concrete application should proceed as quickly as possible after any waterproofing membrane has been installed. This minimises the risk of pressure build up behind the membrane, before it is encapsulated in concrete.

## HPFRCC LININGS

### General Criteria and Properties

The term High Performance Fibre Reinforced Cement Composites (HPFRCC) is the preferred term adopted by the Japanese Society of Civil Engineers in their recently published report (JSCE 2008). Similarly to HPFRCC, Li (2003) uses the term of ECC (Engineered Cementitious Composites). The use of HPFRCC as tunnel linings is an emerging application for high performance composites that can contribute to the sustainability of future tunnels by decreasing the lining thickness and increasing durability.

HPFRCC exhibit a range of improved properties compared to conventional FRC (Fibre Reinforced Concrete) such as high ductility, higher energy absorption capacity and higher toughness. As a result of their design and composition, HPFRCC exhibit the remarkable ability to HHstrain-hardenHH after the first cracking followed by the development of multiple cracking, less than 0.1mm (0.004in) wide, according to Li, Wang and Wu (2001). As a result of the strain-hardening behaviour the crack width is constant over a wide range of tensile strain and therefore can be considered a material property and not structural performance criterion (JSCE, 2008). Because of the large volume of material involved in the inelastic deformation process, energy absorption is significantly enhanced. Another desirable property of HPFRCC is its unit density, which is lower than concrete (on average 10–20%), requiring much less

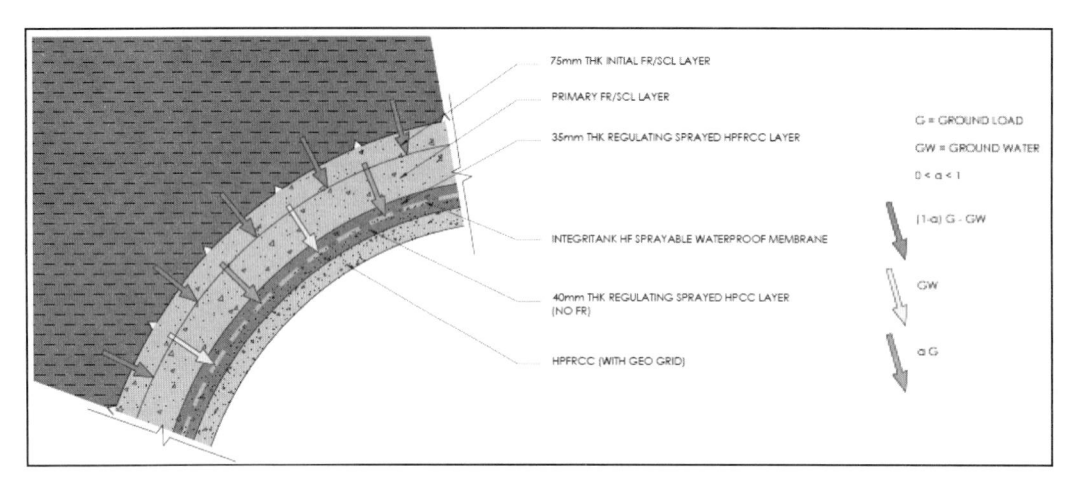

Figure 3. HPFRCC applied as secondary permanent lining

energy to produce, handle and apply, deeming them a more economic building material.

The unique properties of HPFRCC are achieved by a careful consideration of the mixing and preparation of the various constituents. Cement and water should conform to standards for structural concrete. Coarse aggregate is not used since it tends to adversely affect the ductile behaviour. Admixtures and additives are required to enhance the physical properties of the mix. Pumping aid/segregation reducing liquid water-soluble polymers admixtures and curing agents are also considered. One of the main contributors to this remarkable behaviour is attributed to the high tenacity Poly-Vinyl-Alcohol (PVA) fibres. Unlike steel, PVA fibres develop a molecular and chemical bond with the cement during hydration and curing.

The durability can be assessed experimentally by determining the changes in the fibre and fibre-matrix interface properties with specimens exposed to accelerated testing and correlating such changes to changes in the ductility of composites exposed to the same accelerated testing conditions. The accelerated test published is a hot water immersion test simulating a long-term hot and humid environment. It is found by Li et al (2004) that although the PVA fiber-matrix interface chemical bond increases, the apparent PVA fibre strength decreases, when the exposure time reaches 26 weeks. Despite the deterioration, ECC is found to retain high tensile ductility after exposure to an equivalent of 70 years or more of hot and humid environmental conditions. Permeability performance on pre-strained specimen confirmed that HPFRCC loaded at strain-hardening range behave similar to concrete (Lepech and Li, 2005). If the specimen is unloaded before final failure, the micro-cracks are often small enough to prevent the intrusion of water, and may heal if there is sufficient un-hydrated lime available. The self-healing characteristics have been demonstrated (Yang et al 2009) through resonance measurements of samples subjected to tensile load and wetting-drying cycles.

## HPFRCC as Tunnel Lining

HPFRCC is usually applied as cast-in-place but there are also sprayed applications (Kanda et al., 2003). The aforementioned properties make HPFRCC very attractive material for tunnel linings. It is useful to distinguish between the potential applications of HPFRCC within the UltraShell™ lining system:

- As part of the Secondary Lining in a new tunnel
- As a Secondary Lining to repair an existing tunnel liner

It is intended that in cases of the tunnels lining experiencing tension and bending, the fibre will be supplemented by a textile reinforcement. In a new tunnel HPFRCC shall be able to sustain the full ground water load and part of the ground load. The envisaged structural repair application entails only a comparatively thin, and durable layer of HPFRCC, which is designed to yield in a controlled manner (without impairing durability) when subjected to increased loading. The elimination of steel reinforcing ensures that no long-term corrosion risks exist. The HPFRCC material is a ductile material with a high tensile capacity, allowing multiple cracking to be maintained below 100μm (0.004in) and preventing migration of aggressive substances.

In the case of a hydraulic tunnels, the secondary lining is required to withstand the internal test/surge hydraulic pressure. The design practice for hydraulic tunnels in the UK is that no benefit from external geostatic loading is taken into consideration for the internal pressure load case. The functional requirements for the pressure tunnel lining is to be essentially watertightFF[1]FF with no flows associated with infiltration or exfiltration and to have an attainable service life of at least 120 years. To be compliant with Eurocode 2 (BS EN 1992-3), the crack widths have to be maintained below 0.05mm-0.2mm (that is 0.002in to 0.008in, depending on the application), to deliver the requisite durability. This requirement results in a design of heavily reinforced lining incorporating high yield reinforcing bars at close centres. The design aim of using HPFRCC is to satisfy the Service Limit State (BS EN 1990) so that for the 'characteristic combinations of actions' (i.e., operation) the lining works within its elastic range, and when subjected to 'frequent/quasi-permanent combination of actions' yields in a controlled manner forming constant width multiple cracks. The key benefits of adopting HPFRCC can be summarised in the following:

- Reduced construction cost (reduced thickness, elimination of steel reinforcement, reduced material waste and over-excavation)
- Improved construction programme
- Improved safety—elimination of steel fixing
- Improved durability—excellent crack control (less than 0.1mm)
- Improved quality—no concerns regarding cover to reinforcement

**Current Development Program**

In developing appropriate mixes, testing work has been undertaken at Morgan Est's laboratory, three Universities (Warwick, Sheffield, Surrey), UK Building Research Establishment and Sandberg Laboratories. Several of these mixes have also been tested in the field at Morgan Est's R&D facility near Rugby and two full scale shutter trials have been undertaken. Site trials have been carried out to assess the performance of the mix in real scale, including spraying trials and shutter cast in place. The latter application in tunnels require placement (by pumping) behind a semi-mechanical steel shutter. The HPFRCC distribution system, shutter and vibration system need to be purpose-designed and suitable for the particular rheology of the proposed mix. Extensive testing of the sprayable format HPFRCC material has being undertaken, and has been valuable in understanding the material selection issues influencing both the practical handling and placing of the material and also the structural performance. The testing programme is ongoing.

**CONCLUSIONS**

**Permanent Tunnel SCL**

The LaserShell™ method of tunnelling, utilising TunnelBeamer™ delivers unparalleled levels of safety, quality and efficiency for construction of SCL tunnel linings. It provides a comprehensive documentation of "as-built" work for quality and certification reasons. The system has proved successful on a number of major projects in the UK. The LaserShell™ primary lining can be greatly enhanced by the use of a HPFRCC inner lining and an Integritank® HF sprayable waterproof membrane.

**Seamless Sprayable Membranes for Tunnels**

The success or failure of the waterproofing depends not only on the suitability of the product for the task, and the quality of installation, but also on the ability to be able to test the waterproofing for integrity, before it is encapsulated by the secondary concrete.

The choice of waterproofing system depends on the degree of water tightness required, the risk of a waterproofing failure and the costs associated with having to attempt repairs to an encapsulated membrane when the tunnel is complete. Pre-formed sheet membranes with multiple seams are prone to leaking, particularly where complex tunnel geometry is involved and hand seaming is employed instead of automated welders.

Sprayed waterproofing membranes enable the deletion of seaming and the simplification of detailing, removing the biggest risk to tunnel waterproofing. The Integritank® HF membrane in particular enables subsequent testing to prove 100% waterproofing integrity has been achieved on site. The efficacy of the system is dependent upon both the preparation of the substrate onto which the membrane will be sprayed, high quality installation, and post application testing. It is recommended that the supplier should have its own specialised labour to apply their membranes.

**HPFRCC for Permanent Tunnel Linings**

The HPFRCC material offers significant benefits when compared with traditionally reinforced concrete for applications in an underground environment including pressurised water tunnels. The strain compatibility philosophy, which underpins the UltraShell™ tunnel lining system, ensures that substantial reserves of strain capacity are available should they be required. Ultimately, by adopting the HPFRCC in the permanent tunnel lining system, it is possible to install a thinner lining, which is highly durable.

---

1. BTS/ICE, 2009. 'Specification for Tunnelling, Thomas Telford, Clause 508.3

## Sustainability Benefits

In terms of lining performance, sustainability can be defined as the capacity to endure. Sustainable development can be defined as a pattern of resource use that aims to meet human needs whilst preserving the environment. The UltraShell lining system contributes to both of these objectives by:

- Enabling thinner linings using HPFRCC, means less excavation, resulting in lower energy input/$CO_2$ generation during excavation, and less waste being created to be transported and disposed of.
- Thinner linings means lower volumes of virgin materials being extracted, transported and processed to form final materials for the tunnel lining, again reducing $CO_2$ generation from construction activities.
- Integritank® HF spray applied waterproofing is key to delivering a completely waterproof tunnel environment. No water flow through the lining results in greatly enhanced durability and significant longer design life.

In addition to the sustainability benefits and reduced maintenance requirements, the system detailed in this paper also provides a faster build programme and overall a 'whole-life' lower cost than traditional SCL/SEM/NATM linings.

## ACKNOWLEDGMENTS

The authors would like to thank Mr. Christian Neumann at Beton-und Monierbau GmbH for his assistance in developing LaserShell™ and UltraShell™, Mr. Martin Rimes Chief Materials Manager of Morgan Est, and Mr. Dave Helliwell, Chief Chemist at Stirling Lloyd, for his continuing innovation in waterproofing technology.

## REFERENCES

[1] Eddie C. and Neumann C. 2003. LaserShell leads the way for SCL tunnels. *Tunnels & Tunnelling International*, June, p38–42.

[2] Franzen T., Garshol K.F., and Tomisawa N. 2001. Sprayed concrete for final linings: ITA WG Report. *Tunnelling & Underground Space Technology* Vol. 16, p295–309.

[3] JSCE 2008. *Recommendations for design and construction of high performance fibre reinforced cement composite with multiple fine cracks*. Concrete Engineering Series 82, Japan Society of Civil Engineers.

[4] Kanda T., Saito T., Sakata N., and Hirashi M. 2003. Tensile and anti-spalling properties of direct sprayed ECC. *J. Advanced Concrete Technology*, Vol. 1, No. 3, p269–282.

[5] Li, V.C. 2003. On Engineered Cementitious Composites (ECC)—A review of the material and its applications. *J. Advanced Concrete Technology*, 1(3), p215–230.

[6] Li, V.C., Wang, S., and Wu, C. 2001. Tensile strain-hardening behavior of PVA-ECC. *ACI Materials J.*, 98 (6), p483–492.

[7] Li V.C., Horikoshi T., Ogawa A., Torigoe S., and Saito T. 2004. Micromechanics-Based Durability Study of Polyvinyl Alcohol-Engineered Cementitious Composite. *ACI Materials J.*, V101, No.3, May-June, p242–248.

[8] Lepech, M. and V.C. Li, 2005. Water Permeability of Cracked Cementitious Composites. *Proceedings of Eleventh International Conference on Fracture*, Turin, Italy.

[9] Yang Y., Lepech M.D. Yang E.H., Li V.C., 2009. Autogenous healing of ECC under wet-dry cycles. *Cement and Concrete Research*, 39, p382–390.

# Design for Sustainable and Economical Tunnels

## Derek J. Penrice, Bradford F. Townsend
Hatch Mott MacDonald, San Francisco, California

**ABSTRACT:** The concrete industry is one of the planet's largest consumers of natural resources. Cement production results in approximately 7% of the annual global emission of carbon dioxide (CO2). With the continued threat of global warming, and spiraling costs for commodities, it is critical that we as Owners, Designers and Contractors promote the development of sustainable underground structures that make economical use of natural resources.

This process starts with design: the selection of a particular design standard and design concept, and materials specification can significantly influence a project's resource requirements, and hence cost. This paper identifies ways we can develop and promote more sustainable and subsequently more economical practices within our own industry.

## INTRODUCTION

It is well documented that the concrete industry is one of the largest global consumers of natural resources—relatively recent statistics indicate that the concrete industry annually consumes over 10 billion tons of sand and aggregates and 1 billion tons (240 billion gallons) of water, not including water for wash down of mixers or curing (Mehta 2001).

In addition to the depletion of these natural resources, the production of cement in itself expends considerable amounts of fossil fuel and electrical energy. Global annual production of cement amounted to approximately 3.05 billion tons in 2007 (Mehta, 2009). The energy expended on the creation of one ton of cement generates an equivalent weight of $CO_2$, an alleged principal contributor to global warming. The annual production of 3.05 billion tons of $CO_2$ (and rising) corresponds to approximately 7% of the global emission of this gas.

Furthermore, demolition debris constitutes a significant percentage of solid waste disposal. While global statistics, including the markets of China and India, which are now producing and using over 50% of the world's concrete are not available, in North America, Europe and Japan it is estimated that in excess of 1 billion tons of construction and demolition waste is generated each year (World Business Council for Sustainable Development 2009).

Clearly none of these practices are sustainable in the long term. Therefore, in this era of emerging environmental concern it is incumbent on the tunneling industry to develop engineering solutions which are sustainable through best design and construction practices which promote the most economic use of materials.

## SUSTAINABILITY—WHAT IS IT?

In 1987, the World Commission on Environment and Development defined sustainability as *Development that meets the needs of the present without compromising the ability of future generations to meet their own needs.*

In the United States, the cause of sustainability has been championed by the Green Building Council (USGBC), which was founded in 1993. The USGBC is an organization comprised of building industry leaders who have committed to the development of environmentally responsible, cost-efficient residential and commercial buildings.

USGBC developed the Leadership in Energy and Environmental Design (LEED) certification system which provides a framework for assessing building performance relative to achieving sustainability goals. Under the LEED certification system buildings are awarded credits from a series of categories including Sustainable Sites; Water Efficiency; Energy and Atmosphere; Materials and Resources; and Indoor Environmental Quality.

Depending upon how many credits a building gathers from the certification system, a qualifying project can be classified as certified, silver, gold or platinum. The available credits and the credits required for certification are indicated in Table 1.

While the LEED certification system is voluntary, there is considerable prestige associated with obtaining a LEED classification. There is also increased interest in sustainable design from public and private sector clients, to the point where many municipalities now require LEED certification on their new building projects. As an example, the

Table 1. USGBC LEED project certification scorecard

| Category | Credits Available | LEED Classification | Credits Required |
|---|---|---|---|
| Sustainable sites | 14 | Certified | 26–32 |
| Water efficiency | 5 | Silver | 33–38 |
| Energy & atmosphere | 17 | Gold | 39–51 |
| Materials & resources | 13 | Platinum | 52–69 |
| Indoor environmental quality | 15 | | |
| Innovation & design process | 5 | | |
| Total credits | 69 | | |

City of San Francisco has adopted an ordinance that requires all city-owned facilities to at least meet a LEED silver classification.

Some in our industry would suggest that tunnels, by the nature of their function, such as mass transit, or their longevity, are already sustainable solutions. Tunnels are typically designed to have a serviceable life of 100 years, but in reality they can last much longer, as evidenced by the continued use of metro systems from Paris to Boston and beyond. These points are not in dispute. However, it is apparent that there is much more that we as an industry can and should accomplish in terms of providing and promoting a truly sustainable product.

## SUSTAINABILITY FOR TUNNELS

With sustainable design being so prominent in the public arena, it is necessary that our industry takes a proactive role in investigating methods of achieving more sustainable solutions and promotes a more consistent and coordinated approach to sustainability, including the adoption of a tailored project certification system. It is of particular value that we promote our sustainability credentials in an environment where tunnel options regularly compete against other transportation or storage alternatives.

In developing an approach to sustainability, the work undertaken by the USGBC in identifying certification credits and classifications of accreditation provides an excellent starting point. While many of the USGBC LEED certification credits are not particularly applicable to tunnel construction, many are directly relevant—including innovation in design; materials reuse; use of recycled content; and construction waste management.

A key element of a sustainability initiative for the tunnel industry is to significantly reduce the volume of cement we use, which will correspondingly result in the production of less $CO_2$. This paper explores specific areas where we can seek to reduce our cement consumption through innovation in design and use of recycled materials, and suggests a USGBC-type certification system be adopted to benchmark our industry performance relative to achieving sustainability goals.

### Innovation in Design

Design innovation is normally thought of in terms of solving a complex technical challenge. While many underground projects have demonstrated true innovation in design—including the tunnel box jacking completed as part of the Massachusetts Turnpike Authority's Central Artery/Tunnel (CA/T) Project and Kuala Lumpur's Stormwater Management and Road Tunnel (SMART), through its innovative design for dual operation, design innovation for sustainable tunnels translates into economic use of materials.

Two examples of where we as an industry improve our sustainability performance with regard to materials, and in particular cement usage, are through the adoption of consistent design criteria for underground projects, and through increased incorporation of temporary works into permanent structures.

*Design Criteria*

Typically at the outset of a large project where multiple design firms may be engaged, or a single design firm may utilize multiple design offices the designer will be faced with defining specific project design criteria, inclusive of structure loads, combinations of loads and load factors to ensure the consistency of the design.

A significant issue which continues to face the designers and owners of underground structures is that no uniformly applied design standard exists, and that the standards which do exist were not developed with tunnel construction in mind. The design standard selected for a particular project can currently be location-based through the adoption of a city or state building code, can be a material code such as American Concrete Institute, or can be directly related to the function of the tunnel. For example, a highway tunnel design would be in accordance with the American Association of State Highway

**Table 2. Ultimate limit state load factors**

| Design Code | Dead Load | Earth Pressure | Hydrostatic Load |
|---|---|---|---|
| American Association of State Highway and Transportation Officials (AASHTO) | 1.3 | 1.3$\beta_e$* | 1.3 |
| American Concrete Institute 318—Building Code Requirements for Structural Concrete | 1.4 | 1.7 | 1.4 |
| American Railway Engineering & Maintenance of Way Association (AREMA) | 1.4 | 1.4 | 1.4 |
| British Standard (BS) 5400—Design of Bridges | 1.15 | 1.5 | 1.5 |
| British Standard 8110—Reinforced Concrete Design | 1.4 | 1.4 | 1.4 |

*AASHTO lateral earth pressure coefficient = 1.3$\beta_e$, where $\beta_e$ = 1.0 for vertical earth pressure, 1.3 for lateral earth pressure, or 0.5 for lateral earth pressure for checking positive moments in frames.

and Transportation Officials (AASHTO) standard. In other cases, many owners also maintain their own design standards.

Underground structures are subject to many potential loading situations and combinations over their intended service life, all of which must be considered in the analysis and design of the structures. In general, loads on underground structures can be defined in four broad categories:

- Dead Loads: the self weight of the structure and permanent (i.e., non-removable) elements including road or track slabs, walkways etc. Other long-term loads such as mechanical and electrical equipment and finishes can be classified as a 'superimposed' dead load.
- Live Loads: transient loadings imposed by vehicles and/or pedestrians.
- Earth Loads: lateral and vertical earth pressures, hydrostatic loads, and imposed surcharges from adjacent facilities such as buildings or construction equipment.
- Extreme loads: seismic loads, extreme flooding events, and blast forces arising from an explosion.

These loads are grouped into combinations and applied with load factors to generate the structure design forces. The magnitude of the applied load factor typically reflects the uncertainty in the derivation of that particular load and provides a margin of safety for the design to address a number of variables including material strength and density, workmanship, and dimensional tolerances.

Table 2 indicates the differences in minimum load factors required for dominant structural loads for several design standards, the majority of which have previously been used in the design of underground structures.

By inspection of Table 2, it becomes apparent that significantly different forces can be generated from the use of one listed standard versus another. For example, earth pressure factors, which constitute one of the largest loads applied to the structure can vary between 1.3 and 1.7, a difference of approximately 30%. While the adoption of any such standard will result in a functional design, the differences in load factors translate into different applied forces, which in real terms translates into different structure member sizes and hence requirements for cement. Simply stated, project design criteria can have a profound effect on project cost and sustainability.

The adoption of a set of consistent criteria for the design of underground structures will promote more economic use of materials and thereby enhance sustainability by the elimination of overly conservative load factors. With this goal in mind, requirements and recommended load factors for each of the principal design loads are presented.

- Dead loads: These loads can be calculated with a high degree of confidence. Significant variations in member thickness between design and as-built are unlikely, and concrete and steel densities do not vary by more than a few percent. On that basis it would seem that a load factor of approximately 1.2 is appropriate. However, a higher factor of 1.4 should be retained for 'superimposed' dead loads such as systems and finishes whose

requirements may not be known until late in the design process, and which may ultimately be replaced over the lifetime of the structure.
- Earth Pressure: As this load provides the greatest potential for variability, it is recommended that a load factor in the region of 1.3 to1.4 be adopted. Reductions in the load factor for lateral earth pressure for calculating positive moments in frames should be considered.
- Hydrostatic: Design groundwater elevations determined from the results of long-term monitoring of observation wells can give the designer some confidence in the accuracy of this load. However, when projects are located in areas experiencing an extended period of drought, groundwater monitoring results may not always give a true indication of 'normal' groundwater level. While the application of a higher load factor may appear to be onerous there is an inherent risk that the range of potential groundwater elevations may not be captured. For that reason it is recommended that a load factor of at least 1.3 is retained for hydrostatic pressure.

Promoting the use of standardized load factors for underground projects, which may be less than the minimums recommended by another industry code, requires maintaining a high degree of quality control during construction to ensure dimensions and variations in material density are within tolerances. However, the cost for enhanced quality control would be a small price to pay relative to the sustainability benefits and construction economies generated by this design efficiency.

## *Selection of Design Concept*

Considerable effort is expended in the construction of temporary works for underground projects. However, too often these elements are ignored in the design of the permanent structures, and consequently owners derive limited value from this investment. By incorporating 'temporary' works items into the permanent structure, considerable benefits —economic and sustainability—can again be derived for some additional outlay for enhanced quality control during construction.

For cut and cover tunnels the primary temporary works investment is in the shoring system, the requirements for which must be evaluated on a case-by-case basis relative to the specific constraints of each individual project, which may include:

- Location—urban or rural
- Requirements for protection of adjacent buildings or infrastructure
- Excavation depth
- Groundwater regime

These constraints will drive the selection of the shoring system, which may correspondingly influence the construction approach. For instance, for deeper excavations in urban settings where ground movements must be carefully controlled, a rigid shoring system comprising slurry (diaphragm) walls or secant piling is commonly specified. In such cases there is a significant cost associated with the construction of these types of shoring, yet too often the full value of systems is not realized through their incorporation into the permanent structure.

In an urban setting the incorporation of the shoring into the permanent structure has numerous benefits. In addition to the sustainable practice of minimizing materials usage by reducing quantities for excavation, concrete and reinforcement, this approach also minimizes project right of way requirements and lessens the headache of trying to relocate utilities in increasingly congested streets. This approach also promotes the use of 'top-down' construction, which involves excavating to the tunnel roof elevation, casting the roof slab, and backfilling to restore grade, while concurrently completing the tunnel excavation. This approach allows rapid reinstatement of surface roadways and utilities which is highly desirable. In addition, the completed roof slab acts as a strut during subsequent stages of excavation, reducing the shoring wall temporary support requirement. Depending on the nature of the material to be excavated below the roof slab, the excavated material may be directly reusable as backfill on top of the roof slab, which offers some opportunity to reduce the volume of truck traffic, as well as quantities of material disposal and backfill.

From a sustainability perspective, the integration of the shoring into the permanent structure is in theory a win-win situation, but it is one which demands careful attention to detail on the part of the designer and a high degree of quality control in construction to ensure the structure remains in serviceable condition over its intended design life. The use of integral shoring was widely adopted on the CA/T Project. Unfortunately the project and the construction method have become synonymous with leakage. If the method is to be promoted as a durable and sustainable solution, and we as an industry are committed to moving forward in an environmentally friendly manner, then the lessons learned from this and other projects must be shared for the benefit of the industry.

In cases where an owner may retain concerns over durability and be reluctant to fully integrate the shoring into the permanent structure, the designer should still investigate any passive contributory effect that the shoring may have on the design of the permanent structure, as the shoring will continue to sustain a percentage of the ground load over the life of the structure. Software is sufficiently sophisticated to allow the interaction of the shoring system and the permanent structure to be modeled. The designer can thereby determine the impact of the shoring wall in contributing to the required strength and stiffness of the permanent structures, which should promote efficiency in the structure size and cost.

Similarly the contributory effect of mined tunnel initial support is frequently ignored in the design of the final lining, though again considerable time is spent and expense incurred on the installation of the initial support. A non-conservative design assumption that the initial support will continue to sustain ground loads over the design life of the structure and beyond promotes sustainability by again generating efficiencies in requirements for materials and equipment.

The relative amount of the ground load taken by the initial support and final lining can be determined based the ratio of the axial stiffness of the initial support relative to that of the combined initial support and final lining system. However, as initial flashcrete quality and corrosion rates can be difficult to accurately predict, it is recommended that any contribution of support elements including the initial flashcoat of shotcrete, lattice girders and unprotected rock reinforcement including bolts, dowels, and spiles be ignored in the calculation of initial support stiffness.

Provisions for ensuring the durability of the initial support must be defined if the design allows for load sharing between the initial and permanent liners. As the initial support must sustain the applied ground load over the life of the structure, no degradation must occur. This in turn mandates increased quality control requirements during initial support installation. However, this would again be a small price to pay relative to the sustainability benefits.

**Recycling and Reuse of Materials**

In addition to achieving economy of materials usage through the design process, the specification of recycled materials can have significant economic and durability benefits for a project.

The most commonly used material in the underground industry is concrete. Globally an estimated 30 billion tons of concrete were produced in 2006 (World Business Council for Sustainable Development, 2009). In the United States, ACI has recognized the importance of sustainable concrete production through the implementation of its Committee 130—Sustainability of Concrete. This Committee is in the process of developing a publication which will include industry guidance on materials and sustainability tools.

Methods for reducing requirements for virgin materials—aggregates and fresh water—have been in place for many years. The use of cement replacement materials, reuse of concrete wash down water, and the use of recycled aggregates are all fairly common at low rates or for specific applications. However, it is time that a more widespread and consistent usage of these underutilized materials was adopted.

*Cement Replacement Materials*

The use of cement replacement materials such as ground pulverized fuel ash and granulated blast furnace slag has become commonplace in the specification of reinforced concrete. Typically concrete mixes are specified to contain roughly 15%-25% of these cement replacement materials, which are by-products from the power and steelmaking industries respectively.

The use of these replacement materials in reinforced concrete has been demonstrated to offer a number of advantages over ordinary portland cement in terms of the durability and long term performance of the concrete including:

- Improved watertightness through a denser concrete mix
- Reduced heat of hydration during setting and curing
- Increased resistance to chemical attack

These advantages are particularly significant for the durability of underground structures. Lowering the heat of hydration minimizes the temperature difference between ambient air temperature and the peak temperature within the concrete matrix during the setting process. By minimizing the temperature differential the incidence of problematic thermal cracking can be minimized. This is of particular importance for cut and cover and similar structures which feature large concrete pours with onerous conditions of restraint at wall/slab interfaces, as thermal cracks, in theory, will extend through the entire thickness of the concrete section.

Proponents of the use of cement replacement materials in concrete production advocate that the 30% utilization should be significantly higher and a content of 50%-60% cement replacement should be sought. Testing has demonstrated that this volume of replacement has no adverse effects on concrete strength or performance (Mehta, 2004). While the beneficial reuse of cement replacement materials is on the increase, there is considerable room for growth in the use of these materials in all the major construction markets. In 2008, the United States produced 136 million tons of coal combustion products. While approximately 45 percent were used beneficially, nearly 76 million tons were disposed of in landfills (American Coal Ash Association, 2008). It is clear that increased utilization for these materials can and should be supported.

By increasing the content of cement replacement materials to 50%-60%, a significant reduction in cement production—in the order of 25%-35% can be accomplished. This would correspond to a maximum reduction in cement production and $CO_2$ production of over 1 billion tons per year. Concrete mixes using higher percentages of cement replacement materials have also been demonstrated by testing to require 20% less mix water than corresponding mixes which are purely cement-based. In many regions of the world water is a precious resource. Based upon an estimated annual consumption rate of 240 billion gallons, a 20% reduction in concrete mixwater would result in 48 billion gallons of water saved per year.

However, the rate of gain of strength for mixes with high percentages of cement replacement material is slower than for portland cement concrete, which requires forms to be left in place longer. While time is money in our industry, this is a trade-off that must be made.

The sustainability and durability benefits for underground structures arising from the use of cement replacement materials are evident. Because natural resources will become scarcer and more expensive, the specification of 55–60% cement replacement material content in concrete mixes is recommended.

## *Gray-Water*

In addition to the 240 billion gallons of water consumed annually in the production of concrete, it was also estimated in the late 1990s that in the US alone, approximately 1.24 billion gallons of water were used to wash down mixing trucks (Chini & Mbwambo, 1996). It can be assumed on a global basis that the volume of wash-down water, or gray-water, can be currently measured in tens of billions of gallons. While this equates to a relatively small percentage of the total water usage, the reuse of this gray-water is a sustainable practice which should be encouraged to the maximum possible extent.

As gray-water contains cement fines, ultrafine aggregate particles and residual admixtures, it retains a high ph value and therefore must be treated prior to disposal. An alternative to this expensive process is to reuse the gray-water in the concrete batching process. This is already permitted by ASTM C-94 Specification for Ready Mix Concrete, subject to an upper limit on the total mix water solids of 50,000 parts per million, of 5% solids by weight of mix water.

Studies and testing undertaken by the University of Toronto suggest that the use of gray-water at the limit specified by C-94 has no adverse impact on concrete strength or setting time. Conversely, benefits of using concrete mixed with gray-water include improved pumpability and appearance. However, the primary benefit from a sustainability perspective is the reduction in the requirement for fresh water.

To ensure concrete quality is not compromised by using gray-water, it should be treated like any other concrete mix ingredient and tested regularly. Testing for water specific gravity, temperature and density (arising from admixtures) should be conducted and mix proportions adjusted accordingly to ensure specification requirements are met.

For structural pours it is recommended that gray-water use be permitted to the upper limit as specified in ASTM C-94. However, for non-performance mixes—including mud slab concrete, flow fill, pump grout, road bed, and walkways—the use of 100% gray-water is recommended.

## *Use of Recycled Aggregates*

In excess of 1 billion tons of construction and demolition waste are generated globally each year. This debris is traditionally disposed of at landfill sites. However, these material volumes and an increasing lack of available disposal sites demonstrate that this practice is not sustainable.

To date, concrete has most commonly been recycled for use as aggregate in roadway subbase. The use of recycled aggregates for this purpose has been promoted by the Federal Highway Administration (FHWA), resulting in recycled aggregates now accounting for approximately 5% of the total aggregate usage in the US.

However, the use of recycled aggregates in structural concrete has yet to find a widespread market. Its use has potentially been discouraged due to lack of knowledge of the history and properties of the recycled materials. Widespread testing of the properties of concrete using recycled aggregates has been performed in the US, the UK and Australia, which has indicated that up to 30% of recycled aggregates may be reused in structural concrete without any noticeable difference in strength or workability. Both Germany and Switzerland are now marketing structural concrete with recycled aggregates. In the US, 2,305 tons of recycled aggregate were recently used for foundations and tilt-up panels in the Enterprise Park at Stapleton project in Denver, Colorado (Construction Materials Recycling Association, 2009). It is reported that the Contractor noticed little, if any, difference in the recycled material, including the ability to pump and finish, and recorded higher end strengths than those found in traditional mix designs.

The expanded use of recycled aggregates in structural concrete is likely to be an area of continued

Table 3. LEED Certification for Tunnels

| Category | Credit | Credits Available | LEED Classification | Credits Required |
|---|---|---|---|---|
| Sustainable Sites | Site Selection | 1 | Certified | 10–12 |
| | Community Connectivity | 1 | Silver | 13–15 |
| | Alternative Transportation | 4 | Gold | 16–19 |
| | Protect or Restore Habitat | 1 | Platinum | 20–25 |
| | Maximize Open Space | 1 | | |
| | Stormwater Design (Quantity/Quality Control) | 2 | | |
| Water Efficiency | Innovative Wastewater Technology | 1 | | |
| | Water Use Reduction | 2 | | |
| Materials & Resources | Construction Waste Management | 2 | | |
| | Materials Reuse | 2 | | |
| | Recycled Content | 2 | | |
| | Regional Materials | 2 | | |
| Innovation & Design Process | Innovation in Design | 4 | | |
| | Total | 25 | | |

research and development within the concrete industry. However, based on the testing conducted to date it would be prudent to implement the use of recycled aggregates in structural concrete mixes on a limited basis. Therefore it is recommended that structural concrete mixes be specified to contain at least 10%–15% recycled aggregates, and this percentage be increased to 30%, or beyond, to the limits of workability for non-performance mixes.

## LEED Certification System for Tunnels

While specific sustainability measures can be implemented on a project-by-project basis, without some form of industry supported certification system, there is no benchmark against which to measure our performance. We can be reactive, and wait until the American Society of Civil Engineers or other professional body develops a plan for civil infrastructure, or we can be proactive and develop our own criteria that could ultimately be incorporated by ASCE or other entity. Therefore, as a starting point, it is proposed that the existing USGBC criteria be modified as necessary to suit the tunneling industry.

Applying the full range of USGBC LEED criteria to tunnels is obviously inappropriate, as many of the criteria simply do not apply. However, if these inapplicable criteria are removed, then a project checklist can be developed, tailored to the requirements of the tunneling industry. While some credits would appear to favor transportation tunnels and some to favor water/wastewater tunnels, it would appear that based upon the existing LEED system, approximately 25 of the 69 credits would be applicable to tunnel construction under the categories indicated in Table 3.

Using a similar proportion of points accumulated as the USGBC system, an accreditation system for tunnel projects would also be as indicated in Table 3.

The identification of specific credits and determination of credits required for classification would require significant discussion and agreement within our industry, but these aspects of the process are the tip of the iceberg. Larger questions of how any system would be managed, including requirements for certification of projects and professionals, must also be considered.

## SUGGESTED NEXT STEPS

Our current practice is clearly not sustainable. While individual projects may continue to seek best practices on a case by case basis, without a concerted effort by our industry, such efforts may go unnoticed. This paper has recommended specific actions, with the objective of reducing cement usage and thereby $CO_2$ production, and presented an approach to a more consistent approach to sustainable design within our industry. This is the starting point.

To promote best practice for sustainability within our industry the following are recommended:

- The development of a design criteria, including loads, load combinations and load factors, tailored to suit the specific requirements of underground construction
- Increased incorporation of temporary works into permanent structures
- The development of a LEED Certification System for underground structures
- Increased specification and use of recycled materials

It is recognized that some of the recommendations herein—including the development of design criteria and a LEED Certification System may take months or years to implement, as the subject matter is complex and industry-wide agreement is required prior to proper implementation. It is proposed, however, that the development of a design standard and sustainability criteria be accomplished through the Underground Technology Research Council (UTRC).

However, it is evident that the expanded use of cement replacement materials, gray-water and recycled aggregates is a keystone in the development of sustainable concrete. The increased use of cement replacement materials in particular is critical in reducing $CO_2$ production. The increased specification of these items must be supported and implemented now.

## REFERENCES

American Coal Ash Association, 2008. *Coal Combustion Product (CCP) Production & Use Survey Report.* http://acaa.affiniscape.com. Accessed December 2009.

Chini, S.A., and Mbwambo, W.J. 1996. *Environmentally Friendly Solutions for the Disposal of Concrete Wash Water from Ready Mix Concrete Operations.* CIB W89: International Conference in Building Education and Research. Beijing, October 21–24, 1996.

Construction Materials Recycling Association, 2009. *Recycled Concrete Aggregate Ready Mix Used in Structural Applications.* www.concreterecycling.org/histories.html. Accessed October 2009.

Duxson, P., and Provis, J. 2008. *Low $CO_2$ Concrete: Are we Making Any Progress?* Australian Council of Building Environment Design Professionals. 2008.

Mehta, P.K., 2001. *Reducing the Environmental Impact of Concrete,* Concrete International Magazine October 2001 pp 61–66.

Mehta, P.K., 2004. *High Performance, High Volume Fly Ash Concrete For Sustainable Development,* International Workshop on Sustainable Development and Concrete Technology. Beijing, May 20–21, 2004.

Mehta, P.K., 2009. *Global Concrete Industry Sustainability.* Concrete International. February 2009.

World Business Council for Sustainable Development, 2009. The Cement Sustainability Initiative—Recycling Concrete. www.wbcsdcement.org. Accessed October 2009.

## BIBLIOGRAPHY

ASTM C 94/ C 94M. 2009. *Standard Specification for Ready-Mixed Concrete,* ASTM International, West Conshohocken, PA.

United States Department of Transportation, Federal Highway Administration. FHWA-NHI-09-010. 2009. *Technical Manual for the Design and Construction of Road Tunnels—Civil Elements.*

United States Green Building Council. 2009. *LEED 2009 for New Construction and Major Renovations Checklist.* www.usgbc.org. Accessed October 2009.

Vickers. G., 2002. *Grey Water Recycling Basics.* The Concrete Producer. September 2002.

# Sacramento UNWI Sections 1 and 2 Project: Special Tunnel Construction with Plastic-Lined PCC Segments

**Jeremy Theys**
Traylor-Shea Joint Venture, Sacramento, California

**Michael Lewis, Greg Watson**
The Robbins Company, Solon, Ohio

**ABSTRACT:** The UNWI project was undertaken by Traylor Shea, a Joint Venture for the Sacramento Regional County Sanitation District. It features the first use of plastic-lined concrete segments in North America combined with rapid segment delivery and installation, tight curve radii, tunnel conveyor mucking and abbreviated startup. This inner lining prevents deterioration of the segments by corrosive sewer gases. Once the segments are in place, the joints between segments are heat-sealed together to form a complete liner. Unique solutions were devised to protect the plastic lining without sacrificing the high performance objectives of the system or the safety of the crew.

## INTRODUCTION

Plastic (PVC) lined segments have much to offer as long as they can be applied efficiently and effectively. They eliminate the corrosive effects of sewer gases collecting in the crown of the tunnel once in service, reducing the need to periodically drain and flush the interceptor tunnel. Due to their special characteristics, however, new techniques in handling, placement and protection after placement are required. In addition, heat welding of the lining seams within the TBM and backup provides early access for the work, ensuring it can be completed before being blocked by tunnel services.

This paper will outline the experience of adapting this type of segment to present TBM technology, and the techniques that had to be developed to address this challenge. The segments were used on the Upper Northwest Interceptor, Sections 1 & 2 Tunnel project in Sacramento, California, which was the first application of this lining in the U.S. This project was undertaken by the Traylor Shea Joint Venture for the Sacramento Regional County Sanitation District, designed by URS and managed by Hatch Mott McDonald. The geology was clay and running sand. In addition to the 60 inch wide PVC lined segments, the tunnel alignment included 10000-ft radius horizontal curves. A 4.2 m (13.9 ft) Robbins single shield EPB TBM backup and continuous conveyor were selected to excavate the tunnel. The TBM featured tungsten carbide knife edge bits and triple-row disc cutters, as well as active articulation to assist in steering.

## PLASTIC (PVC) LINED SEGMENTS

The PVC lining was embedded into the concrete I.D. of the segments and once the segments were installed the liner was fully heat welded to ensure a continuous "pinhole free" liner. The Traylor-Shea Joint Venture designed the precast concrete segments to accommodate flaps overhanging the edge of the segments. These flaps were incorporated to facilitate welding one segment to the next and to minimize the PVC welding materials and labor needed. Once the segments were erected and grouted in place, the joints or flap edges were joined using a separate welding strip. The welding strips were applied manually using electric heat guns (see Figures 1–2). The strips melted with the application of heat and the molten material created the bond.

The flaps could have been placed on either the radial or circumferential joints or both. Which edges the flaps are placed on is dependent on whether or not the segments are universal, as well as the desired build orientation. For the UNWI project, the flaps were used initially on the radial joints and then eventually also used to the circumferential joint. The flaps were designed to provide the same width as the primary joint strip overlap, so that no loss in sealing capacity was recognized. The size and location of the flaps had also to be considered in the delivery sequence, storage and handling of the segments. The

**Figure 1. PVC lined segments with flaps**

**Figure 2. Sealed radial and circumferential joints**

144" segment ID was relatively small compared to the 60" segment width, which made transport of the segments through the backup already challenging. The overhanging flap on the circumferential edge of the segments added to their width, making transport much more difficult. The flaps also had to be considered throughout the segment delivery and handling processes in the shafts and tunnel. The flaps on the radial joints were less of a problem since they projected fore and aft at that stage while passing through the backup. Alignment dowels at the circumferential joints would also have added to the segment width if they had been inserted before being brought into the tunnel. Therefore, all segment dowels were shipped to the heading and installed in the segment build area immediately before erection at the face.

To maximize efficiency and to satisfy the tight contract schedule, a certain minimum portion of the welding process had to be completed in front of (or somewhere along) the backup. This was in areas that would later be blocked by utilities and services installed to the tunnel wall near the end of the backup. These areas included the continuous conveyor, fan-line, high voltage cable, compressed air, water, dewater, grout and accelerator lines and track.

Workstations for welding were located in proximity to these locations and provision for this had to be considered in the backup design. Well-designed work stations were important because the welding was a continuous process throughout tunneling excavation. Furthermore, the welding process demanded appropriate lighting and power takeoffs for the 220V weld guns.

To permit welding of the PVC joints at all locations, three distinct work areas were provided on the TBM backup gantries. To access the lower portions below springline, the PVC welders had to weld the joints in the area between the segment feeder and the front of Deck #1. This area was often congested as it was the travelway for the precast concrete segments from the TBM backup to the segment feeder.

A dedicated workdeck was also installed in the same area of the TBM, immediately behind the discharge chute of the screw conveyor. This deck, along with numerous welding gun power takeoffs, proper lighting and storage areas to support the welding operation, was used to weld all the PVC joints in the upper portion of the tunnel. It was critical to the process as it allowed the only access to the crown before it was blocked by installation of the mainline conveyor. As it was the only possible location, the welding process had to share the work deck area with the fresh air duct. The duct was made flexible so that it could be pushed from side to side to access the entire crown. In this way, the 15' long deck provided sufficient access to this area of the tunnel to allow efficient welding of the joints as the tunnel advanced (see Figure 3).

The last work area in the process allowed welding of the PVC joints at the approximate springline locations of the tunnel on both sides. Conveniently, the two backup decks designed for installation of the mainline conveyor gear provided adequate access to both springline locations for PVC welding. Adequate lighting and power takeoffs completed the preparations to allow welding in this location.

As the mining progressed, Traylor Shea JV made numerous improvements to the welding process that maximized efficiency, including staging various laborers at each workstation during mining and staging materials in critical areas within easy

Figure 3. Work area with fresh air duct

Figure 4. Heat welding of seams

reach. The Resident Engineer's tunnel inspection team provided oversight of the process as well as proof testing of the welds using rounded-edge putty knives. Traylor Shea provided testing checks of the welds using Tinker Rasor spark testing devices. Final testing and repairs remained to be completed at the time of this writing (see Figure 4).

Traylor Shea JV scheduled the PVC welding to proceed as the mining progressed and planned to achieve at least 50% of the joint welding during the tunnel drive. To make up for lost time in the TBM manufacture/delivery phase, TSJV hired additional labor (certified PVC welders) to increase the production and makeup time on the schedule, eliminating a lot of the welding that was scheduled to take place after the excavation phase was complete. At the 85% excavation point, TSJV had achieved much higher completion percentages than scheduled. Average PVC welding rates were roughly one linear foot of weld per minute.

## TIMING

Traylor Shea was focused on minimizing construction time by maximizing the rate of advance. As such, rapid segment delivery and placement were key factors.

Based on the long tunnel length and use of conveyor belt to remove the tunnel spoils during the excavation, the JV decided it would be advantageous to design the TBM to handle two rings worth of segments on each load. This design resulted in the need to deliver a train full of segments only once every hour, reducing the number of trains in the tunnel and eliminating the need for a passing track. However, two rings worth of segments would cause the segment handling area and trolley to become quite long, making a very long trip from the furthest segment stack to the segment erector. Such a trolley would have to travel so fast to supply the erector during the build that it would not be feasible or prudent.

The remedy was to implement a system which included segment lifters and a segment feeder. The segment lifters allowed the two rings worth of segments to be offloaded and stored on the backup in an instant, allowing the trains to immediately return to pick up another load. The feeder stored enough rings immediately behind the erector to keep it supplied, even though the erector used segments faster than the trolley could bring them. The feeder became depleted at the end of each build but there was enough time to refill it during the push. (see Figure 5)

The time study, shown in Figure 6, was used to determine the speed and capacity parameters for the segment lifters, hoist and feeder. The cycle time was designed to allow both the advance of the TBM and erection of the 6-piece segmental liner in an approximately 30 minute timeframe, which was composed of 15 minutes of ring building and 15 minutes of mining. This cycle resulted in a segment feeder capacity of three segments. To ensure an uninterrupted supply of segments, four sets of segment lifters were included in the backup design to collectively handle 2 complete rings.

## THE FEEDER

Ordinarily, segment feeders run on wheels. Feeders need to be low profile enough so that they can insert the segment below the erector pickup. This limits the possible diameter of the wheels. Given the small diameter and low contact pressure allowed for the wheels, many wheels would have been required. Ensuring that all of these wheels would have equal loading, especially in a tight curve, seemed impractical so the decision was made to suspend the feeder from beams running between the TBM and backup. These beams also served to tow the backup and support walkways along the feeder leading to the TBM.

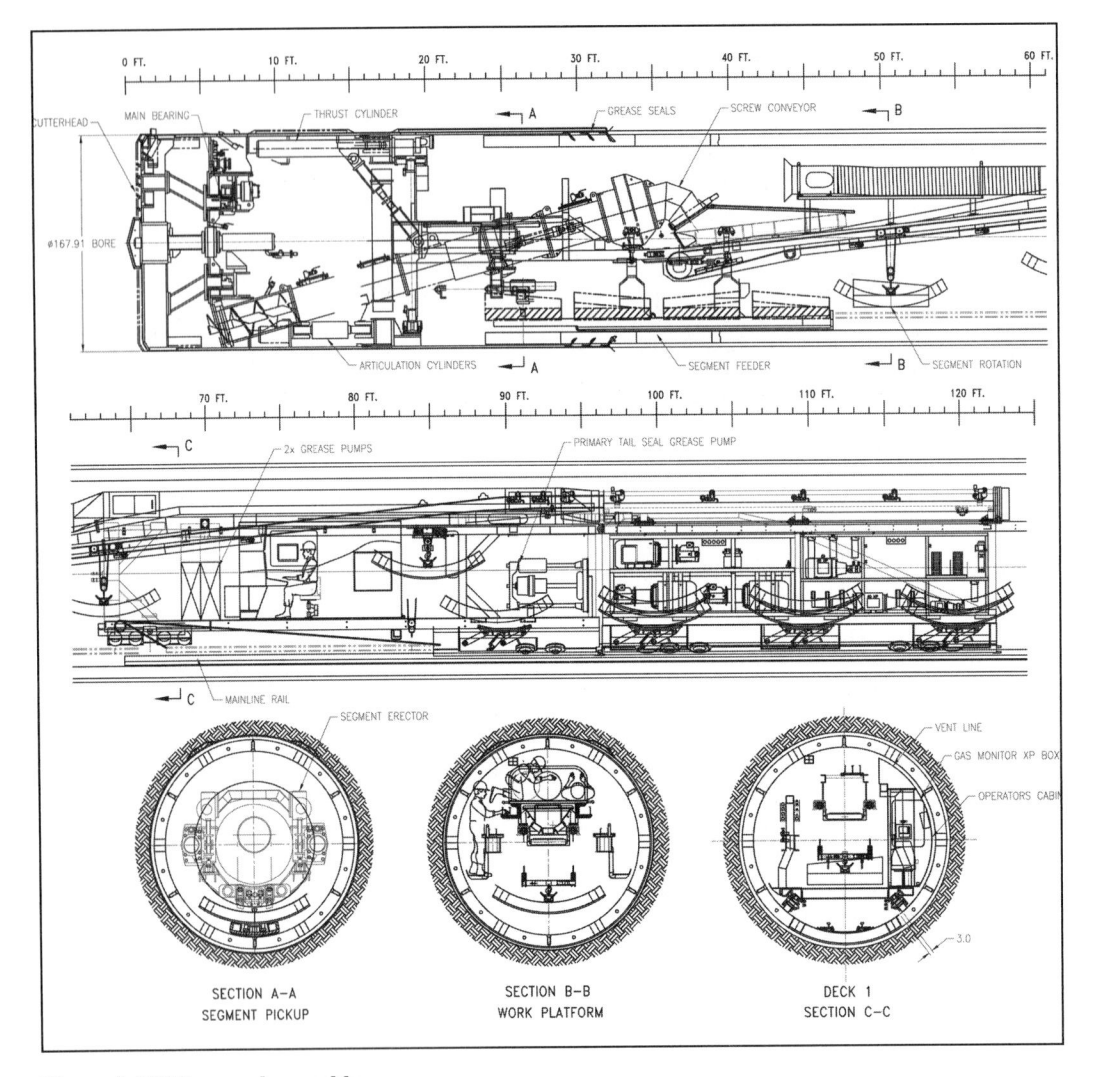

**Figure 5. TBM general assembly**

Because the feeder was much closer to the TBM, most of the feeder load was transferred there. The towing frame of the TBM had to be heavily reinforced to accept this load. This setup was difficult because there was only so much clear space inside the erector and most of this was taken up by the screw conveyor and the many hoses and cables. One consequence of the design was that access into the shield was more limited than it would have otherwise been.

TSJV's design specifications called for a 750 ft horizontal and 1,500 ft vertical curve capability. Clearance both inside and outside of the feeder suspension system had to accommodate these relatively sharp curves as well as the PVC flap on the segment. To provide the maximum clearance for the extended segment with the erector pickup head, the feeder was suspended on four hydraulic cylinders. These lifted the feeder off of the lining for the push but lowered it again prior to feeder extension. This design prevented the suspension system from being damaged in the event that the erector pickup was accidentally extended instead of retracted when picking up the segments (see Figures 7–8).

## ABBREVIATED STARTUP CONFIGURATION

The starting pit contained approximately 130 ft of useable length, meaning that the length of the TBM plus backup, at startup, could not exceed this

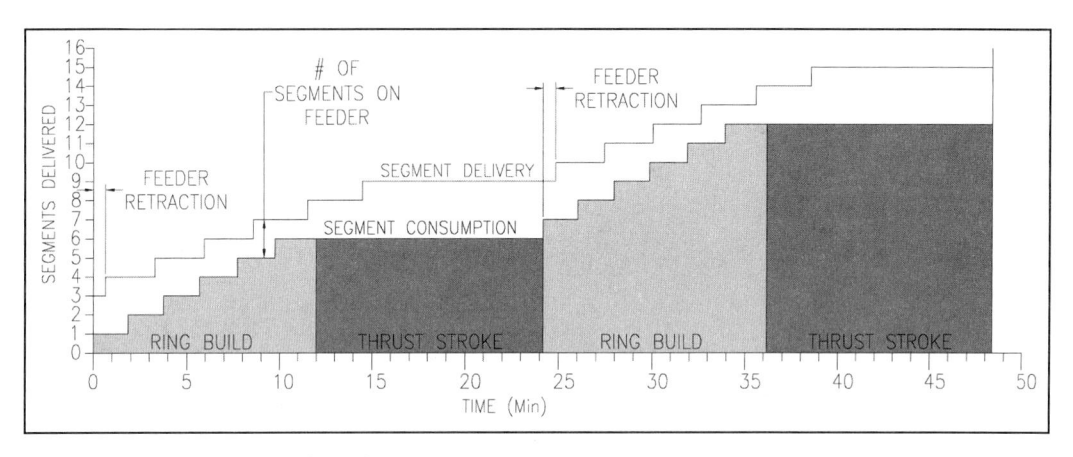

**Figure 6. Segment building time diagram**

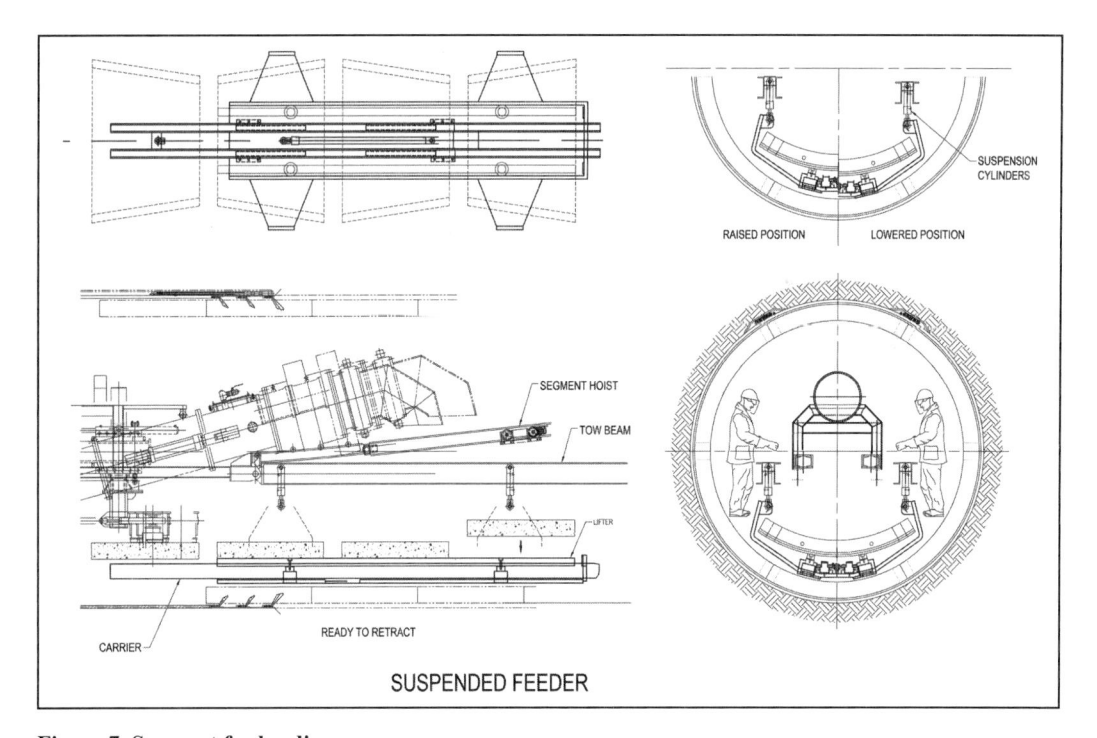

**Figure 7. Segment feeder diagram**

distance. This was initially expected to be sufficient for the bridge and three decks behind the TBM, as there was 25 ft between the TBM and deck 1. Four decks would contain all of the hydraulics but none of the electrical. However, the 25 ft space between TBM and backup was barely sufficient for a short feeder and the installation of half length rails (16.5 ft). Though beneficial for startup, this short arrangement would have been a handicap for the remainder of tunneling. The development of a started tunnel was not practical in the pressurized soil environment and so the most compact starting arrangement had to be devised while still preserving efficient segment storage as well as the ability to set full length (33 ft) rails (see Figure 9).

## DECK 1 STRUCTURE

To minimize roll-back of liquefied spoil, the incline of the belt was limited to 6°. At this angle, to reach

**Figure 8. Installed segment feeder**

**Figure 9. Abbreviated startup configuration**

the elevation of the main run of the conveyor, the bridge needed to be 56 ft long. Given the compressed startup length, this put the rear support point of the bridge near the rear end of deck 1 and caused the bridge to run right through the deck. This meant that the movement of the bridge, due to curves, that normally occurs out in front of deck 1 now had to be accommodated within it. Due to the severity of the curve requirement, the amount of this movement was greater than normal. This movement meant that upper portion of the structure had to be open at the forward end to clear the bridge. The segment hoist, being an integral part of the bridge, also experienced significant lateral movement in a curve. Therefore, the structure had to be contoured to clear the already wide segment when carried along an offset path.

The 33 ft rail requirement, together with the short startup, meant that the rail setting activity would have to take place mostly within the length of deck 1. In order for this to occur, the forward portion of the bottom deck needed to be open. Having both the top and bottom of the deck open at the front would have been ideal from a functional point of view. It was, however found to be impractical from a structural point of view. With an open bottom the front wheels had to contact the tunnel walls at about 4:00 and 8:00. This created additional inward forces, which would have acted to collapse the C shaped deck. In the end a transverse brace was added to the bottom deck at the front. The brace made setting the rail more challenging but still practical. This brace was later lowered to make it less of an obstruction to foot traffic while still allowing rail installation (see Figures 10–11).

## WHEELS IN CONTACT WITH PVC LINING

Since the rails were not in place ahead of deck 1, the front wheels had to run directly on the PVC tunnel

**Figure 10. Track laying process**

lining. Due to the delicate nature of the PVC inner lining and its critical importance, great care had to be taken to prevent damage during installation, or by contact with the backup. Except for the bottom drainage, the PVC lining had to be 100% sealed so any damage had to be fully repaired. Ultimately, the lining was 100% spark tested so quality assurance of welding and prevention of damage was a must. Since some of the backup wheels rolled directly on the lining, the acceptable contact load had to be determined

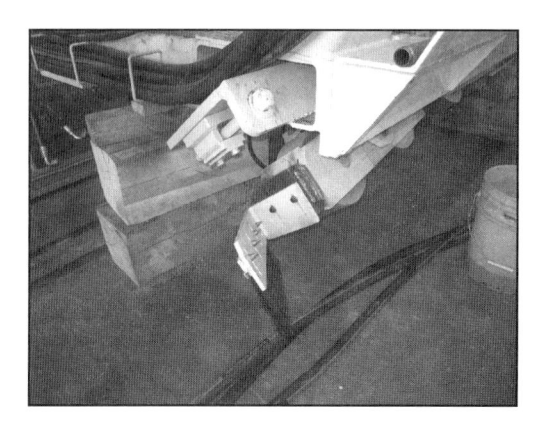

**Figure 11. Elastomeric wheel assemblies with brushes and adjustable alignment**

**Figure 12. PVC liner testing**

for the particular wheel type used. Traylor Shea JV performed loading testing of elastomeric wheels on the Ameron T-lock sheets to determine loading limits and prevent damage. The testing was carried out on site using 60 durometer urethane wheels left over from previous projects. The wheels were tested by pressing the wheel into the PVC sheet, which was then held for up to 24 hour increments using hydraulic cylinders to apply varying loading. The testing resulted in a contact pressure of 1,000 psi, or about 6,000 pounds of force per wheel. Once the critical loading pressure was determined, a series of tests were performed to determine any damage caused by movement of the wheels under the high pressures, namely rolling damage and damage caused by embedded grit from a dirty liner. Traylor Shea determined that the sideways loading on the wheel edges did not damage the PVC liner. However, dirt and debris on the PVC sheet caused very serious damage to the liner as the loaded wheel pressed the grit into the sheet. This caused enough damage to result in a failed spark test. Once this analysis was performed and the results realized, a cleaning system in front of the urethane gantry wheels was deemed to be necessary. TSJV recommended that brushes or air puffers be used to push any such material out of the path of the wheels. Brushes were ultimately supplied. Throughout the tunnel drive the brushes provided adequate cleaning capability to prevent debris from become embedded in the liner (see Figure 12).

The most direct way to protect the lining from the wheels is of course to reduce the load as much as possible. Typically, the front wheels of deck 1 are among the highest loaded wheel positions on the backup because they normally support the rear of the bridge conveyor, segment hoist, segment cars, TBM equipment and half the weight of the deck itself. In this case though, the largest portion of the load was contributed by the full segment feeder. As mentioned earlier, most of the load was supported by the TBM but the rest was supported at the front of deck 1.

The front half of the deck structure was another component of the load but this was already minimal due to the open bottom and top at the forward end. There was very little equipment mounted on the deck and therefore no ballast required either. Only two small grease pumps and the operator cab occupied most of the deck length, and they were not very heavy. The larger tail seal grease pump was located at the rear of the deck but sat directly over the rear wheels so it did not contribute to the front wheel load. Only one set of segment lifters was located on this deck at the rear, so only one segment car would reach that location. It carried 3 segments and weighed 7.5 tons. The car and the lifting mechanisms were located just ahead of the rear wheels and so didn't contribute significantly to the front wheel load.

The rear half of the bridge is normally supported at the front of deck 1 and so is typically a major component of its front wheel load. As noted above, however, due to the 6° maximum belt inclination the rear of this bridge rested near the rear of the deck. Given its minimum length, it would have ended directly over the rear wheels. The bridge, therefore would not have contributed to the front wheel load. Instead of simply allowing the bridge weight to remain neutral, however, the bridge weight was used to counteract the load on the front wheels. The counterbalancing was done by placing the support point as far behind the rear wheels as possible while still remaining on deck 1.

The monorail or segment hoist along with its cargo were supported by the bridge and so their weight also contributed to the force behind the rear wheels, thus acting to reduce the load on the front wheels.

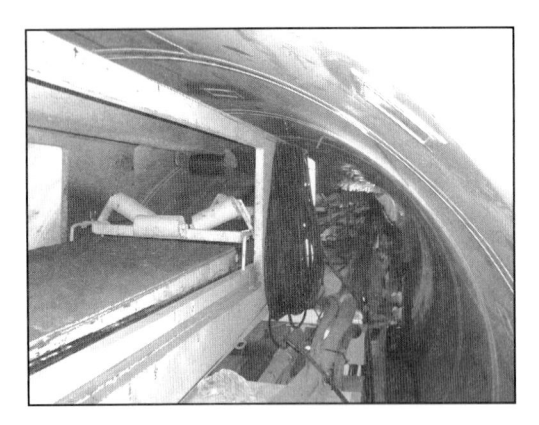

**Figure 13. Rear PVC welding station and continuous conveyor tailpiece**

A downward force is created on the front wheels any time the towing connection at the front of deck 1 is located above the couplers pins between decks. The towing load on a backup can be quite significant particularly if the backup is long. The equipment on long backups is placed on only one side, therefore requiring ballast and bronze bearing wheels– as was the case in Sacramento. This load created a moment based on the vertical distance of the towing point above the deck couplers and was reacted by the front wheels. To remove the load, a hydraulic cylinder was placed at the top of the deck structure between decks 1 & 2. The cylinder passed half the towing load into the deck 2 upper structure where it could be counteracted by the weight of decks 2 and 3. To avoid complex pressure control, the cylinder was simply connected in parallel with the main tow cylinders. In this way, exactly half of the tow load was directed to the top of the structure and half to the bottom. This arrangement removed the moment and thus the vertical component of the wheel load.

Taken together, all the factors above resulted in a wheel load of approximately 2,000 lb per wheel— well below the acceptable load of 6,000 lb determined by testing.

Beyond minimizing the wheel load, bogies were designed to fully equalize the wheel loading. This meant that each two-wheel bogie was mounted on a center pivot. The two bogies were then mounted on a larger bogie which itself had a center pivot connecting it to the deck. In addition, the front wheel assemblies featured adjustable alignment to minimize skidding.

## CONTINUOUS CONVEYOR

A crown-mounted tailpiece was selected for this project. This type of arrangement allows better use of the backup space. With side mounted conveyors, any deck space to the rear and on the same side as the conveyor is essentially useless. In a smaller tunnel, the opposite side of the backup will usually be occupied by a dedicated walkway. This arrangement is used because that space is already occupied by the erected conveyor. The typical solution is to put the tailpiece as far back as possible. This, however, makes for a long transfer conveyor and makes it more difficult to combine the transfer conveyor and continuous conveyor into a single belt.

Combining the two conveyors can be beneficial in that it allows at least one, and possibly two, transfer points to be eliminated. In small tunnels, the head room saved through the elimination of a transfer point can be very valuable. This was the arrangement used on the UNWI project. Due to space constraints, the segment hoist, bridge conveyor and continuous conveyor were integrated into a single unit. A consequence of the design was that the return pulley, just under the TBM discharge, was much larger than it would otherwise be. This cramped the space for bypassing the feeder with the segment but was still utilized successfully.

At the tailpiece, the vent ducts were run downward into deck structure below to allow unobstructed access to the tailpiece. This design also provided for better access to the PVC lining for welding.

## SAFETY

The PVC liner sheets, when wet, were slippery and presented a slipping hazard. Though it was not practical to place walkways and scaffolds in every work area, it was nevertheless a priority in the design criteria. Designers found, however, that the problem could be easily avoided by performing a rough cleaning of the lining after segment installation. Complete cleaning of the joint was performed just before the welding was to take place. Even then, only the area local to the welding location was cleaned so that the remaining dust could provide a safe amount of traction for walking. To further address the slip hazard, the backup gantries were designed to ride on mainline tunnel rail. This design prevented the workers from the constant hazard of walking on the slippery PVC surface. It also prevented damage from the equipment and processes taking place along the TBM backup gear during normal mining operations.

Fire hazards were also of concern and therefore all consumables (greases and oils) used were fire resistant. Also, the conveyor system booster drives were outfitted with temperature sensors. The conveyor system was outfitted with many sensors which were tied into the global PLC control and monitoring network. These sensors included slip sensors, heat sensors, drive amp monitoring, etc. These sensors transmitted the status to the PLC network many times per second. Traylor Shea implemented many alarms monitored on the TBM operator and office monitoring screens in case any of the sensor readings approached levels that required corrective action. All of the sensors were tied to an automatic shut down in case dangerous levels or hazards were indicated.

In addition, before starting the mining operation, Traylor Shea performed rigorous fire testing on the PVC liner to determine the flammability hazard. The testing proved that the fire would not propagate through the PVC sheet.

## CONCLUSION

The PVC lined segments used in the UNWI Sections 1&2 Tunnel Project, along with high performance expectations, tight curves and limited startup length, prompted many new considerations in design, implementation and operation. The solutions and techniques developed on this project have contributed toward what is fully anticipated to be the highly successful first application of PVC lined concrete segments in North America. The high advance rates achieved so far, placing the project well ahead of schedule, have demonstrated the viability of this technology going forward.

# Sustainable Underground Structure Design

Lei Fu
URS Corporation, Gaithersburg, Maryland

**ABSTRACT:** Sustainable design is an integrated design process that complies with the principles of economic, social, and ecological sustainability. The philosophy of sustainability should be applied at various phases of an infrastructure system—the planning, design, construction, and operation. This paper reviews the current practice in sustainable infrastructure design, especially underground structure design, and discusses issues related to the application of the philosophy of sustainability to the design and construction of underground projects.

## INTRODUCTION

The concept of sustainability was first introduced by the Brundtland Report (United Nations, 1987). Sustainable development is "development that meets the needs of the present without compromising the ability of future generations to meet their own needs" United Nations, 1987). Since then, sustainability has been widely accepted as an important consideration for projects world wide. Sustainable design is now generally considered as an integrated design process that complies with the principles of economic, social, and ecological (triple bottom line or three pillars) sustainability. In Figure 1, sustainability is shown as the overlap of the three pillars. Sustainable or green programs such as the Leadership in Energy and Environmental Design (LEED) and the Green Globes system have been developed to promote sustainable practices. These programs focus on buildings and neighborhood development.

## SUSTAINABLE INFRASTRUCTURE

Infrastructure sustainability is a growing area of interest in practice and research. Some agencies of the United States, such as EPA, FTA, FHWA, DOE and DOT, have included sustainability in their mission statements or action plans. FHWA initiated the Green Highway Partnership in 2002 to help sustainable road design. More recently, FTA adapted LEED and the Energy Star systems for the construction of transit buildings, "because of the similarities of transit buildings to other building structures." However, "they may need further analysis and development to be applicable to other transit facilities such as subway stations, tunnels and bridges, considering their unique construction and operational characteristics" (FTA, 2009). There also has been a substantial increase in research on sustainable infrastructure systems. Sahely et al. (2005) proposed a framework for sustainability assessment of urban infrastructure system. Jeon and Amekudzi (2006) evaluated transportation systems through the development of a sustainability index. University of Washington and CH2M Hill developed a sustainable rating system, Greenroads (UW and CH2M, 2010), for sustainable road design.

Underground structures include road tunnels, transit tunnels, water and wastewater tunnels, underground offices, underground storage spaces, etc. On the one hand, underground construction provides an environment friendly alternative to its corresponding on-surface development. Sustainable benefits of tunnels and underground space include low environment impact, efficient use of land parcels, less visual impact, reduced noise and air pollution, etc. On the

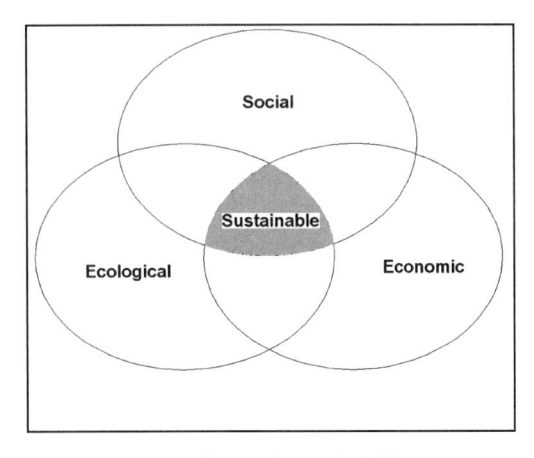

Figure 1. Three pillars of sustainability

other hand, green practices inside the tunnel industry are at an early stage and need to be promoted to keep pace with other industries. Currently, there are very few green programs for underground construction. There are no rating type sustainable frameworks available. However, sustainable practice guidelines/strategies of certain agencies may be used for the underground projects within the corresponding agencies. In the following sections, sustainability methodology is first discussed and sustainability strategies for underground construction are then presented.

## SUSTAINABILITY METHODOLOGY

A number of sustainability frameworks exist. They generally fall into two categories: rating (quantitative) systems, such as LEED (USGBC, 2009), Green Globes (GBI, 2004), and Greenroads, and green guideline or strategy (qualitative) systems, such as GreenGuide (ASHRAE, 2006) and LID National Manuals (PGDER, 1999). In a quantitative system, the performances of sustainable criteria are measured by numbers. In a qualitative system, only practice strategies, guidelines or tips are used to help green practices.

### Rating Systems

A rating system credits a project in several sustainability indicator categories. Each indicator is described by a value, $s$, such as "yes = 1 point" or "no = 0 points." Sometimes, the weights, $w$, are also used to reflect the priorities of indicators. The individual values are added together as shown below to create an aggregate score, $S$, as a measurement of the overall sustainability performance a project.

$$S = \Sigma ws$$

For example, Greenroads credits a project in the following categories (UW and CH2M, 2010):

- Project Requirements
- Environment & Water
- Access & Equity
- Construction Activities
- Materials & Resources
- Pavement Technologies
- Custom Credit

Similar to LEED, a project may be certified to one of the following levels based on total points achieved (UW and CH2M, 2010):

- Certified: All Project Requirements +32–42 Voluntary Credit points (30–40% of total).
- Silver: All Project Requirements +43–53 Voluntary Credit points (40–50% of total).
- Gold: All Project Requirements +54–63 Voluntary Credit points (50–60% of total).
- Evergreen: All Project Requirements +64+ Voluntary Credit points (>60% of total).

A star diagram can also be used to present the performance of an alternative. In the diagram, each sustainable indicator or category can be shown on one finger, giving a bird's eye view depicting the performance of each category or each indicator. The center on star usually designates the minimum allowed score for each criterion or category. The outer unit polygon represents the maximum score. Figure 2 shows a star diagram example using the sustainable categories for underground construction described in this paper later.

### Green Guideline Systems

This type of systems provide practice guidelines to instruct green practices at various stages of a project, including planning, design, construction, operation, and demolition. One example of such systems is the Low-Impact Development Design Strategies: An Integrated Design Approach (PGDER, 1999). The document provides "an overview of LID strategies and techniques and describes how LID can achieve stormwater control through the creation of a hydrologically functional landscape that mimics the natural hydrologic regime." The purpose is to "share some of our experiences, and show how LID can be applied on a national level."

## SUSTAINABLE UNDERGROUND CONSTRUCTION

For an underground construction, sustainability can be assessed in five categories:

- Site and underground space planning
- Structure design
- Construction and resource conservation
- Operation and maintenance
- Retrofit and upgrade

In the following sections, underground environments and resources are discussed first and sustainability strategies are presented later.

### Underground Environments and Resources

Surface environments include lands, surface waters, air, trees, natural habitants, existing developments, and human beings, etc. Underground environments include soil, groundwater, existing underground facilities, etc. Surface developments, such as buildings, bridges, and roads, mainly affect surface environments and potentially impact underground environments. On the contrary, underground

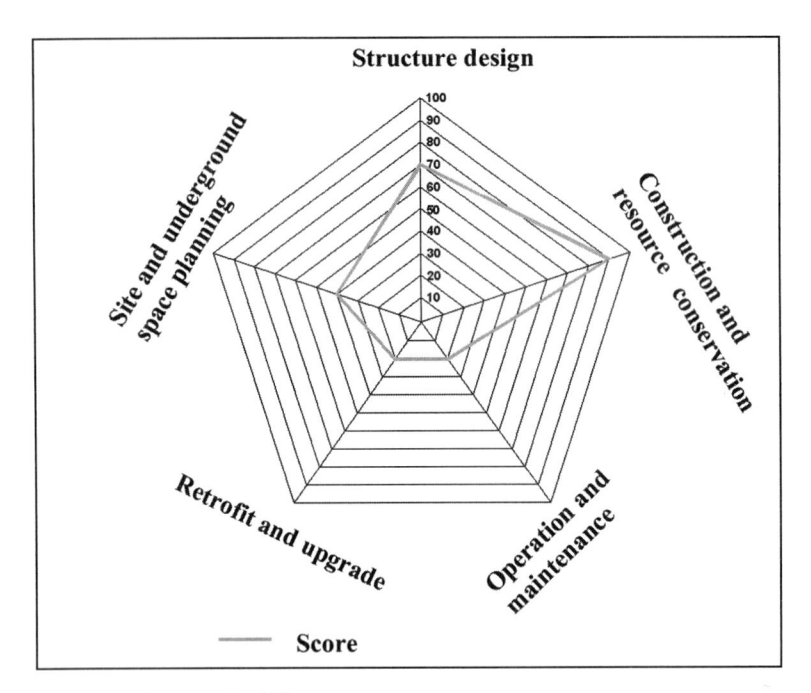

**Figure 2. A star diagram for sustainability assessment**

developments can avoid or greatly reduce the impacts on the surface environments and have potentials of large effects on the underground environments. Sustainable underground construction focuses on protection and sustainable use of the underground resources, including underground space, groundwater, geo-materials, and geothermal energy, etc., and, at the same, minimizing affects on the surface environments from underground construction.

## Site and Underground Space Planning

Like the above ground space, the usage of the underground space needs to be planned to achieve sustainability. The following several aspects need to be considered:

- Selection of sites and alignments
- Protection of underground soil and water resources
- Usage of underground spaces

The alignment and portals of a tunnel project is typically determined by other factors, such as the alignment of the road it connects to for road tunnels, the locations of stations for transit tunnels, the locations of treatment plants for wastewater tunnels. However, whenever possible, the following factors need to be considered: environmental impacts of portals in rural areas, ground settlement from tunneling, and ground conditions along an alignment and their effects on construction costs.

Groundwater is a very important natural resource for human beings and other habitants. Generally, it is in movement status. Hydrogeologic conditions of a project site needs to be clearly understood during the planning stage of the project. The effects of the project on the local hydrogeologic conditions need to be evaluated. Groundwater flow direction should not be changed from the development. Where a hydrogeologic balance is essential to support local and regional environmental regimes, the planned underground construction should not disturb these regimes.

Historically, usage of underground space is poorly planned. Different underground facilities are installed by different agencies or developers at different times. The underground spaces in urban areas are very crowded. Laws or regulations regarding usage of underground space could be established as in Japan—"in 2001, a law was voted to restrict the private use of underground at 40 m ~50 m. "This gave birth to two kind of planning: shallow land planning which result in the major part in trying to improve the existing infrastructures and public project planning in the deep underground" (Maire and Blunier, 2006).

An efficient way of using underground space is to use multi-utility tunnels (MUTs). A multi-utility tunnel is "any system of underground structure containing one or more utility service which permits the placement, renewal, maintenance, repair or revision of the service without the necessity of making excavation; this implies that the structure is traversable by people and, in some cases, traversable by some sort of vehicle as well" (APWA, 1997). In the United States, such systems are used in universities and other large campuses. Legal and other constraints prevent more common usage of MUTs. Figure 3 shows configuration of a MUT.

An innovative MUT project is the Stormwater Management and Road Tunnel (SMART) in Kuala Lumpur, Malaysiais. It is a combination of road tunnel and water tunnel. The 3-level 6- mile long tunnel is to solve the problem of flash floods and also to reduce traffic jams in the project area. It has three operation modes (Figure 4). Under normal conditions, no flood water is in the tunnel and the tunnel works only as a road tunnel (Mode 1). When there is a major storm, the tunnel will be closed for traffic and the tunnel is used to pass flood only (Mode 3).

When there is a moderate storm, the flood water is diverted into the lowest level and the upper two levels are still open for traffic (Mode 2).

## Structure Design

Tunnel structures include temporary supports (anchors, bolts, shotcrete, slurry walls, sheet piles, etc.), final liners (precast concrete segments, cast-in-place concrete, steel pipes, etc.), waterproofing systems, connection joints, and inside structures (road slabs, ceiling slabs, etc.). Structural design strategies include providing information on durability of structural components, considering maintenance measures and costs, and considering renovation or demolition of the structure.

Durability or good long-term performance is a key aspect of structural sustainability for the following reasons:

- Water tunnels, subways, and road tunnels, belong to life lines. Failures or frequent shutdowns for maintenance are highly undesirable.
- It is very difficult to repair, replace, or upgrade underground structures.
- Durable structures may be more cost-effective considering the life- cycle cost of a project.

Fu (2008) discussed typical problems of underground structures and approaches to improve long-term performance of underground structures. Following are some considerations for sustainable underground structural design:

- Use designs that have good long-term performances, such as use cast-in-place (CIP) concrete instead of shotcrete as tunnel permanent tunnel lining.
- Use durable materials, such as use clay or concrete pipe instead of steel pipe as sewer pipe.

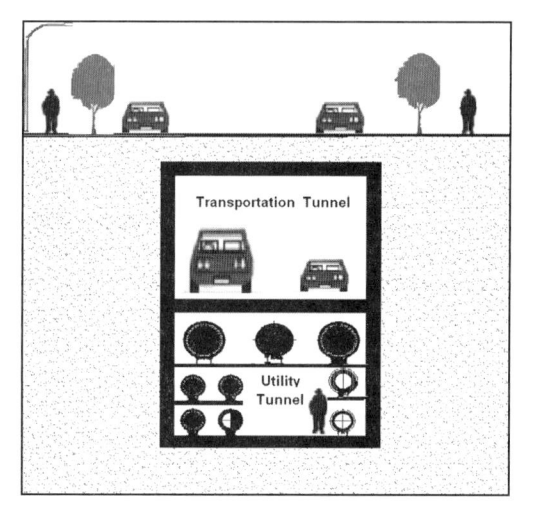

**Figure 3. Sketch of a MUT system**

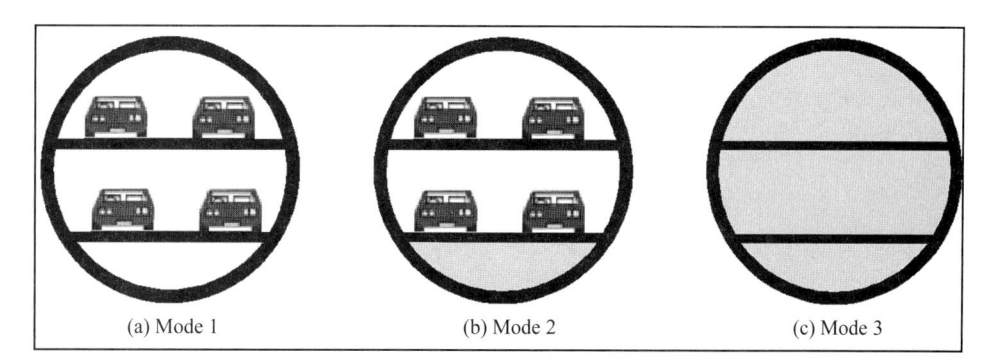

**Figure 4. Operation modes of the SMART**

- Use designs that have relative low expectations on construction quality as the construction quality of the underground structure is hard to be fully inspected.
- Improve the design of tunnel waterproofing systems and construction joints. These are most problematic in existing tunnel projects.
- Reduce operational energy. For example, the tunnel alignment can be designed to take advantage of the natural air flow in a road tunnel to eliminate the need for a mechanical ventilation system.

**Construction and Resource Conservation**

Regarding the sustainable construction, LEED includes the following aspects: materials for construction, recycling and re-use, resource use, and energy efficiency. Additional, there are two important issues related to the underground construction: ground settlement and groundwater contamination. Followings are some considerations for sustainable underground construction:

- Use construction methods that can achieve high construction qualities.
- Use underground resources including geomaterials, groundwater, and geothermal energy. Excavated soils and crushed rocks can be used as aggregates of concrete lining of tunnels and shafts and backfill materials.
- Minimize the affects of construction induced ground settlement on surface structures and underground utilities. Tunnel Boring Machines (TBMs) with positive tunnel face control can effectively control ground settlement.
- Minimize material consumption and energy consumption during construction.
- Avoid or minimize drawdown of groundwater which may lead to ground surface settlement or affect ground habitants.
- Avoid contamination of underground water resources from construction materials or grouting.
- Avoid vertical contamination of water resources from vertical construction and installation of wells.
- Avoid contamination of groundwater from sewer leakage from tunnels into surrounding soil.

**Operation and Maintenance**

Maintenance is costive for underground structure and is very difficult to perform for certain tunnel components. Over the life of a project, operations and maintenance expenses often far exceed the initial cost of the project. Project delivery methods, such as the Design-Build-Operate-Maintain (DBOM) can incorporate operation and maintenance knowledge and requirements into design. Life-cycle cost tools can be used to facilitate decision making. Decisions made in a project design stage have strong affects on the performance and life- cycle cost of a project.

**Retrofit and Upgrade**

For a tunnel project, ideal strategy at the end of the project life is to reuse it. Retrofit or upgrade of a tunnel is an issue need to be taken into consideration in the design phase of a project. Factors need to be considered including potential retrofit and upgrade techniques, reduction of tunnel size from the retrofit, future tunnel dimension requirements, etc. For example, the alignment of a utility tunnel can be designed to be straight to facilitate the application of the pipe basting method for the tunnel retrofit in the future.

## CONCLUSIONS

Concept of sustainability has been widely accepted in various sectors of infrastructure systems. Some agencies already included sustainability in their mission statements or action plans. In the transportation industry, sustainable programs, such as Greenroads and LID National Manuals, have been developed and applied in some projects. The underground construction is a green alternative to its corresponding on-surface development. However, the application of sustainability in underground construction is still at an early stage. There are very few green programs developed for underground construction, not to say an industry-wide accepted program. This paper discussed strategies regarding the design and construction of a sustainable underground project in the following five categories: site and underground space planning, structure design, construction and resource conservation, operation and maintenance, and retrofit and upgrade. Emphasis was put on sustainable use of underground space and sustainable structural design. Many of these are in the frame of green tips rather than quantifiable indicators. Eventually, rating systems including various sustainable indicators for sustainable underground construction need to be developed as the tunnel industry moves forward.

## REFERENCES

American Society of Heating, Refrigeration, and Air Conditioning Engineers (ASHRAE). 2006. *The ASHRAE GreenGuide*.

American Public Works Association (APWA). 1997. *The "How-to" Book on Utility Coordination Committees*.

Federal Transit Administration (FTA). 2009. *Transit Green Building Action Plan*.

Fu, L. 2008. A discussion on design methods to improve long-term performance of tunnel projects. *North American Tunneling 2008 proceedings*. San Francisco, California.

Green Building Initiative (GBI). 2004. *Green Globes v.1 Rating System*.

Jeon, C.M. and Amekudzi, A. 2005. Addressing sustainability in transportation systems: definitions, indicators, and metrics. *ASCE Journal of Infrastructure Systems*.

Maire, P. and Blunier, P. 2006. Underground planning and optimisation of the underground resources' combination looking for sustainable development in urban areas. *Going Underground: Excavating the Subterranean City*, Manchester.

Prince George's County and Maryland Department of Environmental Resources Programs and Planning Division (PGDER). 1999. *LID National Manuals*.

Sahely, H.R., Kennedy, C.A., and Adams, B.J. 2005. Developing sustainability criteria for urban infrastructure systems. *Can. J. Civ. Eng.*, 32(1), 72–85.

United Nations. 1987. Development and International Economic Cooperation: Environment Report of the World Commission on Environment and Development: "Our common future." *Official Records of the General Assembly, 42nd Session*, Supplement No. 25, Brussels, Belgium.

United States Green Building Council (USGBC). 2009. *LEED for New Construction*.

University of Washington and CH2M Hill (UW and CH2M). 2010. *The Greenroads v1.0 Rating System*.

# Use of Underground Space in a Pristine Watershed: Chester Morse Lake Pump Plant and Intake, North Bend, WA

**Joe Clare**
MWH Americas, Inc., Bellevue, Washington


**ABSTRACT:** The design of the Chester Morse Lake Pump Plant and Intake provides a recent example of the use of underground space for lifeline infrastructure. An underground pump plant constructed on the side of a steep slope along Chester Morse Lake provides security, preserves aesthetics, and reduces operation and maintenance costs when compared to an above ground structure. When completed, the pump plant, pipeline and intake tunnel will access dead storage within Chester Morse Lake to provide drinking water to Seattle, and meet in stream fish flow requirements at low lake levels.

## INTRODUCTION

Seattle Public Utilities (SPU) contracted with MWH Americas, Inc. (MWH) to provide engineering services for the design of the Morse Lake Pump Plant (MLPP) project. The project consists of four major components: Pump Plant, Intake, Transmission Pipeline and Discharge Structure.

### Project Background

The Cedar River Municipal Watershed with Chester Morse Lake (Morse Lake) is located on the western flank of the central Cascade Mountains approximately 53 km east of Seattle, and 12 km south-southeast of North Bend, Washington. The 367 square kilometer watershed, owned by the City of Seattle and operated by Seattle Public Utilities (SPU) Department, provides 70 percent of the drinking water needs for 1.4 million people in the greater Seattle area. The outflow of Morse Lake drains into Masonry Pool (historic Cedar Lake) and then passes through Masonry Dam to the Cedar River. The Cedar River then flows to Landsburg Dam where flow is captured for water supply. The remaining water continues as Cedar River to meet requirements for fish, recreation, and operation of the Hiram M. Chittenden Locks where Lake Washington meets Puget Sound. A location and vicinity map of the Cedar River Watershed area is shown on Figure 1.

Pumping during drought conditions is presently accomplished with two barge-mounted pumping plants with a total of 28 pumps and a combined pumping capacity of 908,500 cubic meters per day. Pumps are driven by electric motors that are powered by shore-based mobile 1.5 MW diesel generator sets that are leased. When the potential need for these pumps is anticipated, the barge mounted pumping plant facilities are moved into position and set up for use in summer and taken down again during the fall season. The arrangement and set up of these barge mounted pumping plants is a prolonged, involved and costly process.

The MLPP project would provide the following to replace the above-described barge-mounted pumping plant:

- New lake intake
- Land based pump plant
- Water transmission pipeline
- Discharge structure in Masonry Pool
- Permanent power supply

A site plan showing the location of the proposed MLPP facilities is shown on Figure 2.

### Existing Conditions

Proposed MLPP facilities would all be located in the Cedar River Watershed area. The watershed is closed with access restricted to authorized personnel and to honor tribal agreements. Chester Morse Lake operates under a Limited Alternative to Filtration Agreement with the United States Environmental Protection Agency. Water from the watershed is not filtered in its path from lake to tap. Exceeding the maximum allowable turbidity level could result in the requirement to construct a filtration facility to treat the source water. While the potential for such an event is low during this project all necessary steps must be taken to protect the unfiltered domestic water supply status as well as protect surface water quality.

The 6.2 square kilometer Morse Lake is located on the west slope of the Cascade Mountain range at

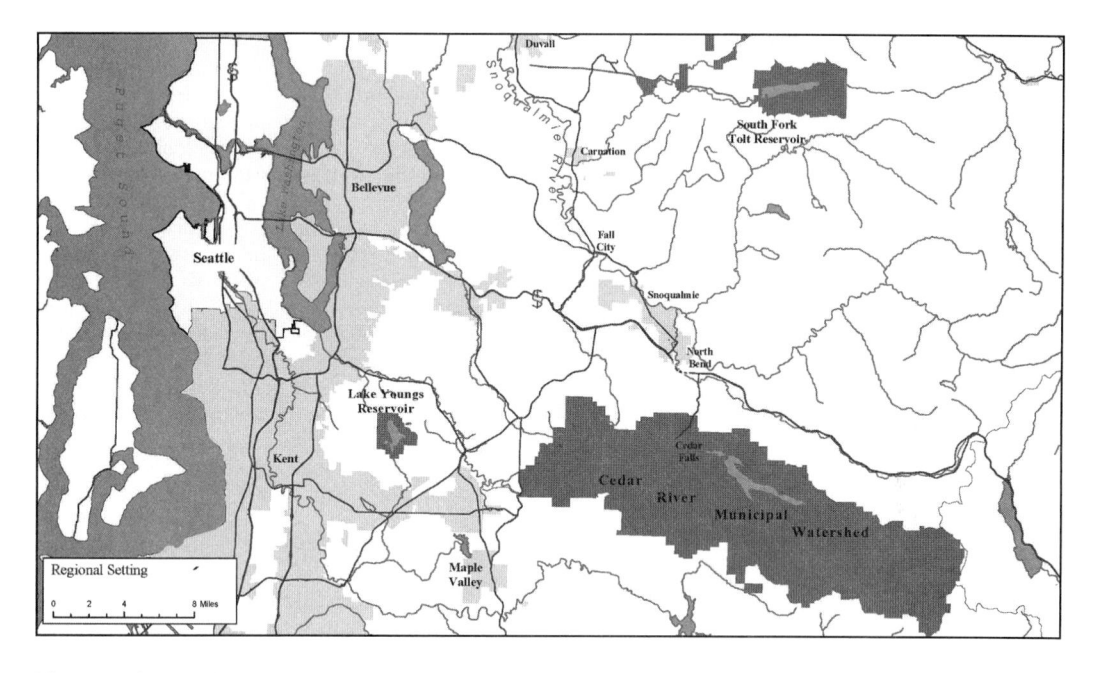

Figure 1. Cedar River municipal watershed

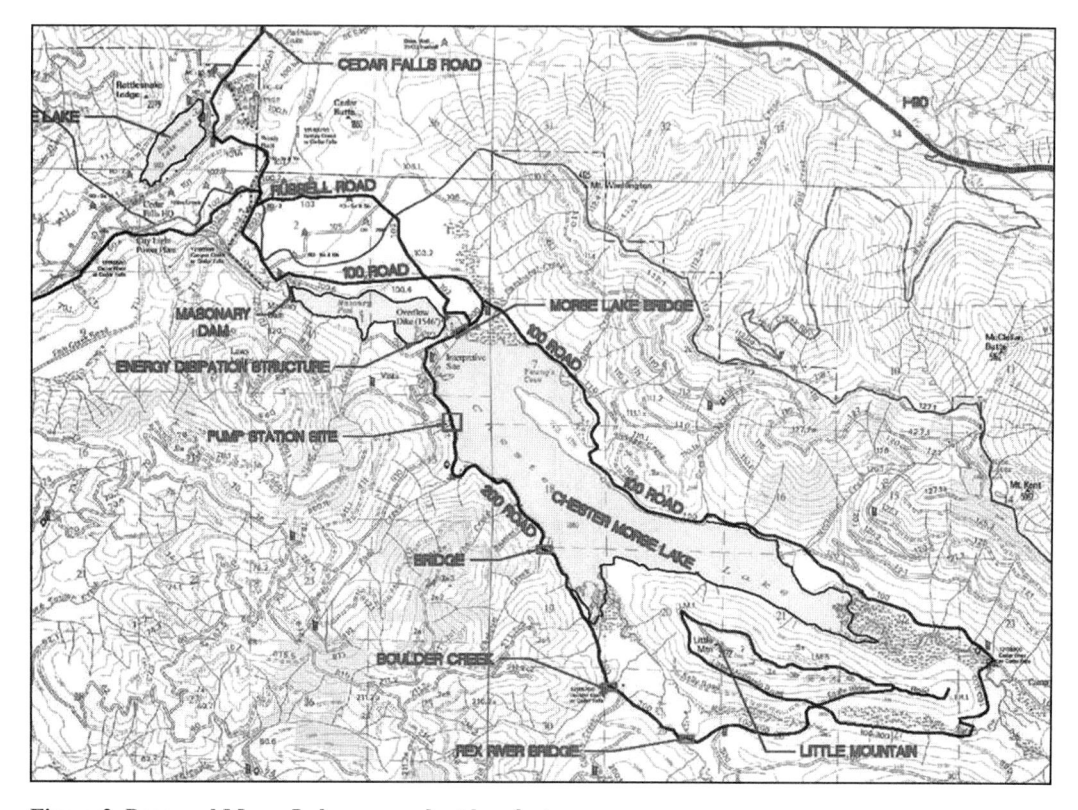

Figure 2. Proposed Morse Lake pump plant key features

approximate an elevation of 488 m in an old and second growth forest and is subject to extreme weather conditions at times (Harris, 2009).

- Elevations from 165m to 1650m
- Annual rainfall amounts 178cm to 280cm
- Snowfall depth 1m
- Periods of hot and dry conditions July-September
- Wind to 130km/hr

The watershed encompasses the head waters of Cedar River at the crest of the Cascade mountains and extend west downstream to rolling foothills at Landsburg Dam. Chester Morse Lake and Masonry Pool represent the reservoir bodies of water within the watershed.

*Geologic Conditions Overview*

This area of the Cascades is largely composed of Tertiary aged volcanic rock with overlying surficial glacial and non glacial deposits. The topography of the area is largely a result of weathering, erosion, glaciation, and regional structural uplift, folding, and faulting. Past glaciation included both continental and alpine glaciers. During the most recent continental ice advance into the area (Vashon stade of the Fraser glaciation) the Puget lobe of the Cordilleran ice sheet covered the project area.

Within the project area, bedrock outcrops, exploration borings, and geologic literature indicate the underlying bedrock to be comprised of the Ohanapecosh Formation. Overlying this formation are Vashon glacial recessional outwash and Holocene age alluvium, lacustrine, colluvium, or modified land (fill) (Tabor, et al, 2000). Vashon-aged glacial deposits, consisting primarily of recessional outwash deposits are present below post-glacial deposits and above the bedrock.

## DESIGN AND CONSTRUCTION CONSIDERATIONS

The pump plant location was selected based on proximity to deep water in the lake, a lack of known cultural resources, and to minimize the length of the discharge pipeline from the pump plant to the discharge structure at Masonry Pool.

The pump plant would be situated within a narrow strip of ground located on and to the side of 200 Road that is presently cut into the hillside above the west side of the lake. Current design included constructing the pump plant as a buried structure within and below the road area with portions of the facility buried into the hillside west the road. From the surface to underground, the pump plant would be composed of an electrical room, pump room, intake shaft with companion drilled shafts, forebay, and intake tunnel.

Partially buried into the hillside, the electrical room would provide all system controls and personnel access for the underlying pump room. The pump room, built directly beneath the road, would house seven motors for the vertical turbine pumps with lift slab access to the road. The intake shaft at 7 m excavated diameter would house two of the pump columns and five drilled shafts would house the remaining pump columns. A mined forebay provides baffling for efficient hydraulic intake of water to the pumps. The intake tunnel and fish screen structure completes the intake side of the system.

Site design at the pump plant facility would have accommodations for snow removal and water runoff from snow melt. The buried facility would have features to protect it from surface operations:

- Recessed lift slabs to allow snow plowing and road grading
- Access protected from snow windrows
- Concrete roof of the electrical room to shed water, snow, and withstand tree fall and debris

The outflow of the facility begins with a 1.4 km long and 1.8 m diameter transmission pipeline buried within 200 Road. A discharge structure would be located at Masonry Pool just downstream of the existing overflow dike.

Major constraints during construction include:

- Maintain single lane road access during construction
- Limit disturbance and removal of vegetation
- Limited area for staging, water treatment, and access
- Wet weather period October–April
- Winter snowfall
- Access and security protocols
- Strict adherence to water quality requirements
- Equipment decontamination procedures
- Fish and wildlife work windows
- Limited site access for deliveries, employee parking, construction offices, etc.
- Steep slopes located above and below the pump plant location
- Remote borrow sites
- Cultural resources and tribal permit conditions

*Pump Plant Arrangement*

The pump plant would be configured to conform to the hillside with the seven vertical line shaft pumps in a linear arrangement parallel to the roadway. The floor of the pump room below the roadway would be located at elevation 478.8 m in order to provide 0.3 m of elevation above the Lake high water elevation of 478.5 m. The total pump room height from the pump room floor to the top of the roof slab is 6 m to accommodate overhead space for the pump head and motors. The pump room would be approximately 32.9 m long and 7.9 m wide.

Construction of the adjacent electrical room would require excavation into the hillside. Preliminary geotechnical explorations describe approximately 6 m of soil overburden overlying bedrock to the west of the proposed structure. Therefore, site excavation would include both rock and soil removal.

### Main Shaft

A 6 m inside diameter (minimum) main shaft would be required to allow construction of the underground works, for installation of two pumps, and serve as access for inspection of the tunnel, forebay and pumps during operation. The shaft would be excavated in rock to 7 m diameter or larger from the ground surface to about elevation 447.4 m and deepened to its full depth after excavation of the intake tunnel or to about elevation 445.6 m (39 m deep).

### Forebay

The forebay would be required to provide a means of connecting and distributing flow from the inlet tunnel to the pumps. The forebay would be a horseshoe shape and excavated using conventional drill and blast methods. The excavated dimensions are approximately 27.1 m long, 8.5 m high, and 9.4 m feet wide.

### Drilled Pump Shafts

In order to make the installation of vertical turbine pumps feasible, each of the remaining 5 pumps would be installed inside individually drilled pump shafts that extend from the pump discharge head down to the forebay. The five pump shafts would be drilled to approximately 1.8 m diameter, then lined with a 1.4 m diameter steel pipe, with the annulus grouted. Drilled shafts would likely be completed using raise bore drilling methods. An isometric view of the proposed pump plant is depicted on Figure 3.

## TUNNELING ALTERNATIVES

Three primary intake tunnel alternatives were initially evaluated for the construction of the intake tunnel for the project. These alternatives were selected based on analyses that identified a variety of preliminary intake layouts that would be feasible at the Morse Lake site. From the preliminary alternatives, tunneling methods and layouts have been compared to develop the optimum design solution for the project. The initial tunnel alternatives comprised a deep rock tunnel, mixed face microtunnel, and a shallow siphon tunnel. Following detailed alternative analyses, a mixed face microtunnel was selected as the preferred alternative as depicted in Figure 4.

### Microtunnel Wet Tap

The microtunnel wet-tap alternative consists of approximately 102.1 m of microtunneling from the shaft to the intake structure. The tunnel would begin at a depth of approximately 36.6 m below ground surface from within a shaft located at the site of the pump station. Microtunneling ends at the location of the intake screens, approximately 6 m below the mud line of the lake.

The microtunneling alternative would begin its alignment in the Ohanapecosh bedrock unit. After approximately 45.2 m, the microtunnel machine (MTBM) would encounter mixed face ground conditions as it breaks out of the rock and transitions to soft ground. The remaining 56.4 m of tunnel are comprised of recessional outwash. Approximately 4.6 to 6 m of very soft lacustrine lake deposits overly the recessional outwash at the location of the intake. As a result, the vertical tunnel alignment was chosen to meet the required intake elevation as well as provide suitable foundation materials for the intake structure and MTBM retrieval. The tunnel is intended to be mined through the recessional outwash materials below the overlying lake deposits.

### Intake Screen Description

Intake screening would comprise four 1.8 m diameter, stainless steel, cylindrical T-screens. An 2.4 m diameter riser from the end of the tunnel forms the base of the screen assembly. It rises from the tunnel to an elevation of about 453.2 m, where it ends in a flange.

## DISCHARGE STRUCTURE

Water pumped from Morse Lake through the transmission pipeline will be discharged along the south bank of Masonry Pool just downstream of the existing Overflow Dike. Since the water surface elevation in Masonry Pool can fluctuate widely over the year, ranging from elevations of 460.2 m to 477.9 m, and pumping is anticipated to occur at water surface elevations ranging from elevation 460.2 m to 473.7 m, a discharge structure will be required to limit effects, such as scour, in Masonry Pool at times when the water level is low. The structure will also be designed to limit effects on fish.

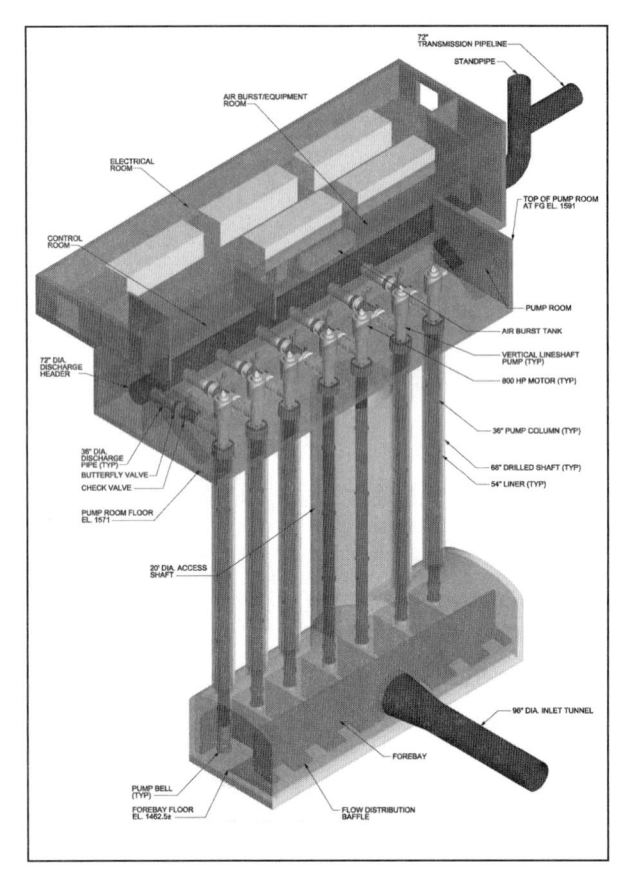

**Figure 3. Pump plant schematic**

## CONSTRUCTION CONSIDERATIONS

Construction at a remote area with strict environmental requirements will be more challenging and expensive than construction in an urban area. Some of the issues that need to be considered during construction are:

- Limited site area
- Evaluation of tree fall hazards during and following construction
- Erosion control and drainage
- Site water treatment and discharge
- Muck hauling and disposal
- Wet weather construction
- Snowfall
- Wildfire danger
- Decontamination of construction equipment
- Security and access protocols in the watershed
- Protection of wildlife
- Wildlife safety awareness

## Staging and Storage

The location of the pump plant is severely limited in area. Contactors would be required to stage equipment and materials at another location within the watershed.

Staging area for the shaft, forebay and tunnel excavation will be along 200 Road. This will be constrained by the need to require single lane access through the work zone for emergency and wildfire uses. Construction is anticipated to be in the following sequence:

1. Site grading and retaining wall
2. Shaft excavation
3. Tunnel excavation
4. Forebay excavation
5. Drilled shafts
6. Pump plant and electrical room
7. Finish grading and restoration

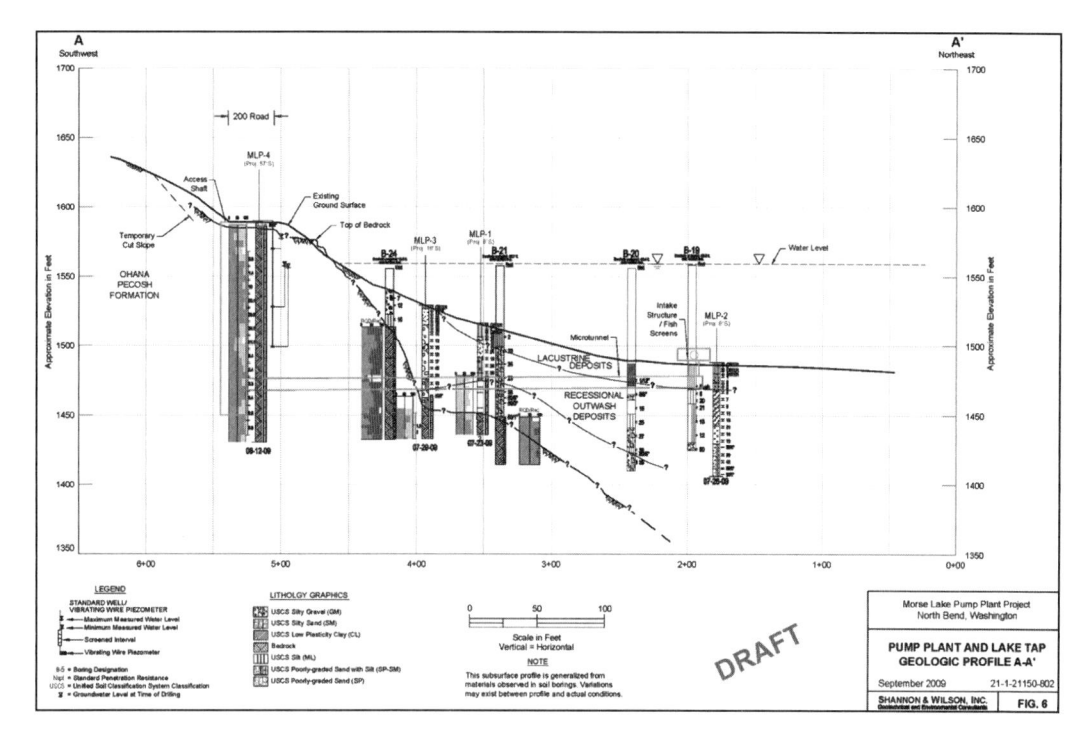

**Figure 4. Intake tunnel vertical alignment**

## UNDERGROUND BENEFITS OF PUMP PLANT

### Pump Plant Appearance

The project facility site will have a footprint of several thousand square feet in building roof and pump station paving. Because the pump plant will be mostly built within and below an existing gravel road (which is essentially an impervious surface) new construction is anticipated to have little increase in additional impervious areas.

The exposed elements of the pump plant that would be visible at the site are the east, north and south walls of the electrical room and its associated doors, ventilation equipment and/or louvers that would be necessary to draw air in to the electrical room and pump room for cooling the electrical equipment and motors, the main standpipe at the connection of the 1.8 m pump discharge header to the 1.8 m discharge pipeline, and the above-grade gooseneck at the end of the access shaft vent pipe. All other pump plant features would be buried.

The exposed elements described above would be treated to blend in with the surroundings. Architectural treatments will be selected to reduce the visibility of these elements. These treatments could incorporate different finishing textures and colors (for example, choosing dark colors that would make the pump plant less noticeable against the vegetated hillside, and/or providing surfaces that have a "rock" appearance). Air handling equipment will be located inside the pump plant to minimize noise as well as visual impact. The exterior walls of the electrical building can be finished with a rockery-type design to blend with the environment, leaving the access doors as the only visible feature of the building.

### *Beneficial Use of Underground Space*

The design of a buried pump plant came easily as a result of meeting major requirements and constraints of the project team such as appearance, protection of the environment, hazards protection, and physical site constraints. The use of underground space for a pump plant minimized the above ground disturbance and appearance which was vital to SPU and cultural stakeholders. The buried pump plant would be anticipated to have a longer lifespan and other benefits than traditional above grade facilities. Compared to above ground facilities, there are limited to no architectural building envelope features that require maintenance and eventual replacement such as siding, windows, and roofs. In addition, the buried facility will offer protection from hazards such as:

- Seismic events
- Tree fall

- Debris flows and avalanches
- Extreme cold and hot weather
- High winds

Other benefits to a buried facility include a reduction in HVAC requirements as a result of greater isolation from temperature extremes. Mechanical equipment is anticipated to utilize shaft intake water for temperature control of motors and electrical equipment.

### *Considerations for Sustainable Construction*

Construction in a remote site often forces the designer and contractor to look for ways to minimize materials and material transport as a means to reduce cost. For the MLPP project, this translates into maximizing the reuse of waste materials generated from construction and reducing the need for imported materials. To provide for the reuse of materials, existing borrow pits within the watershed would be utilized for processing of waste materials as well as a source of quality fill materials for pipeline bedding and backfill. Since the project is not in an urban area, typical strict specifications and requirements for imported high quality structural fill can be relaxed to permit the use of native materials for backfill.

Crushed and screened rock is utilized throughout the watershed for road course, erosion protection, stream, and slope restoration. Normally wasted and exported overburden materials can be stockpiled and reused for restoration of road decommissioning projects. Salvage of native understory vegetation from the pump plant and portions of the pipeline alignment could also provide restoration opportunities within the watershed.

### REFERENCES

Seattle Public Utilities (April 2000), *Cedar River Watershed Habitat Conservation Plan.* http://www.seattle.gov/util/About_SPU/Water_System/Habitat_Conservation_Plan/AbouttheHCP/Documents/index.htm.

Seattle Public Utilties. (2009, January). *Draft Geotechnical Data Report Chester Morse Pump Station, Cedar River Watershed.*

Tabor, R.W., Frizzell, V.A. Jr., Booth, D.B., Waitt, R.B. (2000). *Geologic Map of the Snoqualmie Pass 30×60 Minute Quadrangle, Washington*. U.S. Geological Survey. Map and Accompanying Pamphlet, 57p.

Harris, Gregory, MWH Americas, Inc. (November 2009). *Cedar River Watershed Operations and Supply System, Morse Lake Pump Plant Project.* Presentation to 2009 AWRA Annual Water Resources Conference.

# TRACK 1: TECHNOLOGY

Mike Smithson, Chair

*Session 5: Tunnel Lining and Remediation*

# High-Pressure Concrete Plug Leakage Remediation

**Carlos A. Jaramillo, Camilo Quinones-Rozo**
URS, Oakland, California

**Robert A. McManus, Andrew Yu**
PG&E, San Francisco, California

**ABSTRACT:** The Helms Pumped Storage Project, constructed in one of the California's batholiths, has head of almost 1900 ft at the pump-turbines. A concrete plug sealing the access to the penstock tunnel has leaked since first filling, and a grouting program performed shortly after construction was not successful reducing the leakage. An abrupt increase of leakage from 600 to 1000 gpm prompted an emergency rehabilitation program consisting of implementing a series of high pressure consolidation grout holes using the Grout Intensity Number (GIN) methodology. This paper describes the grouting program, the concrete plug repair, and the behavior of the leakage after filling the tunnel.

## INTRODUCTION

PG&E owns the Helms Pumped Storage Plant (Helms) located in Fresno County, California. This project has one of the largest heads of its kind in the United States (1,630 feet), and the head at the pump turbines is almost 1900 ft. Highly sheared zones (see Figure 1) were intercepted during excavation of the Power House Access Tunnel, the T3 Access Tunnel, the high-pressure tunnels upstream from the steel liners, and the upstream surge shaft. First filling of the tunnels was interrupted by the Lost Canyon pipe failure, and PG&E decided to use the time required for its repair to undertake a grouting program to treat the portion of the shear zone that intercepts the high-pressure tunnel upstream of the penstock access tunnel concrete plug, as very high and unexpected inflows were recorded at the concrete plug. After the repairs were complete, the project was put into operation in June 1984. The T3 Access Tunnel was repaired in 2008 to mitigate the impact of leakage, and deterioration of the shotcrete liner.

On November 2008, leakage from the concrete plug separating the pressure tunnels from the access tunnels increased suddenly. This increase raised concerns about irreversible damage occurring in the high pressure gradient area around the plug. Based on evaluation of the situation, the system was taken off line and the tunnels were drained during the spring of 2009 to make repairs to the plug area. The emergency outage was scheduled between February 28, 2009, and April 23, 2009.

## DESCRIPTION OF THE PROJECT

Helms PSP has an installed capacity of 1,200 megawatts. The project joins Courtright Lake (upper reservoir) and Lake Wishon (lower reservoir) and uses the 1,630 feet (ft) of elevation difference and a design flow of 9200 cfs to generate electricity. Water is pumped to the upper reservoir during the off-peak hours of power demand and it is used to generate power at times of high demand. The main features of the facilities within this project are as follows:

- Intake/Outlet at Courtright Lake
- Tunnel 1: 27-ft diameter, 4,200-ft-long, concrete-lined
- T-1 Gate Shaft: 12.5-ft long, 26.5-ft wide, 237-ft drop, shotcrete-lined
- Lost Canyon Crossing: 22-ft-diameter, 204-ft long, steel pipe
- Tunnel 2: 27-ft-diameter, 9,000-ft-long, concrete-lined
- T-2 Surge Shaft: 47-ft to 60.5-ft diameter with a 27-ft diameter restricted orifice, 578 ft drop, shotcrete-lined
- Inclined Shaft: 27-ft-diameter, 2,500-ft-long, inclined 55° below the horizontal, concrete-lined
- Penstock Tunnels: Three 11.5-ft-diameter, concrete/steel-lined
- Drainage Gallery: 260-ft long, 10-ft wide, 14-ft high, unlined
- Transformer Chamber: 300-ft-long, 41-ft-wide, 41-ft-high, unlined

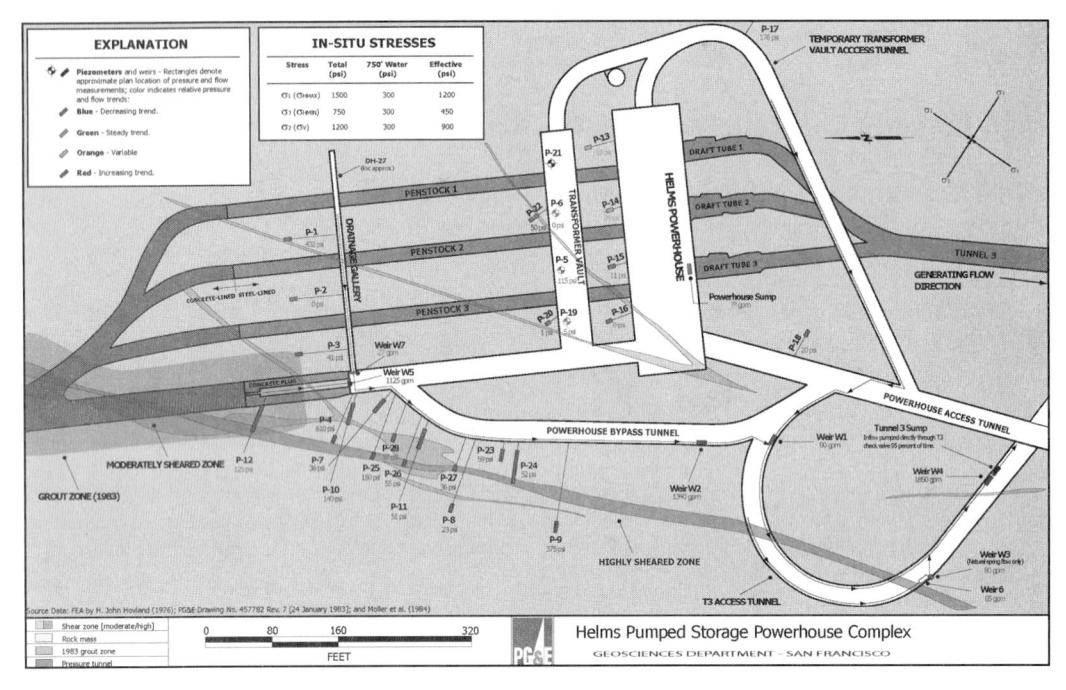

Figure 1. Plan of powerhouse complex

- Powerhouse Chamber: 336-ft-long, 83-ft-wide, 142-ft-high, unlined
- Tunnel 3: 27-ft-diameter, 4,000-ft-long, concrete-lined
- T-3 Surge Shaft: 10-ft to 44-ft diameter with a 27-ft diameter restricted orifice, 971-ft drop, shotcrete-lined
- Powerhouse Access Tunnel: 27-ft-diameter, 3,800-ft-long, with unlined/steel sets plus shotcrete
- Other Auxiliary tunnels and shafts, including the Penstock Access Tunnel, the Powerhouse Bypass Tunnel (PHBT), the T3 Access Tunnel, the Temporary Transformer Vault Access Tunnel, and the Elevator Shaft

## SITE GEOLOGY

The site is in one of the Sierra Nevada batholiths, but it is one with a rather complex history. Four bedrock units are present in the Helms area:

- Dinkey Creek Granodiorite (Kdc), approximately 103 million years (Ma) intrusion age
- Lost Peak Quartz Monzonite (KJlp), 103–98 Ma intrusion age
- Mount Givens Granodiorite (Kmg), approximately 90 Ma intrusion age
- Metaquartzite roof pendant of unknown age (DCRP-qz)

The Lost Peak quartz monzonite is a small pluton intruded into the Dinkey Creek granodiorite along the generally north-south trending shear zones. Tunnel 2 crosses the Lost Peak pluton. The metaquartzite roof pendant lies at the southern end of the Helms project on the north shore of Wishon Reservoir. Tunnel 3 and the Powerhouse Access Tunnel lie within the quartzite to the south and Dinkey Creek granodiorite to the north. The contact between these units was noted in the Tunnel 3 as-built geology log. The western contact between these units follows a north-northeast trending shear zone; however, the regional tectonism believed to have produced the shear zones post-dates the intrusion of the Dinkey Creek granodiorite. The Mount Givens granodiorite lies to the northeast of the Helms area, apparently truncating the shear zones. This implies that the regional tectonism responsible for the shear zones had ended before its 90 Ma intrusion age.

The structural geology of the zone was not well understood before the construction of the project and encountering the shear zones was a surprise to the designers, and impacted the construction progress, and required modifications to the design. The stationing and attitudes of the four shear zones identified during construction are shown in Table 1.

These shear zones were included in a three-dimensional geologic model to identify their surface trace, and obtain a better understanding of the geologic structure of the site. The shear zone crossed by the penstock access tunnel was interpreted not as a single plane, but as the combination of two of surfaces connected to distinctive surface lineaments, implying a large afferent area capable of contributing large flows to the intersecting tunnels.

## HISTORY OF OBSERVED INFLOWS AND GROUNDWATER PRESSURES

PG&E filled the waterways of the project for the first time during September 1982. Initial instrumentation included four weirs (W-1 to 4), and six piezometers (P-1 to 6) to monitor changes in the groundwater regime around the power complex. The maximum pressure registered at the penstock was 650 psi, whereas an unexpectedly high pressure of 580 psi was recorded at piezometer P-4, located immediately downstream of the concrete plug, requiring re-evaluation of the filling process. Throughout the filling, pressures registered at piezometer P-4 were proportional to variation in pressures within the penstock. Inflows in the penstock plug area increased from 76 gpm to a maximum of 450 gpm, consistent with the increase in pressures. Additional sources of inflow were detected along the north wall of the transformer chamber, T3 Access Tunnel, and along the north wall of the powerhouse chamber.

On September 29, 1982, the 22-ft-diameter Lost Canyon Pipe Crossing failed unexpectedly while the project was being placed in operation. Unrelated to the Lost Canyon incident, PG&E decided to conduct a high-pressure grouting program in the penstock access tunnel to take advantage of the standby time provided by the repairs to the Lost Canyon Pipe Crossing, and address the high inflows and high groundwater pressures. The grouting operations concentrated on those areas where the shear zone intersected the Penstock and the Penstock Access Tunnel (PAT). The waterways were refilled on August 1983, and the units were placed in operation in June 1984. The inflows and groundwater pressures measured after the repair were not significantly different to those during the initial filling, resulting only in minor reductions.

Tables 2 and 3 provide a summary of flow rates and groundwater pressures recorded during the period from September 1982 through April 2009 by the instrumentation installed in the vicinity of the PAT plug. Instrumentation records are provided for the following milestones:

- Baseline Reading: Value registered after instrument installation. Note that not all the instruments were installed simultaneously.
- September 1982: First tunnel filling.

Table 1. Location of major shear zones

| Location | Stationing (ft) | Strike and Dip |
|---|---|---|
| Main penstock | Sta 154+26 | N13°E/75°W |
| Penstock access tunnel | Sta 155+21 | N16°E/87°W |
| Tunnel 3 access tunnel | Sta 4+56 | N25°E/86°W |
| Tunnel 3 access tunnel | Sta 5+77 | N22°E/86°W |

- October 1983: Tunnel filling after the 1983 grouting program.
- February 1984: Partial dewatering of T1 and T2 (up to the top of the inclined shaft).
- 1986: Tunnel dewatering and penstock inspection.
- 1988: Tunnel dewatering identified from the instrumentation records
- 1992: Tunnel dewatering and penstock inspection
- October 1997: Tunnel dewatered for penstock inspection
- November 2008: Completion of T3AT repairs works and installation of additional drain on the Powerhouse ByPass Tunnel (PHBT).
- January 2009: Sudden increase in leakage coming out of the PAT plug.
- April 2009: Tunnel filling after the 2009 emergency repairs, including high-pressure grouting and sealing of existing cracks, weep holes, and grout holes.

On November 2008, a sudden increase was noticed in the leakage from the concrete plug separating the PAT from the PHBT. Weir W-5, which records the inflow into the PAT plug, registered a sharp increase. In a period of two weeks, flow measurements at W-5 increased from 600 to 925 gallons per minute (gpm). This increase in flow occurred simultaneously with a 400 pounds per square inch (psi) pulse-like fluctuation in the pressures registered by piezometer P-12, located in the middle of the concrete plug. The inflows measured at W-5 continued increasing until reaching 1200 gpm by mid February. This increase raised concerns about irreversible damage occurring in the high-gradient area around the plug. Based on evaluation of the situation, and the potential impact to the plant and its personnel, it was decided to take the system off line and drain the tunnels during the spring of 2009. An emergency outage took place between February 28, 2009, and April 23, 2009. The repairs performed during the outage consisted of: (1) High-pressure grouting near the plug and (2) Sealing of existing cracks, weep holes, and grout holes in the PAT.

Table 2. Historical record of flow rates measured in the powerhouse complex

| Weir | Baseline Reading (after installation) | Sept. 1982 | Oct. 1983 | Feb. 1984 | Nov. 2008 | Jan. 2009 | Apr. 2009 |
|---|---|---|---|---|---|---|---|
| W-1 | 5 | N/A | 179 | 250 | 80 | 80 | 80 |
| W-2 | 27 | 717 | 448 | 528 | 600 | 1340 | 150 |
| W-3 | 45 | N/A | 76 | 94 | 80 | 80 | 85 |
| W-4 | 152 | N/A | 806 | 1008 | 1500 | 1800 | 600 |
| W-5 | N/A | 250 | 228 | 202 | 600 | 1100 | 20 |
| W-6 | Monitored since 2006 | † | † | † | † | 85 | 90 |
| W-7 | Monitored since 2006 | † | † | † | † | 27 | 50 |
| W-8 | Installed 2009 | † | † | † | † | 100 | 50 |

Notes:
All flows in gpm; gpm = gallons per minute
N/A = Data not available
† = Instrument not installed at the time
Weir flows are not additive (W-4 includes all flows, W-2 includes flows from W-7, W-5, and W-8)

## PREVIOUS GROUTING PROGRAMS

### Grouting Performed During Construction

Construction records indicate that both high-pressure grouting and contact grouting were conducted in the vicinity of the PAT plug during construction. High-pressure grouting was conducted at the PAT and the Penstock Tunnels. Two grout rings were installed immediately upstream of the concrete/steel transition in the penstock tunnels; and two grout rings were placed 8 ft upstream from the PAT plug. Contact grout was performed at the rock/concrete interface of the PAT plug using rings of holes spaced every 10 ft (longitudinally). Grout rings were drilled from within the plug manway, with each ring consisting of eight holes equally spaced around the perimeter of the opening. Each contact grout hole extended to the rock/concrete interface.

### Initial Remedial Grouting Program

As mentioned before, PG&E performed a high pressure grouting program of the PAT tunnel during the repair work on the Lost Canyon Pipe Crossing in 1983. The high-pressure program targeted those areas where the shear zone intersected the Penstock and the PAT. The focus of the 1983 grouting program was to prevent high-pressure water from leaking out of the tunnels.

Grouting operations were performed in the period between March 18 and June, 15, 1983. According to Moller et al. (1984), grout holes were drilled in rings of eight, spaced 10 ft between rings. Holes in adjacent rings were rotated 22.5° to stagger the spacing between the holes. Each grout hole was drilled in stages up to a maximum length of 40 ft into the rock.

A total of 14,300 linear ft were drilled with rotary percussion jack-leg drills (not including re-drilled through previously placed grout). A total of 710 yd3 of grout were injected, corresponding roughly to 110 tons of Ultrafine cement and 410 tons of Type III cement.

## EMERGENCY GROUTING PROGRAM

The emergency grouting program had to be performed following a very tight schedule. Access to the tunnel was possible beginning on 12 March 2009, and all work hat to be done by 23 April 2009.

High-pressure grouting was used to improve the hydraulic properties of the rock mass around the plug, as well as those of the rock/concrete interface. Grouting pressures of 850 psi, exceeding the measured minimum in situ stress of 750 psi, and expected operation pressures were used, so the grout pressure could open and fill tight discontinuities within the rock mass, effectively pre-stressing the rock mass making the discontinuities practically unreachable to seepage during operation.

Grouting concentrated on the upstream section of the PAT plug where most of the inflows originated. This portion of the PAT plug was of great concern because the largest hydraulic gradients are present in this area and the coverage provided by the 1983 grouting program was marginal.

The grouting program included fifteen grout rings/fans distributed evenly along the upstream portion of the PAT plug. Rings/fans were grouped into three sets: V, I, and M, as follows:

- Four V (vertical) rings were located upstream from the PAT plug.

**Table 3 Historical record of groundwater pressures measured in the powerhouse complex**

| Piezometer | Baseline Reading (after installation) | Sept. 1982 | Oct. 1983 | Feb. 1984 | Nov. 2008 | Jan. 2009 | Apr. 2009 |
|---|---|---|---|---|---|---|---|
| P-1 | 0 | 26 | 290 | 230 | 430 | 432 | 200 |
| P-2 | 0 | 500 | 28 | 81 | 0 | 0 | 0 |
| P-3 | 0 | 155 | 20 | 14 | 40 | 41 | 100 |
| P-4 | 20 | 786 | 652 | 662 | 575 | 610 | 650 |
| P-5 | 0 | 315 | 260 | 228 | N/A | 115 | 150 |
| P-6 | 0 | 435 | 338 | 279 | N/A | 0 | N/A |
| P-7 | 15 | † | 72 | 66 | N/A | 36 | 25 |
| P-8 | 10 | † | 128 | 182 | N/A | 23 | 18 |
| P-9 | 20 | † | 210 | 240 | 375 | 410 | 300 |
| P-10 | 15 | † | 50 | 74 | 140 | 140 | 150 |
| P-11 | 15 | † | 88 | 112 | 51 | 60 | 40 |
| P-12 | 65 | † | 526 | 518 | 425 | 125 | N/A |
| P-13 | 10 | † | 20 | 15 | N/A | 18 | 10 |
| P-14 | 15 | † | 22 | 18 | N/A | 24 | 15 |
| P-15 | 0 | † | 12 | 10 | N/A | 11 | 10 |
| P-16 | 10 | † | 22 | 20 | N/A | 0 | N/A |
| P-17 | 50 | † | 122 | 136 | N/A | 176 | 185 |
| P-18 | 90 | † | 110 | 104 | N/A | 20 | 17 |
| P-19 | 0 | † | 0 | 0 | N/A | 5 | N/A |
| P-20 | 0 | † | 10 | 28 | N/A | 1 | 2 |
| P-21 | 0 | † | 0 | 0 | N/A | N/A | N/A |
| P-22 | 0 | † | 22 | 14 | N/A | 50 | 40 |
| P-23 | N/A | † | 84 | 84 | 50 | 59 | 53 |
| P-24 | N/A | † | 78 | 81 | N/A | 52 | 45 |
| P-25 | N/A | † | † | 600 | 150 | 150 | 250 |
| P-26 | N/A | † | † | 96 | 50 | 55 | 54 |
| P-27 | N/A | † | † | 86 | N/A | 36 | 16 |
| P-28 | N/A | † | † | 100 | 50 | 57 | 50 |
| P-29-3 | 0 | † | † | † | † | † | 760 |
| P-30-5 | 40 | † | † | † | † | † | 730 |
| P-31 | 26 | † | † | † | † | † | 20 |

Notes:
Groundwater pressures (psi)
N/A = data not available
psi = pounds per square inch
† = Instrument not installed at the time

- Four I (inclined) fans were used to target the portion of the PAT plug where the 32-inch manhole is located. Inclined holes were used to reach the rock mass left untreated by the 1983 grouting program. Two fans were drilled pointing downwards from the upstream side of the PAT plug, and two more fans were drilled from within the plug manway pointing upstream.
- Seven vertical M (manway) rings were drilled from within the plug manway.

Each ring/fan consisted of 8 radial grout holes (2-inch diameter) spaced evenly around the perimeter of the opening. Grout holes of adjacent rings/fans were staggered to minimize the spacing between holes (22.5°).

The high-pressure grouting was performed as follows:

- Split spacing was used. Each ring/fan consisted of four primary and four secondary holes. For the purposes of drilling and grouting, primary holes were performed first, followed by the secondary holes, and tertiary, as necessary (I2, I3 and I4).
- The intended drilling-grouting sequence was V4, V2, V3, V1, I1, M1, I2, M3, I3, M5, M0, M2, M4, I4, and M6.

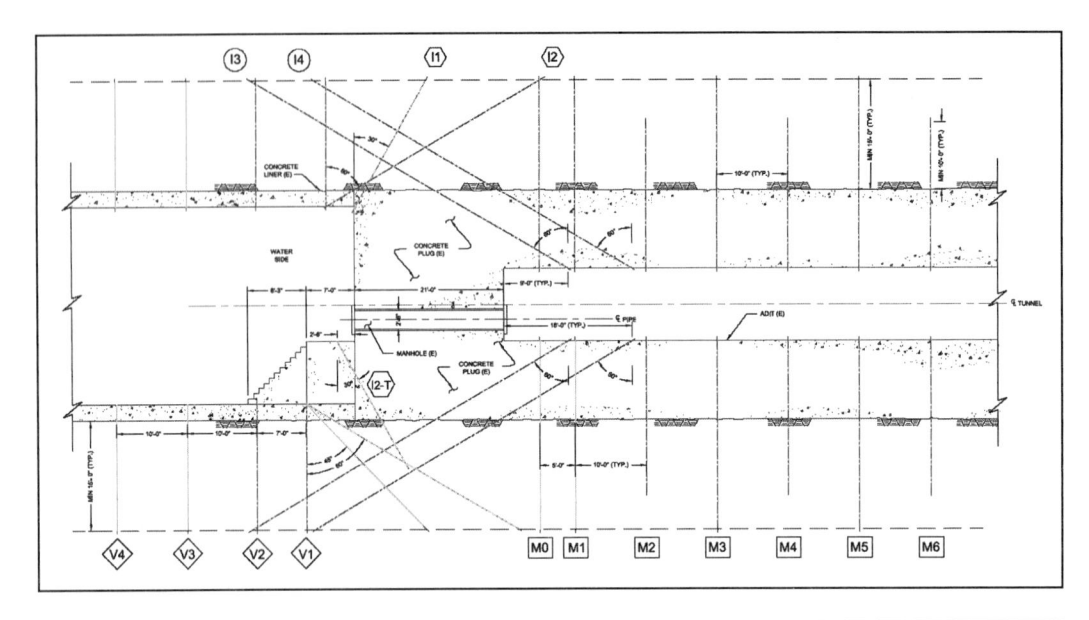

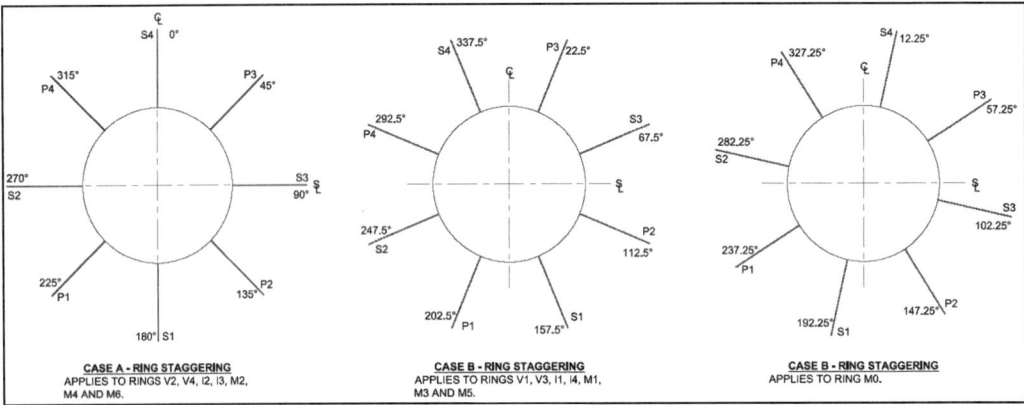

**Figure 2. Location and distribution of groutholes**

- The actual drilling-grouting sequence was as follows: V2, V4, V1, V3, I1, M1, M4, M6, I2, M2, M5, I3, M3, I4, M0, I2 Tertiary, I3 Tertiary, and I4 Tertiary.

Grouting was performed following the Grout Intensity Number (GIN) method (Lombardi and Deere 1993). To determine the grouting intensity during construction, pressure, flow rate, volume injected and penetrability were monitored in real time using a LOGAC G5 recorder manufactured by Atlas Copco. The grouting program was performed using a single stable grout mix throughout the grouting process, regardless of the takes observed. Stable grout mixes usually have water/cement (w/c) ratio between 0.6:1 and 0.8:1 (by weight), and result in less sedimentation of cement grains during low-flow conditions, less porosity, lower permeability, greater bond strength, and less shrinkage (Lombardi and Deere 1993).

Grouting completion based on reaching the Pressure Limit, the GIN curve, or the Unit Volume Limit. The pressure limits were selected based on the grout-hole location, as the liner and the plug had different structural capacities. The V holes were grouted to 500 psi (GIN 2000). The M holes were grouted to 870 psi (GIN 3000).

When the grout path reached the unit volume limit, a decision to do one of the following was made:

- Continue grouting (i.e., if the GIN value was about to be reached)

Table 4. Summary of grout ring/fan configuration

| Ring/Fan | Min. Total Hole Length, ft [Min. Depth into Rock, ft] | Angle with Respect to Tunnel Centerline, ° | Staggering with Respect to Tunnel Crown, ° | Stationing, ft |
|---|---|---|---|---|
| V1 | 17.2 [15.0] | 90 | 22.5 | 156+43.13/156+40.13 (P3, P4, S3 and S4/P1, P2, S1 and S2) |
| V2 | 17.2 [15.0] | 90 | 0 | 156+33.13 |
| V3 | 17.2 [15.0] | 90 | 22.5 | 156+23.13 |
| V4 | 17.2 [15.0] | 90 | 0 | 156+13.13 |
| I1 | 19.9 [17.3] | 60 (downstream) | 22.5 | 156+47.13 |
| I2 | 34.5 [30.0] | 30 (downstream) | 0 | 156+43.13 |
| I3 | 50.0 [30.0] | 30 (upstream) | 0 | 156+77.13 |
| I4 | 50.0 [30.0] | 30 (upstream) | 0 | 156+86.13 |
| M0 | 25.0 [15.0] | 90 | 12.25 | 156+73.13 |
| M1 | 25.0 [15.0] | 90 | 22.5 | 156+78.13 |
| M2 | 20.0 [10.0] | 90 | 0 | 156+88.13 |
| M3 | 25.0 [15.0] | 90 | 22.5 | 156+98.13 |
| M4 | 20.0 [10.0] | 90 | 0 | 157+08.13 |
| M5 | 25.0 [15.0] | 90 | 22.5 | 157+18.13 |
| M6 | 20.0 [10.0] | 90 | 0 | 157+28.13 |

- Stop grouting (i.e., when pressure does not increase, or extensive hydro jacking may occur)
- Discontinue grouting temporarily to allow the grout time to set
- Discontinue grouting and re-drill after the grout has set completely

The project specifications called for the contractor to use a cement-based grout with super-plasticizer meeting the properties listed below:

- Cement Type III
- Bleed < 4%
- Marsh Flow (Apparent Viscosity) <35 sec
- Water/cement (w/c) ratio (by weight) 0.6:1 to 0.8:1
- Super-plasticizer/water (sp/w)ratio (by volume) 0.5% to 1.25%

The selected mix was a 0.7 w/c ratio mix with 0.5 percent Glenium 7500, with a Marsh cone value of about 36 seconds.

Table 5 summarizes the grout takes observed during the high-pressure grouting. In total, 200 cement bags (9 tons) during the high-pressure grouting operation.

## INSTRUMENTATION READINGS

A comparison of instrumentation readings before (Jan 26, 2009) and after (June 10, 2009) the execution of the emergency repairs is presented in Table 6 and highlights the following points:

- Weir 5, measuring the leakage out of the concrete plug, registers a significant decrease.
- The left side of T3 Access Tunnel (looking downstream) registered a continuous increase in the amount of water inflows since the tunnels were refilled.
- The inflows to the drainage gallery increased from 27 gpm before the repairs to 50 gpm after completion of the emergency repairs, but have remained stable.
- The pressures registered by piezometers P-29-3 and P-30-4 (800 and 560 psi, respectively June 10, 2009) are significantly higher than those observed in the past by P-3 and P-12 (41 psi and 125 psi, respectively) in the same area. The pressure readings at P-29 and P-30 correspond roughly to 98% and 69% of the maximum static pressure within the tunnels. It should be kept in mind that pressures can significantly vary over short distances due to the presence of discontinuities and/or preferred paths.

Table 5. Summary of grout takes for the high pressure grouting

| Hole \ Ring | V1 | V2 | V3 | V4 | I1 | I2 | I3 | I4 | M0 | M1 | M2 | M3 | M4 | M5 | M6 |
|---|---|---|---|---|---|---|---|---|---|---|---|---|---|---|---|
| P1 | 8.1 | 0.1 | 0.3 | 20.5 | 0.5 | 1.8 | 3.6 | 1.0 | 1.3 | 3.1 | 0.1 | 1.0 | 0.5 | 0.5 | 0.1 |
| P2 | 0.8 | 15.3 | 0.3 | 0.3 | 0.5 | 4.3 | 4.8 | 5.5 | 1.0 | 1.8 | 0.5 | 5.8 | 1.5 | 0.5 | 0.3 |
| P3 | 0.8 | 0.3 | 0.3 | 0.4 | 0.8 | 0.8 | 1.8 | 1.5 | 0.1 | 0.8 | 0.5 | 2.0 | 0.5 | 7.3 | 0.3 |
| P4 | 0.8 | 0.3 | 0.8 | 2.5 | 1.0 | 1.0 | 4.0 | 2.3 | 0.8 | 1.3 | 0.1 | 0.3 | 0.3 | 3.3 | 0.3 |
| S1 | 0.5 | 0.1 | 0.1 | 0.1 | 1.1 | 7.5 | 1.0 | 3.0 | 0.8 | 0.3 | 0.5 | 0.5 | 0.8 | 3.0 | 0.3 |
| S2 | 0.1 | 0.3 | 1.0 | 0.4 | 0.5 | 1.0 | 6.8 | 2.0 | 0.8 | 0.3 | 0.3 | 3.0 | 1.0 | 5.0 | 0.3 |
| S3 | 0.3 | 0.3 | 1.0 | 0.3 | 1.0 | 0.8 | 0.1 | 1.3 | 0.5 | 0.3 | 1.5 | 0.3 | 0.3 | 0.5 | 0.3 |
| S4 | 0.8 | 1.0 | 1.0 | 0.3 | 0.5 | 1.3 | 1.3 | 1.3 | 0.1 | 2.0 | 0.3 | 0.1 | 0.5 | 0.5 | 0.3 |
| T1 | — | — | — | — | — | 0.1 | 0.5 | 1.0 | — | — | — | — | — | — | — |
| T2 | — | — | — | — | — | 0.4 | 0.3 | 2.0 | — | — | — | — | — | — | — |
| T3 | — | — | — | — | — | 1.0 | 1.0 | 0.3 | — | — | — | — | — | — | — |
| T4 | — | — | — | — | — | 3.5 | 2.0 | 0.3 | — | — | — | — | — | — | — |
| T5 | — | — | — | — | — | 1.3 | — | — | — | — | — | — | — | — | — |
| T6 | — | — | — | — | — | 0.3 | — | — | — | — | — | — | — | — | — |

199.5

Grout take (cement bags)

Table 6. Comparison of instrumentation readings before and after the 2009 emergency repairs

| Instrument | Reading on 01/26/2009 | Reading on 06/10/2009 | Trend^ | Instrument | Reading on 01/26/2009 | Reading on 06/10/2009 | Trend^ |
|---|---|---|---|---|---|---|---|
| P-1 | 432 psi | 320 psi | ↓ | P-29-3 | - | 800 psi | - |
| P-2 | 0 psi | 0 psi | ≈ | P-30-4 | - | 560 psi | - |
| P-3 | 41 psi | 80 psi | ↑ | P-31-1 | - | 28 psi | - |
| P-4 | 610 psi | 720 psi | ↑ | P-32-3 | - | 390 psi | - |
| P-7 | 36 psi | 38 psi | ≈ | P-33-2 | - | 0 psi | - |
| P-8 | 23 psi | 20 psi | ≈ | W-2 | 1340 gpm | 350 gpm | ↓ |
| P-9 | 375 psi | 380 psi | ≈ | W-4 | 1805 gpm | 740 gpm | ↓ |
| P-10 | 140 psi | 170 psi | ↑ | W-5 | 1125 gpm | 25 gpm | ↓ |
| P-11 | 51 psi | 60 psi | ↑ | W-7 | 27 gpm | 50 gpm | ↑ |
| P-12 | 125 psi | - | - | W-8 | - | 70 gpm | - |
| P-23 | 59 psi | 380 psi | ↑ | | | | |

^ ≈ = steady; ↓ = decreasing; ↑ = increasing.

## CONCLUSIONS

The emergency grouting program was a success. The inflows through the PAT were decreased significantly, and the powerhouse system, including T3 access tunnel and PAT plug are currently safe, and the current leakage is not threatening to overwhelm the capacity of the powerhouse pumping system

The grouting program should not be considered a permanent repair as the injected grout is constantly under high pressure and high gradients that could affect its integrity. Additionally, the emergency repair did not address the potential for hydro jacking, which has been considered in the past as a driving factor behind the increase in inflows, or backwards erosion at T3AT and other locations. These factors will remain as potential issues unless pressure water is prevented from entering the rock mass, especially in high conductivity areas such as the shear zone.

The groundwater drainage pattern in the vicinity of the concrete plug was altered by the grouting program, and piezometers close to the face of the excavations are being monitored to identify the presence of high water pressures at shallow depths.

## ACKNOWLEDGMENTS

The emergency work was successful thanks to the participation of numerous persons from PG&E, URS, DTA, and Mitchell Engineering. The participation of Gabriel Perigault, Project Manager was especially valuable.

## REFERENCES

Lombardi, G., and D. Deere. 1993. "Grouting Design and Control Using the GIN Principle." Water Power & Dam Construction.

Moller, D., H. Minch, and J. Welsch. 1984. "Ultrafine Cement Pressure Grouting to Control Ground Water in Fractured Granite Rock." Innovative Cement Grouting: ACI Special Publication. 129–151.

# Structural Inspections of Colorado's Eisenhower Johnson Memorial Tunnel, Hanging Lake Tunnel, and Reverse Curve Tunnel

**Ralph Trapani**
Parsons Corporation, Denver, Colorado

**Jon Kaneshiro**
Parsons Corporation, San Diego, California

**David Jurich**
Hatch Mott McDonald, Phoenix, Arizona

**Michael Salamon**
Colorado Department of Transportation, Idaho Springs, Colorado

**Steve Quick**
Colorado Department of Transportation, Glenwood Springs, Colorado

**ABSTRACT:** The Eisenhower Johnson Memorial Tunnel (EJMT), the Hanging Lake Tunnel (HLT), and the Reverse Curve Tunnel (RCT) in Colorado on Interstate 70 are operated and maintained by the Colorado Department of Transportation. The EJMT has two 40 ft by 40 ft by 9,000 ft long tunnels that cross the continental divide at approximately 11,000 feet above sea level and are subject to extreme geotechnical, weather, and highway traffic conditions. The HLT includes two 40 ft wide by 30 ft high tunnels, approximately 4,000 feet long at an elevation of 6,300 feet above sea level. RC is a single bore two lane tunnel that is about 600 feet long. EJMT was completed in the 1970s and RC in 1984. HLT was completed in 1992. Comprehensive structural integrity inspections were completed in 2007 for the EJMT and in 2009 for HLT and RCT. These projects were some of the first interstate tunnels to be inspected in accordance with the FHWA guidelines described in the Highway and Rail Transit Tunnel Inspection Manual (2005 edition), with emphasis on safety, communication, and standardized written and photographic documentation. Comparisons and contrasts of construction techniques, geologic conditions, and structural condition assessments for tunnels separated by a generation are made.

## INTRODUCTION

The Eisenhower/Johnson Memorial Tunnel (EJMT), Hanging Lake Tunnel (HLT), and Reverse Curve Tunnel (RCT) are located on Interstate 70 in Colorado as shown in Figure 1. The EJMT is located approximately sixty miles west of Denver, Colorado on Interstate 70. It is the highest major vehicular tunnel in the world, located at an elevation of 11,013 feet at the East Portal and 11,158 feet at the West Portal. EJMT traverses through the Continental Divide at an average elevation of 11,112 feet. The facility lies entirely within the Arapaho National Forest and is divided by two counties, Clear Creek County at the East portal and Summit County at the West portal.

The EJMT was originally designed as a twin bore tunnel. Construction on the westbound bore (North Tunnel) began on March 15, 1968 and was completed five years later on March 8, 1973. This bore was originally called the Straight Creek Tunnel, and later was officially named the Eisenhower Memorial Bore, after Dwight D. Eisenhower, "father of the Interstate Highway System." Construction on the second bore began on August 18, 1975 and was completed four years later on December 21, 1979. It was named after Edwin C. Johnson, who served as a state legislator, lieutenant governor, governor, and U.S. Senator, who had actively supported an interstate system across Colorado. Centerline to centerline, the two tunnels are approximately 115 feet apart at the

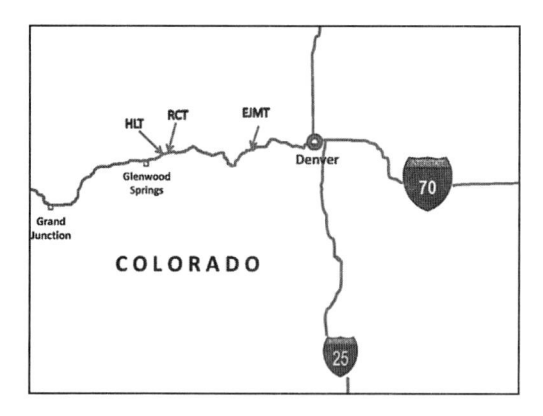

**Figure 1. Location of EJMT, HLT, and RCT on Interstate 70**

east ventilation building entrance, 120 feet apart at the west ventilation building entrance, and some 230 feet at the widest point of separation under the mountain. The length of the westbound (north) tunnel is 1.693 miles, and the length of the eastbound (south) tunnel is 1.697 miles (outside face to outside face of the ventilation buildings).

The HLT and RCT are part of overall improvements to I-70 through Glenwood Canyon for safety, environmental stewardship, and traffic congestion. The 12 mile long corridor was completed in 1992. The HLT is a 4,300 ft long, dual bore highway tunnel carrying I-70 through the mountains bordering Glenwood Canyon, east of Exit No.127 in Garfield County. The HLT houses an earth-sheltered command center, which is accessed from hanger doors inside the bores. The command center monitors cameras through the canyon, and are used to control electronic message signs to slow down or re-direct traffic in the event of an accident, and dispatch tow trucks. The HLT was completed in 1992. The RCT is a 600 ft long, single bore that carries the westbound lanes of I-70 through the north wall of Glenwood Canyon just east of HLT at approximately reference post #127. RCT was completed in 1984 and gets its name from the reverse curve that precedes it.

## GENERAL INSPECTION PROCEDURES

A logging/inspection protocol meeting was held prior to inspections (EJMT in 2007, HLT in 2009) to review how and what forms to fill out, photographing procedures, and other standards with the aim for uniform and objective quality. Two-man inspection teams were directed to map features and record details of special features or concern standard rating inspection forms (a scale from zero to nine) and base maps (an unwrapped 360 degree view of the tunnel), which in general followed the guidelines found in: (1) the FHWA and FTA Highway and Rail Transit Tunnel Inspection Manual (2005) and (2) the American Concrete Institute standards ACI 201.1R, *Guide for Making a Condition Survey of Concrete in Service*. Each day began with a Safety Moment and a brief discussion of the days anticipated activities. At the end of each day, a brief meeting was held to discuss the day's findings.

Two-man teams per area/tunnel heading were used for safety reasons and for more efficient record keeping and inspection. A "walker" was utilized to provide additional communication between the teams and to provide supplemental support. To facilitate inspection, each team was equipped with the tools shown in Table 1.

If any areas requiring Non-Destructive Testing (NDT) were found, they were spray painted and recorded. Where cracks or joint separation were observed, spackling was applied and dated to monitor future cracking or separation. Photographs were taken of areas of concern to illustrate typical conditions found throughout the tunnel. Specifics of the photograph (date, location, and subject) were noted on the manila or grey paper and photographed to serve as a title for the proceeding photographs. Video recordings with narration of key features of the entire length of each tunnel were conducted.

It should be noted that because of the type of construction and nature of tunnels, certain components that make up the structural systems cannot always be observed. Therefore, observations made in the concrete structures must infer if there are any major deficiencies beneath it. For example, hanger rods within the precast divider wall panels are concealed from sight because they are completely encased within the wall. Consequently, deterioration of hanger rods are inspected for, in the form of stress cracks or corrosion on the concrete surfaces or the employment of non-destructive testing (NDT).

## EISENHOWER-JOHNSON MEMORIAL TUNNEL

EJMT are horseshoe shaped tunnels with a maximum excavated height of 48 ft and a width of 40 ft as shown in Figure 2. The main construction method utilized for this tunnel was the Drill and Blast method. The exhaust and supply air ducts are located above a suspended porcelain enamel panel ceiling and a drainage system is provided underneath the roadway surface. The exhaust and the supply air ducts are separated with precast concrete divider walls.

The tunnels were constructed using several initial/primary support systems and a final cast-in-place concrete liner. The three initial/primary support systems utilized consist of:

1. A rock reinforcement system of rock bolts and spiling were used for the majority of

**Table 1. Typical tools employed for Inspection of the EJMT, HLT, and RCT**

| | |
|---|---|
| • A push/pull cart | • Carpenters rulers |
| • A 16 ft portable extension ladder | • Fiber glass 25 ft stadia rod |
| • A digital camera | • Small pocket knife |
| • A digital DVD video camera (and voice recorder) | • Zip lock bags (for sample collection) |
| • Flash lights | • Measuring cup (for recording seeps or drips) |
| • Miners head lamps | • Screw driver |
| • At least one cell phone and CDOT supplied two-way walkie talkie radio | • Wire brush |
| • Shop lamps | • Magnifying glass (loupe) |
| • Power strip/surge protectors | • Field clip board |
| • 50 ft extension cord | • Writing instruments |
| • Inspection forms | • Logging protocol "cheat sheets" with a crack comparator gauge |
| • Base maps | • Spackling and spackling knifes for monitoring crack growth |
| • Photograph logs | |
| • Manila/grey paper for identifying photos | • Personal protective equipment (including hard hats, leather work gloves, protective eye wear, work boots, and safety vests) |
| • Indelible ink markers | |
| • Survey marker spray paint cans | |
| • Sounding hammers | • Particle masks were made available to those wishing to use them. |
| • Tape measures | |

the tunnels (Zone 1, Figure 3). The rock bolts consisted of 20 ft long No. 8 rock bolts (typically on a 5 ft by 5 ft pattern) installed around the perimeter of the arch to create a reinforced rock arch to support ground loads.

2. A ribbed system of horseshoe shaped steel wide flange (12WF106) ribs closely spaced (typically 2 to 4 ft) designed to support increased loads in more difficult ground (Zones III and IV, Figure 3). Korbin (2007) indicates that ribs in the south bore instrumented with load cells indicated only its own dead weight. Thus, at that time, all rock loads were carried by the rock bolt systems. The observations were not made in the stacked drift sections. In some cases, steel ribs (14WF287) were used to close the invert and form continuous steel ribs supports around the tunnel perimeter.

3. A four-type stacked drifts system (Figure 3) of small diameter tunnels advanced around the perimeter of the excavation and filled with concrete to support heavy ground loads in the very difficult ground such as the Loveland Fault. The fault had squeezing ground with loads up to full overburden.

The final liner is cast-in-place concrete that conceals the initial/primary support systems. The final liner provides a uniform finished surface and adds redundant long term ground support, and houses the lighting, signaling, ventilation, protective tiling and fire protection systems.

Precast concrete divider walls are supported by steel rods hanging from the crown of the arch. They support steel angles that carry the precast concrete

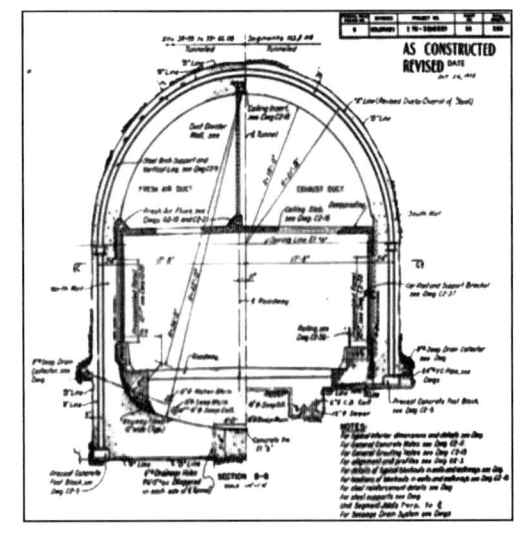

**Figure 2. Typical cross section of EJMT (north bore)**

panels that make up the floor of the plenums/ceiling of traffic tunnels. The steel rods are placed in a precast concrete divider wall that hides, but does not touch the steel hanger rods. The divider wall's function is to separate the supply plenum from the exhaust plenum; the structural loads are the dead weight, nominal maintenance traffic loads, and the wind loads from the fans. The plenums provide for a transverse ventilation system. The plenums and associated divider wall generally consisted of 168 segments and 2 transition areas to the ventilation buildings. Each segment is about 50 ft long, consisting

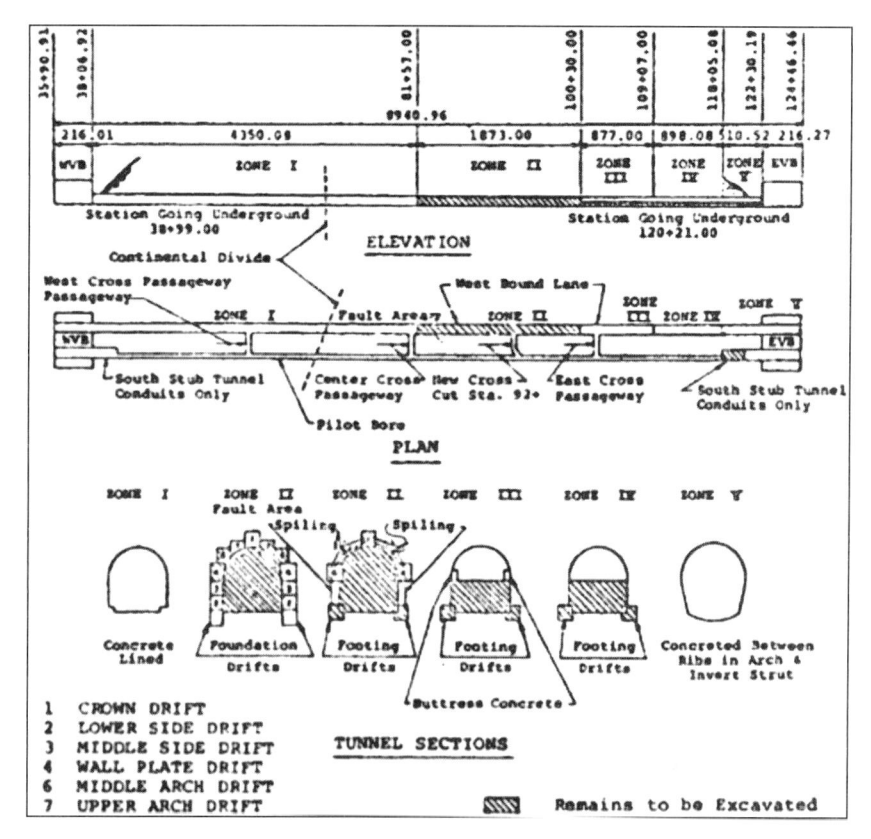

Figure 3. Typical EJMT support (Hooper et al., 1972)

of 7 to 10 panels (precast elements), depending on which tunnel and which side of the plenum being considered (Figure 4). The panels in the north tunnel generally had different lengths than the panels in the south tunnel. Grout is used in the divider walls to seal the precast elements and varies between the different tunnels. Between panel segments, the north exhaust employs neoprene type gaskets employing soft joint material, while the north supply and south exhaust has hard joints, and the south supply has soft joints at every other joint while the others are hard.

At the inverts, the roadway is placed on grade at the bottom of the tunnel structure. Tunnel finishes in the roadway portion of the tunnels are slightly different for the north and south tunnels. The north tunnel has precast panels with ceramic tiles, but the south tunnel has structural glazed tiles.

There are several differences in the construction of the north tunnel and the south tunnel. The north tunnel was built first, and is taller than the south tunnel. The divider wall panels have different length dimensions in the north tunnel vs. the south tunnel; therefore, the spacing of the panels is different. There are differences in the ventilation systems between the north and south tunnels as well. In the north tunnel, the supply vents are in the walls of the tunnel and the exhaust vents are in the floor. In the south tunnel, the supply and exhaust vents are in the floor. The north tunnel supply has ducts that go down the walls. These ducts must cross the exhaust side in the floor, thus the floor is 10" thick with only a 4" precast panel. The remainder of the depth is an air duct with a metal topping.

## EJMT INSPECTION, ANALYSIS, AND FINDINGS

One of the main purposes of the EMJT inspections was to evaluate the structural elements of the ventilation plenums including the divider wall, the plenum floor/traffic tunnel ceiling, and the tunnel's arch or radius wall (the traffic tunnel and ventilation structures, while inspected, are not discussed herein). The divider wall was inspected and analyzed for signs of potential deterioration and structural deficiency of the embedded hangers of which there were a variety of types as exhibited in Figure 5. In addition, the groundwater drainage, mechanical and electri-

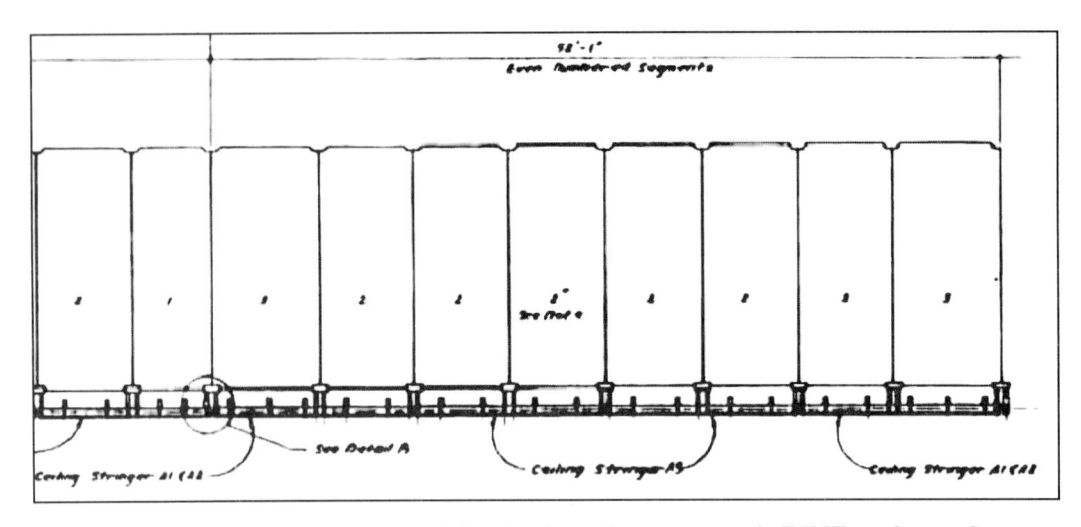

Figure 4. Elevation of precast divider wall showing 8 panels per segment in EJMT north tunnel

cal systems within the plenums were inspected for obvious deficiencies.

All plenums (north bore exhaust and supply, and south bore exhaust and supply) were found to be generally in good to excellent structural condition. The panel walls and floors were generally in very good condition throughout the tunnels. There were no indications of severe deterioration or overstressing. The minor deficiencies found in the radius wall included hairline cracks, pattern cracking and segment joint separation. These deficiencies included varying amounts of efflorescence, mineral deposit buildup, concrete honeycombing, minor concrete spalling, concrete peeling, and concrete leaching. Locally, the waterproofing on the floors required repair. Nuisance groundwater seepage through tunnel walls is an ongoing maintenance issue, primarily in the North Tunnel near each ventilation building. Drainage collection systems installed to handle groundwater seepage in the plenums require repairs or replacement. All plenums were adequately lit for walking, but required supplementary lighting to perform detailed inspection work. All surfaces in the exhaust plenums were coated with a thick layer of black soot and dust. This made detailed inspection difficult since the soot and dust likely masked some shrinkage hairline cracking not readily apparent. The surfaces in the supply plenums were generally clean and free of soot deposits making most cracks, including hairline/shrinkage cracks, fairly clear to identify. The teams recorded all cracks graphically by a freehand sketching; the cracks were easily mapped where water had once been present and grouting repairs had occurred. The presence of autogenously healed concrete, efflorescence, deterioration, or stalactites in the radius wall of the final liner (as shown in Figure 6) correlated with the groundwater inflow recorded in the pilot tunnel (Hurr and Richards, 1965) and the final liner type (i.e., stacked drift supports had less cracks and previous groundwater inflows, Figure 3).

The precast concrete panels that form the divider wall and floor of the ventilation plenums incorporated neoprene type control joints at about every 50 ft. Most control joints fit flush or nearly flush (less than 1 inch of offset) but some joints have greater offset. Moreover, at some joints movement did not occur at the control joint, but adjacent to it, within the precast panel as shown in Figure 7. These locations were marked with spackling and marked with the date for further monitoring.

The hanger rod, hanger rods connections, and steel angle/channels supporting the floor were evaluated and found to be satisfactory for overpressure from the ventilation fans. Furthermore, since steel has a large strain limit, on the order of 3 to 5% before rupture occurs, any movements can be detected with periodic visual inspections and surveys.

Recommendations were prioritized and cost estimates for repairs were prepared according to the FHWA/FTA manual. Recommendations included continued/routine inspection including monitoring of cracks that were spackled and dated, limited testing and NDT at select locations, inspections during periods of higher groundwater, repair of groundwater drainage and floor waterproofing systems, repair of missing mastic at isolated hanger locations, and remedial grouting where required if areas of high groundwater inflow are found during higher groundwater periods.

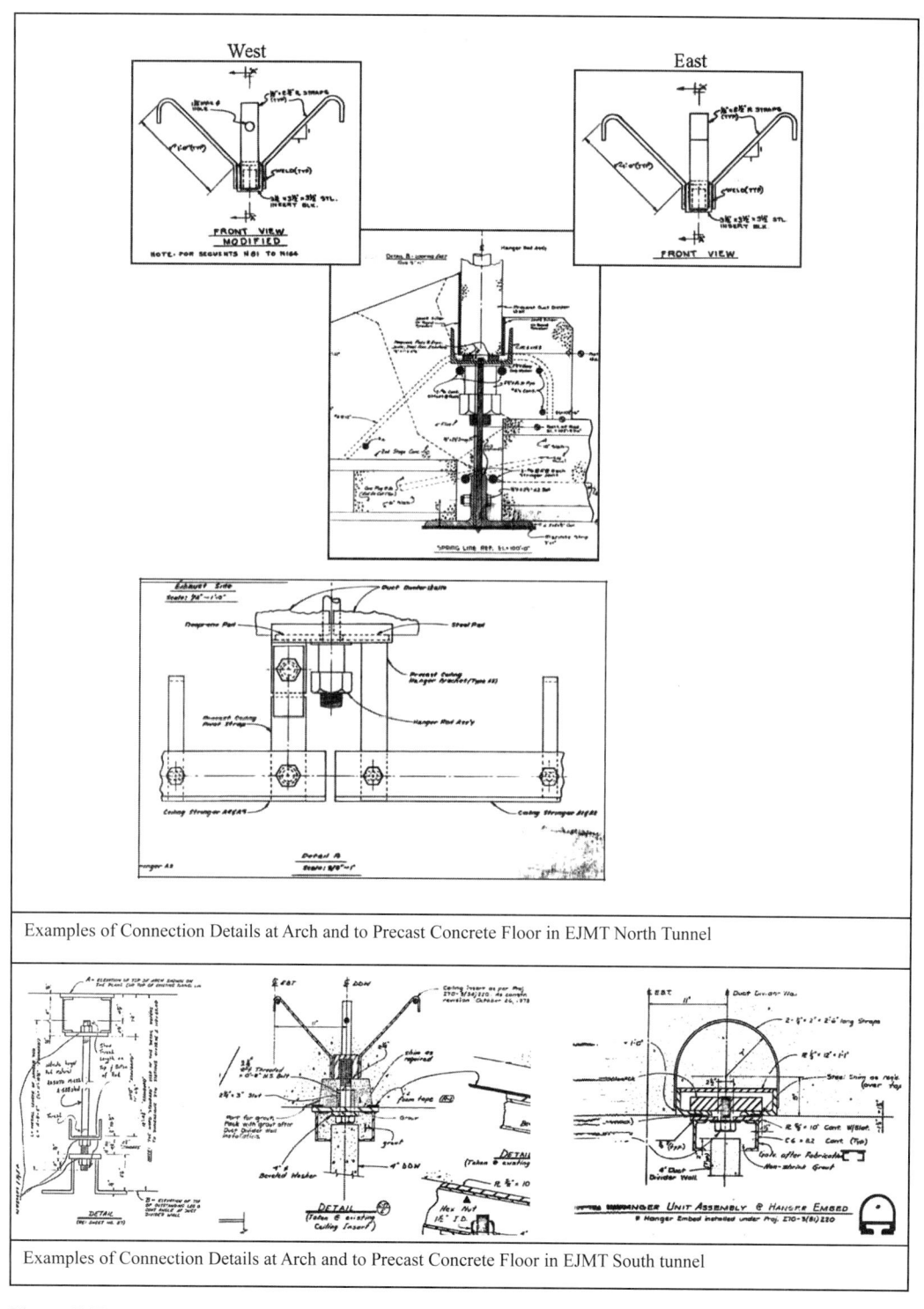

Examples of Connection Details at Arch and to Precast Concrete Floor in EJMT North Tunnel

Examples of Connection Details at Arch and to Precast Concrete Floor in EJMT South tunnel

**Figure 5. Examples various hanger rod and floor connection details in EJMT**

Figure 6. EJMT typical efflorescence and leakage showing drainage pipes in arch/radius wall

## HANGING LAKE TUNNEL AND REVERSE CURVE TUNNEL

The HLT are twin semicircular shaped tunnels as shown in Figure 8. Each tunnel carries two lanes with a concrete pavement road, cast-in-place sidewalls with grouted tile, and suspended aluminum ceiling with porcelain- metal ceiling panels, which provides plenums for the semi-transverse ventilation system. The support for the suspended ceiling is embedded at springline in the final liner as shown in Figure 9. Excavation was performed by drill and blast methods using multiple slash excavation techniques (three top headings and one bench). The primary support system for the HLT and RCT is patterned rock bolts, with wire mesh and steel fiber reinforced shotcrete as required. The final liner is a nominal 15-inch thick cast-in-place concrete liner, cast in 40-ft sections

Figure 7. Crack in pre-cast panel, adjacent to neoprene construction joint, in EJMT

against the PVC waterproofing layer. The ventilation and control structure, called the Cinnamon Creek Control Complex (CCC) is a partially buried structure (up to 30 ft depth), located to the east of the midpoint of the facility. The five-story CCC divides the HLT, such that it was constructed in four bores or headings.

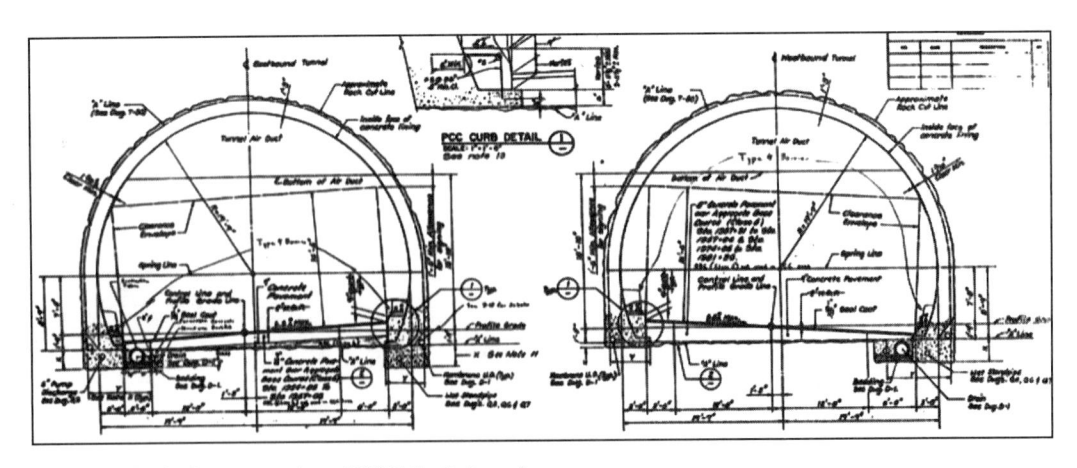

Figure 8. Typical cross section of HLT (both bores)

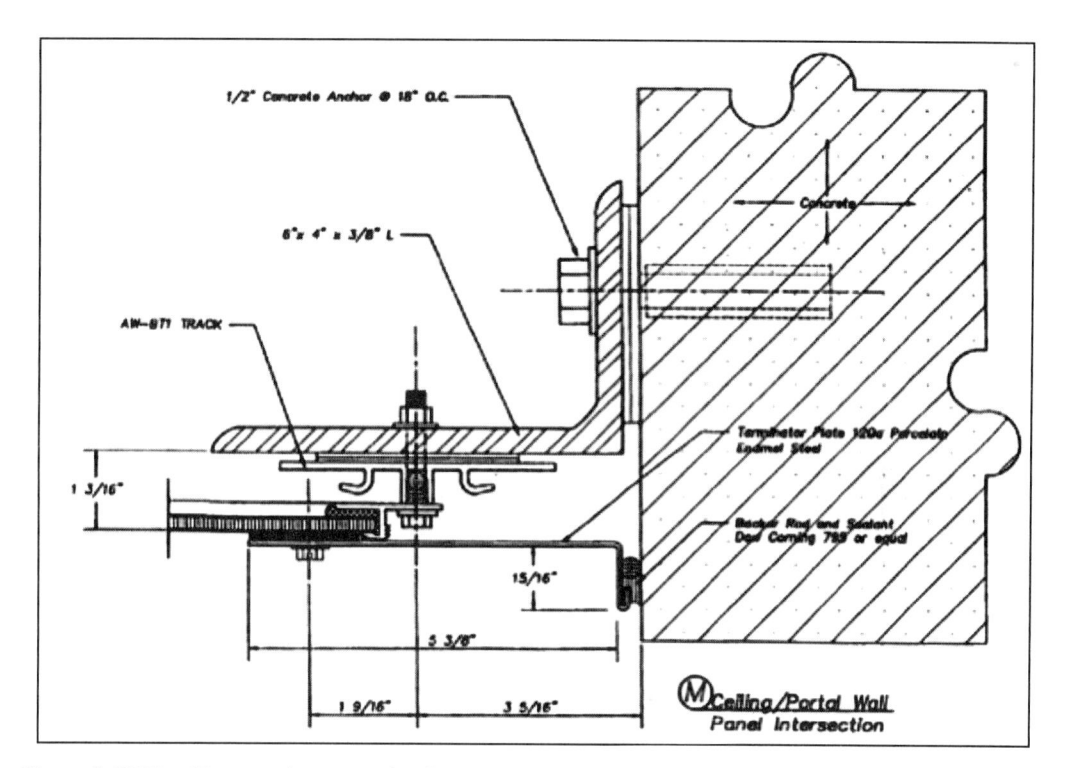

**Figure 9. HLT ceiling panel support detail**

RCT, just east of HLT, is a single bore horseshoe shaped tunnel carrying west bound traffic, while the east bound traffic does not require a tunnel. There is no ceiling or ventilations system in RCT due to its short length. At all ends of both the HLT and RCT, there are portal extensions which form the abutment for approach structures and provide protection from rockfalls.

It should be noted that in 2007, a rock fall damaged the east bound cut and cover transition structure to the HLT tunnel near the CCC, creating a crack for 70 ft (⅔ the length of the roof). A fast track emergency repair was required before the winter and holiday season. The plenum ceiling had to be removed locally for access and a temporary work deck shored from underneath. Rock debris was removed from above with limited access to the roof; rock debris was hauled to the adjacent Union Pacific railroad line. Holes were drilled for # 7 tie rods at two foot centers. A 6-inch thick reinforced shotcrete layer (#8s & #10s) was placed under the soffit while a 12 inch thick reinforced concrete slab (#11s) was placed on the roof. The two layers form a sandwich that is bonded by the tie rods. The structure was then backfilled with over 2500 cy of polystyrene blocks to provide a lightweight fill and cushion protection from rock falls.

## HLT AND RCT INSPECTION, ANALYSIS, AND FINDINGS

The main purpose of the HLT and RCT inspections was to evaluate the structural elements of the ceiling connections, embedded hangers, and beam connections in the arch of the tunnel as well as the condition of tunnel liner (the CCC structure, while inspected, is not discussed herein). In general, the condition of the connections and the concrete arch liner were in excellent shape.. There was distress noted in one cross passage, however, where the shotcrete liner had extensive cracking. Also, tiles in the tunnel space had locations where the grout bond had failed. At the time of writing of this paper, inspection findings, structural analysis, and cost estimates for recommendations are in progress. Similar recommendations as for EJMT are expected for HLT and RCT.

## CONCLUSIONS

The EJMT, HLT, and RCT on the scenic I-70 corridor are subject to harsh winter environments, problematic geotechnical conditions (granitics with shear zones, groundwater inflow, rockfalls), and a demanding tourist and commerce traffic conditions in summer and winter. Advances in tunnel lining approach and highway tunnel design from the late 1960s to the

1980s and 1990s are contrasted in the approach to the final lining systems and the approach to ventilation and systems design. Regardless, the designs are subject to the same harsh environment and require diligence in inspection, monitoring, and repair to keep this asset running smoothly for generations to come.

## ACKNOWLEDGMENTS

The authors wish to thank their respective employers and the Colorado Department of Transportation for the permission to publish this paper. Subconsultants were Hatch Mott McDonald for the EJMT inspections and Yeh & Associates for the HLT and RCT inspections. The inspection teams led by the management of R. Trapani and B. Doyle consisted of: J. Kaneshiro, D. Weir, D. Jurich, T. Mayen, A. Ruis, J. Zeid, R. Brewer, J. Crowder, M. Piek, and D. Neil for EJMT; and consisted of H. Hansson, J. Zeid, T. Mayen, S. White, K. Assay, J. Barker, J. Spanier, J. Braaksma, J. Wehner, W. Egger and R. Phil for HLT and RCT. J. Braaksma was lead structural engineer and D. Neil was lead for NDT for all tunnels.

## REFERENCES

ACI [American Concrete Institute] standards ACI 201.1R, Guide for making condition survey of concrete in service.

ACI, ACI 228.2, Nondestructive test methods for evaluation of concrete structures.

Colorado Department of Transportation (CDOT). CDOT Web Page. Internet. http://www.dot.state.co.us/Eisenhower/description.asp; http://www.dot.state.co.us/Eisenhower/Milestones.asp

FHWA and FTA [Federal Highway Administration and Federal Transit Association], 2005, Highway and Rail Transit Tunnel Inspection Manual.

Haack, A., J. Schreyer, and G. Jackel, 1995, State-of-the-art of non-destructive testing methods for determining the stat of a tunnel lining, Tunnelling and Underground Space Technology, Vol. 10.4, pp.413–431.

Hopper, R.C., T.A. Lang, and A.A. Matthews, 1972, Construction of the Straight Creek Tunnel, Colorado, Rapid Excavation and Tunneling Conference, Vol. 1, Ch. 29, pp. 501–538.

Hurr, R.T., and D.B. Richards, 1965, Ground-water engineering of the Straight Creek Tunnel (pilot bore), Colorado, Association of Engineering Geologists, Engineering Geology Bulletin, Vol. 3, No. 1, pp. 80–90.

Korbin, G.E., Personal communications, Nov. 2, 2007.

McGlothlin, W., Straight Creek Tunnel, Mines Magazine, March 1973.

# Corrosion Protected Systems for Tunnels and Underground Structures

## James B. Carroll, Heather M. Ivory
URS Corporation, Columbus, Ohio

**ABSTRACT:** Wastewater tunnels are continually exposed to highly corrosive environments. With Owner requirements of a 100 year design life, it becomes apparent that these tunnels, traditionally lined with concrete, require protection from the corrosive environments they are deployed in. This paper discusses the available corrosion protection products on the market today for rehabilitation and new constructions, the risk involved with the application of corrosion protection products in underground structures, and the cost associated with these products.

## WHY CORROSION PROTECTION?

The root source of corrosion in wastewater system is hydrogen sulfide ($H_2S$). The mechanism by which $H_2S$ cause corrosion to the concrete and steel of wastewater systems is as follows (Talley, J. and Wallace, G. 2009):

1. $H_2S$ is created by the decomposition of organic materials within the wastewater flow;
2. When turbulence is encountered by a change in flow conditions at an outfall, drop, force main or other change in flow, the $H_2S$ is stripped/released from the wastewater flow;
3. The $H_2S$ settles on the damp concrete surface above the flow of wastewater;
4. The deposited $H_2S$ produces elemental sulfur on the structure or pipe surface;
5. The microbial bacteria Thiobacilus convert the sulfur deposits to sulfuric acid; it is this acid which causes the corrosion within the system.

The amount of hydrogen sulfide produced in wastewater is dependent on several factors; turbulence, retention times, temperature, humidity, terrain, flow strength and flow levels. The limit at which point corrosion by Thiobacilus begins is 2 to 5 ppm of $H_2S$ on the concrete surface (mechanism three above). A reduction in $H_2S$ can be gained with ventilation, controlling wastewater drop structures and removing force mains. However, due to potential changes in a wastewater system over time, it is not feasible to predict long term effects of corrosion with accuracy. Today, engineers and designers of wastewater systems are managing this challenge by incorporating corrosion resistant materials and/or changes in design to aid in extending the design life of sewers.

## DESIGN/CONSTRUCTION CONSIDERATIONS

### Construction Materials (Precast Segments, Cast In-place Concrete Liner, Pipe)

For tunneling applications, corrosion resistant materials can generally be divided into two categories: one-pass systems and two-pass systems. One-pass and two-pass systems contain many of the same corrosion protection products, where the difference is when the material is installed. In a one-pass system, the corrosion resistant material is applied or added during the casting process. In a two-pass system, the corrosion resistant material is installed after tunneling operations are complete, often times extending the total construction time of a project. Corrosion resistant pipe installed using open cut or jacking methods also falls within the one-pass system category; however it can also be placed within the two-pass system category if it is being placed within the tunnel after tunneling operations are complete. Table 1 provides a general listing of available corrosion protection products for one-pass and two-pass systems.

### One-pass Corrosion Resistant Materials

*Anchored Thermoplastic Lining (PVC, HDPE)*

Anchored thermoplastic lining PVC or HDPE, also referred to as cast-in stud liners, have a proven track record in sewer pipe and in cast-in-place concrete structure applications. Available in either t-ribbed

Table 1. Corrosion protection products

| | System | Type | Product Examples |
|---|---|---|---|
| One-pass Systems | Anchored Thermoplastic Lining (PVC, HDPE) | Lining | T-lock, AgruSureGrip |
| | Concrete Faced with Glass Reinforced Plastic | Lining | Combisegments |
| | Precast Polymer Concrete Segments | Lining | U.S. Composite Pipe, Future Pipe |
| | Acid Resistant Concrete Additives | Concrete additive | Consheild |
| | Microsilica (Silica Fume) Concrete | Concrete additive | |
| | FRP/GRP/Polymer Concrete/PVC lined RCP Pipe | Pipe | Hobas, Krah, U.S. Composite, Future Pipe |
| Two-pass Systems | Liquid Applied Polymer Based Protective Lining | Coatings | Tnemec, Sauereisen |
| | Post-Installed Anchored Thermoplastic Lining (PVC, HDPE) | Lining | T-lock, AgruSureGrip |
| | Acid Resistant Gunned Cementitious Lining | Coating | SewperCoat |
| | Deformed Pipe Lining | Lining | InsituGuard |
| | Slip Lining | Lining | Hobas, Krah |
| | Pipe-in-tunnel | Pipe | Hobas, Krah, U.S. Composite, Future Pipe, PVC Lined RCP |

Figure 1. Ameron T-Lock

Figure 2. Agru Suregrip

(Figure 1) or studded (Figure 2) sheet lining materials, these products have been in use since the 1940s to protect new concrete trunk sewers, structures (such as wet wells, manholes and shafts) and monolithic tunnels against hydrogen sulfide attack and other sources of corrosion. The Upper Northwest Interceptor (UNWI) project in Sacramento, California is the first application in the world to utilize a cast-in stud liner with a one-pass precast concrete segmental tunnel lining.

As mentioned above, the UNWI project is the first use of a cast-in-mold anchored PVC lining in conjuction with precast concrete segments. For the UNWI project, the lining was installed by the following steps during the segment casting process:

1. Place pre-cut liner sheets (Figure 3) in the segment molds.
2. Place steel reinforcement cage and other segment parts (grout/lifting port, and dowel sockets) in segment mold (Figure 4).
3. Continue casting operations as per normal casting procedures.
4. Install segments per normal tunneling operations.
5. Complete welding of liner joints/seams (Figure 5).

Once the precast segments have been erected in the tunnel, the gaps in the lining at the longitudinal and radial joints of the segments are welded using plastic welding strips. The welding strips (shown in Figure 5) are fusion welded by welders approved by the manufacturer. These welders use approved methods and techniques. Part of the approval process of welders may be the requirement to pass a

Figure 3. Precut liner sheets

Figure 4. Placement of steel reinforcement cage

Figure 5. Hot air welding 1" weld strip using weld strip roller

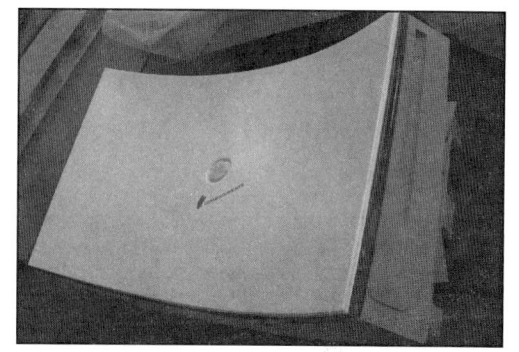

Figure 6. Herrenknecht combisegments

qualification welding test prior to performing the work. The final stage of stud liner application is testing of welds by the contractor, lining manufacturer, a third party inspection group and/or the engineer.

### Concrete Faced with Glass Reinforced Plastic

The glass reinforced plastic face for precast concrete segments is an acid resistant facing. The product is placed in the segment mold during the casting process, similar to the previously discussed anchored thermoplastic lining, trademarked as Combisegments by Herrenknecht AG and shown in Figure 6. These specialized segments are erected in the same manner as a normal segmentaly lined tunnel. Combisegments are advertised as a corrosion resistant one-pass tunnel lining.

Sanitary sewer tunnel Combisegments are constructed of a 3 mm reinforced polyester (Glass Reinforced Plastic) corrosion resistant lining. The GRP corrosion resistant material is added to the segment forms during the casting process, while the position of the segment gasket is moved from the outside diameter to the inside diameter (see Figure 8). The availability to reposition the segment gasket within the same mold makes it possible to change over the manufacturing process from a lined to an unlined tunnel segment without significant delays, as the same mold is used for lined or unlined segments (see Figure 9). Combisegments are designed with a dowel and pin configuration, where bolts are used in the construction of the segmental ring.

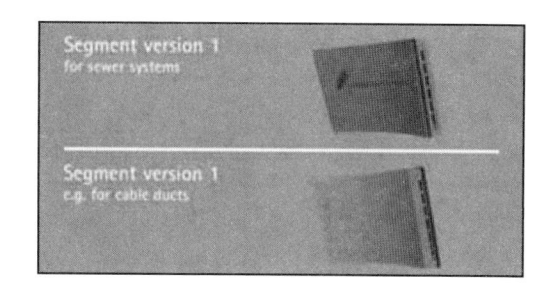

Figure 7. Combisegment versions

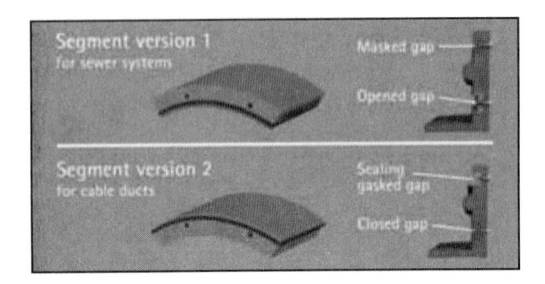

**Figure 8. Combisegment gasket positions**

The molds and the GRP liner for Combisegments are made in Germany and shipped to the job site, to be cast by local precast concrete manufacture familiar with precast concrete tunnel segments. Combisegments were recently installed in a 10-foot diameter tunnel in Moscow, Russia which began in March 2009. It is the first project of its type in the world.

*Precast Polymer Concrete Segments*

Polymer concrete segments have many of the same characteristics as polymer concrete pipe. Although the technology for the segments cast with polymer concrete exists, to date, polymer concrete segments have not been used in the construction of a tunnel to the author's knowledge

Polymer concrete raw material costs are currently around four to six times the cost of Portland cement concrete. Even so, no additional liner for corrosion protection is required with this material, which potentially reduces construction time. It is assumed that the cost of construction of all segment types is comparable and that TBM progress rates will not be significantly altered by the use of one-pass material over the other. To the extent that a design thickness reduction can be achieved, polymer concrete segments will be proportionately lighter than those made from traditional concrete, which makes moving and handling easier. However, the added expense and contractor inexperience with polymer concrete can make it difficult to bid.

There are currently two material suppliers who can manufacture polymer concrete segmental linings in the United States, Meyer Polycrete Ltd. Lüneburg, Germany and Polymer Pipe Technology (PPT) Des Moines, (Figure 10). These two manufacturers use different materials, including agents and resins, based on their own proprietary technology and processes. Manufacturers' materials may include polyester, epoxy, furan polymer, and methyl methacrylate and may or may not include conventional steel reinforcement in the same manner as with conventional RCP as per ASTM C76.

**Figure 9. Finished Combisegment tunnel types, sewer, and cable**

Long-term durability and the mechanical properties of polymer concrete are two aspects that need to be ascertained before a final decision regarding the suitability of this material can be made. Water absorption rates are approximately one order of magnitude less than those of Portland cement concrete. No current standard specifically applies to Reinforced Polymer Concrete segments, as this form of technology is relatively new.

*Acid Resistant Concrete Additives*

Silica fume is a dry powder consisting of micron sized particles of pure silica, specifically amorphous (non crystalline) silicon dioxide, produced as a by product of silicon and ferro-silicon metal production. Silica is typically un-reactive with anything but the strongest acids. Early researchers found that one of silica fume's properties was the ability of silica fume to react with calcium hydroxide, a by-product of concrete. This reaction reduced the volume of weak, calcium hydroxide and replaced it with calcium-silicate hydrate, producing an extremely strong concrete. The protection of calcium hydroxide in concrete was

**Figure 10. Polymer concrete segments**

also found to reduce the permeability of concrete, making the matrix more dense and resistant to liquid penetration.

Silica fume concrete is highly resistant to penetration by chloride ions. Since the silica fume acts as a pore blocker, the rate and quantity of sulfates and chlorides that are able to migrate through the concrete and attack the reinforcing steel and calcium in the concrete, is reduced. These benefits were seen by many transportation agencies currently using silica fume concrete for the construction of new bridges and rehabilitation of older structures. In addition, fine and coarse aggregates which are not susceptible to acid attack (e.g., sand and granite) can be selected in this type of concrete.

There are two ways commonly used to specify silica fume for use with the precast concrete segments:

1. Prescription: 1.5% silica fume by weight or 40 kg/cubic meter
2. Performance: resistivity testing

As an example, on the BWARI Part 1 project in Columbus, Ohio, the precast concrete tunnel segment specification was a high performance concrete that included a percentage of silica fume of 5 to 7 percent and an aggregate per ASTM C33 (consisting of natural sand, gravel, crushed gravel, or crushed rock, ¾" minimum size). Blast furnace slag was not allowed as it slows concrete set up times. Sika super plasticizers were used to improve the workability of the mix. The purpose of this specification was to improve corrosion resistance.

While silica fume will not make concrete acid-proof, it will make it more acid resistant. It increases longevity by slowing the rate of decay due to sulfide attack. Where typical Portland Type II concrete can withstand a pH of 6 continuously without damage, concrete enhanced with 7 percent silica fume is expected to perform similarly at a pH of 4.5 or lower. As such, the use of silica fume cannot guarantee a design life of 100 years, but can result in increases in the design life of a tunnel over an ordinary Portland cement concrete tunnel.

## FRP/GRP/Polymer Concrete/PVC Lined RCP Pipe

Common pipe materials available for use in cut and cover, pipe jacking, microtunneling, and pipe-in-tunnel options include: Fiber Reinforced Pipe (FRP), Polymer Concrete Pipe and PVC lined Reinforced Concrete Pipe (RCP) Pipe.

### Fiber Reinforced Pipe

FRP provides several outstanding qualities for sanitary sewer systems. These properties include exceptional corrosion resistance as well as low maintenance costs. With its high strength-to-weight ratio, FRP can be transported much easier and handled on site using smaller equipment. This high strength is also suitable for jacking and the smooth interior of FRP provides excellent hydraulic characteristics. These properties allow smaller diameter FRP to have a comparable hydraulic capacity to larger sizes of RCP.

Fiber reinforced polymer pipe comes in two main variations: Centrifugally Cast and Non-Centrifugally Cast FRP. Centrifugally Cast Fiber Reinforced Polymer Mortar Pipe (CCFRPMP) by Hobas Corporation is produced within the United States and used extensively nationwide. One major advantage of CCFRPMP is that lower jacking loads are required as a result of constant outside diameters, and consistently straight pipe lengths inherent to the centrifugal casting process. Meyer Pipe manufactures a non-centrifugally cast FRP product, promoted as Glass Reinforced Pipe (GRP). This product is resistant to sulfide attack, and has excellent hydraulic characteristics and performance record; it has been used for about 30 years in wastewater applications.

### Polymer Concrete Pipe

Polymer concrete pipe, such as that produced by Amiantit Pipe Systems has been used in the United States for 25 years. It has a good track record and is used both in open cut applications and for microtunneling and jacking. There are well developed standards for the use of polymer concrete pipe contained in ASTM Standard D 6783. Polymer concrete pipe has excellent corrosion resistance and excellent hydraulic characteristics. Due to the pipe's initial high compressive strength, it is a good choice for pipe jacking and it will allow longer than average drive distances. The largest standard pipe that Amiantit manufactures is 102 inches in diameter. There is some indication in the literature that polymer concrete pipe loses strength over time. However, Amiantit states that this loss of strength has been taken into account during the pipe design.

### Reinforced Precast Concrete Pipe, (RCP) with Liner

RCP has a long history of use in the US for both storm and sanitary sewers. However, for use in sanitary sewers without a liner the concrete is subject to attack by sulfuric acid. Ameron's T-ribbed, poly-vinyl chloride, sheet stud lining material was designed specifically to protect new concrete sewer pipe by providing an embedded barrier between the corrosive atmosphere and the concrete. This material is known as T-Lock and is a PVC cast-in pipe liner. The T-Lock material and pipe system has been used for almost 50 years with a good track record.

Besides Ameron's T-Lock, there are several ribbed or studded liners made from HDPE and Liner Low Density Polyethylene (LLDPE) that have been cast into concrete for pipe and for lining rectangular or cylindrical wastewater basins. These materials have sufficient chemical resistance to be protective in a sewer environment. In service, they are also low maintenance and generally easy to repair.

The main concern with sheet liners is with the attachment/bonding of the liner to the concrete. In a centrifugally cast or other cast-in system this process is completed above ground at the production facility, where quality control can be monitored better than casting underground in-situ linings. When fitted in the tunnel, a cap strip or overlapping joint strip is welded to both sides of the joint. The liner is then tested again with an approved electrical holiday detector for defective welds and pinholes. Studded or ribbed liners can be designed to cover a portion of the pipe circumference consistent with ensuring that dry-weather flow submerges the bottom bare concrete and up to the liner. For pipe sections not in contact with the wastewater flow such as, manhole or structure risers, it is recommended a 360 degree coverage be used since the entire pipe surface is exposed to H2S gasses

As with all liners applied or cast, there is always potential for damage during shipment, installation, and construction activities. It is imperative to select reputable manufacturers for the selected alternative products as well as provide quality performance specifications during final design to ensure a long-lasting quality product with minimal repairs required prior to the acceptance of flow in the pipe.

**Two-pass Corrosion Resistant Materials**

*Liquid Applied Polymer Based Protective Lining (Coatings)*

Spray and trowel applied polymer based protective coatings provide resistance to sulfuric acid and acidic gas by providing an impermeable layer on top of the concrete substrate. There are many available coating systems which fall under this category including Sauereisen, Tnemec, Polibrid, Warren Environmental and many more. While these systems will not provide a 100 year design life as the post applied sheet linings may, properly applied coatings have been shown to provide 10–20 year design life before requiring reapplication or rehabilitation. For proper application, polymer based coatings systems require strict environmental controls and surface preparation during the application and curing of the products. Proper surface preparation includes the cleaning of the substrate while providing a sufficient mechanical anchor pattern. The advantage of coatings over other post applied corrosion protective materials is the reduced cost of materials and application. However, this reduced cost is for the original application only. Applying 4–5 coats of a polymer based protective lining over a 100 year tunnel design life may prove to have a higher life time cost over other systems.

*Post-Installed Polyvinyl Chloride or High Density Polyethylene Studliners*

Post installed studliner systems are commonly used worldwide. These systems include installing a studliner in a final layer of grout or concrete (up to 8–10 inches thick) after the tunnel is constructed. The studliner can be applied to a pipe leaving the invert open, to allow for seepage from groundwater behind the liner to be released. A disadvantage associated with this method is the reduction of the internal diameter of the new or existing tunnel thereby reducing capacity. The reduction in the diameter of the tunnel is caused by the amount of concrete used to embed the studliner. Another disadvantage is the time required to install the liner system and weld the joints after the tunnel has been constructed.

Ameron supplies a T-Hab studliner product to rehabilitate sanitary sewers greater than 72 inches in diameter. This product is installed when raised mechanical anchor is placed over a collapsible form that travels inside the pipe or tunnel. As the collapsible form is expanded the stud liner pushed against the surface of the pipe or tunnel. Self consolidating concrete is pumped into the annular space between the lining and the concrete surface. This highly flowable, special mix of concrete is specially formulated so that it can be pumped long distances through narrow openings in forms and meet the performance requirements of the project. Ameron T-Hab has been used on some large diameter brick sewer rehabilitation projects in California including the Lower North Outfall Sewer which was under construction during the years 2004 through 2008.

The HDPE Agru Sure Grip product was successfully installed on more than 1.5 million square feet of the Singapore Deep Tunnel Sewerage System (DTSS) which included 48 kilometers (29.8 miles) of deep, large diameter sewer tunnels (Figure 11). These tunnels ranged in size up to 7.23 meters (23.8 feet) in diameter. To ensure a 100 year design life, the precast concrete structural lining of this tunnel was protected by an inner corrosion protection lining consisting of a HDPE primary liner 2.5 mm

Aluminates have been in used for more than 65 years with the first U.S. application at a Treatment plant in Southern California in 1959. It is produced by melting limestone (calcium) and bauxite (aluminum) and allowing it to cool; once cooled this very dense and hard calcium aluminate "clinker" is ground to the fineness of normal cement. Additionally, the "clinker" can be milled into different aggregate gradations to replace normal aggregates in the cement. The hydration process of the calcium aluminate cement produces calcium aluminate hydrates and gibbsite. The gibbsite from this cement hydration is not susceptible to an attack by microbial induced corrosion. This is not to say that this cement is corrosion proof, the reaction between the wastewater and the cement produces PH levels below 3.5 contributing to the neutralization of the acid at the cement surface by the consuming hydrogen ions. The resulting neutralization reaction has an inhibitory effect on the metabolism of the bacteria, which create the corrosive acid. This is the manner in which calcium aluminate act as a protective and reactive barrier reducing the corrosion of the concrete.

Generally these products are applied in a two (2) to six (6) inch thick layer on the surface of the segmental lining. However, at this thickness, a geo-textile fabric may be required to assist in the proper application of these cementitious materials as they should not be applied over one (1) to two (2) inches thick per lift. An advantage of these systems is their ability to be applied in a damp and humid environment. Often times the manufactures specification calls for the wetting of the substrate prior to application.

**Figure 11. HDPE Agrusuregrip lining being installed**

(0.10 inches) thick that is exposed to sewage and is anchored to a secondary lining of cast in-situ concrete that is approximately 225 mm (8.8 inches) thick. The corrosion resistant lining was placed inside the tunnel previously constructed with precast segments and installed utilizing steel forms.

*Acid Resistant Gunned Cementitious Lining*

There are numerous products referred to as "acid resistant concrete," generally these products can also be characterized as a gunned cementations lining. For this paper a product which is considered a gunned cementitious product is any acid or corrosion resistant product that is applied with the use of a shotcrete or other spray application. Some manufacturers of these products include: "Permacast MS-10,000" by AP/M Permaform, "Strong Seal MS-2C" by Strong Seal Systems, Inc., "EMACO S88-CI" by Master Builders, Inc., and "SewperCoat" by Kerneos Inc. Unlike other cementitious products, these products rely on one of two corrosion resistant materials; Silica Fume or Calcium Aluminate. EMACO S88-CI by Master builders, Inc uses silica fume as the corrosion limiting agent. Silica fume has been previously discussed in this paper.

The remaining three products utilize calcium aluminate as the corrosion resistant product. Calcium

*Deformed Pipe Lining*

There are two primary examples of deformed pipe linings; compressed and roll down/swagedown. The compressed version, consisting of HDPE or PVC, are compressed to a deformed shape and expanded once placed into the pipe. These systems generally use the heat and pressure of steam to expand the deformed material into the circular shape of the host pipe. This type of deformed pipe lining is used for trenchless rehabilitation projects as they are available up to only 18 inches. The roll down/swagedown lining reduces the diameter of a HDPE pipe allowing it to be installed into the host pipe. This product will return to the original unrolled size creating a secure fit in the host pipe. These products are again generally used for rehabilitation projects as they are limited to pipe diameters less than 36 inches.

*Slip Lining*

Slip lining for corrosion protection consists of placing a corrosion resistant pipe material into an existing

## Table 2. Common sizes of slip lining pipe materials

| Pipe Type | Typical Length | Available Diameter | Joint Configuration | Comments |
|---|---|---|---|---|
| Solid wall polyethylene pipe | 20 feet | Up to 63 inches | Snap-fit | |
| Profile wall polyethylene pipe | 20 feet | Up to 144 inches | Bell and spigot with gasket | Constant ID and OD |
| Spiral rib polyethylene pipe | 20 feet | Up to 144 inches | Bell and spigot with gasket | Constant ID and Variable OD for better anchorage during grouting |
| Fiberglass Reinforced Plastic (FRP) pipe | 20 feet (up to 60 feet) | Up to 169 inches | Bell and spigot with gasket | Constant ID with the OD driven by the bell size |
| Reinforced Plastic Mortar (RPM) | 20 feet | Up to 96 inches | Bell and spigot | Constant OD |

## Table 3. Example evaluation table

| | System | Cost Prohibitive | Size Limitation | Sufficient Product History | Design Life (50 years or more) |
|---|---|---|---|---|---|
| One-pass Systems | Anchored Thermoplastic Lining (PVC, HDPE) | Y/N | Y/N | Y/N | Y/N |
| | Concrete Faced with Glass Reinforced Plastic | Y/N | Y/N | Y/N | Y/N |
| | Precast Polymer Concrete Segments | Y/N | Y/N | Y/N | Y/N |
| | Acid Resistant Concrete Products | Y/N | Y/N | Y/N | Y/N |
| | Microsilica (Silica Fume) Concrete | Y/N | Y/N | Y/N | Y/N |
| | FRP/HDPE/Polymer Concrete/PVC lined RCP Pipe | Y/N | Y/N | Y/N | Y/N |
| Two-pass Systems | Liquid Applied Polymer Based Protective Lining | Y/N | Y/N | Y/N | Y/N |
| | Post-Installed Anchored Thermoplastic Lining (PVC, HDPE) | Y/N | Y/N | Y/N | Y/N |
| | Acid Resistant Gunned Cementitious Lining | Y/N | Y/N | Y/N | Y/N |
| | Deformed Pipe Lining | Y/N | Y/N | Y/N | Y/N |
| | Slip Lining | Y/N | Y/N | Y/N | Y/N |

pipe and grouting the annulus. The pipe materials used for slip lining are the same as those discussed previously (FRP/GRP/Polymer Concrete/T-lock lined RCP Pipe). Slip lining with thermoplastic pipe sections allows for a continuous section by welding individual pipe sections together. The primary limitations with this method of slip lining are the radius bends in the host pipe. In tunneling applications by TBM, the curve radius of the tunnel will generally be sufficient at large diameters for this method. However, applications that contain tight bends or turns may limit this method.

Slip lining using short sections of pipe joined with gaskets, mechanical joints or welding can be completed within the tunnel. Individual sections are brought into a tunnel or carrier pipe and joined to create the continuous pipe. This method may allow for tighter turn radius to be taken as shorter pipes can be manufactured in a variety of sizes and shapes to navigate turns within the carrier pipe. Common sizes and characteristics are given in the Table 2.

## SUMMARY

When considering corrosion resistant materials for an underground construction project, it is useful to evaluate products against each other as each project is individual and each material is variable in their use and cost. Additionally, the availability of materials may change and their use may vary depending upon each application and project specific characteristics. Table 3 is an example of such a comparison for a site specific use and is not applicable in other circumstances.

**Table 4. Cost of several corrosion protection systems**

| | System | Type | Estimated Unit Cost ($/ft²) (2009) | Estimated Product Duration (years) |
|---|---|---|---|---|
| One-pass Systems | Anchored Thermoplastic Lining (PVC, HDPE) | Lining | $25–35 | 20–40 |
| | Concrete Faced with Glass Reinforced Plastic | Lining | $20–30 | 20–40 |
| | Precast Polymer Concrete Segments | Lining | $28–35 | 30–40 |
| | Acid Resistant Concrete Additives | Concrete additive | $15–25 | 15–35 |
| | Microsilica (Silica Fume) Concrete | Concrete additive | $3–5 | 20–40 |
| | FRP/GRP/Polymer Concrete/PVC lined RCP Pipe | Pipe | Varies with pipe size | 25–50 |
| Two-pass Systems | Liquid Applied Polymer Based Protective Lining | Coatings | $12–25 | 10–25 |
| | Post-Installed Anchored Thermoplastic Lining (PVC, HDPE) | Lining | $25–30 | 20–40 |
| | Acid Resistant Gunned Cementitious Lining | Coating | $18–22 | 10–25 |

It is essential that each underground/tunnel project be evaluated on a case by case basis as each may have its own individual qualities and features. Additional criteria which may be used for the selection process include:

- Cost prohibitive
- Size limitation
- Sufficient product history
- Design life (50 years or more)
- Constructability (can be built without modifying tunnel)
- Placement flexibility (shafts, tunnel, steel)
- Engineering confidence level (90% or greater)
- Provides good corrosion protection
- Design life (50 years or more)
- Acceptable performance track record
- Quality of manufactures warranty (length of time and terms)
- Schedule (can be built in 6 to 10 months)
- Surface preparation requirements
- Does not require concrete abrasive blast cleaning
- Easily tied into for future connections (ensuring corrosion protection)
- Resistance to hydraulic scour
- Resistance to hydrostatic pressure should segment gasketed joints leak
- Ease of repair in the future
- Capital maintenance requirements
- Ease of future inspection
- Future tunnel access for repair
- Contractor familiarity
- Contractor ability/product certification

With owners asking for a design life up to 100 years, it is essential to be able to provide adequate life expectancy for the cost. However, it is not feasible to expect to obtain corrosion protection for this period from any product available on the market place today. Presented in Table 4 is the cost of several corrosion protection systems and anticipated design lives found during the design of several tunnel projects in central Ohio. Estimated prices are listed as a unit cost based on dollars per square foot and given as a range by categories as many products were evaluated for each category.

## REFERENCES

Talley, J. and Wallace, G. 2009. Calcium Aluminate Technology and Biogenic Corrosion. Trenchless Technology. 17(6):28-30.

# Lined Concrete Segments: An Alternative Construction Method for a Large Diameter Sewer Tunnel

**Galen Samuelson Klein, Michaele Monaghan, Samuel Gambino**
URS Corporation, Oakland, California

**Dwayne Deutscher**
URS Corporation, Sacramento, California

**Rigoberto Guizar**
Sacramento Regional County Sanitation District, Sacramento, California

**ABSTRACT:** The Upper Northwest Interceptor 1&2 Tunnel (UNWI 1&2) is currently under construction by Sacramento Regional County Sanitation District (SRCSD). This 144-inch inside diameter wastewater tunnel is being constructed in soft ground below the groundwater table using an earth pressure balance tunneling machine. The tunnel is supported with precast concrete segments lined with PVC. This is the first use of precast PVC lined concrete segments to construct a sewer tunnel in the United States. In the past, large diameter sewer tunnels have typically been constructed using pipe jacking, a two pass system, or by adding a liner after construction. Using PVC lined concrete segments to construct sewer tunnels offers the industry an alternative approach.

## INTRODUCTION

The Sacramento Regional County Sanitation District (SRCSD) is expanding its sewer interceptor system for their service areas. The Upper Northwest Interceptor (UNWI) is one of eight major interceptors in the SRCSD Interceptor Program. The UNWI consists of approximately 99,840 feet of pipeline varying in inside diameter from 30 to 120 inches. The interceptor is divided into nine segments including one pump stations (Van Maren), one force main, and seven gravity pipelines.

SRCSD contracted with URS Corporation to design the two downstream segments of the system, UNWI 1 & 2: approximately 3.7 miles (19,400 feet) of large-diameter gravity pipeline and associated transition structures at the upstream and downstream ends of the project, along with access shafts for maintenance. Figure 1 shows an aerial view of the UNWI 1 & 2 Project. UNWI 1 & 2 begins at the New Natomas Pump Station (NNPS) and ends just south of Bridgecross Drive. From the NNPS to the C-1 Canal, the UNWI 1 & 2 alignment parallels the west side of the East Drainage Canal, while the remainder of the interceptor is along the east side of the East Drainage Canal.

## Preliminary Design/Alternative Analysis

A probabilistic risk based life cycle cost analysis was performed to assess the life cycle cost of the selected excavation and tunnel lining alternatives, considering not only construction costs, but also the costs associated with construction risk and with Maintenance and Operation (M&O) (Gambino et al., 2008). This analysis was used to compare pipe jacking a 120-inch diameter pipe and tunneling with 144-inch diameter precast segments for tunnel support. Risk was defined as the probability that the life cycle cost of an alternative would reach or exceed a specified threshold. The life cycle cost of each alternative was dependent on future conditions and events that could not be predicted with certainty. This can result in cost increases beyond those traditionally included in the contingency cost.

The probabilistic risk analysis methodology employed was based on the principles of uncertainty propagation in which the uncertainty in a specified response variable of interest (cost for this project) is analyzed as a function of the uncertainty in the input variables that impact the response variable. In a probabilistic analysis, a range of plausible values is defined for each input variable and the probability of

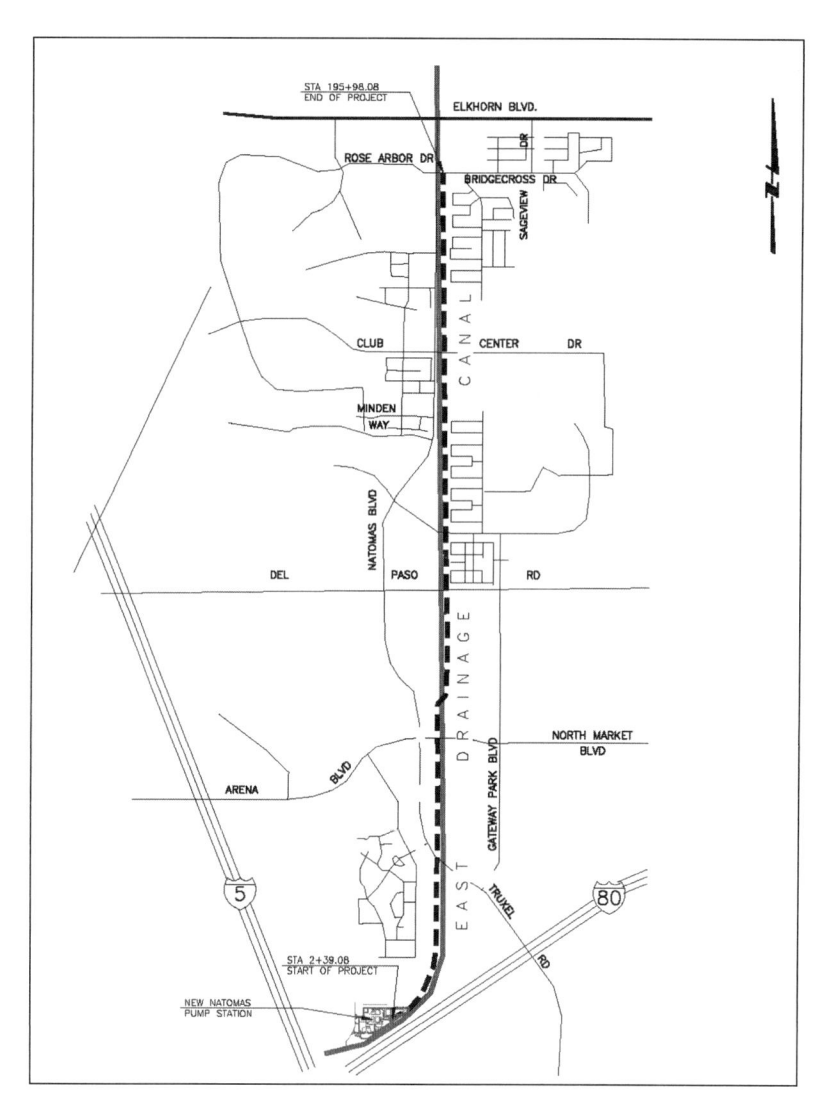

**Figure 1. Aerial view of UNWI 1 & 2**

obtaining each value is specified. The uncertainty in the input variables is quantified in the form of probability distributions.

Discounting methods were used to convert all future cash flows into an equivalent value today, which are termed "present values". In discounted cash flow analysis, a project should be accepted if the present value of the benefits exceeds the present value of its costs. The probability of encountering each risk factor was assessed using an analysis of historic data on similar projects. An extensive literature review was performed and case histories from over 100 tunnel projects were tabulated and categorized according to the types of construction difficulties (i.e., risk factors) encountered. For some of the projects, the data were not explicit, but the occurrence of a risk factor could be indirectly inferred.

The results of the risk analysis indicated that a 144-inch ID precast concrete segmental lining (PCSL), with interior T-lock® lining, installed using conventional tunneling was found to have the highest net present value. The main reasons for the cost difference between the alternatives is that, due to the high cost of shaft construction for the pipe jacking alternative, the construction base cost for the tunneling alternative and the M&O base cost of the tunneling alternative is lower. Substantial costs would have been associated with constructing multiple deep shafts in close proximity to the Reclamation

District (RD) 1000 East Drainage and C-1 Canals for the pipe jacking alternative.

The present worth M&O costs of the two alternatives also differ substantially because of the difference in the number of years during which round-the-clock staff at the New Natomas Pumping Station would be required due to insufficient in-line storage time. The larger pipe increased the storage capacity of the pipe. The increased storage capacity increased the required operator response time should there be a pump station failure. The longer allowable response time extended the time the pump station could function without being staffed full time. For the 144-inch precast concrete segmental lining, increased costs associated with more early inspections and more repairs of joint were more than offset by the benefit of having the additional storage volume and therefore not having to staff the pumping station for the extra 20-year period.

## DESIGN ISSUES AND SOLUTIONS

The selection of tunneling with a PCSL rather than utilizing pipe jacking reduced the risk associated with constructing multiple jacking shafts in very limited working area in close proximity to the drainage canals, levee, and neighboring residences. The tunnel is completely below the groundwater table. Selection of an Earth Pressure Balance Tunneling Machine (EPBTM) eliminated the need to dewater much of the construction along the alignment. However, since it is a wastewater tunnel, there were additional facilities to be constructed and constraints unique to sewer construction.

### Fixed Tunnel Grade and Elevations

The tunnel is relatively shallow, with cover varying from 11 to 32 feet. The relatively shallow cover brought the tunnel into close proximity with existing utilities in many locations. Because of the potential for settlement above the tunnel all utilities sensitive to settlement such as waterlines, sewer lines and storm drains that the tunnel passed beneath were supported by jet grouting. An envelope that extended eight feet on each side was jet grouted in advance of tunneling in areas where settlement was anticipated.

The part of Sacramento in which the tunnel is being constructed has very little elevation variation. As a result the tunnel slope averages less than 0.10 percent grade. The tolerance from theoretical centerline is four inches for alignment and two inches for grade provided that such variation does not result in a reverse sloping invert.

### Construction Below the Groundwater Table

The groundwater levels at the location of the project vary from 5 to 10 feet below the ground surface. Therefore, all elements of the project were designed to be constructed below the groundwater surface. There was also limited capability to construct a dewatering system and dewatering to construct project elements was precluded by the contract documents. The one exception was the main construction shaft where the tunneling machine was launched. It is located at the inlet to a recently constructed wastewater pump station. Groundwater collected by the dewatering system was discharged directly to the pump station wet well and disposed by the SRCSD.

Construction of the tunnel without dewatering along the alignment required development of unique solutions for manhole construction and for connection of smaller sewers to the interceptor:

- The tunnel segments would ultimately support the manhole barrels and a structural collar was designed to restrain the tunnel segments while providing support for the manhole barrels. A jet grout block was installed at each manhole location prior the advance of the tunneling. The EPBTM would then tunnel through the grout block exposing the shaft for subsequent manhole construction. A steel casing was designed to be embedded into grout and was used to shore the excavation and provide a water tight working environment. Figure 2 shows the details of the manhole construction.
- There are three smaller (16-, 24-, and 27-inch) diameter trunk sewer connections. Since open cut without dewatering was not practical, URS design called for construction by use of pilot tube microtunneling. Connections greater than approximately 10 feet in length were constructed using this technique. Short connections were hand mined within the manhole jet grout blocks.
- Dewatering was not allowed for the retrieval shaft. URS specified a water tight secant pile shaft with a concrete base plug that would resist uplift pressure from the surrounding groundwater. The reason for not allowing dewatering was primarily the difficulty associated with disposing of the water.

### Environmental Constraints

The primary environmental constraint to the project was the potential existence of the giant garter snake (GGS), a special status species. The snake is known to inhabit areas near water courses and a strip 200 feet wide from the water line is considered snake habitat. GGS hibernates between October 1st and April 30th. Ground disturbing activities including surface excavations and jet grouting cannot occur during their hibernation, blackout period.

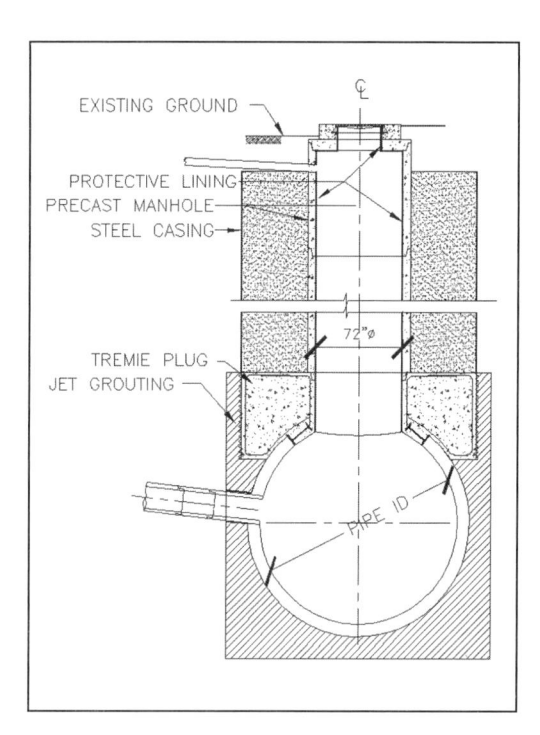

**Figure 2. Manhole construction**

Tunneling was not affected since it was all below the groundwater surface and the snake does not hibernate under water. Surface construction for shafts and other surface excavations had to be scheduled by the Contractor around the GGS blackout periods presenting significant scheduling challenges for the Contractor. Subsequent work that did not disturb the ground could be performed during the blackout period.

**Precast Segment Design**

For the UNWI 1&2 tunnel precast segments were chosen for tunnel support. Preliminary design of the segments, given the ground conditions was developed and 10-inch thick segments with 4-foot wide rings were presented on the design drawings. SRCSD required that the proposed segments be corrosion resistant. Options for corrosion resistance in the segmental tunnel included that PVC or HDPE be mechanically anchored in calcareous concrete or that the segments be constructed of polymer concrete. The use of calcareous concrete was based on SRCSD's experience. They had previously utilized calcareous concrete and found this material to be reliable against corrosion attack.

The use of polymer concrete would have precluded the use of a secondary lining. During design it was determined that polymer concrete had been used in Europe and Japan successfully. The design team evaluated the three proposed systems considering design life, long-term maintenance, bid ability, constructability and cost. The conclusion of this evaluation was to include all three systems in the bid documents and to allow the Contractor to select the most cost effective system and prepare their bid based upon this. Any one of the three proposed systems would have been a unique application in the United States (there were no examples that the design team could identify where sewer tunnels had been previously constructed using these systems).

The project specifications were prepared for both calcareous concrete and polymer concrete. The contract specifications were prepared as performance specifications, directing the Contractor to provide the final design of the segments under the direction of an engineer registered in California. As the winning bidder, Traylor Shea Precast JV based their bid upon using PVC lined calcareous concrete segments. Bids included all three options with the polymer concrete option being significantly higher in cost than the other proposed alternatives. Traylor Shea JV proposed 9-inch thick segments in a 5-foot wide ring. The PVC lining system chosen consisted of T-lock®, an Ameron proprietary product used typically in PVC lined precast pipe. Segment molds were fabricated by CBE of Tours, France and the segment were designed by Halcrow Group Ltd in the United Kingdom. Traylor Shea JV has proven experience with segment production. However, the inclusion of mechanically anchored PVC T-Lock® lining in the segment production process presented an unknown variable to the process.

**TUNNEL CONSTRUCTION**

Tunneling bids were opened in July 2007. SRCSD awarded the contract to TSJV in August 2007 and NTP was given in September, 2007. Tunnel construction began in January 2009 and is scheduled to be completed in November 2009. Substantial completion of the tunnel is required by September 22, 2010. The contractor is on schedule to achieve this. The contractor elected to procure a new tunneling machine which was originally scheduled for delivery 13 months after notice to proceed.

**Surface Facility Construction**

Based on the EPBTM delivery schedule the contractor was planning to begin tunneling activities in November 2008, after the start of the GGS blackout period. While the GGS were not affected by the tunneling operations, all jet grouting to support utilities and for manhole construction had to be completed before the tunneling. The contractor developed a schedule to complete the jet grouting activities

**Figure 3. Blind auger drilling for manhole construction**

**Figure 4. Placement of PVC T-Lock® liner in form**

during the period from May 1 to September 30, 2008. All jet grouting activities were completed according to this schedule with the exception of a reach beneath a paved roadway where construction during the GGS blackout period was allowed.

The Contractor followed the jet grouting by blind auger drilling to set casings at each manhole location. This construction was also completed during the 2008 construction window. The receiving shaft was constructed during 2009 and microtunneling for the trunk sewer connections was completed during the 2009/2010 blackout period. Figure 3 shows the blind auger drilling used to set the casings for manholes.

### Segment Manufacture

Traylor Shea Precast (TSP) constructed a new segment production plant in Stockton, California to produce segments for the UNWI 1&2 project. Production of segments began in early June 2008. Utilizing their previously successful carousel system of production, TSP achieved a plant production level of about 10 rings per shift with two shifts per day. The manufacturing process involved first placing the PVC liner on the bottom of the form with the "T"s facing up. The edges of the PVC were then tucked into the bottom joints of the form. Figure 4 shows placement of PVC T-Lock® liner in the form. Reinforcement cages were then placed on top of the liner and supported on concrete reinforcement chairs or "dobies". There were a total of 6 dobies per cage and these were placed two each at the edges of the cage at the center and at ¼ points on either side of the center of the segment.

The segments were demolded after a 6 hour steam curing cycle. Figure 5 shows the segments being demolded from the forms. Two vacuum suction cups extracted the segment from the mold and the segment then turned over and conveyed to a final station where QC was conducted. Any defects were at this time identified and logged. Test cylinders prepared at the same time as the segments were tested at this point as well. Strength values of the concrete cylinders were reported to be around 7,500psi. Upon leaving the production plant, the segments were stacked as two pallets of 3 each for a total of 6 segments in a stack equivalent to one ring. These were stored in tarp covered storage areas in the yard behind the plant for an additional 5-day minimum moist curing period. These tarp covered structures are used for the moist curing cycle. Humidity was maintained as close to 100% as possible through the use of fogger fans. Overhead sprinklers were also used to maintain the humidity.

### Installation of Segments

Installation of the PVC lined segments raised several concerns during design, including:

- Welding of the T-Lock® joints posed a significant level of effort. There is approximately 68 feet of joints to be welded per ring. There are approximately 3,800 rings in the tunnel, so a total of 260,000 feet of joints required welding.
- Potential damage could occur to the T-lock® through EPBM trailing gear, attachment of utilities, mucking operations and general construction activities.
- The PVC liner needed to be provided for the total 360-degree inside face of the tunnel. It was not possible to leave the T-Lock® off of 1 segment per ring because the joints were staggered from ring to ring. Because of this, the invert of the tunnel is lined and the slipperiness of the walking surface was a safety concern.

**Figure 5. Demolding of segments from forms**

All of these issues proved to be successfully addressed by the Contractor during construction. In an effort to make welding of the PVC joints more efficient, the Contractor had a dedicated platform behind the erector arms on the EPBM built to accommodate a welder for the joints above the spring-line. Below the spring-line, welders work directly behind the erector arms welding over the joints before the track for the EPBM trailing gear and access cars is placed. Trailing gear for the EPBM and access to the tunnel heading occurs on track laid in the invert. Muck is transported on a conveyor system hung from the crown of the tunnel. All utilities, track and conveyor supports are attached to the tunnel through rubber base-plates in order to further protect the PVC lining.

**Line and Grade**

Because the tunnel is a gravity sewer, maintaining grade during tunneling was more critical than alignment. During normal tunneling operations the contractor was able to maintain grade within tolerances, however, there were several instances when the grade was out of compliance with the contract documents. The contractor found that minor adjustments to line and grade were possible during tunneling in the native soils. However, upon entering jet grout blocks steering the machine was difficult. Because the tunnel was a one pass operation, it was not possible to correct the final product for line and grade, as would be done if pipe were being placed inside of initial supports within a tunnel.

**Tunneling Production**

Delivery of the EPBTM occurred behind schedule and the Contractor began tunneling in January 2009. With the successful completion of manhole construction shafts the contractor has been able to attain continuous tunneling progress without being affected by the GGS black out period and regained schedule. Hole through occured on November 18, 2009. Tunneling was being performed during three eight hours shifts, five days a week. The Contractor has frequently achieved over 500 feet per week, attaining as much as 690 feet per week. Most of the welding of the PVC liner has been completed as the machine advances. At the completion of tunneling activities most of the welding will be completed and initial acceptance testing complete. To accomplish this level of completion the Contractor has, at times, delayed tunneling activities so the welding could be performed in designated areas on the trailing gear. Figure 6 shows PVC liner on the concrete segments installed in the tunnel.

**Figure 6. PVC liner on segments in tunnel**

## CONCLUSION

Mining and support for the UNWI 1 & 2 tunnel has been successfully completed. Substantial completion of the project is anticipated to occur ahead of schedule. Segments for the UNWI 1 & 2 tunnel, while of a standard universal configuration are unique in that they have the corrosion protective liner cast into them. This is a first in the United States. Fabrication of the segments was successful resulting in a good quality product. Installation of the lining did not slow production rates relative to installation of conventional segments without a liner.

Since utilizing PVC lined segments was a new approach for sewer tunnel construction, there were understandable concerns to the project team when the tunnel was being design. Now that construction is complete, it has been demonstrated that there are successful methods of both protecting the liner once in place and repairing any damage to the liner that occurs during construction. In addition, welding of the joints was successfully accomplished without significantly impacting the project schedule.

When compared to pipe jacking or installing corrosion protection after construction of the tunnel, the PVC lined segment alternative offered schedule acceleration and additional storage capacity. Successful construction of the UNWI 1&2 tunnel has demonstrated that utilizing PVC lined concrete segments is a viable construction method for large diameter sewer tunnel construction and it offers an alternatives to previously utilized construction methods.

**REFERENCES**

Gambino, S., Klein, G.S., Kulkarni, R., Deutscher, D., Thistlewait, W., and Nguyen, D. 2008. Risk-Based Evaluation of Life-Cycle Costs for the Upper Northwest Interceptor, Proceedings, North American Tunneling, San Francisco, California, June.

# Portal Slope Stability and Tunnel Leakage Remediation

**Carlos A. Jaramillo, Camilo Quinones-Rozo**
URS, Oakland, California

**Robert A. McManus, Andrew Yu**
PG&E, San Francisco, California

**ABSTRACT:** Inspection of the Belden Tunnel 2 after initial water-up revealed a ½ inch wide cracked in the liner. Instrumentation installed after this observation documented the rate of development of this crack, and the apparent acceleration of its opening. Seepage and stability analyses indicated a potential connection between the leakage and the stability. This paper describes the background of the problem, the analyses performed, and the mitigation measures evaluated before deciding on the use of a PVC membrane. The performance of the tunnel and the slope after remediation are also discussed.

## INTRODUCTION

The 15-ft diameter, 1.8 miles long Belden Tunnel 2 is part of the conveyance system connecting Belden Dam to Belden Powerhouse, part of the Upper North Fork Feather River Project of PG&E. Inspection of the tunnel in 1970, during an unrelated outage shortly after initial water-up revealed a ½ inch wide crack in the concrete lining near Portal 3, the beginning of Tunnel 2. Subsequent inspections found additional cracks, prompting installation of instrumentation to monitor the tunnel and the portal. Review of instrumentation data in 2006 indicated that Portal 3 had moved upstream (towards the river valley) approximately 10 inches since monitoring started. This movement resulted in extensive cracking of the concrete lining downstream of Portal 3.

## REGIONAL GEOLOGIC SETTING

The Belden Project is in the northern portion of the Sierra Nevada of California, an actively uplifting, west-tilted structural block that is fault-bounded on the east where the fault scarp forms the precipitous eastern range front. The west-facing range front is moderately inclined with deeply incised, well-developed, west trending drainages such as the North Fork Feather River. The Belden Project is approximately halfway between the eastern range front and the valley floor of the Sacramento Valley. This structural block consists primarily of Jurassic- to Cretaceous-aged granitic plutons, which intrude late Paleozoic- to Mesozoic-aged metasedimentary and metavolcanic rocks. Cretaceous to early Cenozoic Great Valley sediments onlap the western edge of the metasedimentary and plutonic basement rock, and are topped by alluvium, west of the project site. Quaternary deposits include volcanic deposits along the eastern scarp, glacial deposits along the crest of the range and eastern slope, slope debris aprons at the base of precipitous slopes, and isolated terrace deposits along major drainages (Jenkins, 1962).

Regional mapping indicates that the Portal 3 area is underlain by Paleozoic to Mesozoic metamorphic basement rocks of the Central belt, the westernmost member of three regional geologic belts in the northern Sierra Nevada. The rocks of this terrain have undergone regional metamorphism and have well developed schistocity. Rocks of the Central belt are generally less indurated and foliated than the other two belts, and consist primarily of argillite or phyllitic argillite rather than slate. In the vicinity of Portal 3, the west side of the North Fork Feather River canyon is mapped with rock foliation surfaces striking to the north with moderate dips to the west (into the hillside).

The structural fabric of the region is dominated by northwest-trending faults, folds and lithologic boundaries. The northwest-trending Rich Bar Fault traverses across the North Fork Feather River canyon less than ¼ mile upstream of Portal 3. This structural feature is an old tectonic fault (inactive) that separates rocks of the Feather River peridotite belt on the northeast, from Central belt rocks on the southwest.

## SITE GEOLOGY

The geology exposed during excavation of the tunnels was logged and summarized in Table 1.

**Table 1. Tunnel 2 geology beginning at Portal 3**

| From Station | To Station | Length (feet) | Description | General Attitude |
|---|---|---|---|---|
| 254 + 69.78 | 258 + 61.00 | 391 | Highly fractured, partially decomposed, thin bedded phyllite | S25°E/53°SW |
| 258 + 61.00 | 259 + 09.70 | 48 | Phyllite rock is fairly hard, but highly fractured | S52°E/80°SW |
| 259 + 09.70 | 259 + 31.90 | 22 | Hard massive phyllite, thick bedded | S52°E/80°SW |
| 259 + 31.90 | 259 + 31.90 | 0 | At this station 2' to 3' wide shear zone with 6" to 12 " of gouge appears | S33°E/70°SW |
| 259 + 31.90 | | | Fairly hard phyllite | S33°E/ vertical |

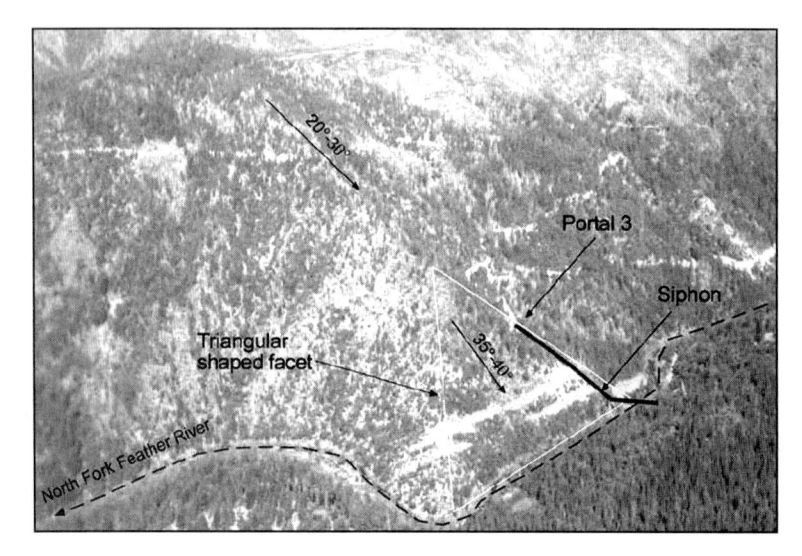

**Figure 1. Belden siphon**

Portal 3 is on the northern edge of a distinctive, triangular-shaped topographic facet in the lower canyon whose east-facing slope dips at 35 to 40 degrees. See Figure 1. The facet appears to protrude into the canyon relative to the sides of the ridge north and south of this area. The shape of this geomorphic feature may be due, in part, to the presence of a very hard and resistant diorite dike mapped at the base of the slope near the siphon. Above this facet, the topography becomes more moderate, with slope angles ranging from 20° to 30°. The lower, steeper slopes appear to reflect the change in incision rate by the North Fork Feather River. The change in rate of down-cutting of the river may have been controlled by the temporary base level controlled by the Spanish Creek, or more likely, the steep lower slopes are related to increased regional uplift of the Sierra Nevada.

The ridgeline above the portal has several distinctive topographic benches. These could be related to toppling movement, but a more common explanation for these features is differential erosion across phyllite units of varying hardness.

Based on a review of aerial photographs and a helicopter aerial reconnaissance, no large deposits of debris suggestive of active rockslide processes were noted in the canyon bottom.

## ROCK STRUCTURE

The rock units exposed in a cut slope near the portal are weathered phyllite, inter-foliated with meta-sandstone, chert and siliceous phyllite. See Figure 2. These descriptions also apply to the metamorphic rocks exposed in the road cuts above the portal all the way to the top of the ridgeline. The quality of the rock varies from poor (Geologic Strength Index, GSI = 15) to good (GSI = 75). Some voids were noted very close to a major set of foliation shears. In general, however, the amount of open voids observed in the portal cut slope, and in the roadway cut slopes above the portal, was not significant, but they affect

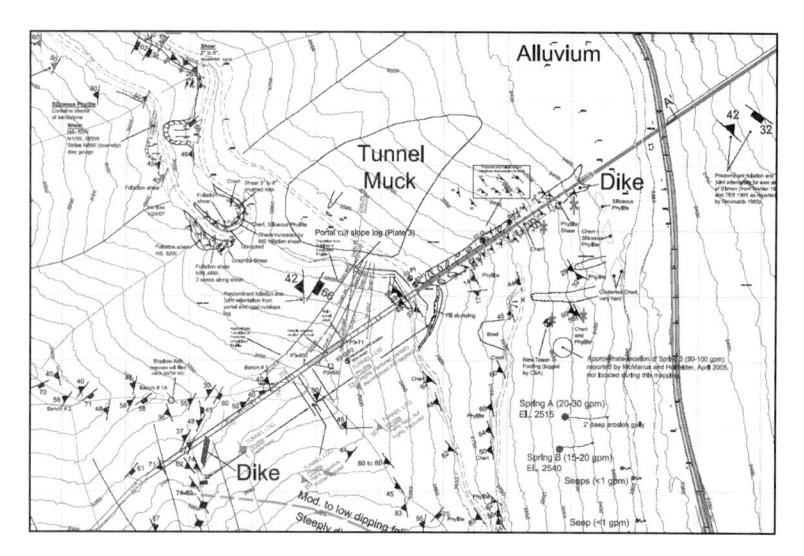

**Figure 2. Belden siphon local geology**

the formation conductivity as interpreted from boreholes drilled for placing instrumentation.

Two major sets of discontinuities were consistently observed from the river to the top of the canyon: 1) foliation surfaces, and 2) a predominant joint set (joint set #1). The foliation surface has an average attitude of N02W 42 SW, a Joint Roughness Coefficient (JRC) of 4 to 14 (generally 8 to 10) and lengths traceable in outcrops generally around 4 to 6 feet. Joint set #1 has an average attitude of N31°E/66°SE, a JRC of 2 to 16 (generally 6 to 8) with lengths traceable in outcrops generally less than 2 feet. In general, the orientation of the foliations mapped at the ground surface match the foliations logged in the weathered rock exposed in Tunnel 2. However, there were areas high on the ridge south of the portal and the lower slopes near the river showing near-vertical foliation, similar to the unweathered rock exposed in Tunnel 2. (Figure 1)

Significant structural features noted in the portal cut slope, and in cut slopes along the roadway, are sheared foliation planes which are characterized by crushed and brecciated rock zones several inches in thickness. In general, the spacing of the shears appears to be on the order of 10 feet to 25 feet. The shears have orientations consistent with the foliation, JRCs from 4 to 14 (generally 10 to 12), and are continuous in outcrop from 6 feet to 50 feet.

During detailed mapping of the siphon cut, a very fine grained, very hard and resistant igneous rock unit (diorite) near the third saddle block above Anchor AB-4 was identified. The hardness of this dike, and the location near the base of the topographic facet discussed in the previous section, suggests the facet may be related to this dike.

Several springs were mapped on the hillside near the base of the topographic facet at elevations similar to the tunnel elevation, down to 300 feet below.

## ANALYSIS OF GEOTECHNICAL INSTRUMENTATION

After finding the crack in the liner, PG&E installed instrumentation to monitor potential slope movements. Figure 3 shows Belden Siphon and Portal 3, and the instrumentation installed around it. The data collected is discussed below.

### Surface Survey Points

The data showed an acceleration trend after 1993, and again after 2003 since monitoring began in 1970. Total displacement measured in survey point A is about 10 inches.

### Extensometers

Extensometers near Portal 3 have not recorded significant movements, implying that these extensometers are in a zone moving as a single block. Extensometers within the Tunnel No. 3 showed an acceleration trend beginning in about 1998.but their rate was only about 70% of the rate measured in survey Point A, likely the result of the extensometers not extending across the full limits of movement in the tunnel.

### Inclinometers

The readings from the inclinometers indicated that there was no relative displacement from the ground

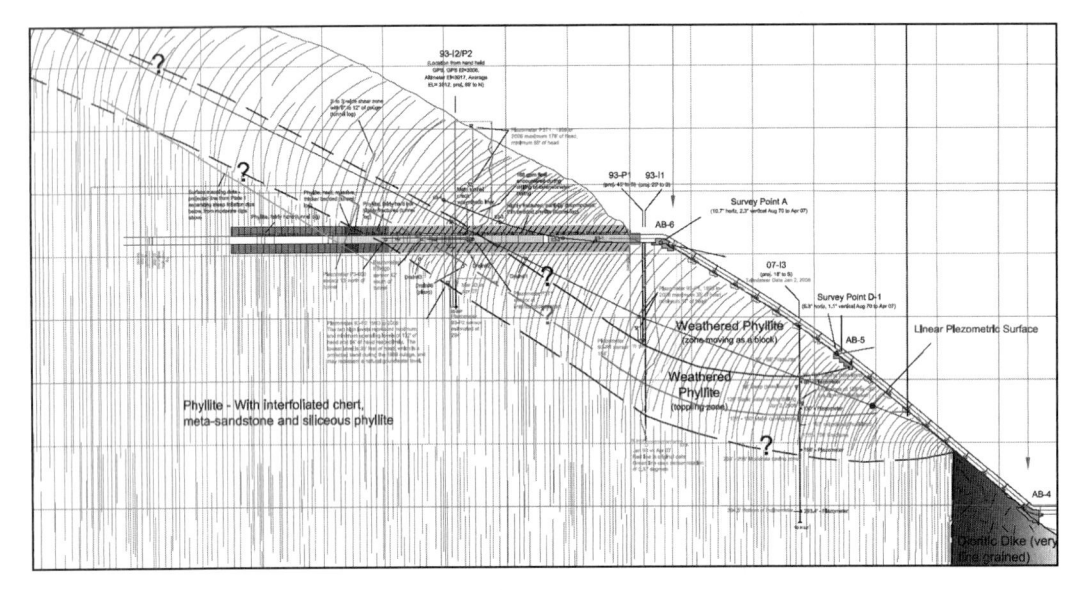

**Figure 3. Belden siphon and Tunnel 2 crosssection**

surface to 164 feet, and significant "leaning" movement is occurring between 164 feet to 274 feet.

Coupling measurements: Monitoring pins installed across the left (south) and right (north) sides of the dresser couplings showed 5.3 inches of contraction across the siphon.

Based on observations of tunnel liner cracks, records of four different instrumentation systems and surface mapping, it was concluded that:

- Portal 3 was moving downslope to the east.
- The movement was slowly accelerating.
- The movement was two to three hundred feet deep and consisted of a lower toppling zone capped by an upper zone moving predominately as a block. Figure 3 shows the estimated limits of the moving zones.
- The limits of the toppling zone correlated with heavy cracking in the tunnel concrete liner, with the limits of moderately dipping weathered and fractured rock noted in the original tunnel logs, and with the projection of surface mapping foliation data.
- The tunnel liner leakage was affecting the local groundwater.

Analysis of available data suggested the existence of three distinct zones within the slope:

- Upper Zone: A band of approximate constant thickness (between 150 and 170 feet) where no or minimal deformation is accumulated.
- Middle Zone: Strip of material where the shearing deformations are concentrated. There is no presence of a well defined surface of sliding, but rather a thick zone, varying between 50 and 100 feet thick, characterized by progressive shearing.
- Lower Zone: Stable region located immediately below the portion subjected to movement.

This interpretation conforms to a particular variation of flexural toppling characterized by the existence of three distinctive zones. The flexural toppling movement was likely a slow, on-going geologic process occurring throughout the formation of the canyon by erosion of the North Fork Feather River, probably accelerated by leakage from the tunnel.

Simplified limit equilibrium analyses of the slope using the geologic model developed above, and using groundwater levels based on instrumentation records extended by two dimensional groundwater modeling, indicated a precarious stability, and insinuated a moderate contribution of the groundwater to the movement.

## ALTERNATIVES STUDY

Water leaking out of the tunnel was considered a driver to the movement so conceptual design alternatives to repair and/or bypass the damaged section of the tunnel were prepared. Five structural alternatives were investigated and are summarized below:

- Installation of a steel liner/pipe with an outside diameter (O.D.) of 13 feet (consistent with the steel lined portion of the tunnel) by fabricating the liner inside the tunnel.

## Table 2. Summary of alternatives for repairing Belden Tunnel No. 2

| Alternative | Description | Advantages / Cost and Schedule | Disadvantages |
|---|---|---|---|
| 1. 13'-0" O.D. Steel Liner Inside Existing Tunnel Section | Collapsible pipe, on saddle supports with internal joints for expansion/rotation | 1. Durability of steel<br>2. No hydraulic head loss relative to existing tunnel<br>$3.5 million to $4.4 million<br>90 days, double shift<br>0.6 ft<br>$ 0.1 million | 1. Limited workspace in existing tunnel<br>2. Intersecting welds may be brittle<br>3. Internal expansion joint requires custom design<br>4. No access to exterior of pipe for maintenance of pipe exterior and supports<br>5. Length of pipe sections limited to rollout section length.<br>6. Cathodic protection required |
| | Pipe constructed in sections, on saddle supports with internal joints for expansion/ rotation – new tunnel adjacent to existing for access to damaged section | 1. Durability of steel<br>2. Past performance of steel liners is excellent<br>3. No hydraulic head loss relative to existing tunnel<br>$3.8 million to $4.7 million<br>192 days, double shift (60 days down time)<br>0.6 ft—$ 0.1 million | 1. Limited workspace in existing tunnel<br>2. Internal expansion joint requires custom design<br>3. No access to exterior of pipe for maintenance of pipe exterior and supports<br>4. Cathodic protection required |
| 2. 13'-0" O.D. Steel Liner Inside Expanded Tunnel Section | Pipe constructed in sections, on saddle supports with external joints for expansion/ rotation – excavate damaged section of tunnel to original unlined size | 1. Durability of steel<br>2. Maintenance of pipe exterior and supports is possible<br>3. Exterior expansion joints are available from various vendors for this diameter pipe<br>4. No hydraulic head loss relative to existing tunnel<br>$4.4 million to $5.5 million<br>180 days, double shift<br>0.6 ft—$ 0.1 million | 1. Limited workspace around pipe in tunnel<br>2. Length of pipe sections limited to rollout section length |
| 3. 11'-0" O. D. Steel Liner Inside Existing Tunnel Section | Pipe constructed in sections, on saddle supports with external joints for expansion/ rotation | 1. Durability of steel<br>2. Maintenance of pipe exterior and supports is possible<br>3. Exterior expansions joints are available from various vendors for this diameter pipe<br>$2.3 million to $3.0 million<br>60 days, double shift<br>3.3 ft—$ 0.7 million | 1. Hydraulic head loss relative to existing tunnel due to smaller diameter pipe<br>2. Limited workspace around pipe in tunnel<br>3. Length of pipe sections limited to rollout section length |
| 4. Geo-synthetic Membrane to Line Existing Tunnel Section | Line damaged section of tunnel with a geo-synthetic membrane with a drainage layer behind to prevent pressure build-up external to membrane. | 1. Geomembrane material can withstand large strains<br>2. Parts are small and or foldable making access to the tunnel easy<br>3. Past performance in hydraulic tunnels has been excellent<br>4. No hydraulic head loss relative to existing tunnel<br>$700,000<br>60 days | 1. Limited history of application to hydraulic tunnels<br>2. Questionable durability when exposed to debris and/ or abrasion from suspended particles |

- Installation of a steel liner/pipe with an O.D. of 13 feet by first removing the damaged concrete tunnel liner.
- Installation of a steel liner/pipe with an O.D. of 11 feet to allow passage through the existing siphon section with the pipe fully fabricated.
- Installation of a geo-synthetic water-tight liner within the damaged portion of concrete liner.
- Installation of a geo-synthetic water-tight liner within the damaged portion of concrete liner, and protect it with a layer of shotcrete.

All five alternatives required an additional external drainage system to prevent the collection of water either in the damaged section of tunnel or behind the geo-synthetic liner. The selected alternative was the geo-synthetic liner, as it represented the lowest installation cost, and required the shortest installation time. An additional advantage of the geo-synthetic liner was the negligible or positive impact on friction losses.

## DESIGN OF THE MEMBRANE SYSTEM

The design of the membrane system was based on the following criteria:

- Watertight attachment to the steel tunnel liner.
- Provision of a continuous geotextile/geonet drainage layer with capacity to carry the seepage around the grout curtain reaching the tunnel.
- Provision of drain pipes to convey drain water to the atmosphere at Portal 3.
- Watertight attachment to concrete lining at the downstream end.
- Provision of the entire geomembrane, support, and drain system to accommodate future concrete liner movements and span open cracks with a maximum opening of 9 inches.
- Differential external head of 80 ft of water for the 100 ft upstream of the end of the liner, and 20 ft in the remaining length.
- Minimum hydraulic capacity of the drainage system established based on a hydrogeology model of the tunnel.

### System Components

The membrane system is composed by several parts that work together to ascertain the performance of the Carpi membrane system.

- Geocomposite membrane liner: The geocomposite membrane consists of a PVC geomembrane heat-coupled during extrusion to a non-woven, needle-punched polyester geotextile.
- Geonet: The geocomposite membrane liner was supported by a geonet. This geonet had multiple functions, not only as a drainage layer, but also as an additional support layer, helping the geocomposite to bridge across cracks.
- Geogrid: The geocomposite membrane liner was supported by a geogrid. The geogrid is the layer that gives continuous support to the geocomposite system. A Mirafi Miragrid 10TX was used all over the tunnel surface below the geocomposite. Additionally, a Miragrid 22TX was used in the areas where large cracking was expected.
- Drainage pipe: The seepage collected by the Tenax GD-7 was transferred to a steel pipe, and discharged at the Portal 3. This discharge pipe was considered essential by PG&E to be able to monitor the effectiveness of the membrane liner.
- Flap valves: Carpi's membrane system design uses flap valves to allow drainage from behind the membrane during tunnel de-watering and prevent membrane bursting. The flap gates were dimensioned based on laboratory tests and past experience.
- Membrane support system. The membrane is attached to the concrete liner using a proprietary system of tensioning strips anchored to the concrete. The tensioning strips consist of two nested stainless steel profiles resembling opposite C sections running upstream-downstream.
- Upstream and downstream stainless steel rings to seal the ends of the membrane against the concrete liner.

## DESIGN OF GROUT CURTAIN AND EXTERNAL DRAINAGE SYSTEM

The Carpi membrane system started in the steel lined section, but its downstream end was in a section exposed to leakage from the unlined tunnel. To protect the membrane system and its drainage system, it was decided to construct a two-row grout curtain formed by eleven 50 ft long holes in the first row, and twelve 50 ft long holes in the second row. The holes were grouted in two stages, using a stable mix, and controlling pressure and volume injected. The goal of the grout curtain was to reduce the conductivity of the rock mass to about $10^{-5}$ cm/s.

The grout curtain criteria were established after evaluating the results from two dimensional

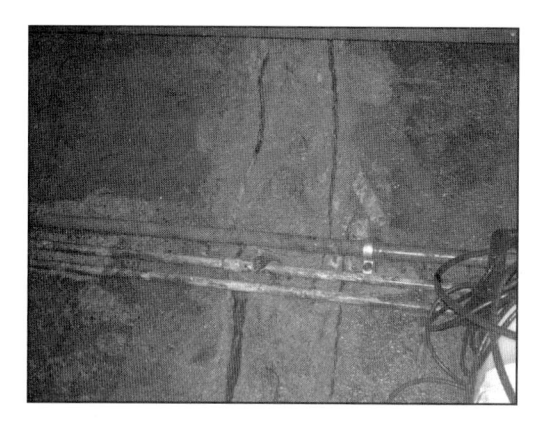

**Figure 4. Existing cracks**

hydrogeology models. Additionally, it was incorporated in a hydrogeology model to estimate the flow that potentially could reach the drainage system behind the membrane. The model used, considered three different scenarios:

- Geomebrane without grout curtain
- Geomebrane with grout curtain—static conditions
- Geomebrane with grout curtain—generation

Additional to the drainage system behind the membrane, it was decided to incorporate a drainage system for the rock mass, as it was expected, based on the results of drilling other boreholes for instrumentation near the tunnel, that some zones of the rock mass could have very high conductivities. The external drainage system selected consisted of horizontal drains drilled from the hillside and extending under the tunnel.

Large diameter drains as those already employed successfully by PG&E in other projects were selected for this location (6-inch diameter holes). The drainholes were drilled using a Casagrande C6 rig and a downhole hammer. The orientation of the drainholes was monitored frequently as their position was close to the tunnel. Some of the longer holes deviated from their original position but were still considered useful. The Appendices show the borehole logs, and the updated geologic plans based on their interpretation.

## INSTALLATION OF CARPI MEMBRANE

Installation of the Carpi membrane is better described in photographs, as shown below. The process consisted generally on the following steps:

- Cleaning of the surface
- Marking the main cracks
- Erection of scaffolding

**Figure 5. Installation of base layers**

- Installation of the base layers on the walls and crown of the tunnel
- Installation of the membrane on the walls and crown of the tunnel
- Welding of sections of the membrane
- Installation of tensioning profiles on walls and crown
- Sealing of the walls and crown
- Installation of support layers on the invert
- Installation of membrane on the invert
- Welding of section of the membrane
- Installation of tensioning profiles on the invert
- Sealing of the invert
- Installation of the closure sections upstream and downstream

Several photographs illustrating installation of the membrane are shown in Figures 4–7.

## CONCLUSIONS

The installation of a Carpi membrane in the initial 600 ft of Belden Tunnel 2 changed the hydrogeologic conditions around the tunnel, as evidenced by readings on piezometers installed around the tunnel as part of the rehabilitation program. Main conclusions and recommendations derived from the installation works, and the observations after initial tunnel filling are:

- The rock mass is very tight at the location of the constructed grout curtain.
- The natural groundwater is depressed, and below the tunnel level, as observed during the tunnel outage, as most of the tunnel was quite dry a few days after being drained, and the springs identified before installing the liner have not reappear afterwards.

Figure 6. Installation of membrane

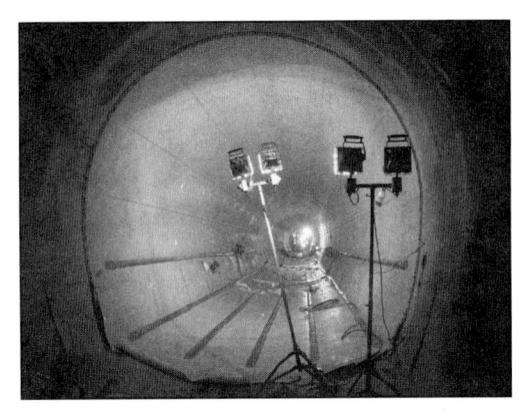

Figure 7. Membrane almost ready

- The rock mass fractures reflected as cracks on Belden Tunnel 2 liner communicate with many other fractures in the rock mass. The drilling of the drain holes resulted in air intrusion on the tunnel in the area these cracks were present.
- The drainage system installed behind the liner is a fundamental part of the liner, and protects it during operation of the project and drainage of the tunnel.
- The liner support system will require additional engineering when planning its application to a higher pressure system. The pressure level in the Belden Tunnel 2 was enough to deform the pieces.

- As demonstrated by the inspection after initial water up of the tunnel, even very small perforations of the liner could result in important leakage from the tunnel.
- The Carpi liner worked as expected, and its installation was accomplished in a very short outage.
- Monitoring of the instrumentation installed at the site should continue to ascertain performance of the liner and its impact on stability.
- The groundwater and rock mass conditions of the Belden Tunnel 2 upstream end.

# Inspection and Rehabilitation of Heroes Highway Tunnel in Connecticut

## Mohammad R. Jafari, Larry Murphy, Michael Gilbert
Camp Dresser and McKee, Cambridge, Massachusetts

**ABSTRACT:** The Heroes Tunnel, the only highway tunnel in the state of Connecticut, is approximately one-quarter-mile long and consists of two barrels and a ventilation shaft. Severe and extensive cracking and spalling, particularly at tunnel portals, along with significant deterioration of the tunnel lining and radial section joints are currently evident. The seepage of water during winter creates an extensive icing condition that can be a danger to traffic and requires constant maintenance. This paper presents details of an extensive site inspection of the tunnel and shafts along with geophysical investigation, evaluation, and proposed rehabilitation program.

## INTRODUCTION

Heroes Tunnel is the only highway tunnel in the state of Connecticut and is part of the Wilbur Cross Parkway, a scenic connection between Hartford and New York. Originally constructed and opened to traffic in 1949, the tunneled portion of the parkway passes through the West Rock Ridge. The tunnel was originally named West Rock Tunnel, but was re-designated as Heroes Tunnel in 2003.

Connecticut Department of Transportation (CTDOT) recognized the risk involved with general aging of the infrastructure, which is compounded by potential safety issues to the vehicular traffic caused by groundwater and New England winter weather. During the winter temperatures, ice build-up occurs, forming icicles that drop and create icy patches on the traveled way. To responsibly manage the risk and financial implications of tunnel rehabilitation, CTDOT retained CDM to conduct an inspection of the tunnel.

## PURPOSE OF INSPECTION

The main goals of this inspection are to assess the condition of various elements of the tunnel, including lining, portals, ventilation shafts, lighting, drainage system, HVAC, and communication systems. This will provide a prioritized rehabilitation program and associated costs to enhance and upgrade the tunnel performance and safety.

## SITE SETTING

Heroes Tunnel is located within the town of Woodbridge and the city of New Haven (shown in Figure 1). The tunnel is approximately 1 quarter mile long and consists of two 28-foot wide and 19-foot high barrels passing through the West Rock Ridge. West Rock Ridge, or West Rock of south-central Connecticut, with a high point of 700 feet, is a 7-mile long rock ridge located on the west side of New Haven. The ridge forms a continuous line of exposed cliff visible from New Haven (shown in Figure 2). West Rock Ridge is located in the municipalities of New Haven, HHamdenH, HWoodbridgeH, and HBethany, ConnecticutH, and is 1 mile wide at its widest point, although steepness of the terrain make the actual square mileage much larger. Notable peaks on the ridge include the high point, alternately called High Rock or York Mountain, approximately 700 feet at the north terminus of the ridge, and the southern prominence, usually referred to as West Rock, 400 feet.

## GEOLOGICAL SETTING

Geotechnical field investigation consisting of 21 borings (boreholes 1 through 5, B through D, G, H, J, L through N, P, Q, R through U, and W shown on Figure 3) was made before construction of the tunnel in 1941 to investigate the subsurface condition of the site. These borings were drilled along both north and south bound tunnel alignments, having depths that varied from 15 to 236 feet. The boring and subsurface profile is shown in Figure 3. As shown on the geological profile, the subsurface ground consists of four major formations of loam, sand and gravel, sandstone, and trap rock. The geological profile also indicates that the north and south portals have been constructed through soil formations. A brief description of each geological earth unit known to exist within the site vicinity is provided below.

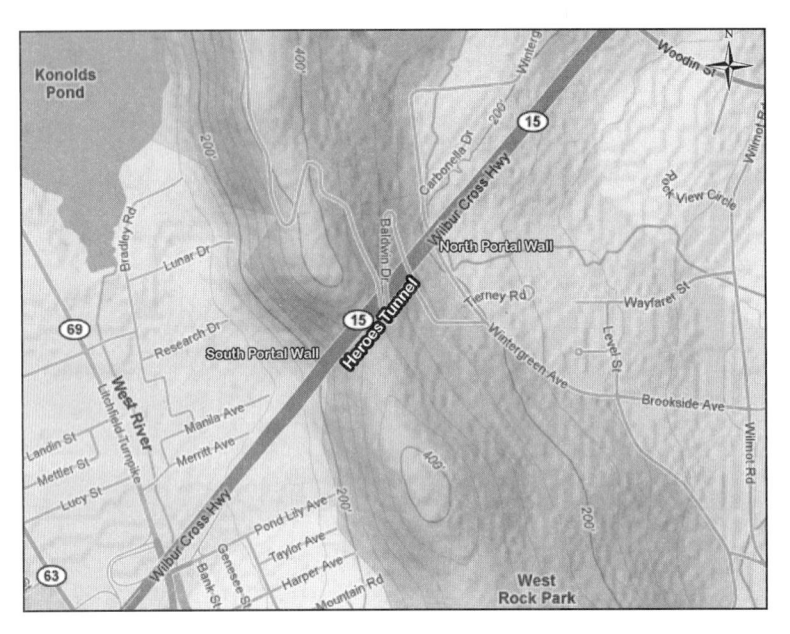

**Figure 1. Tunnel location and topography**

**Trap Rock:** The trap rock is typically a dark-green, fine-grained, hard igneous rock, and very homogeneous throughout with a few exceptions. The geological term for trap rock is "diabase" or "dolerite," and the main minerals forming this rock are plagioclase, feldspar, and augite. It is hard, having a hardness of 6 on the geological scale, or slightly softer than quartz. The unconfined compressive strength (UCS) tests performed on three samples indicated a range of strength ranging between 3100 to 8200 pounds per square inch (psi). The trap rock forming the West Rock Ridge was molten lava under high pressure, which forced its way between layers of sedimentary sandstone (brownstone). The erosion of the softer sandstone leaves the trap rock exposed as a ridge, as shown in Figure 4. The trap rock had a number of fine seams due to shrinkage, cooling, and shearing caused by fault movements. The borings report for Heroes Tunnel stated that the seam concentration, open fissures, and weathering are local deficiencies, and as far as it could be established, do not affect the general excellent condition of the rock, which was characterized by the high recovery and state of most samples. This report indicated that measured recovery for rock samples were above 90 percent, with an average of 95 percent.

**Sandstone:** Borings 1, B, C, D, G, H, J, and Q consist partially of sandstone (brownstone) stratum. The brownstone is coarse-grained sandstone (or fine-grained conglomerate) and shale. The brownstone is sound rock, which possessed fissures slightly to the horizontal in the same proportion as the trap

**Figure 2. South prominence of West Rock from New Haven**

rock. No strength tests were performed on brownstone samples; however, the boring report stated that the general impression the brownstone produced is favorable for tunneling.

**Soils (overburden):** The soil samples obtained from several borings indicated that it consisted of loam, sand and gravel, and sand and gravel with cobbles and boulders.

**Groundwater:** Based on historical data collected for the borings in close vicinity of the tunnel's north portal, the groundwater table is approximately 3 to 5 feet below ground surface. This data showed a high groundwater, which required the installation of a drainage system behind the tunnel and portal wall to lower the groundwater table. Currently there is no

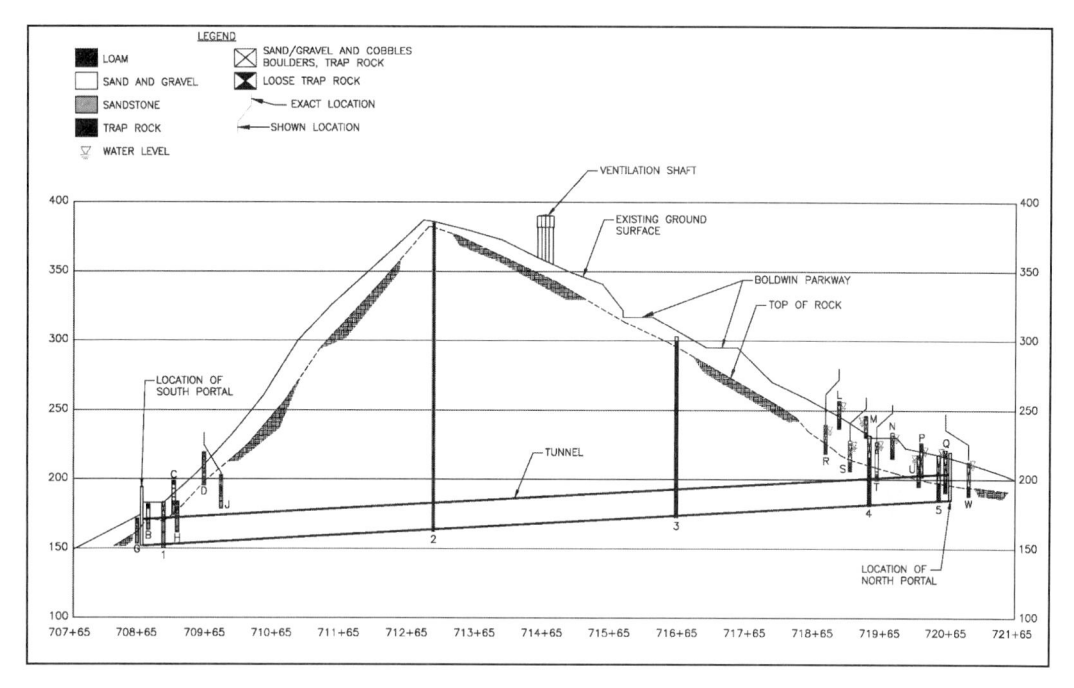

Figure 3. Geological profile and boring location

updated information on groundwater table along the tunnel alignment since the tunnels were constructed.

## TUNNEL STRUCTURE DESCRIPTION

The Heroes Tunnel is comprised of two main roadway tunnels. Each tunnel barrel has interior dimensions of approximately 28 feet wide by 18 feet 6 inches high at the center crown. The tunnels can be generally described as horseshoe shaped with straight sides and an arched ceiling. Each tunnel barrel has two 11-foot-6-inch travel lanes and a 2-foot-6-inch wide pedestrian walkway along the east and west side walls; the total interior width of each barrel is 28 feet.

The tunnels were constructed using drill and blast and cut-and-cover techniques. For drill and blast segments, there are steel support ribs around the circumference of the tunnel, spaced at intervals between 2 and 6 feet o.c., which were placed after blasting to support the rock mass above. A cast-in-place concrete liner was then placed over the ribs for the final support of the tunnel. There is no evidence indicating that the contact grouting has been used to fill any remaining annular spaces between the liner and rock. The structure consists of four 6-foot diameter shafts grouped together. Each shaft is approximately 200 feet in height. The octagon shaped ventilation structure is located atop West Rock Ridge and serves as the outlet.

Figure 4. Exposed trap rock close to tunnel ventilation shaft

## ISSUES RELATED TO HEROES TUNNEL

Results of field inspections for Heroes Tunnel indicated that groundwater infiltration is a major cause of tunnel deterioration, especially for the reinforced concrete lining. Figure 5 shows the deteriorated tunnel roof joints and the growth of icicles at these locations. A review of cross sections for the cut-and-cover and rock tunnel segments reveals the following issues:

- The tunnel portions constructed in rock lacks either a continuous or partial waterproofing system, and the reinforced concrete tunnel liner is in direct contact with the surrounding rock and with groundwater due to the nature of trap rock.
- The cut-and-cover section of the tunnel is partially waterproofed by the installation of a waterproofing membrane on the roof. However, the waterproofing membrane at the cut-and-cover section does not eliminate the possibility of water leakage through construction joints, since groundwater can seep under the waterproofing membrane where the membrane is terminated. The deterioration of existing waterproofing membrane by aging and the interaction of membrane and the chemical and biological agents in the surface water can also be the cause of seepage of groundwater through concrete defects.
- Tunnel does not have continuous longitudinal land drains behind the tunnel lining. The drainage system consists of a 2.5-inch diameter and 1-foot long holes drilled perpendicular to the tunnel direction every 30 feet. However, the effectiveness of a drainage system depends on the ability of the system to maximize the connection between the source of seepage (rock fractures) and the drainage system. For a tunnel constructed by drill-and-blast, the thickness of fractured rock zone around the tunnel perimeter could be at least 3 to 6 feet. Permeability of this zone is particularly increased in the longitudinal direction. Therefore, the ground near the tunnel may act as a longitudinal drain, and an effective drainage system should be in the longitudinal direction (parallel to tunnel direction).

The existing drawings indicate a drain area with variable thickness and height behind the tunnel wall for the northbound tunnel between Station 709+00 and Station 710+122. The nature of this drain area is not clear, and it is possible that the previous repairs performed on tunnel walls using grouting techniques may have suppressed the ability of this drain area to drain and direct the water to 2.5-inch diameter pipes.

The existing drawings also indicate that a drainage system has been installed behind the portals at both ends of the tunnel. The drainage system is parallel to the portal wall and consists of 6-inch diameter pipe. However, a field investigation performed after a heavy rain indicated that the water is seeping through the portal wall at the northern end of the southbound tunnel. Considering the age

**Figure 5. Icicles forming at deteriorated tunnel joints**

of tunnel and difficulty to access the drainage system for frequent cleanup, the seeping water could be an indication of a clogged drainage system.
- Field investigation performed above the tunnel structure at the south and north portals indicated that the surface water runoff collection system, designed through grading, was ineffective due to lack of maintenance and gradual changes in surface grades caused by soil settlement and erosion. Although the existing drawing indicates an interception ditch for south portal, the field investigation could not verify such a ditch was constructed. Vegetation occupied the majority of the area, which could potentially increase the amount of water infiltration into the soil and reduce the amount of runoff water reaching the collection system. The gutters are covered with vegetation and leaves.
- Field inspection performed on tunnels and shafts revealed that various types of structural defects, such as cracks, spalls, and delaminations, have developed in the tunnels and shafts. These defects are also a result of the adverse effects of water leakage (infiltration) through the tunnel and shaft liners.

## INSPECTION PROGRAM

Because this was the state's first real tunnel inspection, CTDOT did not have formal procedures. Therefore, a multi-phase inspection and evaluation program was developed. The initial step was to review existing data. This was followed by conducting field studies to identify tunnel characteristics, constraints, and limitations that could affect the inspection process. An intensive testing process included two forms of geophysical testing—ground penetrating radar and ultrasonic measurements—as well as detailed visual inspection and hammer sounding to determine tunnel liner integrity, was performed.

Prior to conducting tunnel inspection, a mobilization period of planning and organizing for the inspection was necessary to conduct an efficient inspection. During this time CDM coordinated with the tunnel owner to determine available access times for inspecting within the road area, where vehicles can be parked, communication procedures for shutting down electrical systems for testing, discussion of known problem areas, coordination among various inspection groups (e.g., mechanical, electrical, structural) to minimize tunnel closure. The health and safety plans, where confined space entry is deemed necessary, were completed to ensure that the inspectors were knowledgeable of their responsibilities.

Prior to the field work, an internal workshop was held to identify the probable risk items that would be present. This was based on the review of the existing data. The results of this evaluation were that water leakage is the major concern and source of most of the troubles in the tunnel. Leakage through tunnel lining in conjunction with frost would cause:

- Reduction of the size of the tunnel opening by the formation of ice barriers;
- Icing of the road pavement;
- Obstruction of ventilation and other service ducts and shafts;
- Hazard from icicles forming in the tunnel roof;
- Frozen drains that can cause groundwater to find or create a new path to enter the tunnel; and/or
- Frequent freeze thaw cycles at joints and cracks that can accelerate the concrete lining deterioration.

Water leakage would also adversely affect the tunnel lining. It was projected that leakage would:

- Cause a loss of cement by effervesce and thereby reduce the strength of the concrete lining;
- Increase the permeability of the lining and accelerate the deterioration of the lining;
- Trigger the corrosion of reinforcement in reinforced concrete lining resulting in cracking and spalling;
- Transport fine dust that could result in dust traps. Certain combustion gases can result in high acidity of the water and be very corrosive of the ambient moisture in the tunnel;
- Cause corrosion of fixations, destruction of lighting cables.

## RESULTS OF FIELD INSPECTION

### Geophysical Survey and Non-Destructive Testing Results

A geophysical investigation (non-destructive) using both sonic/ultrasonic and ground penetrating radar (GPR) was performed by NDT Corporation to:

- Determine the spacing and depth of cover of the reinforcing within the concrete tunnel lining;
- Identify the location of steel ribs;
- Assess the condition between the lining and rock surface;
- Assess in-situ dynamic strength of the concrete lining;
- Identify areas of weakened concrete due to delamination and cracking.

The non-destructive testing was performed and data was acquired from portal to portal along predetermined survey lines (five lines for each tunnel). Each line measures approximately 1,185 feet at the tunnel wall, spring line, quarter point, and crown.

The sonic/ultrasonic method uses a projectile impact energy source, which produces stress waves in the concrete. These stress waves are detected by the array of sensors that measure the amplitude of the stress wave in time. Sonic/ultrasonic compression and shear wave transmission velocities are used to determine mechanical characteristics of concrete, such as compressive strength, elastic modulus, shear modulus, and Poisson's ratio.

### *Sonic/Ultrasonic Test Results—Northbound Tunnel*

Compression and shear wave velocity data indicate that the tunnel lining strength is variable, ranging from less than 1,000 to 8,700 psi with an average strength of approximately 6,000 psi. For most test locations, the strengths range from 4,000 to 7,500 psi. The sonic/ultrasonic compression and shear wave velocity measurements indicate that approximately 16 percent of the test measurements have poor or no transmission of compression and/or shear wave, which is an indication of weakened concrete due to open cracking. These conditions are more prevalent at the crown with approximately 23 percent of the crown sonic/ultrasonic measurements having poor or no wave transmission. The sonic/ultrasonic measurements also indicated that approximately 2 percent of the measurements at the spring lines and quarter points, and 5 percent of measurements at the crown have a "ringing" character, which is a signal characteristic associated with delamination. Most of these locations are adjacent to measurement indicative of open cracking discussed above, suggesting that delamination may be associated with open cracking.

Resonant frequency/impact echo data indicate that the concrete tunnel lining thickness varies from approximately 1.2 to 3.2 feet, with an average of approximately 2.0 feet. Thickness resonant frequency measurements are adversely affected by poor sensor coupling due to dirt or rough surface conditions, internal micro/macro cracking, irregular/rough back of lining surface, and well bonding of concrete lining to the bedrock with similar velocity and density values.

### *Ground Penetrating Radar (GPR) Test Results—Northbound Tunnel*

The GPR results indicate the locations of moisture entrapment at the lining-bedrock boundary and in bedrock as shown in Figure 6. Moisture entrapment behind the tunnel lining and in the bedrock is indicative of debonding and voiding at this boundary. These areas are concentrated on the west spring line between Stations 709+85 and 714+65, and the crown line between stations 717+65 and 719+65. The GPR test also provided information for reinforcing element, such location and distance between steel ribs, location and distance between reinforcing bars, and concrete thickness.

### *Sonic/Ultrasonic Test Results—Southbound Tunnel*

Compression and shear wave velocity data indicate that the tunnel lining strength is variable, ranging from less than 1,000 to 8,700 psi, with an average strength approximately 6,300 psi. For most test locations, the strengths range from 4,000 to 7,500 psi. The sonic/ultrasonic compression and shear wave velocity measurements indicate that approximately 28 percent of the test measurements have poor or no transmission of compression and/or shear wave, which is an indication of weakened concrete due to open cracking. These conditions were relatively evenly distributed on each of the lines of coverage. The sonic/ultrasonic measurements also indicated that less than 1 percent of the measurements at the spring lines and inside quarter point, 6 percent of measurements at the outside quarter point, and 10 percent of measurements at the crown have a "ringing" character, which is a signal characteristic associated with delamination. Most of these locations are adjacent to measurement indicative of open cracking as discussed above, suggesting that delamination may be associated with open cracking.

Resonant frequency/impact echo data indicate that the concrete tunnel lining thickness varies from approximately 1.2 to 3.2 feet, with an average of approximately 2.0 feet.

### *Ground Penetrating Radar (GPR) Test Results—Southbound Tunnel*

The GPR results indicate the locations of moisture entrapment at the lining-bedrock boundary and areas of possible loose fill between the lining and bedrock, as shown in Figure 6.

### Structural Inspection Results

### *Tunnels*

A structural inspection was performed following the geophysical investigation. The inspection consisted of both visual and hands-on inspection methods of the east and west walls of each tunnel barrel, the tunnel ceiling, the north and south exterior portals, and the ventilation shafts. Figure 7 shows crews performing the structural hands-on and visual inspections.

In general, the hands-on and visual inspection indicated that the concrete tunnel liner is in fair to poor condition, and noted cracks, spalls, hollow areas, areas of prior repairs, areas of patch failures, and other concrete surface defects. These concrete surface defects were classified according to the criteria described in Federal Highway Administration (FHWA) manual for the type of defect, ranging in severity from minor to severe. Areas of moisture, leakage, and efflorescence were also noted. A tabular summary of the concrete surface defects noted for each tunnel barrel, listed by the type of defect and the severity is provided in Table 1. Figure 8 illustrates a typical spalling at tunnel crown.

In general, the structural inspection results tended to verify the geophysical investigation findings in the following respects:

- There is a high correlation between the locations noted in GPR investigation results, as indicative of having moisture/water behind the concrete liner, and the locations where leakage, icicles, efflorescence, spalling, and hollow areas were observed during the structural inspection.
- There is a high correlation between the areas noted in the sonic/ultrasonic testing results, as indicative of delaminations, and similar areas where hollow areas and spalls were observed during the visual inspection.
- There is a high correlation between the areas noted in the sonic/ultrasonic testing results, as indicative of fractured/weakened concrete, and areas where cracking and patch failures were observed during the visual inspection results.

### *Portal Walls*

Inspections were performed on both the north and south tunnel portals. The portals consist of reinforced

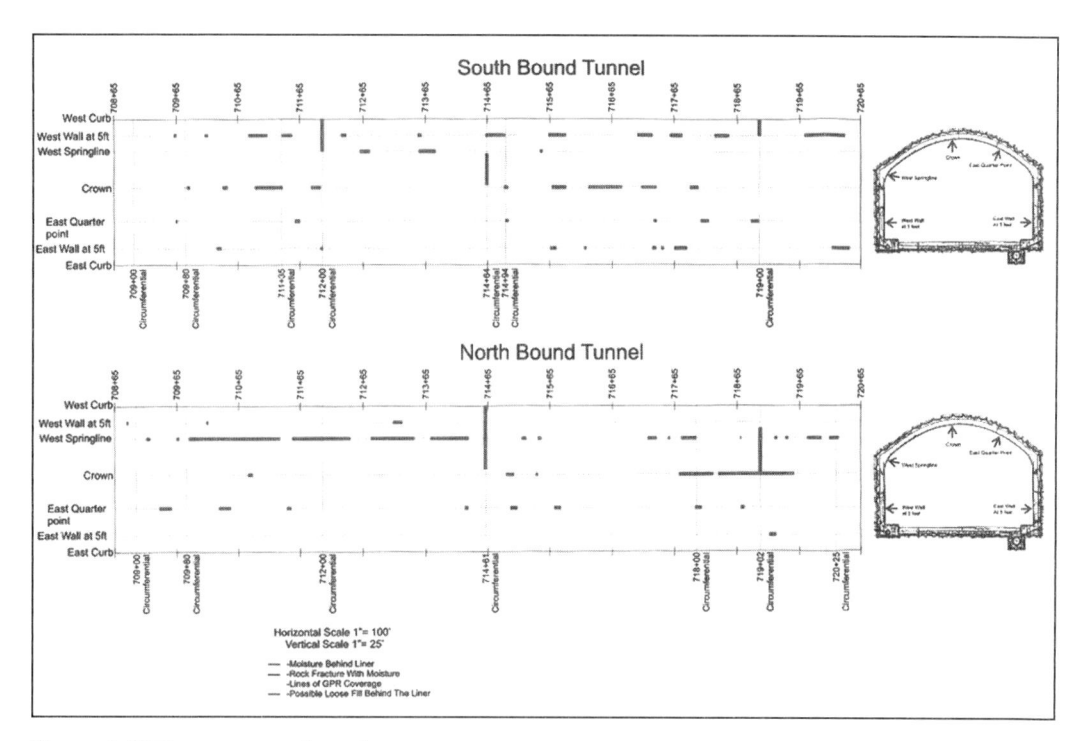

**Figure 6. GPR coverage and results**

**Figure 7. Crews performing structural hands-on and visual inspection**

concrete counterfort retaining walls with an ashlar stone masonry veneer. Only a small percentage of the reinforced concrete was accessible to the inspectors, which included the areas at the very top of the back sides of the portals. Defects in the stone masonry veneer, locations of deteriorated or missing mortar, areas of vegetation and moss growth between the mortar joints, and areas of efflorescence and surface staining were noted in the inspection. The inspection also noted areas of missing and deteriorated joint filler.

The inspection of the north and south tunnel portals observed that these portals appear to be in fair structural condition, and noted isolated areas of missing mortar, efflorescence, and surface staining of minor to moderate severity. Some areas of at the very top of the portals also showed vegetation and moss growing through and between the mortar joint beds. Areas of vegetation growth and missing mortar occur predominately at the top three to four courses of the stone masonry. Some of these isolated areas of missing mortar exceeded 6 inches in depth, but no loose stones were observed.

Table 2 presents the summary of the defects for both north and south portals.

*Ventilation Shafts*

A hands-on inspection was performed within each of the four vertical ventilation shafts. The shafts are not currently operational, and the outlets at the bottom of the shafts into the main tunnels have been closed with steel doors. Access into the shafts was gained from within the ventilation structure atop West Rock Ridge (See Figure 9).

Each shaft was inspected hands-on, and the concrete surfaces were hammer sounded for detection of hollow areas. Areas of spalls, leakage, and

**Table 1. Summary of tunnel defects**

| Tunnel Barrel | Type of Defect | Minor | Moderate | Severe |
|---|---|---|---|---|
| Northbound | Cracks (CR) | 2171 LF | 1799 LF | 346 LF |
| | Spalls (SP) | 294 SF | 679 SF | 356 SF |
| | Scaling (SC) | 488 SF | 170 SF | 219 SF |
| | Hollow Areas | — | — | 2548 SF |
| | Patch Failures | — | — | 3408 SF |
| Southbound | Cracks (CR) | 3231 LF | 837 LF | 22 LF |
| | Spalls (SP) | 399 SF | 326 SF | 260 SF |
| | Scaling (SC) | 1757 SF | 437 SF | 111 SF |
| | Hollow Areas | — | — | 1903 SF |
| | Patch Failures | — | — | 2966 SF |

**Figure 8. A typical spall at tunnel crown**

**Table 2. Summary of portal wall defects**

| Tunnel Barrel | Type of Defect | Minor | Moderate | Severe |
|---|---|---|---|---|
| Northbound | Missing Mortar | — | 61 LF | — |
| Southbound | | — | 60 LF | — |

**Table 3. Summary of defects**

| Type of Defect | Minor | Moderate | Severe |
|---|---|---|---|
| Leaking construction joints/patch failures (PF) | — | — | 2000 SF |
| Cracks (CR) | 80 LF | 0 LF | 0 LF |
| Spalls (SP) | 10 SF | 10 SF | 0 SF |
| Scaling (SC) | 0 SF | 0 SF | 5200 SF |

efflorescence were photographed during the inspection. The shafts were observed to be in fair condition with only isolated areas of a few minor to moderate spalls and few minor cracks. No hollow areas were detected by the hammer sounding. Moderate to severe surface scaling was also observed in the lower portions of the shafts.

The ventilation shafts appear to have infiltration of groundwater, with inflow most likely between the construction joints (23 to 25 per shaft) of the concrete shaft liner. These joints appear to have been previously repaired, but no information was available from the CTDOT regarding the timing or method of repair. These repairs appear to be experiencing continued leakage and groundwater infiltration. The concrete shaft liners of all shafts were observed to be wet, and heavy efflorescence has accumulated at these construction joints.

Interviews with the CTDOT maintenance staff during the inspection indicated that when the shafts were operational, heavy amounts of ice would accumulate inside the shafts during the winter months. The accumulated ice would often fall down the shaft damaging the fans, and ultimately exit out the bottom of the shaft onto the roadway surface. The falling ice caused both maintenance and safety issues for the CTDOT. Several attempts were made to install heaters inside the shafts to keep the ice from accumulating, but ultimately, these attempts proved to be costly and time consuming for the maintenance staff.

Table 3. presents the summary of the defects for ventilation shafts.

**Inspection Results for Other Disciplines**

As previously mentioned, inspections were also conducted to assess the condition of various elements of the tunnel, including lining, portals, ventilation, lighting, drainage system, HVAC, and communication systems. Results for these inspections were provided in final reports and are not provided here.

## REHABILITATION ALTERNATIVES

Results of field inspections for Heroes Tunnel indicated that groundwater infiltration is a major cause of the tunnel deterioration, especially for the reinforced concrete lining. The major causes for water infiltration could be related to:

- Lack of continuous or partial waterproofing system around tunnel and ventilation shaft structures;
- Lack of continuous longitudinal groundwater drains behind the tunnel lining;
- Possible clogging or ineffectiveness of existing drainage system;
- Lack of insulation sheet between rock and concrete lining to prevent freezing of groundwater;
- Progressive deterioration of existing defects.

These issues were considered and addressed for developing and recommending rehabilitation techniques for Heroes Tunnel.

An extensive literature review was conducted on the existing rehabilitation techniques for the tunnels with similar issues. In general, waterproofing membranes (watertight linings), effective drainage systems, concrete crack injection, and soil/rock grouting are the most popular waterproofing methods. It is also possible to utilize one or more of these techniques in combination to stop or reduce water leakage into the tunnel and ventilation shafts.

Several options to address the management of the groundwater were considered, and the most viable and effective options are recommended as presented in following sections.

**Tunnel Rehabilitation Alternatives**

*Option #1—Install New Deep Fan Drains, Tunnel Drainage Membrane, and Insulating Liner System*

This option consists of the following operations:

- Concrete surface repairs, including repair of spalls, hollow areas, patch failures, joints, and severe cracks.
- Installation of new deep fan drains by drilling through the existing drain into tunnel crown and side walls from inside the tunnel. This will drain the groundwater through the liner and direct it to a waterproofing membrane/drainage layer.
- Installation of a waterproofing membrane/drainage layer.

**Figure 9. Exterior view of ventilation structure atop West Rock Ridge**

- Installation of a new longitudinal drainage system on each side of the tunnel to direct the collected groundwater catch basins or other system outside the tunnel.
- Installation of new insulation panels.
- Installation of final protection and fireproofing layer of shotcrete.

*Option #2—Soil/Rock Grouting (Back-wall Grouting)*

This option consists of the following operations:

- Concrete surface repairs, including repair of spalls, hollow areas, patch failures, joints, and severe cracks.
- Drilling from within the tunnel through the existing liner at a certain interval spacing and depth (to be determined) and then grouting the soil/rock. This would fill the rock discontinuities (joints) and eliminate the contact between the tunnel lining and groundwater.
- Installing additional longer and more closely spaced transverse drains through the side walls of the existing tunnel liner. This would drain the groundwater from behind the grouted zone of the soil/rock and prevent the accumulation of hydrostatic pressure.

*Option #3—Surface Applied Waterproofing System*

This option consists of the following operations:

- Concrete surface repairs, including repair of spalls, hollow areas, patch failures, joints, and severe cracks.
- Installation of a surface applied waterproofing system, such as crystalline waterproofing system, or other equal system.

- Installation of additional longer and more closely spaced transverse drains through the side walls of the existing tunnel liner to drain the groundwater from behind the tunnel liner.
- Installation of a new main drain under the roadway to collect the groundwater from the transverse drains and direct out the tunnel.

**Ventilation Shaft Rehabilitation Alternatives**

*Option #1—Soil/Rock Grouting (Back-wall Grouting)*

This option is similar to the alternative introduced for tunnel rehabilitation.

*Option #2— Polyurethane Joint and Crack Injection*

This option consists of the following:

- Concrete surface repairs including the repair of spalls.
- Polyurethane injection grouting of the horizontal construction joints of the ventilation shaft liner.

## CONCLUSIONS

CTDOT recognized the risk involved with the general aging of the Heroes Tunnel, which is compounded by potential safety issues to the vehicular traffic caused by groundwater and New England winter weather. CDM performed multi-phase field inspection to assess the condition of various elements of the tunnel, including lining, portals, ventilation shafts, lighting, drainage system, HVAC, and communication systems and provided a prioritized rehabilitation program and associated costs to enhance and upgrade the tunnel performance and safety.

## ACKNOWLEDGMENT

The authors would like to acknowledge the collaboration of several individuals for preparing this paper. Particular thanks are due to Scott MacDonald, Mathew Ash, and Jeff Cabrera.

## REFERENCES

U.S. Department of Transportation, Federal Highway Administration and Federal Transit Administration, Highway and Rail Transit Tunnel Inspection Manual (2005).

U.S. Department of Transportation, Federal Highway Administration and Federal Transit Administration, *Highway and Rail Transit Tunnel Maintenance and Rehabilitation Manual* (2005).

# TRACK 2: DESIGN

Matt Fowler, Chair

*Session 1: Design Validation by Instrumentation, Monitoring, and Mapping*

# Instrumentation of Freight Tunnel in Chicago

**Alireza Ayoubian**
STV Incorporated, New York, New York

**Richard J. Finno**
Northwestern University, Evanston, Illinois

**ABSTRACT:** The Block 37 development in downtown Chicago involved construction of a high rise building, two tunnel connections and a new platform. During excavation of the tunnel connections, cracks were noted in the freight tunnel underlying one of the excavation zones. The tunnels at that location were horseshoe-shaped, approximately 7.7-foot wide and 9.7-foot high and were instrumented. The freight tunnels house sensitive utilities vital to the City of Chicago. The objective of this paper is to evaluate the instrumentation program employed for freight tunnels at Block 37, analyze the data and help design more effective instrumentation plans for Chicago freight tunnels for future projects.

## INTRODUCTION

The network of freight tunnels in Chicago consists of about 62 miles of horseshoe-shaped tunnels built in early 1900s beneath the downtown area of the city. The tunnels were hand-excavated and supported temporarily with steel ribs and wood laggings. The 10-inch-thick final liner consisted of unreinforced concrete forced by hand between the wood laggings and wooden forms. The freight tunnels were constructed to carry telephone and telegraph cables, but later were used to transport merchandise and solid waste from buildings (Moffat, 2002). Currently, the freight tunnels are used to house sensitive utilities such as power and fiber-optics cables. Any future excavation in downtown Chicago will likely be in the proximity of this network of tunnels. Damages to freight tunnels can be quite costly and must be avoided.

During the development of Block 37 at downtown Chicago, a segment of the active freight tunnel located below Randolph Street was damaged and a 17-foot longitudinal crack was developed in the crown and walls of the tunnel. The cracked segment of the tunnel was properly repaired. The incident can be used to evaluate the instrumentation records and help understand the tolerable deformations freight tunnels can sustain without being damaged.

## BLOCK 37 PROJECT IN DOWNTOWN CHICAGO

Major excavations and construction activities were undertaken by several contractors at the Block 37 site including:

- Constructing a high rise building with multiple levels of basement.
- Connecting the blue and red transit lines by constructing two tunnel connections at northwest (Dearborn Street Connection) and southeast (State Street Connection) corners of the project site.

A plan view of the project site is presented in Figure 1. Construction of the high-rise building required a large excavation (approximately 320 by 380 feet in plan) that covered the majority of the Block 37. The excavation was made using the top-down technique with reinforced concrete slurry and secant-pile walls installed around the perimeter of the excavation to a depth of 85 feet below ground surface. These walls also served as permanent building walls.

Construction of the two tunnel connections and a future new station at Block 37 site were parts of a bigger construction plan to build express train services linking O'Hare International and Midway Airports to downtown Chicago and creation of a transportation hub at Block 37. The new subway station will occupy subsurface space within Block 37 located between the southeast (State and Washington Streets) and northwest (Dearborn and Randolph Streets) corners of the site. The tunnel connection to the red line was constructed at the intersection of State and Washington Streets and the connection to the blue line was built at the intersection of Dearborn and Randolph Streets (Figure 1).

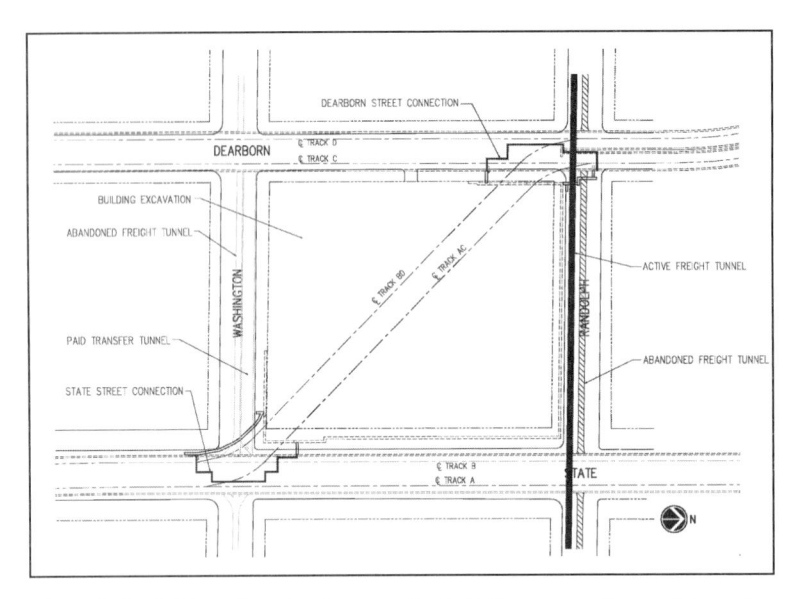

**Figure 1. Plan view of Block 37 project site showing the building excavation and tunnel connection excavations along with active and abandoned freight tunnels**

Construction of tunnel connections was performed using the cut-and-cover method. A mixed of SPTC (Soldier Pile Tremie Concrete) walls and Soldier Pile & Lagging (SPL) walls were used for each connection. The excavation at the intersection of State Street and Washington Streets (referred to as State Street Connection) is about 150 ft × 80 ft in plan with a curved SPTC wall as shown in Figure 1. The excavation at the intersection of Dearborn and Washington Street (referred to as Dearborn Street Connection) is approximately 150 ft × 50 ft in plan (Figure 1). The SPTC walls were taken to about 80 ft below ground surface. The SPL walls were shallow and used for retaining the ground above the existing red and blue line tunnels. The soldier piles were installed such that their tips were close to the top of the existing tunnels. Lateral support for the excavation system was provided by deck beams installed at ground surface, one row of cross-lot braces and the diaphragm slab poured above red and blue line tunnels. Excavation continued below the diaphragm slab to the bottom of a new invert slab to create space for the future tunnel connections. The new diaphragm and invert slabs were components of the permanent structure of the new tunnel connections.

This paper focuses on the Dearborn Street Connection covering blue line tunnels and the active freight tunnel which suffered cracking during construction activities. Figure 2 shows a plan view of the Dearborn Street Connection. The active freight tunnel below Randolph Street runs in east-west direction at the northern part of the Dearborn Street Connection. The blue line tunnels run perpendicular and above the freight tunnel in north-south direction below Dearborn Street. The blue line tunnels below Dearborn Street were constructed with double-tube and triple-arch formations and the transition point from double-tube to triple-arch formation occurred just above the active freight tunnel. The Block 37 excavation and slurry walls were to the east of the Dearborn Street Connection and just south of active freight tunnel. Figure 3 presents an elevation view of the Dearborn Street Connection through cross section 1 (shown in Figure 2). The damaged zone at the freight tunnel is also depicted in Figure 3. Figure 3 shows the double-tube formation of blue line tunnel above the active freight tunnel. This figure also illustrates the three lateral supports elements for Dearborn Street Connection: deck beams at ground surface approximately at elevation 14.5 CCD (Chicago City Datum), cross-lot braces at elevation –4.0 CCD and diaphragm slab above the blue line tunnels. The bottom of the diaphragm slab beyond the blue line tunnels was approximately at elevation –11.0 CCD.

After installation of SPTC and SPL walls around the perimeter of Dearborn Street Connection, excavation of soil started in early February 2008. Installation of deck beams, cross-lot braces and diaphragm slab was completed as excavation was advancing. After placement of diaphragm slab, excavation below and backfilling above the slab began. The excavation below the diaphragm slab reached the bottom of the invert slab at elevation –37.0 CCD, about 5 feet below the active freight tunnel crown, in late August 2008. The space created between

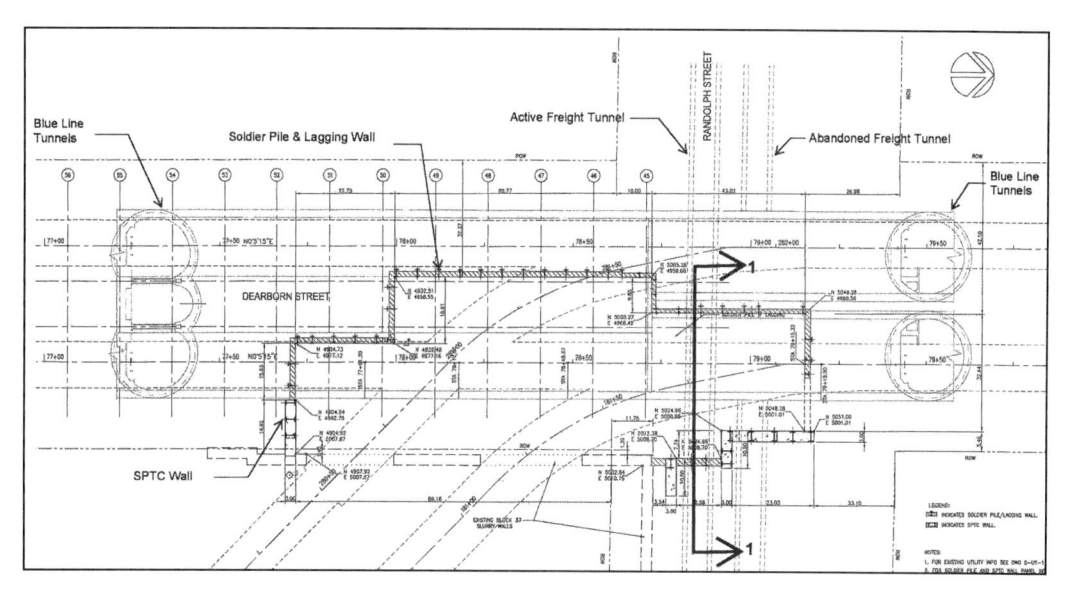

Figure 2. Plan view of Dearborn Street Connection (Source: Kiewit-Reyes, 2007)

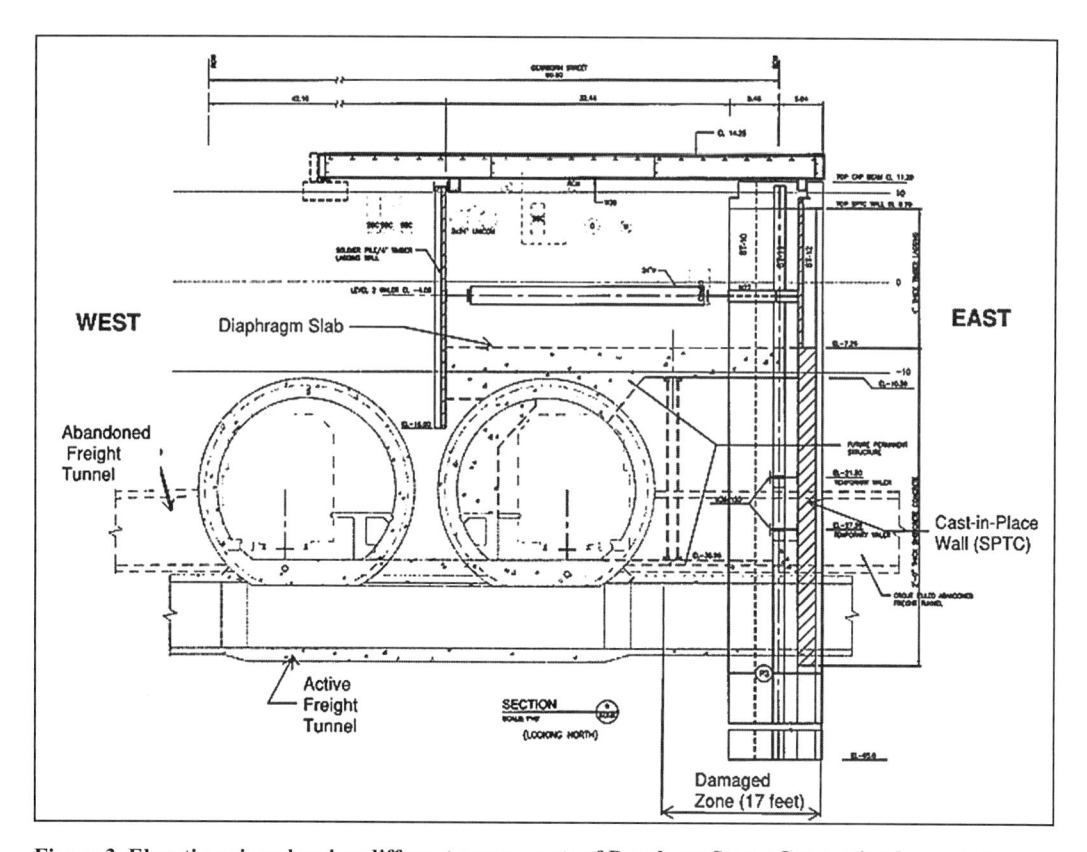

Figure 3. Elevation view showing different components of Dearborn Street Connection located above the freight tunnel (Source: Kiewit-Reyes, 2007)

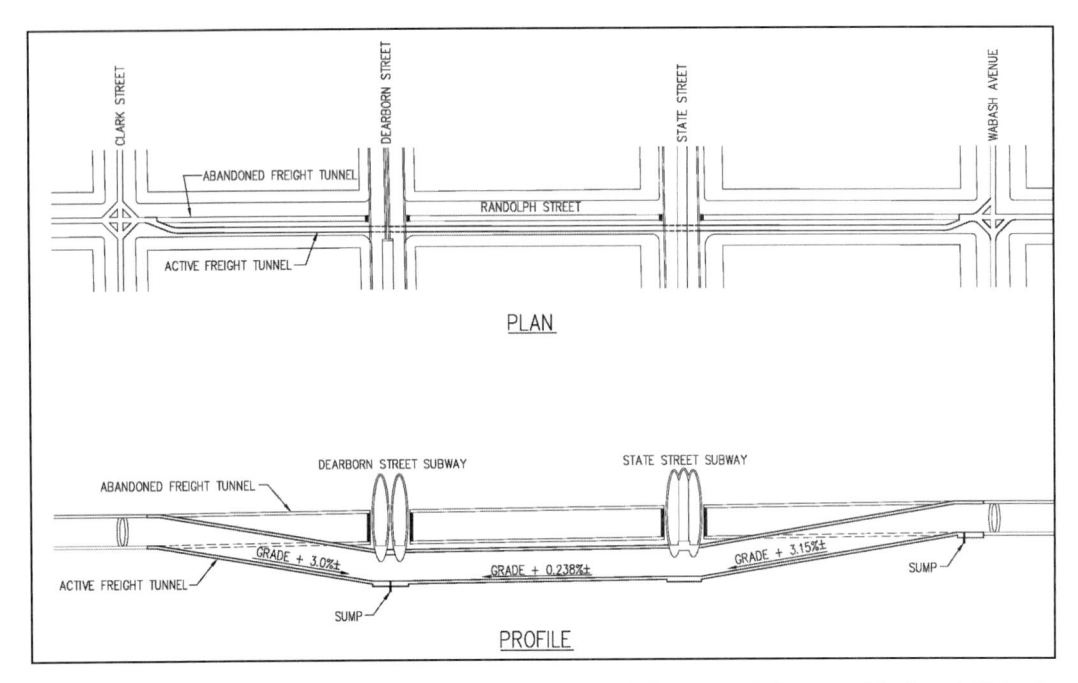

Figure 4. Plan and elevation views of abandoned and active freight tunnels between Clark and Wabash Streets (scale is distorted) (Moffat, 2002)

diaphragm and invert slabs will be used in the future for connection to blue line tunnel. Lightweight flowable backfill was placed above the diaphragm slab up to elevation 5.0 CCD and granular backfill was used from elevation 5.0 CCD to the ground surface. New pavement was placed followed by restoration of the Dearborn Street.

Excavation levels for the Block 37 building reached an approximate elevation of –35 CCD at its deepest location. Most of the excavation within Block 37 at the northwest corner of the site (in the vicinity of Dearborn Street Connection) was completed by the end of March 2008.

## ACTIVE FREIGHT TUNNEL AT BLOCK 37 SITE

The active freight tunnel in the vicinity of Block 37 construction site is located below Randolph Street. Figures 1 and 2 show the location of freight tunnel relative to Block 37 building excavation and Dearborn Street Connection.

Record drawings show that before construction of the blue and red line tunnels in 1940s, the freight tunnel below Randolph Street was blocked off and abandoned to the east of the red line and west of blue line tunnels as shown in Figure 4. After construction of red and blue line tunnels, a new segment of the active freight tunnel was constructed in 1940s between Clark and Wabash Streets, parallel to, below and to the south of the abandoned freight tunnel as illustrated in Figure 4. The active freight tunnel has a downward (easterly) slope of 3% from Clark Street to Dearborn Street and an downward (westerly) slope of 3.15% from Wabash Street to State Street. The middle section of the active freight tunnel between State and Dearborn Streets has a gentle downward (westerly) slope of 0.238% from State Street to Dearborn Street such that all the running water in that part of the freight tunnel network is being collected at a sump located below the blue line tunnel at Dearborn Street. This segment of the active freight tunnel was constructed using steel ribs installed every 3 feet with wood laggings between the ribs. The unreinforced final liner was constructed by hand packing concrete between the forms and the wood laggings. A cross section of the active freight tunnel below Randolph Street is presented in Figure 5. For the remainder of this paper, this active freight tunnel will be simply referred to as the freight tunnel. The freight tunnel was approximately 7.7 feet wide and 9.7 feet tall with its top and bottom located at elevations –32.0 and –41.7 ft CCD, respectively.

### Installation of ST-13 Pile

SPTC walls were used for support of excavation for tunnel connections beyond the limits of red and blue line tunnels. However, SPTC wall panels could not be construction at the northeast corner of the

Dearborn Street Connection because the freight tunnel was well above the proposed bottom of the SPTC wall (Figures 2 and 3). Therefore, a W24×229 pile, designated ST-13, was installed from ground surface and rested on top of the freight tunnel between 3-foot-thick P2 and P3 panels as shown in Figure 6. The support of excavation between panels P2 and P3 was provided by wood lagging between these panels and ST-13. Wood lagging was also installed between the pile installed in P2 panel and the Block 37 wall for support of excavation. The significance of the ST-13 pile will be discussed later.

## Temporary Bracing of Freight Tunnel at Block 37 Site

To protect the freight tunnel during construction of P2 and P3 panels and to brace the tunnel against additional pressures, longitudinal walers with lateral struts were installed inside the freight tunnel before construction of these panels. The length of the walers was about 10 feet and they were supported laterally by 4 timber struts. The centerline of the braced length of the freight tunnel coincided approximately with the center of the P3 panel as shown in Figure 7. Figure 7 also shows the longitudinal crack at the roof and walls of the freight tunnel and the blue line tunnel wall. Figure 8 shows the cross section A-A (shown in Figure 7) of the freight tunnel with an elevation view of the temporary braces installed inside the freight tunnel

## Cracking of Freight Tunnel at Block 37 Site

Cracks and damages at the crown and walls of the freight tunnel during excavation of Dearborn Street Connection were first noted on August 29, 2008 by surveyors who were working in the tunnel. At this time, excavation at the Dearborn Street Connection had reached the bottom of the invert slab at elevation –37 CCD, which was about 5 feet below the top of

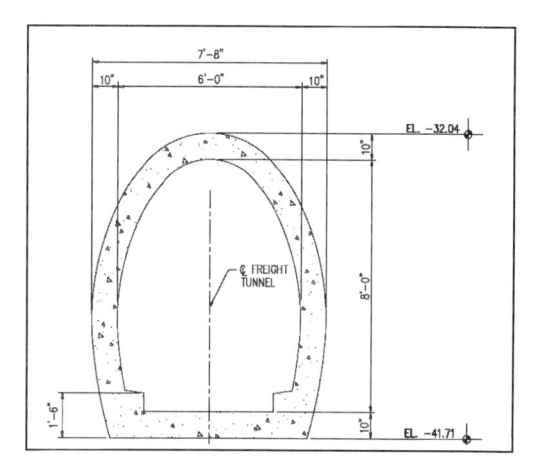

Figure 5. A typical cross-section of active freight tunnel at Dearborn Street Connection

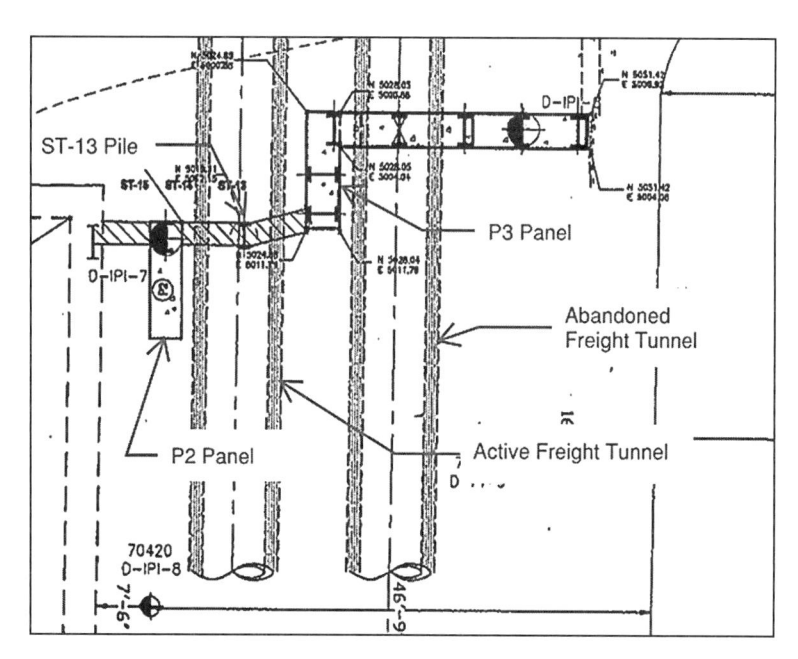

Figure 6. Location of ST-13 (plan view) resting on top of the freight tunnel below Randolph Street (Source: Kiewit-Reyes, 2007)

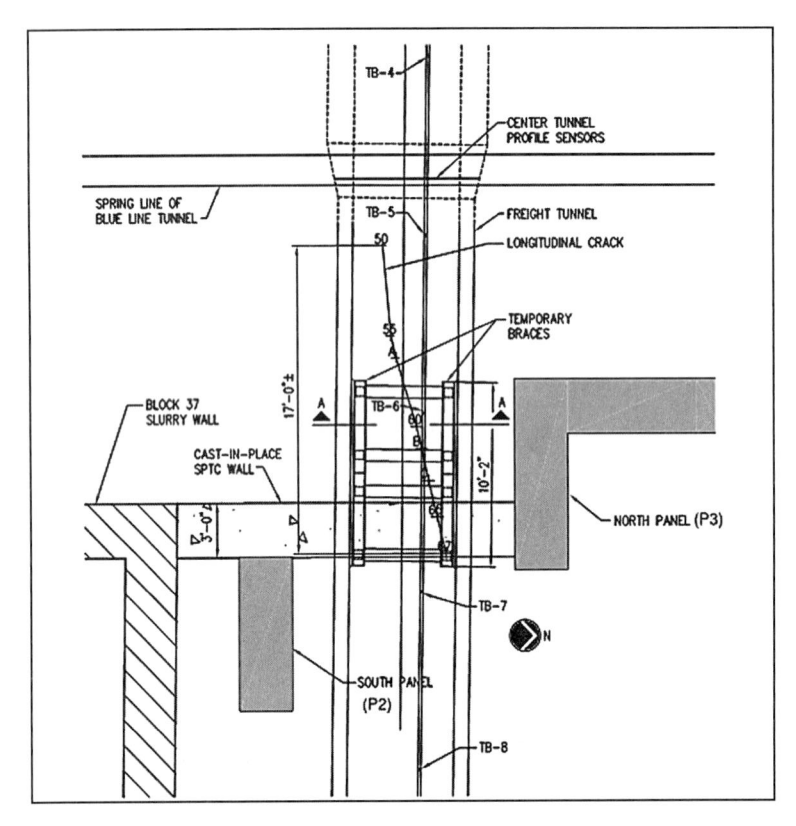

**Figure 7. Plan view of the damaged zone of the freight tunnel at Dearborn Street Connection relative to the temporary braces, blue line tunnel, and tilt beam sensors**

the freight tunnel crown. Picture 1 shows the freight tunnel at the Dearborn Street connection when all the soil above it has been excavated. Also shown is the blue line tunnel and the P3 Panel. Later, when the wood laggings were removed from the exterior of the freight tunnel, cracks were visible on the outside face of the tunnel. The damaged zone in the tunnel was between the soldier pile and lagging wall and blue line tunnel and contained one major longitudinal crack along the crown and walls of the freight tunnel as depicted in Figure 7. The length of the crack was about 17 feet and it started just below the soldier pile wall segment of the excavation support system on its east end and continued west to about 5 ft of the blue line tunnel. Crack width data suggest that the width of the crack in the locations measured from the inside of the freight tunnel was approximately 0.04 to 0.06 in. (about 1 to 1.5 mm).

## GEOTECHNICAL SUBSURFACE CONDITIONS AT BLOCK 37 SITE

Results of subsurface investigation undertaken at the Block 37 site showed that the site was underlain by a 15-foot-thick layer of fill and three layers of glacial clays (Blodgett, Deerfield and Park Ridge) which ranged in strength from soft to medium stiff. The clay layers were underlain by a thick layer of hardpan overlaying the bedrock at the site. The red and blue line tunnels were located in Blodgett and Deerfield layers and the freight tunnel was constructed within the Deerfield clay in the Block 37 area. Perched groundwater table was encountered at about 14 feet below ground surface.

## INSTRUMENTATION OF FREIGHT TUNNELS AT BLOCK 37 SITE

Tilt beam and tunnel profile sensors were installed inside the active freight tunnel below Randolph

Street at Dearborn Street Connection. These instruments were installed prior to the start of the construction at Block 37. Automatic data collection was used for collecting data from the tilt beam and tunnel profile sensors.

## Tilt Beam Sensors

Tilt beams (TB) were used to measure differential movements at the crown of the freight tunnel in its longitudinal direction. Each 10-foot long fiberglass tilt beam contained a sensor mounted inside and approximately in the middle of the beam which records the rotation of the beam. The rotation of the beam recorded by the TB sensor is translated to the movement of one end of the beam relative to the other end. The beam is attached to the structures with anchor bolts. Ten TB sensors were installed approximately every 10 feet on the interior face and close to the crown of the freight tunnel in longitudinal (east-west) direction and were used to measure vertical movements of the tunnel with time. The recorded movements at all the TB sensors starting from TB-1 at the west end were added to obtain the cumulative vertical displacement at the east end of any tilt beam bar. The tilt beam sensors were not installed exactly at the tunnel crown and had an off-set of about 1.1 to 1.5 feet. The locations of the tile beam sensors along the longitudinal axis of the freight tunnel are shown in Figure 9. The sensors were installed below the blue line tunnel tubes, excavation zone at Dearborn Street Connection and past the excavation support

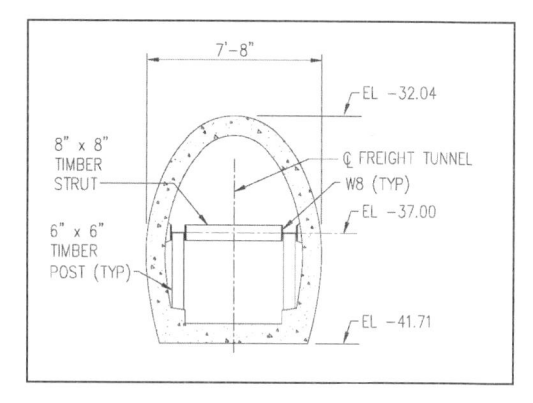

**Figure 8. Cross section A-A showing the location of the temporary braces installed inside the freight tunnel prior to the construction**

**Picture 1. Freight tunnel at Dearborn Street Connection relative to Blue line tunnel and the SPTC wall (looking north and taken on 8/19/08)**

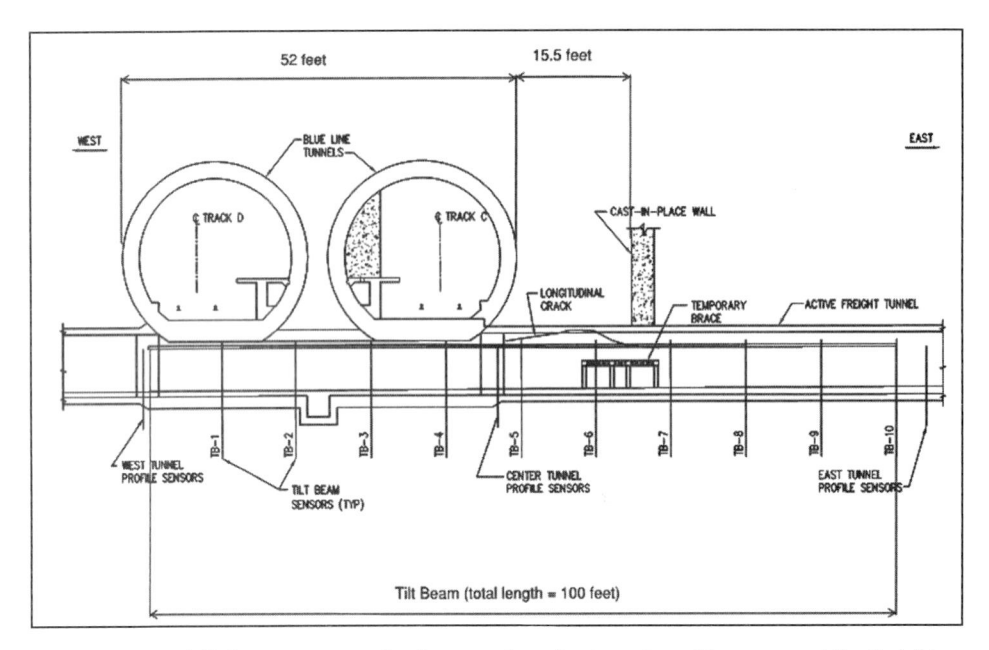

**Figure 9.** Location of tilt beam sensors, the three sections for tunnel profile sensors at the freight tunnel, with the longitudinal view of the crack and temporary braces

wall. Figure 7 shows the location of TB-4 through 8 sensors relative to the longitudinal crack. What is referred to in Figures 7 and 9 as TB sensors is the location of the "east end" of each tilt beam bar where the vertical displacements are recorded for each tilt beam and not the actual location of the sensors in the middle of each tilt beam bar. TB readings were relative to the west end of the tilt beam. The intent was to install the TB-1 sensor beam far enough away from the excavation zone such that it would not be effected significantly by the construction activities within the Dearborn Street Connection. The manufacturer-specified resolution of the data collected by the TB sensors is about ±5arc second (0.0003 inch/1 ft). Since the length of the tilt beams used for the freight tunnel was 10 feet, the expected resolution in recording vertical movement of the tunnel for each tilt beam was about 0.003 inches.

**Tunnel Profile Sensors**

Tunnel profile (TP) sensors were installed at three sections (east, center and west) along the length of the freight tunnel as illustrated in Figure 9. TP sensors were used to record movements in horizontal (north-south) and vertical directions and monitor distortions (differential movements) in the tunnel in transverse direction. Figure 10 presents a view of the center cross section (looking east) which shows the location of the TP sensors mounted inside the freight tunnel. The movements recorded by each TP were relative to the movements of the TP anchor point at the bottom north corner of the freight tunnel at each section. Tilt beam is also shown schematically in Figure 10 at the crown of the tunnel, however the actual locations of the tilt beams had an offset of 1.1 to 1.5 feet to the north. Figure 10 also shows the utility conduits schematically within the tunnel.

**Surveying**

To supplement the data obtained from tilt beam and tunnel profile sensors, manual survey of additional survey points installed inside the freight tunnel was conducted three times a week between July 7 and September 23, 2008. The locations of these survey points were close to the TP anchor points and were installed at three sections along the longitudinal axis of the freight tunnel. The west and east sections were located approximately at the western and eastern ends of the tilt beam at stations 0+00 and 1+02, respectively, and the center section was located at station 0+45, approximately between TB-4 and TB-5 (Figure 9). The west end of the tilt beam was assumed at station 0+00. Four to five survey points were added at each section on the interior face of the freight tunnel. A cross section of the freight tunnel showing the locations of survey points 10, 22, 23, 24 and 25 at center section is presented in Figure 11.

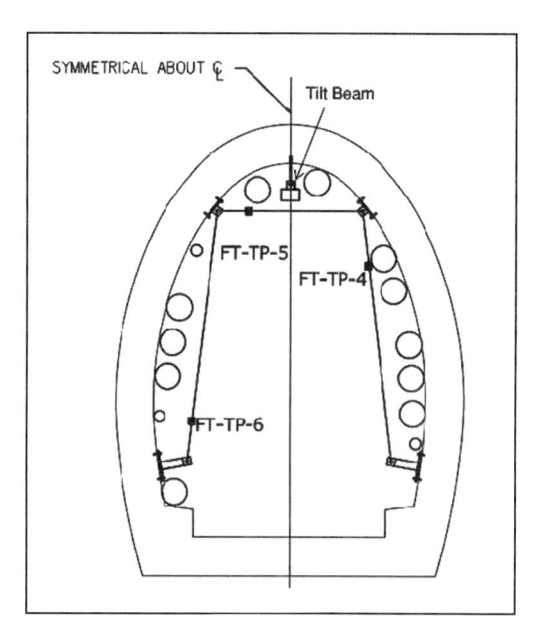

**Figure 10.** A cross section of freight tunnel showing the location of tunnel profile sensors at "center" section, looking east (Source: GZA, 2008)

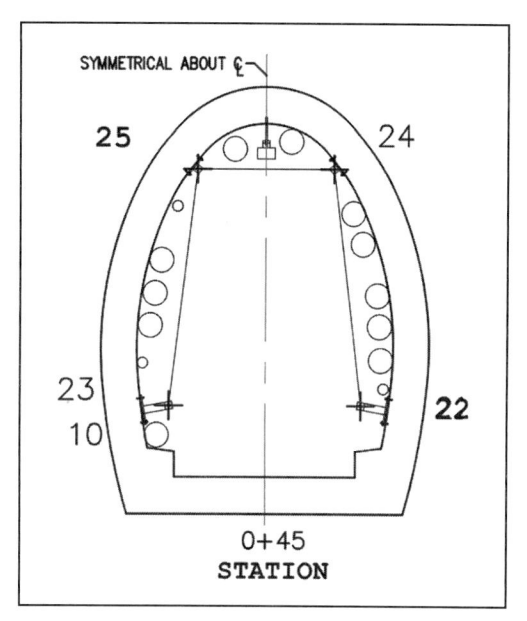

**Figure 11.** A cross section of freight tunnel showing the locations of survey points at "center" section, looking east (Source: GZA, 2008)

## EVALUATION OF INSTRUMENTATION RECORDS

Instrumentation records obtained from tunnel profile sensors (TPs) showed multiple erratic jumps which could not be explained. Such results could have been caused by workers in the tunnel accidentally hitting and disturbing the sensors or instrument malfunctioning. TP sensors were mounted inside the tunnel and could easily be disturbed by outside interference. As a result, the data did not provide reliable information and were not included in analyzing the instrumentation records at the freight tunnel. The instrumentation data were collected and reported by GZA GeoEnvironmental Inc. (2008).

### Excavation Records at Dearborn Connection

To properly evaluate the instrumentation records obtained at the freight tunnel, it is imperative that the construction records be evaluated with the instrumentation data. Figure 12 shows the excavation and backfilling elevation changes with time above the freight tunnel in Dearborn Street Connection. Excavation of soil started at the ground surface (elevation 14.5 CCD) around 2/8/08 and reached elevation −3 CCD around 4/28/08. Then there was a quiet period with little excavation, during which other construction activities were underway at the excavation zone. The excavation reached the bottom of the diaphragm slab at elevation −11 CCD around 6/24/08. After pouring the diaphragm slab and allowing enough time for the concrete to reach acceptable strength, excavation below the diaphragm slab started around 7/28/08 and reached elevation −31 CCD (close to the crown of the freight tunnel at El. −32 CCD) around 8/12/08. Excavation then continued and reached elevation −37 CCD (bottom of the invert slab) around 8/19/08. During excavation of soil below the diaphragm slab, backfilling with light flowable fill was in progress above the slab as shown in Figure 12.

### Tilt Beam Records

Vertical movements recorded by tilt beam sensors at the freight tunnel crown during the excavation period (2/8/08 to 8/19/08) are presented in Figures 13 and 14 for TB-1 through TB-7 and TB-8 through TB-10, respectively. Upward movements are positive and downward movements are negative in Figure 13 and 14 and the movements presented are cumulative movements at the east end of each TB bar as shown in Figure 9.

All TB sensors (1 through 10) showed zero or close to zero movements prior to the start of Dearborn Connection excavation. TB-1 and TB-8 recorded very small vertical movements (less than 0.05") throughout the entire excavation period. Figure 13 shows that as excavation proceeded, TB-1 through

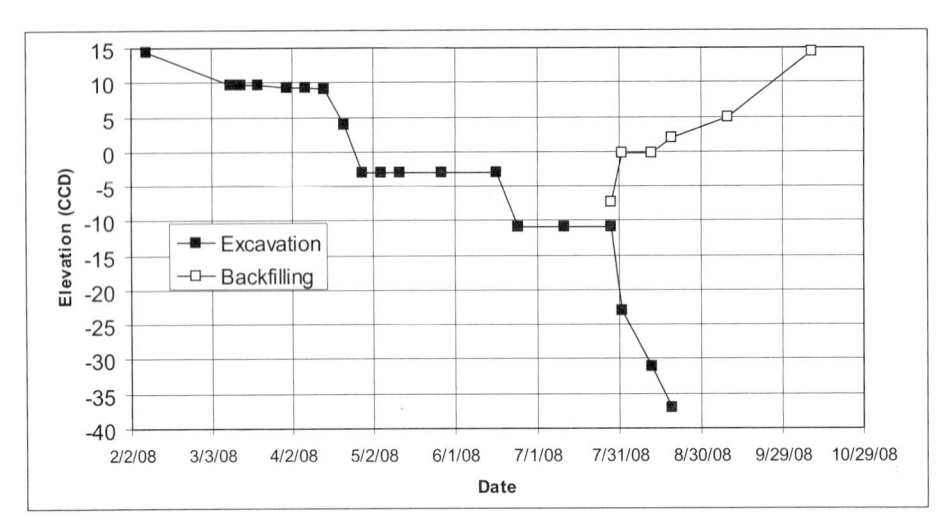

**Figure 12. Excavation and backfilling elevation changes above freight tunnel**

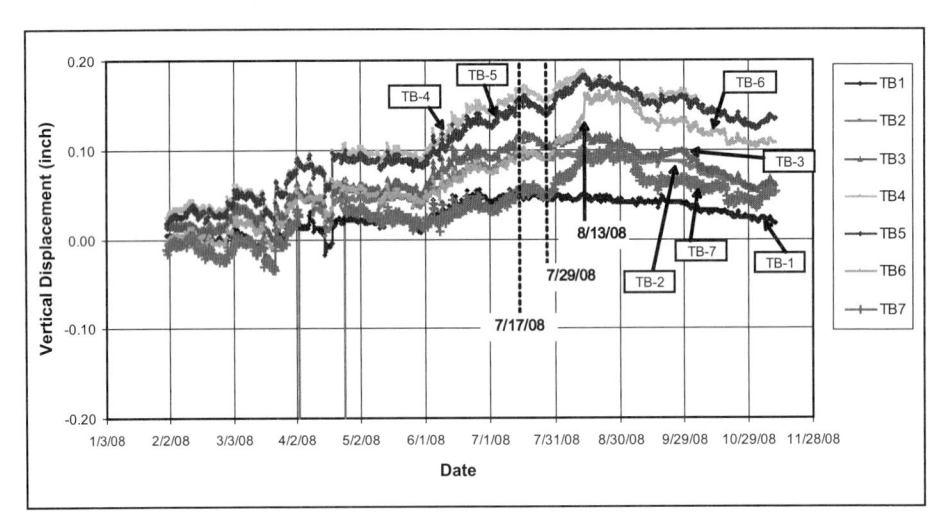

**Figure 13. Vertical movements of TB-1 through TB-7 sensors during excavation (Data from GZA, 2008)**

TB-7 showed increasingly upward movement due to unloading of the freight tunnel until about 8/12/08 when excavation reached close to the top of the freight tunnel. In Figure 13, TB-1 and TB-7 showed the least amount of upward movement. The maximum upward movements were recorded by TB-4 and 5, about 0.19 inches on 8/12/08 when the excavation had reached elevation −31 CCD (about a foot above the freight tunnel crown). Figure 14 shows that TB-9 and 10 recorded downward movements. Figures 13 and 14 illustrate that vertical movements of the tilt beam installed within the freight tunnel were directly impacted by excavation of soil above the tunnel. The sensors to the west of TB-8 showed upward movements and sensors to the east of TB-8 exhibited downward movement during excavation period. Once backfilling started around 8/1/08, this pattern of movements began to reverse gradually in most TB sensors. The deformations in the crown along the longitudinal axis of the freight tunnel can be visualized by plotting the displacement of sensors with distance over time as shown in Figure 15. Figure 15 illustrates the upward movement of the freight tunnel between TB-1 and TB-8 and downward movement between TB-8 and TB-10.

Figure 13 shows that the vertical movements recorded by TB-6 showed an abrupt increase on 8/13/08 from 0.14 to 0.16 in. This abrupt vertical displacement is interpreted as the formation of the crack in the crown of the tunnel. Figure 7 shows

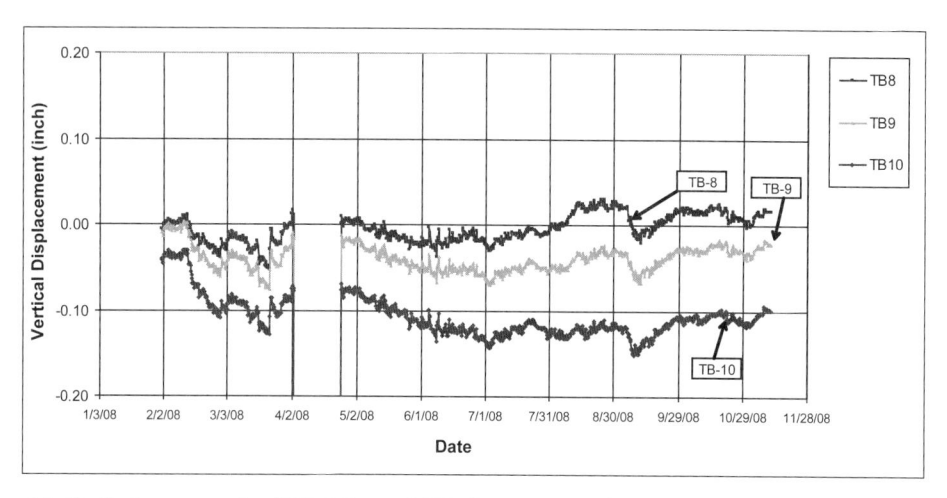

Figure 14. Vertical movements of TB-8 through TB-10 sensors during excavation (Data from GZA, 2008)

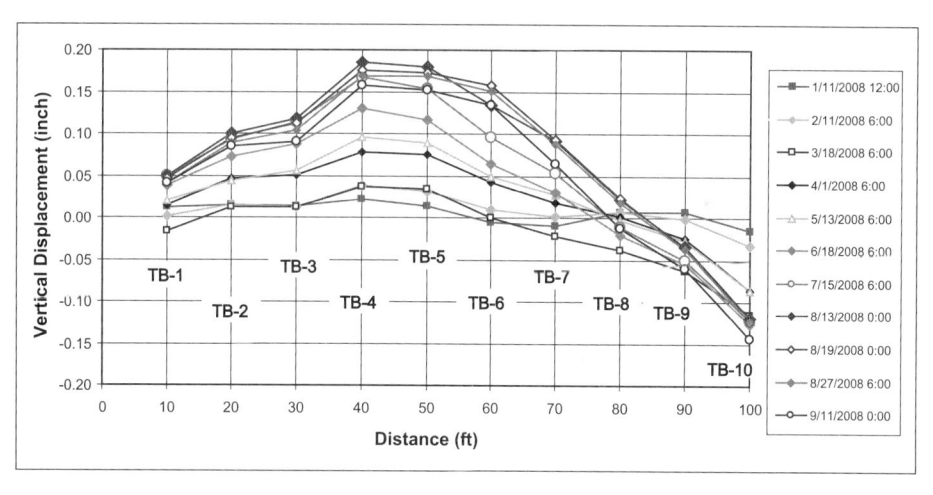

Figure 15. Vertical movements of TB-1 through TB-10 sensors during excavation (Data from GZA, 2008)

that TB-6 was the closest TB sensor to the crack and explains why crack formation was only recorded by this sensor.

It can also be noted from Figure 13 a rather abrupt downward movement recorded by TB-3 through TB-7 from about 7/17/08 to 7/29/08. The magnitude of this movement ranged from 0.01 to 0.02 inches. Diaphragm slab was poured on 7/12/08 and the period from 7/17/08 to 7/28/08 corresponded to the period waiting for the diaphragm slab to cure and start of the excavation below the diaphragm slab. It appears that during this period, additional loads could have been imposed on the freight tunnel by the ST-13 pile, causing downward movements at the crown of the tunnel. This additional load could be due to the transfer of some of the diaphragm slab weight to ST-13 after curing. Another potential source of additional load applied to the top of the freight tunnel was start of excavation of soils along the length of the ST-13 from bottom of the diaphragm slab, reducing skin resistance of the pile and transferring more load to the tip of the pile and ultimately to the crown of the freight tunnel. Excavation below diaphragm slab started on 7/28/08.

### Surveying Records

Manual survey of the additional points inside the freight tunnel was performed from 7/17/08 to 9/24/08. The readings recorded the position of each survey point in terms of station, offset from the centerline of the freight tunnel and elevation. Thus the survey readings recorded the movement of each point in east-west, north-south and vertical directions. The

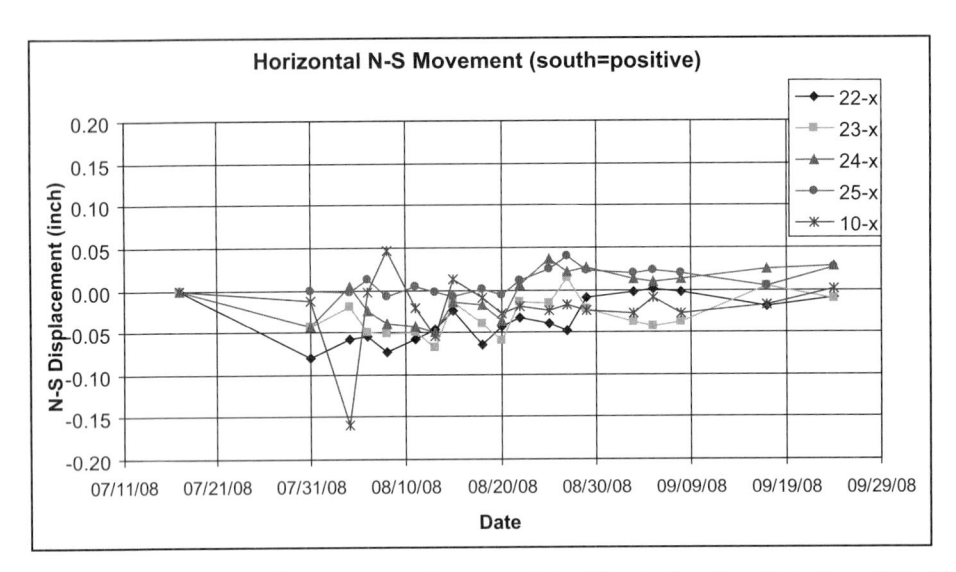

**Figure 16. Horizontal north-south movements of survey points at "center" section (Data from GZA, 2008)**

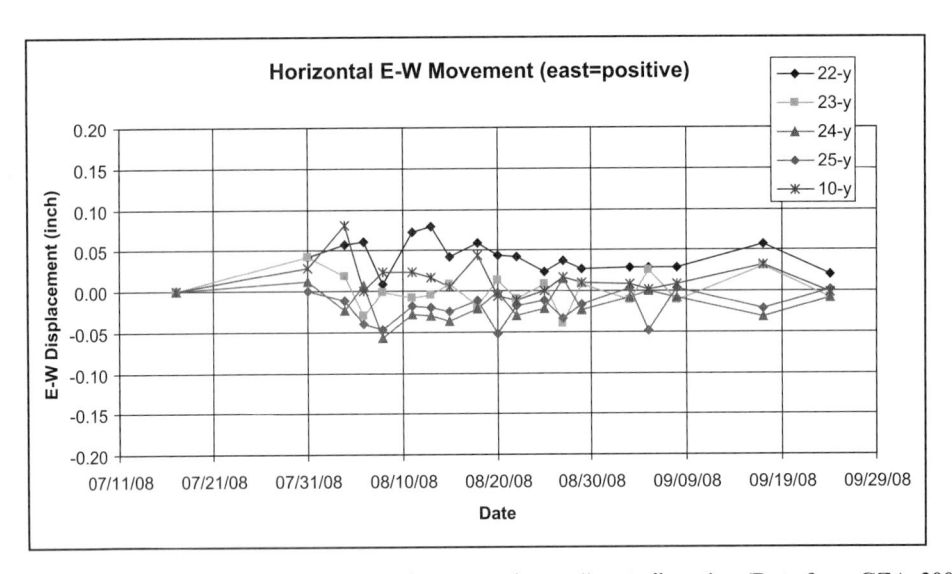

**Figure 17. Horizontal east-west movements of survey points at "center" section (Data from GZA, 2008)**

data collected from west and east sections did not show any significant movements throughout the survey period (mostly less than 0.05 in.) and are not presented herein. The movements recorded at the center section in north-south, east-west and vertical direction are presented in Figures 16, 17 and 18, respectively. It can be seen from these figures that the recorded movements in north-south and east-west directions were also small and generally less than 0.08 in. However, movements recorded in vertical direction were larger and are discussed herein. It must be noted that the movements recorded by the survey points are relative to their initial readings taken at the beginning of the survey period. Initial readings for all survey points at the center section were taken on 7/17/08 except for survey point 25. The initial readings for survey point 25 were taken on 7/31/08. Therefore, displacements for survey point 25 were shown in Figures 16, 17 and 18 starting on 7/31/08.

The locations of the survey points at the center section are shown in Figure 11. Figure 18 shows that all survey points recorded vertical upward movements with the start of excavation below the

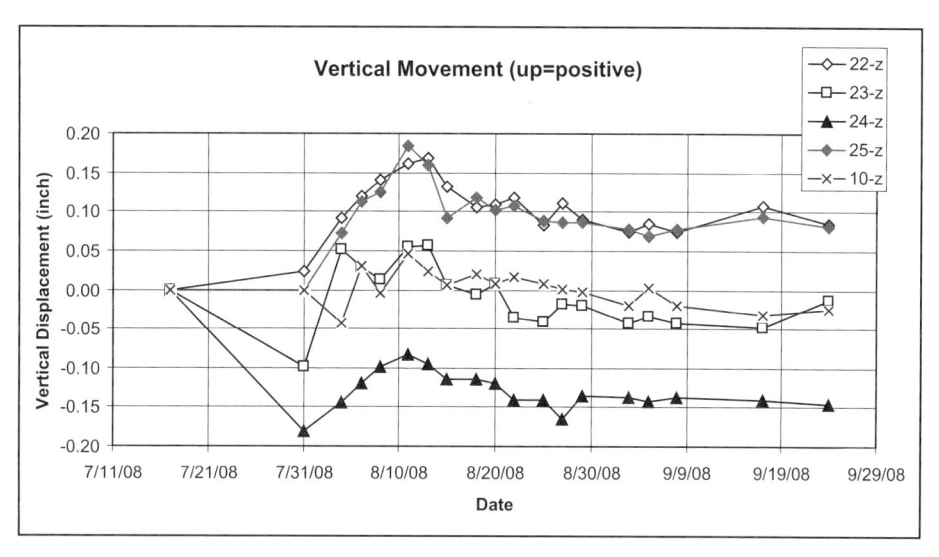

Figure 18. Vertical movements of survey points at "center" section (Data from GZA, 2008)

diaphragm slab beginning around 7/28/08, as a result of reducing the overburden stress acting on the tunnel. The recorded vertical upward movements reached peak values around 8/11/08 at all survey points when the excavation reached close to the top of the freight tunnel. This observation is consistent with data collected from TB sensors.

Another observation from Figure 18, is the development of differential vertical movements recorded at the survey points at the center section during 7/17/08–7/31/08 period, which persisted throughout the entire survey period. It can be seen that survey point 24 showed a downward movement of about 0.18 inches whereas other survey points showed less downward or slight upward movements. As discussed in the previous section, downward movements were also recorded by TB-3 through TB-7 during the same period (7/17/08 to 7/29/08). The development of differential vertical movements between survey points at the center section suggest that the vertical load exerted on the freight tunnel by ST-13 could have applied not at the centerline of the freight tunnel, but at an offset with respect to the tunnel centerline. The offset of the additional load imposed by ST-13 on the freight tunnel could also be the cause for the skewed crack that was formed at the crown and walls of the tunnel (Figure 7). The magnitude of vertical downward movements recorded by the survey points at the center section during 7/17/08–7/31/08 period were much higher than the vertical downward movements recorded by the TB sensors during the same period. Such differences show that the data recorded by the TB sensors revealed only a fraction of the vertical movements that took place at the crown of the tunnel and did not disclose the differential movements at the crown in the transverse direction.

The impact of the Block 37 excavation on the development of the crack in the freight tunnel in Dearborn Street Connection appears to be negligible. Almost the entire excavation for the Block 37 building was completed by the end of March 2008. The vertical movements recorded by TB-1 through TB-7 sensors at the end of March 2008 were generally less than 0.07 (upward) inches, which were believed to have occurred due to excavation at Dearborn Street Connection. Some downward vertical movements were recorded by TB-9 and TB-10 around the end of March 2008 which were generally less than 0.13 and 0.08 inches, respectively, and could have occurred due to the Block 37 excavation. However, the timing of cracking and the magnitude and nature of tunnel movements prior to cracking suggest that the impact of Block 37 excavation on the freight tunnel cracking was very small.

## CONCLUDING REMARKS

The instrumentation data collected in the freight tunnel located in Dearborn Street Connection at the Block 37 site was used to explain the cracking that occurred at the crown of the tunnel during excavation. Results from tilt beam sensors and conventional optical survey showed that a combination of upward movement of the tunnel due to excavation and additional load imposed on the freight tunnel by the ST-13 pile could have led to cracking near the crown of the freight tunnel. The imposed load from ST-13 pile appeared to have been applied not at the centerline of the tunnel but with some off-set with respect to the crown centerline. Such off-set of the additional load from ST-13 pile could explain the skewed pattern of the cracking. Additional load from ST-13 pile might have developed as a result of transferring some of the diaphragm slab load to the pile and loss of skin resistance along the length of the pile due to excavation of the soil below the diaphragm slab. Such a point load acting on the crown of the tunnel with off-set is a plausible reason for the developing of differential movements at the crown of the tunnel which could have led to cracking of the tunnel. The vertical movements recorded by the tilt beam sensors did not fully disclose neither the magnitude of vertical movements at the tunnel crown nor the differential movements in the transverse direction. Timing of the cracking was also estimated from tilt beam and it was estimated to have occurred on 8/13/08. Bracing of the freight tunnel did not seem to have a direct impact on the damages to the freight tunnel. However, it could have contributed to the stiffening of the tunnel at that point.

For future projects, it is important to predict the movement of the tunnel and possible mechanisms that can lead to non-symmetric deformations. Freight tunnels in Chicago are unreinforced and brittle. This was confirmed by sudden development of cracks in the tunnel without any prior warning. Enough redundancy should be built into the instrumentation program in case some instruments malfunction or data become unreliable.

**REFERENCES**

Moffat, B.G. 2002. The Chicago tunnel story: Exploring the railroad "forty feet below." Bulletin 135 of the Central Electric Railfans' Association.

GZA GeoEnvironmental Inc. 2008. Internal instrumentation reports.

Kiewit-Reyes 2007. Internal documents.

# Field Mapping and Photo Documentation of the Southern Nevada Water Authority's Lake Mead Intake No. 3 Project, Saddle Island, Lake Mead, Nevada

**Scott Ball, Richard Coon**
MW/Hill, Las Vegas, Nevada

**Erika Moonin**
Southern Nevada Water Authority, Las Vegas, Nevada

**ABSTRACT:** The Southern Nevada Water Authority's (SNWA) Lake Mead Intake No. 3 Project is a water supply program on and near Saddle Island in Lake Mead, Nevada. This program has three underground construction contracts (Table 1): Contract No. 1 includes a deep shaft and a bored tunnel; Contract No. 2 includes a surface excavation, a deep surge shaft, and a drill and blast connector tunnel; Contract No. 5 includes a deep gate shaft and a drill and blast connector tunnel (Figure 1). The Contract No. 2 and No. 5 excavations are the subject of this presentation. These contracts involve drill and blast excavation in Precambrian age mylonitized amphibolite and quartz-feldspar gneiss in the Lower Plate metamorphic rock of Saddle Island. Documentation of the work described in this paper includes (1) geologic maps, (2) photo-geologic maps and (3) annotated photographs of ground conditions and construction operations. The main purpose of the geologic mapping and field documentation performed for these contracts is to verify the geologic and geotechnical conditions as presented in the Geotechnical Baseline Reports (GBR), identify locations for Engineer directed (as required by the contracts) rock reinforcement (e.g., rock bolts and shotcrete), evaluate overbreak, and provide a record of geologic conditions. This paper describes the geologic mapping methodology, development of a photo-geologic record of the exposed rock, and provides examples of formats to present these data.

## PUBLISHED GUIDANCE AND METHODS FOR ENGINEERING GEOLOGIC MAPPING

Recommendations and guidelines for geologic and geotechnical mapping, logging, and documentation of rock type, structural features, and rock discontinuities in surface and underground excavations for civil engineering projects have been discussed and presented at conferences, in technical manuals, papers, and books for at least the last 65 years. Several of the most significant contributions to this literature are discussed below.

Terzaghi (1946), Robb (1951), Cooper (1968), and Jack (1969) provided some of the first recommendations regarding geologic techniques for mapping, evaluation, and documentation of tunnels and underground structures. During the 1970s through the 1980s the U.S. Army Corps of Engineers (1970 and 1975), Proctor (1971), Houghton (1977), McCrae et.al. (1979), McFeat-Smith (1982), Hatheway (1982), and Klassen et.al. (1987), further developed engineering geologic mapping and geologic hazard evaluation procedures and guidance documents specifically targeted for application in trench, tunnel, shaft, and underground civil construction projects.

In the late 1990s, the U.S. Bureau of Reclamation (1997 and 1998) developed an underground mapping guidance document and an engineering geology field manual for use by engineering and geologic staff. This guidance document and field manual are used and referenced by most applied engineering geology practitioners today. In 2002, the American Standards for Testing and Materials developed a guide (ASTM: D 4879-02, Reapproved 2006) for geotechnical mapping of large underground structures in rock that is a reference for applied engineering geologist working on large underground civil projects.

While the recommendations, procedures, and guidance contained in these references are significant and appropriate for the authors' intended purposes, they were applied and adapted to the Lake Mead Intake No. 3 Project based upon specific criteria including, the geologic environment, technical objectives, desired level of geologic/geotechnical documentation, budgetary constraints, and construction means and methods. Recommendations by McFeat-Smith (1982), Hatheway (1982), U.S. Bureau of Reclamation (1997 and 1998), and ASTM: D 4879-02 (Reapproved 2006) were used on this

**Figure 1. Photo of Southern Nevada Water Authority Contract No. 5 connector tunnel**

**Table 1. Summary of SNWA Lake Mead Intake No. 3 Contracts**

| Contract | Shaft | Tunnel |
|---|---|---|
| No. 1 | 600-foot deep access | 3-mile long by 22-foot diameter bored |
| No. 2 | 450-foot deep surge | 2,700-foot long by 14-foot wide by 16-foot high drill and blast connector |
| No. 5 | 360-foot deep gate | 560-foot long by 14-foot wide by 16-foot high drill and blast connector |

project because they provided more recent information for shaft and tunnel mapping techniques. At a minimum, the level of detail of the engineering geologic mapping is dependent upon the features being mapped and the proposed use of the data. In general, sufficient engineering geologic data and photographic documentation should be obtained to potentially facilitate further interpretation of the observed ground conditions, provide a comparison to the GBRs, provide a permanent record of actual geologic/geotechnical conditions encountered during construction, and provide a record of construction activities. The GBRs provided baselines for the orientation of joints and the number, orientation and thickness of faults.

Since 2000, three-dimensional ground-based laser mapping and stereo-photographic mapping of cut slopes, mines, quarries, and tunnels has become more common. The quality and quantity of engineering geologic data and evaluation that can be produced by these two mapping methods is impressive. However, these mapping methods require an up-front cost that many engineering geologic practitioners, consulting engineering firms, and clients typically have difficulty accepting for single projects. The authors expect these new methods will be used on more projects in the future if equipment and data processing costs are reduced. The photographic documentation methods presented in this paper are an important part of future project documentation.

## PROJECT GEOLOGIC MAPPING AND DOCUMENTATION

The main objective of the geologic mapping and field documentation of the Contract No. 2 Surface Excavation and Surge Shaft and the Contract No. 5 Connector Tunnel was to compare the encountered ground conditions to the anticipated geologic and geotechnical conditions described in the GBRs. The Contract No. 2 GBR stated that final ground support of the surface excavation slopes be directed by the owner based upon the geologic mapping. Final ground support for the Contract No. 2 and No. 5 shafts and tunnels would consist of cast-in-place concrete lining and patterned rock bolts and shotcrete, respectively. Initial ground support for both contracts was designed and installed by the contractor. The rock mass classification that was utilized for the geologic mapping was based upon the Geologic Strength Index values developed by Marinos, Marinos, and Hoek (2005).

The geologic and photo-geologic maps for this project contain general statements about rock type, a general evaluation of rock quality, and the location and inclination of faults, joints and igneous intrusions which typically extend more than 20 feet. The detailed variation in rock structure from massive, to gneiss to schist and short discontinuities are not mapped because these changes rarely affect the stability of the underground openings or surface excavations in this strong and relatively massive bedrock. The detail of geologic mapping performed during the Contract No. 2 and No. 5 projects has been commensurate with that performed during two existing intake tunnels also constructed in the Lower Plate metamorphic rock of Saddle Island with the exception that photo-geologic maps of surface excavation and shafts have been prepared during the recent construction work.

Project size and Contractor's means and methods do affect the quality of geologic mapping. Restricted and noisy environments where potential hazards such as loose or falling rock, uneven and wet ground, and heavy equipment traffic do have an impact on the detail of geologic mapping that is performed. In addition, if tunnel excavation is conducted on a two or three shift per day schedule and final shotcrete or cast-in-place concrete is placed soon after excavation, it may be difficult to complete geologic mapping.

## Contract No. 2 Surface Excavation Geologic Mapping

Contract No.2 required a hill-side surface excavation from which the surge shaft is being excavated. This Surface Excavation included an upper 26-foot high 1:1 rock slope through overburden and weathered rock, a 15-foot wide 5:1 bench and drainage ditch, and a 43-foot high 1:4 rock slope in slightly weathered to fresh rock. Rock bolting of the 1:4 rock slope was directed by the Engineer based upon a field evaluation of rock discontinuity orientations mapped along the slope face and the potential that these discontinuities may define rock blocks which loosen after excavation. Ground support of the 1:4 rock slopes will also include partial shotcrete. Ground support recommendations were presented to the Construction Manager on these photographs.

Photo-geologic mapping for Contract No. 2 Surface Excavation was easier than the underground geologic mapping for the Contract No. 5 Connector Tunnel because of the setting and the construction method. The excavation was performed only on day shift and in vertical stages with the work area changing frequently. Placement of shotcrete was deferred until the completion of the excavation. Measurements of faults and joint discontinuities were recorded on photo-geologic maps. Figure 2 is one of the photo-geologic maps of the north facing 1:4 rock slope on the south side of the Surface Excavation. This map is a mosaic of several photographs and shows the traces and orientations of some prominent faults (solid lines) and joint discontinuities mapped along the slope; the traces of quartz/feldspar veins (dotted lines) along the slopes; the locations of rock bolts (double triangle) installed in the slope; the top and bottom of the cut slope (dashed lines); the approximate toe-ditch stationing and elevations, and the estimated Geotechnical Strength Index (GSI) values of the rock. The GSI values shown on this map are based upon visual observations of the rock utilizing Marinos, Marinos and Hoek's (2005) methodology for estimating GSI values in jointed rock. Boundaries for these GSI values were not drawn on the photo-geologic maps as the rock characteristics change over a given area; rather these GSI values are general ranges for observed rock strength across the rock slope. This mapping methodology provides a convenient tool for evaluating locations for installing rock reinforcement (rock bolts and shotcrete) and a record of the rock conditions encountered during excavation of the rock slope prior to installation of shotcrete. Table 2 provides explanations for symbols and data that are included on the photo-geologic maps.

## Contract No. 2 Surge Shaft Geologic Mapping

Geologic mapping of the Contract No. 2 Surge Shaft excavation posed more difficulties than mapping the surface excavation because it was difficult to schedule mapping and photography. The ideal time to perform mapping is after completion of the blast-hole drilling and before the blast holes are loaded. Generally, the geologic mapping was performed during the loading of the explosives and the loading operations limited the engineering geologist's access to make measurements.

After each shaft round blast the rock is mucked out and the engineering geologist inspects the bottom of the shaft taking overlapping photographs of the lower shaft wall, sketches the geologic conditions, measures strike and dip of major discontinuities and the width of faults, and evaluates rock quality. In the office the photographs are used to make a circumferential photo-mosaic of the shaft wall and the field measurements are entered as notes on the photo-mosaic. Figure 3 is a photo-geologic map of the Surge Shaft from elevation 1,217 feet to 1,226 feet above mean sea level (msl). Figure 3 is a mosaic of seven photographs (limits shown) showing the locations and strike and dip measurements of faults, joint, and foliation discontinuities, the traces of faults (red solid line), rock contacts (white dashed line), and quartz/feldspar veins (white dotted lines) around the shaft, the approximate azimuth and distance around the shaft, and the estimated rock GSI values. This photo-geologic mapping provides a

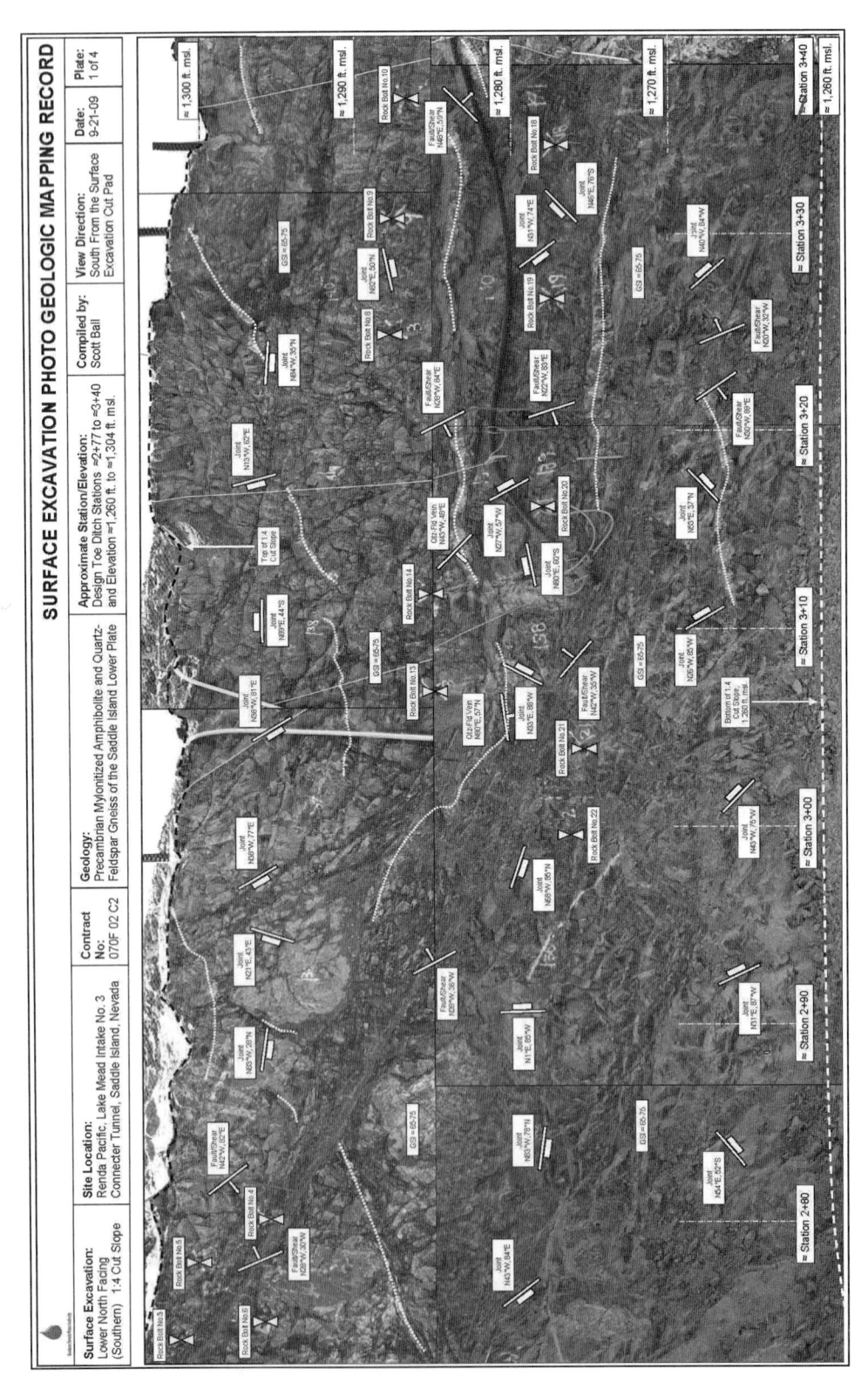

Figure 2. Photo-geologic map of 1:4 rock slope for contract no. 2 surface excavation

Table 2. Explanations for symbols and notes shown on photo-geologic maps

**Qualitative Estimation of Rock-Mass Properties:** Numerical values shown on the photo-geologic maps are based upon Geological Strength Index (GSI) values as presented by Marinos, P. and Hoek, E. (2000, *GSI: a geologically friendly tool for rock mass strength estimation*, Proceedings GeoEng 2000 Conference), and the application and limitation of these GSI values as discussed by Marinos, V., Marinos, P and Hoek, E. (2005, *The geological strength index: applications and limitations*, Bulletin of Engineering Geology and the Environment V 64: Pg. 55-65).

**Notes:** Photo-geologic maps were created with a mosaic of photographs. Maps show strikes and dips of joints, faults, quartz/feldspar veins and foliation. Geologic symbols shown on the photo-geologic maps are based upon standard symbols used in geologic investigations: ⊥⊥ , bedding; ⊏⊐ , joint; ▽ , foliation; and ⊥⊥ , fault. Solid lines (———) represent faults, dashed lines ( ⌐ _ _ · ) represent a contact or limit of a feature, and a dotted line ( ·········· ) represents a quartz/feldspar vein. Rock bolts are represented by a double triangle symbol (▷◁).

detailed record of the subsurface rock conditions encountered during shaft excavation.

## Contract No. 5 Connector Tunnel Geologic Mapping

Hand drawn geologic mapping logs were prepared during construction of the Contract No. 5 Connector Tunnel. Since the tunnel was a modified horse-shoe shape, the logs show the walls and crown of the tunnel from invert-to-invert in a typical tunnel mapping format. These logs include data for contract number, tunnel heading, excavation progress, tunnel bearing, stationing, discontinuity information (i.e., type, orientation, aperture, roughness, filling, etc.), groundwater inflow, lithology, crown overbreak, and rock reinforcement. Figure 4 shows a geologic map of the north heading from Stations 27+66 to 27+96. This map shows two faults, a mafic dike, and a zone of quartz/feldspar dikes and sills encountered in the north heading that cross the tunnel alignment. Geologic mapping was typically performed while the tunnel crews were drilling or placing ground support. Photo-geologic maps were not prepared for the tunnel excavation.

## GUIDANCE FOR UNDERGROUND PHOTOGRAPHY

The authors have found that the literature on photography in surface or underground excavations is limited. Susan Bednarz (2008) published an article titled *Underground Construction and Inspection Photography* in the December issue of Tunneling Magazine and a paper on the same issue at the 2008 North American Tunneling Conference. These papers recommend several techniques for successful photography of underground excavations and tunnels. The recommendations include uniformly illuminating the subject with lights and/or an external camera flash to reduce shadows, using wide angle lenses to capture broader images of small areas, using a mono-pod or standard tripod, bouncing the flash off the ceiling or wall of the excavation or tunnel for better illumination, using a telephoto lens for dramatic perspectives, using water resistant equipment, and using a digital single lens reflex (SLR) camera. The primary author has used both a digital point-and-shoot camera with only a built-in flash and the equipment recommended by Ms. Bednarz and concluded that point-and-shoot cameras rarely provide adequate photographs. A digital SLR camera (including lens) and a high quality external flash, although more expensive than a point-and-shoot camera and more difficult to carry underground, provide dramatically better images.

Additional recommendations for photography in surface and underground excavations include:

- Use the scene to provide scale for the image or carry a painted range pole to lean against the excavation.
- Take photographs of the subject from several different camera locations and angles, use different lighting effects, and use the photos with the best perspective and shading and which best represents the site.
- Always take extra photographs to increase the coverage of the documentation.
- Reduce snow-like reflection spots in the image caused by dripping, running, or spraying water by using a shutter speed greater than 1/125 of a second, an external flash held away from the camera, and pointing the flash at the top of the excavation or tunnel.
- Keep the camera and flash covered until you are ready to take a photograph.
- Be prepared to clean the camera equipment while in the field and completely clean the equipment when you return to the surface.

It is recommended that photographs be taken using the camera raw image file format (RAW). The

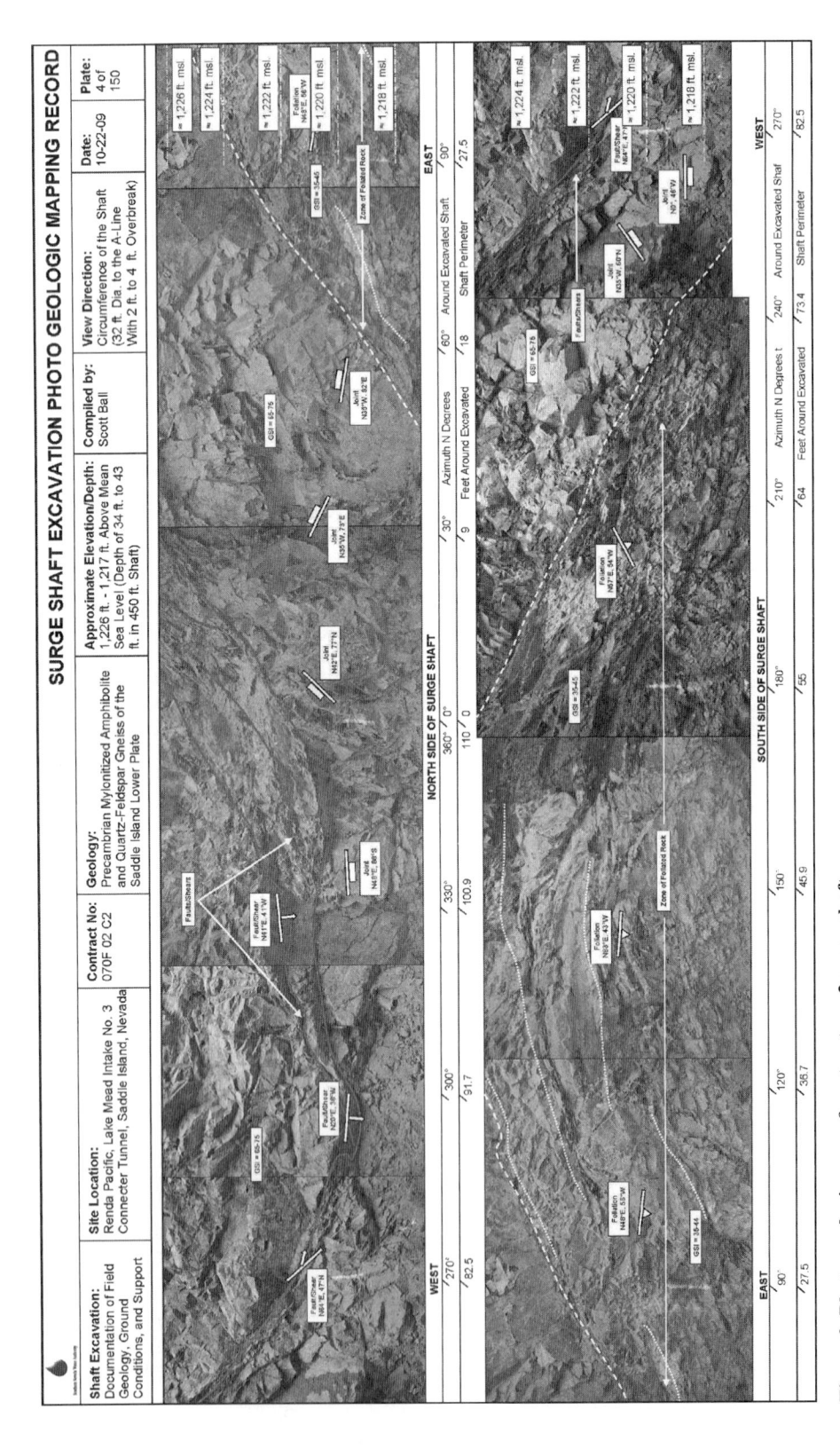

Figure 3. Photo-geologic map of contract no. 2 surge shaft

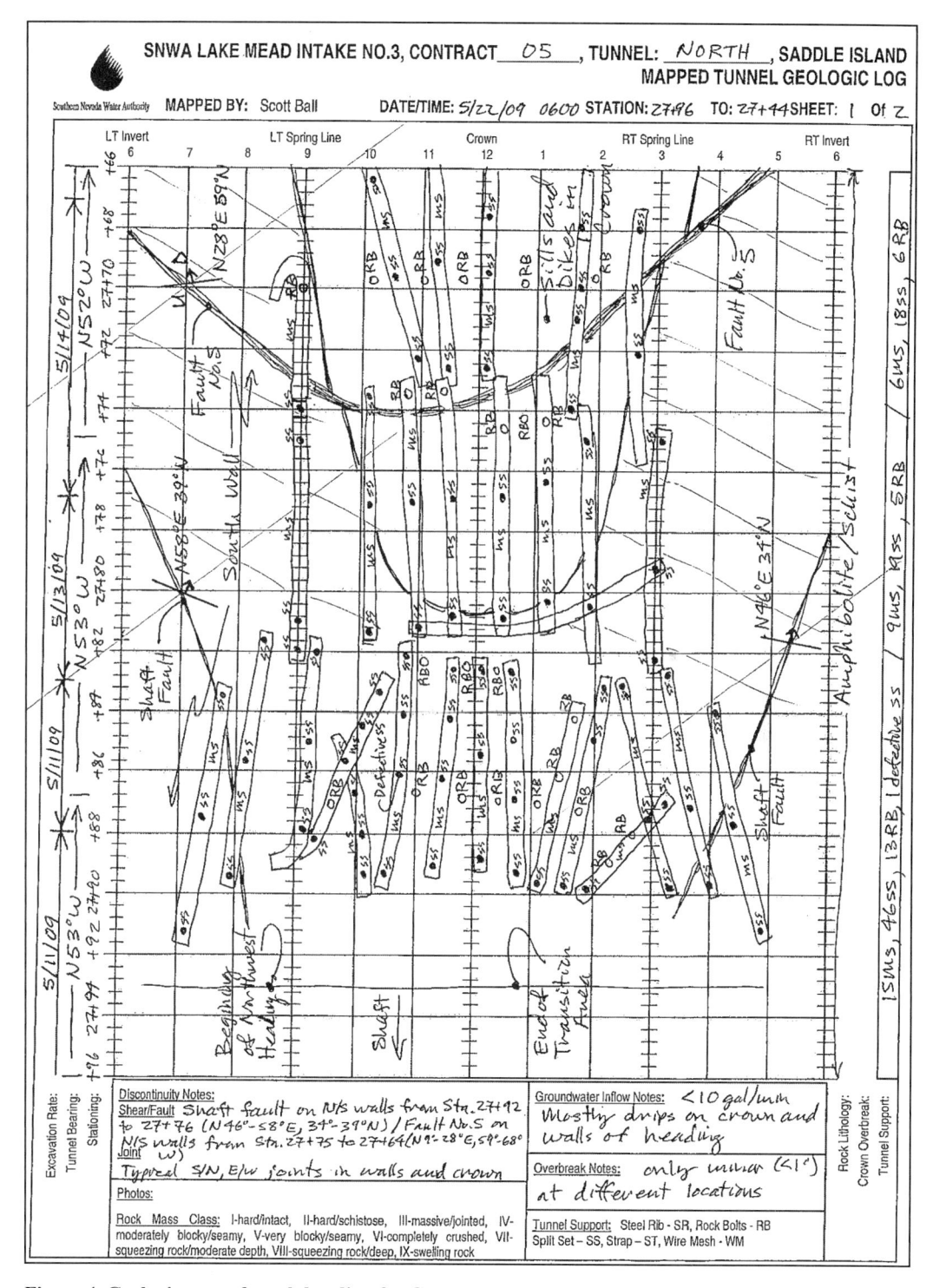

Figure 4. Geologic map of north heading for Contract No. 5 connector tunnel

RAW file format contains minimally processed data from the camera image sensor and thus contains more pixel and color data. This image format is useful in low light conditions and for documenting surface texture detail. Software provided with the SLR camera, and Adobe® Photoshop Elements was used for cataloging and processing the RAW format photographs. These programs allow the user to convert RAW format photographs into a "positive" file format such as TIFF or JPEG for storage, printing, or further manipulation.

## PROJECT PHOTOGRAPHIC RECORDS

Photographic records of surface or underground civil infrastructure projects can provide a record of ground conditions, construction operations and the progress of the work. Photographs of construction activities during the Contract No. 2 and No. 5 projects are presented in a Microsoft Word® template that includes a title block, spaces for two landscape images per page, image descriptions, image location and shooting direction, and image data.

The Contract No. 2 Surface Excavation was inspected after each 10-foot excavation lift to take photographs, make geologic observations and recommendations for ground support. Approximately three field hours were required to obtain the geologic mapping and final slope photographs shown in this paper. An additional four hours of office time were required to format the photographs, add geologic notes and prepare the reports. The Contract No. 2 Surge Shaft and Contract No. 5 Connector Tunnel were inspected approximately every one-and-a-half to three days to take photographs requiring between two to three hours of field time per photographic session including the time to access the work areas. Office time to format and annotate the photographs and prepare the reports was about four to six hours per photographic session.

### Contract No. 2 Surface Excavation Photographic Records

Figures 5 and 6 show photographic records of construction operations in the Contract No. 2 Surface Excavation. Photographs EW119 and EW120 document the direction and inclination of drilling for rock bolts on the upper portion of the 1:4 rock slope. Images from two perspectives of the same activity provide better spatial reference. Photographs EW195 and EW196 document the fractured nature of quartz/feldspar veins and a set of three main joint discontinuities (out-of, into, and across slope) exposed in the 1:4 rock slope. Perspective for these geologic documentation photos was important as the light, shadows, and contrast for these two features changed based on orientation. These surface excavation photographs provide a record of permanent rock slope support and ground conditions.

### Contract No. 2 Surge Shaft Photographic Records

Figures 7 and 8 show photographic records for the Contract No. 2 Surge Shaft excavation that provide documentation of blasting and the ground conditions encountered in the shaft. Photographs S17 and S18 document the ground conditions at the top of the surge shaft prior to the next blast, a portion of the blasting pattern for the next shot, placement of blast mats, and the results of the blast. Photographs S53 and S54 document the faults, the foliation in the rock, and the placement of the steel "curb ring" form in preparation for placement of the upper part of the surge shaft lining. These surge shaft excavation photographs provide a record of ground conditions, and initial rock support that can be compared to the contractor's excavation work plan.

### Contract No. 5 Connector Tunnel Photographic Records

Figures 9 and 10 show photographic records for the Contract No. 5 Connector Tunnel that provide documentation of the engineering geologic conditions encountered in the tunnel, installation of initial tunnel support, and the construction of final lining within the underground excavation. Photographs 98 and 99 document two faults that were encountered near the south end of the north tunnel heading just beyond the transition excavation below the shaft. The location and character of these faults was documented in these photographs (and on tunnel geologic logs, see Figure 1) to provide a clear record for future inspection or adjacent underground construction efforts as four inches of shotcrete was placed over these features. Photograph 126 documents drillers installing initial tunnel support within the north heading. Photograph 127 documents installation of steel rebar within the transition excavation below the shaft. This record documents the equipment and methods used by the contractor to install initial rock support and construction of the final lining in the transition area.

The Contract 2 Surge Shaft and the Contract 5 underground photography required about two hours of field time per photographic session including the time to access the work areas. Office time to format and annotate the photographs and prepare the reports was about two hours per photographic session.

## CONCLUSIONS

Engineering geologic mapping and photographic documentation can be an important part of construction records for underground civil infrastructure projects. Annotated photographs can be a permanent

| | SURFACE EXCAVATION GEOLOGIC MAPPING PHOTOGRAPHIC RECORD | |
|---|---|---|
| **Rock Bolts and Surface Excavation Support:** Cut Slope Earthwork | **Site Location:** Lake Mead Intake No. 3 Connecter Tunnel, Saddle Island, Nevada | **Contract No.:** 070F 02 C2 |

**Photo No. EW119**

**Location of Photo:** Renda Pacific Intake No. 3 Connector Tunnel Site

**View Direction:** Southwest

**Date of Photograph** September 1, 2009

**Description:** Photo shows drill jumbo drilling rock bolt holes near Stations 2+70 and 2+80 between the north facing 1:4 cut slope (right side photo) and the west facing 1:4 cut slope (photo background).

**Photo No. EW120**

**Location of Photo:** Renda Pacific Intake No. 3 Connector Tunnel Site

**View Direction:** South

**Date of Photograph** September 1, 2009

**Description:** Photo shows drill jumbo drilling rock bolt hole 1 near Station 2+70 on the west facing 1:4 cut slope. Rock bolt hole 7 was drilled below bolt hole 1. All drill holes are inclined 15° downward from horizontal.

Figure 5. Photographic record of Contract No. 2 surface excavation

| | SURFACE EXCAVATION GEOLOGIC MAPPING PHOTOGRAPHIC RECORD ||||
|---|---|---|---|
| **Rock Bolts and Surface Excavation Support:** Cut Slope Earthwork | **Site Location:** Lake Mead Intake No. 3 Connecter Tunnel, Saddle Island, Nevada || **Contract No.:** 070F 02 C2 |

**Photo No. EW195**

**Location of Photo:** Renda Pacific Intake No. 3 Connector Tunnel Site

**View Direction:** Southwest

**Date of Photograph** October 14, 2009

**Description:** Photo shows a highly fractured 14 to 18 in. thick quartz/feldspar vein exposed in the north facing (south) 1:4 cut slope. The upper surface of the vein dips 10° to 30° out of slope while the lower surface dips 5° to 15° into the slope.

**Photo No. EW196**

**Location of Photo:** Renda Pacific Intake No. 3 Connector Tunnel Site

**View Direction:** Southwest

**Date of Photograph** October 14, 2009

**Description:** Photo shows typical small (6 in.) to large (2 to 4 ft.) wedges/blocks of rock exposed in the north facing (south) 1:4 cut slope. These wedges/blocks of rock are created by three typical joint sets identified in the GBR.

Figure 6. Photographic record of Contract No. 2 surface excavation

| | SHAFT EXCAVATION GEOLOGIC MAPPING PHOTOGRAPHIC RECORD | |
|---|---|---|
| **Shaft Elevation:** 1,260 – 1,244 ft. mean sea level | **Site Location:** Lake Mead Intake No. 3 Connecter Tunnel, Saddle Island, Nevada | **Contract No.:** 070F 02 C2 |

**Photo No. S17**

**Location of Photo:** Renda Pacific Intake No. 3 Surge Shaft

**View Direction:** Northeast

**Date of Photograph** October 12, 2009

**Description:** Photo shows Sanders workers completing placement of the blast mats over the shaft area prior to setting off the charges.

**Photo No. S18**

**Location of Photo:** Renda Pacific Intake No. 3 Surge Shaft

**View Direction:** East

**Date of Photograph** October 12, 2009

**Description:** Photo shows the resultant blast of the second charge for excavation of the surge shaft from approximate elevation 1,252 ft. msl. to 1,244 ft. msl.

Figure 7. Photographic record of Contract No. 2 surge shaft

| SHAFT EXCAVATION GEOLOGIC MAPPING PHOTOGRAPHIC RECORD | | |
|---|---|---|
| **Shaft Elevation:** 1,218 – 1,208 ft. mean sea level | **Site Location:** Lake Mead Intake No. 3 Connecter Tunnel, Saddle Island, Nevada | **Contract No.:** 070F 02 C2 |

**Photo No. S53**

**Location of Photo:** Renda Pacific Intake No. 3 Surge Shaft

**View Direction:** South

**Date of Photograph** October 27, 2009

**Description:** Photo shows the rock from approximately 1,218 ft. to 1,208 ft. msl. in the surge shaft, the foliated zone in the rock, and the collar form (curb ring) that will be used as the base for the upper 49 ft. of the surge shaft lining.

**Photo No. S54**

**Location of Photo:** Renda Pacific Intake No. 3 Surge Shaft

**View Direction:** West

**Date of Photograph** October 27, 2009

**Description:** Photo shows the rock from approximately 1,218 ft. to 1,208 ft. msl. in the surge shaft, the fault and foliated zone in the rock, and the collar form (curb ring) that will be used as the base for the upper 49 ft. of the surge shaft lining.

Figure 8. Photographic record of Contract No. 2 surge shaft

| NORTH TUNNEL GEOLOGIC MAPPING PHOTOGRAPHIC RECORD |||
|---|---|---|
| Tunnel Heading and Stationing: North Heading / 27+50 – 27+90 | Site Location: Lake Mead Intake No. 2 Connection and Modifications, Saddle Island, Nevada | Contract No.: 070F05C1 |

**Photo No. 98**

Location of Photo: Southwest side of North Heading

Station: 27+70 to 27+90

View Direction: South

Date of Photograph: May 22, 2009

Description: Photo shows the two northeast trending, northwest dipping faults encountered near the south end of the North Heading and just beyond the north end of the Transition. Some minor iron staining can be observed adjacent to the faults and on joint surfaces.

**Photo No. 99**

Location of Photo: Southwest side of North Heading

Station: 27+70

View Direction: Southwest

Date of Photograph: May 22, 2009

Description: Photo shows a close-up of the northeast trending (N9° to 28°E), steeply dipping (68° to 89°) northern fault/shear shown in Photo No. 98. The fault is the three inch thick dark material to the right of the white paint line and is composed of clay and fractured rock

Figure 9. Photographic record of Contract No. 5 north connector tunnel

# SOUTH TUNNEL GEOLOGIC MAPPING
# PHOTOGRAPHIC RECORD

| Tunnel Heading and Stationing: Transition and South Heading / 28+00 – 30+22 | Site Location: Lake Mead Intake No. 2 Connection and Modifications, Saddle Island, Nevada | Contract No.: 070F05C1 |
|---|---|---|

**Photo No. 126**

**Location of Photo:**
Face of South Heading

**Station:**
30+22

**View Direction:**
South

**Date of Photograph**
June 9, 2009

**Description:**
Photo shows drillers installing initial ground support, including 6-foot long split sets and 10-foot long mine straps, in the crown of the south heading. Half of the muck pile has been removed and drilling of the blast holes for the next blast will be performed after muck has been removed.

**Photo No. 127**

**Location of Photo:**
Transition at base of shaft between headings

**Station:**
28+00 – 29+00

**View Direction:**
Southeast

**Date of Photograph**
July 23, 2009

**Description:**
Photo shows rebar being set and tied on top of the rat slab for the concrete and steel gate structure inside of the transition below the shaft.

Figure 10. Photographic record of Contract No. 5 south connector tunnel

record of geology, geotechnical conditions and construction operations. This documentation can be of great future value in the evaluation of construction claims, planning future inspections or facility repairs, or future projects. The scope and details of the project geologic mapping and photography will vary with geologic environment, ground behavior, and the level of geologic/geotechnical documentation required by the Owner, available funds, and the Contractor's means and methods. The approach to geologic mapping and photography may be changed during construction to better meet the exposed ground conditions and the issues identified by the parties. In order to obtain detailed photographic documentation a digital SLR camera is recommended.

Many factors can impact the level of documentation possible on any given project; however the construction means and methods employed by a contractor typically have the greatest impact on engineering geologic mapping during underground civil infrastructure projects.

**REFERENCES**

American Standards for Testing and Materials, *Standard Guide for Geotechnical Mapping of Large Underground Openings in Rock*, Designation: D 4879-02 (Reapproved 2006).

Bednarz, Susan L., 2008, *Underground Construction and Inspection Photography*, Tunneling Magazine, December 2008 Feature Story.

———. 2008, *Underground Construction Photography: Documenting a Success Story*, 2008 North American Tunneling Conference, pp. 101–107.

Cooper K.R., 1968, *a Rapid, Accurate Method of Mapping Tunnels of Circular Cross Section*, Bulletin of the Association of Engineering Geologists, Vol. V, No. 2, pp. 63–68.

Hatheway, Allen W., 1982, *Trench, Shaft, and Tunnel Mapping*, Bulletin of the Association of Engineering Geologists, Vol. XIX, No. 2, pp. 173–180.

Houghton, D.A., 1977, *Application of Engineering Geological Mapping in Hazard Evaluation for Mining and Other Underground Structures*, Bulletin of the International Association of Engineering Geology, Vol. 19, pp. 166–175.

Jack, H.A., 1969, *Discussion: Tunnel Mapping Method*, Bulletin of the Association of Engineering Geologists, Vol. VI, No. 2, pp. 151–156.

Klassen et al., 1987, *Engineering Geological Mapping*, 1987 RETC Proceedings, Vol. 2, pp. 1311–1323.

Marinos, V., Marinos, P. and Hoek, E., 2005, *The geological strength index: applications and limitations*, Bulletin of Engineering and the Environment, 64, pp. 55–65.

McCrae, R.W., Powell, D.B., and Y, H.T., 1979, *Engineering Geological Mapping of Large Caverns at Dinorwic Pumped Storage Scheme, North Wales*, Bulletin of the International Association of Engineering Geologists, Vol. 19, pp. 182–190.

McFeat-Smith, I., 1982, *Logging Tunnel Geology*, Tunnels and Tunneling, May, pp. 20–25.

Proctor, R.J., 1971, *Mapping Geological Conditions in Tunnels*, Bulletin of the Association of Engineering Geologists, Vol. VIII, No 1.

Robb, G.L., 1951, *Geologic Techniques in Civil Engineering*: P. 1120–1149 *in* Subsurface Geologic Methods, L.W. Leroy, ed., Colorado School of Mines, 1156 p.

Terzaghi, Karl, 1946, *Introduction to Tunnel Geology in* R.V. Proctor and T. L. White, Rock Tunneling with Steel Supports: The Commercial Shearing and Stamping Co., Youngstown, Ohio, 271 p.

U.S. Army Corps of Engineers, 1970, *Engineering and Design, Geologic Mapping of Tunnels and Shafts by the Full Periphery Method*, Office of the Chief of Engineers, Engineering Technical Letter No. 1110-1-37; Washington, DC.

———. 1975, *Geologic Mapping Procedures, Open Excavation*, Office of the Chief of Engineers, Engineering Technical Letter No. 1110-2-203; Washington, DC.

U.S. Bureau of Reclamation and U.S. Geological Survey, 1997, *Underground Geologic Mapping*, Yucca Mountain Project, Technical Procedure YMP-USGS-GP-32, R1, SCP No. 8.3.1.4.2.2;4 WBS 1232212, for work funded by the U.S. Department of Energy, Effective Data, July 14, 1997.

U.S. Department of the Interior, Bureau of Reclamation, 1998, *Engineering Geology Field Manual*, Second Edition, Vol. 1, Reprinted 2001.

# Tunnel-Induced Surface Settlement on the Brightwater Conveyance East Contract

## Michael A. Lach, Farid Sariosseiri
CDM, Bellevue, Washington

**ABSTRACT:** This study presents the results of an investigation on surface settlement due to construction of a 5.87-meter (m) (19.3-foot [ft.]) diameter bored tunnel project in King County, Washington. During the design phase of the project, a field investigation was conducted and basic classical theories, as well as finite element and analytical methods, were used to estimate the surface deformation with respect to overburden pressure and geologic setting. Surface deformations were monitored by a series of optical surveys as the earth pressure balance tunnel boring machine advanced. This study compares the settlements measured during construction to the various prediction models.

## INTRODUCTION

In response to rapidly increasing population in the Seattle area, King County is constructing the new Brightwater wastewater treatment facility and conveyance system. The Brightwater system includes the treatment plant, tunneled influent and effluent conveyance lines, connections to existing sewers, and a marine outfall. Upon its completion, this new facility will treat sanitary sewage from northern King and southern Snohomish counties, and transfer the treated wastewater to the Puget Sound.

The Brightwater system includes 20.4 kilometers (km) (12.7 miles [mi.]) of large-diameter tunnels constructed in four segments, and 1.4 km (0.87 mi.) of microtunnels designed and constructed in six segments. The microtunnels transfer wastewater from existing trunk sewers in the Swamp Creek Valley and North Creek Valley to the North Kenmore Portal and North Creek Portal, respectively. In order to minimize the impact on roads and structures along the conveyance alignment, tunnel boring machines (TBM) and micro-tunnel boring machines (MTBM) were used. The location of each component with respect to the county line is shown in Figure 1.

The east-west Brightwater tunnel alignment has been excavated through complex geologic conditions. The general geology of the area consists of glacial and inter-glacial deposits, which together comprise a series of aquifers and aquitards. Due to the complex geology, depth of the tunnels, and groundwater head along the tunnel alignment, the Brightwater tunnels are considered to be one of the most challenging projects in the United States (Newby et al. 2007).

The tunneling portions of the project were divided into three contracts: East, Central, and West. At the time of this writing, the East Contract is the only tunnel section that has been completed. This paper will cover that portion of the project only.

### East Contract Tunnel (ECT)

The main tunnel drive for the ECT is 4,285 m (14,050 ft.) long and was driven from west to east. The contract specified that the selection of a TBM and the design of the initial segmental lining system was the contractor's responsibility. To allow the contractor some flexibility in these decisions, a minimum inside diameter of 5.08 m (16.7 ft.) was specified with no limits placed on the outside diameter. The contractor elected to complete the main tunnel drive using a 5.87 m (19.3 ft.) diameter Lovat earth pressure balance TBM.

Two sections of the tunnel were shallow enough to warrant close scrutiny for potential settlements: the North Creek Valley, at the west end of the alignment, and the Little Bear Creek Valley at the east end. The depth of the tunnel (measured from ground surface to the springline) was as low as 15.5 m (51 ft.) in the North Creek Valley, and as little as 7.6 m (25 ft.) in the Little Bear Creek Valley.

## GEOLOGIC SETTING

The Quaternary geologic history of the Puget Sound area is dominated by a series of at least six continental

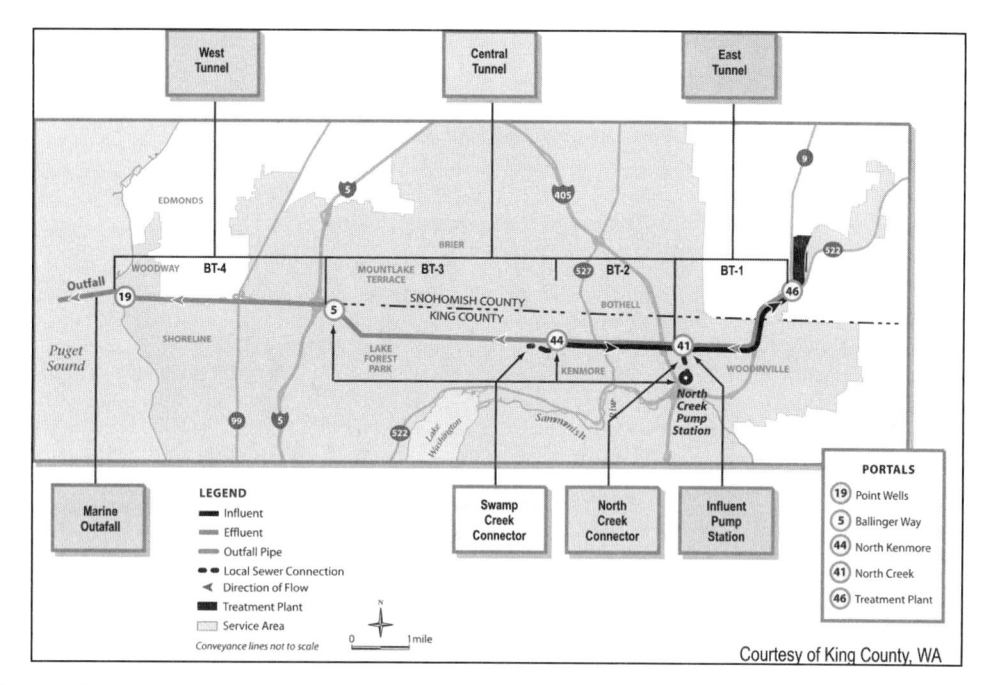

**Figure 1. Overview of Brightwater conveyance system**

glaciations. During these glacial periods, continental glaciers advanced southwards from Canada, covering many of the low-lying areas of northern North America in glacial ice. In the Puget Lowland, the most recent continental glacier was present as a lobe of ice that reached its maximum extent just south of Olympia, Washington, approximately 97 km (60 mi.) south of the Brightwater tunnel alignment. In the project area, the ice lobe exceeded 910 m (3,000 ft.) in thickness (Troost and Booth 2003). As a result of this ice loading, many deposits in the Puget Lowland have been highly over consolidated. Within the Little Bear Creek Valley, the ECT was constructed in these over consolidated soils.

In some locations, additional sediment has been deposited on top of the over consolidated materials since the most recent glacial period. This has occurred due to natural processes, such as alluvial deposition, landsliding, and lakebed deposition, as well as by human activities, such as filling and grading. Within the North Creek Valley, the ECT was constructed in normally consolidated alluvial soils. These soils consisted predominantly of sand, with varying amounts of gravel and silt. In addition, the tunnel alignment crossed a deep trough filled with highly compressible organic silt.

## ANALYSIS

During the pre-design phase of the project, empirical and semi-empirical methods were used to estimate the order of magnitude of the potential surface settlement in the shallow sections of the tunnel alignment. During final design, finite element studies were used to further refine these predictions. In general, the analyses were conducted for a wide range of anticipated ground loss percentages. This range was chosen to give the designers an appreciation of the importance of limiting ground loss through such measures as tail void grouting. It should be noted that in these analyses, ground loss percentage is defined as the volume of ground lost divided by the total volume of the excavation. In addition, only ground losses caused by the inward movement of the soil around the TBM and liner were considered. In other words, ground losses due to over or under pressurization of the face leading to movements in front of the machine were not considered.

Generally, it was assumed that in tunnel sections with a depth-to-diameter (Z/D) ratio of greater than five, surface settlement would be negligible. On the ECT alignment, all locations with a Z/D ratio of greater than five were to be mined in soils with relatively high strength and low compressibility characteristics. In these conditions, the team's lead engineers felt that the development of arching stresses in the soil above the tunnel crown would limit deformation. Simple empirical correlations, such as those cited by Leca et al. (2000), were used to validate this assumption.

The methods used to estimate settlement are described briefly in the following sections.

### Geometric Analysis

Cording and Hansmire (1975) have presented a method of estimating surface settlement based on geometric considerations. In this method, it is assumed that the volume of ground loss around the tunnel opening will equal the volume of the settlement trough at the surface. Their method includes an empirical relationship between the general soil type and the angle at which the settlement propagates from the tunnel opening to the surface, or β. The angle β is the only measure of soil behavior in this model. We assumed that the soils on this project would exhibit β-angles ranging from 33 degrees for stiff clays to 50 degrees for saturated sands.

### Closed-form Solution

Bobet (2001) presented a closed-form analytical solution for shallow tunnels in homogeneous ground. During the pre-design phase of the project, this method was primarily used to estimate loading on the final tunnel liner. One side benefit of this analysis is that it also includes a solution for ground surface settlement.

This model is more complex than the empirical method described above. Input parameters include the volume of ground loss, stiffness parameters for both the soil and the liner, density of the soil, and the at-rest earth pressure coefficient.

### Finite Element Method

During the final design process, the 2-dimensional load-deformation finite element analysis software package SIGMA/W (GeoSlope 2004) was used to further refine the estimates generated during pre-design.

Sands and gravels in the vicinity of the tunnel heading were modeled using a non-linear (hyperbolic) elastic model. The initial modulus and unload-reload modulus were assumed to be equal and were taken as the average from pressure meter tests carried out during the exploration phase of the project. The modulus value was also assumed to have a lower bound of 14.4 megapascal (300,000 pounds per square foot), equivalent to loose sand (Bardet 1997). Sand and gravel strata that were more than 6.1 m (20 ft.) above the tunnel crown were modeled as linear-elastic materials since strain levels at that distance from the tunnel opening would likely be insignificant. Clay soils were also modeled as linear-elastic materials. Peat deposits in the North Creek valley were modeled using the modified Cam-Clay method based on one-dimensional consolidation testing results.

## COMPARISON OF PREDICTIONS

The results of two of the three analysis methods were very compatible. However, the closed-form solution predicted smaller settlements, especially at larger ground loss percentages. For two critical stations (station 138+50 in the North Creek Valley and station 267+00 in the Little Bear Creek Valley), the predicted maximum settlements are plotted on Figure 2 as a function of ground loss percentage. The depths to the tunnel springline for these stations are 16 m (53 ft.) and 7.6 m (25 ft.), respectively, equating to Z/D ratios of 2.7 and 1.3.

Based on these predictions, the design team established specifications for the TBM. Most critically, tail void grouting was specified. Based on local tunneling experience, such as the Alki Tunnel (Webb and Breeds 1998), it was anticipated that with appropriate tail void grouting, ground loss could be limited to 2 percent or less. The tail void grouting specifications included both minimum and maximum volume limits to limit ground movement during mining.

## SETTLEMENT MONITORING PROGRAM

To verify that the specified mining and grouting procedures were effective in preventing excessive settlement and potential damage to surface structures, settlement was monitored using a series of instruments at the surface. Three action limits were set for each monitoring point; designated "trigger level," "level 2," and "maximum allowable." If the trigger level or level 2 are reached, monitoring frequency is increased, and planning begins for implementation of mitigation strategies. The contractor was made responsible for mitigation strategies, but they could include adjustment of operational parameters, slowing of the TBM advance, or adjustment of the tail grout composition or injection methodology.

In order to monitor the settlement, optical survey points were installed along the alignment, both on the center line and offset, typically 6 m (20 ft.) to each side. The monitoring points were generally installed at 30-m (100-ft.) intervals within the North Creek Valley and at 91-m (300-ft.) intervals in the Little Bear Creek Valley. Additional points were installed in areas where the alignment crossed major utilities, structures, or highways. The optical points were surveyed daily when the TBM cutterhead was within 30 m (100 ft.), and weekly afterwards.

### Monitoring Results

Within the North Creek Valley, 26 of the 44 installed settlement points exceeded the defined trigger level, and three exceeded the maximum allowable level. The location of the instruments that exceeded the defined trigger level within the North Creek Valley and the date at which maximum settlement occurred

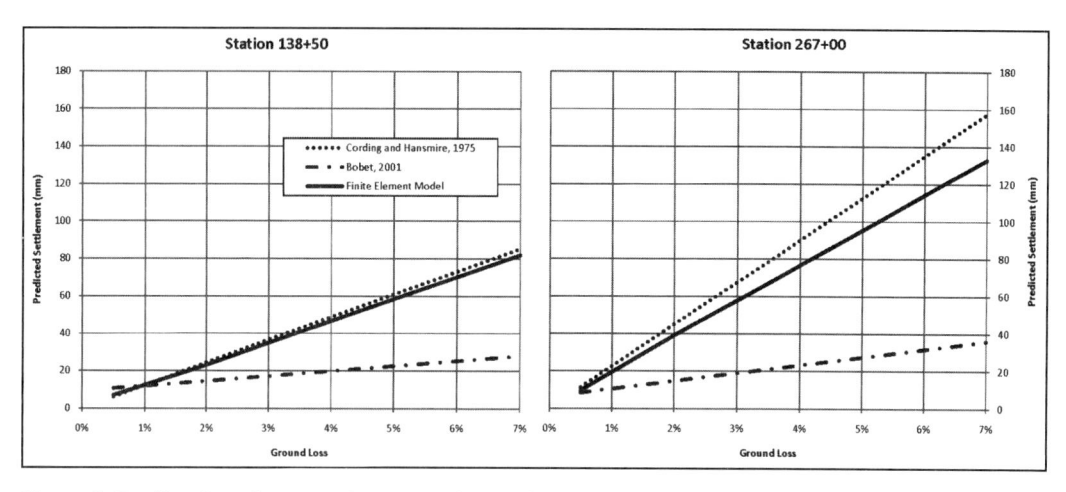

Figure 2. Predicted settlement values at stations 138+50 and 267+00

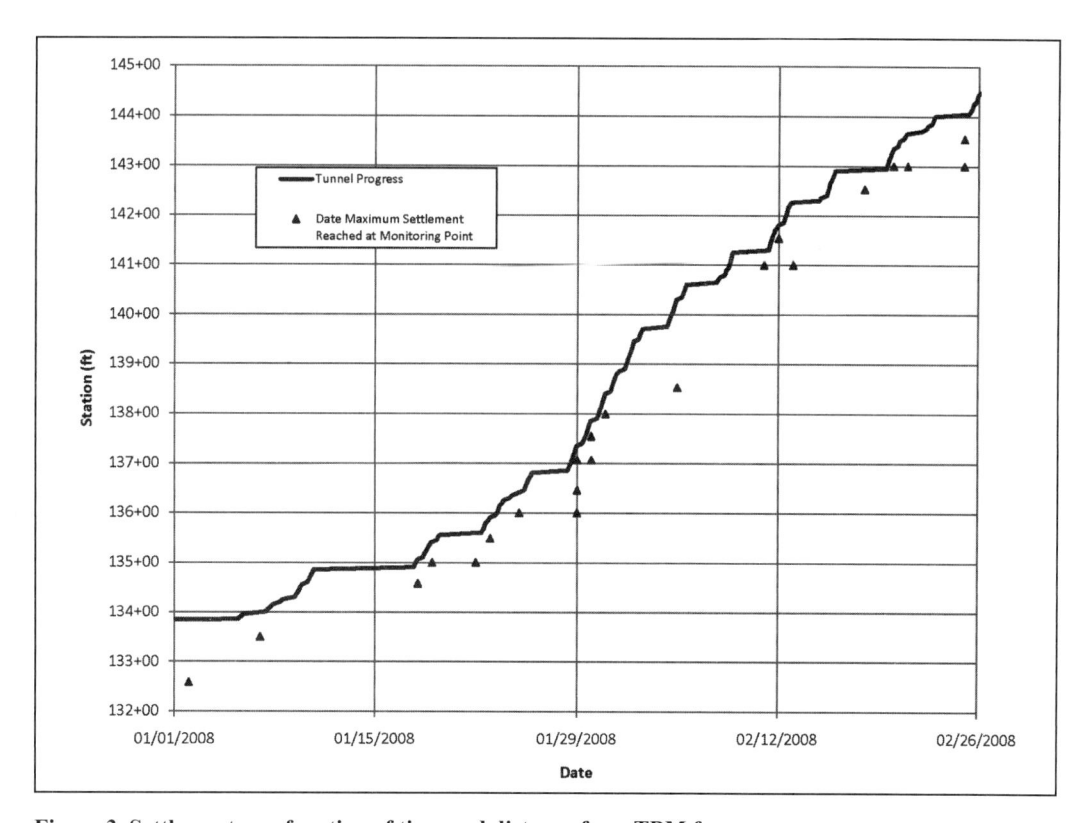

Figure 3. Settlement as a function of time and distance from TBM face

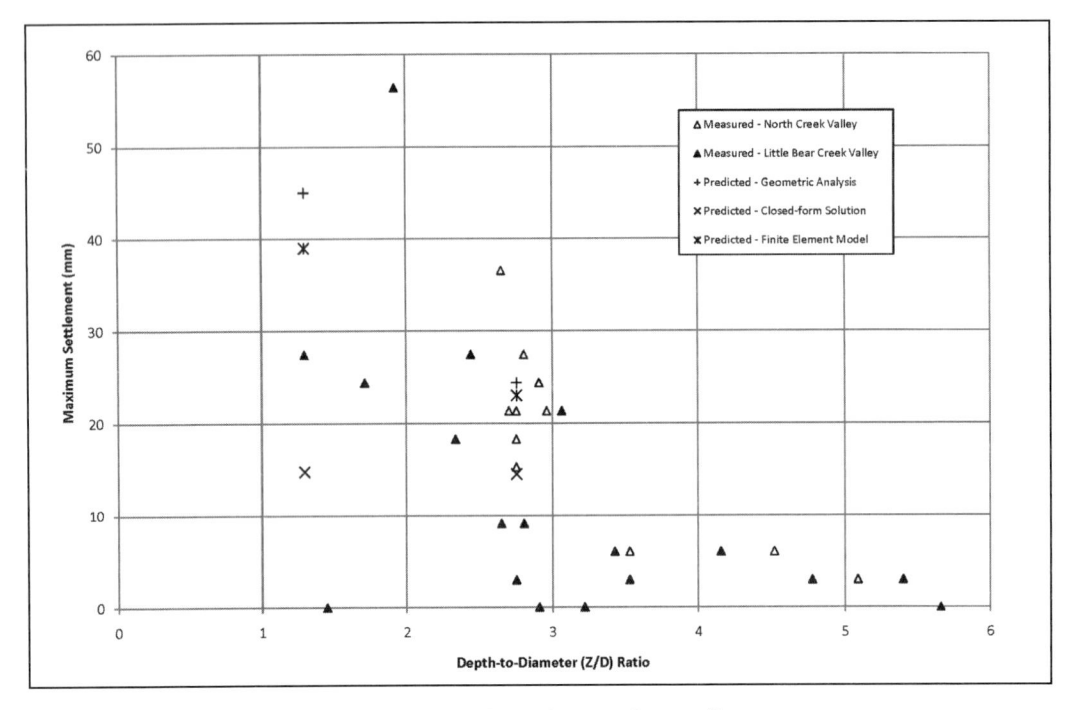

**Figure 4. Measured vs. predicted settlements above the tunnel centerline**

are superimposed over TBM progress, as shown in Figure 3. It can be observed that generally, the maximum settlement occurred after the TBM passed and the horizontal distance between the settlement point and theTBM cutterhead was between 7.5 and 30 m (25 and 100 ft.). The back of the tail shield was approximately 9.0 m (29.5 ft.) behind the cutterhead. Since the settlement did not generally continue after passage of the TBM, it can be inferred that the tail void grouting was very effective in preventing further movement of the soils around the tunnel. Although the database for the Little Bear Creek Valley is not as comprehensive, a similar pattern was observed for this section of the tunnel.

Maximum settlement versus Z/D ratio is plotted on Figure 4 for all monitoring points directly above the centerline of the tunnel. It can be seen clearly that settlement decreases with increase in overburden, as would be expected. For comparison, the predicted values of settlement at stations 138+50 and 267+00 (for an assumed ground loss of 2 percent) are also included in the figure. It can be seen that in general, the predictions closely matched the measured settlements.

It can also be seen in the figure that in general, for similar Z/D ratios, the settlements measured in the North Creek Valley were slightly higher than those measured in the Little Bear Creek Valley. The differences are minor, but not negligible. There are two factors that could be influencing this. First, the additional settlement could be as function of the difference in soil density due to glacial over consolidation. Second, the contractor mined through the Little Bear Creek Valley much later in the project and had the benefit of learning from earlier experiences on the project.

Finally, after a certain height of overburden, settlement is minimal. In this case, settlement was less than 10 millimeters (mm) (0.4 inches [in.]) for all Z/D ratios of greater than about 3.5, and less than 5 mm (0.2 in.) for Z/D ratios greater than about 4.5, confirming the initial assumption that settlement would be negligible at Z/D ratios of greater than 5. However, settlement was highly variable in the Little Bear Creek Valley for Z/D ratios of less than about 2, indicating that localized variability within the glacial geology has a greater effect on surface settlement in shallow sections.

In the locations where offset settlement points were also used, the data clearly indicated that the settlement took the shape of a trough as predicted by all models. Although the derivation is beyond the scope of this paper, it should be noted that the average β-angle was estimated to be 43 degrees through a back-calculation. This is slightly lower than the value of about 50 degrees that Cording and Hansmire (1975) correlated for saturated sands.

## CONCLUSION

In general, the specified mining and grouting procedures in conjunction with proper execution by the contractor resulted in tolerable settlements along the alignment. In addition, measured settlement closely matched various predictive models. This indicates that when good data, good engineering, and good judgment are applied, valid predictions can be made, even in the most challenging geologic settings. However, it should be noted that the closed-form solution tended to under-predict the settlement. This reinforces the notion that in complex problems, engineers should not rely solely on one analysis method.

## REFERENCES

Bardet, J.P. 1997. *Experimental Soil Mechanics*. Prentice Hall, Inc. Upper Saddle River, NJ.

Bobet, A. 2001. Analytical Solutions for Shallow Tunnels in Saturated Ground. *Journal of Engineering Mechanics.* 127(12):1258–1266.

Cording, E.J. and Hansmire, W.H. 1975. *Displacements Around Soft Ground Tunnels*. General Report of Session IV, 5th Panamerican Conference on Soil Mechanics and Foundation Engineering, Buenos Aires. 571–632.

GeoSlope. 2004. *SIGMA/W.* Software package. Version 5.

Leca, E., Leblais, Y., and Kuhnhenn, K. 2000. Underground Works in Soils and Soft Rock Tunneling. *International Conference on Geotechnical and Geological Engineering.* Volume 1. 220–268

Newby, J.E., Gilbert, M.B., and Maday, L.E. 2007. Brightwater Conveyance System Will Expand Seattle's Wastewater Treatment. *Tunneling and Underground Construction.* 1(2):31.

Troost, K.G. and Booth, D.B. 2003. *Puget Lowland Geologic Framework.* Training Course for King County Wastewater Treatment Division. Unpublished material.

Webb, R. and Breeds, C.D. 1998. Soft Ground EPBM Tunneling—West Seattle, Alki Tunnel. *Proceedings of the 1998 Conference on Underground Tunneling.* London.

# Strategic Tunnel Enhancement Programme, Abu Dhabi, UAE: An Overview of Geology and Anticipated Geotechnical Conditions Along the Deep Sewer Tunnel Alignment

**Mubarak Obaid Al Dhaheri**
Abu Dhabi Sewerage Services Company (ADSSC), Abu Dhabi, United Arab Emirates

**Mark Marshall, Panagiota Asprouda**
CH2M HILL Inc., Abu Dhabi, United Arab Emirates

**Suresh Parashar**
CH2M HILL Inc., Chantilly, Virginia

ABSTRACT: Abu Dhabi Sewerage Services Company has developed the Strategic Tunnel Enhancement Programme (STEP) to improve sewerage systems in Abu Dhabi. STEP will include a deep sewer tunnel, link sewers and an underground pumping station. The deep tunnel runs approximately 41 km from Abu Dhabi Island to the pumping station on the mainland. A preliminary geotechnical investigation was carried out along the tunnel alignment through urban areas, desert environment and a wetlands reserve. The investigation comprised 85 boreholes and over 6 km of geophysical surveys. This paper introduces aspects of STEP, describes the preliminary geotechnical investigation for the deep tunnel, outlines some factual geological information and summarizes encountered soil, rock and groundwater conditions.

## INTRODUCTION

Abu Dhabi is the capital of the Emirate of Abu Dhabi and the federal capital of the United Arab Emirates (UAE), a country of seven Emirates on the edge of the Arabian Gulf – see Figure 1. Abu Dhabi is the largest of the Emirates and is the federal seat of Government. Most of the city is on Abu Dhabi Island but it also has suburbs on the mainland to which it is linked by three bridges.

The city is experiencing steady growth, with residential and commercial construction sectors witnessing a major boom occurring on an aggressive timescale. Additionally, the population is expected to increase from the 2007 baseline figure of 930,000 to 1.3 million in 2013, 2 million in 2020 and to 3.1 million in 2030 (Abu Dhabi Urban Planning Council 2007). Given the projected growth, the need for an efficient wastewater management system is vital. Abu Dhabi Sewerage Services Company (ADSSC), as the service provider for sewerage services throughout the Emirate of Abu Dhabi, has developed a comprehensive plan for the wastewater management system. The Strategic Tunnel Enhancement Programme (STEP) forms a key part of the plan. CH2M HILL has been appointed by ADSSC as the STEP programme manager for the US$2.0 billion project.

CH2M HILL will manage the delivery of STEP using six design-build contracts (three deep tunnel contracts, two link sewer contracts and one pump station contract). CH2M HILL has undertaken preliminary design of STEP to 30 percent completion and has also produced the bid documents for each of the six contracts. Bid evaluation and construction supervision of the design-build contracts will be performed by CH2M HILL.

To characterise the anticipated geology and the groundwater regime for the tunnelling methods and to facilitate the preliminary design and preparation of bid documents, a preliminary geotechnical investigation was undertaken along the deep tunnel alignment, as described below.

### Project Details

Components of STEP include the deep sewer tunnel, several link sewers and a large underground pumping station. The existing sewerage network on Abu Dhabi Island consists of gravity lines with internal diameters ranging from 100 mm (4 in) to 2200 mm (7.2 ft) and pumping mains with internal diameters ranging from 100 mm (4 in) to 1200 mm (4 ft). The

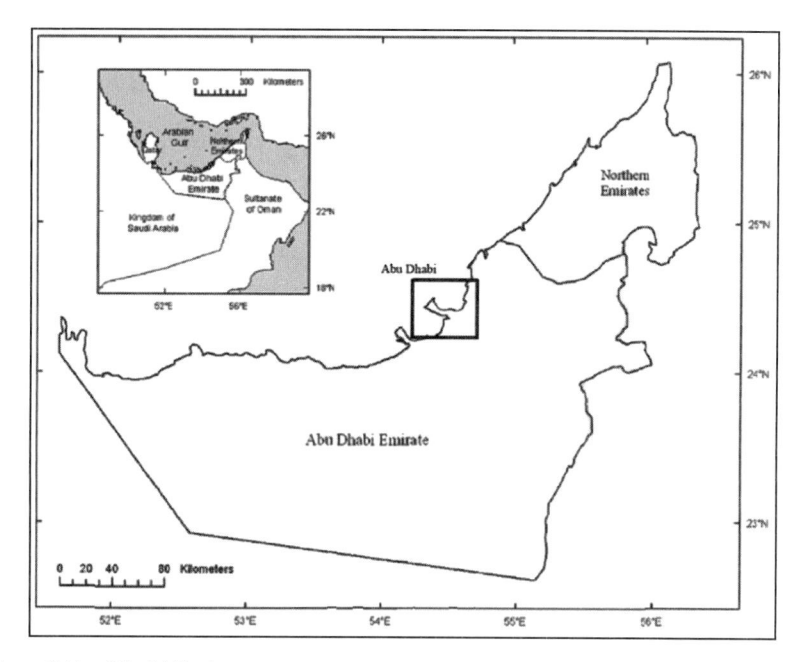

**Figure 1. Plan of Abu Dhabi Emirate**

network also includes 50 pumping stations currently in service. Wastewater is collected at several main pumping stations and pressure main systems and pumped to Mafraq Wastewater Treatment Plant (MWTP) located on Abu Dhabi mainland. When implemented STEP will intercept flows from these existing gravity sewers at the existing pump station influent lines and convey them by gravity to a new pumping station at Al Wathba. The pumping station will lift the sewage to the Al Wathba Independent Sewage Treatment Plants (ISTPs). The sewage will then be processed and reclaimed as treated effluent for irrigation purposes. STEP will also allow ADSSC to decommission 35 of the existing pumping stations, which will dramatically reduce operating and maintenance costs, as well as releasing valuable real estate for other development purposes.

The deep sewer tunnel alignment (shown in Figure 2) extends over 41 km (25 miles) from an existing pumping station at WS1 on Abu Dhabi Island, to Al Wathba on Abu Dhabi mainland. The depth of the tunnel below ground at the upstream end is approximately 20 m (66 ft) and downstream at Al Wathba, it is approximately 80 m (262 ft) deep. To meet the hydraulic design criteria the tunnel alignment will fall at a gradient of 1 in 1300.

The design-build contracts for the deep tunnels include access shafts and ancillary structures associated with the trunk sewer (vortex chambers, drop shafts and adits). There will be nine work shafts along the tunnel alignment (WS1 to WS9 in Figure 2) of diameters large enough for Tunnel Boring Machine (TBM) launch and/or recovery. The work shafts will be converted into smaller diameter permanent access shafts when all construction work in the tunnel has been completed. Eight additional permanent access shafts (AS1 to AS8 in Figure 2) are also required and will be up to 6 m (20 ft) finished internal diameter. They are located approximately halfway between the work shafts and since they are off the line of the tunnel, short adits will be constructed (using NATM techniques) to connect the shafts to the main sewer. At four of the shafts, vortex chambers and drop shafts will be required for the permanent works.

A double lining system will be employed for the deep tunnel. A primary lining of segmental and gasketed pre-cast concrete tunnel lining (PCTL) will provide the permanent ground support. After completion of the structural primary lining, a corrosion protection lining (CPL) will be installed. The CPL will comprise a mechanically anchored HDPE membrane fixed into a cast in-situ secondary concrete lining. This lining system is illustrated in Figure 3.

The deep tunnel will be implemented through three design-build contracts: T-01, T-02 and T-03. Delineation of the three contracts is indicated in Figure 2. The T-01 tunnel will be approximately 16.1 km (10 miles) long, will begin at WS1 (adjacent to an existing pumping station) and end at WS4, and will mainly be in urban areas. The final internal diameter of T-01, after installation of the CPL, will be 4 m (13 ft). This tunnel section also crosses Maqta

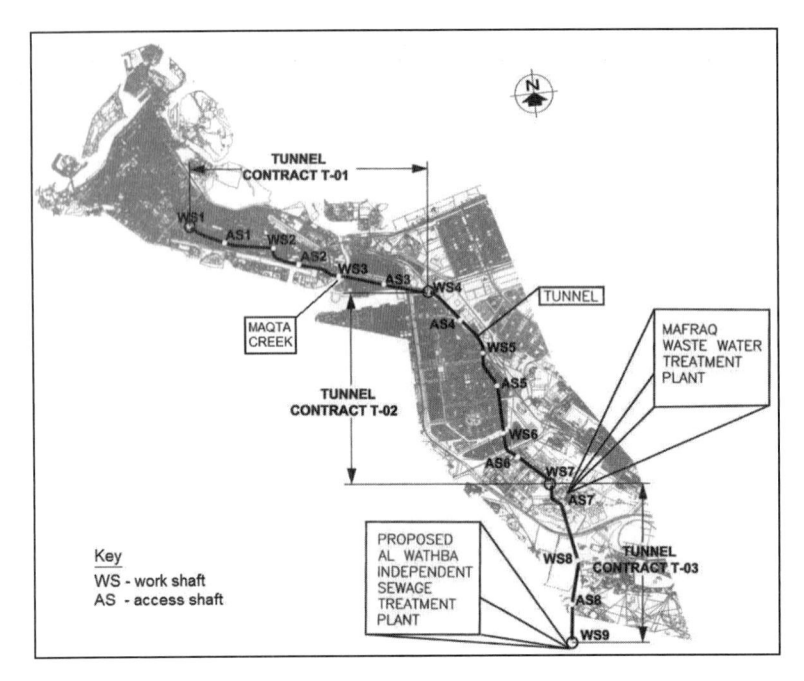

Figure 2. STEP deep tunnel alignment

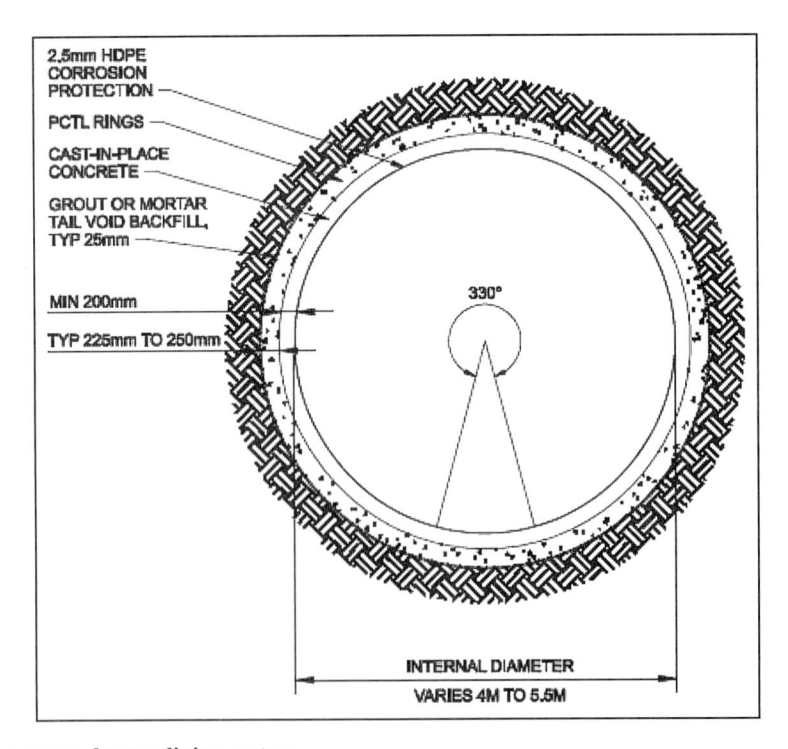

Figure 3. Deep tunnel sewer lining system

Creek, the channel separating Abu Dhabi Island and the mainland. The T-02 tunnel will be approximately 15.5 km (9.6 miles) long, will begin at WS4 and end at WS7, close to the existing MWTP. The alignment will follow the utilities corridor route through areas that are not yet fully developed, but are zoned for both residential and industrial developments. The final internal diameter of T-02 will be 5 m (16 ft). The T-03 tunnel will be approximately 9.7 km (6 miles) long, starting at WS7 and will end at the pumping station, close to WS9, and will have a final internal diameter of 5.5 m (18 ft). The T-03 alignment will mostly be in open desert areas but will pass close to the existing MWTP and beneath the environmentally sensitive Al Wathba wetland.

The tunnels will be excavated by pressurized face TBMs using slurry pressure or earth pressure to prevent uncontrolled movement of ground and groundwater and to minimise surface settlements. Eight TBMs will be required for the three contracts to excavate the tunnels and install the PCTL. Three TBMs will be used on each of the T-01 and T-02 tunnels and two TBMs on the T-03 tunnel.

Construction on the deep tunnel contracts commenced in September 2009 and is expected to be completed by May 2013.

## Geology of Abu Dhabi

Most of the bedrock in the Abu Dhabi area is Tertiary Miocene age weak sedimentary rocks overlain by Quaternary Pleistocene and Holocene formations. The geological conditions in Abu Dhabi are known to support karst formation and a number of developments within Abu Dhabi have encountered karst features (Tose & Taleb 2000).

Geological conditions in Abu Dhabi comprise a linear coastline dissected by ancient channels and creeks. Superficial deposits consist of marine sands and silts in the coastal zones. In addition, wind erosion, capillary action and evaporation has led to some extensive Sabkha deposits (flat areas between the desert and ocean, characterized by a crusty surface of evaporates), notably around the creeks forming hypersaline deposits. Sabkhas form primarily through the evaporation of sea water that seeps upward from a shallow water table and through the drying of windblown sea spray.

These superficial deposits overlay alternating beds of cemented sands, carbonate mudstone, carbonate siltstone, calcarenite, carbonate sandstones and gypsum.

## GEOTECHNICAL SITE INVESTIGATION

### Introduction

The preliminary geotechnical investigation for the three deep tunnel contracts was carried out during a nine-month period from June 2008 to February 2009. A separate investigation was carried out for each contract using two geotechnical investigation contractors. The geotechnical contractors were required to perform the fieldwork, prepare the boring logs, carry out laboratory testing and provide geotechnical data reports. CH2M HILL provided partial oversight during the fieldwork but did not participate in any independent logging.

The investigations comprised the drilling of exploratory boreholes, disturbed sampling of soils, rock coring, in-situ testing and surface based geophysical surveys. The geophysical surveys comprised seismic refraction, electrical resistivity tomography (ERT) and time electromagnetic method (TEM).

### Drilling

Eighty-five (85) exploratory boreholes were drilled along the tunnel alignment. The planned interval was one borehole every 500 m (1640 ft) along the alignment but the actual intervals varied due to drilling rig access or permitting requirements. The borehole interval achieved was an average of 490 m (1607 ft) but varied from approximately 200 m (655 ft) up to a maximum of 1820 m (5970 ft) at the Al Wathba wetland, where drilling was not permitted due to environmental restrictions. Twenty-one (21) boreholes were drilled in the urban areas of Abu Dhabi Island and 64 on Abu Dhabi mainland. Information from this preliminary geotechnical investigation will be augmented by the design-build contractors during the detailed design stage by additional exploratory drilling requirements. The additional requirements include, but are not limited to, exploratory boreholes at the shafts and along the tunnel alignment to close the interval to approximately 250 m (820 ft) between boreholes.

For the drilling in the urban areas of the T-01 tunnel, 31 boreholes were drilled. The borehole depths varied between 35 m (115 ft) and 50 m (164 ft), with an average depth of 41 m (134 ft). The typical depth of the borehole below tunnel invert was 10 m (33 ft). The depth of bedrock varied between 4.2 m (14 ft) and 15.5 m (51 ft) below ground surface (bgs), with an average depth of 12 m bgs.

Thirty-three (33) boreholes were drilled for the T-02 tunnel in the less confined conditions of the residential and industrial development areas. Depth of the boreholes varied from 45 m (148 ft) to 70 m (230 ft) along the alignment, with an average borehole depth of 54 m (177 ft) and typically 10 m (33 ft) below tunnel invert. The bedrock was encountered between 1.5 m (5ft) and 12.2 m (40 ft) bgs, with an average depth of 7.4 m (24 ft) bgs.

Drilling for the T-03 tunnel, where some of the surface conditions were desert-like (see Figure 4) but open and with few utilities, saw 21 boreholes drilled.

**Figure 4. Drilling for a T-03 borehole**

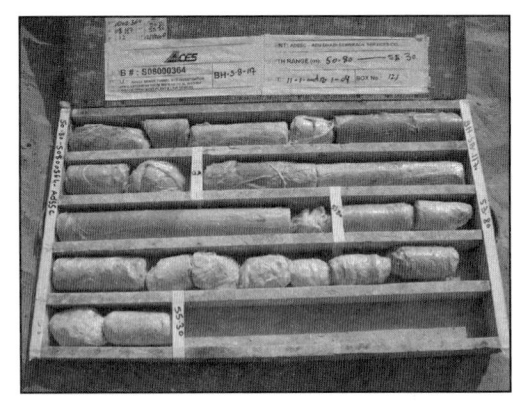

**Figure 5. Core box containing plastic wrapped core from T-03 borehole, SB-117**

The borehole depths, along the deepest alignment of the three tunnel contracts, varied from 82 m (259 ft) to 101 m (331 ft), with an average depth of 90 m (295 ft)—generally 10 m (33 ft) below tunnel invert level. Bedrock was encountered between 4.6 m (15 ft) and 25.0 m (82 ft) bgs, with an average depth of 10.5 m (34 ft) bgs.

All the boreholes were drilled using rotary drilling techniques, with bentonite drilling fluid for retention of the borehole sides. Steel casing was installed in many of the boreholes for additional support in the weaker strata. SPT sampling and testing was carried out in the overburden soils at 1.0 m (3 ft) intervals. Rock coring was performed using 76mm (3 in) diameter double tube core barrels.

**Logging and Supervision**

The geotechnical contractor's field supervision comprised a geologist per two rigs (since several boreholes were typically drilled concurrently), one field operations foreman and one senior geologist who supervised the drilling and logging procedures. CH2M HILL's geotechnical engineers provided limited oversight during the drilling.

Once recovered, the soil samples were placed in plastic bags and marked. Rock core samples were wrapped in clear plastic wrap for moisture protection (see Figure 5) and placed in wooden boxes. The geotechnical contractor's field geologists prepared draft logs on site, which included details on drilling times, fluid loss and rod drops, if encountered. Subsequently, the rock core and samples were transported to the contractor's laboratory for further evaluation, logging, laboratory testing and storage.

**In-Situ Testing**

In-situ tests included permeability packer tests, high pressure dilatometer (HPD) tests, downhole camera televiewer, acoustic televiewer and installation of standpipe piezometers for groundwater testing and monitoring.

*Groundwater Measurement*

Standpipe piezometers were installed at selected borehole locations to monitor the groundwater levels along the tunnel alignment at varying depths and the tunnel horizon. Fifteen (15) standpipe piezometers were installed along the alignment, many of them at or near shaft locations. More specifically, four piezometers were installed along the T-01 alignment, at an average depth of 35 m (115 ft), and seven piezometers along T-02 at an average depth of 36 m (118 ft). In the T-03 alignment, four piezometers were installed, two of which were screened at the soil-rock interface, one was screened at the deep tunnel invert level, and one screened at the midpoint level.

Water levels were recorded for five to ten days after installation; a number of piezometers were however monitored over a period of several months after installation.

*Packer Permeability Tests*

Both single and double packer permeability tests were conducted in rock formations over a 1.0 m (3 ft) test section. Up to ten such sections were tested per borehole to gain information on rock mass permeability for the different strata along the alignment. For the T-01 alignment, double packer tests were conducted in 30 of the 31 boreholes; along the T-02 alignment, double packer tests were conducted in 27 of the 33 boreholes; and for the T-03 alignment, single packer tests were conducted in 7 of the 21 boreholes (typically every fifth borehole and at shaft locations).

*High Pressure Dilatometer*

High pressure dilatometer (HPD) testing was used to gain a measure of deformation moduli of the rock mass at selected borehole locations. The tests were performed in a secondary borehole drilled adjacent to the primary hole. The purpose of the secondary borehole was to achieve the correct bore diameter for the test probe and obtain a freshly cored rock surface with minimal disturbance.

The HPD tests were carried out in 13 boreholes, mainly at shaft locations: eight sections were tested in each of four boreholes in the T-01 alignment; five test sections were tested in each of two boreholes in T-02; and an average of three sections were tested in seven boreholes along T-03.

*Televiewer*

The acoustic televiewer uses high resolution acoustic scanning of borehole walls to generate digital images from which fracture orientation and frequency can be determined. Additionally the televiewer provides a continuous log over sections of ground where core loss may have occurred. The acoustic televiewer logging tests were mainly carried out at shaft locations and included six boreholes on the T-01 alignment and eight boreholes on the T-02 alignment.

Downhole camera logging, recording video files of the borehole walls, was also used as it offered visual observation of borehole walls including any cavities or voids of karst solution features. This method was used to log six boreholes along the T-03 alignment: four of the boreholes at shaft locations. The results from this method varied since the video camera was submerged in groundwater and image clarity was affected by rock particles, loosened by the camera equipment, suspended in the water.

**Laboratory Testing**

To determine the physical properties of overburden soils a suite of laboratory tests was performed that included particle size analysis, natural moisture content, specific gravity and Atterberg limits. Tests performed on the rock core samples to determine physical and mechanical properties included natural moisture content, unconfined compressive strength (UCS), Young's modulus, point load test, splitting tensile strength (Brazilian) test, slake durability test, Schmidt rebound hardness and abrasivity testing. Chemical analyses were performed on select soil and water samples, which were tested for sulphate content, carbonate content, water soluble chloride content, pH and organic matter content.

Petrographic analyses and X-Ray Diffraction (XRD) were conducted to determine the mineralogical composition of the encountered rocks. This information was required to better describe the rock and its approach to weathering; its swelling potential; its aptitude for sticking; and from the quartz content, the expected wear on cutting tools. Abrasivity testing of rock samples—using the Cerchar Abrasion Index (CAI), Abrasion Value (AV) and Abrasion Value Cutter Steel (AVS) tests—was performed for consideration of TBM drive performance and the frequency of cutterhead tool replacement.

**Geophysical Survey**

Geophysical surveying was carried out along 6124 m (20090 ft) of the alignment to identify potential karst features and determine the top of bedrock where drilling was not possible. Surveying methods included seismic refraction, ERT and TEM. One or more of these methods were employed at four surveyed segments.

The geophysical investigations were performed in areas where karst solution features were suspected during drilling, either due to losses in drilling fluid circulation, or where wash out zones and small cavities (less than 1 m, 3 ft) were reported. The targeted areas included ERT along a 370 m (1214 ft) long section of the T-01 alignment and a combination of ERT, seismic refraction, and surface wave surveys along 4000 m (13120 ft) of the T-02 alignment.

For the T-03 alignment, geophysical surveys were undertaken at the Al Wathba wetland reserve. The wetland is an environmentally protected area and drilling boreholes was not permitted. Permission for non-intrusive geophysical surveying was gained and consequently seismic refraction and TEM surveys were carried out over a length of 1900 m (6234 ft). The tunnel alignment through the wetland traverses an 800 m (2620 ft) stretch of shallow water (less than 1 m (3 ft) deep). The TEM survey was performed by floating a 5 m (16 ft) transmitter loop on buoys with the TEM receiver contained in an inflatable raft as shown in Figure 6. The equipment and loop were temporarily anchored every 5 m (16 ft) to take a set of measurements. The seismic refraction survey was carried out by driving a series of steel bars into the marine bed with the geophones supported above the water and the seismograph placed in the raft.

# RESULTS OF PRELIMINARY GEOTECHNICAL INVESTIGATION

## Anticipated Subsurface Conditions Along the Tunnel Alignment

This section presents some of the engineering properties of the encountered overburden soils and rocks. Since the geological conditions were found to be generally uniform across the project site, data from the T-02 alignment is summarized here.

**Figure 6. TEM surveying across Al Wathba wetland**

*Overburden Soils*

The thickness of overburden soils along the deep tunnel alignment was found to vary from 2 m (6 ft) to 12 m (39 ft), with a typical thickness of approximately 6 m (23 ft) in an average 55 m (180 ft) deep borehole. The overburden soils were generally brown or grey silty sand. Gravel, shell fragments, and rock fragments were encountered in most of the boreholes with clay, clayey silt and organic materials encountered in a few boreholes. Boulders, rock pieces generally greater than 0.3 m (1 ft), were also encountered.

The SPT N-values in the overburden soils tested varied between 0 to 50 blows, with an average of 20 blows; the equivalent relative density of very loose to dense, with an average relative density of medium dense. The consistency of the clay or silt materials encountered was from very soft to medium stiff.

*Rock Formation*

A typical stratagraphic column is shown in Figure 7. This represents actual subsurface conditions encountered at borehole TB-133, located between WS4 and WS5 in the T-02 alignment. The rock encountered comprised interbedded layers of mudstone and gypsum (as demonstrated in the column of Figure 7) but also included sandstone, calcarenite, siltstone and claystone. The rocks were weak to very weak with the massive bedded gypsum layers being the hardest and strongest. Some of the rock crumbled under finger pressure to dry silt and clay.

A geological description of the encountered rocks is described below.

**Claystone and Mudstone.** The claystone and mudstone encountered was massive, laminated and thinly bedded. It was generally weak to very weak and the weathering varied from very slight to highly weathered. Significant multiple layers of claystone and mudstone were encountered in every borehole and the typical cumulated total thickness was approximately 30 m (98 ft) out of an average 55 m (180 ft) deep borehole.

**Siltstone.** The siltstone encountered was typically very weak to weak, massive, moderately to highly weathered, light grey pinkish and brown, and fine grained. It had very closely to medium spaced fractures and had gypsum inclusions. Multiple layers of siltstone were encountered in many boreholes and the average cumulative total thickness was approximately 5 m (16 ft) out of an average 55 m (180 ft) deep borehole.

**Gypsum.** The gypsum encountered ranged from thin discrete nodules to massive bedded layers.

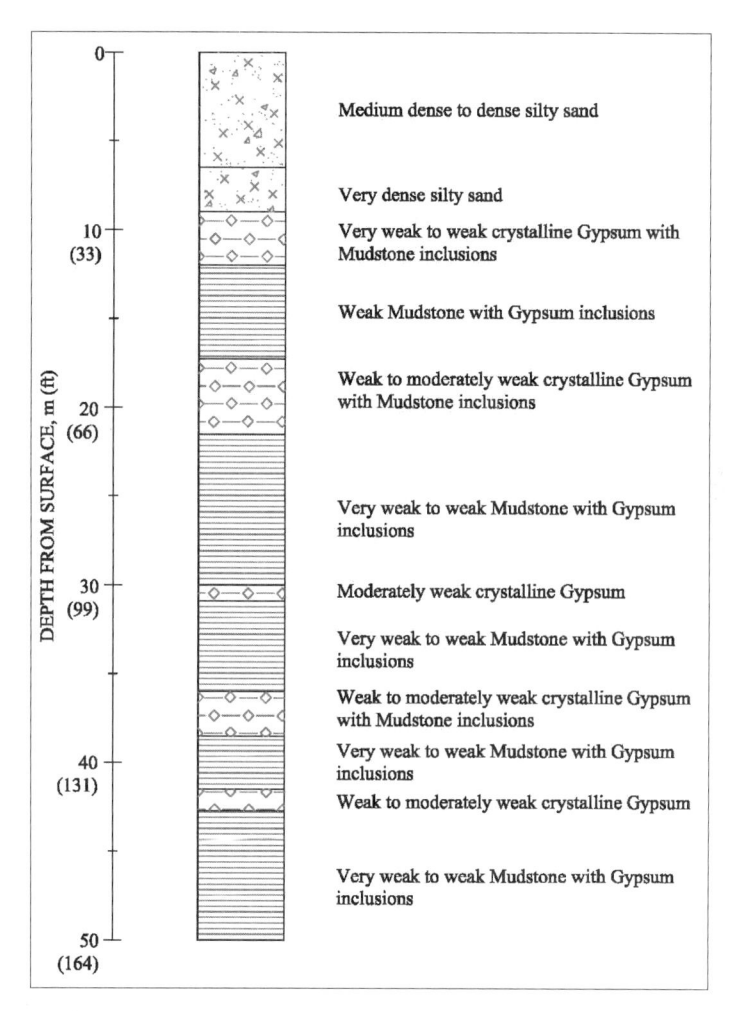

Figure 7. Stratagraphic column of borehole TB-133

The cumulated total thickness of the gypsum was approximately 12 m out of an average 55 m (180 ft) deep borehole. Occasionally gypsum was found mixed with mudstone or claystone and appears to have been deposited that way.

The depositional environment seemed gradational with gypsum nodules being present in the uppermost meter or two of the underlying mudstone or claystone and then transitioned to a pure gypsum layer. Another gypsum occurrence was individual gypsum crystals disseminated in a mudstone or claystone matrix.

**Calcarenite.** Calcarenite is a limestone consisting mainly of detrital calcite of sand size particles and is a calcium carbonate-rich material. It is often a conglomerate varying from a little shell material to nearly all fossil shells with little sand. The thickness of encountered calcarenite layers was generally less than 2 m (6 ft) and the cumulated total thickness of calcarenite was approximately 2 m (7 ft) in the average 55 m (180 ft) deep borehole.

**Sandstone.** The sandstone encountered was generally weak to moderately weak, laminated, moderately weathered, brown, fine to medium grained and closely to medium spaced. Sandstone was encountered only in few boreholes and had thicknesses of less than 2 m (6ft). The average cumulated total thickness of the sandstone in one borehole was generally less than 1 m (3 ft) out of an average 55 m (180 ft) deep borehole.

## Petrographic and XRD Examination

The petrographic and XRD examination on rock samples from the preliminary geotechnical investigation concluded that the sedimentary rocks were predominantly dolomitic in nature with a range

Table 1. Unconfined compressive strengths of samples from T-02 boreholes

| Rock Type | Minimum UCS MPa (psi) | Maximum UCS MPa (psi) | Average UCS MPa (psi) |
|---|---|---|---|
| Gypsum | 1.4 (205.9) | 26.8 (3886) | 15.6 (2262) |
| Mudstone/claystone | 0.1 (14.5) | 33.9 (4915.5) | 3.8 (551) |
| Siltstone | 0.3 (43.5) | 6.3 (913.5) | 2.2 (319) |
| Calcarenite | 0.7 (101.5) | 15.7 (2276.5) | 5.8 (841) |
| All | 0.1 (14.5) | 33.9 (4915.5) | 7.7 (1116.5) |

in porosity from 7% to 27%, which indicates high porosity and potential development of solution cavities. The analyses also showed a presence of anhydrite (that hydrates to gypsum with an associated volume increase of up to 63%) from 1% to 4% by weight in the gypsum, and palygorskite in mudstone (up to 24% by weight), sandstone (3% by weight) and calcarenite (up to 12% by weight). Palygorskite is a clay with a fibrous nature with a high surface area and a porosity that gives it the properties of sorption and gelling. It has swelling properties and will swell in salt water.

**Intact Rock Properties**

*Unconfined Compressive Strength (UCS)*

A summary of the UCS of intact rock cores is provided in Table 1.

Gypsum was the strongest rock encountered. Using the UCS values in Table 1 and the rock strength classifications adopted in BS 5930: 1999, it can be described as varying from a very weak to a moderately strong rock with an average classification of moderately strong. Mudstone and claystone were the major rock types encountered in the geotechnical investigation and can also be described as varying from a very week to a moderately strong rock but with an average classification of weak. Siltstone varied from very weak to moderately weak with an average classification of weak and calcarenite varied from very weak to moderately strong with an average classification of moderately weak. No particular trend for any of the rock samples was observed in the UCS strengths versus sample depth, elevation, or borehole location as shown in Figure 8.

*Permeability*

Packer permeability tests were typically conducted at ten sections in the 27 borings tested along the T-02 tunnel alignment. The rock mass permeability derived from these packer tests are summarised in Table 2.

The results show a mean permeability of all the rocks as 1.2E–06 m/s (3.93E–06 ft/s). The maximum measured permeability of 9.66E–06 m/s (3.17E–05 ft/s) is indicative of 'Class K3—Highly Permeable Rock' according to AFTES 2003. The mean permeability could be considered at the upper end of moderate permeability or low end of highly permeable rock based on AFTES 2003.

Estimates of groundwater infiltration were made for the three tunnels. T-02 tunnel comprises the three drives WS4 to WS5, WS5 to WS6 and WS6 to WS7. Infiltration rates for each of these three drives, using an empirical method described by Heuer 2005, are included in Table 3. The infiltration rates have been calculated for an unlined tunnel to provide estimations of water inflows at open excavations such as the adits and at tunnel junctions. For each of the three tunnel drives a different permeability distribution was used based on water pressure tests conducted in borings adjacent to the drives.

*Abrasiveness*

Cerchar abrasion tests (CERCHAR 1986) were conducted on a number of samples for each of the rock types. The CAI results for the T-02 samples are summarized in Table 4. Based on these test results and the CAI classification of Table 5, the rocks are classified as "not abrasive" to "not very abrasive."

A number of tests were also performed for determination of abrasion value (AV) and abrasion value cutter steel (AVS). The results from AV and AVS tests were classified as extremely low to very low for AV and extremely low for AVS.

**Groundwater Conditions**

Observations made at the piezometers showed the groundwater table to be roughly at ground surface. There were several borehole locations where groundwater under artesian conditions was encountered. The artesian flows were estimated to be around 0.3 l/min (0.07 gallon/min). These borings were located between WS6 and MWTP. Usually the overflow of water from these boring ceased within 24 hours, indicating perched limited water source conditions but at two locations, the overflow of water continued for several months after first observed. These conditions

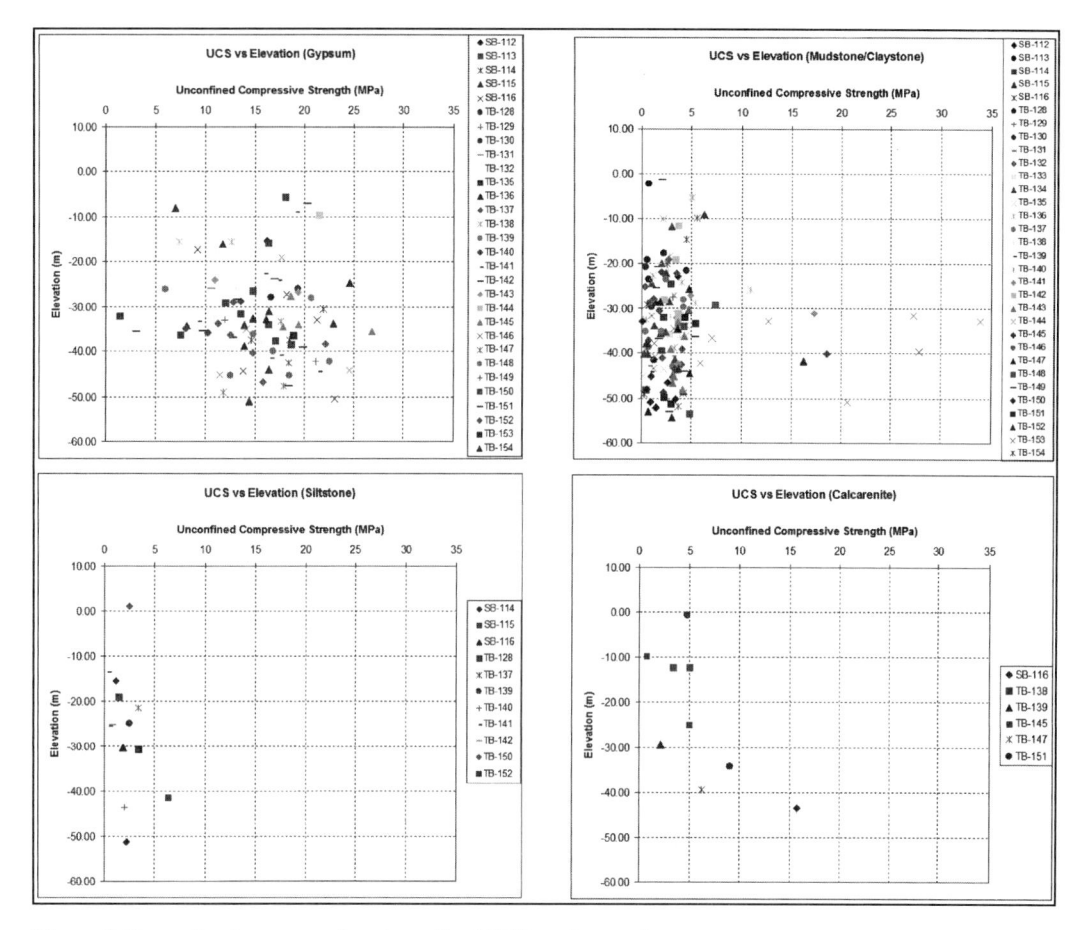

**Figure 8. Unconfined compressive strength of T-02 rock samples**

**Table 2. Summary of rock mass permeability from the T-02 packer tests**

| Rock Type | Minimum Permeability, K: m/s (ft/s) | Maximum Permeability, K: m/s (ft/s) | Mean Permeability, K: m/s (ft/s) |
|---|---|---|---|
| Gypsum | 6.14E–09 (2.01E–08) | 8.37E–06 (2.74E–05) | 1.12E–06 (3.67E–06) |
| Mudstone/claystone | 3.81E–08 (1.24 E–07) | 8.45E–06 (2.77E–05) | 1.11E–06 (3.64E–06) |
| Siltstone | 9.99E–08 (3.28E–07) | 9.66E–06 (3.17E–05) | 1.79E–06 (5.87E–06) |
| Calcarenite | 3.86E–07 (1.27E–06) | 3.22E–06 (1.05E–05) | 1.37E–06 (4.49E–06) |
| Sandstone | 6.14E–09 (2.01E–08) | 9.66E–06 (3.17E–05) | 1.20E–06 (3.93E–06) |
| All | 6.14E–09 (2.01E–08) | 9.66E–06 (3.17E–05) | 1.20E–06 (3.93E–06) |

will be further evaluated during the detailed design stage by the design-build contractor.

## Soil Chemistry

Selected soil samples were tested for sulphate and chloride content (sulphate content as $SO_4$ and chloride content as Cl). These samples were also tested for the pH values. Of 38 samples tested the sulphate content as $SO_4$ of the tested soil samples ranged between 0.013 g/litre to 4.34 g/litre. The chloride content as Cl of the tested soil samples ranged from 0.016% to 8.84%. The pH value of the tested samples ranged from 7.85 to 9.35.

Table 3. Summary of groundwater infiltration rates for the T-02 tunnel

| Tunnel Drive | Segment Length: m (ft) | Groundwater Infiltration Rate l/min (gallon/min) | | | |
|---|---|---|---|---|---|
| | | Reasonably Expected | | Worst Expected* | |
| | | Stabilized† | Heading‡ | Stabilized† | Heading‡ |
| WS4 to WS5 | 5326 (17469) | 8146 (1792) | 602 (132) | 16293 (3584) | 1204 (265) |
| WS5 to WS6 | 5297 (17374) | 3273 (720) | 137 (30) | 6547 (1440) | 275 (30) |
| WS6 to WS7 | 4907 (16095) | 8123 (1787) | 824 (181) | 16247 (3582) | 1648 (363) |

Notes:
* Based on reasonable expected case with an uncertainty factor of 2 applied
† Longer term steady-state inflow
‡ Initial 3D inflow at the head and the first 100 m of tunnel

Table 4. CAI results for the T-02 samples

| Rock Type | Minimum CAI | Maximum CAI | Mean CAI |
|---|---|---|---|
| Gypsum | 0.01 | 0.36 | 0.12 |
| Mudstone/Claystone | 0.01 | 0.40 | 0.11 |
| Siltstone | 0.03 | 0.40 | 0.22 |
| Calcarenite | 0.03 | 0.31 | 0.16 |
| Sandstone | 0.13 | 0.13 | 0.13 |
| All | 0.01 | 0.40 | 0.12 |

From this soil chemistry the classification of the ground conditions using the British Cement Association 2001 guidelines indicates that the design sulphate class varies between DS-1 and DS-4 and the Aggressive Chemical Environment Class (ACEC) ranges from AC-1 to AC-4.

## CONCLUSIONS

Following implementation of the STEP system in 2013, ADSSC will have the capability to convey and discharge approximately 1.7 million m³ per day (449 million gallons per day) of sewage to the new Al Wathba ISTPs. It will allow the removal of 35 existing pumping stations, will reduce ADSSC operating and maintenance costs and release valuable real estate for other development purposes.

The preliminary geotechnical investigation for the deep sewer tunnel allowed sufficient information to be included in the tender documents. The contractors are required to perform a minimum number of additional exploratory boreholes for detailed design. The boreholes are required at shafts, along the tunnel alignment to close up borehole intervals to 250 m (820 ft) and to investigate potential karst features and artesian conditions.

Information from the preliminary geotechnical investigation indicates that the deep tunnel excavation will primarily encounter weak to very weak

Table 5. Classification of CAI (CERCHAR 1986)

| Category | CAI |
|---|---|
| Not abrasive* | < 0.3 |
| Not very abrasive | 0.3–0.5 |
| Slightly abrasive | 0.5–1.0 |
| Medium abrasiveness to abrasive | 1.0–2.0 |
| Very abrasive | 2.0–4.0 |
| Extremely abrasive | 4.0–6.0 |
| Quartzitic | 6.0–7.0 |

* The specialist testing laboratory, SINTEF, has added the category of "not abrasive" for samples showing results of less than 0.3 (SINTEF 2009)

interbedded mudstone and gypsum, with some relatively thin layers of calcarenite and sandstone. It is likely that the design-build contractor's choice of TBM for these anticipated conditions will be earth pressure balance machines.

The cavities encountered during the preliminary geotechnical investigation included voids, loss of drilling fluid, core losses and wash out zones. These are typical indications of potential karst features – geohazards known to be present in some areas in Abu Dhabi. Mitigation measures for cavities include requirements for additional borings and geophysical investigations to try find cavities along the alignment and forward probing and a means for ground treatment from the TBMs. Other geohazards that exist in this area are high sulphate and chloride contents that will affect tunnel lining durability.

Although the presence of anhydrite in the gypsum samples was low, there is the possibility of swelling of the gypsum rocks when the excavation works provide an easy path for groundwater to the gypsum. It was recommended in the tender documents therefore that the design-build contractors perform swelling tests on gypsum with anhydrite to determine the swelling potential. The presence of palygorskite in mudstone samples, and to a lesser extent in the sandstone and calcarenite samples, also offers swelling potential in saline groundwater. This was also highlighted in the tender documents.

## ACKNOWLEDGMENTS

The authors wish to acknowledge the support of ADSSC and CH2M HILL senior management in supporting the publication of this paper. The encouragement and support of Shahzad Orakzai (ADSSC) and Robert Marshall (CH2M HILL), Patrick Doig (CH2M HILL) and Emad Farouz (CH2M HILL) is greatly appreciated.

## REFERENCES

Abu Dhabi Urban Planning Council 2007. Plan Abu Dhabi 2030: Urban Structure Framework Plan. http://www.abudhabi.ae. Accessed October 2009.

AFTES 2003. *Characterisation of Rock Masses Useful for the Design and the Construction of Underground Structures*. AFTES http://www.aftes.asso.fr. Accessed December 2009.

British Cement Association 2001. *Specifying concrete to BS EN 206-1/BS 8500: Concrete resistant to chemical attack*. British Cement Association, Camberley, UK.

BS 5930:1999. *British Standards Institution Code of Practice for Site Investigations*. BSI, London

CERCHAR, 1986. *The CERCHAR abrasiveness index*. Centre d'Etudes et Recherches de Charbonnages de France, Verneuil, S (ed).

Heuer, R.E. 2005. Estimation Rock Tunnel Water Inflow—II, In *Proceedings of the 2005 Rapid Excavation and Tunneling Conference*, pp 394–407.

SINTEF, 2009. *Determination of CAI* [for STEP rock samples]. SINTEF Building and Infrastructure Rock Engineering, Trondheim, Norway. Unpublished report.

Tose, J.A., and Taleb, A. (2000). Khalifa city "B" ground conditions program. In *Geoengineering in Arid Lands*, Mohamed A & Al Hosani (eds), Balkema, Rotterdam. pp 75–81.

# Geotechnical Variability and Uncertainty in Long Tunnels

Jack Raymer
Jordan, Jones, and Goulding, Norcross, Georgia

**ABSTRACT:** Geotechnical analyses for long tunnels should account for both variability and uncertainty. Variability is a natural condition of the ground. Uncertainty involves limits in knowledge about the ground. Variability can be described geologically and statistically using models based on the bell curve. Ground problems typically involve the extreme conditions at the tails of the bell curve. Uncertainty comes from having a limited amount of data, models that are imperfect and the change in scale between boreholes and the tunnel. This paper uses examples to show how baselines can be developed to account for both variability and uncertainty.

## INTRODUCTION

Long tunnels are geotechnically different from other types of civil construction. Long tunnels typically pass through a variety of ground conditions along their lengths. They lie deep under ground in geology that is typically very different from what can be seen at the surface. Test borings are expensive, few and far between. The geotechnical challenge for long tunnels is to predict the ground conditions as they vary along the entire length of the tunnel in spite of the high degree of geotechnical uncertainty.

The geotechnical uncertainty in a long tunnel creates considerable business risk, especially for the Contractor. Most of the risk comes from the overall rate of advance, and most of that risk comes from the extreme conditions that occur in a few places along the alignment. A single zone of particularly difficult ground can reduce the advance to a crawl or even stop it altogether, turning what should have been a profitable project into a financial disaster.

Geotechnical Baseline Reports (GBR's) have been a major advancement in the management of geotechnical uncertainty. Instead of giving the Contractor the data and having him assume the risk of interpreting it, the GBR tells the Contractor what geotechnical conditions he is entitled to expect by contract. But writing clear and reasonable GBR's is not as easy as it sounds. Owners fear the criticism that comes with change orders, engineers are conservative by nature, and the ground is anything but straightforward and clear. Business forces push the Contractor into being as optimistic as possible in order to win the bid and then using claims of differing site condition to make up for the reality. Political forces at the Owner's end and liability considerations at the engineer's end push towards unrealistic conservatism in order to shed risk and avoid embarrassing changes. And who can know what is "realistic"? It is easy to look at the ground after it has been opened and say what "realistic" should have been; it is much harder to look at a few boring logs and some field mapping and guess what "realistic" ought to be.

## VARIABILITY AND UNCERTAINTY

Variability is a condition of the ground. Uncertainty is a condition of our knowledge about the ground. More data reduces uncertainty but has no effect on variability. Realistic GBR's should describe the variability as accurately as possible and then manage the uncertainty with reason and moderation.

The bell curve is a familiar way to illustrate variability (Figure 1). The peak of the curve indicates the average condition and the tails indicate the extreme conditions. Most of the problems in long tunnels occur in the extreme ground in the tails of the bell curve. If the ground conditions in a long tunnel could be described as a bell curve, then the probability of encountering bad ground could be calculated. And since the tunnel is long, even a low probability condition could result in a substantial length of difficult ground.

Uncertainty arises from many sources. Some can be quantified and some cannot. Some involves a simple lack of data and others involve biases in the data due to scale, orientation, the test method and location of data points. Managing uncertainty requires an understanding of the various ways in which the data is to be used, because an approach that is conservative for one application can be grossly optimistic for another. If the engineer baselines a substantially higher intact strength than indicated by the data, he has been conservative with respect to TBM boreability but optimistic with respect to ground stability.

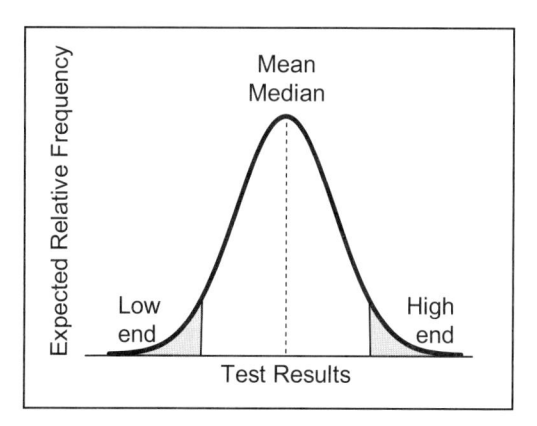

**Figure 1. Standard normal curve, or "bell curve"**

The goal of this paper is to show how the mathematical model of the bell curve can be used as a predictive tool to quantify variability and manage uncertainty so that baselines can be made more realistic with confidence. The general approach is to use the data to build a statistical model and then apply that model to the tunnel. The model indicates the variability of the ground and, by inference, what percentage of the tunnel is likely to have a certain ground condition. The model does not indicate where in the tunnel those conditions should occur.

## NORMAL DISTRIBUTION

The normal distribution, or "bell curve" is the mathematical model. It can be described by two parameters: the mean ($\mu$) and the standard deviation ($\sigma$). The mean marks the center of the curve and the standard deviation describes the spread, which is the variability. The equation for the bell curve is in the appendix but is actually unnecessary except for illustration purposes. A simpler and more powerful equation can be made by taking $\sigma$ and $\mu$ as the slope and intercept of the line

$$Y = \sigma z + \mu \qquad (1)$$

where $Y$ is a data value and $z$ is the transformed probability of that value occurring. (In statistics $z$ is called the *standard variable*). Values of $z$ for various probabilities are listed in Table 2 and can be calculated in Excel® using the function =NORMSINV(). The probability $P$ is interpreted as the percent chance that any given point in the tunnel will have a data value less than the corresponding value of $Y$. The probability $1 - P$ is the percent chance that $Y$ will be greater.

If the relationship between a geotechnical property and probability can be adequately described by Equation 1, then the property is normally distributed. If a property is normally distributed, then the mean of the property equals the median equals the mode. In Equation 1, $\mu$ is actually the median because at $z = 0$ there is an equal chance that the next data point will be greater or less than $\mu$.

## BASIS

The normal distribution arises from the central limit theorem. This theorem holds that the sum of a large number of independent random variables will tend to be normally distributed, which is to say, fit the bell curve (Bulmer, 1979). Key words are "sum" and "independent."

Rock strength is normally distributed because the strength of a block of rock is the sum of the strengths of any smaller blocks of rock into which it might be divided. Furthermore, the strength of each of those smaller blocks is independent of the strength of the others. From a geological perspective, the fact that the small blocks of rock came from the same parent rock would likely make them similar but from a statistical perspective, the strength of one block does not depend on the strength of another.

Fracture spacing, grain size and permeability are not normally distributed because they are not independent and additive. Each of these properties is actually a function of block size. Fractures are the boundaries between blocks of rock. Grains are each tiny blocks themselves. Permeability is a function of the spaces between the blocks or grains. These properties are not independent because they are subdivisions of a fixed volume of rock. If one block is large, then the other blocks must be small so that the overall volume is maintained. The sizes of each block become ratios of the whole and of each other, which is not additive. But they can be made additive by taking their logarithms, since $\log (a/b) = \log a - \log b$.

Fracture spacing, grain size and permeability are log-normally distributed. They can be made normally distributed so they fit Equation 1 if the data are first transformed by taking the logarithm of each data value. The values of $\sigma$ and $\mu$ are now in terms of the logarithms of the data and $\mu$ is the logarithm of the geometric mean. The results of any analysis must be reverted to their original units taking their antilogs before they are reported.

## DOMAINS

This method requires a model domain and a corresponding data domain each time it is used. Both should be defined explicitly and both are typically based on geological criteria. The model domain consists of the ground along the entire tunnel or specifically defined parts of the tunnel. The data domain consists of test results that are considered representative of the ground in the model domain. The model

## Table 1. Values of z for given probabilities

| z | P | 1 – P | z | P | 1 – P |
|---|---|---|---|---|---|
| –3.0 | 0.13% | 99.87% | 0.0 | 50.00% | 50.00% |
| –2.5 | 0.62% | 99.38% | 0.5 | 69.15% | 30.85% |
| –2.0 | 2.28% | 97.72% | 1.0 | 84.13% | 15.87% |
| –1.5 | 6.68% | 93.32% | 1.5 | 93.32% | 6.68% |
| –1.0 | 15.87% | 84.13% | 2.0 | 97.72% | 2.28% |
| –0.5 | 30.85% | 69.15% | 2.5 | 99.38% | 0.62% |
| 0.0 | 50.00% | 50.00% | 3.0 | 99.87% | 0.13% |

$P$ = probability that the parameter will be less at a given point.
$1 - P$ = probability that the parameter will be greater at a given point.

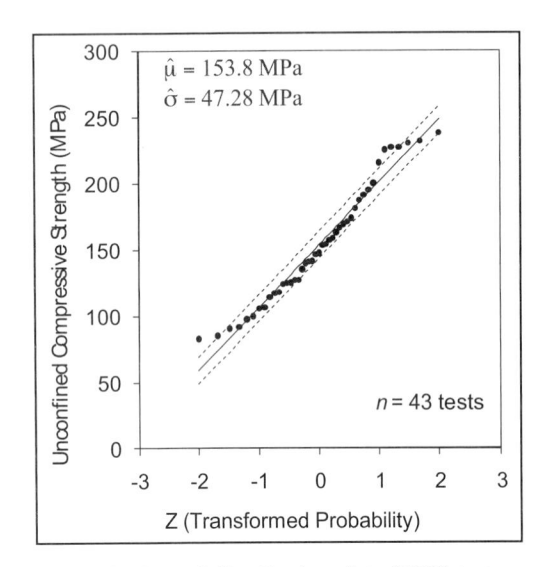

**Figure 2.** Normal distribution plot of UCS test results showing best-fit line (solid) and bounds (dashed)

is used to correlate the data domain to the model domain.

For example, if a tunnel passes through limestone and shale, then it would be reasonable to have one data domain consisting of the results from the limestone and another consisting of the results from the shale. The corresponding model domains would be those parts of the tunnel in each type of rock.

## BUILDING THE MODEL

Figure 2 is a normal distribution plot of unconfined compressive strength (UCS) tests from a long tunnel in metamorphic rock. The vertical axis ($Y$) represents the data domain; the horizontal axis represents the model domain. The best-fit line through the points is the estimated correlation between the data domain and the model domain.

Each point on Figure 2 represents one test result from the data domain. The results are ranked so they always increase in value from left to right. The z-value for each test result is calculated from its ranking as described in Table 2. The z-value of a given test result represents the probability that the next new result would have a lower data value. For example, Figure 2 is based on 43 test results. The highest point has a data value of 238.5 MPa and a z-value of 2.00. The probability that the 44th test result would be less than 238.5 MPa is 97.72 percent (see Table 1). Conversely, the probability that the 44th test result would be greater than 238.5 MPa is 2.28 percent.

The normal distribution is a straight line on this graph; if the data points follow a straight line, they are normally distributed. The equation for the best-fit straight line is

$$\hat{Y} = \hat{\sigma}z + \hat{\mu} \tag{2}$$

where the slope is $\hat{\sigma}$ and $\hat{\mu}$ is the intercept at $z = 0$. The slope ($\hat{\sigma}$) indicates the estimated variability; the intercept ($\hat{\mu}$) indicates the estimated median. The "hats" over $\sigma$ and $\mu$ indicate that those parameters are *estimates* of the unknown true values of $\sigma$ and $\mu$. Because the line is straight, the median equals the mean equals the mode, all of which can be regarded as the "average condition." A flatter line indicates lower variability and a steeper line indicates greater variability. A horizontal line indicates uniform conditions.

The method of calculating $z$ has some important consequences. If a point is removed or added, then $z$ values for all the remaining points will change. It is not valid to simply delete a point from the line without recalculating $z$ for all of the points. The points toward the high and low ends of the line are much more likely to be far off the best-fit line than points

**Table 2. Procedure for making normal distribution plots**

| Step | Procedure | Variables |
|---|---|---|
| 1 | Rank the data from lowest to highest. Assign index numbers starting with 1, 2, 3 ... n. | $i$ = index number of each data point<br>$n$ = total number of data points<br>$y_i$ = value of each data point |
| 2 | Calculate the probability of each data point $i$ | $P_i = \dfrac{i}{n+1}$ |
| 3 | Transform the probability to $z$ values using Excel® function =NORMSINV() | $z_i$ = NORMSINV($P_i$) |
| 4 | Make a scatter plot chart with $z$ as the horizontal axis and $y = Y$ as the vertical axis. | $z_i$ on horizontal axis<br>$y_i$ on vertical axis |
| 5 | Find the best-fit line through the points. The equation is $\hat{Y} = \hat{\sigma}z + \hat{\mu}$ | $\hat{\sigma}$ = estimated population standard deviation<br>$\hat{\mu}$ = estimated population mean = median |

nearer the middle because the horizontal distance between points is much greater at the ends of the line than in the middle. This means that large deviations toward the ends could easily be due to random scatter. Humps in the line, as in Figure 2 near $z = 1$, occur when several points have nearly the same test result.

## DATA EVALUATION

Uncertainty arises from how well the data fits the line. Random scatter about a best-fit line is assumed to be caused by having a limited number of data points. As more and more data is added, the scatter should diminish and the data should converge about a clear line.

If a large number of points systematically deviates from the line, then it indicates that the data might not be normally distributed as it is presently plotted. Systematic deviation typically appears as a curve or change in the slope of the line. If there are not many data points and the slope change is subtle, then there is always the question of whether the slope change is real or just a random manifestation of scatter.

If the slope change is real and the line of points is concave upward and flattens to the left, then the data is skewed to the left. If the data is concave downwards and flattens to the right, it is skewed to the right. If the data forms two line segments with a steeper segment in the middle, then it is bimodal. These patterns typically require about 12 to 24 data points in each segment to be confident that they really exist and are not just random patterns in the data.

Skewness can occur if some secondary process has affected the data, either by adding points to one of the tails of the bell curve or removing points from one of the tails. (Recall that $z$ is recalculated each time points are added or removed.) Some secondary processes are geological, some are artifacts of the testing procedure, and some are caused by selective sampling or culling of the data. Permeability reduction by hydrothermal cementation can result in a skewed permeability distribution if the hydrothermal fluids favor the more permeable zones and avoid the less permeable zones. If test results go out of the calibrated range of the equipment, then the high or low end tails of the bell curve will be truncated, which will make the line of points flatten to the right or left. If someone throws out data points based on the results of the tests themselves, then the results will take on characteristic patterns of skewness. There are various methods to account for skewness but these are beyond the scope of this paper.

Bimodalism and skewness can occur if there are two very different populations in the data domain, such as limestone and shale. The more correct way to solve this problem is to go back to the specimens and separate them into two data domains based on geological criteria. This would produce two model domains that have to be specifically defined in the GBR. The result would be one distribution for the shale and another distribution for the limestone. Another way to solve the problem is to develop a non-linear correlation between the model domain and data domain. If a non-linear correlation is used, the intercept $\mu$ is the median but not the mean, the slopes of each segment are not the standard deviation of the whole, and the method of quantifying uncertainty by using bounds become much more complicated.

## BOUNDS

Uncertainty can be managed by using bounds. The bounds place constraints on where the true normal distribution model can be for a given level of confidence.

The best-fit line with slope $\hat{\sigma}$ and intercept $\hat{\mu}$ is only an approximation of the unknown true conditions in the model domain. Assuming the property is normally distributed, its variability and average can be represented by a line with slope σ and intercept μ, (hats removed). As the number of data points increases, then the best-fit line should come closer and closer to indicating the true average ($\hat{\mu} \to \mu$) and true variability ($\hat{\sigma} \to \sigma$) of the ground in the model domain.

The true line should be close to the best-fit line. Bounds can be placed around the best-fit line to constrain the probable location of the true line for a given level of confidence. The level of confidence is described as the probability that the true line is between *both* bounds. More data or a lower level of confidence pull the bounds together; less data or a higher level of confidence push the bounds apart. The bounds shown in Figure 2 were calculated on the basis of an 80 percent two-sided level of confidence.

There are various methods of calculating bounds; some are quite involved. The simplest method, which the author has found to be adequate for evaluating geotechnical data and establishing geotechnical baselines, is to calculate straight-line prediction intervals that are parallel to the best fit line (see Figure 2). The upper and lower bounds have the same slope σ as the best-fit line but different values of μ. These bounds only extend left and right as far as the data.

The difference in μ is calculated from the distribution of residuals about the best-fit line. The residual $R_i$ at each data point $i$ is the difference between the data value of that point ($y_i$) and the value of the best-fit line opposite that point ($\hat{Y}_i$), where $R_i = y_i - \hat{Y}_i$. The residuals are normally distributed themselves with a mean of zero and some standard deviation ($s_r$). The intercept values for the upper and lower bounds ($\mu_{UB}$ and $\mu_{LB}$) are calculated as:

$$\mu_{UB} = \hat{\mu} + (s_r)(t_{\alpha,df}) \quad (3)$$

$$\mu_{LB} = \hat{\mu} - (s_r)(t_{\alpha,df}) \quad (4)$$

where ($t_{\alpha,d}$) is the value of the t-distribution. The input parameters for the t-distribution are alpha (α = 1 − CI) and degrees of freedom ($d_f = n - 2$), where CI is the confidence interval, $n$ is the number of samples, and 2 is because two points are needed to define a line. The value of ($t_{\alpha,df}$) can be calculated in Excel® using the function =TINV(α,df). The value of CI in Figure 2 is 80 percent; the value of $n$ is 43 tests.

There are a number of other methods for calculating bounds. The more sophisticated methods produce hyperbolic bounds that extend past the data to infinity. Strictly speaking, some of the methods produce confidence intervals and some produce prediction intervals; all have certain nuances. This author has found through experimentation that prediction intervals are much better suited for developing

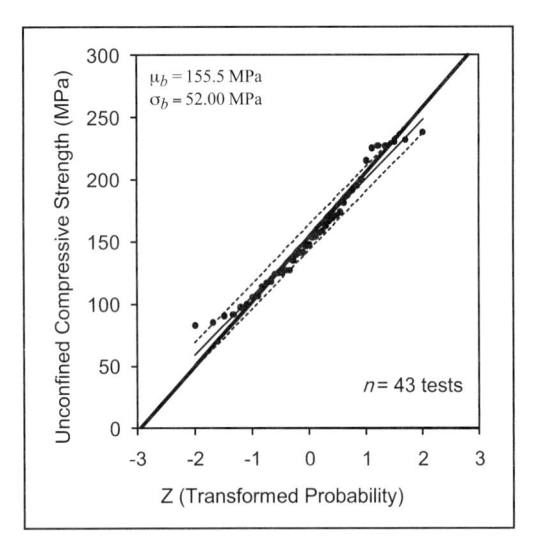

**Figure 3. Baseline distribution selected within the bounds to maximize variability**

baselines than confidence intervals, and that the straight-line method described above is adequate for the types and amounts of data typically associated with geotechnical designs. If one uses statistical software packages, it is important to understand the method being used, including the types of bounds being produced.

## SELECTING THE BASELINE

The engineer can select as the baseline any line that fits between the bounds without crossing them (Figure 3), recalling that straight bounds only extend as far as the limits of the data. This is reasonable because the true condition of the ground is unknown but constrained by the bounds. There is no reason to consider that the best-fit line through the data is any more accurate than any other line that could fit between the bounds. The engineer's line can be described by Equation 1 with a slope of $\sigma_b$ and average of $\mu_b$. The subscript indicates that it is the engineer's baseline estimate for the true but unknown condition of the ground. Note that the baseline extends past the data to infinity in both directions.

Allowing the engineer to choose the baseline anywhere within the bounds might seem arbitrary but is actually quite well constrained. If there is a decent amount of data and the confidence limits are reasonable, then the bounds are likely to keep the baseline estimate fairly close to the data itself, probably within two significant figures. If the engineer is more concerned about the range of variability of a property rather than just the average, he can choose the baseline as the line of maximum slope, such that it just touches upper bound on the right side and the lower bound on the left side. If infinite hyperbolic bounds were used, the baseline thus described would be tangent to them.

## APPLYING THE BASELINE MODEL

The baseline maximum strength $Y_b$ expected to occur in a given percentage of ground $P$ is calculated using Equation 1, except that $\sigma_b$ and $\mu_b$ are used instead of $\hat{\sigma}$ and $\hat{\mu}$. The word "maximum" is used because of the way the calculations are performed: if $P = 50\%$, then half the ground should have strength *less than* $Y = \mu$. Baseline values for certain percentages of the model domain can be calculated and tabulated as shown in Table 3. A GBR would typically only include the first and last columns of Table 3.

Figure 4 is a synthetic histogram generated from the model used in Figure 3. Evenly spaced values of $Y$ are chosen ($Y_1$, $Y_2$...) and the probability ($P_1$, $P_2$...) is calculated for each. The first column has boundaries at $Y_1$ and $Y_2$, and a height of $P_2 - P_1$, and so forth. The height of the columns depends on the width chosen for the columns.

## CHALLENGING THE BASELINE

If the contractor wants to show that the baseline is wrong he must demonstrate that, with more data, the baseline distribution cannot be contained within a reasonable confidence interval. (Recall that the bounds become closer together as more data is collected.) There is no reason why he should have to get all new data, but could supplement the existing data with more tests. If the Contractor can demonstrate that the baseline falls outside a reasonable confidence interval, then the he can calculate the difference between the confidence interval and the baseline. The result will be a distribution of the differences versus percent of tunnel, which can be integrated to calculate the quantum.

## NON-NATURAL DATA

The normal distribution applies to measurements of natural properties in the range where the test method is accurate. Index measurements, such as Rock Quality Designation (RQD), are commonly not normally distributed and are not easily transformed to become normally distributed. The problem with RQD is that the results are bounded by 100 and 0 and limited by the length of the coring run. This can make the points pile up at one end or the other. Further work needs to be done on transforms to make RQD and other index measurements fit a normal distribution model. Until then, these indexes are best represented by data summary histograms.

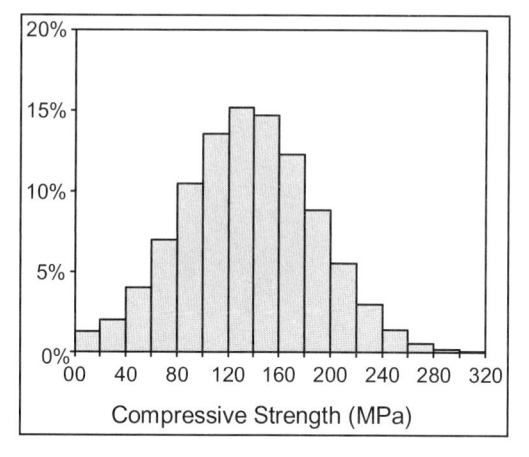

**Figure 4. Synthetic histogram of baselined strength from Figure 3. Vertical axis is interpreted as percentage of the model domain.**

Table 3. Baseline maximum strength in MPa Based on Figure 4

| Amount of Tunnel | | Calculated Results $Y$ | | Rounded Results $Y$ | |
| --- | --- | --- | --- | --- | --- |
| Percentile | Z | Best Fit | Baseline | Best Fit | Baseline |
| 0.001 | −3.09 | 7.68 | −5.19 | 7.7 | 0 |
| 0.05 | −1.64 | 76.02 | 69.97 | 76 | 70 |
| 0.1 | −1.28 | 93.20 | 88.86 | 93 | 89 |
| 0.2 | −0.84 | 114.00 | 111.74 | 110 | 110 |
| 0.5 | 0.00 | 153.80 | 155.50 | 150 | 160 |
| 0.8 | 0.84 | 193.59 | 199.26 | 190 | 200 |
| 0.9 | 1.28 | 214.39 | 222.14 | 210 | 220 |
| 0.95 | 1.64 | 231.57 | 241.03 | 230 | 240 |
| 0.999 | 3.09 | 299.91 | 316.19 | 300 | 320 |

## DATA SUMMARIES

Summarizing the data in a histogram is different from building a statistical model. A data summary only reflects what was actually collected. It does not provide a systematic method to evaluate uncertainty and it gives only a limited representation of the variability because the tails of the bell curve will always be inadequately represented. (Full representation occurs as the number of data points approaches infinity.) Histograms of the data should not be used to establish baselines unless the number of data points is large enough to make the statistical uncertainty trivial. Where that is the case, the baseline should be based directly on the histogram with no padding for uncertainty.

## CONCLUSIONS

Ground conditions tend to follow the bell curve. The width of the bell curve indicates the variability of the ground. How well the data fits the bell curve indicates the statistical uncertainty. The typical ground falls in the middle of the bell curve but it is the extreme ground in the tails of the bell curve that cause most of the problems in tunnels.

The bell curve can be reduced to a line as defined by a slope and intercept. The line is based on the test data from the geotechnical investigation. The slope indicates the variability and the intercept indicates the median. Upper and lower bounds can be established around the data points to constrain the uncertainty. More data points move the bounds together. The true condition of the ground is expected to lie somewhere between the bounds.

The engineer can select the baseline distribution as any line that fits between the bounds. This approach keeps the baselines close to the data but also gives the engineer flexibility to be reasonably conservative.

## APPENDIX

Equation for the ordinates of the bell curve (Bullmer, 1979):

$$\phi = \left(\frac{1}{\sigma\sqrt{2\pi}}\right)\exp\left[-\frac{1}{2}\left(\frac{y-\mu}{\sigma}\right)\right]$$

where $\phi$ is the probability density on the vertical axis, $y$ is the data value, $\mu$ is the mean, and $(y-\mu)/\sigma = z$.

## REFERENCES

Bullmer, M.G., 1979, *Principles of Statistics,* Dover Publications, New York.

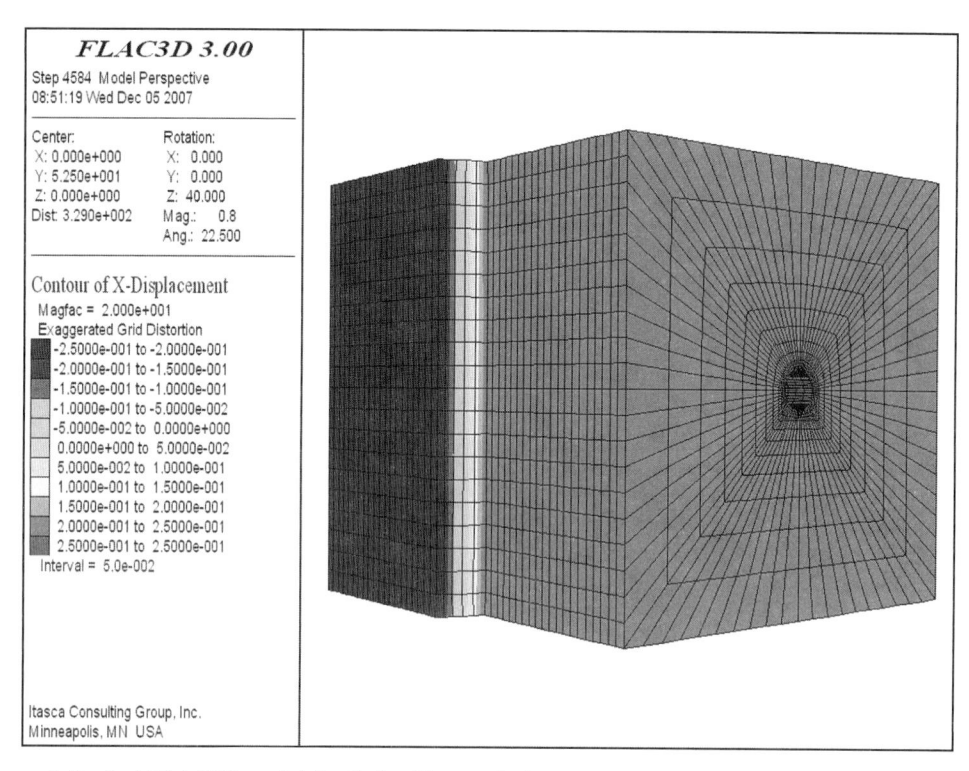

**Figure 4. Typical FLAC3D model for fault offset analysis**

on a seismic hazard and ground response analysis (URS 2007).

Depending on distance from the active faults, the estimated peak ground accelerations (PGA) along the tunnel alignment for the design earthquake vary from 0.85 g to 1.02 g at the ground surface level, and from 0.45 g to 0.58 g at the tunnel level. The peak ground velocities (PGV) range from 148 to 160 cm/s (58.3 to 62.9 in./sec) at the ground surface level, and from 47 to 58 cm/sec (18.5 to 22.8 in./sec) at the tunnel level.

Although the NIT alignment does not cross any seismically active faults, up to 150 mm (6 in.) of sympathetic displacement over a width of 1.5 m (5 ft) is considered possible on any or all of the four mapped secondary faults (Pirate Creek, Sheridan Creek, Unnamed, and Mill Creek faults). Sympathetic displacement could occur in response to a significant earthquake event on either the Hayward or Calaveras fault (WLA 2007).

In order to minimize the potential impact of sympathetic displacements on the NIT final lining, the design included steel pipe final lining in areas where the NIT crosses the four mapped fault zones. The steel pipe is more ductile than concrete and can tolerate much higher deformations or strains without rupture or collapse. However, because the design displacement occurs over a very short length, high shear strains in the steel pipe are expected. To investigate the effects of potential fault offset on the steel pipe final lining and finalize the pipe design, numerical analyses were performed using the three-dimensional finite-difference program FLAC3D (Fast Lagrangian Analysis of Continua in 3 Dimensions) Version 3.0 (Itasca, 2005). The FLAC3D analyses were used to calculate deformations and stresses induced in the steel pipe final lining by the design. A typical model used in the FLAC3D analyses is illustrated in Figure 4. In the analyses, two blocks of rock, one on either side of a 2.7-m (9-ft) wide fault zone were offset in the opposite directions along the fault plane. The resulting total relative displacement between these two blocks of rock masses was equal to the design displacement of 150 mm (6 in.) considered.

Key parameters were varied in the analysis to optimize the design and assess the impact of potential uncertainties. The key parameters include the stiffness of backfill concrete, wall thickness of steel pipe, magnitude of fault offset, and deformation moduli of rock mass and fault zone. Figure 5 shows a typical deformed shape (magnified by 20 times) of the tunnel steel pipe final lining following a 150-mm (6-in.) fault offset. Results of the analyses indicated

# TRACK 2: DESIGN

Matt Fowler, Chair

*Session 2: Challenging Conditions and Site Constraints*

# New Irvington Tunnel Design Challenges

## Glenn Boyce, Steve Klein, Yiming Sun
Jacobs Associates, San Francisco, California

## Theodore Feldsher
URS Corporation, Oakland, California

## David Tsztoo
San Francisco Public Utilities Commission, San Francisco, California

**ABSTRACT:** The San Francisco Public Utilities Commission (SFPUC) is designing the 5.6-km-long (3.5-mi-long) New Irvington Tunnel from the Sunol Valley to the City of Fremont in Alameda County, California, with a minimum finished diameter of 2.6 m (8.5 ft). The tunnel will have a two-pass lining system—an initial support system (such as steel sets) and a final lining consisting of steel pipe, concrete pipe, or cast-in-place concrete. Ground conditions are anticipated to be difficult and highly variable, with groundwater heads of 113 m (370 ft) and potential inflows as high as 95 L/sec (1,500 gpm). This paper discusses some of the issues faced during design of the tunnel, including potential risks associated with construction.

## PROJECT DESCRIPTION

The existing Irvington Tunnel (EIT) was constructed between 1928 and 1931 and extends approximately 5.6 km (3.5 mi) from the Sunol Valley to Fremont in Alameda County, California. The tunnel has a finished diameter of 3.2 m (10.5 ft). The east end connects with the Alameda Creek Siphons and the west end connects with the Bay Division Pipelines. The portals are the Alameda West Portal (east end) and the Irvington Portal (west end). Flow through the tunnel is from east to west.

The EIT and the Alameda Creek Siphons are critical lifeline components of San Francisco's Hetch Hetchy Water System, carrying approximately 85 percent of all of the water delivered to the City of San Francisco's customers. The tunnel and siphons are located between two active faults—the Hayward fault and the Calaveras fault. Movement on either of these faults during a major earthquake could seriously damage the tunnel and siphons and disrupt flow to the City's customers.

Because of the need to maintain continuous operations, the tunnel has not been taken out of service for inspection and maintenance in over 43 years (since 1966). To provide seismic reliability and ensure reliable delivery of high quality water to all of its customers, the SFPUC plans to construct the New Irvington Tunnel approximately parallel to the existing tunnel (Figure 1).

The NIT is expected to encounter difficult and highly variable ground conditions. The rock mass is generally composed of weak, intensely fractured and sheared sedimentary rocks (mainly sandstone, siltstone, interbedded siltstone/sandstone, and shale), and also includes some sections of stronger and more massive rock. Along the proposed alignment, the tunnel will also intercept a number of fault zones with abundant clay gouge. The EIT encountered running, caving, flowing, raveling, and squeezing ground in a number of areas along the alignment. Heavy groundwater inflows were also encountered in the EIT during excavation, with reported portal flows that ranged from 500 to 2,000 gallons per minute (gpm).

The concrete and steel final lining for the new tunnel will serve as the water conduit and steel pipe sections will connect to a combination of new and existing steel pipelines at each portal. The new facilities will include an overflow shaft to control the maximum hydraulic grade line in the tunnel and in the downstream pipelines. The tunnel extends about 5.6 km (3.5 mi) from a new Alameda West Portal to a new Irvington Portal. On its east (Alameda Creek) end, the new tunnel will connect to Siphons 1, 2, and 3, and a new Siphon 4. On its west (Irvington) end, the new tunnel will connect to the existing Bay Division Pipelines (BDPLs) Nos. 1, 2, 3, and 4, and a new pipeline No. 5.

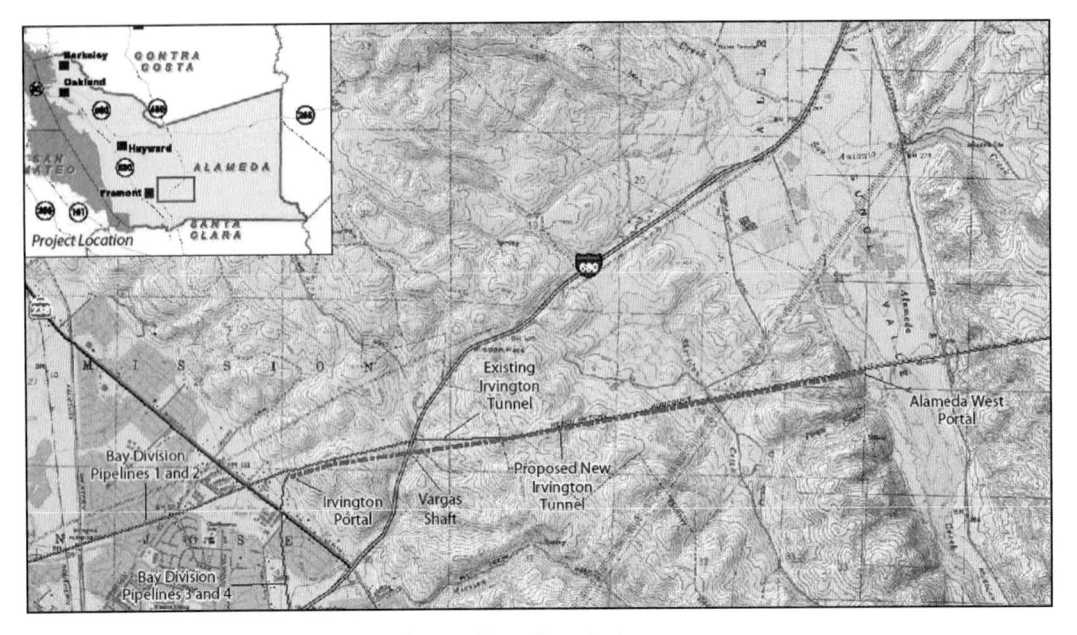

**Figure 1. Location map of the New Irvington Tunnel project**

The NIT alignment is about 58 m (190 ft) south of the EIT from the Alameda West Portal (Sta.41+40) to Sta. 154+00, for a distance of about 3,432 m (11,260 ft). From Sta. 154+00 to Sta. 183+50, the horizontal separation between the NIT and the EIT alignments increases to a maximum distance of about 204 m (670 ft). To the west of this point, near where the tunnel crosses I-680, the separation between the NIT and EIT gradually decreases to zero at Sta. 224+00, where the NIT alignment crosses below the Irvington Portal of the EIT. The NIT extends along this bearing north of the EIT for the remaining 122 m (400 ft) of the tunnel.

The NIT vertical alignment has two slopes. From the Alameda West Portal face (Sta. 41+40) to Sta. 200+00, the design slope is 0.00125. West from this point, the design slope is 0.029. The design invert elevation of the NIT varies from about El. +93 m (305 ft) at the new Alameda West Portal to about El. 62 m (202 ft) at the new Irvington Portal. The NIT will be lower than the EIT for its entire length. The vertical separation between the EIT and NIT alignments ranges from about 9 m (30 ft) at the Alameda West Portal to 37 m (120 ft) at the Irvington Portal.

The Alameda West Portal (AWP) provides access for constructing one of the NIT headings, connecting to the Alameda Siphon Mixing Manifold, and constructing the NIT portal access structure. In addition, the Alameda West Overflow Shaft will be constructed on the hillside above the portal. The NIT will connect to the Alameda Siphon Mixing Manifold via a connecting pipeline, intersecting the new tunnel portal pipe with a wye section.

A temporary construction shaft, referred to as the Vargas Shaft, will be constructed on the east side of I-680 at Vargas Road (Figure 1). The purpose of this shaft is to provide access for the excavation of two tunnel headings in the NIT. One heading will be mined to the west towards the Irvington Portal. The other heading will be mined to the east towards the new Alameda West Portal. The shaft size will be determined by the contractor, and is anticipated to be about 10.7 to 12.2 m (35 to 40 ft) in diameter. The shaft depth will be about 36.6 m (120 ft).

The Irvington Portal provides access for construction of an eastward tunnel heading and to connect the tunnel with the Bay Division Pipelines. The Irvington Portal is located at Sta. 228+00, where the tunnel has an invert elevation of about El. 62 m (202 ft). The Irvington Portal is adjacent to private homes, and most work will be limited to daytime hours in order to minimize impacts. The maximum length of tunnel that can be driven from this portal will be limited to 168 m (550 ft) to reduce the amount of truck traffic in the area.

## GEOLOGIC CONDITIONS

The ground surface above the tunnel generally consists of rolling hills covered with grass, brush, and trees. A number of privately owned parcels will be crossed by the tunnel alignment. These parcels have domestic water supply wells, livestock and irrigation wells, and natural springs that indicate the local availability of groundwater in the vicinity of the tunnel alignment.

Ground cover above the tunnel varies along the alignment, ranging from about 6.1 to 228.6 m (20 to 750 ft). Ground elevations along the tunnel alignment vary from about El. 72 m (235 ft) at the Irvington Portal to about El. 320 m (1,050 ft) at the high point in the central portion of the alignment.

The rocks along the NIT alignment consist of marine sedimentary sandstone, shale, siltstone, and chert ranging in age from Miocene to Cretaceous (5 to 144 million years old). Younger Quaternary-period deposits include alluvium and colluvium, which are present at the Alameda West Portal, Sheridan Valley, in the vicinity of I-680, and at the Irvington Portal. Geologic units crossed by the tunnel alignment include (in order of increasing age) unconsolidated Quaternary-period deposits and artificial fill (Qoa, Qc, af), Briones Formation (Tbr), Tice Shale Formation (Tt), Oursan Sandstone Formation (To), Claremont Formation (Claremont Formation Chert and Shale Member [Tcc] and Sandstone Member [Tcs]), and the Cretaceous Sandstone and Shale (Ks). The geologic unit names are consistent with USGS geologic maps, but in some cases, the names do not match the bedrock lithology present within the tunnel. For example, the Ks unit includes extensive siltstone and sandstone but relatively little shale.

The NIT lies in the western ridges of the Diablo Range. This range has been subjected to significant faulting and folding, which has fractured and sheared the rocks along the tunnel alignment. Regional tectonic compression has uplifted the range and created folds that form at least one anticline and one large syncline (Niles Syncline) in the site area.

Fault-bounded blocks are formed by four mapped faults that cross the tunnel alignment: the Pirate Creek, Sheridan Creek, Unnamed, and Mill Creek faults. At least three additional faults and shear zones without surface expressions were observed during the EIT construction. Strata within the fault-bounded blocks tend to dip to the southwest between the AWP and Sheridan Creek fault, and are folded into a syncline between the Sheridan Creek fault and the Unnamed fault east of I-680. Strata between the Unnamed fault and the Mill Creek fault are folded to form a northwest-trending anticline. West of the Mill Creek fault, the strata dip to the northeast. Bedding dip inclinations range from moderate to vertical within these folds and are commonly steepest in the vicinity of faults. The largest fold feature is the Niles Syncline, with a near vertical axial plane mapped at approximately Sta. 140+50. The Claremont, Tice, and Briones formations will be encountered two to three times along the alignment because of the folding involved with the regional structure.

An extensive program of geotechnical investigation was completed during design. The investigation objectives were to characterize the soil and rock conditions, lithologic and fault contacts, and groundwater conditions along the tunnel alignment. The investigations included geologic mapping, exploratory borings, downhole testing and logging, surface geophysics, and laboratory testing.

Because of the complexity of the conditions encountered, the subsurface investigations program ultimately grew to include 38 exploratory borings, ranging from 15.2 to over 228.6 m deep (50 to over 750 ft). The total length drilled was over 2,652 m (8,700 ft), nearly double the originally anticipated footage. Twenty-five of the borings were drilled vertically, and 13 were inclined. Standpipe piezometers and/or vibrating wire piezometers were installed in 28 borings. The drilling was primarily performed using LF-70 and CS-1000 skid-mounted rotary rigs equipped with HQ-3 wireline core barrels. Sonic drilling equipment was also used in one area in an effort to improve sample recovery.

Water pressure (packer) testing was performed in a total of 17 core borings. Downhole geophysical surveys, including caliper logging and televiewer logging, were performed in 15 borings. Seismic velocity surveys (OYO suspension) were performed in 8 borings. The televiewer logging results were used to characterize the in situ frequency and orientation of rock discontinuities, fractures, bedding, and shear zones.

Surface seismic refraction surveys were completed at each portal area to assist in characterizing the overburden depths and bedrock properties. Surface wave geophysical surveys were performed to investigate the deeper overburden in the Vargas Road area. Aquifer pumping tests were carried out at the Vargas Road and the Sheridan Valley sites to investigate the hydrogeologic formation properties.

Laboratory testing was performed on selected rock core samples. The tests included uniaxial compressive strength, point load strength, indirect (Brazilian) tensile strength, punch-penetration, slake durability, Cerchar abrasivity index, thin-section petrographic analysis, specific gravity, modulus and uniaxial compression, triaxial compression, and multistage direct shear tests on joint samples.

## GROUND CHARACTERIZATION

An extensive program of geotechnical investigations was completed as part of the design process (URS 2009). Based on the results, the tunnel alignment has been divided into eight reaches. The reach boundaries were established at the estimated contacts between the geologic formations (Figure 2). The locations of the geologic contacts were estimated based on the available geologic data, interpolation

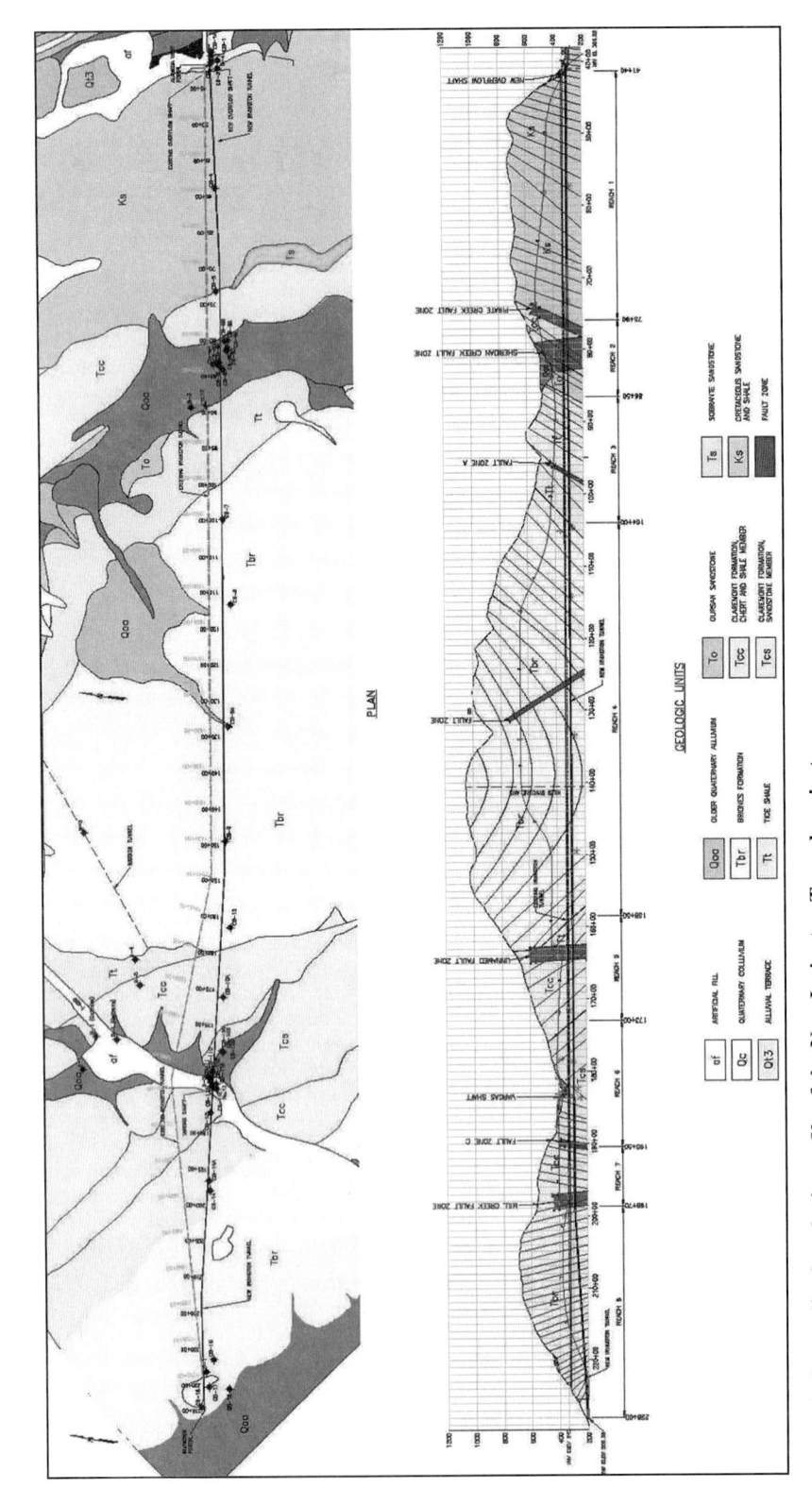

Figure 2. Generalized geologic profile of the New Irvington Tunnel project

## Table 1. Summary of the tunnel reaches

| Reach (Stationing) | Length, m (ft) | Geologic Formations | Anticipated Fault Zones |
|---|---|---|---|
| 1 (Sta. 41+40 to Sta. 75+90) | 1,052 m (3,450 ft) | Cretaceous Sandstone and Shale (Ks) | None |
| 2 (Sta.75+90 to Sta. 86+50) | 323 m (1,060 ft) | Claremont Chert and Shale (Tcc); Oursan Sandstone (To) | Pirate Creek Fault; Sheridan Creek Fault |
| 3 (Sta. 86+50 to Sta. 104+00) | 533 m (1,750 ft) | Tice Shale (Tt) | Fault A |
| 4 (Sta. 104+00 to Sta. 158+50) | 1,661 m (5,450 ft) | Briones Formation (Tbr) | Fault B |
| 5 (Sta. 158+50 to Sta. 173+00) | 442 m (1,450 ft) | Tice Shale (Tt); Claremont Chert and Shale (Tcc) | Unnamed Fault |
| 6 (Sta. 173+00 to Sta. 190+50) | 533 m (1,750 ft) | Claremont Sandstone (Tcs) | None |
| 7 (Sta.190+50 to Sta. 198+70) | 250 m (820 ft) | Claremont Chert and Shale (Tcc) | Fault C; Mill Creek Fault |
| 8 (Sta. 198+70 to Sta. 228+00) | 893 m (2,930 ft) | Briones Formation (Tbr) | None |

between borings, evaluation of surface geologic mapping data, and correlation with the EIT construction records. The reaches are shown in Table 1.

Ground conditions that will be encountered along the tunnel alignment have been divided into four ground classes to aid in the selection of tunnel excavation and support methods. Ground classes were defined based on the physical characteristics of the ground and its anticipated behavior during the tunnel excavation. The ground assigned to a particular class is expected to perform similarly in the tunnel excavation, and to require similar support methods.

Each ground class will be encountered multiple times throughout the tunnel in all tunnel reaches. The ground class definitions, predominant ground behaviors, and key characteristics associated with each ground class are described in Table 2. Potentially unstable ground conditions will be encountered throughout the tunnel, including but not limited to, raveling/caving, squeezing, swelling, running, and flowing conditions. The sheared nature of the rock mass, weak rocks, abundant clay infilling materials, intensely fractured rock mass, and high groundwater levels all will contribute to the instability of the tunnel excavation if not properly controlled. Control of groundwater inflows into the tunnel by pre-drainage and/or pre-excavation grouting will be necessary to minimize adverse effects on unstable ground and excavation progress.

The anticipated ground conditions along the tunnel were estimated based on evaluation of the geologic and geotechnical data collected for this project, along with review and correlation of the available EIT construction records. Rock mass quality evaluations were performed utilizing the Rock Quality Designation (RQD) (Deere and Deere 1988) and the Rock Mass Rating (RMR) system (Bieniawski 1988).

During the review process, we discovered that the EIT construction records contain some inconsistencies in their descriptions of the ground conditions and lithology encountered in the tunnel. In addition, the records provide no definitions of the descriptive or geologic terms used, and it appears that some of the terms used differ from current practice. Therefore, the EIT records were interpreted based on our understanding of site geology and the construction methods employed at the time of construction for each reach of the NIT. As an example, "running ground" identified in the EIT in areas of high groundwater inflow was concluded to be "flowing ground" in current tunneling terminology.

Significant shearing and many shear zones were observed in the geotechnical investigation borings completed for the NIT. Similar conditions are reported in the EIT construction records. The estimated amount of sheared/faulted rock in the NIT based on the EIT records ranges up to 90 percent over some reaches (in terms of the tunnel length impacted by sheared/faulted rock).

## TUNNEL EXCAVATION AND SUPPORT

The NIT will be constructed in variable ground conditions, ranging from strong, massive rock to very weak and intensely fractured and sheared rock

Table 2. Definitions of ground classes

| Ground Class and Definitions | Typical Rock Characteristics | Typical Discontinuity Characteristics | Ground Behavior |
|---|---|---|---|
| I. Massive to Moderately Fractured Rock | Sandstone, siltstone, and interbedded siltstone/sandstone; weak to strong rock; slightly weathered to fresh | Very rough to rough; fresh to slightly weathered surfaces | Structurally controlled block instability; spalling |
| II. Highly Fractured Rock | Sandstone, siltstone, interbedded siltstone/sandstone, and shale; weak to moderately strong rock; highly to slightly weathered | Rough, smooth, or slickensided surfaces or bedding planes; moderately to highly weathered/altered surfaces with infillings of clay and/or sand | Slow raveling; fast raveling where flowing groundwater is encountered |
| III. Intensely Fractured Rock | Sandstone, siltstone, interbedded siltstone/sandstone, and shale; thinly bedded to laminated rock structure; very weak to moderately strong rock, may be friable, poorly cemented; highly to slightly weathered/altered | Smooth, slickensided surfaces; highly weathered/altered with occasional moderately wide clay/sand-filled joints, shears, and shear zones | Fast raveling/caving; potentially flowing ground |
| IV. Heavily Sheared/Faulted Rock with Clay Gouge/Infilling Materials | Heavily sheared rock including fault gouge, shattered rock, all with abundant clay; extremely weak to very weak rock; moderately to completely weathered/altered | Slickensided surfaces; highly weathered/altered with wide clay-filled joints, shears, and fault/shear zones | Squeezing; swelling; caving; fast raveling |

including fault gouge. The tunnel will encounter high groundwater inflows and difficult ground conditions including raveling, running, flowing, caving, and squeezing ground. Conventional tunneling methods, including the use of roadheaders, drill-and-blast techniques, and hydraulic excavators, are expected to be adaptable to the anticipated wide range of ground conditions, including running, caving, flowing, raveling, and squeezing ground in a number of areas along the alignment.

The potential for use of a tunnel boring machine (TBM) was evaluated and concluded to be unacceptable for several reasons. Given the anticipated highly variable and difficult ground conditions, high groundwater inflows, and high groundwater pressures present along portions of the alignment, the use of a TBM, especially a machine of the size required for this project, was concluded to present excessive risks in terms of the cost and schedule impacts to the project. Use of a TBM would also restrict access to the tunnel face and would hamper the use of ground improvement techniques needed on this project. In addition, the seismic reliability goal of the project requires a steel pipe lining across zones of potential sympathetic fault offset. The use of a TBM (with a continuous full-perimeter lining) would restrict the face inspection and mapping needed to detect and delineate the lengths and locations of these zones.

The tunnel excavation methods, initial support systems, and groundwater control measures for the tunnel will be determined by the contractor. The EIT was mainly excavated using drill-and-blast methods, although "hand spades" were used in two sections of the tunnel to excavate the Claremont Formation in the vicinity of Reach 5. Several excavation methods could be applied to the NIT, including roadheaders, drill-and-blast methods, and mechanical excavation. The selected approach will depend mainly on economics, equipment availability (for a tunnel of this size), and the skills and experience of the contractor's crew. Key geotechnical issues include rock types, strength and degree of fracturing of the rock mass, and groundwater conditions.

Roadheaders are expected to be capable of excavating most of the tunnel reaches where the rock mass is moderately weak or weaker and highly fractured. The performance of a roadheader will depend on the machine size and weight, and the rock strength (UCS), fracture spacing, and abrasivity.

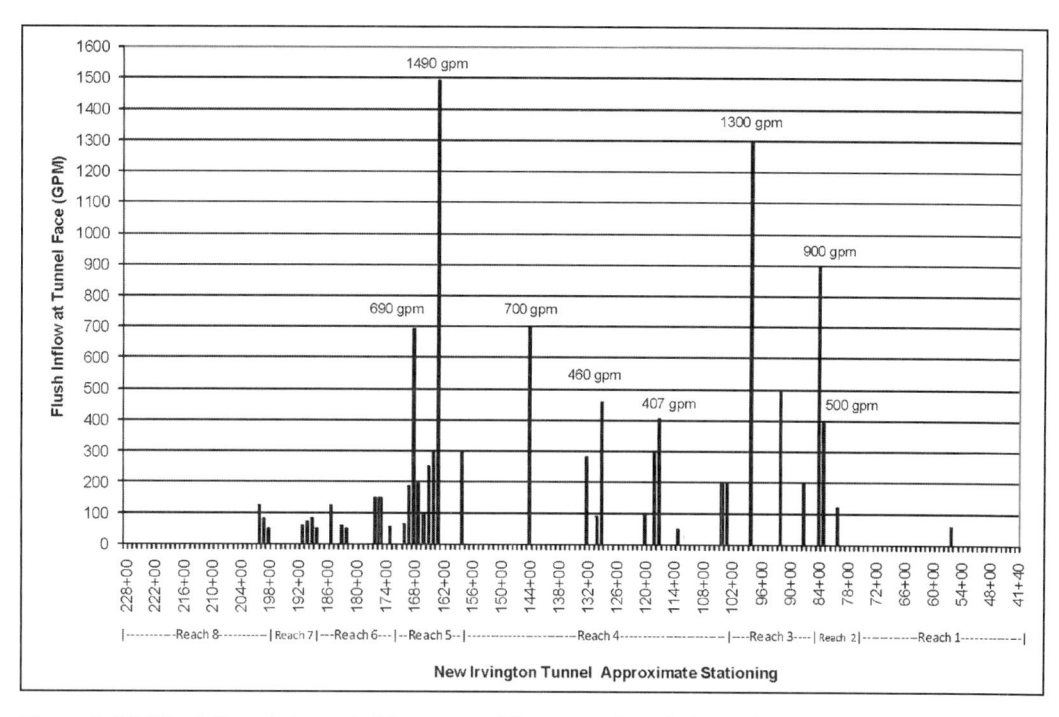

**Figure 3. EIT flush flows interpreted from portal flow records and shown based on the NIT stationing and reaches**

Drill-and-blast excavation methods were used extensively for the EIT and are applicable to most of the NIT, the exception being zones of intensely fractured, sheared rock with significant clay content and the fault zones. The maximum round lengths for blasting will depend on face stability, rock mass quality, initial support requirements, and other factors. Wherever blasting is used, appropriate blast designs and vibration and noise monitoring will be required to control and minimize potential impacts of blasting on the existing tunnel, existing pipeline facilities, and residences adjacent to the portals and above the tunnel alignment.

The NIT will have a two-pass support system consisting of an initial support system and a final lining. The initial support requirements will vary along the tunnel due to the range of ground conditions that will be encountered during construction.

Presupport using spiling and/or forepoling will be required to control raveling, caving, and crown instability, primarily in tunnel reaches with Ground Class III and IV conditions. Presupport may also be required to prevent structurally controlled block instability, expected primarily in Ground Classes I and II. Face support in conjunction with pre-support is expected to be required to control block instability, overbreak, raveling, running/flowing, slaking and caving behaviors at the tunnel face in Ground Classes II, III, and IV.

## GROUNDWATER CONTROL

The anticipated groundwater levels above the tunnel crown range from zero to 112.8 m (370 ft). Groundwater inflows are anticipated throughout a significant portion of the tunnel. The heaviest flows will occur where the ground is highly fractured and where fault and shear zones are encountered. Estimates of the maximum potential groundwater flush flows and sustained flows were made for each of the NIT reaches. The estimates were based on the results of the groundwater modeling for present-day conditions and on interpretation of the EIT construction records (Figure 3).

Implementation of effective groundwater control measures will be required to limit uncontrolled inflows into the tunnel and to reduce the impact of inflows on tunnel construction. Additional inflow control measures will be required where needed to protect groundwater wells and resources, as directed by SFPUC.

Predrainage of the rock mass ahead of the face is expected to be feasible from within the tunnel in many areas. At selected locations, predrainage from the ground surface is also feasible. The contractor must implement pre-excavation grouting, predrainage, and/or a combination of both measures, as necessary, to reduce sustained groundwater inflows to within workable limits. Predrainage from within the tunnel is considered feasible when the probe holes indicate the potential for significant groundwater inflows into the tunnel. Predrainage has the following objectives:

- Reduce high groundwater inflow potential,
- Improve the efficiency of the pre-excavation grouting, and, if necessary,
- Improve ground behavior.

Drain holes drilled ahead of the tunnel face can reduce the groundwater head and aid in the control of heading inflows. Depending on the fracture openings, fracture spacings, and storativity of the rock mass, the effectiveness of drainage will vary. Typically, drainage would be done ahead of the advancing face to improve the ground behavior at the tunnel walls, roof, and face.

Predrainage from the ground surface using dewatering wells is planned to supplement dewatering from within the tunnel in two areas along the alignment: Sheridan Valley and the Vargas Road/I-680 corridor. At each site, the contractor will be required to design and install a dewatering system to lower the groundwater table in advance of tunnel excavation, in areas where problematic ground conditions and/or high water inflows are expected.

Pre-excavation grouting requires the injection of grout to fill open fractures in the rock mass. Typical pre-excavation grouting will not penetrate intact rock or joint infillings with low porosity. In zones of completely weathered/altered rock, clay-filled shears and clay fault gouge, or intensely fractured rock, grout penetration is expected to be limited because of low hydraulic conductivity, so the effectiveness of pre-excavation grouting for groundwater control and ground improvement will also be limited. In areas where the rock exhibits high hydraulic conductivity, treatment of the fractured rock mass through grouting is expected to be more effective. The performance objectives for pre-excavation grouting and drainage are as follows:

- Limit the groundwater inflows at the tunnel face to a rate compatible with the selected tunnel construction means and methods.
- Mitigate adverse ground behavior caused by heavy groundwater inflows as necessary to allow adequate installation of initial support measures.

The contractor's drilling, casing, and grouting equipment and methods must be capable of staged grouting in unstable rock formations. The probe and grout holes are expected to encounter hole stability problems in weak rock that includes highly to intensely fractured rock and clayey shear and fault zones. Due to uncertainties associated with the characteristics of groundwater flows in a fractured rock mass, the required probe and verification holes are not expected to detect all potential inflows. The actual inflows encountered at the tunnel face may vary substantially from estimates based on probe hole flows. The contractor must be prepared to adjust pre-excavation grouting and drainage techniques, criteria, and procedures during construction to accommodate the expected rapidly varying ground conditions.

Inflows into the tunnel may negatively impact existing water wells and nearby springs. If monitoring data indicate unacceptable impacts are occurring, the owner may direct the contractor to implement additional control measures. Feasible additional measures include additional pre-excavation grouting, the installation of a built-up shotcrete lining for controlling water inflows in conjunction with controlling ground stability, and other effective measures proposed by the contractor.

**SEISMIC DESIGN**

The NIT is located in a seismically active region of California, dominated by the San Andreas fault system. The nearby Hayward fault on the west and the Calaveras fault on the east are capable of generating large (Magnitude ≥7.0) earthquakes. In addition to these two faults, numerous other active faults are located within 48.3 km (30 mi) of the tunnel site. However, the tunnel alignment does not cross any faults that have been zoned as Alquist-Priolo Earthquake Fault Zones by the state of California. The Alquist-Priolo Earthquake Fault Zones are defined based on evidence of Holocene surface rupture (i.e., within the last 11,000 years). Surface mapping completed as part of the geotechnical investigations (URS 2009) found no evidence for Holocene surface rupture on any of the mapped faults crossing the tunnel alignment.

According to the SFPUC's General Seismic Design Requirements (SFPUC 2006), the NIT is specified as a Seismic Performance Class III facility. The design earthquake for the Seismic Performance Class III facilities has a 5 percent probability of exceedance in 50 years (975-year approximate return period). The SFPUC's objective for seismic performance of the tunnel is to deliver winter day demand of water within 24 hours of a major earthquake (SFPUC, 2006). In order for this objective to be achieved, catastrophic damage to the facilities during the design earthquake must be avoided. The seismic design parameters for the NIT were developed based

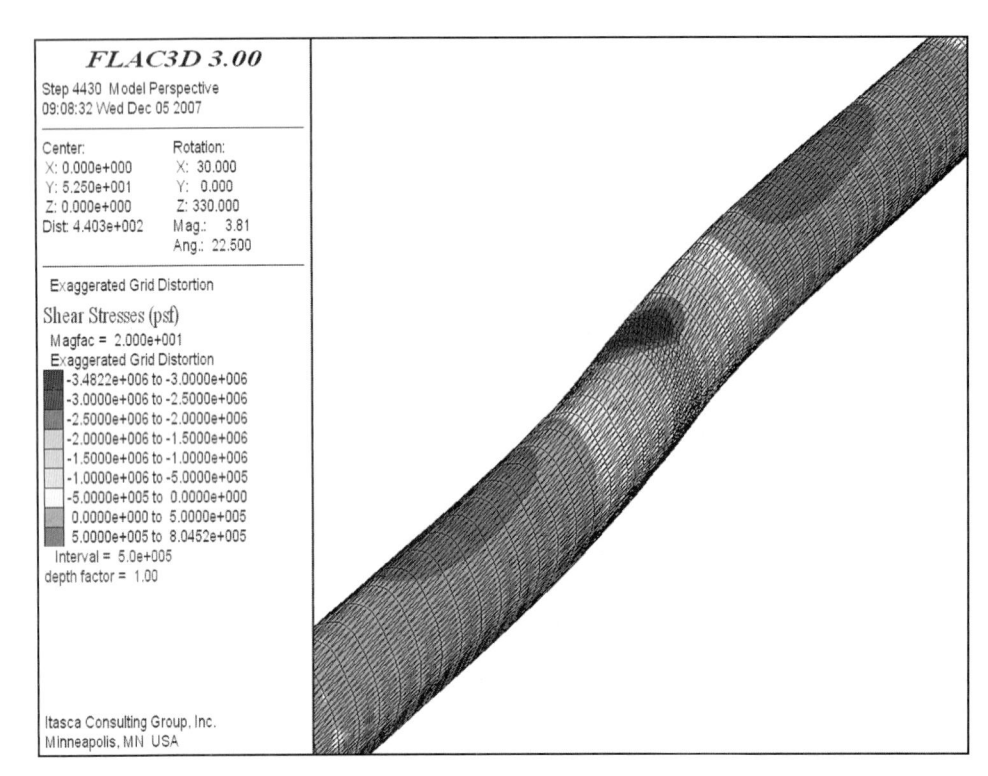

**Figure 5.** Contours of shear stresses and deformed shape of the steel pipe lining caused by fault sympathetic displacement

that the maximum stress in the steel pipe increases with stiffness of the backfill concrete. In order to control the maximum stresses in the steel pipe to within tolerable limits, use of a special low-density backfill concrete to fill the annular space between the initial support and the steel pipe will be required. This special backfill will consist of cellular concrete with an unconfined compressive strength between 1.37 and 2.07 MPa (200 and 300 psi).

## PORTAL EXCAVATION AND SUPPORT

Development of the portals for tunnel construction and installation of pipeline connections will be a critical element of the project. The preparatory work will include access development, installation of erosion and sedimentation control measures, and protection of existing SFPUC pipelines and portal structures.

The two portal excavations will require both soil and rock excavation techniques. Rock near the portals is expected to be highly to intensely fractured, with an average fracture spacing of less than one foot. The use of appropriately sized excavating and earth-moving equipment will be suitable for most of the portal excavations. Drill-and-blast methods or impact hammer methods will be required to break up harder, more massive rock in localized areas. However, such excavation methods and related construction activities may disturb nearby residents and affect the stability of adjacent pipelines, slopes and shoring systems. The project includes very tight noise and vibration criteria and required measures to reduce and mitigate potential impacts.

To achieve slope stability, all portal excavation cut slopes will require temporary support measures, such as fully grouted soil nails, shotcrete, rock dowels, rock bolts, and/or other measures as appropriate for the subsurface conditions encountered. In addition, to achieve a stable tunnel excavation at the portal, reinforcement of the rock mass above the tunnel crown will be required, by installation of portal spiles or forepoling before starting the tunnel. Typical minimum required ground support measures for the portal excavation sidewalls are shown in Figure 6.

## PROJECT SCHEDULE

The design of the project was completed in December 2009 and advertised for bidding in January 2010. Bid opening was March 2010. Notice to proceed is anticipated in June 2010. The project has a 42-month

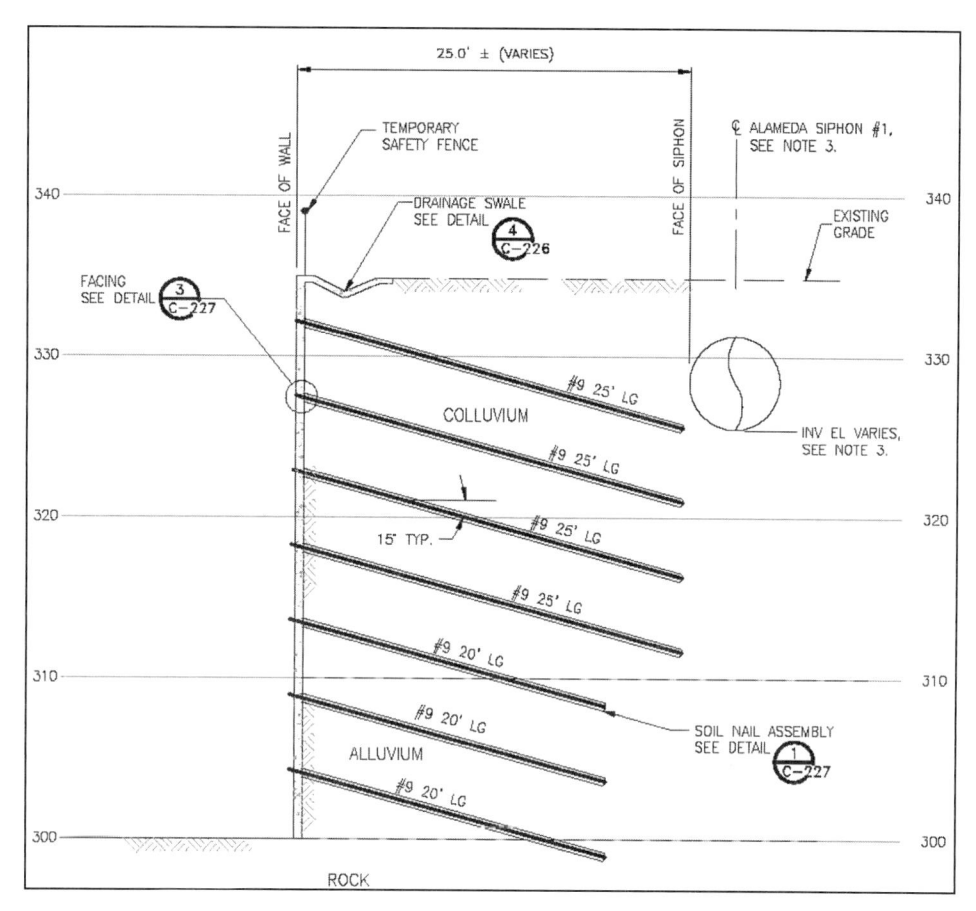

Figure 6. Typical portal excavation support

construction schedule and completion is scheduled for November 2013.

## CONCLUSION

The first tunnel was built with very little advanced information. The miners in 1928 drove the tunnel blindly, with no advanced exploration, no probe hole drilling, limited ability to grout, and used timber sets. Information collected during construction of the existing tunnel has been valuable in the design of the new Irvington Tunnel project to help limit some of the challenges to be faced. That information combined with today's tunneling equipment, technology, materials, increased exploration, advanced planning, and the ability to probe, drain, grout and dewater from the surface, should allow the New Irvington Tunnel to be completed with much fewer construction problems and much greater ability to survive the next major earthquake in the area.

## REFERENCES

Bieniawski, Z.T. 1988. The Rock Mass Rating (RMR) system (geomechanics classification) in engineering practice. In *Rock Mass Classification Systems for Engineering Purposes*. Edited by L. Kirkaldie, 17–101. Philadelphia, Penn.: American Society for Testing of Materials, ASTM STP 984.

Deere, D.U., and Deere, D.W. 1988. The Rock Quality Designation (RQD) index in practice. In: *Rock classification Systems for Engineering Purposes*, L. Kirkaldie, ed., 91–101. Philadelphia, Penn.: American Society for Testing of Materials, ASTM STP 984.

Itasca. 2005. *Fast Lagrangian Analysis of Continua in 3 Dimensions (FLAC3D)*, Version 3.0, Itasca Consulting Group, Minneapolis, Minnesota.

SFPUC (San Francisco Public Utilities Commission). 2006. *General Seismic Requirements for Design of New Facilities and Upgrade of Existing Facilities*.

URS (URS Corporation). 2007. *Seismic Hazard Analysis, New Irvington Tunnel Project.* Draft Technical Memorandum 6-06, prepared for the San Francisco Public Utilities Commission, June.

URS. 2009. *Geotechnical Data Report, New Irvington Tunnel Project.* Final Report prepared for the San Francisco Public Utilities Commission, November.

URS and Jacobs Associates. 2009. *Geotechnical Baseline Report, New Irvington Tunnel Project.* Final Report prepared for the San Francisco Public Utilities Commission, November.

WLA (William Lettis & Associates, Inc.). 2007. *Geologic Investigation of Possible Deformation Expected on Secondary Faults Along the New Irvington Tunnel Corridor, Alameda County, California.* Technical Memorandum prepared for URS Corporation, October 26.

# The Sunnydale CSO Tunnel: Dealing with Urban Infrastructure

**Renée L. Fippin, Heather E. Stewart, Richard M. Nolting III**
Jacobs Associates, San Francisco, California

**Manfred M. Wong**
SFPUC, San Francisco, California

**ABSTRACT:** The Sunnydale Auxiliary Sewer Project is a CSO tunnel designed to prevent storm events from flooding residences near San Francisco Bay. Although this 2.44–3.35 m (8–11 ft) diameter sewer pipeline is less than 1,219 m (4,000 ft) long, its installation will require a single-pass, segmentally lined EPB drive, a microtunneled drive, a jacked-shield tunnel, and a short cut-and-cover section. Undercrossings include a major four-track commuter rail station, deeply buried gas and high-voltage electric utilities, and a culvert carrying ten-lane State Highway 101 that will require jet-grouting for support. In addition, the EPB tunneling will encounter a plume of contaminated soil and groundwater currently undergoing remediation. This paper discusses the reasons for the multiple construction methods, including variable subsurface conditions and infrastructure crossings.

## INTRODUCTION

The existing 100-year-old Sunnydale Sewer Tunnel, owned by the San Francisco Public Utilities Commission (SFPUC), transports wastewater and storm water from an approximately 720-acre service area in San Francisco's Visitacion Valley to the Sunnydale Transport/Storage Structure and Pump Station at Harney Way near Candlestick Park. The service area includes primarily residential sources, open space area in McLaren Park, and some paved roadway and parking areas. The system also accepts dry weather flows from the Bayshore Sanitary District and the Brisbane/Guadalupe Valley Municipal Improvement District in San Mateo County.

The existing 1910s-era sewer system is unable to fully accept storm flows. As a result, it is often overwhelmed during significant wet weather events, and temporary flooding occurs within portions of the service area. The proposed auxiliary tunnel will serve as a wet-weather overflow. That is, it will be empty during the dry-weather season of May through October; and only see flows of combined storm water and sewage during significant storm events. Once constructed, the Sunnydale Auxiliary Sewer Tunnel (SAST) (Figure 1) will provide capacity to prevent flooding in the Visitacion Valley community.

The Phase 1 project area, located between the San Francisco Bay and Talbert Street, is the main focus of this paper. Phase 1 will consist of the new 2.44 to 3.35 m (8 to 11 ft) diameter SAST. At Harney Way, adjacent to the San Francisco Bay, the SAST will connect to the existing below-grade Sunnydale Transport/Storage Structure and Pump Station. As the alignment heads west along the county line, the proposed SAST will run approximately parallel to, but north of, the existing tunnel. Near Bayshore Boulevard, the tunnel will be installed within approximately 6.1 m (20 ft) of the existing 1.7 m (5.5 ft) diameter Sunnydale Sewer Tunnel. The 1220-m-long (4,000 ft) SAST is a gravity sewer and relatively shallow, having an average of 7.6 m (25 ft) of cover. The SAST downstream invert is fixed based on the invert of the transport box.

## GEOLOGY

### Site Investigation

The final excavation methods and support systems were recommended in part based on anticipated ground conditions in the project area (Figure 2). Preliminary site investigations were performed in 1996, with the final site investigations being completed by Arup in 2008. The investigations included a total of 45 borings. In addition to the drilling and sampling, exploration included: cone penetration tests, downhole seismic suspension logging, installation of vibrating wire piezometers, and packer tests at selected boring locations.

The subsurface conditions vary widely from soil deposits of sand and clay to Franciscan Complex bedrock. Portions of the site between Bayshore

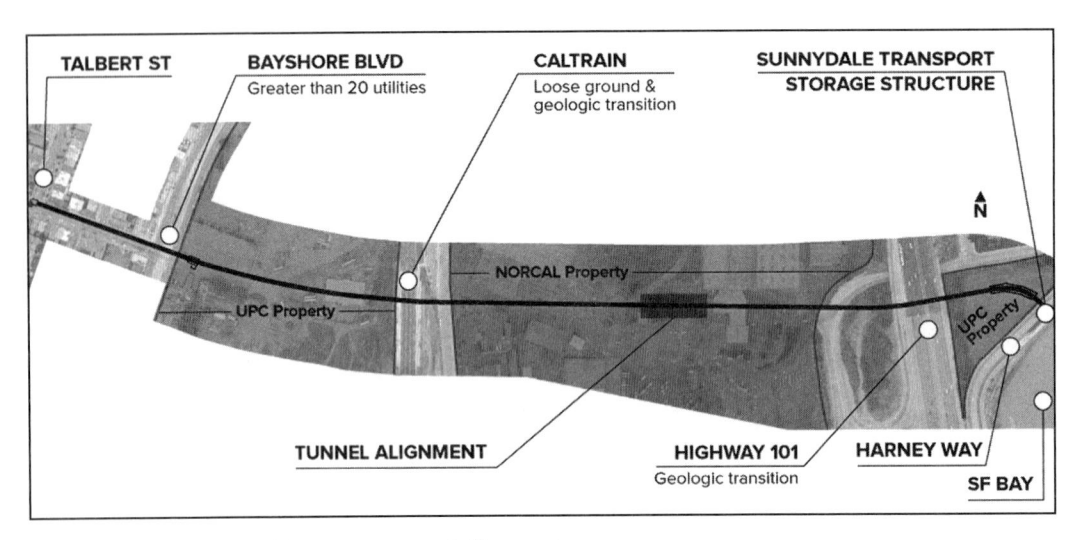

Figure 1. Sunnydale auxiliary sewer tunnel alignment

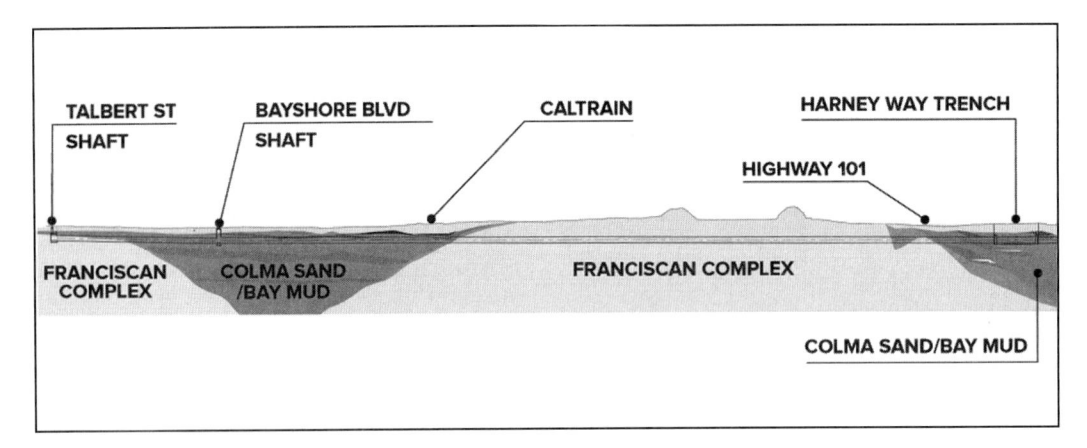

Figure 2. Geologic profile along the alignment

Boulevard and Tunnel Avenue and between Highway 101 and the Sunnydale Transport/Storage Structure at Harney Way are located outboard of the historic (1848) shoreline.

**Regional Geology**

The project site is located in the Coast Ranges geomorphic province, which is characterized by northwest-southeast trending valleys and ridges. These are controlled by folds and faults that resulted from the collision of the Pacific and North American plates and subsequent strike-slip faulting along the San Andreas Fault Zone. This highly active zone of faulting has been the source of numerous moderate to large-magnitude historical earthquakes that have caused strong ground shaking in the project area; however, none of these faults cross the project alignment.

The bedrock, the Franciscan Complex, consists mainly of a chaotic tectonic mixture of variably sheared shale and sandstone containing resistant rock masses largely of greenstone, chert, graywacke, and serpentinite. The degree of shearing ranges from gouge (mélange matrix) to unsheared rock. Fresh, relatively unsheared rock is hard, the larger resistant rock masses are pervasively fractured, and smaller resistant rock masses are commonly tough and relatively unfractured.

This mixture of rock materials exhibits a range of characteristics, including massive, closely jointed, completely crushed, and conditions resembling soft clay. The main consequence of this geologic mixture, especially as related to the prediction of ground conditions, is that the Franciscan Complex typically lacks spatial continuity and, therefore, exhibits frequent and abrupt variations in geomechanical characteristics. Similarly, weathering is highly variable between and within geological units.

### Sediments

The Quaternary sediments include residual soil at the rock/soil interface, discontinuous units of sand, sandy alluvium, and clayey to sandy colluvium, overlain in many places by Bay Mud with discontinuous units of peat. The main sediments include Colma Sand characterized as medium dense to very dense, medium to fine-grained sand with variable amounts of silt and clay. At the east end of the alignment, the Colma Sand decreases in thickness from about 27.4 m (90 ft) at Harney Way to where it pinches out below the center of Highway 101. The Colma Sand is also present between Tunnel Avenue and the western extent of the alignment. In this reach, the Colma Sand increases in thickness to a maximum of greater than 30 m (100 ft) and decreases to a 3-m (10-ft) thickness near Talbert Street.

The Bay Mud consists of marine clay and silt that was deposited in the San Francisco Bay and adjacent marshlands, tidal sloughs, and tidal mudflats when the Bay was inundated by rising sea levels through the Holocene epoch. Along the project alignment, the Bay Mud overlies the Colma Sand and is up to 3 m (10 ft) thick. Bay Mud includes silty clay, lean clay, sandy lean clay, elastic silt, silt, and clayey sand. The unit also contains occurrences of fat clay and occasional sand layers, shell deposits, peat deposits, and organic silts and clays of varying thicknesses and extent.

The fill material overlying these sediments is composed of interbedded deposits of gravel, sand, and clay, and generally covers the entire project area. The fill is generally loose to medium dense varying in thickness from less than 0.3 m (1 ft) up to 4.6 m (15 ft).

Groundwater levels throughout the alignment vary from 1.5 to 4.9 m (4.9 to 16 ft) below the ground surface.

## PROJECT CONSIDERATIONS

The proposed SAST alignment crosses under Highway 101; Union Pacific Railroad (UPRR) spur tracks; Caltrain Railroad Station and main tracks; structures located at the San Francisco Recycling & Disposal Inc. property (Norcal); property being developed by the Universal Paragon Corporation (UPC); and Bayshore Boulevard, a busy city street with numerous deep utilities and surface light rail facilities. Efficient permitting and agreements with affected agencies and property owners have been a major challenge of the project.

### Structures/Easements

The structures under which the alignment crosses include the eight-lane Highway 101, a major artery serving San Francisco. The alignment crosses directly beneath a 1950s-era strutted-bottomed culvert that was originally developed as a railroad access to the Bay. There is currently no public access through the 6-m-wide (20-ft) culvert. It is maintained by the California Department of Transportation (Caltrans), and the crossing required extensive coordination with Caltrans to obtain an encroachment permit.

This crossing is particularly complex as the ground transitions from Colma Sand into residual soil and finally into the Franciscan Complex, all within the Caltrans right-of-way. The major concerns expressed by Caltrans included the potential for settlement and/or damage to the culvert or the travel way.

Furthermore, the alignment crosses beneath existing buildings within the Norcal property and proposed buildings both on Norcal and UPC properties. The City of San Francisco (City) had numerous meetings and conversations with both property owners in order to resolve the easement requirements for the proposed sewer.

The major concerns expressed by both property owners included settlement induced by tunneling and the effect of the tunnel's presence to future development, such as increased structural complexity to future foundations and the impact on their development costs.

### Railroad and Light Rail

West of the Norcal property, the alignment crosses beneath a UPRR spur and a four-track railroad station operated by Caltrain. Caltrain serves commuters on the peninsula going into and out of San Francisco to San Jose. Near the UPRR spur, the ground transitions out of rock and into the Colma Sand and Bay Mud deposits overlain by a thick loose fill.

The primary consideration at this location is the capability to excavate beneath the tracks without disturbance to train service. A stoppage in service for Caltrain would result in a significant revenue loss and passenger inconvenience. With severely limited work windows, the project team evaluated several alternatives to mitigate the potential for settlement.

At Bayshore Boulevard the San Francisco Metropolitan Transportation Agency (SFMTA) operates a light rail station, which serves the local community. Although SFMTA is also a City agency, proper compliance and coordination procedures still apply to the project. The major concerns expressed by SFMTA included settlement of the tracks and damage to their facilities.

**Other Agencies—DTSC and BCDC**

The undeveloped land west of Caltrain was formerly the site of the Southern Pacific Railroad Yard and more recently the Schlage Lock Factory. The soils and groundwater in this region are known to be contaminated with solvents and volatile organic compounds (VOCs). The area has been and is currently undergoing an environmental cleanup program, which is overseen by the California Department of Toxic Substance Control (DTSC). The land is currently being prepared for multiuse development; hence, the clean-up efforts have been expedited. Because the proposed SAST passes through this zone, the project requires coordination with DTSC and compliance with its protocol. The project team has also extensively coordinated with the property owner and its environmental consulting firm.

The major concerns expressed by DTSC were whether the tunnel construction could negatively impact the cleanup efforts. In addition, because the western staging portal for the tunnel is located in this area, UPC was concerned that the project may interfere with the cleanup program. The cleanup includes injections of soluble vegetable oils in a grid pattern, which must be completed on schedule to avoid delays to the multiuse development project.

Lastly, all projects within 30 m (100 ft) of the San Francisco Bay fall under the jurisdiction of the Bay Conservation and Development Commission (BCDC), whose mission is to protect and enhance the Bay and encourage its responsible use. The permit requirements include accessing the impact of the construction activities occurring within its jurisdiction near Harney Way and the impact on public access.

**Utilities**

Because of its urban location, an estimated 54 utilities cross over or run parallel to the alignment. Typically, a tunnel of this size would be deep enough to avoid urban utilities; however, due to the tie-in elevation constraints at the Sunnydale Transport/Storage Structure and a minimum slope necessitated by the hydraulic requirements, this tunnel will be relatively shallow with a minimum cover of 4.9 m (16 ft) and average cover of 7.6 m (25 ft). Additionally, where the alignment crosses Bayshore Boulevard, a former route for Highway 101 and main artery into San Francisco, the existing Sunnydale Sewer runs parallel to and approximately 6.1 m (20 ft) to the south of the proposed tunnel. This sewer comes within 0.6 m (2 ft) of the ground surface, which forced utilities built along Bayshore Boulevard to dive below it, resulting in deeper than usual utilities in this area. Therefore, the proposed tunnel may be within 3 feet of some major electric and gas transmission lines.

*Potholing Program*

During the initial design stage, utility maps were collected from utility companies. These utilities included electric, gas, water, sewer, and fiber optic/telecommunication lines. The majority of the collected information provided no depth information. Since the tunnel will be shallow and mostly excavated via closed-faced microtunneling machines, it was imperative to know the elevation of utilities before tunnel excavation to avoid damaging existing utilities or creating a hazard during construction. Also, because the vertical alignment of the tunnel is fixed, any utilities found to be in the path of the proposed tunnel would need to be relocated. Early identification would allow utility owners enough notice to complete relocation projects prior to the start of construction. As a result, a potholing program to locate the deeper utilities was developed. After reviewing the utility maps and the initial field markings, 20 utilities were identified for potholing, the majority of which are located within the intersection of Bayshore Boulevard and Sunnydale Avenue.

**Summary of Risk Factors**

Risk factors along the alignment are shown in Table 1.

**SELECTION OF EXCAVATION METHODS**

During the early stages of design, the project team evaluated several potential alignments and construction methods for the proposed SAST. The methods were recommended both to accommodate the variable geology and also to address the concerns of the multiple stakeholders. The primary focus during preliminary stages was mitigating risks by approaching each geologic unit with what was viewed as the best-suited construction method. The excavation methods considered included open-trench near the Bay tie-in, conventional excavation with a roadheader within the rock reaches, and microtunneling in the contaminated ground. These options were pursued through the 35% design level.

However, increasing project costs forced the re-evaluation of alternate construction approaches. The most viable economic approach was found to be an earth pressure balance machine (EPBM) erecting a single-pass segmental lining for the majority of the alignment. A short pipe jack and microtunnel section was also recommended at the opposite ends of the project for reasons described in the following paragraphs. This approach reduced project costs by eliminating a shaft and combining the initial support

Table 1. Risk factors along the alignment

| Reach | Anticipated Ground Conditions | Primary Stakeholders | Risk |
|---|---|---|---|
| Harney Way to Highway 101 | Bay Mud | Universal Paragon | • Disruption to Harney Way |
| Highway 101 eastward to Tunnel Ave. | Colma Sand<br>Residual Soil<br>Franciscan Complex | Caltrans<br>Norcal | • Settlement at Highway 101<br>• Mixed Face Conditions<br>• Future impact on Norcal development plans |
| Tunnel Ave. to Bayshore Blvd. | Colma Sand<br>Bay Mud<br>Artificial Fill | Universal Paragon<br>Caltrain<br>UPRR | • Disruption to Caltrain/UPRR service and settlement of tracks<br>• Disturbance of contaminated plume and interference with soil/groundwater remediation program<br>• Future impact on UPC development plans |
| Bayshore Blvd. to Talbert St. | Colma Sand<br>Artificial Fill<br>Franciscan Complex | City of San Francisco<br>PG&E | • Numerous utility crossings beneath Bayshore (including critical transmission lines)<br>• Settlement of light rail tracks |

with the final lining; finally, the hydraulic design criteria were re-evaluated, which resulted in a smaller-diameter tunnel.

**Ground Support**

The ground support system must be compatible with the TBM operations. For this reason, a TBM will be used in conjunction with a prefabricated ground support system, which commonly consists of precast concrete segments that are bolted and gasketed to form a watertight lining.

Because the SAST is a CSO-type tunnel that only carries storm overflow, and the City of San Francisco has not historically had issues with hydrogen sulfide attack, the proposed support method for the tunnel is a single-pass segmental lining. This precast concrete lining will be watertight, bolted, doweled, and gasketed. The support will be relied upon as long-term support designed for hydrostatic and ground loads as well as seismic loading conditions.

**Harney Way to East of Highway 101**

This 130 m (426 ft) stretch of the alignment comprises an overflow parking lot for San Francisco 49er's games that will be utilized as a staging area, and Harney Way. Harney Way is a heavily trafficked street serving as an on/off ramp for Highway 101 leading to Candlestick Park. The area is within old fills, Bay Mud, and Colma Sand. The Harney Way trench will serve as the launching trench for the EPBM. The trench length was expanded in order to aid in the set up of trailing gear from within the trench.

Because of the need to tie into the below-grade Sunnydale Transport box at an invert elevation of –7.9 m (–26 ft), the design originally considered the open trench extending to the side wall of the transport box. However, two key considerations led to reconsideration: heavily trafficked Harney Way and BCDC jurisdiction.

The staging of the construction to provide appropriate traffic flow to/from Highway 101 proved to be considerably costly, particularly with regard to two large sewer lines located within Harney Way, requiring support in place. In addition, the short stretch across Harney Way is within the jurisdiction of the BCDC. Limiting the impact to this zone and having flexibility in staging is to the benefit of the City. Therefore, a 23.1 m (76 ft) portion of the alignment is proposed to be constructed using pipe jacking from the Harney Way trench, breaking into the transport box below grade (Figure 3). The inner dimensions of the transport box are approximately 6.1 by 6.1 m (20 by 20 ft), with a large access opening on the surface located nearby. Because the sewer will be tying into the wet-weather compartment, sewage flows will not be an issue during the dry-weather season. The cost of this pipe jacking was less than the open-trench option.

***Deep Utility Within Harney Way Construction Staging Area***

Within the proposed Harney Way staging area, of particular interest is the 0.76 by 1.14 m (2.5 by 3.75 ft) reinforced concrete egg-shaped sewer on piles (Figure 3). This sewer is of interest for two reasons: the piles will interfere with pipe jack excavation and will need to be cut, and the as-builts indicate that the

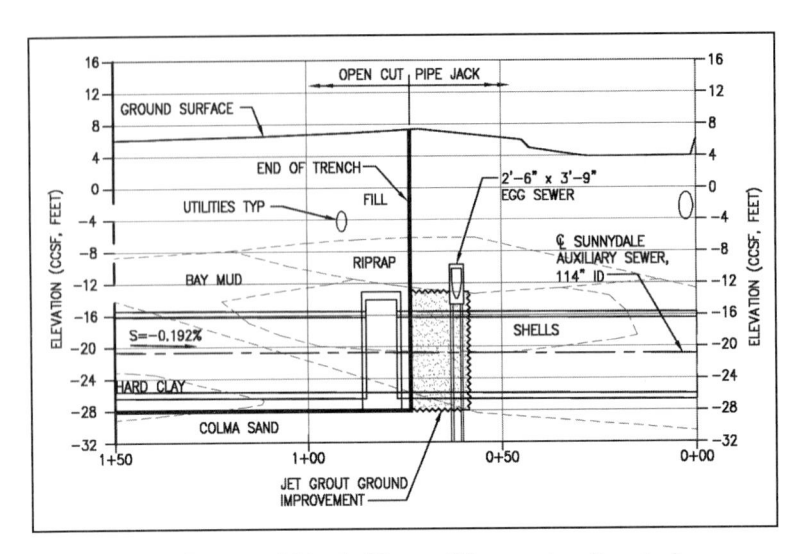

Figure 3. Profile of egg-shaped sewer within the Harney Way construction staging area

bottom of this sewer is less than 0.3 m (1 ft) above the top of the proposed tunnel excavation. The depth of this sewer was confirmed based on invert measurements in adjacent manholes and as-builts. Due to the limited clearance, the contractor will be required to pothole and daylight this sewer prior to pipe jack excavation to visually confirm its plan and elevation. Should the bottom of this sewer be located within the area of planned tunnel excavation, the contractor will be required to develop a plan outlining its proposed methods to construct the pipeline in this area. Because of the traditional design-bid-build contract on this job, this potential work will be considered as change-order work and will be compensated as such. Should the sewer be found where expected, the ground surrounding the utility will be jet grouted. The jet grouting will provide stability and prevent the ground (a combination of shells and riprap) from raveling during pipe jacking, which could ultimately leave the sewer unsupported.

## Highway 101 to Bayshore Boulevard

### Highway 101

The alignment beneath Highway 101 crosses through ground transitioning from fill, Bay Mud, and Colma Sand to Franciscan Complex bedrock, where it reaches a pinnacle then transitions again into a valley of residual soil and weathered rock. Although the soils and rock beneath Highway 101 are suitable for tunneling as individual units, the numerous transitions in relative hardness were viewed as complications for the TBM excavation. Although numerical models predicted low levels of settlement, experience dictates that the inability to adequately condition the soil and pressurize the face could lead to significant settlement. To mitigate these concerns, the project team recommended jet grouting a tunnel envelope from the surface. The addition of ground improvement will reduce the amount of expected settlement and angular distortion and create a consistent transition for the TBM beneath the highway. In addition, the jet grouting will create a zone for potential retooling of the cutterhead.

### Norcal

For the subsequent 610 m (2,000 ft) west of the Highway 101 undercrossing, tunneling will continue through greenstone, sandstone, siltstone, and chert within the Franciscan Complex bedrock beneath the Norcal property. Overall, the rock mass within this reach segment varies; conditions include limited zones of intensely fractured rock, but on average are moderately blocky and blocky. As the reach approaches Tunnel Avenue, the rock degrades to residual soil and slopes downward, where Colma Sand is present in transitions throughout the remainder of the excavation. Ground cover throughout the reach averages approximately 6.1 m (20 ft) above the tunnel excavated crown.

Although the alignment passes beneath existing buildings, their shallow foundations, built on the rock, are not anticipated to be impacted. The tunneling machine will be required to have the necessary capabilities to excavate the widely variable rock. The major design consideration for this stretch is to provide a final tunnel lining with adequate capacity to carry heavy foundation loads from future solid waste transfer station facilities. The tunnel lining design is capable of meeting the loading requirement of the future development. In addition, the specifications require design submittals to address the potential loading.

## *Caltrain*

As the tunnel passes beneath Tunnel Avenue and approaches the UPRR spur, the ground transitions out of rock and into the Colma Sand and Bay Mud deposits overlain by a thick loose fill. Settlement through ground loss is a significant concern to the project team and Caltrain. Without ground improvement, settlements could potentially exceed the limit of 13 mm (0.5 in.) set by the railway engineers. For this reason, the project team investigated different ground improvement methods for the loose fill above the tunnel crown, including chemical and jet grouting.

The width of the station platform would have required ground treatment to be performed from the ground surface, thereby obstructing the tracks. With only a maximum four-hour window on most days between continually serviced routes, the scheduling of any productive ground improvement work from the surface would prove to be impractical. In addition, Caltrain engineers expressed concerns about the potential for track heave and ballast contamination. In the end, a joint decision was made to not pre-treat the ground, but to instead perform the tunneling work during a 72-hour weekend work window, when trains are less frequent, and to utilize a stand-by reballasting crew. This will require the contractor to tunnel around the clock until out of the railroad right-of-way. In addition, extensive instrumentation and monitoring of the track was specified for this crossing.

## *Universal Paragon*

The tunnel alignment will pass through mostly Colma Sand in this stretch. Because the site is unoccupied, settlement is not a major concern. The largest concern is the contaminated soils/groundwater and the impact on the ongoing DTSC cleanup program. Environmental analysis of the soils and groundwater within the former factory site indicated that contamination at the depth of the tunnel was less than originally anticipated. It is anticipated that the soils will be hauled to a California Class 2 disposal facility, while the groundwater will require treatment for VOC contamination prior to disposal into the City sewer system. The site will, however, require an extensive clean-up program for the upper soils; which is being performed by the developer.

Although an open-cut was evaluated for this stretch, it was seen as having a potentially negative impact on the remediation program. Not only would this have obstructed the developer's activities, but also it would have generated large volumes of water to treat and would have potentially obstructed the localized groundwater flow. The environmental analysis showed no impact to the program from the tunnel construction; however, the specifications include provisions for the tunneling contractor to closely coordinate with the clean-up contractor because the staging area at Bayshore is primarily within land owned by the developer with on-going remediation activities.

In addition, it is anticipated that the site will be developed with single-story and multistory buildings. The multistoried buildings will require pile foundations. The tunnel lining was redesigned to accommodate increased loading from shallow foundations. Where pile foundations are required, the property owner will be compensated for the increased cost of spanning the tunnel.

## Bayshore Boulevard to Talbert Street

As the alignment heads west from the Bayshore Boulevard shaft, it will closely parallel the shallow existing Sunnydale sewer. The diameter required to meet the project hydraulic criteria along this is segment of the tunnel is 2.44 m 8 ft). As discussed above, the 46-m-long (150 ft) crossing beneath Bayshore Boulevard will encounter numerous utilities running north-south (Figure 4) as well as the SFMTA light rail system. Once across Bayshore Boulevard, the shallow alignment will end about 152.4 m (500 ft) west at the Talbert Street intersection.

Potholing above the water table was very effective, and utilities above this depth were easily located. However, potholing for deep utilities was substantially more difficult. The potholes were excavated through variable loose to medium dense fills and medium to very dense Colma Sand. Below the water table, the holes had a tendency to ravel and flow and then collapse. Because of the quickly recharging groundwater table, water could not be removed or blocked from the holes fast enough to obtain visual observation.

This was a major concern due to the presence of four Pacific Gas & Electric (PG&E) 110 to 115 kV high voltage electric lines and one 0.61 m (2 ft) gas line in a 0.76 m (2.3 ft) casing crossing the alignment within Bayshore Boulevard. Although all of these lines were marked, after the initial potholing, only one of these utilities was able to be located. The located line was felt at a depth within 1.5 m (5 ft) of the excavated diameter of the tunnel. For the remaining locations, the utility locators probed to a depth of approximately 3.2 m (10.5 ft). The failure to find these lines resulted in concerns that the remaining utilities could be even deeper, which was confirmed after further communication with PG&E and the discovery of additional as-builts.

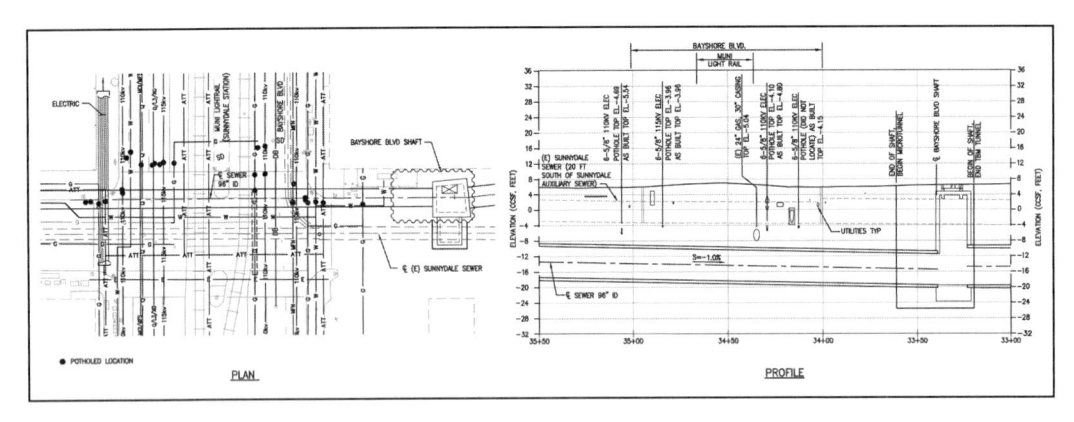

**Figure 4. Utilities beneath Bayshore Boulevard**

Because of the extreme consequences associated with hitting any of these utilities during tunnel excavation, a second potholing program was completed. Initial field marks for the PG&E utilities had been placed by as-builts. For the second program, PG&E entered nearby vaults and connected equipment directly onto the gas and electric lines to provide a more accurate location. This proved effective as two of the electric lines were discovered to be in significantly different locations than originally thought, and the gas line was discovered to be under the edge of the Muni light rail concrete slab, which meant that potholing would have to occur on an angle.

The angled potholing and ultimate depth of the gas utility meant that a visual would not be feasible, although it was attempted. With PG&E assistance, the top of the gas line was probed at a depth of 3.4 m (11.2 ft), with the bottom of the cased gas line anticipated to be within 0.64 m (2.1 ft) above the top of the proposed tunnel excavation. Two more electric lines were felt by probing at depths of approximately 3 m (9.8 ft) below ground surface. The fourth electric line was never found, despite three separate potholing attempts. The City has requested that PG&E locate this line. Because of relatively good correspondence between the as-builts of the other lines and their potholed depths, the depth of this line is also estimated to be at a depth of 3 m. Although none of these electric lines appear to be in direct conflict with the tunnel, they are all within 1 to 1.5 m (3.3 to 4.9 ft) of the proposed excavation.

Although probing in conjunction with coordination with PG&E was adequate for the design stage, these lines will need to be daylighted prior to tunnel excavation. Because the utilities are both deep and within different traffic lanes of a very busy intersection, this will present its own series of challenges. In addition, the option remains for the contractor to coordinate with PG&E to schedule a gas line outage during construction.

While it seems intuitive to have continued excavation with the EPBM machine to Talbert Street, these four high voltage transmission lines and gas transmission line dive deep beneath the existing sewer. This resulted in inadequate clearance to continue with the larger diameter machine.

A deep open-trench with dewatering was viewed as too disruptive to the residential community, impractical to cross the utilities within Bayshore Boulevard, and having the potential for extending the contaminated plume through dewatering. For this reason, several trenchless methods were evaluated for this stretch; in the end, microtunneling was cost comparable and advantageous to reduce settlements to tolerable limits at the Muni light rail station and for impacted utilities.

## SUMMARY/CONCLUSION

### Transition Zones

The SAST alignment crosses beneath several transit structures, two of which are critical: Highway 101 and the Caltrain Bayshore Station. The crossings are also the transition zones of geology—transitioning from fill, Bay Mud, and Colma Sand into residual soil and Franciscan Complex bedrock. The material with the lowest strength and compaction is the fill that overlies Bay Mud near the crown of the tunnel. The project team recommended ground improvement at the Highway 101 transition zone. However, the risk of heaving the track or contaminating the ballast led to the decision to perform the Caltrain railroad crossing during a low traffic weekend work window with a reballasting crew on standby.

### Utilities

To aid in the design and planning phase, an extensive potholing program was undertaken and several deep utilities were located. Depths indicated by probing closely coincided with as-built locations, providing

**Table 2. Summary of reach construction methods**

| Reach | Reach Length | Anticipated Ground Conditions | Pipe Inner Diameter | Construction Method |
|---|---|---|---|---|
| Harney Way to Highway 101 | 130 m (426 ft) | Bay Mud Colma Sand | 2.9 m (9.5 ft) | • Open trench<br>• Pipe jack beneath Harney Way |
| Highway 101 to Bayshore Blvd. | 895 m (2,935 ft) | Colma Sand Residual Soil Franciscan Complex bedrock | 2.9 to 3.4 m (9.5 to 11 ft) | • EPBM excavation<br>• Single-pass, bolted, gasketed reinforced concrete segments |
| Bayshore Blvd. to Talbert Street | 196 m (643 ft) | Colma Sand Artificial Fill Franciscan Complex bedrock | 2.4 m (7.9 ft) | • Microtunnel |

relative certainty that the lines are all located outside of the tunnel's path, even though some utilities are close. As with any project, there always remains a chance that unknown utilities will be encountered. An allowance item for all work related to the protection, maintenance, or relocation of unconfirmed or unknown existing utilities is included in this contract.

Recommendations for future pothole programs include an evaluation of anticipated utility depths, especially when obstructions or deep utilities are anticipated, and a workable plan for probing several feet below the anticipated depth. If there are critical utilities, it must be ensured that the personnel performing the field marking have taken the time to connect to the line at a nearby vault. It is also recommended that an owner's representative be on site during the potholing to make decisions on where to pothole and what amount of additional effort should be spent on locating a particular utility.

**Excavation Methods**

Based on the final hydraulic analysis, the inner diameter of the EPBM tunnel has been set at a hydraulic minimum of 2.9 m (9.5 ft). At the same time, it is beneficial to choose a tunnel size that is familiar to U.S. contractors and that will allow the use of machines that are widely available, hopefully resulting in a lower bid price. Given these considerations, the project team set a minimum tunnel inner diameter at 2.9 m, allowing for an inner diameter up to 3.4 m (11.2 ft) in the contract documents, should the contractor be better equipped to supply a larger tunnel.

With the goal of balancing the cost benefit of mitigating risks with completing the envisioned project, the project team recommended that the project be constructed as an EPBM tunnel, erecting a single-pass, gasketed segmental lining, with supplemental work at either end of the alignment by microtunneling and pipe jacking. Although other options were pursued, they ended up making the project cost impractical. The recommended methods result in a cost-effective solution, which allows the SFPUC to achieve all the goals of this project.

**BIBLIOGRAPHY**

Deere, D.U., Peck, R.B., Monsees, J.E., and Schmidt, B. 1969. *Design of Tunnel Liners and Support System, Final Report for Office of High Speed Ground Transportation*. U.S. Department of Transportation, Washington, D.C.

Heuer, R.E., and Virgens, D.L. 1987. Anticipated behavior of silty sands in tunneling. *Proceedings of the Rapid Excavation and Tunneling Conference*. Littleton, CO: Society of Mining Engineers, Inc.

Kuesel, T.R. 1972. Soft ground tunnels for the BART project. *Proceedings of the Rapid Excavation and Tunneling Conference*. Littleton, CO: Society of Mining Engineers, Inc. Vol. 1, pp. 287–313.

O'Carroll, J.B. June 2005. *A Guide to Planning, Constructing and Supervising Earth Pressure Balance TBM Tunneling*. Parsons Brinckerhoff.

Sutter, D.A., Frobenius, P. and Wu, C. 1995. Tunnel Design of MUNI Metro Turnback Project. *Proceedings of the Rapid Excavation and Tunneling Conference*. San Francisco, California. Vol. 1, pp. 703–722.

Thon, J.G., and Amos, M.J. 1968. Soft ground tunnel for BART. *Civil Engineering, ASCE*, pp. 52–59.

# New York Harbor Siphon Project

**Colin Lawrence, David Watson**
Hatch Mott MacDonald, New York, New York

**Michael Schultz**
Camp Dresser and McKee, Cambridge, Massachusetts

ABSTRACT: The New York Harbor Siphon project involves construction of a 1.8 mile long 12-foot diameter tunnel beneath the New York Harbor, using a pressurized face tunnel boring machine. This subaqueous crossing will be the first of its kind using pressurized face tunneling technology in New York City. A steel water main inside the tunnel will connect water distribution systems on Staten Island with Brooklyn. The project is being managed by the New York City Economic Development Corporation on behalf of the owner, the New York City Department of Environmental Protection, and the Port Authority of New York and New Jersey. This paper outlines the complex geological conditions and highlights key design and construction considerations.

## INTRODUCTION

The Port of New York and New Jersey is one of the most heavily used transportation arteries in the world, handling nearly 40 percent of the North Atlantic shipping trade and directly providing nearly 230,000 jobs to the local economy. In 2004, $100 billion worth of consumer goods moved through the port.

In order to accommodate future cargo volumes in the port, which are expected to double over the next decade and possibly quadruple in 40 years, deeper shipping channels are needed to provide access for a new generation of cargo megaships with drafts exceeding 45 feet when loaded. Current channels within the harbor range in depths up to 45 feet, thus preventing carriers from using these larger ships, or requiring significant reductions in cargo to achieve lesser drafts to operate safely within the harbor.

Under Section 101(a)(2) of the Water Resources Development Act of 2000 (P.L. 106–541), Congress authorized the deepening of a number of channels in the New York/New Jersey Harbor, including the Anchorage Channel, which extends from the Verrazano-Narrows Bridge to its confluence with the Port Jersey Channel. In 2006, PANYNJ, in cooperation with the USACE, began dredging operations to allow larger ships to access the port.

As part of the Harbor Deepening Project, the Anchorage Channel will be deepened to 50 feet below mean low water (MLW), for a length of 19,000 feet. This dredging will expose and/or potentially impact two existing siphons that are situated at an insufficient depth to be protective of their continued use.

The New York City Department of Environmental Protection (NYCDEP) owns, operates, and maintains the two existing water siphons in the Harbor. Due to their shallow depth, both existing siphons must be relocated before dredging of the Anchorage Channel can be completed.

The siphons were constructed across the Anchorage Channel of the Upper New York Bay using the wet-trench excavation method and are 36-inch diameter (Siphon No. 1) and 42-inch diameter (Siphon No. 2) constructed circa 1917 and 1925, respectively.

To replace the existing siphons, a new 9,440-linear-foot, 72-inch-diameter pipeline to be installed within a 12-foot diameter tunnel. The existing and proposed siphons are shown in Figure 1. The NYCDEP have previously crossed this reach of the harbor to the north with the construction of the Richmond Tunnel in the 1960s, although at depths of over 900-feet, the tunnel crossing of the harbor was excavated through rock.

A pressurized face tunnel boring machine will be used to excavate the tunnel through the predominantly soft ground soil conditions from a launch shaft in Staten Island to a receiving shaft in Brooklyn. The tunnel will be lined with gasketted, precast concrete segments. Following completion of the tunnel drive, a welded steel pipeline will be installed and the void between the steel pipeline and the tunnel will be backfilled with grout.

The harbor siphon tunnel is scheduled to be the first tunnel built under the Hudson River or the New York Harbor in many decades. The pressurized face

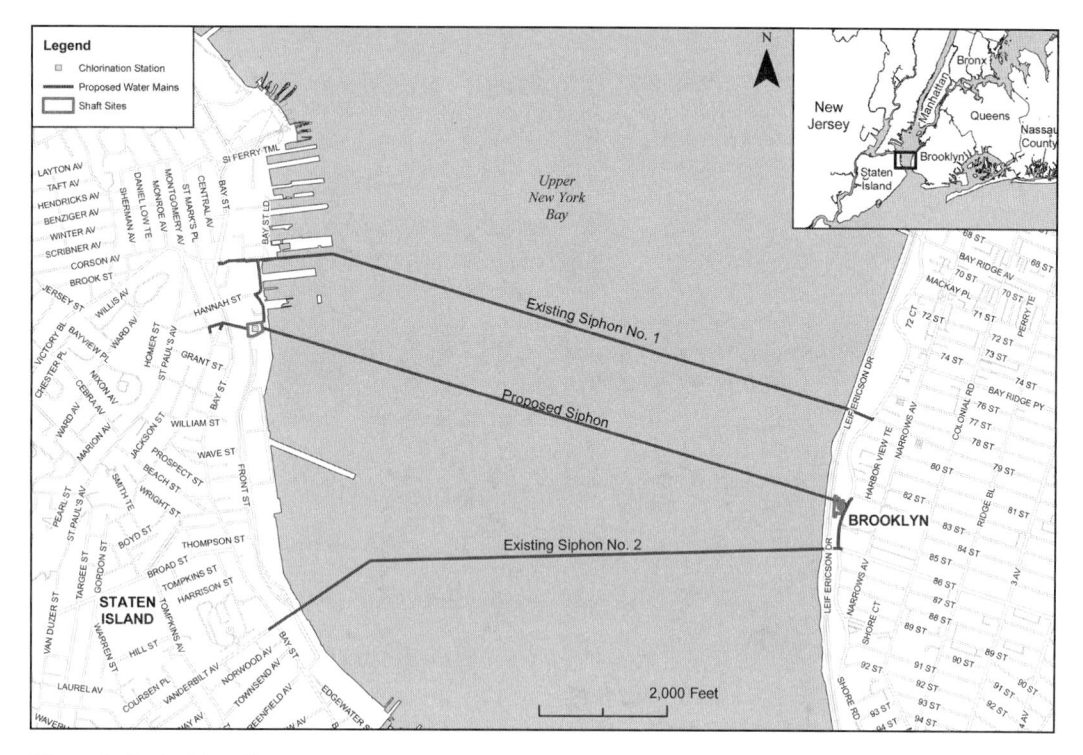

**Figure 1. Project location**

tunneling methodology that will be used to construct the harbor siphon will be the first application of this technique for a sub-aqueous crossing in New York City.

On behalf of the New York City Department of Environmental Protection (NYCDEP) and the Port Authority of New York and New Jersey (PANYNJ), the New York City Economic Development Corporation (NYCEDC) is managing a project. The CDM/Hatch Mott MacDonald Joint Venture (JV) has been retained by the NYCEDC to perform engineering design services for this project.

This paper discusses the development of the design and the various challenges faced by the design team in the first project of its scope to be constructed in New York, focusing on the design of the siphon tunnel beneath the New York Harbor.

**Project Overview**

The project comprises the following primary components:

1. Bored Tunnel (Siphon): 12-foot nominal diameter bored tunnel to be constructed using a pressurized face tunnel boring machine and lined with a precast concrete, gasketed, segmental lining system. The tunnel boring machine will be launched from a shaft in Staten Island and driven over a distance of approximately 9,440 feet, beneath the Anchorage Channel, to a receiving shaft in Brooklyn. A 72-inch diameter steel transmission pipeline will be backfilled inside the tunnel.

2. Shafts: The Staten Island launching shaft will be located near the intersection of Front Street and Murray Hulbert Avenue on Staten Island. The Brooklyn receiving shaft will be located in Shore Road Park near the intersection of Shore Road Land and Shore Road. The shafts are to be constructed using either slurry wall or ground freezing methods.

3. Trenchless Crossings: Two trenchless crossings will be constructed beneath the Staten Island Railroad (SIR). The crossings are to be constructed using a microtunnel boring machine (MTBM) up to 84-inch in diameter and are approximately 325 and 120 feet in length.

4. Staten Island and Brooklyn Land Piping: Water transmission mains constructed in open cut to connect the new infrastructure with the existing water distribution system.

5. New Chlorination Station: A new chlorination station is required to boost the chlorine residual in the new siphon water supply.
6. Abandonment of Existing Siphons and Metering Chambers: The existing siphons and metering chambers will be abandoned in place following successful commissioning of the new siphon.

**Project Purpose**

The purpose of this project is to replace the existing water siphons between Brooklyn and Staten Island, New York. The project involves the abandonment of the two existing water siphons and their replacement with a new water siphon across the New York Harbor and the Anchorage Channel, between Brooklyn and Staten Island. The general locations of the existing and proposed siphons are presented in Figure 1.

As part of this project, water mains on both the Brooklyn and Staten Island sides of the crossing will be installed to connect the new siphon to existing water transmission mains. In addition a new chlorination station will be constructed on Staten Island to serve the new siphon. Various sewers will also need to be constructed or relocated to accommodate the project.

The proposed siphon will be located at a depth of at least 85 feet below MLW within the channel limits. This depth was selected in order to place the new siphon at a depth that will not be affected by the currently proposed dredging of the Anchorage Channel, as well as any reasonably anticipated future dredging within the channel. The scheduling of the new siphon construction is also important to allow the harbor deepening efforts within the Anchorage Channel and associated port activity to continue as planned.

**PRELIMINARY ENGINEERING**

During the Preliminary Engineering phase, an evaluation of feasible construction methods for the harbor crossing resulted in horizontal directional drilling, micro-tunneling, and pipe jacking being eliminated from further consideration owing to the length and diameter of the crossing. The evaluation focused on two primary alternatives: a dredged trench; and a bored tunnel. The alternatives were evaluated against the following criteria:

- Schedule
- Constructability
- Construction risk
- Environmental impacts

Configurations of the number and diameter of pipes, pipeline materials (ductile iron, fiberglass reinforced polymer, high density polyethylene, pre-stressed concrete, and welded steel), and shaft construction methods were also included in the evaluations.

It was agreed that a bored tunnel was preferable to a dredged trench for this project, even though the estimated construction cost of the bored tunnel alternative was slightly higher. The bored tunnel alternative was selected primarily for the following reasons:

- Substantially less environmental impact
- No impact on navigation in the harbor
- More predictable permitting process and timeline
- Less risk of weather-related construction delays
- Less risk of unexpected costs for disposal of contaminated soils

The final recommendation included the establishment of the design concept and preliminary design parameters for the project. These included:

- A 12-foot nominal diameter bored tunnel, constructed primarily in soft ground, using a pressurized face tunnel boring machine (TBM). Tunnel length will be approximately 9,440 feet.
- A gasketted, precast concrete segmental lining erected by the TBM.
- Tunnel launching and receiving shafts constructed by either slurry wall or ground freeze technologies.
- A 72-inch (6-foot) nominal diameter welded steel water transmission pipeline installed within the tunnel.

**GEOLOGY AND HYDROGEOLOGY**

**Regional Geology and Hydrogeology**

The project site lies within the Hudson River Basin on the border between Staten Island and Brooklyn, near the confluence of the Manhattan Prong of the New England Uplift, the Newark Basin Physiographic Province, and the Atlantic Coastal Plain Physiographic Province. The region is generally characterized by thick glacial sediments overlying sedimentary and metamorphic rock. The bedrock at the project site is judged to be of the Hartland/Manhattan Schist Formations. Following the glacial retreat and subsequent sea level rise, sediments including poorly graded sand, silty sand, slightly organic silt, clay, and peat have been deposited in and adjacent to New York Harbor. The surficial geology of the land sides of the proposed tunnel alignment are dominated by glacial soils overlying bedrock.

The hydrogeologic setting in the project area is dominated by the presence of the New York Harbor and the glacially deposited soils. The groundwater flow generally trends from the interior of Brooklyn and Staten Island toward the New York Harbor. The data also show a general trend of higher groundwater elevation with increasing distance from the New York Harbor. The glacial soils encountered at the landside portions of the project generally consist of water-bearing coarse-grained soils with occasional layers of silt and clay, and exhibit tidally influenced groundwater elevations.

**Site Geology and Groundwater Conditions**

A phased approach to the site investigation was adopted to allow a limited number of borings to be taken prior to deciding on the preferred construction method for the siphon tunnel. In total, 38 marine borings, 16 borings in Brooklyn, and 30 borings on Staten Island were drilled. A range of in-situ and laboratory tests were carried out, including: in-situ vane shear; in-situ pressuremeter; slug/bail testing; mini-vane; triaxial; consolidation; and frozen soil testing. A total of 26 observation wells were installed, and a geophysical survey program was conducted across the harbor.

The available subsurface data gathered during the multiphase geotechnical investigation program indicated high variability in the subsurface conditions along the proposed tunnel alignment, the shaft locations and water transmission mains.

The subsurface soil stratigraphy in the project vicinity (land and marine), as encountered in the geotechnical borings, can be categorized into stratigraphic units as shown in Figure 2, with typical descriptions as follows:

- Fill:
  - Loose to dense, miscellaneous sand and gravel with variable amounts of silt and clay, debris (brick, asphalt, glass, wood etc.) and occasional cobbles and rock fragments.
- Glacial Soils:
  - Silty Sand and Gravel (SSG): Medium dense to very dense, fine to coarse sand with variable amounts of silt and gravel, and occasional layers of stiff to hard silt with variable amounts of sand and gravel. Within the stratum, there are occasional cobble and boulder zones.
  - Fine to Medium Sand (FMS): Medium dense to very dense, fine to medium sand, trace to some silt and sporadic traces of fine gravel, with occasional layers of stiff to hard, fine-grained soils consisting primarily of silt with little to some sand and variable amount of gravel. Also present are occasional zones of gravel, cobbles and boulders.
- Recent Marine Sediments:
  - Plastic Silt and Clay (PSC): Very soft to medium stiff, slightly organic, plastic, Silty clay, Clayey silt and clay.
  - Marine Sand with Silt (MSS): Very loose to loose, fine sand with trace to some silt, and occasionally silt with little to some fine sand.
  - Interlayered Clay, Silt and Sand (ICSS): Interlayered zones of: 1) soft to medium stiff, low plasticity silt and clay with little to some fine sand and occasional silty clay, and 2) very loose to loose, non-plastic silty sand and sandy silt.
- Lower Deposits: Likely of Pleistocene age or older and may be either coastal or glacially reworked deposits:
  - Lower Silt and Clay (LSC 1 and LSC 2): Medium stiff to hard, low plasticity clay and silt to silty clay with trace to some sand.
  - Lower Sand and Silt (LSS): Dense to very dense, fine to medium sand with little to some silt.
  - Lower Sand and Gravel (LSG): Very dense and typically consists of various combinations of fine to coarse sand and gravel, with trace to little silt.
- Bedrock:
  - Predominantly consists of decomposed to fresh, soft to moderately hard, mica schist and gneiss with zones of pegmatite. Based on the subsurface exploration data from this project, the bedrock slopes down toward the east, from about El. −67 at the Staten Island shaft site, to be between El. −270 and −320 near the Brooklyn seawall.

Groundwater elevations were recorded at a number of observation wells located on Staten Island and in Brooklyn:

- Staten Island Shaft Site: Measured groundwater elevations range from El. 2.5 to El. 8.0 and indicate between 1 to 2 foot higher piezometric head levels in the deeper wells screened in the bedrock, compared to the shallower wells screened in the glacial soil.
- Brooklyn Shaft Site: Measured groundwater elevations range from approximately El. 2 to El. 6 and indicate a horizontal hydraulic gradient from east to west toward the harbor, and from north to south.

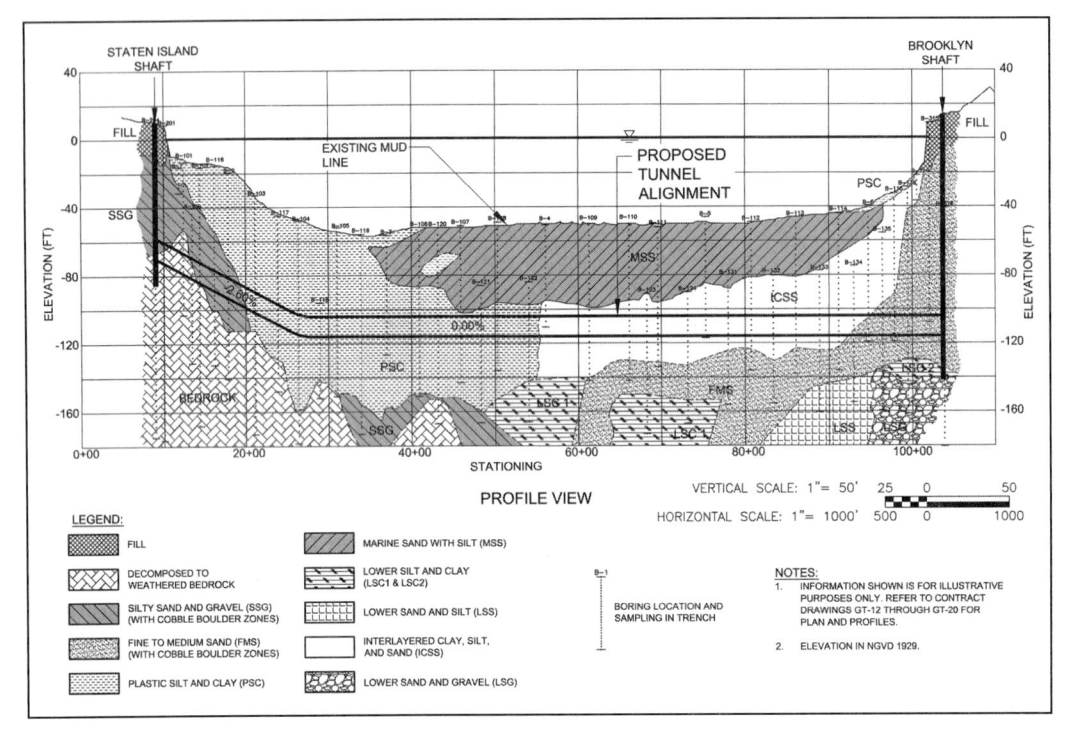

Figure 2. Generalized subsurface profile

The groundwater levels exhibit tidally influenced fluctuations of 1 to 1.5 feet that lag behind the tide cycle by between 1 and 3 hours. The tidal variation in the deeper wells shows generally less variations and greater lag time than in the shallower wells. There is more variation (3.5 to 5.5 ft) in groundwater levels due to seasonal and rainfall changes than due to the tidal variation. At the tunnel horizon, the groundwater pressure will be primarily controlled by the water level within the harbor. Toward the land side portions of the alignment, the groundwater head at the tunnel horizon will be greater than the hydrostatic pressures from the harbor tide level, as indicated by observation well data.

## Geotechnical Baseline Report

A Geotechnical Baseline Report (GBR) was developed for the bored tunnel, microtunnel and shaft components of the project. The principle purpose of the GBR was to set clear, realistic baselines for conditions anticipated to be encountered during subsurface construction.

Some of the primary considerations addressed in the GBR included:

- Soil/Rock Parameters. Due to the mixed face (soil and rock) and variable soil conditions (soft marine sediment to dense glacial deposits) rock and soil parameters were baselined in concise summary tables. In addition, the baseline parameter tables included statistical evaluation of the available test data including the minimum, maximum, mean and standard deviation of the laboratory and field data, providing an overview of the statistical variability of the parameter data.
- Tunnel Face Conditions. Tunnel face conditions were defined by one or a combination of the project strata to be encountered along a tunnel alignment. The tunnel face conditions were baselined by providing a range of drive length for each face condition as well as a face diagram for each face conditions. The face diagrams provide the baseline grain size distributions of the face conditions to be encountered within a specific face condition (see example diagram in Figure 3).
- Soil Abrasion. Based on recent project experience in dense glacial soils the abrasivity of the soil strata on the tunnel cutting tools was addressed in addition to the abrasivity of the

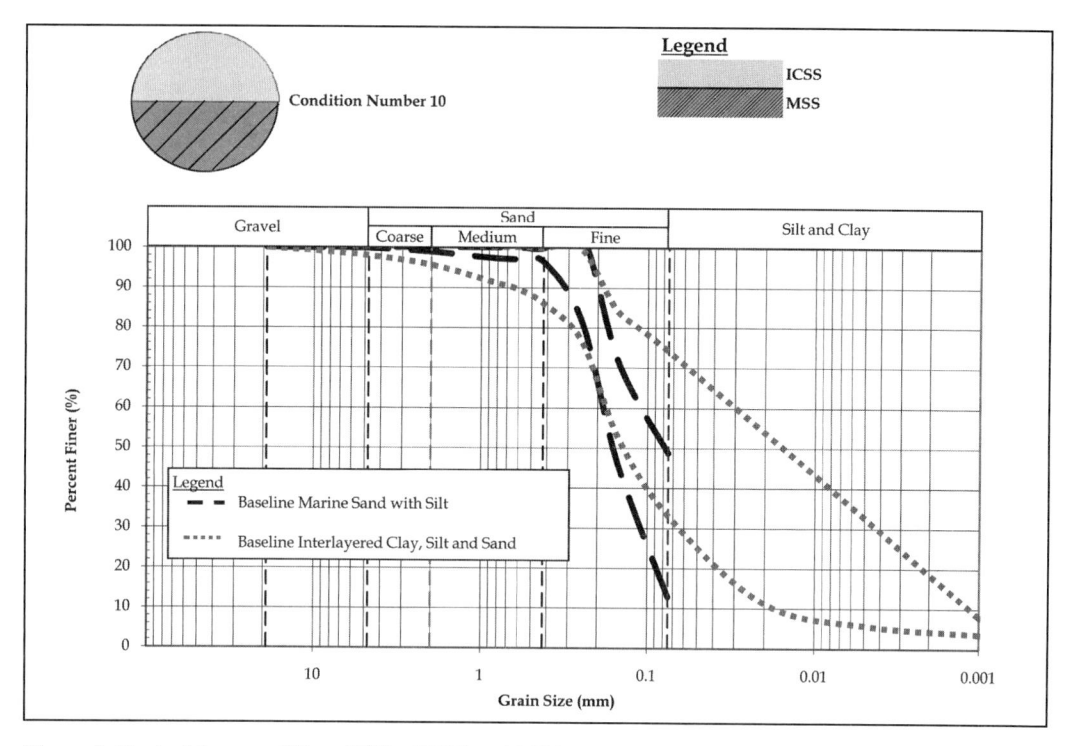

Figure 3. Typical face condition (#10) – ICSS and MSS

bedrock. Previous project experience indicates that tool wear can be excessive in both slurry and EPB TBMs in soil conditions with similar quartz content and density as the glacial soils that will encountered for the Siphon project.
- Cobbles and Boulders. The presence of multiple zones of nested boulder and cobbles in the glacially deposited project strata (SSG and FMS) is anticipated and was baselined. Each cobble/boulder zone was assumed to extend up to 50 feet along the bored tunnel alignment.
- Stickiness/Clogging. The stickiness/clogging potential of the soft recent marine sediment on the TBM tools and muck conveyance was evaluated through the use of methodology developed by Thewes (2004) which defines zones of stickiness/clogging potential based on Atterberg limit laboratory data. Based on this approach PSC and ICSS strata baselined as low stickiness/clogging potential while the LSC stratum as medium to high stickiness/clogging potential.
- Design Considerations. The criteria and methodologies used for the design of shaft and tunnel ground support and ground stabilization including tunnel lining was addressed. In addition, the basis for the selected horizontal and vertical tunnel alignment, which was influenced by the possible presence of abandoned piling supporting historic Pier Nos. 8 and 9 on Staten Island, was addressed.
- Construction considerations. Key construction considerations addressed in the GBR include the anticipated ground behavior in response to construction operations in the variable soil/rock conditions including tunnel face stability, anticipated face pressures and the handling and disposal of excavated material. Interventions and inspections of cutterhead and forward shield were anticipated due to the variable ground conditions (mixed face) and the anticipated TBM tool wear. It was recommended that planned interventions be avoided in the ICSS stratum due to the instability of the soils comprising that stratum.

## SHAFTS

### Shaft Locations

The locations of the two shafts were established based on identifying connections with the existing water distribution system, availability of suitable land, environmental impacts, and a suitable site for launching and servicing of a tunnel boring machine.

The City of New York owns a vacant site in an industrial waterfront area on Staten Island that will be used to construct the launch shaft and to stage the tunnel boring machine operations. The site is adjacent to the former Navy Homeport site that is currently in planning to be redeveloped as part of the Stapleton Waterfront Project.

The limitation of available land on the Brooklyn side of the tunnel alignment has resulted in the receiving shaft being located within a park between the Belt Parkway and Shore Road, near Fort Hamilton High School. This park is owned and maintained by the New York City Department of Parks and Recreation. The exact location of the shaft within the park has been refined in coordination with the New York City Department of Parks and Recreation (NYCDPR) to ensure that temporary and permanent impacts on the park are minimized to an acceptable level, with a number of trees requiring protection during construction.

The proposed pipeline will connect to the existing distribution system in Brooklyn at the intersection of Shore Road and 86th Street. On Staten Island, the proposed pipeline will connect to the existing distribution system at two points, one at Van Duzer Street Extension and the other at Victory Boulevard.

**Shaft Design**

The internal diameter of the Staten Island Launching Shaft is 28-feet with the base slab approximately 88-feet below existing ground level. The Brooklyn receiving shaft is 24-feet internal diameter and the base slab is approximately 140-feet below existing ground level. Shaft diameters were selected to accommodate water main piping and appurtenances, TBM launch at Staten Island, and TBM reception at Brooklyn.

Two alternative methods of shaft construction were recommended in the preliminary engineering phase: ground freezing; or slurry walls. Although slurry wall construction has been designed and depicted on the contract drawings, ground freezing will be allowed as an acceptable shaft construction method.

The slurry wall panels are to be excavated through predominantly granular soils with decomposed and weathered rock on the Staten Island side. The use of a hydromill (hydrofraise) was considered to be suitable for the anticipated ground conditions and to provide the required verticality tolerance (1 in 200) for the construction of the slurry wall panels.

The panel geometry was set out to accommodate a three-bite primary panel with a one-bite secondary panel. The secondary panel has an overlap of at least 6-inches with the adjacent primary panel concrete to form a watertight joint between panels. A shear key is provided to enhance shear transfer between panels and improve the watertightness of the joint.

Watertightness of the shaft is defined in the contract specifications by an allowable inflow criteria of 0.07 gallons per minute overall, and 0.0125 gallons per minute from any single source, with no running water from the wall permitted. The panel arrangement for the Staten Island Shaft is shown in Figure 4.

The reinforcement at the tunnel eyes for the break-out and break-in was specified using glass fiber reinforced polymer (GFRP) to allow the TBM to mine through the shaft wall without encountering steel reinforcement. The GFRP reinforcement design was carried out in accordance with ACI 440.

**TUNNEL ALIGNMENT**

The bored tunnel will consist of a nominal 12-foot excavated diameter, precast concrete, segmental lined tunnel, extending from the Staten Island shaft beneath New York Harbor to the Brooklyn shaft. A 72-inch welded steel pipeline will be installed in the tunnel to convey water between Brooklyn and Staten Island. The annular void between the steel pipeline and the bored tunnel lining will be backfilled with concrete. The typical tunnel cross section is shown in Figure 5. The alignment was selected to meet a number of construction and operational considerations described below.

**Staten Island Bulkhead and Demolished Piers**

The tunnel vertical alignment near the Staten Island shaft has been located at a depth to provide clearance beneath the timber piles of both the existing bulkhead wall and the demolished Pier No. 8. The timber piles of the demolished pier were either pulled or cut at mudline. The pile tip elevations have been assessed based on available historic drawings.

The results of the site investigation indicated that the Pier No. 8 and Pier No. 9 piles toward the harbor end of the piers were likely driven deeper than indicated on the historic drawings. As part of the risk management approach adopted on the project, the horizontal alignment was amended to avoid the plan location of the harbor end of the demolished piers.

**Harbor Dredging**

The proposed siphon must be constructed at a depth sufficiently below the proposed channel depth of El. −50.0. The NYCDEP has also expressed an objective of constructing the proposed siphon deep enough to accommodate possible future harbor deepening programs. Based on the constraints defined for the project, the proposed siphon will be installed with the top of pipe at or below El. −75.9. This depth will be

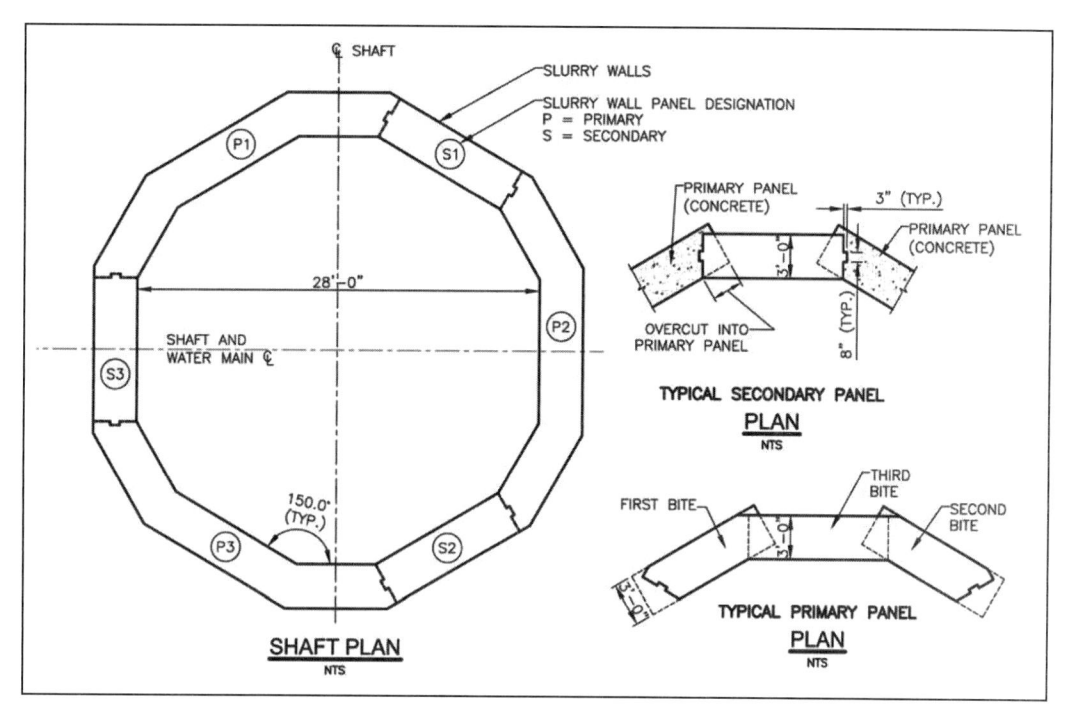

Figure 4. Staten Island shaft slurry wall panel arrangement

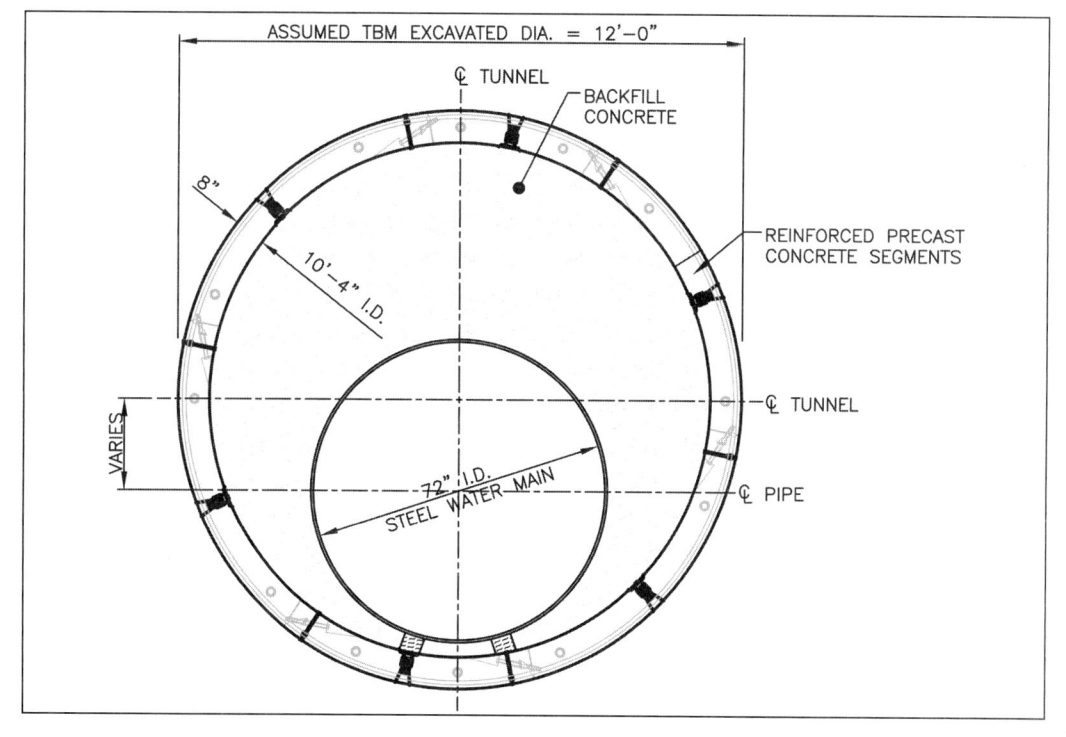

Figure 5. Typical tunnel cross section

maintained across the Anchorage Channel, the Bay Ridge Channel, and the Stapleton Anchorage.

**Brooklyn Seawall**

The Belt Parkway, a six lane highway, and the adjacent Promenade at Brooklyn are protected from the harbor by a seawall. Historic design drawings and bathymetric data show that the seawall is founded on a relatively shallow riprap foundation placed on the previously existing mudline. The tunnel vertical alignment was set to provide sufficient cover below this structure.

**Gravity Drainage of the Tunnel**

The vertical alignment of the siphon tunnel has been subject to a number of changes. The final alignment is based on the requirement to provide gravity drainage of the tunnel to one of the shafts and to position the vertical alignment with due consideration of the ground conditions. The final vertical alignment provides a slope from the Staten Island Shaft toward the Brooklyn Shaft.

**Water Transmission Main Operations**

- Provide key connection points to the existing water main distribution systems on Staten Island and in Brooklyn.
- Connect with a suitable new Chlorination Station for the project.

## TUNNEL LINING DESIGN

The lining rings have an internal diameter of 10'-4" and an outer diameter of 11'-8". The rings consist of four 67.5° parallelogram segments and two 45° trapezoidal segments. The nominal width and thickness of the lining rings are 56.0 inch and 8.0 inch, respectively. The rings are tapered 0.5 inch, with a minimum width of 55.5 inch and a maximum width of 56.5 inch.

Several different cross sections were analyzed to study the effect of rock and the various soil strata along the project alignment on the tunnel lining. The effect of dredging in reducing the load on the lining after construction was also investigated.

Each segment is fitted with ethylene-propylene-diene monomer (EPDM) gaskets to resist water ingress into the completed tunnel. Dowels are provided at the circumferential joints, with a typical pitch of 22.5°, and two bolts are provided at each radial joint.

The bolts, dowels, and the inserts keep the gaskets compressed. The gaskets are designed for a maximum hydrostatic pressure of 9.0 bars, including gap and offset, providing a factor of safety of two in relation to the actual hydrostatic pressure.

The material properties for the segments are:

Compressive strength of concrete, $f_c' = 7,500$ psi

Yield strength of reinforcement, $f_y = 75,000$ psi

## CURRENT STATUS OF THE PROJECT

In October 2009, the NYCDEP issued a Negative Declaration determining that the proposed project will not have a significant effect on the environment. An Environmental Assessment Statement was prepared by the JV to provide supporting evidence to the determination.

The NYCEDC issued a Request for Qualifications (RFQ) for the provision of tunnel construction services on August 19, 2009. Contract Award and Notice to Proceed are anticipated to be in the second quarter of 2010.

# Rehabilitation of Rail Tunnels with Widening the Cross-Section While Maintaining the Regular Rail Traffic

**Christian Wawrzyniak**
CDM Consult GmbH, Stuttgart, Germany

**Heiner Fromm**
CDM Consult GmbH, Alsbach, Germany

ABSTRACT: This paper deals with the renewal of existing railroad tunnels while maintaining the running railroad traffic. Due to technical reasons of modernization, in Germany many new tunneling projects as well as reconstruction of existing tunnels are underway at present. Depending on tunnel conditions and traffic density, two different construction methods may be applied. In stable ground with moderate rail traffic, the tunnel may be renewed while maintaining the traffic flow. Alternatively a new tunnel may be built next to the existing one. Different methods of ground investigation and design of the tunnel lining for reconstruction of existing tunnels are described.

## INTRODUCTION

The German railroad company Deutsche Bahn AG is currently maintaining about 640 tunnels with a total length of more than 450 km. In addition, 35 railroad tunnels with a length of 90 km are presently under construction. A large number of existing tunnels were built during growing industrialization in the second half of the 19th century (see Figure 1).

Regardless of ongoing traffic growth in recent years, the stressed financial situation of the Federal Republic of Germany has lead to a halt of financial investment. However, as a result of the good economical situation before the recent global financial crisis, a number of new tunnels as well as modernizations of existing tunnels are presently underway. Depending on tunnel condition and traffic density, two different construction methods may be applied. In stable ground with moderate rail traffic, the tunnel may be renewed while maintaining the traffic flow. Alternatively a new tunnel may be built next to the existing one. This paper focuses on the renewal of existing tunnels.

## RECONSTRUCTION OF EXISTING TUNNELS

When an existing tunnel is reconstructed under running traffic, as a first step the inner lining has to be removed. Moreover, due to requirements of larger clearance and wider track separation distance a widening of the tunnel cross-section often has to be implemented. As a result parts of the rock mass have to be excavated.

As temporary supporting means an anchored and reinforced shotcrete lining is applied. For the final lining cast-in-place concrete with reinforcement is used. In addition to the lining the portals also typically have to be reconstructed. For the latter typically the slopes have to be supported by retaining walls or the tunnel has to be lengthened.

As a basis of planning the reconstruction of existing tunnels, the existing tunnel structure and the surrounding ground have to be analyzed with respect to their configuration and mechanical properties. Typically the following procedure is followed in this regard:

Figure 1. Portal of Buedenholz Tunnel

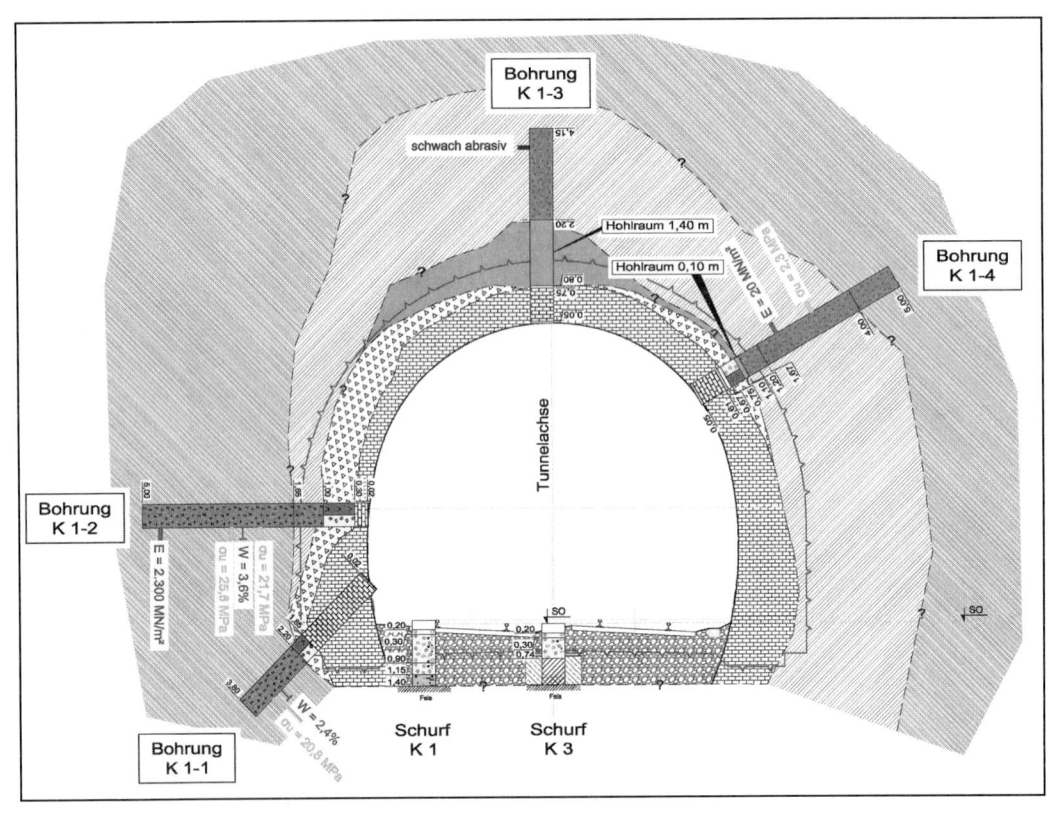

Figure 2. Example of a cross-section for investigation with core hole drillings

- Survey of on-site conditions
- Core hole drillings
- Field and laboratory tests
- Geophysical investigations
- Environmental investigations
- Characterization of ground and ground water conditions
- Investigation of existing lining with regard to brickwork thickness, backfilling and voids
- Determination of mechanical properties of existing lining and rock mass
- Classification of the tunnel alignment into homogeneous regions
- Technical recommendation for excavation and supporting means as a basis of blueprint planning
- Estimation of loading on the new inner lining
- Specifications for the monitoring program

## CONSTRUCTION METHOD OF HISTORICAL TUNNELS

A large number of railroad tunnels were constructed within the second half of the 19th century. Generally these tunnels have been built using traditional construction methods such as the German Core Construction Method, the Belgian Construction Method with excavation of a crown gallery or the Austrian Construction Method using invert gallery and crown trench. Depending on rock quality, the excavation was carried through using the drill and blast method or with hand tools. Most often tunnels were built in solid rock. With respect to the state of technology, often a wider cross-section than needed was excavated (see Figure 2). In order to shorten construction time for longer tunnels, very often shafts were excavated from the ground surface along the tunnel alignment. The shafts were used as intermediate construction access or ventilation and were refilled after construction of the tunnel.

The lining of these historical tunnels most often consists of natural stone with various thickness or brick masonry covering the rock mass from the invert bearing up to the tunnel crown. In sections with strong and stable rock often no masonry was used. Most historical tunnels were constructed without an invert arching and with a variety of different profiles for the inner lining. Most often a horseshoe shaped profile was used; however profiles with shallow arching or door frame profiles were used.

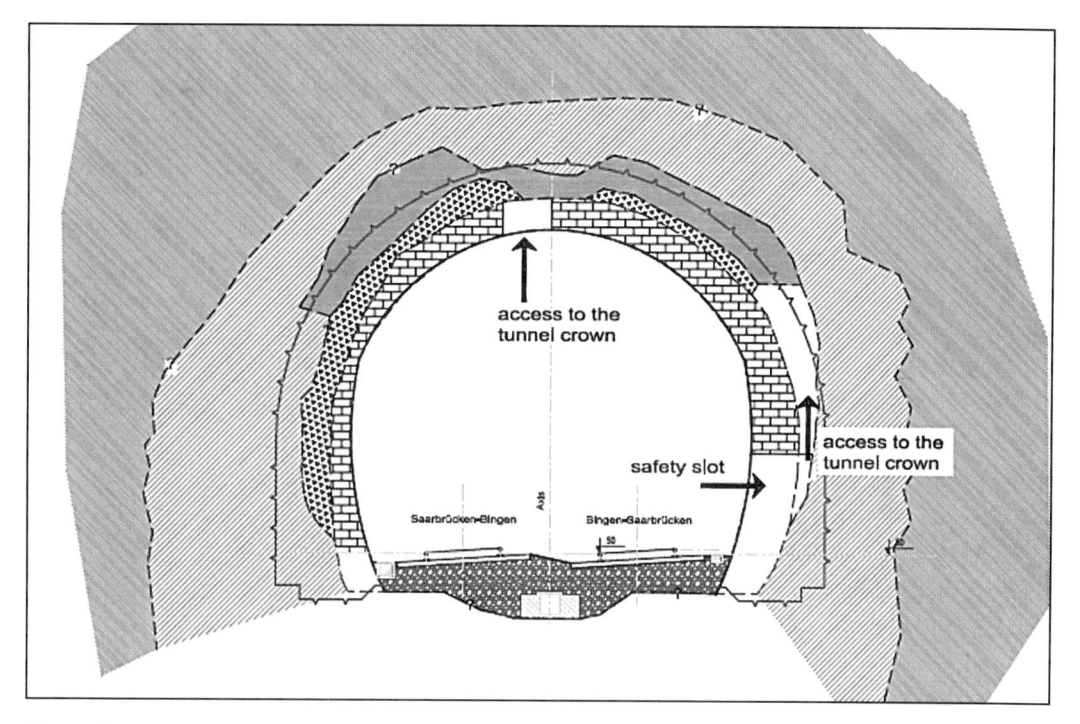

**Figure 3. Access to the tunnel crown**

Many tunnels include a number of safety slots along both sides of the lining having access shafts to the tunnel crown. Furthermore there are access holes distributed along the tunnel crown (see Figure 3).

The inner lining sometimes is sealed at the outside against seepage ("umbrella sealing") using tar pitch with cardboard, which require special treatment of disposal. In the invert drainage passages have often been installed.

## RESTORATION AND DAMAGE PATTERN

In the course of their up to 150 years operating time, the historical tunnels of the Deutsche Bahn have been maintained applying several construction methods, e.g., additional dewatering channels in the middle of the invert or improvement of the invert bearing and masonry. For the latter, poor conditioned masonry has been partially removed and replaced using additional measures such as sealing plaster or if necessary reinforced shotcrete. In addition, often masonry joints have been renewed or anchors have been applied. The safety slots have been sealed with concrete.

The current damage pattern of historical railroad tunnels typically includes a large number of voids in and behind the existing tunnel lining with longitudinal as well as transverse cracks. In regions with water leakage, calc-sinter cracks, sintering and efflorescence in the masonry are present. Furthermore flaking and burst with loose bricks of the masonry has often occurred.

In many cases the portal walls are also in poor condition. The masonry of the portals often shows bursts, weathered joints and efflorescence. The slopes above the tunnel portals often show several loosened rock blocks.

## GROUND INVESTIGATION

To investigate the existing tunnel lining and the adjacent ground conditions in the vicinity of the tunnel, direct exploration methods like core hole drillings, test pits and soundings are applicable. The investigations are typically conducted from inside the tunnel, preferably at nightly traffic breaks. As shown in Figure 2, exploratory drilling is conducted at different inclinations into the nearby rock mass. The exploration drillings are usually between 2 m and 6 m long. Due to the limitation of traffic breaks, the drilling is usually of relatively short length. In addition exploration drillings can be carried through from the ground surface down to the tunnel. At the slopes in front of the tunnel portals outward oriented core hole drillings are recommended.

Additional borehole field testing such as side pressure tests for investigation of ground stiffness or camera inspection for identification of fractures are

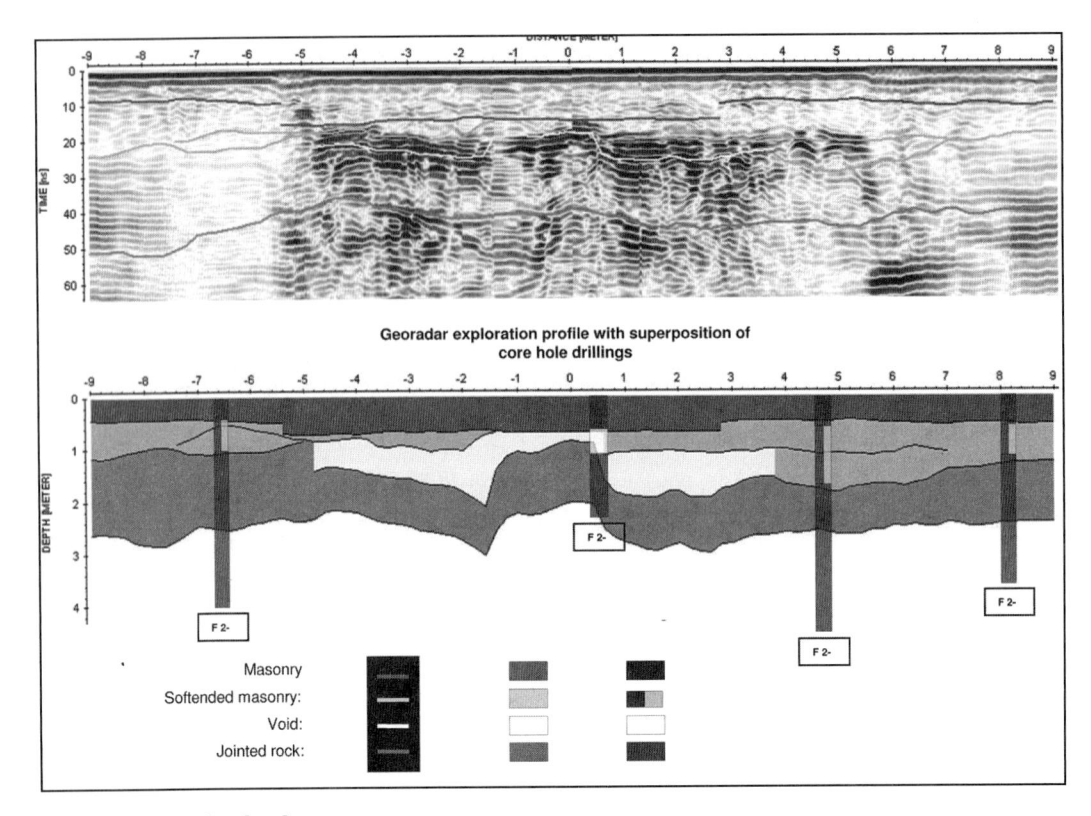

**Figure 4. Example of radargram**

typically conducted. Selected specimens from boreholes are used for standard laboratory testing. For investigation of thickness and condition of the track ballast, hand excavation and soundings are typically conducted within the tunnel. The hand excavations are also used to investigate the bearing condition of the tunnel lining as well as the location and condition of the dewatering system. In some existing tunnels environmental testing with respect to chemical contamination of the track ballast and the air-sided surface of the lining masonry are required. Likewise the tar pitch sealing material on the ground side of the tunnel lining has to be examined. It has to be verified whether or not the different stones and other material may be excavated separately or whether a mix of different materials has to be accounted for.

In addition to the above mentioned investigation methods, geophysical investigation methods are particularly well suited for the examination of existing tunnels. The advantage of geophysical methods over bore hole drilling is obvious. Using these methods, the detection of voids may be obtained not only at point locations but also throughout the lining.

A well suited geophysical method is the electromagnetic reflection method (also called geo radar). The electromagnetic reflection method (EMR-Method) in many ways is similar to echo sounding: runtime and amplitude of reflected or scattered waves are registered and visualized as a so-called radargram (see Figure 4). The EMR equipment consists of a central send-and-receive device and the antennas. During measuring procedure the antennas are moved along the profiles of interest and electromagnetic waves are sent as high frequency impulses into the material. Depending on dielectricity constants of the material as well as objects or structures, the waves are reflected or diffracted. Thereafter the waves are registered at the receiver and are digitally processed with their amplitudes and runtimes.

The reflection of waves requires that the contrast of the particular physical parameter (dielectricity constant, DC) between structures and objects and the surrounding material be large enough for detection. In general a difference of one between DC's of two neighboring layers is large enough to cause reflection of electromagnetic waves at the layer boundary. This implies also that within one geological layer changing water content or changing chemical property may cause reflection. The detection of small layer thickness depends on the frequency used as well as on dc of the layer and is ¼ of the wave length of the electromagnetic waves. Accordingly air filled spaces can be detected starting from a dimension of about 20 cm, backfill material starting from a dimension of about 7.5 cm.

The vertical axis of the radargram shows runtime of electromagnetic waves, the horizontal axis displays length of survey along which the antenna was moved during measuring.

During measuring, graphical information is available on a monitor, showing the ground displayed as a radargram. According to this, it is possible to optimize registration parameters with respect to the material properties encountered. As a result, it is possible to obtain insight into material structure or existence of voids in real-time in the field. Detailed information, however, can only be obtained after a computer based post processing of data. In order to obtain good quality data, the characterizing signals are enhanced while the interfering signals are filtered using suitable software. Depending on the task, antennas may be used with a center frequency from 15 MHz to 2.5 GHz. The detection depth of these frequencies (the higher the frequency the lower the detection depth) ranges from >>10 m to about 0.2 m (depending on the ground and its physical properties, e.g., concrete, rock, sand, etc.) with a resolution in the range of meters to millimeters.

After calculating a velocity-depth-model for the electromagnetic waves traveling inside the material, one is able to obtain layer depths and void sizes from the radargram.

## DESIGN OF INNER TUNNEL LINING

In contrast to construction of new tunnels, reconstruction of existing tunnels requires building measures within previously excavated rock mass. The loads on the existing tunnel structure have typically achieved an equilibrium state, depending on constitutive behavior of rock mass and tunnel lining, effects of geometry including voids, backfilling, etc. This equilibrium state of loads can only be simulated in a numerical model by approximation, because the interaction of rock mass and tunnel lining is highly influenced by the irregular contour of the excavation boundary as well as the irregular and partial backfilling of the lining. In the following example the acting load will be characterized by means of numerical analyses. The numerical model used takes into account the existing conditions. However, due to economical reasons some simplifications have to be applied.

The two-dimensional analyses were conducted using the program Tochnog. The symmetrical half of the Finite-Element (FE) mesh comprises 2,900 6-noded elements with quadratic interpolation and has a width of 60 m. The number of degrees of freedom is around 17,000. To take into account construction stages and supporting means, the mesh is refined around the tunnel (see Figure 5). The input parameters in Table 1 are based on results of laboratory and field testing.

Because of high rock strength there is no indication of plasticity in the rock mass or in the tunnel lining. Therefore it may be assumed that plastic yielding has minor impact on the results of the analyses and may be neglected. Based on this assumption all analyses are carried through using a model of elasticity.

The analysis of the new shotcrete lining and the new inner lining is carried through in several steps. First the construction of the existing lining is simulated to take into account the present state of the tunnel. Thereafter calculation steps are performed taking into account the planned construction phases. For this analysis the following steps were performed:

Step 1: Calculation of initial stresses before construction of existing tunnel.
Step 2: 60 % load reduction of initial stresses within tunnel cross-section to take into account stress release in the vicinity of the unsupported tunnel heading.
Step 3: Excavation of existing tunnel and installation of masonry lining, taking into account voids and backfilling (present state of tunnel).
Step 4: Excavation of new tunnel cross-section and installation of shotcrete lining.
Step 5: Installation of new inner reinforced concrete lining.
Step 6: Simulation of softening of shotcrete lining to take into account time dependent degradation of the shotcrete. Calculation of loads on the inner lining.

Some important results for the evaluation of stability and loading on the new lining are highlighted below. The calculated displacements of the existing conditions in Step 3 are relatively small. The largest displacements of millimeters were obtained at the tunnel crown. At the invert limited heave is obtained.

The stresses of Step 3 are shown in Figures 5 and 6. As a result of the irregular bedding of the masonry lining, the stresses are also distributed irregularly around the tunnel contour. Basically, the model indicates that the voids between inner lining and rock mass cause substantial stress release at the tunnel crown (see Figure. 7). At the tunnel sides a considerable stress increase with respect to the initial stress state takes place. The vertical stresses are transferred through the backfill onto the masonry at the tunnel sides.

In Step 4 the stresses are changed only gradually. Marginal stress change takes place at the tunnel crown. As a result of the improved interaction

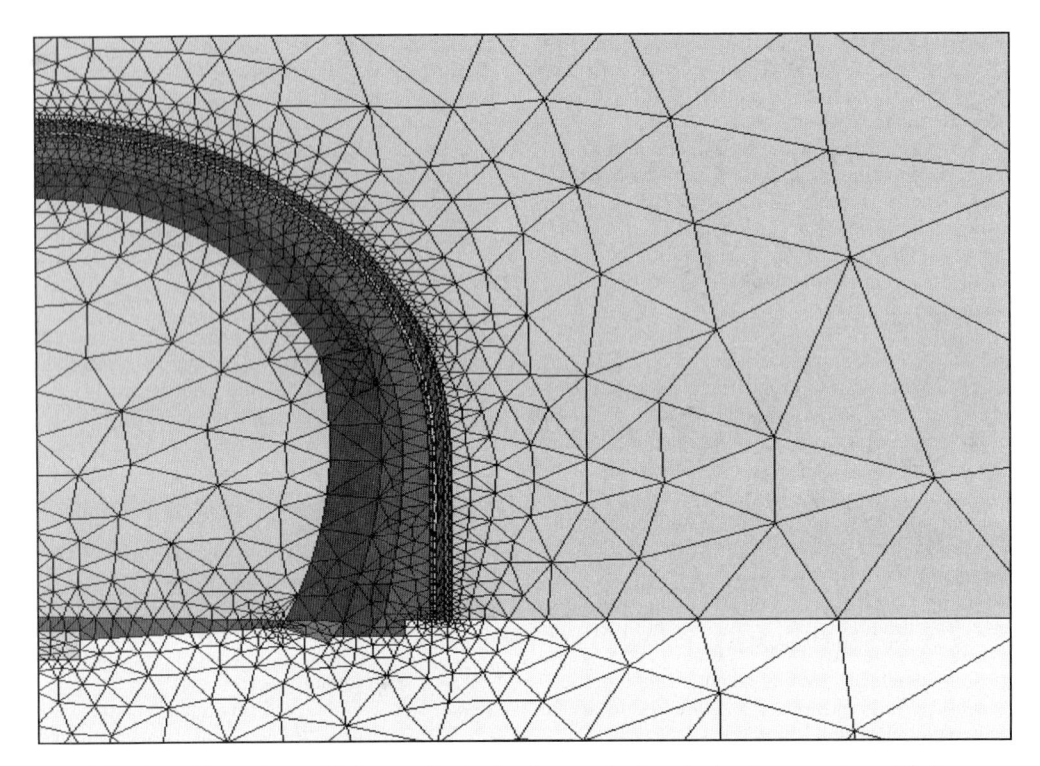

**Figure 5.** Design of inner tunnel lining on the basis of numerical analysis, close-up view of finite-element mesh

of rock mass and shotcrete lining, the inhomogeneous stresses at the tunnel sides are smoothed to a more homogeneous stress state. On the other hand, the widening of the tunnel cross-section leads to stress redistribution with a substantial loading of the shotrcete lining. As a result vertical and shear stress increase is obtained at the toe of the shotcrete lining (See Figure. 8).

The displacements due to widening of the cross-section are relatively small, in the range of some millimeters. In the present example, crown displacements are estimated to be on the order of about 6 mm. Furthermore the analyses show that softening of the shotcrete lining leads to no significant change of stresses and displacements. This implies that the loading of the shotcrete lining is transferred onto the inner lining without any considerable change of the loading situation.

Altogether it may be concluded that as a result of the planned construction measures, more homogeneous stresses are to be expected in the rock mass. Due to the widening of the cross-section and the change of the bedding conditions, stress redistribution takes place which leads to displacements in the range of some millimeters. Local stress concentrations are to be expected at the toe of the shotcrete lining as well as the new inner lining. Yielding of the rock mass due to stress concentration are limited to relatively small regions. This has been tested taking into account plasticity of the rock mass at critical regions, obtaining no significant change of results. These modeling results indicate that for strong and stable rock the use of an elastic stress-strain relationship may be justified.

## TUNNEL PLANNING AND CONSTRUCTION METHOD

For the renewal of existing tunnels, railroad traffic has a direct influence on the planning of reconstruction works. Therefore it is very important that consulting engineers are experienced in requirements of both tunnelling and railroad operations. The method is particularly well suited for relatively short two lane railroad tunnels.

As a first measure in the presented method, the two lanes are merged into a new centre-lane within the existing tunnel. On this centre-lane the railroad traffic is then scheduled to operate in one-way traffic mode during the projected reconstruction time.

Excavation of the new widened tunnel cross-section is achieved through first removing the existing masonry lining. Thereafter the drill-and-blast

Figure 6. Horizontal stresses Step 3

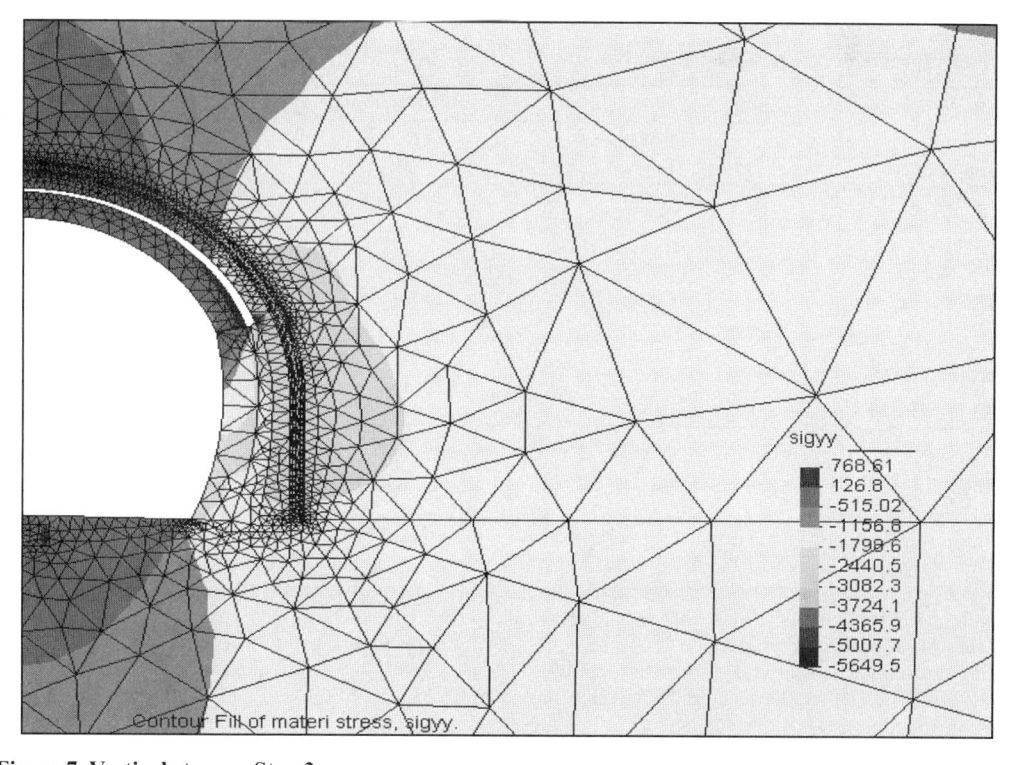

Figure 7. Vertical stresses Step 3

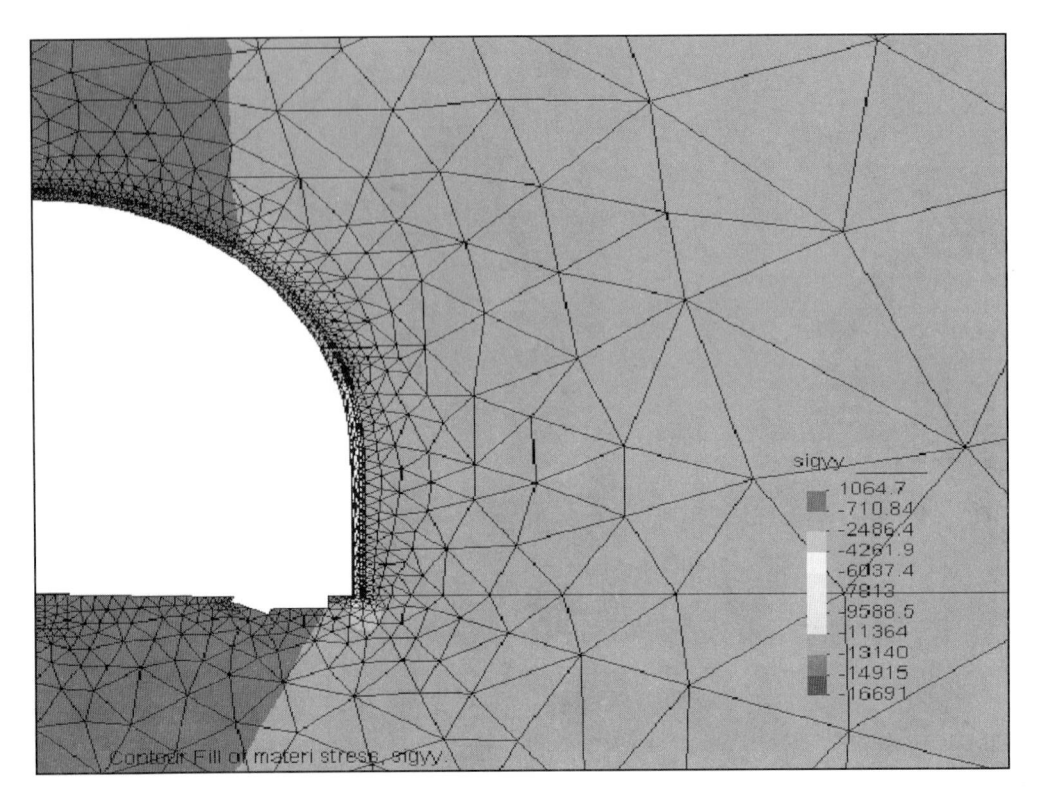

**Figure 8. Vertical stresses Step 4**

**Table 1. Parameters used in the FE-analyses**

|  | Input Parameters | | |
| --- | --- | --- | --- |
| Ground Layer | Self Weight $\gamma$ [kN/m$^3$] | E-Modul [MN/m$^2$] | $\nu$ [–] |
| Rock | 26 | 2,000 | 0.25 |
| Existing masonry lining | 25 | 4,000 | 0.25 |
| Backfill | 24 | 500 | 0.30 |
| New shotcrete lining | 25 | 15,000 | 0.20 |
| Decayed shotrete lining | 20 | 100 | 0.35 |
| New inner lining | 25 | 30,000 | 0.35 |

method is used to excavate the rock mass to achieve the planned new profile.

Starting from one tunnel portal a support-casing is built. On the outer side of the casing, equipment needed for the construction works (cutting tools, anchor boring equipment, shotcrete manipulator, etc) are installed. On the inner side of the support-casing the railroad traffic is continuously operating (see Figure 8). The support-casing has to take into account requirements of railroad clearance. For the reconstruction works the support-casing provides hydraulic support-plates which allow to temporarily support the existing masonry lining before it is removed, as shown in Figure 9. As also shown in this figure, the subsequently excavated areas are supported with anchors and reinforced shotcrete according to principles of the NATM.

After completion of tunnel excavation, the support-casing is removed and the formwork carriage is put in place. Again the design of the formwork

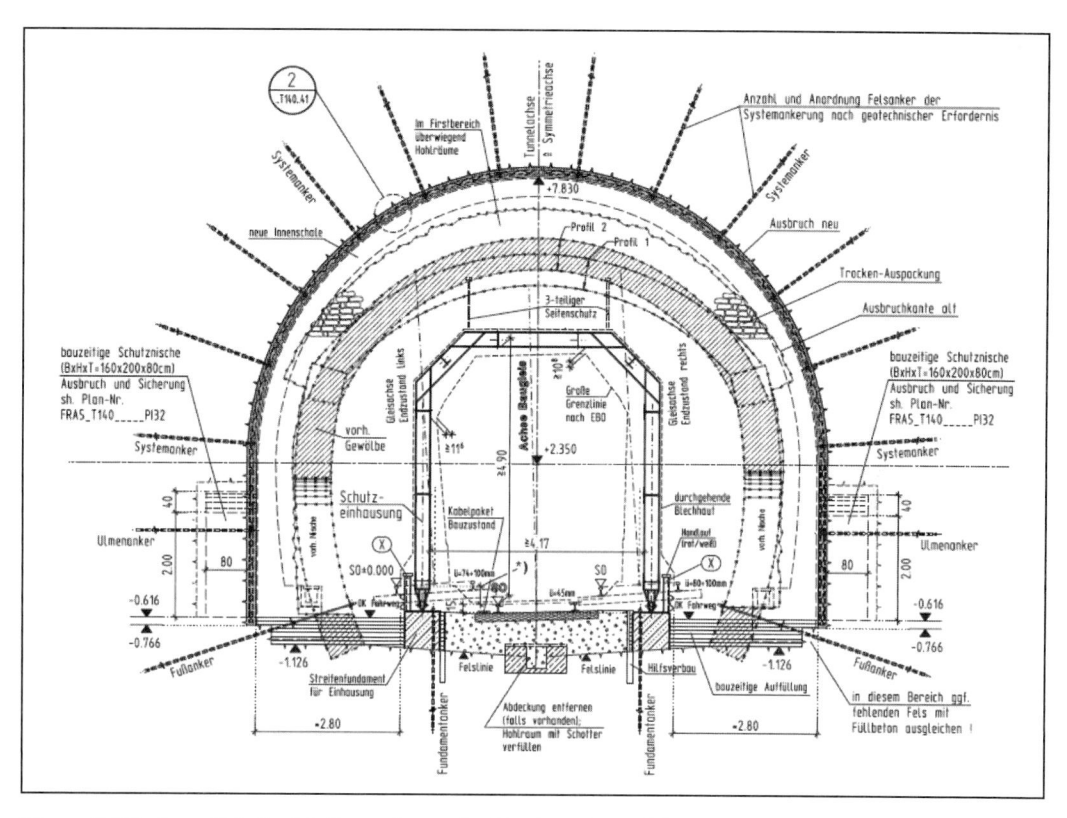

Figure 9. Display of Method 1. Renewing and widening of the old tunnel at ongoing railroad traffic

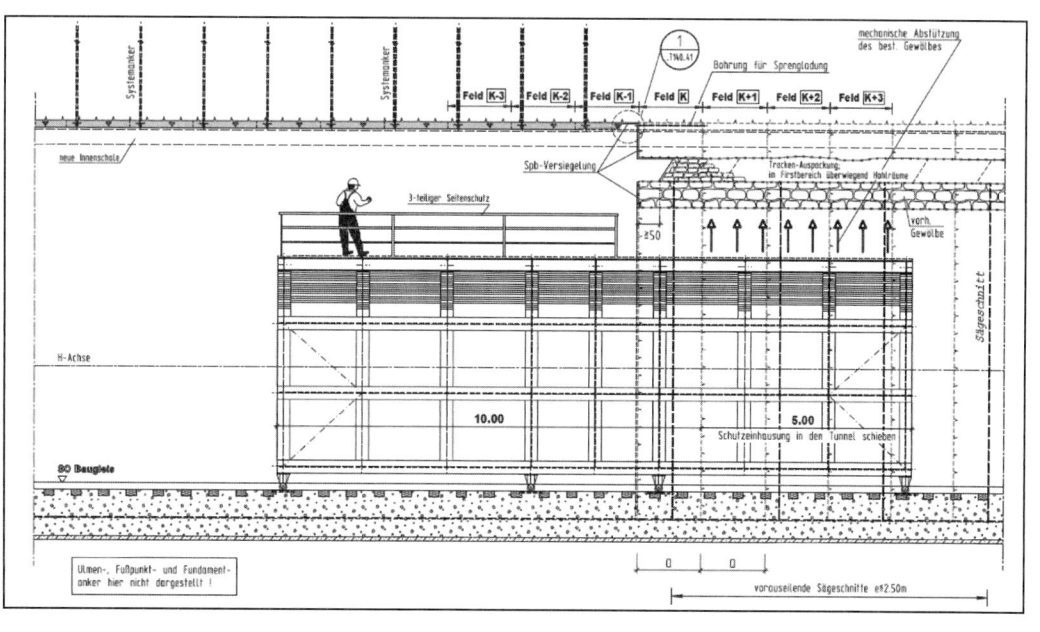

Figure 10. Removing of old masonry lining and excavation and support of new tunnel cross-section with support-casing

carriage must take into account the railroad clearance. The latter condition also requires special structural solutions to carry horizontal loads caused by concrete pressure.

After installation of the inner lining, formwork and curing carriages as well as the existing middle-lane are removed. Subsequently two final new lanes are installed.

## CONCLUSIONS

This paper discusses the renewal of old railroad tunnels, considering the ongoing railroad traffic.

For relatively long tunnels, requirements of tunnel operation, logistical reasons as well as safety regulations for one-lane traffic with emergency exits, demand for the construction of a new tunnel next to the existing one.

For relatively short tunnels, the described method of reconstruction under ongoing railroad traffic is both economical and time efficient. Moreover, requirements of tunnel operation, relatively high traffic density and security regulations are being satisfied by this method. Only two short breaks of railroad traffic are needed in order to merge the tracks.

The described method has been successfully applied to existing tunnels with non-electrified tracks. Further development of the method is aimed at reconstruction of existing tunnels with electrified tracks.

## LITERATURE

[1] Breidenstein, Matthias: New tunneling method for modernization under ongoing traffic. (In German); Tunnel 26 (2007), 2

[2] Deutsche Bahn AG: Directive of railroad tunnel planning, construction and maintenance, Ril 853, 01.01.2007 (in German).

# Development of Seismic Design Criteria for the Coronado Highway Tunnel

**David Young**
Hatch Mott MacDonald, Pleasanton, California

**Anil Dean**
Hatch Mott MacDonald, Sacramento, California

**James R. Gingery**
Kleinfelder, San Diego, California

**Tomas Gregor**
Hatch Mott MacDonald, Mississagua, Ontario

ABSTRACT: Seismic design has advanced significantly in the last 20 years, yet there is no formalized procedure for seismic design of tunnels. This paper presents the preliminary seismic design criteria for the SR 75/282 Project in Coronado, California. The project is currently in the planning/environmental phase, and may include a one-mile long, 36.5 foot diameter twin bored tunnel in soft ground below the water table. The tunnel crosses the active Coronado Fault and could experience a magnitude 7 earthquake. This paper presents an overview of site characterization, seismic hazards evaluations, tunnel design criteria for seismic loads and deformations due to shaking rupture and liquefaction, and seismic performance criteria.

## INTRODUCTION

The City of Coronado has initiated the State Route 75/282 Transportation Corridor Project to improve traffic flow between the San Diego-Coronado Bay Bridge and the Naval Air Station North Island (NASNI), a distance of approximately 2.25 km (Figure 1). The project is in the Project Report/Environmental Document (PR/ED) phase. As such, several alignments and construction method options are under consideration. The two major project configuration options are: a cut-and-cover tunnel alignment, and a twin-bored tunnel alignment. Schematic cross-sections for these options are shown on Figure 2 and Figure 3 (HMM, 2007).

Hatch Mott MacDonald (HMM) performed tunnel design services for Parsons Brinckerhoff Quade & Douglas (PBQ&D) and the City of Coronado (the City). Geotechnical data, seismic hazard assessments and fault capability studies were conducted by Kleinfelder, Inc. of San Diego.

Criteria presented herein were reviewed by Caltrans, the City, PBQ&D, and an external Technical Advisory Panel (TAP) convened by Caltrans. However, this paper presents preliminary seismic design criteria limited to the current Caltrans

Figure 1. Project location map

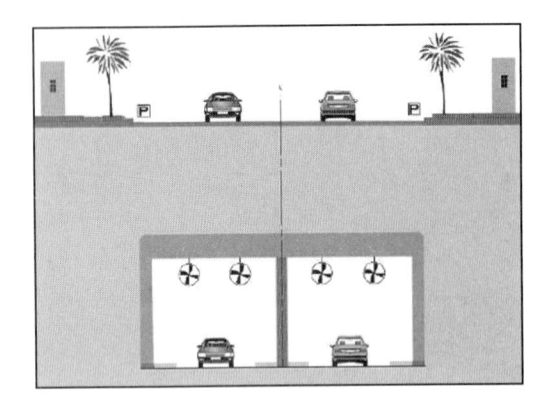

**Figure 2. Cut-and-cover option**

**Figure 3. Twin-bored tunnel option**

Project Report phase only, and the data may be revisited for use during final design.

## GENERALIZED SOIL AND GROUNDWATER CONDITIONS

Subsurface conditions for Phase I of the project were characterized with seven rotary wash borings, two resonant sonic borings, and twelve cone penetration test (CPT) holes, a seismic reflection survey to assist with interpretation of fault locations within the earthquake fault hazard zone for the Coronado Fault, and laboratory testing of selected samples recovered from the borings. Additional exploration was conducted in the vicinity of the Coronado Fault, as part of the fault rupture hazard characterization study described below.

The alignment proceeds through soils consisting of interbedded clays and sands of the Quaternary-age Bay Point Formation overlain by younger deposits of dune sand and fill. Bed thickness in the Bay Point Formation typically varies from a few inches to tens of feet. The natural water contents for the Bay Point Formation clays are typically at or near the plastic limit, and the over consolidation ratio of these materials is typically in the range of 2.5 to 2.9. The Bay Point Formation sands are medium dense to very dense, with an effective friction angle typically ranging from 33 to 38 degrees. Groundwater exists 3.0 to 4.5 m below ground surface, and depth to invert varies from 12 m for the cut-and-cover option to 30 m for the bored tunnel option.

## SITE SEISMICITY AND GEOLOGIC HAZARDS

### Seismic and Tectonic Setting

The project is situated in an area of southern California that is structurally and tectonically controlled by a system of northwest-trending active faults. The San Andreas, San Jacinto, and Elsinore Faults lie northeast of the site at distances of 150, 105 and 70 km, respectively. Southwest of the site, the Descanso, Coronado Bank, and San Diego trough faults are at distances of 11, 18 and 35 km, respectively. The site is located near the southern terminus of the Rose Canyon Fault Zone (RCFZ), which extends approximately 66 km north of the site, then transitions into the Newport-Inglewood Fault Zone. The RCFZ is considered active (Lindvall & Rockwell 1995), and capable of generating a moment magnitude 7 to 7.2 earthquake (Mualchin 1996, Cao et al. 2003), which dominates the seismic hazard at the site.

### Fault Rupture Hazard Characterization

The RCFZ trends southeast through La Jolla to downtown San Diego. In the downtown area the RCFZ branches off toward the south in three well defined segments: the Spanish Bight, Coronado (CF), and Silver Strand Faults, and distributed faulting in the southern portion of the San Diego Bay. Figure 4 shows the tunnel alignment in relation to regional faults.

Evaluation of the hazard posed to the project by the CF was considered critical to the planning and feasibility evaluations. Prior to this study, the location of the CF on land had been inferred based on offshore seismic reflection surveys performed in the San Diego Bay (Kennedy & Clarke 2001) and the Pacific Ocean (Kennedy & Welday 1980), and based on an apparent scarp feature that crosses Coronado and aligns with the offshore lineaments. Previous trenching on land in the scarp area had not detected the CF (Artim & Streiff 1981).

A detailed study was undertaken that utilized aerial photo analysis, high resolution seismic reflection, cone-penetrometer test (CPT) profiling, downhole logging of large diameter bucket auger boreholes, and fault trenching to locate and study the displacement characteristics of the CF. The investigation was performed in a phased manner. After initial seismic reflection survey data suggested that offset marker beds were present at depth below the site, a row of CPTs with an initial spacing of nine (9) meters was performed across the potential fault area. After review of the initial CPTs, additional CPTs spaced as closely as 2.3 m were performed to provide better resolution in the suspected fault zone. The interpreted CPT data were then used to locate an exploratory trench focused on the suspected fault zone. Careful logging of offsets in the pedogenic soil horizon revealed that 29 cm of vertical throw had occurred on the CF in one or more events within the past several hundred years. Mismatching of pedogenic B-laminations was observed in the trench suggesting strike slip offset, but the absence of piercing points made it impossible to estimate the magnitude and sense of slip along strike. Coupling of the trench and CPT data provided a clear picture of the location and orientation of the active fault trace (Figure 5) and the trench data suggested the fault was capable of 29 cm of vertical throw per event and some amount of strike slip.

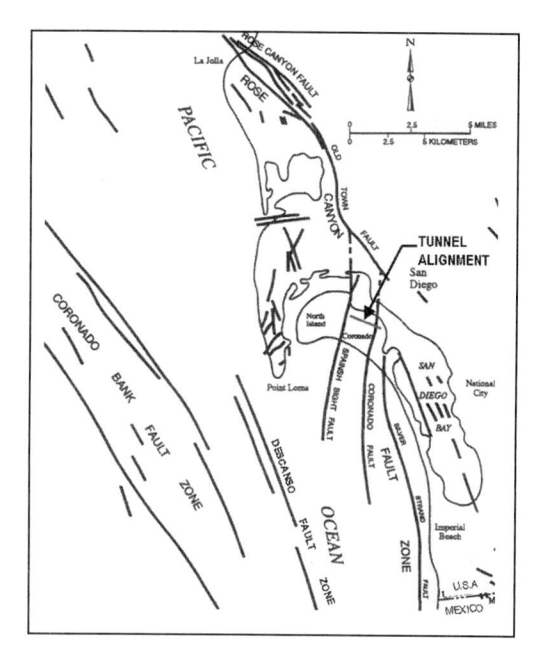

Figure 4. Regional fault map

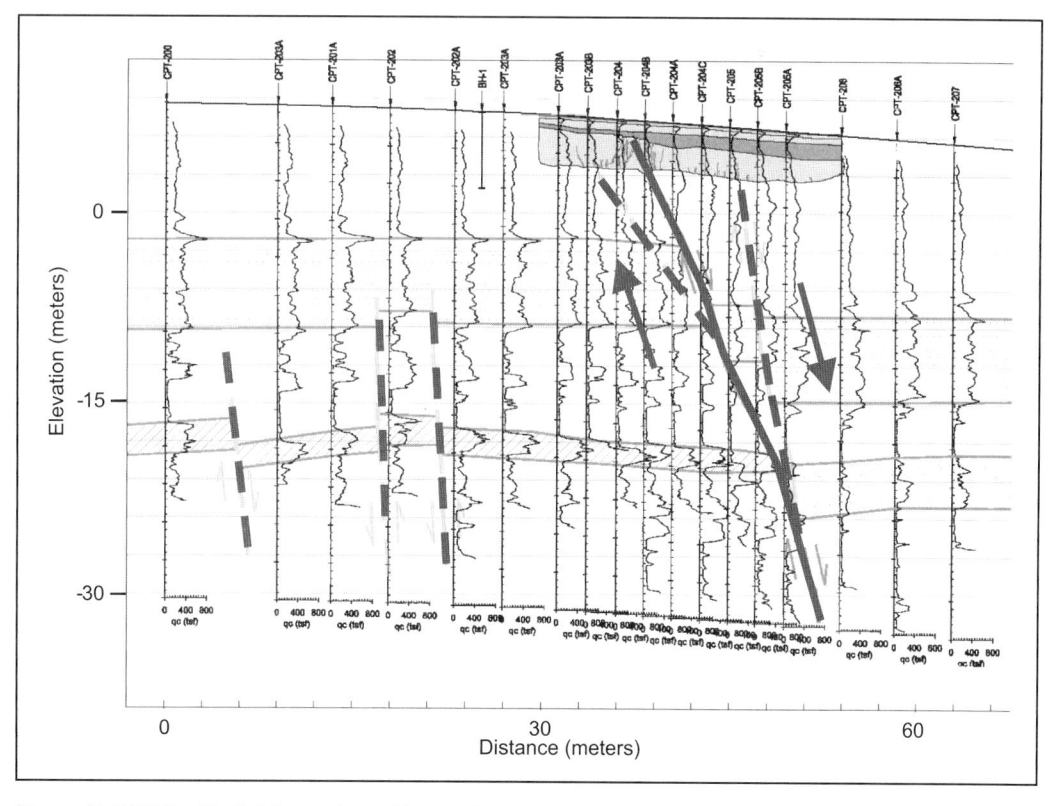

Figure 5. CPT Profile (with stratigraphic and faulting interpretations. Solid line is the active fault trace and dashed lines are potentially active traces.)

**Table 1. Summary design fault rupture offsets**

| | Displacement Linearly Distributed Within a 15.2 m Wide Secondary Zone in Both the Hanging Wall and Footwall | | Displacement Linearly Distributed Within a 0.6 m Wide Main Fault Trace Zone | |
|---|---|---|---|---|
| | Strike Displacement | Vertical Throw | Strike Displacements | Vertical Throw |
| Right lateral: oblique scenario | 28 cm right lateral | 15 cm | 56 cm right lateral | 29 cm |
| Left lateral: oblique scenario | 18 cm left lateral | | 36 cm left lateral | |

**Table 2. Summary of ground motion parameters**

| Ground Design Motion Level Parameter | Earthquake FEE with 150-Year Return Period | SEE |
|---|---|---|
| Peak horizontal ground acceleration | 0.17g | 0.57g |
| Peak horizontal ground velocity | 20 cm/sec | 67 cm/sec |
| Peak horizontal ground displacement | 15 cm | 69 cm |
| Free-field peak shear strain | 0.04% | 0.28% |

The strike slip offset used in design was developed considering four deterministic models. Based on the normal offset data from the field investigation and the strike slip data from modeling, a right lateral-oblique scenario and a left lateral-oblique scenario were developed for design. These design cases, which include fault offset in the main fault trace zone and secondary zones in both the hanging and footwalls, are summarized in Table 1. The location of the next sfault rupture on the Coronado Fault is assumed to occur within 10 m of the location of the previous rupture mapped in the fault trench.

The strike of the fault has an Azimuth of approximately 15 degrees. With a tunnel bearing of approximately 117 degrees, the fault strike intersects the tunnels at an angle of 78 degrees. The dip of the Coronado Fault was observed to be 70 degrees to the east.

**Design Ground Motions**

The design ground motion criteria were developed based on input and agreement from the owner, project design team, Caltrans, and the TAP. The Functional Evaluation Earthquake (FEE) was defined as the earthquake having a probability of exceedance of 28 percent in 50 years (150-year return period). The Safety Evaluation Earthquake (SEE) response spectrum was developed by enveloping (i.e., taking the greater spectral acceleration) the site-specific Probabilistic Seismic Hazard Assessment (PSHA)-based 975-year ground motion spectrum and the site-specific deterministic-based median response spectrum.

Considering the proximity of the site to the Rose Canyon Fault Zone, rupture directivity was accounted for by increasing spectral accelerations at higher periods. Peak ground velocities (PGV) were developed using the PGV/PGA ratios recommended by Hashash et al. (2001) and peak ground displacements (PGD) were based on an unpublished correlation with the four (4) second period spectral acceleration and earthquake magnitude suggested by TAP member Dr. Norm Abrahamson. Translation of ground motion parameters from the ground surface to the tunnel depth was performed in accordance with ratios presented in Power et al (1998). Free-field peak shear strains due to vertically propagating shear waves were developed for ovaling and racking analysis of the tunnel using a variation of the simplified technique of Gingery (2007) which is similar to the shear strain estimation performed for seismic compaction settlement analyses (Tokimatsu and Seed, 1987). A summary of ground surface ground motion parameters is presented in Table 2.

**Liquefaction Susceptibility**

Liquefaction susceptibility was evaluated using the simplified procedure of Youd, et al. (2001), Idriss and Boulanger (2004), and Bray and Sancio (2006). The sands and silty sands of the Bay Point Formation were generally non-liquefiable under the FEE, and only isolated, discontinuous "pockets" of liquefiable material were predicted under the SEE at tunnel

### Table 3. Proposed minimum performance level for SR75/282 project

| Ground Motion | Minimum Performance Level |
|---|---|
| Functional Evaluation Earthquake (FEE) | Immediate Full Service: Reparable damage within 90 days. Lane closure allowed outside peak hours. Minor concrete spalling and joint damage. |
| Safety Evaluation Earthquake (SEE) | No Collapse: Significant damage requiring closure. No rapid inundation with soil and groundwater. No sinkhole development. Significant damage may require closure to the public. Repairs, if possible, will require a complete evaluation. |
| Fault Rupture (with SEE ground motion) | No Collapse: More extensive damage than SEE event requiring closure. No rapid inundation with soil and groundwater. No sinkhole development. Repairs, if possible, will require a complete evaluation. |

### Table 4. Engineering parameters used in analyses

| Material | Model Unit | Elevation at Analyzed Section (m) | Total Unit Weight (kN/m$^3$) | Effective Friction Angle Ø' (degrees) | Shear Modulus, MPa (ksf) | K$_o$ | Poisson's Ratio | Undrained Shear Strength, Su, kPa (psf) |
|---|---|---|---|---|---|---|---|---|
| Bay Point formation SP, SP-SM, SM, SW, non-plastic ML | S-1 | 3.05 to 6.1 | 18 | 36 | 7.2 MPa (150 ksf) | 0.7 | 0.3 | NA |
| | S-2 | −15.8 to 3.05 | 20.4 | 36 | 19.2 MPa (400 ksf) | 0.7 | 0.3 | NA |
| | S-3 | −19.8 to −18.3 | 20.4 | 34 | 24.9 MPa (520 ksf) | 0.7 | 0.3 | NA |
| | S-4 | −25 to −23.5 | 20.4 | 36 | 28.7 MPa (600 ksf) | 0.7 | 0.3 | NA |
| Bay Point formation CL, CH, MH, plastic ML | C-1 | −18.3 to −15.8 / −23.5 to −19.8 | 19.6 | | 10.1 MPa (210 ksf) | 0.7 | 0.49 | 196 kPa (4,100 psf) |
| | C-2 | −30.5 to −25 | 19.6 | | 13.4 MPa (280 ksf) | 0.7 | 0.49 | 247 kPa (5,150 psf) |

depth. The plastic silts and clays of the Bay Point Formation generally had plasticity indices between 15 and 35, and therefore do not meet the compositional criteria for liquefaction susceptibility per Bray and Sancio (2006).

## SEISMIC PERFORMANCE CRITERIA

Performance criteria are described in Caltrans' Memo to Designers 20-1 (Caltrans, 1999). The performance criteria are dependent on the maximum level of damage and post-earthquake service level required. Post-earthquake service levels are categorized as either "Ordinary" or "Important." The "Ordinary" designation is used unless the structure has been formally designated as "Important" by Caltrans. This project is currently categorized as "Ordinary" because existing surface roads serve as the emergency lifeline. The Coronado Tunnels must have immediate serviceability and reparable damage in the FEE event. Limited serviceability and significant damage are acceptable for the SEE for structures classified as ordinary. Further definitions of the level of damage and level of service were prepared for the PR/ED phase, based in part on criteria developed by the San Diego-Coronado Bay Bridge Seismic Retrofit, which addressed fault rupture. Project specific seismic performance criteria are presented on Table 3.

## SEISMIC TUNNEL DESIGN

Seismic deformations are combined with external ground and hydrostatic loads for tunnel design. Some highlights of the structural analysis are provided below. For readers interested in more information on the structural analysis, refer to Gregor et al (2007). Soil parameters used in the analyses were provided in Kleinfelder (2006) and are presented on Table 4.

### Cut-and-Cover Option

A 3-D model of the cut-and-cover box was built using StaadPro software. Soil was represented by compression-only springs that were placed at every exposed model node of the structure. Spring coefficients were developed through 2-D finite difference

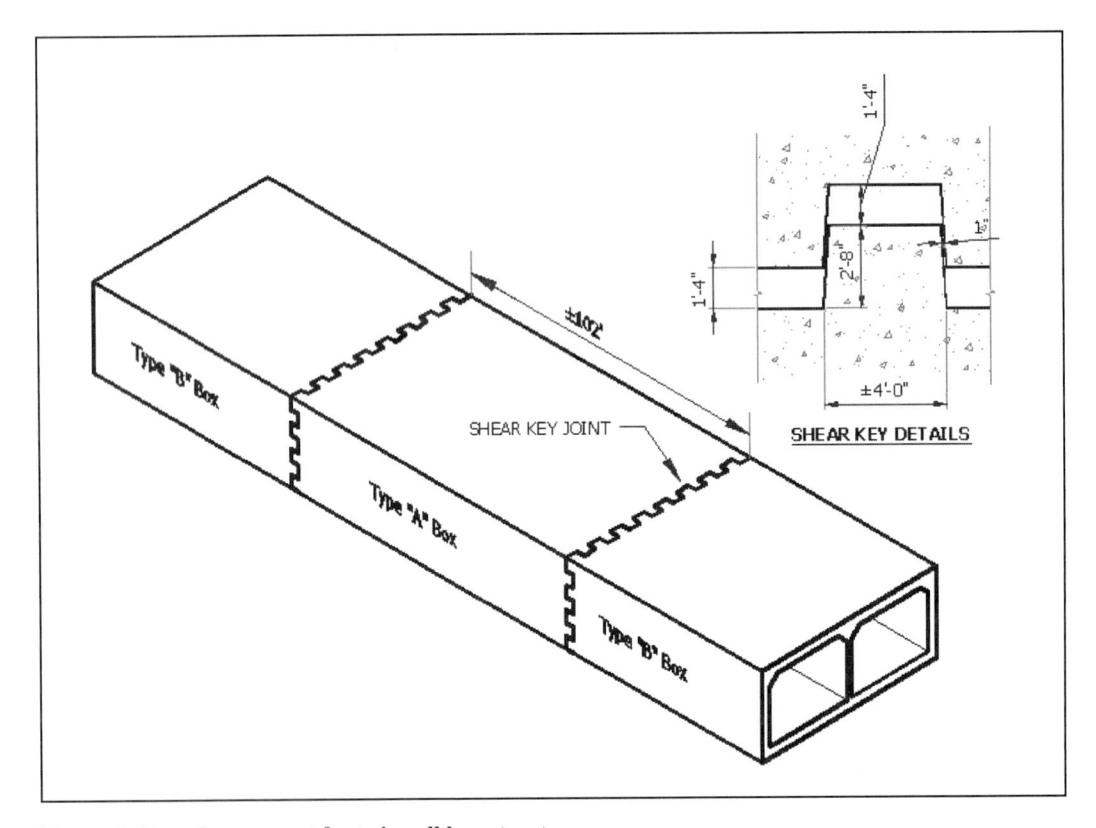

**Figure 6. Shear key concept for twin-cell box structure**

modeling with FLAC, and have variable stiffness that is defined by multi-linear curves.

In trial runs for the fault rupture case, a seamless continuous box was found to produce very high compressive and tensile forces along the box length that required heavy reinforcing. For that reason, shear key joints were introduced in the structure so that axial stresses are relieved during a fault rupture event. A conceptual sketch of a shear key joint is presented in Figure 6. In the model, these joints are simple hinges that are only capable of carrying in-plane shear forces only along the joint edges. The two shear key joints were placed 31.1 m (102 feet) apart. Consideration was given to use of engineered compressible materials, such as EPS Geofoam or low density cellular concrete, outside of the box and was shown to have some benefit in terms of reducing the reinforcement necessary to meet the required SEE performance criteria.

**Twin-bored Tunnels Option**

The twin-bored tunnels were modeled using FLAC3D software by ITASCA Inc. Only the primary slip zone displacements were modeled for the bored tunnels, since the gradient of displacement is much more severe in the primary rupture zone in comparison to the secondary zone. Therefore, it seems reasonable to assume that if the structure can withstand the primary zone displacement, it can withstand the secondary zone displacements at the same time.

The precast concrete tunnel lining (PCTL), shown in Figure 7, is a segmental lining, therefore its flexural stiffness in the hoop direction will be less than that of a solid tube. The Muir Wood (1975) methodology was used to estimate the stiffness of the segmental lining in the model. A variety of possible longitudinal connection scenarios between adjacent PCTL rings were evaluated, including yielding bolt connections.

**CONCLUSION**

Historically, underground structures have survived strong ground motions better than surface structures for various well-documented reasons (i.e., Young and Dean, 2006). However, some underground structures have experienced significant damage, particularly when they cross a fault that may rupture. The anticipated horizontal PGA for the SEE ground motion puts this project at moderate risk of damage. Potential rupture of the CF raised questions about

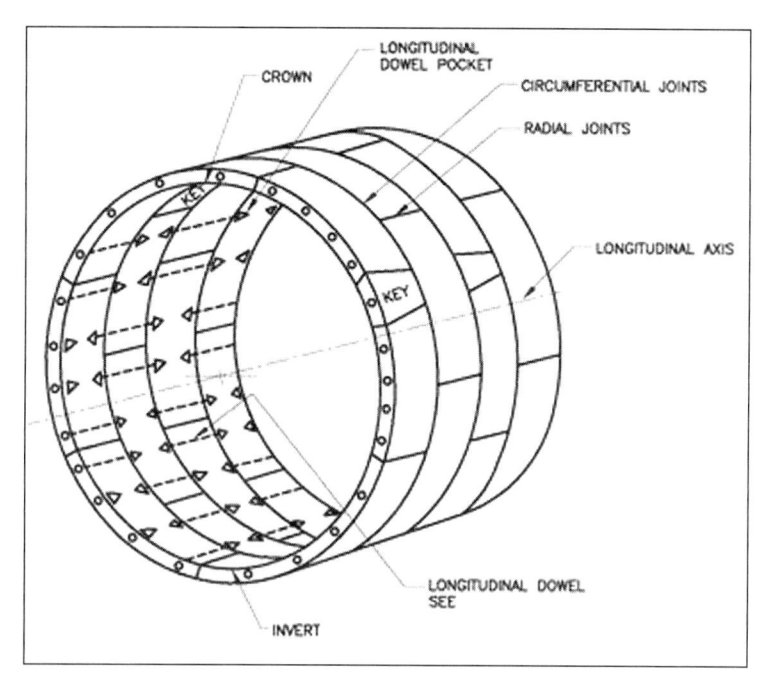

**Figure 7. Segmental precast concrete tunnel lining (PCTL)**

project feasibility; therefore, fault rupture had to be analyzed at the planning stage. As a result, the project team decided to allocate project funds in the planning stage on fault studies and to employ pseudo static analysis methods to confirm project feasibility. Site response analyses and generation of time histories were not conducted, as they were not critical to confirm project feasibility.

It was obvious that the CF rupture would impose extreme loads on both the precast concrete tunnel lining and cut-and-cover options so it was most important to characterize the fault rupture. The analyses show that it is unlikely that damage to either structure could be avoided. However, we believe that with a carefully detailed design, which would consider structural stability as well as ductility requirements, it is feasible to develop a design for both options that is capable of withstanding fault rupture, while meeting the performance criteria for life safety.

Other significant findings of the preliminary design analyses for the SR 75/282 Transportation Corridor Project are:

- Characterizing the geometry and width of the fault displacement zone was essential to the structural analyses. The CF was located on land in Coronado for the first time as part of this study.
- The soil surrounding buried structures helps to distribute the displacement away from the main trace of the fault rupture, making soil structure interaction analyses essential for analyzing structural response to fault displacement.
- Although the thickness of the bored tunnel lining is controlled by construction loads, the reinforcing and segment connection details will be influenced by seismic loads.
- For the box structure, the special joint detail and amount of reinforcing are a result of the seismic load cases analyzed.

# REFERENCES

Artim, E.R. and Streiff, D. 1981. Trenching the Rose Canyon Fault Zone, San Diego, California, *U.S. Geological Survey Contract No. 14-08-0001-19118*.

Bray, J.D. and Sancio, R.B. 2006. "Assessment of the Liquefaction Susceptibility of Fine-Grained Soils," *JGGE*, Vol. 132, No. 9, September 2006, pp. 1165–1177.

Cao, T., Bryant, W.A., Rowshandel, B., Branum, D., and Wills, C.J. 2003. *The Revised 2002 California Probabilistic Seismic Hazards Maps*, California Geological Survey, June 2003. www.consrv.ca.gov/cgs/rghm/psha/index.htm.

Caltrans, "Caltrans Memo to Designers 20-1 Caltrans, 1999.

Gingery, J.R. 2007. "A simplified method for estimating shear strains for ovaling and racking analysis of tunnels," *4th International Conference on Earthquake Geotechnical Engineering*, June 25–28, 2007, Thessaloniki, Greece,

Gregor, T., Garrod, B., Young, D.J. 2007. "Analyses Of Underground Structures Crossing An Active Fault In Coronado, California," *International Tunneling Association Conference,* June 25–28, 2007, Prague, Czech Republic.

Hashash Y.M.A., Hook, J.J, Schmidt, B., Yao, J.I-C. 2001. "Seismic Design and Analysis of Underground Structures," *Tunneling and Underground Space Technology 16* (2001), pp 247–293, Elsevier Science Ltd.

Hatch Mott MacDonald 2007. Advance Planning Study (APS), SR 75/282 Transportation Corridor – Fourth Street Tunnel Alternatives. April 2007.

Idriss, I.M. and Boulanger, R.W., 2004, "Semi-empirical Procedures for Evaluating Liquefaction Potential During Earthquakes," *Proceedings of the 11th International Conference on Soil Dynamics and Earthquake Engineering and 3rd International Conference on Earthquake Geotechnical Engineering*, Volume 1, University and California, Berkeley, January 2004.

Kennedy, M.P. and Clarke, S.H. 2001. "Late Quaternary Faulting in San Diego Bay and Hazard to the Coronado Bridge," *California Geol*ogy, July/August.

Kennedy, M.P. and Welday E.E. 1980. Character and Recency of Faulting Offshore Metropolitan, San Diego, California, *CDMG map sheet 40.*

Kleinfelder, 2006. Final Preliminary Engineering Geotechnical Investigation, Volume I. State Route 75 and 282 Transportation Corridor Project, Coronado, California. April 12, 2006.

Lindvall, S.C. and Rockwell, T.K. 1995. "Holocene activity of the Rose Canyon Fault Zone in San Diego, California," *Journal of Geophysical Research*, vol. 100, no. B12, pp. 24,121 – 24,132.

Mualchin, L. 1996. A Technical Report to Accompany the Caltrans California Seismic Hazard Map 1996, *California Department of Transportation Engineering Service Center Office of Earthquake Engineering*; July.

Muir Wood, A M. 1975. The circular tunnel in elastic ground. *Geotechnique*, 1, 1975, 115–127.

Power, M.S., Rosidi, D., Kaneshiro, J., Gilstrap, S.D., and Chiou, S-J., 1998. Draft Report, Summary of Evaluation of Procedures for the Seismic Design of Tunnels (unpublished).

Tokimatsu, K. and Seed, H.B. 1987. "Evaluation of Settlements in Sands Due to Earthquake Shaking." *J. of Geotech. Engr. Div.*, ASCE. Vol. 113, No. 8. 1987.

Youd, T.L. et al. 2001. "Liquefaction Resistance of Soils: Summary Report from the 1996 NCEER and 1998 NCEER/NSF Workshops on Evaluation of Liquefaction Resistance of Soils," *JGGE*, Volume 127, no. 10, October 2001, pp. 817–833.

Young, D.J. and Dean, A, 2006. "Seismic Performance of Precast Concrete Tunnel Linings," *Proceedings Tunnelling Association of Canada, Nineteenth National Conference, 2006,* Vancouver, B.C. Canada.

# TRACK 2: DESIGN

Matt Fowler, Chair

*Session 3: Managing Risk, Safety, and Security Through Design*

# Performance-Based Design Using Tunnel Fire Suppression

**Fathi Tarada**
Mosen Ltd., Crawley, West Sussex, United Kingdom

**James Bertwistle**
WSP Group plc, London, United Kingdom

ABSTRACT: This article reports on a quantitative risk assessment, supported by Computational Fluid Dynamics calculations, to compare the fire risk levels of two design options for the Yas Island Southern Crossing Tunnel in Abu Dhabi: a 5-cell tunnel with two escape galleries, compared to a 3-cell solution with no escape galleries, but with a fire suppression system installed in the highway cells. The risk levels for the two designs were shown to be broadly similar, and the 3-cell design was accepted as meeting the required level of fire safety by Abu Dhabi Civil Defence. The 3-cell tunnel benefited from a reduced construction cost compared to the conventional 5-cell design, as well as reduced risks to meeting the target date for construction completion.

## INTRODUCTION

Fire safety is a key issue that can influence the configuration, structural design, mechanical/electrical/traffic control systems and the operation and management of tunnels. Recent high-profile fires in tunnels include the three fires in the Channel Tunnel between the UK and France (in 1996, 2006 and 2008), and the Burnley tunnel fire in Melbourne, Australia (2007), which was successfully controlled with a deluge system. On mainland Europe in the previous decade, there have been road tunnel fires with multiple fatalities at Mont Blanc (1999), Tauern (1999) and Gotthard (2001).

A number of guidelines, design standards and statutory instruments have been written or updated as a result of these fires, including NFPA 502 "Standard for Road Tunnels, Bridges, and other Limited Access Highways—2008 Edition," the European Union's Directive 2004/54/EC on "Minimum Safety Requirements for Tunnels in the Trans-European Road Network," and the World Road Association's report on "Road Tunnels: An Assessment of Fixed Fire Fighting Systems" (2008). These standards allow a certain degree of flexibility in defining performance-based alternatives to standard prescriptive measures, as long as engineering analysis can demonstrate that an equivalent level of fire safety is maintained. The analysis usually takes the form of a qualitative and/or quantitative risk assessment, as recommended for example by BS 7974:2001 "Application of fire safety engineering principles to the design of buildings—code of practice."

The risk assessment process normally comprises hazard identification, scenario development, consequence analysis, probability assessment and risk reduction (Figure 1). The ideal outcome of the risk assessment is a design that is agreeable to all stakeholders, including the Authority Having Jurisdiction. In the authors' experience, this is best achieved by obtaining agreement at an early stage on the methodology of the risk assessment, as well as the primary data and assumptions that are fed into that assessment.

## PROJECT OVERVIEW

The Southern Tunnel links Yas Island with the mainland of Abu Dhabi, allowing access for public and private vehicles to the hotels, residential and commercial areas on the island (Figure 2 and Figure 3). The tunnel is 698m long, and therefore under the NFPA 502 standard has a category C rating. This determines the minimum level of fire protection equipment the tunnel must have. However, the fire safety specifications for the tunnel also call for consistency with the requirements of the United Kingdom's Design Manual for Roads and Bridges (BD78/99), which require a maximum distance of 100m between escape doors in the tunnel. For comparison, NFPA 502 prescribes a maximum distance between exits of 300m.

During the initial stages of the conceptual design, the tunnel was conceived as a five-cell structure, as per Figure 4. This comprised two 3-lane highway cells, a central Light Rail Transit (LRT) cell, and two emergency escape galleries on either side of the LRT cell.

In order to minimise the project construction costs, and to keep to the demanding construction programme, an alternative 3-cell concept (Figure 5) was proposed. This involved installing a low pressure deluge fire suppression system in the two highway cells to mitigate the risks of fire, and the construction of exits on both sides of the creek, which kept the maximum escape distances down to 294m (Figure 6 and Figure 7).

Although the 3-cell design conformed to the 300m maximum exit distance requirement of NFPA 502, it did not satisfy the more stringent 100m maximum exit distance of BD78/99. It was therefore decided to undertake a quantitative risk assessment to ascertain whether the overall fire risks of the 3-cell option were comparable to those of the 5-cell concept.

## EVACUATION ROUTES

A longitudinal ventilation system with jetfans was designed for the highway cells. In normal (non-congested) traffic scenarios, traffic downstream of any fire will be able to drive out of the tunnel, while traffic that is stuck behind a fire incident will be located in fresh air. A traffic management system ensures that priority is given to tunnel traffic in case of an incident, to reduce the likelihood of congested traffic in the tunnel. Once an incident has been confirmed, the traffic lights at entry to the affected tunnel will be switched to red and the traffic barriers will be lowered, in order to reduce the number of vehicles entering the tunnel.

Civil Defence will, in principle, be able to access the tunnel from both portals, since bi-directional

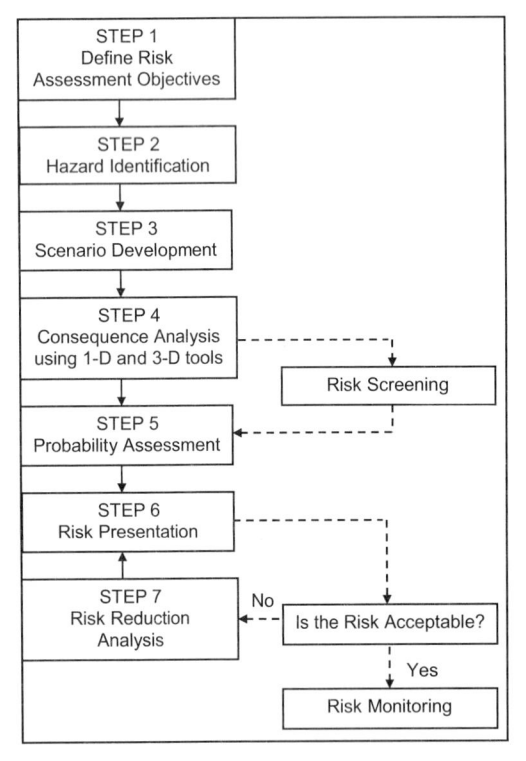

**Figure 1. Basic fire safety engineering process (adapted from Barry, 1995)**

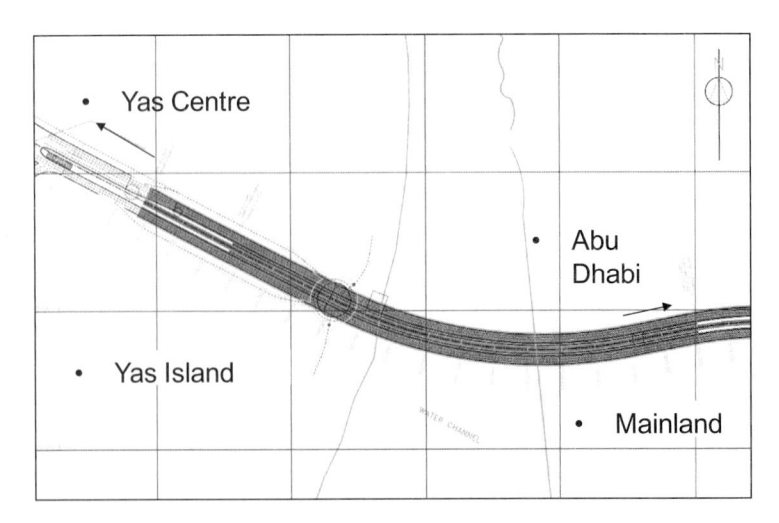

**Figure 2. Horizontal alignment of Yas Island southern crossing tunnel**

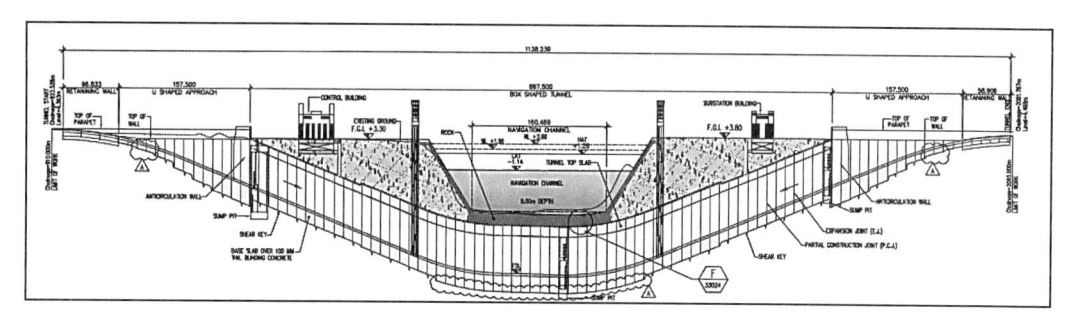

Figure 3. Vertical alignment of the Yas Island southern crossing tunnel

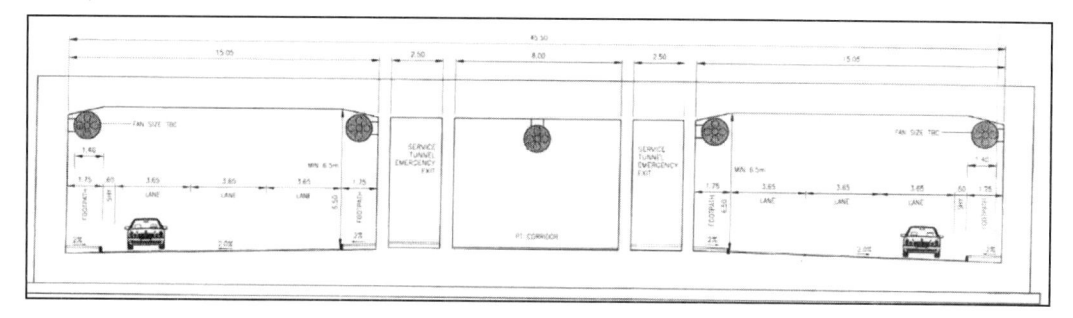

Figure 4. Original 5-cell tunnel cross-section (concept)

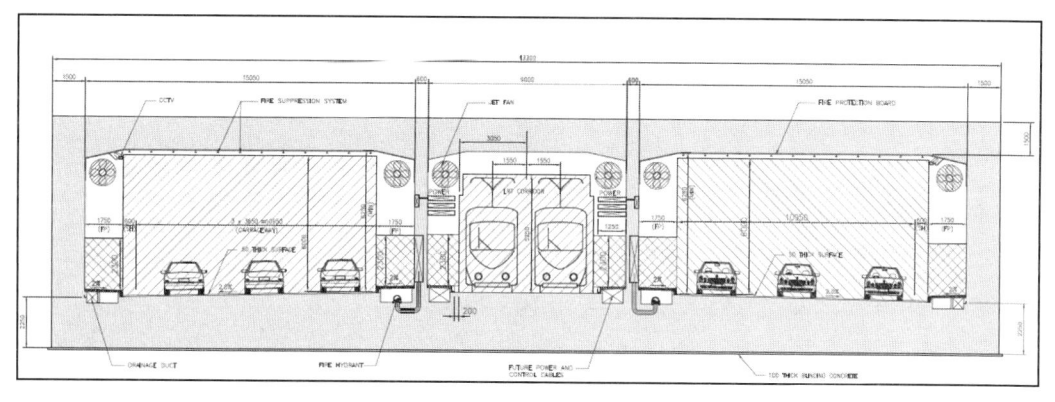

Figure 5. 3-cell tunnel cross-section

jetfans will be specified. At the initial (evacuation) stages, the airflow direction will be the same as the traffic direction. The airflow direction can be reversed by Civil Defence if required, to enable them access to the source of fire from the opposite direction.

## FIRE RISK ASSESSMENT

### Fire Size

The maximum heat release rate for a design fire within the Yas Island Southern Crossing Tunnel is determined by the type of vehicles permitted in the tunnel. Petrol tankers and other dangerous goods vehicles are banned from using this tunnel, but there are no restrictions on heavy goods vehicles, hence it was considered that a maximum design fire heat release rate of 150MW should be assumed for the tunnel design (Opstad, 2005). The access for flammable good vehicles is to be via the Shahama, Saadiyat freeway.

There is experimental evidence that fire heat release rates are substantially reduced by fixed fire suppression systems. From a review of experimental

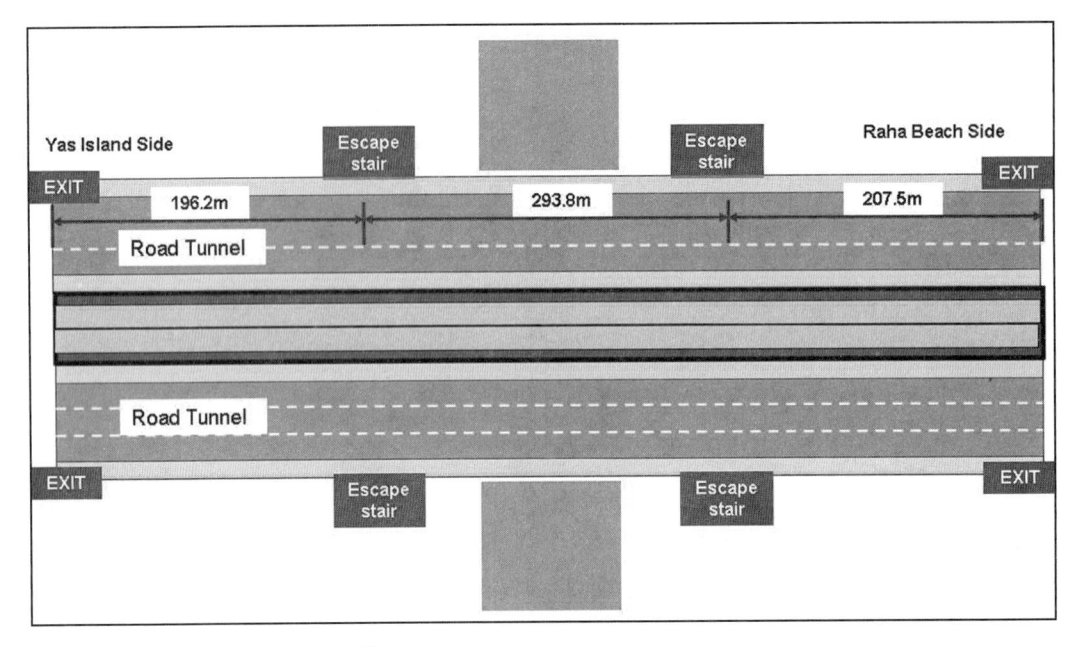

**Figure 6. Escape distances in 3-cell tunnel**

**Figure 7. 3-cell tunnel cross-section at location of emergency exits**

evidence, a design fire heat release rate of 30 MW is considered appropriate for the 3-cell option if a fixed fire suppression system were installed in the Yas Island Southern Crossing Tunnel.

Therefore assumed maximum fire heat release rates for this project were:

3 Cell Option:    30 MW (HGV) – with fire suppression
5 Cell Option:    150 MW (HGV) – without fire suppression
Car Fire:    4 MW – with fire suppression

Car Fire:    8MW – without fire suppression

**Evacuation Scenarios**

Evacuation scenarios under both congested and non-congested traffic conditions were considered. For the purposes of our risk analysis, it has conservatively been assumed that there will be congestion downstream of an accident in 12.5% of incidents (and hence that there is no traffic congestion in 87.5% of incidents). This corresponds to expected periods of congestion for 1.5 hours in the morning and evening peak periods. The likelihood of congestion

Table 1. Breakdown of evacuation procedure times

| Variable | Meaning | Assumed value |
|---|---|---|
| $t_1$ | Detection time | 1 to 2 minutes from fire growth (1 to 5 minutes from start of incident) |
| $t_2$ | Operator or automatic decision time | 1 to 2 minutes |
|  | Stop traffic | 1 to 2 minutes |
| $t_3$ | Instruction time | 1 to 2 minutes |
| $t_4$ | Motorists' decision time | 1 to 2 minutes |
|  | Fire suppression activated | 3 minutes after instructions to evacuate are given |
| $t_5$ | Walking time | Distance/38m/min |
|  | Total time to empty cell | 8 – 13 min (5-cell option) 11 – 16 min (3-cell option) |

downstream is minimized since the traffic plans will allow for priority to tunnel traffic in an emergency. The provision of the 3-cell option will only differ from that of the 5-cell option in the length of time required to evacuate the tunnel, due to the increase in distance and the environment in which people will have to evacuate in the 3-cell option.

*No Congestion Downstream*

In the scenario where there is no congestion downstream of the fire, vehicles in front of the incident may continue to drive out. Therefore, smoke can be ventilated downstream, as there will be no people evacuating their vehicles in this area. The visibility will be reduced downstream of the incident for both the 3 and 5 cell options, but this should not normally increase the risk to the motorists driving out of the tunnel.

Drivers evacuating upstream of the fire can be expected to be located in a fresh air environment, since the piston effect of moving vehicles and the thrust of the jet fans should help drive the smoke downstream. Therefore, although the 3-cell option involves an additional distance to a place of safety, there are no strict time limits on the time required for evacuation to a place of safety.

*Congestion Downstream*

In the scenario where there is traffic congestion downstream of the fire, vehicles in front of the incident will not be able to drive out. All drivers will therefore need to leave their cars and evacuate to the nearest exit point. Drivers upstream of the fire will be in a similar situation to the non-congested scenario and therefore have no immediate limits on their evacuation time due to the conditions.

For the first few minutes of the evacuation time, any fire suppression system will not be active. Therefore until the incident has been detected, confirmed and the delay time of the fire suppression system exceeded, the only difference between the 3 and 5 cell options is the distance required to exit the tunnel. After this time, the fire suppression and longitudinal ventilation systems are activated (not necessarily simultaneously). The fire suppression system reduces fire growth, lowering temperatures in the tunnel and reducing the risk of fire spread.

*Timelines*

A typical fire scenario in the Yas Island Southern Crossing Tunnel is described below and has been assumed as part of quantitative risk model [with times in parenthesis]:

1. Fire breaks out in a vehicle [T=0 s]
2. The fire is detected by the tunnel operators via the CCTV-based Automatic Incident Detection system [$T = t_1$]
3. A decision is made by the tunnel operators to order a full evacuation of the tunnel [$T = t_1 + t_2$]
4. The tunnel operators stop the traffic and instruct motorists to leave their vehicles and evacuate the tunnel [$T = t_1 + t_2 + t_3$]
5. The fire suppression system is activated 3 minutes (NFPA 502) after the instructions to evacuate the tunnel are given
6. The motorists make a decision to abandon their vehicles and start to evacuate [$T = t_1 + t_2 + t_3 + t_4$]
7. Motorists walk to the points of safety (portals and evacuation shafts), until the tunnel is empty of all people [$T = t_1 + t_2 + t_3 + t_4 + t_5$]

Table 1 indicates a likely range of overall evacuation times for the 5 and 3 cell options, with the maximum time taken assuming the fire is blocking an exit.

For an HGV fire with fire suppression, the effective fire size is limited to 30MW since the fire suppression is assumed to be activated 8 minutes after the start of the incident. Without fire suppression the fire is assumed to grow to 129MW after 25 minutes, based on a 'fast' growth rate as per PD 7974-1 (British Standards Institution, 2003).

The evacuation timeline has been represented as a fault tree using PrecisionTree software, based on the fire growth rate assumptions for the suppressed and unsuppressed fires. Fires directly in front of an exit (full evacuation distance) and those between exits (half evacuation distance) have been included in the quantitative risk analysis. The probability of the incident blocking an exit has been predicted using various fire sizes and therefore changes with the incidents.

**Table 2 Estimated fatalities resulting from different fire types**

| Fatality Rates | 5-Cell Option | 3-Cell Option |
|---|---|---|
| Damage only (up to 5MW)—car fires | 0 | 0 |
| Minor fire (up to 10MW and 30min burning time)—car fires | 0 | 0 |
| Severe fire (up to 50MW)—car fires | 1 | 2.5 |
| Very severe fire (up to 100MW)—goods vehicle fires | 3 | 7.5 |
| Catastrophic fire (greater than 100MW)—Goods vehicle fires | 20 | 50 |

## Consequence Analysis

Table 2 shows estimates of the fatalities resulting from different fire types. As a result of the higher evacuation times, the consequences of a severe or catastrophic fire are worse for the 3-cell option than the 5-cell option. However, very severe and catastrophic fires can only occur if the fire suppression system fails during a HGV fire. The probability of these two events occurring simultaneously is very low. The fault tree in Appendix A assumes 1% of incidents result in an HGV fire and that the fire suppression system is unavailable (either due to maintenance or failure) for 1.8 days per year. Based on these assumptions, such circumstances could be expected to occur in approximately 1 out of every 20,000 incidents.

## Monte Carlo Simulations

In order to ascertain the overall fire risk profile of the two tunnel options, Monte Carlo simulations were undertaken of a range of fire scenarios, with each scenario being defined by 17 parameters that are allowed to vary within Gaussian distributions, each with a defined mean value and standard deviation. A typical result of these calculations is shown in Figure 8. While car fires define the highest frequency/low consequence fires, HGV fires correspond to low frequency/high consequence events. Fire suppression is seen to shift the 'secondary peak' of the frequency/consequence diagram towards lower consequences in terms of fatalities, almost by an order of magnitude. It is seen that the 3-cell option with fire suppression provides a similar risk profile to that of the 5-cell option without fire suppression.

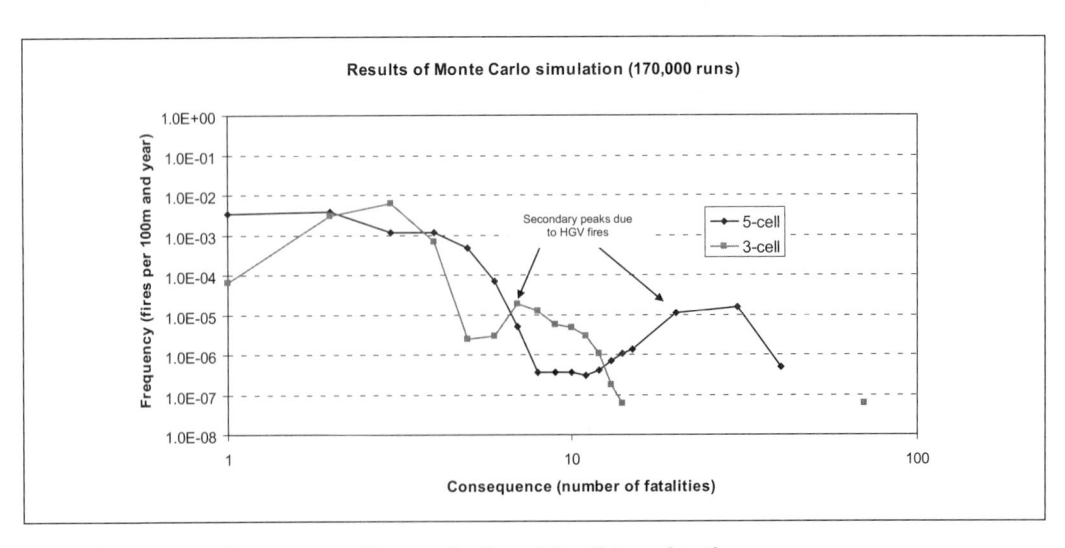

Figure 8. Frequency/consequence diagram for 3- and 5-cell tunnel options

In practice, there will be a range of light goods vehicles (LGVs) that have an intermediate heat release rate and accident probability between those of passenger vehicles and HGVs. Such vehicles have not been included in the present analysis. However, the authors consider that the conclusions of the study remain valid.

**Results of Quantitative Risk Assessment**

Although the 3-cell option increases the distance between exits and therefore the evacuation time for motorists, installation of a fire suppression system is likely to reduce the overall life safety risk to that of a 5-cell option, which has escape accesses every 100m along the tunnel.

If the fire suppression system should fail, the fire may grow to a size where the jet fans cannot fully ventilate the tunnel, as the system has been designed assuming a maximum of a 30MW fire. The probability of this situation occurring is very low (approximately 1 out of every 20,000 incidents), as indicated by the quantitative risk assessment described above.

The only other scenario where the 3 cell and 5 cell options differ significantly is that of a heavy goods vehicle fire in the presence of congested traffic (due to the higher evacuation times for the 3-cell option). Since such circumstances will only account for a small proportion of incidents any increase in fatality rates is likely to be marginal.

Two parameters within the risk assessment significantly affect these results: the pre-movement time (i.e., the time elapsed prior to the movement of motorists towards points of safety) and the time during which the fire suppression system is unavailable (either due to planned maintenance or failure). Hence, the installation of public address systems to encourage motorists to leave their vehicles and evacuate the tunnel in an emergency, and rigorous standards for maintenance and reliability of the chosen fire suppression system are required.

## COMPUTATIONAL FLUID DYNAMICS (CFD) CALCULATIONS

In order to provide a better insight into the safety consequences of the 3-cell and 5-cell designs, a number of CFD calculations were undertaken for fire scenarios with 150MW heat release rate (unsuppressed HGV fire for 5-cell design) and 30MW heat release rate (suppressed HGV fire for 3-cell design). The calculations were undertaken using the Fire Dynamics Simulator (FDS) software from National Institute of Standards and Technology.

For the proposed sprinklered tunnel, the worst credible fire scenario is a HGV fire within the central portion of the road tunnel. Based on NFPA 502, World Road Association and experimental data, this has been modelled as a fast growth rate 30MW fire. This was defined as a prescribed $t^2$ fire growth curve and therefore the sprinklers will not be used to suppress the fire but will not cool the gas layer. This method of defining the fire is more conservative and will allow for the fact that shielding of a fire can occur.

For the non-sprinklered tunnel a fast growth 150MW fire will be used in the same location designed to simulate a large HGV fire with additional vehicle involvement. A 10m × 2m block the same size as a HGV was modelled in the tunnel with the prescribed fire applied to this object.

The reaction used in both these fires is cellulosic with a soot yield of 0.1g/g. The CO yield was set as 0.05mol/mol. These values were chosen to conservatively represent the type of fuel loads that would be present in a large HGV fire.

A steady airflow velocity of 2.3m/s imposed by the jetfans was assumed in the analysis for both the 30MW and 150MW fire scenarios.

Owing to the size and geometrical complexity of the tunnel, it is meshed with 0.5m × 0.5m × 0.5m grids throughout, giving a total cell count of 3.5 million. Grid sensitivity checks were carried out on the calculated air velocities in the tunnel with a range of grid sizes (Figure 9). These checks confirmed that the choice of the 0.5m mesh size was reasonable for the purposes of this analysis.

Figure 11 shows a graph of visibility for the 150MW fire scenario along the entire length of the tunnel at four minute intervals. Four minutes after the fire has started, visibility is better than 10m in all parts of the tunnel. After five minutes, visibility deteriorates below 10m, at a distance 50m downstream of the fire up to the tunnel exit. After 16 minutes the visibility is untenable at all points downstream of the fire. The upstream visibility distance is maintained at greater than 10m at all times.

Shown in Figure 12 is a graph of temperature along the entire length of the tunnel at four minute intervals. Conditions are tenable until 14 minutes after the fire has started. By 19 minutes, conditions have become untenable up to 220m downstream of the fire. At twenty three minutes the whole of the tunnel downstream of the fire has become untenable. Upstream of the fire, temperatures are maintained tenable at all times.

Figure 13 shows a graph of CO concentration during the tunnel fire. The levels of CO continue to rise until 32 minutes where steady state conditions are observed in all areas of the tunnel apart from immediately around the fire source. The peak level is 402ppm. Around the fire source levels continue to increase and after 60 minutes the level is 1150ppm. CO levels are not shown to increase upstream of the fire location.

Figure 14 shows the expected visibility distances for the 30MW fire scenario. Visibility is better than

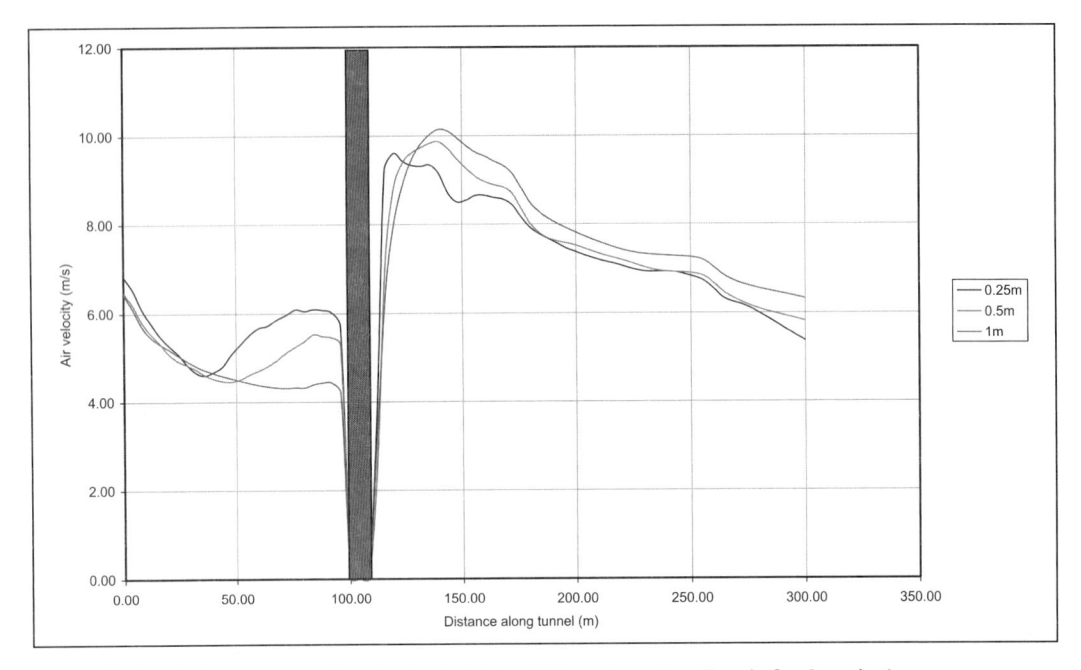

Figure 9. Graph showing the air velocity (m/s) for different meshes (bar is fire location)

10m, up to seven minutes after the fire has begun. After seven minutes, visibility conditions deteriorate in the tunnel. By 16 minutes, visibility is worse than 10m at all locations downstream of the fire.

Figure 15 shows the temperatures along the entire length of the tunnel at four minute intervals. The temperature is tenable in all parts of the tunnel up until twelve minutes. After this time conditions are tenable in all parts of the tunnel apart from a 30m section 50m downstream of the fire source. This remains the case for the rest of the simulation.

Figure 16 shows the CO concentration in ppm for the 30MW fire scenario. Upstream of the fire there is no increase in concentration. Downstream of the fire the levels increase until sixteen minutes where steady state conditions are observed. The highest concentration in any part of the tunnel is 102ppm, which is well within the tenability limit of 800ppm.

In summary, the CFD calculations show that the longitudinal ventilation system with jet fans works well to maintain tenable conditions upstream for escape and for fire service access. It is also noted that a significant improvement in tenability conditions (temperature, CO) can be obtained downstream of an HGV fire due to fire suppression, although the visibility conditions are still expected to be poor.

## CONCLUSIONS

On the basis of the quantitative risk assessment presented in this paper, the Yas Island Southern Crossing

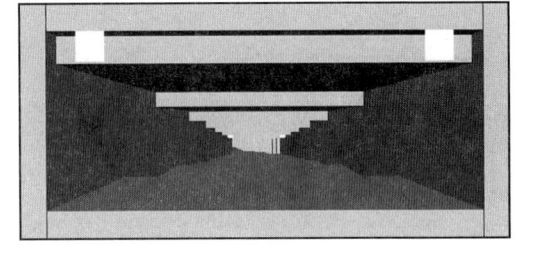

Figure 10. FDS Model of Yas Island southern crossing tunnel (end view)

Tunnel was constructed as a 3-cell structure, with the agreement of the Abu Dhabi Civil Defence, and in time for the Formula 1 Grand Prix that was held in Abu Dhabi between 30 October and 1 Nov 2009. The 3-cell structure with fire suppression in the highway cells and with a maximum distance to exits of 293.8m was shown to have a similar risk level to a 5-cell structure with 100m between exits. The performance-based fire safety design of this tunnel thus provided significant construction and programme advantages to this project.

## REFERENCES

Barry, T.F., 1995. *An Introduction to Quantitative Risk Assessment in Chemical Process Industries,* in the SFPE Handbook of Fire Protection Engineering, 2nd Ed.

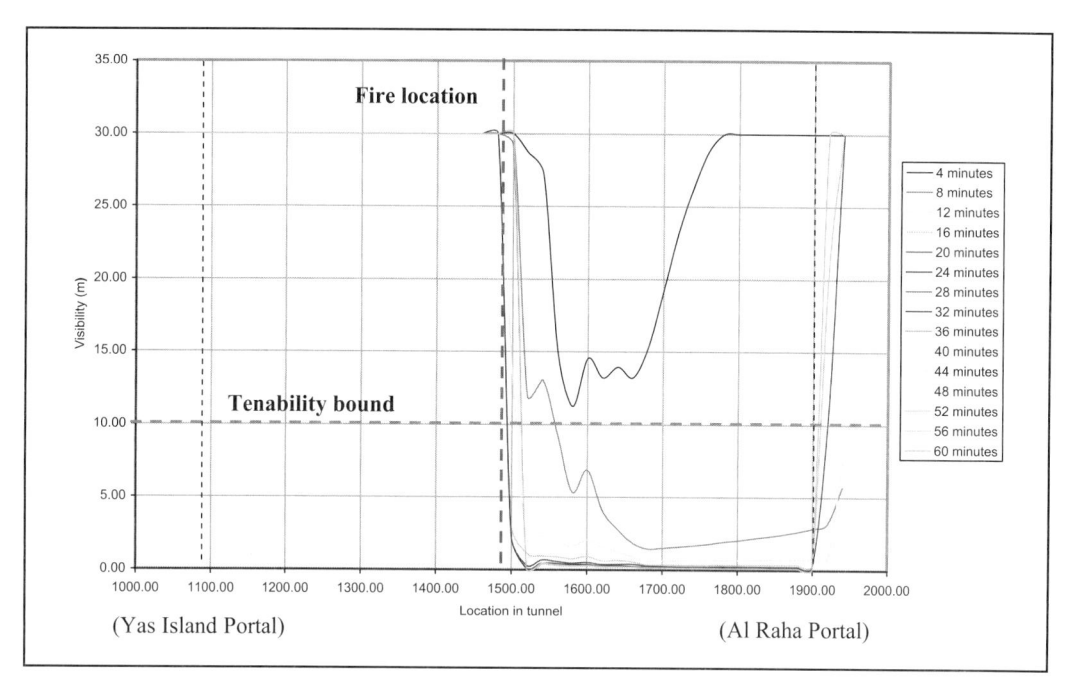

Figure 11. Visibility distances during the 150MW tunnel fire

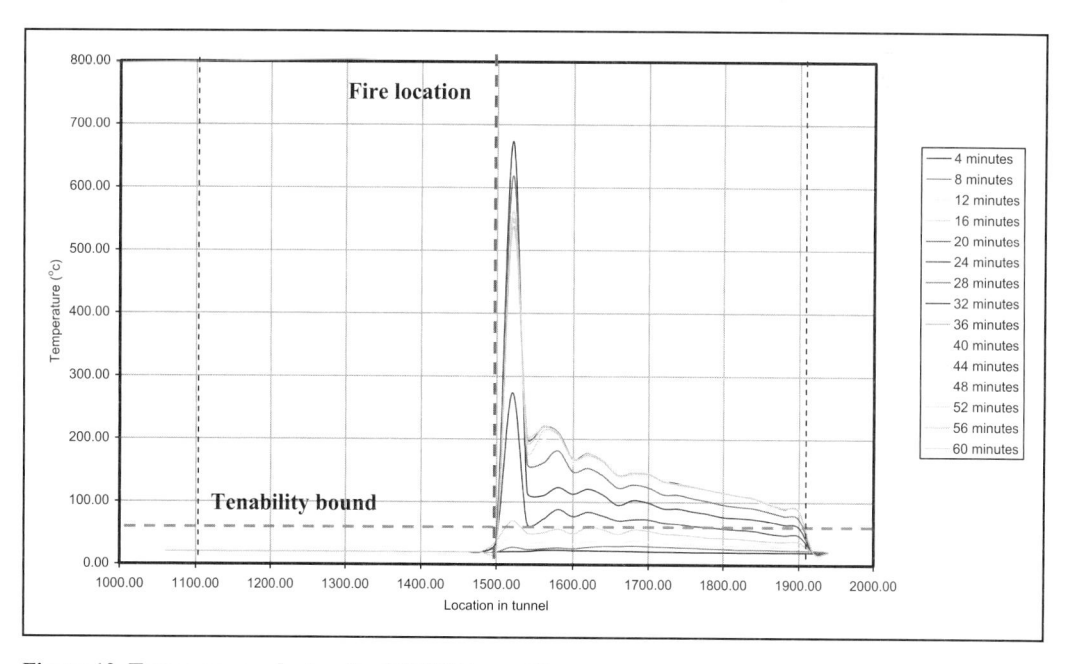

Figure 12. Temperatures during the 150MW tunnel fire

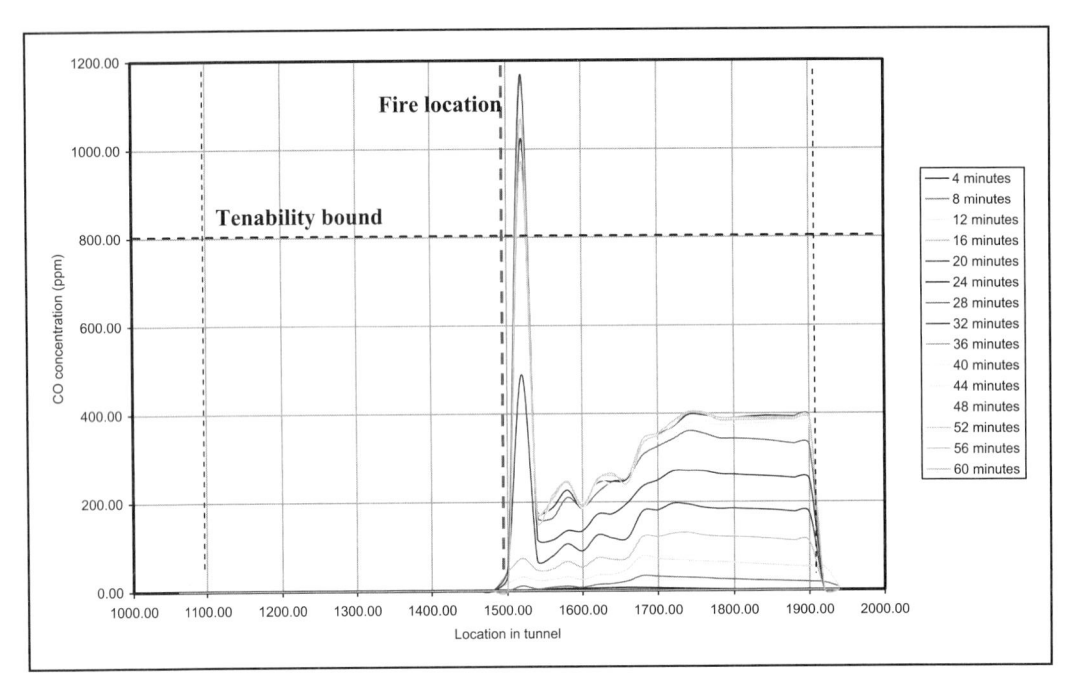

Figure 13. CO concentrations during the 150MW tunnel fire

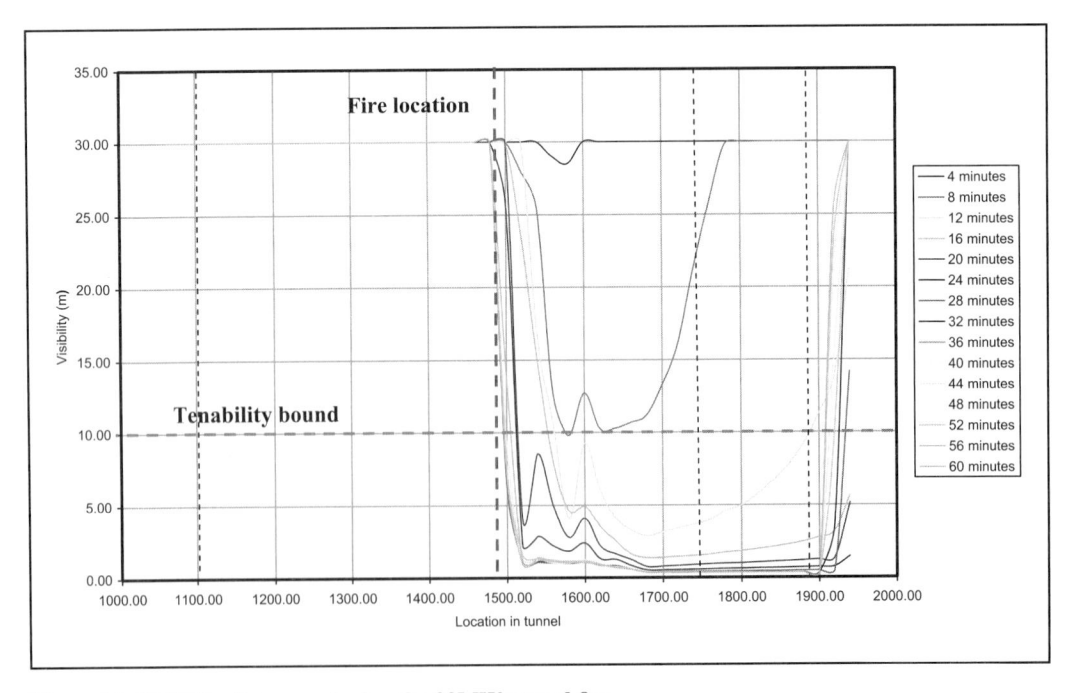

Figure 14. Visibility distances during the 30MW tunnel fire

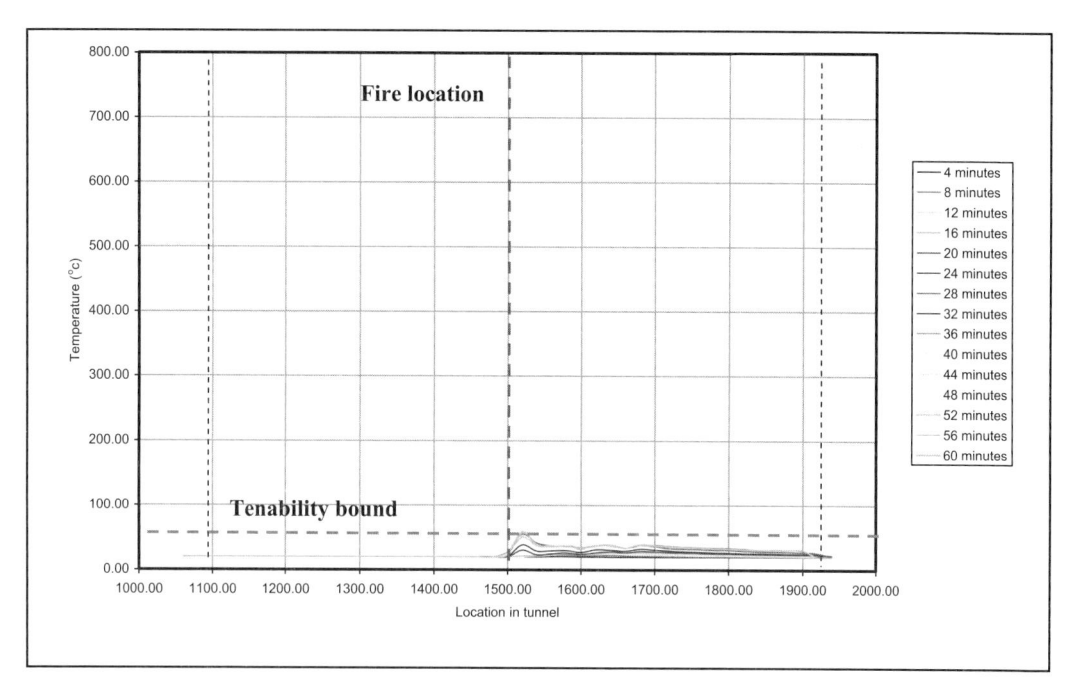

Figure 15. Temperatures during the 30MW tunnel fire

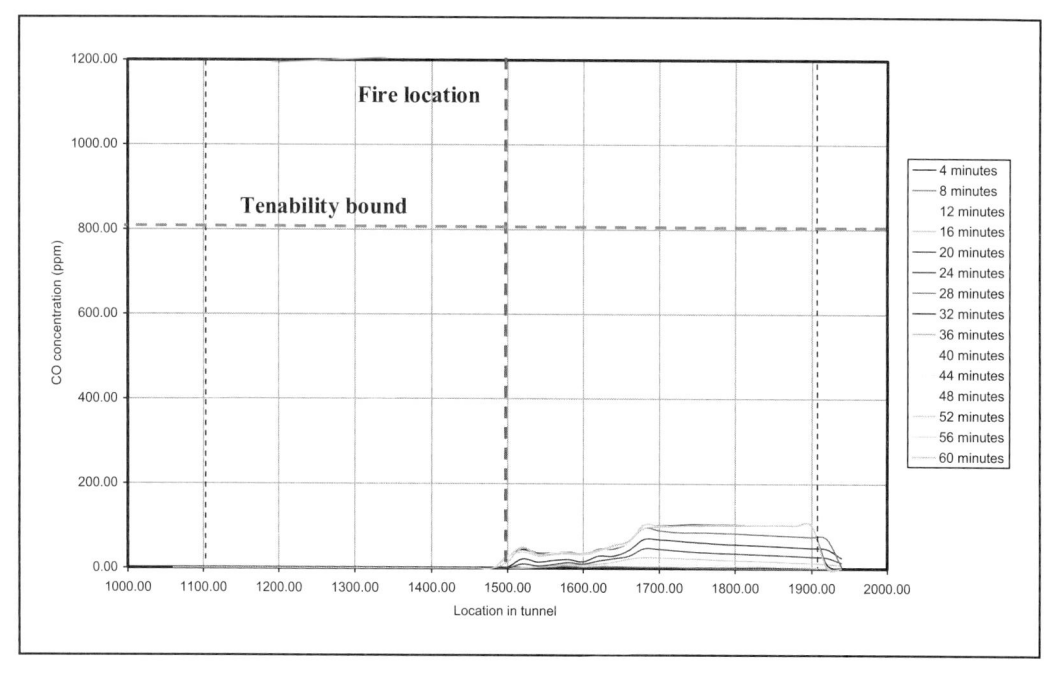

Figure 16. CO concentrations during the 30MW tunnel fire

British Standards Institution, BS 7974:2001. *Application of fire safety engineering principles to the design of buildings—code of practice*

British Standards Institution, PD 7974:2003. *Application of fire safety engineering principles to the design of buildings—Part 1: Initiation and development of fire within the enclosure of origin (sub-system 1).*

European Union, 2004. *Directive 2004/54/EC of the European Parliament and of the Council on Minimum Safety Requirements for Tunnels in the Trans-European Road Network.*

Highways Agency, United Kingdom, 1999. *Design Manual for Roads and Bridges, Volume 2—Highway Structures Design (Substructures and Special Structures) Materials—Section 2, Special Structures—Part 9, BD78/99—Design of Road Tunnels.*

National Fire Protection Association, 2008. *NFPA 502 Standard for Road Tunnels, Bridges, and other Limited Access Highways.*

Opstad, K, 2005, *Fire scenarios to be recommended by UPTUN WP2 Task leader meeting of WP2, Minutes from a meeting in London.*

World Road Association, 2008. *Road Tunnels: An Assessment of Fixed Fire Fighting Systems, Technical Committee C3.3 on Road Tunnel Operations.*

# First Comprehensive Tunnel Design Manual in the United States

**James E. Monsees**
Parsons Brinckerhoff, Orange, California

**Nasri A. Munfah, C. Jeremy Hung**
Parsons Brinckerhoff, New York, New York

ABSTRACT: The increased use of underground space for transportation systems prompted Federal Highway Administration (FHWA) to commission the development of a design manual for road tunnels. The manual, which was issued in 2009, is the first comprehensive publication by a US federal agency for the design of tunnels. It is intended to be a single-source manual providing guidelines for planning, design and construction of road tunnels, and it encompasses all tunnels types including bored tunnels, sequential excavation (SEM), cut-and-cover, and immersed tubes; and all ground conditions. Subsequently, AASHTO's Technical Committee on Tunnels (T-20) adopted the manual and is in the process of publishing it under its domain. This paper presents a summary of the manual and provides general guidelines for the design and construction of highway tunnels.

## INTRODUCTION

The increased use of underground space for transportation systems and the increasing complexity and constraints of constructing and maintaining above ground transportation infrastructure have prompted the need to develop this technical manual. In 2006 FHWA Commissioned Parsons Brinckerhoff to develop a design manual that will establish design guidelines for road tunnels. This FHWA road tunnel manual (FHWA 2009) is intended to be a single-source technical manual providing guidelines for planning, design, construction and rehabilitation of road tunnels, and encompasses various types of tunnels as shown in Figure 1.

The scope of the manual is primarily focused on the civil elements of road tunnels with the intent of future development of a manual for tunnel systems. It includes the following sixteen (16) chapters:

- Planning
- Geometrical Configuration
- Geotechnical Investigations
- Geotechnical Reports
- Cut-and-Cover Tunnels
- Rock Tunneling
- Soft Ground Tunneling
- Difficult Ground Tunneling
- Sequential Excavation Method (SEM) Tunneling
- Tunnel Lining
- Jacked Box Tunneling
- Immersed Tunnels
- Seismic Considerations
- Tunnel Construction Engineering
- Geotechnical and Structure Instrumentations
- Tunnel Rehabilitations

A summary and few highlights for the above chapters are presented hereafter.

This manual focuses primarily on the civil elements of design and construction of road tunnels,

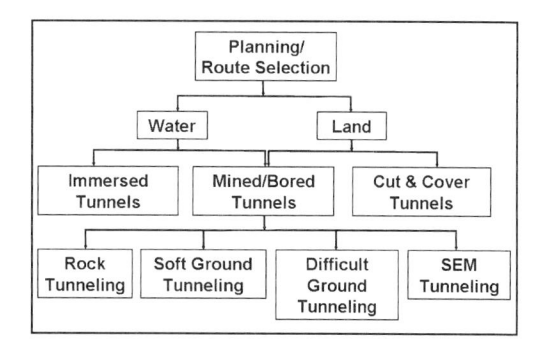

Figure 1. Types of road tunnels and tunneling methods included in the *FHWA Road Tunnel Manual*

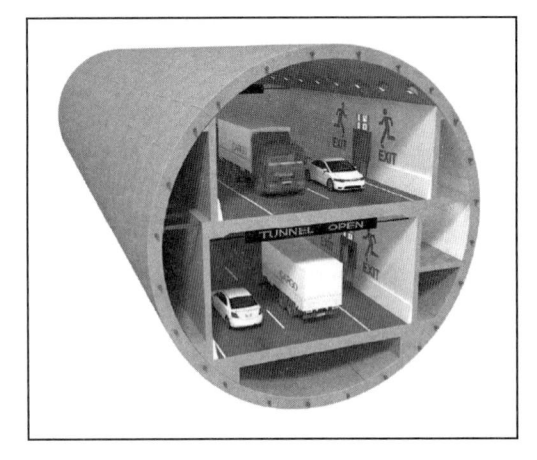

Figure 2. Chongming Tunnel, Shanghai (left) and Alaskan Way Tunnel, Seattle (right)

thus it provides only limited guidance on the system elements and fire-life safety issues when appropriate. It is the intent of FHWA to collaborate with AASHTO to further develop manuals for the design and construction of other key tunnel elements, such as, ventilation, lighting, fire life safety, mechanical, electrical and control systems. FHWA also intends to work with road tunnel owners in developing a manual on the maintenance, operation and inspection of road tunnels.

## PLANNING

The planning chapter provides a general overview of the planning process of a road tunnel project including alternative route study, tunnel type study, operation and financial planning, and risk analysis and management. It recommends that, in addition to the capital construction cost, a life-cycle cost analysis should be considered. Life cycle cost estimate will more appropriate to account the longer life expectancy of a tunnel. This will provide a better cost comparison with other transportation alternatives such as a bridge or a surface (at grade) facility. In evaluating the life cycle cost of a tunnel, costs should include construction, operation and maintenance, and financing (if any) using Net Present Value. It also recommends that values for potential air right developments should be taken into consideration. Furthermore, although assessing monitory values is difficult, the environmental benefits and sustainability aspects of underground facilities should be considered in the comparison.

As the technology advances, tunnel boring machines have been developing with a major increase in diameter, better ground control, and improved reliability. Advances in technology, as well as lessons learned from previous projects, would help planning larger, deeper and longer road tunnels to accommodate growing traffic demands in the future. The maximum size of a circular TBM existing today is about 51 ft (15.43 m) for the construction of Chongming Tunnel, a 5.6 mile (9-kilometer) long tunnel under China's Yangtze River, in Shanghai (Figure 2). Presently, the Alaskan Way Tunnel Project is being planned in Seattle to replace the Alaskan Way Viaduct with a double-decker of 54 ft diameter. When implemented, it will be the largest soft ground road tunnel.

As early as in the planning phase, it is important to understand fire-life safety issues of a road tunnel and make provisions for them in the alignment, cross section, location of emergency exits, ventilation provisions, geometrical configuration, right-of-way, and cost estimates. Space planning of the tunnel systems is of critical importance for efficient design.

## GEOMETRICAL CONFIGURATION

The manual provides the geometrical requirements and recommendations of new road tunnels including horizontal and vertical alignments and tunnel cross section requirements.

Based on the general requirements for the cross section elements provided in the AASHTO's *Green Book: A Policy on Geometric Design of Highways and Streets*, the manual recommends considering minimum geometrical requirements to economize the overall size of the tunnel yet maintain a safe operation through the tunnel, particularly considering the high costs of mined and bored tunneling and increasing restrictive right-of-way. The manual recommends the use of modified shoulder widths (comparable to other European road tunnels) to reduce the overall tunnel diameter.

Furthermore, geometrical design for road tunnels must consider tunnel systems such as fire life safety elements, ventilation, lighting, traffic control,

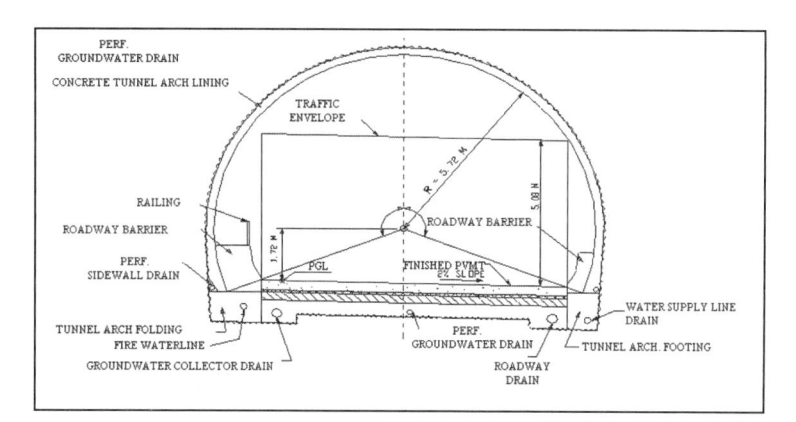

Figure 3. Typical two-lane road tunnel cross section and tunnel elements (Cumberland Gap Tunnel)

fire detection and protection, communication, and the like. The manual recommends that planning and design of the alignment and cross section of a road tunnel must also comply with the most recent National Fire Protection Association (NFPA) 502 —Standard for Road Tunnels, Bridges, and Other Limited Access Highways.

## GEOTECHNICAL INVESTIGATIONS AND REPORTS

A chapter in the manual covers the geotechnical investigative techniques and parameters required for planning, design and construction of road tunnels. In addition to extensive discussions on subsurface investigation techniques and requirements, the chapter also includes discussions for investigation methods including information study; survey; site reconnaissance, geologic mapping, instrumentation, and other investigations made during and after construction to obtain a broad spectrum of pertinent topographic, geologic, subsurface, geo-hydrological, and structure information and data. The techniques and procedures focus on the specifics for tunnel and underground projects, and their variation related to the tunneling methods.

A geotechnical investigation program for a tunnel project must use appropriate means and methods to obtain factual information about the distribution and engineering characteristics and properties of the ground as a basis for planning, design and construction. The investigation focus will be on the anticipated behavior of the ground during excavation and to identify potential construction risks, and to establish realistic engineering cost estimate and construction schedule. The manual recommends phasing of the geotechnical investigations to provide an economical and rational approach for adjusting to anticipated changes to a road tunnel project. It also suggests a geospatial data management system be used

Figure 4. Water boring investigation from a barge for the Port of Miami Tunnel, Miami, FL

so data collected throughout the various phases and time can be managed and utilized efficiently.

The extent of the investigation should be consistent with the project scope (i.e., location, size, and budget), the project objectives (i.e., risk tolerance, long-term performance), and the project constraints (i.e., geometry, constructability, third-party impacts, aesthetics, and environmental impact). In some cases, such as road tunnels in mountainous areas or for water crossings, the cost for an extensive subsurface

**Figure 5. Voest-Alpine AM 105 Roadheader, Australia**

investigation may be prohibitive. Therefore, the challenge to geotechnical engineers is to develop an adequate and diligent geotechnical investigation program that can improve the predictability of ground conditions within a reasonable budget and acceptable level of risk. It is important that all involved parties have a common understanding of the limitations of geotechnical investigations, and be aware of the inevitable risk of not being able to completely define all existing geological conditions, or to fully predict ground behavior during construction.

The manual recommends the use of the updated ASCE "Geotechnical Baseline Reports for Construction – Suggested Guidelines" publication (ASCE, 2007) to provide improved risk sharing mechanism. It recommends using three general types of geotechnical reports throughout the planning, design and construction phases of a road tunnel including: Geotechnical Data Reports (GDR) which present all the factual geotechnical data; Geotechnical Design Memorandum (GDM) which presents interpretations of the geotechnical data and other information used to develop the designs; and Geotechnical Baseline Report (GBR) which defines the baseline conditions upon which contractors will base their bids.

It is important to note that, although the GBR reflects the findings of the geotechnical investigations and design studies, a GBR is not intended to predict the actual geotechnical and geological conditions at a project site, or to accurately predict the ground behavior during construction. Rather, it establishes the bases for delineating the financial risks between the owner and the contractor.

## ROCK TUNNELING

The rock tunneling chapter presents various excavation methods and temporary support elements in rock and focuses on the selection of temporary support of excavation and input for permanent lining design. It describes the basic rock failure mechanism and common rock mass classifications including Terzaghi's system, Barton's Q System, and Bieniawski's Rock Mass Rating (RMR) System. It presents various rock tunneling methods including drill and blast, tunnel boring machine (TBM), and roadheaders. Sequential excavation method (SEM) is discussed in a separate chapter. The use of rock TBM is discussed extensively in Appendix D of the manual. It provides descriptions and uses of various types of rock TBMs including open gripper main beam, single shield, and double shield TBMs.

The chapter also discusses recent uses of roadheaders in road tunnels. Their advantages and disadvantages (Figure 5) and offers the general guidance when roadheaders may be considered. It is recommended that roadheaders be used for rock strength below about 20,000 psi—preferably below 15,000 psi—with low abrasivity. They are best used for self supporting rock with little water intrusion. They can be used for short runs, non circular in cross sections such as connections or cross passages or cross sectional enlargements.

Subsequently, the chapter presents the common types of rock reinforcement and initial support system as show in the Table 1, and discusses the design methodologies of rock tunnel support systems.

The chapter also addresses groundwater control measures during excavation including:

Table 1. Typical initial support used in the current practice (after TRB, 2006)

| Ground | Rock Dowels/ Bolts | Rock Bolts with Wire Mesh | Rock Bolts with Shotcrete | Steel Ribs and/ or Lattice Girder with Shotcrete | Precast Concrete Segments |
|---|---|---|---|---|---|
| Strong rock | • | • | | | |
| | | • | • | | |
| Medium rock | • | • | • | • | |
| | | | • | • | |
| Soft rock | | | | • | • |
| | | | | • | • |
| Residual soil | | | | • | • |

- Dewatering at the face
- Drainage ahead of face by probe holes
- Drainage from pilot bore/tunnel
- Grouting
- Freezing
- Pressurized closed face machine
- Other measures

Lastly, the chapter addresses permanent rock tunnel lining design issues including rock and hydrostatic loads, waterproofing systems, and final liner details.

## SOFT GROUND TUNNELING

The soft ground tunneling chapter addresses analysis, design and construction issues specific for tunneling in soft ground covering cohesive and cohesionless soils. It addresses soft ground classification and general behavior and focuses on the latest shield tunneling techniques such as Earth Pressure Balance and Slurry Face Tunnel Boring Machines. Appendix D of the manual demonstrates the components and excavation sequences of both types of TBMs, and presents some of the unique machines currently available such as the hybrid EPB-Slurry TBM for the A-86 SOCOTOP Road Tunnel in Paris (Figure 6).

The chapter addresses ground load and ground support issues for LRFD design of precast segmental lining, and provides general guidance for continuum numerical analysis using Finite Element Method (FEM) and Finite Difference Method (FDM).

The tunneling induced settlement issues and their impact on the surrounding surface buildings and structures are also discussed. Mitigation and protection of the impacted structures are summarized subsequently. Lastly, the soft ground tunneling chapter touches on ground control and stabilization techniques as listed in Table 2.

## DIFFICULT GROUND TUNNELING

Factors that make tunneling in difficult ground are generally related to instability, which inhibits timely placement or maintenance of adequate support at or behind the working face; heavy loading from the ground which creates problems of design as well as installation and maintenance of a suitable support system; mixed face conditions, natural and man-made obstacles or constraints; and physical conditions which make the work place untenable unless they can be modified. The following challenging conditions are discussed:

- Non-cohesive running sand and gravel
- Flowing ground
- Soft clay
- Blocky rock
- Adverse combinations of joints and shear zones
- Faults and alteration zones
- Excessive groundwater
- Mixed face conditions
- Shallow covers
- Squeezing and swelling rock
- Presence of boulders
- Unchartered obstacles such as abandoned foundations, utilities, etc.
- Karstic limestone
- Methane, hydrogen sulfide, and other gassy ground
- High temperatures

The chapter discusses and makes recommendations to deal with these conditions.

## SEQUENTIAL EXCAVATION METHOD (SEM) TUNNELS

Sequential Excavation Method (SEM), commonly known as the New Austrian Tunneling Method (NATM), is becoming increasingly popular in the

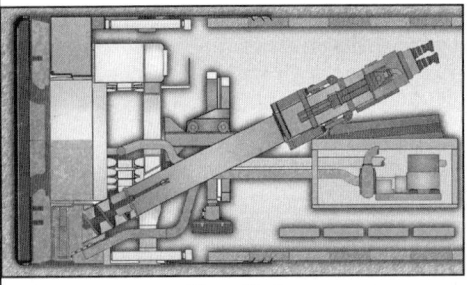

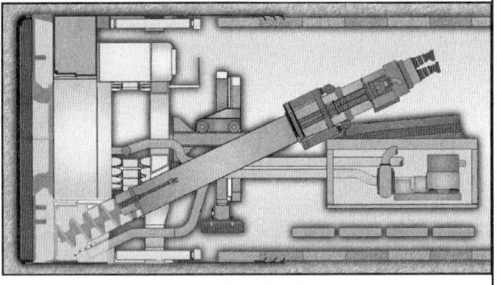

Slurry Mode | EPB Mode

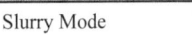

**Figure 6. Hybrid EPB—Slurry TBM for A-86 road tunnel, Paris (courtesy of Herrenknecht)**

US for the construction of tunnels, cross passages, subway stations, and other underground structures. A typical SEM cross section for a road tunnel involves generally a curvilinear shape as shown in Figure 7 to promote smooth stress distribution in the ground around the newly created opening. By subdividing a tunnel cross section into multiple drifts (depending on the quality of the ground), and adjusting the construction sequence expressed mainly in round length, timing of support installation and type of support the SEM allows for tunneling through rock, soft ground and a variety of difficult ground conditions.

This chapter introduces the history, principles, and recent development of Sequential Excavation Method (SEM), and addresses analysis, design and construction issues for SEM tunneling including ground classification; SEM excavation and support classes; ground support elements; ground improvement techniques; structural design considerations; instrumentation and monitoring considerations;

**Table 2. Ground treatment methods**

| Challenging Ground Conditions | Treatment Method(s) |
|---|---|
| Weak soils | • Vibro compaction<br>• Dynamic compaction<br>• Compaction grouting<br>• Permeation grouting<br>• Jet grouting |
| Ground water | • Dewatering<br>• Freezing<br>• Grouting |
| Unstable face | • Soil nails<br>• Spiling<br>• Soil doweling<br>• Micro piles |
| Soil movement | • Compensation grouting<br>• Compaction grouting |

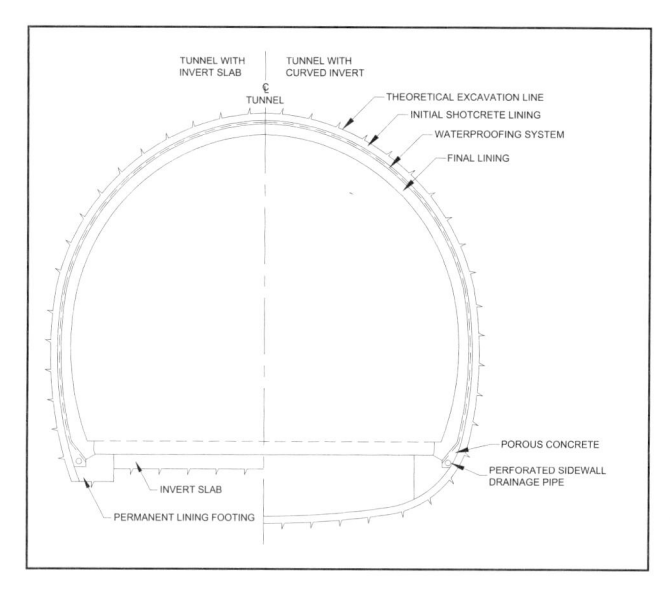

**Figure 7. Typical SEM tunnel cross section**

and construction and contractual considerations. Numerous examples of SEM excavation and support Classes in rock and soft ground are provided. Appendix F of the manual presents a calculation example which involves the tunneling analysis and lining design of a typical two-lane SEM highway tunnel using the finite element method. A paper entitled "Design Guidelines for Sequential Excavations Method (SEM) Practices for Road Tunnels in the United States" by V. Gall and N. Munfah in this conference addresses this topic in more details.

Application of the SEM involves practical experience, knowledge of engineering sciences, good QA/QC, and skilled execution. The SEM tunneling chapter addresses:

- Ground and excavation and support classification based on a thorough ground investigation
- Definition of excavation and support classes by:
  - Round length (maximum unsupported excavation length)
  - Support measures
  - Subdivision of the tunnel cross section into multiple drifts or headings as needed (top heading, bench, invert, side wall drifts)
  - Ring closure requirements
  - Timing of support installation
  - Pre-support by spiling, fore poling, and pipe arch canopy
  - Local, additional initial support by dowels, bolts, spiles, face support wedge, and shotcrete
- Instrumentation and monitoring
- Pre-support and ground improvement measures prior to excavation

## LRFD DESIGN OF TUNNEL LINING

The tunnel lining chapter covers considerations for the LRFD structural design (AASHTO, 2008), detailing and construction of various types of tunnel linings for mined and bored road tunnels. Although the focus of this chapter is mostly on the final liner, the discussions in the chapter can be applied to initial stabilization of the excavation, permanent ground support or a combination of both. The materials for tunnel final linings covered in this chapter are cast-in-place concrete lining, precast segmental concrete lining, and shotcrete lining. The final architectural finishes are not specifically addressed. A design example for precast segmental lining is presented in Appendix G of the manual.

Special topics such as fire protection measures (e.g., using specialty concrete, polypropylene fibers, etc.), steel fiber reinforced concrete, corrosion protection, lining selection criteria, etc. are also discussed.

## IMMERSED TUNNELS

The chapter discusses immersed tunnel design and construction. It identifies both steel and concrete types of immersed tunnel and their construction techniques. It also addresses the structural design approach and provides insights on the construction methodologies including fabrication, transportation, placement, joining and backfilling. It

**Figure 8. Chesapeake Bay Bridge tunnel**

addresses the water tightness and the trench stability and foundation preparation requirements. It also provides guidance in the interpretation of the AASHTO specifications, and the design of items related to immersed tunnels not specifically addressed in AASHTO.

The chapter provides basic descriptions of immersed tunnel construction methodology including trench excavation, foundation preparation, tunnel element fabrication, transportation and handling of tunnel elements, lowering and placing, element placement, backfilling, locking fill, general backfill, protection blanket, and lastly anchor release Protection. Lastly, loads, load combinations, structural analysis, design and water tightness issues are discussed.

### CUT AND COVER TUNNELS

The manual presents the construction methodology and excavation support systems for cut-and-cover road tunnels for bottom-up and top-down construction. It describes the structural design in accordance with the AASHTO LRFD Specifications (AASHTO, 2008). Other considerations dealing with support of excavation, maintenance of traffic and utilities, and control of groundwater and how they affect the structural design are discussed. A comprehensive design example is included in Appendix C of the manual.

### SEISMIC CONSIDERATIONS

Tunnel structures, are constrained and supported by the surrounding ground. During a seismic event they cannot be excited independently of the ground movement or be subject to strong vibratory amplification similar to the inertial response of a bridge structure during earthquakes. Therefore, seismic considerations of tunnel design is different than any other structure. This chapter provides general procedure for seismic design and analysis of tunnel structures, which are based primarily on the ground deformation approach (as opposed to the inertial force approach used for above ground structures); i.e., the structures should be designed to accommodate the deformations imposed by the ground.

The main factors influencing tunnel seismic performance addressed include (1) seismic hazards such as ground shaking and ground failure (i.e., fault rupture, tectonic uplift and subsidence, landslide, and soil liquefaction)., (2) geologic conditions, and (3) tunnel structure. Simplified screening processes are provided to identify if potential seismic risks exist that may require more detailed evaluations. Seismic risks considered are:

- An active fault intersecting the tunnel alignment
- A potential landslide intersecting the tunnel alignment
- Liquefiable soils adjacent to the tunnel
- History of static distress to the tunnel (e.g., local collapses, large deformations, cracking or spalling of the liner due to earth movements, etc.)

The chapter presents simplified seismic evaluation procedures and design recommendations for ground failure effects and ground shaking effects.

### TUNNEL CONSTRUCTION ENGINEERING

The tunnel construction engineering chapter focuses mostly on mined/bored tunnel construction engineering issues; the engineering that must go into a road

**Figure 9. Highway tunnel lining falling from tunnel crown: 2004 Niigata earthquake, Japan**

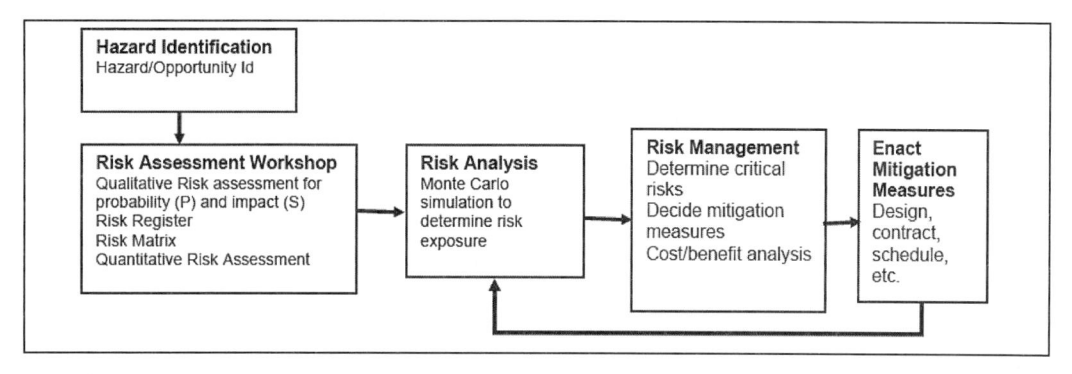

**Figure 10. Risk management process**

tunnel project to make it constructible. This chapter examines various issues that need to be addressed during the design process including project cost drivers; construction staging and sequencing; health and safety issues; risk management, and logistics such as muck transportation and disposal.

The chapter discusses risk management approach for tunnels. Risk assessment, analysis, and management are required to assure that the project is kept on schedule and within budget, and to provide greater accuracy in the application of project contingency. A comprehensive risk management process includes the use of risk management workshops, development of an "actionable" risk register, risk analyses and the development of risk management action plans are discussed. What's important is early identification and communication of potential risk factors that might create delays and bottlenecks, followed by proactive management of threats to cost and schedule adherence and to identify opportunities for improvement. Figure 10 provides a flow chart of risk management approach.

## GEOTECHNICAL AND STRUCTURAL INSTRUMENTATION

The primary purpose of geotechnical and structural instrumentation program is to monitor the performance of the underground construction process and the impacts on surface facilities and structures in order to detect, avoid, or mitigate problems. The chapter provides the types of measurements typically made to monitor the ground behavior:

- Ground movements: Vertical and lateral deformation
- Movement of existing buildings and structures within the zone of influence
- Deformation and movement of the tunnel
- Dynamic ground movement and vibration
- Groundwater movement and pressure change

Subsequently, the chapter presents over thirty (30) commonly available instruments and their applications to monitor each of the tunnel behavior.

Guidelines for selection of instrument types, numbers, and locations are provided, and commentary for issues such as remote (automated) versus manual monitoring, establishment of warning/action levels, and division of responsibility are discussed.

## TUNNEL STRUCTURAL REHABILITATIONS

Tunnel inspections and rehabilitations require a coordinated multi-disciplinary approach to deal with various functional aspects of a tunnel including civil/structural, mechanical, electrical, drainage, and ventilation components, as well as operational aspects such as signals, communication, fire-life safety and security components.

Recognizing that tunnel owners are not mandated to routinely inspect tunnels and that inspection methods vary among entities that inspect tunnels, the FHWA and the Federal Transit Administration developed guidelines for the inspection of tunnels known as "Highway and Rail Transit Tunnel Inspection manual" available at www.fhwa.dot.gov/bridge/tunnel/inspectman00.cfm (FHWA, 2005). Note that at the time of preparing this manual, the FHWA is proposing to create a regulation establishing National Tunnel Inspection Standards (NTIS) which would set minimum tunnel inspection standards that apply to all federal-aid highway tunnels on public roads.

This chapter focuses on the civil/structural aspect of tunnel condition assessment and rehabilitation including identification, characterization and repair of typical structural defects. It addresses the following:

- Groundwater leakage repairs
- Structural concrete and shotcrete repairs
- Segmental linings repairs
- Steel/cast iron repairs

**Figure 11. Rock tunnel with shotcrete wall repair and arch liner (I-75 Lima, Ohio)**

- Masonry repairs
- Unlined rock tunnel repair

## CONCLUSIONS

This paper summarizes the contents and provides some highlights of the newly published FHWA Manual for Design and Construction of Road Tunnels—Civil Element (FHWA-NHI-09-010). The manual ambitiously attempts to cover most common tunnel types and tunneling methods, and to provide guidelines for planning, design, construction and rehabilitation of road tunnels. The FHWA Road Tunnel Manual is now available for download from the FHWA Office of Bridge Technology web page at http://www.fhwa.dot.gov/bridge/tunnel/pubs/nhi09010/index.cfm.

In July 2009, AASHTO Subcommittee on Bridges and Structures and its Technical Committee on Tunnels (T-20) had approved to adopt this road tunnel manual in 2010. It is the intent of FHWA to collaborate with AASHTO to further develop manuals for the design and construction of other key tunnel elements, such as, ventilation, lighting, fire life safety, and mechanical, electrical and control systems.

## ACKNOWLEDGEMENTS

The development of *FHWA Road Tunnel Manual* has been funded by the National Highway Institute, and supported by Parsons Brinckerhoff, as well as numerous authors and reviewers from Parsons Brinckerhoff and Gall Zeidler Consultants, LLC.

The authors would like to especially acknowledge the continuous support of Louisa Ward of National Highway Institute, Firas Ibrahim of FHWA, and Jesus Rohena of FHWA, and reviews by the members of AASHTO T-20 Tunnel Committee.

## REFERENCES

American Association of State Highway and Transportation Officials (AASHTO) (2008) "AASHTO LRFD Bridge Design Specifications 4th Edition," Washington, D.C.

American Association of State Highway and Transportation Officials (AASHTO) (2004) "A Policy on Geometric Design of Highways and Streets," Washington, D.C.

American Society of Civil Engineers (ASCE) (2007). "Geotechnical Baseline Reports for Construction—Suggested Guidelines," NY.

Bickel, Kuesel and King, 2nd ed. (1996) "Tunnel Engineering Handbook"; Chapman & Hall, N.Y.

Federal Highway Administration (FHWA) (2009) "Technical Manual for Design and Construction of Road Tunnels—Civil Element," FHWA-NHI-09-010, Washington, D.C.

Federal Highway Administration (FHWA) (2005a) "Highway, Rail and Transit Tunnel Inspection Manual," FHWA-IF-05-002, Washington, D.C.

Federal Highway Administration (FHWA) (2005b) "Highway, Rail and Transit Tunnel Maintenance and Rehabilitation Manual," FHWA-IF-05-017, Washington, D.C.

# Geotechnical Investigations for the Anacostia River Projects

**Amanda Morgan**
Jacobs Associates, Washington, DC

**Kevin Fu**
URS Corporation, Gaithersburg, Maryland

**Ronald E. Bizzarri, Carlton M. Ray**
District of Columbia Water and Sewer Authority, Washington, DC

**ABSTRACT:** The Anacostia River Projects (ARP) is the major component of the Long Term Control Plan (LTCP) for the District of Columbia Water and Sewer Authority (DC WASA). The ARP consists of an approximately 20.4-km-long (12.7 mile) tunnel system, including 18 large-diameter deep shafts and supporting structures. This paper presents the ongoing geotechnical investigations used to characterize the subsurface for the ARP. Drilling methods include sonic and conventional, and boring spacing is about 190 m (600 ft). Field testing includes pressuremeter, vane shear, and crosshole seismic. Laboratory testing includes index, triaxial, consolidation, soil abrasion testing (SAT), soil chemistry, and water quality. Estimated cost for the investigations is $6.5 million.

## INTRODUCTION

The District of Columbia Water and Sewer Authority (WASA or Authority) provides wastewater collection and treatment for the District of Columbia, and wastewater treatment for surrounding areas, including parts of suburban Virginia and Maryland, at the District's Advanced Wastewater Treatment Plant at Blue Plains (Blue Plains). Like many older cities, the District of Columbia's storm water and sanitary conveyance system is combined in many geographic areas. During heavy storms, this results in direct discharge of untreated combined sewer overflows (CSOs) into rivers or streams. Hence, the Authority negotiated an agreement with the U.S Environmental Protection Agency (EPA) to improve water quality in the city's waterways by implementing both short- and long-term plans to control storm water discharges contaminated with sewage and other pollutants. Currently, WASA is in the process of implementing their Long Term Control Plan (LTCP) of facilities, infrastructure, and system improvements needed to enhance the quality of the receiving waterways and achieve (and maintain) the water quality standards in accordance with WASA's National Pollutant Discharge Elimination System permit (NPDES).

## PROJECT OVERVIEW

The LTCP includes the construction of several miles of storage and flood relief tunnels in addition to other components such as shafts, diversion chambers, and overflow facilities. Cumulatively, these facilities are considered the Anacostia River CSO Control Project (ARP). The tunnels system and the major ARP Contract Divisions are shown on Figure 1. The term Contract Division refers to the separation of each design and construction contract. The LTCP will capture, store, and convey the combined sewer flow of existing CSO outfalls along the Anacostia River. The flow captured in the tunnels will be treated at Blue Plains and flows in excess of the tunnels storage capacity and Blue Plains treatment capacity will overflow to the Potomac and Anacostia Rivers at locations C and D shown on Figure 1. Geotechnical investigations will be performed to characterize the ground conditions for the final design of the LTCP structures.

There are approximately 13 miles of tunnels to be constructed for the LTCP and 18 shafts. The tunnels and shaft inverts will be constructed at depths to invert between 18.3 and 48.9 m (60 and 160 ft) below existing ground elevation. The diameters of

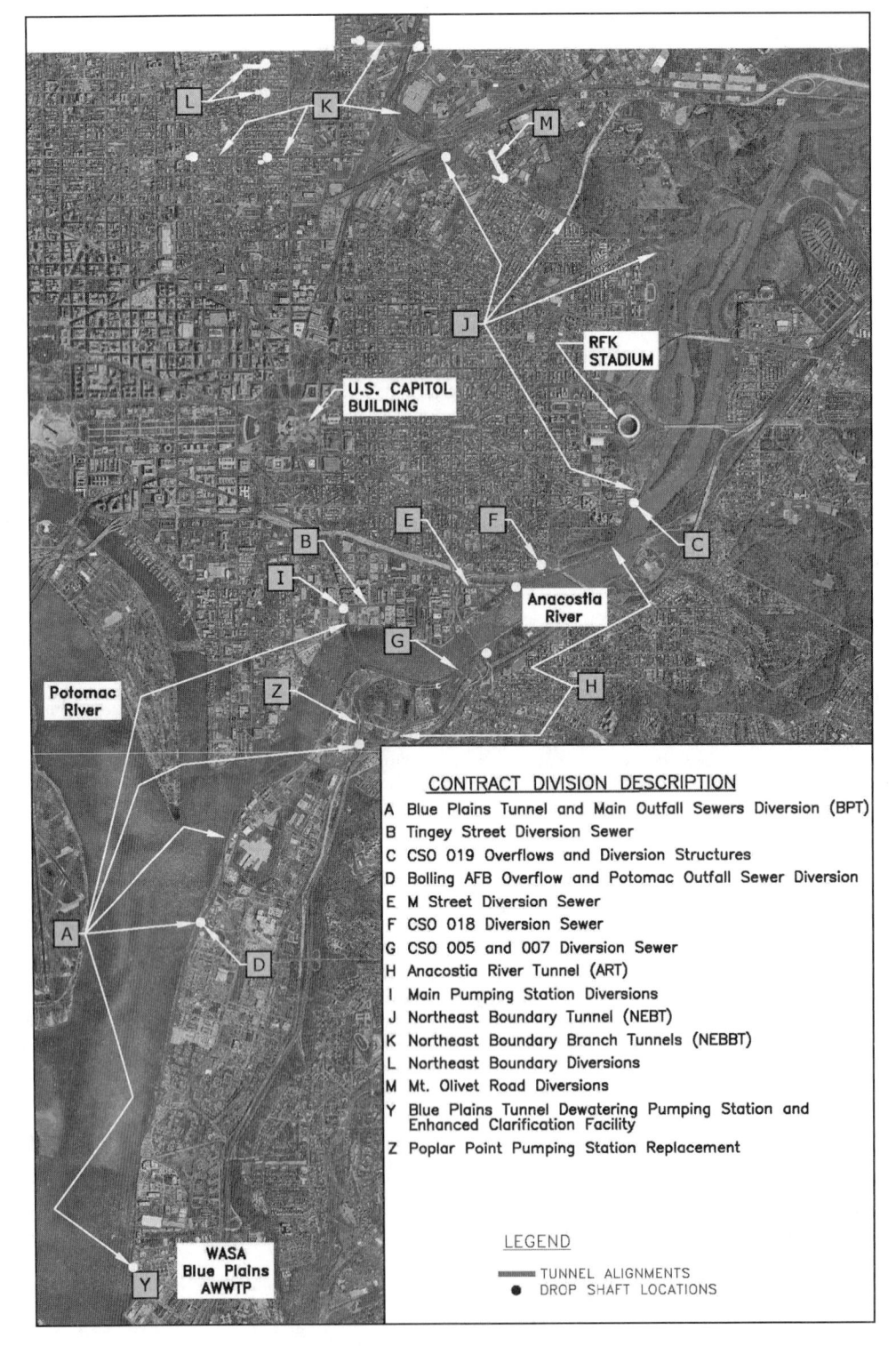

Figure 1. DC WASA LTCP contract divisions for the ARP

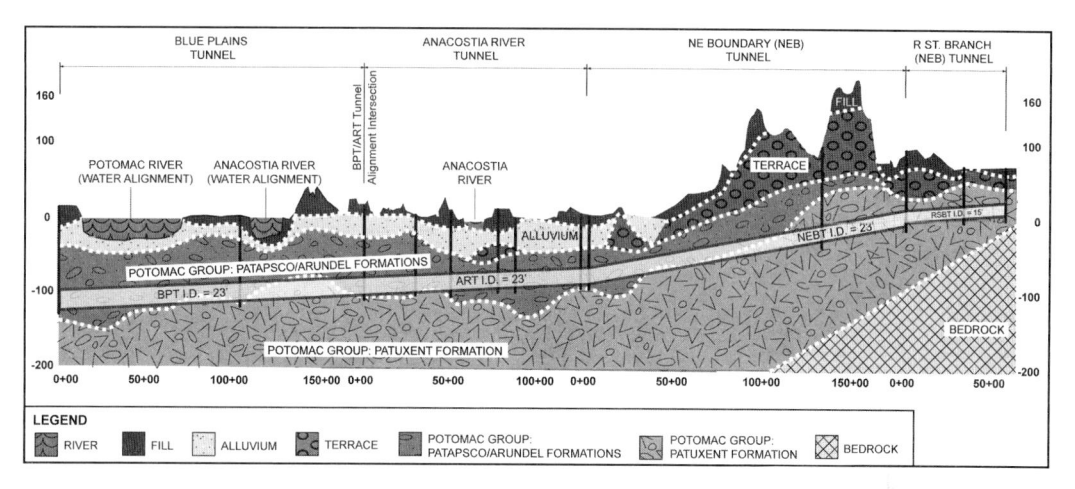

Figure 2. General geologic profile for the ARP

these shafts range from 9.1 to 33.5 m (30 to 110 ft). The principal tunnels that comprise the LTCP are show on Figure 1 as Contract Divisions A, H, J and K. The tunnels for Contract Divisions A, H, and J will have an inside diameter of 7 m (23 ft), and Contract Division K will have an inside diameter of approximately 3.7 m (12 ft).

To capture and convey flows from the existing combined sewer system to the respective drop shaft facilities, diversion chambers will be constructed at the points of diversion, and diversion sewers will be constructed from those points to the nearest drop shafts. The invert of the diversion chambers and sewers are typically about 9.1 to 12.2 m (30 to 40 ft) below the existing ground surface. The most significant diversion sewer alignments can be seen on Figure 1 as Contract Divisions B, E, M and L. The diameters range from 914 to 2,438 mm (36 to 96 in.) and have the potential to be larger based on hydraulic requirements.

This paper discusses project information available during the early stages of the LTCP's 30% design. As of the writing of this paper, the final phase of the geotechnical investigation is approximately 75% complete for the first tunnel contract, the Blue Plains Tunnel (BPT).

## GEOLOGIC SETTING

The Blue Plains facility lies within the Atlantic Coastal Plan physiographic province, which is a broad belt of sedimentary soils that were deposited on older bedrock. Coastal Plain formations in the vicinity of the site include, from oldest to youngest: Cretaceous period Potomac Group sediments, Pleistocene epoch terrace deposits, and relatively recent Quaternary period alluvium. The Potomac soils were deposited in relatively shallow seas from steams flowing eastward out of the continental interior. Pleistocene Terrace sediments were carried in braided streams charged by glacial melt water and were deposited on top of Potomac Group soils. More recently, river alluvium was deposited over Terrace soils. The uppermost soils at the site consist of existing fill that is believed to be associated with previous development.

The fill deposits frequently contain fragments of construction debris, metal, cinders, and/or trash in varying amounts. The alluvial deposits consist of loose/soft organic silt, clay and fine sand. Sand and gravel are also present at some locations, usually at depths underlying the fine grained material. The Terrace deposits consist of older alluvial sand and gravel that are often yellow or orange in color The Potomac Group consists of sediments that have been subdivided into the Patapsco/Arundel formations and the underlying Patuxent Formation. The Patapsco/Arundel formations typically consist of hard, reddish brown silt and clay with minor sand. The Patuxent generally consists of silty and/or clayey sand, locally with minor gravel. The bedrock generally consists of crystalline schist and gneiss that is more than 450 million years old.

Figure 2 shows a general geologic profile of the ARP for the DC WASA LTCP. The majority of the tunnels will be excavated in the Potomac Group soils. This geologic profile will be continuously updated based on the findings of the additional investigations of the LTCP, which are presently ongoing and which are the focus of this paper.

## GEOTECHNICAL CONSIDERATIONS

The structures for the LTCP will be designed and built by a variety of construction methods. Table 1 lists the main design and construction considerations

Table 1. Geotechnical design and construction consideration

| Proposed Structure | Design | | | | | | | Construction | | | | | | | | | |
|---|---|---|---|---|---|---|---|---|---|---|---|---|---|---|---|---|---|
| | Foundation Bearing Capacity /Settlement | Lateral Earth and Water Pressure | Global Slope Stability | Buoyancy and Uplift Resistance | Ground Improvement | Liner and Wall Loads | Backfill Materials | Site Preparation And Grading | Ground Deformation | Subgrade Preparation | Obstructions | Mixed-Face Tunneling | Gas/Ventilation | Soil Disposal Or Reuse | Groundwater Control And Disposal | Soil Abrasiveness | Soil Stickiness Potential | Support of Excavation | Temporary Dewatering | Ground Improvement |
| Tunnels | | | | | • | • | | | • | | • | • | • | • | | • | • | | | • |
| Shafts | • | | | • | • | • | | • | • | • | • | | | | • | • | | | • | • |
| Overflow & Chamber | • | • | • | • | • | • | • | • | • | | • | | | | • | • | | | • | • |
| Sewers | | | | | • | • | | | • | | • | • | • | | • | • | | | | • |

for each of the main LTCP structures. These considerations have been used in the development of the geotechnical investigation program for the LTCP. The following sections discuss the details for the design and construction considerations for each of the proposed LTCP structures.

## Tunnels

The main design and construction considerations for tunneling include the liner design, machine selection and design, obstructions along the tunnel alignment, tunneling-induced ground settlements, and ground improvement. The tunnel liner will be designed to resist static and dynamic ground loads as well as hydrostatic loads. Other loading such as structure loads above the tunnel on the tunnel will also be considered. Other geotechnical considerations for liner design include chemical testing of the soil and groundwater to ensure there are no constituents in the ground that could adversely affect the concrete. The selection of a tunnel boring machine (TBM) mostly depends on the anticipated soil types and groundwater pressure.

Obstructions are objects that are encountered within the planned tunnel such as utilities and piles. The most common obstruction that has been found along the BPT is piles. Forensic studies are being performed to determine the length of piles where as-builts and constriction records are missing, which requires stratigraphy at the site in question and defined soil properties.

Ground settlement caused from tunneling and excavations is also being considered. The magnitude of the settlement will depend on the subsurface conditions, construction methods, equipment utilized, structure geometry, and the contractor's means and methods. The impact of the estimated settlement on structures is being evaluated and ground improvement techniques may be required to reduce the effects of settlement on utilities and structures. Properties such as grain-size distribution, moisture content, and soil strength are important factors when selecting ground improvement techniques.

## Shafts

The design and construction issues for the large-diameter, deep shafts include excavation support, ground deformation, water inflow, excavation bottom stability, and break-in and break-out of TBMs. Geotechnical information at each shaft site will play a critical role in its design. Subsurface profiles will be used to determine the lateral forces that will act on the shafts. Information concerning the groundwater head and permeability of the soil at the base of the shafts will be required. Water-tight excavation support systems, such as slurry walls, secant pile walls, steel sheet piling, or ground freezing, will most likely be required.

## Overflow and Diversion Chambers

Structures, including overflow facilities and diversion chambers, are expected to be built by open-cut excavation methods. Considerations for such construction include excavation support, ground deformation, dewatering, and excavation bottom stability. Relatively stiff excavation support systems, such as slurry walls, anchored or braced sheeting systems, etc., are required if ground deformation is a concern. Groundwater drawdown outside the excavation may result in the settlement of ground surface, nearby structures, and utilities due to the consolidation of soils caused by dewatering. Measures to stabilize excavated bottoms will also need to be considered.

**Sewers**

Many of the design and construction considerations discussed in the previous tunnel section are also applicable for the diversion sewers. The diversion sewers may be constructed using microtunneling, pipe jacking, and/or hand mining methods. Open-cut excavation methods are also being considered for some of the shallower sewer locations.

**GEOTECHNICAL INVESTIGATION PROGRAM**

The investigations discussed in this paper are being performed to support the final design of the LTCP tunnels, shafts and shallow structures. All geotechnical data collected will provide a basis for understanding the subsurface conditions and the geotechnical parameters for the final design and construction of the proposed LTCP facilities. Both the physical characteristics and engineering properties of the soil are important for the design and construction of the LTCP structures.

To simplify the soil profile in terms of soil behavior for tunneling, a system for grouping of soil has been developed. The tunnels are anticipated to be constructed primarily within the Potomac Formation. For this reason, the Potomac Formation was further divided into five soil groups (G1 through G5). The engineering properties of the five soil groups are being defined by field and laboratory testing data. The following sections describe the drilling, in situ testing, and laboratory testing being used to characterize the ground conditions for the LTCP.

**Geotechnical Drilling and Sampling**

The location and spacing of the borings, as well as the depths, are structure specific. The spacing of the borings along the tunnel and sewer alignments is approximately 183 m (600 ft). The depth of the borings along the tunnel and sewer alignments is about two tunnel diameters below the tunnel invert. At shaft locations, borings are being drilled to approximately 1 to 1.5 shaft diameters below shaft inverts, depending on the size of the shafts and the ground conditions encountered. At shaft locations, generally two borings will be drilled, depending on the size of the shaft. At structure locations, generally two borings will be drilled, depending on the structure size, the accessibility of the site to obtain borings, and the ground conditions encountered during drilling.

Drilling is being performed using both mud rotary and sonic drilling methods. Borings are being performed from land and water using barges. Standard penetration test (SPT) sampling is being performed and hammer energy testing will be performed for all drill rigs using hammers. The SPT sampling zone will be continuous starting one diameter above the tunnel, continuing within the tunnel, and going one diameter below the tunnel. The remaining depths will be sampled on 1.5-m (5-ft) spacing. The sampling zone for a structure and shaft boring spans from the ground surface to the depth of the boring. SPT sampling in structure borings will use 1.5-m (5-ft) spacing. Undisturbed soil samples will be obtained and are being collected using a thin-walled Shelby tube sampler, Pitcher Sampler, or Denison sampler.

**In Situ Testing**

The objectives of in situ testing are to gather additional information regarding site geology, hydrogeology, and physical and engineering properties of soils. In situ tests include those for the determination of the strength, stiffness, and permeability of major strata, as well as quality. Table 2 summarizes the in situ testing methods planned for the geotechnical investigations. The field testing methodologies are discussed in detail below.

*Standard Penetration Tests (SPTs)*

The SPT blow counts provide a qualitative indication regarding soil density or consistency. Soil samples collected by the SPT samplers are used for soil classification and index testing. SPT testing has been generally easy to obtain; however, there has been an occasional borehole stability issue when drilling in the deep coarse-grained Patuxent Formation. Bottom instability of the borehole has been encountered, leading to inaccurate blow counts. This issue has been corrected by ensuring the drilling mud is thick enough and there is enough mud in the hole.

*Pressuremeter Tests*

Pressuremeter tests are being performed using a Menard pressuremeter. Test results are used to estimate in situ lateral earth pressures, soil stiffness, and shear strength for the design of tunnel and other structures. The tests are applicable for sandy and cohesive soils; however, they are more difficult to perform in the sandy soils because of borehole stability issues. No tests are being performed for gravelly

Table 2. In situ testing methods

| Geotechnical Design and Construction Issues | SPT | Pressuremeter | Vane shear | Groundwater Monitoring | Soil gas screening | Groundwater quality test | Slug test | Geophysical surveys |
|---|---|---|---|---|---|---|---|---|
| **All Underground Facilities** | | | | | | | | |
| Shoring/under pinning | ● | ● | ● | | | ● | | |
| Soil/muck reuse/disposal | | | | | | | | |
| Ground water level/pressure | | | | ● | | | | |
| Corrosion | | | | | | ● | | |
| **Tunnels** | | | | | | | | |
| Tunnel liner earth load | | ● | ● | ● | | | | |
| Obstructions | ● | | | | | | | ● |
| Abrasiveness | ● | | | | | | | ● |
| Stickiness | | | | | | | | |
| Gas | | | | | ● | | | |
| Face stability | ● | ● | ● | ● | | | | |
| Groundwater inflow & disposal | | | | ● | | ● | ● | |
| Ground surface subsidence | ● | ● | ● | ● | | | | ● |
| **Shafts** | | | | | | | | |
| Water inflow/dewatering | | | | ● | | ● | ● | |
| Excavator resistance | ● | ● | ● | | | | | ● |
| Ground movement | ● | ● | ● | ● | | | ● | ● |
| Excavation bottom stability | ● | ● | ● | ● | | | ● | |
| Lateral earth pressure | ● | ● | ● | ● | | | | |
| Backfill materials | | | | | | | | |
| Gas | | | | | ● | | | |
| **Overflows & Chambers** | | | | | | | | |
| Foundation capacity/settlement | ● | ● | ● | ● | | | | |
| Obstructions | ● | | | | | | | |
| Lateral earth pressure | ● | ● | ● | ● | | | | |
| Ground movement | | ● | ● | ● | | | | |
| Buoyancy & uplift resistance | | | | ● | | | ● | |
| Water inflow/dewatering | | | | ● | | ● | ● | |
| Excavation bottom stability | ● | ● | ● | ● | | | ● | |
| Backfill materials | | | | | | | | |
| **Sewers** | | | | | | | | |
| Tunnel liner earth load | | ● | ● | ● | | | | |
| Obstructions | ● | | | | | | | ● |
| Abrasiveness | ● | | | | | | | |
| Stickiness | | | | | | | | |
| Gas | | | | | ● | | | |
| Face stability | ● | ● | ● | ● | | | | |
| Groundwater inflow & disposal | | | | ● | | ● | ● | |
| Ground surface subsidence | ● | ● | ● | ● | | | | ● |

materials. The tests are generally being performed in borings along the tunnel alignments and at shaft locations. Undisturbed samples are being obtained adjacent to test locations to determine if there is agreement between the in situ tests and laboratory tests. Approximately four tests are being performed in tunnel borings and six tests are being performed shaft borings. Tests for the tunnel borings are generally located in the tunnel zone, and tests performed for shaft borings are along the shaft profile.

*Field Vane Shear Tests*

The field vane shear test is used to determine in situ undrained shear strength within soft to medium stiff (SPT blow counts <9) clayey or silty soils. This test is not applicable for testing granular soils or stiffer clays. Soft soils are anticipated to be encountered in the alluvial materials at locations adjacent to the rivers and areas where old marshes and tributaries to the rivers existed. The vane shear test requires the apparatus to be calibrated to account for the friction along the rods connecting to the vane. The vane shear test has been performed in three areas thus far in the ongoing geotechnical investigation: the main pumping station; the CSO-019 area; and the second river crossing, adjacent to the Naval Annex. Approximately five tests are performed in one boring, and undisturbed samples are obtained adjacent to the test for comparison purposes.

*Groundwater Monitoring, Permeability Testing, and Quality*

Two kinds of groundwater monitoring methods are being used: monitoring wells and vibrating wire piezometers (VWPZs). Monitoring wells are being placed in coarse-grained materials in the tunnel zone, shaft inverts, and gravel layers within surface structures. Vibrating wire piezometers (VWP) are being placed in the fine-grained materials in and near the tunnel zone and within shafts. Slug testing is being performed in all monitoring wells to determine the permeability of the material. These test results are the basis for defining the groundwater conditions for all LTCP facilities. Groundwater quality testing is being conducted in a few wells within the tunnel zone to evaluate the chemistry of the groundwater. Parameters such as alkalinity, total dissolved solids, calcium content, sulfate, pH, etc., are being obtained through laboratory and field testing.

*Soil Gas Screening*

The potential for encountering explosive and toxic gases during construction of the LTCP facilities is being assessed by testing both the soil and groundwater. The soil is being screened using a photoionization detector (PID) and confined space monitor, which detects both hydrocarbons and methane. Groundwater samples are being collected from monitoring wells within the tunnel zone and are being tested for concentrations of hydrogen sulfide and methane. Gas is not anticipated to be encountered within the tunnel zone.

*Geophysical Investigations*

Geophysical survey methods being used include seismic reflection and cross-hole seismic surveys to gain additional information about the characteristics of subsurface layers. Seismic reflection is being used in the three water crossing for the BPT to delineate the Potomac clay contact between borings. Cross-hole seismic surveys are being performed at most shaft sites to determine the shear wave velocities of the soil strata within the shafts, which can be used to determine geotechnical properties of soil, including Poisson's ratio and elastic moduli.

**Laboratory Testing**

The objectives of geotechnical laboratory testing are to measure physical and engineering properties of soil and water samples obtained during the field investigation. The test results, combined with the data collected by field tests, provide a basis for understanding subsurface conditions and soil parameters for design and construction. In general, there are two types of soil samples that can be tested in a laboratory: disturbed and undisturbed. Disturbed soil testing generally yields the physical properties of the soil, while the undisturbed testing yields the engineering strength properties of the soil. Table 3 presents a matrix of proposed laboratory tests and the geotechnical design and construction issues the tests will be used to support. These laboratory testing methodologies are discussed below.

*Disturbed Testing*

Disturbed soil tests are being performed to characterize the physical properties of the soils, such as amount of natural moisture. These disturbed properties are being used to provide insight into the anticipated behavior of the soil.

**Moisture content, Atterberg limits, and grain-size distribution.** The majority of the disturbed soil tests performed are moisture content, Atterberg limits, and grain-size distribution. Grain-size analyses include sieve analysis and hydrometer analysis and are used to determine the percentages of various soil grain sizes for the purposes of USCS classification. Properties gained from these three types of disturbed tests are being correlated with other physical and engineering properties such as

## Table 3. Laboratory testing methods

| Geotechnical Design and Construction Issues | Unit weight | Moisture Content | Atterberg Limits | Gradation | Strength | Compressibility | Compaction | Permeability | Organic Content | Soil Corrosion | Water Quality | Dissolved Gas | Rock/Boulder Hardness and Strength | Mineralogy | Soil Abrasion |
|---|---|---|---|---|---|---|---|---|---|---|---|---|---|---|---|
| **All Underground Facilities** | | | | | | | | | | | | | | | |
| Shoring/under pinning | ● | ● | ● | ● | ● | ● | | | | | | | | | |
| Soil reuse/muck disposal | ● | ● | ● | ● | | | ● | | | | | | | | |
| Groundwater level/pressure | | ● | | | | | | | | | | | | | |
| Corrosion | | ● | | | | | | | | ● | ● | | | | |
| **Tunnels** | | | | | | | | | | | | | | | |
| Tunnel liner earth load | ● | ● | ● | ● | ● | ● | | | | | | | | | |
| Obstructions | | | | | | | | | | | | | ● | | |
| Abrasiveness | ● | | | ● | | | | | | | | | ● | ● | ● |
| Stickiness | | ● | ● | | | | | | | | | | | ● | |
| Gas | | | | | | | | | | ● | | ● | | | |
| Face stability | ● | ● | ● | ● | ● | | | | ● | | | | | | |
| Groundwater inflow & disposal | | ● | | | | | | ● | | | | | | | |
| Ground surface subsidence | ● | ● | ● | ● | ● | ● | | | | | | | | | |
| **Shafts** | | | | | | | | | | | | | | | |
| Water inflow/dewatering | | ● | | ● | | | | ● | | | ● | | | | |
| Excavator resistance | ● | | ● | ● | ● | | | | | | | | ● | | |
| Ground movement | ● | ● | ● | ● | ● | ● | | | | | | | | | |
| Excavation bottom stability | ● | ● | ● | ● | ● | ● | | | ● | | | | | | |
| Lateral earth pressure | ● | ● | ● | ● | ● | | | | | | | | | | |
| Backfill materials | ● | ● | ● | ● | | | ● | | | | | | | | |
| Gas | | | | | | | | | | ● | | ● | | | |
| **Overflows & Chambers** | | | | | | | | | | | | | | | |
| Foundation capacity/settlement | ● | ● | ● | ● | ● | ● | | | ● | | | | | | |
| Obstructions | | | | | | | | | | | | | | | |
| Lateral earth pressure | ● | ● | ● | ● | ● | | | | | | | | | | |
| Ground movement | ● | ● | ● | ● | ● | ● | | | | | | | | | |
| Buoyancy & uplift resistance | | ● | | ● | | | | | | | | | | | |
| Water inflow/dewatering | | ● | | ● | | | | ● | | | ● | | | | |
| Excavation bottom stability | ● | ● | ● | ● | ● | | | ● | | | | | | | |
| Backfill materials | ● | ● | ● | ● | | | ● | | | | | | | | |
| **Sewers** | | | | | | | | | | | | | | | |
| Tunnel liner earth load | ● | ● | ● | ● | ● | ● | | | | | | | | | |
| Obstructions | | | | | | | | | | | | | ● | | |
| Abrasiveness | ● | | | ● | | | | | | | | | ● | ● | ● |
| Stickiness | | ● | ● | | | | | | | | | | | ● | |
| Gas | | | | | | | | | | ● | | ● | | | |
| Face stability | ● | | ● | ● | ● | | | | | | | | | | |
| Groundwater inflow & disposal | | ● | | ● | | | | ● | | | | | | | |
| Ground surface subsidence | ● | ● | ● | ● | ● | ● | | | | | | | | | |

soil stickiness and liquidity index and permeability and strength. On average, five of each type of test are being performed in each boring. For tunnel borings, testing is focused in the tunnel zone. For shaft borings, testing is being performed throughout the boring, targeting the shaft invert and any permeable layers.

**Unit weight and specific gravity.** Unit weight and specific gravity tests are generally being performed in conjunction with other tests, such as consolidation tests and triaxial tests. Both tests are used in geotechnical calculations. A few tests are being performed without an undisturbed test; however, this is not typical of the investigation.

**Organic content.** The purpose of the organic content test is to determine the percentage of organic materials in a soil sample if the sample has been classified as organic or if the sample has any notes of organic materials. This test is being performed on peaty soils or soils having high moisture contents within shaft and shallow structure areas.

**Soil abrasion and mineralogy**. The purpose of soil abrasion tests is to provide an understanding of the abrasivity of the soil particles and help in the selection of soil conditioners for reducing the wear on cutter tools of the TBMs. Currently, there are no standard methods to measure and evaluate the impacts of abrasive soils on cutter tools for soft ground excavation. Two types of tests are being performed to quantify the abrasivity of the soil for the LTCP: X-ray diffraction testing, and the Norwegian University of Science and Technology (NTNU) Soil Abrasion Test (SAT). These tests are being performed on coarse-grained soils and sandy and gravelly fine-grained soils obtained from the tunnel zones.

X-ray diffraction testing is being performed to identify the minerals in the representative soil samples and quantify their relative abundance. The NTNU SAT is an extension of the existing NTNU Abrasion Value (AV) test and the Abrasion Value Cutter Steel (AVS) test for rock. The SAT result is a value calculated as the mean value of the measured weight loss in milligrams (mg). Approximately 10 of each type of test have been performed on coarse-grained material in the BPT tunnel zone.

**Soil corrosion tests.** Soil corrosion tests are being performed to determine concentrations of soil parameters such as sulfate, chloride, and sulfide. Other properties obtained during this test are pH, moisture content, electrical resistivity, and redox potential. The test results will be used to determine the potential adverse impacts of soil on concrete and steel during design. The corrosion tests are distinct from and not intended to substitute for appropriate environmental chemical analyses for evaluation of potential contamination. Corrosion tests are being performed at various locations at the tunnel level and near the ground surface at all shafts and shallow structure locations.

### Undisturbed Testing

Undisturbed soil tests are being performed to characterize engineering properties of the soils such as the undrained shear strength. These engineering properties, as well as the physical properties, are being used to provide insight into the anticipated behavior of the soil. Disturbed tests for moisture content and Atterberg Limits are also performed for each undisturbed sample that is tested. All undisturbed samples are being examined by X-ray radiography to determine their suitability for testing. The target is to obtain five tubes from each boring; however, it is not always possible to get all five tubes because of the difficulty in sampling the Potomac clays. In addition, it is anticipated that undisturbed testing will not be performed on all tubes obtained. Some of the tubes will be disturbed, and some will be kept for backup.

**X-ray radiography.** Undisturbed soil samples are being transported vertically in racks to from the field to storage and from storage to the two laboratories. All tubes are radiographed upon arrival to the laboratory in order to assess the sample quality and any potential disturbance, general material type, presence of inclusions, and variation in macro-fabric. Based on the radiography results, suitable zones are identified to assign samples for testing; therefore, relatively high-quality samples are used for strength and deformation tests. The biggest challenge in performing X-ray radiography was finding a facility that performed the testing and was close to one of the two undisturbed testing laboratories. The radiography lab used on this project had never tested soil before, only concrete. Laboratory technicians from the soil lab trained the concrete radiography lab to test the soil using the ASTM standard.

**One-dimensional consolidation/swelling.** One-dimensional constant rate of strain (CRS) consolidation tests, one-dimensional incremental consolidation tests, and swelling tests are being performed on fine-grained soils. The consolidation tests measure the coefficient of consolidation for estimating the rate of soil consolidation and provide an estimate of the maximum past pressure. These parameters will be used to evaluate strength-deformation properties and the degree of overconsolidation (OCR) of fine-grained soils. Determination of the maximum past pressure provides a better understanding of the strength-deformation behavior obtained from triaxial test results.

The swelling tests are anticipated to provide swell pressures in the Potomac clays that should be considered for the tunnel liner and TBM machine design, as well as the shaft design. Approximately 10 swell tests have been performed for the BPT alignment.

**Triaxial tests.** Triaxial tests are being performed to measure the undrained shear strength (Su), effective soil strength parameters (c', ϕ'), and deformation properties (E) of cohesive soils. There are currently three types of triaxial tests being performed to define the strength properties of the fine-grained soils: SHANSEP CK$_0$U, Recompression CK$_0$U, and CIU. SHANSEP refers to the Stress History and Normalized Soil Engineering Properties technique (Ladd and Foott 1974) for estimating strength properties of cohesive soils. CK$_0$UTC refers to K$_0$-Consolidated Undrained Triaxial Compression and CIU refers to Isotropically Consolidated Triaxial Compression. The recompression technique (Bjerrum 1973) involves reconsolidating specimens to in situ vertical stresses, and then the specimens are sheared to failure. In SHANSEP tests, test specimens are reconsolidated to a normally consolidated state (to stresses in excess of maximum past pressure) and then unloaded to varying OCRs, where required. The purpose of the SHANSEP technique is to minimize the effects of sample disturbance and to develop relationships between normalized undrained shear strength properties as a function of OCR.

SHANSEP testing is more expensive than recompression testing, and it is time consuming. Also, since the Potomac clays are highly overconsolidated materials, the test requires triaxial equipment that can handle high pressures, which is not common for the typical soils laboratory. A small number of SHANSEP tests are being performed to evaluate the usefulness of the testing for this project. Based on test results, additional tests may be assigned. The majority of the strength tests being performed are Recompression CK$_0$U. Approximately 25 tests have been performed along the BPT alignment. Additional testing may be required based on the results for each fine-grained soil group.

## SUMMARY

For the Anacostia River, one of the Districts's receiving waterways, the LTCP includes the construction of several miles of storage and flood relief tunnels in addition to other components such as shafts, diversion chambers, and overflow facilities. There are numerous design and construction considerations for each of the main LTCP structures, and these considerations have been used in the development of the geotechnical investigation program for the LTCP. Geotechnical investigations are currently underway to support the final design of these structures, starting with the BPT. Both the physical characteristics and engineering properties of the soil are important for the design and construction of the LTCP structures and are determined using field and laboratory testing methods. The types of field and laboratory testing performed for each structure depends on the construction and design issues for each and are often site specific.

## REFERENCES

Bjerrum, L. 1973. Problems of Soil Mechanics and Construction on Soft Clays and Structurally Unstable Soils (Collapsible, Expansive and Others). *Proceedings of the Eighth International Conference on Soil Mechanics and Foundation Engineering, Moscow*. Vol 3. pp 111–159.

Camp Dresser & McKee Inc. and Hatch Mott McDonald. 2009. *Facility Plan—ANACOSTIA RIVER PROJECTS*. Washington, D.C. District of Columbia Water and Sewer Authority.

Ladd, C.C., and R. Foott. 1974. New design procedure for stability of soft clays. *Journal of Geotechnical Engineering, ASCE*. 100(7):763–786.

# TRACK 2: DESIGN

Matt Fowler, Chair

*Session 4: Strength, Stresses, and Stability Assessment Selection*

# Lining Design Issues Associated with the Storage of Cryogenic Fluids in Rock Caverns

## Christopher Laughton
Fermi Research Alliance, Batavia, Illinois

**ABSTRACT:** Achieving a robust solution for cavern storage of large volumes of cryogenic fluids such as Liquid Natural Gas (LNG) and Liquid Argon (LAr) is important to both the oil and gas industry and the particle physics research community. For the gas industry, cavern storage can better address life safety issues as well as environmental concerns associated with the surface operation of tank facilities. For the physics community, underground experiments based on the use of LAr as a detection medium, can support the development of improved particle tracking systems. Designing hard rock cavern facilities that provide for the safe, long-term storage of cryo-liquids, offers new opportunities to improve the supply and distribution of natural gas supports a new generation of research tools for the experimental physicist.

This paper discusses general design issues associated with the storage of cryogenic fluids in hard rock caverns with a particular emphasis on the development of options for a new particle physics experiment, the Long Baseline Neutrino Experiment (LBNE). The LBNE will be sited at the Deep Underground Science and Engineering Laboratory (DUSEL). DUSEL is to be constructed within the footprint of the Homestake Mine, South Dakota, US.

## INTRODUCTION

A new underground particle physics experiment, the Long Baseline Neutrino Experiment (LBNE) is under design for deployment at the Deep Underground Science and Engineering Laboratory (DUSEL). The experiment will study the properties of neutrino particles. Concept development for a new underground facility to house a Liquid Argon-filled detector is drawing heavily on recent design experience gained in the LNG industry and at physics laboratories. The DUSEL site and some LAr Cavern design concepts are discussed below.

## THE DEEP UNDERGROUND SCIENCE AND ENGINEERING LABORATORY (DUSEL)

### DUSEL at the Homestake Mine

DUSEL is sited within the boundaries of the recently closed Homestake Gold Mine in the town of Lead, South Dakota. Lead is located in the Black Hills of South Dakota, some 60 km north, north-west of Rapid City.

Prior to its closure in the early 2000s, gold ore was extracted from the Homestake Mine for over a hundred years. Within the mine footprint, there is a network of over 500 km of tunnel. Vertical shafts, winzes, and ramps extend down to a depth of approximately 2,400 m. Figure 1 presents a schematic view of the main permanent excavations developed over the mine life. The Ross and Yates shafts identified on the section have been rehabilitated by the State of South Dakota to provide temporary access to the 4850 level. In Figure 1, the levels are designated as feet below the approximate collar elevations of the Yates and Ross shafts. The shaft and tunnel network will be further developed to provide access to new DUSEL facilities, built adjacent to existing excavations.

Preliminary design and re-entry work is being managed by a team of physicists and engineers under the leadership of the South Dakota Science and Technology Authority (SDSTA) and the University of California (UC). The SDSTA is responsible for re-opening the mine to the 4850 level and supporting an initial phase of laboratory operation as the Sanford Laboratory. The full DUSEL will be designed and constructed by the UC with funding from the National Science Foundation (NSF). The full DUSEL includes plans to develop laboratory space, campus sites, and "research outposts" down to the 7400 Level. Shallow research sites (100–300 levels) will be accessed and serviced by drive-in portal. Intermediate and deep levels will be accessed via shafts, winzes, and ramps and serviced by new and refurbished infrastructure systems.

The start-up research program planned for the 4850 Level will accommodate experiments that will

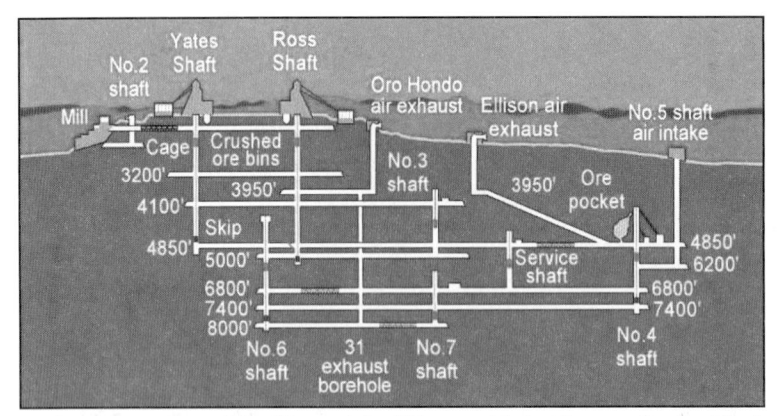

Figure 1. Schematic cross-section of the Homestake Mine workings (Source: Sanford Laboratory)

begin operation early in the decade. The experiments that will constitute the full DUSEL program have not yet been selected, but will likely include a core physics program and research initiatives in the fields of geo-science and engineering. The National Science Foundation (NSF) is currently planning to provide construction funding for the facility, including access to deeper levels, starting in 2013.

## Homestake Geology

The main rock units mined at Homestake were the Poorman, Homestake and Ellison. These geologic units are meta-sedimentary in origin, largely consisting of schists, phyllites and amphibolites. The units have been subject to significant deformation, as can be surmised from the cross-section shown in Figure 2.

The mine units are heavily folded, and contain faults and cross-cutting rhyolite dikes. Joint sets and fracture zones are present across the mine site. During mine development, some fracture zones yielded hot water (45 to 85°C) under pressure.

The complexity of the host geologic structure suggests that particular attention will need to be paid to the characterization of the rock mass at candidate excavation sites during the initial site investigation period. Intact strength, stress, fracture and water conditions may all be expected to vary markedly within and between rock units.

## Rock Mass Characterization

Extensive information on geologic structure, rock mass properties, and in situ stress regime were gathered over the operating life of the mine. During the latter stages of mine operation, a computer-based 3-D model of the mine geology was also developed. The historical data sets and model are currently

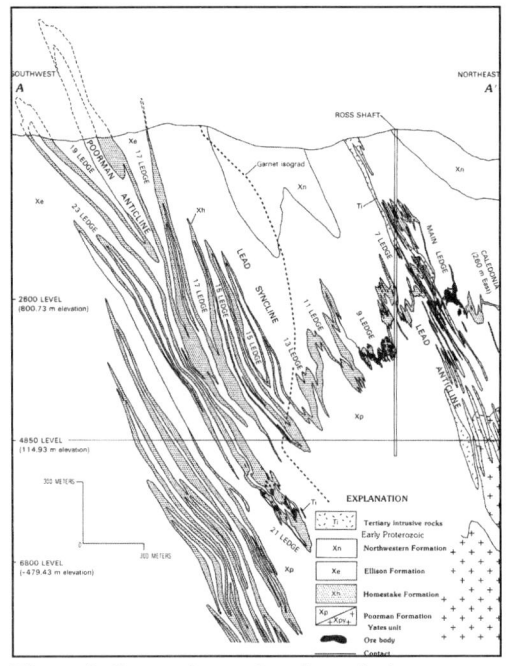

Figure 2. Composite section through the Homestake Mine geology (Source: Sanford Laboratory).

supporting the selection of areas for geo-research. The model and mine reports on seismic and mining-related geotechnical data are also proving of great value to the engineers as they offer key insights into the potential range of rock mass behaviors to expect in situ.

Preliminary Q-System parameters (Barton and Grimstad, 1993) for the rock units present on the 4850 Level (P, Poorman and Y, Yates) are shown in

**Table 1. Preliminary rock mass Q parameters for the Poorman and Yates formations**

| Q Parameters | Factor | Yates | Poorman |
|---|---|---|---|
| Block Size | RQD | 85 | 75 |
| | Jn | 6 | 4 |
| Inter-block Strength | Jr | 4 | 1 |
| | Ja | 0.75 | 1 |
| Water Pressure & Strength:Stress | Jw | 1 | 1 |
| | SRF | 5 | 12 |
| Q-Index | | 15.1 | 1.6 |

Table 1 (Steed and Cavahlo, 2006). These empirical parameters were used to evaluate preliminary ground support requirements. The engineering properties of the rock units are currently being investigated within the context of an on-going site investigation program. The program will support further work relative to the siting and design of the major excavations.

## CAVERN FACILITIES FOR PARTICLE PHYSICS EXPERIMENTS AND CRYOGENIC FLUID STORAGE

### Overview

The physics community has significant experience building caverns to accommodate particle detectors. The community has also gained significant experience in the use of cryogenic liquids in underground environments; in particular, a number of national and international laboratories operate cryogenic cooling systems in support of on-going particle accelerator operations. However, the new LAr detector systems being proposed for LBNE require containment of larger volumes of cryo-liquids than have previously been deployed underground. To evaluate the design implications of building and operating such facilities, underground designers are able to reference construction experiences and operating practices successfully applied elsewhere to achieve the safe containment of cryogenic liquids, such as helium, nitrogen, and particularly LNG.

Since the 1990s the LNG industry has conducted numerous detailed studies and constructed several pilot projects to test the concept of storing large quantities of methane underground, in either a gaseous or liquid state. Of particular relevance to the physics community are Lined Rock Cavern (LRC) concepts developed for construction in hard rock formations similar to those of the DUSEL site.

Drawing on laboratory and industry precedent and an early characterization of the host rock mass, engineering guidance can be provided to support the development of a preliminary set of LAr cavern design criteria. The basis for developing an early set of cavern design criteria is outlined below.

### Laboratory Design Criteria

During the later stages of the design, the LAr Cavern facilities will be optimized to meet functional requirements with a focus on cost-effectiveness. However, at a pre-conceptual level, a preliminary set of design criteria is needed to establish the affordability of the endeavor. A preliminary list of design issues are noted below:

- Excavation dimensions and shape, including a consideration of users' orientation and shape preferences.
- Mechanical/electrical systems, including handling devices, ventilation, power, and communications.
- Environmental systems, including air conditioning dust, humidity, radon dilution), and water control.
- In-cavern monitoring systems, including leak detection, oxygen deficiency, climate, smoke, and fire.
- Access, egress and refuge provisions, to service the facility under normal and emergency conditions.

Exposed permanent materials will also need to meet criteria for fire resistance and corrosion durability. For the LAr facilities, specific problem scenarios associated with filling, operating and emptying cryo-vessels and LAr leaks will need to be addressed within the context of the underground design. LAr released within a "warm" underground enclosure will evaporate and expand in volume. Argon gas is colorless, odorless and heavier than air. Argon gas will displace oxygen from the base of underground enclosures create. Detailed evacuation modeling of all occupied areas of the underground facilities will be required to safeguard against loss of life under a comprehensive range of fire and LAr release scenarios. Released cryo-fluids will also subject liners, drain water and the rock mass surrounding the excavation, to extreme cold temperatures. To specifically mitigate against LAr release scenarios, the cavern design may incorporate a number of additional design measures, including:

- LAr leak monitoring system(s).
- Cold-resistant materials to withstand cold temperature (e.g., cryogenic concrete and reinforcements).
- Insulation materials to protect surrounding structures against freeze damage.
- LAr drain and sump provisions to collect and contain LAr leaks and spills.

**Figure 3. LAr and neutrino nucleon and nuclear decay detector LANNDD concept (Source: Cline, 2001)**

- Groundwater drain and sump provisions to collect and remove groundwater from the cavern.
- Rock mass grout and drainage provisions to limit groundwater inflow into the cavern.
- Dedicated systems for ventilation of emergency egress ways and exhaust of Argon gas.
- Bulkheads to ensure that LAr leaks are contained within the design space.

If the LAr facilities are to be connected to the larger DUSEL facility, emergency plans will need to be coordinated with the site-wide DUSEL safety program.

**DUSEL Site Design Criteria**

Building LAr detector facilities at DUSEL will present the LBNE Project with opportunities for lifecycle economy. In-place infrastructure can service a range of LBNE's construction and operational needs. However, cost savings must be balanced the added costs, distributed site-wide, needed to upgrade the twentieth century mine to meet the standards of a twenty first century National Laboratory. The implementation of the laboratory program may also place added engineering constraints on individual experiments. In particular, the concurrent performance of research and construction activities may limit the capacity of infrastructure to fully service experimental or contract work and require the incorporation of costly measures to mitigate against the deleterious impacts of dust, noise, heavy traffic and blast overpressure and vibration.

**Geotechnical and Constructability Design Criteria**

Selmer-Olsen and Brock (1982) describe four basic steps to follow in selecting sites and designing caverns in hard rock. These step are outlined below:

1. Site the cavern at a place that represents the conditions of optimum stability.
2. Align the long axis of the cavern in a direction that minimizes stability problems and overbreak.
3. Shape and size the cavern to optimize stability relative to rock properties, fracture, and in situ stresses.
4. Configure the cavern and auxiliary structures (cavern, portals, tunnels and junctions, shielding, bulkheads, auxiliary chambers) to improve facility constructability and overall economy.

Steps one, two and three are associated with stability optimization, whereas step four focuses on issues of practicality and cost-effectiveness. The data to support these early steps is being obtained through the acquisition of site investigation data and the solicitation of constructability input from an experienced underground team.

## DESIGN OPTIONS UNDER CONSIDERATION

**Overview**

The oil and gas industry has identified a number of options for storing methane as either a cryogenics liquid or compressed gas. Prototype physics detectors under development have adopted the liquid storage option and use insulated, slightly-pressurized, vessels and refrigeration plants to maintain purified Argon in a liquid state (approx. −192°C) through the circulation of liquid nitrogen. Three of a number of options under consideration by the physics community are briefly discussed below.

**A Single Vessel Option**

A single freestanding vessel option developed for deployment at the Waste Isolation Pilot Project (WIPP) site located in Carlsbad, New Mexico, is shown in Figure 3.

The LANNDD is based on the use of a single cylindrical steel container, similar in concept to those used for LNG storage on surface. As laid-out underground, an assembly hall would be located to the side of a pit structure. A single overhead crane would service both hall and pit. The lined pit structure would serve to fully contain LAr in the event of a leak. The

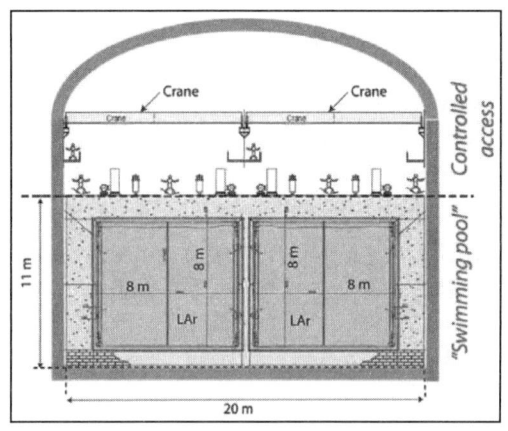

**Figure 4. Modular liquid argon imaging chamber concept (Source Baibussinov et al., 2007)**

cavern facility would be equipped with a dedicated Argon gas exhaust system.

**A Multi-Vessel Option**

Figure 4 shows multiple rectangular vessels or tanks embedded in an insulation material (Baibussinov et al, 2007). The upper section of the cavern is equipped to provide for installation, operation, and maintenance of the facility. This concept has some advantages and drawbacks compared to the single vessel solution shown in Figure 3 above. The use of a thick, external insulating layer offers improved protection of the rock mass against cold-damage under normal and leak conditions. The use of modules limits the maximum LAr volume released in a worst case scenario and improves the practicality of keeping a spare, empty module underground to transfer and store LAr if repair and maintenance work were to call for the evacuation of an adjacent module. Module drawbacks are likely to include a higher capital cost and a loss of active detector volume when compared to the single vessel option. A volume of LAr adjacent to the vessel walls cannot be instrumented for particle detection purposes; the smaller the vessel the greater the percentage of the contained LAr that has no experimental value.

**Membrane-Lined Cavern Option**

In addition to the drawbacks noted above, both the single and modular concepts dedicate a significant fraction of the excavated cavern volume to non-physics functions. If it could be proven economic to build, operate, and maintain, a preferred detector option would be a single, insulated, membrane-lined vessel that completely filled the excavated space. The membrane-lined option being considered by LAr proponents draws on recent experiences in the LNG

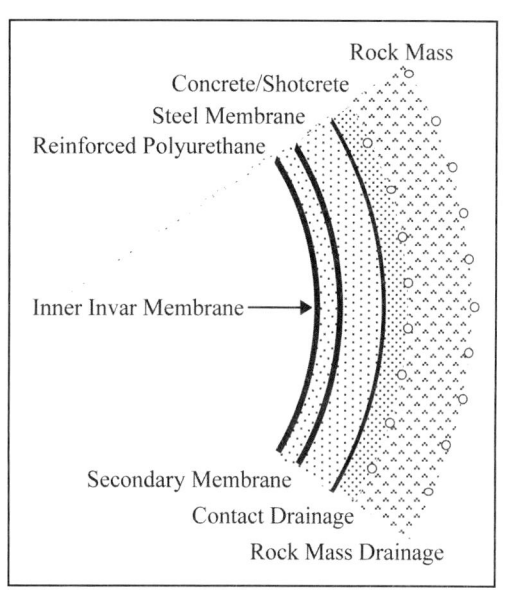

**Figure 5. Pre-conceptual composite liner for liquid argon containment in a LRC (After: Bromberg 2006)**

industry. Here, flexible membrane liners, which were originally developed to line the hulls of ocean-going LNG tanker vessels, have been adapted to serve as liners for LNG storage caverns. A demonstration project using the adapted liner system was recently constructed and put into operation in Daejon, South Korea (Amaniti and Chamfreas, 2004).

The prototype facility was constructed in fractured granite. A multi-layer system placed in direct contact with the rock was used to contain liquid methane. The composite liner incorporates concrete, polyurethane foam panels and a flexible inner membrane of invar. This composite liner serves to cool, insulate and seal the cryo-liquid. A further barrier, external to the excavation profile, is provided by a freeze curtain formed adjacent to the cavern wall. This curtain functions as an additional barrier to prevent leaked methane gas migrating to surface. A drained zone is maintained around the ice curtain. The drained zone serves to control ice wall thickness and groundwater pressures.

Figure 5 shows a cut-away schematic section of a liner considered for the containment of LAr at the DUSEL site. For the purposes of this discussion, the rock mass at the cavern site is assumed to be relatively impermeable and above the water table. The system is similar in concept to that used for LNG storage, but does not include a freeze curtain. Unlike methane, Argon gas is heavier than air and will not migrate towards surface. Laboratory testing and modeling work performed in the early

2000s indicated that long-term (months-years) exposure to cold temperatures, such as those associated with LNG or LAr storage, will result in the development of radially-oriented tensile cracks (Inada & Kinoshita, 2001). If water is present in the rock mass, ice will build-up in these fractures. The impact of cracks and freeze-thaw damage in the rock mass surrounding an embedded cryo-vessel was investigated during recent field studies at the Daejon pilot cavern. Gatelier (2008) found that the "thermo-geomechanical effects are fully acceptable for the rock mass and have no detrimental effect on the containment system." Based on the successful performance of the pilot cavern, a commercial-scale LRC facility is now being developed for construction at Taean, South Korea.

Although the Daejon studies have validated a LRC concept for LNG storage, site- and application-specific studies will be necessary before a LAr membrane-type storage concept can be confidently selected for deployment at DUSEL. Rock data need to be collected and thermo-geomechanical and hydro-geologic modeling performed. This work is needed to develop a detailed understanding of site-specific ground conditions and behaviors. The LAr-LRC is an attractive option for physics end-user. It maximizes the amount of mined space filled with LAr. However, the benefits of more efficient space use need to be balanced against the added challenges and costs that building a membrane-lined detector may incur. Most notably, a membrane design may require the use of more sophisticated controls and monitoring systems to ensure that functional requirements are reliably met.

## SUMMARY

The oil and gas and particle physics communities are studying options for the underground storage of large volumes of cryogenic fluid. For the gas industry, the underground storage of LNG is an attractive option as it offers important opportunities to reduce community impacts and improve grid performance. For the physicists, the containment of large volumes of LAr in an underground environment offers significant opportunities to probe more deeply and efficiently into the fundamental properties of matter. In developing designs for a new generation of detectors, the physics community can draw on experience constructing large-scale underground laboratories and benefit from reference to progress being made in the gas industry relative to the underground storage of LNG. In this regard, of particular interest is the successful development of an LRC option for LNG storage at Daejon, South Korea. Selection of a preferred option for LaR project will focus on a consideration of life cycle costs, risk mitigation and, most importantly, life-safety as the underground facilities will be occupied during operation.

## REFERENCES

Amaniti, E., and Chanfreau, E. 2004. Development and construction of a pilot lined cavern for LNG underground storage. LNG 14 Conference.

Baibussinov, B. 2008. A new, very massive modular Liquid Argon Imaging Chamber to detect low energy off-axis neutrinos from the CNGS beam. *Astroparticle Physics* 29:174.

Barton, N and Grimstad, E. 1994. The Q-System following twenty years of application in NMT support selection. *Felsbau* 12(6):428.

Bromberg, C. 2006. Deep Underground Cryostat for a LArTPC. Michigan University. Unpublished Report.

Cline, D.B., Sergiampietri, F., Learned, J.G., and McDonald K. 2003. LANNDD—A Massive Argon Detector for Proton Decay, Supernova and Solar Neutrino Studies, and a Neutrino Factory Detector. *Nuclear Instrumentation Methods* A503.

Gatelier, N. 2008. Underground Storage of LNG in Mined Rock Cavern. International Gas Union Research Conference, Paris, France.

Inada, Y. and Kinoshita, N. 2001. A few remarks on thermal behavior of rock mass around openings due to low temperature materials storage. Rock Mechanics—a Challenge for Society.

Selmer-Olsen, R., and Broch, E. 1982. General design procedures for underground openings in Norway. Publication of the Norwegian Soil and Rock Engineering Association 1:11.

# Design Guidelines for Sequential Excavations Method (SEM) Practices for Road Tunnels in the United States

**Vojtech Gall**
Gall Zeidler Consultants, Ashburn, Virginia

**Nasri Munfah**
Parsons Brinckerhoff, New York, New York

ABSTRACT: In the last 20 years the Sequential Excavation Method (SEM) or the New Austrian Tunneling Method (NATM) has been gaining popularity and use in the United States. Its use is versatile, in various ground conditions and at various depths. Although many of the projects were successfully completed, the lack of design guidelines for underground construction and in particular for the SEM construction, in which it relies on observational method and assessment of the ground behavior at the face, has negatively impacted the tunneling industry. Recognizing the need to develop design guidelines for underground construction, in 2007 FHWA awarded a contract to Parsons Brinckerhoff to develop and publish a design manual for road tunnels. As a result of this contract, "Technical Manual for Design and Construction of Road Tunnels—Civil Elements" was published in November 2008. The Manual provided specific guidelines for SEM construction. This paper provides a summary of the guidelines for the design and construction of tunnels using Sequential Excavation Method with emphasis on its technical aspects, contractual issues, and practices in the US based on the recommendations and guidelines made in the above stated publication.

## INTRODUCTION

The increased use of underground space for transportation systems and the increasing complexity and constraints of constructing and maintaining above ground transportation infrastructure has prompted the United States Federal Highway Administration (FHWA) to recognize the need to develop a technical manual for the design and construction of road tunnels in the US. In 2007 it awarded a contract to Parsons Brinckerhoff to develop and publish design guidelines for road tunnels. As a result of this contract the "Technical Manual for Design and Construction of Road Tunnels—Civil Elements" was published in November 2008 and placed on FHWA website www.fhwa.dot.gov . The manual included a dedicated chapter on the design and construction of tunnels using the Sequential Excavation Method (SEM) or the New Austrian Tunneling Method (NATM) or Conventional Tunneling using the nomenclature of Working Group 19 of the International Tunneling Association (ITA).

It is important to recognize that the manual consists of guidelines and not code provisions and its use by the highway and road authorities of each state is not mandatory. However, the lack of any other authorities' guidelines or codes renders this manual to be an invaluable source of information for the design and construction of tunnels in the United States.

The authors of this paper (being the main author of the SEM chapter and the principal investigator of the manual) provide their insight on the practices of SEM tunneling in the US relying on their experiences, knowledge, and recommendations made in the above stated manual.

## SEQUENTIAL EXCAVATION METHOD GUIDELINES

Sequential Excavation Method (SEM) design practices in the US rely on the development of ground classification and support classes based on extensive geotechnical investigations, the establishment of excavation support classes and initial support, and the use of supplemental measures (tool box) for tunnel excavation, pre-support and ground improvement measures coupled with a comprehensive monitoring and instrumentation program.

### Ground Classification and Support Classes

A series of qualitative and quantitative rock mass classification systems have been developed over the years and are implemented on tunneling projects worldwide including the Q system and the Rock Mass Rating (RMR) system and are used on rock tunneling projects to establish a geotechnical baseline and basis for the derivation of excavation and support classification.

Rock mass classification systems aid in the assessment of the ground behavior and ultimately lead to the definition of the support required to stabilize the tunnel opening. While the above quantitative classification systems lead to a numerical rating system that results in suggestions for tunnel support requirements these systems cannot replace a thorough design of the excavation and support system by experienced tunnel engineers.

All classification systems have in common that they should be based on thorough ground investigation and observation. The process from the ground investigation to the final definition of the ground support system can be summarized in three models:

- Geological Model
- Geotechnical Model
- Tunnel Support Model

### *Geological Model*

A desk study of the geological information available for a project area forms the starting point of the ground investigation program. Literature, previous projects, maps and published reports (e.g., from the US Geological Survey) form the basis for a desk study. Subsequently and in coordination with initial field observation and mapping results, a geotechnical investigation program is developed and carried out. The geological information from the geotechnical investigation, field mapping, and the desk study are compiled in the geological model.

### *Geotechnical Model*

With the data from the geological model in combination with the test results from the ground investigation program and laboratory testing, the ground response to tunneling is assessed. This assessment takes into account the method of excavation, tunnel size and shape as well as other parameters such as overburden height, environmental issues, proximity of adjacent structures and facilities, and groundwater conditions. The geotechnical model assists in deriving zones of similar ground response to tunneling along the alignment and Ground Response Classes (GRC) are defined. These GRCs form the baseline for the anticipated ground conditions during tunneling. Typically, the ground response to an unsupported tunnel excavation is analyzed in order to assess the support requirements for the stabilization of the opening.

### *Tunnel Support Model*

After assessing the ground support needs, excavation and support sequences, subdivision into multiple drifts, as well as the support measures are defined. These are combined in Excavation and Support Classes (ESCs) that form the basis for the Contractor to develop a bid as well as to execute the tunnel work.

## Excavation and Support Classes (ESC) and Initial Support

Excavation and Support Classes (ESCs) contain clear specifications for excavation round length, subdivision into multiple drifts, initial support and pre-support measures to be installed and the sequence of excavation and support installation. They also define means of additional initial support or local support or pre-support measures that augment the ESC to deal with local ground conditions that may require such additional support. They also define supplemental support if needed.

Initial support is provided early on. In soft ground and weak rock it directly follows the excavation of a round length and is installed prior to proceeding to the excavation of the next round in sequence. In hard rock tunneling initial support is installed close to the face. The intent is to provide structural support to the newly created opening and ensure safe tunneling conditions. Initial support layout is dictated by engineering principles, and risk management needs.

The amount and design of the initial support was historically motivated mainly by the desire to mobilize a high degree of ground self support and therefore economy. This was possible at the outset of SEM tunneling applications in "green field" conditions where deformation control was of a secondary importance and tolerable as long as equilibrium was reached. Nowadays, however, safety considerations, risk management, robustness and conservatism, design life, and the need for minimizing settlements in urban settings add construction realities that ultimately decide on the layout of the initial support.

Initial support is provided by application of a layer of shotcrete to achieve an interlocking support with the ground. Shotcrete is typically reinforced by steel fibers or welded wire fabric. Plastic fibers are used for reinforcement only occasionally although its application appears to become more frequent. With higher support demands of the ground and with shotcrete thicknesses of generally 150 mm (6 inches) or greater lattice girders are embedded within the shotcrete to provide for the structural requirements. Occasionally and if needed by special support needs rolled steel sets are used in lieu of, or in combination with lattice girders.

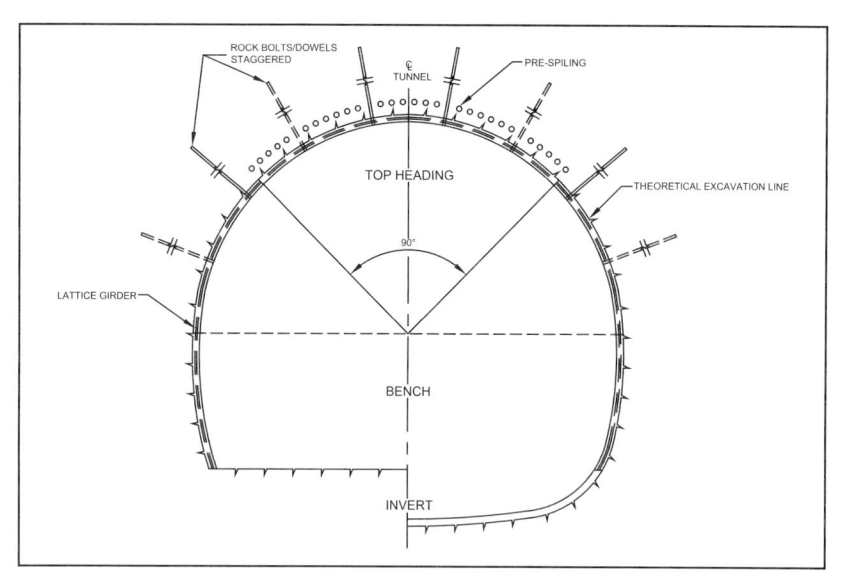

**Figure 1. Prototypical excavation support class (ESC) cross section**

Figure 1 and Figure 2 show a prototypical ESC cross section and longitudinal section respectively. Figure 1 displays a cross section without a closed invert on the left side and ring closure on its right side. Invert closure is typically required in soft ground and weak rock conditions and in squeezing ground. Figure 2 includes elements of typical initial support including rock bolts/dowels, initial shotcrete lining and tunnel pre-support. The arrangement of rock bolts/dowels is typical and varies depending on the excavation and support. The table in Figure 2 provides details of initial support measures for a prototypical ESC Class IV. In that sense, conventional tunneling is a prescriptive method which defines clearly and in detail tunnel excavation and initial support means.

## Tunnel Profile and Distribution of Excavation and Support Classes

Contract documents contain all Excavation and Support Classes (ESCs) assigned along the tunnel alignment in accordance with the Ground Response Classes (GRCs) and serve as a basis to estimate excavation and initial support quantities. A summary longitudinal section along the tunnel alignment shows the anticipated geological conditions, the GRCs with the relevant description of the anticipated ground response, hydrological conditions and the distribution of the ESCs. Figure 3 displays a prototypical longitudinal profile with an overlay of GRCs and corresponding ESCs, which form a baseline for the contract documents.

Geological data, Ground Response Classes, Excavation and Support Classes, the Longitudinal Tunnel Profile as well as design assumptions and methods are described and displayed in reports that become part of the contract documents. When defining the reaches and respective lengths of GRCs and corresponding ESCs it is understood that these are a prognosis and may be different in the field. Therefore contract documents establish the reaches as a basis and call for observation of the ground response in the field and the need for their adjustment as required by actual conditions encountered. Actual conditions must be accurately mapped in the field to allow for a comparison with the baseline assumptions portrayed in the GRCs.

## Tunnel Excavation, Support, and Pre-Support Measures

The use of most common initial support measures, along with excavation and support installation sequencing frequently associated with conventional tunnels depending on the basic types of ground encountered, i.e., rock and soft ground were summarized and presented in the FHWA Technical Manual for Design and Construction of Road Tunnels. These tables indicate basic concepts to derive Excavation and Support Classes (ESCs) for typical ground conditions portrayed.

Table 1 in the manual addresses rock tunnels and builds on the use of Terzaghi's Rock Mass Classification. It distinguished between the following rock mass qualities:

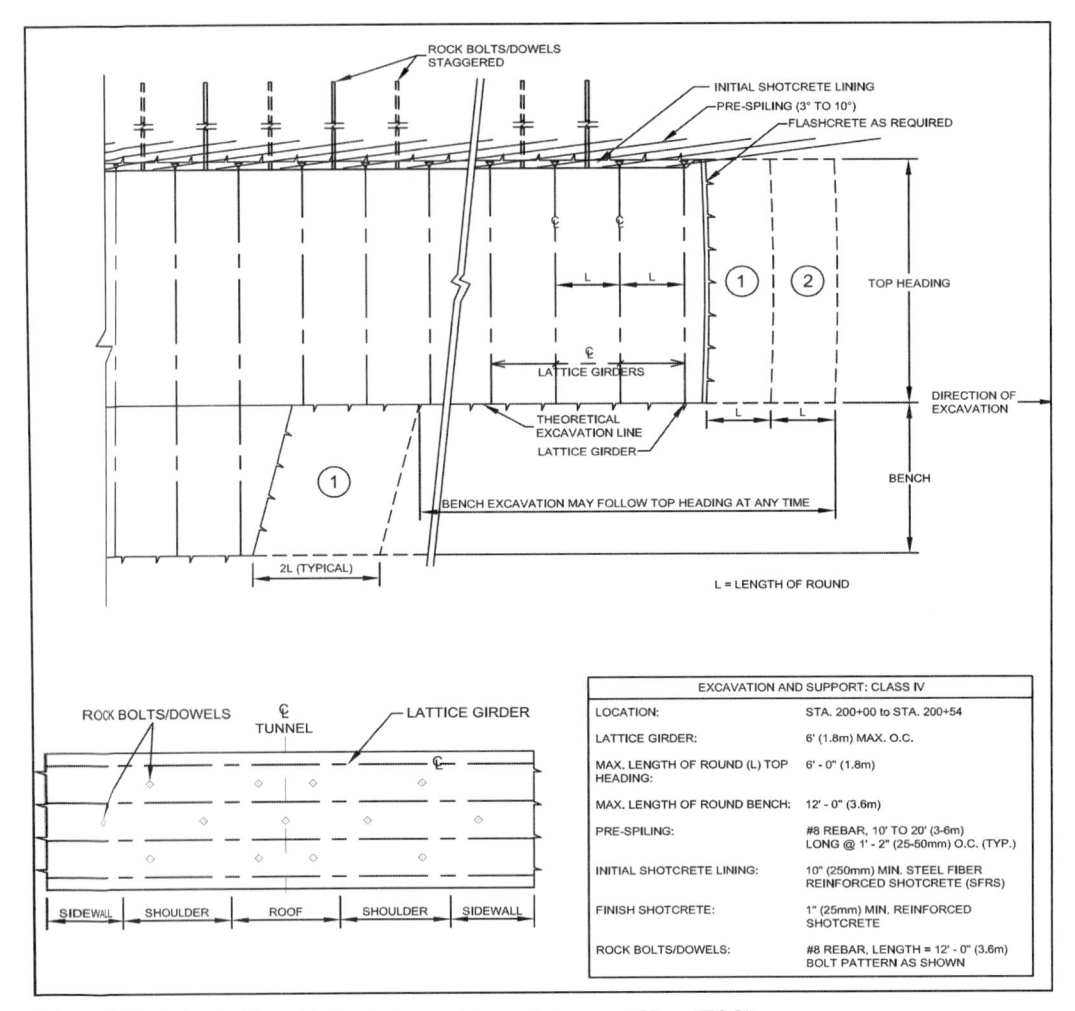

**Figure 2.** Prototypical Longitudinal Excavation and Support Class (ESC)

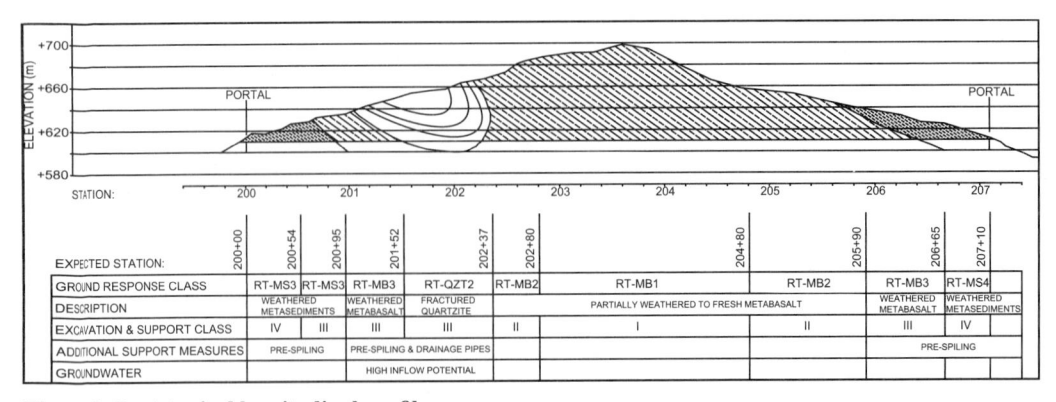

**Figure 3.** Prototypical longitudinal profile

- Intact Rock
- Stratified Rock
- Moderately Jointed Rock
- Blocky and Seamy Rock
- Crushed, but Chemically Intact Rock
- Squeezing Rock
- Swelling Rock

Table 2 in the Manual shows elements commonly used in excavation and support classes for soft ground. It distinguishes among the various ground types and the ground water conditions as follows:

- Stiff cohesive soil—above groundwater table
- Stiff cohesive soil—below groundwater table
- Well consolidated non-cohesive soil—above groundwater table
- Well consolidated non-cohesive soil—below groundwater table
- Loose non-cohesive soil—above groundwater table
- Loose non-cohesive soil—below groundwater table

The tables are not meant to be a "cook-book" but rather guides to the engineers to determine the potential excavation and support classes to be used.

**Pre-support Measures and Ground Improvement: Tool Box Measures**

With the significantly increased use of conventional tunneling in particular in soft ground and urban areas over the past decades, traditional measures to increase stand-up time were adopted and further developed to cope with poor ground conditions and to allow an efficient initial support installation and safe excavation.

These measures are installed ahead of the tunnel face. They include ground modification measures to improve the strength characteristics of the ground matrix including various forms of grouting, soil mixing and ground freezing, the latter for more adverse conditions. Most commonly methods include mechanical pre-support measures such as spiling installed ahead of the tunnel face often with distances of up to 18 to 30 m (60 to 100 feet) referred to as pipe arch canopies or at shorter distances, as short as 3.6 m (12 ft) utilizing traditional spiling measures such as grouted solid bars or grouted, perforated steel pipes. Ground improvement and pre-support measures can be used in a systematic manner over long tunnel stretches or only locally as required by ground conditions.

Local non-systematic use of not only pre-support measures, but additional measures including temporary Shotcreting the face, subdivision into smaller excavation faces (multiple faces), face support earth wedges, etc. form what is often referred to as the "tool box" measures applied as required by ground conditions.

**Instrumentation and Monitoring**

An integral part of the SEM tunneling is the verification by means of in-situ monitoring of design assumptions made regarding the interaction between the ground and initial support as a response to the excavation process.

For this purpose, a specific instrumentation and monitoring program is laid out. The SEM instrumentation aims at a detailed and systematic measurement of deflection of the initial lining. While monitoring of deformation is the main focus of instrumentation, stresses in the initial shotcrete lining and stresses between the shotcrete lining and the ground are monitored to capture the stress regime within the lining and between the lining and the ground. Reliability of stress cells, installation complexity and difficulty in obtaining accurate readings have nowadays led to the reliance on deformation monitoring only in standard tunneling applications. Use of stress cells is typically reserved for applications where knowledge of the stress conditions is important, for example where high and unusual in-situ ground stresses exist or high surface loads are present in urban settings.

Monitoring data are collected, processed and interpreted to provide early evaluations of:

- Adequate selection of the type of initial support and the timing of support installation in conjunction with the prescribed excavation sequence
- Stabilization of the surrounding ground by means of the self-supporting ground arch phenomenon
- Performance of the work in excavation technique and support installation
- Safety measures for the workforce and the public
- Long-term stress/settlement behavior for final safety assessment
- Assumed design parameters, such as strength properties of the ground and in-situ stresses used in the structural design computations

Based on this information, immediate decisions can be made in the field concerning proper excavation sequences and initial support in the range of the given ground response classes (GRC) and with respect to the designed excavation and support classes (ESC).

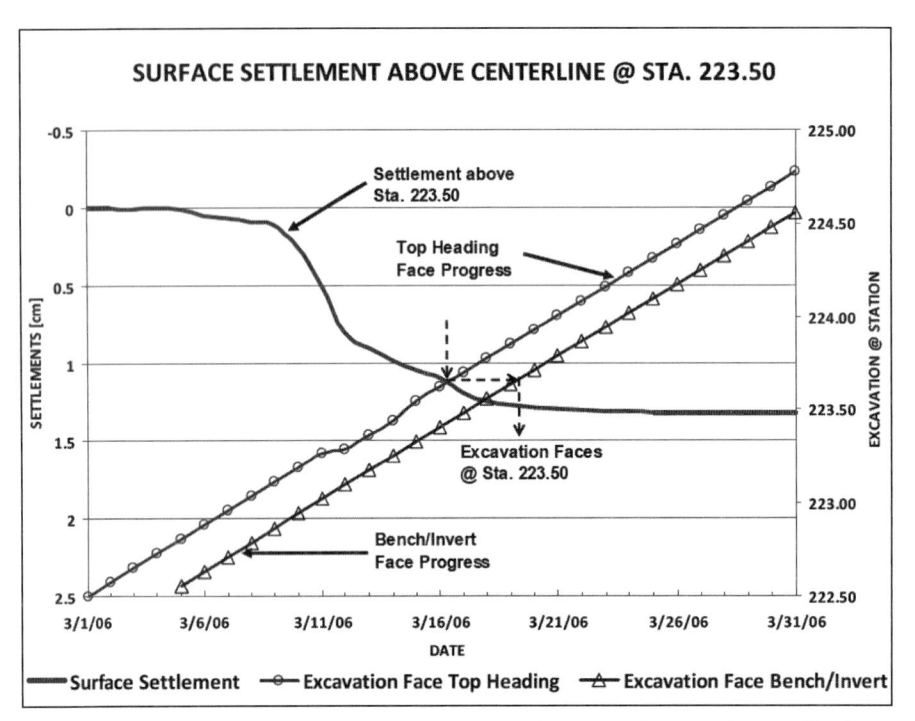

Figure 4. Prototypical monitoring of a surface settlement point above the tunnel centerline in a deformation vs. time and tunnel advance vs. time combined graph

## Interpretation of Monitoring Results

All readings must be thoroughly and systematically collected and recorded. An experienced tunnel engineer, often the tunnel designer, should evaluate the data and occasionally complement it by visual observations of the initial shotcrete lining for any distress such as cracking. To establish a direct relationship between tunnel excavation and ground behavior, it is recommended to portray the development of monitoring values as a function of the tunneling progress. This involves a combined graph showing the monitoring value (i.e., deformation, stress or other) vs. time and the tunnel progress vs. time. An example is shown in Figure 4. As can be seen from this graph, the surface settlement increases as the top heading and later bench/invert faces move towards and then directly under the monitored point and gradually decrease as both faces again move away from the location of the surface settlement monitoring point. The settlement curve shows an asymptotic behavior and becomes near horizontal as the faces are sufficiently far away from the monitoring point indicating that no further deformations associated with tunnel excavation and support occur in the ground indicating equilibrium and therefore ground stability.

The evaluation of monitoring results along with the knowledge of local ground conditions portrayed on systematic face mapping forms the basis for the verification of the selected excavation and support class (ESC) and the need to make any adjustments to it.

## CONTRACTUAL ISSUES

SEM tunnel construction requires solid past tunnel construction experience and personnel skills. These skills should relate to the use of construction equipment and handling of materials for excavation, installation of the initial support including shotcrete, lattice girders, pre-support measures, and rock reinforcing elements and even more importantly observation and evaluation of the ground as it responds to tunneling. It is therefore important to invoke a bidding process that addresses this need formally by addressing contractor's qualifications, personnel skills, and making payment provisions on unit prices basis.

### Contractor Pre-qualifications

It is recommended that the bidding contractors be pre-qualified to assure a skilled tunnel execution.

This pre-qualification can occur very early on during the design development, but at a minimum it should be performed as a separate step prior to soliciting tunnel bids. On critical projects the owner may solicit qualifications from contractors as early as the preliminary design stage. This early involvement also ensures that contractors are aware of the upcoming work and can plan ahead in assembling a qualified work force. Pre-qualification documents identify the scope of work and call for a similar experience gained on past projects by a tunneling company and for key staff including project manager, tunnel engineers, and tunnel superintendents.

**Unit Prices**

To suit the method's observational character and support its flexibility, it is recommended that SEM tunneling be procured within a unit price based contract. Unit prices also suit the need to install initial support in accordance with a classification system and amount of any additional initial or local support as required by field conditions actually encountered. The following is bid on a unit price basis:

- Excavation and Support on a linear meter (foot) basis for all excavation and installation of initial support per Excavation and Support Class (ESC). This includes any auxiliary measures needed for dewatering and ground water control at the face.
- Local support measures including:
  – Shotcrete per cubic meter (cubic yard) installed.
  – Pre-support measures such as spiling, canopy pipes and any other support means such as rock bolts and dowels, lattice girders, and face dowels are paid for each (EA) installed.
  – Instrumentation and monitoring is paid for either typical instrument section (including all instruments) or per each instrument installed. Payment is inclusive of submitted monitoring results and their interpretation.
  – Ground improvement measures per unit implemented, for example amount of grout injected including all labor and equipment utilized.
- Waterproofing and final lining installed to complete the typical dual lining structure may be procured on either lump sum basis or on a per tunnel meter (foot) basis.

The anticipated quantity of local support (additional or supplemental initial support) measures should be part of the contract to establish a basis for bid.

**Experienced Personnel**

Because SEM tunneling strongly relies on experience and personnel skills, it is imperative that experienced personnel be assigned from the start of the project, i.e., in its planning and design phase. The SEM tunneling design must be executed by an experienced designer.

The SEM tunneling contract documents must identify minimum contractor qualifications regardless whether the project is executed in a design-bid-build, design-build or any other contractual framework. For example, if the project uses the design-build framework then it is imperative that the builder take on an experienced SEM tunnel designer.

The construction contract documents must spell out minimum qualifications for the contractor's personnel that will initially prepare and then execute the tunnel work. This is the case for field engineering, field supervisory roles and the labor force that must be skilled. Contract documents call for a minimum experience of key tunneling staff by number of years spent in the field on SEM tunneling projects of similar type. Experienced personnel include Senior Tunnel Engineers, Tunnel Superintendents and Tunnel Foremen. All of such personnel should have a minimum of ten (10) years SEM tunneling experience. These personnel are charged with guiding excavation and support installation meeting the key requirements of conventional tunneling:

- Observation of the ground
- Evaluation of ground behavior as it responds to the excavation process
- Implementation of the "right" initial support

Face mapping including all ground exposed should occur for every excavation round and be formally documented and signed off by both the contractor and the owner's representative. Knowledgeable face mapping, execution of the instrumentation and monitoring program and interpretation of the monitoring results aid in the correct application of excavation sequencing and support installation.

The senior tunnel engineer is generally the contractor's highest authority for the tunneling and supervises the excavation and installation of the initial support, installation of any local or additional initial support measures and pre-support measures in line with the contract requirements and as adjusted to the ground conditions encountered in the field. As a result the ground encountered is categorized in accordance with the contract documents into ground response classes (GRCs) and the appropriate excavation and support classes (ESC) per contract baseline. Any need for additional initial support and/or pre-support measures is assessed and implemented. This task is carried out on a daily basis directly at the active tunnel face and is discussed with the owner's representative for each round. The outcome of this process is subsequently documented on form sheets that are then signed by the contractor's and owner's representatives for concurrence.

This frequent assessment of ground conditions provides for a continuous awareness of tunneling conditions, for an early evaluation of adequacy of support measures and as needed for implementation of contingency measures that may involve more than additional initial support means. Such contingency measures may include heavy pre-support and face stabilization measures or even systematic ground improvement measures.

To be able to support this on-going evaluation process on the owner's behalf the construction management (CM) and inspection team must also include relevant experience in conventional tunneling. It is recommended that the field representation includes a designer's representative who is familiar with the basis of the design. Represented in the field, the designer is able to verify design assumptions, will aid in the implementation of the design intent, and will make design changes on the spot if needed.

**Risk Management**

It is recommended that owners should initiate at the beginning of every SEM tunneling project a risk management plan. The plan should continue throughout the design and construction phases. The risks should be documented and managed with the best available tools at each phase. A risk register should be established and maintained throughout the life of the project. It is a living document that should be updated regularly and the effectiveness of the mitigation measures should be reassessed as more information becomes available. Risks should be allocated or shared among the parties on the basis of who has better control of the risk. The allocation of risks should be clearly documented and proper cost of accepting the risk should be included in the contract value.

**CONCLUSION**

This paper highlights the recommendations made by the FHWA "Technical Manual for Design and Construction of Road Tunnels: Civil Elements" for SEM tunneling construction and it includes the authors experience and insights on this subject. The American Association of State Highway and Transportation Officials (AASHTO), under its Technical Committee T-20 on tunnels, adopted the manual and is in the process to publish it under its domain. It is important to recognize that the manual consists of guidelines and not code provisions and its use by the highway and road authorities of each state is not mandatory. However, the lack of any other authorities' guidelines or codes renders this manual to be an invaluable source of information for the design and construction of tunnels in the United States.

**REFERENCES**

American Society of Civil Engineers (ASCE) (2007). "Geotechnical Baseline Reports for Construction – Suggested Guidelines," NY.

Federal Highway Administration (FHWA) (2009) "Technical Manual for Design and Construction of Road Tunnels – Civil Element," FHWA-NHI-09-010, Washington, D.C.

ITA Working Group 19 (2008) "Conventional Tunnelling in the United States—Contractual Issues" V. Gall and N. Munfah.

Underground Construction Association of Society of Mining, Metallurgy, and Exploration, Inc (UCA of SME) (2008) "Recommended Contract Practices for Underground Construction."

# Continuum and Discontinuum Modeling of Second Avenue Subway Caverns

**Verya Nasri, William Bergeson, Nils Pettersson**
AECOM, New York, New York

**ABSTRACT:** In recent years, several major underground projects with shallow, large-span caverns have been in the design phase in Manhattan. The construction area under consideration mainly consists of Manhattan schist, which has been subjected to intense faulting and folding. The defining feature of these design projects is the consideration of placing wide caverns with shallow rock cover in close proximity to tall buildings. While previous similar projects employed massive steel sets and thick concrete support, the current design philosophy is based upon rockbolts and shotcrete. New approaches to design, including verification of numerical modeling techniques must be employed under these circumstances.

Early tunnel design methods relied heavily upon the use of continuum modeling techniques. More recently, however, attention is being paid in the design stage to the use of discontinuum modeling. Where shallow caverns must be designed in jointed rock masses, it is critically important to consider the effects of the jointing characteristics on the development of rock loadings in the cavern linings. These rock loads are a function of the rock quality, the cavern geometry, and the type of lining that is installed.

Continuum versus discontinuum modeling at the design and excavation/support stages in hard rock tunneling has been discussed. A brief review of fundamentals in both numerical modeling approaches has been presented. The results of two- dimensional finite element modeling and distinct element modeling have been used to analyze several large shallow rock caverns. With the finite element method an equivalent continuum approach including the influence of major discontinuities was applied. With the distinct element method a fully deterministic discontinuum approach was used. Results generated from the discontinuum modeling confirm the empirical design approach employed by the Norwegian Geotechnical Institute's Q system.

## INTRODUCTION

The Second Avenue Subway Project is a major capital expansion project of the New York City subway that will provide a dedicated line for the east side of Manhattan with a link to the existing subway network. The proposed alignment runs from Harlem in the north to the financial district in the south with possible extension to Brooklyn. The project is approximately 13.7 km long including 16 stations, and its estimated cost is about $ 16 billion (Figure 1). Under the current design of the whole subway route, 10 stations will be cut-and-cover and 6 will be mined caverns which will be constructed through vertical shafts within the right-of-way of Second Avenue. In addition, there are numerous multi-track tunnels, crossovers and connections that will be constructed in caverns. The excavated diameter of the bored tunnels is 6.6 m and the caverns span ranges from 12.0 m to 21.0 m. All caverns have rock cover less than their span. As the geology of Manhattan varies along its length, the subway will pass through both hard rock and soft ground and there will be multiple rock/soil interfaces along the alignment.

The philosophy behind the construction methodology was to minimize the impact at street level on neighboring communities and businesses during the construction period. With most of the work being done within the right-of-way of Second Avenue, the largest impacts will be related to the maintenance and protection of traffic and street restoration as travel lanes are reduced from six to four during the construction. Because of the nature of the work, close proximity of high rise buildings, critical nature of adjacent utilities, the characteristics of the ground along alignment, and the visibility of this project, strict performance criteria and limitations were imposed and comprehensive instrumentation and monitoring programs were designed to ensure compliance with action and trigger levels to protect third parties for noise, vibration, subsurface movements and protection of overlying utilities and structures. The final engineering is being undertaken for the

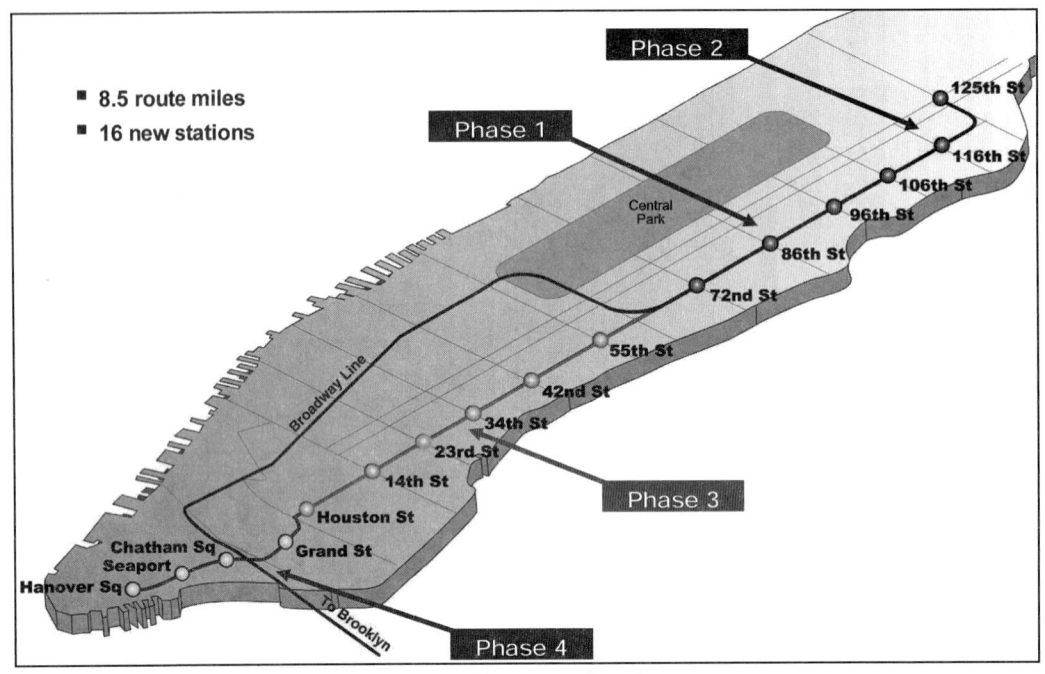

**Figure 1. Second Avenue subway alignment and construction phases**

New York City Transit Authority by a joint venture of AECOM and Arup.

The Second Avenue Subway project has been broken into four construction phases, which could potentially overlap, to make funding of this mega project more manageable. The project is being funded by a combination of State and Federal contributions. The budget for phase 1 is $3.8 billion, in year of expenditure dollars, and it is scheduled for completion by the end of 2016. Phase 1 includes 3.9 km of twin TBM rock tunnels, double-track 21.0 m span mined rock cavern stations at 72nd Street, and 86th Street, and a double-track cut-and-cover station at 96th Street. The overall configuration of the stations aimed to achieve as shallow a cavern as feasible to minimize passenger access time between entrances and platforms and to avoid interaction with existing subway underground structures. This had to be balanced against the need to provide an adequate rock cover for the caverns. Phase 1 of the Second Avenue Subway provides early revenue service, with ridership expected to be over 200,000 weekday riders when operational. The 3D models of two mined stations and their connecting tunnels are shown in Figures 2 and 3.

## GEOLOGICAL SETTINGS

The geological setting of New York City has posed many challenges to the construction industry and particularly to subsurface projects. The rock types encountered ranges from Precambrian to Devonian in age. The Pleistocene glaciation has added further complications with subsequent active erosion in Holocene times. The erosion and deposition has accumulated vast glacial till, modified glacial drifts, sand and gravel and glacio-lacustrine silt, clays and marshland.

The project area mainly consists of the Manhattan schist rocks, calcareous rocks of the Inwood marble and Fordham gneiss. Manhattan schists are typically crystalline variations of essentially quartz and mica composition with quartz and feldspar rich zones, garnetiferous biotite and muscovite mica schist, quartz-hornblende-mica-garnet schists, and chlorite schists. Numerous pre and post to late thrust kinematic pegmatite intrusions of varying size have been emplaced within these schists typically along and occasionally across the foliation and along other fractures.

The underlying bedrock geology of New York City is highly complex. The crystalline rocks of New York City are divided into two major units separated by Cameron's thrust fault. This regional feature has been classified as a suture of the proto-american plate. The Cameron thrust faulting has affected both these units and imparted various structural features such as faults, shears and joint systems. The rocks of Manhattan area have undergone multiple deformation events causing three identifiable foliations. The

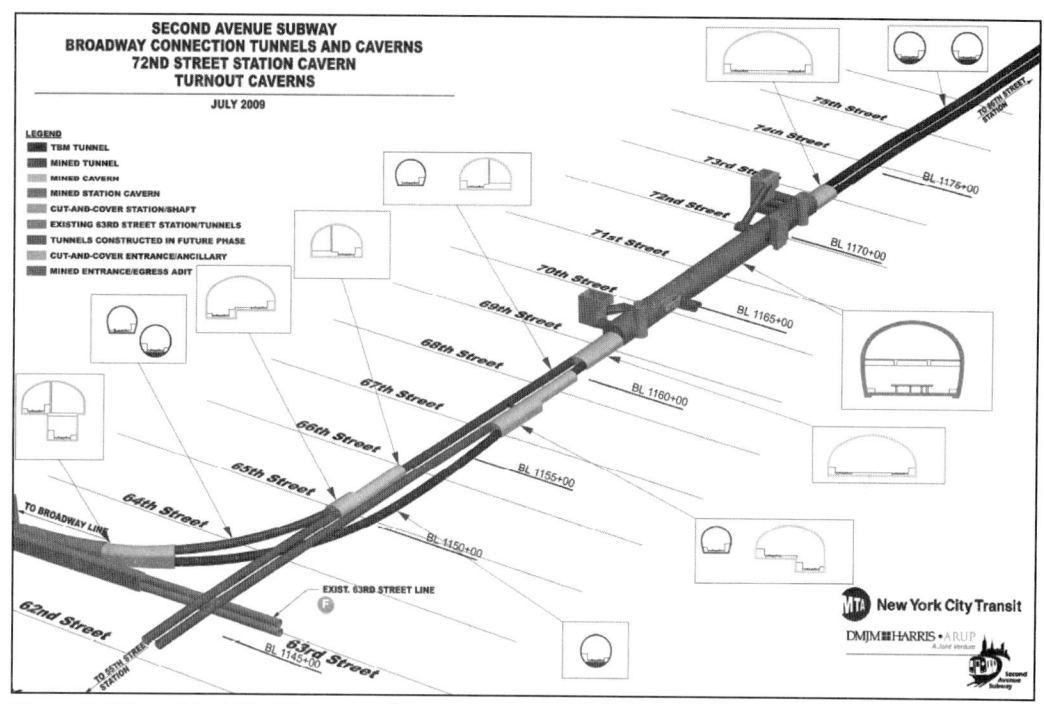

Figure 2. 3D model of 72nd Street station and connecting tunnels

rock mass is characterized by three principal joint sets with sub-sets and the dominant joint set is parallel to the foliation.

## SUBSURFACE INVESTIGATION

It is very difficult to distinguish between folded, faulted and unfaulted ground using conventional methods of core logging even if the full range of ground types and rock mass conditions are intercepted by the boring. It is possible to make general interpretations of the structural geology with rock outcrops to supplement the borings but these are rare in Manhattan. The conventional methods of fracture logging provide basic spacing and dip angle data but it is not possible to make direct interpretation of these data into structural groups or joint sets because the dip direction is unknown. Therefore, the detailed fracture logging of the core was enhanced by imaging the borehole wall with an acoustic televiewer. This data was used to approximate the thickness of the faults and shears and their relative orientation to the proposed tunnel alignment.

In addition, fabric and petrographic logging was used to classify the rocks by stratigraphy, genesis and deformation event. Petrographic analysis by thin section helped to determine the proportion of hard minerals, the degree of alterations and decomposition, and the extent of mineral segregation. These are critical concepts for understanding the engineering properties of the intact material where anisotropy may influence the behavior of the specimen under load. This examination can identify alteration and weathering associated with the faulting and hydrothermal action.

Because the quality of the rock can change dramatically in a very short distance, major features can be overlooked (Figure 4). This image presents a major fault with hydrothermal alteration. It shows a zone of extremely fractured and degraded rock. The rock has been reduced to silty sandy gravel in places and the intact pieces are friable. There is alteration in the form of secondary mica and distortion of the schistosity. These are characteristics of faulted ground. If there is three or more borings in close proximity that intercept this feature then it is possible to estimate the orientation of the fault.

Groundwater generally follows the interface between soil and rock or stands approximately 15 feet below ground surface. The rock mass permeability is generally very low with local high permeability associated with fracture zones, faulting and alteration.

The subsurface investigation included microscopic to regional geological studies of the ground conditions. The investigation started with collection of existing information such as old maps and construction records showing geomorphology, geology,

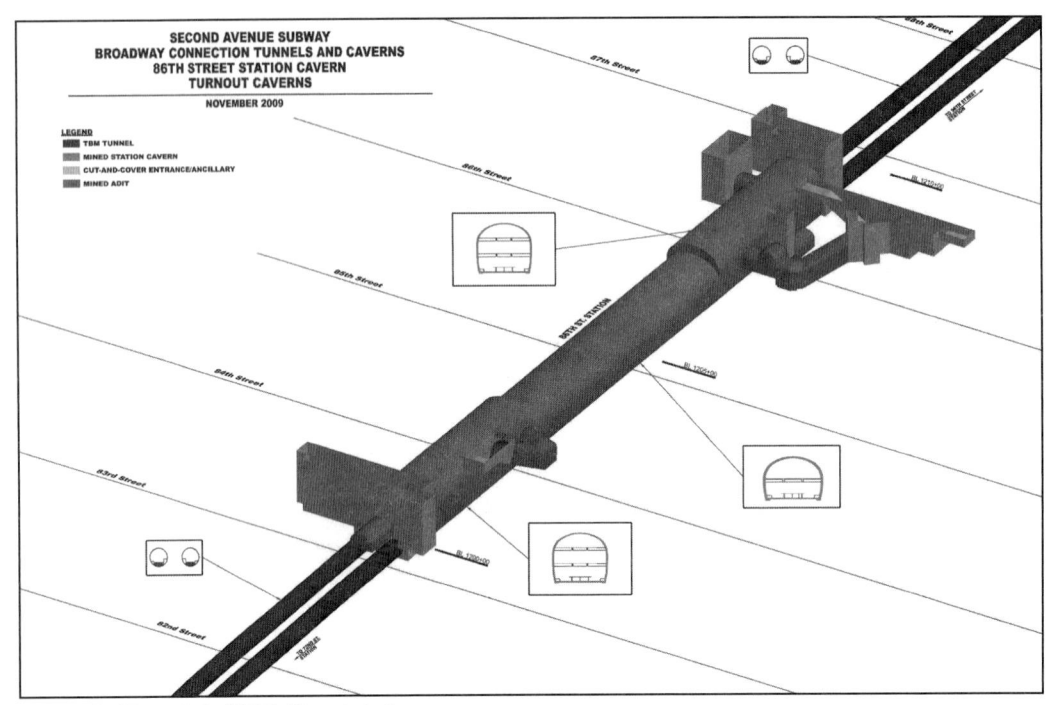

Figure 3. 3D model of 86th Street station

land-use and more than 600 historic borings. During preliminary engineering over 350 new borings were taken along the Second Avenue Subway corridor to determine and/or verify ground conditions. In addition, over 200 environmental borings were taken in the soil overburden at locations where present or prior activities may have resulted in hazardous or industrial soil contamination. The investigation included not only the basic soil sampling and rock coring for laboratory testing and classification, but also oriented core drilling, cone penetration tests, geophysical surveys of boreholes, installation of monitoring wells and vibrating wire piezometers, observation wells, packer testing in bedrock, cross-hole seismic testing, seismic refraction testing and in-situ stress testing. Total number of borings along the Phase 1 alignment during preliminary and final engineering was more than 180.

The key was to understand the rock at the most fundamental level and build a credible geological model. This required detailed investigation to meet the major objectives. A solid model was needed for the orientation of discontinuities, their properties, discrete features that may have a local influence on behavior and the potential risks from major failure. To achieve this goal, subsurface exploration and testing program included orientation and frequency of fractures, shear strength properties of fractures, abrasivity of rock, faults and shear zones, intrusions and alteration, rock material properties, rock mass properties, and soil-rock interface profile and condition.

The variable quality of the rock mass along the cavern alignment required the development of multiple models representing the zonal differentiation of the rock mass in terms of foliation, jointing, and the presence of joint swarms and fractured zones. The methodology adopted, which can be described as "deterministic" was based on:

- The geometrically exact projection of the main rock mass features (e.g., shear zones, etc.) found in adjacent boreholes onto the section of analysis.
- The inclusion of the sets of joints onto the plane of analysis on the basis of statistically derived spacing and dip angles as determined from adjacent boreholes to the section of analyses. The low bound spacing values were selected in all cases, whereas all joints in the sets projected were inferred to be through-cutting.

## DESIGN APPROACH

Large excavation spans, low rock cover, variable geotechnical conditions, relatively large and complex intersections, and dense urban environment characterize the design challenges of the Second

Avenue Subway caverns. Based on the operational requirements and geotechnical information, the station and crossover configurations were developed for two track alignment (Figure 4). In response to constructability concerns, various drill and blast sequencing scenarios were developed to analyze the impact on the required initial and final lining systems. The mined cavern excavation sequence and support system were designed to ensure the stability of the rock mass and adjacent structures. Therefore, maximum allowable vertical ground movement in crown was limited to 2 inches and maximum allowable differential settlement for historical buildings near cavern to less than 1/1000.

Large cavern and crossover sections require multiple drill and blast drifts. The design of drift sizes and shapes was governed by excavation rate, different drifts and cavern stability, and ground settlement and vibration concerns. Various possible cavern excavation sequence including center out drift, side in drift and their combination was considered and their pros and cons were studied through numerical modeling. The analyses show that given the nature of the rock mass (generally competent) and the tendency for gravity induced rock mass stability mechanisms, a center out sequence of excavation may be potentially more beneficial than an equivalent side in approach (Figure 5).

The center out sequence involving the opening of a central heading followed by lateral extension and excavation of the side headings to form the full top heading through a three stage sequence for the larger cavern, will allow continuous dissipation of the induced stresses away from the excavation profile and will facilitate the gradual formation of a rock arch over the crown. The side in sequence can be considered to initiate an increasing concentration of stresses in the central pillar, which will add to the gravity loadings released upon pillar removal during the final development of the top heading. The critical top heading excavation drifts need to be separated longitudinally to allow optimum stress redistribution to occur as well as to facilitate parallel excavation and stabilization activities in the different headings. A minimum distance of one cavern span would be appropriate in our case.

In addition to excavation sequence and support system impact on the cavern stability, size of various drifts (cross section and round length) was adjusted in order to limit the amount of charge per delay for each blasting cycle to satisfy the strict vibration limit of 0.5 inch/sec peak particle velocity under the historical buildings. The Phase 1 construction schedule requires that the TBM tunnels be excavated prior to drill and blasting of station caverns, which imposes some restriction on the excavation sequence configuration and mucking process. The side in drift incorporating the TBM tunnel provides some advantages in terms of unconfined blasting and temporary muck storage.

Empirical data shows that there is a breakdown of the natural arching concept below some minimum cavern rock cover to span ratio. Underground rock engineering practice sets a limiting cover to span ratio of $\geq \frac{1}{3}$. The $\frac{1}{3}$ rule has long been used as a rule of thumb in the mining industry. To avoid heavy support requirements and allow conventional construction methods in hard rock, the cover to span ratio over all of the cavern length was kept above $\frac{1}{3}$. High arches lead to favorable compressive stress distribution in the rock mass around the tunnel and in the tunnel primary and final linings. However, high arches increase the cost of excavation and reduce the thickness of rock cover. Shape of the caverns and high or low arch configurations were investigated using continuum and discontinuum analysis methods and optimum shape was selected for each cavern. Because of limited expected long-term groundwater infiltration, drained invert concept was used for the caverns resulting in a relatively thin and flat invert slab.

## Design Based on Q Empirical Method

The cavern design features represented by large, shallow, non-circular openings, jointed rock masses, random shear zones, and variable rock covers required a robust design procedure including a combination of empirical methods, continuum and discontinuum analyses. Barton's Rock Tunneling Quality Index empirical method, Q, was employed to ensure that the designed support system was compatible with successful existing and similar rock caverns. In the Q method, the rock mass is divided into different categories of quality, and initial support systems are derived for various corresponding rock quality classes. The Q system was developed with a view to determining the mechanism and mode of failure in the rock mass based roughly on the block size, inter-block shear strength, and the active stress regime, with the aim of evaluating stability as one of the first steps in designing an underground excavation. The Q rating was used to provide a first indication of initial ground support.

The raw Q values were developed for each core run from more than 50 deep borings encompassing a zone that extended at least $\frac{1}{4}$ cavern span above and below the crown. From these raw Q values, the weighted average over the crown zone was taken to obtain representative Q values. Using these representative values, along with the northing and easting coordinates for each of the borings, an input file was generated to plot Q contours across the cavern plan and the centerline Q values were obtained by cutting a longitudinal section across the contours.

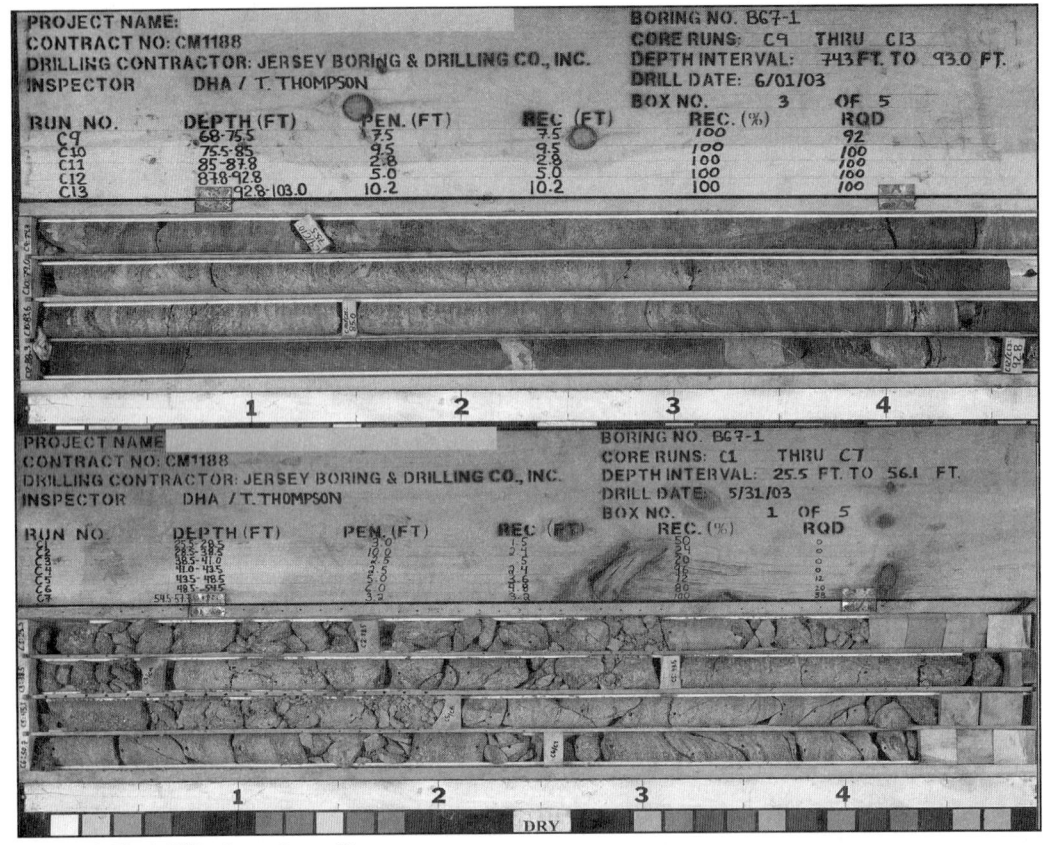

Figure 4. Variability in rock quality

In zones influenced by penetrations and portals, the centerline Q values had to be reduced by a factor of 3.0 and 2.0, respectively, after the contours had been generated and the longitudinal section had been cut. For this analysis, the Q reduction was taken over a zone of the station cavern roughly equal to one-half the width of either the penetration or the portal, as appropriate.

**Discontinuum Analysis**

The existence of low rock cover within a jointed rock mass led the designers to consider a block interaction problem rather than a stress strength one. Discontinuum analysis was used to ensure that the presence of joints and faults in the rock mass around the cavern does not result in unacceptable bolt loads or displacements in the cavern structure. The Universal Distinct Element Code, UDEC, was employed to perform the discontinuum analysis and calculate the ground response, and rock bolt and shotcrete forces. Basic UDEC input parameters including cavern geometry, rock cover thickness, joints pattern, rock mass and rock joint parameters, and material properties of shotcrete and bolt structural elements were determined by the geotechnical investigation program or through literature review.

The first step in the design process was to divide the cavern into different ground class zones. For each ground class zone two deterministic jointing patterns (expected worst condition and expected typical condition) and a support class obtained from empirical methods were assigned and the available data for intact rock, rock joints and soil properties were interpreted and best estimate and lower bound values were determined. UDEC was used to evaluate the global stability of each excavation drift and the entire cavern after each drift excavation and before and after its support installation. The analysis aimed at optimization of the design in terms of excavation sequence and type and quantity of support. Intrinsic stability mechanisms of the caverns were studied by excavating each drift and the entire cavern without installing the support. This was critical to the identification and interpretation of the range of rock mass responses resulting from key physical attributes.

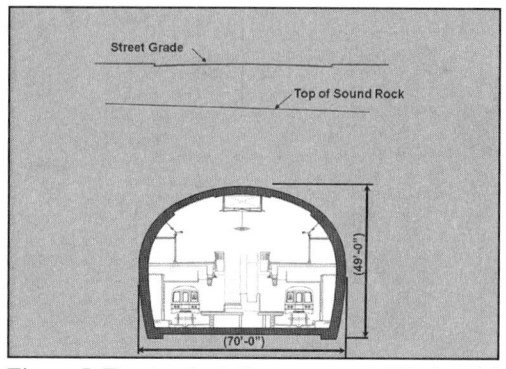

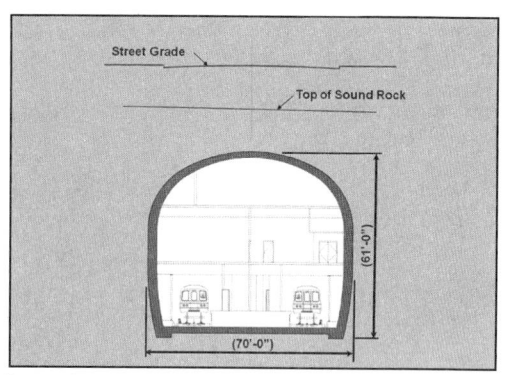

Figure 5. Two track station caverns at 72nd and 86th Street Stations

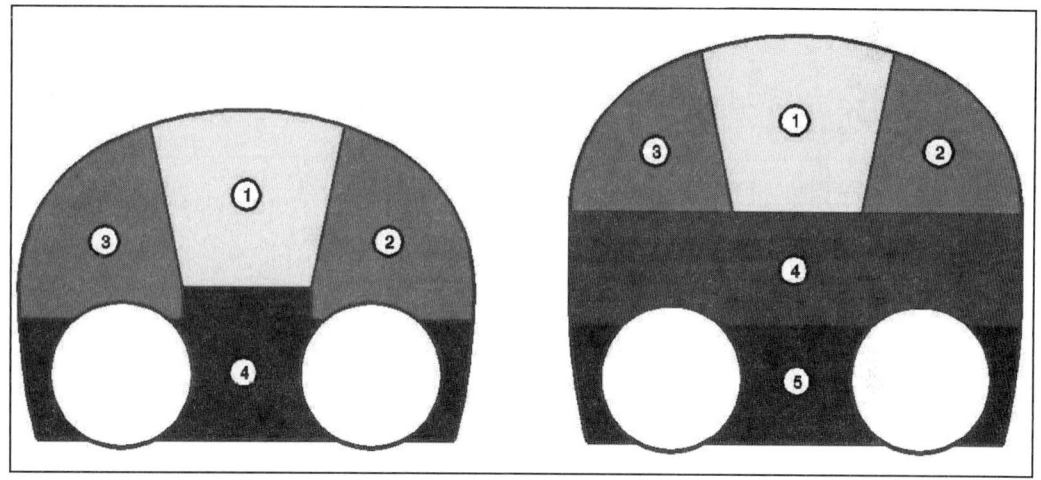

Figure 6. Excavation sequence analyzed

For intact rock the Hoek-Brown criterion, for the foliation and cross foliation joints the Barton-Bendis joint behavior model, and for the shear zones the Mohr-Coulomb shear failure criterion was used in the UDEC analysis. The convergence-confinement analysis method was used to account for the three dimensional effects of the excavation face and a relaxation of 50% of the initial stress was applied after the excavation of each drift and prior to the installation of the initial liner. Three different shotcrete strengths (1, 7, 28 days) were used according to the timing of different excavation stages. The effect of groundwater flow was not included in the modeling of the cavern excavation and support. Adequate drainage during construction was assumed to relieve hydrostatic pressures on the initial lining.

The main modeling steps consisted of:

- Development of a rock mass model representing physical and mechanical characteristics of the ground, which was the principal factor controlling the structural behavior (Figure 7),
- Initialization of the primary stress state through model consolidation under rock, soil and buildings gravity loading,
- Excavation of various drifts and entire cavern without support to assess intrinsic stability state,
- Installation of the primary support in line with the appropriate excavation sequencing.

The evaluation of the results included:

- Evaluation of the principal stability mechanisms,
- Review of the induced stress-displacement fields,
- Assessment of supporting function of various rock reinforcement systems, in terms of tunnel profile deformation control and their load capacity requirements,

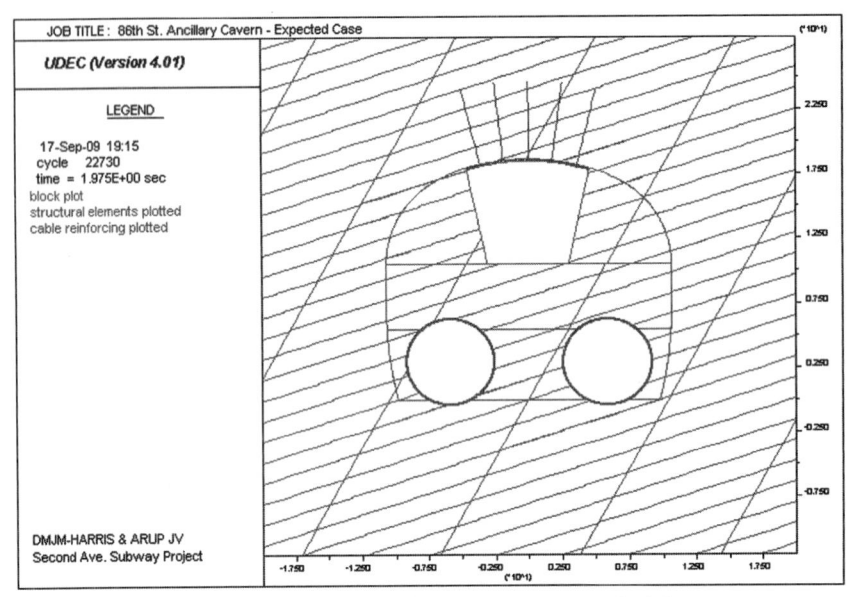

Figure 7. UDEC model for analysis of the two track cavern station at 72nd Street

- Overall engineering evaluation of the modeling results.

In shallow discrete structure model, global stability may be associated with the unique geometrical combination of the joints. In addition, adverse through-cutting structure can be likely, while the crown arching capability is very limited. Given the nature of the rock mass and the potential stability mechanisms, it was considered unlikely that a controlled pre-failure deformation response would allow support upgrading during construction.

**Continuum Analysis**

Continuum analysis was used to ensure that the design does not result in adverse stress strength condition in the rock mass around the cavern opening. Rock mass parameters were determined using RockLab and the excavation sequence and support installation of the cavern was modeled using Phase$^2$.

RockLab was used to determine the Generalized Hoek-Brown strength parameters as well as the rock mass deformation modulus. The input parameters comprise uniaxial compressive strength, intact rock parameter (mi), geological strength Index (GSI), disturbance factor (D), and the intact rock deformation modulus.

The uniaxial compressive strength and the deformation modulus of the intact rock were obtained from rock core laboratory tests. The disturbance factor (D) was assumed to be 0.8 based on the expected rock blasting quality. The intact rock parameter (mi) was considered to be 10 as recommended for schist. The structure of the rock mass was expected to be blocky with fair to good joint surface conditions. The expected typical condition assuming good joint surface condition resulted in a GSI value of 60 while the expected worst condition assuming fair joint surface condition resulted in a GSI value of 50. Based on these input parameters, Hoek-Brown strength parameters and rock mass deformation modulus were calculated using RockLab for the expected typical condition and the expected worst condition.

The excavation sequence and support installation of the cavern was analyzed using Phase$^2$. Two 2D continuum models with 6'×6' bolt spacing for the expected typical condition and 5'×5' bolt spacing for the expected worst condition and their corresponding sets of rock mass properties were developed. The caverns were excavated in 3 top heading drifts and one or two benches and the corresponding support systems were installed after each excavation stage and a relaxation coefficient of 50% was applied.

Based on the analyses performed, it can be concluded that this type of continuum analysis for jointed hard rock cases results in very small deformations and a very low level of stress in bolts and shotcrete. Therefore, continuum analysis in this kind of jointed rock cases fails to detect the local and global failure mechanisms generated by the joint sets and cannot be used in the design or verification of the design of excavation sequence and support systems.

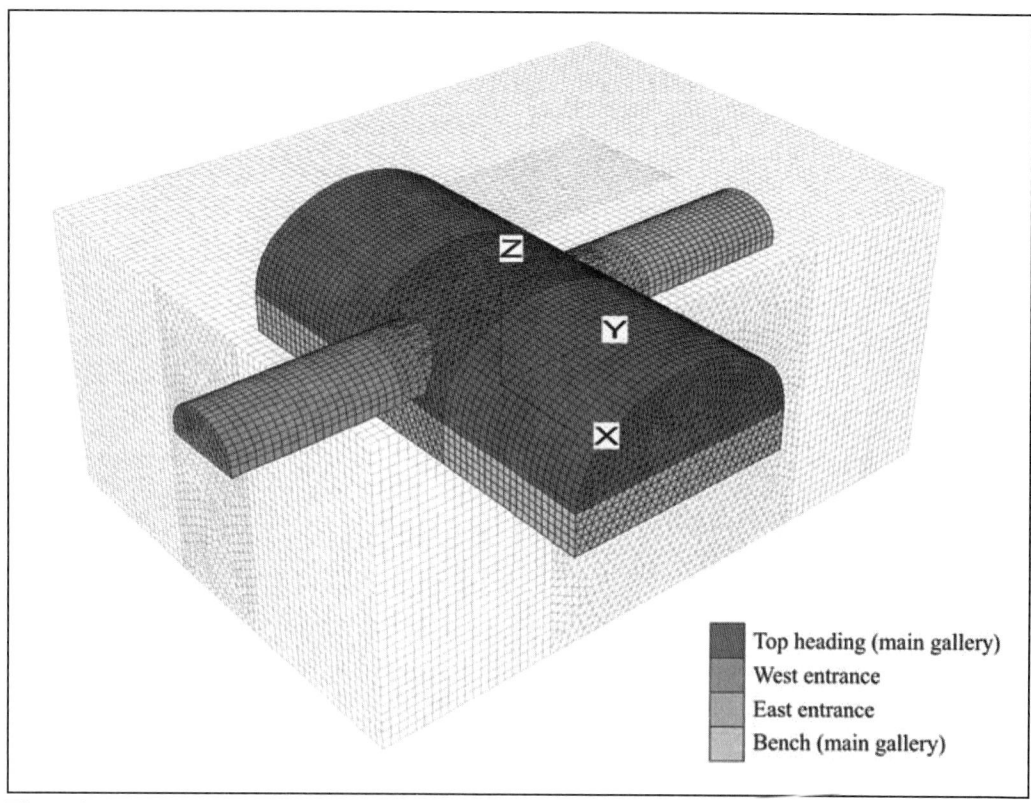

**Figure 8. FLAC3D model for analysis of the effect of penetrations on station behavior**

### 3D Analysis for Penetrations

A 3D continuum analysis with FLAC3D software was used to evaluate the effect of the excavation of entrance and ventilation penetrations on the deformation and stress distribution in crown and sidewalls of the station cavern in the vicinity of these penetrations (Figure 8). This analysis provided the required information for designing the penetrations and determining the additional initial support needed to reinforce the station cavern excavation at the proximity of these penetrations. First the station cavern and then the penetrations were excavated incrementally based on their specified round lengths and following the projects construction schedule. To make the 3D analysis practical, the division of cross section into multiple top heading and bench drifts was ignored and the entire section of the cavern and penetrations were excavated in the same step over their specific round length and then the initial liner including bolts and shotcrete was installed for that particular round length.

### Initial Liner Design

The design approach to the stabilization of the caverns relied on the use of phased excavation by drilling and blasting of the relatively competent but jointed rock mass generally unaffected by significant weathering or alteration coupled with the installation of patterned rock bolt support and the application of shotcrete lining to the exposed rock surface in order to achieve temporary stability during excavation. The use of tensioned reinforcement integrated with reinforced shotcrete to form a composite support system (rock mass + rock reinforcement + reinforced shotcrete) was considered suitable for the good or fair rock mass conditions anticipated in most areas. Long term stability was assumed to be assured by the construction of cast in place concrete lining, once excavation is completed.

The main primary support element for the larger cavern with 21 m span included 6 m long 32 mm diameter 13.6 tons tensioned bolts at 1.8 m × 1.8 m grid for crown and shoulders combined with 18 cm of shotcrete. Similar passive bolts at 1.8 m × 3.6 m spacing were used for sidewalls. Fiber reinforced shotcrete was recommended for the initial liner of all the caverns due to the increased capacity provided by its ductile behavior. The satisfactory performance of the support system was indicated also by the favorable redistributed stresses in the pre-stressed arch and the level of shear strength mobilization. Use of passive reinforcement resulted in a similar level of cavern stabilization, but with less effective radial restraint over the crown arch and an excessive overloading for a number of bolts. Therefore tensioned bolts were used as the main support for the cavern arch.

## CONCLUSIONS

The Second Avenue Subway Project is one of the largest and complex construction projects in the United States and a critical part of the success for the project will be the safe and optimum design of its large and shallow rock caverns. Best known design tools with fundamentally different approaches including empirical methods and two and three dimensional continuum and discontinuum analyses were used in cost effectively and conservatively designing the excavation sequence and initial support system. Understanding each method's differences and limitations, and comparison of their results provided a comfortable margin of safety, which compensated for the unknowns in the design process.

Construction of the Second Avenue Subway was started in 2007, over 80 years after the line was first planned. There are significant construction difficulties, many resulting from the mass of subsurface utilities, the need to maintain traffic flows and the densely populated neighborhoods. Construction of two large and shallow rock caverns will be significant challenges that will need to be overcome in the next few years.

## REFERENCES

Bandis, S.C., et al., 2004. Modeling-aided design of a very large span underground excavation. 1st International UDEC/3DEC Symposium, Bochum, Germany, pp. 57–63.

Bennett, C.K., 2006. Design challenges of New York's largest public works project of the decade – Second Avenue Subway. Proceedings of North American Tunneling Conference, Chicago, IL, June 10–12, 2006, pp. 21–28.

Desai, D., Naik, N., Rossler, K., and Stone, C., 2005. New York Subway caverns and crossovers – A tail of trials and tribulations. Proceedings of Rapid Excavation and Tunneling Conference, Seattle, WA, June 27–29, 2005, pp. 1303–1314.

Grimstad, E., et al., 2003. Q system advance for sprayed lining. Tunnel and Tunneling International, Vol. 35, No. 3, pp. 46–48.

Snee, C.P., Ponti, M.A., and Shah, A.N., 2004. Investigation of complex geologic conditions for the Second Avenue Subway tunnel alignment in New York City, New York. Proceedings of North American Tunneling Conference, Atlanta, GA, April 17–22, 2004, pp. 357–362.

## AKNOWLEDGMENTS

The authors would like to thank Anil Parikh, the Program Manager of Second Avenue Subway Project for the MTA Capital Construction Company and Chris Bennett, the Project Manager of the Joint Venture for their permission to publish this paper.

# Shaft, Cavern, and Starter Tunnel Construction for Lake Mead Intake No. 3: Temporary Support and Permanent Lining Solutions

**Luis Piek, Rob Harding**
Arup, San Francisco, California

**Jeff Hammer**
Brierley Associates, Littleton, Colorado

## INTRODUCTION

In March 2008 Vegas Tunnel Constructors (VTC), a joint venture of Impregilo SpA and SA Healy were awarded a US$ 447M Design-Build contract by the Southern Nevada Water Authority (SNWA) for a section of the Lake Mead Intake No. 3 Project. Arup, supported by Brierley Associates, is the Design Engineer for VTC.

The VTC contract scope of new facilities for Intake No. 3 includes three major components: a deep tunnel access shaft and cavern, a tunnel beneath Lake Mead, and a submerged intake structure. The first major construction phase of the project is nearing completion. Construction of the 9 m diameter 185 m deep access shaft was completed using a drill-and-blast top-down construction method, with a 450 mm thick unreinforced concrete final lining placed as excavation was completed. The shaft passed through a major fault zone and discontinuity swarms requiring extensive pre-excavation grouting for water control in portions of the shaft.

At the base of the shaft, a large cavern is excavated, containing the TBM erection chamber, a backshunt tunnel for future expansion with steel bulkhead, and the TBM starter tunnel. Due to differential hydrostatic loads as high as 17 bar in the dewatered condition during maintenance and inspection, the shaft and cavern linings are designed as drained structures for economy. An unreinforced plain concrete liner was adopted for the shaft structure; the cavern is lined with steel fiber reinforced permanent shotcrete and fibre reinforced plastic (FRP) rock bolts. Design considerations and construction practicality for the structures are discussed. Further background of the project purpose and its description can be found in Hurt, et al, RETC 2009, and Feroz, et al, RETC 2007.

## ACCESS SHAFT DESIGN

The access shaft forms only a part of the larger Intake No. 3 system. The lake water will be drawn out through the intake tunnel into the shaft, out through the IPS-3 Stub tunnel and into to the Alfred Merritt Smith Water Treatment Plant prior to distribution to the city of Las Vegas. The contract allowed limited groundwater inflows at the construction completion; however the design challenge was to withstand the high hydrostatic loads on the shaft liner during the temporary maintenance and inspection condition under a maximum design lake level of 1234 ft amsl. This was accomplished by designing the shaft lining as a drained structure.

The design approach for a drained liner is to provide a lining permeability higher than that of the surrounding rock such that the water pressure on the lining is significantly reduced. However, for construction of the shaft to occur a rather intense pre-excavation grouting operation must be performed below the water table. This forms a thick ring of rock with a lower permeability than the host rock; thus, in order to lower the permeability of the grouted region, drainage holes must be installed through the cast in place liner which penetrate into the grouted rock mass. The drainage holes should be sufficiently long to effectively grade the pressure head across the grout curtain, but not so long as to allow excessive flows into the excavated shaft, which would complicate any future inspection or maintenance activities. To provide this, the drain holes were designed on an effective 3 m by 1.5 m grid. These drainage holes extend a minimum of 0.6 m into the rock mass. This provides a lining permeability higher than the grouted rock permeability such that the water pressure on the lining is significantly reduced. For design purposes, a seepage analysis was performed and the resulting pressure (calculated as a percentage reduction of the initial hydrostatic head) used for the design of the cast-in-place final lining.

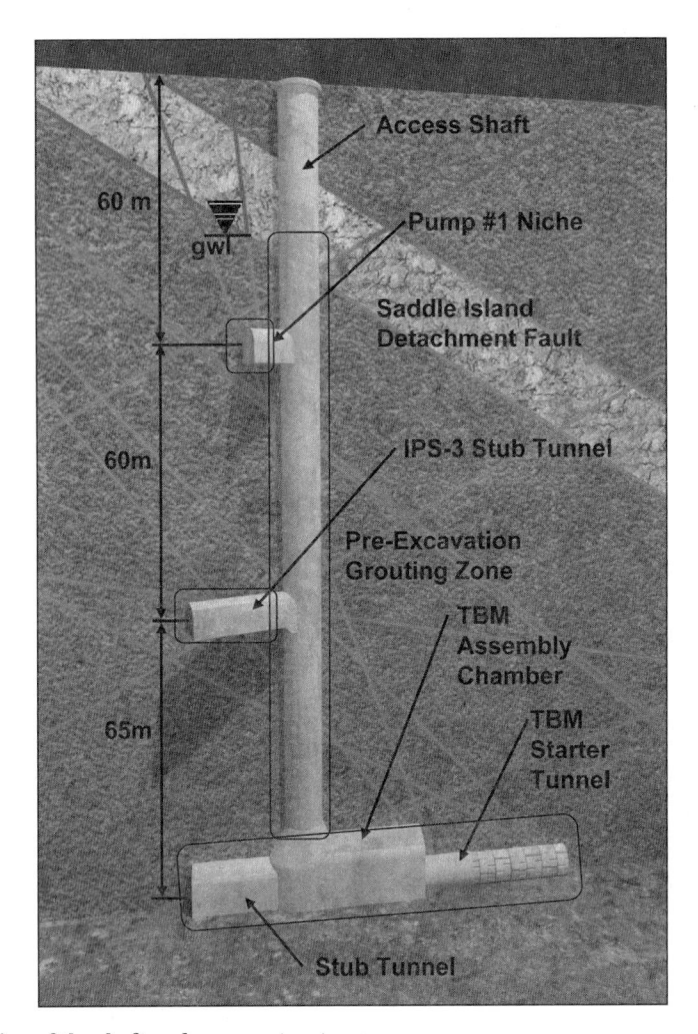

**Figure 1. Elevation of the shaft and cavern structures**

The final lining was modeled as a thick concrete cylinder that resists lateral rock, ground water, surcharge, and seismic loads. For the final lining, lateral rock loads were evaluated based on the estimated zone of plastic deformation around the shaft, which is a function of the rock mass strength.

Structural design was performed in accordance with ACI 318-08 to consist of plain (unreinforced) concrete throughout the shaft, reducing durability concerns that may arise with the use of reinforcement, such as spalling due to corrosion. At the two junctions (the temporary niche and IPS-3 stub) a steel rib and post frame was provided within the concrete lining for additional rigidity. A concrete cover of 75 mm was included to address durability concerns. The shaft final lining was required to be a minimum of 450 mm thick with minimum 28-day strength of 32 N/mm$^2$ (4,500 psi).

## GROUNDWATER DRAWDOWN MODEL

As stated previously, the major design load for the permanent works of the shaft and cavern structures was the external water load that develops when the shaft is dewatered for inspection and maintenance. To model this condition, a groundwater drawdown model was established to reflect a realistic prediction of the effect of dewatering through the shaft and intake tunnel. This model allowed for a minimum period of 14 days for the dewatering of the entire system.

The model reduced the internal water pressure inside the intake shaft and tunnels from the maximum water table level (Elev. 1234 ft) to tunnel invert level (Elev. 647 ft) by applying a "rapid drawdown" function (change in water level head vs. time). Seepage analyses were performed to assess the rate of water

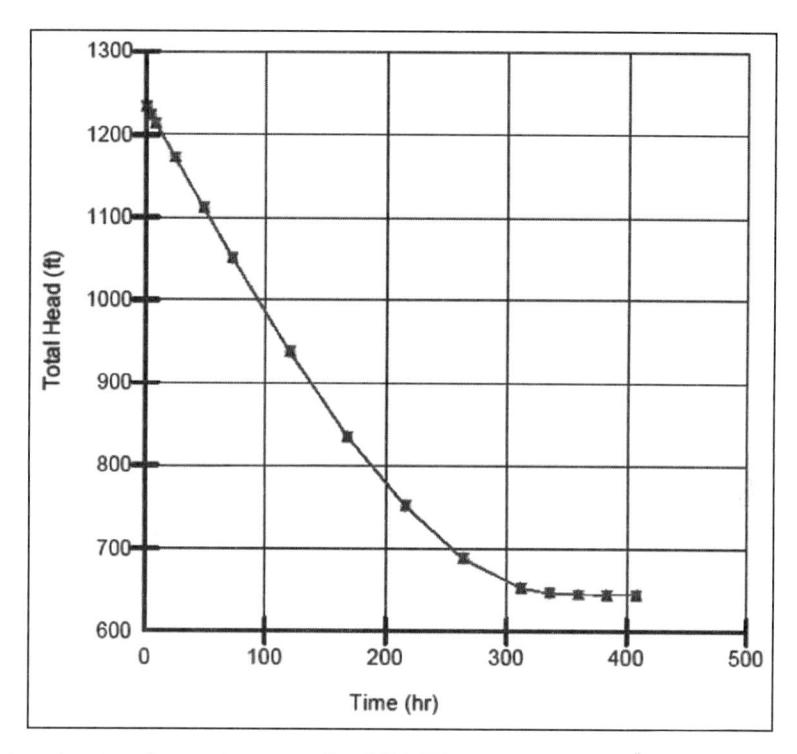

Figure 2. Internal system dewatering, water level (AMSL) vs. time

inflow into the shaft and cavern excavations, measuring the reduction in external ground water pressure in respect to the reduction of the internal water pressure in the intake system.

An axi-symmetric model, using the program SEEP/W was used to analyze two cases; an undrained (without weep holes) and a drained design case, (with weep holes). The size and effectiveness of the weep holes were varied to assess the sensitivity and the resulting drop in external groundwater pressure. It should also be stressed that the model assumes a uniform permeability, which may not necessarily be the case where groundwater inflows are dependent on secondary flows e.g., through rock discontinuities. The authors stress the use of sensitivity analyses in order to account for such an effect. The pressure differential between the internal and external pressure was used to identify the maximum pressure acting on the cavern lining at any point in time. The resulting pressure differential was applied to the lining and to rock wedges, depending on their size and shape.

The SEEP/W analysis was verified using an analytical solution for seepage forces in an Elasto-Plastic tunnel medium (Barbosa, 2003). This verification analysis considered the same two cases (drained and undrained), taking into consideration the fact that low permeability zones are expected to have a major impact on the time required to reach the steady state condition. The internal drawdown process considers an initial hydrostatic water pressure, with dewatering of the system controlled by a "Rapid Drawdown Head vs. Time boundary function" representing a transient flow with a time dependent hydraulic boundary condition and a change in porewater pressure. For the analysis the minimum estimated time to dewater the intake system was taken as 14 days.

## SHAFT PRE-EXCAVATION GROUTING

Due to the presence of water-bearing features, such as the Saddle Island Detachment Fault and several anticipated discontinuities, pre-excavation grouting was required to allow shaft construction to proceed, and was the primary means of limiting water inflows.

The grouting work plan was based on using a top-down method. This consisted of pouring a concrete slab at the base of the excavation as a working surface (if required by ground conditions), setting 57mm ID casings for the grout holes in the excavation base, and grouting the casings in place using a heavy cement mix. 48mm grout holes were then drilled from the casings down into the rock using a three-boom Tamrock drill jumbo.

Prior to the start of grouting, water inflow was recorded with all holes open and from each hole with all the other holes closed. Based on the inflow of the holes, either grouting was implemented or the drilling was advanced. The production process was to drill, test for water, grout, drill again, and repeat the process until, based on the final test, the curtain had been advanced such that the water inflow fully met the post grouting inflow criteria.

During implementation of the first grout curtain the drilling was advanced in 3.0m intervals for the first 24.4m. A complete set of 22 primary and 22 secondary holes were drilled to depth with a surface hole spacing of 0.9m to 1.2m on center. At about 24.4m the drilling program was changed to drill in larger 6.1m intervals. In total, the first grout curtain depth was 36.6m.

Successive grout curtains modified the grouting plan by drilling the shaft grout holes in two 3m and 6m rings on the outside of the shaft excavation line on 0.9m to 1.2m spacings, respectively. The grout casings were placed using the same general method as previously described. The primary holes were drilled to 20 meters and grouted using BASF Rheochem 650 microfine cement at a 1:1 water cement ratio with a superplasticiser additive (2%-3%). The water cement ratio was increased if difficulties in pumping the grout were encountered. Refusal criteria used an approximation on the GIN (Grouting Intensity Number) method. The plan specified either a pumping volume of 751 L/m or a pressurization of the grout hole from 31.0 bar to 32.7 bar, though these parameters varied depending on the depth of the drilling operations.

Each grout hole was re-drilled and tested for water inflow. Holes were then re-grouted using the same criteria of volume or pressure if the inflow of water exceeded post grouting criteria. The secondary holes were then drilled to 18.3m, tested, and if necessary grouted using the same criteria. The holes were finally advanced to 40 meters and grouted, primary first and then secondary. As a final test, four test holes were placed in the eye of the curtain to measure the effectiveness of the grout curtain.

The grout was mixed in a specialized plant that included colloidal mixers suspended from a steel platform just above the shaft invert and pumped into the grout holes from a header. The pressure and quantity of grout in gallons were recorded by a pressure gauge and a flow meter on the grout plant. Although this equipment is often properly designed for the harsh construction environment, a secondary method of measuring the volume of grout, such as counting the number of tanks of grout, should be used in the event of equipment failure. Such a method was used to identify an error in the flow meter during the first grouting operation.

## ACCESS SHAFT INITIAL SUPPORT AND CONSTRUCTION

Design of the initial support system employed two rock mass classification systems: Rock Mass Rating (RMR) and Geologic Strength Index (GSI). These systems were used to evaluate rock mass strength and deformation parameters which were in turn then used to design the initial support (Hurt, et al RETC 2009). The Shaft Design Engineer's representative (SDER) and contractor's representative independently assessed the excavated rock face, assigning GSI Values, before meeting to agree on the initial support type. Six types of initial support were specified:

- Type 1A—75mm shotcrete
- Type 1B—75mm shotcrete with a single layer of steel welded wire fabric or mesh Spot rock reinforcement (dowels or bolts) and shotcrete
- Type 1C—75mm shotcrete with a single layer of steel welded wire fabric or mesh and 3.0 m long pattern No8 hollow bar, spaced 1.5m on centre vertically and horizontally
- Type 2A—Spot rock reinforcement (dowels or bolts) and a flash coat of shotcrete as needed
- Type 2B—Spot rock reinforcement (dowels or bolts) and shotcrete
- Type 2C—1.8 m long pattern SS-46 Split Sets, spaced 1.5m on centre vertically and horizontally, and shotcrete with a single layer of steel welded wire fabric or mesh

Generally, installed support consisted of Type 2C with a layer of chain link mesh placed behind the dowels. Shotcrete was placed in depths up to 100mm. In some areas where the rock mass was exceptionally good only a flash coat of shotcrete was applied. In terms of ground support, the predicted amount of rock support in vertical meter lengths in the shaft can be summarized as follows:

| Ground Type | Predicted Rock Support | Rock Support Constructed |
|---|---|---|
| Type 1A | 37 m | 35 m |
| Type 1B | 3 m | 3.3 m |
| Type 1C | 2 m | 3.3 m |
| Type 2A | 110 m | 130 m |
| Type 2B | 14 m | 0 m |
| Type 2C | 6 m | 0 m |

Rock support installed in the access shaft was substantially less than predicted due to two main

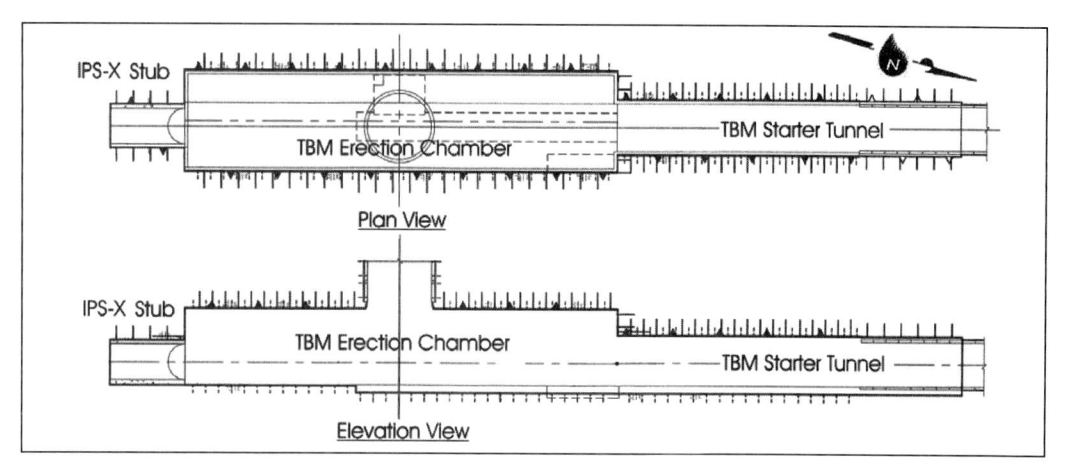

**Figure 3. Access shaft cavern general arrangement**

advantages of the top-down construction method. First, the installation of the final lining within 6 meters of the floor of the excavation provided a minimal amount of exposed ground. The second advantage was the pre-excavation grouting performed, which reduced the amount of time and materials needed for support. Nearly all the excavation in the class 2A region of the Access Shaft did not require any support beyond the shotcrete flash coat.

The construction of the top-down lining for the Access Shaft followed two stages. In the first stage the shaft lining form was split into a lower and upper portion. The lower portion, identified as the curb ring, was approximately 0.6 meters in height and had a tapered bottom to allow the joining of subsequent concrete castings. The curb ring was hung from the previous pour using rebar couplers. Expanded sheet metal sheet and scribing pins were placed in the curb ring to fill the gap between the form and the rock surface. Concrete was then poured into the curb ring form and allowed to set. During that time, the second stage was started. The second part of the shaft form, identified as the wall form, was lowered into placed on top of the curb ring form. Concrete was then poured into the 3.04 meter tall form until filled. Joints between concrete pours were later contact grouted with a cementitious grout without a waterstop. Subsequent contact grouting of these joints was performed.

Vibrations due to blasting could potentially cause damage to young concrete. Based on the requirements of CIRIA Technical Note 142, "Ground-borne vibrations arising from piling," appropriate limits to avoid damage are a minimum concrete strength of 600 psi and a maximum peak particle velocity of 2 inches per second. Due to the sequencing of the excavation cycle, there was sufficient time to meet this minimum requirement. The blast-proof forms remained in place over the previously cast ring during blasting to reduce any potential damage. Stripping was also a concern for the contractor. An unconfined compressive strength greater than 1200 psi was required before the concrete forms could be stripped, lowered, and set up for the next pour. Again, due to the excavation sequencing, there was sufficient time to meet this requirement. The strengths were confirmed by testing in accordance with ASTM C39 prior to striking the forms. In general the concrete mix design preformed well and exceeded 6200psi at 28 day strength in testing.

Water infiltration into the shaft was managed throughout shaft construction, though the water inflows below the water table averaged 400 L/m. Pumping of infiltrating water was essential to construction operations in the shaft. The shaft bottom was flooded on a few occasions due to pumping problems.

The average actual advance rate of shaft excavation was 9m per week. This advancement rate does not include all stoppages for grouting and for driving the two short adits.

**CAVERN INITIAL SUPPORT**

The cavern excavation at the base of the access shaft was conducted using drill-and-blast techniques with a heading and bench configuration. Collectively referred to as the Access Shaft Cavern, it consists of three distinctly individual elements, with their designation and excavated sizes as follows:

- IPS-X Stub Tunnel (TBM Backshunt)—26 m long, 5.8 m wide, by 6 m high
- TBM Cavern—61.5 m long, 13.7 m wide, by 10.5 m high

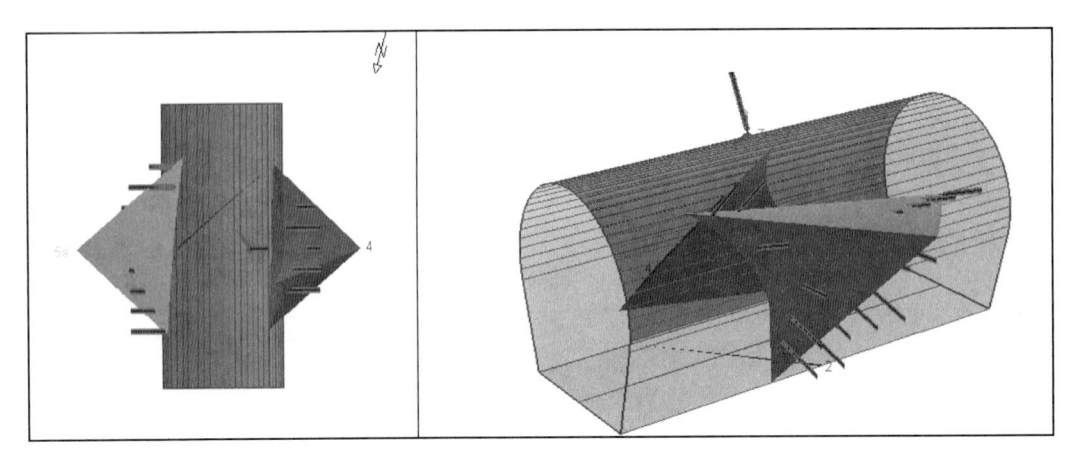

**Figure 4. Example of wedge analysis for one cross section**

- TBM Starter Tunnel—106 m long, 7.6 m wide, by 7.9 m high

The size of the cavern is driven by a large rail mounted gantry crane that will be installed in the TBM Erection Chamber to enable VTC to erect the TBM and move large mechanical and support components in the cavern. The TBM Starter Tunnel was extended to allow lowering and assembly of the prefabricated parts of the TBM into essentially the full configuration at the time of launch. The general arrangement of the Access Shaft Cavern is provided in Figure 3.

Design of initial support employed two rock mass classification systems: Rock Mass Rating (RMR) and Geologic Strength Index (GSI). These systems were used to evaluate rock mass strength and deformation parameters and to evaluate rock reinforcement spacing, lengths, and size with empirical, structural, and numerical evaluations in general accordance with applicable codes.

Discontinuity (wedge stability) analyses were performed to assess wedge stability and to evaluate and design initial support systems for the Access Shaft. Data from major rock joints, fractures and fault planes that were collected during site investigation, which consisted of rock core drilling, surface mapping, and acoustical televiewer surveys, were used to establish the design joint sets and major discontinuity planes. The discontinuity planes were combined in UNWEDGE 3.0 to execute the stability analyses. These major planes produce several combinations of rock wedges which were examined for individual wedge stability with each of the cavern design cross sections. To meet a minimum factor of safety of 1.3, a combination of rock reinforcement and shotcrete to the design sections was applied. Figure 4 shows a representation of the analysis performed. It is important to note that where the initial support incorporated the final lining, the presence of water pressure on the wedges during an 'inspection and maintenance' condition was also included in the design.

Finite element analyses (FEA) were conducted for a total of five cavern cross sections using Phase2 to evaluate the performance of the proposed initial support of the Access Shaft Cavern. Rock mass strength and deformation parameters were varied in these analyses to evaluate stresses and deformations as part of a parametric study to confirm the stability of structural elements that were designed using empirical methods. Given that the majority of the rock mass surrounding the Access Shaft Cavern possess GSI values greater than 60, three GSI values (60, 70, and 80) were evaluated in the parametric study. In general, the predicted deviator stress was less than rock mass unconfined compressive strength, indicating the rock mass when supported with an initial support system will behave elastically.

In addition to the two-dimension parametric study, a three-dimensional stress analyses encompassing the construction from existing ground surface (EL. 1230ft) to the cavern base (EL. 620 ft) was performed to evaluate stress concentrations within the rock mass during construction of the Access Shaft, Temporary Adit, IPS-X Stub Tunnel, and Access Shaft Cavern. This analysis was conducted using Examine 3D, an elastic boundary element program developed for visualizing stresses associated with rock excavation. The analysis was conducted using the strength and deformation parameters for a Geological Strength Index (GSI) of 30 and a unit weight of 0.026 MN/m$^3$. Predicted rock mass deviator stresses were compared to rock mass unconfined compressive strength ($\sigma_{cm}$) values for various GSI values to assess whether the rock mass will behave in an elastic or plastic manner during construction.

The analysis confirms satisfactory performance of the selected initial support in the TBM Erection Chamber, consisting primarily of 3 m long K60-32 post-tensioned corrosion resistant fiberglass bolts fiberglass installed on a 1.8 m by 1.8 m pattern in the crown and 2.4 m by 2.4 m in the sidewall with a 250 mm fiber reinforced shotcrete layer in side walls and crown. The first 100 m of the TBM starter tunnel will incorporate the final support into the initial support through the use of 2.4 m long K60-32 post-tensioned corrosion resistant fiberglass bolts fiberglass installed on a 1.8 m by 1.8 m with a 150 mm fiber reinforced shotcrete layer. As the initial support is also the final support for these sections, the bolts and shotcrete have been designed with higher capacity to meet the more stringent requirements for a permanent lining required in the contract documents.

The remaining portion of the TBM Starter Tunnel will only require temporary support designed as 1.8 m long SS-33 split sets dowels installed on a 2.4 m by 2.4 m pattern from springline to springline across the crown and a 150 mm thick layer of plain shotcrete all-around. A secondary final lining utilizing the TBM segments will be placed later.

A 100 mm thick mud slab is cast on the cavern floor. In similar fashion to the Access Shaft liner, the final lining for the TBM Erection Chamber is designed as a drained structure and will be provided with drainage holes. Also as explained herein, pre-excavation grouting will take place.

## CAVERN FINAL LINING

As mentioned previously, approximately the last 15 m of the TBM Starter Tunnel will incorporate a secondary final lining. The end of the starter tunnel will be lined with precast segments, forming an undrained lining.

The IPS-X stub tunnel will also incorporate a final undrained lining. The contract documents require a 26 m length of stub tunnel to be constructed at the far end of the cavern to serve as a future expansion connection, should SNWA wish to increase the capacity of the intake system. All of the required 26 m length of this tunnel will be used during construction to provide additional length for locomotive operations, and lined with cast in-situ concrete after tunnel completion. At the juncture of the stub tunnel and main chamber, a steel bulkhead will be placed to seal off the IPS-X stub tunnel from the rest of the Intake No.3 Project. The completed 26 m stub tunnel length has been designed as an undrained concrete lining. This undrained design of the IPS-X Stub Tunnel will allow a future contractor to connect to the existing intake system without having to dewater the entire system. It is envisaged that a future contractor would provide a mined connection to the IPS-X stub by first providing a ground seal (perhaps via either grout or ground freeze) around the stub tunnel and then proceed to pump all water out of the stub tunnel before making connection. Once this connection has been made, full hydrostatic load will be exerted onto the bulkhead and the adjacent undrained tunnel lining. A robust design of both the tunnel lining and bulkhead has been provided to resist this worst case loading.

Full hydrostatic loading and ground loads (based on Hoek-Brown plastic loading assessment and Examine 3D models) were applied to the final undrained lining. The IPS-X Stub Tunnel was designed as an unreinforced section, based on the ACI 318-08 code with a phi value equal to 0.6 used to produce a simplified 4-point plain concrete capacity envelope. Reinforcement has been provided to give temperature and shrinkage control.

The final design solution for the bulkhead consists of a 4826 mm (15'-10") external diameter hemispherical shell, of constant thickness 20.6 mm (13/16") Grade 70 carbon steel with improved notch toughness for pressure vessels, as per ASTM A516. The design of bulkhead, fabrication tolerances, and non-destructive testing requirements are in accordance with American Society of Mechanical Engineers Boiler and Pressure Vessel Code (ASME BPVC) Section VIII Division 2 Alternative Rules, 2007 Edition. The bulkhead was analyzed for an internal pressure of 267psi (1.84MPa) which is a combination of the differential hydrostatic pressure and the design surge pressure. Handing loads during installation and future removal were not considered as the governing case for design.

SNWA requirements also request a provision to be made for removal of the bulkhead. The IPS-X bulkhead design contains details of how this bulkhead could be removed using current technology. For example a valve has been provided in the design to allow equalization of pressure (once IPS-X has been filled to operating capacity).

Corrosion resistance of the steel bulkhead is provided using a two-part system of epoxy coating expected to provide corrosion protection for at least the first twenty years with the remaining eighty years ensured by a 6.35 mm (¼ inch) sacrificial steel thickness. The thickness of the sacrificial layer is based on a corrosion rate the range of 0.025 to 0.050 mm/year, based on slow moving oxygen saturated water.

As mentioned above, the TBM segments will be installed in the last section of the TBM Starter Tunnel to form the final lining. The interface between the shotcrete lining and the segments will be grouted and cured prior to launch of the TBM. In addition to resisting the thrust via this interface, a thrust frame will be assembled to provide the jacking resistance to advance the TBM.

## CAVERN CONSTRUCTION

Prior to cavern excavation, an extensive grouting regime was conducted similar to that in the access shaft. The combination of grout hole and test hole inflows and conditions were used to implement a pre-excavation grouting stage (grout cover) for each lift. The on-site Design Engineer's Representative (SDER) conferred with the superintendent to select a pre-excavation grouting geometry, grout mix and grouting procedure system. In the end, an observational approach was employed in the evaluation of ground water inflows and the application of pre-excavation grouting.

The main cavern was constructed using a heading and bench approach with temporary support installed as the excavation progressed. The design engineer's representative on site monitored the probe drilling and pre-excavation grouting prior to advancing the excavation face. During excavation and prior to the placement of the initial support system, the SDER mapped the excavated face and assessed the Geologic Strength Index (GSI) value for each lift. The SDER then conferred with the superintendant to select an initial support system in accordance with the contract drawings.

Due to construction sequencing, the heading was advanced the full length with the bench excavated thereafter. Wire mesh and split sets were installed after each advance round, with flash coats of steel fibre reinforced shotcrete applied after several advance rounds. Better ground conditions influenced the decision to install only temporary support during excavation, with the final lining placed in a series of consecutive days after the completion of the cavern.

## TBM LOWERING AND ASSEMBLY

The TBM will be assembled lowered in pre-assembled portions consisting of the cutterhead, forward shield, middle shield, and tail shield. The screw auger will be lowered as one long 18m continuous piece. In order to accomplish this, the headframe used to construct the shaft will be removed and two specialty gantry cranes brought in to lower the pieces of the TBM. The gantry crane in the TBM Cavern will be pulled back and out of the lowering path. A skid system will be used to move the TBM components into the starter tunnel once delivered to the bottom of the shaft.

The long backup system required will be constructed within the limits of the cavern and portions of the tail tunnel and starter tunnel using the cavern gantry crane. The tail tunnel also facilitates locomotive logistical operations at the base of the shaft during construction of the tunnel. The 105m length of TBM Starter tunnel is required to construct the full length of TBM and back-up systems before mining and installation of pre-cast segments.

After the TBM has been fully assembled, it will be inched forward into place. A thrust frame will be erected and a bulkhead between the shotcrete liner and first ring installed. The TBM will then be advanced with a few free strokes before beginning mining.

## CONCLUSION

The first of three major phases of the Lake Mead Intake No.3 Project has been completed, consisting of a 10m diameter 185m deep access shaft with a large cavern at its base to house the TBM erection chamber, backshunt tunnel, and TBM starter tunnel. Excavation of the shaft was completed using a top-down construction methodology which installed the 450mm thick final drained liner as excavation progressed. Design of the drained and undrained linings for the shaft, IPS-X and TBM starter tunnel has been discussed, as well as technical construction challenges including extensive pre-excavation grouting, blasting near fresh concrete, controlling large water inflows, and quality control.

At the time of publication, the TBM components have been lowered into the cavern and are being assembled for launch.

## REFERENCES

Barbosa, R, 2003, Analytical Solution for a Cylindrical Tunnel Excavated in a Porous Elasto-plastic Material Considering the Effect of Seepage Forces, Canadian Rock Symposium, Toronto, Canada

Hurt, J., McDonald, J., Sherry, G., McGinn, A.J., and Piek, L., Design and Construction of Lake Mead Intake No. 3 Shafts and Tunnel, Las Vegas, Nevada, USA, RETC 2009, Las Vegas, US

Feroz, M., Jensen, M. and Lindell, J.E., 2007, The Lake Mead Intake 3 Water Tunnel and Pumping Station, Las Vegas, Nevada, USA, RETC 2007, Toronto, Canada.

Feroz, M., Moonin, E. and McDonald, J., 2009, Project Delivery Selection for Southern Nevada's Lake Mead Intake No. 3, RETC 2009, Las Vegas, NV.

Bieniawski, Z.T. (1989) "Engineering Rock Mass Classifications," Wiley-Interscience, 272pgs

Littlejohn, A (2007) "A review of Glass Fiber Reinforced Polymer (GFRP) tendon for rock bolting in tunnels," Ground Anchorages and Anchored Structures, Thomas Telford, London

D.J. Rockhill, M.D. Bolton and D.J. White, CIRIA Technical Note 142 (1992), "Ground-borne vibrations due to press-in piling operations" Cambridge University Engineering Department.

# Methodology for Structural Analysis of Large-Span Caverns in Rock

## Sanja Zlatanic, Patrick H.C. Chan, Ruben Manuelyan
Parsons Brinckerhoff, New York, New York

**ABSTRACT:** The objective of this paper is to present general design criteria and a design methodology for structural analyses of large-span caverns in rock beneath dense urban environments. A design approach to structural analyses of the cavern liner and interior structures including non-linear finite element analyses incorporating ground-structure interaction is addressed. Structural design criteria including the rationale for determining cavern loads, load factors and load combinations are discussed. The application of the presented design approach for major caverns in New York metropolitan area resulted in practical solutions and economical designs.

## INTRODUCTION

As the initiatives for more efficient public transportation systems dominate the headlines and surface availability becomes scarce especially within context of major cities, design and construction of tunnel structures is receiving unprecedented attention from engineers of all disciplines. In particular, planning and designing complex underground structures housing large commuter railroad stations in form of large-span reinforced concrete caverns beneath dense urban environments provide a unique engineering challenge and require specialized methods amid the absence of a national code in tunnel design. For structural engineers, analysis of complex underground structures including large-span caverns in rock comes hand in hand with design, which starts with practical and constructible layout of all their components. Clearly stated methodology for structural analyses at very onset of the project is crucial and becomes instrumental for performing a consistent and economical structural design, which benefits the successful execution of the project. Since the methodology herein is presented for reinforced concrete caverns in rock constructed in US to house railway trains, references are made to the AREMA 2009 code with ACI 318-08 code as a supplement.

This paper begins with a discussion on the basis for structural design of cavern structures. Structural analyses considered use beam-spring method; design procedures are presented with illustration by a flow-chart, and include a brief discussion on finite element modeling followed by a detailed discussion on material and section properties of reinforced concrete liner. Various types of loads and their combinations to be considered in the analysis of cavern structures are presented, followed by a discussion on structural design of reinforced concrete liners and serviceability check.

## BEAM-SPRING VERSUS CONTINUUM ANALYSES

At first, it shall be understood that the structural methodology presented supplements the geotechnical analyses since, after all, the subject structures are situated in the ground and structure-ground interaction can best be simulated by continuum analyses. These are usually finite element, finite difference or discrete element analyses, which generally produce valuable results that primarily contribute understanding the behavior of both smaller and larger underground openings. Structural beam-spring analyses cannot and shall not replace the continuum analyses; moreover, the former could easily be misleading if not properly implemented or interpreted. It shall also be noted that geotechnical continuum analyses allow for a more realistic interaction between structural linings and rock mass than those using a beam-spring model. However, it is generally recognized that there is an issue regarding design of underground structures, especially as it relates to application of Load Factor Design method (which assures that all elements of the structure are properly dimensioned to account for type, duration and severity of the 'loads'). Also, structures design service life, their serviceability requirements, or special types of loadings such as fire or blast need to be provided for. Therefore, regarding design of major complex underground structures, conventional continuum models shall be supplemented with structural analyses, while still serving as a primary tool for evaluation of structural input parameters (such as magnitudes of tangential and radial spring stiffness used in the beam-spring analyses). Structural analyses using beam-springs shall be coordinated with continuum numerical models, particularly when it is suspected that the beam-spring liner models are providing unrealistic results.

## STRUCTURAL UNDERGROUND DESIGN CONSIDERATIONS

At the onset of any major construction project containing large cavern structures, it is customary to establish project-specific design criteria establishing ground rules for structural and geotechnical designs for all underground structures. The attempt to summarize those fundamental requirements is outlined below.

Generally, principles outlined in ACI, AISC, ITA, US Army Corp of Engineers EM 1110-2-2901 and AREMA (for railroad applications) are drawn upon in preparing the project-specific design criteria. There are no national codes specifically addressing tunnel and underground cavern design. For each project, a compilation of existing guidelines or standards is provided by following the applicable and widely recognized industry standards. This compilation usually takes a form of a project-specific design guidelines and criteria which shall be approved by the owner of the project. The guidelines shall (a) recognize unique owner requirements and requirements of other owners, agencies, or stakeholders the project is interfacing with, and (b) reconcile those requirements.

Performance and service life of the final lining, interior structures, services and utilities in tunnels and especially in deep station caverns are an important project consideration. Usually, a project goal is to provide owner and user of the facilities a safe and dry underground space for 100 to 120 years of its service life and reduce maintenance and operation costs. The right selection of construction materials with consideration of the underground environment, positive control of the water leakage into the structure and corrosion control measures are among the most important design considerations. Among the service requirements for underground cavern structures, the following are the most important.

### Fire Protection

Cavern structures should be protected in accordance with Fire/Life Safety design criteria so that integrity of the structure would not be compromised during a fire event. Project-specific fire curves or fire curves based upon the ITA Fire Guidelines are used. Integrity of structure should be checked taking into account degradation in material properties due to exposure to high temperature. For such extreme events, it is acceptable that the structure may require repair, but it is imperative that collapse or major structural failure be prevented. Explosive spalling of concrete and the formation of toxic fumes are not permitted. All materials should have a certified classification of non-combustibility.

### Watertightness

In terms of watertightness, the underground structures, especially passenger cavern stations, should be watertight as per defined criteria within the specified design service life, taking into account the proposed construction methods, relative movements across joints during the service life of the structures and the waterproofing methods specified. The project shall be to provide owner and users with a friendly, safe and dry underground space and reduce maintenance and operation costs.

## STRUCTURAL ANALYSES AND DESIGN PROCEDURES

### Application of Structural Analyses for Design of Underground Structures

During the recent decades, application of analytical methods based on the Theory of Elasticity, followed by the Finite Element method of analyzing structures and continua, together with advances in the science of rock mechanics, permitted the prediction of interaction between underground structures and the surrounding rock with increasing precision and confidence in structural integrity. This process, in turn made possible the realization of increasingly larger size structures in rock, and predominant among such structures are tunnels and stations of underground transportation networks which are inherent to densely populated metropolises. These applications often require the creation of controlled waterproofed environments, which is achieved by lining the tunnels and caverns with cast-in-place (CIP) concrete, and in special cases, by precast concrete segments or shotcrete. In current practice, concrete lining constitutes the permanent 'support' of the excavated spaces, ensuring long term durability of the internal structural components and the structure watertightness. It is constructed immediately following application of temporary support systems (rock bolting, shotcreting, etc.) whose sole purpose is to maintain stability of the excavated openings in the short term. Temporary support systems are not designed to ensure watertightness, and are usually considered to be subject to deterioration in the long term. Consequently, the waterproofed concrete lining constituting the permanent support system is required to withstand long term, short term and special loading conditions.

## Types of Analyses

In performing structural design of reinforced concrete cavern structures in rock, several types of analyses have to be considered:

- Analysis under normal conditions with permanent or frequently occurring loads such as water load and rock load
- Analysis under natural extreme events such as earthquake
- Analysis under human-caused extreme events such as blast and fire

The methodology for analysis in this paper is limited to the first type of loading conditions listed above. Analyses under extreme events are out of the scope of this paper.

Steps in structural analysis and design of reinforced concrete cavern structures under normal conditions can be summarized in a flowchart as shown in Figure 1. It has to be noted that liner thickness and reinforcement ratio of the cavern liner and interior structures should also be verified by further analyses for adequacy under earthquake, blast and fire conditions, each considered separately.

Once the layout, configuration and the minimum interior dimensions of the cavern have been determined in consideration with the surrounding geology, hydrology, constructability, functional requirements of the facility, and economy, an initial selection for the thickness of the liner and the size of interior structures is made based on preliminary computations. Preliminary section and material properties of structural components are then checked by geotechnical continuum analyses. Soil-structure interaction is considered in subsequent analyses through the use of non-linear ground springs (provided through the previous process of continuum analyses). After determining various loads and assembling load combinations, non-linear finite element structural analyses can be performed by means of finite element software. It is important to examine and evaluate results of the analyses to detect and eliminate numerical input and modeling inconsistencies, thereby avoiding unduly discrepancies when compared with geotechnical continuum analyses. Final output in the form of force and moment envelopes on the structures would be used for design purposes.

## FINITE ELEMENT MODELING

Usually, a typical section of the cavern structure can be identified as representative of cavern configuration along the longitudinal profile of the cavern structure and a 2-dimensional finite element model, as shown in Figure 2, can be developed containing all the significant components of the structure considered. Hence, it is essential during the initial analytical stage to determine key locations of interest and assemble a representative numerical model. Final design shall include number of 3D numerical models as well to account for three-dimensional features as tunnel/cavern intersections, excessive surcharge under limited rock cover, etc. Basic modeling principles, guidelines and considerations that should be taken into account during the structural modeling process are as follows:

- The liner and interior floor system can be modeled by using 2-node beam elements. Nodes of the beam elements are developed by using the centroidal axis of each structural component. The beam elements can be visualized as a series of chord members connecting all the nodes. Each node has 3 degrees-of-freedom.
- If beam elements are too long, fictitious moments may result. Conversely, beam elements that are too short may result in longer computation time. In general, a beam element length that approximates the liner or slab thickness will suffice and give reasonable results.
- There are certain nodes that are essential to be included in the modeling process so that force and moment demands from the analysis can be used directly in design. The nodes include those at the face of a support of a structural member (for flexural design) and those located an effective depth away from the face of a support ('effective depth' represents a distance from extreme compression fiber to centroid of longitudinal or vertical reinforcement). In addition, some other essential nodes include nodes at the crown and at the springline of the cavern.
- Rigid links are used to model monolithic connections between the liner walls and the interior floor system (slab/transverse beams), or corner joints between liner walls and invert slab (in the case of a 'tanked' structure designed under undrained conditions); therefore, there is no relative rotations expected between structural elements at these connections. The moment of inertia of the rigid links can be taken as 100 to 1,000 times of the moment of inertia of the adjoining structural components. This factor can vary and engineers should determine it by verifying that results of analysis have converged.
- Longitudinal width of the 2-dimensional finite element model is determined by considering the overall layout of the structural

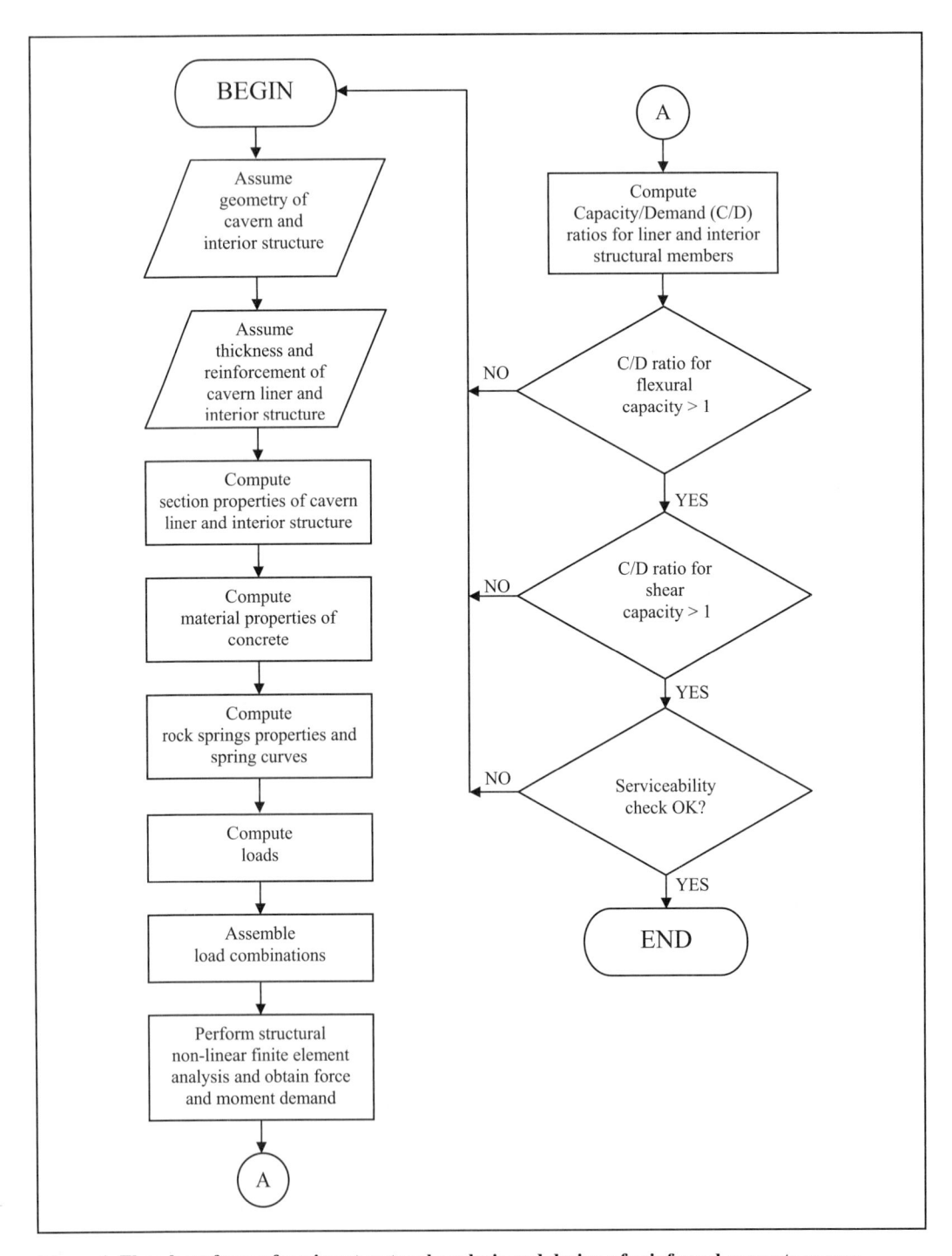

**Figure 1. Flowchart for performing structural analysis and design of reinforced concrete cavern structures in rock under normal conditions**

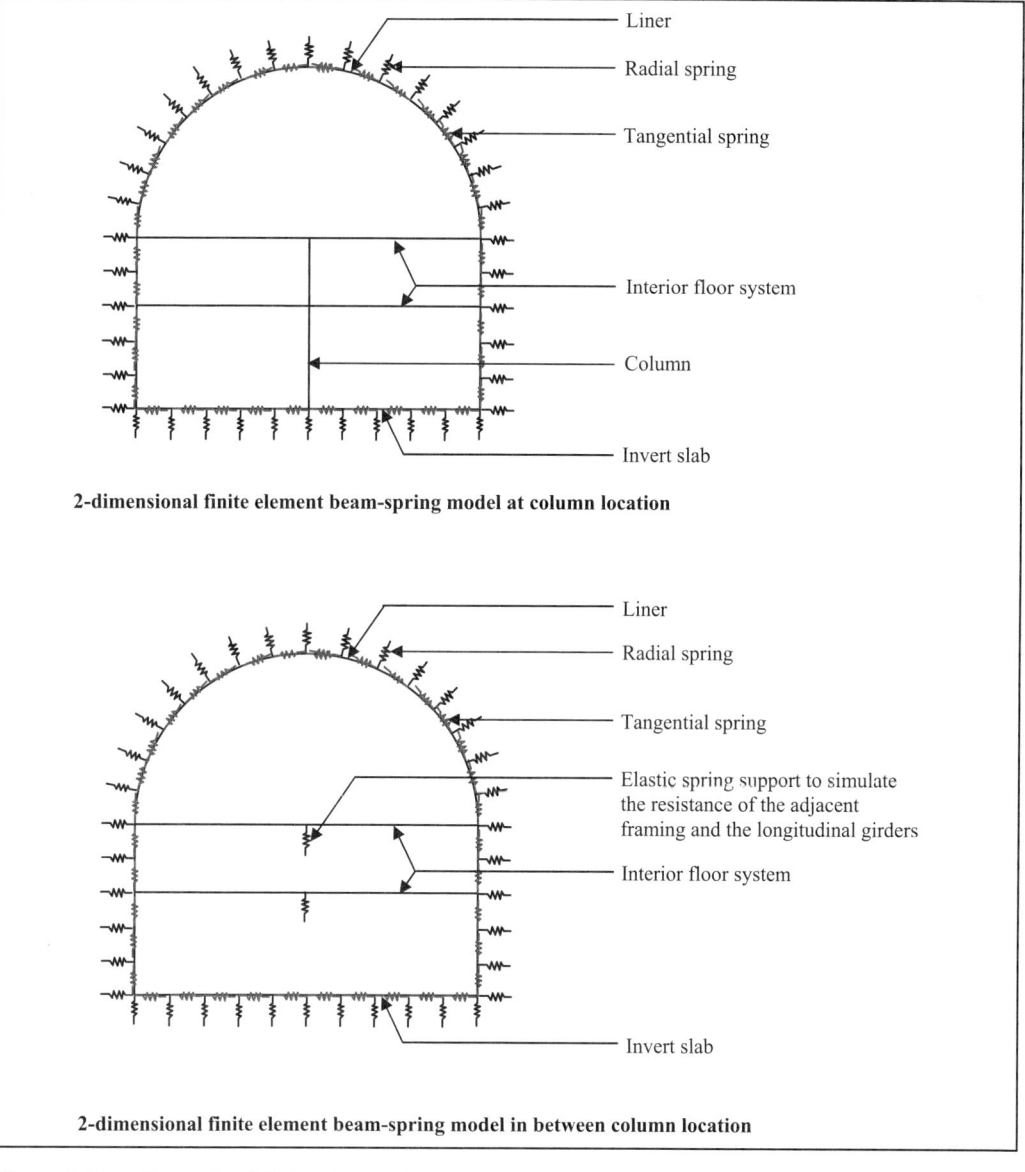

**Figure 2. Two-dimensional finite element beam-spring model**

framing system of interior structures. The spacing between the transverse beams within the representative portion of the structural cross-section can be taken as the width of the finite element model. Some analysts prefer to use a unit width model by converting section properties, stiffness and loads on all the structural components within a finite width of the cavern structure into a per unit width basis.

- If the interior floor system is supported on columns, a section of the cavern at the column location is modeled. The section of the cavern that is not directly supported on columns (i.e., between columns) is also considered by replacing the column supports with elastic spring supports to simulate the resistance of the adjacent framing and the longitudinal girders (See Figure 2).
- In order to simulate interaction between the liner and the rock, rock springs surrounding

the liner in radial direction are modeled by using 2-node spring members. Non-linear spring curves are used to simulate the non-linear stiffness characteristics of these radial springs. Similarly, tangential springs are also incorporated into the model by using 2-node spring members. Stiffness properties of rock springs and spring curves are presented later in the paper.

## MATERIAL PROPERTIES OF REINFORCED CONCRETE LINERS

Per AREMA 2009 Chapter 8 Section 2.23.4 and ACI 318-08 Section 8.5, modulus of elasticity of concrete, $E_c$ (in MPa or psi), is given by:

$$E_c \text{(in MPa)} = w_c^{1.5} \, 0.0043 \sqrt{f_c'}$$

$$E_c \text{(in psi)} = w_c^{1.5} \, 33 \sqrt{f_c'} \qquad (1)$$

where:
$w_c$ (in kg/m³ or pcf) = unit weight of concrete
$f_c'$ (in MPa or psi) = compressive strength of concrete

For normal weight concrete, we have:

$w_c$ = 2,300 kg/m³ (145 pcf)

$E_c$ (in MPa) = 4,700 $\sqrt{f_c'}$

$E_c$ (in psi) = 57,000 $\sqrt{f_c'}$ (2)

Per AREMA 2009 Chapter 8 Section 2.23.5, thermal coefficient for normal weight concrete can be taken as 0.0000105 per °K (or 0.000006 per °F).

In general, steel reinforcement should conform to the requirements of ASTM A615 Grade 60.

## SECTION PROPERTIES OF REINFORCED CONCRETE LINERS

Tension cracks regularly form in the concrete liner; their depth depends on the combination of moment and axial load (thrust) on the liner. Hence, the use of the moment of inertia of gross section of the liner, $I_g$, in computing liner section properties is likely overly conservative. ACI 318-08 Section 10.10.4.1 suggests a value of $0.7 I_g$ as the moment of inertia for columns or uncracked wall. It also suggests a value of $0.35 I_g$ as the moment of inertia for a cracked wall, which is defined as one that would crack in flexure (based on the modulus of rupture developed through initial analysis, with the wall moment of inertia equal to $0.7 I_g$). As ACI 318-08 Section 14.4 indicates that walls subjected to axial load or combined flexure and axial load shall be designed as compression members, the rational approach is to use the value of $0.7 I_g$ for the reinforced concrete liner in the Load Factor analysis.

During service load analysis performed to account for serviceability check, moment of inertia of a reinforced concrete liner should be representative of degree of cracking at the various service load levels investigated. In the Commentary of ACI 318-08 Section 10.10.4.1, it is suggested to use 1.43 times the moment of inertia used in Load Factor analysis. Therefore, for serviceability check, section properties of reinforced concrete liner should be computed based upon the value of $I_g$ ($1.43 \times 0.7 I_g = I_g$).

## PROPERTIES OF ROCK SPRINGS AND SPRING CURVES

Radial and tangential springs are considered around perimeter of cavern structure at the nodal points of the beam elements forming the concrete liner to simulate ground-structure interaction (refer to Figure 2).

The stiffness of radial and tangential springs ($k_r$ and $k_t$ respectively) is computed by multiplying the moduli of subgrade reaction ($K_r$ and $K_t$) with the tributary area at the corresponding node, $\Delta A$, as indicated in equations (3) and (4) below. Moduli of subgrade reaction can be obtained either by using equations in the US Army Corp of Engineers EM 1110-2-2901 that involve the properties of rock and geometry of cavern, or by performing a stiffness analysis on the rock mass using the geotechnical continuum model.

$$k_r = K_r \, (\Delta A) \qquad (3)$$
$$k_t = K_t \, (\Delta A) \qquad (4)$$

where:

$$\Delta A = \Delta L \, w \qquad (5)$$

$\Delta L$ = Tributary length at the node considered
$w$ = Longitudinal width of the 2-dimensional finite element model

Units of $K_r$ and $K_t$ are in (force/length³) whereas units of $k_r$ and $k_t$ are in (force/length). Figures 3 and 4 show the force-displacement curves for the radial and tangential springs.

In Figure 3, the slope of the force-displacement curve for radial springs in the compression region (3rd quadrant) is given by $k_r$. At the interface between liner and rock, no tension force can be developed on the radial springs. This is reflected by the infinitesimally small slope of the force-displacement curve for the radial springs in the tension region (1st quadrant). A very small slope is used instead of a theoretical zero slope to ensure numerical stability during the iterative process of non-linear analyses. The engineer

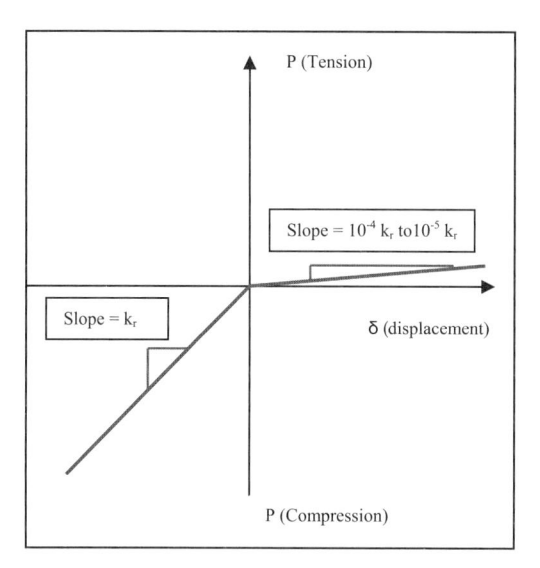

Figure 3. Force-displacement curve for radial springs

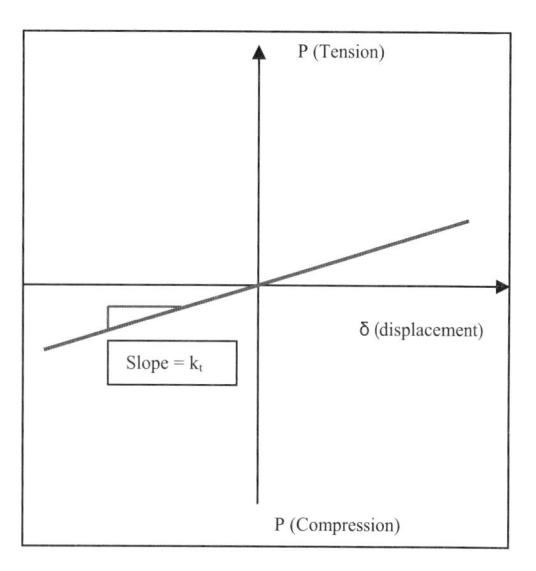

Figure 4. Force-displacement curve for tangential springs

should verify that tensile forces, if developed in the radial springs, should be close to zero.

In Figure 4, the slope of the force-displacement curve for tangential springs in both the compression and tension regions (3rd and 1st quadrants) is given by $k_t$. Unlike radial springs, tangential springs can act in both compression and tension, since the resisting tangential force from the rock can act in either direction. Effects of waterproofing membrane and compressibility of the drainage material on both radial and tangential springs shall be modeled and determined in advance through geotechnical continuum analyses.

## LOADS

### Dead Load (DL1)

Dead load consists of the self-weight of the concrete liner. Density of concrete and section area of the liner and interior floor system are used to compute dead load of the structure.

### Superimposed Dead Loads (DL2)

Superimposed dead loads consist of the weight of permanently installed trackwork, pipes, conduits, utilities, services, partitions, finishes, service walks and platforms, effects due to adjacent or overlying structures and all other permanent construction and fixtures. Superimposed dead loads can be applied as concentrated or/and uniform loads.

### Live Loads (LL)

These include uniform and/or concentrated live loads applied on pedestrian areas such as station platforms, stairways, pedestrian ramps, mezzanines, and station support service areas. In addition, loads at equipment rooms, storage spaces, escalator and passenger conveyors also have to be considered. Magnitude of these loads should be specified in the design criteria of the project.

### Train Load and Impact Load (TL+IM)

As indicated in AREMA 2009 Chapter 8 Section 2.2.3, for reinforced concrete structures, such as the interior floors of the cavern structure, if a project recommended train load per track is the Cooper E80, loading with axle loads and axle spacing are shown in Figure 5. If Cooper E60 loading is to be used per project design criteria, loads indicated below in Figure 4 should be multiplied by a factor of 0.75 (60/80 = 0.75).

Impact factors have to be applied to the train loads defined above. Theoretical computation of impact factors is complex, and is related to the speed of vehicle and the dynamic characteristics of structural members, which are function of their stiffness, mass and boundary conditions. Considering the structural floor system inside the cavern to be similar to a bridge superstructure, the formula provided by AREMA 2009 Chapter 8 Section 2.2.3 (d) can be used to compute the impact factor due to train movement.

In general, these impact factors result in a conservative estimation of train loads because of the

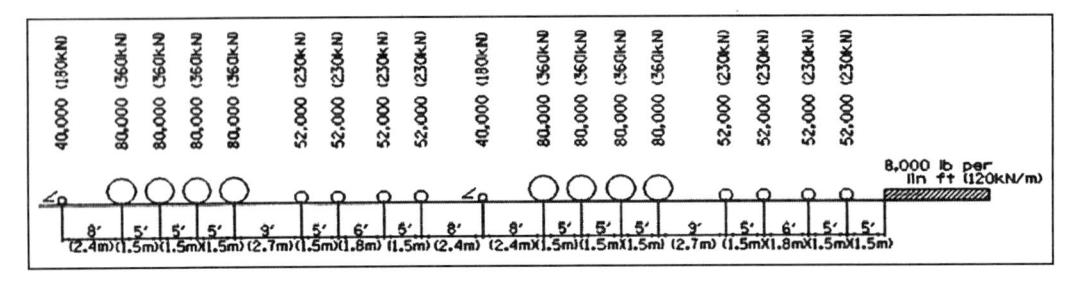

Figure 5. Cooper E80 axle load diagram (from AREMA Manual for Railway Engineering, 2009)

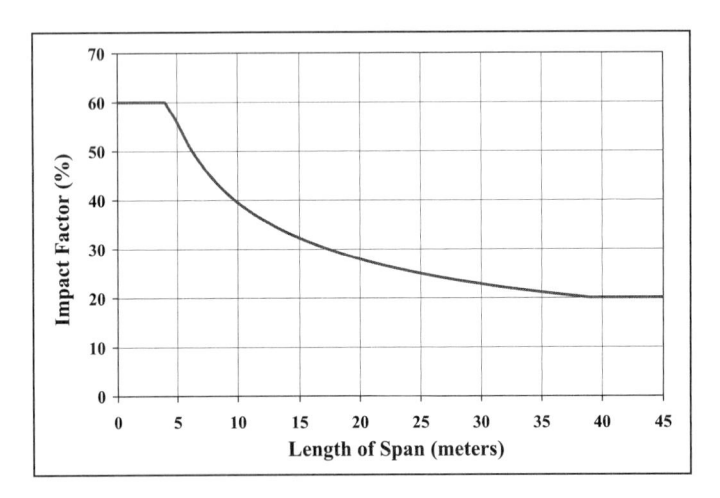

Figure 6. Impact factor versus length of span

relatively slow speed of train movement in a station. Lower values for impact may be used if speed limits for train movement can be assured operationally.

Impact force is applied at the top of rail and is as follows:

$$IM = TL \times I \qquad (6)$$

where:

I = Impact factor given in terms of percentage of the train load and is a function of the span length L (distance between supports on the transverse beams) in meters or feet
= 60% when $L \leq 4$ meters (14 feet)
= $125/\sqrt{L}$ (in %) when 4 meters $< L \leq$ 39 meters or $225/\sqrt{L}$ (in %) when 14 feet $< L \leq 127$ feet
= 20% when $L > 39$ meters (127 feet)

A plot of impact factor (I) versus length of span (L) is shown in Figure 6.

Where several tracks on a given floor are occupied by trains, consideration must be given to the constraints such as signalization limiting the number of moving trains, and impact load must be applied accordingly.

In calculating the maximum train loads on a structural member due to simultaneous loading on two or more tracks, operational characteristics of the station need to be considered; in general, the following proportions of the specified train load usually suffice:

- For two tracks—full train load
- For three tracks—full train load on two tracks and one-half on the other track
- For four tracks—full train load on two tracks, one-half on one track and one-fourth on the remaining track

Under certain circumstances, trains on some tracks are stationary and trains on other tsracks are moving when impact factor need to be applied. Coordination with systems engineers is essential to obtain an understanding of the positions of stationary and moving trains. The tracks selected for full live load are those tracks which will produce the most critical design condition.

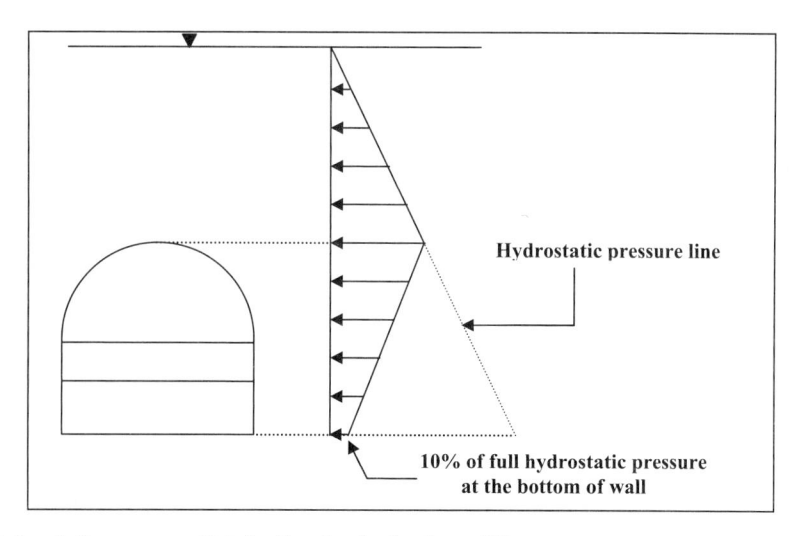

**Figure 7. Hydrostatic pressure distribution for drained condition**

### Water Load (WL)

Cavern structures, for transit applications, have to be waterproofed. Two waterproofing systems can be considered: a closed (undrained, tanked) system and an open (drained) system.

### *Drained System*

In an open (drained) system, hydrostatic pressure on the cavern structure is reduced by using a pressure relieving drainage system. Thus, more economical excavation outline and concrete liner can be designed. In general, the magnitude of reduction of hydrostatic pressures achievable by the pressure relief system is determined by hydraulic analysis. Alternatively, a simplified water load diagram can be used as shown in Figure 7. Full hydrostatic pressure is assumed on the crown, and decreases linearly to 10% of the full hydrostatic pressure at the bottom of wall.

An open (drained) system may not be appropriate and a closed (undrained) system may be warranted under the following conditions:

- When there is a potential to cause surface settlements by drawing down the groundwater. Ground water draw-down related settlements may occur when ground water elevation is above top of rock and within soil layer that may be prone to settlements due to dewatering.
- In high inflow and/or contaminated ground conditions, it is undesirable to collect and dispose of large groundwater volumes and/or contaminated groundwater. Also, moving the contaminated plume due to dewatering should be avoided. These scenarios may cause relatively high operation and maintenance cost due to special pumping and discharging requirements.

### *Undrained System*

In a closed (undrained) system, the waterproofing is applied around the entire cavern structure and the liners are designed for full hydrostatic pressures as indicated in Figure 8. Absent a pressure relieving system, cost saving can be achieved through the elimination of pressure relief system related maintenance and operating costs.

### Rock Load (RL)

Rock loads can be evaluated by empirical or analytical methods.

The empirical methods could be based on the Terzaghi's method, RSR-system, RMR-system, Q-system, and RMI-system which have their limitations (details of these methods are out of the scope of this paper).

Selection of analytical methods is based on characteristics of the rock mass, especially jointing, weathering, identification of the week zones, bedding, etc. A three-dimensional limit-equilibrium wedge stability analysis program such as UNWEDGE is often used for this purpose. Different combinations of rock wedges and blocks are evaluated to determine the most critical combination in ascertaining the rock 'load'. Rock bolts used to hold loose key blocks in place are usually assumed to corrode and become ineffective in the long term.

For purpose of structural analysis, two types of rock "loads" are considered:

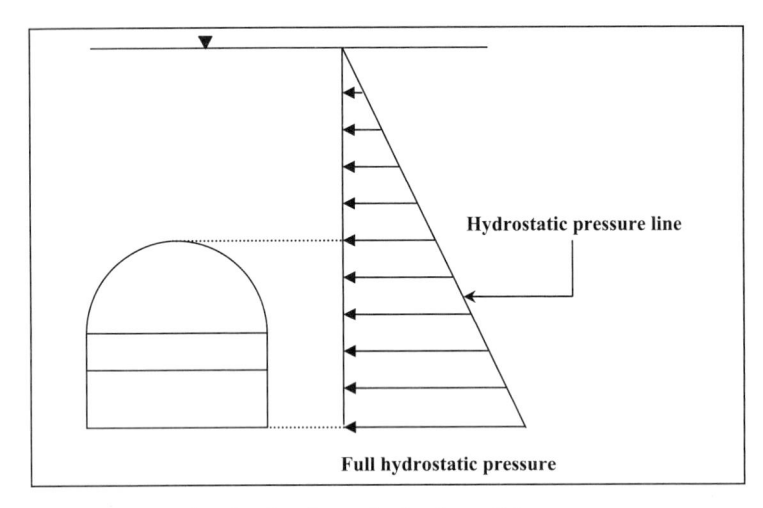

**Figure 8. Hydrostatic pressure distribution for undrained condition**

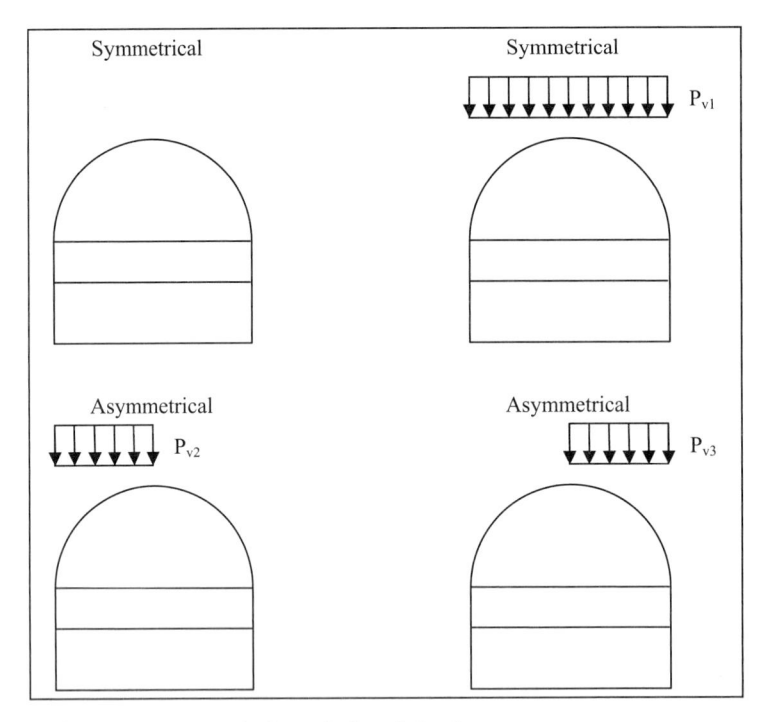

**Figure 9. Symmetrical and asymmetrical vertical rock loads**

- Long term rock load – computed in terms of effective stress and to be applied with water load
- Short term rock load – computed in terms of total stress and not to be applied with water load

Short term rock load corresponds to the scenario when the final liner is in place but a draw down of water level may occur (i.e., as a result of construction of adjacent structures).

Figures 9 and 10 indicate typical rock load patterns obtained from UNWEDGE analyses, which apply for both long term and short term conditions,

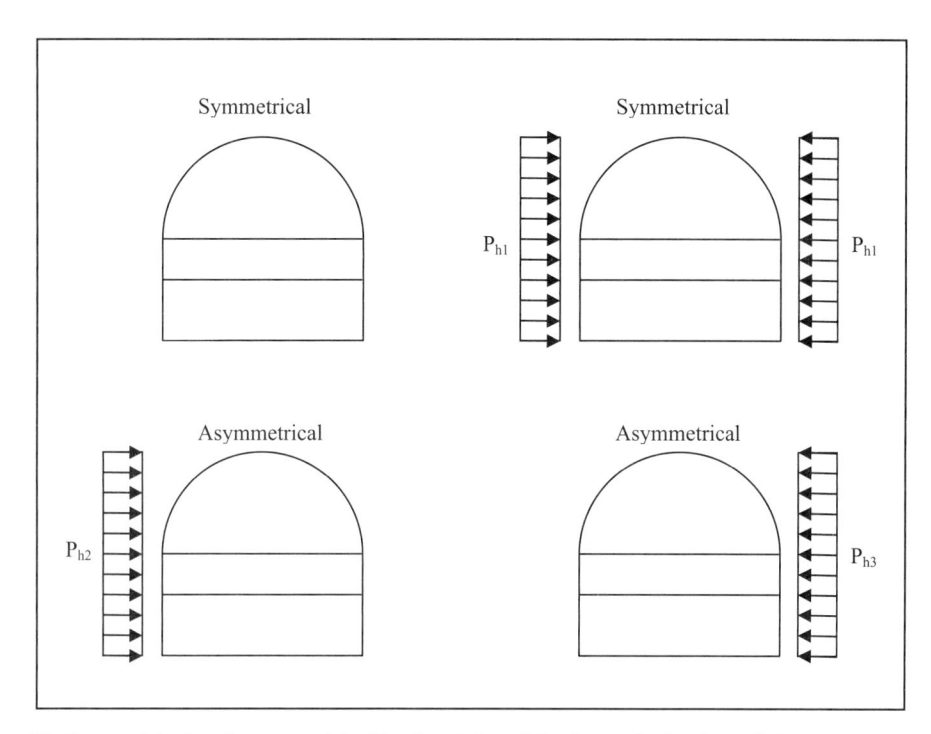

**Figure 10. Symmetrical and asymmetrical horizontal rock loads, undrained condition**

but with differences in the values of rock loads $P_{v1}$, $P_{v2}$, $P_{v3}$, $P_{h1}$, $P_{h2}$ and $P_{h3}$.

For both long term and short term rock loads, two independent sets are considered. The first set consists of vertical rock loads as shown in Figure 9. The second set consists of horizontal rock loads as shown in Figure 10.

In addition, in each of the two independent sets of loads (i.e., vertical and horizontal), usually four load patterns are possible (Symmetric, Asymmetric on the left side, Asymmetric on the right side and Nil).

Hence, because of the inherent uncertainty of location of the rock wedges, for each of the two types of rock loads (long term and short term), the number of possibilities of rock load patterns is 4×4=16. Nevertheless, among the 16 rock load patterns, there is one load pattern in which both the vertical and horizontal rock loads are nil. Hence, the total number of rock load patterns is 15 for each of the two types of rock loads (long term and short term).

Figures 11 and 12 show different possible patterns of rock loads on the cavern that are considered for the case of long term rock load (Load Cases RL101 to RL115) and short term rock load (Load Cases RL201 to RL215) respectively. It must be noted that values of rock loads $P_{v1}$, $P_{v2}$, $P_{v3}$, $P_{h1}$, $P_{h2}$ and $P_{h3}$ in Figure 11 (computed in terms of effective stress) are different from those in Figure 12 (computed in terms of total stress).

### Temperature Load (T)

The temperature inside a cavern structure may vary during its service life. In coordination with ventilation engineer the cavern designs temperatures are determined and a heat transfer analysis (to compute temperatures on the interior and exterior surfaces of the concrete liner) is performed. Stresses resulting from thermal loads are incorporated into design.

Two methods can be used to obtain the temperature distribution across the thickness of a concrete liner for two types of solutions as described below.

#### *Transient Solution*

This is a more complex analysis and will not be discussed in particular in this paper. In general, thermal properties of concrete and the surrounding medium have to be input into heat transfer analysis software. The reinforced concrete liner is divided into finite elements. At each time-step of the time-history analysis, a number of iterations are specified to ensure numerical convergence of the model. Temperature distributions can be obtained at different time steps.

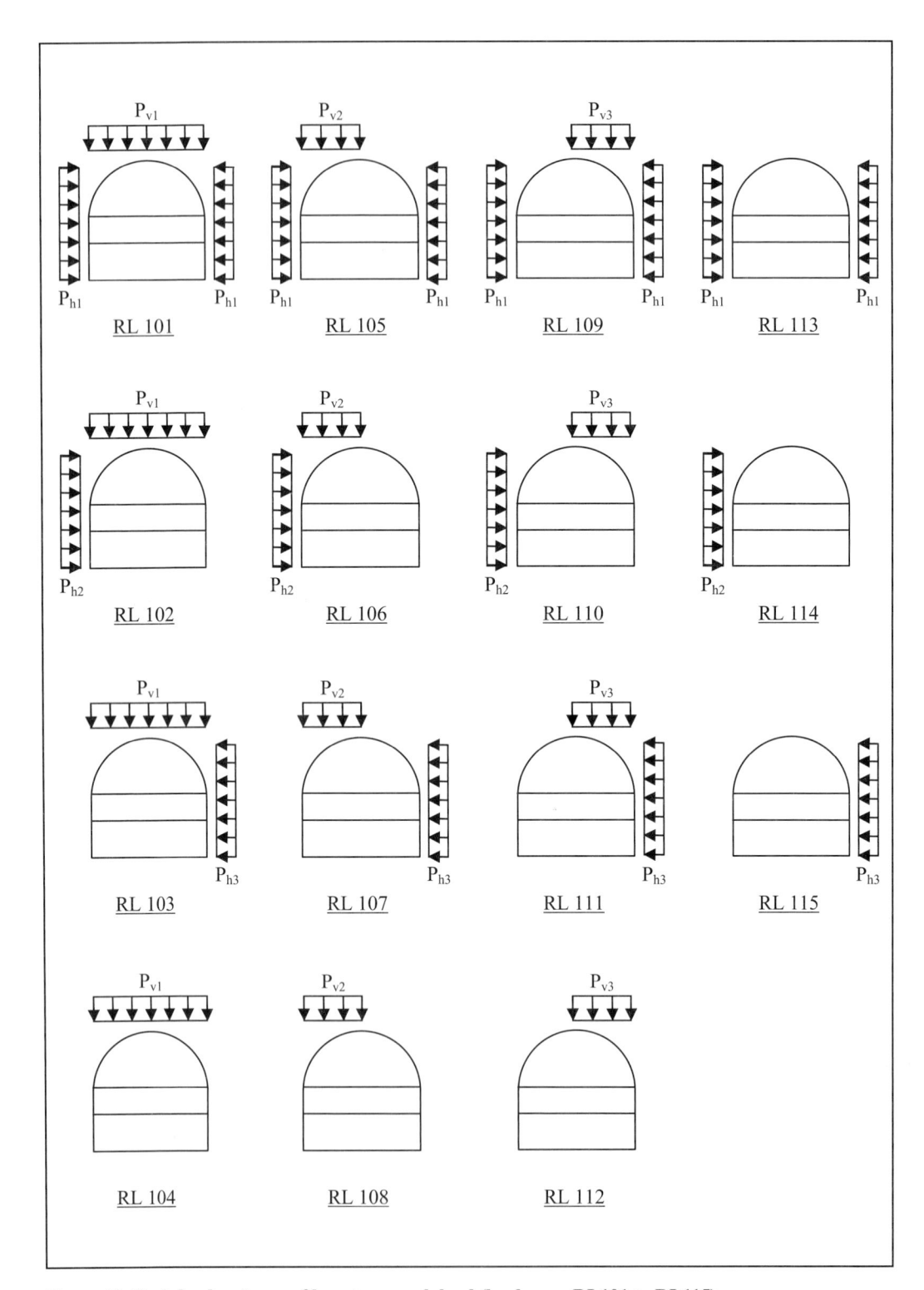

**Figure 11. Rock load patterns of long term rock load (load cases RL101 to RL115)**

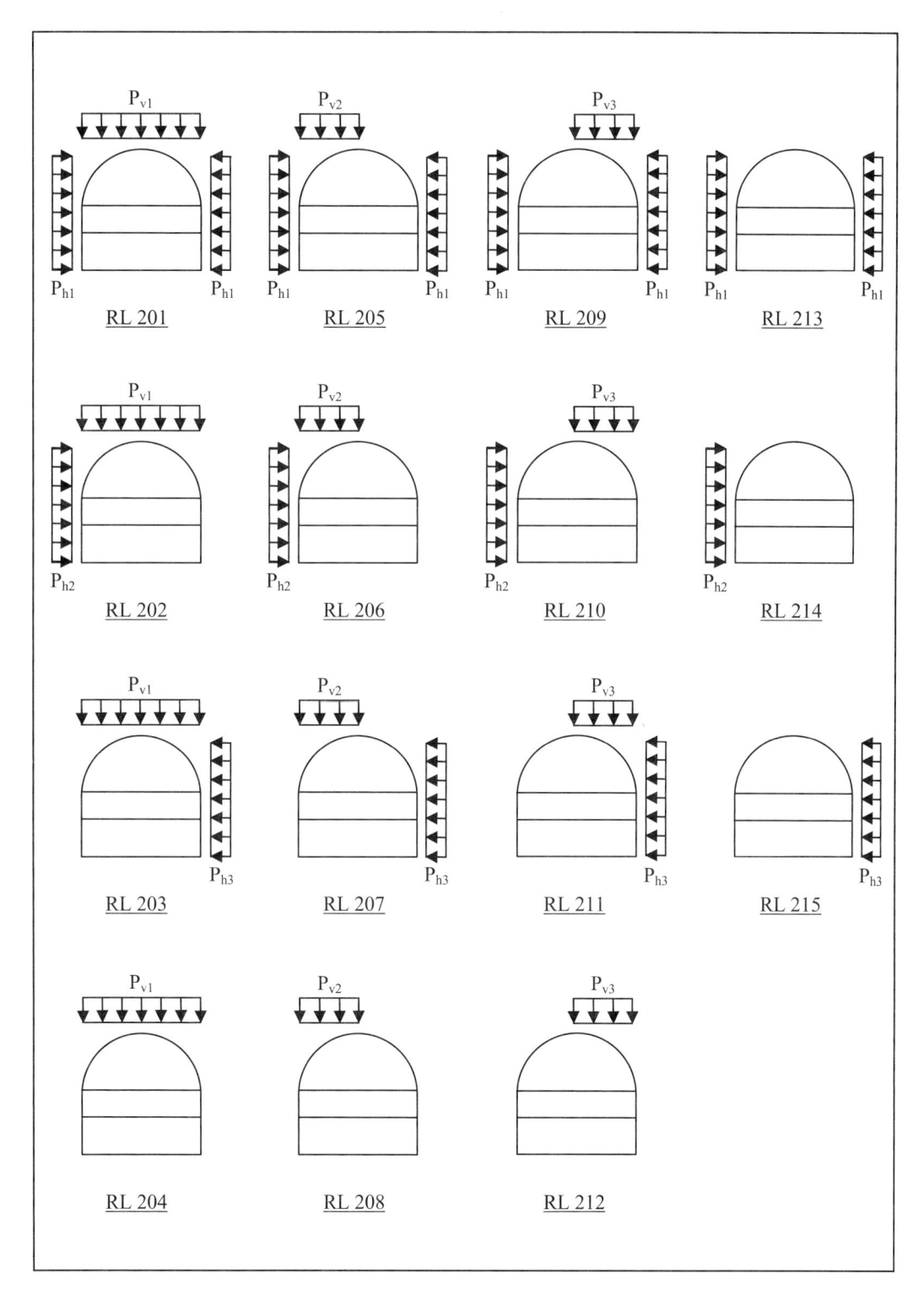

Figure 12. Rock load patterns of short term rock load (load cases RL201 to RL215)

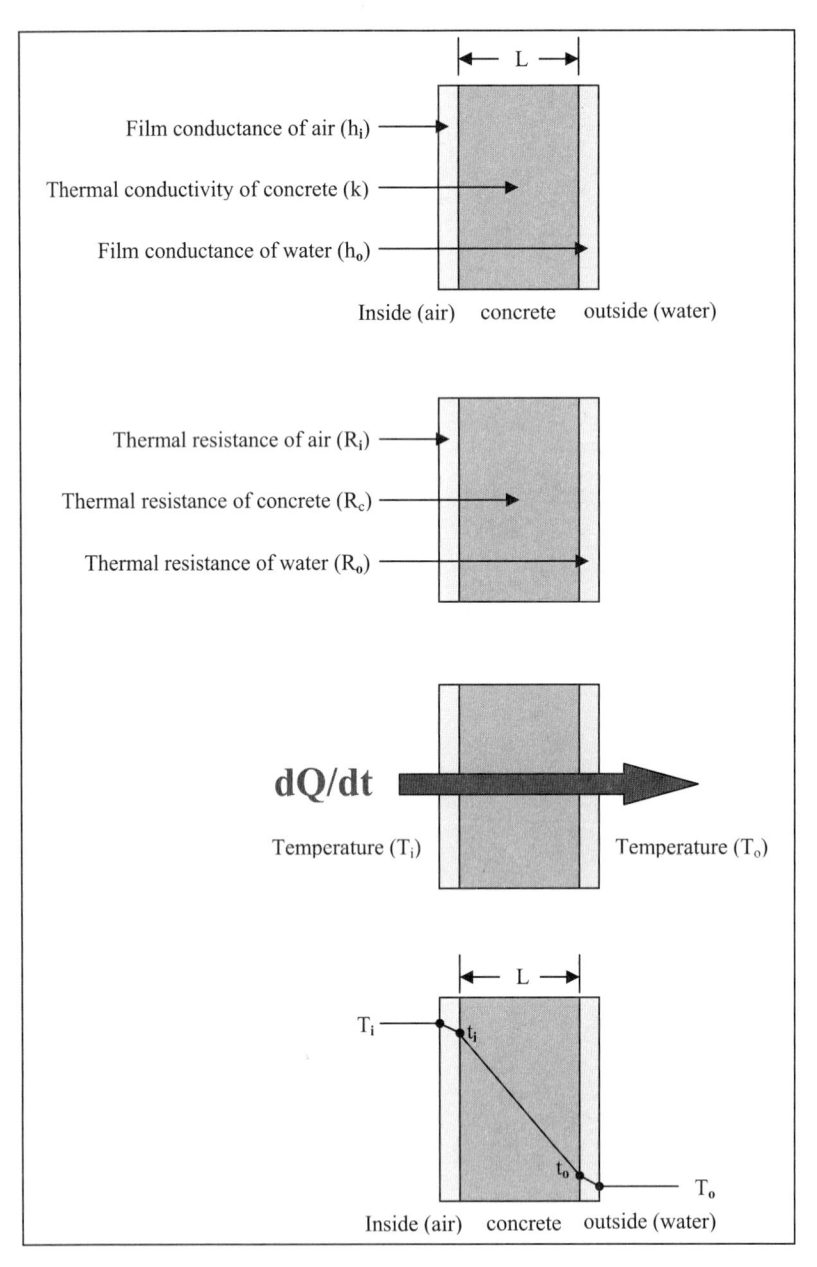

**Figure 13. Heat transfer analysis for steady state solution**

*Steady-state Solution*

A more practical (although more conservative) methodology is to consider the steady-state solution. Methodology of the heat transfer analysis (summer conditions) is presented below. Note that methodology for the winter conditions is similar. (Refer to Figure 13)

During summer condition, assume:

$T_i$ (in °K or °F) = Temperature inside the cavern structure (i.e., air temperature)

$T_o$ (in °K or °F) = Temperature outside the cavern structure (i.e., rock temperature)

$h_i$ (in W/m$^2$ °K or Btu/hr ft$^2$ °F) = Film conductance of air

k (in W-cm/m² °K or Btu-in/hr ft² °F ) = Thermal conductivity of concrete
h_o (in W/m² °K or Btu/hr ft² °F) = Film conductance of water

Therefore, thermal resistances of air, concrete and water over an exposed area A (in m² or ft²) can be given as below:

$R_i$ (in °K/W or hr °F/ Btu) =
Thermal resistance of air = $1/h_i A$ (7)

$R_c$ (in °K/W or hr °F/ Btu) =
Thermal resistance of concrete = $L/kA$ (8)

$R_o$ (in °K/W or hr °F/ Btu) =
Thermal resistance of water = $1/h_o A$ (9)

where:

L (in cm or inch) = Thickness of concrete liner (waterproofing layer ignored)

The total thermal resistance, $\Sigma R$, will be the sum of thermal resistances of air, concrete, and water and is given by:

$$\Sigma R = R_i + R_c + R_o \quad (10)$$

Given the difference between the temperature inside the cavern and temperature outside the cavern as $\Delta T$, the heat flux, dQ/dt, flowing across the concrete liner over an exposed area A can be computed by relating $\Delta T$ and the total thermal resistances, $\Sigma R$, as:

$$dQ/dt = \Delta T/\Sigma R = (T_i - T_o)/\Sigma R \quad (11)$$

Using heat flux and thermal resistance of each layer, the temperature on the interior face of concrete, $t_i$, and the exterior face of concrete, $t_o$, can be computed as:

$$t_i = T_i - (dQ/dt)(R_i) \quad (12)$$

$$t_o = T_i - (dQ/dt)(R_i + R_c) \quad (13)$$

If the base construction temperature, $T_c$, is assumed as the temperature of rock at the time of pouring the concrete liner, then the difference between the temperature on the interior surface of the concrete and the base construction temperature ($\Delta t_i$), and the difference between the temperature on the exterior surface of concrete and the base construction temperature ($\Delta t_o$) should be used to calculate thermal stresses, and thereby the forces and moment demand on the concrete liner. Therefore:

$$\Delta t_i = t_i - T_c \quad (14)$$

$$\Delta t_o = t_o - T_c \quad (15)$$

It is useful to observe the analogy between the following two equations:

Heat Flux (dQ/dt) = Temperature Differential ($\Delta T$)/Thermal Resistances ($\Sigma R$) (16)

Electrical Current (I) = Voltage Drop ($\Delta V$)/Electrical Resistance ($\Sigma R$) (17)

In a series of several electrical resistances, a larger resistance causes a larger voltage drop. Similarly, in the heat transfer through a series of materials, temperature drop occurs across the material with higher thermal resistance (lower conductivity).

## LOAD COMBINATIONS

Load combinations based on dead load, superimposed dead loads, live loads, train load, impact load, water load, rock load and temperature load are shown below for Load Factor analysis and design of reinforced concrete liner using AREMA 2009 as a reference. For serviceability check, load factors for all the load combinations shown below are taken as 1.0. See Table 1.

Several issues regarding the use of the load combinations shown in Table 1 are noted:

- A load factor of 1.4 for water load is suggested in AREMA 2009, which is prudent under the drained condition due to reliance of such system on regular maintenance. However, when cavern structure is designed under undrained condition, a lower load factor of 1.2 is suggested since hydrostatic pressure acting on the liner is more predictable.
- Despite the fact that the load combinations that have a load factor of 0.9 for dead load and superimposed dead load (as per ACI318-08) are not addressed in AREMA 2009, these load combinations are recommended to be included. Because they may result in a minimum axial load on the liner.

## STRUCTURAL DESIGN OF REINFORCED CONCRETE LINERS

Based upon the load combinations described in the previous section, non-linear finite element analyses are performed using the analytical model. The liner and interior structures are subjected to combination of axial forces and bending moments. It is a good practice to divide the axial force and moment demand from the analysis by the longitudinal width w of the model to obtain axial force, $P_u$, and moment, $M_u$, demand in a per unit foot basis.

**Table 1. Load combinations**

| | | | | | |
|---|---|---|---|---|---|
| (1) | 1.4 (DL1+DL2) | | | + 1.4 (or 1.2) WL | |
| (2) | 1.4 (DL1+DL2) + 1.4 T | | | + 1.4 (or 1.2) WL | |
| (3) | 1.4 (DL1+DL2) | | | + 1.4 (or 1.2) WL | + 1.4 RL' |
| (4) | 1.4 (DL1+DL2) + 1.4 T | | | + 1.4 (or 1.2) WL | + 1.4 RL' |
| (5) | 1.4 (DL1+DL2) | + 2.33 (TL+IM) | + 1.6 LL | + 1.4 (or 1.2) WL | |
| (6) | 1.4 (DL1+DL2) + 1.4 T + 1.4 (TL+IM) | | + 1.6 LL | + 1.4 (or 1.2) WL | |
| (7) | 1.4 (DL1+DL2) | + 2.33 (TL+IM) | + 1.6 LL | + 1.4 (or 1.2) WL | + 1.4 RL' |
| (8) | 1.4 (DL1+DL2) + 1.4 T + 1.4 (TL+IM) | | + 1.6 LL | + 1.4 (or 1.2) WL | + 1.4 RL' |
| (9) | 1.4 (DL1+DL2) | | | | |
| (10) | 1.4 (DL1+DL2) + 1.4 T | | | | |
| (11) | 1.4 (DL1+DL2) | + | | 1.4 RL | |
| (12) | 1.4 (DL1+DL2) + 1.4 T | | | | + 1.4 RL |
| (13) | 1.4 (DL1+DL2) | + 2.33 (TL+IM) | + 1.6 LL | | |
| (14) | 1.4 (DL1+DL2) + 1.4 T + 1.4 (TL+IM) | | + 1.6 LL | | |
| (15) | 1.4 (DL1+DL2) | + 2.33 (TL+IM) | + 1.6 LL | | + 1.4 RL |
| (16) | 1.4 (DL1+DL2) + 1.4 T + 1.4 (TL+IM) | | + 1.6 LL | | + 1.4 RL |
| (17) | 0.9 (DL1+DL2) | | | + 1.4 (or 1.2) WL | |
| (18) | 0.9 (DL1+DL2) | | | + 1.4 (or 1.2) WL | + 1.4 RL' |
| (19) | 0.9 (DL1+DL2) | | | | |
| (20) | 0.9 (DL1+DL2) | | | | + 1.4 RL |
| (21) | 0.9 DL1 | | | + 1.4 (or 1.2) WL | |
| (22) | 0.9 DL1 | | | + 1.4 (or 1.2) WL | + 1.4 RL' |
| (23) | 0.9 DL1 | | | | |
| (24) | 0.9 DL1 | | | | + 1.4 RL |

DL1 = Self-weight of concrete lining
DL2 = Superimposed dead loads
LL = Live loads
TL = Train load
IM = Impact load due to TL
WL = Water load
RL = Rock load (computed in terms of total stress, not acting with water load)
RL' = Rock load (computed in terms of effective stress, acting with water load)
T = Temperature load

Figure 14 shows a typical interaction diagram for a reinforced concrete section. The exterior curve is the nominal curve which is developed based upon nominal capacity of the section. Point A represents pure axial compression. Point B corresponds to a strain distribution with a maximum compression strain 0.003 on one side of the section and a tension strain of $\varepsilon_y$ (the yielding strain of the reinforcement), at the level of the tension steel. This represents a balanced failure in which crushing of the concrete happens simultaneously with yielding of the tension steel. Point C represents pure bending. It must be noted that for all the points lying on the nominal curve, compression strain 0.003 occurs on one side of the section.

The design curve is developed by applying a strength reduction factor $\phi$ to the nominal curve. Per AREMA Chapter 8 Section 2.30.2, strength reduction factors are given as below.

For flexure, $\phi = 0.90$
For tied reinforced compression members with or without flexure, $\phi = 0.70$

Axial force, $P_u$, and moment, $M_u$, are input into a software program, such as pcaColumn, which develops the interaction diagram for reinforced concrete sections, corresponding to the interior curve in Figure 14. The interaction diagram displays the envelope of acceptable combinations of bending moment and axial force, for a given cross sectional area, reinforcement area, material properties of concrete and reinforcement and strength reduction factor. Each pair of axial force $P_u$ and moment $M_u$ constitutes a point to be plotted in the interaction diagram computed by the pcaColumn program. If the point lies within the interaction envelope, the Capacity/Demand (C/D) ratio is larger than 1, which means the section and reinforcement are acceptable. If the point falls outside the interaction envelope, the

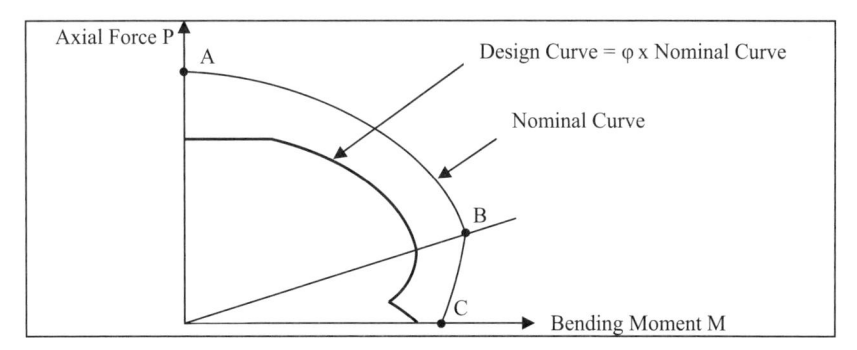

**Figure 14.** Interaction diagram

C/D ratio is smaller than 1, the section and reinforcement are unacceptable and need to be revised.

Shear capacity of concrete liner is computed based on AREMA 2009 Chapter 8 Section 2.35 that takes into account the effect of axial compression, axial tension on ascertaining shear capacity of concrete. If the Capacity/Demand (C/D) ratio is larger than 1, the section is acceptable. Strength reduction factor, φ, for shear is 0.85 per AREMA 2009 Chapter 8 Section 2.30.2.

## SERVICEABILITY CHECK

Per AREMA 2009 Section 2.37, requirements for distribution of reinforcement (crack control), deflection and fatigue need to be considered to account for serviceability check. Procedures for checking serviceability requirements for deflection and fatigue control per AREMA 2009 and ACI 318-08 codes can be applied on interior reinforced concrete floor structures, and will not be presented in this paper. Serviceability aspects of concrete liner need to be addressed separately to satisfy structures functional and aesthetic (where necessary) requirements.

As for crack control, tension cracks can form in the liner as well as interior floor structures. The requirement for distribution of reinforcement for crack control has to be satisfied by both liner and interior floor structures. The requirement per AREMA 2009 Chapter 8 Section 2.39 is similar to the requirement per ACI 318 (1995 edition), in which a limit is put on the value of Z where Z is given by:

$$Z = f_s (d_c A)^{1/3} < 30 \text{ kN/mm (170 k/in)} \quad (18)$$

where:
$f_s$ (in MPa or ksi) = Tensile stress in tension reinforcement at service load and can not be greater than $0.5 f_y$
$d_c$ (in mm or inch) = Distance from the extreme tension fiber to the center of the reinforcement located closest to it

A (in mm$^2$ or inch$^2$) = Effective tension area of concrete surrounding the tension reinforcement, and having the same centroid as that reinforcement, divided by the number of bars

The limiting value of Z (30 kN/mm or 170 k/in) corresponds to structural members in moderate exposure conditions since the cavern liner is waterproofed.

It should be noted that the ACI formulation subsequent to its 1995 edition regarding rebar distribution have not been incorporated into AREMA.

## CONCLUSIONS

Amid the lack of a national code for tunnel design, this paper attempts to provide a systematic approach for structural analyses of large openings in rock utilizing the available analytical tools while following referenced guidelines. The structural methodology presented supplements the geotechnical analyses and it should be carefully executed since it could be misleading if not properly implemented and interpreted. Structural analyses using beam-springs shall be coordinated with continuum numerical models, particularly when it is suspected that the beam-spring liner models are providing unrealistic results.

It is generally recognized that there is an issue regarding design of underground structures, especially as it relates to application of Load Factor Design method. This method assures that all elements of the structure are properly dimensioned to account for type, duration and severity of the 'loads' and cannot be executed in context of geotechnical continuum analyses. Also, structures design service life, their serviceability requirements, or special type of loading such as fire or blast need to be provided for. Therefore, regarding design of complex underground structures, conventional continuum models shall be supplemented with structural analyses, while still serving as a primary tool for evaluation of structural input parameters.

**REFERENCES**

*AISC Steel Construction Manual Thirteenth Edition.* American Institute of Steel Construction Inc.

*AREMA Manual for Railway Engineering 2009.* American Railway Engineering and Maintenance-of-Way Association.

*Building Code Requirements for Structural Concrete (ACI 318-08) and Commentary.* American Concrete Institute.

*EM 1110-2-2901 Tunnels and Shafts in Rock.* Department of the Army, US Army Corp of Engineers.

*Guidelines for Structural Fire Resistance for Road Tunnels.* International Tunneling Association.

*pcaColumn v3.64 Design and Investigation of Reinforced Concrete Column Sections.* Portland Cement Association.

*UNWEDGE Key Block Analysis Program v3.0.* RocScience, Inc.

# TRACK 2: DESIGN

Matt Fowler, Chair

*Session 5: Design Optimization and Alignment Selection*

# Factors Influencing Tunnel Design in Mass Transit Applications

## Irwan Halim, Marci Benson
URS Corporation, Boston, Massachusetts

**ABSTRACT:** Given the choice, tunnel engineers would rather design tunnels which are in the best subsurface condition, with the shortest alignment, and smallest geometry to save cost. However in mass transit applications, there are many other design factors than geotechnical, including right-of-ways, impacts to adjoining properties, operation and maintenance, and passenger experience. This paper uses as an example the preliminary design of the Silver Line tunnel project in Boston, Massachusetts. Compromise must often be reached amongst sometimes conflicting factors to reach design conclusions. However it is the tunnel engineer's responsibility to ensure a project that is buildable with an acceptable risk level.

## INTRODUCTION

Nowadays, new tunnel projects are typically performed to either build underground connections between two or more well established points, or underground extensions from existing points to some future developments. Water/wastewater tunnels are being built either as conveyance or storage to expand and connect to the existing facilities. New sanitary sewer tunnels sometimes need to be built to accommodate future population growth and development around the existing ones. Transit tunnels are being built either as capacity enhancement and improvement to the existing systems, or as new connection between the existing systems. Many of these tunnels are necessitated by the urban development in the area, some of which have to be built in the most congested part of major cities and towns.

Some of the most important aspects of tunnel design are establishing horizontal and vertical alignments. In order to save costs, the tunnel would need to be built in the best geotechnical condition possible, with the shortest distance between the connecting points. The tunnel geometries are the smallest possible and optimally shaped to suit the in-situ conditions. This is the main reason why many of the large diameter water conveyance/storage tunnels have been built deeper in hard rock, rather than in soft ground soils. Their alignments were often tweaked to avoid potentially bad ground. However in mass transit applications, due to the much closer passenger interface, there are other important factors to be considered such as the system operational attributes and passenger stations experience. These tunnels are usually built in congested urbanized areas, therefore their construction impacts on public facilities and adjoining properties would need to be addressed. Finally, the construction cost would have to be weighed against the potential user's benefit from the project. For instance, a tunnel alternative with the higher construction cost would not necessarily be a bad choice if it can bring in much more in user's benefit. This paper will describe the tunnel design process for a mass transit application that considers the various aspects described above, using the preliminary design of the Silver Line Phase III Bus Rapid Transit (BRT) Tunnel Project in Boston, Massachusetts (Yates, Ainsley, and Gallagher 2009) as an example.

The Silver Line Phase III segment is approximately 1 mile, and runs through one of Boston's oldest and most congested downtown areas. It is flanked on both sides by many historic buildings and landmarks. It is the final underground segment linking the existing Phase I and II Services, resulting in a continuous line from Dudley Square to Logan Airport and South Boston as shown in Figure 1. The interim preliminary engineering (30%) design of the project was completed in October 2008, with the project alignment shown on Figure 2. This alignment consisted of twin "stacked tunnels" under Essex and Boylston Streets to limit the tunnels within the narrow public right-of-ways, and two cut-and-cover stations connecting to the existing Orange Line at Chinatown Station and the existing Green Line at Boylston Station. In order to do this, the project alignment had to go under a major sewer interceptor pipe in South Street (the 72-inch NESI), a major interstate highway tunnel (I-93 South/CASB), and the existing subway Orange Line at Chinatown and Green Line at Boylston (see Figure 3).

All the mined tunnels are to be constructed using the Sequential Excavation Method (SEM or

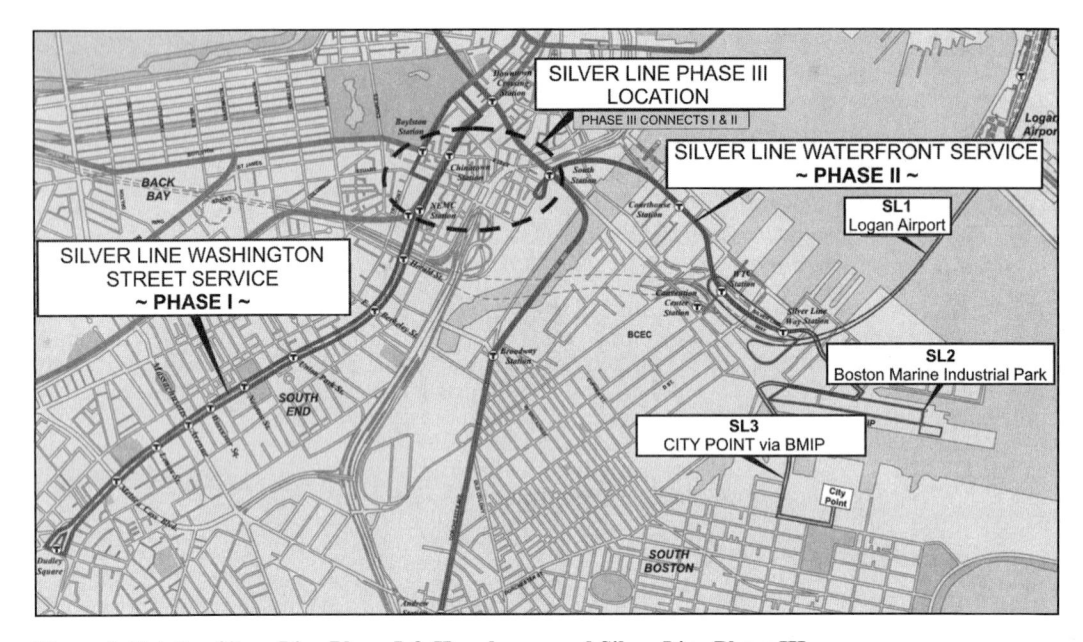

Figure 1. Existing Silver Line Phase I & II and proposed Silver Line Phase III

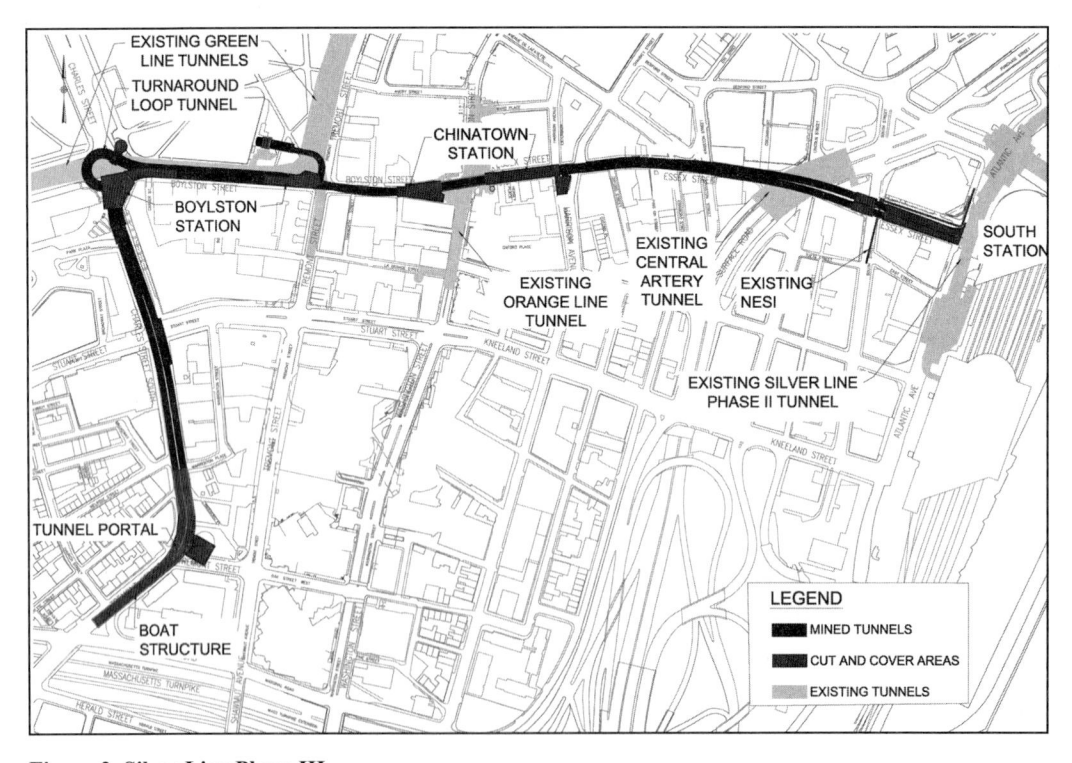

Figure 2. Silver Line Phase III

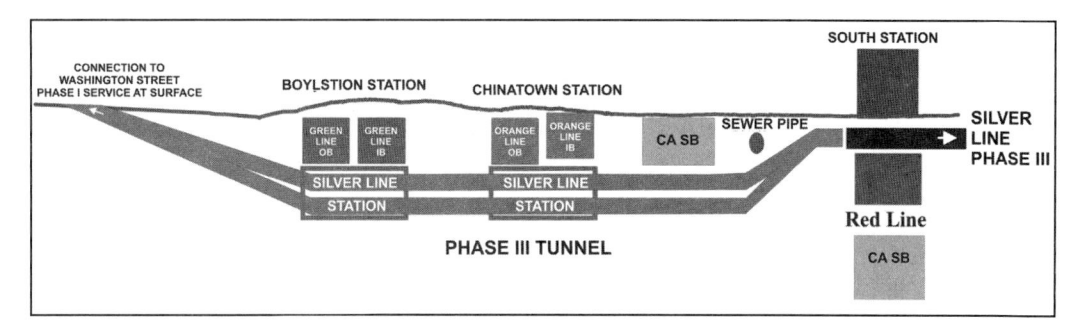

**Figure 3. Silver Line Phase III alignment sketch**

NATM) and vary in size, depending on their function either as running tunnel, blister tunnel, or station platform tunnel. Shotcrete will be used as the initial liner, with a waterproofing membrane and a cast-in-place final liner to complete the tunnel structure. The cut-and-cover stations are to be constructed using slurry walls serving as excavation support and permanent structural walls. An interior cast-in-place wall will be poured against the slurry wall with a waterproofing membrane placed in-between the walls. Detail descriptions of the NATM tunnel design approach for the project are given in Nasri, Vrouvlianis, and Halim 2009.

In November 2008 a value engineering study was conducted, that resulted in the re-evaluations of alternative alignments and configurations. Among the alternatives, a single side by side "binocular" tunnel rather than "stacked" tunnels, and a single station (mid-block) connection to both the Orange and Green Lines rather than the two separate stations, were considered as potentially cost-saving (URS – DMJM/Harris Joint Venture 2009). This paper presents evaluations of the different non-tunneling factors during examination of these alternatives, and how they impacted the tunneling conditions and support design in general. The decision making process in coming up with the most optimum alternative tunnel/station configuration will be summarized as a conclusion.

**GEOLOGICAL AND TUNNEL SUPPORT CONSIDERATIONS**

The subsurface profile along the project alignment is shown on Figure 4. Starting at the ground surface, the profile consists of the following layers: fill, organic soil (in limited areas), marine clay and sand, glacial till, and bedrock. Marine clay and sand layers were typically encountered beneath the fill and/or organic soil (when present). This layer ranging from about 11 feet to 117 feet thick, generally became thinner as the tunnel alignment progressed from its starting point at Marginal Road to its end point at Atlantic Avenue. In general, the individual sub-layers of clay and sand were discrete and relatively homogeneous. However, finely stratified clay and sand, consisting mostly of clay but with sand layers a few millimeters thick up to several inches thick, were encountered in several areas.

Glacial till was encountered beneath the marine sand and clay, and overlying bedrock, at most locations. The maximum till thickness was about 48 feet, and generally the top of the till layer rose in elevation, and the till thickness increased, as the tunnel alignment progressed from its starting point at Marginal Road to its end at Atlantic Avenue. The depth to bedrock ranging from 72 to 134 feet, generally decreased as the tunnel alignment progressed from its starting point at Marginal Road to its end point at Atlantic Avenue. Except for the short section near the tunnel portal on the Charles Street South end of the alignment, the remaining tunnels are completely below the groundwater table.

The currently stacked tunnel and alternate binocular tunnel alignments are also shown on the subsurface profile in Figure 4. In order to maintain the necessary clearance between the two tunnels and also to the existing underground structures, the overall depth from the ground surface to the lowest tunnel roadway grade and station platform is approximately 100 feet in the stacked configuration. By constructing the tunnels side-by-side in a binocular configuration, the tunnels and stations would not have to be as deep, with the platform elevation approximately 80 feet below the ground surface.

The mined tunnels under both the stacked and binocular alignments will mainly be excavated in the glacial tills, marine clay and sand, or stratified clay and sand. A small portion of the inbound (lower) tunnel under the stacked alignment will be excavated in the argillite bedrock. The different subsurface profiles will determine the tunneling ground condition or behavior, excavation sequence, and initial support required to maintain the tunnel stability. Since the glacial till and bedrock are relatively compact and very dense, standard initial tunnel support consisting

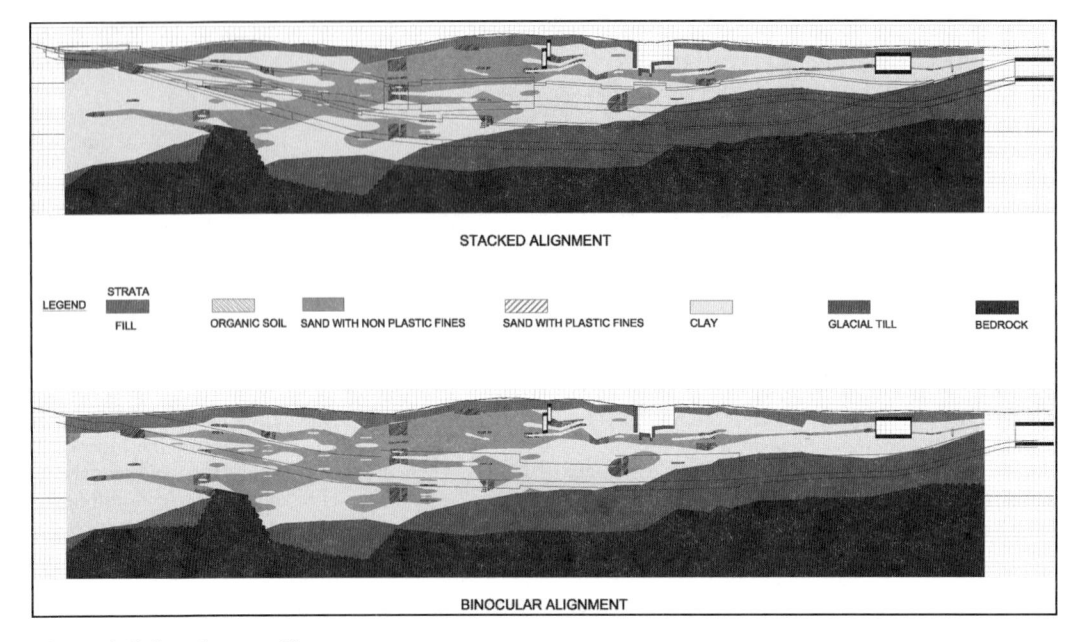

**Figure 4. Subsurface profiles**

of lattice girders and shotcrete is typically adequate, without additional ground improvement or pre-support requirements. On the other hand, tunnel excavation in the clay will invariably encounter 'squeezing' ground condition, where the severity of the squeeze depends on the clay strength and tunnel depth. Therefore, tunnel presupport using face reinforcement and forepoling will be required in this case. Without any ground improvement prior to excavation, tunneling in the sand below groundwater will encounter a 'flowing' condition. Full pre-support of a horizontal jet grout umbrella arch will be required in this case, in combination with face reinforcement.

Therefore from the tunneling perspective, it was determined that based on the ground conditions, four different classes of initial support would be required for the tunnels, as illustrated in Figure 5 and described below in the order of increasing capacity:

- Class I: Standard lattice girders and shotcrete support—For tunnels in glacial till and bedrock in general.
- Class II: Standard support plus forepole presupport—For tunnels with predominantly till face or in very stiff to hard clays in general.
- Class III: Standard support plus forepole and face reinforcement—For tunnels in other clays or with minor stratified sands in general.
- Class IV: Standard support plus horizontal jet grout umbrella and face reinforcement—For tunnels in sands or interlayered sand/clay in general.

The most optimum construction with the least cost would be if the tunnels are located deeper and as much as possible in the more competent glacial till and bedrock, which would require the minimum support. However, the start and end point connections with the existing Phase I and Phase II Services, and the maximum allowed profile grade for vehicle operation, would require significant portions of the tunnels to be built in the less desirable marine clay and sand formations.

**Stacked Tunnel Alignment**

From Figure 4, it is evident that the upper outbound tunnel, located mostly in the clay and sand layers, will require heavier Class III and Class IV support, whereas the lower inbound tunnel in the glacial tills will require minimal support. Both the outbound and inbound tunnels for the Charles Street South segment will require the heaviest Class IV support since they are located within the sand or stratified sand and clay layers. Excavation sequence for the smaller running tunnels will typically consist of a single top heading and bench or invert (see Figure 5), whereas the larger blister and station platform tunnels will be excavated using a single sidewall drift with top heading and bench sequence. In areas where continuous horizontal jet-grout umbrella support is used (i.e.,

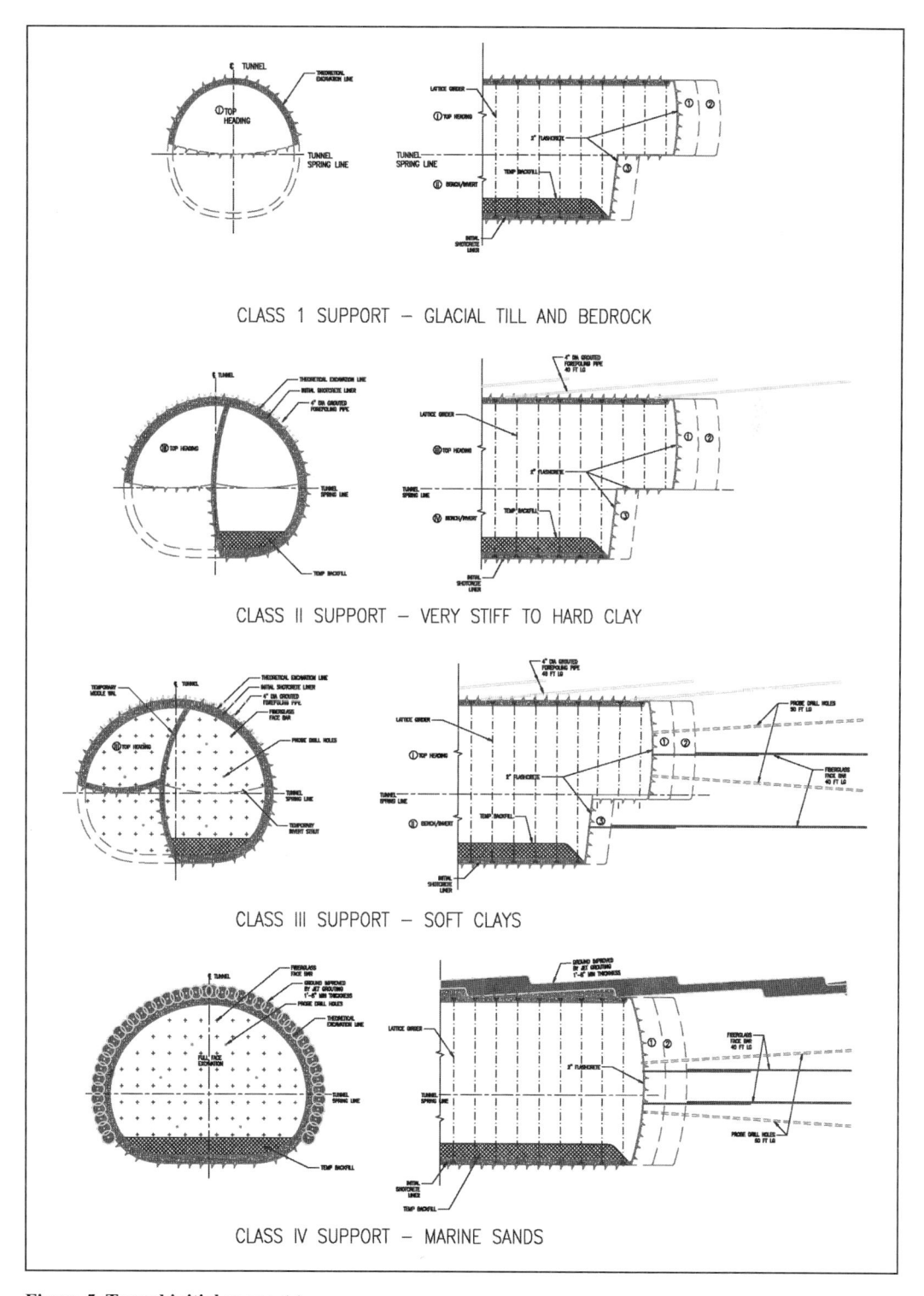

Figure 5. Tunnel initial support types

Class IV), the tunnel will be excavated using a full face heading.

**Binocular Tunnel Alignment**

The single mined binocular tunnel will be higher in elevation than the inbound tunnel but lower than the outbound tunnel in the stacked alignment due to its larger size (see Figure 4). East of Chinatown Station, the tunnel will mainly be excavated in the glacial tills, which will require relatively light initial support system (i.e., Class II). The remaining tunnel will be excavated in marine clay and sand, or stratified clay and sand. By combining the outbound and inbound lanes into a side by side tunnel configuration, the total tunnel length will be about half of that in the stacked alignment. However, the binocular tunnel cross section will be much larger than the individual outbound/inbound tunnel in the stacked alignment. Therefore, the excavation sequence, rate of advance, and initial support installation will be much different, i.e., more elaborate excavation sequence, slower advance rate, and more initial support per LF of tunnel. Excavation sequence for the binocular running tunnel will typically consist of a single sidewall drift with top heading and bench sequence; whereas the larger station cavern will need to be excavated using double sidewall drifts with a central section, and top heading and bench sequence. Again in areas where continuous horizontal jet grout umbrella support is used (i.e., Class IV), the tunnel may be excavated using a full face heading.

## PASSENGER STATION EXPERIENCE

One of the most important aspects of station design is the passenger experience of moving through the station on the way to or from the platform. It is helpful to compare tunnel configurations and the related station options with the passenger experience in mind, specifically comparing each option on the basis of wayfinding (number of directional changes, vertical circulation elements or VCEs, and clarity of circulation), transfer times, egress, and platform width etc.

The passenger paths of travel from point of origin (into the station system) to point of destination (out of the system) is an important area for comparison, especially the number of directional changes and decision making points in each path. In general, an increase of either of the two is a negative impact on the passenger experience: the fewer, the better the experience. For passengers, knowing quickly which path to take to reach the desired destination enhances the efficiency of their travel, improves the passengers' orientation and greatly improves their experience and sense of place.

Because of the generally great depth of the station platforms below the surface, the two main modes of vertical circulation within the station are escalators and elevators. Elevator connections to the platforms are direct from the mezzanine; all decision-making (where required) is at the mezzanine level. Escalator connections are also direct but due to the depth of the stations, they require switchbacks. As described previously, changes in direction or switchback conditions along the path of travel can be detrimental to passenger orientation.

The various station options differ substantially in how they impact the various transfer routes and transfer times for passengers, depending on the station layouts, location of VCEs, and designed circulation paths. The projected transfer volumes between the existing Orange Line inbound and outbound (IB/OB), and the Silver Line IB/OB, and also between the existing Green Line IB/OB and the Silver Line IB/OB, are different for the different transfer routes and times of day (peak versus non-peak hours). Therefore, one good measure for comparison purpose could be the average transfer time during the day weighted by the passenger volumes over the different transfer routes.

The station platform width is determined by a combination of factors. These factors include: the required Level of Service (LOS), the forecasted station capacity, the requirements of the VCEs, and the nature of the vehicle berthing (dynamic vs. static – to be explained later). The length of the platform is determined by the number of vehicles that berth at the platform. In evaluating the simplicity of the platform functions for the various options, the design must take into consideration both the regular system users, persons with disabilities, and people unfamiliar with the system.

### Stacked Versus Binocular Tunnel

Figure 6 shows comparisons of the Chinatown Station layouts and cross-sections between the stacked and binocular tunnel configurations. The binocular station platform would be less deep and on a single level to be more passenger-friendly. To minimize the platform tunnel width and number of VCEs, the binocular stations would be constructed using a single center platform rather than double side platforms. Figure 7 shows the passenger circulation diagram for the respective stacked and binocular Chinatown Station configurations. From the figure, it is clear that stations with binocular alignments need fewer VCEs than stations with stacked alignments because the separate inbound and outbound platform levels are integrated into a single level center platform.

With regard to the transfer times, generally the binocular alignment has shorter inbound transfer times, due partly to the fact that inbound and outbound platforms are now on the same level, and

**Stacked Tunnel Alignment**

**Binocular Tunnel Alignment**

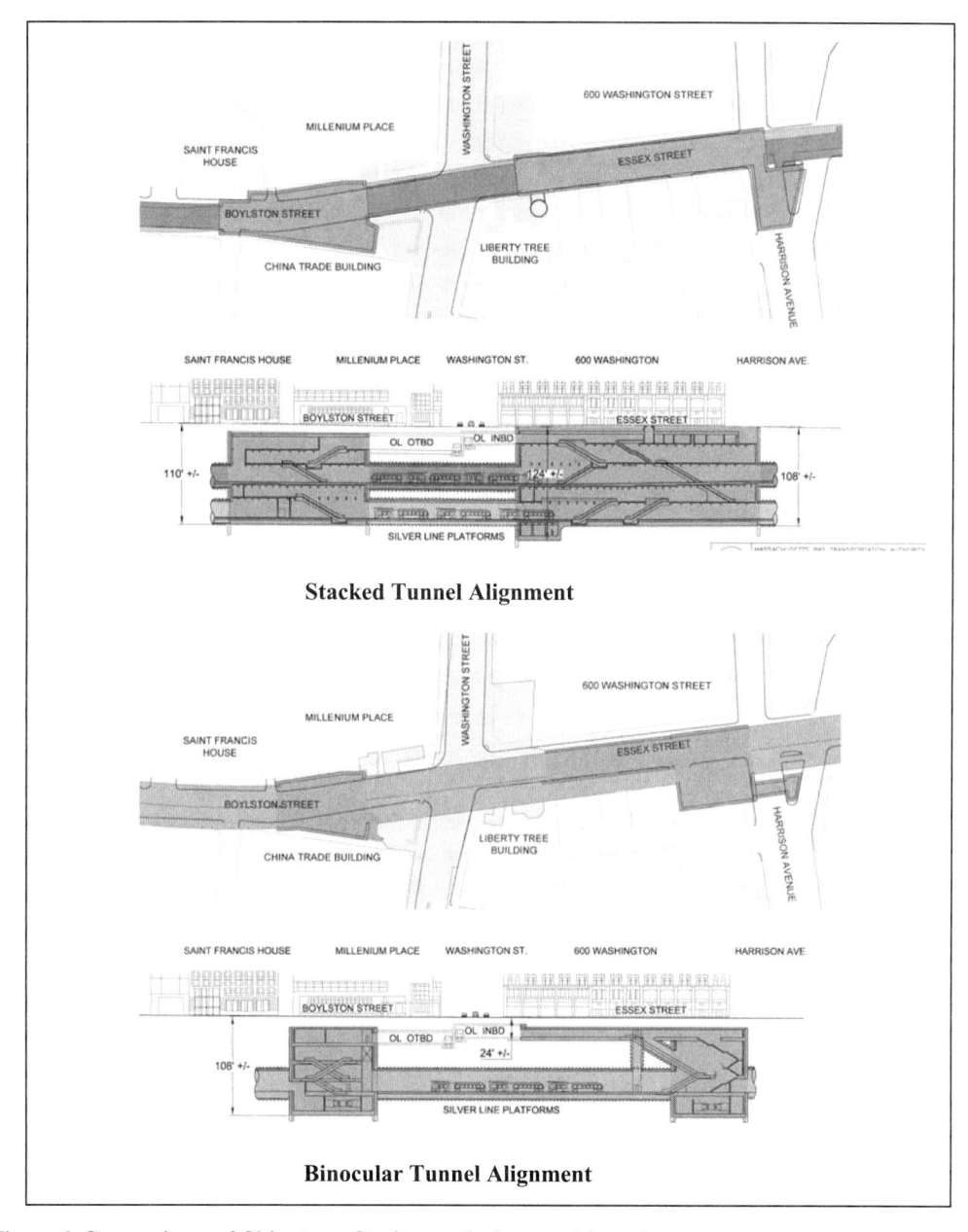

**Figure 6. Comparisons of Chinatown Station stacked versus binocular**

at a higher elevation, than the inbound platform in the stacked alignment. Conversely, the outbound transfer times are longer due to the fact that the outbound platform is now at a lower elevation than in the stacked alignment. Due to the depth of the inbound platform in the stacked alignment, a separate emergency egress structure or shaft would also be required. On the other hand, the wider binocular tunnel would likely require the presence of columns or encumbrance on the center platform that would in turn necessitate a wider platform for the passenger service requirements.

### Two Station Versus Mid-block Station

Figure 8 shows comparisons of the two station and mid-block station layouts and cross sections for the

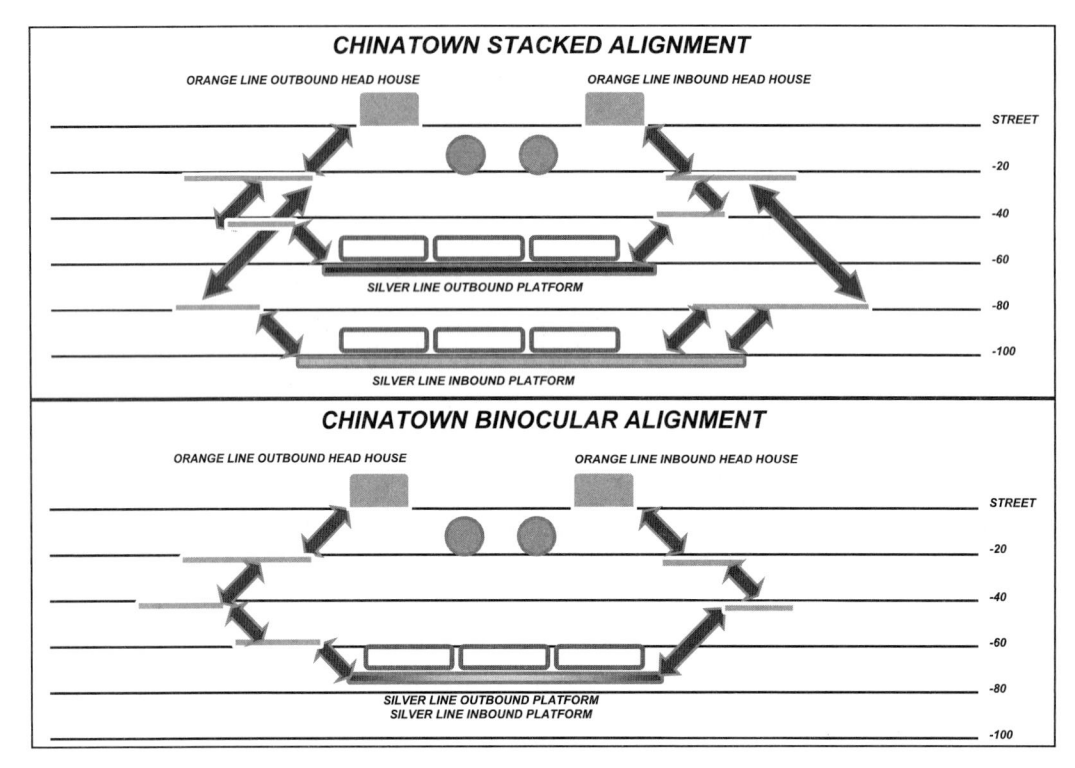

**Figure 7. Comparisons of Chinatown Station transfer diagrams**

stacked tunnel alignment. Figure 9 shows the passenger circulation diagrams for the respective two station and mid-block station options. With regards to wayfinding, the two station options provide a simpler passenger experience because each Silver Line station connects with only one existing station. This is especially beneficial to first time users and tourists. On the other hand, for both the mid-block station options (with stacked or binocular tunnel alignments) the location of the platform between the existing Boylston Green Line and Chinatown Orange Line Stations means that at the Silver Line platform level, passengers must be able to quickly determine which direction to take to their connection, and therefore more switchbacks. However, due to combining the two stations into a single station platform, the mid-block stations would generally require fewer VCEs.

The two station options would require a minimum platform length of 220 feet for three vehicles berthing. Due to the combined passenger capacity from the two stations, the mid-block station options would require a minimum platform length of 360 feet for five vehicles berthing. The longer platform for the mid-block station will require more walking time, and thus generally longer transfer times due also to its required location between the two existing stations. The mid-block option with the binocular alignment especially presents the most challenging environment for passenger circulation due to the fact that this platform has inbound and outbound passengers on the same platform, more vehicle berths, and columns in the center of the platform, while providing clear access to four routes leading to the existing Green Line and Orange Line stations.

## OPERATIONAL ATTRIBUTES

The operational viability of the various tunnel alignment and station options need to be evaluated in light of the targeted design ridership and service plan. This evaluation was used to determine the number of vehicle berths required for each option, and whether a bypass lane would be required for the mid-block station options which must accommodate the passenger loads of two stations simultaneously. Currently, the Silver Line stations have three vehicle berths with the restriction that the route to Logan Airport uses only the last berth in the inbound direction (static berthing).

### Stacked Versus Binocular Tunnel

As mentioned previously, the binocular tunnel configuration has many benefits in terms of shallower

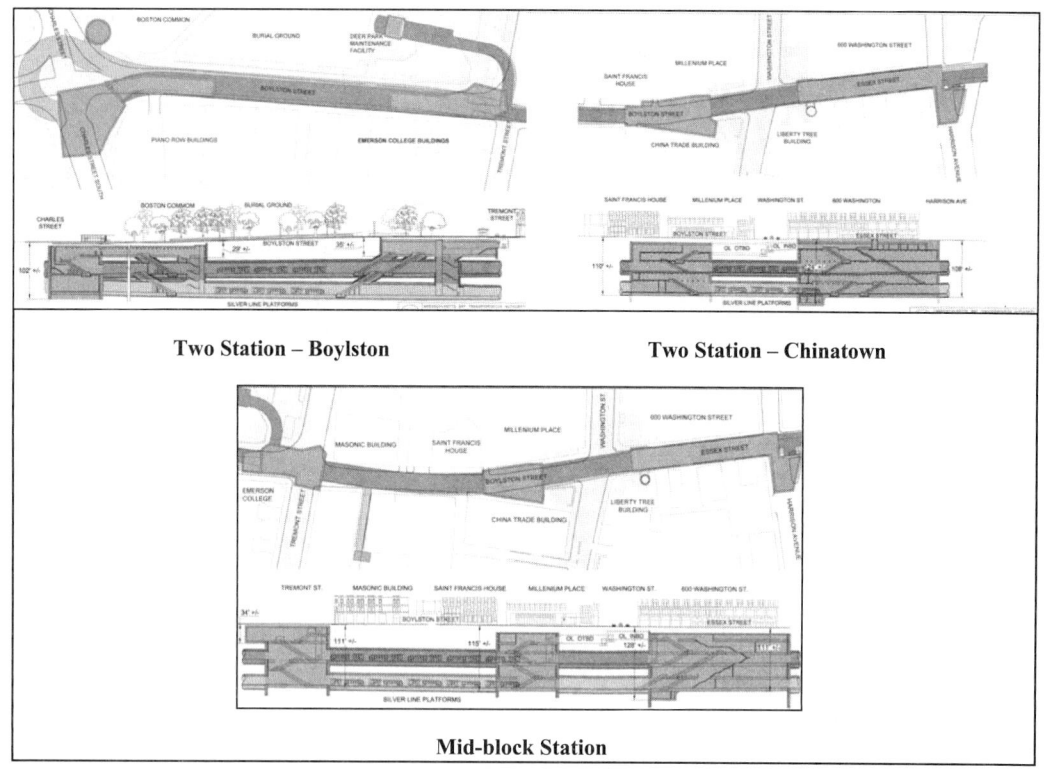

Figure 8. Comparisons of two station and mid-block station stacked tunnel

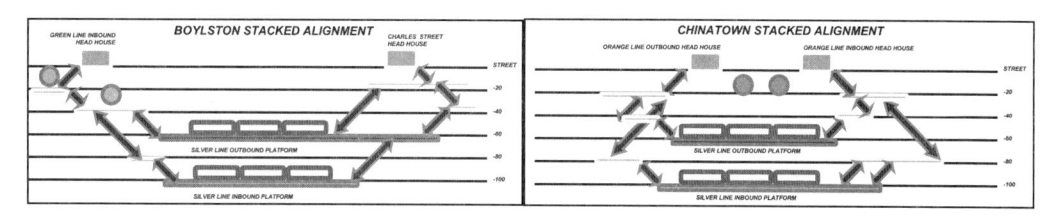

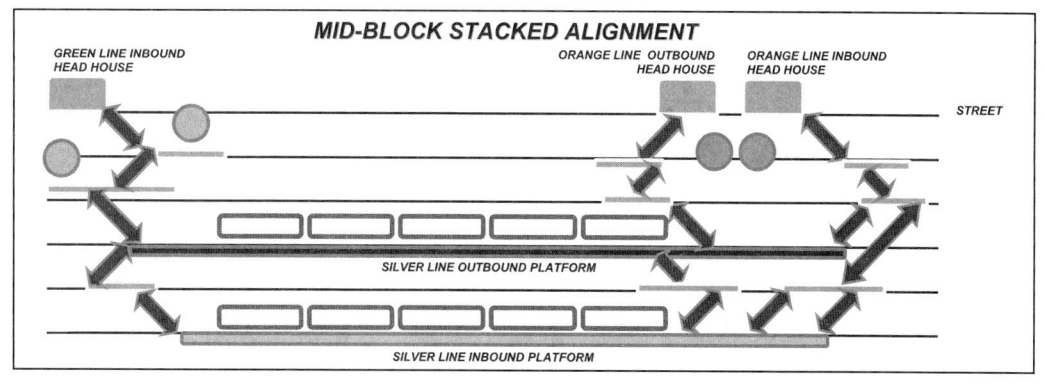

Figure 9. Comparisons of two station versus mid-block transfer diagrams

**Figure 10. BRT vehicles with two driver-side doors**

depths of construction, more openness of stations, simplified wayfinding, and the efficiency of a center platform. The vehicle travelways at the same elevation provide for additional passenger safety by being able to transfer passengers from the tunnel in which an emergency situation may exist into the other tunnel for safe evacuation, and thus less extra emergency egress requirements. However the use of a center platform also brings to light one significant issue that must be addressed – the Silver Line vehicles only have doors for alighting and boarding on the right-hand side of the vehicle. While it may be possible to procure the new fleet of vehicles with doors on both sides of the vehicle, there will only be two sets of the doors on the left-hand side of the vehicle, instead of three door sets typical on the right-hand side of the vehicle, thus increasing the dwell times and the overall travel time. Figure 10 shows typical bus rapid transit vehicles having two doors on the left-hand side or driver-side of the vehicle.

The other operating scenario that was examined and modeled was one in which the vehicles would actually "crossover" and operate through the stations in the Phase III section using a left-hand driving style. This scenario actually would result in quicker travel times due to the fact that now the vehicles would have their full compliment of three doors on the right-hand side for boarding and alighting passengers.

By having a center platform on a single level, rather than two separate platforms and tunnels for the outbound and inbound directions, an added potential operational benefit with the binocular alignment is the reduction in the number of systems requirements for the stations and within the tunnels. This includes the mechanical/electrical systems, tunnel ventilation, the overhead catenary system (OCS), traction power, signals, communications, intelligent transportation systems (ITS) and security.

**Two Station Versus Mid-block Station**

In the mid-block station options, a bypass lane would be required in order to allow following vehicles to go around longer dwelling Logan Airport-bound vehicles. However, adding the bypass lane onto the binocular station in particular would make the station become unattainably wide and un-constructible. Alternatively, it was determined that a bypass lane would not be required if dynamic berthing was used; i.e., dynamic berthing eliminates the operational restriction for Logan Airport-bound vehicles, and allows them to stop at any available berth. With dynamic berthing, vehicles would be connected to an ITS which would provide passengers with information regarding the destination of the next vehicle arriving at each berth. This would allow passengers, including passengers with luggage heading to Logan Airport, to maneuver to the proper berth. But this would also require a wider platform width for the extra maneuvering room. And again by having a single station rather than two separate stations, there may be a reduction in the number of duplicated systems requirements for the station, and thus costs savings.

**CONSTRUCTION IMPACTS**

The biggest impact of the project construction is anticipated due to the cut and cover portions of the tunnels and stations, where open cut excavations from the surface will be required. Potential impacts to adjoining properties in the cut and cover construction areas include: detouring of vehicular and pedestrian traffic, maintaining building access and parking, relocating utilities and utility service connections, construction of slurry walls, scheduling of construction activities to minimize disruption to adjoining properties, monitoring and controlling of construction noise and vibration, and settlement impact caused by excavation and dewatering.

For mined tunnels, most work is done underground and materials and equipment are brought into the tunnel from access shafts. Excavated material from the tunneling operation is removed from the tunnel through the access shaft and hauled away. Construction impacts for all tunnels include monitoring and control of groundwater and mitigation of

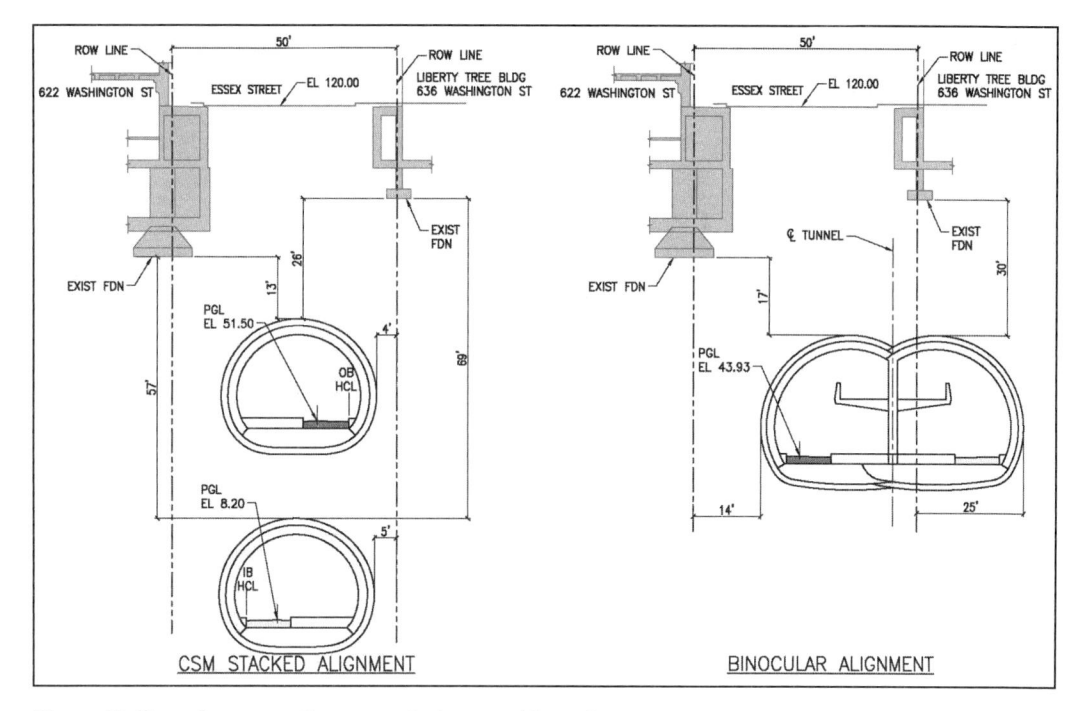

Figure 11. Tunnel cross sections – stacked versus binocular

settlements. It is anticipated that all buildings along the alignment will have instrumentation installed and will be continuously monitored during construction for impact mitigation purposes.

**Stacked Versus Binocular Tunnel**

Depth of open cut excavation is one of the major construction cost drivers of each option. The binocular tunnel alignment generally would not require as deep of excavations as with the stacked tunnel alignment, and therefore is a better option. However, since the binocular tunnel is wider, it would extend outside of the public right-of-way and into properties of adjacent buildings in certain areas, requiring permanent subsurface easements to varying degrees along the tunnel alignment. Figure 11 shows a comparison between the stacked and binocular tunnel sections with respect to adjacent buildings. Additional building mitigation must also be performed in terms of either additional soil stabilization and/or direct building underpinning. These present additional risks to the tunnel construction.

**Two Station Versus Mid-block Station**

The mid-block station options would generally require less open cut excavations than the two station options. However, the mid-block stations would require wider station platform tunnels and cavern, which in this project case happens to occur at the narrowest point of the public right-of-way, and in the area with some of the worst ground conditions, i.e., soft clays, along the project alignment. As shown in Figure 12, this forces the binocular mid-block station cavern to extend up to 20 feet into the existing properties on the south side of Boylston Street between Washington and Tremont Streets. Even the stacked mid-block station tunnels would encroach much closer onto the building properties, such that the requirements for soil improvement and building mitigation measures would increase. The construction risks would therefore increase under the mid-block station options.

**IMPACTS ON ADJOINING PROPERTIES**

One of the criteria established for pursuing a stacked configuration of tunnels was to stay within the public right-of-way and away from adjacent buildings to the greatest extent possible. While there may be some impacts to the adjacent buildings for the stacked alignment, they are considered relatively *minimal*. Where the impacts are minimal, instrumentation would be provided for all buildings along the alignment which would be continuously monitored during construction. Basic mitigation costs are included for protection, i.e., cosmetic repair to buildings such as crack sealing and painting, including an allowance for compensation grouting to limit settlements, but

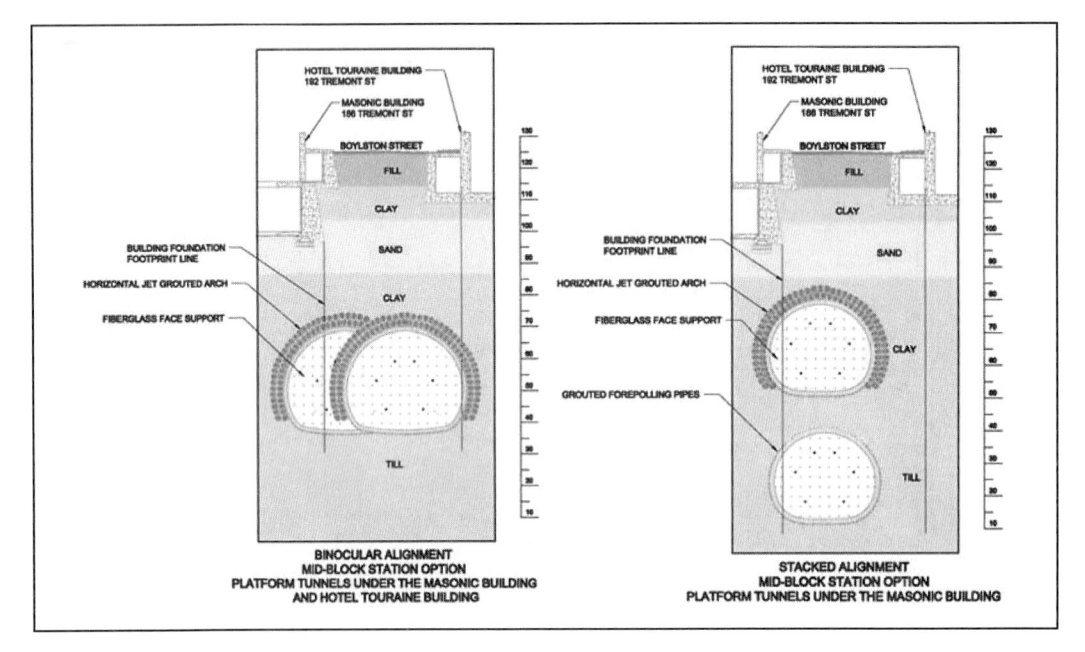

**Figure 12. Tunnel cross sections – mid-block options**

not considering direct underpinning of the building foundation.

As depicted in Figures 11 and 12, the alignment change from stacked twin tunnels into a single binocular tunnel is anticipated to change the settlement patterns due to the tunneling at the ground surface. Due to the relatively shallower and larger binocular tunnel, the settlement impacts on the adjacent buildings are expected to be more significant, requiring more mitigation measures. Furthermore, the tunnel drifts outside the street right-of-way and beneath building foundation footprints in some areas also requiring additional underpinning or ground improvement measures to protect the buildings, in addition to the standard allowance for building mitigation where the tunnels remain completely within the street right-of-way.

Cases where the binocular tunnel alignments encroach upon the right-of-way limits and slightly under a building foundation, the impacts are considered to be *moderate*. Moderate impacts would require additional ground improvements such as grouting and/or building stabilization. Again costs are included for any cosmetic repairs that may be necessary upon the completion of construction. Finally, where binocular tunnel alignments fall outside the public right-of-way and are significantly into the private right-of-way, *substantial* ground improvements and/or building stabilization would be required. The actual mitigation measures that would address the *minimal*, *moderate* and *substantial* encroachment into private rights-of-way and under adjacent buildings would depend on the building conditions, and have not been determined during the preliminary design. However, increasing degrees of costs and risks have to be included in the evaluations of each options considered.

## COST CONSIDERATIONS

Cost is one of the most important factors in determining project viability. Accordingly, the Year of Expenditure (YOE – estimated date or year when incurred including escalation) costs were estimated for each option. The interim preliminary engineering (30%) design cost estimate for the stacked tunnel two station configuration was used as the basis, and costs for all the other options were estimated using the same cost estimating methodology based on their conceptual designs. Based on the total YOE costs, the binocular alignment was found to be less expensive than the stacked alignment with two station option. However, the mid-block station option was deemed to be more expensive than the two station option mainly due to the much larger impacts on adjacent properties. The binocular alignment with mid-block station option was the most expensive of all the options due to its far encroachment underneath existing buildings.

The Federal Transit Administration (FTA) also requires calculation of Cost Effectiveness for projects being considered for Section 5309 New Starts discretionary funding provided through the FTA

| SIGNIFICANT RISK ISSUES | Stacked Alignment - Two Station (Core Only) | Stacked Alignment - Mid BlockStation (Core Only) | Binocular Alignment - Two Station (Core Only) | Binocular Alignment - Mid BlockStation (Core Only) |
|---|---|---|---|---|
| Groundwater stabilization due to different soil conditions | ○ | ● | ● | ● |
| Groundwater control | ○ | ○ | ○ | ○ |
| Utility construction and relocation | ○ | ◐ | ◐ | ◐ |
| Known and unknown obstructions | ○ | ○ | ● | ● |
| Potential need for direct underpinning | ○ | ● | ● | ● |
| Acquisition of permanent easements for tunnel construction | ○ | ○ | ◐ | ● |
| Acquisition of staging and lay down sites | ○ | ○ | ● | ● |
| Support of excavation construction | ○ | ◐ | ◐ | ◐ |
| Construction interfaces with existing structures | ○ | ◐ | ◐ | ◐ |
| Contractor interfaces | ○ | ○ | ○ | ○ |

○ Base Case   ● Worst than Base Case   ◐ Better than Base Case

**Figure 13. Risk comparison for different options**

such as in this case. Cost effectiveness is defined as the annualized capital cost and operations and maintenance cost divided by the transportation system user benefit (TSUB). TSUB represents the changes in mobility for individual travelers that are induced by a project, and used by the FTA to compare projects throughout the U.S. They are measured in hours of travel time savings and summed over all travelers who use the system. Based on this Cost Effectiveness factor, although the binocular alignment with mid-block station had the highest cost, it still fared better than all the stacked tunnel options due to the more efficient operation and passenger's benefit. Based on this exercise, the option with both the lowest total YOE cost and best Cost Effectiveness was found to be the binocular tunnel alignment with the two station option.

## SUMMARY AND CONCLUSIONS

This paper discussed the preliminary engineering for the Silver Line Phase III BRT Tunnel Project in Boston, Massachusetts. Several different alternative configurations for the tunnels and stations were evaluated during the design value engineering process, which is typically required for projects of this significance. Other than the aspect of tunneling and ground conditions, other important factors impacting the design in this case include the system operational requirements, passenger's station experience, and construction impacts to adjoining properties. Often times these factors are in conflict with each other. For instance, the tunnel support requirements would prefer locating the tunnels deeper in the better glacial till and bedrock layers; however, the deep tunnels would cause more costly stations construction. Or, a single level station would be preferred for more efficient passenger circulation in and out of the station; but the resulting wider tunnel would then significantly impact the adjacent properties. During the design process, difficult compromises sometimes must be achieved amongst the various disciplines, including tunnel engineering, to achieve a reasonable and constructible project.

The bottom line construction and operation/maintenance costs for the project would typically serve as the main deciding factor. In mass transit applications such as this case, the costs must be factored against the benefit to the user in terms of travel time savings and cost effectiveness. The risk of underground construction for the different alternatives should also be incorporated into the decision making process. Unfortunately, these risks as well as the costs are sometimes difficult to quantify accurately during the preliminary design process, when the important decision of selecting an alternative is being made. Therefore, the real value of the cost estimating and risk identification exercises during this stage of the design probably lies more in their comparative nature of the results rather than their absolutes. They would need to be fine tuned during final design of the selected alternative. Figure 13 shows the risk comparison for the different alternatives or options considered in the project. It is important then when comparing the costs, project assumptions are being made to even out the risk levels amongst the various options.

This paper illustrates the fact that tunnel design for mass transit applications involves a lot more than designing a safe and stable opening in the ground. It is a negotiation process whereby tunnel engineers may be asked to take the backseat and let others run the show. But in the end, it is the tunnel engineer's responsibility to ensure a project that is buildable with an acceptable level of risk.

## ACKNOWLEDGMENT

The authors would like to acknowledge the following principals of the Silver Line Phase III Project, which is used as example in this paper:

- *The Massachusetts Bay Transpor*tation Authority (MBTA), the project owner.
- AECOM, joint-venture project design partner with URS.
- IBI/DHK, subsonsultant to the JV as the project architect/operational system designer.

## REFERENCES

Yates, G., Ainsley, M., and Gallagher, W. 2009. MBTA Silver Line Phase III – Completes Boston's Newest Transit Line. *Proc. 2009 Rapid Excavation and Tunneling Conference.* Las Vegas, Nevada.

Nasri, V., Vrouvlianis, K., and Halim, I. 2009. Design of NATM Tunnels and Stations of Silver Line Phase III Project in Boston. *Proc. 2009 Rapid Excavation and Tunneling Conference.* Las Vegas, Nevada.

URS – DMJM/Harris Joint Venture 2009. *Binocular Tunnel & Mid-Block Station Concept Report.* MBTA Silver Line Phase III Technical Memorandum. Boston, Massachusetts.

# Transbay Transit Center Program Downtown Rail Extension Project

**Meghan M. Murphy, Derek J. Penrice, Bradford F. Townsend**
Hatch Mott MacDonald, San Francisco, California

**ABSTRACT:** The Transbay-Downtown Rail Extension (DTX) Project involves the construction of approximately 1.6 miles of underground structures that will extend Caltrain commuter service and the future California High Speed Rail system into downtown San Francisco. The extension will be constructed by mined tunnel and cut-and-cover methods and will encounter challenges including difficult ground conditions, low rock cover, the presence of historic buildings along the alignment, and a large tunnel span of greater than 50 feet. This paper provides a summary of the preliminary engineering phase and contract packaging approach and provides an overview of the coming design phases and potential challenges.

## INTRODUCTION

The Transbay Transit Center Program in the San Francisco Bay Area will result in the creation of the largest multi-modal transit facility in the United States west of New York City. The Program will enhance regional public transportation service by improving the bus and rail connectivity of eight major transit providers at one multi-modal facility in downtown San Francisco. The Transbay Joint Powers Authority (TJPA) is responsible for the completion of the design, construction, and the subsequent operation and maintenance of the Program. The TJPA represents a collaboration of Bay Area government and transportation agencies. A full description of the Program can be found on the TJPA's website (www.transbaycenter.org) (TJPA 2010).

The Downtown Rail Extension (DTX) is an essential aspect of the Program. The DTX project is composed of approximately 1.6 miles of underground structures that will extend the existing Caltrain commuter rail service and the future California High Speed Rail (CHSRA) system into the new Transit Center (TC) in the heart of downtown San Francisco (Figure 1).

The rail extension will be constructed using a combination of mined tunnel and cut-and-cover methods largely under city streets in the public right-of-way. The construction of the tunnel will encounter challenges such as difficult ground conditions including low rock cover, extensive utilities, and the presence of historic buildings along and above the alignment. These challenges are further exaggerated by the magnitude of the proposed tunnel structures.

This paper discusses these challenges and the solutions developed to overcome them by the project team during the development of the preferred project configuration, and outlines the proposed approach to the contract packaging, procurement and schedule for the forthcoming tunnel construction.

## PROJECT DESCRIPTION

The DTX project will wind northeast from the existing Caltrain tracks at Seventh and Common streets then turn northwest and continue along the existing Caltrain Depot and Townsend Street. A new proposed underground station will extend from Fifth and Townsend to Fourth and Townsend streets. The alignment will turn beneath existing structures at Townsend and Second streets and proceed northwest along Second Street before turning northeast again north of Howard Street to enter the beneath-grade trainbox at the TC (Figure 2). The TC will provide three platforms and six tracks for Caltrain and future CHSRA service.

The DTX alignment consists of six primary structure types: a retained cut, cut-and-cover tunnels, a cut-and-cover station box, a mined tunnel, ventilation shafts, and cut-and-cover tailtracks. The mined tunnel stretches approximately 0.6 mile and varies in dimension to a maximum of 42 feet high by 60 feet wide. The cut-and-cover excavations are similarly massive, stretch nearly 0.7 mile in length and reach depths of up to 70 feet and widths of up to 175 feet. The retained cut portion of the underground construction is approximately 0.2 mile in length and will span approximately 40 feet in width. The retained cut

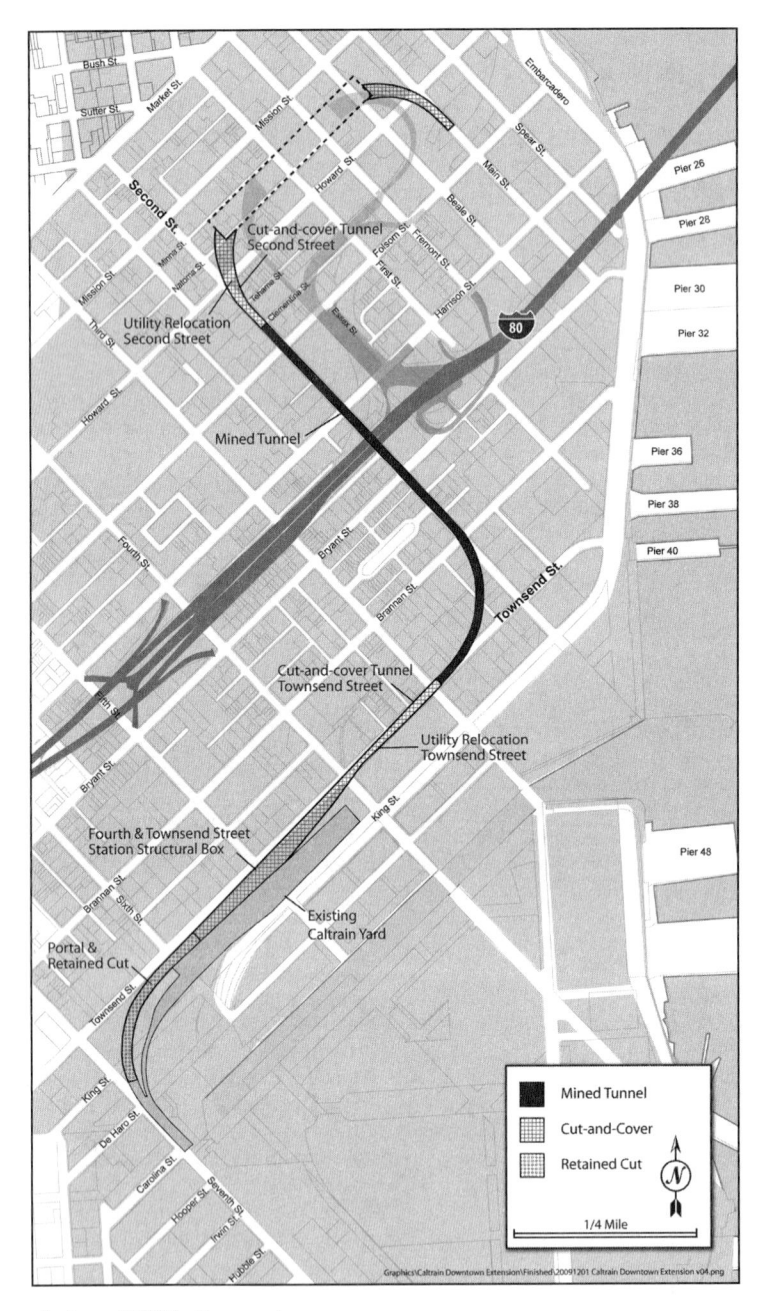

**Figure 1. General plan of DTX alignment**

will contain a two-track approach to the Fourth and Townsend Underground Station which will expand to three tracks as it enters the station. The remaining 0.1 mile of underground construction comprises a section of cut-and-cover tailtracks which extend past the TC down Main Street.

There are two ventilation shafts proposed along the alignment. The ventilation shafts will be approximately 25 feet by 40 feet in dimension and will be located at the intersection of Harrison and Third streets and Townsend and Third Streets. The shaft at the intersection of Harrison and Third streets is approximately 100 feet in depth and is proposed

**Figure 2. Cross-section of the transit center**

as the start point of a third tunneling heading during construction. The ventilation shaft at Townsend and Third streets will be approximately 65 feet in depth and will act as a staging area during construction.

Along with the TC, there will be a new underground station at Fourth and Townsend streets which will extend under the existing Caltrain Yard and into Townsend Street. The station will have a mezzanine and a platform level dedicated to Caltrain service. The station will include a bypass track to allow high-speed trains to proceed to the TC without stopping.

Two tracks will be provided in the tailtracks structure to provide operational advantages for Caltrain. The tailtracks will wrap south on Main Street from the TC from the southernmost two tracks in the TC.

## FUNCTIONAL AND LOGISTICAL CHALLENGES

External challenges on the DTX project result from the need to meet the operational and functional requirements of multiple operators: Caltrain and CHSRA. The primary challenge lies in the design schedule which has advanced significantly ahead of the operators' capital programs. Therefore, required design parameters, operational procedures and rolling stock have not yet been determined by the operators and may not become available until after the DTX design is completed. The operational procedures affect assumed headways and platform dwell times, while the rolling stock choice impacts the required clearance envelopes, platform and walkway heights and allowable radii for curves along the alignment. As the operators are stakeholders in the Program, but not the client, the project schedule has required that assumptions be made until information becomes available from the operators. For example, clearances were developed using an envelope of potential vehicles for Caltrain and CHSRA. Platform and emergency walkway heights have been based on this fabricated clearance diagram.

The project team has coordinated with the operators to obtain information as soon as it becomes available for issues such as fire-life safety requirements and emergency response procedures. The operators are issued project design submittals as they become available for their review and comment to facilitate the coordination effort.

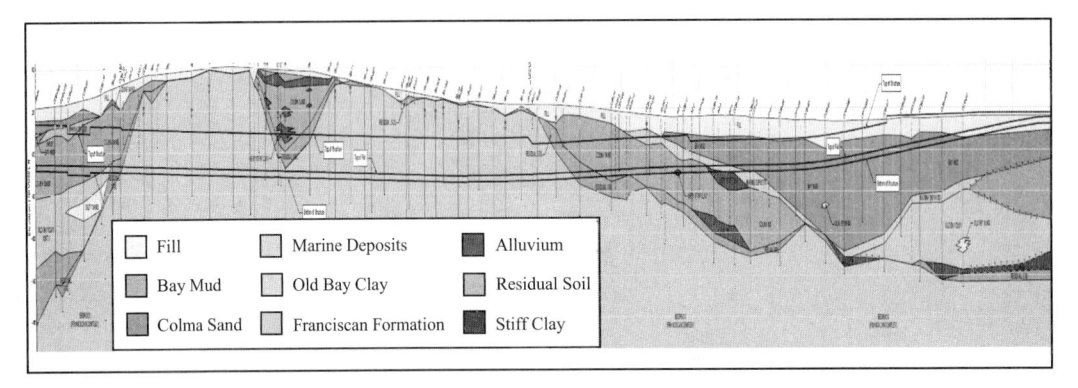

**Figure 3. Subsurface profile along the DTX alignment (Arup 2009)**

## TECHNICAL CHALLENGES

The DTX project is constrained by ground conditions, low rock cover, extensive utilities, and fragile historic buildings adjacent to and above the alignment. These challenges have shaped the engineering design as well as the proposed construction methods along the project alignment. These challenges and their proposed solutions are described in the following sections.

### Ground Conditions

The ground conditions vary greatly along the DTX alignment as is shown in Figure 3. The southern portion of the alignment (between the start of the project alignment near Common Street and the intersection of Townsend and Third streets approximately) is located bayward of the historic 1848 shoreline and is adjacent to an inlet of the San Francisco Bay. The southern portion of the alignment (on the right portion of Figure 3) consists of loose surficial fill that ranges in thickness from 10 feet to 30 feet which is primarily composed of debris from the 1906 San Francisco Earthquake and is partially classified as California hazardous waste. The fill is underlain by a 10-foot to over 110-foot-thick layer of compressible clay, locally termed Bay Mud, under which is a thin layer of marine deposits. The marine deposits overlay layers of dense Colma Sand or Old Bay Clay as the alignment moves east along Townsend Street. Both the Colma Sand and the Old Bay Clay are approximately 60 feet to 80 feet in thickness and are underlain by thin layers of Alluvium and residual soil, which rest above the highly weathered bedrock of the Franciscan Formation. The bedrock ranges in depth from near ground surface to over 200 feet beneath grade. A relatively shallow groundwater table and the potential for amplification of seismic waves through the Bay Mud during a seismic event cause further design and construction challenges.

The Franciscan Formation rises to near the ground surface near the intersection of Townsend and Third streets as the alignment turns onto Second Street and continues to be near the surface until approximately Second and Harrison streets. The bedrock is interrupted for approximately 650 feet between the I-80 crossing of Second Street to past the intersection of Second and Bryant streets by an approximately 100-foot-deep Paleolithic valley, which is predominantly filled with Colma Sand, interbedded with stiff clay, which is underlain by approximately 20 feet of residual soils above the Franciscan Formation.

Along the northern portion of the alignment between the intersection of Second and Folsom streets and the TC, the bedrock returns from near ground surface to over 200 feet beneath grade. The fill remains approximately 10 to 20 feet thick and is underlain by thin layers of dune sand and marine deposits followed by a layer of Colma Sand which ranges in thickness from 10 feet to 60 feet. The Colma Sand is in turn underlain by residual soil to approximately Second and Tehema streets and Old Bay Clay above bedrock for the remainder of the northern portion of the alignment to the TC (Arup 2009).

### Historic Buildings

The majority of the alignment is planned to be in the public right-of-way; however, as the DTX alignment turns northwest from Townsend Street to Second Street, the alignment passes beneath eleven existing buildings. The buildings were constructed in the early 1900s and range from one to six stories in height, some with brick facades. All the buildings form a part of the Rincon Point/South Beach Industrial District, and may be eligible for the National Register of Historic Places. The mined tunnel will pass beneath these buildings with approximately 20 to 35 feet of rock cover, a small depth compared to the tunnel span. Additionally, numerous existing buildings line the street adjacent to the cut-and-cover excavations. Most of these buildings range from two to seven stories; however, there are several high-rise buildings along the northern portion of the alignment (Klein 2007). An extensive monitoring program is planned for the structures along the alignment including advance survey work and real-time vibration and ground movement monitoring. Analysis to identify structures susceptible to damage by tunneling is ongoing. At-risk structures will be underpinned or otherwise strengthened prior to tunnel construction.

**Utilities**

The DTX alignment is entirely in an urban environment in downtown San Francisco. As with most urban environments, an extensive network of dry and wet utilities from various stages of the city's development cross and run parallel to the alignment. The utilities range from small diameter pipes to large duct banks which must either be relocated or supported in place during cut-and-cover operations. On Second Street there is an extremely large communications duct bank which is over 20 feet by 15 feet in dimension. This duct bank will be supported in place during cut-and-cover construction. The project team is currently working on a design for supporting the massive structure for which no structural information is available from the communications company. Additionally, old brick sewers cross the alignment which must be replaced with more contemporary materials then supported in place. One significant sewer relocation along the southern portion of the alignment will require the jacking of a 10-foot-diameter pipe approximately 300 feet under active Caltrain tracks at a depth of 20 to 30 feet. Though the pipe will be buried in the Bay Mud stratigraphic unit, it is not currently anticipated that underpinning will be required for the new sewer pipe.

**Construction Methodology**

The technical challenges for the DTX project have shaped the decisions on tunneling methodology as well as the application of these methodologies along the alignment. Limiting ground movements and associated surface settlements to avoid damage to the adjacent and overlying buildings is a critical concern for the project. The ground conditions played a large role in determining the sections of the alignment where cut-and-cover methods and mined tunneling methodologies would be used. The project team strived to maximize the mined tunnel segments to minimize the surface disruptions caused by cut-and-cover operations. Figure 1 shows the zones along the alignment where cut-and-cover and mined tunnel methodologies will be employed.

*Tunneling*

It was decided to pursue mined tunnel in sections of the alignment which are in bedrock and are relatively uniform in cross-section. The extent of the mined tunnel generally aligns with the extent of the Franciscan Formation bedrock in the tunnel horizon, as indicated in Figure 3. Several tunneling methodologies were evaluated: the Stacked Drift Method, the New Austrian Tunneling Method/Sequential Excavation Method (NATM/SEM), and Tunnel Boring Machines (TBM). The number of tracks and locations of track crossovers combined with the available right-of-way precluded this use of a TBM. In comparison with the stacked drift method, NATM/SEM offered cost, schedule and risk benefits. NATM/SEM was determined to be the appropriate tunneling method for the DTX alignment as it allows for relatively small ground movements to occur around the tunnel while still mobilizing the strength of the surrounding materials. These characteristics of the NATM/SEM tunneling method were particularly attractive to the project team given the presence of historic structures as well as the relatively soft rock materials and the low rock cover compared with the large tunnel cross section.

With NATM/SEM, materials are excavated in drifts which are shown in the typical tunnel cross section in Figure 4. At the conceptual design stage it was anticipated that five drifts would be required for the mined tunnel, including three top heading drifts plus bench and invert. However, more recent analysis by the project team has demonstrated that ground movements and corresponding surface settlements can generally be controlled to tolerable limits using only three drifts, as indicated in Figure 4. This provides a tremendous advantage to the project in terms of the ability to use larger equipment and achieve better production. Four drifts will be used only when ground conditions require the second top heading such as in the mixed face conditions near the northern portal of the mined tunnel.

Four excavation support types are proposed for the mined tunnel to accommodate the anticipated variability in rock quality over the length of the tunnel. Each of the support types makes use of a combination of auxiliary support measures including tube spiles, rock dowels, a pipe canopy, sloping core, fiberglass face dowels, and fiber reinforced shotcrete. Figure 4 is an example of one support type which includes a single pipe canopy, rock dowels, fiberglass dowels, and fiber reinforced shotcrete. This support type will be used in zones of tunneling through blocky, disturbed, and seamy sandstone, siltstone and shale of the Franciscan Formation. The

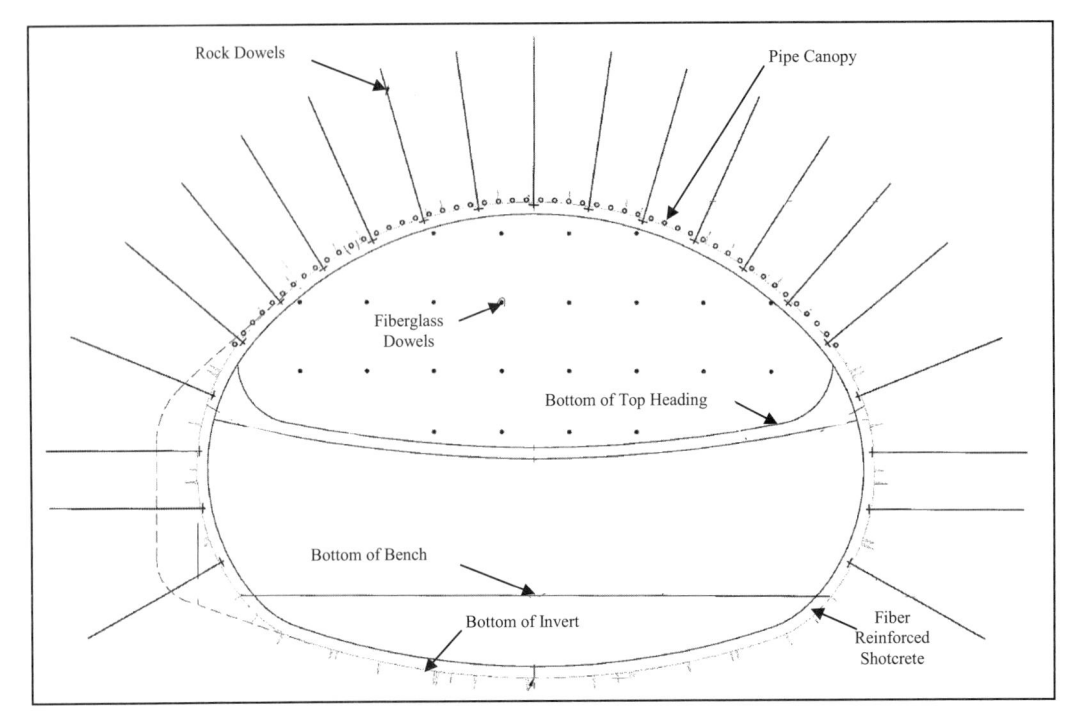

**Figure 4. Typical tunnel cross section for NATM/SEM (Jacobs Associates 2009)**

tunneling method offers the flexibility to adjust the proposed support measures to suit the actual in situ conditions encountered, thereby helping to mitigate ground movements during construction.

In the Paleolithic valley on Second Street and at the north portal of the mined tunnel Colma Sand extends into the tunnel horizon. At the Paleolithic valley site, the tunnel profile is sufficiently deep, and the material has sufficient clay content to enable tunneling to be completed without the use of ground improvement. Preliminary analysis has indicated that ground movements and surface settlements in the Paleolithic valley are tolerable. However, due to the risks associated with portalling, it is currently anticipated that ground improvement by means of jet grouting will be required to stabilize the area around the tunnel portal.

*Cut-and-Cover*

The project team sought to minimize the extent of any cut-and-cover construction along the alignment. As the excavation will occur in an urban environment, street closures and work hours for construction will affect the overall construction schedule and cost. Additionally, the project team has focused on minimization of ground movements, and thus impacts on adjacent structures, as a result of the cut-and-cover excavations. The cut-and-cover construction will begin with the installation of deep soil mix shoring walls with embedded steel soldier piles placed under their own weight. The excavation will then proceed with traffic decking placed over the street to allow surface traffic to be maintained during the construction. Steel pipe struts and walers will be placed as temporary excavation support as the excavation deepens. An illustration of the cut-and-cover method is shown in Figure 5.

There are two individual sections along the DTX alignment where the cut-and-cover construction method will be employed. The first of which is along Townsend Street between Sixth Street and Clarence Place, at the southern limit of the mined tunnel. This section also includes the proposed Fourth and Townsend Underground Station. This section will be fairly uniform in cross section and will be primarily constructed in the public right-of-way of Townsend Street. As described previously, the ground conditions on this portion of the alignment are comprised of softer materials. As the tunnel alignment is relatively shallow in this area, mining would require extensive and expensive ground treatment. In this area, cut-and-cover construction poses less risk.

The second section of cut-and-cover construction works will occur between the northern portal of the mined tunnel at the intersection of Second and Clementina streets and the interface with the TC. This segment of cut-and-cover allows for the

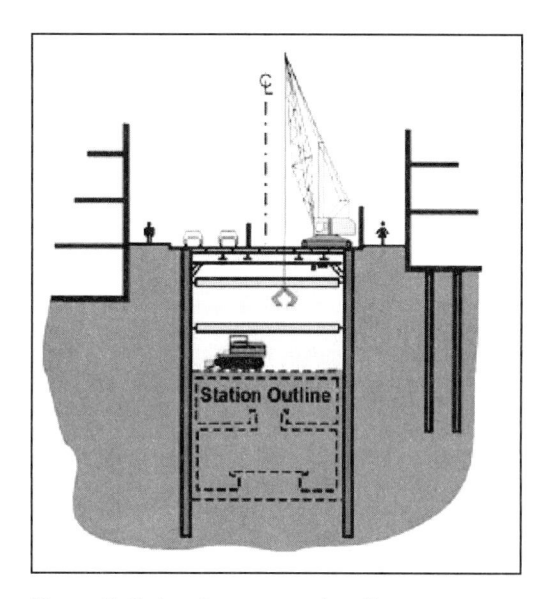

**Figure 5. Cut-and-cover construction methodology**

three-track tunnel to expand to six tracks on a curved entry into the TC, resulting in a structure that varies in width to a maximum of 175 feet. The variation in cross section and excavation size demands that cut-and-cover construction be adopted. The Second Street cut-and-cover tunnel is particularly constrained as the alignment turns from the public right-of-way towards the TC and the number of tracks must increase from three to six. Within the public right-of-way, the limited width precludes the use of internal structural walls, as these would increase clearance requirements and hence structure width. In addition, major communications utility infrastructure above the tunnel constrains the structure depth. Therefore, to overcome these issues, a cast-in-place post-tensioned roof is proposed.

The portion of the alignment past the intersection of Sixth and Townsend streets to Common Street will be constructed as an open retained cut, which will be constructed in a manner similar to that described for the cut-and-cover structures.

## CURRENT STATUS

The DTX project is currently in the Preliminary Engineering (PE) Phase of Design, which is scheduled to be complete in June of 2010. The objective of this design phase is to define the structural size and limits and the configuration. At the conclusion of PE, the DTX engineering design will be approximately 30 percent complete.

## Contract Packaging

The development of the DTX Contract Packaging Strategy (CPS) is the responsibility of Hatch Mott MacDonald as part of the Program Management/Program Controls (PMPC) team. The task has not been without challenges and was discussed in great detail in a paper presented at RETC 2009 (Penrice 2009). The overall approach of the CPS has been to develop a strategy which allows the DTX project to be constructed in the shortest timeframe possible. In addition to allowing the DTX project to open for revenue service earlier, benefits of this approach include the minimization of impacts of escalation, minimization of time-dependent costs such as program and construction management, and importantly, the minimization of impacts to Caltrain operations.

The physical construction cost for the DTX project contained in the Program Baseline Budget was established in March 2008 at $1.22 billion. This figure is inclusive of contingencies, but exclusive of soft costs and right-of-way acquisition. An update to this figure is expected at the end of the PE Design Phase in June 2010. The project team considered the current estimate too large to be procured as a single construction contract, and thusly endeavored to divide the scope into several discrete and manageable construction contracts. The project team worked to optimize the number of contracts as well as to limit the maximum value of any individual contract to approximately $250 million, which has become a North American industry rule-of-thumb figure. As with any urban setting, space for staging construction is at a premium. A primary consideration for the development of the DTX contract packages is the scarcity of construction staging areas and how the available areas could best be utilized to support concurrent construction of the major civil works. The contract has thusly been broken into several construction packages which include 1) Townsend Street cut-and-cover, 2) mined tunnel, Second Street cut-and-cover, and ventilation shafts, 3) track, 4) systems for the tunnel, 5) systems for the buildings, and 6) finishes.

## Schedule

The project is structured as a design-bid-build project allowing for funding to be obtained gradually. The current PE, 30 percent design phase is anticipated to be completed in summer of 2010 with the Final Design Phase completion in 2012. Advance contracts for utility relocation and survey work will be issued prior to the conclusion of final design and are anticipated to commence in late 2011. Contracts for the primary civil works are anticipated to be awarded for construction in 2012, with systems contracts flowing once civil construction is substantially complete. Revenue service for Caltrain is scheduled for 2018.

## CONCLUSION

This paper summarizes the functional, logistical and technical challenges the DTX project has encountered. The DTX design is proceeding on schedule and is being refined as more information becomes available from the operators. However, in the absence of operator input, the DTX designers can still effectively progress the design by making a variety of assumptions based on industry standards and the designers' past experience. As the DTX design progresses and coordination continues with adjacent projects (i.e., the Central Subway project, the Moscone Center East project, and the electrification of the Caltrain Peninsula Corridor with the California High Speed Rail Project), the necessity of the DTX project to provide a regional transportation hub which is centrally located in downtown San Francisco becomes increasingly apparent.

## ACKNOWLEDGMENTS

The authors would like to thank the Transbay Joint Powers Authority for the permission to publish this paper. In addition, the contributions of the other key members of the project team are gratefully acknowledged: Parsons Transportation Group—prime consultant, rail design, and open cut construction; Jacobs Associates – tunnel design; and Arup – geotechnical investigations.

## REFERENCES

Arup. 2009. *Final Geotechnical Data Report*, Prepared for the Transbay Joint Powers Authority, August 14, 2009.

Jacobs Associates. 2009. *Numerical Analysis of Tunnel Excavation and Support Methods Preliminary Design Phase,* Prepared for the Transbay Joint Powers Authority, May 26, 2009.

Klein, S., Hopkins, D., Townsend, B.F., Penrice, D.J., and Sum, E. 2007. Evaluation of Large Tunnels in Poor Ground – Alternative Tunnel Concepts for the Transbay Downtown Rail Extension Project, *Rapid Excavation and Tunneling Conference.*

Penrice, D. and Townsend B. 2009. Proposed Contracting Practices for the Caltrain Downtown Extension, *Rapid Excavation and Tunneling Conference.*

Transbay Joint Powers Authority. 2010. Official website of the Transbay Joint Powers Authority. www.transbaycenter.org. Accessed January 2010.

# A Tale of Two Capitals: Modeling Helps Designers Manage Strong Surge and Pneumatic Forces in Deep Combined Sewer Storage Tunnels

**Daniel J. Lautenbach, John K. Marr, Peter Klaver**
LimnoTech, Ann Arbor, Michigan

**John F. Cassidy**
Greeley and Hansen, Washington, DC

**David Crawford**
London Tideway Tunnels Delivery Team, London, England

ABSTRACT: Rapid advancements in tunneling technology and the need to avoid surface disruptions have made deep tunnels popular options for dense urban areas to store and transport large volumes of untreated combined storm and waste water. The District of Columbia and Thames Water Utilities (London) have selected large branching tunnels to capture combined sewage for treatment. Rapid filling is a given for these types of systems and presents many design challenges due to the massive forces involved. Wastewater geysers, blown manholes- even structural failures can result if not adequately considered. Advanced computer modeling used to help designers avoid these problems in Washington and London is showcased.

The District of Columbia Water and Sewer Authority (DCWASA) and Thames Water Utilities (TW) are each designing large tunnel systems to meet pollution control requirements and reduce flooding potential. Plans for both national capitals are to rapidly design, construct and put the tunnel systems in operation. An innovative computer model, based on research by Vasconcelos and Wright, has been simulating rapid filling, transient surges and pneumatics as an integral part of the evaluation of the tunnel designs, particularly in managing or controlling adverse conditions. The tunnel model, called SHAFT, simulates both open-channel and pipe-filling bores, the flows and forces at work and predicts locations and volumes of air entrapment and displacement. The model uses a novel shock capturing technique to simulate both flow regimes and the transitions from open channel to closed conduit flow. The model further predicts where air will be trapped in otherwise unexpected portions of the interconnected tunnels due to the piston-like action of hydraulic surges, the timing of different surges and wave reflections. Model results for a matrix of hydrographs, initial fill levels, control alternatives and tunnel profiles have been used to adjust the location and sizing of drop shafts and vents to avoid dramatic sewage geysers and loud air releases in both capitals. Some general lessons learned are presented and discussed that will be informative to tunnel designers elsewhere.

## INTRODUCTION

Combined sewer overflows are major sources of contamination to urban rivers and streams (U.S. Environmental Protection Agency 2004a; CIWEM, 2004). Combined storm water and sanitary sewer systems are especially prevalent in older cities of the Northeastern and Midwestern U.S. and in England. During wet weather events, total inflows to these systems exceed wastewater treatment capacity and are discharged untreated to surface waters. U.S. CSO policy requires cities to implement Long-Term CSO Control Plans (LTCPs) sufficient to meet water quality standards or meet presumptive criteria, which include overflow frequency and/or percent of flow captured and treated (U.S. Environmental Protection Agency 2004b).

Sewer separation is expensive, often leads to an increase in pollutant loadings, and is disruptive for urban communities; most communities look to storage of wet weather flows to meet the regulatory requirements and CSO policy. If storage facilities are properly sized, storm flows for many rain events can be captured and routed to wastewater treatment after elevated storm flows have subsided. Sizing improvements to minimize overflows and maximize capture can be based on watershed modeling

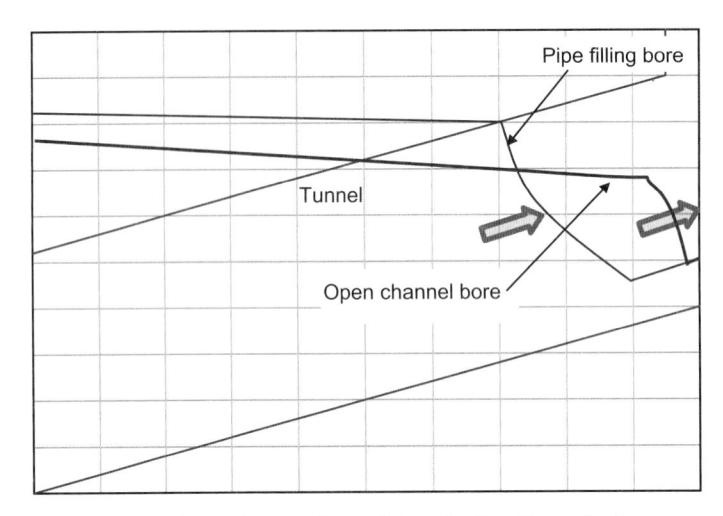

**Figure 1. Comparison of pipe filling and open channel bore hydraulic grade lines**

of representative events to simulate flows entering the system, combined with hydraulic modeling to route flows through the sewer system. Computer programs to perform these simulations are widely used and well developed. Many communities (including Chicago IL, Minneapolis MN, Portland OR, and Toledo, OH) have installed storage tunnels, and others like Washington, DC and London, England are including tunnels in their CSO control plans. Attractive attributes of tunnels include economies of scale and generally lower community impacts.

As storage and conveyance tunnels fill, it is also important to consider the potential effects of transient surges, including high hydraulic grade lines (HGLs), geysering caused by trapped air, and the forces created by rapidly moving bores. Predicting these transient impacts is beyond the capabilities of the watershed and collection system software packages mentioned above. Accurate simulations of rapid filling dynamics are important to prevent accidental venting of captured sewage, to ensure the safety of tunnel operators and the public, and to prevent damage to sewer system and tunnel infrastructure.

In collaboration with scholars at the University of Michigan and the University of Brasilia, LimnoTech developed a software program entitled SHAFT (*Surge and Hydraulic Analysis for Tunnels*). SHAFT simulates all stages of the tunnel filling process, including the creation of and transitions between open channel and pipe filling bores as the tunnel fills and the locations prone to air entrapment.

The District of Columbia Water and Sewer Authority (DCWASA) is designing a large tunnel system to capture and store combined sewage generated in the City of Washington for later treatment, as is the Thames Water Utilities (TW) serving London, England. The SHAFT model is being used in both capitals to simulate transient hydraulic grade lines and air pocket formation for planned tunnel geometry, using a variety of tunnel filling and dewatering scenarios. These simulations support and test surge control strategies and the adequacy of design modifications to prevent adverse impacts of tunnel-filling surges, insure the needed CSO control and avoid costly oversizing.

This paper briefly describes the SHAFT modeling framework, contrasts its structure and performance with pre-existing transient surge models, and presents sample results and lessons learned from transient surge modeling performed to support designs for the Washington and London tunnel systems.

## SURGE MODEL FRAMEWORK

### Overview of the Rapid Filling Problem in Tunnels

Non-steady flow during rapid filling of sewer system storage tunnels produce shock fronts that stress physical structures and can result in discrete pockets of entrapped air, often associated with bore reflections from tunnel transitions. These fronts can take the form of either pipe-filling bores (PFBs) or open channel bores. These fronts are flow discontinuities that move through the tunnel system like a moving hydraulic jump. The discontinuity can reach from open water in the tunnel to the crown of the pipe (a PFB) or can create a jump in the water surface itself, without pressurizing the tunnel (an open channel bore). These bores, illustrated in Figure 1, can strike the end of a tunnel with tremendous force, and drive air and entrained sewage upwards into dropshafts encountered along the length of the tunnel. If there is a sufficient surge in the resulting hydraulic grade line (HGL), flooding to grade and backups at low points in the collection system can occur (Guo and Song, 1990).

An open channel bore, as shown by the thicker line in Figure 1, can cause the problems described above, and in addition can trap pockets of air when this bore reflects off a tunnel transition point or meets another bore within the system. When air pockets are not ventilated, they reduce the volume of CSO storage available, and also reduce the conveyance capacity of the tunnel. Air is compressible, so a rapid increase in pneumatic pressure can occur when the air pocket is squeezed by water in the tunnel (Zhou et al., 2002). In addition, trapped pockets of air rise rapidly when they migrate to points in the tunnel that are accessible to the surface. The rapid rise of air at access points that are already partially filled with captured sewage can push captured sewage upwards, resulting in buoyancy-driven geysering. This geyser effect is different than the phenomenon of high peak HGLs at dropshafts mentioned above, and depends both on the size of the trapped air pocket and the geometry of structures that would provide venting. Vertical shafts that have a relatively small cross sectional area can inhibit the downflow of water past the escaping air as it rises, and these are more prone to geysering. Therefore it is important to determine both the sizes of pockets and locations where they form, and evaluate venting points near these locations.

**Model Development**

In collaboration with scholars at the University of Michigan and the University of Brasilia, LimnoTech developed a surge modeling tool named SHAFT (*Surge and Hydraulic Analysis for Tunnels*). SHAFT is based on a program of research by Vasconcelos and Wright (all citations) into methods of numerically simulating surge behavior in tunnels. SHAFT simulates both open channel and pipe-filling bores, and predicts locations of air entrapment. The model utilizes a shock capturing solution technique that decouples hydrostatic pressure due to water depth in the conduit from surcharge pressures occurring only in pressurized conditions, and takes advantage of the structural equivalence between unsteady incompressible flow equations in elastic pipes and unsteady open channel flow equations in the model governing equations. These two concepts allow SHAFT to simulate both flow regimes using the same generalized set of equations, and to readily model flow transitions (i.e., from open channel flow to closed conduit flow) (Vasconcelos et al., 2006). Predictions from the more widely applicable method used in SHAFT have even been compared for pressurized flow situations fitting the more narrowly applicable Method of Characteristics models and the results are favorable where the two methods should agree. (Vasconcelos and Wright, 2007).

SHAFT implements a non-linear finite volume numerical scheme that utilizes an approximate Riemann solver applied to the Saint Venant unsteady flow equations. This approach was adapted from finite volume method schemes that have successfully used Riemann solvers for other highly dynamic fluid flow problems. The use of a nonlinear numerical scheme minimizes problems associated with smearing of flow discontinuities associated with shock phenomena such as hydraulic bores. The benefits of using a non-linear scheme such as the one employed in SHAFT have been demonstrated by Machione and Morelli (2003).

Air is currently represented in SHAFT as void space; that is, volume not occupied by water. A PFB tends to expel all trapped air at upstream vent shafts, assuming there is sufficient venting capacity, but the reflection of the bores from ends of tunnels can result in discrete trapped air pockets, and both phenomena are simulated by SHAFT. This identifies locations and sizes of vent shafts needed at intermediate locations, in addition to ventilation that may be needed at the ends of tunnels and thereby avoids buoyancy-driven geysering of water in the path of migrating air pockets.

**Laboratory Studies**

The SHAFT model formulation was developed and favorably compared to data from extensive laboratory testing of scale models in the University of Michigan hydraulics lab, which included both PFB and sub-PFB filling conditions. The approach of that work was to generate a wide range of filling behaviors in a physical model of a tunnel system and then develop a numerical model that could accurately reproduce them (Vasconcelos and Wright, 2005; Vasconcelos et al., 2006). The ability of the SHAFT model to predict a flow regime transition from free surface flow to pressurized flow that is not coincident with the movement of an open channel bore gives the model the ability to realistically define tunnel dynamics under a variety of inflow scenarios and to predict large air pockets that can cause geysers if not vented appropriately (Wright et al., 2007). The model framework has also been compared to laboratory and numerical studies by others. The SHAFT model framework simulates the full range of conditions and shows good agreement with the results of these laboratory and numerical studies that were focused each on narrow transient flow situations (Vasconcelos and Wright, 2007). This model can therefore be used confidently and can negate the need for more expensive physical models.

## DISTRICT OF COLUMBIA SYSTEM

The DCWASA LTCP (2002) includes a range of controls to reduce the volume of combined sewage overflow (CSO) to receiving waters in an average year from approximately 2.5 billion gallons (with current in-system controls fully operational) to 138 million gallons per year; the controls will also reduce the number of overflow events from 75/74/30 overflows per year in the Potomac, Anacostia and Rock Creek receiving waters (currently) to 2/4/4 overflows per year under the LTCP. Capture of combined sewer flows in storage tunnels includes a 49 million gallon storage tunnel from Poplar Point to the Northeast Boundary Outfall and a 77 million gallon storage and conveyance tunnel parallel to the Northeast Boundary Sewer. These tunnels were also designed to relieve flooding in sections of the city and provide a higher level of control for design storms not reflected in the average year CSO statistics. The proposed tunnel system has since been extended in a plan supplement (DCWASA, 2007) to include a tunnel from Poplar Point to the DCWASA Blue Plains advanced wastewater treatment plant. The expansion increases total tunnel storage from 126 million gallons to 157 million gallons, and the apportionment of that volume among the various tunnel sections is to be determined in the course of detailed facilities planning. The additional tunnel storage was added to reduce total nitrogen discharge and meet a new requirement that was not in effect at the time the plan was finalized.

### Tunnel Layout and Inputs

Several vertical and horizontal tunnel alignments were developed to intercept 14 CSOs that would otherwise discharge to the Anacostia River and to provide flooding relief to sewers in the Northeast Boundary Drainage area. Factors causing geometry changes included the addition of the Blue Plains Tunnel (BPT) that will connect the Anacostia River Tunnel (ART) to the Blue Plains WWTP, geotechnical conditions encountered in the field, tunnel constructability, siting and accessibility concerns, desire to maximize storage of the system built in 2025 below an elevation of 3 feet above sea level, high HGLs predicted during modeling of early designs, and simulated tunnel responses to extreme events. Of these factors, the simulated HGL responses to peak events required the greatest modifications to model geometry. The geometry simulated in the SHAFT model of the DCWASA tunnel system evolved throughout the modeling process and guided tunnel design. Tunnel geometries naturally also varied depending on the route of the BPT. Modeled tunnels included the Northeast Boundary Tunnel (NEBT) and associated spur tunnels, the ART, and the BPT. As modeled the Complete Tunnel System to be built by 2025 consisted of approximately ten miles of 23 foot diameter tunnel and three miles of spur tunnels ranging from 8 to 15 feet in diameter. Simulations were performed for two basic configurations: the "Complete Tunnel," consisting of all three component tunnels, and the "Intermediate Tunnel," consisting only of the ART and BPT. The Intermediate Tunnel will be constructed first, and is expected to be operational on its own while the NEBT and associated spurs are constructed in a second phase. Horizontal alignments for the proposed system are shown in Figure 2.

Flows used to drive the SHAFT filling scenarios were generated using a hydrologic/hydraulic model of the DCWASA collection system. Simulated flow into and out of dropshafts was regulated in several ways. Flows to the tunnel system were modeled as free-flowing outfalls in the collection system model; that is, no backwater effects produced by a full tunnel were simulated. However, inflows to SHAFT were restricted where applicable, either by flap gates that closed at specific elevations, or by head-based diversion curves that were developed independently. Flow out of drop shafts (and from the tunnel) were simulated either by free-flowing weirs, or by head-discharge curves that were developed for various overflow points, based on river water surface elevations; several outfall conditions were simulated including the 10-year flood and 100-year flood river surface elevations. SHAFT scenarios used hydrographs of storm frequencies ranging up to the 100-year, 6-hour event, including possible future connections and inflows to the tunnel system that may be added when the Complete Tunnel System begins operation in 2025.

### Conditions Simulated

The conditions simulated could be varied in three essentially independent ways: tunnel layout/geometry, filling scenario, and river surface elevation. Simulation conditions that were evaluated in coordination with the DCWASA tunnel designers as part of the surge modeling process included the following:

- several storms taken from 1988 to 1990 that generated the largest CSO response in the collection system;
- tunnel filling for the 15-year, 6-hour storm inflow hydrographs for present and future wastewater flows;
- tunnel filling for the 100-year, 6-hour storm hydrographs for present and future wastewater flows;
- partial fill antecedent conditions i.e., the tunnel is incompletely de-watered from a previous rainfall event;

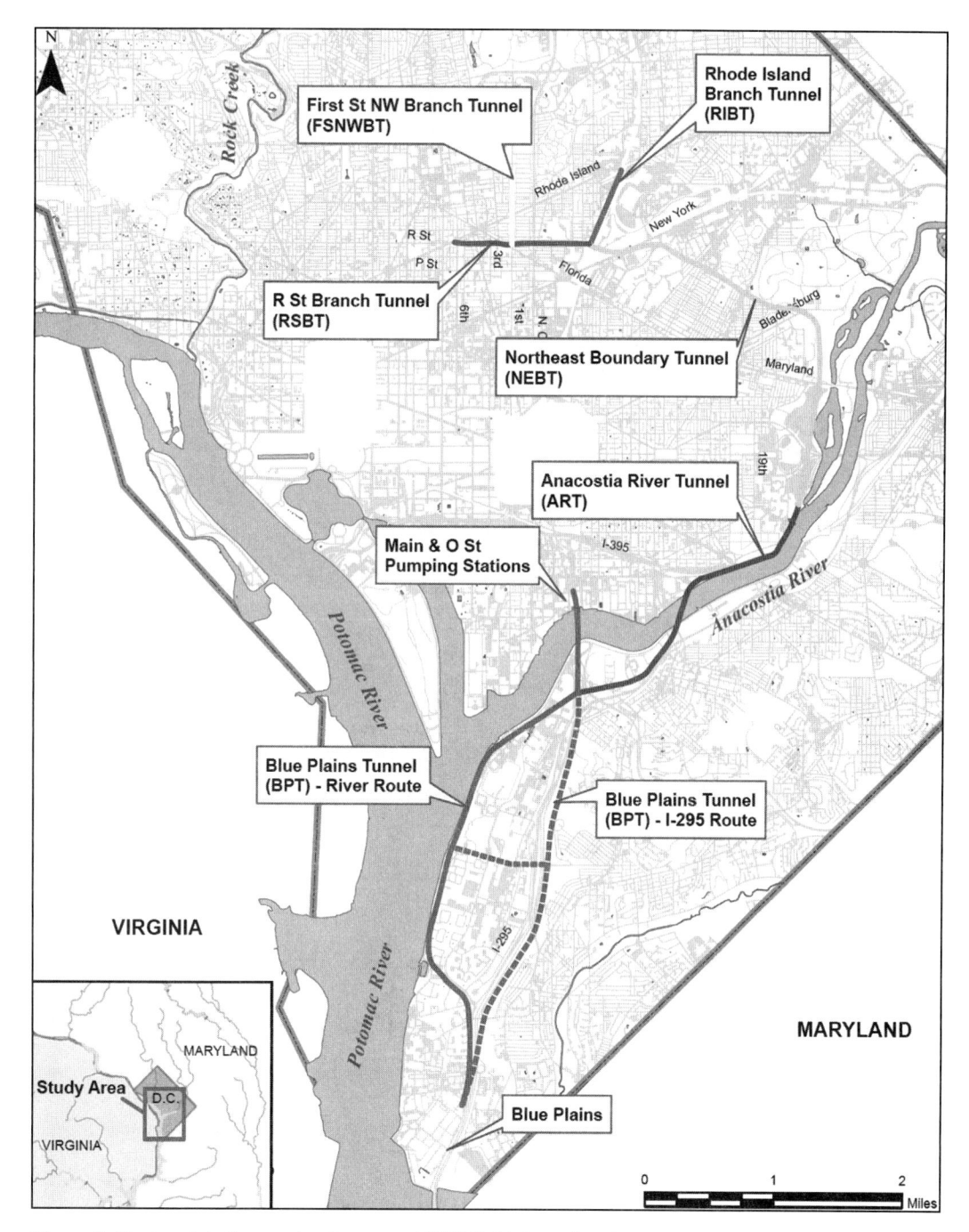

Figure 2. Horizontal alignments of proposed DCWASA tunnel system

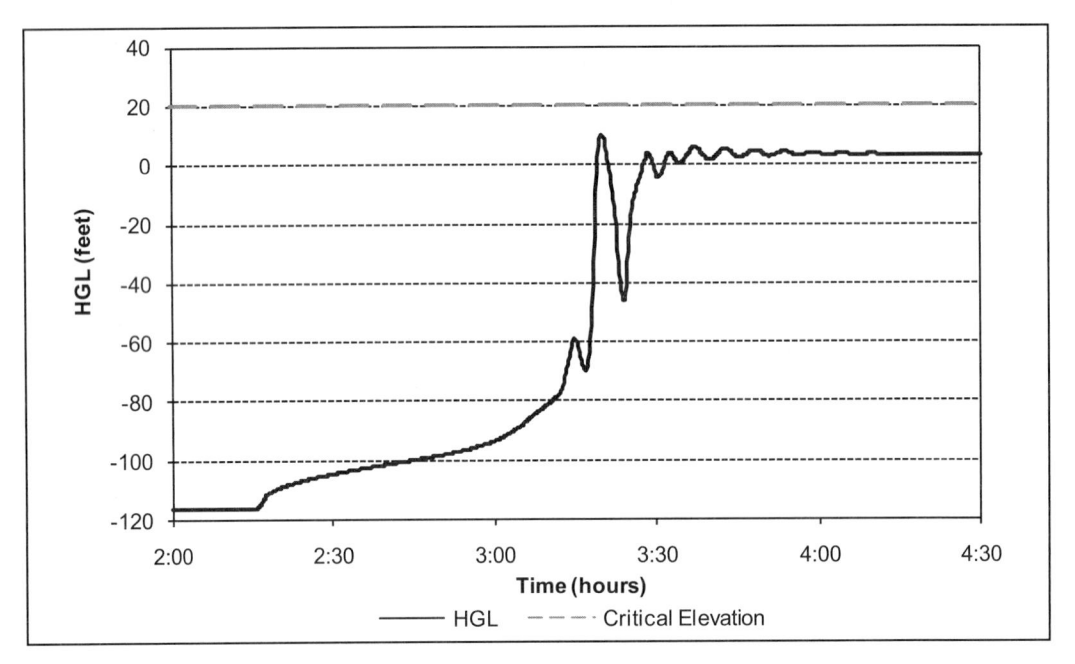

**Figure 3. Hydraulic grade line as a function of time at the Blue Plains WWTP pump station during the 100-year 6-hour storm**

- scenarios using the 10-year and 100-year Potomac and Anacostia river surface elevation;
- simulations of a large storm moving west to east across the drainage area; and
- BPT tunnel route following I-295 and one following the Potomac River.

Among all of the resulting simulations, the 100-year, 6-hour storm events for the Complete Tunnel System with future flows caused the largest concern in predicted tunnel performance, with initial predictions of flooding to grade at several dropshafts at the upstream end of the tunnel system. Consequently, several different changes to tunnel geometry were evaluated to reduce peak HGLs. These changes in tunnel geometry can be broken out into the following categories:

- changes in the slope of the NEBT or other branch tunnels;
- changes in the vertical alignment of the NEBT tunnel and/or other branch tunnels;
- addition of offline storage at or increasing the sizes of various shafts at various elevations; and
- increasing the sizes of selected upstream spur tunnels to increase linear storage.

**Results**

The tunnel surge model was used to develop economical and passive design features that are predicted to successfully control hydraulic surge and pneumatic challenges in the DCWASA tunnel system in both the intermediate and full service area build-out situations. Steepening the NEBT to a slope of 0.0045 ft/ft and crown-to-crown vertical alignments of spur tunnels produced the greatest benefit in reducing peak HGLs in the upper sections of the tunnel system, and also reduced excessive HGLs due to reflections in other sections of the tunnel system as well.

Comparison of peak HGLs within the BPT and ART tunnels showed a potential for more severe HGLs while the Intermediate Tunnel is in operation and prior to the completion of other tunnel sections in 2025. These peak HGLs are short duration, and are primarily caused by the reflection of surge waves off the upper end of the ART tunnel before the CSO 019 overflow structure is engaged. Once the Complete Tunnel System is in operation, upstream storage within the NEBT and other high spur tunnels and the initiation of overflow to the Potomac and Anacostia Rivers before surge waves reach the upper end of the Complete Tunnel System serve to minimize the height of surge generated peak HGLs. Figures 3 and 4 show example model predicted hydraulic grade lines at the Blue Plains Pump station and another tunnel overflow location as a function of time for the

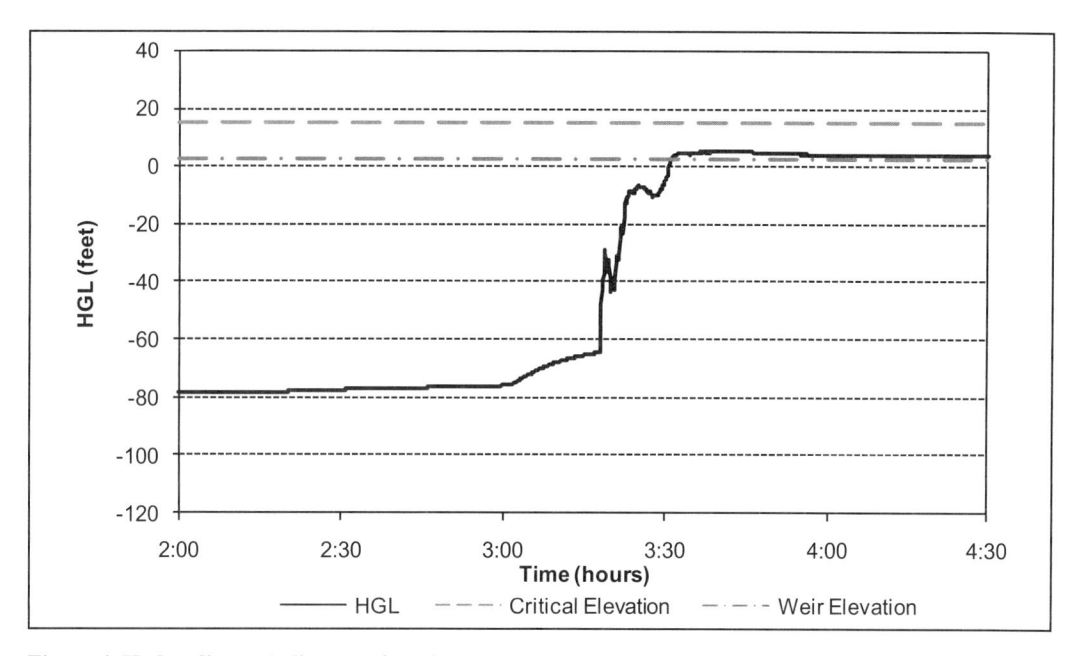

Figure 4. Hydraulic grade line as a function of time at the southerly CS0 019 overflow during the 100-year 6-hour storm

Complete Tunnel System filling under the 100-year, 6-hour storm design scenario. Figures 5, 6, and 7 show the tunnel filling process for the same storm at two different stages before the tunnel reaches steady state conditions, and also approaching steady state after the tunnel has filled and is discharging to the Potomac and Anacostia Rivers.

Figure 5 show two distinct bores: an open channel bore near the CSO 019 dropshaft and a second, pipe-filling bore just upstream of the CSO 007 dropshaft. Figure 6 shows that ten minutes later in the scenario, the second bore has caught up with first, and there is now a single pipe-filling bore approaching the Mt. Olivet dropshaft. In Figure 7 it is seen that inflows have receded, especially in the upper reaches of the NEBT (at the right of the figure), and the system is overflowing at CSO 019.

Other scenarios evaluated included various combinations of the filling scenarios and river surface elevations listed previously. None of these additional conditions resulted in adverse HGLs or surge conditions. The entire suite of tunnel-filling scenarios listed above was also evaluated using the BPT I-295 route for both the Intermediate and Complete Tunnel System.

**Maximum HGLs Between Drop Shafts**

Maximum HGLs between shafts were also determined using the SHAFT model simulation of various tunnel filling scenarios. Figures 8 and 9 show the maximum HGLs for the 100-year storms in both the Intermediate Tunnel System and Complete Tunnel System for the BPT River Route. Predicted peak HGLs within the tunnel rise higher than predicted peak dropshaft HGLs. This is primarily because dropshafts have storage available to mitigate pressure spikes due to hydraulic transients caused by the filling process, whereas tunnel sections do not have extra storage if they are already filled when the transients occur. Similarly, peak tunnel HGLs rise higher in the Intermediate Tunnel simulations than in the Complete Tunnel simulations within the common sections since the steeply sloping section of the NEBT dampens peak HGLs for the entire system. In the Intermediate Tunnel System, surge waves that reach the upstream of the tunnel under extreme storms reflect and cause pressure spikes in the tunnel.

**Minimum Gage Pressures**

The SHAFT model was used to predict negative pressures in the tunnel system, which can occur when surge waves are reflected through reaches that are already surcharged. Although highly transient, the magnitude of these pressures is of interest because of the stresses they impose on the tunnel walls. Figure 10 shows the composite maximum vacuum pressures in the Intermediate Tunnel System for the 100-year event. The strong negative gage pressure in the section between CSO 019 and CSO 018 occurs after an open channel bore reaches the

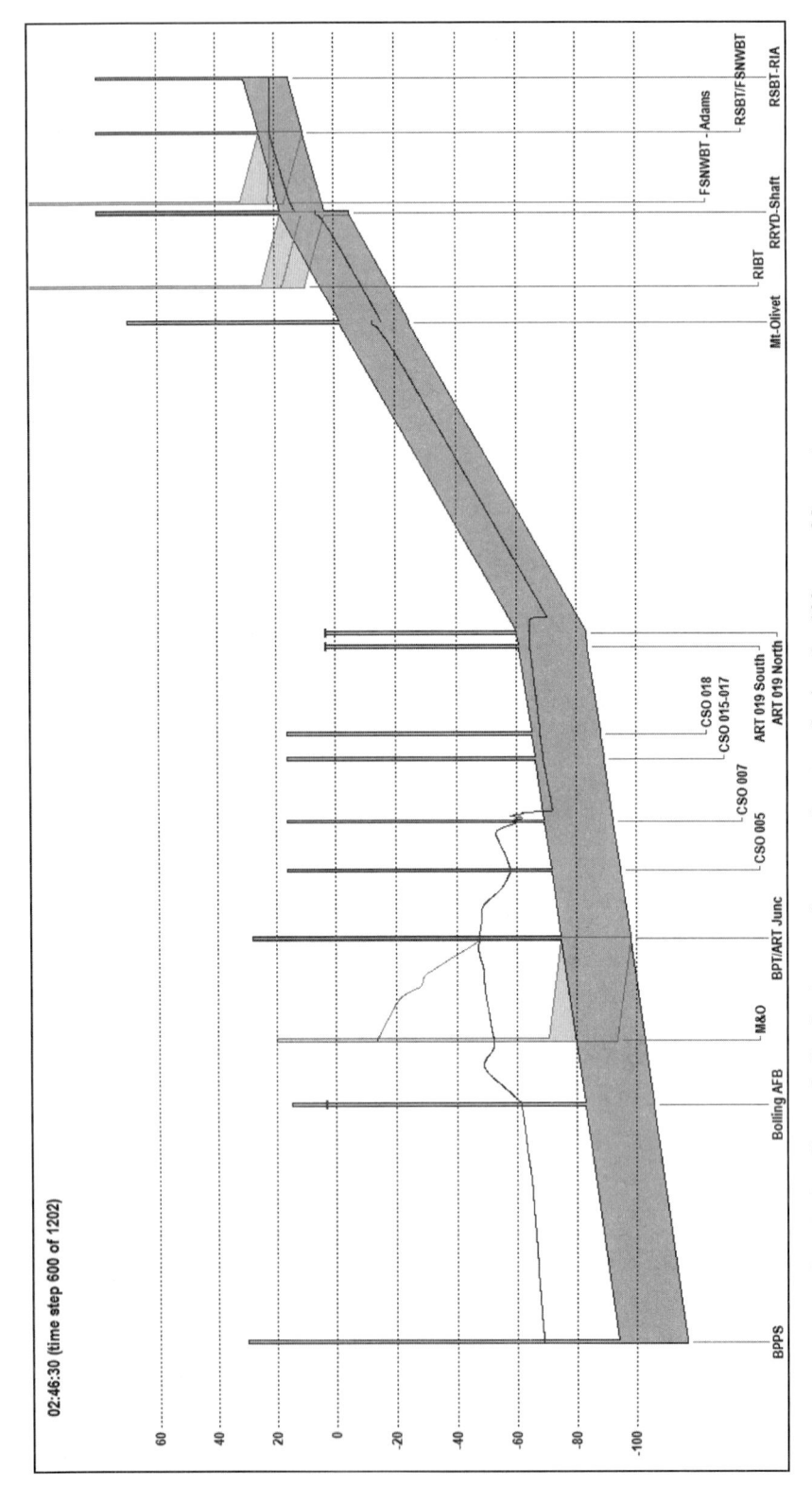

Figure 5. Snapshot of the hydraulic grade line in the complete tunnel system during the 100-year 6-hour storm

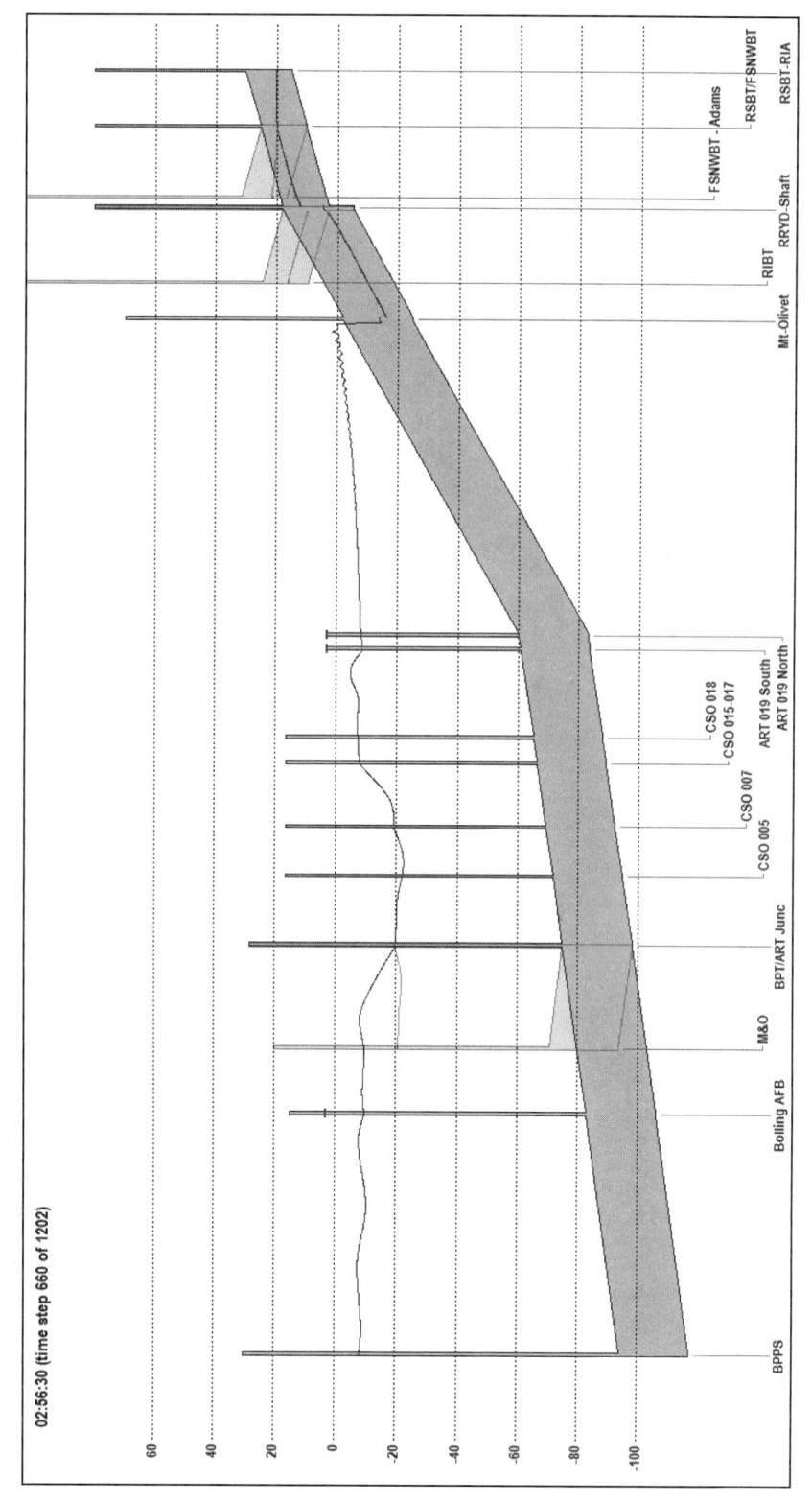

Figure 6. Later snapshot of the hydraulic grade line in the complete tunnel system during the 100-year 6-hour storm

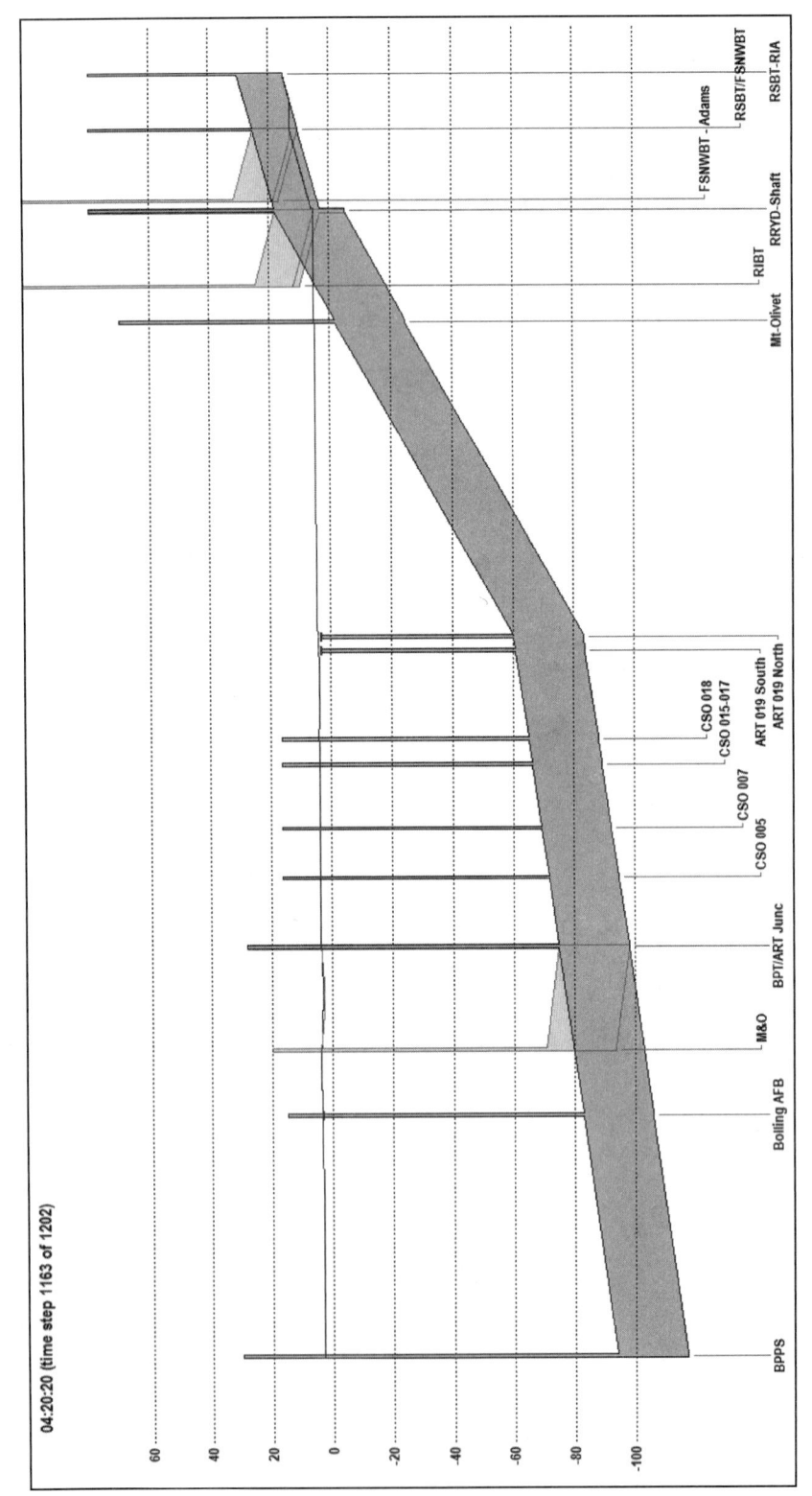

Figure 7. Snapshot of the hydraulic grade line in the complete tunnel system during the 100-year 6-hour storm approaching steady state conditions

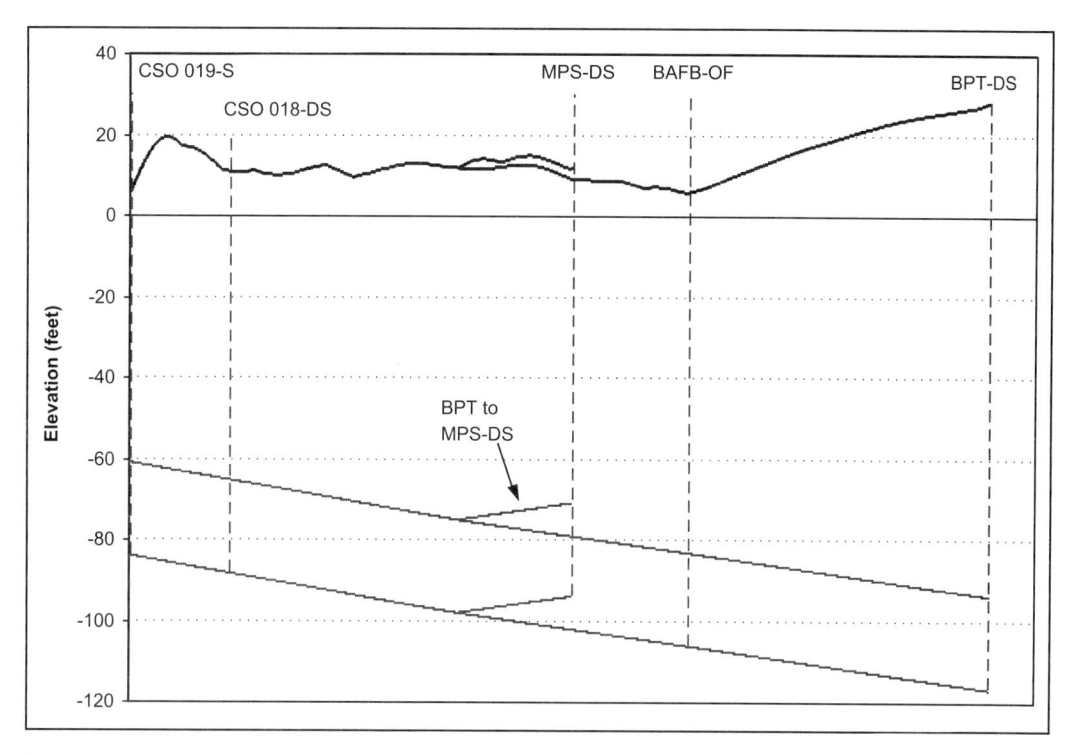

Figure 8. Maximum HGL for intermediate tunnel BPT river route: 100-year event

upstream end of the tunnel, reflects off the upstream end and travels back downstream as a pipe filling bore. Meanwhile, a second pipe filling bore develops, traveling up towards CSO 019, and these two bores collide in the tunnel between CSO 018 and 019, creating a large pressure spike in the now completely surcharged tunnel. When this surge wave reaches CSO 018 it is reflected back up the tunnel toward CSO 019 as a negative pressure surge. The negative pressures shown in the figure below do not occur at the same time and do not take up a large portion of the tunnel section between CSOs 018 and 019 at any one time. Nor do they occur for very long; the passage of the negative surge takes less than 10 seconds to pass from dropshaft to dropshaft. This figure also shows that the Main & O Pump Station branch experiences negative surge waves in these simulations as well.

In general, the most extreme minimum gage pressures calculated by the model occur when two bores meet within a tunnel segment away from a dropshaft. Model simulations showed that spur tunnels were more likely to experience negative gage pressures during extreme filling events, as well as the section of tunnel bounded by the CSO 018 and 019 dropshafts when the Intermediate Tunnel system is in operation.

SHAFT simulations have shown that surges and transients generated as part of the DCWASA tunnel filling process during extreme storms can be successfully mitigated through the selection of appropriate geometry. Current tunnel geometry successfully mitigates the possibility of the creation of large tunnel HGLs at most dropshafts during the tunnel filling process for the extreme events and critical conditions considered. Where model-predicted peak HGLs do temporarily rise above critical elevations, mitigation measures have been evaluated at selected dropshafts for both the Intermediate and Complete Tunnel systems, and solutions have been found using the SHAFT model. The following passive mitigation measures were considered or employed:

- Design the shaft to withstand the hydraulic pressure—The shaft structure and all openings are designed to take the predicted hydraulic pressure, plus a safety factor.
- Extend shaft above predicted surge elevation—For this option, the top slab of the shafts was extended above the predicted surge elevation, plus a freeboard allowance.
- Add storage at the shaft sites—This option involved adding storage to contain the predicted surge volume. When CSO influent rises during an extreme surge event, it

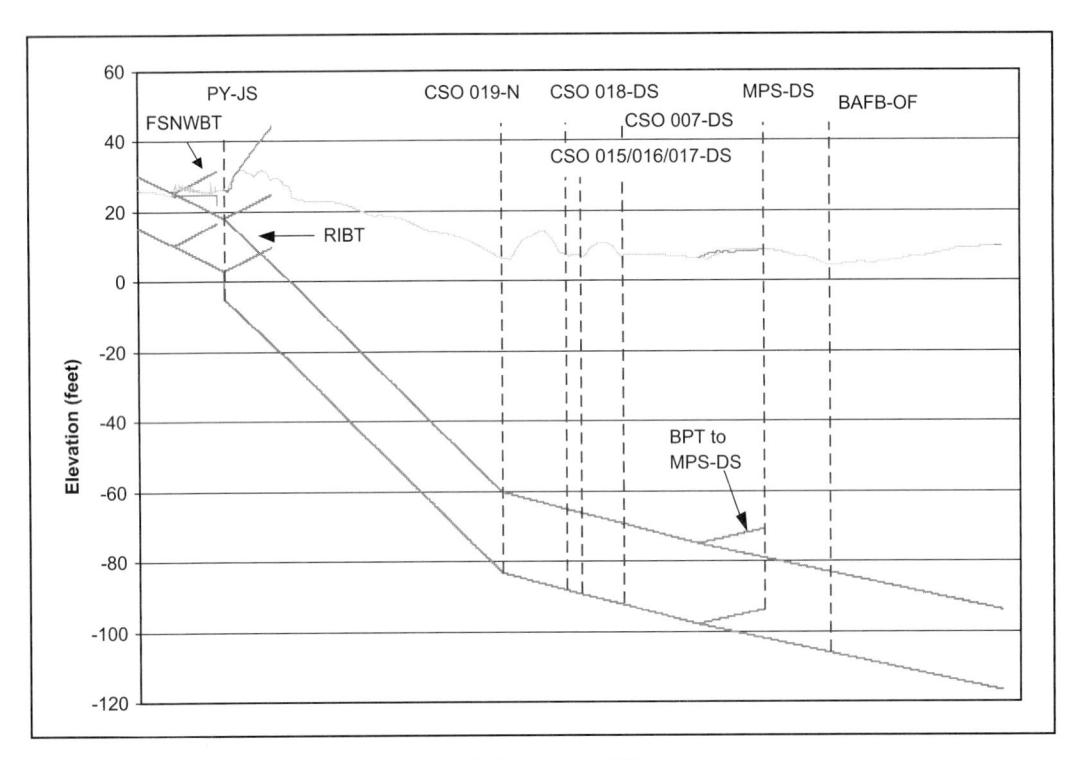

Figure 9. Maximum HGL for complete tunnel river route: 100-year event

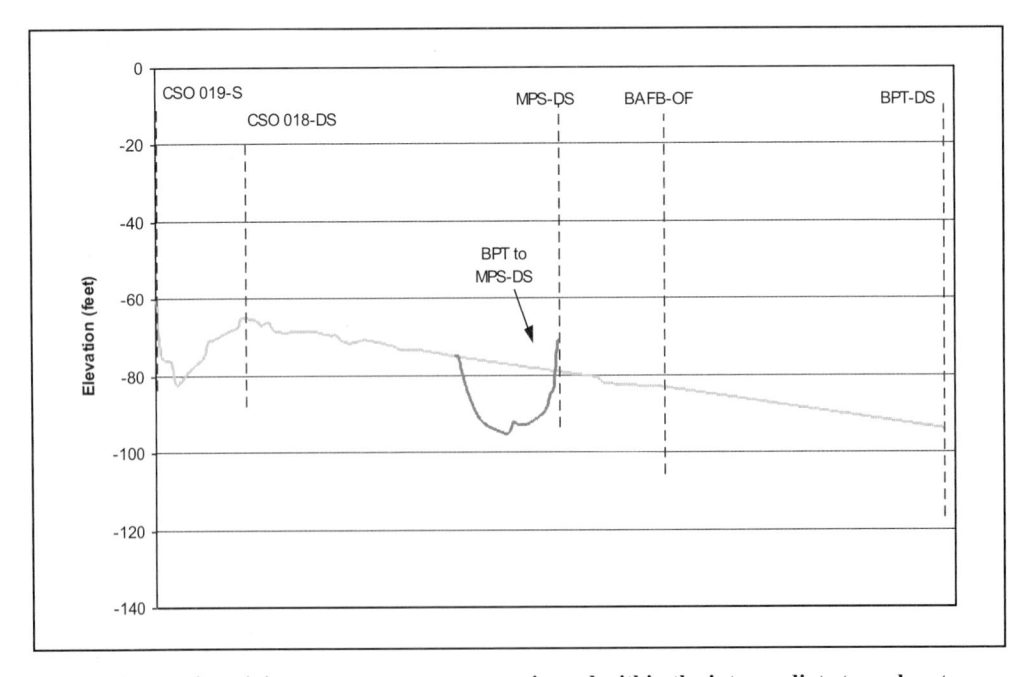

Figure 10. Composite minimum gage pressures experienced within the intermediate tunnel system during the 100-year 6-hour storm (river route)

overflows an annular weir and is contained in the surrounding temporary storage as part of the drop shaft. Any stored volume flows by gravity back to the tunnel.
- Planned surge relief to the river or to existing CSOs—The approach involved adding an overflow weir near the top of the shaft. When the surge overflows the weir, it is conveyed to an existing CSO overflow (if capacity were available) or to a new outfall to the river.
- Combinations—The above alternatives were also combined at some locations to get the most effective solutions.

## Venting

Tunnel venting facilities must not only handle the large quantities of air displaced by water rapidly filling the tunnel, but must also provide for the venting of air pockets trapped by reflecting surge waves. In addition, water falling in the drop shafts will entrain air that also needs to be controlled. All predicted air pockets were reasonably small, and the oversized design of the drop shaft structures were predicted to provide sufficient relief for both trapped air pockets and for venting in general. Air management evaluations, including calculation of peak venting and inflow rates to the tunnel, as well as determining locations of large air pocket development, were performed. In the Intermediate Tunnel System, air entrapment was predicted to occur primarily in the upper end of the tunnel system in the ART and the BPT. When the Complete Tunnel System is in operation, smaller pockets of air may be trapped in the same locations. The model also predicted that the NEBT could develop air pockets during filling at two locations: a short distance up-tunnel from the transition between the ART and NEBT, and a short distance downstream of the upstream end of the NEBT. Minimum vent and air intake areas were calculated using a peak acceptable air velocity of 3,000 ft/minute; the largest cross sectional area needed for venting at any vertical shaft was approximately 160 ft$^2$.

## THE LONDON TUNNEL SYSTEM

Like the DCWASA tunnels, the London tunnels is a CSO storage and conveyance (to treatment) system. The London tunnels will also be constructed in phases, with the first phase (the Lee Tunnel) being completed by 2015 and the second phase (the Thames Tunnel) targeted for completion by 2020. The objective of these tunnels is to reduce the frequency and volume of combined sewer overflow into the tidal River Thames and its tributaries within Greater London. SHAFT modeling is being conducted to ensure that the Lee and Thames Tunnels will not experience excessive HGLs, causing flooding to grade or backups in the existing collection system and to predict venting rates at relief points and locations where trapped air pockets may develop.

### Tunnel Layout and Inputs

A layout of the Lee Tunnel operating alone, when it is completed in 2015, and several layouts of the Lee and Thames tunnels operating in concert were evaluated for this study. The various combined Lee and Thames layouts were designed to evaluate how different tunnel lengths, and arrangements of tunnel connections, would affect the filling process, storage volume captured and cost of construction. The London Tideway Tunnels (LTT) system will consist of about 29 kilometers of 7.2 meter main tunnel and 9 kilometers of connection tunnels ranging in size from 2.2 meters to 4.5 meters. The LTT, in conjunction with significant capacity expansions of two major treatment works, will capture about 96% of the typical year CSO volume and reduce spills to less than 4 events per year.

The transient and pneumatic performance of the Lee Tunnel (operating alone) and the complete tunnel systems were tested using output from a hydrologic/hydraulic model of the greater London collection system (large portions of which are Victorian brick sewers that are and will be continued in use). Most simulations were based on inflows produced from rainfall corresponding to a 15-year, 2-hour event, which, when applied to the entire catchment area at the same time, was equivalent to a 50-year event at a more local scale. The simulated flows delivered to the tunnel system include both pumped flows and flows routed to the tunnels via collection tunnels and CSO consolidation structures. Parts of the existing collection system are capacity-limited, and events larger than this event are not expected to deliver significantly greater peak flows to the tunnels. In addition, a 5-year, 2-hour event and large historical storms were used to evaluate tunnel performance under less severe conditions.

Inflow hydrographs were produced by simulating the existing London collection system with various tunnel and collection tunnel alternatives. Within the hydraulic model, the tunnel system was linked to the existing system so that interaction between the existing system and tunnels and real time control of flows to the tunnel system could be simulated. An integrated system model was developed for selecting where CSO discharges would occur first and how often.. The hydrographs included flows delivered to over twenty inlets to the tunnel system, the exact number depended on the arrangement of tunnel geometry under each scenario. Flap gate closure was also simulated based on simulated water levels in drop shafts rising above collection basin elevations. The majority of CSO discharges are pumped to the rivers as the existing collection system near the rivers is below mean tide levels. The relative ground elevation is a key consideration in determining critical water levels in the tunnel system as the value of land and adverse public opinion would preclude structures above ground level.

## Results

Surge modeling identified issues with the various tunnel geometry arrangements during the 15-year storm event, including premature pressurization and excessively high HGLs at locations along the Thames. Premature pressurization, in which intermediate portions of the tunnel become surcharged before the most downstream section of tunnel surcharges, generally results from the cumulative effect of multiple inflow rates that, in sum, exceed the conveyance capacity of the tunnel. Surcharging of upstream tunnel sections in conjunction with bores moving upstream towards the surcharged upstream section can cause large pockets of air to become trapped in the tunnel, increasing the risk of geysering. An example of premature pressurization within the tunnel system is shown in Figure 11 (elevations shown are project datum which are 100 m above ordnance datum). Note the large air pocket that has formed at the lower end of the tunnel, due to surcharging in the vicinity of Abbey Mills Shaft F.

Recommended options for resolving these issues included restriction of peak inflows and total inflow volume to the tunnel, addition of storage and/or relief overflow points to reduce the magnitude of surges, and upsizing of the main Thames tunnel diameter. The first two options were investigated through a series of model runs using several tunnel layouts; upsizing of the tunnel was deemed impractical, primarily because of cost and potential for oversizing the tunnel system solely for transient conditions. This section presents a modification of one alternative, including both inflow controls and offline storage, sufficient to eliminate premature pressurization for both the 5-year and 15-year events, and to reduce peak HGLs to practical and implementable levels.

The first approach taken was to reduce the peak composite rate of inflow to the tunnel to prevent premature pressurization, since a moderation of peak inflow would reduce both the surges and the risk of geysers occurring within the system. Model simulations did not predict premature pressurization in the 2-year event. Based on this observation, controls of inflow were set at all locations to exclude sufficient volume over critical times during a 15-year event such that the total peak inflow was similar to the 2-year rates. Peak HGL levels within the Thames were restricted by collection system concerns and constructability issues at selected dropshafts, so inflow controls were amended to ensure that the final system HGL was below desired system elevations. There was confidence that a workable inflow control scheme could be implemented within the system, since much of the inflow was delivered to the system by pumping stations. Additional offline storage at selected shafts was sized to mitigate transient HGLs caused by the quickly filling tunnel. Another method used for controlling surge was to place overflow relief points at various locations, which has the effect of locally restricting the HGL to the level of the relief point. However, peak Thames River tide levels prevented this method from being used effectively while limiting HGLS below desired elevations. In addition, peak surge pressures at locations between relief points could still exceed the elevation of the points themselves, but these effects were minimized.

Figure 12 is a snapshot of the tunnel HGL during simulated filling with inflow controls in place, which shows that premature pressurization is avoided: a pipe-filling bore advances up the tunnel, but nowhere does the tunnel become surcharged before the bore arrives. For the most part, simulated peak HGLs are kept below critical elevations, with some exceptions occurring at the upstream ends of certain connection tunnels. Figure 12 also shows that the model predicts high HGLs in the smaller diameter portion of the tunnel at the upstream end of the Thames Tunnel between the Acton CSO connection tunnel and the upstream end of the main tunnel.

Figure 13 shows the predicted volumetric venting rates at selected shafts from a 15-year storm simulation with inflow controls. As air pressurization is not explicitly modeled in SHAFT, a simplified methodology for computing venting was employed, based on the reduction in volume available for air within each reach of the tunnel as it fills. The peaks in venting rate are greater than 200 cms and generally occur along the main tunnel. The predicted peak inflow rates at each shaft rise to a peak rate in sequence, moving from downstream to upstream in the Thames Tunnel. These peak rates, which are associated with the movement of bores up the tunnel, are likely to be conservatively high, because it is assumed that all air moving ahead of the advancing bore is exhausted through the first shaft it encounters. In reality, high-rate air flows would be attenuated somewhat by a build-up of pressure, so that some of the air flow would be diverted to shafts located farther up-tunnel.

Figure 14 is a lower envelope showing the minimum gage pressure experienced at each London tunnels location. The greatest negative pressures are seen in upstream reaches of the main tunnel, and in certain connecting tunnels, where the model predicts the collapse of air pockets. The largest negative pressure predicted by SHAFT is approximately negative 3.5 meters at its most extreme point. Overall, the negative pressure resulting from surges is not a

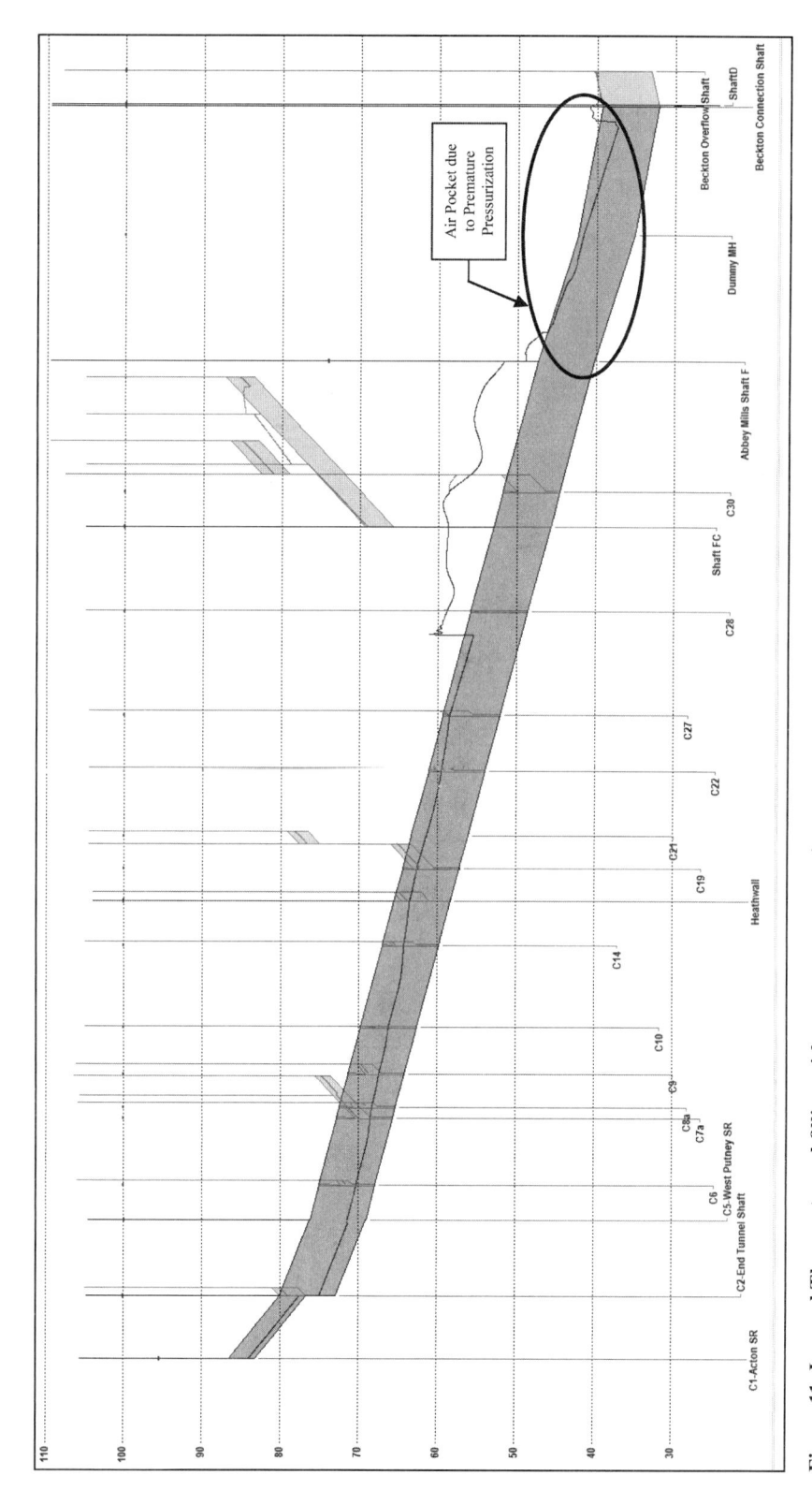

Figure 11. Lee and Thames tunnel filling with premature pressurization

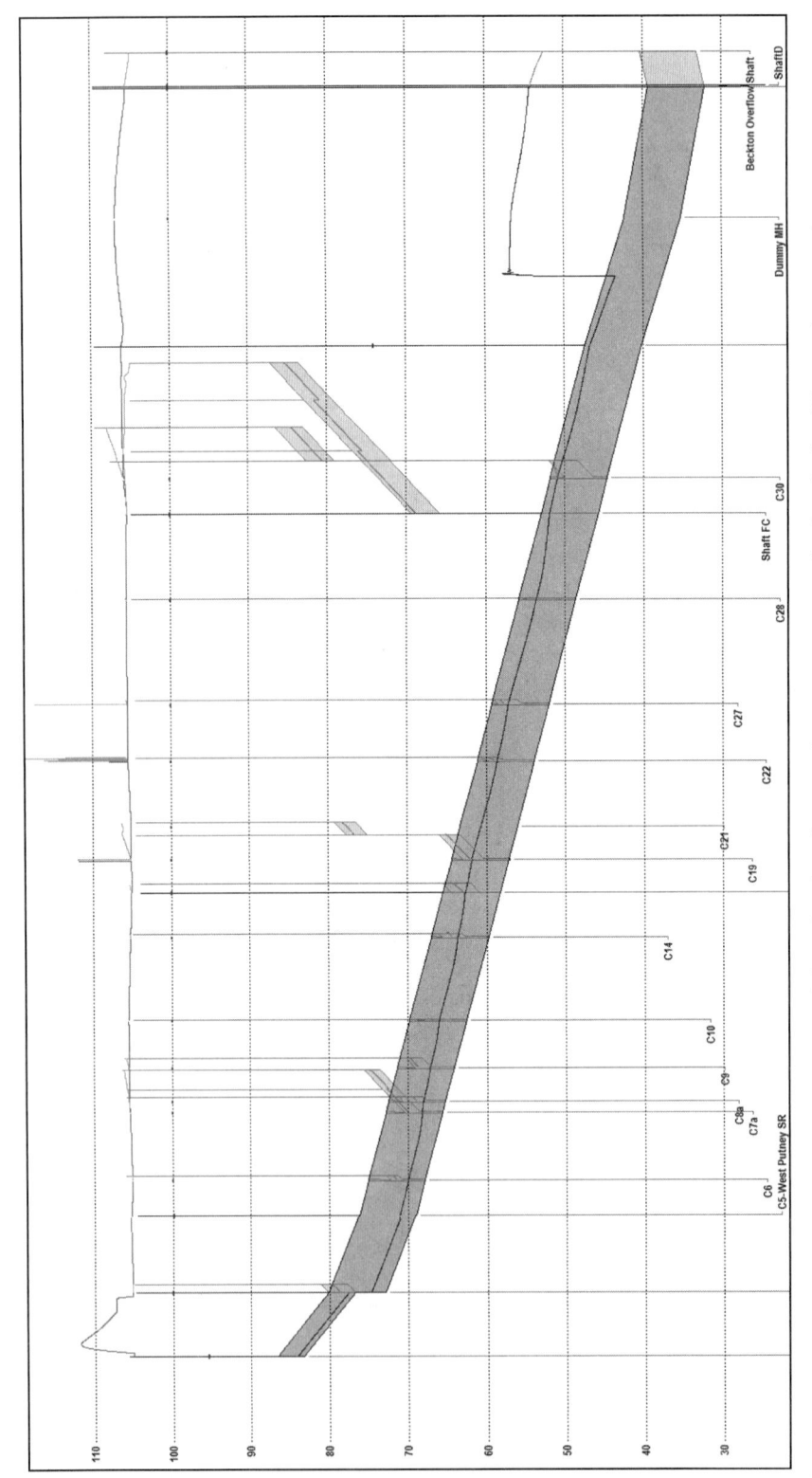

Figure 12. Lee and Thames Tideway tunnel mid-filling, showing absence of premature pressurization during 15-year storm event

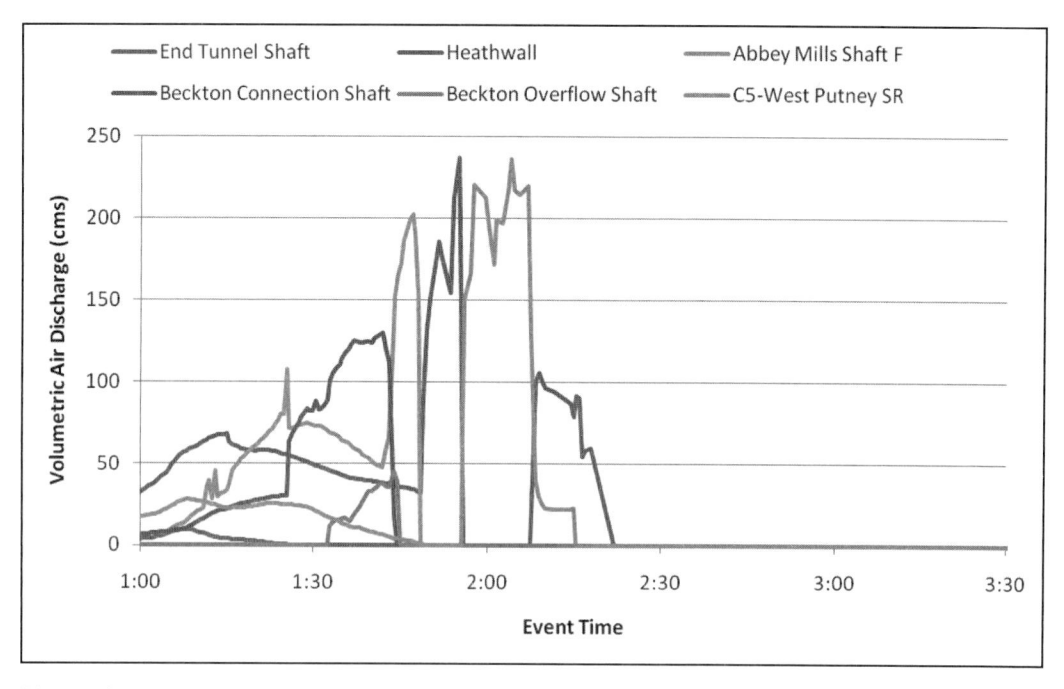

**Figure 13. Venting rate versus time for the End Tunnel, Heathwall, Abbey Mills, Beckton Connection, Beckton Overflow, and West Putney Shaft**

significant issue, with most reaches of the London tunnel experiencing no negative pressure.

The London simulations reported above demonstrate the feasibility of the Lee and Thames Tideway Tunnels, with the inclusion of inflow controls and selected local offline storage. Inflow control and offline storage was essential in the simulations to minimize transient hydraulic grade line fluctuations. The addition of offline storage at the upstream end of the tunnel produced large reductions in peak HGLs in the complete system. Even with offline storage in place, however, inflow controls to the London tunnel are also essential to prevent premature pressurization and the formation of air pockets in the tunnel system. With inflow controls in place, the simulated maximum negative pressure due to air pocket collapse was acceptably low, and the probability of geyser formation was minimized.

## CONCLUSIONS

The SHAFT model is a very useful computational tool for the evaluation and evaluation of alternatives to mitigate surges in large diameter tunnel systems. The SHAFT model's innovative approach for the calculation of PFBs, open channel bores, hydraulic transients and location of trapped air enables users to accurately determine potential problem areas in proposed or existing tunnel systems and evaluate solutions. The simulations performed for the DCWASA and London Tideway Tunnels systems have enabled designers to estimate peak and minimum HGLs, peak venting rates, and potential locations of large air pockets; they can also adapt their designs to minimize the effects of surges and trapped air. This analysis reduced the potential risk of failure of the proposed tunnel system under extreme events, and increased confidence that expensive retrofit solutions can be avoided.

## REFERENCES

The Chartered Institution of Water and Environmental Management (CIWEM), 2004. *The Environmental Impacts of Combined Sewer Overflows*, a Position Paper, London, England.

District of Columbia Water and Sewer Authority (2002) *Combined Sewer System Long Term Control Plan*; Final Report; District of Columbia.

District of Columbia Water and Sewer Authority (2007) *Blue Plains Total Nitrogen Removal/ Wet Weather Plan, Long Term Control Plan Supplement No. 1*; Final; District of Columbia.

Guo, Q.; Song, C.S.S. (1990) Surging in Urban Storm Drainage Systems. *J. Hydraul. Eng.*, 116 (12), 1523–1537.

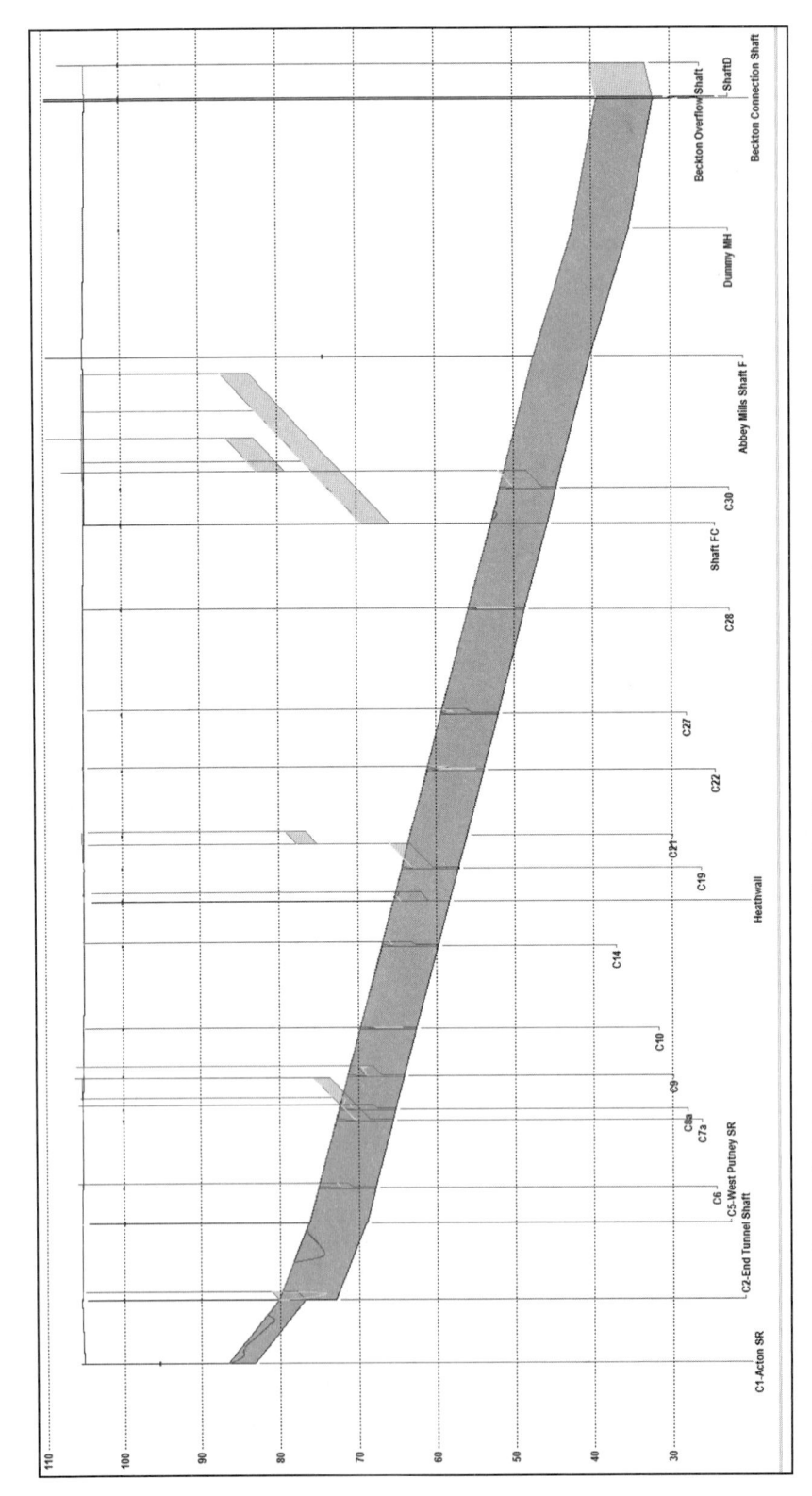

Figure 14. Composite picture of minimum gage pressures in London Tunnels during 15-year storm event

Macchione, F.; Morelli, M.A. (2003) Practical Aspects in Comparing the Shock-Capturing Schemes for Dam Break Problems. *J. Hydraul. Eng.*, 129 (3), 187–195.

U.S. Environmental Protection Agency (2004a) *Report to Congress, Impacts and Control of CSOs and SSOs*; EPA-833-R-04-001; Washington, D.C.

U.S. Environmental Protection Agency (2004b) *Report to Congress, Impacts and Control of CSOs and SSOs, Appendix A.2 Combined Sewer Overflow (CSO) Control Policy*; EPA-833-R-04-001; Washington, D.C.

Vasconcelos, J.G.; Wright, S.J. (2005) Experimental Investigation of Surges in a Stormwater Storage Tunnel. *J. Hydraul. Eng.*, 131 (10), 853–861.

Vasconcelos, J.G.; Wright, S.J.; Roe, P.L. (2006) Improved Simulation of Flow Regime Transition in Sewers: Two-Component Pressure Approach, *J. Hydraul. Eng.*, 132 (6), 553–562.

Vasconcelos, J.G.; Wright, S.J. (2007) Comparison Between the Two-Component Pressure Approach and Current Transient Flow Solvers. *J.Hydraul. Res.*, 45 (2), 178–187.

Wright, S.J.; Vasconcelos J.G.; Creech, C.T.; and Lewis, J.W. (2007) Mechanisms of Flow Regime Transition in Rapidly Filling Stormwater Storage Tunnels, Proceedings, *5th International Symposium on Environmental Hydraulics*, Tempe, Arizona.

Zhou, F.; Hicks, F.E.; Steffler, P.M. (2002) Transient Flow in a Rapidly Filling Horizontal Pipe Containing Trapped Air. *J. Hydraul. Eng.*, 128 (6), 625–634.

# McCook Reservoir Main Tunnel Connection Marks Another Significant Milestone in Chicago's TARP

**Faruk Oksuz, Megan Puncke**
Black & Veatch, Chicago, Illinois

**Jeffrey Bair**
Black & Veatch, Pittsburgh, Pennsylvania

**Paul Headland**
Black & Veatch, Gaithersburg, Maryland

**ABSTRACT:** The Metropolitan Water Reclamation District of Greater Chicago's (District's) Tunnel and Reservoir Plan (TARP) and the McCook Reservoir will further reduce flood damages and CSOs for the city of Chicago. The McCook Reservoir will receive over 10 billion gallons of water via the 33-ft dia. Main Tunnel Connection. The USACE tasked Black & Veatch to design the system. The paper presents design and constructability considerations for 86 ft dia and 300 ft deep construction/gate shaft; 33 ft dia and 1,600 ft long rock tunnel; wet-well shaft arrangement to house six high head and large (14.5 ft by 29.5 ft) wheel gates; tunnel bifurcation, steel and concrete lining; live tap connection to existing Mainstream TARP tunnel; energy dissipation and portal work; risk management and construction sequencing.

## HISTORY

The District began dealing with the problem of combined sewer overflow (CSO) related problems in the late 1960s and formally adopted the Tunnels and Reservoir Plan (TARP) in 1972. Phase I of TARP included 109 miles of deep tunnels located up to 350 feet below grade with diameters up to 33 feet. These tunnels provide for storage and conveyance of CSO discharges. Construction of Phase I is complete and the tunnels are on-line an accepting CSOs. The result has been a substantial improvement in the river system and water front quality with continued improvement expected as Phase II comes on-line.

Phase II includes a series of above ground storage reservoirs to increase the flood storage capacity and further reduce the impacts of CSO discharges with major new storage capacity projected to come on-line over the next several years.

## MCCOOK OVERVIEW

Originally authorized in the Water Resources Development Act of 1999, the McCook Reservoir Project is a key component of Chicago's ongoing TARP Project. McCook will provide approximately 10 billion gallons of additional CSO and flood water storage to the District. In order to minimize flood damages (during periods of peak flow) to the City of Chicago and 36 surrounding suburban communities, the reservoir will store excess CSO and flood water from TARP's Mainstream and Des Plains Deep Tunnel systems until the floodwaters levels recede. This stored volume will then be pumped to the water treatment plant at Stickney, Illinois for treatment and appropriate discharge to TARP network.

McCook Reservoir is currently being excavated in a limestone quarry. The walls are unfinished limestone, nearly vertical. The McCook Reservoir is part of the larger Chicagoland Underflow Plan (CUP) and Phase II of TARP and includes tunnels for stormwater storage and conveyance, and reservoirs for stormwater storage.

The McCook Reservoir will allow CSO from the Mainstream Tunnel to be stored until sufficient capacity is available at the water reclamation plant. The subject of this paper is the McCook Reservoir Main Tunnel System and the connection of TARP Mainstream system with the McCook Reservoir.

## SUMMARY OF CURRENT PROJECT

The McCook Reservoir Main Tunnel is characterized by the following components:

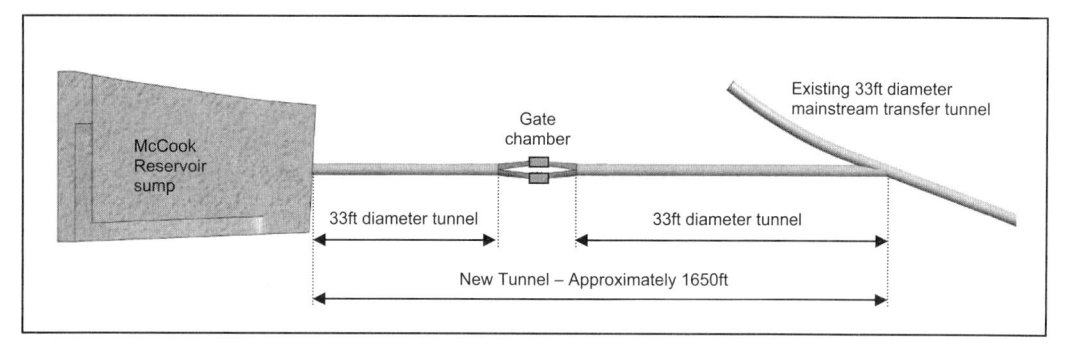

**Figure 1. Layout of new tunnels and gate chamber**

- A 1,200 foot long, 33 foot finished diameter (ID) Main Tunnel extending from the Mainstream Tunnel to the McCook Reservoir, bifurcated through the Gate Chamber (central shaft location).
- An 88 foot ID central shaft, located near the mid-point of the main tunnel alignment, extending some 284 feet below grade for construction of tunnel and to install the gate assembly.
- Three steel wheel gates and gate control structures on each of the bifurcated trunks (6 total) along the Main Tunnel alignment; and
- Connections to both the Mainstream Tunnel and the McCook Reservoir.

The Main Tunnel design, construction, operation, and commissioning will be coordinated with overall McCook Reservoir water control plan as well as other activities such as the reservoir excavation, high wall stabilization, groundwater protection system construction, and connection of the Distribution Tunnels to the reservoir.

The Main Tunnel includes a live connection to the Mainstream Tunnel. Mainstream Tunnel operation disruptions will have to be minimized as part of the live and final connection construction and all other reservoir facilities must be completed and ready to receive water. This connection will be one of the more challenging aspects of the construction project.

The project is on an accelerated (fast-track) schedule with Bid-Ready Contract Documents to be completed within roughly 15 months of Notice to Proceed. To facilitate the fast-track, the project was divided into 10 sub-projects with each completed somewhat independently. Frequent design meetings and three formal design review workshops were scheduled to facilitate coordination between the various disciplines completing the work.

## GEOLOGIC CONDITIONS

The site is located in Willow Springs, Illinois and lies within Lyons Township, T38N R12E in western Cook County. It is bordered by the Stevenson Expressway (I-55) and the Des Plaines River to the northwest, District operated sludge drying lagoons lie to the northeast, railroad tracks and the McCook Reservoir to the southwest and the Sanitary and Ship Canal to the southeast.

The project area is covered by glacial drift which in turn is underlain by about 4,000 feet of sedimentary rocks, ranging in age from Cambrian to Silurian. The bedrock surface is an undulating plain on which valleys have been incised by glacial and pre-glacial erosion. The uppermost part of the bedrock surface is generally fractured with the frequency of fracturing decreasing with depth. Some fractures have been enlarged by solutioning and some vuggy porosity is present.

The site is located on a substantial veneer of surficial deposits comprising fill material (presumably associated with ongoing dredging activities), poorly sorted recent alluvium and lake sediment derived from a prehistoric proglacial lake that existed from approximately 19,000 to 15,000 years ago, near the end of the Pleistocene.

Underlying these deposits is a series of massive, relatively homogenous Silurian and late Ordovician dolomites, including less prevalent interbedded shales that formed during the early stages of diagenisis during this time. These rocks form a relatively uniform 300 feet thick plus sequence across the site and incorporate the Racine Formation, Sugar Run Formation, Joliet Formation, Kankakee Formation, Elwood Formation and Wilhelmi Formation.

## HYDRAULIC ANALYSIS

The layout of the system as generally modeled in this study is shown in Figure 1. The goals of the hydraulic analyses were as follows:

- Develop designs for the mainstream connection and the bifurcation which give acceptable hydraulic performance under the maximum design flow condition of approximately 26,000 cfs into an empty reservoir.
- Assess the performance of the system under various possible operating states which could impose more severe hydraulic conditions and determine whether operational restrictions would be required.
- Provide data on the velocities and pressures expected within the tunnel system for use in the design of the tunnel lining and assess the risk of cavitation.

The main concerns are the potential for damage due to high velocity flows and differential pressure heads and resultant cavitation, abrasion and erosion. Although, we were unable to simulate these processes directly, we were able to predict velocities and pressures which can be used to assess the risk(s).

## Cavitation

Cavitation is the formation of vapor bubbles and their subsequent condensation. It will occur where the local absolute pressure approaches the vapor pressure of water (approximately 0.25psi) causing the water to boil. Small obstructions or surface irregularities in high velocity flow (greater than about 40ft/s) can also cause cavitation. Vapor bubbles produced by cavitation will be carried downstream. As these bubbles reach a location of higher pressure, the vapor will condense and the bubbles collapse suddenly. This implosive shock can cause severe structural damage. Of particular concern in this system is whether geometric details at the connection with the mainstream tunnel, the gate chamber or the bifurcation could be improved to reduce regions of low pressure.

It is important to recognize that cavitation is not necessarily restricted to the regions in which the model predicts zero absolute pressure. An informed judgment should also be based on the maximum velocities and the occurrence of any regions with a sharp velocity gradient between adjacent flow streams.

## Abrasion

A storm water system will convey significant quantities of sand, gravel, rock and other debris all of which can be highly abrasive and cause damage to the tunnel lining in high velocity flows. It is therefore important to minimize the peak velocities at the tunnel walls and ensure that the tunnel lining can survive the abrasive affect of sediment. The maximum velocities at the tunnel walls, as predicted by the CFD modeling, will allow the lining to be designed to withstand these abrasive affects.

## Modeling Method

The hydraulic analysis presented is based on the use of Computational Fluid Dynamics (CFD). All models were solved using Ansys CFX which is one of the leading commercially available CFD codes. The method involves setting up a mesh which splits the water into a large number of small elements. The software then calculates the predicted flow by solving iteratively a series of equations for conservation of mass momentum and energy. Free surface simulations have been run where necessary.

## System Operation and Design Conditions

This modeling was based on the understanding that the McCook Reservoir Main Tunnel system will normally be operated with the Main Gates fully open. The design condition is a peak flow of approximately 26,000 ft$^3$/s as derived from a 58-year of record modeling using a one dimensional hydraulic model of the entire system by others.

The tunnel and gates are designed to handle the hydrostatic pressure condition experienced when the gates are shut with the tunnel full and reservoir empty, or the reservoir full and tunnel empty. However, extreme hydraulic conditions would occur if the gates were opened under this loading condition. The severity of the velocities and pressures is such that damage should be expected if the gates were to be opened under these extreme conditions.

## Connection with Mainstream Tunnel

The new McCook Tunnel will connect into the existing mainstream transfer tunnel approximately 1,600 feet north east of the reservoir. During storm events, the majority of the flow will be routed through the connection into the McCook tunnel. The design flow gives an average velocity in excess of 30 ft/sec in the 33 foot diameter tunnel. This relatively high velocity will be difficult to turn and there is a risk of local high velocities and low pressures which could cause problems due to abrasion and cavitation. From a hydraulic perspective, it would therefore be preferable to have a wide radius swept connection with the mainstream tunnel. However, this will increase construction difficulties and risks. We have therefore assessed a range of different connection details with an aim of giving a reasonable compromise between constructability and acceptable hydraulic performance. We have considered two main options for the connection detail:

- Option A—Connecting into an existing construction shaft: There is an existing shaft on the mainstream tunnel about 400 feet east (upstream) of where the new McCook Tunnel would ideally connect. If the new tunnel were connected into this shaft, it would avoid the hazards associated with breaking into the existing tunnel directly.
- Option B—New Connection: We have assessed a series of connection details which involve breaking a new connection into the existing tunnel.

*Option A—Connection Into Existing Shaft*

The existing construction access shaft has a diameter of 25 feet. Two different designs have been modeled for this option as shown in Figure 2 and 3.

Both layouts J and K were simulated with a fixed flow passing though the connection into the McCook tunnel. The shaft is open to the atmosphere and so the models were therefore run as multiphase free-surface simulations which simulate the flow of both the water and the air phases.

The models predicted that both layouts would be unacceptable from a hydraulic perspective. There are three main causes for this poor performance:

- The cross-sectional area of the 25 foot diameter shaft is nearly half that of a 33 foot tunnel and so the velocity will be at least doubled as it passes through the connection.
- The existing shaft connection is perpendicular to the tunnel and there are no wide fillets or radii at the connection to aid the turning of the flow.
- The open shaft allows air to be drawn into the system.

These limitations cannot be overcome without breaking into the tunnel and so we conclude that connecting into the existing shaft is not a viable option.

*Option B—New Connection*

A series of models were analyzed to test different connection details direct into the existing tunnel. The connection will inevitably weaken the existing tunnel, but the stresses imposed on the tunnel can be significantly reduced if the opening is cut into the tunnel below the crown and/or above the invert. The connections modeled can be divided into three categories:

- Full bore connections (aligned invert and crown): These are geometrically the simplest layouts with a mitered circular cross-section for the branch. Because these are aligned at both the invert and the crown, they will be more difficult to construct without weakening the existing tunnel.
- Slot connections: These involve breaking an elliptical, oval or rectangular slot into the existing tunnel. They avoid breaking into the crown and invert region which should give structural advantages. Since the invert is raised at the connection, the depth to which the reservoir can be drained via the McCook tunnel will be reduced by 5 to 7 feet. The reduced crown level at the connection could also give venting issues as the tunnel fills.
- Invert aligned but not crown: This will allow the reservoir to be fully drained to the same depth as a full bore connection, but could have similar venting issues to the slot connection.

Eighteen different models were run to test the different connection types and to develop better performing designs. Five of the layouts tested represent key steps in our design iterations and are presented in Figure 4. The velocity and maximum absolute head predicted full design flow into an empty reservoir are shown in Figure 5.

An absolute pressure head of 34 feet indicates atmospheric pressure and absolute pressure head of 0 feet indicates full vacuum. If the absolute pressure head falls below 10 feet, there is a serious risk of cavitation. Maintaining a higher head is beneficial as it reduces the risk of cavitation damage and will provide some protection against higher flows than those analyzed.

The development of the different designs is discussed below:

*Model A—20° Full Bore Miter*

This was the base model and was recommended in previous studies. This gave moderate hydraulic performance and would be marginally acceptable. The minimum absolute pressure head of 15 feet is at the lower end of acceptable values and any increases in design flow would raise concerns about cavitation. The layout causes some construction difficulties as it leaves a narrow crotch and requires an opening in excess of 96 feet wide to be broken into the existing tunnel.

*Model C—45° Full Bore Miter*

Using a 45° miter gives an opening in the mainstream tunnel which is half the width of a 20° miter and is therefore easier to construct. However, the geometry does not enable the flow to turn smoothly

| Details | Views |
|---|---|
| Model J<br>• Break into existing 25 ft diameter shaft 5 ft above crown of tunnel<br>• Circular to rectangular transition between McCook tunnel and connection<br>• Swan-neck to drop invert level of McCook tunnel | 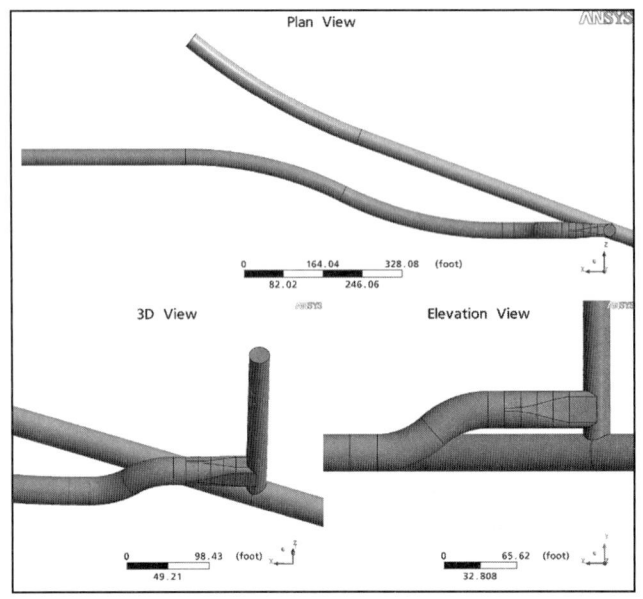 |
| Model K<br>• Enlarge shaft to 66 ft diameter<br>• Swan-neck to drop invert level of McCook tunnel | 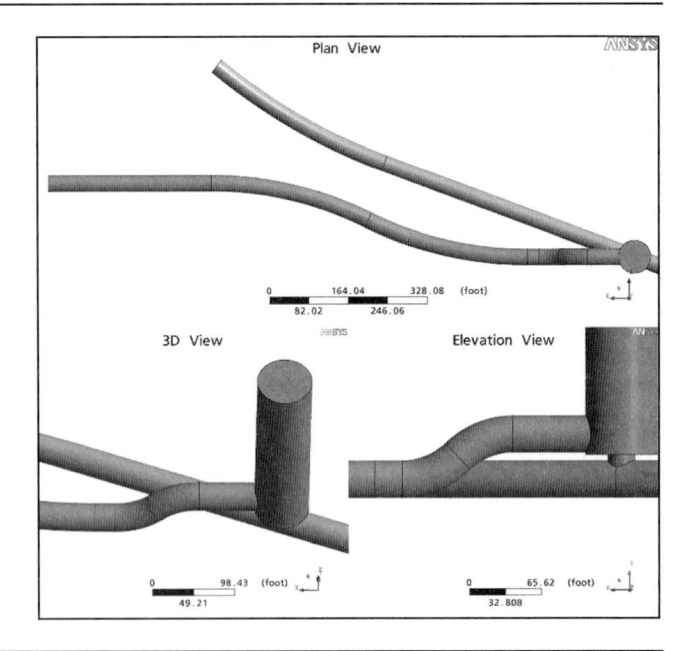 |

Figure 2. Details for connections into existing shaft

| Details | Predicted velocity on free surface |
|---|---|
| Model J<br>• Maximum velocity 103 ft/s<br>• Extensive regions below zero absolute pressure (cavitation)<br>• Unstable air core vortex<br>• Extensive air entrainment into McCook tunnel | 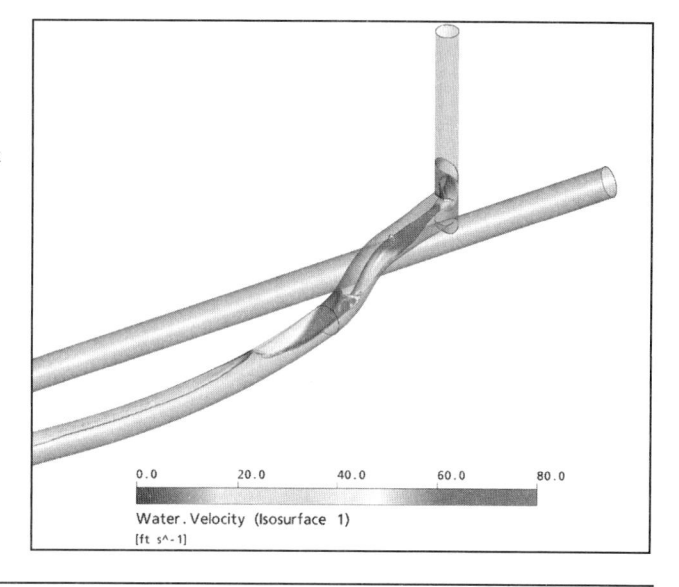 |
| Model K<br>• Maximum velocity 95 ft/s<br>• Extensive regions below zero absolute pressure (cavitation)<br>• Air entrainment into McCook tunnel | |

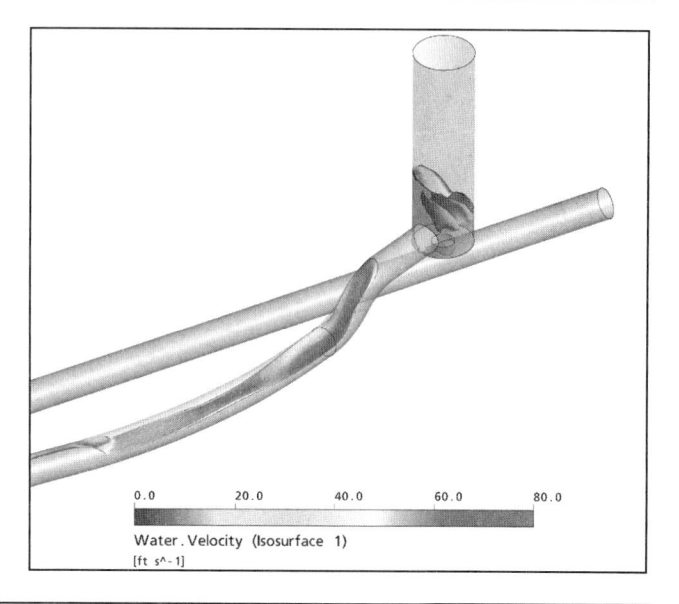

Figure 3. Hydraulic performance for existing shafts at 26,500 ft$^3$/s

| Model | 3D-View | Description | Break in dimensions Height [ft] | Width [ft] |
|---|---|---|---|---|
| A | 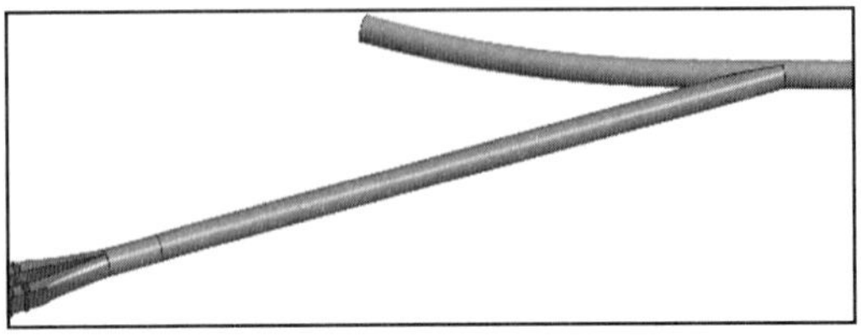 | • Full bore connection<br>• 20° branch angle<br>• 33 ft diameter circular cross-section | 33.0 | 96.5 |
| C | 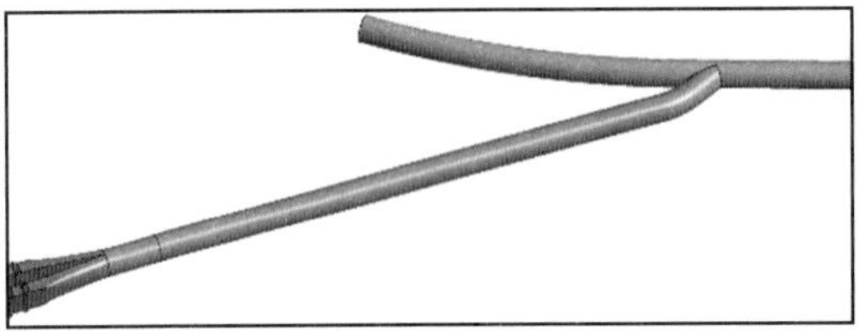 | • Full bore connection<br>• 45° branch angle<br>• 33 ft diameter circular cross-section | 33.0 | 46.7 |
| L | 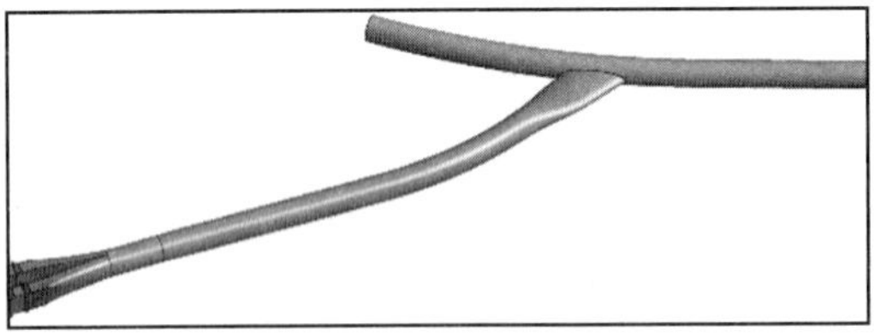 | • Slot connection<br>• Section: elliptical<br>• 45° branch angle<br>• 3 ft radius on all edges | 21.4 | 81.4 |
| W | 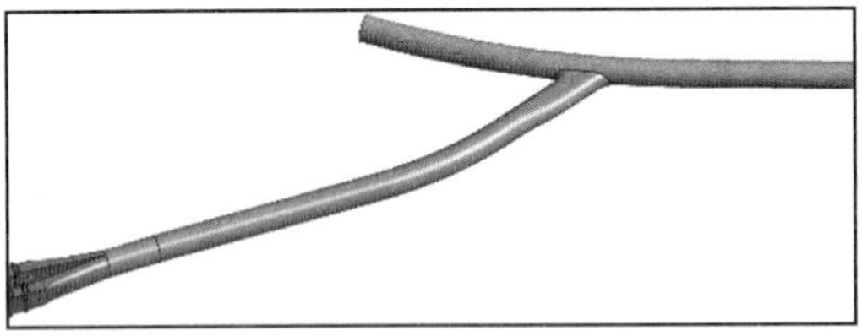 | • Invert aligned connection<br>• Slot connection<br>• Section: Rectangular with semi-circular sides<br>• 45° branch angle<br>• Swept branch and crotch<br>• 1.5 ft radius on all edges | 27.0 | 68.2 |
| Y | 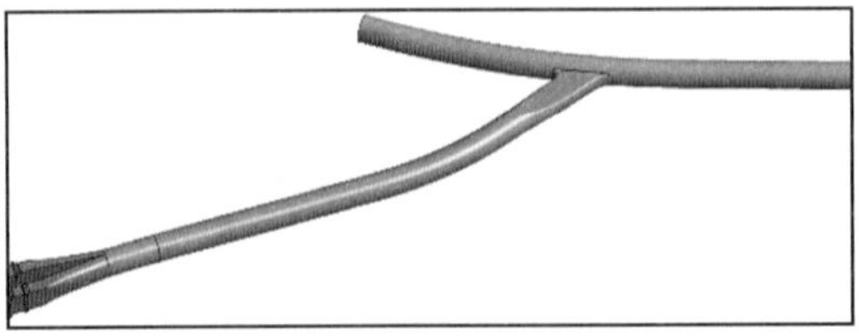 | • Invert aligned connection<br>• Section: parallelogram with 16.5 ft radius arched sides<br>• 45° branch angle<br>• Swept branch and crotch<br>• 1.5 ft radius on all edges | 27.0 | 70.6 |

Figure 4. Details for different connection geometries tested

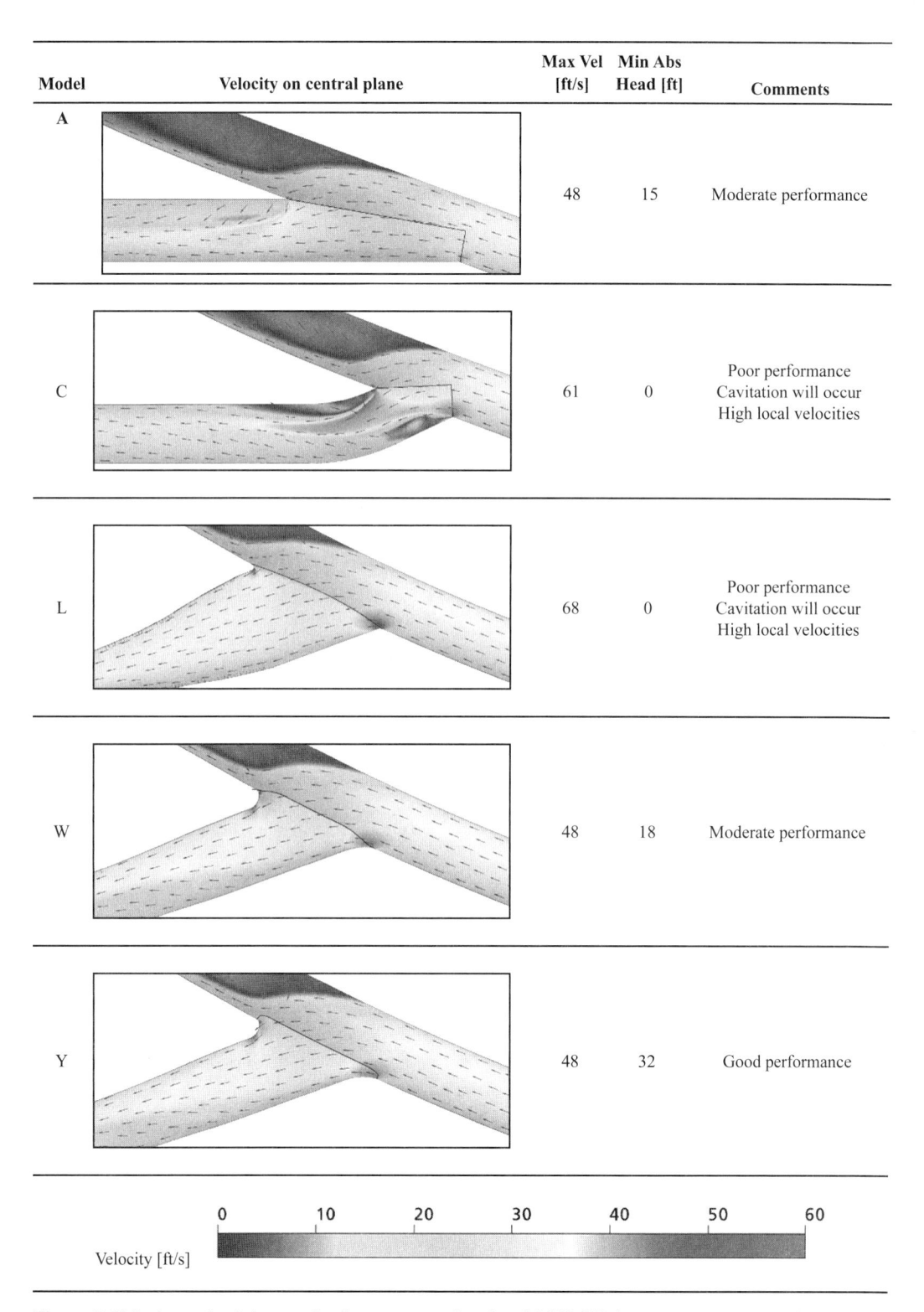

| Model | Velocity on central plane | Max Vel [ft/s] | Min Abs Head [ft] | Comments |
|---|---|---|---|---|
| A | | 48 | 15 | Moderate performance |
| C | | 61 | 0 | Poor performance<br>Cavitation will occur<br>High local velocities |
| L | | 68 | 0 | Poor performance<br>Cavitation will occur<br>High local velocities |
| W | | 48 | 18 | Moderate performance |
| Y | | 48 | 32 | Good performance |

Velocity [ft/s]: 0 — 10 — 20 — 30 — 40 — 50 — 60

**Figure 5. Velocity and minimum absolute pressure head at 26,500 ft³/s into an empty reservoir**

into the branch. This results in high velocities as the flow enters the McCook tunnel and gives regions of very low pressure which would cause cavitation.

### Model L—45° Miter Elliptical Slot Connection

Visual inspection of the velocities shows that the general flow pattern through this connection detail is good. The slot connection gives a flared end which allows a smoother change in direction of the flow. However, closer inspection of the results shows that there are small regions of high velocity and low pressure. These are located at the sharp mitered edge, and although the region affected is small, the absolute pressure drops to zero and so cavitation will occur.

### Model W—45° Invert Aligned Connection with Rounded Edges

Model W has a rounded rectangular rather than an elliptical cross section. This gives a reduced width for a given cross-sectional area which is more efficient for construction. In addition all the edges of the mitered connection have been smoothed. This rounding of edges was very effective and resulted in a peak velocity of 48 ft/s and a minimum absolute pressure of 18 feet.

### Model Y—45° Invert Aligned Connection with 16.5 ft Radius Arched Sides

Model Y is similar to Model W, but the sides of the branch are radiused at 16.5 feet instead of 10 feet. This gives a perfectly smooth connection where the sides sweep into the 33 feet diameter mainstream tunnel. This minor change increases the minimum absolute pressure from 18 feet to 32 feet. This would give a significant reduction in the risk of cavitation for higher flows during more extreme events. This was the best performing design. The various features of this design are discussed below:

### Section Shape

The geometry of the connection is shown in Figure 6. The selected profile was found to have several benefits over alternative profiles:

- The 16.5 foot radius sides enables the branch to be swept smoothly into the cross-section of the 33 foot diameter mainstream tunnel.
- The flat top maximizes the height of the profile across the section. This reduces the extent to which the flow from the crown of the mainstream tunnel is forced downwards.
- Reducing the height, but increasing the width tends to be beneficial as it results in a flare (in plan view) which enables a smoother turning of the flow into the McCook Tunnel.

### Sweep Radius

The model demonstrated a significant improvement in performance if the branch was swept horizontally into the main tunnel. Models without a swept connection all predicted that the absolute pressure would fall below zero and so cavitation would occur. We therefore conclude that a swept connection should be used.

### Edge Radius

All edges at the connection have been smoothed to give a minimum radius of 1.5 feet. Testing of alternative edge details found that it was beneficial to include a radius of this size or greater to reduce separation of flow and avoid highly localized regions of low pressure.

### Opening of Gates

The designs were developed considering the hydraulic behavior with the gates fully open at the start of a storm event. If the gates were opened under pressure, flow will jet out under the lip. This will give very high velocities and low pressures downstream of the gates. Previous analyses predict velocities of up to 120 ft/s and localized regions of zero absolute pressure if the gates were opened under pressure. Velocities in the region of 120 ft/s could occur over the full length of the tunnel downstream of the gates. For this reason, we recommend that the gates not be operated when the differential head across the gate exceeds 30 feet.

## TUNNEL DESIGN

Deep tunneling in the Chicago area has been on-going for some 30 years and significant data and information exists. The tunnel was designed using all available rock test data from previous studies and investigations and three additional borings drilled by the USACE.

Detailed analyses were performed using the Universal Distinct Element Code (UDEC) to explore the various excavated profiles as well as for both temporary and permanent lining. UDEC is a two dimensional numerical program based on the distinct element method for discontinuum modeling. UDEC simulates the response of discontinuous rock mass subjected to either static or dynamic loading.

The purpose of this analysis is to provide the design basis and section forces for final lining design. Several areas of the tunnel system were analyzed to sufficiently describe the design load cases for structural purposes, including the following:

1. Standard Main Tunnel Section (Circular and Horseshoe shape).

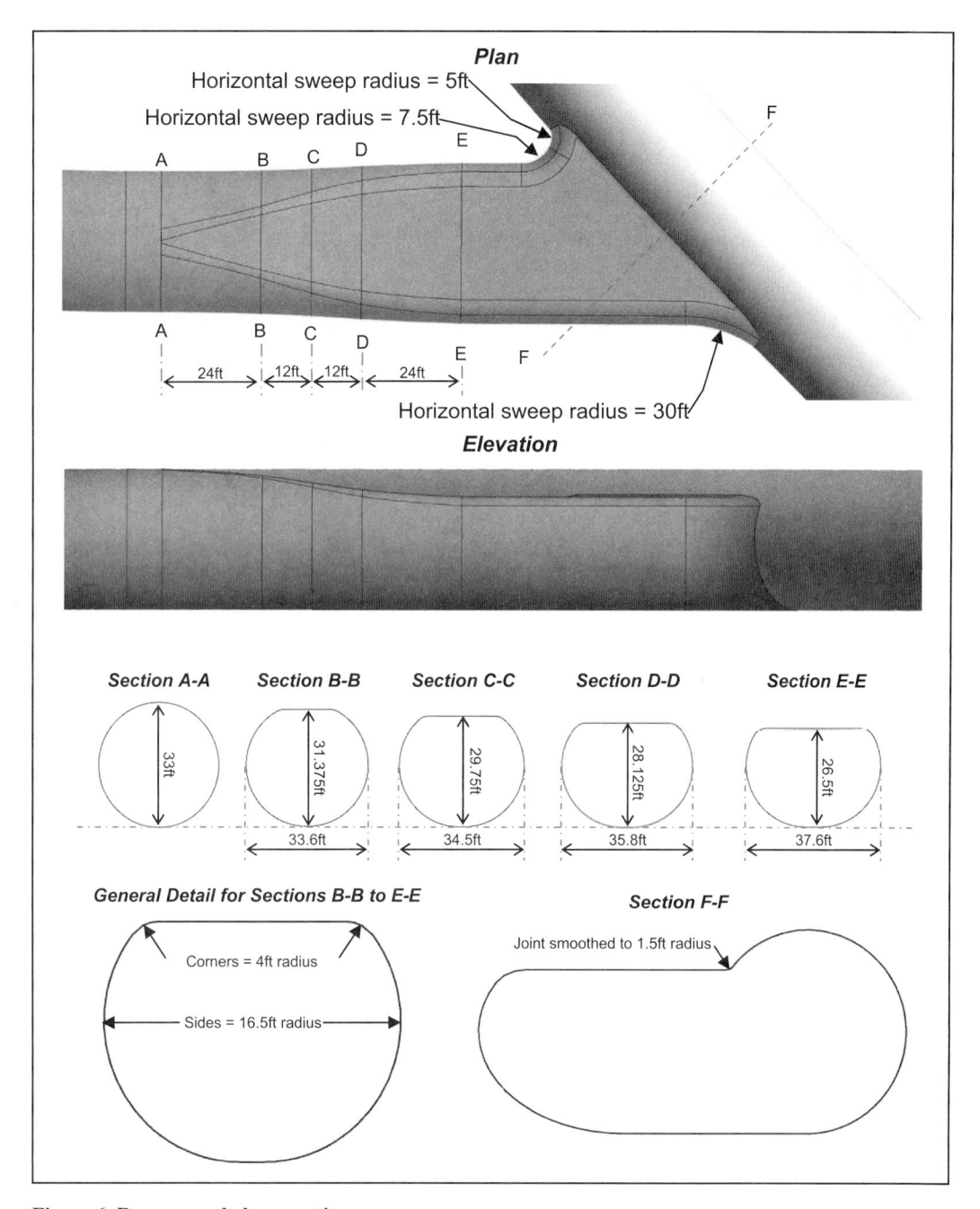

**Figure 6. Recommended connection**

2. Main Gate Shaft Geometry.
3. Main Tunnel Bifurcation at the intersection of the Main Gate Shaft and Tunnel.
4. Construction Shaft Geometry.

Based on the subsurface conditions and to ensure that operational requirements for the project are satisfied, the design criteria for the Main Tunnel includes 33 foot diameter circular tunnel with 0.26% slope towards the portal from the Gate Shaft and 0.11% slope from the Gate Shaft to Mainstream

Tunnel to keep the gates in dry conditions when the water is not flowing in the tunnel. Transition zones on both side of Gate Shaft will be leveled with no slope for 90 feet on each side.

## Initial Support

The initial tunnel support is placed to provide a safe and stable opening in the ground of such dimensions that the permanent structure can be constructed. The initial support elements can be made up of several components including rock bolts, steel straps, wire mesh, and shotcrete, if necessary.

## Final Lining

The final lining or permanent structure is designed to resist both stresses from rock loading as well as internal stresses from the function and use of the tunnel as a conveyance for water which will be, at times, under considerable pressure head with high velocity flow and debris. The erosion, cavitation, dynamic pressure and static load resistance are all of interest to the durability of the permanent structure.

## Geologic Conditions for Tunnel Design

Two major joint sets are visible in the quarry rock faces of the future McCook Reservoir. The orientations of the joint sets are consistent within the region. The primary joint set strikes at about N45°W to N55°W and steeply dipping at about 80° to 90° to the northeast or southwest. This joint set is often filled with clayey shale up to approximately 6-inches thick, but mostly it is less than 1-inch thick. Typical spacing perpendicular to strike of the joint set is average about 50-ft but range from 5-ft to more than 200-ft.

The secondary set is orthogonal to the primary set, striking at about N45°E to N55°E and steeply dipping at about 80° to 90° to the northwest or southeast direction. The continuity of the joint set is 60 to 120 feet and the average spacing among joints ranges between 100 to 250 feet. Joint apertures are open over 6-inches near the surface and progressively get tighter with depth. Another major rock discontinuity is the bedding. The apparent dip of the bedding is sub-horizontal.

## UDEC Background

UDEC (Universal Distinct Element Code) is described as "distinct element program," and falls within the classification of discontinuum analysis technique. A discontinuous medium is distinguished from a continuous one by the existence of contacts or interfaces between discrete bodies. Discontinuum numerical methods can be further categorized by the way they represent the contacts and discrete bodies.

Therefore a numerical model must contain two types of mechanical behavior in a discontinuous system: 1) contacts or discontinuities; and 2) solid material or rock mass.

UDEC analysis simulates the behavior of jointed rock mass subject to either static or dynamic loading. Joint sets and discontinuities in rock mass are treated as contacts and interfaces between the neighboring discrete blocks in the model system. Discrete blocks in UDEC can be modeled either as rigid or deformable material. Deformable blocks are further subdivided into a mesh or zone of finite-difference elements. Each element responds according to a prescribed constitutive model with linear or non-linear stress-strain relationship. Linear (elastic) or non-linear (non-elastic) load-displacement relationship in both normal and shear directions govern the behavior of the discontinuities and rock mass.

## Objective of UDEC Analysis

The main objective of the numerical analysis of the Main Tunnel is to obtain in-situ stress conditions of rock mass upon excavation of the tunnel. Results of the numerical analysis will be used to design the final tunnel lining (future work). From UDEC analysis, the major principal stress, x- and y-direction stresses and displacements, axial load and maximum bending moment in lining structure were obtained as input to the final lining design.

Model execution for all sections on main tunnel, bifurcation, and shafts involves the following steps which are further described in the following sections:

1. Model set-up and definition of constitutive model
2. Material property
3. Boundary and in-situ stress conditions
4. Tunnel excavation and concrete lining installation
5. Final equilibrium

The purpose of the analysis is to determine the mechanical behavior of rock mass and development of stress field around the tunnel. This behavior is highly dependent on the constitutive model used for the blocks and discontinuities.

### *Model Setup*

Individual joints are modeled in addition to the rock material to better represent the behavior of a rock mass. For the rock mass in the project area, two high angle joint sets exist in addition to horizontal bedding. When the rock mass properties are satisfactory, the geometry of the desired excavation is drawn onto the mesh for later removal or "excavation."

The corners of each discrete block are rounded to avoid infinite concentrated load and strength due to zero contact area and to allow adjoining blocks to smoothly slide past one another when two corners interact. Blocks are modeled as either quadrilateral or triangle. Quadrilateral zones are kite-shape parallelograms that contain two opposing triangular subzones. This zone shape is applicable only to blocks that contain 4 or 5 edges. Blocks that do not meet this requirement utilized triangle zones. For the McCook tunnel and shaft models, blocks within an area of 80 feet wide by 80 feet high adjacent to the tunnel profile are quadrilateral in shape (less than 4 edges), and outside of this region to the boundaries of the model, zone shapes are triangular. This region is created with different block densities. Horizontal bedding thickness of 0.5, 1.0, and 2.0 feet are used within this region, and bedding thickness of 10.0 feet is used outside of this region.

None of the models extend to the ground surface and rock mass is expected to behave beyond the elastic limit, therefore only the Mohr-Coulomb model is used. Continuously yielding constitutive model was selected for discontinuities. This model considers residual strength of the contacts in terms of friction angle and joint roughness parameters.

*Material Property*

A total of four sets of properties are used in the modeling—rock mass, joint sets, concrete lining, and rock-structure interface properties. Numerical values for each of these sets are summarized in Tables 1 to 4.

*Description of Numerical Modeling Steps*

The modeling is carried out in discrete stages to increase understanding of the rock mass behavior on excavation. Each of the four primary stages of analysis is described below:

**Table 1. Rock mass properties**

| Rock Mass Properties | Values | Units |
|---|---|---|
| Density | 172.3 | lb/ft$^3$ |
| Elastic bulk modulus | 4.03E+08 | lb/ft$^2$ |
| Elastic shear modulus | 2.30E+08 | lb/ft$^2$ |
| Internal friction angle | 50 | Degree |
| Cohesion | 3.86E+05 | lb/ft$^2$ |
| Tensile strength | 1.76 | lb/ft$^2$ |
| Horizontal to vertical stress ratio | 4 | — |

**Table 2. Joint and bedding properties**

| Joint and Bedding Properties | Unit | Vertical Joint | Bedding |
|---|---|---|---|
| Conditions | — | Clay Filled | Clean |
| Normal stiffness | lb/ft$^3$ | 7.20E+06 | 1.73E+08 |
| Shear stiffness | lb/ft$^3$ | 7.20E+06 | 1.73E+07 |
| Initial friction angle | degree | 13.5 | 52 |
| Intrinsic friction angle | degree | 13.5 | 35 |
| Cohesion | lb/ft$^2$ | 1.28E+03 | 1.93E+03 |
| Tensile strength | lb/ft$^2$ | 1.76E+01 | 1.76E+02 |
| Dip angle with horizontal | degree | 80 | 0 |
| Spacing | ft | 15 | 5 |
| Roughness | ft | 1E-2 | 7E-3 |

**Table 3. Concrete properties of the tunnel lining**

| Concrete Properties for Lining | Values | Units |
|---|---|---|
| Unit weight | 150.0 | lb/ft$^3$ |
| Compressive yield strength | 4.90E+05 | lb/ft$^2$ |
| Poisson's ratio | 0.15 | — |
| Yield strength (reinforcing steel) | 4.80E+04 | lb/ft$^2$ |
| Young's modulus | 5.20E+05 | lb/ft$^2$ |

**Table 4. Rock and concrete interface properties**

| Rock and Concrete Interface Properties | Values | Units |
|---|---|---|
| Tensile strength | 1.25E+04 | lb/ft$^2$ |
| Friction angle | 35 | degree |
| Cohesion | 0 | lb/ft$^2$ |
| Interface normal stiffness | 1.59E+09 | lb/ft3 |
| Interface shear stiffness | 6.23E+06 | lb/ft$^3$ |

1. Initial equilibrium is the most important step of analysis as it confirms that the mesh and geometry as well as the initial stress conditions and boundary conditions are correct and consistent. All displacement values are reset to zero once equilibrium is attained so that the "present day situation is the zero basis for all movement.
2. Excavation stage is modeled by removal of the desired tunnel profile and the application of a reaction force for one single cycle that establishes initial required support pressure. This support pressure is then allowed to dissipate by 50% to model the relaxation of the rock mass and the residual reaction pressure is applied on concrete lining to equilibrium. This approach essentially models the effect of an initial support lining.
3. Installation of final concrete lining is wished in place immediately after the chosen support pressure step. The remaining support pressure is then released to model the degradation of the initial lining and transfer of previously supported rock pressure to the permanent lining.
4. Run the model to equilibrium that allows the full rock pressure to develop. At equilibrium, the results displayed in the results tables for the various runs is taken from the model and transferred to structural design calculation inputs for final design detailing.

*Standard Tunnel Profile*

The final shape of the Main Tunnel will be cast to a circular shape with the permanent concrete lining. Since the length of the tunnel is less than 1,000 feet on either side of the Gate/Access shaft it is envisioned that the tunnel will be excavated by the conventional drill and blast method. To evaluate and compare the conditions that will affect the tunnel lining due to excavation method, two scenarios with circular and a horseshoe shape excavation have been modeled.

Sensitivity of excavation stability has been checked with bedding plane spacing being reduced from 2 feet to 1 foot and 6 inches. Each scenario is further evaluated using a 1-foot and 2-feet thick concrete lining to determine section forces for design.

The model for each of the above scenarios considers two critical conditions, one where a clay-filled joint (of the J2 set and orientation that is sub-parallel to the direction of tunneling) lies directly in the crown and the other where two of these joints exist within the 30-foot span of the tunnel.

*Bifurcation and Approach to Main Gate Shaft*

We have assumed in our analysis that the proposed shape of the gate approach and the final required structure of a bifurcation to the Main Tunnel will result in construction of a single large-span tunnel that the concrete and steel bifurcation will be built inside.

While two meshes were generated for analysis, the larger span cavern was analyzed first as the worst case. This section is immediately adjacent to the breakout from the shaft. As part of the initial stress field assumptions in the analysis, the already high horizontal stress was further increased as a result of stress concentration around the already excavated 88 foot diameter shaft. This stress concentration over the field stress (determined during the analysis of the shaft structure) was applied to the large cavern analysis as an additional factor on horizontal stress. It was found from the shaft analysis that stress field at the base of the shaft after excavation is twice the stress before excavation, and maximum stress is acting in the direction approximately perpendicular to the tunnel axis. Therefore a factor of 2.0 is applied to the horizontal in-situ and boundary stresses for bifurcation near the gate shaft only. This factor is applied in addition and after the factor of 4.0 mentioned previously.

The analysis was then performed in the same way as for the standard profiles above. Once results were obtained that indicated that there was no need to increase lining thickness and that no change in rock mass behavior was observed between the standard profile and the largest profile adjacent to the shaft, no further analysis took place on the intermediate profile generated.

*Shaft Lining Analysis*

There are two shafts on the project, the Main Gate Shaft and another working shaft anticipated adjacent to the main-line tunnel to facilitate safer working practices. A horizontal mesh was carried out on both structures using vertical features and no bedding. The mesh was oriented so that the major and intermediate stresses could be applied to the mesh and results (section forces) were obtained for a 1-foot and 2-foot thick concrete lining.

Table 5. Parametric study on bulk and shear modulus of rock mass

| Run | Bulk Modulus (Psf) | Shear Modulus (Psf) | Max. Axial Force (Kips) | Max Moment (Kips-ft) | Max. Structural Displacement (In.) | Max Block Displacement (In.) |
|---|---|---|---|---|---|---|
| 1 | $4.03 \times 10^8$ | $2.3 \times 10^8$ | $6.1 \times 10^2$ | $-1.6 \times 10^5$ | 0.38 | 0.40 |
| 2 | $4.03 \times 10^7$ | $2.3 \times 10^7$ | $7.8 \times 10^2$ | $-1.8 \times 10^5$ | 4.87 | 3.29 |
| 3 | $4.03 \times 10^6$ | $2.3 \times 10^6$ | $4.0 \times 10^2$ | $-2.7 \times 10^5$ | 44.3 | 24.9 |

*Sensitivity Analysis for a Robust Model*

The critical parameters for this analysis have been identified as the shear and bulk modulus of rock mass. To verify the validity of UDEC model and determine the behavior of rock mass, the bulk and shear modulus are reduced 10 and 100 times from the original values. All other parameters have remained the same, and the tunnel excavation profile is horseshoe shape. In this sensitivity study, no relaxation is allowed between tunnel excavation and lining installation.

Reducing the bulk and shear modulus of rock mass effectively results in a weaker rock. Such weakening is expected to increase loadings on tunnel lining thus increasing structural requirements of the lining. The results of the study are summarized in the following Table 5.

The effect of rock weakening is apparent from the recorded block and lining displacement in Run 2. The increased displacement of the rock blocks has caused both axial and moment of the lining to increase. For Run 3 where bulk and shear moduli are reduced 100 times, structural lining has moved more than 3 feet. This magnitude of displacement has effectively altered the structural integrity of the tunnel lining beyond the scope of analysis of this study. However, this is consistent with the expected behavior of weaker rock mass in terms of shear and bulk modulus. The model sensitivity to these parameters has been checked and the variance is considered within reasonable bounds to determine that the model is robust and providing reasonable results.

**Horizontal and Vertical Alignment**

The horizontal and vertical alignment of the Main Tunnel is developed in accordance with the following design criteria:

- The tunnel bifurcation shall be as far as possible from the reservoir for hydraulic reason.
- The tunnel length will be straight with large radius bends.
- The slope of the tunnel will be minimized.
- The tunnel will work under gravity flow in both directions driven by the head of the tunnel system or the reservoir.
- The Gates will be the highpoint and therefore will not be standing in water under dry tunnel conditions—even when kept closed.
- McCook reservoir portal is fixed with an invert at elevation –264.84 CCD.

*Connections*

The Main tunnel will be connected to the existing Mainstream Tunnel at 45 degree angle for hydraulic requirements. For the stability of the existing tunnel lining and constructability of the connection, an elliptical opening will be excavated to coincide with the invert of the Mainstream Tunnel. The area and hydraulic properties of the elliptical excavation are equivalent to a much narrower angle of intersection that would be preferred hydraulically but would be difficult to construct.

*Tunnel Support*

Chicago TARP construction has over 30 years history of successful, large span tunnel excavations, chambers and crossovers in the dolomites within which the Main Tunnel System is located. Immediately west of the McCook Reservoir there are similar pump house and distribution tunnels and chamber structures in the same geology that confirm the feasibility of the proposed Main Tunnel System drill & blast excavations. Conventional rock bolts, welded wire mesh, and shotcrete will provide safe and stable openings that further validate the approach here to allow for some convergence that re-distributes rock loading around the excavation—reducing the load seen by the permanent concrete lining.

*Bifurcations*

For high flow velocities, high performance concrete would be required to reduce concerns with scour and deterioration of the lining material over time. Further, to reduce the risk of cavitations in the bifurcation zone, steel linings will be utilized in the vicinity of the gate structure. This will also prevent unacceptable leakage from the pressurized waterway into the gate slots downstream of the closed gate. In addition, steel to concrete connection details will follow industry practice with seepage collars and provisions for cutoff grout rings drilled through collars in the steel liner to minimize longitudinal flows between the steel and concrete liner. Measures to ensure constant contact between the steel lining and the surrounding structural concrete with unreinforced concrete shall be a requirement in the contract specifications.

*Drill and Blast Excavation*

Based on the length and diameter of the Main Tunnel, it is envisioned that tunnel will likely be excavated by conventional drill and blast method of excavation in two lifts. Since the Main Tunnel excavated diameter will be minimum 35 feet and maximum of 37 feet in a high stress rock environment, the tunnel will be excavated using the principles of NATM, installing a relatively flexible lining that allows controlled deformation and reduces the stresses in the tunnel lining. Either full face excavation or a top-heading and bench sequence of excavation could be allowed depending on the contractor's proposed design, means and methods and equipment.

*Tunnel Constructability*

Main tunnel construction access will be primarily via the 88-foot diameter gate shaft. The shaft is sufficiently large for tunnel construction purposes. Clearly, the access restrictions would affect cost and duration of construction. If possible, access via a portal at the McCook Reservoir sump may reduce costs for construction since the mucking operation will be more efficient.

For this project, given the short lengths of tunnel and variable cross section at bifurcations, transitions, through the gate waterways and at bulkheads, it is almost certain that drill and blast excavation techniques will be used. With heading spans up to 56 ft, tunnel excavations are likely to be undertaken using a top-heading and bench sequence of operation.

An additional 25-foot diameter shaft is proposed between the rock bulkhead and Mainstream Tunnel connection to increase worker safety as an ability to access the connection area while allowing the gate area to remain isolated in case of inundation.

Tunnel muck volume and removal is a key variable when it comes to advance rate. For maximum flexibility, rubber tired equipment is the likely choice for excavation and mucking. After each blast the blasting fumes will be dissipated, the face, crown and sidewalls made safe and a loader and trucks or specialized underground load haul excavator may be used to transport the muck to the bottom of the shaft where a crane or vertical conveyor can be used to hoist the material to the surface. The dolomites have intrinsic commercial value, and the possibility of blending the tunnel spoil material with quarry material for aggregate production should be evaluated for disposal.

*Tunnels Risk Management Strategy*

The following are the most critical issues that can affect the progress and success of the project during and after construction the Main Tunnel:

- Operation of the existing Mainstream tunnel system during the construction.
- Safety concerns due to flooding of the work during Main Tunnel and Mainstream connection.
- Lack of coordination between the contractors of this project and the already bid shaft construction.
- Impact to existing Mainstream Tunnel structures due to connection.
- Excessive groundwater inflows in the shaft and tunnel during construction.
- Excessive over break along the bedding and joints due to blasting during the construction.
- The reservoir excavation is not completed on schedule to daylight the tunnel.
- The gates do not fit properly or are damage by large debris.

## CONCLUSION AND ACKNOWLEDGMENT

The design of the McCook Main Tunnel System is a culmination of many years of effort by the U.S. Army Corps of Engineers (USACE) and the MWRDGC as the project local sponsor. The McCook (and Thornton Composite) Reservoirs will mark the completion of Phase II of TARP and collectively represent another milestone achievement for protection of Chicago's waterways and providing flood control benefits to many communities in Chicago and Cook County.

Black & Veatch, in cooperation with the USACE and MWRDGC, has been designing and providing construction phase engineering services for the various components of the reservoir projects. The authors acknowledge the guidance, support and cooperation of the staff of the USACE and MWRDGC to this project, and look forward to a successful completion of the design and subsequent construction."

## REFERENCES

Nov 99 Design Documentation Report, Chicagoland Underflow Plan, McCook Reservoir, Illinois.
Oct 94 Hydraulic Transient Study of Mainstream Tunnel System and Control System.
ER 1110-2-1150 31 Mar 94 Engineering and Design for Civil Works Projects.
2008 McCook Reservoir Gates Open Shaft Design CFD Modeling Results Technical Report No. 361.

# Integration of Operations and Underground Construction: Sound Transit University Link

**John Sleavin**
Sound Transit, Seattle, Washington

**Peter Raleigh, Samuel Swartz, Phaidra Campbell**
Jacobs Associates, Seattle, Washington

**ABSTRACT:** Connecting an operating transit system to a new extension is always a challenge. This paper discusses the decision-making methodology and design solution of the bored tunnel connection at the Pine Street interface between the Central Link and the University Link tunnels currently being constructed for Sound Transit in Seattle, WA. The design incorporates a unique combination of ground improvement, shaft construction, and underground excavation aimed at avoiding disruption of an important arterial street and the transit operations within the Downtown Transit Tunnel. Lessons learned from Pine Street have translated into the design of the University of Washington Station, where a reception shaft for tunnel boring machines (TBMs) has been incorporated within the permanent works.

## INTRODUCTION AND BACKGROUND

As part of the Link Light Rail Project in Seattle, WA, work was completed at the Pine Street Stub Tunnel (PSST) in early 2007 for the Central Link Project. This tunnel was excavated using cut-and-cover construction within the limits of Pine Street. The stub tunnel provides both a turn back via double crossover for light rail trains running in the Downtown Transit Tunnel (DTT), as well as a connection point for the next phase of the project the University Link: (see Figure 1).

Designing the new University Link tunnel connection to the existing PSST was a tricky task. The on-site geotechnical conditions, buried obstructions, geometry, requirements for construction and balancing operational considerations and neighborhood stakeholder concerns associated with the recently completed PSST combined to create a uniquely challenging assignment.

Site geotechnical challenges included rubble fill, landslide deposits, and possible contamination soils. Soldier piles and tiebacks in the path of the both the Northbound (NB) and Southbound (SB) running tunnels, an electrical vault, duct bank, and a vent shaft partially above the tunnels, in addition to an existing deep sewer above the Southbound (SB) tunnel rounded out the buried obstructions we had to contend with. The geometry of the PSST connection was originally designed for a tunnel alignment towards First Hill and during the preliminary engineering design the alignment was changed to Capitol Hill. The recent history of construction in the area presented us with a further challenge of making the connection without causing disruption of traffic on Pine Street. Operational considerations within the PSST dictated limited access and work hours impacting the construction stage where complicated connections to the PSST for waterproofing, electrical, mechanical, temporary ventilation, and systems components needed to be made.

### Preliminary Engineering Design

The preliminary engineering (PE) concept for the connection shown in Figure 2 involved the excavation of two tunnel boring machine (TBM) retrieval shafts (one for the SB tunnel and one for the NB tunnel), the construction of tunnels excavated using the sequential excavation method (SEM), between these shafts, and the PSST headwall. The proposed short SEM tunnels, which were between 27 and 35 m (90 and 115 ft) in length, would have included the removal of tiebacks and soldier piles that intersect the proposed tunnel alignment adjacent to the PSST headwall.

During final design, an alternative approach was developed that would fulfill the tall order of:

- reducing the overall costs of the connection
- facilitating access to the existing PSST headwall

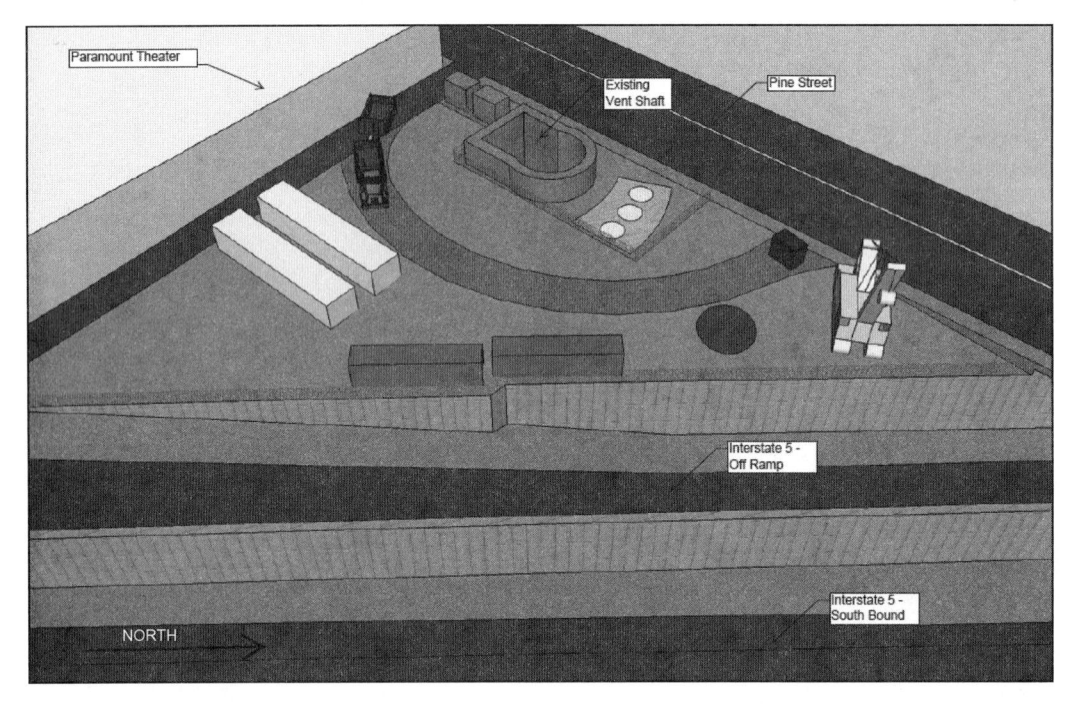

Figure 1. Conceptual layout of the Pine St. site showing temporary access shaft and adjacent to PSST ventilation shaft

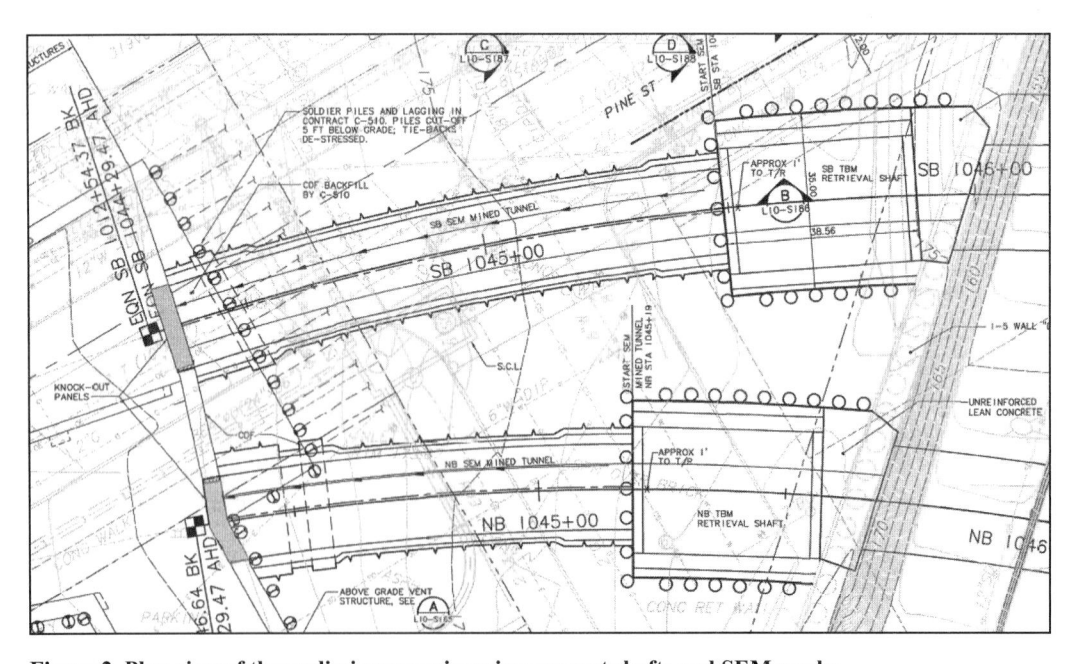

Figure 2. Plan view of the preliminary engineering concept shafts and SEM works

- avoiding excavation of retrieval shafts in close proximity to the existing I-5 Freeway
- removing the soldier piles and anchors within the tunnel envelope
- preparing for TBM excavations up to the face of the PSST headwall
- removing the heavily reinforced concrete tunnel "eyes" without undue disturbance to the ongoing transit operations

## ALTERNATIVE DESIGN

A few alternative approaches were originally considered, in addition to the PE design. One of the initial ideas was to determine whether any of the TBM breakthrough preparation work could be carried out under the PSST contract, which was still underway at the beginning of the University Link design period. This preparation work would have consisted of the removal of the partially exposed soldier piles and part of the anchors that intersect the proposed tunnel alignment from the surface prior to the restoration of Pine Street, as well as removal of the break-out panels in the PSST headwall. However, it was quickly ascertained that this idea would be difficult to implement given the necessity for a very late change to the scope and schedule of the PSST contract, which was near completion at that time.

The alternative discussed in this paper was developed as part of the Capitol Hill Tunnel contract (U230), and eliminated the need for any further construction work within Pine Street, minimizing disruption to residences and businesses in the area. This alternative was used in the final design, and involved the following activities:

- Ground treatment to facilitate tieback removal through the TBM cutterhead for the SB tunnel, and stabilizing TBM break-ins for both tunnels.
- Installation of "demising wall" bulkheads within the PSST to facilitate removal of the NB and SB break-out panels and installation of utility connections and light rail within an agreed length of the PSST.
- Temporary access/retrieval shaft construction for the NB tunnel only, taking advantage of the PSST headwall and the existing Controlled Density Fill (CDF) backfill on two of the four shaft sides.
- Access drift from the temporary shaft to a temporary chamber, constructed within the safety of the CDF located between the existing soldier pile wall and the PSST headwall.
- Removal of soldier piles from the temporary chamber and replacement with CDF backfill.
- In-tunnel disassembly of the SB tunnel TBM.
- Removal of soldier piles from NB retrieval shaft.

Figure 3 shows a general layout of the alternative used in the final design.

### Ground Treatment

Due to the presence of both recent alluvium deposits and landslide deposits below the groundwater and overlying the over-consolidated glacial soils, a limited ground treatment zone was determined to be required for both tunnels. The ground treatment zones, as shown in Figure 3, vary for each tunnel. For the NB tunnel, the zone is large enough to provide a stable face to allow for bottom removal of the east soldier piles, which are used for support of the retrieval shaft and later be removed from the path of the TBM. For the SB tunnel, the zone also needed to provide a stable face for the east soldier pile removal, but also provide stability for the tunnel heading to allow removal of tiebacks from within the face of the TBM, to be carried out under atmospheric pressure. The SB tunnel geometry was also dictated by an existing sewer that needs to stay in operation throughout the tunnel construction phase.

Ground treatment to stabilize the tunnel crown and improve the soil standup time was designed as jet grouting because of the high silt content of the in situ soils, and used to create a consolidated block of material in the zone of landslide debris between the alluvium and over-consolidated glacial soils. This work has been planned to be carried out from the Sound Transit staging site shown in Figure 1 next to Pine Street and extend at an angle below the street to prevent further surface disruption and minimize any potential traffic impacts.

### Operational Considerations and Neighborhood Stakeholder Concerns at the PSST

Early in the design of the connection design it was made clear to us that disruption of the Sound Transit and King County Metro operations within the PSST had to be held to a minimum. After some reflection on all of the construction activity that could not be avoided within the PSST and the risks this posed to ongoing transit operations the concept of "demising walls" was developed. The "demising walls" are fixed bulkheads fitted out with roller and personnel access doors constructed between 15 and 20 m (50 and 65 feet) from the PSST headwall in order to create a construction exclusion work zone. These bulkheads have been designed to prevent the communication of dust and noise from the zone, control personnel access into the active transit operations area, and maintain the integrity of the existing FLS (Fire Life Safety) ventilation. Installation of the

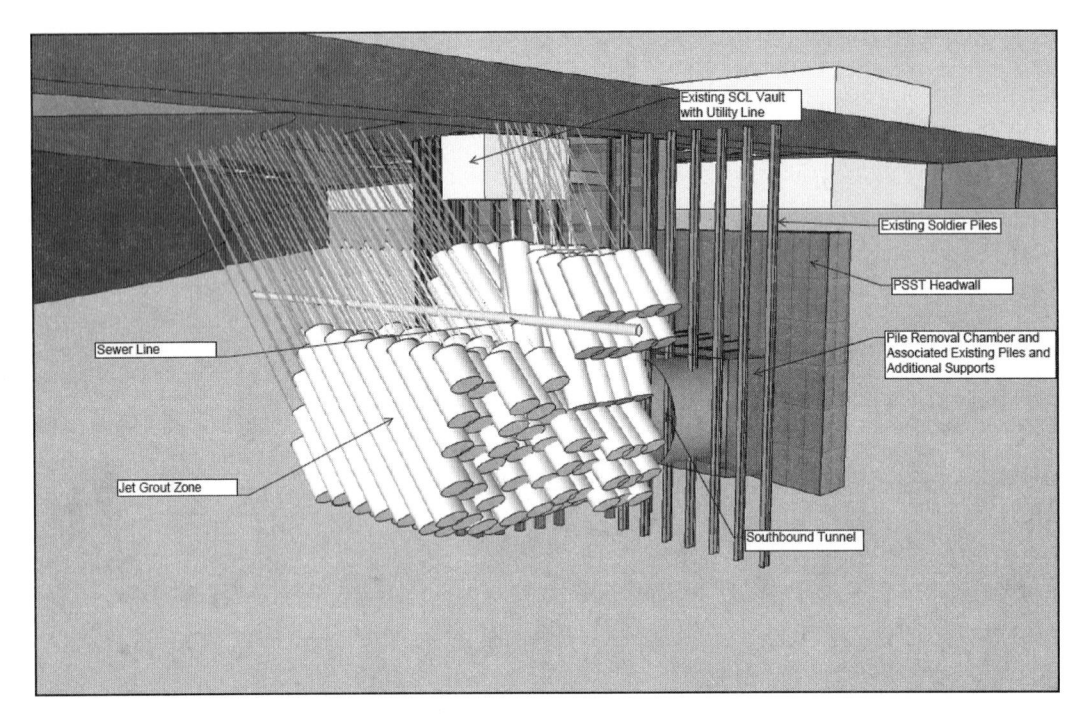

**Figure 3. Jet grouting below Pine St.**

bulkheads could not avoid impacts to the PSST. Relocation of the light rail "bumper posts" reduces the available storage length for light rail vehicles by approximately 60 feet and restricts Sound Transit to two-car travel. However, the advantages of the bulkheads outweighs this temporary inconvenience to operations and Sound Transit will not need a three-car service until the completion of University Link in 2016. Once the bulkheads have been installed they will remain in place until all systems and other finishing works have been carried so that the seamless integration of the U-Link with the PSST can be completed.

To address the concerns that site neighbors and other stakeholders would have of further construction being carried out that would disrupt traffic on Pine St. the design team came up with a feasible approach that would ensure that for the most part construction activities would take place within the site boundaries, only stepping outside into the side walk areas for very specific operations such as the angled jet-grouting below Pine St as shown in Figure 3.

**Temporary Shaft Support**

Construction of the temporary access/retrieval shaft for the NB TBM tunnel has been designed to proceed according to the following steps:

- A roughly rectangular shaft will be constructed so that the PSST headwall and additional temporary soldier piles will support the shaft excavation from elevation 52 m (170 ft) to the base of the PSST structure. The layout of the piles avoids the electrical duct bank and the overhang of the existing vent structure.
- The 22 m (70 ft) shaft will rely on the temporary soldier piles, wales at 2.4 to 3.7 m (8 to 12 ft) level intervals, and timber lagging are anticipated, similar to the successful model used for temporary excavation support of the PSST.
- Temporary soldier piles will be installed in order to safely excavate the shaft to the level of the access drift and provide access for removal of the existing soldier piles within the shaft.
- Upon completion of the works for the temporary pile removal chamber, the shaft will be excavated to a point where the NB tunnel headwall break-out panel can be removed.
- Existing soldier piles that were used as temporary support for the PSST and are within the Temporary Access Shaft will be removed.
- Once this work has been completed, all the soldier piles within the tunnel envelope will be cut or extracted after bracing the existing

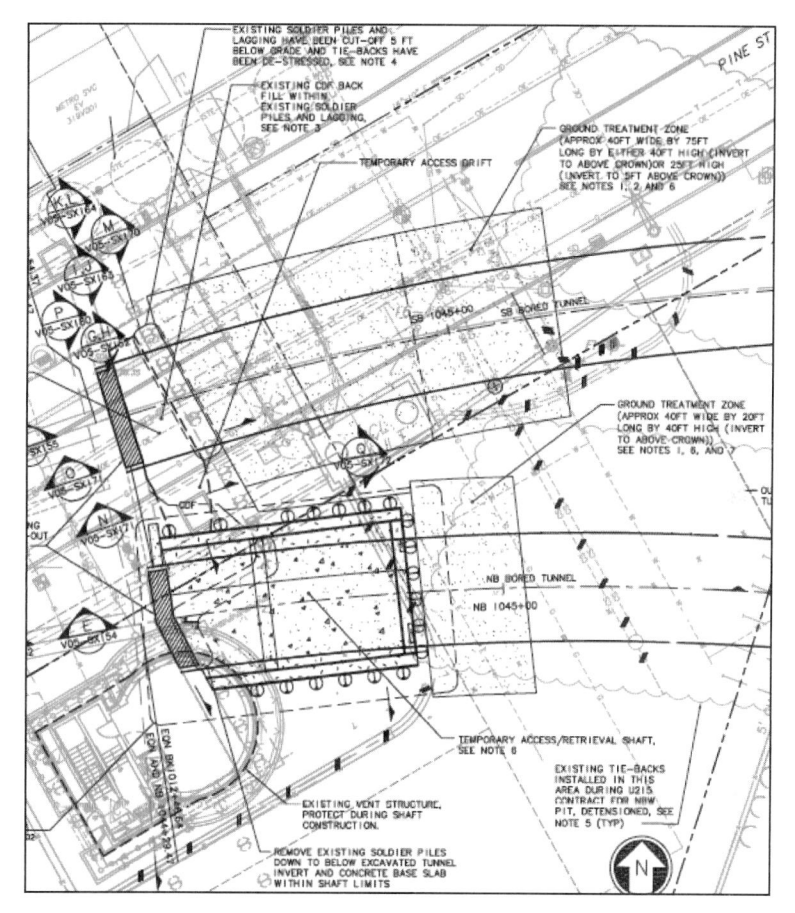

**Figure 4. Plan view showing the adopted final design alternative connection (section arrows provide some idea of the detailed engineering required to make the design work)**

piles in lifts, with removal carried up to 0.6 m (2 ft) above the crown.

- The shaft will then be partially backfilled in lifts corresponding to the pile removal sequence above, with CDF material to allow for the NB tunnel TBM to mine into the shaft.

Figure 5 shows an isometric view of the shaft.

**Access Drift and Pile Removal Chamber**

In order to avoid surface disruption to Pine Street a 3×3 m (10×10 ft) access has been designed to be driven from the shaft above tunnel axis level within the CDF material between the NB and SB tunnels. This access drift will take advantage of the existing PSST soldier piles on the east side for support. The drift excavation and chamber top bench will be supported by partial steel sets, with lagging or shotcrete to ensure ground stability, placed in line with the existing piles. At this stage it will be possible for the upper part of the SB tunnel break-out panel to be exposed and removal of the concrete will begin. Subsequent benches will be excavated from the top down, exposing the entire break-out panel for removal and the complete length of piles and laggings to be removed from the tunnel envelope.

Beginning from the bottom bench, laggings, piles, and 0.9 to 1.2 m (3 to 4 ft) of existing tieback will be removed after stabilizing the soldier piles. The lower portions of the pile removal chamber are expected to be in the over-consolidated Qpgm and Qpgl materials, which are stiff to very stiff clays. The upper portion of the chamber will be within the zone of ground treated soils, which should not become unstable during the short period that they are left unsupported. Figure 6 illustrates a section through the fully developed access drift and pile removal chamber which is larger than required to accommodate the tunnel envelope (SB tunnel profile shown) because of the presence of tieback anchor points that

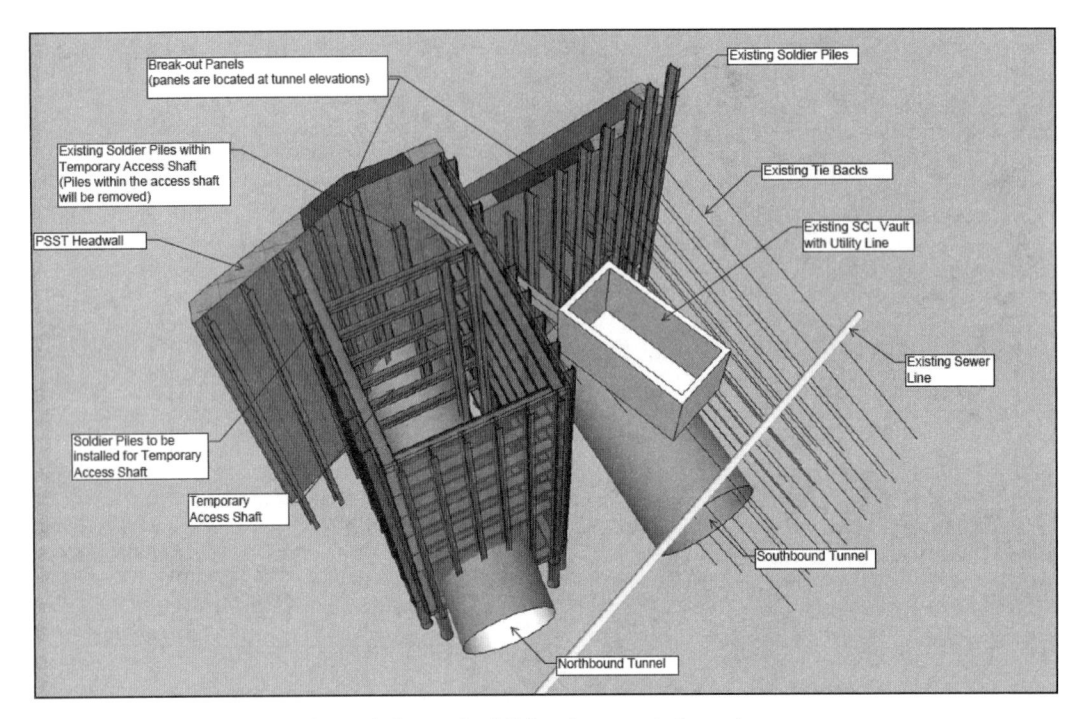

**Figure 5. View of shaft showing existing and additional support elements**

connect with tiebacks intersecting the tunnel horizon, as well as to allow waterproofing, mechanical, electrical, and systems connections to the existing PSST structure. To facilitate TBM excavation, the tiebacks will be disconnected from their associated piles within the pile removal chamber.

### Tieback Removal Through the TBM

During construction of the PSST headwall structure, the temporary excavation support soldier pile wall running northwest was supported by a tieback anchoring system. The tieback system was arranged in five rows at intervals of 3.4 to 3.7 m (10 to 12 ft), which intersect the proposed SB tunnel envelope as shown in Figure 6. The tiebacks consist of steel cables anchored over a minimum 4.6 m (15 ft) length at the cable terminus, and intersect the SB tunnels to varying degrees. The TBM is likely to encounter tiebacks over a 13.7 m (45 ft) long interval, starting approximately 16.8 m (55 ft) before the PSST headwall. In accordance with the specification an earth pressure balance (EPB) TBM will excavate in closed mode (pressurized face) up to this position and then convert to open mode (non-pressurized face) while excavating under the cover of the jet-grouted tieback zone. Following each of the 7 to 8 ring excavation sequences required to mine through the tieback zone, interventions are to be carried out as necessary to cut the cables engaged by the cutterhead or exposed in the face. This is anticipated to ensure that at no time there will be more than 1.5 m (5 ft) of cable exposed which could become entangled in the TBM cutterhead. Stability of the crown during these interventions will be provided by the ground treatment zone. Figure 7 shows a perspective view of the intersection of tiebacks with the SB tunnel envelope.

### TBM Drives from I-5 to PSST

Once the temporary excavation supports have been removed from the tunnel envelope, both NB and SB tunnel TBMs should be able to proceed up to the PSST headwall without difficulty. The NB TBM will be driven up to the PSST headwall and removed via the temporary access shaft. The cutterhead and shield components will be hoisted out of the shaft and loaded onto a flatbed trailer in easily transportable pieces, to be reassembled at the Capital Hill Station for the SB tunnel drive.

The SB TBM will pass through the anchors (as described above) and then through the CDF, aligning roughly perpendicular with the PSST headwall. Once the SB TBM shield is in position it will be grouted and the internal elements disassembled, leaving the shield carcass as temporary support for the tunnel. The gap created following removal of the cutterhead between the shield and the PSST headwall will be temporarily supported by bracing around the shield in order to ensure ground stability. The shield

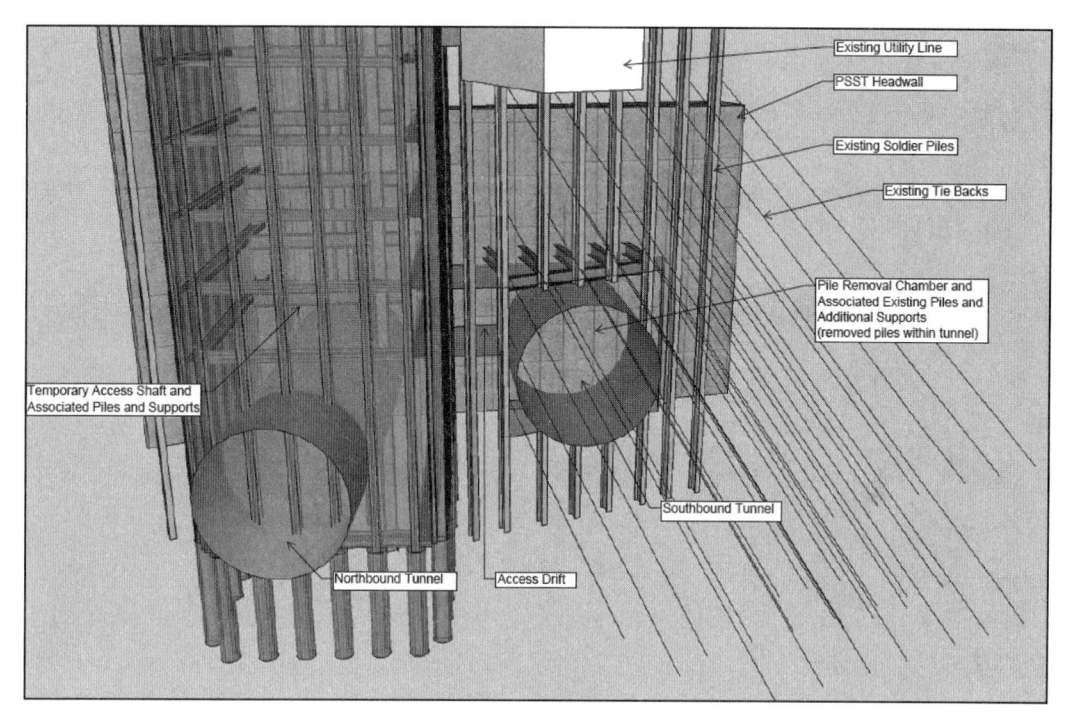

Figure 6. Transverse section showing access drift and SB pile removal chamber

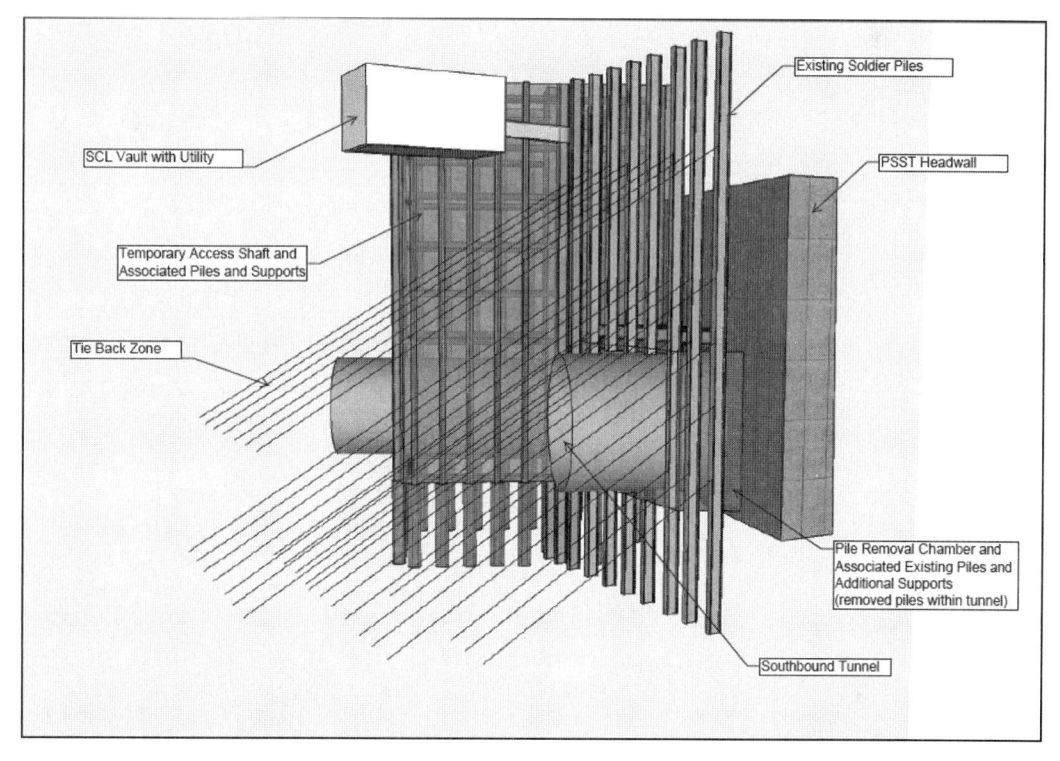

Figure 7. Perspective of tiebacks intersecting the SB tunnel envelope (ground treatment not shown for clarity)

diaphragms will be removed, and waterproofing, rebar, and concrete or shotcrete will be used to complete the circular cross section of the tunnel up to the PSST headwall. Connections for mechanical, electrical, and systems components will be made prior to placing the final lining.

**Final Lining and Connections**

After completion of the tunnel drives, the temporary shaft will be left open to allow subsequent contractors to transport materials to tunnel level without requiring access from the existing PSST. As a final step, a cast-in-place concrete lining will be installed to bridge the gap between the precast concrete segmental lining installed in the tunnel and the PSST headwall, including connections for waterproofing, mechanical, electrical, and systems components. The shaft will then be backfilled to the ground surface and the existing site restored.

**LESSON LEARNED: DESIGN OF THE NORTHLINK CONNECTION INTERFACE**

As illustrated above the conditions in and around the PSST were less than ideal for reception of the TBMs and considerable design was required to address the unique challenges of the site. As part of the University link both Sound Transit and their designer wanted to "think ahead" and avoid the difficulties encountered in designing the Pine Street connection. To address this issue, the north end of the University of Washington station (UWS) has been designed to incorporate a reception area and shaft for TBM removal. A number of design elements were incorporated into the north end of UWS to ease future construction. These elements include:

- Access rights have been worked out with the University of Washington to allow for the removal of TBMs from the north end of UWS.
- A TBM retrieval shaft has been built into the permanent works of the north end of the UWS, to allow removal of the TBMs as they mine into the station.
- Fiberglass reinforcing bars have been incorporated into the final design of the headwall at the north end of the UWS, to allow easier removal of the concrete headwall for the TBM break-ins.
- A block of treated ground will be created at the break-in points to the shaft headwall.
- The north headwall was designed to be perpendicular to the direction of the anticipated TBM drives.

**CONCLUSIONS**

The design of the connection of the University Link tunnels to the existing infrastructure at the Pine Street Stub Tunnel (PSST) presented many challenges. Preliminary engineering (PE) concepts anticipated retrieval shafts and short tunnels excavated using Sequential Excavation Methods (SEM) for this connection, as shown in Figure 2. However, limited access within the existing PSST structure for the SEM tunnels required an alternative approach. Use of a retrieval shaft adjacent to the existing box structure was designed to accommodate the Northbound Tunnel, and a short access drift to the Southbound Tunnel will allow construction to be performed with only limited impact on operations within the existing PSST structure via the use of a demising wall. This method also limits impacts to adjacent Pine Street; eliminates SEM works; allows tunneling to be performed by TBM for the entire tunnel alignment which has both schedule and cost advantages; ensures safety and security in the PSST; minimizes interference with existing or ongoing transit operations to reduce risks from both safety and contractual points of view; reduces schedule risk by performing preparatory works at PSST prior to the arrival of the TBMs; and gives some additional flexibility for making connections to the existing structure for waterproofing, as well mechanical, electrical, ventilation, and systems components.

Finally, the lessons learned from the PSST connection have been directly put to use at the north end of the U-Link project where UWS ties in with the future running tunnels expansion to the North. Future running tunnels coming into the station will be provided with a TBM retrieval shaft built into the permanent works of the station, greatly reducing the impact of the future expansion on the operations of the University Link light rail.

# TRACK 3: PLANNING

Bradford Townsend, Chair

*Session 1: Project Cost Estimating/Finance*

# Size Matters If You're a Tunnel

Lee W. Abramson
Hatch Mott MacDonald, Pleasanton, California

ABSTRACT: The varieties of mined tunnel sizes and ground characteristics being used on recent infrastructure projects have reached unprecedented levels. For a number of possible reasons, the project teams on these tunnels are frequently asked about incremental costs, durations and risks for incrementally larger tunnels. The cost answer is embedded in the fundamentals of contractor-style tunnel cost estimating. Similarly, schedule impacts are dependent on sophisticated scheduling techniques. Risk scaling is based on a proper risk evaluation, mitigation and management program. This paper discusses issues related to tunnel size and some recent projects that performed studies on alternative tunnel sizes and resultant conclusions that can be used for future studies.

## INTRODUCTION

One of the first decisions often made on mined tunnel projects is what size the tunnel will be. Because of the costly nature of tunnels, minimizing the size is usually a necessary requirement. Size requirements are based upon the type of tunnel, function, capacity, internal installations, geometrics, shape, access/egress, safety, operations, maintenance and a multitude of other considerations. At the earliest phases of a project, many of these requirements are not well-known initially especially during feasibility, planning and conceptual design. However, to gain necessary consensus, approvals, funding, etc. so the project can move forward expeditiously, an inordinate amount of specificity is demanded, often more than can be reliably given. Occasionally, this early work may even be conducted without the advice of engineers and geologists with extensive experience in tunneling. Invariably, the project gets known to be a tunnel of a specific size and it is extremely difficult to increase during subsequent engineering when final requirements are better and more reliably known. One tempting direction to take initially would be to start the size very large. But this could have very negative effects on impacts, constructability, cost, risk and schedule and could very well kill a feasible and worthwhile project. Also, without extensive engineering studies done, the information may be insufficient to scale these up or down reliably. The flip side of starting too large is starting too small and having, costs grow sometimes exponentially, decreasing acceptance of the project and fueling a notion in and outside of the tunneling industry that tunnel costs always spiral out of control. This paper discusses issues related to tunnel size and some recent projects that performed studies on alternative tunnel sizes and resultant conclusions that can be used for future studies.

## TUNNEL SIZE CONSIDERATIONS

Even though there is some overlap, tunnel size considerations are strongly affected by the ultimate use of the tunnel. From this standpoint, tunnels can be grouped into three broad size categories. Smaller utility tunnels are used for electric, telecom, gas, oil, water, wastewater, combined sewer overflow (CSO), etc. Larger transportation tunnels are used for highways, rail, transit, airport people movers, etc. The third category includes other unique applications of tunneling for high energy physics, the military, etc. Some of the considerations used to size these tunnels are listed in Table 1.

The general topic of flow capacity is meant to mean number of wires, cables, liquid or gas flow, people, train cars vehicles and other requirements. Related to this in transportation tunnels is number of lanes, tracks, etc. Control structures and connections include pull boxes, junctions, branches, intersections, shafts, weirs, gates, valves, sumps, screens, portals, stations, crossovers, cross passages, etc, In recent years, particularly for CSO tunnels, in-line storage is designed into the project to manager flow rates, treatment requirements, pump out scheduling, etc. Length affects size largely due to the tunneling methods that may be required and the room for equipment, lining and spoil removal. Similarly, curvature both horizontal and vertical may affect size for the same reasons. Another important consideration is whether human access is required during construction and/or later during the design life. If humans are expected to be in the tunnel, fire life safety must be taken into consideration including access/egress,

Table 1. Recommended considerations for sizing tunnels

| Size Considerations | Electric/Telecom | Gas/Oil | Water | Wastewater | CSO | Rail/Transit | Highway | Physics/Military |
|---|---|---|---|---|---|---|---|---|
| Flow capacities | • | • | • | • | • | • | • | • |
| Flow control structures | | | • | • | • | • | • | • |
| In-line storage | | | | | • | | | • |
| Length | • | • | • | • | • | • | • | • |
| Curvature | • | • | • | • | • | • | • | • |
| Connections/junctions | • | • | • | • | • | | | • |
| Human access | | | | | | • | • | • |
| Screens/debris pits | | | • | • | • | | | |
| Size of vehicles | | | | | | • | • | |
| No. of tracks/lanes | | | | | | • | • | |
| Public fire/life/safety | | | | | | • | • | • |
| Lighting/signage/comm. | | | | | | • | • | • |
| Unique applications | | | | | | | | • |

Table 2. Typical tunnel diameter ranges

| Typical Diameters (meters) | Electric/Telecom | Gas/Oil | Water | Wastewater | CSO | Rail/Transit | Highway | Physics/Military |
|---|---|---|---|---|---|---|---|---|
| 0.1–0.5 | • | • | • | • | • | | | |
| 0.5–1.0 | | • | • | • | • | | | |
| 1.0–3.0 | | | • | • | • | | | |
| 3.0–6.0 | | | • | • | • | • | | • |
| 6.0–9.0 | | | • | • | • | • | | • |
| 9.0–12.0 | | | | | | • | | • |
| 12.0–15.0 | | | | | | • | • | • |
| 15.0–18.0 | | | | | | • | • | • |
| 18.0–21.0 | | | | | | | • | • |
| >21.0 | | | | | | | | • |

ventilation, lighting, communications, signals and sensors. Finally, some unique tunnels as for physics and military uses require some of the most challenging sizes and shapes encountered underground apart from ore mining applications.

## TYPICAL TUNNEL DIAMETERS

The considerations listed above often result in some typical tunnel sizes. As seen in Table 2, electrical and telecom tunnels are often circular and less than 0.5 meters in diameter. Gas and oil pipelines are usually also circular and up to 1.0 meter in diameter. Water supply, wastewater and CSO tunnels are also circular usually but diameters range widely from 0.1 to 12.0 meters. Tunnels for hydropower use are often larger. Rail tunnels vary in size depending on vehicle type and number of tracks. Most are between 6.0 and 12.0 meters except for high speed rail tunnels and can be circular or oval/horseshoe-shaped. Highway tunnels are frequently 12.0 to 18.0 meters wide and oval/horseshoe-shaped except for very recently where large (15.0 to 20.0 meter) circular tunnels have been constructed or are under design. Much larger tunnels have been constructed or are being planned for the military and high energy physics research communities.

Table 3. Typical tunneling methods for various types of tunnels

| Trenchless Method | Electric/ Telecom | Gas/ Oil | Water | Waste-water | CSO | Rail/ Transit | Highway | Physics/ Military |
|---|---|---|---|---|---|---|---|---|
| HDD | • | • | • | • | • | | | |
| Auger | • | • | • | • | • | | | |
| Ramming | • | • | • | • | • | | | |
| Microtunnel | • | • | • | • | • | | | |
| Hand Mine | | | • | • | • | | | |
| Digger Shield | | | • | • | • | • | • | • |
| EPB/Slurry | | | • | • | • | • | • | • |
| Rock TBM | | | • | • | • | | • | • |
| Drill & Blast | | | • | • | • | • | • | • |
| SEM/NATM | | | | | | • | • | • |
| Jacked Tunnel | | | | | | • | • | |
| Immersed Tube | | | | | | • | • | |

## TRENCHLESS METHODS USED IN A VARIETY OF GEOTECHNICAL CONDITIONS

As shown in Table 3, gas and telecom tunnels are almost always excavated using remote tunneling methods in soft ground. They can be installed in rock if necessary but unit costs are quite high. Water, wastewater and CSO tunnels can be constructed using every method in every subsurface condition and are almost always circular unless they are very short making it uneconomical to use tunnel boring machines. Transportation tunnels are similar except that the tunnels are too large to use remote tunneling methods. Large physics and military caverns are most often excavated in bedrock. Long connecting tunnels such as that started for the super conductor super collider in Texas used tunnel boring machines.

## COST ESTIMATES

Cost estimates for tunneling projects should be prepared the same way contractors put together their bids. Often this is referred to as a "bottoms-up" estimate. There are no short cuts to this and urges to scale other projects and/or use rates from published cost estimating manuals should be avoided. Tunnel cost estimates involve very detailed information and extensive underground experience. A cost estimate that is systematically prepared and organized will provide an excellent platform to examine and evaluate incremental costs. Within narrow ranges, incremental variations in tunnel diameter have a significant impact on muck volumes and lining quantities but production rates and crew sizes (labor costs) are not as sensitive.

A session held at the xxx Transportation Research Board meeting in 2009 included a specific session on getting tunnel cost estimates right. Jim Peregoy (2009) described many features of a proper tunnel cost estimate. First of all, it is important to review all documentation including geotechnical information, special conditions of the contract, specifications, plans/drawings and make a site visit. Next, it is very important to estimate production rates based on general underground work conditions, means and methods, general production (and cost) groups, quantity take-offs, crewing, materials, equipment and subcontractors. Then the actual cost estimate can be prepared consisting of appropriate cost categories, labor costs, material costs, service costs, subcontractor costs, and equipment costs. Finally, an overall review of the estimate should be made and any anticipated or required escalation added on. It is absolutely essential to have this detail if one is to evaluate incremental costs of larger and/or longer tunnels. Estimates done in this way can be very reliable as shown in Table 4.

On average, the described methodology tends to work out quite well. However, every tunnel is different and the relationships between engineer's estimates and bid prices do vary on a case by case basis. Reasons for this depend considerably on factors such as market pressures, contractor workload, risk sharing, owner-agency records with contractors, permitting and regulatory issues, potential third party issues, environmental constraints, quality and accuracy of contract documents, project location, etc. A worthwhile target is for the Engineer's Estimate to be within 5 to 10 percent of the low bid price.

Table 4. Cost estimating history for selected recent tunnel projects

| Project | Eng. Estimate ($ millions) | Low Bid ($ millions) | % Difference of Low Bid |
|---|---|---|---|
| **Transit tunnels** | | | |
| • Minneapolis LRT | 53 | 52 | −1.89% |
| • San Diego State LRT | 14 | 15 | +7.14% |
| • Sheppard Subway, Toronto | 107 | 100 | −6.54% |
| • Dulles APM Tunnels – Phase I | 450 | 440 | −2.22% |
| • Beacon Hill Project, Seattle | 250 | 290 | +11.96% |
| **Water/Wastewater Tunnels** | | | |
| • Sacramento LNW Interceptor | 44 | 44 | +2.07% |
| • Atlanta West Area CSO | 260 | 210 | −19.23% |
| • 9th Line Sewer, Ontario | 45 | 50 | +11.11% |
| Totals | **$1,223** | **$1,201** | **−1.83%** |

## SCHEDULING

A tunnel construction schedule should be prepared in detail based on the activities and production rates in the cost estimate. There are various ways of depicting this including the Critical Path Method using various popular scheduling programs. Often the project activities are broken down into a list of work commonly referred to as Work Breakdown Structures. These usually have a close correspondence to the plans and/or specifications. For long linear projects such as tunnels, it is also very useful to prepare what is referred to as a sloping line diagram that depicts project stationing, time and specific activities. Potential conflicts and interfaces become abundantly apparent utilizing this approach. On one recent project, this approach alerted the team to a "bust" in a transit tunnel schedule prepared by others that had rails being installed in a tunnel before the tunnel had been excavated and lined! Project schedule is not super sensitive to incremental variations in tunnel size usually.

## RISK MANAGEMENT

Assessing, controlling and mitigating risk is a crucial element to the successful delivery of any tunnel project. Risk management activities include:

- Development and implementation of the risk management process.
- Risk analyses including identification, ranking and mitigation.
- Development of Risk Management/Risk Mitigation Plan.
- Coordination with the master schedule.
- Tracking and management of contingencies.
- Periodic updates of risk management plan.
- Recommendations and coordination related to program insurance/OCIP/Wrap-up.
- Development of strategy and implementation of claims avoidance plan.

Tunneling is an inherently risky undertaking. Risk, however can be controlled by a proactive plan to identify risks early in the project development cycle. The risk management approach should be woven into every task – from data gathering and documentation, through design and evaluating potential environmental impacts, to construction cost estimates and schedules. There should be no surprises and the unexpected should be expected. Potential concerns should be anticipated in advance, there should be contract line item cost and schedule provisions to address these circumstances, and the contract should clarify how such provisions will be implemented during the work. As an example, for underground construction, while the "owner owns the ground" through which the facility is to be built, the owner only pays for the conditions the contractor actually encounters. If the adverse ground conditions are not actually encountered, the provisions are not implemented, and the owner does not pay for those conditions.

Owners contemplating underground construction projects are increasingly skeptical of projected costs and schedules. Innovative techniques allow quantification of risks associated with budgets and schedules, and the development of strategies to deal with them. Risk Registers are created specifically for projects and become an everyday tool to monitor and anticipate events with significant outcomes on project performance. To ensure that risk continues to be proactively and continuously monitored and reviewed during the entire design and construction

Table 5. Possible cost and schedule impacts of risk on tunnel projects

| Risk | Possible Schedule Impact | Possible Cost Impact |
| --- | --- | --- |
| Damage to adjacent facility | Weeks–Months | $10ks–100ks |
| TBM failure | Weeks–Months | $10ks–100ks |
| Excessive settlement/sink hole | Months | $100ks–Millions |
| Worker injury/death | Weeks–Months | $100ks–Millions |
| Obstruction | Weeks–Months | $100ks–Millions |
| Differing site conditions | Weeks–Year(s) | $100ks–Millions |
| Slow production/progress | Months–Year(s) | $100ks–Millions |
| Shaft/portal failure | Months–Year(s) | $100ks–Millions |
| Major collapse/flood/explosion | Months–Years | $10s–$100s Millions |
| Major collapse/flood/explosion | Months–Years | $10s–$100s Millions |
| Failure to get record of decision* | Years | $10s–$100s Millions |
| Failure to get NOD (FFGA)* | Years | $10s–$100s Millions |

* Federally funded projects

process, the Risk Register should be formally reviewed on a regular basis. Each review should identify new risks and reconsider the severity of the previously identified risks, based upon the incidents of the previous period. This ongoing process allows the project owner to carry firm contingency budgets as the project proceeds.

Are all risks created equal? Not really. Certainly very frequent risks that have a high degree of severity should be mitigated to the extent possible. Infrequent risks that don't have a high degree of severity can probably be tolerated on most projects. Also, project risk severity must be looked at ultimately in respect to cost and schedule. Some relative comparisons are made in Table 5.

Often, cost and schedule impacts are referred to the base cost and schedule and contingencies planned for in the program planning and approvals. It is wise to plan for 10 to 20 percent variances (contingencies) in cost and schedule ahead of time should certain anticipated risks manifest themselves. It is very difficult and cumbersome on a program to obtain additional funding to deal with a realized risk after a firm project or program cost and schedule has been approved by governing bodies and/or the public. In this sense, tunnel size does not matter except the larger and longer the tunnel, the more severe cost and schedule impact magnitudes will be.

## EXAMPLES

The following examples provide a brief glimpse of where size mattered on recent tunnel projects or will matter on future ones.

### Santa Cruz Landfill Water Bypass Tunnel

The City of Santa Cruz, California owns and operates a Class III landfill that was constructed in canyons close to the Pacific Ocean. For several years, the landfill acted as a barrier to fresh storm water drainage in these canyons. Surface runoff ended up flowing into the landfill and becoming polluted by landfill contaminants. The State of California Regional Water Quality Control Board imposed a Cease and Desist Order on the landfill until the city constructed a system to intercept fresh water flow above the landfill, conduct it around the landfill and discharge it downstream of the landfill back into the creek bed that the landfill occupies. The solution provided for the city included two 0.9-m-diameter microtunnels to convey the fresh water around the landfill and soil-cement-bentonite slurry cutoff walls to prevent subsurface migration of fresh water into the landfill (Abramson et al, 1996). The two microtunnels were approximately 400 and 600 meters long in soft bedrock. One of three tunneling technologies could be bid: Microtunneling (MTBM), Tunnel boring machine (TBM) or Horizontal directional drilling (HDD). The horizontal directional drilling option was the lowest bid at $4.1 million.

On this project size mattered, not during excavation of the tunnel but after backfill grouting of the HDPE pipeline. Unbeknownst to the contractor, water had leaked out of the pipeline during grouting of the annulus outside of the pipeline causing a partial collapse. The collapse had to be reamed and a smaller pipe installed. Luckily, the hydraulic design could tolerate the smaller pipe diameter. Also, the initial pipe sizing took into consideration the possible need for human access which did become necessary to assess the conditions albeit with one brave soul riding down the pipeline on a skate board tethered with rope!

## MCUA Tunnel and Force Mains Under the Raritan River, New Jersey

The Middlesex County Utility Authority in New Jersey required two new force mains to provide a redundant means for sewage conveyance from MCUA's Edison Pump Station on the northern shore of the Raritan River to the Central Wastewater Treatment Plant on the southern shore (Rautenberg et al, 2009). Rather than drive two separate tunnels through low strength estuarine deposits, sand and gravel with limited cover below the Raritan River and close proximity to diabase bedrock, one single larger tunnel was constructed. The tunnel is 1,192 meters long and 4.09 meters inside diameter. The tunnel liner consists of gasketed concrete segments and two 1.5 meter diameter wastewater force mains to be installed within the tunnel. The construction also included two 8.5 meter diameter shafts by permanent slurry wall construction as well as installation of valves, piping and provisions for future tunnel access. Size mattered because this configuration accomplished the installation of the required force mains, worker access for operations and maintenance as well as the added feature of allowance for future installation of additional utilities across the river.

## MARTA's Peachtree Center Station

Peachtree Center station is the deepest station in the Metropolitan Atlanta Rapid Transit Authority (MARTA) rail system, at 36 meters below Peachtree Street in the heart of Atlanta's business district (Kuesel and King, 1979). Initial proposals contemplated cut-and-cover construction but were ruled out due to the expected negative impact on commerce, traffic, etc. The station is 183 meters long with a center platform and one trainway on each side. The resulting cavern is 18 meters wide and 14 meters high. The cavern "walls" are carved out of solid gneiss rock. During geotechnical explorations including a pilot tunnel, it was disclosed that the depth of rock weathering increased and the general quality of the rock deteriorated at the northern end of the platform. Size mattered because the station was sufficiently sized to accommodate a reduction in cross section and additional concrete lining in the poorer rock sections without significant changes to alignments, configurations, access/egress points, ventilation and other features that would have been very difficult and costly to change during final design and construction.

## Los Angeles Metro Red Line North Hollywood Extension

The 10-kilometer Metro Red Line North Hollywood Extension (Segment 3) included three new underground stations and two new line sections (Albino. High capacity passenger vehicles powered by an electric third rail system operating through a twin-tunnel subway. The contract was financed through a grant from the USDOT, the FTA, and with funds from the State of California and local sources. The tunnels were 6.7 meters in diameter and constructed through faulted rock and mixed ground with advanced grouting for ground water mitigation while crossing a major fault line. The line tunnels driven with two full-face TBMs through a variety of rock conditions ranging from soft shales to fresh granodiorite. At the south end of the drives, an oversized tunnel or special seismic section was incorporated into the tunnels to cross the active Hollywood Fault. This was the terminus of the TBM tunnels that were being mined from the north and was originally to be mined by drill and blast methods. However, a horizontal borehole confirmed the ground to consist of intensely sheared, brecciated and decomposed granodiorite and the tunnels were enlarged and shortened to include lattice girders and mining by NATM using roadheaders.

## California High Speed Rail Tunnels

California is setting the pace for transportation in the U.S. with its landmark California High Speed Rail (CAHSR) system that will span nearly 1,300 kilometers from San Francisco/Sacramento in the north to San Diego in the south (Abramson and Crawley, 1995). The proposed system will be the first of its kind in the U.S. and will connect with existing rail, air and highway networks. It is projected to transport 100 million passengers by 2030. Operating at speeds up to 220 mph, the trains will be electrified steel-wheel-on-steel rail. A passenger boarding an express train in San Francisco can step off in Los Angeles in 2½ hours, which is three to four hours less than traveling the same distance by car without traffic. Many sections will require tunneling. Normally for rail and transit tunnels, the cross section is optimized for the car size, dynamic clearances, internal fixtures, track bed, etc. For high speed rail trains however, air resistance is quite significant and tunnels must be sized larger to reduce air drag and heat generation especially traveling uphill. For a single track tunnel, up to a 50 percent increase in cross section might be required depending on grade and speed through the tunnel. This would also need to be considered for double track tunnels. Also in California, some of the tunnels will cross active faults and allowances for fault slippage and rupture might need to be taken into account when sizing the tunnels.

## Interstate 90/93 Interchange on the Central Artery Tunnel Project (Jacked Tunnels)

The purpose of the Central Artery project in Boston was to cure massive traffic congestion in the heart of the historic city by replacing the elevated Central Artery section (opened in the 1950s) with a new eight to ten lane highway running mainly underground. The I-90/I-93 interchange links the Central Artery to the Ted Williams immersed tube tunnel leading to Logan Airport. This multi-level interstate interchange is one of the most complex sections of the project and involves three crossings beneath eight active railway tracks providing Amtrak and MBTA service into South Station. Jacked tunnels form the deepest section of the multi-level interstate interchange that also includes an open boat section, at-grade roadways, several levels of viaduct that tie into the adjacent roadways, and immersed tube tunnels (Taylor and Winsor, 1998). Early design concepts would have made interruption of rail traffic inevitable. Alternatively, a tunnel jacking solution was developed that allowed construction without interrupting train service, saving millions of dollars of railway operating revenues. This part of the project included three jacked tunnel sections, three jacking pits where the to-be-jacked tunnels were constructed on-site and jacked from, and 366 meters of cut-and-cover tunnels as approaches to the jacked tunnel segments. The jacked tunnels are the largest and most complex ever constructed in the world. The largest of the three was 24 meters wide by 11 meters high by 113 meters long. The deepest tunnel section had 7.3 meters of cover. Combinations of dewatering and ground treatment methods including ground freezing and jet grouting were used to control and minimize settlement during excavation. Soil conditions consisted mostly of reclaimed land, Boston Blue Clay, thin layers of fine sand, and glacial till overlying Cambridge Argillite bedrock. It is unlikely that other types of tunneling could have been used for such large cross sections and little cover the to the commuter rail tracks above without having major disruptions to Boston's busiest train terminal.

## Mission Valley East Light Rail Extension

The Mission Valley East Light Rail extension (Green Line) extends the San Diego Trolley from Mission San Diego to the Orange Line at Baltimore Drive. The project was separated into a guideway contract and a tunnel contract. The tunnel segment, located where the line crosses the San Diego State University campus, originally called for the design of twin-bore tunnels running under SDSU and an underground station servicing the campus (Field et al, 1995). During the design phase, the client altered the alignment, which changed both the length and profile of the tunnel. Changes were made expeditiously to change from conventional mined tunnels to NATM tunnels. The preliminary design for included twin tunnels with connecting cross-passages and a mined underground station to be constructed by conventional mining. Preliminary drawings and specifications incorporated tunnels that were 1,100 meters in length and 6.1 meters in diameter. The presence of boulders required the specification of an open-faced shield for tunnel excavation. Lining was to be either one-pass, bolted and gasketed concrete segments, or temporary expanded segments with a final cast-in-place liner. Final design for the revised alignment led to the need to locate both tracks within a larger but shorter single tunnel. This section was 11.3 meters wide, 9.8 meters high and 305 meters long. The main geological formation at the tunnel horizon was Stadium Conglomerate, a competent, partially cemented mixture of gravel and sandy clay, containing occasional large boulders. The water table varied up to 6.1 meters above the tunnel roof. Surface settlement was of particular concern, and settlement monitoring of historic buildings and other structures and utilities was an essential element of the excavation process. The twin-bored, gasketed segmental lining would have enabled the inherent flexibility of the lining to accommodate seismic movements and loadings. The change to the more monolithic NATM lining resulted in detailed seismic design of both the lining itself and the watertight movement joints at the two ends of the tunnel where it abuts onto adjacent structures.

## Seattle Sound Transit Beacon Hill Station and Tunnels

Beacon Hill Station and Tunnels is a segment in the new 23-kilometer Central Link light rail line, a critical component in Sound Transit's long-term regional transportation network. The station, one of 12 new stations on the line, is expected to serve approximately 3,000 people a day by 2020. The twin 1,280-meter-long tunnels, one northbound and one southbound, were excavated using an earth pressure balance tunnel boring machine and lined with one-pass precast segmental linings measuring 5.7 meters internal diameter (Varley et al, 2007). The design included three cross passages to meet the National Fire Protection Association's 130 egress requirements. Tunnel construction was in soft ground that primarily consists of firm to hard clays, but also includes water bearing sand and silt zones.

The Beacon Hill station was mined from two shafts: one 14 meters in diameter, the other 7.9 meters in diameter. The shafts are 56.4 meters deep and act as both entrance and exit and ventilation structures. The 116 meter-long platform tunnels plus connector tunnels, concourse adit, ventilation adits and cross-passage tunnels were constructed using the sequential excavation method (SEM) using shotcrete initial linings that take advantage of ground relaxation to reduce loading. The large diameter concourse adits measure 13.7 meters in diameter. Stage-grouted barrel vault pipes and grouted pipe spiles formed the presupport for the SEM tunnels. Additional tool box support items were used where necessary. The support methods were selected onsite as ground conditions were assessed. The project included multistage excavation sequences, including twin sidewall drifts and single sidewall drifts. This station is one of the deepest soft ground SEM stations in the world.

## WSDOT-Project SR99-Alaskan Way Viaduct and Seawall Replacement

The SR99 Alaskan Way Viaduct in Seattle was damaged during the Nisqually earthquake in February 2001. Prior to this, the seismic vulnerability of the viaduct had already been recognized. After numerous studies and evaluations, the decision was made early in 2009 by the Washington State Department of Transportation (WSDOT) to realign SR 99 as a limited access four-lane divided highway through downtown Seattle in a deep bored tunnel alignment along 1st Avenue (WSDOT, 2009). This large-diameter bored tunnel will have an approximate interior diameter of 16 meters and an approximate length of 2.8 kilometers (not including the cut-and-cover sections at each end). WSDOT is in the process of procuring a design-builder for this project that will include twin deck roadways each with two 3.66 meter lanes, a 0.61 meter and 2.44 meter shoulder, and a 4.88 meter vertical clearance cross section. The South portal structure is expected to be located in the vicinity of First Avenue South between Charles and Dearborn Streets and serves as the starting location for the tunnel boring machine (TBM). The current proposed tunnel alignment will then extend along First Avenue South, passing under a rail tunnel, to the intersection of Pike Street where it will make a sweeping turn to the east going beneath a sewer tunnel and numerous buildings and ends at the north portal on highway SR 99, in the vicinity of John Street. The tunnel will be constructed using a pressurized face TBM and supported with a bolted, gasketed, pre-cast concrete lining. As an integral part of the tunneling operation, comprehensive ground movement and building settlement monitoring and mitigation action plans will be required. Considerable work will be required at the North and South portals to make these areas ready for the tunnel work, including: the design and construction of permanent and temporary retaining walls, relocation of utilities, removal of unsuitable materials and/or soil improvements, removal of adjacent temporary building tie-back supports, providing temporary and permanent power supplies, muck disposal operations, design and construction of ventilation buildings, and design and construction of connecting cut-and-cover work at both the North and South ends of the tunnel.

## The Large Water Neutrino Detector Project for DUSEL

The former Homestake gold mine in Lead, South Dakota is the site for the Deep Underground Science and Engineering Laboratory (DUSEL). One of the projects planned there is to construct very large caverns nearly 1.6 kilometers below ground for very large multipurpose detectors for proton decay and neutrinos from many different natural sources (Diwan, 2007). These caverns will be constructed within the Yates Member of the Poorman Formation consisting of metamorphosed tholeiitic basalt and schist. The caverns are each expected to be approximately 50 meters in diameter and 50 meters high with several various connecting ramps, drifts and smaller caverns. Once excavated, these large test caverns will be supported, lined, fitted with 50,000 photo multiplier tubes and then filled with ultra-pure, deionized water. Size matters!

## CONCLUSIONS

1. For small tunnels, it is very desirable for sizing to consider human access should that be necessary.
2. For larger utility tunnels, particularly circular ones, excess space may provide opportunities for additional usages such as future access and additional utility crossings.
3. For mechanized tunnel (TBM) drives that require workers inside the tunnel to carry out the work (i.e., not microtunnels), there is an optimum size for adequate productivity and safety. TBM tunnels that are smaller than approximately 2 meters in diameter make it very difficult to maneuver and in the long run, tunnels larger than 3 meters in diameter might be less expensive because they can be constructed faster.
4. Transportation tunnels ranging in diameter between 6 and 15 meters have become quite common place in recent times. Tunnels of this size often pose greater risks to the environment. A formal risk assessment and mitigation program is extremely important to avoid excessive delays and cost overruns.
5. Larger and larger tunnels are being contemplated and built in marginal ground conditions around the world. Unprecedented equipment, linings, methods and expertise are required for these tunnels. Extensive

evaluations and analyses must be undertaken to minimize risk. More progressive contract mechanisms and bidding strategies should be developed.

6. Cost, schedule and risks can not and should not be scaled linearly for incrementally larger tunnels. There are sometimes fundamental differences between tunnels of similar size and these differences need to be addressed in detail from the very basic building blocks that proper estimates have been prepared.

## ACKNOWLEDGEMENTS

The author wishes to thank his clients and colleagues who have assisted in making these projects successful and sharing their knowledge and experiences with him freely in the preparation of this paper.

## REFERENCES

Abramson, L.W. and Crawley, J.E. 1995. "High Speed Rail Tunnels in California," Proceedings of the Rapid Excavation Tunneling Conference, San Francisco, pp. 575–594.

Abramson, L.W., Fowler, M.E. and Boyce, G.M. 1996. "Santa Cruz Landfill Water Bypass Tunnel," Proceedings of the North American Tunneling Conference, Washington, D.C., pp. 17–25.

Albino, J.H., Roy, G. and Monsees, J.E. 1999. "Crossing the Hollywood Fault Using Special Section," Proceedings of the Rapid Excavation Tunneling Conference, Orlando, pp. 39–44.

Diwan, M. 2007. "The Large Water Detector Project for DUSEL," Presentation to Fermilab.

Field, D.P., Hawley, J. and Phelps, D. 2005. "The North American Tunneling Method—Lessons Learned," Proceedings of the Rapid Excavation Tunneling Conference, Seattle, pp. 871–881.

Kuesel, T.R. and King, E.H. 1979. "MARTA's Peachtree Center Station," Proceedings of the Rapid Excavation Tunneling Conference, Atlanta, pp. 1521–1544.

Peregoy, J. 2009. "Tunnel Cost and Scheduling—A Contractor's Perspective," Annual Meeting of the Transportation Research Board, Washington, D.C., 11 pp.

Rautenberg, B., Prada, J., Perrone, F. and Tanzi, D. 2009. "Construction of the MCUA Tunnel and Force Mains Under the Raritan River, New Jersey: A Case History," Proceedings of the Rapid Excavation Tunneling Conference, Las Vegas, pp. 824–833.

Taylor, S. and Winsor, D. 1998. "Developments in Tunnel Jacking," Proceedings of ASCE Geo-Congress '98, Boston, pp. 1–20.

Varley, Z., Martin, R., Robinson, R., Schmall, P. and Parmantler, D. 2007. "Beacon Hill Station Dewatering Wells and Jet Grouting Program," Proceedings of the Rapid Excavation Tunneling Conference, Toronto, pp. 346–359.

Washington Department of Transportation. 2009. WSDOT-Project SR99-Alaskan Way Viaduct and Seawall Replacement, www.wsdot.wa.gov/Projects/Viaduct.

# Setting the Owner's Budget: A Guideline

## Paul T. Gribbon
Bureau of Environmental Services, City of Portland, Oregon

## Julius Strid
EPC Consultants, Portland, Oregon

### THE SKY *IS* FALLING

Many major underground public works programs have had problems with original published budgets compared with actual final costs (Flyvbjerg et al. 2009). The cause of this in most cases is a poorly considered initial budget. The problem is the public expects an accurate project budget to be published before there is any site investigation, engineering or design; let alone a bid from a contractor, and not to mention the inevitable changes that occur during all phases of an underground public works program.

The responsibility for establishing the budget falls squarely on the owner. This usually occurs when the owner has no avenue for assistance from contractors or designers familiar with the type of work since they cannot hire someone to help them until program has begun. Some owners are lucky enough to have experienced personnel on staff familiar with major underground projects, but most do not since these projects are not a regular occurrence for owners.

Additionally, the benefits of a project are not necessarily fully explored prior to the go-ahead to offer a contrast between the anticipated public investment in the project vs. what the economic/environmental/community benefits are estimated to be over its life.

This paper presents an approach to setting a realistic Owner's budget and preparing a benefits analysis for an underground public works program. This paper will attempt to define the problem and give some answers.

- **Appendix A** provides a **Program Budget Guideline Checklist.**
- **Appendix B** provides a **Benefits Checklist**.
- **Appendix C** provides some **Rules of Thumb for Predicting Budgetary Items.**

These lists have grown to be quite lengthy since we have invited everyone in our industry to provide input on them. There isn't room in this paper for the full listings. We are in process of setting them up on the internet for free access. The access address will be http://www.ownersbudget.com/.

### A Published Estimate at the Outset Determines If a Project Begins

Unfortunately, the initial budget amount announced for the program is the one the media and public will remember and will always be used as a yardstick. This has been referred to as "anchoring" (Flyvbjerg et al. 2009). Getting the initial budget amount in the correct range at the start makes or breaks your program. If the budget amount can be compared to an expected life cycle benefit, it may be more palatable during initial public review. If instead it results in public "sticker shock," the public will reject it immediately. This would not help the situation, since a realistic budget could bring the perceived amount within a range acceptable to the public. If the budget amount is significantly under the costs as construction bids begin to come in, the program may be delayed until additional funding can be arranged. This is unfortunate because at that stage, a large amount of effort and treasure has been expended on site investigations and design work. If the project goes forward anyway, an owner may already have a perceived failed program because they are over budget before they begin. We have to note that if a proposed project is too expensive for the public to fund, perhaps some other solution should be considered rather than going too far and getting into a position where all of the funding has run out and the project is only partly completed. In our opinion, this situation is often the result of inexperience, unwarranted optimism, or wishful thinking. This issue was addressed at the 2001 Northwest Regional Conference of the American Underground Construction Association *Planning for Successful Underground Projects: Applying the Lessons Learned:*

…The key issues identified in the first morning's presentations and discussion had to do with an attempt to quantify or at least identify the true costs/benefits of what they referred to as "the utilization of underground space." The benefits, usually not quantified, include longer life cycles, greater opportunities for the use of aboveground space, lower long-term maintenance costs, social and economic benefits, including less disruption during construction, maintaining the connection between neighborhoods…, and environmental/aesthetic...

One of the biggest challenges identified was how to obtain quantifiable statistics of the true cost/benefits of major underground projects that could be used to provide to political entities and planners early on in the process, since public acceptance and political will were both mentioned as crucial to getting a major project off the ground. It was stated that most crucial cost issues are resolved between the time of policy formation and implementation planning. Another approach mentioned was to attempt to quantify the costs of not doing a program in comparison with the cost of the program (e.g., is there a way to quantify the cost to Portland residents of not pursuing the CSO program?) (Gribbon 2001)...

**The Actual Investment Measured Against the Published Budget Defines Success**

The initial budget yardstick will always be used to compare your actual investment to the original estimate. A revised budget in the middle of the project is often not an option, or at least, it is a poor option from the public's point of view. The public should also be shown the effect of annual escalation on long-term projects and consequently escalation over the project life should be included in the original budget. An initial estimate in a given year for a long-term project without reference to expected annual cost escalation will almost assuredly result in unnecessary public skepticism when the initial stated estimate begins to gradually rise in the public's eyes.

**There Is No Reason for Budget Optimism at the Beginning of the Project**

There is an inclination in our industry toward optimism in estimating a project/program's initial cost. This is out of concern that if the estimate appears too high, the project will not be authorized or public resistance will occur. This treats the public dishonestly and sets the project up for public failure once the actual investment amounts begin to accrue. The public has the right to know what the true expected investment is from the outset.

There are almost always unexpected items that come up during the design and the construction of an underground project. The public has the right to expect that the original budget contains contingencies for these unforeseen issues (Edgerton 2008). A project that is under-funded also usually has problems during construction in resolving change orders. These can spiral out of control and result in disruption and extended overhead claims. Claims of this type are difficult to resolve and result in years and, sometimes, decades of litigation.

History shows that an estimated investment required for an underground project almost never goes down. Escalation and unforeseen items take care of that.

Therefore, at the beginning of the project, behave as Chicken Little, i.e., *The Sky Is Falling*. On a public works project that has been mandated by a regulatory agency or has universal appeal in the area, there is no downside to being conservative at this moment of time in the program. If you are required to do something, then you can be as conservative with the initial estimates of cost as you want. Realism in the acceptance of the many things that can go wrong during the delivery of a major capital project can help to counter common myths that government cannot successfully manage capital programs without significant cost and schedule overruns.

In a presentation on the Boston Harbor Project, the Massachusetts Water Resources Authority was faced with massive rate increases to fund what was potentially a $6 billion program to clean up Boston Harbor from sewage pollution. A conscious decision was made to release the worst-case cost to the public early to "…avoid the hemorrhaging of credibility that comes from repeated bad news…" The thought was to inform everyone of the worst case, then attempt to beat it. The public outrage of rates was also used to get politicians to find a way to procure state or federal funding. The strategy eventually worked, with the procurement of almost $1.3 billion in state and federal funding and the announcement that with favorable inflation rates, funding, and cost saving measures, the final cost was projected to come closer to $4 billion (the actual cost is approx. $3.8 billion). (McBride 2001)

However, on a major project that is not mandated, there has historically been a tendency toward optimism where the inevitable unknowns are downplayed and initial schedule and cost estimates are presented that may allow the project to look more attractive than it should. There is a natural tendency for unwarranted optimism where a project is being "sold" or a constituency needs to be convinced that a project is necessary or beneficial. The downside to being conservative in this case is the public or politicians may reject it as too expensive or time consuming and your program/project never gets off the ground. The resulting tendency toward early optimism has often led to the inevitable attention in the media and public opinion that government project cost and schedule overruns are the norm. Our main point of this paper is that the initial cost estimate for a public works project should fully include all likely costs based on historical record of projects of similar size and scope so that in theory the estimate is the same whether the project is mandated or not.

> ...Former Amsterdam city alderman Tjeerd Herrema resigned earlier this year in frustration at the project. None of the companies involved would testify in public to the committee, but Herrema told a parliamentary committee last month the city had deliberately underestimated the costs to win backing when parliament approved it in 2002. This was the reality that welcomed him to office in 2006. "Many times I have cursed this decision," Herrema said. Estimates already have the North-South line running at a future loss. The 2 billion euros the new metro is expected to yield to the city in its 100-year lifetime falls short of its 3 billion euro construction cost. Explaining the string of accidents, deadline extensions and budget overruns, Amsterdam mayor Job Cohen blamed business after a government commission in June ruled construction had literally reached the point of no return. "We all are just amateurs in this perspective and we have to rely on professionals," Cohen told the commission. ... "It is almost systematic that the costs are underestimated for metro systems," said Jack Short, chief of the International Transport Forum (ITF), a global platform for transport, logistics and mobility,... "We definitely have seen a tendency of what you might call 'project optimism'," Short said, attributing this to the prestige involved in such projects. "They have to make the numbers look good, the revenues a bit higher and the costs a little bit lower than what they really are. (RPT Feature 2009)

One factor that is usually left out of the project budget and risk analysis is the impact of the highly improbable. This is addressed in the book *The Black Swan* by Nassim Nicholas Taleb. Mr. Taleb includes a skeptical empiricism and the a-Platonic School way to approach randomness in his book. One point is that we have a tendency to be too focused on the internal workings of a project to attempt to quantify the effects of events "straying from our mental projections." Part of his approach is to consider that we cannot easily compute probabilities and that we should seek to be approximately right across a broad set of eventualities. He also makes the point that we should look outside of our projects to compare to others histories to check to see if the budgets are within the same range for similar work (Taleb 2007). A similar point is also made in an article called "Delusion and Deception in Large Infrastructure Projects" (Flyvbjerg et al. 2009).

**A Beneficial Public Works Project Has No Cost**

It's not too much of a stretch to say that a public works program that has a beneficial purpose and is constructed for a reasonable price has no net cost (if it's clearly shown upfront that the life cycle benefits exceed the initial price). It is an asset and deserves the investment of the public's money. If a project cannot pass this test, it should not be built.

An example of this is the Portland, OR Aerial Tram that was recently constructed. This project exceeded the original published cost by almost a factor of four and consequently was pilloried in the local press. What was lost from the discussion after the extent of the cost overruns was published was the ultimate benefit of the project. As summarized in a risk analyses report conducted after the project became embroiled in financial difficulty:

> ...The Portland Aerial Tram, when complete, will be a dramatic, one-of-a-kind facility that will become a Portland landmark—easily overshadowing its earlier history of budget and schedule problems (Busch 2006)...

**An Approach to Analyzing the Investment/Benefit for an Underground Public Works Program**

Analyzing a conceptual project to see if it makes investment sense is not new. The method of Cost Benefit Analysis is used by many private organizations and public agencies and is well documented (*Mn/DOT Benefit Cost Analysis Guidance* 2005; *FAA Airport Benefit-Cost Analysis Guidance* 1999; Scott 2006; *Cost Risk Assessment (CRA) and CEVP®*, n.d.). In fact, all of us make these types of analyses on a daily basis in our private lives, just not on the same scale as for a large program and certainly not in as complex a manner. We would prefer to change the terminology of Cost Benefit Analysis a bit and use the word *investment* rather than *cost*, since what we are trying to determine is whether a project has a net benefit or not. If the project results in a cost, it should not go forward. If the analyses show a net benefit, it can be presented to the public for investment.

The Investment/Benefit Analysis Steps have been defined (Galambos and Schreiber 1978) as:

- Identify investment and benefits of the project
- Measure investment and benefits in dollars
- Consider investment and benefits over life of the project
- Make a decision

Elements (McKenna 1980) within the steps:

- Objectives
- Alternatives
- Benefits
- Investment
- Model
- Criterion

Since the method of analysis has been defined by others and to describe it thoroughly is out of scope of this paper, we are going to concentrate on the two main parts of the puzzle, the investment and the benefits. The result is three simple, but extensive, and we hope comprehensive, lists of items to be considered by an Owner at the conceptual stage of a project. These lists will provide the basis for establishing what the benefits of an underground project would be and what investment will be required to achieve it.

**Estimate the Benefits of the Program**

Compare the capital investment against:

- Lower long term maintenance
- Longer life cycles
- Greater opportunities for use of above-ground space
- Public's use: social and economic benefits (less disruption, maintaining connection between neighborhoods, environmental/aesthetic

**An Approach to Setting a Realistic Owner's Budget for an Underground Public Works Program**

There are numerous ways to approach the establishment of a budget at the outset of the project. We are presenting just one method to provide a guideline for the owner. Appendix A contains a full checklist of items. Some of the items may not apply to a particular project and can be deleted, as the case may be. The purpose is to alert the owner to items that can slip through the cracks and become surprises during the design or construction.

**CONCLUSION**

Our goal with this paper and the referenced appendices is to provide a guideline for an owner just starting out on a project and needs some basis for a realistic budget. We recommend that owners focus on the benefits of the project and present the level of investment required to achieve the benefits. The budget should consider what realistically has historically gone wrong on other major underground projects and contain a healthy dose of project pessimism. The budget for either a project you are forced to do by mandate or to sell a project should be the same. Then a realistic investment can be compared against the realistic benefits. A project should not go forward if the comparison results in a net cost. It should only be built if there is a net benefit.

**BIBLIOGRAPHY**

Administration for Children and Families and Health Care Finance Administration— Department of Health and Human Services. July 1993. Feasibility, Alternatives, And Cost/Benefit Analysis Guide. http://www.acf.hhs.gov/programs/cb/systems/sacwis/cbaguide/index.htm

Asset Management Economic Analysis Primer— U.S. Dept. of Transportation. 29 August 2007. Cost-Benefit Analysis. http://www.fhwa.dot.gov/infrastructure/asstmgmt/primer05.cfm

Asset Management— U.S. Dept. of Transportation. 20 March 2009. Evaluation and Economic Investment Team. http://www.fhwa.dot.gov/infrastructure/asstmgmt/invest.htm

Busch, Pinnell. 1 February 2006. *Portland Aerial Tram Risk Assessment, Prepared for the Portland Development Commission.*

Cost Risk Assessment (CRA) and CEVP®. Washington State Department of Transportation. http://www.wsdot.wa.gov/Projects/ProjectMgmt/RiskAssessment/default.htm

Edgerton, W.W., ed. 2008. *Recommended Contract Practices for Underground Construction.* Littleton, CO: SME. page 52.

Flyvbjerg, Garbuio, and Lovallo. Winter 2009. *Delusion and Deception in Large Infrastructure Projects: Two Models for Explaining and Preventing Executive Disaster.* California Management Review Vol. 51, No. 2.

Gribbon, P., Ryan, W. 19–20 March 2001. *Northwest Regional Conference of the American Underground Construction Association "Planning for Successful Underground Projects: Applying the Lessons Learned."*

Guess, George M., and Farnham, Paul G. 2000. *Cases in Public Policy Analysis.* Washington, D.C.: Georgetown University Press. http://books.google.com/books?id=KGFnn4KJHN8C&pg=PA304&lpg=PA304&dq=%22Flood+Control+Act+of+1939%22&source=bl&ots=Dc5oLG0mkm&sig=tXTtcTe9hPO9xYIBTseb9LO2Ndw&hl=en&ei=bjawSdavN4_HtgeNmvnaBQ&sa=X&oi=book_result&resnum=6&ct=result

McBride, Michael. 2001. *Boston Harbor Project.* The Massachusetts Water Resources Authority.

Mn/DOT. June 2005. *Mn/DOT Benefit Cost Analysis Guidance.*

Office of Aviation Policy and Plans Federal Aviation Administration and others. 15 December 1999. *FAA Airport Benefit-Cost Analysis Guidance.*

Planning—U.S. Dept. of Transportation. 6 November 2007. Case Study: Portland, Oregon. http://www.fhwa.dot.gov/planning/toolbox/portland_overview.htm

Planning—U.S. Dept. of Transportation. 6 November 2007. Cost-Benefit Case Studies.http://www.fhwa.dot.gov/planning/toolbox/costbenefit_forecasting.htm

Scott, Elizabeth A. 2006 *A Benefit-Cost Analysis of the Wonder World Drive Overpass in San Marcos, Texas.* Texas State University-San Marcos, Political Science Department, Public Administration.

Taleb, Nassim Nicholas. 2007. *The Black Swan, The Impact of the Highly Improbable.* New York: Random House. page 284.

Transportation Planning—California Dept. of Transportation. 2007. Benefit-Cost Analysis. http://www.dot.ca.gov/hq/tpp/offices/ote/benefit_cost/index.html

Wikipedia. 2009. Cost-benefit analysis. http://en.wikipedia.org/wiki/Cost-benefit_analysis

Wolde, Harro ten. 26 October 2009. *In a hole, Amsterdam tunnellers just keep digging.* Reuters.

# Show Me the Money: The Real Savings in Tunnel Contract Payment Provisions

**John M. Stolz**
Jacobs Associates, San Francisco, California

**ABSTRACT:** Bidders read the contract general conditions for projects under consideration for bid before committing to the expense of preparing a bid. This is a smart practice because it oftentimes reveals that the payment terms dictate that the bidder must either add significant financing costs or front-load the bid to obtain a reasonable cash flow. While the industry has made qualitative recommendations to reduce financing costs and discourage unbalanced bids for common contract clauses relating to retention, capped mobilization, the timing of their payment, and the use of equipment mobilization items, there has been no attempt to quantify these savings. By use of a "typical" example tunnel contract, this paper first guides the reader through these payment provisions while quantifying their savings. It then concludes by reinforcing the message that when owners do not consider contractor cash flow in the contract language, they are ultimately either subsidizing significant financing costs or receiving unbalanced bids, and calls for engineer's estimates to have the ability to make these analyses to owners' benefit.

## INTRODUCTION

It is generally accepted that tunnel construction costs more and is generally riskier than other kinds of construction. It is therefore in the owners' interests to reduce costs and risks when they have the ability to do so. Contractor financing costs are the low-hanging fruit that can be picked with minimal effort and risk by simply tailoring tunnel construction payment methods and provisions to reduce contractor financing costs. Added benefits include a measure of certainty that bidders are less likely to unbalance their bids—or at least unbalance them to a significantly lesser extent—while reassuring owners that they are not paying a disproportionately high amount relative to contract work completed. This in turn tends to reduce the quantification of disputes since it gives owners more confidence that the contract costs they can analyze are based on some measure of reality

While many owners have employed these recommended practices, they have done so using qualitative arguments to justify their rationale. These qualitative arguments may be sufficient, but for those who demand a more rigorous accounting, some method must be used to quantify the savings. Owners who do not have the resources to make these analyses during the design process should find the following of interest as a rough measure of the possible amount of financing costs that may be included in a balanced bid.

## A TYPICAL TUNNEL CONTRACT

The basis for this discussion is a tunnel construction contract recently prepared for a client. The figures presented are censured scans taken from the estimate itself.

The project cost-loaded schedule is shown in Figure 1. The work involves the construction of a long tunnel with shaft access using a hard-rock tunnel boring machine (TBM) and requiring a final lining. The $135M project requires about 57 months of construction, including a 10-month TBM procurement period, time to develop the shaft access for two TBM drives (taking 3 months and 15 months, respectively), followed by 17 months of final lining. Progress payments average about $3M per month. In addition to the $4M in fleet equipment, the contractor purchases an additional $18M of equipment for the job, including $13M for a new main-beam TBM and trailing gear, of which 90% is written off to the project because of the long length of tunnel. About $15M is devoted to shaft and permanent surface works. The contractor has included about $20M, or about 15% of cost in markup. Cost items are grouped into bid items, and the bid items themselves are balanced in accordance with contract requirements.

Figure 2 shows the cash flow for this balanced bid and provides the details of the base project payment provisions. Refer to the glossary at the end of this paper for an explanation of parameters.

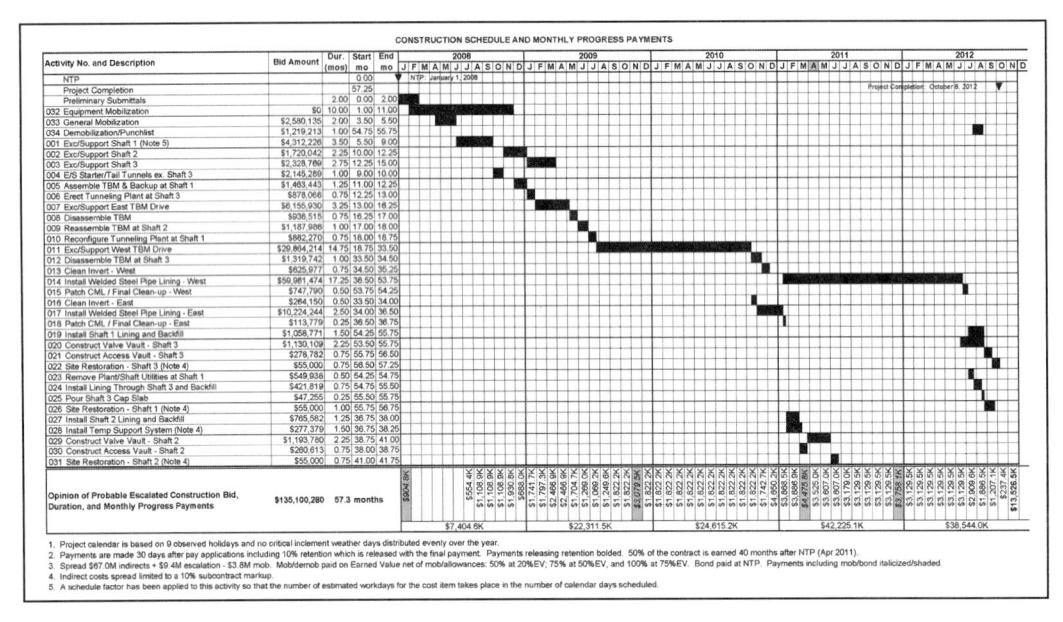

Figure 1. Cost-loaded schedule for a typical tunnel project

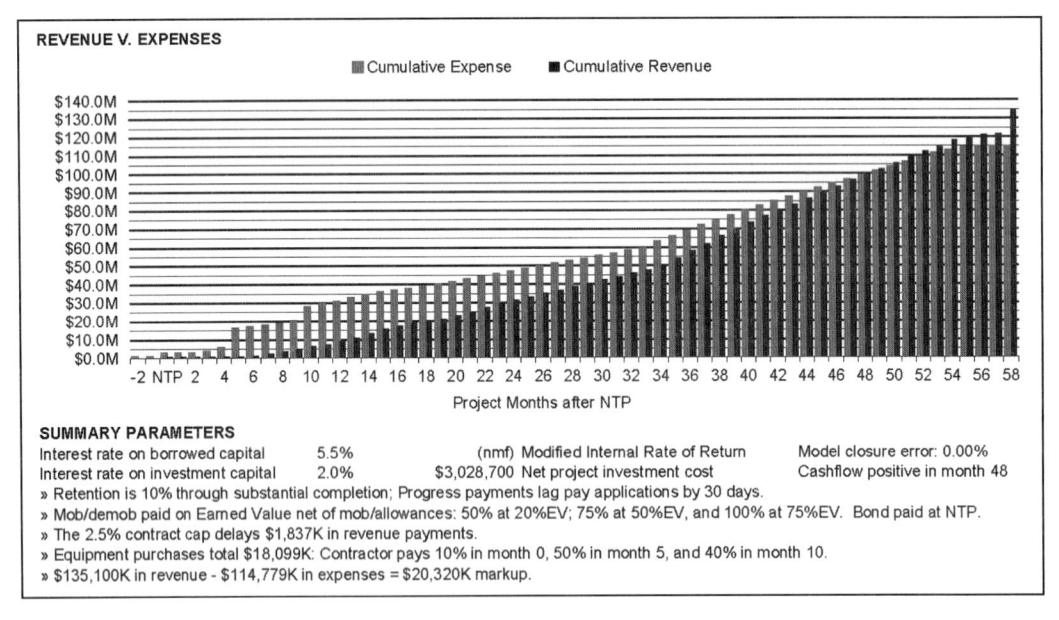

Figure 2. Cash-flow base scenario

The financing costs are the costs of money for the difference in the timing of revenue relative to expenses. The base scenario reflects various payment conditions identified by the chevron bullet items. This includes a requirement for withholding 10% of earnings, capping mobilization payment to 2.5% of the contract value, making payments for mobilization at 20%, 50%, and 75% of earned contract value (EV), and the absence of an equipment mobilization bid item.

Looking at the cumulative expense plot, one can see there is an initial disbursement at the time of award (taken to be 2 months prior to issuance of Notice to Proceed) for replacement of the bid bond

**REVENUE V. EXPENSES**

■ Cumulative Expense   ■ Cumulative Revenue

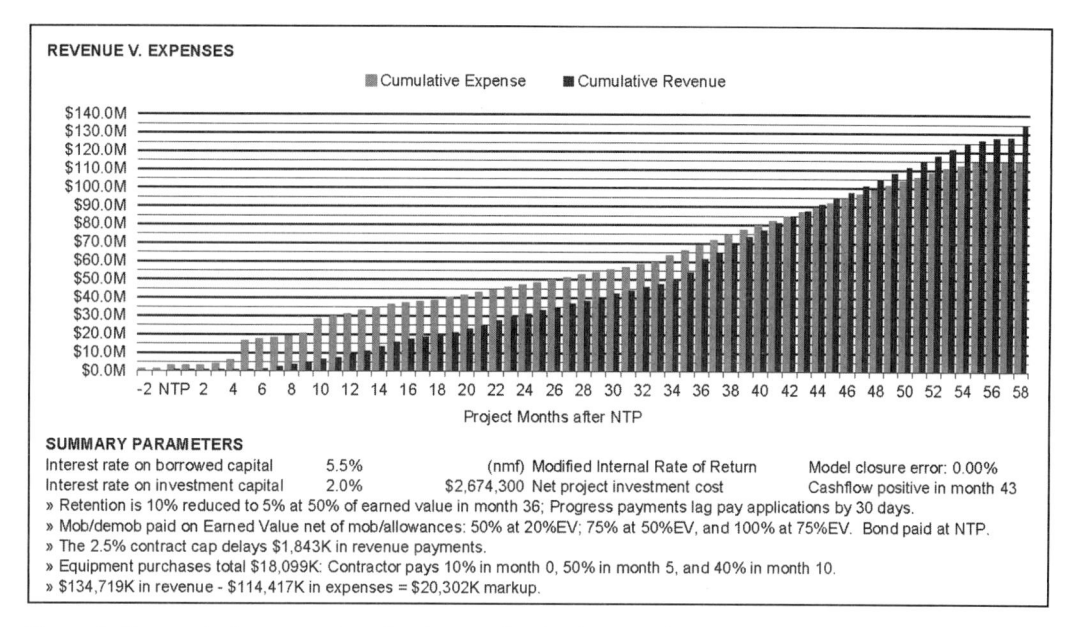

**SUMMARY PARAMETERS**
Interest rate on borrowed capital    5.5%         (nmf) Modified Internal Rate of Return    Model closure error: 0.00%
Interest rate on investment capital  2.0%         $2,674,300 Net project investment cost     Cashflow positive in month 43
» Retention is 10% reduced to 5% at 50% of earned value in month 36; Progress payments lag pay applications by 30 days.
» Mob/demob paid on Earned Value net of mob/allowances: 50% at 20%EV; 75% at 50%EV, and 100% at 75%EV. Bond paid at NTP.
» The 2.5% contract cap delays $1,843K in revenue payments.
» Equipment purchases total $18,099K: Contractor pays 10% in month 0, 50% in month 5, and 40% in month 10.
» $134,719K in revenue - $114,417K in expenses = $20,302K markup.

**Figure 3. Base cash-flow scenario reflecting a reduction in retention of 5% from 10% at 50% EV**

with a performance bond. This is followed by a short period of relatively low expenditures during the contract submittals and planning phase. After approval of the work plan submittals, there is a significant increase in spending for equipment procurement and mobilization.

On the revenue side, there is very little activity in the beginning months of the project since project payments are limited to tangible contract progress. Thus, 10 months into the work the contractor has received about $6M in progress payments but has expended almost $30M to gear up for the work ahead. This huge deficit is maintained over the next 4 or 5 months and then slowly eroded over the course of the work until in month 48 it is finally erased, and the contractor is now operating in the black for the remaining 10 months. Note that with the release of final retention, the contractor receives its profit on the job.

This is a dramatic display of what results from a balanced bid requirement coupled with an ideal contractor not concerned with bidding climate: a significant financing cost to the owner of just over $3M, or 2.2% of the project bid. Keep in mind that the relative fractional contribution of financing cost to project cost is a function of many variables, not the least of which is the magnitudes of and relative differences in the interest rates for borrowed and invested capital.

**CASH-FLOW SCENARIOS**

The following scenarios are made to provide an indication of relative savings that can be expected by modification of the above base contract payment language. Again, keep in mind that the amount of savings for each scenario is a function of many variables that are likely to change with both project specifics and commercial terms but which nevertheless provide a useful gauge of approximate savings.

**Retention**

All construction contracts call for retention on earnings. In many cases, this requirement can be satisfied by the substitution of securities in lieu of cash retentions, and it should be noted that the models presented herein are based on this assumption. Figure 3 shows the effect of modifying the base cash-flow scenario to reflect a reduction in the amount of retention at the 50% EV milestone. At 50% of earned contract value, retention is reduced from 10% to 5%.

As can be expected, the effect of this modification is less successful at reducing initial negative cash flow since it does not take effect until after 50% of the contract is earned. Nevertheless, it does generate a modest spike in revenue at month 36. It also moves the transition from negative to positive cash flow by 5 months, or about 10% of the contract time, resulting in a decrease in financing costs of some $350K, or almost 0.3% of the contract price.

**REVENUE V. EXPENSES**

■ Cumulative Expense ■ Cumulative Revenue

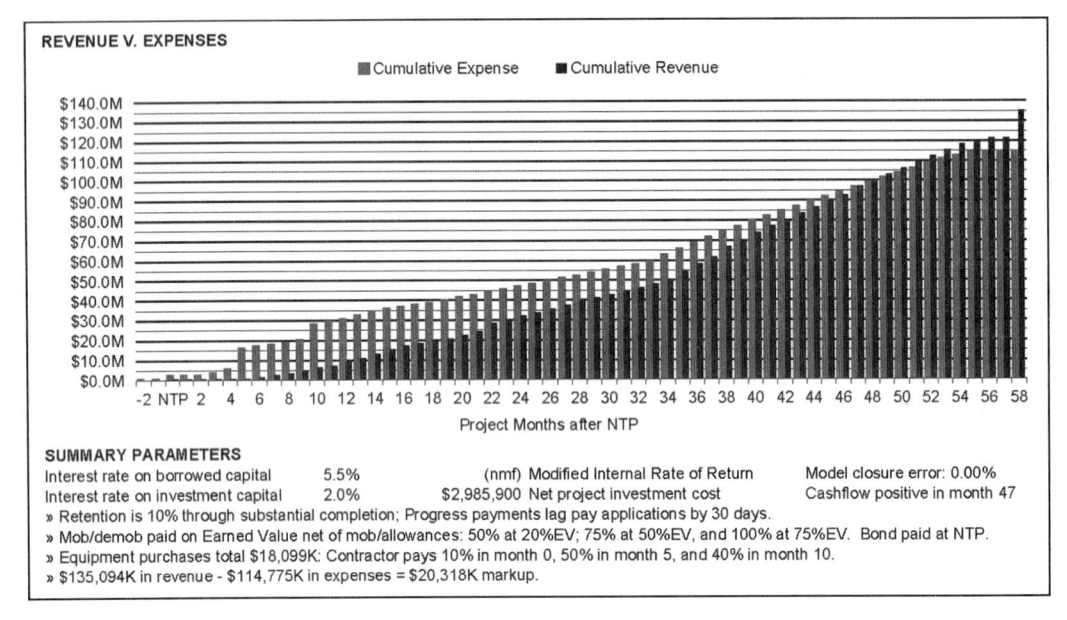

**SUMMARY PARAMETERS**
Interest rate on borrowed capital 5.5% (nmf) Modified Internal Rate of Return Model closure error: 0.00%
Interest rate on investment capital 2.0% $2,985,900 Net project investment cost Cashflow positive in month 47
» Retention is 10% through substantial completion; Progress payments lag pay applications by 30 days.
» Mob/demob paid on Earned Value net of mob/allowances: 50% at 20%EV; 75% at 50%EV, and 100% at 75%EV. Bond paid at NTP.
» Equipment purchases total $18,099K: Contractor pays 10% in month 0, 50% in month 5, and 40% in month 10.
» $135,094K in revenue - $114,775K in expenses = $20,318K markup.

Figure 4. Base cash flow scenario reflecting a mobilization cap that reflects actual cost

### Mobilization Cap and Payment Schedule

Many owners specify contract caps on the amount bid for mobilization as a means of assuring that contractors do not front-load this item. Usually demobilization is included in the mobilization item. There may be some recognition that mobilization for smaller projects has a higher amount of mobilization relative to contract price, and so these caps may be graduated based on contract price. In the base example, mobilization was capped at 2.5% of the bid in recognition of the $120M to $150M contract value, but this cap did not reflect the contractor's actual cost for mobilization and demobilization. Figure 4 shows the scenario for a 5% contract cap, which more accurately reflects the actual costs for mobilization and demobilization without regard for equipment procurement costs.

While the effect of this modification is to more accurately reflect the actual cost of mobilization, its effect on curtailing financing costs is negligible: the positive cash flow point changes only marginally, reducing financing costs by a negligible amount. The reason for this is that the *timing* of the mobilization payments in Figure 4 is not a better fit to the actual stream of mobilization expenditures. Figure 5 shows that if more attention is paid to the timing of mobilization payments, the reduction in financing costs is more significant at $415K, or 0.3% of the contract value.

### Separate Equipment Mobilization Item

The largest influence on financing cost is the cost of specialized tunnel equipment, as evidenced by the large expenditure spikes starting in month 10 of the base scenario cash-flow profile. Figure 6 shows how the effect of these spikes can be softened by use of an equipment mobilization item that makes timely payments to the contractor in response to these expenditure surges.

Note that equipment mobilization only covers the main-beam TBM, the trailing gear, and an allowance for their erection and commissioning in an amount of $13.3M. The actual equipment procurement costs are about $18M in addition to $4M in book value for fleet equipment. Nevertheless, the contractor receives payments in some structured fashion over the 10-month equipment procurement period that result in a significant $1.7M reduction in financing costs, or 1.3% of the contract price. Note that it is not necessary to cover all equipment purchases to generate this magnitude of savings in financing costs.

### Cumulative Effects

The cumulative effect of the above-described modifications to retention, general mobilization language, and the use of an equipment mobilization item is shown in Figure 7.

**REVENUE V. EXPENSES**

■ Cumulative Expense　　■ Cumulative Revenue

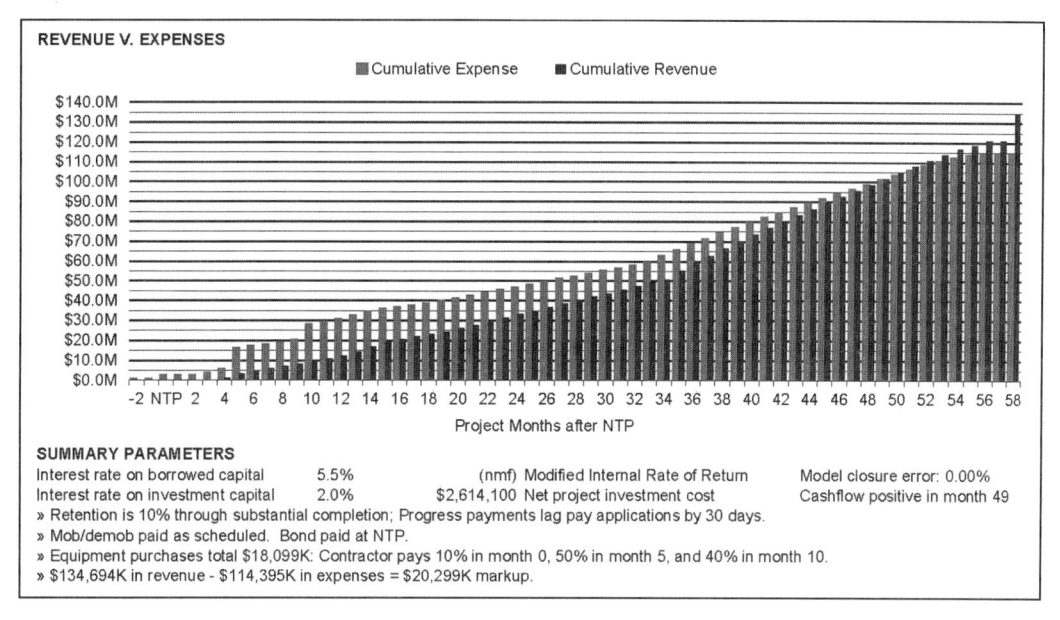

[Bar chart showing cumulative expense and revenue from month -2 NTP to 58, values from $0.0M to $140.0M]

**SUMMARY PARAMETERS**
Interest rate on borrowed capital　5.5%　　(nmf) Modified Internal Rate of Return　　Model closure error: 0.00%
Interest rate on investment capital　2.0%　　$2,614,100 Net project investment cost　　Cashflow positive in month 49
» Retention is 10% through substantial completion; Progress payments lag pay applications by 30 days.
» Mob/demob paid as scheduled. Bond paid at NTP.
» Equipment purchases total $18,099K: Contractor pays 10% in month 0, 50% in month 5, and 40% in month 10.
» $134,694K in revenue - $114,395K in expenses = $20,299K markup.

Figure 5. Base cash-flow scenario reflecting a better tuned mobilization payment schedule

**REVENUE V. EXPENSES**

■ Cumulative Expense　　■ Cumulative Revenue

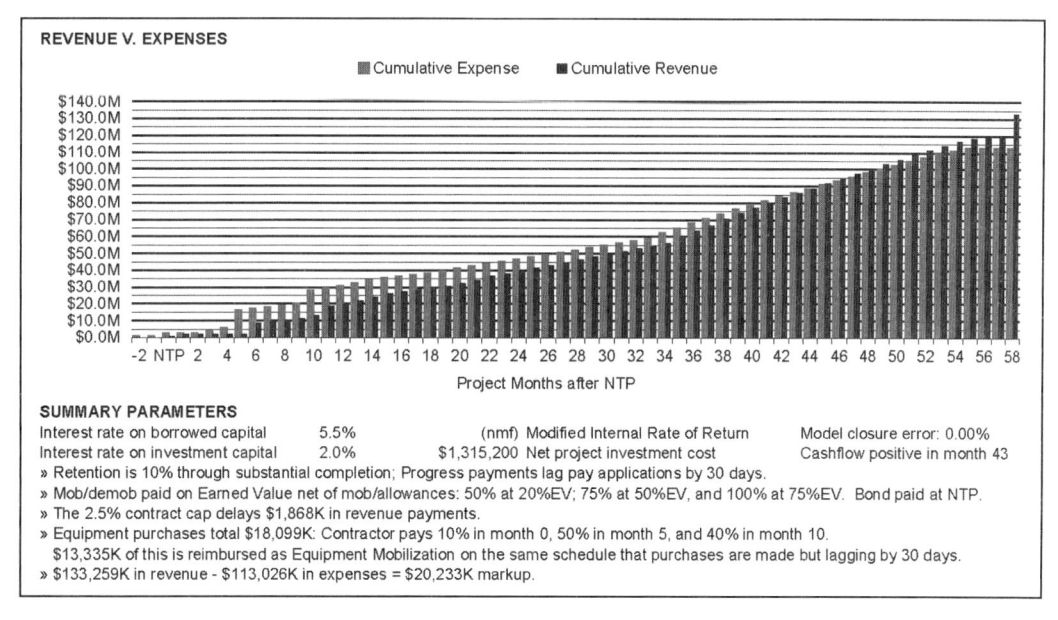

[Bar chart showing cumulative expense and revenue from month -2 NTP to 58, values from $0.0M to $140.0M]

**SUMMARY PARAMETERS**
Interest rate on borrowed capital　5.5%　　(nmf) Modified Internal Rate of Return　　Model closure error: 0.00%
Interest rate on investment capital　2.0%　　$1,315,200 Net project investment cost　　Cashflow positive in month 43
» Retention is 10% through substantial completion; Progress payments lag pay applications by 30 days.
» Mob/demob paid on Earned Value net of mob/allowances: 50% at 20%EV; 75% at 50%EV, and 100% at 75%EV. Bond paid at NTP.
» The 2.5% contract cap delays $1,868K in revenue payments.
» Equipment purchases total $18,099K: Contractor pays 10% in month 0, 50% in month 5, and 40% in month 10.
　$13,335K of this is reimbursed as Equipment Mobilization on the same schedule that purchases are made but lagging by 30 days.
» $133,259K in revenue - $113,026K in expenses = $20,233K markup.

Figure 6. Base cash-flow scenario reflecting an equipment mobilization item

Financing costs have been reduced by an impressive $2.4M—almost 2% of the contract value—from $3.0M to $0.6M. The turning point to positive cash flow—the point at which the project can sustain itself—has been shifted from the 83% completion point to the 62% completion point. Note that the sum of the parts is not equal to the whole because the measures that improve cash flow earliest dilute the cash-flow imbalance for those that improve cash flow later.

**REVENUE V. EXPENSES**

■ Cumulative Expense ■ Cumulative Revenue

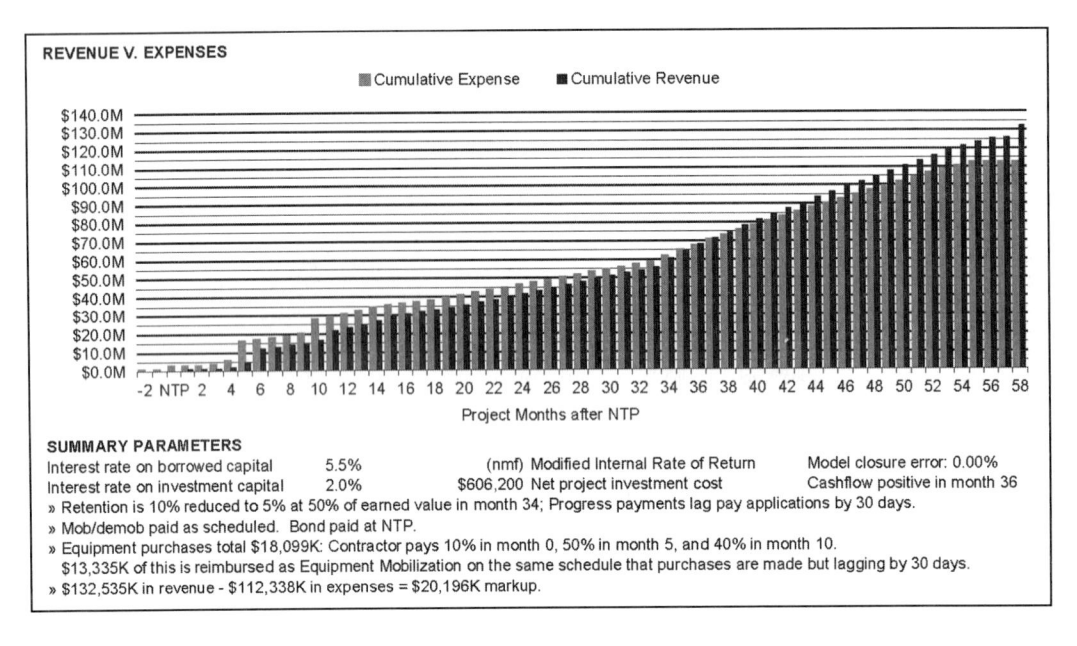

**SUMMARY PARAMETERS**
Interest rate on borrowed capital     5.5%          (nmf) Modified Internal Rate of Return      Model closure error: 0.00%
Interest rate on investment capital    2.0%          $606,200 Net project investment cost       Cashflow positive in month 36
» Retention is 10% reduced to 5% at 50% of earned value in month 34; Progress payments lag pay applications by 30 days.
» Mob/demob paid as scheduled. Bond paid at NTP.
» Equipment purchases total $18,099K: Contractor pays 10% in month 0, 50% in month 5, and 40% in month 10.
   $13,335K of this is reimbursed as Equipment Mobilization on the same schedule that purchases are made but lagging by 30 days.
» $132,535K in revenue - $112,338K in expenses = $20,196K markup.

Figure 7. Base cash-flow scenario reflecting all measures

## SUMMARY

Real savings are possible by minimizing contractor financing costs, encouraging more balanced bids, and providing a more accurate measure of earned value. The forgoing example of a typical tunnel construction project shows that these measures can reduce financing costs by about 80%. Since financing costs can contribute in excess of 2% to project costs, these measures are a relatively easy and risk-free way to reduce project costs a significant amount.

When owners do not have the ability to determine these savings in justification of amending standard contract payment language, the example presented herein provides a rough order of magnitude of the dollars involved. For those engineer's estimates that calculate these costs, a project-specific analysis can easily be made to quantify these savings for the owner's benefit. However, considering the size of these financing costs, it is recommended that owners require their engineer's estimates to be capable of making these assessments.

## GLOSSARY

*Cumulative expenditure:* For any month, the sum of all expenditures charged to the project, including the interest expense for prior months.

*Cumulative revenue:* For any month, the sum of all contract payments to the project, including the interest income for prior months.

*Interest rate on borrowed or invested capital:* Shown as a nominal interest rate. The period interest is 1/12th of the nominal rate. When the cumulative project cash flow for any month is negative (positive), the investment cost is calculated using the borrowed (invested) rate.

*Model closure error:* This identifies how well the cash-flow model independently arrives at total project cost. The error is typically zero; however, slight errors can be generated by the assumptions in the timing of expenditures for escalated estimates relative to where the escalation period rate date begins.

*Modified Internal Rate of Return (MIRR):* The discount rate that makes the net present value of all cash flows from a particular project equal to zero. This differs from the conventional method of calculating Internal Rate of Return (IRR) because with IRR, both the borrowing/investing rates are the same, whereas MIRR allows these borrowing and investment rates to be different.

The MIRR metric by itself does not identify the desirability of undertaking a particular project (other than in general, the higher a project's internal rate of return, the more desirable it is to undertake the project). Rather, MIRR should be used to rank different prospective projects that a contractor is considering with the rationale being that if all other factors are equal between two projects, the one with the highest MIRR is preferable.

Since MIRR is intended to return a single discount rate, the metric is useful only for short-term projections, since the discount rate can vary over time. If cash flow alternates between negative and positive more than once of the project life, there will be multiple project MIRRs with the number of MIRRs equal to the number of times cashflow changes sign. The complexity of the cash-flow model can render the calculation of MIRR meaningless and is not an indication of a flaw in the analysis itself.

*Net project investment cost/income:* The algebraic sum of the monthly project investment costs (this tabulation is omitted in the above figures in the interest of brevity). The project investment cost for any month is the difference between cumulative expense and cumulative revenue multiplied by the appropriate interest rate. A positive number identifies the total amount that the contractor finances over the life of the project. A negative number indicates that more income is derived from the financing than is expended.

*Positive cash flow:* The month in which cumulative cash flow becomes positive. This is a coarse metric used to rate the financial attractiveness of a project.

# Planning Level Tunnel Cost Estimation

**Hamid Javady**
AECOM, Toronto, Ontario

**Jamal Rostami**
The Pennsylvania State University, State College, Pennsylvania

**ABSTRACT:** Preparation of preliminary cost estimation for planning is difficult challenge because of the uniqueness of these structures relative to their application, unknown nature of the site geology, and risks and uncertainties involved with tunneling in various ground conditions. Any tunneling operation will include certain categories of activities and therefore the cost estimation can be broken into the following activities and component: 1-Site preparation and mobilization, 2- Excavation and support of launch and exit shaft 3-Excavation and support of tunnel 4-Dewatering, ventilation, utilities, and other services, 5-Fit outs and finishing work within the tunnel, and 6- Demobilization and clean up. Typical tunnel ground support methods include: 1-One pass method involving Segmental lining 2- Two pass method with temporary lining and Cast in Place (CIP) concrete lining. Different systems can be used for temporary lining depending on the ground conditions and tunnel sizes. This paper will discuss available methods for tunnelcost estimation to determine project feasibility and to compare alternative tunnel sizes and routes during the planning stage.

## INTRODUCTION

Tunnel Advance depends on the ground condition, water infiltration to the tunnel, skills and experience of the crew and specially the operators, mucking and time to erect support. In order to optimize the operation, all activities should be properly coordinate to minimize delay. Different types of support systems can be used in tunneling operation and each has a major cost implication for the planning level estimation purposes. This includes Steel rib and lagging, shotcrete and wire mesh, and rock bolts and steel straps. The mucking system which transport spoil from the face to disposal area can be used train or conveyor belts, and in some larger tunnel diameters, off rubber tire trucks. Any small variance in production creates large variations in construction costs because of high hourly cost of labor and equipment. Similarly, efficiency of equipment and its performance will affect productivity. Additionally, advance rate and hence cost of the tunnel depends on skill of the crew, capability of equipment, and ground conditions. There are a lot of parameters such as labor cost, Bond cost, Machine depreciation cost, Insurance cost, Management cost, Staff cost, general expenses, and etc.

## COST ESTIMATION

Cost estimating is one of the most important steps in project management as it relates to justification of the projects, especially if there are alternative means of achieving a certain foal objective. A cost estimate establishes the base line of the budget and funding requirements at different stages of the project development. It represents the predicted project cost on the basis of available data from similar project in the area or other locations at certain time in the past.

Tunnel cost usually constitutes only a fraction, though a substantial portion of the total project cost. However, it is the part of cost which typically involves highest level of risk and variation during the planning, design, and construction of the project. The required levels of accuracy of construction cost estimates vary at different stages of project development. The cost estimates made at the early stage are less accurate and it depends to available data at that time. The type of cost estimate could be classified to different categories as follows:

- Prefeasibility and feasibility analysis estimates
- Planning stage
- Design stage and engineer's estimates
- Construction estimates (bid) done by the contractors
- Post construction estimates including all the adjustments and claims.

The objective of this paper is to introduce cost estimation formulas, which can allow for estimation of tunnel cost for feasibility and planning stages.

These formulas can be used by project mangers to estimate the cost of given alternative plans by changing some of the design variables and examining the potential impacts on the overall cost.

The formulas should be improved by using some "Engineering Judgment" to consider some of the project specific issues and also issues related to local construction market.

The design stage cost estimate could be, 30% design, 60% design step, 90% or 100% design which is the engineers estimate for the contract documents. Winning bid usually is lowest bid that is typically between 5–30% lower than engineer's estimates, but from time to time it could be up to 20% higher. This paper presents the preliminary level costs estimate formulas for tunnel. All costs presented herein reflect potential cost in 2008 dollars, although various projects have been used for its initial development. The preliminary opinion of probable costs were developed using bid information, experience, and cost provided by some equipment manufactures.

## BASIC ASSUMPTION

In order to develop a cost estimate certain assumptions needs to be made. One of the important assumptions at this level of work is that the project is going to be a normal operation without any particular complications due to weather, unknown geological problems, social and political changes that can impact the project, and certainly no surprises on the environmental issues such as contaminations that need special treatment (handling, disposal, mitigation etc.) There should be some provisions for site mobilization depending on the working conditions of the area and availability of the utilities. This issue is not covered in more details here, due to the fact that for the most part the operations needed to mobilize the site can be estimated using standard construction cost estimate systems such as Means and HCSS. This includes the site clearing, setting up the fence, and perhaps noise barriers, earth work and leveling for the shaft sites or portals, cost of electric line and substations, cost of road into and out of the site, rental of the temporary mobile homes/containers and related accessories for the offices, shops, storages, locker rooms, etc. even building small buildings for such purposes, water/gas/phone/cable hookups, surface material handling facilities, truck wash out areas, setting up the cranes or gantries for lifting of heavy loads, and other facilities as needed for the projects. Most if not all these items can be estimated from typical cost estimation methods. The same applies to the demobilization and clean up operation after completion of the tunnel.

Another set of costs that is not covered in this paper is the cost of project completion including fit outs, and connection to other structures and final finish of the tunnels. This could include the cost of architectural finishes and decorations, permanent moving devices such as elevators, escalators, conveyors, in transportation tunnels and installing additional pipeline, weirs, pumps, deairation chamber and swirls, wet wells and related structures in water and sewer projects. Also, permanent utilities in mining applications are not included in these estimates. Thus the estimates offered here is for the construction of the tunnels.

Tunnel cost was calculated base of different components such as:

1. Shaft excavation
2. Tunnel excavation
3. Ground support
4. Mucking

## Tunnel Construction Utility

In estimating the cost of shaft construction, the diameter of shaft is considered to be 2.5 times of tunnel diameter. Tunneling construction cost like any other cost can be divided into direct and indirect. However, for the ease of application in this paper, it is assumed that the costs estimated for any of the operations will include the direct and indirect cost of tunneling as it is borne by the contractor. In other words, it is assumed that the prices listed for each of the above items are the most likely prices for doing the work by a contractor and thus it includes all the direct and indirect costs to the contractors. However, there are some indirect costs that the owner should bear and for the purposes of this paper we call the contract cost direct and the additional costs as indirect cost. The indirect cost to the owner includes the cost of engineering and design, (typically in 10–12% range of the direct), cost of construction management or CM (~5% of direct), cost of financing, which depends on the sources of funds and duration of the project, cost of bonds and insurance that has to be carried by the owner depending on the project type and complexity, and finally the contingency and reserve costs. Although the construction contingency of 10–15% has been considered for basic components of the tunnel construction activities, higher values of contingency in the range of about 30–35% should be considered at planning level to cover the potential changes in the design during the course of progression of the project. As one expert put it, "the estimates are rough and order of magnitude, but they will be treated as gospel as the project moves forward" (Redmond 2009). This is to say that at planning level, it is recommended to use contingency of about 30–35% and as the project and design moves forward, this contingency will be decreased to normal level of 10–15% construction contingency when the project is fully designed and it ready for bidding. Obviously, most clients/owners

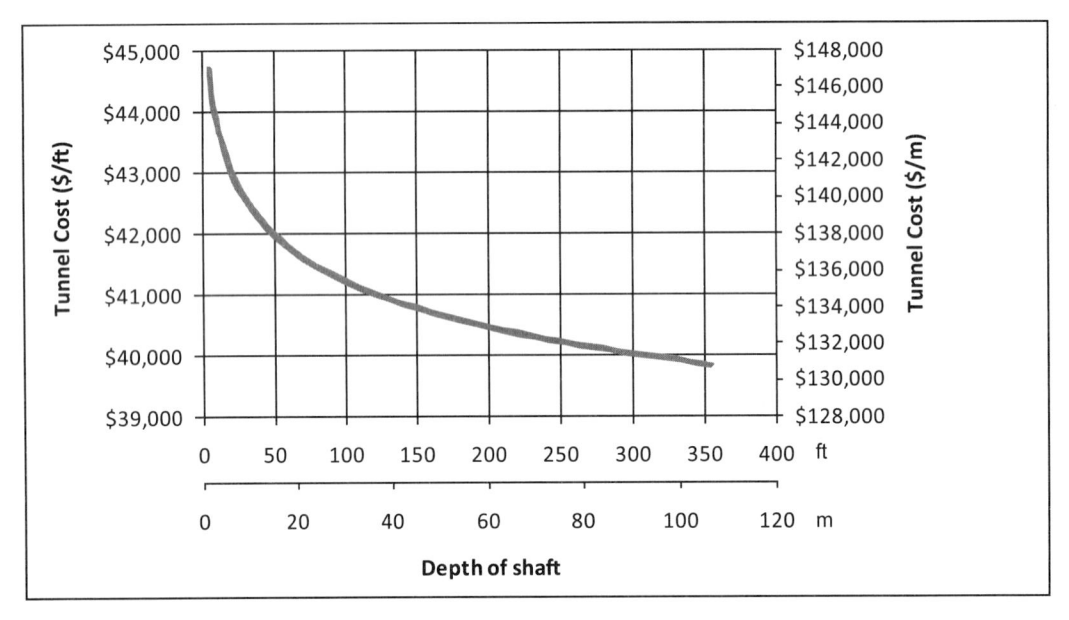

**Figure 1. Shaft cost versus depth for 30 ft diameter**

can live with reduced project cost, but not many can cope with elevated costs if after design, the estimated costs of the project exceeds their initial budgets. This is specially important since in many cases allocated budgets are based on with the preliminary estimates. If this means that the project is no go at the beginning, that should be accepted as a natural consequence of preliminary estimates to relieve the owners of the scrambling to find money for the construction and claims towards the end. In reality, the higher estimated budgets means that either scope of work will be cut to fit available financial sources or project may be delayed for a while to get the proper sources of funding. Either way, additional contingency at the early stages are justifiable.

The magnitude of each of these cost components vary depending on the nature, size, and location of the project.

In each component some main variable has changed to see how the affect the cost. For that component best fit graph has been drawn and base of these graphs the best mathematical formula has calculated.

## TUNNEL COST COMPONENTS

This section will explain the estimated tunnel cost components for the above mentioned breakdown of cost items.

### Shaft Excavation Cost

As you can see in Figures 1–3, cost of shaft change accordingly as tunnel diameter change so base of this graph we could estimate the formula as following:

$$SC = H(1134D + 12529)\{H^{0.03(LnD-0.1269)}\}$$

Where:
  $SC$ = Shaft cost ($/Lf)
  $H$ = Depth of shaft (ft)
  $D$ = Tunnel diameter (ft)

### Tunnel Excavation Cost

In this component, sub components of ventilation cost, mucking cost, dewatering cost, lighting cost, machine operating cost such as generator, machine assembly, cutter cost, compressor, and pump has been included. As you can see, tunnel excavation cost for longer tunnel reduced. Base of Figure 4 we could estimate the formula as following:

$$TEC = L\{11.5D - 678Ln(L) + 8045\}$$

Where:
  $TEC$ = Tunnel excavation cost ($)
  $D$ = Tunnel diameter (ft)
  $L$ = Tunnel length (ft)

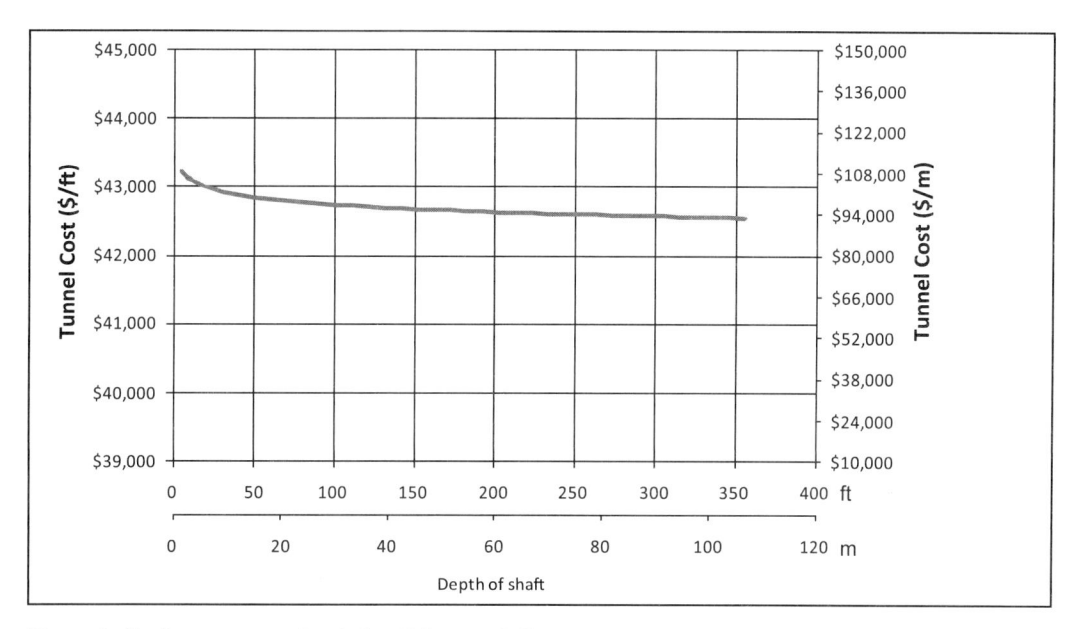

Figure 2. Shaft cost versus depth for 20 ft tunnel diameter

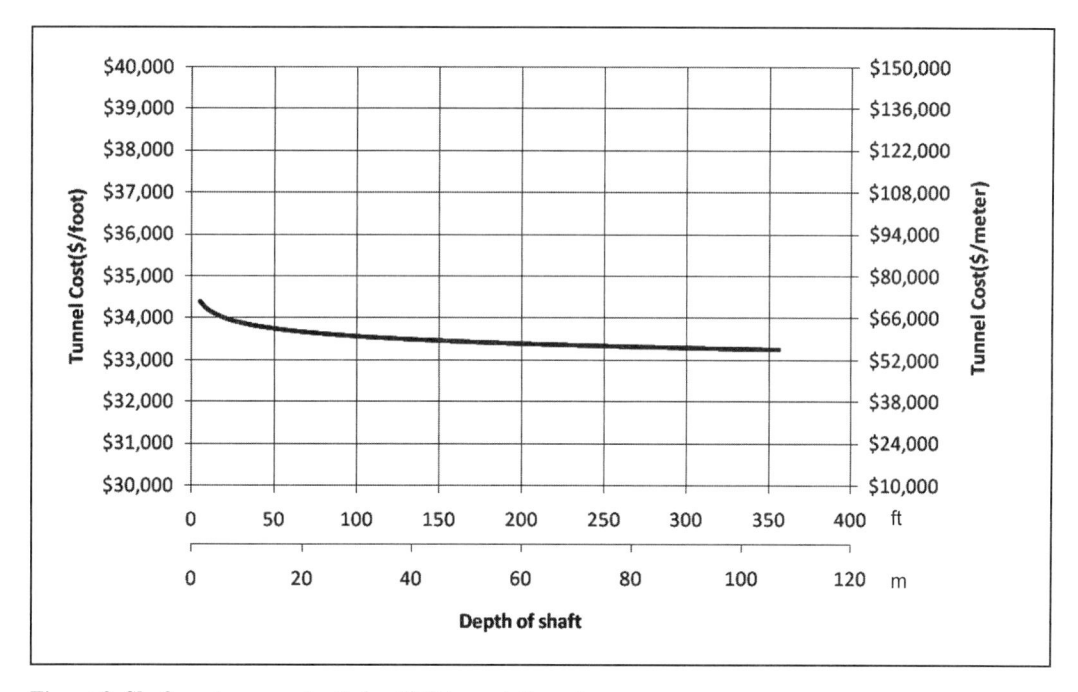

Figure 3. Shaft cost versus depth for 10 ft tunnel diameter

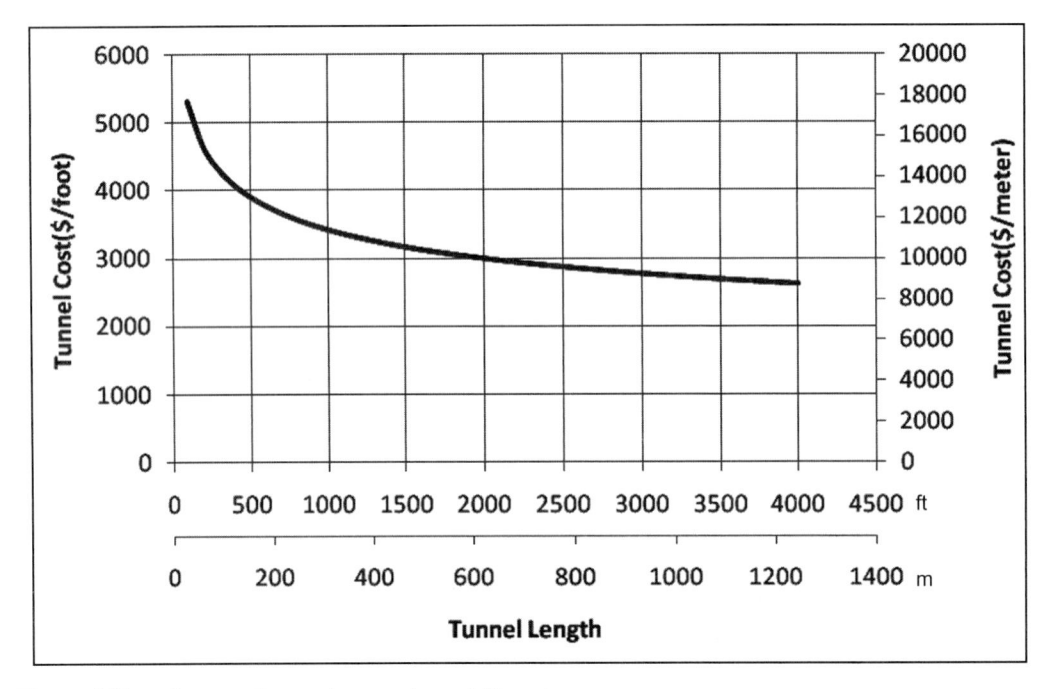

Figure 4. Tunnel excavation cost versus tunnel diameter

## Ground Support Cost

In order to define formula for support cost, type of support and amount necessary. Beside increase number of rock bolt, thickness of shotcrete, and steel ribs spacing will reduce rate of advance.
 Cost of support could be break down to:

1. Cost of segment and CIP
2. Cost of shotcrete
3. Cost of steel liner plate
4. Cost of steel ribs
5. Cost of rock bolt

### *Cost of Segment and Cast in Place*

Based on Figures 5–6, we could estimate the formula as following:

$$CS = A(124D+623)$$

$$CCIP = 59B$$

Where:
 CS = Cost of segment ($)
 CCIP = Cost of cast in place
 A = Length of segment installed
 B = Total length of cast in place installed in total length of tunnel
 D = Tunnel diameter

### *Total Cost of Shotcrete*

$$CSH = 1.743EF(D+F)$$

Where:
 CSH = Total cost of shotcrete ($)
 E = Thickness of shotcrete (ft)
 F = Length of shotcrete sprayed
 D = Tunnel diameter

### *Total Cost of Steel Liner Plate*

$$CLP = 553.5DJ$$

Where:
 CLP = Total cost of liner plate installed ($)
 D = Tunnel diameter (ft)
 J = Length of liner plate installed (ft)

### *Cost of Steel Ribs*

$$CSR = (3.14MD/S)\,V$$

Where:
 CSR = Total cost of steel ribs ($)
 M = Total length of steel ribs installed in total length of tunnel (ft)
 D = Tunnel diameter (ft)
 S = Spacing of steels ribs
 V = Cost of unit foot of steel ribs ($)

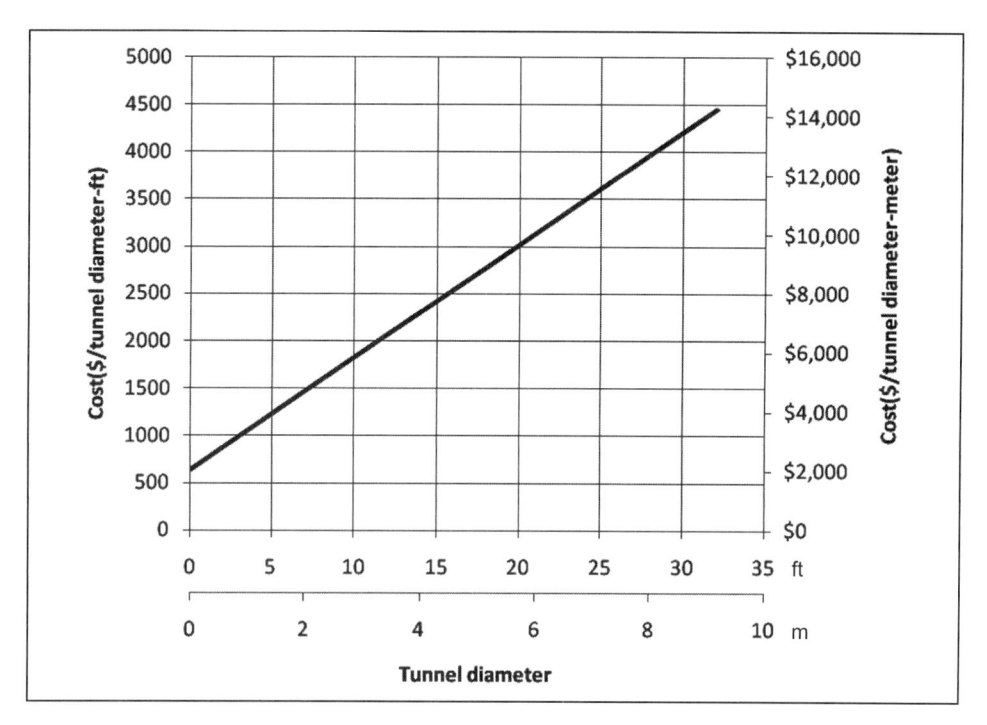

Figure 5. Segment cost versus tunnel diameter

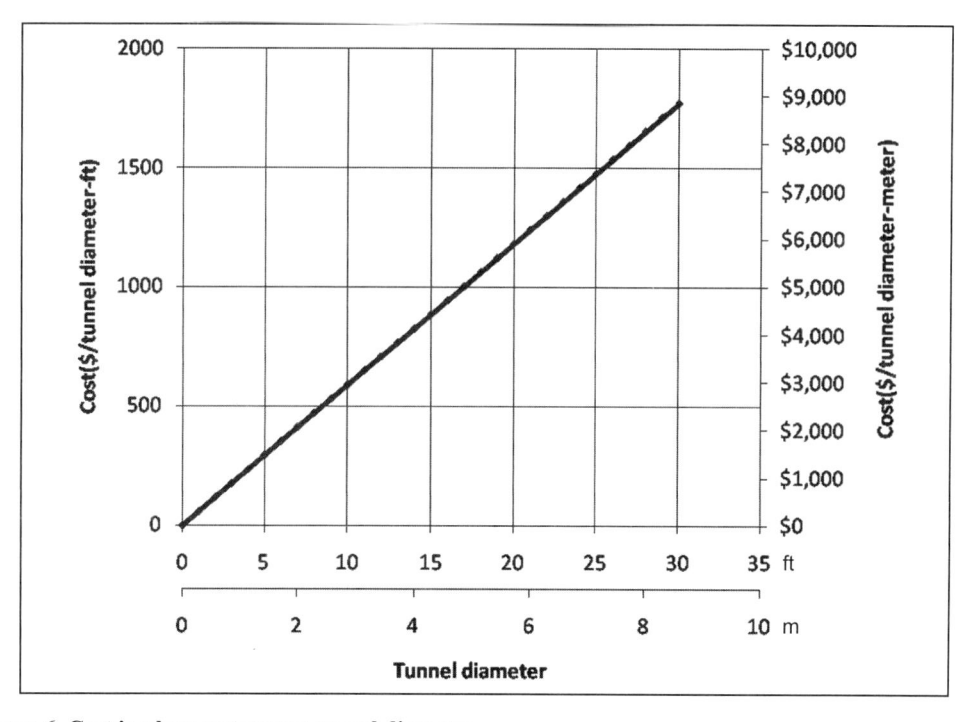

Figure 6. Cast in place cost versus tunnel diameter

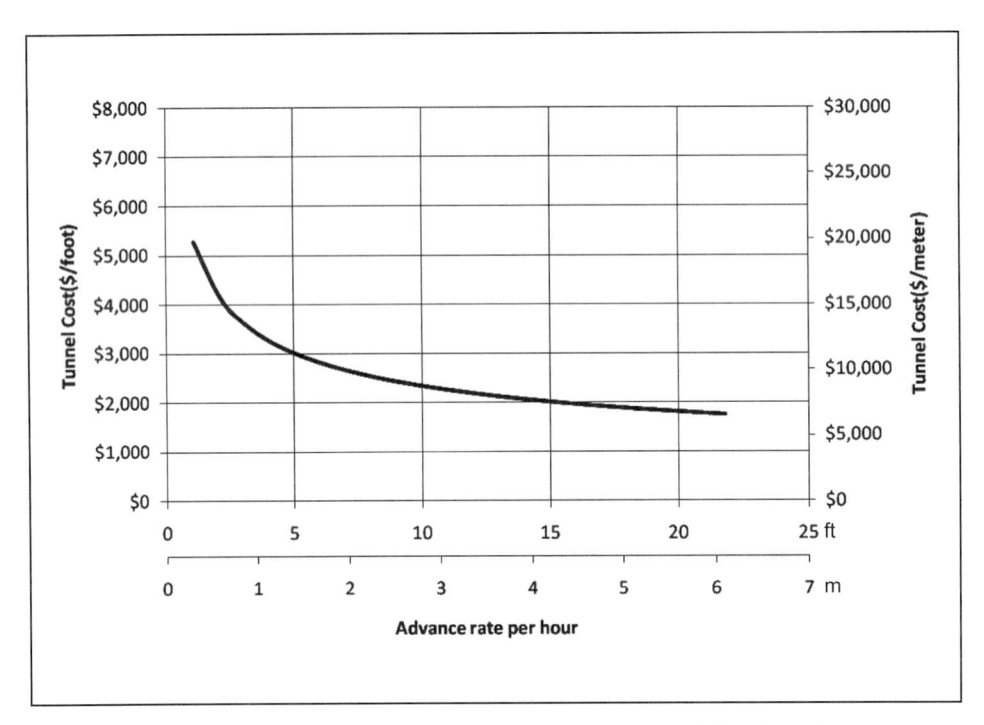

Figure 7. Tunnel cost versus advance rate for 9 ft tunnel diameter and 1,000 ft length

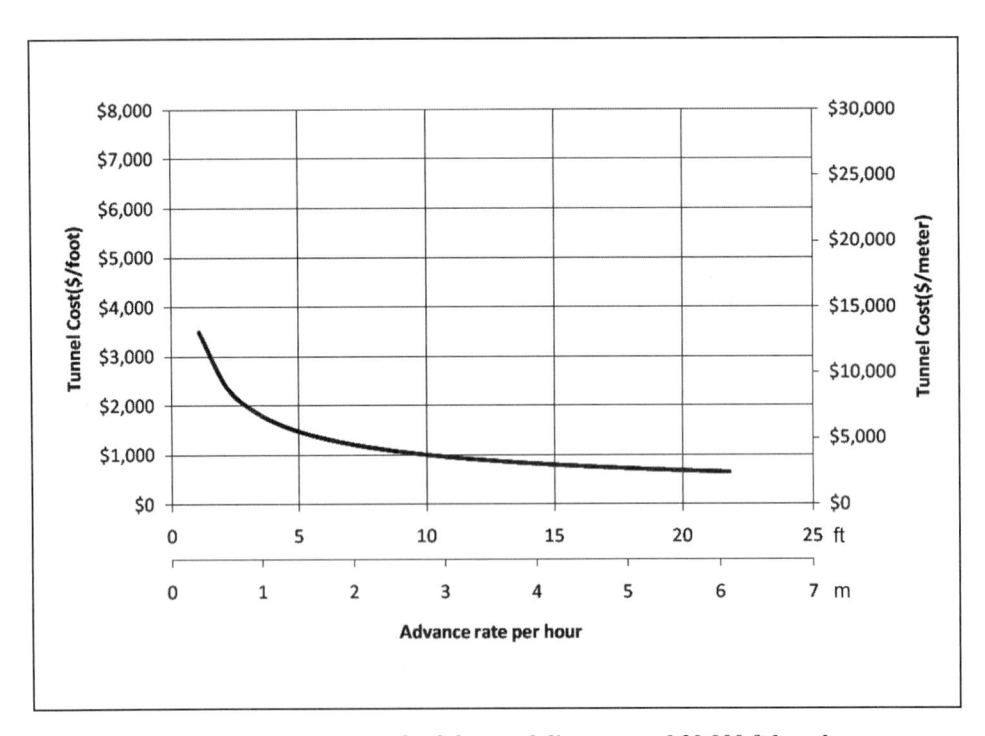

Figure 8. Tunnel cost versus advance rate for 9 ft tunnel diameter and 30,000 ft length

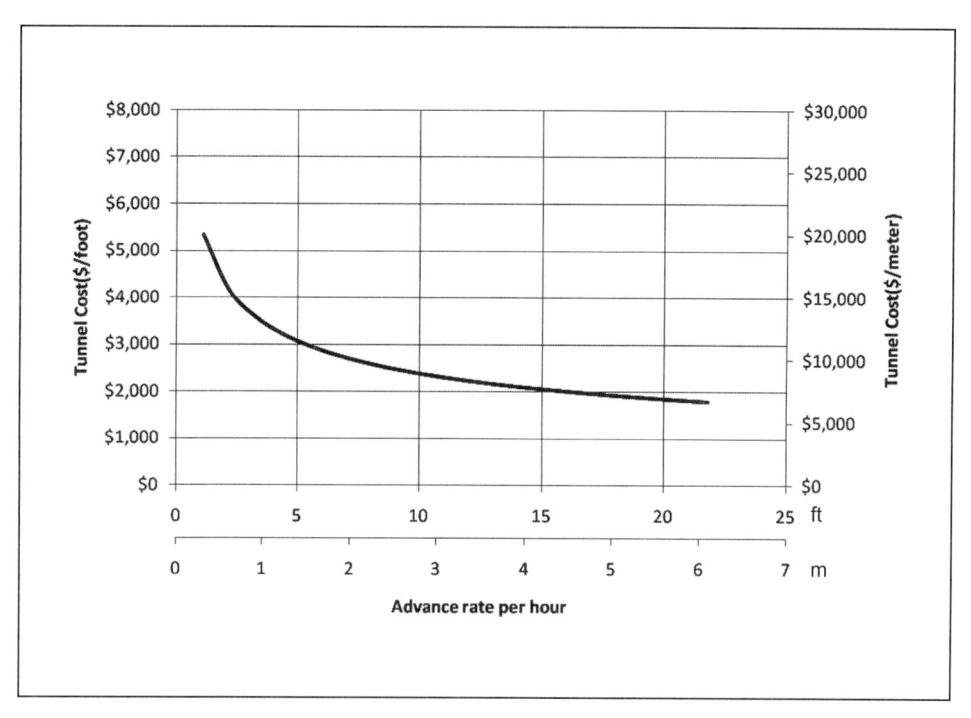

Figure 9. Tunnel cost versus advance rate for 18 ft tunnel diameter and 1,000 ft length

*Cost of Rock Bolt*

$$CRB = 20NWK$$

Where:
 CRB = Cost of rock bolt ($)
 N = Number of rock bolt installed per linear tunnel foot
 K = Length of tunnel which rock bolt installed
 W = Depth of rock bolt (ft)

**Mucking Cost**

To determine cost of mucking this formula has been used:

$$CM = 19.78LD^{0.69}$$

Where:
 CM = Total cost of mucking for whole tunnel length ($) (disposal on site)
 D = Tunnel diameter (ft)

**Cost of Tunnel Base of Tunnel Advance**

Cost of tunnel effect excessively by rate of tunnel advance. If rate of advance for TBM is less than 5 ft per hour then cost of tunneling tremendously increase as you can see at Figure 7–9.

Time related cost index is a very important parameter that impacts realistic cost value. Variation order can cause a delay and it affects total cost. Type of the machine for excavation is very important and directly affect on total cost. Such cost need to be incurred by engineer or contractor to make sure it working out on acceptable rate. The consequences of variations could be significant. All necessary precautions need to be considered to prevent the occurrence of delay.

At the end two case study add to check accuracy of the formula. All the cost is in US dollars.

**CONCLUSIONS**

The cost estimate for tunnel project is one of the most important steps in project management. The formula need more improvement by try and error method which can be achieved by more case study.

Table 1. Case history for YDSS interceptor sewer project, Toronto, Canada, final closure, 2009 cost calculated from formula

| Description | Leslie-Bayview 2285m | Bayview-young 1308m | Leslie South 534m | Total ($) | Factors |
|---|---|---|---|---|---|
| Tunnel Excavation Cost | $15,555,307 | $10,527,416 | $5,362,009 | $31,444,732 | Mobilization |
| Cost of Segment | | | | $20,133,576 | 0.00% |
| Cost of CIP | | | | | $0 |
| Cost Of Liner Plate | | | | | Esc. Factor |
| Cost of Shotcrete | | | | | 0% |
| Cost of rock bolt | | | | | $0 |
| Cost of Steel Rib | | | | | Contingency |
| Cost of Mucking | $561,356 | $321,336 | $131,188 | $1,013,880 | 25.0% |
| Cost of Shaft | $637,765 | $1,512,394 | $1,302,332 | $3,452,491 | $14,011,170 |
| | | | | $56,044,679 | |
| Total | $16,754,428 | $12,361,146 | $6,795,529 | | |
| | | | | $70,055,849 | |

Table 2. Comparison for YDSS interceptor project in Toronto

| Project | Description | Cost $ | Deviation from Real cost | Deviation Percentage | Deviation from 100.0% |
|---|---|---|---|---|---|
| YDSS Interceptor | By Formula | $56,044,679 | -$13,305,632 | -19.2% | -$1,150,307 |
| | By Engineer : 100% Design | $57,194,986 | -$12,155,325 | -17.5% | |
| | Final Paid to Contractor | $69,350,311 | | | |

# TRACK 3: PLANNING

Bradford Townsend, Chair

*Session 2: Project Delivery*

# Lake Mead Intake No. 3, Las Vegas, NV: A Transparent Risk Management Approach Adopted by the Owner and the Design-Build Contractor and Accepted by the Insurer

**Michael Feroz**
Parsons Water Infrastructure, Las Vegas, Nevada

**Erika Moonin**
Southern Nevada Water Authority, Las Vegas, Nevada

**James Grayson**
Vegas Tunnel Constructors, Boulder City, Nevada

ABSTRACT: The results of shared risk management used by Owner and Design-Build Contractor are discussed. The potential for high contingency to be included in the Owner's budget and Contractor's lump sum bid was eliminated by conducting partnering meetings to address potential issues and to jointly develop mitigating solutions. The paper discusses the approach to Risk Management which complied with the intent of ITIG. The joint risk management goals were to: minimize cost to the Owner, maximize profit for the Contractor, demonstrate risk management process to the Insurers, and successfully design and build the project on schedule and budget.

## PROJECT OVERVIEW

Southern Nevada (Clark County and Greater Las Vegas Valley) gets more than 90 percent of its water supply from Lake Mead on the Colorado River. The Colorado River system has experienced the worst drought in its recorded history. The drought is expected to continue and is significantly impacting water storage in Lake Mead. The Southern Nevada Water Authority (SNWA) operates two water-intakes in Lake Mead, located 20 miles east of Las Vegas, Nevada. Severe drought has caused decline in lake level by 110 feet since 2000 and future declines are expected. Due to severe drought conditions future operation of both the existing intakes is at risk and water supply to Clark County and the Las Vegas valley will be adversely impacted.

In order to sustain existing water capacity SNWA is constructing a new deep-water intake (Intake No. 3) in Lake Mead against the potential inoperability of Intake Nos. 1 and 2 should lake levels continue to fall. SNWA has undertaken the task of building a new project, Lake Mead Intake No.3 (Project), to overcome the severe drought challenges and construct water facilities to sustain community viability in severe drought conditions in Southern Nevada. The Project is located in Lake Mead, in the Lake Mead National Recreation Area. Program costs are estimated at $817 million with completion slated for first quarter-2013.

This paper describes expected construction challenges and associated design and construction risks for the new project. The results of shared risk management used by Owner and Design-Build Contractor are discussed. The potential for high contingency to be included in the Owner's budget and Contractor's lump sum bid was eliminated by conducting proprietary meetings with the potential bidders to address potential issues and to jointly develop mitigating solutions. The paper discusses the use of the transparent risk management approach and the development of the Risk Register that complied with the intent of the international code "International Tunneling Insurance Group" (ITIG). The joint risk management goals were to: minimize cost to the Owner, maximize profit for the Contractor, demonstrate risk management process to the Insurers, and successfully design and build the project on schedule and budget.

## PROJECT DESCRIPTION

The major components of the Lake Mead Intake No. 3 project were divided into six major contract packages:

**Figure 1. Planned location**

1. Design-Build of a 20-ft diameter, 15,000 ft long intake tunnel driven beneath the Lake and a portion of Saddle Island and a deep-water intake riser and inlet structure (Contract No. 070F 01 C1)
2. Design-Bid-Build of a Pumping Station UndergroundIPS-3 (Contract No. 070F 02 C1)
3. Design-Bid-Build of a Pumping Station Superstructure IPS-3 (Contract No. 070F 02 C2)
4. Design-Bid-Build Power Supply Facilities (Contract No. 070F 03 C1)
5. Design-Bid-Build of a discharge pipeline from IPS-3 connecting to the AMSWTF (Contract No. 070F 04 C1)
6. Design-Bid-Build of a shaft and tunnel (Intake 2 Connection and Modifications) connecting the IPS-3 facilities with the existing IPS-2 (Contract No. 070F 05 C1)

Packages 2, 3 and 5 have been put on hold for indefinite period. A new construction package has been introduced and awarded as Design-Bid-Build of a Connector Tunnel between Intake No. 3 Shaft and existing IPS-2 tunnel.

The overall Lake Mead Intake No. 3 project is currently being constructed under four different contracts.

**070F 01 C1: Intake No. 3 Design-Build Shafts and Tunnel**

- Construction awarded to Vegas Tunnel Contractors JV (Impregilo and S.A. Healy) on March 20th 2008. Duration: 4-years. Bid amount: $450 Mil, approximately.
- 30-ft diameter and 600-ft deep tunnel construction access shaft.
- 3-miles long, 20-ft internal diameter tunnel with pre-stressed concrete segmental liner (bolted and gasketed).
- 1,200 MGD intake structure in 250-ft of water with connection to the new tunnel.

The intake tunnel will connect the new intake riser. Elements that determine the location and alignment of the intake tunnel are:

- Location of the intake riser
- Location of the proposed pumping station (IPS-3)
- Hydraulic requirements
- Topography
- Ground conditions
- Tunneling technology

The intake will be north of the Las Vegas Wash, in the Boulder Basin of Lake Mead, and will be designed for a capacity of 1,200 million gallons per day (mgd) at an intake elevation of 860 feet amsl. The intake will be a lake tap structure with a riser structure above the lake bottom and a shaft below the lakebed that connects the riser to the intake tunnel.

The riser structure will provide for a future extension to deeper waters northeast of the planned location (see Figure 1).

The tunnel will be constructed from a shaft near IPS-3 with an uphill gradient to the connection with the intake riser. The topography between the intake riser and pumping station is highly variable, as confirmed by pre-lake impoundment photographs and the results of the geophysical, bathymetric, and geotechnical field investigations. The most significant topographical feature between the intake riser and pumping station is the submerged continuation of the Las Vegas Wash. The intake tunnel needs to cross the submerged wash with adequate ground cover, creating a vertical constraint for the intake tunnel alignment.

The tunnel will be constructed from a shaft on Saddle Island (future location of IPS-3) with an uphill gradient to the connection with the intake riser. The topography between the intake riser and pumping station is highly variable, as confirmed by pre-lake impoundment photographs and the results of the geophysical, bathymetric, and geotechnical field investigations. The most significant topographical feature between the intake riser and pumping station is the submerged continuation of the Las Vegas Wash. The intake tunnel needs to cross the submerged wash with adequate ground cover, creating a vertical constraint for the intake tunnel alignment.

The invert of the low end of the tunnel will be approximately 600 feet below the maximum water elevation of Lake Mead and approximately 500 below Lake Mead's current water surface. Geological conditions for the tunnel construction, intake riser construction and the connection of the two are very challenging.

### 070F 02 C2: Design-Bid-Build Connector Tunnel

- Construction awarded to Renda Pacific dba KW Pipeline on June 2009, Duration: 3.5-Years. Bid amount: $30 Mil., approximately
- 30-ft diameter and 600-ft deep tunnel construction access shaft:
- The IPS-2 connector tunnel will extend from the distribution shaft at the IPS-3 site to the existing IPS-2 intake tunnel, a distance of approximately 2,500 feet. The drill-and-blast connector tunnel will be modified horseshoe-shaped, with a width of 14-feet and a height of 16-feet, which is the same configuration as the IPS-2 intake tunnel.
- 2,500-feet long, 20-ft internal diameter tunnel with pre-stressed concrete segmental liner (bolted and gasketed).

### 070F04 C1: Design-Bid-Build Permanent Electric Power Supply

- Construction awarded: 1st quarter 2009. Construction Duration: 18-months. Bid Amount: $8 Mil., approximately.
- 69/13.8-kV power transformer, 69-kV power circuit breakers, a three-cubicle relay switchboard, and duct bank.
- The material will be received and installed by an installation contractor under a separate contract.

### 070F 05 C1: Intake 2 Design-Bid-Build Connection and Modifications

- Construction awarded May 19, 2008. Duration: 2 Years. Cost estimate: $45 Mil., approximately.
- 270-ft long stub tunnel, 14-ft wide by 16-ft high modified horseshoe section, upstream of isolation gate shaft.
- 250-ft long stub tunnel, 14-ft wide by 16-ft high modified horseshoe section, downstream of the isolation gate shaft.
- 45-ft long rock plug excavation and connection to the existing Intake Pumping Station No. 2 intake tunnel.
- Modifications to the existing Intake Pumping Station No. 2 intake structure.
- 22-ft diameter × 380-ft deep shaft and isolation gate.

## TRANSPARENT RISK MANAGEMENT APPROACH USED BY THE OWNER

SNWA used a shared risk management approach. The results of shared risk management used by owner and design-build contractor are discussed. The potential for high contingency to be included in the owner's budget and contractor's lump sum bid was eliminated by conducting proprietary meetings to address potential issues and to jointly develop mitigating solutions. The purpose of the proprietary meetings was to receive feedback from the Finalist bidders on the RFP requirements, including the contract agreement and allocation of risk issues considered during RFP Process. The following issues were discussed:

- Appropriate and fair allocation of risk between D-B and owner.
- Technical requirements.
- Use of allowances to compensate contractor for unforeseen conditions for specific Unit Items.
- Differing site conditions clause applied after allowance item is fully used for specific Unit Items.

- Providing more flexibility for Finalist to determine appropriate means and methods.
- Allowing the finalist to establish the contractual completion date by specifying the number of days in proposal.
- Providing additional time for finalists to complete and submit proposal.

The paper discusses the transparent risk management approach and the development of the Risk Register that complied with the intent of the international code "International Tunneling Insurance Group" (ITIG).

The joint risk management goals were to: minimize cost to the owner, maximize profit for the contractor, demonstrate risk management process to the insurers, and successfully design and build the project on schedule and budget.

This project will be the deepest sub-aqueous tunnel constructed with a pressurized-face TBM in the world to date. The current lake level is 1,128 feet amsl; however, the lake has the potential to rise to 1,229 feet. With a tunnel elevation of approximately 650 feet amsl, this project will have a current hydrostatic pressure of approximately 14.3 Bar, but with a potential pressure of approximately 17.4 Bar. Tunneling in pressures of this magnitude in closed (pressurized-face) mode presents significant risks. Mitigation of these risks involves developing alignments and profiles that minimize this pressure, and tunneling in geologic units that minimize the amount of pressure necessary for stability and water control. These issues are discussed later in this paper. Based on current evaluations, the West Tunnel Corridor appears to offer the best potential to minimize the potential number of cutter-head interventions in hyperbaric conditions. The West Tunnel Corridor also appears to offer the least adverse geological conditions and shortest construction period. Consequently, the West Tunnel Corridor should offer the lowest relative risk and lowest construction cost. This should not be interpreted that there will be no risk or challenges during construction.

## RISK MANAGEMENT PROCESS CONTRACTOR'S VIEW

### Overview of Risk Management Process

For the Lake Mead Intake No. 3 project, the Southern Nevada Water Authority (SNWA) and Vegas Tunnel Constructors (VTC) have agreed to a transparent risk management approach. The risk management approach was conceived through voluntary compliance with the International Tunneling Insurance Group's Code of Practice for Risk Management of Tunnel Works (Code), and minimizes the risks of physical loss or damage and associated delays. As a transparent approach, the management of the anticipated risks of the project is openly shared between SNWA and VTC. This includes the brainstorming of the anticipated risks as well as determining what suitable mitigating measures would be to reduce the negative effects of those risks.

Having engaged in an open risk management process with SNWA, VTC has strengthened their partnership with the SNWA. Although formal partnering was part of the original bid documents, VTC proposed a more informal partnership approach. This partnership allows for open communication between persons from both organizations enabling decisions and actions to take place at the lowest possible levels in the organizations. With this approach, collaboration is necessary on a daily basis between the parties of the partnership. This approach not only allows for transparency during construction of the project, but it paves the way for a transparent risk management partnership.

In keeping with the spirit of the Code, VTC has incorporated the objectives of the code into its Risk Management Program. As part of the provisions of the Code, representatives of the insurance group audited the project and had determined that SNWA and VTC were in compliance with the code of practice and that it was their determination that it would be highly unlikely that the Code would not be used.

Although, the main tool for the Risk Management Program is the Risk Register, the Code also stipulates that several other documents are to be produced. These include an overall site organization chart, a training plan, an audit plan, and construction method statements. These documents when applied in conjunction with a realized risk management program can help to reduce the risk ratings of individual risks.

Risk management allows us to address potential issues and to jointly develop mitigating solutions with SNWA. In doing so, we can prepare for anticipated risks and reduce costs due from unforeseen events. This cost savings helps to decrease overall cost for both parties.

### Risk Management Plan

On this project, the sections of the risk management plan cover the purpose and scope of risk management, organization and roles, the approach and method to risk management, the risk breakdown structure, and include the schedule for implantation of the plan. SNWA reviewed the VTC Risk Management Plan and accepted the plan for the project.

The risk management approach is shared between the owner and the contractor, but SNWA is not made owner of any of the risks of the project nor are they transferred any risks. It was determined by VTC that ownership of the risks of the project during construction are the responsibility of VTC to avoid, mitigate or accept. SNWA, although sharing in the risk management is not held responsible for the risks and their avoidance, mitigation or acceptance specifically in the Risk Register. Rather SNWA is kept well informed of the risk management on the project and regularly participates and engages with VTC regarding the risk management of the project. As part of the partnership, SNWA also aids VTC in the mitigation of risks when necessary and appropriate. For instance, during the design stage, the VTC design engineers had asked for additional boring data for the tunnel alignment to verify subsurface conditions were as described in the GBR. SNWA agreed that the cost of the additional investigation was a benefit to the project to prevent unnecessary risk and to prevent increase cost and time to the project.

## IMPLEMENTATION OF RISK MANAGEMENT

To implement risk management on the project, a schedule of the risk management activities was created. The activities included issuance of the Risk Management Plan, the Risk Workshops, and Project Risk Reporting. SNWA and VTC jointly engage in the implementation of the risk management program. This ensures a thorough understanding of the projects risk management by all parties.

### Risk Management Team

VTC's risk management team is made up of the project director, risk manager, risk manager assistant, and risk owners.

### Risk Facilitator

In addition to the contractor's risk management team is the Risk Facilitator. Mr. John Reilly was hired to help facilitate risk management on the project. His services are used to explore the existing risk management program and to introduce aspects of risk management that would better enable a high level of risk management on the project.

### Collection of Risk Data

The major collection of the risk data is through the brainstorming activities during the Risk Workshops. These workshops are held to discuss specific portions of the work. The workshops allow for the intimate discussion of anticipated risks and the consequences of those risks between the owner of the project and the contractor. Thus the workshops become a platform for ensuring risk and management of risk is fully transparent to all parties associated with the project.

For this project the workshops have been broken down into the main topics, access shaft, stub tunnels and cavern, tunnel, and intake.

These workshops are attended by the risk management team of the contractor as well as persons from the owner's project team and the risk facilitator. The workshops have been held in a couple of different formats. At the first workshop, the group of attendees was broken down into smaller groups of approximately 8 persons. Each group went through the existing Risk Register and made edits to it. Any risks that were not on the register that were deemed as valid were then subsequently added to the register. The next workshops were held where the whole group discussed the risks.

In addition to the Risk Workshops, any person that determines that there is a risk that has not been addressed by the Risk Register can have the risk added. The information is given to the risk manager Assistant who in turn updates the Risk Register and distributes to the others for comment.

## RISK REGISTER

The ITIG Code suggests that the Risk Register be a document that identifies and clarifies ownership of risks and how the risks are to be allocated, controlled, mitigated and managed. The Risk Register is a living document and it is revised and updated as necessary. We have created a Risk Register that focuses on the qualitative analysis of the risks. The qualitative analyses are done for the anticipated risks prior to mitigation and then again for the risks after mitigating measures are put into place. The analysis of each risk on the register gives us a risk rating before and after mitigation. This rating is used to determine the effectiveness of mitigation that has been performed.

For the Lake Mead Intake #3 project, the Risk Register has undergone several iterations since the RFP stage. The initial Risk Register that was submitted met the requirements of the RFP and the Code, but did not include all of the information that was recommended by the risk facilitator. Following the advice of the risk facilitator, the Risk Register was modified to include triggers that would cause the anticipated risk to occur and links between risks. These changes address the statement of "If this then that…"

It has been widely accepted that responses to risk include avoidance, mitigation, acceptance and transferring. We have determined that transferring of risk is not an acceptable method to respond to an anticipated risk. Instead we have determined that the only acceptable responses to risk are to avoid, mitigate, actively accept, or to passively accept.

Avoidance of the risk is to eliminate the potential risk by eliminating the cause, such as changing the construction method. Mitigation of the risk

## Table 1. Probability matrix and scale

| Probability (P) | | Severity ($, T) | | |
|---|---|---|---|---|
| Expected | P>75% | Catastrofic | $>10M | T>6m (month) |
| Probable | 51%<P<75% | Major | 2M<$<10M | 3m<T<6m |
| Likely | 26%<P<50% | Moderate | 0.5M<$<2M | 1m<T<3m |
| Unlikely | 6% <P<25% | Minor | 0.1M<$< 0.5M | 1w<T<4w (week) |
| Remote | P<5% | Negligible | $< 0.1M | T<1w |

Scales are then converted into a numerical as follows: high =1, medium=3, low =5.
The probability and impact ratings are then combined to give a score for each risk.
The criticality is simply the product of probability and impact as shown on the 5×5 Score Matrix in Table 2.
If the impacts on cost, schedule, and performance are different, then the greater impact of the three is used to rate the risk (i.e., the highest criticality is retained).

## Table 2. Risk score matrix

| | | LAKE MEAD INTAKE Nr. 3 PROJECT | | | | |
|---|---|---|---|---|---|---|
| | SEVERITY | 1 | 2 | 3 | 4 | 5 |
| PROBABILITY | | CATASTROFIC | MAJOR | MODERATE | MINOR | NEGLIGIBLE |
| 1 | EXPECTED | RR1 | RR2 | RR3 | RR4 | RR5 |
| 2 | PROBABLE | RR2 | RR4 | RR6 | RR8 | RR10 |
| 3 | LIKELY | RR3 | RR6 | RR9 | RR12 | RR15 |
| 4 | UNLIKELY | RR4 | RR8 | RR12 | RR16 | RR20 |
| 5 | REMOTE | RR5 | RR10 | RR15 | RR20 | RR25 |

Unacceptable
Acceptable with mitigating measure
Acceptable

reduces either the probability of the risk or the impact of the risk. Active acceptance of the risk includes implementing a contingency plan upon an identified trigger event. Passive acceptance is to not react to the risk, but to accept the consequences regardless of the cost.

**Information Organization**

The Risk Register is divided into separate sections and subsections and then by anticipated risk. The risks are uniquely numbered and are removed from the register when the work is complete and the risk is no longer applicable to the project.

**Qualitative Analysis of Risk Data**

Although, risk management processes can be complex and complicated, it was decided that a more simplified approach to risk management was appropriate for this project. In this way more emphasis is placed on the qualitative analysis of the risks. The risks are ranked by determining the likelihood of occurrence and the severity of an occurrence based on time and cost. The qualitative analysis uses a 5 × 5 table where 1 is the most severe or the highest probability and 25 is the least severe or the lowest probable. The outcome of this analysis provides us a Risk Rating that is used to determine the most severe risks and where high effort is needed to mitigate the anticipated risks. Using a mostly qualitative approach removes the complexity of other Monte Carlo simulations from distorting or distracting the Risk Owners from fully understanding where their efforts are best served (see Table 1).

Based on Table 1, risks will be categorized on Table 2 by their criticality or score to enable project to prioritize where the most urgent attention is required.

Based on this classification of the risk criticality, the action planned are as follows:

**Low risk** (criticality 16 to 25): Monitor periodically to follow evolution.

**Medium risk** (criticality 5 to 15): Requires action by the risk owners, either to move as far as possible:
- To the "Low" risks area for risks,
- To the "High" risks area for opportunities.

**High risks** (criticality 1 to 4): Requires immediate action either:
- To reduce criticality to acceptable level for risks,
- To increase the chance of the opportunity to happen.

Extra resources may be necessary and shall be determined to achieve the mitigation/enhancement. Fallback plan in case of event occurrence shall be envisaged and described.

## CONCLUSIONS

### Proactive Risk Management

Rather than being reactive, the Risk Management approach in accordance with the Code causes a proactive management process. With the transparent risk management approach, SNWA benefits in the fact they are informed of the potential risks of the project. This allows them to review certain aspects of the design as it is developed to ensure mitigating measures are incorporated by the engineers. SNWA also gets to include their ideas and wishes regarding mitigation of anticipated risks. This allows them to have a higher level of comfort in regards to the final work.

### Cost Savings

Cost savings are realized by VTC with this risk management approach due to the fact the risks are managed proactively allowing for smooth continuation of construction rather than stopping and starting or redoing work. Additionally, the "Over-the-shoulder" approach that was proposed by VTC during the RFP stage allows for smooth continuation of work especially for those tasks that are on the critical path. Ensuring that tasks on the critical path are well executed mitigates risks to time and cost, which benefits the SNWA and VTC.

# Tunneling MegaProjects: They Are Different

**David J. Hatem, David H. Corkum**
Donovan Hatem, Boston, Massachusetts

## INTRODUCTION

No one familiar with the design and construction industry can legitimately or genuinely dispute that megaprojects—large, complex civil construction projects having a cost of $1 billion or more—typically result in controversy. A megaproject's cost, its multitude of stakeholders with competing interests, and its high public profile, assure "mega" opportunities for substantial risk, conflicting and disappointed project participant expectations, disputes, and claims.[1] Concomitant with the cost, public interest, and public impact comes an intense involvement from elected officials. When you add to the megaproject profile that the project involves tunneling or other major subsurface components the realization of all of the foregoing "opportunities" is all the more probable and the consequences all the more potentially negative. To be sure, megaprojects and tunneling, separately and especially combined, are not for the timid or faint of heart.

For the majority of individual participants within the owner, consulting engineer, and constructor organizations, a megaproject will be a once in a lifetime opportunity and experience. Similarly, these owners and private firms that undertake megaprojects rarely have a chance to take on a second megaproject with their same project team and accumulated intuitional knowledge. The learning on megaprojects, especially subsurface megaprojects, is in its relative infancy. In part, this state of development is attributable to the fact that for most project owners a megaproject is a once in a more than decade experience. It requires an uncharacteristically far-sighted public owner to perceive the value of, and thus commit the resources to, reflecting upon and capturing lessons learned from such experiences. Simply put, the benefit of such an exercise is often considered too remote and not likely to contribute to more pressing immediate organization goals. For the consulting engineers and constructors involved in the megaprojects, the last phase of that experience is typically fraught with disputes, conflicts and claims, and reflecting on that experience is often seen as painful, unconstructive and useless: "things are not going to improve in any significant way and the pain that we are experiencing is an inherent part of the megaproject experience."

Things are changing, however, including efforts to capture, transmit and learn lessons—what works and what does not—on megaprojects. This is a positive development and is a critically important step in the process of improving the planning and delivery of these projects. At the same time, there have also been some negative trends especially with respect to risk allocation for subsurface conditions in certain project delivery approaches, such as design-build and public-private partnerships. As the backlog of conceptualized subsurface megaprojects increases and public owner exploration of and experimentation with alternative project delivery approaches also increases, now is the appropriate time to reflect upon all of these developments with a view toward improving and learning from the lessons of the past.

## MEGAPROJECTS: WHAT MAKES THEM CHARACTERISTICALLY DIFFERENT FROM MORE CONVENTIONAL PROJECTS

Megaprojects are, by definition, large and often involve complex and innovative design and construction methodologies. It is not, however, the simple scaling up of the size of the organization's internal management structure that poses the challenge. Megaprojects generally require these newly formed and generally untested internal organizations to coordinate with a variety of design consultants, and successfully manage numerous interfaces between multiple prime contractors. The projects often entail substantial risk for all project participants, along with the corollary of significant opportunity for seriously disappointed contractual, economic and other important project objectives and expectations. The megaproject participants are also challenged by public perception of their efforts by the often less than charitable, or even objective observations and reporting of the press.

The complexity of megaproject planning starts with some very basic questions. Who is the client? Whose expectations are we trying to satisfy? Will the project owner who starts the project be the owner or primary user at the end of the project? Megaprojects involve many stakeholders, often with contradictory objectives and unpredictable powers to influence decisions. It is not at all unusual for the identity of the project owner to evolve over the typically decade-plus duration of the project period. Certainly the individuals will change as will the political leadership that oversees the owner's organization. This diversity of interest, the evolutionary character of project owner identity and the fact that the ultimate project owner (responsible for operations and maintenance of the completed project) may be different from the project owner during the planning stages—all significantly contribute to the possibility, if not probability, of lack of clarity, conflict in project priorities, and disappointment in the expectations.

The amount of public funds—typically federal, state and local in nature—required for the completion of a megaproject justify the significant degree of public oversight on megaprojects. Equally important, the potential impact and disruption to the public caused by the megaproject construction process and the fact that megaproject funding may preclude other state or local projects with important political constituencies also justify and rationalize intense public scrutiny and oversight over all aspects of megaproject planning and delivery.[2]

Public and governmental interest is yet further drawn to megaprojects given the rather serious and highly publicized and political experiences on some of those projects in recent years. For decades, megaprojects have been plagued by several serious problems:

- Artificially low and strategic underestimations of project cost
- Exaggerated benefit to the public by project proponents
- Unrealistic project performance expectation
- Underestimation of risk
- Optimism bias to promote project approval and funding
- Lack of clear accountability of project participants
- Lack of transparence and reporting to the public
- Asymmetric information and strategic information disclosure, non-disclosure or misinformation

In significant part, because of these experiences the "rules of engagement" for megaproject delivery (planning, funding, design, construction and oversight) have significantly and positively evolved in several important respects over the last five years. In the past, optimism, not realism, was the rule and guiding principle for megaprojects. Risk registers, independent costs and schedule validation; transparency with respect to project costs/schedule reporting; and accountability in the performance of all project participants (owner included)—were, to say the least, not universally embraced. The FTA and FHWA have taken a significant leadership position in addressing these problems. One need only visit the agencies' websites and peruse the Major Projects' links to understand their commitment to improving megaproject delivery. Mega transportation projects—at least those in which federal funding is implicated—has fundamentally changed in the last 5 years and, again, significantly for the better.

## TUNNELING MEGAPROJECTS: WHAT MAKES THEM CHARACTERISTICALLY DIFFERENT FROM CONVENTIONAL TUNNELING OR OTHER SUBSURFACE PROJECTS?

This entire inquiry may, in some respects, be answered in an abbreviated way by saying that tunneling megaprojects are "megaprojects, plus." Clearly, subsurface projects pose the potential of encountering conditions different from those anticipated (so-called differing site conditions) that can (and often do) significantly increase the estimated cost and/or time for completion of the project. And just as clearly these differing site conditions have the potential to seriously disappoint programmatic, planning, schedule, cost, contractual risk and other expectations of major project participants.[3] Often these differing site conditions are encountered early in the project schedule and result in substantial reduction or exhaustion of the planned or available contingency funds. On tunneling or other major subsurface megaprojects—by virtue of their typically linear and interdependent components—these differing site conditions can have consequential impacts that manifest themselves early in the project work and have a significant effect on the critical path. Unfortunately, when the subsurface project is halted by a differing site condition the next steps toward resolving the problem are not immediately obvious, and there is often open disagreement as to financial responsible that hampers the effective cooperation and efficient resolution of the problem. In these instances the leading stakeholders often come to believe that they were seduced into supporting or funding a project that was destined—only shortly after the "first shovel is in the ground"—to be significantly over budget, behind schedule and prone to disputes and claims among project participants and perhaps threaten the viability of the project in its entirety. Simply put, subsurface projects pose the risk of serious disappointment in costs and schedule expectations and, on megaprojects, that risk (like virtually all aspects of megaprojects) is significantly intensified and magnified.

The public rarely sees any value added in exchange for the additional money or time is consumed to address and equitably adjust for differing site conditions. While addressing those conditions is necessary to implement and progress the project, the public rarely sees those steps as providing the public with any benefit. Given the magnitude of funding involved in megaprojects, the intense public oversight over the expenditure of funds on those projects, differing site conditions quickly gain public attention and are fertile ground for not only disputes and claims but also intense public criticism of the project and all of its participants. In many respects that criticism is indiscriminate.

In some respects, however, implementation of improved contracting practices on megaprojects, as on any subsurface project, can and does have the potential for significantly mitigating this potential for disappointed expectations and other collateral problems such as the degradation of the public's confidence in the project.[4] The planning, definition, and implementation of an adequate pre-construction subsurface investigation program is essential to minimizing and mitigating the occurrence and impact of encountering unanticipated subsurface conditions during the construction process. The design and implementation of both permanent project work and construction means and methods depends on reasonably accurate and reliable information about subsurface conditions expected to be encountered during construction. Project owners should recognize the value in authorizing and implementing an adequate and sufficiently comprehensive subsurface investigation program, especially on tunneling megaprojects. The products of such investigation should be disclosed to constructors who are bidding the project.

An important adjunct to the authorization and implementation of an adequate subsurface investigation program is fairness in risk allocation, especially between the project owner and constructor with respect to equitable adjustment due to the encountering of subsurface conditions which are materially different from those indicated in contract documents. On the tunneling and other major subsurface megaprojects, owner decisions regarding risk allocation and the mechanisms to implement them are among the most important decisions on the planning of the project. Risk allocation decisions should be grounded in fairness, expressed in contract documents, and administered in a clear and consistent manner. Fairness and clarity in risk allocation decision making will reduce the likelihood of contentious claims and disputes, increase the likelihood that dispute resolvers adhere to those risk allocation decisions, increase the likelihood that the constructor will achieve contractual performance, and maximize the likelihood that the project will be completed on schedule and within expected cost.

Recently, project owners have expressed increasing interest in exploring and implementing alternate project delivery methods, such as design-build and public-private partnerships for their subsurface megaprojects. The choice of delivery method in subsurface project work depends upon a variety of factors and considerations requiring an evaluation of project-specific factors, interests, objectives and considerations.[5] While these alternate project delivery methods may be relatively untested in the major subsurface project delivery context, it is fair to state that the successful utilization of those alternate delivery methods depends upon proper implementation of many of the same "improved contracting practices" as had been recommended for use in the traditional design-bid-build delivery method. While many responsible and prudent project owners have recognized that basic principle, some project owners have demonstrated a tendency to transfer complete or disproportionate degrees of subsurface condition risk to the constructor on design-build projects, or to the concessionaire and members of the concessionaire's design-build team on public-private partnerships.[6]

The failure in design-build and public-private partnership delivery approaches to adopt "improved contracting practices" has several negative consequences including: reduction in the pool of responsible respondents/bidders; undisclosed contingency; substantial increase in risk of disputes and claims; substantial increase in the risk of performance failure or default; and significant increase in risk of professional liability claims against the consulting engineer. There is no reason why the "improved contracting practices should not be implemented in the context of alternate project delivery approaches such as design-build and public-private partnerships."[7]

## CONCLUSION

Megaprojects, in themselves, and tunneling or other major subsurface megaprojects, pose substantial risk for all project participants. Prudent decisions by project owners regarding subsurface investigations, subsurface data disclosure and sharing of risk—as well as other long-accepted improved contracting practices—should serve to manage that risk to the benefit of all project participants.

## REFERENCES

[1] Megaprojects: Challenges and Recommended Practices, Part II, "Risk Management and Professional Liability Issues for Engineering Consultants." 2010 ACEC.

[2] R. Capka, in "Megaprojects: They Are a Different Breed," *Public Roads* 68 (July/Aug. 2004): 7.
[3] *Megaprojects: Challenges and Recommended Practices*, Part II, "Risk Management and Professional Liability Issues for Engineering Consultants." 2010 ACEC.
[4] W. Edgerton E.D. "Recommended Contract Practices for Underground Construction." *Society for Mining, Metallurgy, and Exploration, Inc.*, 2008.
[5] Brierley & Hatem Design Build Subsurface Projects, Second Edition (2010).
[6] *Megaprojects: Challenges and Recommended Practices*, Part II, "Risk Management and Professional Liability Issues for Engineering Consultants." Section 4.7. 2010 ACEC; and D.J. Hatem Presentation: "Public-Private Partnerships: Opportunities and Risks for Consulting Engineers Involved in Subsurface Projects" at Geo Halifax 2009 (September 23, 2009; The Anatomy of Construction Risks: Lessons from a Millennium of PPP Experience." Standard & Poor Report, April 2007.)
[7] The Florida Department of Transportation ("FDOT") in its Contract Documents for the first major United States Public-Private Subsurface Project—Port of Miami Tunnel, utilized a technical baseline report in conjunction with a differing site condition provision to address the issue of fairness in risk allocation for differing site conditions. These contractual provisions were augmented with a Geotechnical Contingency Fund to address and fund added project costs arising out of differing subsurface conditions encountered during construction. For further discussion, see Wern-Ping Chen, Port of Miami Tunnel Update: A View from Design-Builder's Engineer, RETC Proceedings (2009) P. 687.

# Alternative Contracting and Delivery Methods

**Janette Keiser**
The Keiser Group, Poulsbo, Washington

**John J. Reilly**
John Reilly Associates International, Framingham, Massachusetts

**ABSTRACT:** Owners and contractors need better contractual and other tools to respond to issues such as diminishing competition, enhanced risk and highly complex project design and delivery associated with very large projects. Traditional contract methods, including design-bid-build and design-build rely on the principal of risk allocation as their foundation. This foundation is ill-suited to highly complex projects where the risks are high and costly, such as most underground projects. Contracts that are based on a foundation of risk sharing are better suited to manage the risks of underground construction because such contracts facilitate the collaboration that is necessary to engage in effective problem solving. The ability to use alternative contracting tools, based on risk sharing, would overcome the fatal flaws of traditional contract methods and enhance our ability to deliver complex projects on time and on budget with better results.

## TRADITIONAL CONTRACT METHODS HAVE A FATAL FLAW

In the United States, the traditional contract method used on underground public works projects, or any type of public works project for that matter, is design-bid-build. Most states require public owners to use design-bid-build as the standard practice. In the past 10–15 years states have allowed their public works agencies to experiment with non-traditional methods such as design build, general contractor-construction manager, public-private partnerships and so forth. But, design-bid-build is still the most common approach, at least on the state and local scene. The problem with design-bid-build contracts is that they depend on a system of risk allocation; that is, risks are allocated to one party or another. The enlightened principal is to allocate risk to the party that is in the best position to control or manage the risk. While this method appears to be a rational method that should result in well-reasoned outcomes, it is still risk allocation; that is, risks are owned by one party or another. That is the fatal flaw of design-bid-build contracts.

There's lots of literature about the difference between design-bid-build contracts and design-build contracts. The federal government conducted surveys of agencies that have used both methods, trying to find out if using design-build saves time or money. Many papers and books have been written about this topic and yet, the jury is still out. Some of the material concludes that design-build saves time but not money. Some conclude that design-build saves time and money but reduces quality. Some conclude that quality is improved but the cost is higher. There is no definitive conclusion because the outcomes differ from project to project because each project, particularly a complex project, such as an underground project, has too many variables from which meaningful comparisons can be drawn.

Plus, the research has missed an important point. It is difficult to make a legitimate comparison between design-bid-build and design-build because both contract mechanisms suffer from the same fatal flaw. This fatal flaw distorts the conclusions because it hides a key element of a project's success. This fatal flaw is that traditional design-bid-build and traditional design-build both depend on the traditional method of risk allocation; that is, risk is owned by one party or another. As long as this element is the foundation of the contract method, the outcomes will not really change.

## THERE IS A FUNDAMENTAL DIFFERENCE BETWEEN RISK ALLOCATION AND RISK SHARING

There are other contract mechanisms that do not rely on risk allocation. They rely on risk sharing; that is, the risks are jointly owned by all the parties. While this sounds simple, it creates a fundamental difference in the way contracts are procured, managed, and administered. With risk allocation, the party to whom the risk is allocated is responsible for managing the risk and paying for the consequences if the risk occurs. With risk sharing, both parties are responsible for managing the risk and sharing the costs if the risk occurs. Traditional contract mechanisms such as design-bid-build and design-build can be effective if the project is low risk and if the costs of the risk are reasonable because one party or another can afford to bear the risk. However, if the project is high risk and the costs of the risks are very high, the "risk allocation" method breaks down because one party cannot easily bear the costs of the risks. It is a fact of commercial reality that a party will only perform well if it is in that party's best interests to do so. This fundamental fact of human nature is true regardless what the black and white letter of the contract says. This fundamental fact is demonstrated all the time in construction contracts.

For example, typically a construction contract starts out with everyone feeling excited about getting to work and everyone is hopeful that the project will go well. The parties work together to use their technical skills to solve problems. But, at some point, there could be a breakdown. There's a problem that can't be solved without compromise. If the parties can't communicate well and aren't willing to find a compromise that is somehow mutually beneficial, the outcome will be a solution that is lopsided. It might be a good solution for one party, but the other party is going to feel like they got the short end of the stick. That party is going to feel that the out come was not in their best interests and is no longer going to feel motivated to continue communicating effectively to solve the problems that will inevitably continue to arise on the project. This scenario is a project headed for trouble. This is a typical outcome is a project that uses traditional risk allocation principals. The party to whom the risk is allocated often feels like they got the short end of the stick and is left feeling like this contract is no longer in their best interests. This happens all the time and we have seen it frequently in our own experiences. This is the natural and inevitable consequence of risk allocation.

On the other hand, with risk sharing, both parties are sharing the costs of the risk and thus, have a greater vested interest in effectively managing the risk. It is now in the best interests of both parties to continue to perform well. This is the natural and inevitable consequence of risk sharing. This is a proven, demonstrated fact in projects where principals of risk sharing are used.

Other problems with risk allocation are that (1) it happens too late in the game and (2) it is done, for all practical purposes, unilaterally. Typically, owners and/or their consultants are the ones who allocate project risk. Sometimes the contractor community is consulted and sometimes not. The contractor may not agree with the risk allocation mechanisms set forth in the contract documents, but when they place a bid and sign the contract, they have little or no opportunity to comment on, or change the manner in which the risks were allocated in the contract. This is true whether the design-bid-build or design-build methods of contracting are used. Traditionally, regardless of the contracting method, the public owner will not negotiate the risk allocation principals.

The contractor can do one of two things, pass on this contract and not bid; or submit a bid, which includes contingency factors that account for the contractor's perception of the project risks, which have been allocated to it. If the contractor guesses wrong and estimates the cost of the risks too high, the contractor can lose the bid. If the contractor is awarded the contract and discovers during the course of construction, it has guessed wrong, the contractor can lose money or go into default. With risk allocation, the risks are unilaterally assigned by the owner and this assignment is for all practical purposes part of a contract of adhesion. The contractor has little negotiating leverage to discuss or change this assignment. This is another symptom of both design-bid-build and design-build contracts. It is another fatal flaw.

With risk sharing, the risks are not merely allocated to one party or another; they are shared by both parties. The contractor's burden of predicting what the risks are and what the costs of managing the risks are, is substantially easier. This is particularly true in cases where the contractor is guaranteed to be paid its costs.

With risk sharing, there is no need for finger pointing. Decisions can be made on the basis of "what is best for the project" regardless of fault. Because of this "no fault" principal, active and transparent collaboration is crucial so that the parties work together to find the most cost effective solution that is in the best interests of the project. If both parties share the costs of the risks, both parties must be fully engaged in addressing the risks; that is, both parties must share responsibility for decision making, communications and financial accountability. These human factors become as critical to the project's successful outcome as the technical factors. Leadership that teaches, facilitates, and encourages these factors rises to the surface and is built into the fabric of the project.

## RISK SHARING FACILITATES EFFECTIVE LEADERSHIP

High risk projects require strong leadership, which can effectively resolve the many issues that inevitably arise. This need for leadership can arise in many forms. It is manifested in the manner in which the project owner addresses risk. A highly risk-adverse contract relationship sends a strong message to the industry that the owner is not interested in a collaborative relationship. This message will ultimately impact price, competition and schedule. Leadership is required to effectively manage the expectations of, and relationships with, elected officials, regulatory agencies and the public.

Leadership is key in the tender process, to balance the need for a fair public forum with the need to protect the confidentiality of proprietary information. In fact, it will be the leadership message sent by the project owner which will determine the willingness of competing contractors to submit ideas for innovation or cost savings. Leadership is required to ensure that the decision-making processes employed throughout the development and execution of the project will truly be in the best interests of the project. At the end of the day, it is leadership which determines the success or failure of the project.

Many decisions affecting large, high-risk projects, such as decisions relating to cost, schedule, permitting and risk, are dependent on the means and methods the contractor will employ to build the project. The project owner and its consultants can creatively speculate on which means and methods a contractor will use to build a project, which methods are most technically viable and which methods are most cost effective. However, these efforts, well-reasoned as they may be, are still speculative because neither the project owner nor its consultants are required to commit to building the project within a stated price and schedule. The only way to get definitive answers to the questions regarding means and methods is to have meaningful discussions with the actual company who is contractually bound to build the project. This is why on a large, high-risk project; it is not only beneficial, but crucial, that the contractor, who will commit its money and resources to building the project, be engaged as early as feasible.

With risk allocation, effective leadership depends on the personalities, training and experience of the personnel involved as well as the "chemistry" between the key players. It is a synergistic combination of factors that when it works, it is magic and when it doesn't work, it can be catastrophic; either way, it is almost accidental. With risk sharing, leadership is built into every project system as a deliberate and conscious effort and process. It has to be this way because everyone must take responsibility for effective problem solving. This is another fundamental difference between risk allocation and risk sharing.

You see this all the time in underground projects. For example, the papers in the RETC 2009 Proceedings, and every Proceeding before that, have many stories about difficult underground projects where the risks were high and something went wrong. The happy stories are the ones where the parties come together under strong leadership, and collaborate on finding a solution. Where this doesn't happen, the project often fails. You usually don't read about these stories in the Proceedings.

The risk sharing concept also affects how dispute resolution is handled. For example, traditional contract law is based on traditional rules of risk allocation. If you aren't relying on traditional rules of risk allocation, traditional premises of contract law don't always apply. This is why the parties on a risk sharing project agree to "no recourse to litigation."

## HOW DO YOU BUILD PRINCIPALS OF RISK SHARING INTO A CONTRACT?

First, the easiest way to build risk sharing into a contract is to start with a non-traditional form of contracting, such as design-build. This is because, while there are many different kinds of design-build contracts, the basis of selection is usually cost and other factors, rather than just price alone. This basis of selection introduces flexibility into the procurement and contracting process that can be highly leveraged.

There are several forms of contracts where variations of risk sharing are being used. The most prominent form is used in Australia and New Zealand, and is called the alliance method. Here in the United States, there is also a risk sharing method being used, right here in Portland, aptly known in the industry as the "Portland Method."

### The Alliance Contract

*Characteristics of Alliance Contracts*

Alliance contracts were created by the oil and gas industry for the development of offshore oil platforms in the North Sea. They have been successfully implemented on public works in Australia, and are of recent interest in the United Kingdom. Alliance contracts differ from project to project but share the following four characteristics:

- The partners to the alliance (owner and contractors) are collectively responsible for performing the work as well as owning and managing risk on the project.
- The alliance agreement contains terms which provide for the following means of compensation:
  - The contractor is paid for direct costs (project specific costs and overheads) subject to "open-book" accounting and verification;
  - The contractor is paid indirect costs (a fee for corporate overhead and normal business profit);
  - The partners share cost overruns and savings in accordance with a project-specific formula.

- The partners make decisions based on "the best interests of the project."
- The partners agree to resolve issues within the alliance with no recourse to litigation, except in limited cases, as set forth in the alliance agreement.

Collectively, these characteristics set alliances apart from design-bid-build, design-build, and general contractor-construction manager forms of contract.

An alliance contract helps achieve a greater degree of cost certainty as well as more effective risk management. This is because with an alliance contract the success or failure of the project is a joint responsibility because risk is shared. More on this later.

Alliance contracts are tendered under a negotiated, qualifications-based procurement process which is concluded by the successful negotiation of a target cost. The target cost, usually validated by an independent estimator, is intended to be a reasonable estimate of what it should take to deliver the agreed scope of work, including schedule, quality, and any other specifications of value to the owner (e.g., traffic control, safety, environment, public outreach). Agreements hinge on the expectation of risks and the willingness of each party to accept their share of the consequences of risks, including sharing in any cost savings or cost overrun which occurs when the risks materialize.

### *Alliance Contracts Have Been Successful*

The case studies from Australia demonstrate that alliance contracts generate a wide spectrum of tangible and intangible benefits, of which one of the most visible is the ability to achieve greater degree of cost certainty. Lack of cost certainty is a pervasive problem on complex public works projects. Some agencies have developed tools to help them achieve greater cost certainty. For example the Washington State Department of Transportation ("WSDOT") has developed a process, known as the Cost Estimate Validation Process ("CEVP"), to, in part, help identify (and quantify) the risks that cause a project's high cost escalation. By combining elements of statistical probability with elements of risk management, a budget developed using CEVP provides a more complete picture of a project's expected cost. Still, the cost estimates developed using CEVP are based on the principal of risk allocation. Further, the costs associated with those risks are quantified by one party, WSDOT, with input from consulting experts, but with no input from the actual contractor selected to perform the work. As a result, as careful and comprehensive as the CEVP process may be, it provides an incomplete picture of the expected cost to complete the project.

Alliance contracts help solve the cost escalation problems because in an alliance contract, both parties are engaged in the process of identifying and evaluating the cost consequences of risk. This occurs in an open dialogue which starts during the selection process and continues throughout the project, holding both parties accountable to pre-determined measures for sharing cost increases, and supporting transparency in estimating and expenditures with independent estimate verification and audits of expenditures. Further, the parties work as an integrated team, which is sure to be more cost effective than if both parties had separate but equal staffs working on similar, or identical, issues. This minimizes the transaction costs associated with the project.

Another feature of alliance contracts, which affects cost certainty, is the use of open-book accounting and independent audits. These features offer financial incentives to enhance owner and contractor performance while ensuring both that their needs will be met. In alliances, the recovery of direct costs is guaranteed irrespective of the outcome of the project. Indirect costs (any and all margins) are payable on performance relative to the target cost. Performance by the alliance that exceeds targets should lead to superior returns, while outcomes that fall short of targets should place margins at risk. There should be no circumstances that allow either party to gain at the expense of the other. As a whole, savings and overruns are shared amongst all alliance participants: target costs and actual costs are compared, and savings or overruns are distributed according to pre-determined commitments from each party to complete the scope of work.

### *Alliance Contracts Can Manage Change More Effectively*

Contracts that are based on the principal of risk allocation tend to give one party or another controlling interest over change. The way parties manage change is fundamentally affected by the principal of risk sharing. Alliance contracts make both parties commercially responsible for change, and set out organization and communication structures for the purpose of supporting the mutual definition and management of change. This structure is equally useful for facing and sharing foreseen and unforeseen risk. In alliance contracts, decision-making is a shared responsibility. As a result, alliance projects should have a better chance of maintaining commercial consensus, and a greater chance of demonstrating transparency and accountability over cost, schedule, and other performance parameters.

### *Alliance Contracts Facilitate Collaboration*

Involving the contractor in the design development process is crucial because this enables the project owner, its consultants, and the contractor, to work together to plan the project, finalize permitting efforts, and otherwise answer the many project development questions which are unanswerable until the contractor's construction methodology is known. Achieving these objectives takes active leadership. A design-build project involves the contractor in the design development process, but a design-build project that is based on the principal of risk allocation doesn't have the same vested interest in collaboration during this design process as a project based on the principal of risk sharing. For example, contracts based on risk allocation will usually give the contractor very little right to rely on the design documents developed by the owner. Further, the contracts will contain volumes of exculpatory language warning the contractor of the liability that will be attached to the contractor's design. The alliance contract, based on risk sharing, avoids this. Further, it builds effective leadership into the design development process, which lays the groundwork for encouraging innovation and managing cost. This has been the experience in Australia and New Zealand.

*Alliance Contracts Facilitate Leadership*

An alliance contract builds leadership into the fabric of the project from the beginning by focusing as much energy and attention on building an effective business culture as is spent on the technical aspects of the project. Alliance contracts employ facilitators whose mission is to ensure that the relationships are working, that the communications systems are flowing and that the decision-making processes are resulting in "best of project" decisions. All project participants undergo training on how to improve their communications and team building skills. As result, many alliance team members report much greater levels of satisfaction from working on an alliance contract than from working on a traditional contract. Under an alliance contract, attention is paid to installing effective communications and decision-making systems into the project as a whole, as well as in upgrading the communications and decision-making skills of the individual team members. An entire industry has developed in Australia to conduct conferences and training programs to address this aspect of alliance contracts. Good projects managers will work hard to include these types of features into projects using traditional means. But, the success or failure of the features is dependent on the strength of the individual's leadership style. A good project manager can make a bad contract better but a bad project manager can kill even a good contract. In alliance contracts, no single individual has that kind of impact because good leadership is built into the project systems.

*Alliance Contracts Increase Competitiveness*

The decision to use an alliance contract affects the competitive field of contractors who would respond to the project's tender documents by demonstrating the project owner is committed to effective project execution. This is important because today's mega-projects, due to their sheer size, raise the potential for a limited field of competition, regardless of contract packaging, construction means/methods, or any other parameter. In all likelihood, a winning contractor to a high-risk mega-project is a joint venture of national or global players. Project owners need to employ an aggressive marketing and educational campaign to reach these players and attract their attention to their procurements. Further, the quality of the winning contractor team will be directly proportional to the extent to which the contracting community believes the project owner will be a fair, reasonable owner who is truly interested in developing and executing a meaningful collaboration in the project's development, equitable risk sharing, and effective decision-making. The project owner needs to demonstrate its commitment to these values in order to attract the best contractor team. Choosing the alliance contract method helps demonstrate this commitment. Initially, the contractors in Australia were skeptical about alliance contracts. But, as they saw the results, they became believers. Now, the contracting industry actively supports, and is an advocate for, the use of alliance contracts.

**The "Portland Method"**

There's been a lot of buzz in the industry over the past few years about what's going on in Portland. As almost everyone in the industry knows, the City of Portland's Bureau of Environmental Services successfully built one project and is in the process of building a second using a contract mechanism, authorized by Oregon statutes, called the Construction Manager General Contractor method. The results have been impressive.

The West Side tunnel was completed on schedule. The price at bid time was $293 million and escalated to $308 million at completion, with no contingency used. The East Side is now over seventy per cent complete, ahead of schedule and under budget, with very little contingency used. These were not easy projects. Both projects faced a number of challenges that would have normally caused serious impacts to the budgets and schedule. For example, on one of the contracts they had severe problems building a pump station shaft, comprised of one of the world largest slurry walls, and they experienced over 175 interventions under difficult conditions for repair/maintenance of the TBMs.

What made these contracts different? The contracts were based on a foundation of risk sharing. For example, the contractor was guaranteed to be paid its costs. The owner agreed to pay all of the direct costs for craft labor, equipment and materials, at cost, and to pay the costs of overhead, profit and salaried employees under a fixed fee. This is a mechanism of risk sharing. It allowed the owner to attract highly qualified, world-class tunnel contractors during the procurement process and empowered the project team to focus on problem solving rather than finger pointing. Each party to the contract was able to focus on what was important to them. For example, the owner maintained control over project controls for purposes of project documentation and audit. Plus, the owner received 100% benefit of any cost reduction. The contractor maintained control over construction means and methods. For example, the contractor was able to modify the originally very conservative settlement mitigation measures as they got more experienced with the ground.

Some of the pertinent contract provisions, reflecting the risk sharing approach include:

a. Team building. Contractor is required to *"Facilitate team building with owner representatives and other construction team participants. Recommend methods of conflict resolution and dispute avoidance."*
b. Liquidated damages
   i. There's a 2-tiered liquidated damage process. $13,115 per day before Dec 1, 2011 and $36,098 after Dec 1, 2011
   ii. There's a cap of $10,000,000 on liquidated damages
c. Contractor is entitled to extra fees for:
   i. Extra work
   ii. Differing site conditions
   iii. Excusable delay
   iv. If owner suspends the work, through no fault of the contractor
   v. If a subcontractor performs extra work
d. Contract payments:
   i. Owner pays prime twice a month
   ii. Prime is required to pay subcontractors twice a month
e. Dispute resolution process:
   i. Negotiation at working level and then referred to persons with authority to resolve the matter. These folks shall set a meeting to try to resolve the dispute.
   ii. Mandatory mediation
   iii. Voluntary final arbitration, using AAA's Large Complex Construction Cases rules and panelists or litigation
f. Costs of materials testing and compliance documentation:
   i. These are reimbursable costs

Many of the provisions of the Portland contract are fairly traditional, but the differences identified above took away the contractor's burden to guess about risk and cost. This transformed the contract into a risk sharing contract and has made all the difference.

## CONCLUSION

The principal of risk allocation is a fatal flaw in tradition public works contracts, whether design-bid-build or design-build. This is because the principal of risk allocation can reduce a party's motivation to perform well on a project because it's not always in their best commercial interests to do so. The principal of risk sharing, on the other hand, increases a party's motivation to perform well on a project because managing the costs associated with risks are shared between the parties. Contracts based on the principal of risk sharing experience better problem solving through more effective leadership, greater levels of cost certainty, and increased levels of project satisfaction. Examples of contracts founded on the basis of risk sharing include the alliance contract method used in Australia and New Zealand and the so-called "Portland Method" used in Portland, Oregon.

# Design of the Waller Creek Tunnel, Austin, Texas

**Tom Saczynski, Nieves Alfaro**
Kellogg Brown and Root, Austin, Texas

**Prakash Donde**
Jenny Engineering Corporation, Springfield, New Jersey

ABSTRACT: The main tunnel portion of Waller Creek Tunnel (WCT) project in Austin, Texas, is in the last stages of final design. This paper focuses on the Main Tunnel component of the project. It specifically addresses the design of the initial support and final lining that resulted from the predicted behavior of the rock mass surrounding the excavated opening. The discussion centers on the more critical geologic features of the ground through which the contractor will drive a large diameter tunnel, up to 9.1 m (30 ft) excavated.

## INTRODUCTION

Situated below Waller Creek, the Waller Creek Tunnel (WCT) will be 1,626 m (5,335 ft) long and located between and roughly parallel to I-35 and Red River Street (Figure 1). The project in Austin, Texas, is in the last stages of final design. Preliminary engineering and planning for the project were presented at the 2009 RETC in Las Vegas (Jackson, Evans, Saczynski, Jewell, & Donde, 2009).

The purpose of the project is to capture a 100-year flood event and safely divert it to Lady Bird Lake, improve the water quality in Waller Creek, and improve recreational opportunities along and near the Creek. For more detail, see the RETC paper (Jackson, et al, 2009).

The main tunnel is one of four separate contract packages that comprise the project. The other three components, that is, the Inlet Structure, Outlet Structure and 8th Street Shaft, are not included in this paper as the designs are less advanced at this time. This paper addresses the main tunnel contract only, which includes the 4th Street shaft.

The lined tunnel will consist of three reaches, each with a different diameter to accommodate additional flows from inlet shafts at 8th Street and 4th Street. There are also two segments comprising smooth transitions between the differing diameters, as well as a connecting tunnel from the access shaft at 4th Street to the main tunnel. The tunnel components and dimensions of the tunnel are shown in Table 1.

The tunneling community recognizes Austin limestone as excellent ground for tunneling as the relatively low compressive strength of the rock type of approximately 10 mPa, (1,500 psi) affords high production. The reputation is, however, based on relatively the relatively small diameters of the tunnel openings. Because of the larger tunnel diameters of the WCT project, geologic considerations are more influential. The discussion turns to the effects of geology on the design.

## DESIGN CHALLENGES

### Site Geology

WCT will be constructed through sedimentary rocks of the Austin Group, Atco Formation Limestone (AFL) and the Eagle Ford Shale (EFS). The project is located in the Balcones Fault Zone, a belt of inactive faults, which trends generally southwest to northeast through central Austin. Several small displacement faults have been identified in the project area and one large displacement fault traverse the alignment. These faults commonly include a series of fault breaks in stair stepped "echelon" formation fault zones. The large displacement fault zone located near 11th Street, and the several small displacement faults to the south of 1st Street, have created conditions which will result in excavation through contacts between limestone and shale at certain locations. Several other faults cause offsets within the limestone.

**Table 1. Variations in tunnel diameter and reach length**

| Structural Feature | Beginning Station | Reach | Excavated Diameter | Finished Diameter | Increment Length |
|---|---|---|---|---|---|
| Outlet to 4th St. | 1+65 | 1 | 9.1 m (30 ft) | 8.1 m (26.5 ft) | 687 m (2,255 ft) |
| Transition | 22+40 | — | 9.1 m to 7.5 m (30 ft to 24.5 ft) | 8.1 m to 6.9 m (26.5 ft to 22.5 ft) | 30 m (100 ft) |
| Connecting Tunnel | 22+84 | — | 5.8 m (19 ft) | 5.2 m (17 ft) | 66 m (215 ft) |
| 4th St. to 8th St. | 25+20 | 2 | 7.5 m (25 ft) | 6.9 m (23 ft) | 438 m (1,436 ft) |
| Transition | 38+56 | — | 7.5 m to 6.9 m (24.5 ft to 22.5 ft) | 6.9 m to 6.2 m (22.5 ft to 20.5 ft) | 30 m (100 ft) |
| 8th St. to Inlet | 39+56 | 3 | 6.9 m (22.5 ft) | 6.2 m (20.5 ft) | 440 m (1,444 ft) |
| End of Contract | 55+00 | — | — | — | — |

The AFL consists of light gray to white chalk, marly limestone and limestone. The limestone varies from hard-fine grained limestone to chalky and clayey limestone. The degree of weathering ranges from the gray to white un-weathered intact limestone to tan slightly weathered limestone.

The Eagle Ford formation, composed of shale and clayshale members, consist of dark gray calcareous clay with sandy and silty "flaggy" (thin-bedded) limestone layers. The formation includes four members, from top to bottom: 1) South Bosque, 2) Bouldin Flags, 3) Cloice and 4) Pepper shale.

*Discontinuities*

The design considered rock joints and fractures in the rock mass, which are generally closed and tight but typically not "healed" by mineralization. Joint surfaces are commonly clay coated with slickensides frequently observed. Also included in the analysis were the many faults found by seismic refraction/reflection (Figure 2).

Fault breaks with small off-sets of less than 3 m (10 feet) have generally been noted as tight with narrow fracture zones, with many as low as a few cm (inches). These features have not significantly impacted tunnel construction when encountered in past tunnel projects. Faults with larger off-sets are associated with wider fault zones including series of fault breaks. Voids have been observed in open excavations on other projects. Faut gouge has been encountered in the downtown Austin area in association with the faulting, but without significant impact on tunnel construction.

The geologic longitudinal section of the project (Figure 3) shows smaller offsets about 1 m (3 ft) or less towards the Outlet shaft. The true dip is typically about 60 degrees with a range of 45 to 75 degrees towards the east. The strike, however, is roughly parallel to the tunnel alignment from station 8+00 to +20+00.

Larger offsets of 3 to 5 m (10 to 15 ft) are present closer the Inlet. Fortunately, the fault strikes are more normal to the alignment. Unfortunately, borings have not penetrated any of the fault zones, thereby rendering the character of the faults as largely unknown. At the time of this writing, additional investigation to determine the character of the faults is under consideration.

*Analysis*

In preparing the design for the project, the design team considered three systems to predict the behavior of the rock mass as the tunnel opening is excavated and supported: the Rock Quality Designation (RQD; Deere & Deere, 1988); the Rock Mass Rating system (RMR; Bieniawski, 1989); and the Q system (Q; Barton, Lien & Lunde, 1974). RQD is a critical component of the RMR and Q systems, but by itself it is valuable in envisioning the beam action in horizontally laminated rock. Of course, one cannot represent the rock mass as a continuum of structural beams as other discontinuities (e.g., faults and joints) intersect the lamina, thereby creating additional planes of weakness that influence the so-called beam action. Furthermore, the intact strength of the rock, water and stress regimes are also significant. The RMR and Q systems incorporate all of these: RQD, rock strength, joint spacing and conditions, water, and stress.

Obviously, the rock mass under consideration is not exposed until the opening is excavated. Although the investigation entailed a considerable number of holes (geotechnical exploratory borings) into the rock, sampled and tested them, the samples represent only a tiny fraction of the volume to be excavated. Additionally, the aggregate of the borings cannot give any valuable information concerning the pervasiveness of the discontinuities (e.g., are the joints continuous and how are they spaced?).

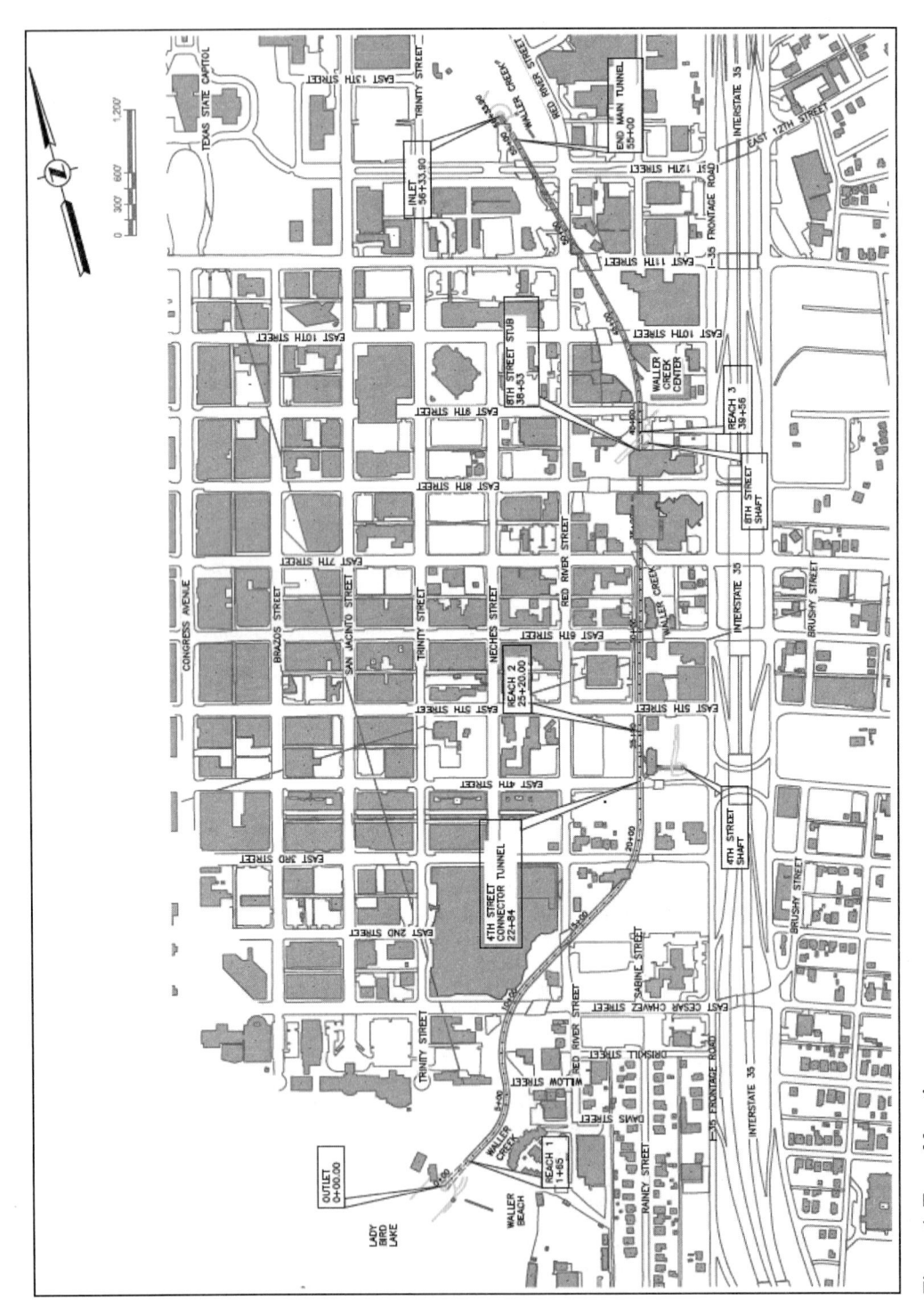

Figure 1. Tunnel location

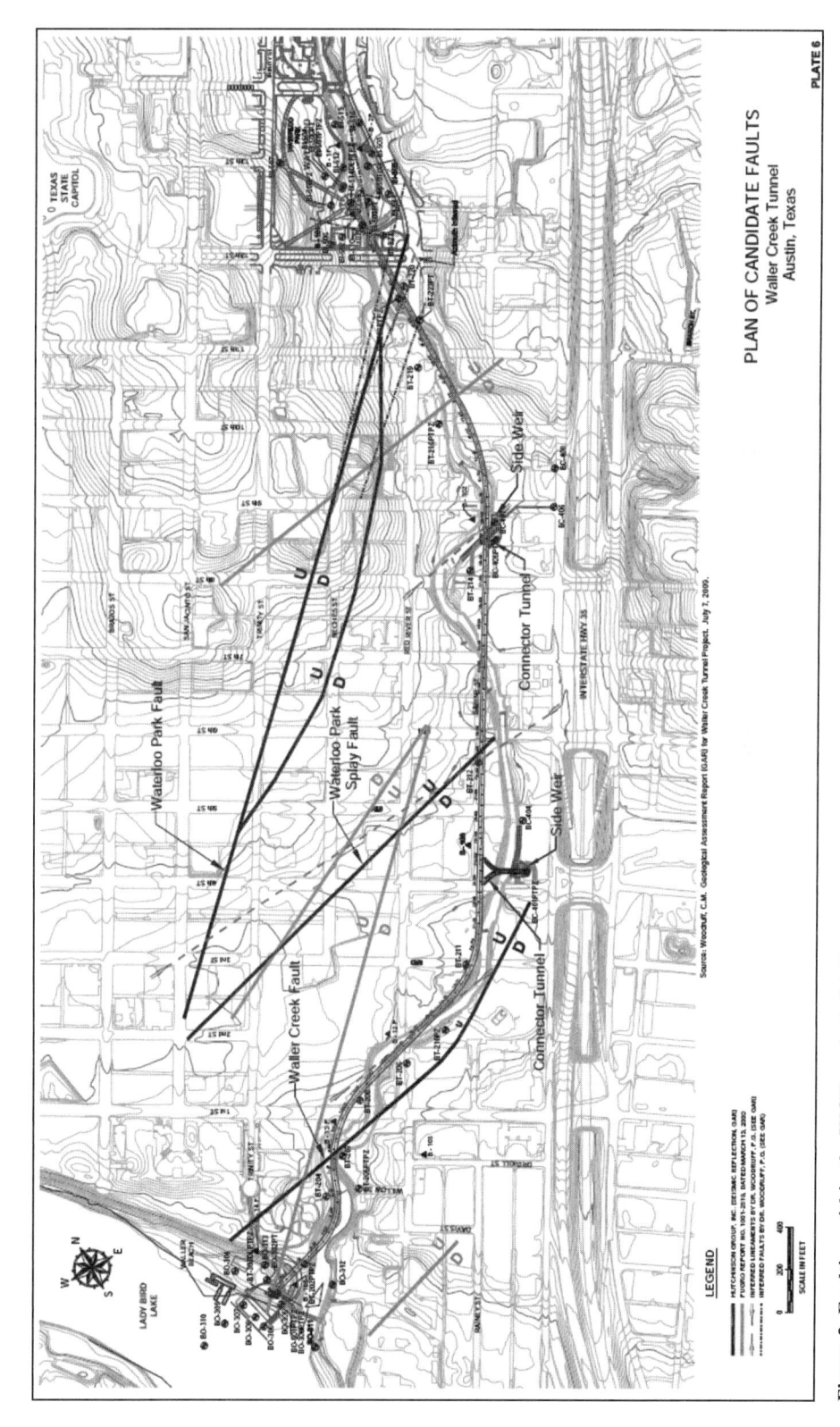

Figure 2. Faulting within the WCT project vicinity

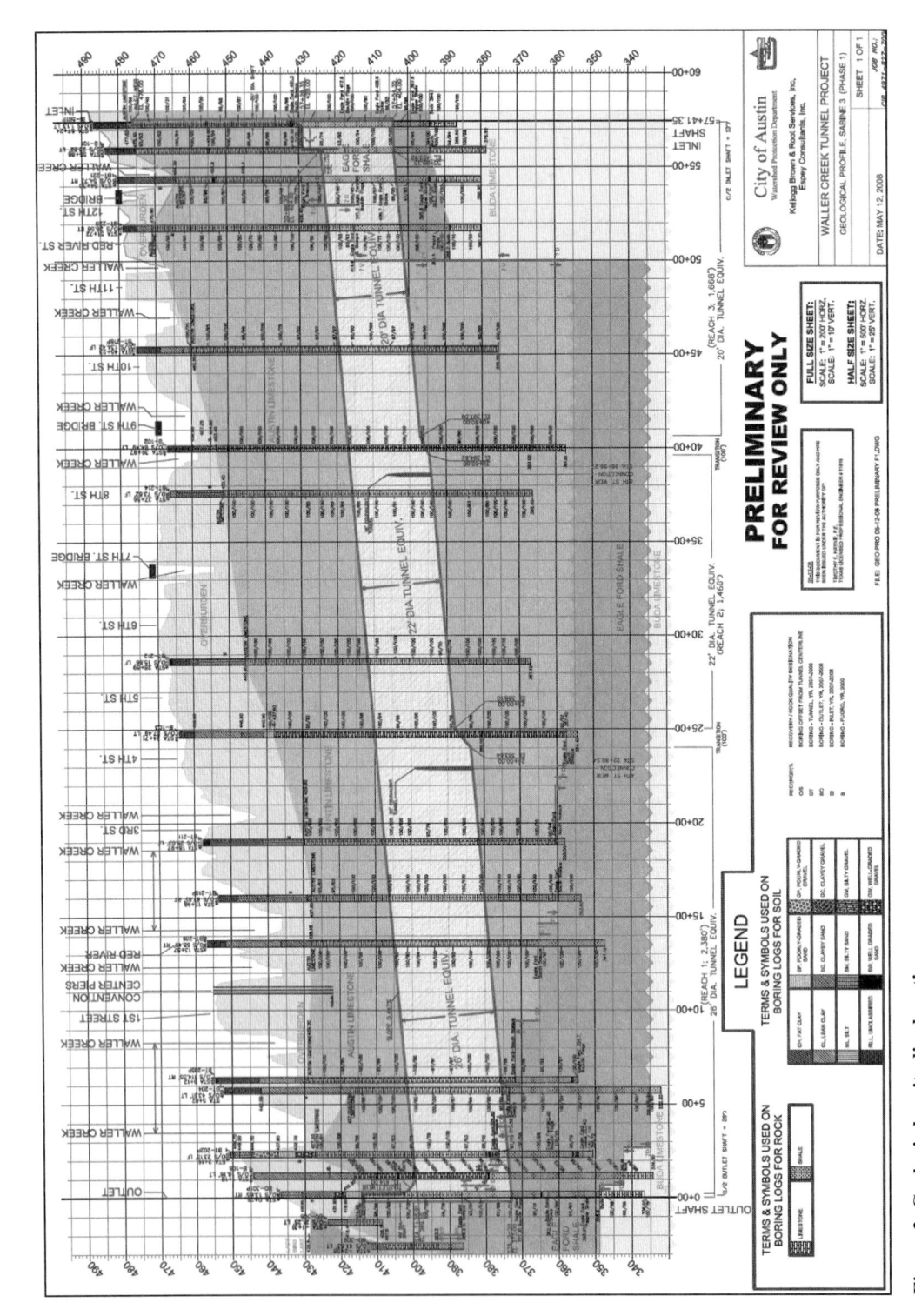

Figure 3. Geologic longitudinal section

Geotechnical characterization for tunnels often relies upon the rock mass that exposed at the surface. Unfortunately, the amount of exposed rock over this tunnel alignment is scant. Observations of the limestone exposed in the Waller Creek bed and sidewalls suggest massiveness. However, information recently developed by the team's geotechnical consultants (Holt Engineering and Fugro) suggests otherwise, as follows:

> The proposed tunnel alignment…between 9th and 10th Streets shows some evidences of multiple small-displacement faults and joints. Based on review of boring logs for nearby existing structures, observations of exposed massive limestone in Waller Creek bed south of 10th Street, and previous observations of the excavation for the existing condominium structure located at 901 Red River, the limestone in this sub-reach appears to have multiple small displacement local faults, inclined at roughly 45 degrees, with displacements of 1 ft or less, yet slickensides are very pronounced and many fracture planes are curved (and intersect). It is suggested that the limestone within Reach 3a may be influenced by the nearby fault displacement within Reach 3b coupled with local bedrock that is less indurated. This seemingly friable and less durable rock appears to have formed slickensides with minimal stress.

The conditions and spacing of discontinuities remain difficult to define. Nonetheless, to determine the RMR and Q ratings, the team has relied on the ages-old practice of rendering assumptions based on engineering judgment. The ratings displayed thus represent the best judgment of the team.

Table 2 shows statistical parameters for RMR, RQD and Q for both AFL and EFS. Upon examination of the table, one see readily observe that the dispersion is considerably greater for Q than for RMR and RQD. Q ratings are not consistent, as the table suggests by the coefficient of variation, V. The following statistical are show for the project, in which:

$V = \sigma/\mu$, in which
$\mu$ = the mean of the sample and
$\sigma$ = the standard deviation

The project team considered the Q values to be unreliable reliable, so that Q was not used as a *quantitative* determinant in the design. Notwithstanding such a finding, the aggregate of Q values indicates *qualitatively* that the interaction of the relatively weak rock, large clear span, joint water, and plasticity

**Table 2. Rock mass parameters statistics**

| Statistic | RQD (%) | RMR | Q |
|---|---|---|---|
| $\mu$ | 90 | 57 | 10 |
| $\sigma$ | 16 | 12 | 13 |
| N | 234 | 226 | 226 |
| V | 17% | 22% | 80% |

Notes:
1. The variability for both AFL and EFS are quite similar, so they are grouped as a whole.
2. Because Q has a semi-logarithmic base, $V = 10^{\log(\sigma/\mu)}$.
3. N = sample size.

of the shale render the rock mass less stable than initial impressions would suggest.

*Swelling Rock*

The north and south ends of the Waller creek tunnel will be constructed partially or entirely in the Eagle Ford Shale. The total combined length of the tunnel passing through the shale in those two reaches is approximately 700 feet.

Certain members of the Eagle Ford Shale exhibit significant swelling potential (ISRM, 1989, 1994) when exposed to water. To determine those characteristic properties of the shale stratum, sixteen (16) swell pressure tests were performed on samples taken from the stratum within those reaches of the proposed tunnel. Swell pressures in the tested samples ranged approximately from a low of 1 to a high of 70 kPa (1 to 70 ksf), averaging about 16 kPa (16 ksf).

The design team decided to allow for the shale to expand to some degree, otherwise the size of support would be too large to install. A ground support system was, therefore, developed to support the ground in those reaches in shale to allow for to expansion. Such a design requires the supports to resist a portion of the swell pressure rather than the entire theoretical load.

Thus an initial support system design for the shale reaches consists of W8 × 31 ribs at 3.05 m (4 ft) on center, with special yielding rib joints. When exposed to high swelling pressures, the yielding rib joints will be able to contract between bolted plates. Once full swelling has taken place, miners will tighten the bolts in the yielding joint. The ribs can then carry the rock loads without having to resist the entire theoretical swell pressure.

Furthermore, to reduce the potential for swell pressure, the design prescribes installing an immediate flash coat of shotcrete upon excavation, thereby limiting the exposure of the shale to air and moisture. The cycle then requires rib installation with an additional 5 inches of shotcrete initial lining between

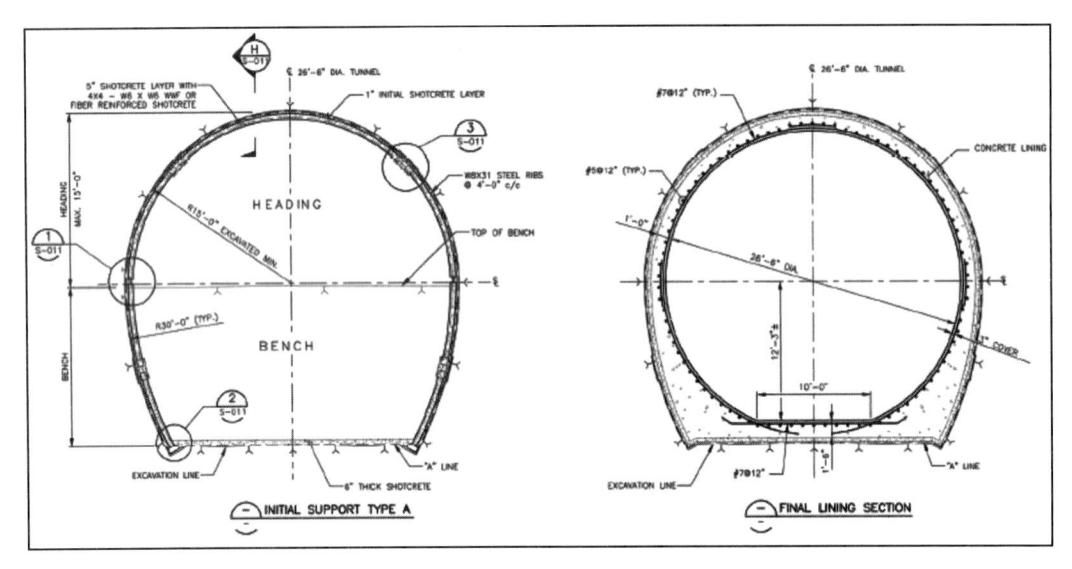

**Figure 4. Initial support and final lining for Type A ground**

the ribs. The steel rib support will be monitored periodically for contractions at the joints to determine swelling movements of the shale. It is anticipated that after a period of time the swelling of shale will subside for the lack of water as well as the restraint. Further adjustment of the support joints will not therefore be necessary.

In addition to swell potential, Eagle Ford shale is also susceptible to slake. As the rock mass dries out and disintegrates, joints and bedding planes become wider and stability problems develop. Therefore rapid application of shotcrete is necessary to preclude slaking, as well as to provide immediate support of the opening.

*Groundwater*

Based on the investigation and analysis to date, the rock would appear to be *tight*, as with experience on other tunnel projects in Austin. Packer tests indicate very low permeability of the rock mass. Additionally, although there are some faults along the alignment, experience has demonstrated that faulting in the area does not transmit water.

A challenge stems from the fact that, in spite of the exploration, the design team recognizes the heterogeneity of the rock masses, especially with respect to the jointing and faults. For example, even though one can envision flow through shale as highly unlikely, two of the piezometers at the Inlet Structure site show fairly rapid recharge rates. There is a concentration of faults around the Inlet Structure, which could contribute potential for groundwater inflow at this end of the project.

The team used field testing and geotechnical analysis to arrive at predicted inflow values. The analysis also led to the determination initial transient flows when the ground is first opened (i.e., flush flow) would be seven times the estimated inflow for the steady state condition.

The successful bidder will be required to treat all water that emanates from the worksite, prior to discharge. They will have to determine how much water they can accommodate without disruption of operations, as well as the quantity that requires treatment.

The Geotechnical Baseline Report (GBR) will contain values of baseline inflow rate to the tunnel. Additionally, in view of the uncertainties involved, the contract will include bid items for grouting: linear feet drilled, hookups made, and bags mixed. The bidders will determine the balance between the extent of water treatment and grouting quantities.

**Tunnel Design**

The differing characteristics of the rock masses and tunnel dimensions form the basis for seven different categories of ground and support types for the main tunnel. The narrative continues with elaborations on the more pertinent characteristics and challenges that will affect the excavation and stability of each type.

*Types A and B*

Types A and B extend from the open cut construction for the transition structure of the outlet facility at Station 1+65 to Station 12+50. The tunnel envelope is in limestone except for up to about 1 m (3 ft) of

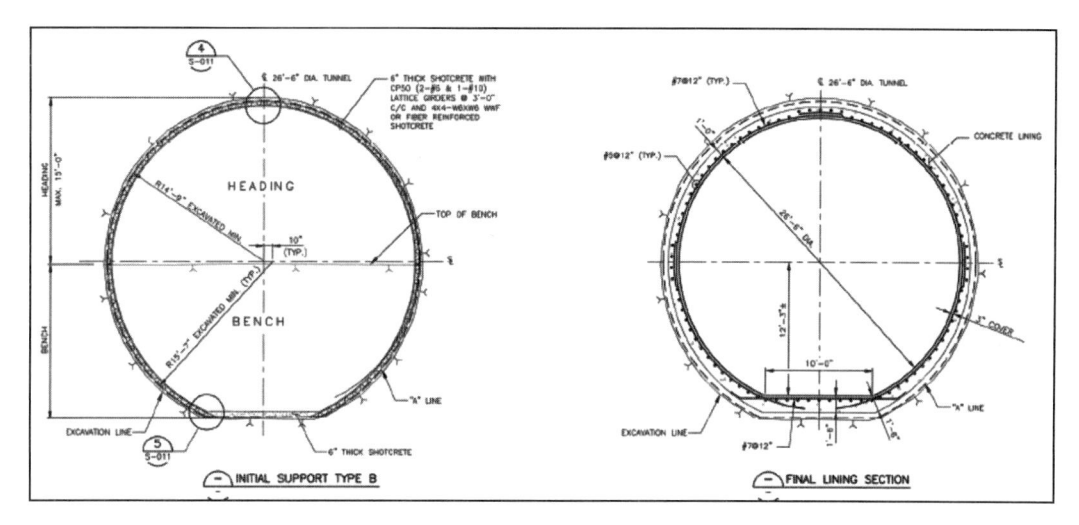

**Figure 5. Initial support and final lining for Type B ground**

invert in shale. The close proximity of the Lady Bird Lake provides potential for direct recharge into the overburden and rock discontinuities above the tunnel. Additionally, the rock mass includes a series of faults. Further, the rock cover is only 0.43 diameters in thickness. Nonetheless, RMR is Fair and RQD is Good with groundwater inflow estimated at 0.6 l/m flow (0.5 gpm/ft).

Type A ground requires the steel sets (W8 × 31 ribs with yielding joints) and 150 mm (6 in) of shotcrete. Figure 4 shows the initial support and final lining for Type A ground.

Type B ground requires CP50 lattice girders with 150 mm (6 in) of shotcrete. Figure 5 shows the initial support and final lining for Type B ground.

### *Type C Ground*

Type C ground begins at station 12+50, includes Junction 1 (Station 22+70 to 23+30).and runs to station 24+20. The tunnel envelope is entirely within limestone, which includes a series of faults. The shale is about 0.7 m (2 ft) below the invert. The RMR is Fair, RQD is Excellent and the estimated groundwater inflow is 0.6 l/m flow (0.5 gpm/ft). The rock cover is about 0.75 diameters.

Initial support for Type C ground consists of #9 rock dowels with 75 mm (3 in) of shotcrete. Figure 6 shows the initial support and final lining for Type C ground.

### *Type D Ground*

Type D ground comprises all of Reach 2, station 24+20 to 40+56 and contains Junction 2 (station 38+50 to 39+00). The tunnel envelope is entirely within limestone with RMR and RQD rated as Good.

The rock cover over the tunnel is about 1.1 diameters and the approximated groundwater inflow is 0.2 l/m (0.2 gpm/ft) with no flush flow. Initial support and final lining is similar to Type C ground support, as shown in Figure 7.

### *Type E Ground*

Type E ground extends from station 40+56 to 49+50 and is comprised entirely of limestone. The tunnel excavation conditions will consist of blocky and seamy limestone. The excavation may encounter faulted limestone, which might render the rock mass as very blocky and seamy in localized zones. The rock mass is good (RMR and RQD are Good). The rock cover over the tunnel is approximately 1.5 diameters. Groundwater inflow is estimated to be less than 1 l/m (1 gpm/ft). The ground support for Type E is similar to Type D ground support (see Figure 8).

### *Type F Ground*

Type F Ground extends from Station 49+50 to Station 50+75. There is a fault plane with large offset of about 3 m (10 ft) located between stations 49+50 and 51+50. Before the fault is encountered the tunnel excavation will be entirely within the limestone. Beyond the fault, however, the limestone comprises the upper half and shale the lower half.

Tunnel construction in shale will require special precautions including immediate shotcrete application to prevent slaking and deterioration of the shale, as well as to provide a workable traffic surface. The shale is also most likely to stick to the equipment and cause increased handling difficulties in comparison to limestone. Excavated fragments of shale will also

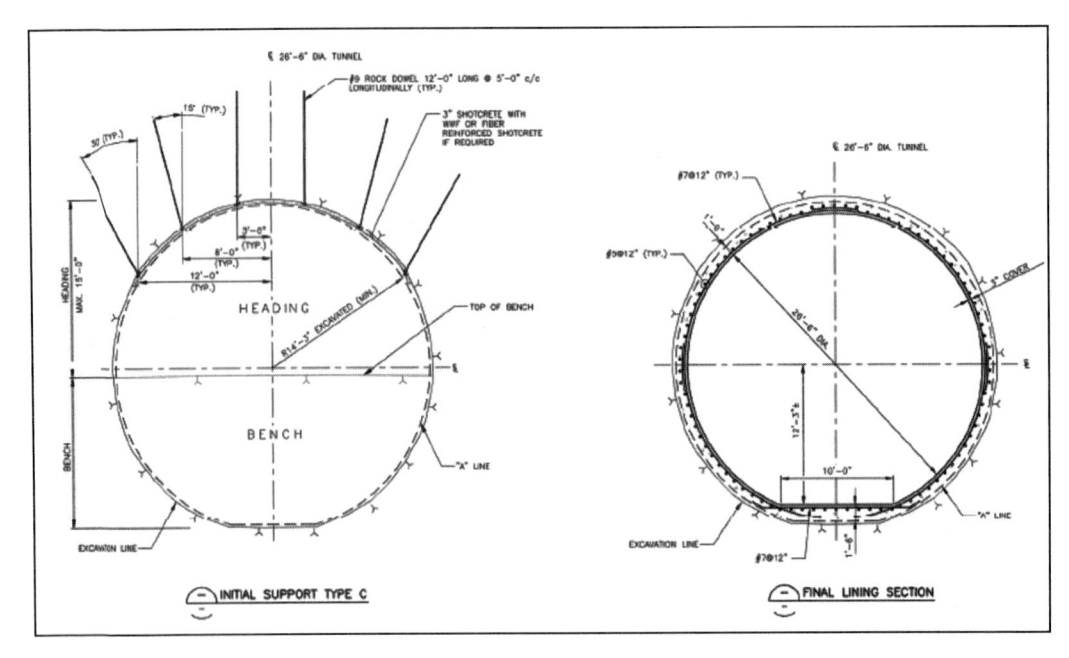

**Figure 6.** Initial support and final lining for Type C ground

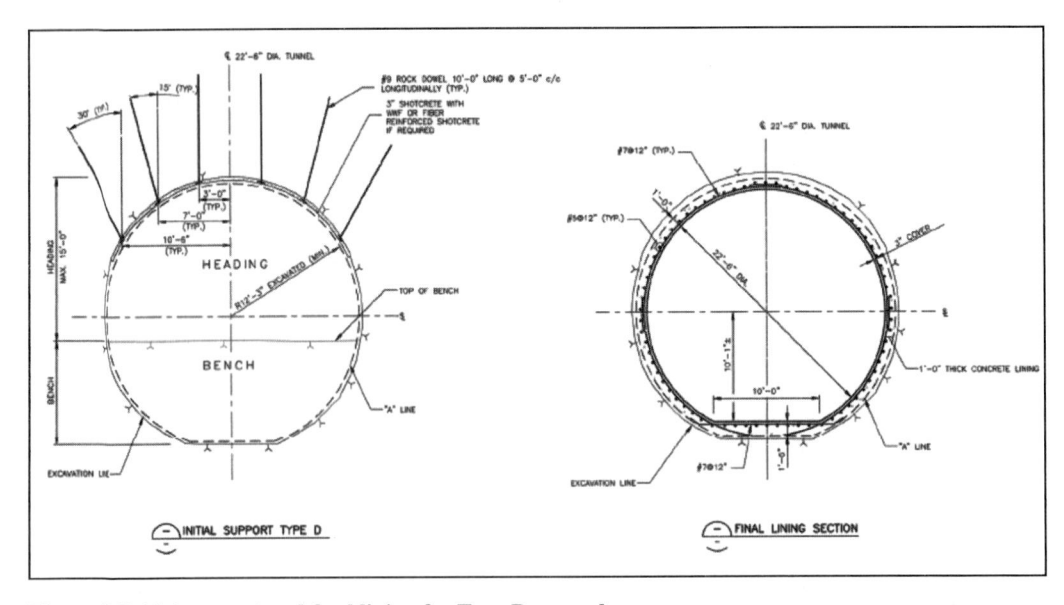

**Figure 7.** Initial support and final lining for Type D ground

increase the sediment content in the tunnel drainage water, which will require additional sediment control measures before discharge.

Groundwater inflows of about 3.5 l/m (3.0 gpm/ft) are expected in the faulted areas. The inflow for the unfaulted zones should be less than 1 l/m (1 gpm/ft). Ground support for Type F is similar to Type B, that is, CP50 lattice girders with 150 mm (6 in) of shotcrete, as shown in Figure 9.

### Type G Ground

Type G ground extends from Station 50+75, which is interpreted as the north limit of the major fault zone, to the open cut construction for the transition

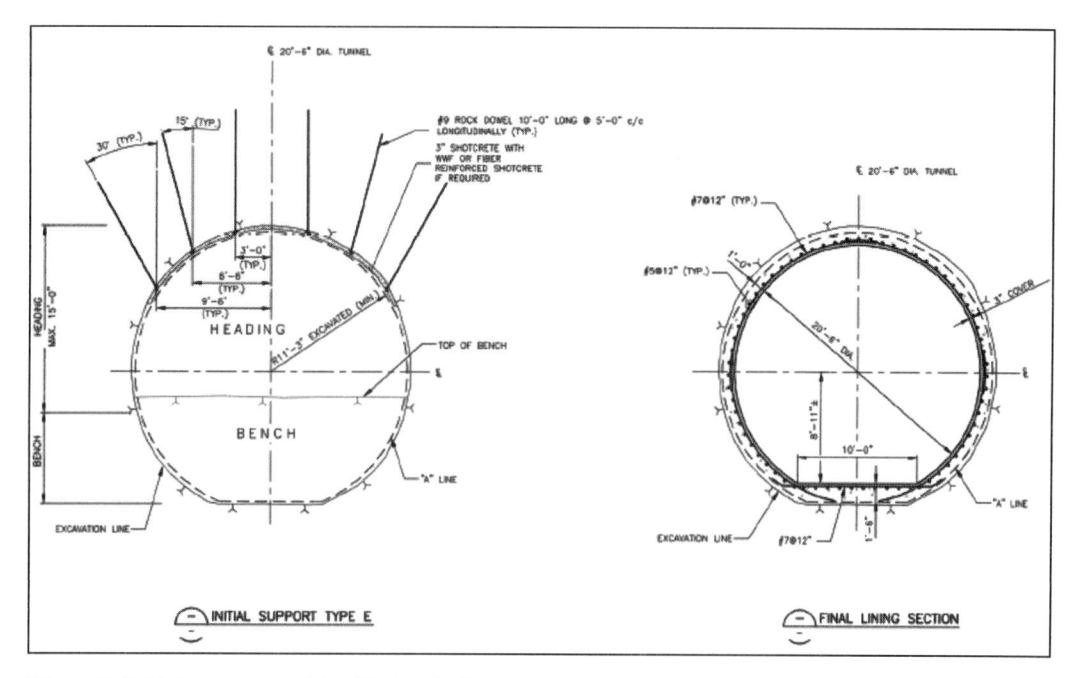

**Figure 8. Initial support and final lining for Type E ground**

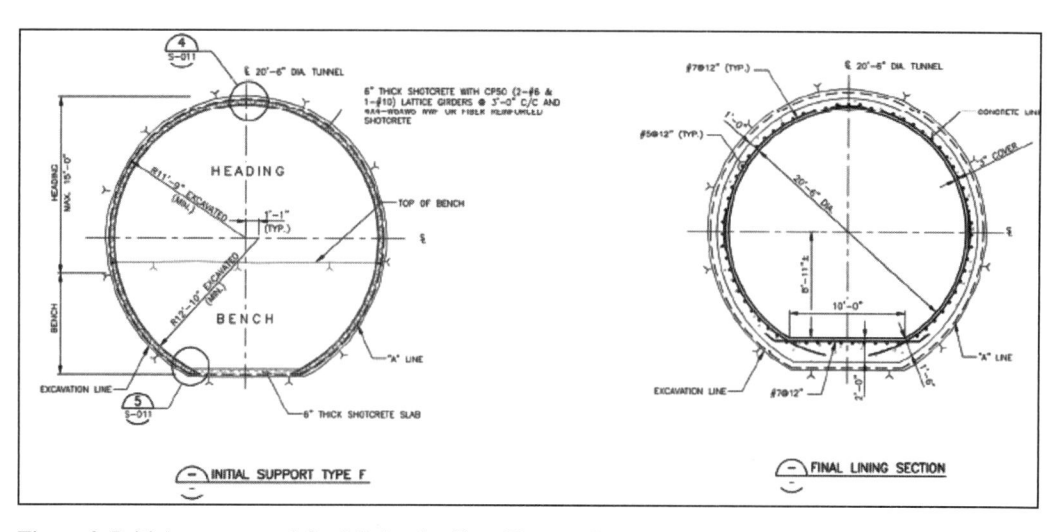

**Figure 9. Initial support and final lining for Type F ground**

structure of the inlet facility at Station 55+00. There is a fault plane offset of about 3 m (10 ft) located between stations 52+00 and 54+00, after which the tunnel envelope will be entirely within the shale. The RMR is Fair, in spite of Excellent RQD. The rock cover over the tunnel is about 1.7 diameters.

Groundwater inflows of 3.5 l/m (3.0 gpm/ft) are expected in the faulted areas. The inflow for the unfaulted zones should be less than 1 l/m (1 gpm/ft).

Ground support for Type G is similar to Type A, that is, steel sets with 150 mm (6 in) of shotcrete, as shown in Figure 10.

The ground types and more pertinent parameters are summarized in Table 3.

## SUMMARY AND CONCLUSIONS

The main tunnel contract of the Waller Creek Tunnel project will be 1,626 m (5,335 ft) long with variable

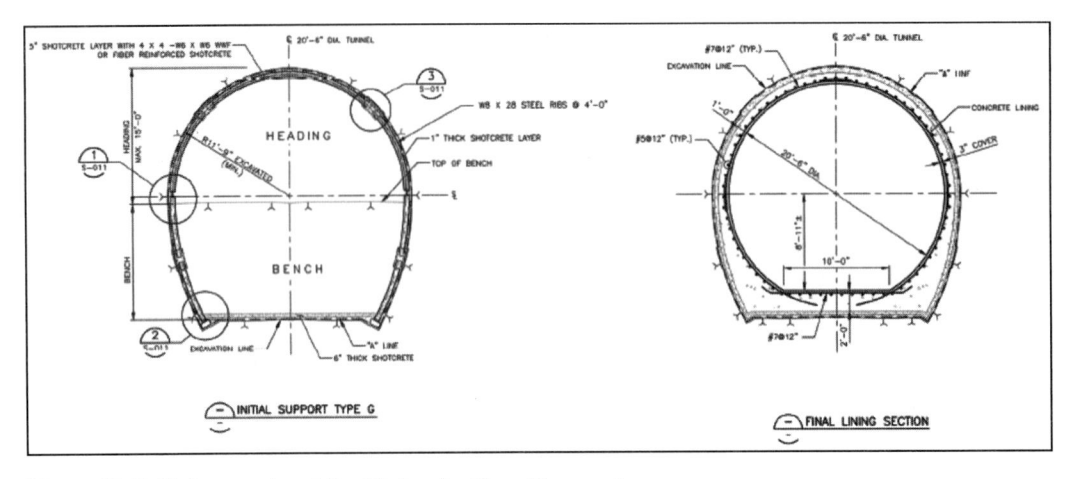

Figure 10. Initial support and final lining for Type G ground

Table 3. Summary of ground types, rock mass quality, and groundwater inflow

| Type | Segment Length (ft) | Station Limits From | Station Limits To | RMR Rating | RMR Quality | RQD Rating | RQD Quality | Groundwater Inflow Flush | Groundwater Inflow Steady State |
|---|---|---|---|---|---|---|---|---|---|
| A & B | 331 m (1,085 ft) | 1+65 | 12+50 | 54 | Fair | 90 | Exc. | 3.5 l/m (3.0 gpm/ft) | 0.6 l/m (0.5 gpm/ft) |
| C | 357 m (1,170 ft) | 12+50 | 24+20 | 55 | Fair | 94 | Exc. | 3.5 l/m (3.0 gpm/ft) | 0.6 l/m (0.5 gpm/ft) |
| D | 499 m (1,636 ft) | 24+20 | 40+56 | 65 | Good | 88 | Good | 0.20 l/m (0.20 gpm/ft) | 0.20 l/m (0.20 gpm/ft) |
| E | 272 m (894 ft) | 40+56 | 49+50 | 64 | Good | 89 | Good | 0.7 l/m (0.6 gpm/ft) | 0.3 l/m (0.3 gpm/ft) |
| F[2] | 38 m (125 ft) | 49+50 | 50+75 | 53 | Fair | 81 | Good | 6.9 l/m (6.0 gpm/ft) | 3.5 l/m (3.00 gpm/ft) |
| G[2] | 130 m (425 ft) | 50+75 | 55+00 | 50 | Fair | 90 | Exc. | 6.9 l/m (6.0 gpm/ft) | 3.5 l/m (3.00 gpm/ft) |

diameters of 9.1 m (30.0 ft), 7.5 m (24.5 ft), and 6.9 m (22.5 ft). The rock mass surrounding the tunnel will be the Atco Formation (Limestone), but the north and south ends of the main tunnel will be located within Eagle Ford Shale.

Although the RQD values computed of the limestone and shale showed essentially "good to excellent" rock, the RMR values consistently reflected lower quality, that is, "fair to good." The downgrade in apparent quality suggested by RQD owes to the presence of faults and joints with detrimental characteristics. Furthermore, swell tests on shale samples resulted in relatively high swelling pressures.

Although taking advantage of some of the qualities that contributed to reputation of the limestone as the ideal tunneling environment, WCT presents some challenges. The relatively large diameters coupled with the less than optimal rock characteristics generate potential for some difficulties that can impact stability of the opening as well as production. Such hindrance is certainly not insurmountable.

REFERENCES

Barton, N., Løset, F., Lien, R. and Lunde, J. 1980. Application of the Q-system in design decisions. In *Subsurface space,* (ed. M. Bergman) 2, 553–561. New York: Pergamon.

Bieniawski, Z.T. 1989. *Engineering rock mass classifications.* New York: Wiley.

Deere, D.U. & Deere, D.W. 1988. The rock quality designation (RQD) index in practice. In *Rock classification systems for engineering purposes,* (ed. L. Kirkaldie), ASTM Special Publication 984, 91–101. Philadelphia: Am. Soc. Test. Mat.

ISRM Commission on Swelling Rock 1994. Comments and Recommendations on Design and Analysis Procedures for Structures in Argillaceous Swelling Rock. *Int. J. Rock Mech. Min. Sci. & Geomech. Ahstr.* 31(5), pp. 535–546. London: Pergamon.

ISRM Commission on Swelling Rock 1989. Suggested Methods for Laboratory Testing of Argillaceous Swelling Rock. *Int. J. Rock Mech. Min. Sci. & Geomech. Abstr,* 26(5), pp. 415–426. London: Pergamon.

Jackson, G., Evans, S., Saczynski, T. Jewell, P. & Donde, P. 2009. Planning and Design Features of the Waller Creek Tunnel, Austin, Texas.. In *2009 Rapid Excavation and Tunneling Conference, Proceedings*. Edited by J. Almeraris, B. Mariucci. Littleton, CO: SME.

# Tunneling as a PPP-Project: Risks from the Viewpoint of the Insurer on a Case Study of a Tunnel Collapse

**Christian Wawrzyniak**
CDM Consult GmbH, Germany

**Winfried Luig, Achim Dohmen**
Zurich Insurance, Germany

**ABSTRACT:** With regard to the new construction of big infrastructure projects more and more tunneling projects are carried out based on functional tenders or as construction projects according to the model of Public Private Partnership (PPP). The special risks are presented from the insurer's point of view using the example of a tunnel collapse. A number of four tunnels consisting of two tubes each were excavated during the new construction of a motorway section in Hungary. In July 2008 a collapse occurred in the tunnel heading with the result that both tubes collapsed on a length of 200 m. In case of PPP projects risks are especially seen in the comparatively high degrees of freedom in the process of interpretation of foundation models as a basis for a further planning. Furthermore, designers of the 100% design and the supervision both are acting on behalf of the construction firm. Moreover there is an increased degree of freedom in the construction itself which also leads to a higher risk potential.

## INTRODUCTION

### Principles of a Public Private Partnership Model

The construction industry in Europe has been characterized by a considerable price competition for many years. In the offer phase very favorable prices are often calculated which have to be balanced by claim-management in the phase of construction. For the owners this phenomenon leads to an increased incertitude that the costs may not be covered. For this reason big infrastructure projects and especially as well tunnel constructions are increasingly awarded on the basis of functional tenders or as construction projects according to the model of a Public Private Partnership (PPP).

In case of PPP projects an economic realization of infrastructure projects is based on an interdisciplinary cooperation and on partnership. Thereby an optimization of the individual project phases (planning, financing, construction, operating and realizing) shall lead to a more effective cooperation. The following advantages are especially mentioned in this type of project realization:

- Quick realization
- Increase in efficiency
- Use of private capital
- Use of private know-how

A further principle of the PPP model is the fact that the contractor is designing and constructing almost on his own responsibility. These basic conditions form an integral part of the contract.

With regard to the underground conditions this means that the underground investigation is down under the responsibility of the owner. The results of the site investigation are a part of the tender documents. The interpretation of these results and the constructional realization are however subject to the responsibility of the contractor. Beyond that, in many cases, especially in case of PPP contacts abroad, the ground risk is also transferred to the responsibility of the contractor.

In projects according to the PPP model there is a higher degree of freedom in interpretation of the results of the site investigation and moreover in drawing conclusions from these results that lead to the development of underground models which are the basis for the later design and construction. This freedom of interpretation is much higher than in projects common contract in which the construction is paid by prices per unit. This phenomenon may lead to additional risks in the state of construction.

### Underground Risk

Tunnel construction always means construction with the material *subsoil*. In spite of the best methods of investigation the subsoil remains to a certain degree an unknown material due to its inhomogeneity in consequence of its natural genesis. The risks combined with the unknown part in the process of realization of tunnel construction projects are the so-called underground-risks.

According to the standards the underground risk is defined as "an inevitable risk which can lead to unforeseeable effects and difficulties, for example construction damages or delays, although the person who provides the subsoil (owner) has completely fulfilled the obligations of site investigation and description of the subsoil and groundwater conditions according to the technical standards."

As a basis of the design and in order to minimize the ground risk, usually geotechnical explorations are carried out during the state of preliminary design in order to determine the structure and characteristics of the subsoil. The results are summarized in a report: the geotechnical report.

The main conclusion derived from the geotechnical report is the understanding of the underground structure and mechanical behavior (underground model) which is defined more or less accurately depending on the quality of the expertise.

The model of the subsoil basically consists of the following information:

- Structure of the subsoil, for example structure of layers and joints
- Groundwater conditions, for example groundwater table and aquifers
- Mechanical characteristics, for example strength and young's modulus
- Hydraulic characteristics, for example permeability and water infiltration capacity
- Determination of homogeneous sections, for example in case of tunnels or dams
- Description of special inhomogenities, for example strongly weathered layers or fault zones in the solid rock, cavities

The model of the underground is the basis for the design of the tunnel and therefore is the foundation for a successful realization of the project.

As in the literature (Raabe, 2008) explained, 12% of all construction damages in the civil engineering are due to insufficient geotechnical investigation. With regard to the quality of the geotechnical investigation the risks often go back to the following circumstances:

- Information of the expert regarding the design
- Quantity of bore holes
- Quantity of laboratory and field tests
- Execution of the laboratory and field tests
- Interpretation of the test results
- Specific experience of the expert with similar projects

Further risk factors consist in the interface between the geotechnical investigation and the design of the construction. In other words: The designer does not realize, only partly realizes the ground model or with faults. The reasons for it can be on both parts of the interface.

A specific situation arises in case of construction contracts with design-built specifications as for example those of PPP projects. In this case planning and execution are to a large extent in the responsibility of the constructor. The geotechnical investigations are usually initiated in advance by the owner. The geotechnical expert delivers geotechnical report and is often not included into the process of further design and construction.

With the regard to save costs the freedom of interpretation in the underground model is often misused to design and carry out simpler and more cost-efficient constructions methods that reduce the safety of the construction. Depending on the contract situation, in case of a damage-event the insurance company should pay for the loss. In the following example a tunnel collapse is shown, in which the designer estimated too high mechanical parameters of the subsoil.

## EXAMPLE OF DAMAGE

### Tunnel

The construction of a 40 km long section of the Motorway M6 in Hungary was commissioned by the Federal State Government as a PPP project. In this project several bridges, dams and cuttings as well as four tunnels had to be constructed. The tunnel constructions consist of two tubes each with a full cross-section of about 100 m² (Figure 1).

The tunnels have a length between 400 m and 1,400 m and an overburden of up to 45 m. The distance between the two tunnel tubes is about 12 m, that approximately one diameter of the tunnel. All four tunnels are constructed by the Shotcrete-Method (NATM).

### Geotechnical Site Investigation

During the geotechnical site-investigation a number of drillings have been carried out. In chosen bore holes field tests have been executed. However, special procedures as for example pressiometer tests to determine the young's modulus of the subsoil have not been done.

In order to determine the mechanical characteristics of the subsoil several geotechnical laboratory tests have been carried out at. Grain distribution,

**Figure 1. Tunnel portal Tunnel A**

determination of the natural water content and the plasticity, void ratio, degree of saturation, sheer tests and determination of the rigidity module belonged to the tests.

According to the results of the site-investigation the tunnels are lying in quaternary layers of loess silt with predominantly stiff, locally also smoother consistency (Figure 2). The quaternary silt is underlying by a clay with mostly stiff consistency. In single soil layers consolidations of lime have been determined. The groundwater table lies under the tunnel invert.

Regarding the cohesive soil layers comparatively high shear strengths were derived from the laboratory tests. For example for the loess an angle of friction of 27.5 ° and a cohesion of up to 70 kN/m² were indicated as characteristic shearing parameters.

### Planning

The tunnel was constructed by the shotcrete method (NATM). The ground was removed by an excavator und the walls of the tunnel were supported by reinforced shotcrete. To ensure to the stability of the excavation face the cross section of the tunnel was divided into crown, bench and invert.

For the design of the tunnels the soil parameters were taken from the geotechnical report without an additional of check of plausibility. Thus the high parameters for the shear strengths of the subsoil became basis of the design and the different excavation and support types.

All in all eight different excavation and support types named "A" to "H" were designed. The support

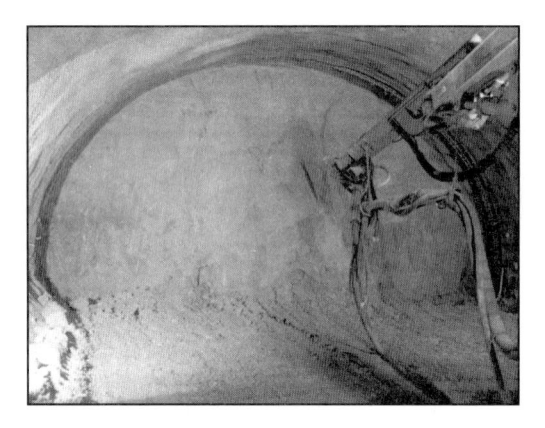

**Figure 2. Tunnel face of the tunnel in loess clay**

types vary between the thickness of the shotcrete-lining, the additional stabilization of the excavation face or additional spiles to support the working area of excavation. In very soft ground a temporary invert in the crown-section was planned. In case of instabilities at the excavation face an additional subdivision of the crown heading should be carried out.

### Collapse of the Tunnel

In the tunnel heading the excavation and support types were chosen by the contractor on his own responsibility. Predominantly the types C and D were used, in which a stabilization of the excavation face with shotcrete was supposed. The type E which

Figure 3. Cracks in the shotcrete lining between the first and second collapse

contains in addition to the stabilization of the excavation face spiles in the top of the crown. In some parts of the tunnel the displacements measured during the tunnel heading have exceeded the maximum permissible values according to the planning. The change into a higher excavation type has however not been realized.

During the heading of the longest tunnel of the motorway section with a length of 1,400 m the center pier between the two tunnel tubes collapsed in the area of the maximum overburden of about 45 m in July 2008. On the insides of the tunnel tubes cracks occurred in the shotcrete-lining which led to a shearing of the wall and to a collapse of both tunnel tubes on a length of 113 m. The collapse continued up to the ground surface and led to a syncline of 60 m width and up to 6 m depth.

Emergency measures in order to stabilize the tunnel in the area next to the collapse have for safety reasons not been permitted by the public supervisory authority. Due to the new distribution of the loads in the ground new cracks occurred after approximately two weeks on the insides of the tunnel tubes (Figure 3) and a second collapse of a length of 72 m as well as a corresponding syncline on the ground surface (Figure 4) took place.

## Reconstruction

In order to investigate the reason for the collapse additional drillings and laboratory tests have been carried out. Moreover the available geotechnical data have been re-evaluated. It became clear, that some of the laboratory tests have not been executed according to the standards. For example the triaxial shear tests in order to determine the shearing strength have been carried out as undrained tests without consideration of the pore water pressure. The shearing strengths derived from these tests correspond to those of a soil in an undrained condition and cannot be applied as an effective shearing parameters. In the geotechnical report this small difference was not considered. In the report the shearing strengths investigated in the undrained condition were indicated as effective shearing parameters.

For the further planning in order to reconstruct the tunnel in the area of collapse now the new shear parameters derived from the additional laboratory tests were taken. In form of a feasibility study several possibilities to reconstruct the tunnel have been discussed. In this context the mined reconstruction as well as the cut-and-cover method have been considered.

In order to reconstruct the area of the collapse first the cavities remained in the underground had to be filled by low pressure injections. Moreover, the soil beside the tunnel tubes was stabilized by Jet grouting columns. The following re-excavation of the tunnel was carried out with the support of spiles out of 15 m long steel pipes filled with concrete and a temporary stabilization of the invert of the crown heading.

The insurance company has continuously provided technical support and supervision regarding the reconstruction of the tunnel in the area of collapse. Moreover, the heading works in the other tunnels have been accompanied by the technical consultant of the insurance company. The costs of the collapse amount to several millions Euro and shall be paid by the insurance company. The headings in the

**Figure 4. Area of collapse on the surface**

other tunnels have temporarily been interrupted after the collapse and were subject to a safety review. The further heading works have uniformly been carried out in type G with a stabilization of the tunnel face with shotcrete and anchors as well as a temporary invert of the crown heading.

## CONCLUSIONS

In tunnel projects which are awarded at a fixed price and/or according to the PPP model there are special degrees of freedom for the contractor as the above-mentioned example shows. Consequently, risks regarding design and construction are generated.

However, these degrees of freedom must not lead to the fact that the current technical standards are not adhered to in order to save costs in the process of design and construction. In order to take steps against this process the insurance company thinks it would be helpful to accompany the process with technical support and supervision. So risks can be identified prematurely and damages avoided.

## LITERATURE

Raabe, E.-W. (2008): Investigation, Planning, Execution—Handling with conflict potential dependent on geotechnical engineering in the heavy construction, publication on the 15. Darmstadt Geotechnical Engineering Colloquium on March 13, 2008, page 175–190.

# DCWASA's Project Delivery Approach for the Washington DC CSO Program

**Ronald E. Bizzarri, Carlton Ray**
DCWASA, Washington, D.C.

**William W. Edgerton**
JA Underground, PC, San Francisco, California

**ABSTRACT:** To comply with the National CSO Policy, DCWASA completed a Long-Term Control Plan in 2002 and entered into a consent-decree agreement with the United States and DC government, establishing an implementation schedule for a number of projects to control CSOs into the Anacostia River. Several of these projects are tunnels. Previously, DCWASA used the traditional design-bid-build delivery process for capital projects. Now it is studying alternative project delivery methods to ensure contractor interest in individual projects. This paper summarizes contract packaging approach, anticipated project delivery methods, and risk allocation techniques that will be used for the DCWASA CSO program.

## BACKGROUND

The District of Columbia Water and Sewer Authority (DCWASA) is a multijurisdictional regional utility that provides drinking water distribution and wastewater collection and treatment to more than 500,000 residential, commercial, and governmental customers in the District of Columbia, and also collects and treats wastewater for 1.6 million customers in Montgomery and Prince George's counties in Maryland and Fairfax and Loudoun counties in Virginia. DCWASA operates the Blue Plains Advanced Wastewater Treatment Plant (Blue Plains), the largest advanced wastewater treatment plant in the world, which covers 150 acres and has a dry-weather flow capacity of 370 million gallons per day (MGD) and a peak pumping capacity of 1.076 billion gallons per day. To collect this wastewater, DCWASA operates 1,800 miles of sanitary and combined sewers, 22 flow-metering stations, 9 off-site wastewater pumping stations, and 16 stormwater pumping stations within the District. Some of this construction dates back to the early 1800s. As shown in Figure 1, separate sanitary and storm sewers serve approximately two-thirds of the District area; the remaining one-third is served by combined sewer systems. The combined systems discharge an estimated 2 billion gallons of combined sewage into local waterways annually. The Anacostia River receives most of this, roughly 1.3 billion gallons, the Potomac River almost 640 million gallons, and Rock Creek averages 50 million gallons.

Section 402(q) of the Clean Water Act enacted in 2001 (P.L, 106–554) requires communities with combined sewer systems to prepare long-term plans for control of combined sewer overflows (CSOs). DCWASA began CSO control planning in the late 1990s and finalized a Long Term Control Plan (LTCP) in July 2002. This LTCP includes a system of tunnels for the Anacostia River, Rock Creek, and the Potomac River (Figure 2), which will capture combined sewer flow for treatment at Blue Plains.

The schedule for completing the LTCP is included in a federal court consent decree among the United States, the District Government, and DCWASA. Scheduled for completion in 2025, the LTCP will play a significant role in restoring DC waterways. As part of this ambitious plan, the Anacostia River Projects (ARP) will reduce the overflows and improve water quality of the Potomac and Anacostia rivers. The project includes $2.2 billion for construction of deep tunnels, diversion structures, and a pumping station to divert, store, and convey combined sewer overflows to Blue Plains, reducing sewage overflows to the Anacostia River by 98 percent. These projects are the highest priority in the court-ordered schedule, and the portion

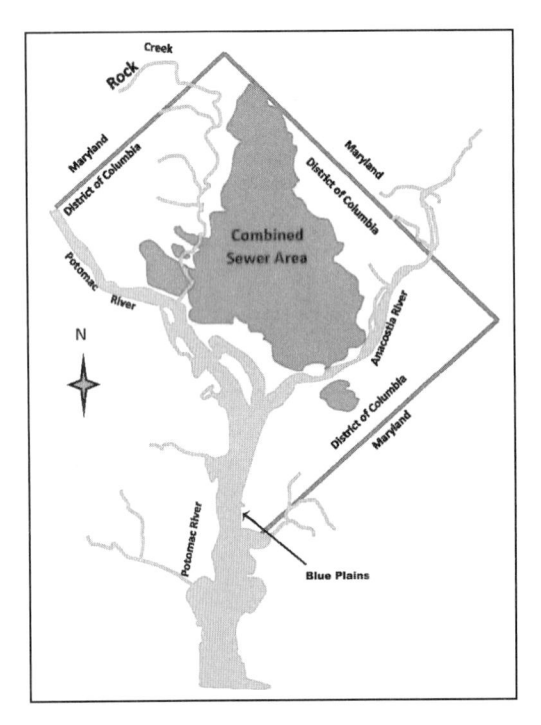

**Figure 1. Metropolitan DC area served by separate sanitary/story and CSO systems**

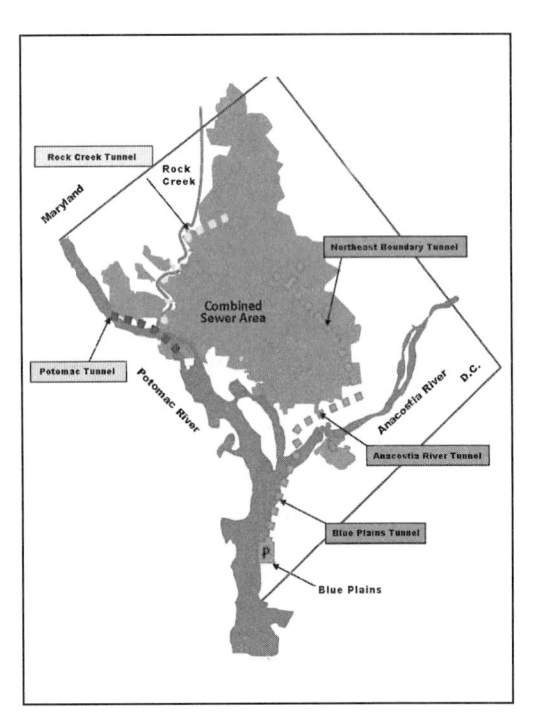

**Figure 2. LTCP tunnel system**

of the LTCP for the Anacostia River projects, from Blue Plains to just south of RFK Stadium, must be in operation by March 23, 2018.

The Blue Plains Tunnel (Figure 3) is the first tunnel project in this program. This tunnel will launch from the Blue Plains Advanced Wastewater Treatment Plant, run approximately 18.2–36.5 m (60–120 ft) deep under the eastern shore of the Potomac and Anacostia rivers, and traverse north to the retrieval shaft near the new Nationals' Baseball Park. The ground is primarily clay, silt, and sand. The tunnel will be constructed using an earth pressure balance machine. It will be supported using a single pass precast segmental liner with a finished inside diameter of 7 m (23 ft).

Five deep, large-diameter shafts will be built along the alignment to service tunneling operations and provide space for future permanent facilities to be constructed under separate contracts. Shaft construction techniques being considered include slurry wall and ground freezing. It will be up to the contractor to select the best method for the anticipated ground conditions and physical constraints at each shaft site. Shafts will be constructed to accommodate installation of a final liner meeting minimum geometric requirements. Shaft final liner diameters will range from 15.2 to 30.5 m (50 to 110 ft). When complete, these shafts will function as hydraulic facilities, and will include a pumping station, a grit and screening facility, drop shafts, and an overflow shaft.

This tunnel project involves multiple challenges, including varied soil conditions, potentially contaminated high groundwater, tunneling under the Potomac and Anacostia rivers, and controlling/mitigating movements of existing facilities and utilities during excavation and tunneling. Existing site constraints need to be addressed above and below ground. The alignment runs under U.S. and District of Columbia government property, and it is critical that the design team work closely with various government agencies and private developers to coordinate staging and mitigate the negative impacts of construction to the surrounding communities.

## CONTRACT PACKAGING

In order to meet the consent decree dates, the LTCP was further developed and a facility plan prepared. The facility plan established a detailed implementation schedule, which divided the overall Anacostia River Program into a series of contract packages, with the underlying objective being to achieve the best value for DCWASA, using the philosophy that the best bid prices can be obtained through enhancing economic competition by ensuring the maximum number of qualified bidders for each contract package. Packages were designed to be large enough that well-qualified and experienced contractors would be encouraged to bid, and small enough that they would not be overly restrictive (due to bonding, insurance, management of subcontractors, etc.). The dividing line between "large enough" and "small enough" in monetary terms is highly variable, and depends on the type of work and industry practices for bonding and sharing financial risk. To the extent possible, contracts were also structured to match market conditions and the availability of labor and equipment at the time of bidding.

The determination of contract packaging on the ARP applied these principles in the following manner:

1. **Type of Work:** Establishing separate contract packages for (1) tunnels, which would likely be constructed by national and/or international tunneling firms; and (2) surface diversion facilities, which would be constructed by local sewer contractors familiar with local practices.
2. **Contract Value:** Limiting the total size of the contract so that it would not exceed the anticipated bonding limits of qualified firms. Based on the most recent bidding results in the underground industry, tunnel contracts in the range of $300 million seem to strike a good balance of qualifying without restricting.
3. **Geographical:** Minimizing the contract interface between adjacent contractors.

In addition to these factors, the contract packaging strategy considered DCWASA's policy of meeting EPA's fair share objectives for MBE/WBE participation. The resulting contract packaging approach is shown in Figure 4.

In summary, this plan includes four deep, large-diameter, soft ground tunnels with a combined total length of 20.6 km (12.8 mi). The four main tunneling contracts are the Blue Plains Tunnel, the Anacostia River Tunnel, the Northeast Boundary Tunnel, and the Northeast Boundary Branch Tunnels. In addition to the four tunnel contracts, there are seven separate surface diversion structure contracts, two overflow structure contracts, two pumping station contracts, and one demolition/site preparation contract—for a total of sixteen separate contracts. The schedule for these contracts is shown in Figure 5.

## PROJECT DELIVERY METHODS

DCWASA considered two basic approaches for project delivery methods for contracts related to the tunnels and near-surface structures:

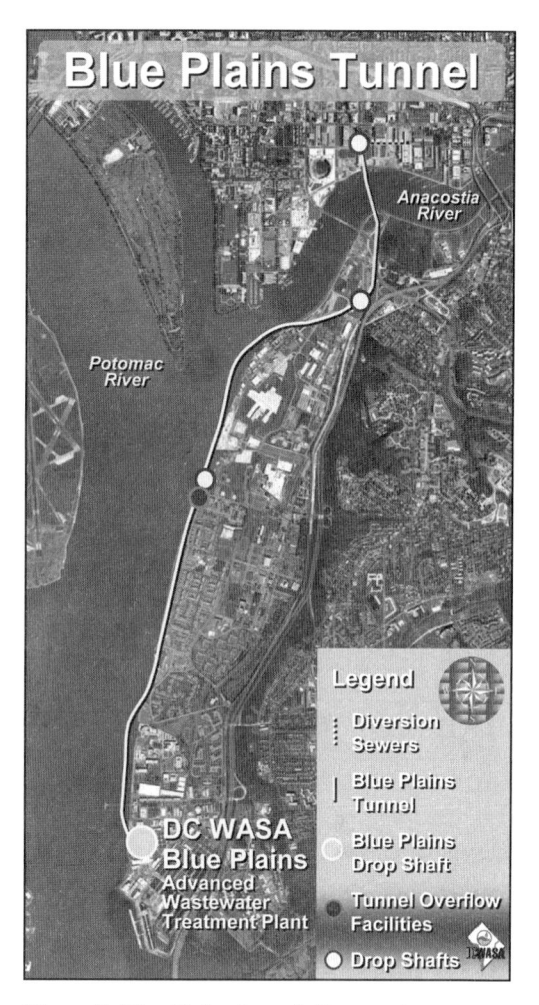

**Figure 3. Blue Plains tunnel alignment**

1. *Design-Bid-Build (DBB)*, which has been DCWASA's traditional method for project delivery, and
2. *Early Contractor Involvement (ECI)*, an umbrella term that can include various methods associated with design-build (DB), construction manager at risk (CMAR), cost reimbursement fixed fee (CRRF), and public-private partnership (PPP).

## DBB: A Traditional Approach Considered

### Advantages

DCWASA found several advantages to the traditional DBB approach. It has a long history of use, the terms and conditions are well defined and understood, the legal aspects are well established, and the

**Contract Divisions: Anacostia River Projects**

| Division | Description |
|---|---|
| A | Blue Plains Tunnel |
| B | CSO 013/014 Diversion Sewer |
| C | CSO 019 Overflow & Diversion Facilities |
| D | Bolling AFB Overflow & Diversions |
| E | CSO 015, 016 and 017 Diversion Sewer |
| F | CSO 018 Diversion Sewer |
| G | CSO 005/007 Diversion Sewer |
| H | Anacostia River Tunnel |
| I | Main Pumping Station Diversion Facilities |
| J | Northeast Boundary Tunnel |
| K | Northeast Boundary Branch Tunnels |
| L | Northeast Boundary Area Diversion Sewers |
| M | Mt Olivet Road Diversion Sewer |
| W | Blue Plains Tunnel Site Preparation |
| Y | Tunnel Dewatering Pumping Station and Enhanced Clarification Facility |
| Z | Replace Poplar Point Pumping Station |

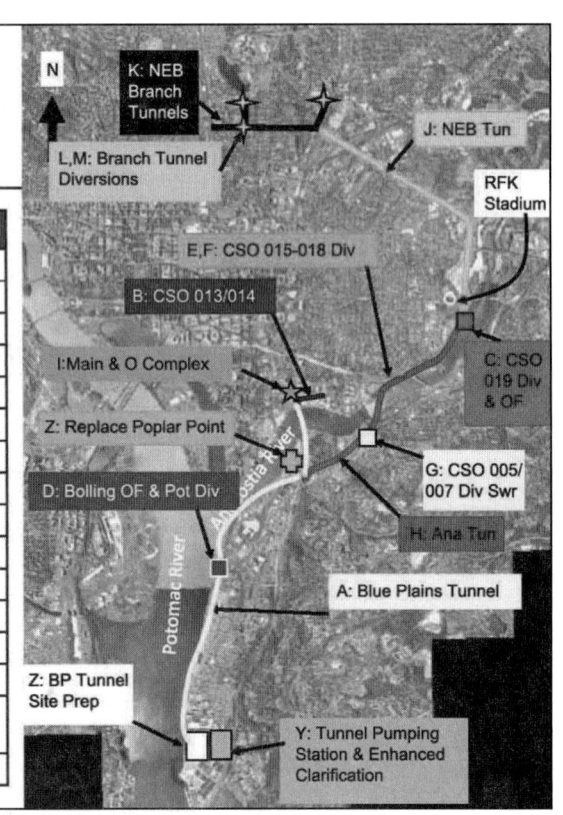

**Figure 4. Contract packaging approach for Anacostia River projects**

lowest price can be obtained through competitive bidding. In addition, this approach can also contain ECI-like aspects (i.e., some degree of collaboration) through informal contactor discussions prior to bid, and it gives the owner the maximum control over the design.

*Disadvantages*

For the owner, there are a number of disadvantages to the DBB approach. The owner has multiple contacts to manage (architect/engineer [A/E], contractor, construction manager [CM]); the contractor considers the owner's design to be warranted and is entitled to base its bid on "best" assumptions; and there are uncertainties as to what the "lowest responsive bid" includes. In addition, given the nature of the roles and responsibilities, the relationships among the parties tend to be adversarial as opposed to collaborative. There is a proclivity for cost growth through design changes needed to resolve errors and omissions, which can result in claims and litigation; there are difficulties in obtaining the combination of best value, quality, and lowest price; and changes can result in extended schedule. Also, although the owner does have maximum control over design, it should be kept in mind that it does not necessarily have maximum control over the budget.

**ECI: An "Umbrella" Alternative Approach Considered**

*Advantages*

The ECI umbrella approach provides the flexibility to obtain the desired end of the day results. The schedule can be shorter than DBB, and can provide reliability and predictability. There is the opportunity to maintain the competitive lowest price feature of the DBB. There are opportunities to reduce uncertainties for all parties: owner, engineer, contractor, sureties, and third parties; decisions don't have to be made in a vacuum, given the collaborative atmosphere; the approach promotes equitable risk sharing and risk management; and there are opportunities to obtain risk sharing and risk management input from all parties. The ECI approach recognizes that design and construction is an integrated process, so it provides a forum to capture creativity, innovations, and preferences from the engineering and contracting communities. There are opportunities to obtain

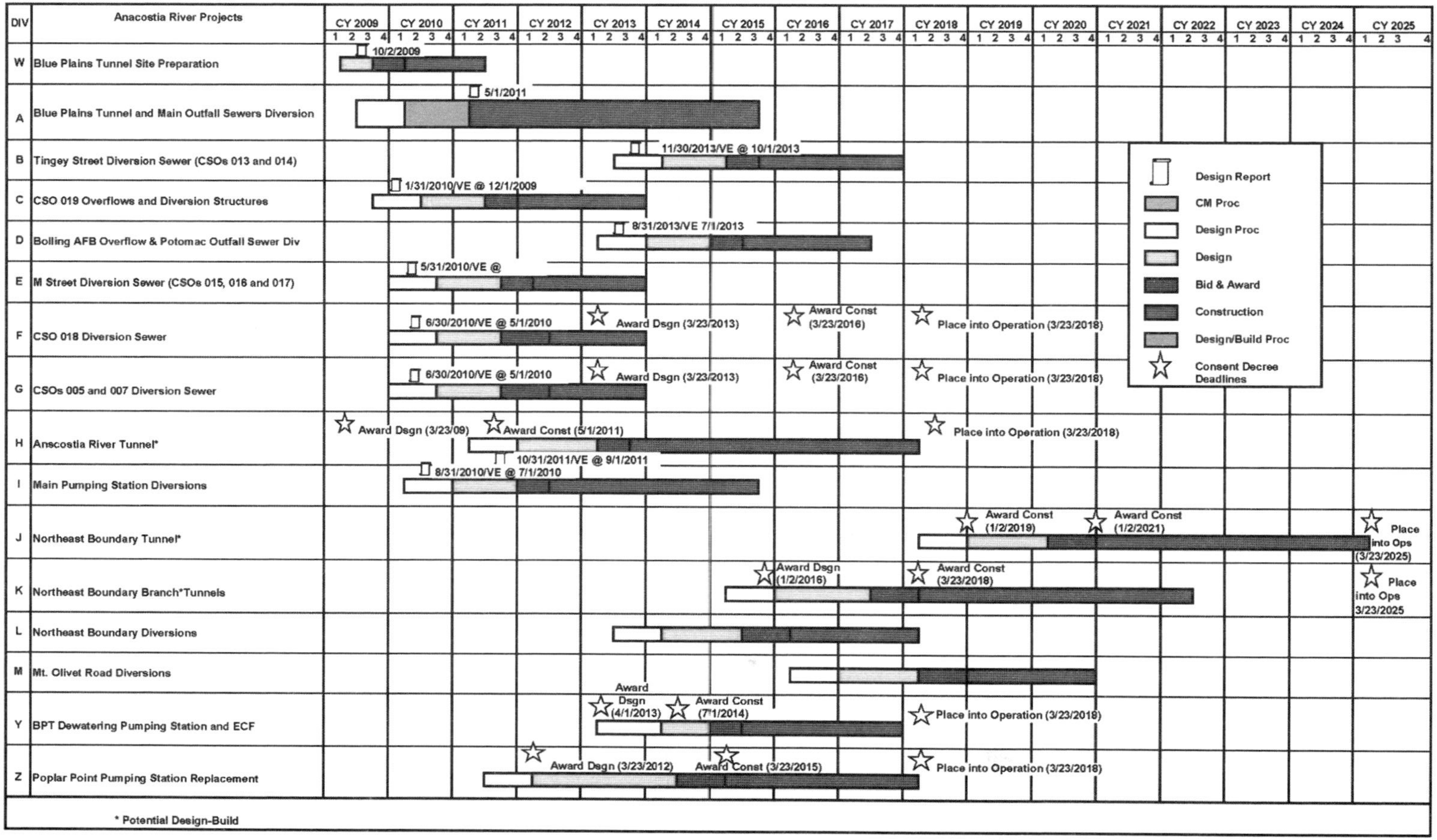

**Figure 5. DC WASA LTCP master schedule**

construction expertise before the design is complete, and the contractor "buys-in" to the design solution. In addition, this approach offers settings to obtain the best quality and value, and the owner can obtain single-source accountability.

*Disadvantages*

However, when considering this approach, there are some things to be aware of. ECI umbrella options are not well proven. The owner relinquishes some control over value, quality, function, and other objectives; and the owner depends on qualitative information in selecting the parties (e.g., prices are often absent in the criteria used to select a team under various ECI umbrella methods). Finally, this approach requires the owner to maintain transparency throughout the process.

**Traditional and Alternative Approaches**

In July 2009, DCWASA established new procurement regulations that provide for the use of both the traditional DBB approach and the various alternative project delivery options included under the ECI umbrella. Specific delivery methods considered for application on the Anacostia River Program included the following:

*Traditional Design-Bid-Build (DBB)*

An A/E firm would be engaged to prepare final contract documents for competitive public bidding. A CM would then conduct a constructability review prior to bidding, and that firm would monitor the construction. Depending on the complexity of the project, an expert panel or project review board (PRB) might provide an independent review of the design process.

*Design-Build (DB)*

Using this model preliminary, plans would be prepared, and up to three DB teams would be short-listed by qualifications. The short-listed teams would be asked to submit technical/price proposals based on proposal and contract requirements issued in a Request for Proposal (RFP) document. During preparation of the technical/price proposal, proprietary meetings would be held with the short-listed teams to achieve the Early Contractor Involvement. The final selection would be based upon a combination of price and technical factors. Subsequent to the final selection, the unsuccessful teams would be compensated with a stipend for submission of responsive technical/price proposals.

*Construction Management At-Risk (CMAR)*

An A/E firm and a CM firm would be engaged to provide design and construction monitoring services. A solicitation for qualifications would then be issued to CMAR teams or firms. Short-listed firms would submit technical proposals that would contain information on the team, preconstruction deliverables, the construction process, ability to self perform, a plan for MBE/WBE utilization, development of a Guaranteed Maximum Price (GMP), and a cost proposal for overhead, profit, staff cost, sharing savings, and other cost-related items other than construction-specific costs.

The CMAR team would be selected based on a "best value" scoring system, whereupon an A/E agreement would be executed with the selected CMAR team to provide preconstruction services. The CMAR team would then participate in the development of the project design, typically entering the process at 30 to 50% design. At some point in the design (e.g., 80 to 90%), the CMAR team would submit a GMP that would be either accepted or rejected. If accepted, a construction Notice to Proceed (NTP) would be issued. If rejected, the owner would be free to use the information obtained in the preconstruction phase and seek competitive bids, or select another CMAR team.

**Determining Project Delivery Systems**

During a six-month period beginning in December 2008, DCWASA investigated which project delivery system would be most likely to result in a successful project, considering the technical features of the project, an internal evaluation of agency characteristics, and how the expected bidders would be likely to react to the various delivery systems. This investigation began with a two-day workshop, and concluded with an all-day presentation to the General Manager, with recommendations for which delivery system to use on which contracts.

During the internal and external evaluation the following conclusions and criteria were reached:

1. Early contractor involvement during design completion would significantly improve the chances of success.
2. A competitive procurement process was necessary to ensure the support of DCWASA staff and its board of directors.
3. Whatever delivery system was used had to encourage contractor participation by addressing concerns related to risk allocation and procurement confidentiality.
4. The proposed process had to have been used successfully in other jurisdictions.

Comparing the various delivery systems considered against these criteria resulted in the following recommendations.

*Tunnel Projects*

For the tunnel projects, and in particular for the Blue Plains Tunnel, the design-build (DB) process was determined to be the best approach. The most efficient way to involve the contractor and designer together in addressing design decisions is for the designer to have a contract with the contractor. This would also minimize DCWASA's design risk. In addition, the DB procurement process can use a combination of price and technical factors, with a defined evaluation system for the more subjective elements, thereby retaining the competitive nature of the procurement process. Also, during the proprietary meetings, the contractor's concerns about risk allocation could be addressed, and changes made in the solicitation document if warranted. A communication protocol could be established that would satisfy any lingering contractor anxiety related to the confidentiality of information exchanged during these meetings. This DB process has been used successfully on other tunneling projects, and the contractors expected to be interested in this project are familiar with it.

The success of the design-build process used on the Blue Plains Tunnel will be evaluated before its use is recommended for the follow-on tunnel contracts. The master schedule has sufficient time for this evaluation to be completed.

*Near-Surface Projects*

For near-surface projects, the design-bid-build (DBB) process was determined to be the best approach. DCWASA could control the design process through separate procurement of AE firms, which was thought beneficial for near-surface structures. In addition, contractor input on constructability need not be integral with the final design, and thus can be provided during the bidding period. The DBB procurement process selects the lowest responsive, responsible bidder. The apparent low bidder is determined through public opening. Contractors who would be interested in the near-surface structures would, for the most part, be local and familiar with the risk allocation philosophy contained in standard DCWASA construction contracts. DCWASA typically uses the DBB procurement process for all construction contracts.

## RISK MANAGEMENT

DCWASA recognizes that it faces a number of challenges, both technical and commercial, in order to achieve a cost-effective project within the time period required to meet the consent-decree date. To address these challenges, DCWASA is committed to using a balanced and fair contracting approach that appropriately allocates risk among the parties. In furtherance of this goal, the agency is seeking to modify certain contracting practices by implementing commercially viable risk management provisions.

The DCWASA legal department is currently modifying existing contract language to be more appropriate for underground projects and to incorporate this risk management policy. At the time of this writing, a draft of the revised General Conditions is not yet available, but it is expected that contract terms will include the following:

### Geotechnical Baseline Report (GBR)

A draft GBR will be provided to the short-listed firms to use in the preparation of their proposals. This GBR will establish baselines for ground parameters, based on the investigations completed. As part of the proposal process, each bidder will interpret the various baselines expressed in this draft GBR, and consider those baselines in the development of its design and construction approaches. At the proposer's request, the GBR could be modified to incorporate additional agreed-upon geotechnical risk allocation measures, and in such a case, a revised GBR would be issued by addendum to all bidders.

### Liquidated Damages

Project completion time is an important factor in achieving project success. DCWASA intends to incorporate liquidated damage provisions for substantial completion, as well as interim milestones that would allow follow-on contractors access to certain work sites. At present, there have been no decisions made on either the use of incentive provisions for finishing early, or on instituting a project cap on the amount of liquidated damages that could be assessed.

### Performance and Payment Bonds

Because the largest contract is expected to be in the $300 million range, it is expected that contractors and sureties will be able to comply with DCWASA standard performance and payment bonds in the amount of 100% of the contract price.

### Retention Practices

DCWASA's retention practice is to withhold 10% until the contract is 50% complete and then, given satisfactory progress, not subject further pay applications to any withholding, thus leaving a retention balance of 5% at project completion.

**Payment Provisions**

Recognizing that tunnel construction contracts impose special financial burdens on the contractor, DCWASA expects to include a provision allowing payment for materials stored off-site that would apply to precast concrete segments. In addition to the standard mobilization provisions, there may be another provision that would allow progress payments for engineering, fabrication, and delivery of the tunnel boring machine (TBM). Because most permanent materials are expected to be purchased relatively early in the contract term, it is not expected that a cost escalation provision for commodities will be included.

**Partnering**

There will be a partnering provision allowing for quarterly meetings, similar to those found on most underground projects.

**Dispute Resolution Board**

DCWASA anticipates having a three-person dispute resolution board to make recommendations on disputes. Details of the specification and the dispute resolution process are not yet final.

**Hazardous Materials**

Excavation at some shaft sites may encounter contaminated and/or hazardous materials. DCWASA is in the process of undertaking environmental investigations to identify both the quantity and quality of any such material. If, based upon these investigations, some of this material is expected, then a unit price will be included in the bidding schedule that will allow separate payment for hauling and disposing of such material at a licensed landfill.

**Escrow Bid Documents**

The contract documents will require that detailed bidding preparation documents be provided and held in escrow for use in future change negotiations.

**Owner Controlled Insurance Program (OCIP)**

DCWASA has used a Rolling OCIP for years on its projects. It is expected that such an OCIP will be in place on the Blue Plains Tunnel project, although the details of what coverages will and will not be included are still being negotiated with the carriers.

**SUMMARY AND CONCLUSIONS**

The DCWASA CSO program is a large system with a number of underground structures that is just beginning the detailed design and construction phase. The contract packaging established using the criteria discussed herein has resulted in 16 separate contracts, 4 of which are tunnel contracts. The delivery methods to be used are a combination of design-build and design-bid-build, with the first tunnel contract, for the Blue Plains Tunnel, to be issued as a design-build. Risk allocation provisions intended to create a balanced and fair contract that allocates risk reasonably among the parties is in the process of being developed. It is anticipated that the success of this CSO program will be determined in part by the results of these contract decisions.

# How to Deliver Your Project On Time: An Owners Procurement Strategy

Wayne Green
The Regional Municipality of York, Newmarket, Ontario, Canada

## INTRODUCTION

The Regional Municipality of York, located north of the City of Toronto in Ontario Canada has experienced an unprecedented growth over the past decade with new home construction averaging 15,000 units per year. This rate of growth is planned to continue well into the future with a projected growth of over 700,000 new residents by year 2031 combined with a target of over 300,000 new jobs to provide a sustainable economic base for the region. It should be noted that York region has been designated as one of the key growth centres by the Province of Ontario aimed in part to meet the housing and employment needs of Canada's growing population. The region is a mix of urban and rural areas and is one of the most desirable areas to live in Southern Ontario. The region offers an attractive lifestyle with its 9 local communities and the amenities created by the many rivers, streams and protected greenbelt areas which comprise over 69% of the region's area. The existing wastewater system referred to as the York/Durham Sewage System was constructed in the 1970s and is comprised of over 200 km of large diameter trunk sewers (2.4 to 3.0m) extending from a treatment plant on Lake Ontario to the most northerly community, a distance of approximately 80 km from the treatment plant (Figure 1).

Parts of this system and in particular a 15 km length of the Southeast Collector trunk sewer will reach it's hydraulic capacity by 2012 thus necessitating a major program of twinning and trunk system expansion to accommodate the servicing needs for the planned future growth in York region.

Growth pressures combined with the need to protect and sustain the region's many natural and heritage features are the key challenges for planning and expanding the capacity of the Southeast Collector trunk sewer portion of the region's wastewater system. The region has undertaken an extensive environmental planning study and developed a unique strategy for procuring equipment, materials and labour that will meet the challenges of providing new trunk wastewater capacity in a timely way while at the same time protecting the many natural and cultural features of the area. A number of strategies are underway to achieve these goals.

## EXTENSIVE ENVIRONMENTAL PLANNING

The region has undertaken for the first time in Ontario an Individual Environmental Assessment to expand the capacity for the Southeast Collector trunk sewer (Figure 2). This study examined a range of alternative solutions including an assessment of 13 alternative routes for the trunk sewer expansion along with an extensive program of public consultation with agencies, stakeholders and property owners. This comprehensive environmental planning study has included the collection of baseline data on soils, surface and groundwater conditions as well as data on natural environmental features related to terrestrial and aquatic species of the area. A detail geotechnical and hydro geological investigation was carried out at an early stage of the project planning. This thorough knowledge of soil conditions has been used to set the sewer profile, alignment and location of drop structures to maintain the tunnelling activity in very competent till material regionally referred to as the Newmarket Till deposit. This geological data base will be used for the development of a Geological Baseline Report (GBR) for construction purposes (Figure 3). Similarly, groundwater pump tests have been conducted along the pipe alignment to confirm that minimal dewatering will be required at the construction shaft locations.

The study has recommended the use of Earth Pressure Balance Machine (EPBM) technology using a single pass segmental liner system. Further, sealed shaft construction has been recommended for the construction shafts. The study concludes that the use of this equipment and construction methods will ensure a minimal impact on the communities and natural environment along the construction route. This early planning study and community engagement program will provide a solid framework for the design, approvals and construction phases of the project to follow.

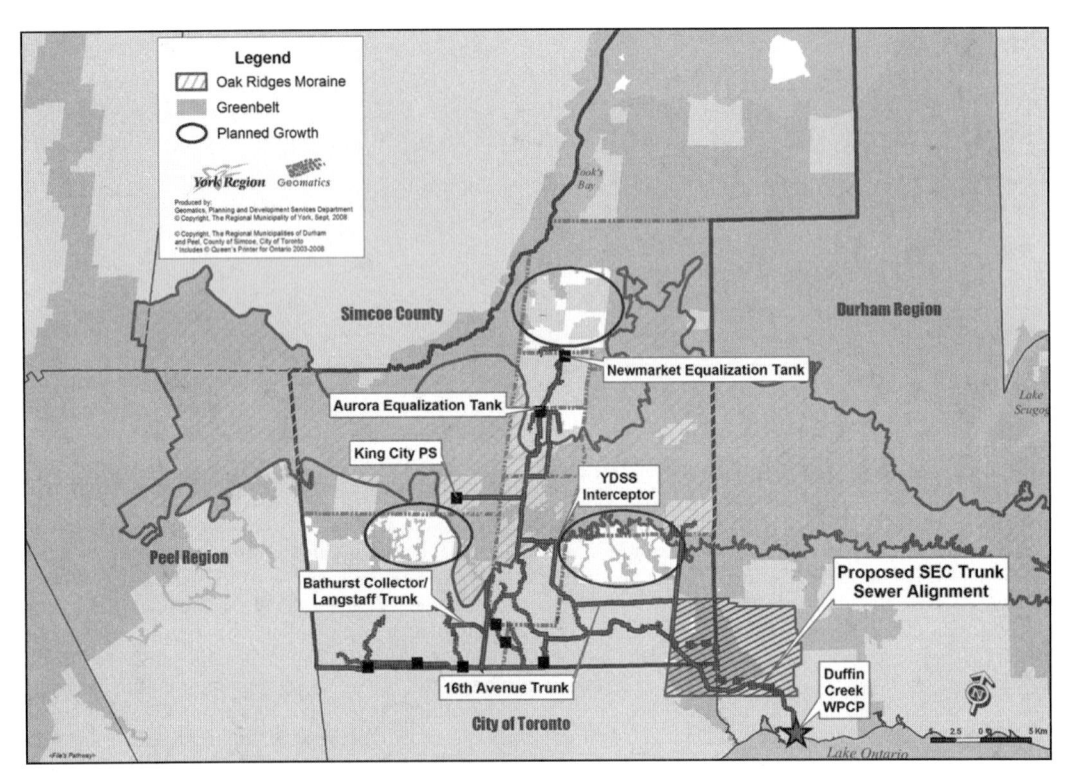

Figure 1. Future growth, greenbelt, and watershed areas

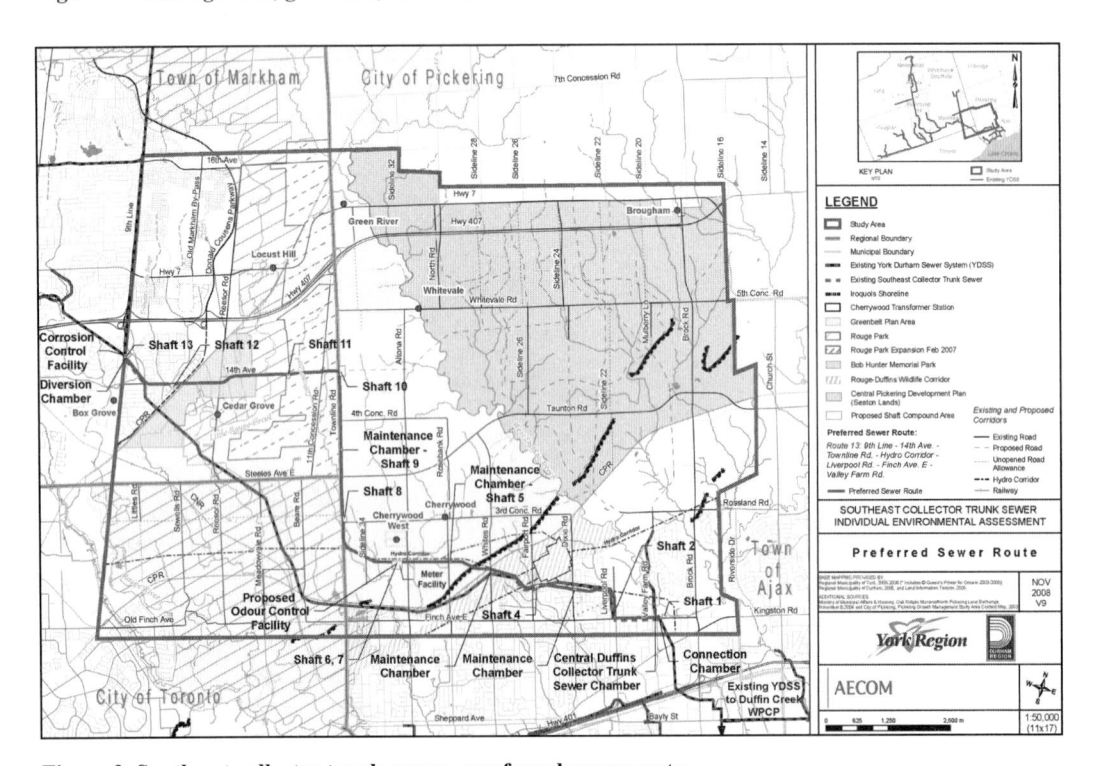

Figure 2. Southeast collector trunk sewer—preferred sewer route

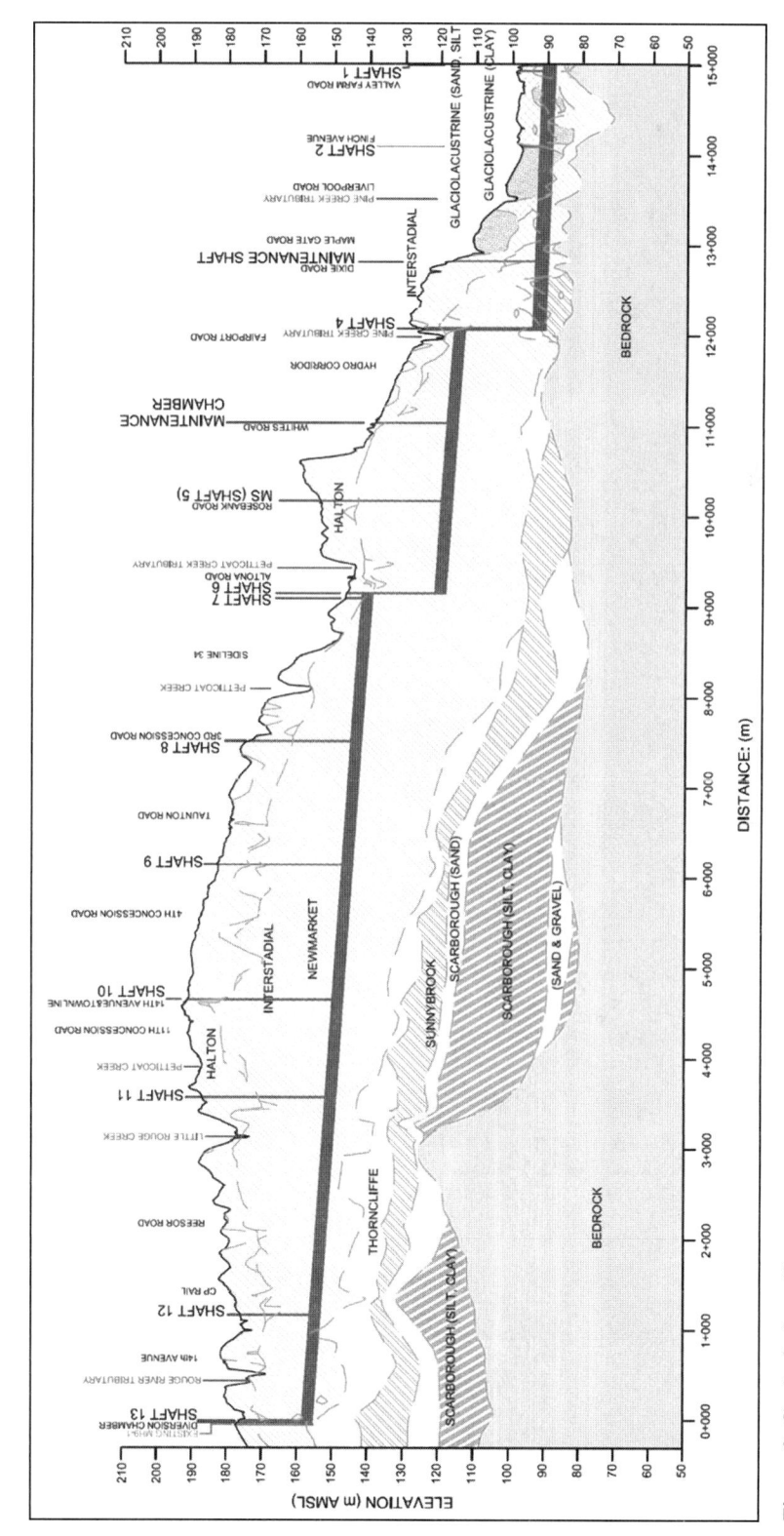

Figure 3. Geological profile

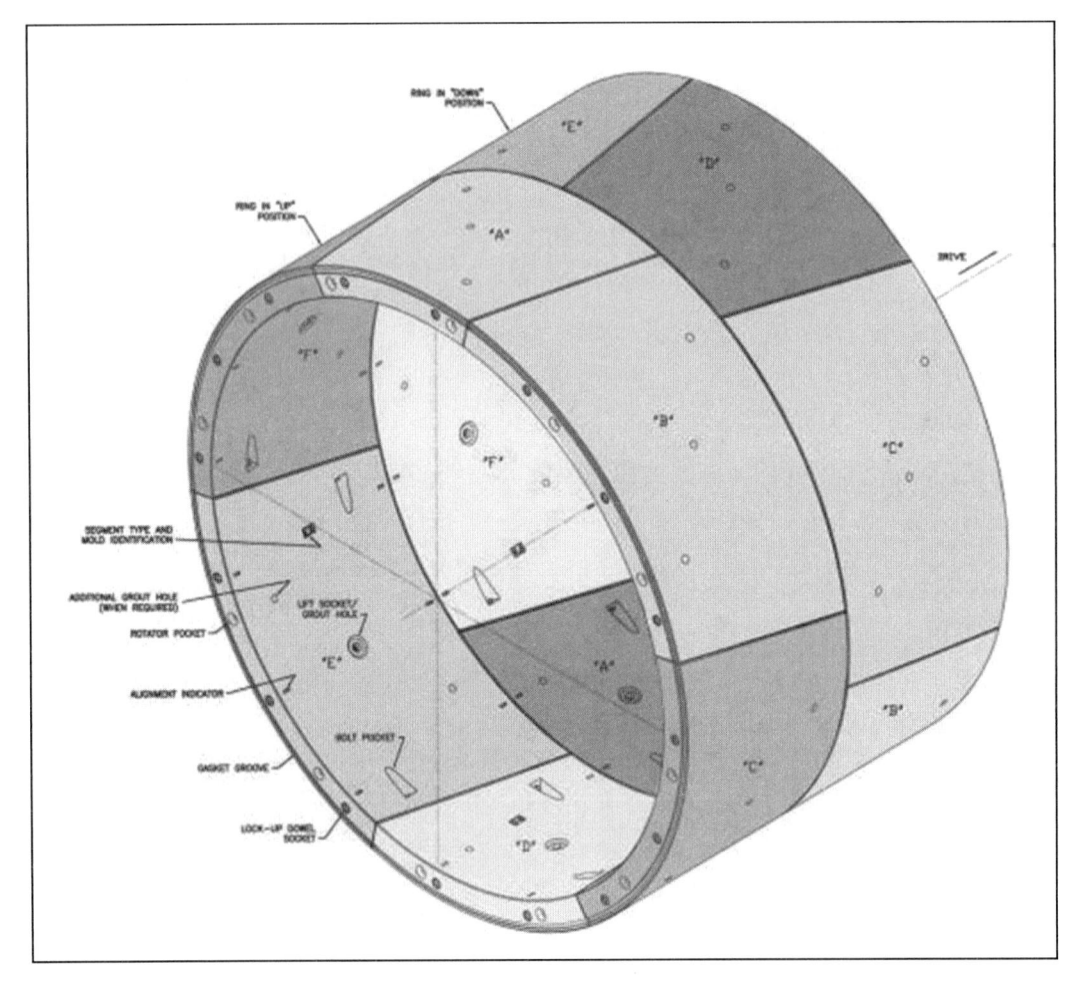

Figure 4. Segment shape—trapezoidal

## ADVANCED PROCUREMENT OF EQUIPMENT AND MATERIALS

### Procurement of TBM Equipment

The region has entered an agreement with Lovat Inc to prepurchase four (4) Earth Pressure Balance Machines with a machine specification to undertake the tunnelling activity in the Newmarket Till materials. Given that the Lovat Inc. local office and assembly plant is located in Ontario Canada it was a logical choice for York region to sole source this equipment manufacturer for the 4 TBM machines. Appropriate due diligence was carried out to confirm the pricing structure and with the assistance of KPM Management Consulting who were hired by the region as a fairness consultant, a satisfactory price was negotiated with Lovat Inc. for the purchase of tunnelling equipment.

This early procurement of equipment by the region will allow the chosen contractor to begin the tunnelling activity at or about the same time using all 4 machines. The total project length of 15 km will be split into contracts each of approximately equal length. Two machines will be launched in each contract and generally work towards a centre exit shaft.

### Procurement of Segmental Liner Materials

The region has developed tender documents to issue for the supply of segmental liner materials to all 4 Tunnel Boring Machines (TBMs) (Figure 4).

It is anticipated that a supply of up to 400 liner segments will be required on a daily basis to meet the pipe installation progress of the 4 Tunnel Boring Machines (TBMs). Given this volume, the contract specification for the liner materials will require the successful bidder to assemble a dedicated plant to maintain this liner supply rate.

## ADVANCING PROJECT FUNDING

One key aspect of this project involves the owner's commitment to allocate significant funds in advance of the project approval by the Province of Ontario to undertake the purchase of long delivery items such as TBM equipment and the setting up of a plant for the delivery of segmental liner materials. This advanced funding of materials and equipment demonstrates the owner's commitment to the successful commissioning of the new Southeast Collector trunk sewer on schedule in 2012. It should be noted that the capital cost of this trunk infrastructure is funded through a per lot charge levied against the future developments that will be serviced by the trunk system. In this way a "user pay" principal is maintained and the burden for payment is allocated against the future home owners who will benefit from the expanded wastewater system.

## EARLY PREQUALIFICATION OF CONTRACTORS

A prequalification process is currently underway to select 3–5 firms to act as general contractors for the two projects. It is planned that the prequalified firms will also participate as an advisory committee for the supply of the tunnel boring machines and segmental liners and where appropriate advise the design team on constructability issues. The owner recognizes the importance of having contractor input and advice during the manufacture and specifying of contract materials and equipment in order to effectively transfer ownership of these contract components to the general contractor. Further, contractor input early in the design phase of the project will lead to an improved contractor/owner relationship and avoid where possible need for dispute resolution processes.

### Project Marketing

Conventional tendering allows the market place to assess the opportunities for tenders based on a number of factors relating to current workload, proximity of work to home base, owner/consultant reputation and early knowledge and awareness of project details. Owners can influence some of these factors by making an effort to inform the construction industry of project details and seeking feedback on design, construction and tendering issues. Workshops, trade journals, conferences and bidders information packages are all useful techniques to keep the industry informed and prepared for the tender packages when released.

### Assessing Bidders Risk

One key factor that limits bidder's interest in project tendering may relate to the exclusive transfer of risk to the successful contractor through general conditions clauses, performance bonds, liquidated damages and the absence of dispute resolution methodologies. In certain circumstances owners may consider some degree of risk sharing clauses to increase bidders' interest in the competitive bidding process.

## PREAPROVALS BY OWNER

Schedule certainty can be increased through the early procurement of project approvals and permits. Given that the Southeast Collector trunk sewer is "time driven," the owner will undertake to acquire many of the key project approvals, permits and permissions in advance of project tendering. Where appropriate, the conditions of approval will be assigned to the general contractor as part of the contract specifications. This early procurement of permits will be a further step in clearing the way for an early construction start and successful completion of the project Some of the key project approvals that will be secured in advance of tendering will include the permission to pump groundwater at shaft locations, local approvals of site plans for compounds and above ground buildings and structures related to meters, odour control facilities. Further, it is planned acquire noise bylaw exemptions for extended construction hours to permit two shifts of tunnelling activity.

## ADVANCED SITE PREPARATION CONTRACTS

It is the owner's objective to prepare to the extent possible the construction project areas in advance of the general contractor mobilizing tunnelling contacts. In this regard, a number of early works are planned to prepare the work sites for the general contractor.

### Haul Road Construction

It is planned to undertake the upgrading of haul roads for spoil removal in advance of launching the tunnel boring machines and undertaking the tunnel mining. This proactive improvement of haul roads will reduce impacts from spoil removal on local community residents and will further provide a positive legacy on the local communities impacted by the construction activity. Such road improvement will also ensure the uninterrupted access of heavy loads relating to construction equipment and segmental liner delivery.

### Securing of Spoils Disposal Site

Several options will be considered to secure a satisfactory spoil site(s) for the approximate 20,000 cm of spoils material. Spoils sites selected in advance of the tunnelling activity will ensure that all local and environmental approvals are in place for site filling thus ensuring continuous and uninterrupted haulage of spoils from the construction shaft locations. Alternative sites will be considered based on the net benefit that can be realized from clean fill to enhance communities or to improve areas for future development activity.

### Property Acquisition

The acquisition of all property needs including purchases, property access rights as well as temporary and permanent easements will be secured by the owner in advance of contract awards. Any conditions resulting from property access agreements will be transferred where appropriate into tender documents for compliance by the contractor(s). The early resolution of property issues for the use of lands for construction compounds, shafts, material and spoils storage will ensure the uninterrupted progress of construction activity particularly where private lands are involved.

### Local Utility Relocation

The size of construction launch shafts particularly within developed areas along the sewer alignment will necessitate the relocation of local utilities to accommodate 10–12 meter diameter construction shafts. The relocation of local utilities will be undertaken in advance of the general contractor's site mobilization. This early work will clear the way for an early and continuous mining operation.

### First Nation Clearance of Compound Areas

Up to 13 First Nations communities have inhabited the planned work site areas over a 300 year period leading up to the early 1800s. Past cultural and burial sites are unrecorded and can only be determined through onsite stripping of top soil and involvement of First Nations communities and archaeologists to confirm there are no significant historical remains. This type of advanced contract work will be carried out to clear the compound and shaft locations of any such cultural artefacts and confirm that there is no evidence of burial sites thus ensuring construction continues uninterrupted.

## CONCLUSIONS

The need for additional sanitary servicing capacity by a firm date of 2012 to allow the continued development of resident, commercial and industrial growth in York region has necessitated an owner initiated procurement of materials and equipment including a number of site preparation contracts all aimed at ensuring final deliver of increased sewer capacity on schedule. The owner's risk of undertaking expenditures of up to 30% of the project capital cost in advance of the project approvals and general contract awards is offset by the increased certainty of delivering the project schedule on time.

# TRACK 3: PLANNING

Bradford Townsend, Chair

*Session 3: Project Planning and Implementation I*

# Sustainability Drives Jollyville Transmission Main Tunnel Design

**Clay Haynes**
Black and Veatch, Austin, Texas

**Tom Knox**
Black and Veatch, Kansas City, Missouri

**ABSTRACT:** Sustainability principles regarding environmental preservation and community impact mitigation were incorporated in the alignment and construction method selection for the Jollyville Transmission Main in Austin, Texas. The City of Austin (COA) and the Black & Veatch (B&V) design team used the Criterium Decision Plus® (CDP) model in accordance with agreed-upon selection criteria. Tunneling was the selected construction method for installation of this critical segment of Austin's treated water supply infrastructure.

## BACKGROUND

The City of Austin, Texas is consistently ranked as one of the most livable cities in the United States. As a result, population growth has imposed heavy demands on the existing treated water supply infrastructure. To address the ever-increasing water demands, the COA plans to construct a raw water intake facility, a large water treatment plant (WTP4) and two major treated water transmission mains (TMs) to serve their growing population in the northwest side of the city. The TMs include the Forest Ridge Tunnel and the Jollyville Tunnel (JV) which is the subject of this paper.

The JV TM will begin at WTP4 which will be constructed on a 92-acre parcel to the south of RM 2222/Bullick Hollow Road and to the west of RR 620. The terminus of the Jollyville Transmission Main will be at the Jollyville Storage Reservoir at the intersection of Highway 183 and McNeil Dr as depicted on Figure 1.

Approximately ten alternative alignments were developed and subjected to a fatal flaw analysis. Four alternative alignments survived the fatal flaw analysis and were subjected to a detailed selection analysis using CDP software. Alignment alternatives 1 and 3 included approximately 10,000 feet of trenched construction on the northeast segment of the alignment. Alignment alternatives 2 and 4 used tunnel construction exclusively.

## SELECTION CRITERIA

The four criteria used in the decision-making process were environmental impacts, community impacts, constructability and cost. The project team, consisting of the COA and the Black & Veatch design team, agreed that the criteria should be weighted equally (i.e., each criterion would make up 25 percent of the decision).

### Environmental Impacts

Much of the JV TM in Reach B and C as depicted on Figure 1 crosses beneath the Balcones Canyonland Preserve (BCP) which was created as a tool for private landowners in this section of Austin to comply with the Endangered Species Act (ESA). The BCP functions as a land mitigation bank where landowners within the area in endangered species habitat can purchase land to set aside as a preserve for perpetuity. The BCP provides critical habitat for two neotropical migratory songbirds, six karst invertebrates and 27 species of concern.

The Golden-cheeked Warbler (Dendroica chrysoparia) and Black-capped Vireo (Vireo atricapilla) are neotropical songbirds listed by the United States Fish and Wildlife Service (USFWS) as endangered. Six karst invertebrates including the Tooth Cave Spider (Neoleptoneta myopica), Bone Cave Harvestman (Texella reyesi), Tooth Cave Ground Beetle (Rhadine Persephone), Kretschmarr Cave Mold Beetle (Texamaurops reddelli), Tooth Cave Pseudoscorpion (Tartacreagris texana), and Texella reddelli.

The Jollyville Plateau Salamander (Eurycea tonkawae) JPS is a lungless salamander that occurs in the BCP that the USFWS considers to be a candidate for listing as an endangered species.

With the comprehensive list of rare and endangered species in the BCP, environmental issues were of paramount concern to the City of Austin.

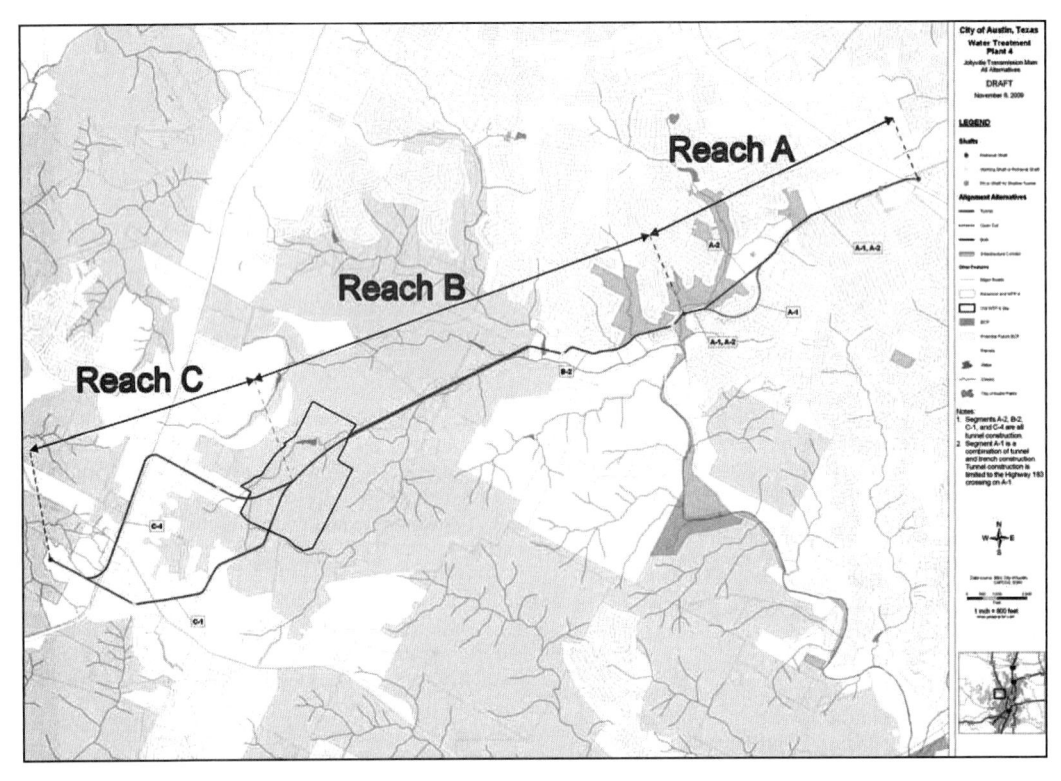

**Figure 1. Alignment segments that depict the four alternatives subjected to detailed evaluation**

Environmental issues included impacts to groundwater resources, surface water resources, flora/fauna and challenges associated with acquiring new permits or modifying or complying with existing permits.
Environmental subcriteria included:

- Number of shafts required
- Linear feet of alignment disturbing the Edwards Aquifer
- Linear feet of alignment within a specific spring's contributing area
- Total area of disturbance within a specific spring's contributing area
- Proximity of shafts to protected caves
- Acreage of total disturbance
- Area of disturbance within critical water quality zone
- Proximity of shaft location to Bull Creek or tributaries
- Disturbance to Jollyville Plateau Salamander habitat
- Disturbance to karst invertebrate habitat
- Disturbance to Golden Cheek Warbler or Black-capped Vireo habitat
- Area of vegetation or tree loss
- Need for permit amendments

**Community Impacts**

Community issues included impacts to residents and businesses during construction.
Community subcriteria included:

- Traffic disruption
- Dust and exhaust emissions
- Noise emissions
- Hours of operation
- Number of residents directly impacted
- Number of businesses directly impacted
- Emergency access restrictions
- Proximity to schools/daycare

**Constructability**

Constructability subcriteria included:

- Schedule impacts
- Right of way requirements
- Shaft site constraints
- Utility relocations
- Turning radii for tunnel

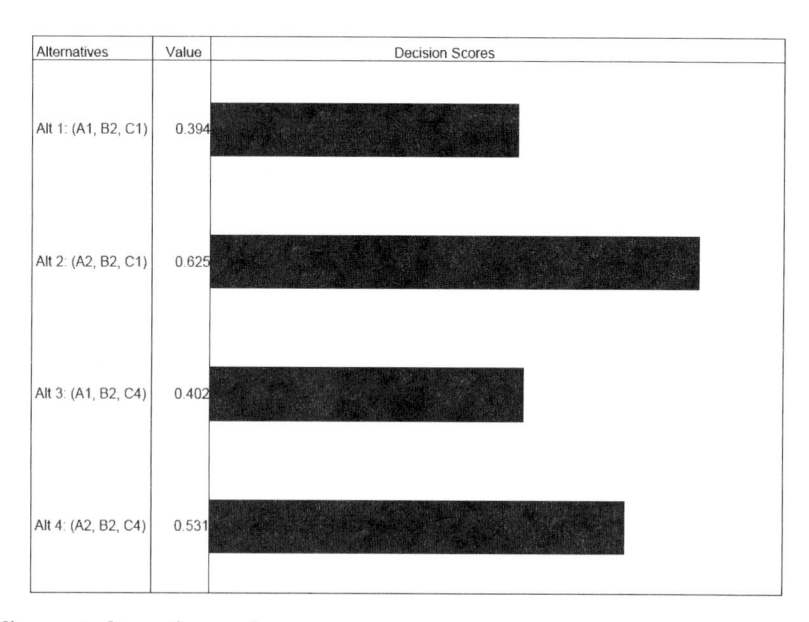

Figure 2. Alignment alternative scoring

## Cost

Cost subcritera included both construction and easement cost.

## SCORING THE ALTERNATIVES

Each alternative alignment was scored at the subcriteria level. Where possible, subcriteria were scored on a quantitative basis. This was possible for subcriteria such as construction cost, easement cost, number of shafts required, etc.

For the remaining subcriteria, a qualitative scoring system was used with scores ranging from the integers of 5 (most favorable) to 1 (least favorable).

The decision model results are shown in Figure 2.

## SENSITIVITY ANALYSIS

A sensitivity analysis was conducted to determine what changes to criteria weighting would result in a change to the model results. For example, the cost criterion weighting would need to be increased to 40.8 percent for Alternative 3 to outscore Alternative 2. Similarly, the community impacts criterion would need to be reduced to 3.2 percent for Alternatives 1 and 3 to outscore Alternative 2. It was noted that changing the weighting for environmental or constructability weighting would not change the model scoring.

Another sensitivity analysis was conducted to determine the model's sensitivity to cost fluctuations by construction method (see Figure 3). Due to uncertainties associated with overestimating trenching costs in an urban setting, the trenching cost estimate was reduced by 50 percent for Alternatives 1 and 3. The revised model results indicated that the cost criterion would have to be weighted at 39.1 percent for Alternative 3 to outscore Alternative 2. Thus, the decision model was considered to be relatively insensitive to trenching costs.

## SUMMARY AND RECOMMENDATIONS

Considering the acceptable alignment alternatives available for consideration, Alternative 2 outscored the other three alternatives using equal weighting among environmental impacts, community impacts, constructability and cost criteria. Sensitivity analysis indicated that cost weighting would have to increase substantially to change the decision model results. Considering the superior scoring of Alternative 2 in relation to environmental and community impacts, the COA and the B&V design team selected Alternative 2 as the preferred alignment alternative.

## REFERENCES

Black & Veatch. 2009. Preliminary Engineering Report. Austin, Texas.

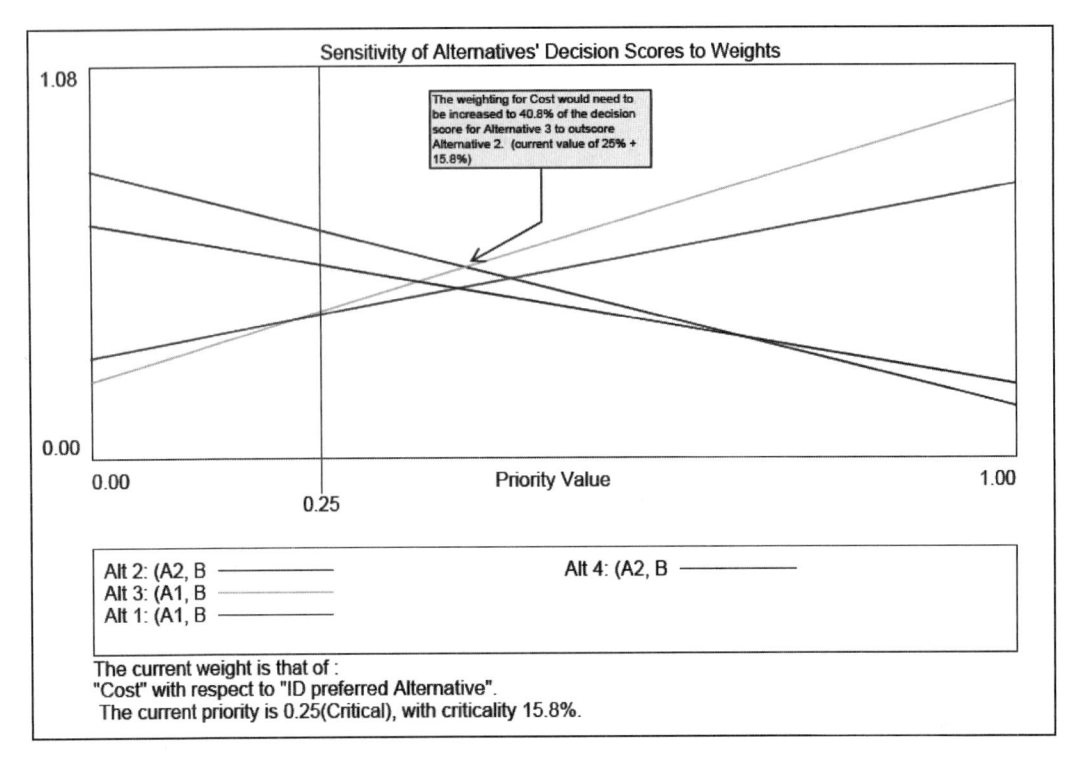

Figure 3. Sensitivity of alternatives to cost

# Sustainable Underground Solutions for an Above Ground Problem

**Laura S. Cabiness, Steven A. Kirk**
Department of Public Service, Charleston, South Carolina

**Stephen A. O'Connell, Jason T. Swartz**
Black and Veatch Corp., Charleston, South Carolina

**ABSTRACT:** City Engineers have turned to underground solutions to alleviate flooding problems at the surface, recently completing the design of an approximately $146M deep tunnel conveyance system. The new system will achieve sustainability through reducing freshwater inundation into saltwater wetlands and by increasing the quality of water discharged into the Ashley River. The Project consists of 8,500 linear ft of 12-foot diameter soft ground tunnel, collection system, wet well, pump station, and an outfall into the Ashley River. The City is currently seeking funding for the project and this paper will outline the extraordinary steps taken to secure such funding (funding outcome will be known in February 2010) in addition to describing key aspects of the design.

## INTRODUCTION AND BACKGROUND

Charleston has deep roots in the historic fabric of the United States. First founded in 1670, Charleston was a significant economic and cultural colony in the early years of the county. Many historic events occurred around Charleston such as the first shots of the Civil War which were fired off the coast at Fort Sumter. Today many people recognize Charleston for its historic significance, unique architecture, Southern charm, and coastal beauty, while to others it is home to their families and businesses. The historic Charleston Peninsula is home to nearly a third of Charleston's 124,000 residents and the nucleus of Charleston's economy.

### History of Flooding on the Charleston Peninsula

Charleston's coastal environment was a key to its success over the years. The coastal environment and inner harbor provided ideal protection from seaward and landward invaders while facilitating ease of travel and commerce. The same geographic characteristics that make Charleston so appealing are also the cause of some its biggest challenges. Charleston is located in the heart of what is regionally known as the Lowcountry, a designation that quite aptly describes the landscape of this historic city. With a surface elevation of only a few feet above sea level and a near table top topography, flooding has been a major obstacle for all Charlestonians and its many visitors. Parts of Charleston are prone to flooding due to daily high tides. The most severe flooding occurs when rain events coincide with high ocean tides as the drainage has nowhere to go and can result in hours of standing water (Photograph 1).

Charleston was among the first American cities to construct separate sanitary and stormwater drainage infrastructure. This concept has roots back to its founding and was a credit to the foresight of its founders. However, implementing a feasible stormwater solution was another matter. One of Charleston's early mayors struggling with providing proper stormwater drainage in 1837 offered a small reward in the form of a $100 gold coin to anyone who could solve Charleston's drainage issues. A solution was eventually born from many of the ideas submitted. The early solution was to construct a network of interconnected brick arches that discharge to the two rivers on either side of the peninsula. Gates were installed on the outfalls to control the tidal water. The piping was slightly undersized with the intention that scouring velocities would be achieved through proper management of the outfall gates and tidal exchanges. However, over the years flooding problems continued and many of the gates were removed. Siltation eventually further clogged the already undersized system.

To make matters worse, a six-lane "expressway" was constructed across the peninsula in 1968. This expressway is part of U.S. Highway 17 and is locally known as the Septima Clark Parkway but often referred to as the *Crosstown* by locals. From the moment the Crosstown was constructed across the Peninsula to connect the Ashley River bridges to the Cooper River bridges, a jagged, indelible scar was slashed across the City of Charleston, dividing neighborhoods, separating friends and families, and creating a tear in the fabric of the City. The six-lane expressway's sole focus was on quickly moving vehicles giving little thought or planning to improve stormwater management. The asphalt expressway added additional stormwater runoff to an already undersized system. As a result, during times of moderate to heavy rainfall that are as common as mosquitoes in Lowcountry summers, the Crosstown becomes impassable to vehicles, oftentimes for many hours, cutting off access to vital entities.

## SUBSURFACE CHARACTERIZATION AND HISTORY OF THE CHARLESTON PENINSULA

### Regional Geology

The project site is located in the Coastal Plain geomorphic province of South Carolina. A thick sequence of Cretaceous and Quaternary age, seaward-dripping sediments form the Coastal Plain province and the continental shelf throughout the area. This wedge of sediments lies on Paleozoic and Mesozoic crystalline rock and increases in thickness seaward from the Fall line at Surry Scarp about 110 miles inland. Progressively older formations are exposed at the surface from the coast to the Fall Line. Through the Charleston region, the sedimentary wedge is about 0.75 miles thick.

The composition of the Tertiary through Quaternary age sediments varies greatly. The Paleocene Mingo Group consists predominately of mudstone, sandstones, and limestone. Overlying this are the predominately calcareous Eocene and Oligocene age Santee and Cooper Groups. The Miocene through Holocene age sediments are primarily non-calcareous clastic sediments.

### Project Site Geology

Across the project site, the surficial soils (soils above the Cooper Group) consist of approximately 30 to 70 feet of loose, fine to medium grained sand, organic clayey silts and clays. Underlying the surficial soils is the Cooper Group, locally known as the Cooper Marl. An irregular errosional contact separates these two stratigraphic units. The Cooper Marl is a relatively massive homogenous olive green, highly calcareous, phosphatic, fossiliferous, clayey sand and silt which exhibits sufficient standup time for erection of initial support behind the excavation face.

Coarse-grained surficial soils exhibit *Flow* like to *Running* behavior and the fine-grained soils *Raveling* to *Squeezing* behavior according to the Tunnelman's Ground Classification system (Terzaghi, 1950; modified by Huer, 1974). Behavior of the Cooper Marl is generally described as *Slow Raveling*.

**Photograph 1. Typical flooding on the Charleston peninsula**

### Previous Underground Experience

Over 40 miles of tunnel have been excavated in the Charleston area and in particular within the Cooper Marl since 1928 to convey water, stormwater, and wastewater. Many of the tunnels were constructed using no temporary or initial support and in the case of one subsequent inspection have been classified in "most excellent condition," being less than 0.5 yard of material that sloughed or slipped off.

## PROJECT DESCRIPTION, DESIGN, AND SYSTEM COMPONENTS

The US17 Septima Clark Parkway Transportation Infrastructure Reinvestment Project is a significant city effort to alleviate frequent flooding within a large portion of the Charleston Peninsula. The project is designed to improve the mobility, efficiency, emergency preparedness, community livability, and to alleviate many of the flooding problems by reinvesting in the infrastructure. The project consists of constructing improved and additional surface collection systems throughout parts of the basins, drilling several shafts from the surface down as much as 150 ft, boring 12-ft diameter tunnels connecting the shafts, constructing a new pump station on the Ashley River, and constructing an 550 linear foot outfall from the pump station to the Ashley River. In addition, transportation advancements incorporate safer travel lanes for vehicles, improved intersections for pedestrian safety and vehicle efficiency, Intelligent Transportation Systems (ITS), and new energy-efficient traffic signals.

The project area makes up approximately 20% of the historic peninsula of Charleston that experiences regular, significant flooding (Figure 1).

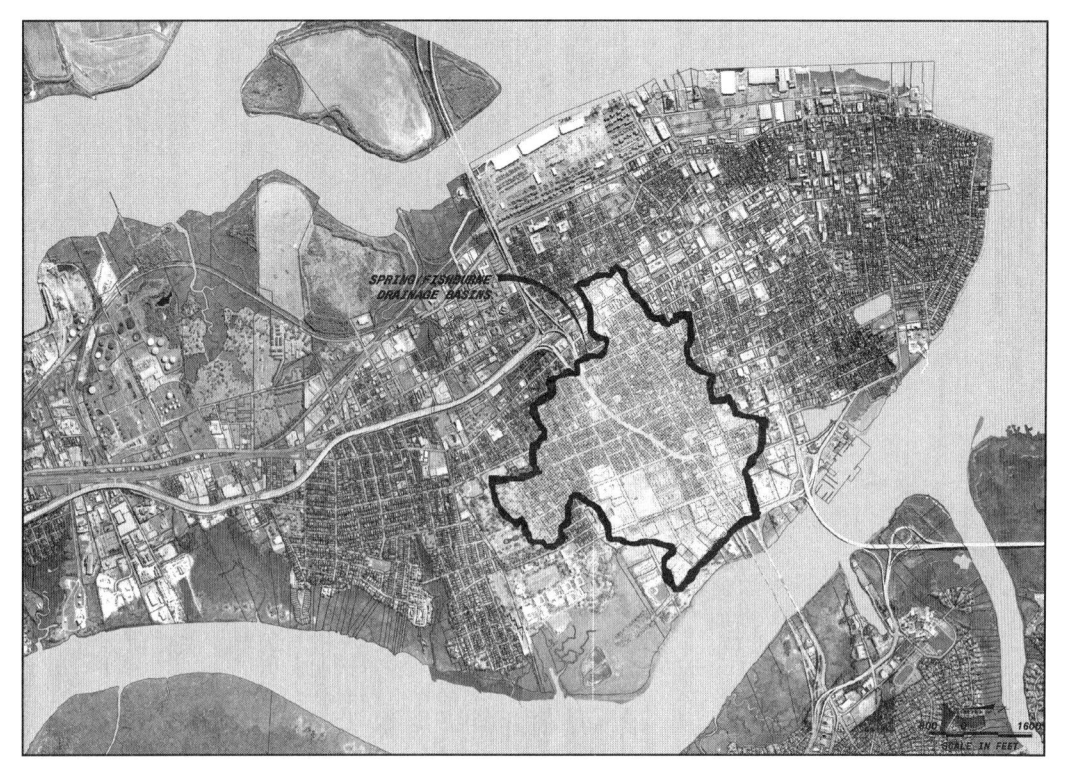

**Figure 1. Spring and Fishburne drainage basins**

Affected entities include the Medical University of South Carolina (MUSC), Roper Hospital, the VA hospital, the Citadel, Burke High and Middle Schools, Mitchell Grade School, US 17 (which facilitates interstate travel, commerce, and emergency vehicles), the City of Charleston police, the City of Charleston fire department, the Army Corps of Engineers, Berkeley-Charleston-Dorchester Council of Governments, the South Carolina Department of Transportation, several churches, and many businesses and residents in the area. This project will make U. S. Highway 17 (Septima Clark Parkway) functional and beautiful, ensuring that whether it is heavy rains or an approaching hurricane, the highway is passable, its adjacent emergency response and critical medical facilities accessible, and the surrounding community repaired.

**Collection System**

A conventional near surface collection system consisting of 12,300 ft of new near surface piping and 350 drainage structures is proposed along approximately 8.3 miles of state, county and city owned streets. The system will convey stormwater from the surface to the deep tunnel conveyance system via 8 drop shafts located within the drainage basins. Piping, inlets and drop shafts were strategically located within the project area to minimize disruption to private properties and public streets while maintaining effective service to the lowest lying areas in the basins.

This component also includes modifications and additions to the conventional stormwater drainage system to direct flows to drop shaft positions throughout the Spring Fishburne drainage basin.

*Design*

Vortex structures were implemented into the design of the collection system at each of the 8 drop shafts to limit the amount of entrained air entering the deep tunnel conveyance system. As air enters and accumulates in the conveyance system it has detrimental affects on the systems overall hydraulic performance by limiting capacity and reducing efficiency. A physical 1:11 scale hydraulic model (Photograph 2) of the vortex chamber and drop shafts was performed by Clemson Engineering Hydraulics (CEH) to determine the flow characteristics of the conveyance system and allow the design team to optimize the structure.

## Tunnels and Shafts

Drop shafts and a deep tunnel system responsible for conveying stormwater from the surface to the pump station were designed beneath the highly urban streets of the Charleston peninsula, minimizing disruptions to private, commercial, and residential properties. The system is comprised of 8,420 ft of cast-in-place 6 to 12 foot finished diameter concrete lined tunnels, 4 large diameter working shafts (20–30 ft ID) and 8 drop shafts (48–54 inch ID). The system was designed approximately 120 to 160 ft below the surface within the Cooper Marl geological formation (Figure 2)

### *Design*

Determination of main working shaft, drop shaft, and tunnel locations were based on stormwater service needs in the basins while limiting disruptions to private property owners and existing utilities. Existing wastewater tunnels within the horizontal tunnel alignment determined the final vertical alignment of the tunnels. A minimum clearance of 20 vertical feet was maintained between the existing wastewater tunnels and the proposed stormwater tunnels. The selection of tunnel and shaft sizes were based on the capacity requirements for handling a 10 year storm event while maintaining proper hydraulic conditions set forth by the pump station design.

As stormwater generally has low corrosive potential compared to wastewater, final lining requirements for the shafts and tunnels were based on an economical solution designed for a service life of 100 years. Cast-in-place reinforced concrete is proposed as the final lining for the main working shafts and tunnels while lengths of welded steel casing with epoxy coating will form the final lining for the drop shafts.

Charleston was near the center of the highest intensity and greatest damage from an earthquake on August 31, 1886. With an estimated Magnitude 7.3 (Modified Mercalli) this event is the largest historical earthquake that has occurred in the southeast United States and it devastated Charleston. Structural damage was reported up to several hundred kilometers away, with about 90% of the buildings in Charleston damaged and 60 people killed. Three other major earthquakes have occurred within the last 3,000 to 3,600 years, and the center of the area of highest intensity during the 1886 earthquake is on a zone of continuing micro-seismicity. Structures whose components extend through the surficial soils into the Cooper Marl are the most susceptible to damage and were therefore designed with seismic deflections in mind. The final lining of reinforced concrete for the main working shafts was designed to withstand a heavy seismic event without catastrophic failure to the structure. It was recognized that some minor repair work would be necessary but that the system would continue to provide some level of service. Due to the relatively low ground accelerations anticipated in the Cooper Marl during a seismic event no additional design efforts were undertaken for the final lining of the tunnels.

**Photograph 2. Physical model of vortex chamber**

### *Components*

The main working shafts will be constructed by the sinking caisson method of construction. Construction by caisson method involves assembling lifts of reinforced concrete above grade as removal of soils is ongoing at the excavation floor inside the caisson. The caisson is then left to sink under its own weight and the process is repeated until final invert depth is reached. Large diameter shafts previously constructed in Charleston have utilized the sinking caisson method through the surficial soils (coarse-grained and fine-grained materials) above the relatively self supporting Cooper Marl. Once in the Cooper Marl a simplistic system of steel ribs and timber lagging is often utilized.

Drop shafts are anticipated to be constructed utilizing a vertical auger drilling method in the wet. A vertical drill rig with helical augers and a temporary steel casing is installed as the drill progresses through the soil to final invert depth. A final lining of High Density Polyethelene (HDPE) or stainless steel is then placed inside the sacrificial casing and the annulus space grouted. The process is completed in the wet to aid in maintaining wall stability.

Construction of the tunnels is expected by means of a soft ground Tunnel Boring Machine (TBM), roadheader and hand mining methods with temporary support provided by steel ribs and timber lagging and a final liner of cast-in-place concrete. Two main spines of the deep tunnel compose approximately 7,400 linear ft of the overall system

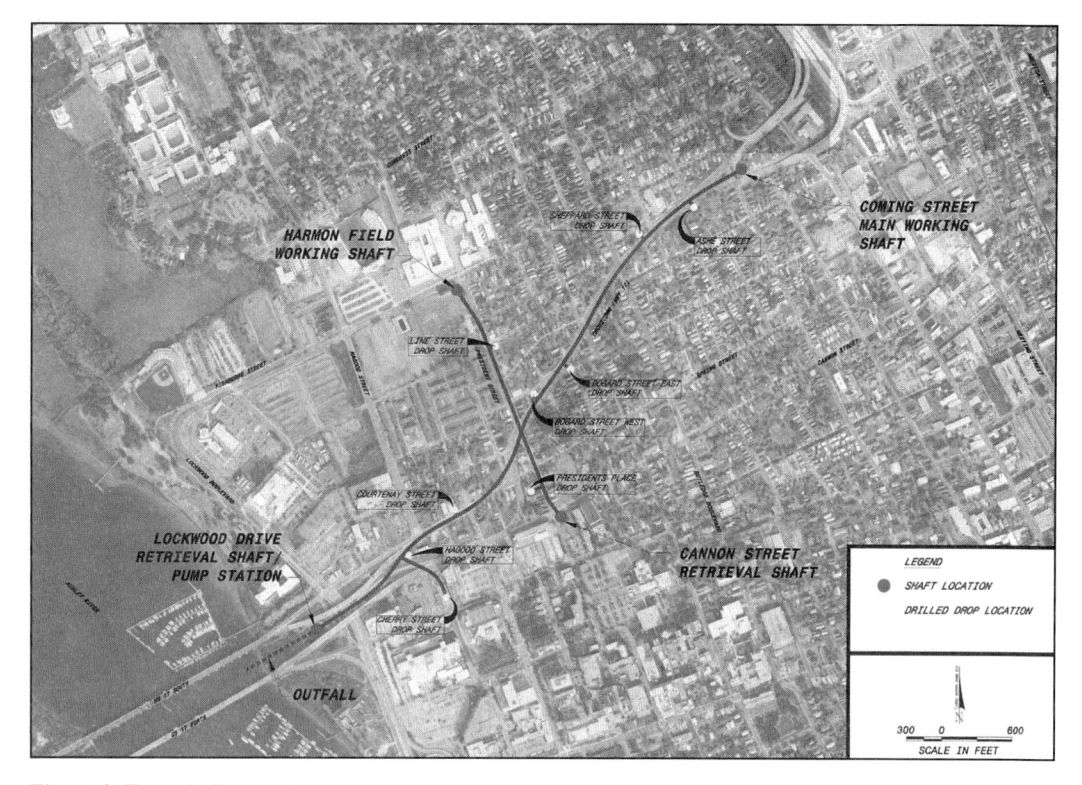

**Figure 2. Tunnel alignment**

and are anticipated to be constructed by means of soft ground TBM with the remaining 1,000 ft of tunnel constructed by roadheader and/or hand mining techniques. The full 8,420 linear ft of tunnel will be constructed through the geologic formation known as the Cooper Marl. Finished diameters of the tunnel conveyance system will range from 6 ft in the branch tunnels, and 8 to 12 ft in the main spines.

## Pump Station

The pump station building will be set on a cast-in-place concrete wet well structure and extend to grade at both ends. The structure is cast-in-place concrete where it extends below wet well height and structural steel above the wet well deck. Architectural features were added to exterior of the pump station including brick/CMU veneers, cast stone ornamentation, and landscaping to add aesthetic appeal to the industrial use building.

The base of the pump station structure (wet well) will be a 55 ft by 135 ft cast in place concrete structure approximately 33 ft below grade to house the three 66-inch diesel pumps with the top slab of the structure approximately 10 ft above grade. A foundation of prestressed concrete piles will support the wet well structure Completion of the wet well will coincide with construction of the outfall as described in the outfall section below.

### Design

Hydraulic modeling was carried out by CEH to determine the efficiency and performance of the three 66 inch diesel driven pumps. Adjustments to the design including the addition of a baffle wall in front of the pumps were added based on recommendations from the model.

## Outfall

The stormwater will be pumped into the Ashley River by way of an approximately 550 foot long cast in place concrete outfall consisting of three parallel 8 ft × 10 ft box culvert sections. Designed to be supported on prestressed concrete piling the outfall will be entirely below grade following completion of construction. The construction site is located between the two Ashley River Bridges (US 17 North and South) and is within the tidally influenced zone.

### Design

Design of the outfall was determined by the peak pumping capacity of the pumps, limiting the

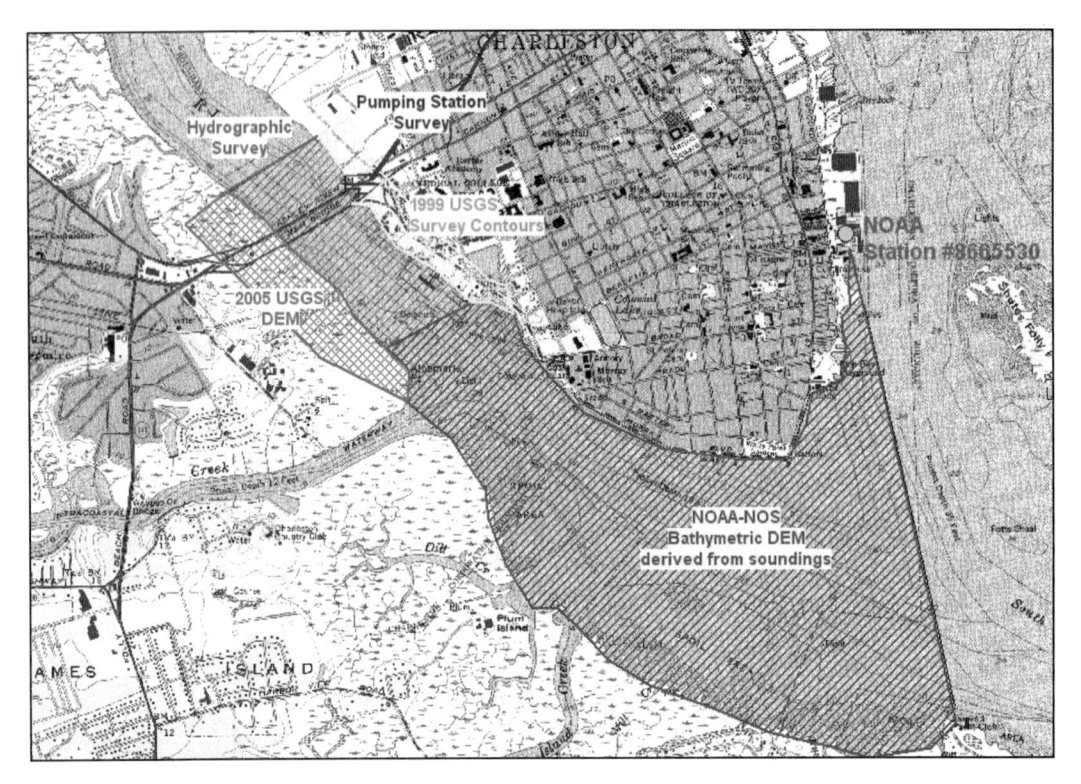

**Figure 3. Outfall model**

discharge velocities into the Ashley River, and mitigating disruptions to the marshland through which the outfall will be constructed.

Each box culvert section is dedicated to one of the three pumps and is capable of handling flow up to the maximum possible output of approximately 100,000 gpm. The outfall will be constructed below the Lower Low Level Water (LLLW) mark to ensure they are completely filled with water at all times during operation.

Both a 2D grid model and a Computational Fluid Dynamic (CFD) model were performed at the discharge point of the outfall to determine its affects on the existing bridge piers, shipping channel and the marinas both upstream and downstream of the system. The model was run at both low and high tide conditions with all three pumps running to establish a worst case scenario (Figure 3). It was determined that the addition of an outfall to the Ashley River would have very minimal impacts to the bridge piers and marinas and no impact to the shipping channel. To further dissipate the discharge velocities and reduce scouring of the river bottom an enlarged box culvert section and a mat of rip-rap and gabion mattresses will be constructed at the terminal point of the outfall.

As part of the physical hydraulic model performed by CEH a 1:11 scale model of the outfall was included to determine if the amount of entrained air being pumped out of the culverts would hamper hydraulic conditions or cause massive bubbling of the river at the point of discharge. Based on the physical model, modifications to the system were made and an air vent just downstream of the pumps was added for air release.

Construction of both the outfall and pump station wet well will be completed simultaneously as both require the use of a temporary cofferdam for support and occupy the same site between the Ashley River Bridges. The steel sheet pile temporary cofferdam will support the side walls of the excavation and reduce the risk of damage to the adjacent bridges from slope failures of the extremely weak predominately cohesive soils. Excavation is expected to take place in the dry dewatered state with the option to excavate in the wet still available to the contractor.

### Roadway and Transportation Improvements

Transportation improvements and enhancements along US 17 will include resurfacing of the six lane section, improved pedestrian and disabled access, visually enhanced streetscaping including landscaping and lighting, and more efficient traffic control through updated signalization and intelligent traffic systems.

## SUSTAINABILE SOLUTIONS THROUGH DESIGN

### Reducing Fresh Water Inundation

Charleston's proximity to the Atlantic Ocean produces saline conditions in the nearby tidally influenced rivers, estuaries, tidal flats and marsh lands. Saline waters are home to a vastly different ecosystem than freshwater environments and require a percentage of brackish water for survival. Stormwater flooding paired with the current inadequate drainage system endangers these saline habitats by diluting the waters. By designing a new system capable of handling the frequent flooding the saltwater habitats can return to there previous state and thrive.

### Enhancing the Quality of Discharged Water

The poor quality of the existing drainage system, including undersized piping as well as poorly placed and insufficient inlets, leads to delayed gravity drainage of the stormwater into the Ashley River. As stormwater rises on the peninsula and becomes stagnant on streets and residential properties for hours after an event, garbage, animal waste, fertilizers, oil and gas from the under carriages of cars, and other debris become incorporated into the ponding waters and are eventually discharged into the regions natural waterways. By incorporating a design which provides sufficient capacity, number and strategic placement of inlets in the basins the proposed pump system will be capable of removing stormwater from the streets and residential properties more expeditiously and efficiently. Removing these ponding and stagnant waters will reduce the amount of debris and detritus in the discharged water and inherently increase water quality in the Ashley River.

A desiltation chamber has also been incorporated into the design of the pump station to remove silt, sands, and other small materials that make their way to the pump station through the near surface and deep tunnel conveyance systems.

As a final measure of increasing the quality of water discharged to the Ashley River a screening mechanism inside the pump station will remove any larger items that have made there way into the system. The mechanical screens have a 2 inch opening which allows water to pass through unimpeded while capturing large debris such as plastic bottles, buckets, and other items larger than 2 inches. The screens have a mechanical scraper which will remove the debris after each storm event and place it, by conveyor belt, into a dumpster on the outside of the building to be hauled off by truck when filled.

By virtue of reducing the time floodwaters reside at the surface and through desiltation and screening mechanisms at the pump station, the quality of water discharged into the Ashley River will be vastly improved.

### Reducing the Use of Fossil Fuels and Lowering Greenhouse Gas Emissions

When complete, the project will reduce the use of fossil fuels by eliminating idling motorists and removing ancillary vehicles including vehicles supporting public safety during a flood event. By reducing fossil fuel consumption greenhouse gas emissions will be reduced.

## SECURING PROJECT FUNDING

### The Beginning

The City of Charleston currently has two main sources of revenue for addressing the many stormwater challenges. The first is a property tax levy set aside for ongoing City wide stormwater management and operations. In 2008, this tax levy generated $1,662,010. The second is a stormwater utility fee on City wide sewer and water bills associated with the user's property area and use. In 2008, this stormwater utility fund generated $5,709,263. The 2008 expenses were $3,333,566 (City wide stormwater management, repair and operations costs) therefore $2,375,697 was set aside for long term maintenance, capital improvements and design services/permitting for the stormwater system.

The magnitude of this project is more than the residents and rate payers of Charleston can manage alone. With the City's mean household income at 84% of the national average and 95% of the state average, the City is not in a position to add a significant tax or fee increase. If the City attempted to fund the entire Project through debt financing, the annual interest and principle payments would run about $8,000,000/year (based on issuing a 20 year, 5.00%, $145,000,000 principle municipal bond). This payment would require the City to increase its city-wide stormwater revenues by more than 2½ times.

The City of Charleston has looked for alternate funding sources over the years. Repeated requests to legislators have resulted in a number of Authorizations for funding. The most recent was a $3M Authorization as part of the Water Resources Development Act of 2007. Unfortunately, the funding only becomes available once the funds have been appropriated and officially receive final passage into law, this has yet to occur. Other opportunities that the City of Charleston is currently working on are an effort to get funds appropriated from a $150K EPA STAG grant. The City's objective is to receive funding from as many sources as possible to supplement the little money already available. The city used a large portion of their available funding to fund the design efforts of the project with anticipation that funding would eventually be secured.

**Federal TIGER Grant Application**

In an effort to secure additional funding, the City of Charleston has submitted an application for funding through the American Recovery and Reinvestment Act. Specifically the city targeted the Federal Transportation Investment Generating Economy Recovery (TIGER) program. This program provides an opportunity for the City of Charleston to address the flooding issues within the Spring and Fishburne basins while improving the aesthetics and mobile efficiency of the US 17 Septima Clark Expressway.

Charleston is competing with projects from across the country for $1.5B available through the TIGER Grant program. The objectives of the TIGER Grant program include preserving and creating jobs, and promoting economic recovery. The grants will be awarded on a competitive basis to projects that have demonstrated a significant impact on the Nation, region, or metropolitan area. The most grant money any one state can be awarded is $300M. Awarding of the TIGER Grant money is expected to be announced in February 2010. To support the City's application, an extensive and highly-detailed Benefit Cost Analysis (BCA) was performed to clearly quantify the benefits and associated costs.

## CONCLUSION

The Charleston peninsula's low relief and proximity to tidally influenced bodies of water exacerbates frequent flooding during storm events rendering many of the streets impassable to vehicles and pedestrians, causing damage to residential and commercial property, reducing the quality and number of saltwater habitats, and polluting the waterways. Engineered solutions designed to alleviate stormwater flooding in these highly urban areas is essential for economic growth, community livability, and environmental sustainability.

# Large Diameter TBM Solution for Subway Systems

Verya Nasri
AECOM, New York, New York

ABSTRACT: This paper describes the use of a single TBM tunnel of approximately 12 to 13 m diameter for transit systems to house both light rail or subway tracks in a stacked configuration. This advanced concept allows significant savings in construction cost and time and reduces to a large extent the negative effects on the environment.

This solution, similar to the recently completed Barcelona Metro Line 9, integrates station platforms, train storages, crossovers, bypass tracks, ancillary areas and utility corridors inside a single tunnel. Each of the underground stations would consist of a cut-and-cover entrance structure and ventilation shafts, which would also serve as emergency exits. The excavations for the station structures and shafts would be done on one side of the street with nominal impact on traffic and utilities.

## INTRODUCTION

The North American transit agencies are familiar with the application of the twin tube tunneling approach, as most of their recent subway extensions used this method of construction. At the same time, large bore tunneling has been used as the preferred method on multiple projects over the last two decades all over the world including Europe, Asia and elsewhere.

The primary feature of the single tube tunnel configuration for a transit line is that station platforms, crossovers and tail tracks are all accommodated within the tunnel cross section. As such, the location of each of these major elements can be adjusted along the entire corridor to maximize design efficiency and minimize construction impacts. Also, another key feature is that station structures would be located on either side of the street. This would allow for such structures, which are constructed using cut-and-cover method, to be built with minimal impact on traffic along the street. This is critical to residents, businesses and the general public.

The new transit lines are introduced as a means of alleviating traffic congestion and the associated noise and air pollution. But the traffic impacts, noise, dust and business disruption that would be generated by cut-and-cover construction along a highly dense corridor may be counterproductive to the objectives of these projects. Part of this disruption would also affect existing public transportation services during several years construction phase of the underground structures.

## PLACING THE STATIONS INSIDE THE RUNNING TUNNEL

Since the beginning of the 90s in many parts of the world, the design of new transit lines generally follows a typical configuration: first excavation of cut and cover stations as rectangular boxes using slurry walls as support of excavation, and then connecting these boxes by a single tube TBM tunnel housing two side by side tracks.

The twin tube system which was preferred during the 80s since it was considered to be easier to build and to generate less settlement, is nowadays abandoned. Aside from major geometrical constraints, the experience shows that the twin tube is clearly more expensive than the single tube. It should be noted that at the present time, a TBM with a diameter large enough to place two side by side subway tracks within it (external diameter of about 9.5 m) can be well controlled. The conventional mining method for subway tunnel construction is now almost abandoned because of the risks involved with construction crew and adjacent structures which is not totally controllable.

After multiple use of large diameter TBM for building double track side by side transit tunnels, the Spanish engineers decided to go farther and put the stations inside the tunnel by slightly increasing the TBM diameter. This revolutionary concept was applied to 28 km of tunnel of the Line 9 of Barcelona Metro using a 12 m diameter TBM housing two stacked tracks (Figure 1) (Della Valle 2003).

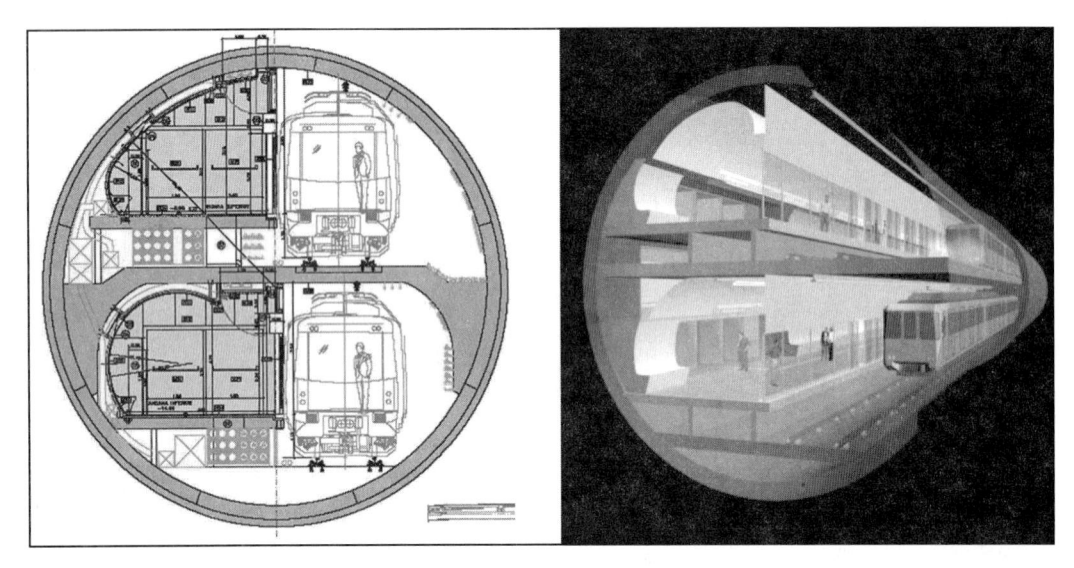

Figure 1. Stacked platform station within a single large diameter TBM tunnel (Line 9 Barcelona Metro)

The conventional twin tube tunnels with cut and cover stations construction is disruptive to vehicular and pedestrian traffic and requires significant construction costs to provide for temporary decking, special staging and phasing of the project and usually extends the construction duration due to the restrictiveness of worksite below the surface.

For the single tube option, since the platform is integrated within the tunnel, the remaining underground work at the stations is limited to access shafts and short connector tunnels. Station configuration can be developed to satisfy the transit agencies design criteria and requirements for vertical circulation, ventilation and fire/life safety. Each of the underground stations consists of a cut-and-cover entrance structure and ventilation shafts, which also serve as emergency exits. These excavations will be performed in one side of the street with no impact on the traffic and utilities.

The single tube solution will promote operation safety; in fact the internal slab will seal off the two platforms which will be behaving like two separate tubes with the possibility of closely spaced emergency stairs equipped with fireproof doors between the two tracks. Each platform acts as a refuge area for the other, with connection possibility at a much higher frequency than the actual standard safety requirement.

Platform edge doors solution can be adopted similar to other major cities like Paris, Singapore, London, Bangkok, Hong Kong, and Barcelona. The doors are physically separating the tracks from the platform enhancing safety in case of fire and reducing the psychological effect of the incoming train.

The access shafts can be equipped with high capacity elevators and emergency stairs. The elevators may be synchronized with train arrival, to minimize waiting time at the shaft bottom. The number of elevators per each shaft varies in function of the expected number of passengers.

Based on the studies performed by the author on several North American projects, the duration of construction is clearly shorter for the single tube option. The advance rates for both single and twin alternatives are comparable based on reported cases. In the case of single tube alternative, the platform structure can be completed just after the passage of the TBM. These studies also show that the estimated construction cost for the single tube alternative is clearly lower.

Some of the other advantages of the single tube option compared to twin tube include:

- Minimal construction disruptions on the street, as building the stations, entrances and other structures would be done outside the street right-of-way. Especially considering the large construction areas and muck disposal activity that would occur at each station location, associated with the cut-and-cover that would be required for the twin tube.
- Cost of disruption to businesses as a result of construction, including muck disposal, traffic and pedestrian reductions/restrictions. Cost of construction, especially considering the additional costs that would be associated with the hoarding, piling, decking, detouring, dewatering, utilities relocation and others at the stations.

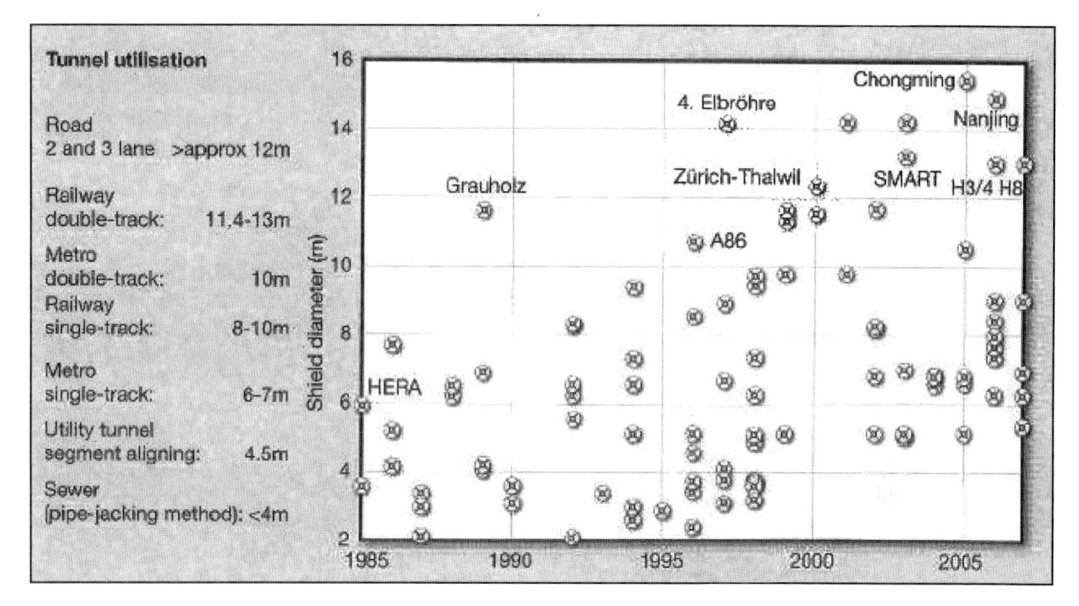

Figure 2. Increase in Mixshield diameter manufactured by Herrenknecht in the past two decades

## TBM SELECTION

The worldwide growing need for efficient infrastructure systems encourages tunneling solutions, particularly by mechanized methods. The mechanized tunneling process offers a safe, settlement controlled alternative to other tunnel excavation methods. Innovative solution like the multipurpose use of the tunnel profiles is increasingly demanded. This requires large diameter tunneling technology. Mechanized tunneling with diameters of up to 16 meters is reliably controlled and compared with conventional tunnel construction is substantially faster.

The TBM method requires a more comprehensive and thorough geological investigation, mapping and testing during the planning stage in order to evaluate the overall TBM viability and possible machine types that could be applicable. Considerations to use the TBM method have to take place during the early planning stage to establish tunnel alignment and profile, and possibilities to reduce adits and job sites while still maintaining the overall construction deadline.

Planning the tunnel projects with diameters exceeding 10 m is a common practice now. The large diameter TBMs can be designed for soft soil, hard rock, or mixed face conditions. Mechanized tunneling with diameter even larger than 15 m is today the state of the art and can be dealt with safely. Compared to conventional construction methods, the mechanized shield tunneling even with larger diameters is considerably faster and its limits are set by logistical issues such as mucking rather than by construction safety or financial concerns (Herrenknecht, and Bäppler 2008).

Today, it is possible to use the TBM method for just about any purpose and ground condition including Open Hard Rock TBMs and Shielded TBM designs like Single Shields, Double Shields, Mix Shields, Earth Pressure Balance Machines, and Slurry Shields in diameters from about 2–3 meters and up to approximately 16 meters. Figure 2 shows the increase in TBM diameter in recent years of one particular machine type (Mixshield) produced by one of the main TBM manufacturers.

Typically a TBM consists of a rotating head which excavates the material and from there the spoil enters into a chamber from which the material is transported to the surface. The complete operation requires a crew driving and running the cutting head, an excavation handling crew, and a segmental liner installation and storage crew. The diameter of the cutting head is selected based on the required tunnel geometry and total thickness of the assembled segmental liner ring and annular grout ring behind it.

Three types of TBM technology are predominantly used in current practice. They are referred to as Open Face, Earth Pressure Balance (EPB), and Slurry Shield. The open face shield method cannot effectively control the inflow of water or support the face of the poorly graded sand and gravel, and water pressure anticipated at most soft ground sites.

The EPB has been traditionally used in finer grade materials such as silts and clays but with the development of foams and polymers can now be used in a much wider range of soils and is even being

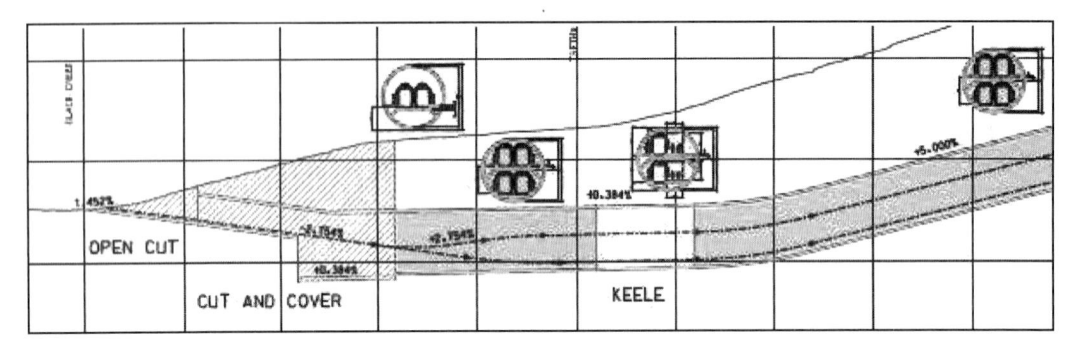

**Figure 3. Change in track configuration from side by side to stacked**

used in rock tunneling. Advantages for the EPB over the Slurry TBM includes: more flexible in the chosen mode of operation, easier access to the face for cutter changes, potentially cheaper machine costs, faster rate of advancement for tunnels in mixed ground conditions. Disadvantages include the difficulties associated with maintaining a good quality spoil in highly-permeable, coarse-grained soils.

Advantages for the slurry TBM include: operating well in sands and gravel soils; tunnel spoil is pumped to the surface avoiding use of mucking cars or conveyor belt system on steep grades. Disadvantages for the slurry TBM include: higher operating and slurry handling equipment costs, potential for slurry "blowout" causing environmental pollution.

Recent developments in TBM technology have brought about effective methods of building tunnels in various types of soil, rock and mixed face conditions. A dedicated TBM is usually designed for the specific subsurface materials and conditions expected to be encountered on each project. The alignment is selected to provide at least one tunnel diameter of cover to the extent possible.

The development of underground technology using tunnel boring machines in recent years has reduced the potential differences between the EPB and Slurry TBM systems. A pressurized face machine is often recommended for use on the project. The contractor will have the option to choose either an EPB or a Slurry TBM. Figure 3 shows the change in tracks configuration from side by side at the portal to stacked at the station.

For transit projects, a large diameter single tube tunnel (12 to 13 m TBM diameter with about 15 cm annular grouting) can be considered to house both light rail or subway tracks in a stacked configuration. This solution which is similar to the recently completed Barcelona Metro Line 9 integrates station platforms, train storages, crossovers, bypass tracks, and ancillary rooms inside the single tube tunnel (Figure 4). Based on the existing experience, the construction cost and time and environmental impact of this concept is lower than the twin tube option.

## STATION CONFIGURATION

Stations for single tube tunnel consist of 3 separate functional elements:

- Stacked side platforms
- Vertical circulation (entrances and emergency exit buildings)
- Fire ventilation units

The platforms are located within the single tube tunnel itself, one at each level, to one side of the running tunnel. Essentially the station consists of side platforms stacked one atop each other within the tunnel structure to provide consistent access to the vertical circulation elements (Figure 5). As such, a single point of entry/egress to/from both platforms can be located anywhere along the length of the platform(s). In addition, the stacked side platforms can be located to the either side of the running tunnel within the tunnel depending on the preferred location of the main entrance building.

Vertical circulation structures include a main entrance building and at minimum, one emergency exit building, each located at either end of the station. These structures are separate entities connected to the single tube tunnel and platforms via individual pedestrian adits.

The main vertical circulation can include two sets of escalators (one up, one down, 1.6 m each), stairs (2.4 m) and an elevator (2.5 m). Due to the depth of the station platforms, an additional escalator can be included to accommodate faster passenger access between the platforms and the street (Figure 6).

The emergency exit building provides egress from each of the platforms and would include fire doors at each platform level to secure a safe access route to the surface.

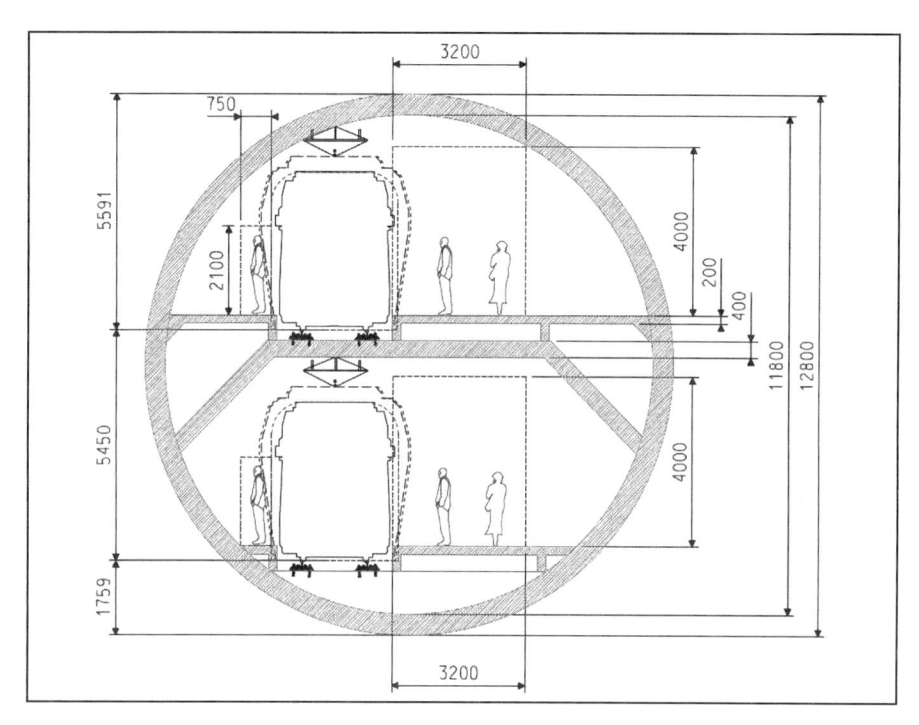

Figure 4. Example of cross section of the single tube tunnel for an LRT project (lengths shown in mm)

Figure 5. Single tube station platform

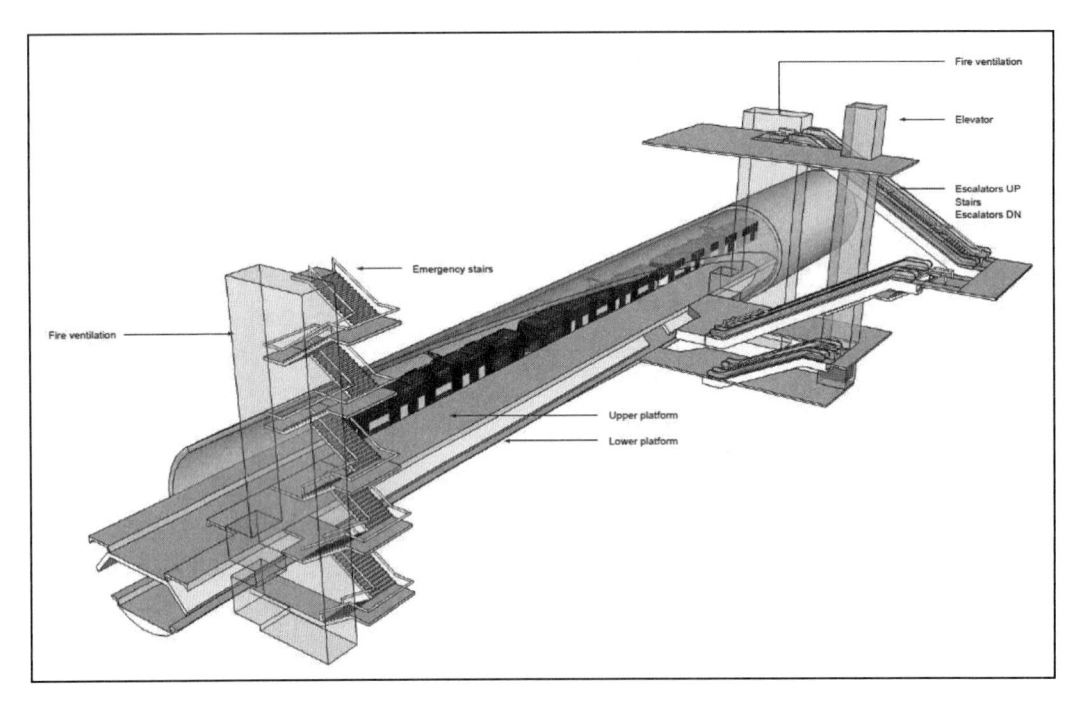

**Figure 6. Single tube station 3D view**

Fire ventilation units (FVU's) are required, one at each end of the station beyond the platforms. For the purpose of this study, standard FVU's applied in the transit systems have been used, stacked atop one another beyond either end of the platform. This allows for all vertical circulation and fire ventilation elements to be consolidated into single or adjoining structures reducing the number of footprints required (Figure 7). As such, ventilation shafts can be raised higher, beyond street level to avoid impacts during smoke discharge.

Station design initiatives from other single tube tunnel studies had investigated options for staggering the side platforms within the tunnel, each to opposing sides of the subway or LRT running tunnel. The staggered platform layout was intended to provide direct access from each platform to both sides of the street above. The design investigations revealed that staggered side platforms within the single tube tunnel would require a complex combination of mezzanines and pedestrian tunnels outside of the main tunnel to connect the platforms with one another and any vertical circulation to the street. This complexity would not only necessitate far more elevators and escalators but also lead to disorientation amongst passengers attempting to connect between the platforms and the street.

The ability to combine FVU's with vertical circulation into a single or adjoining structures reduces the number of structural components required. Albeit larger in size, ideally the station requires only two vertical structures:

- Main entrance building with adjoining FVU
- Emergency exit building with adjoining FVU

These structures exist independently of the single tube tunnel (horizontal structure) and can be located anywhere along the length of the station platform but must always be located on the same side of the tunnel as the platforms. The platforms also exist independent of any site constraints since they are within the single tube tunnel and can be located anywhere along the horizontal alignment limited only by natural topography (slopes) and associated track work. In addition, due to the larger diameter size, the stacked platforms can be located on either side of the single tube tunnel. This versatility in sitting individual elements allows the main entrance building to be located at a preferred location, typically at an intersection allowing direct connection to surface transit routes. The location of the entrance building will determine the location of the platforms. Opportunities exist for satellite entrances to be located in the immediate vicinity (either side of the street) and connect via pedestrian tunnels to the main vertical circulation building.

Construction of the platforms occurs within the tunnel structure which has no impact on adjacent properties or surface activities. In addition,

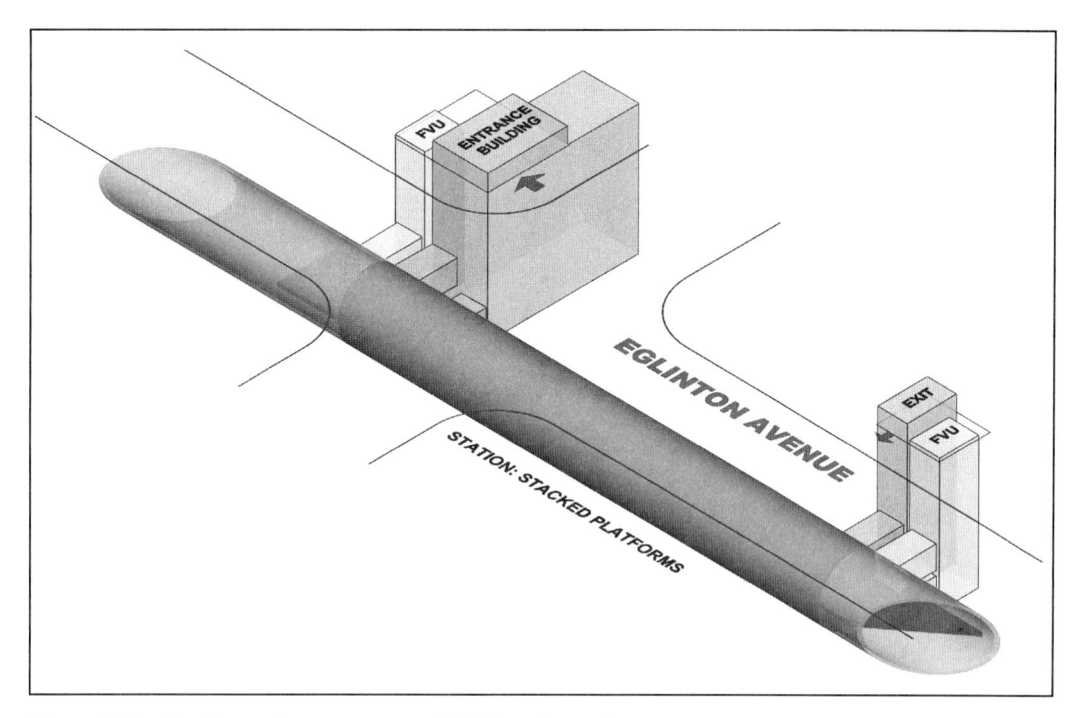

**Figure 7. Single tube station entrance and FVU configuration**

the continuity and independence of the tunnel from the vertical structures allows for increased capacity with ability to lengthen the platforms without additional impacts. As an example, the platforms could be constructed at 64 m lengths for a 2-car train and then in the future extended to 96 m for a 3-car train. Additionally, should demand warrant, these platforms can be extended even further, limited only by the vertical alignment constraints and special track work.

The combined structures of vertical circulation and fire ventilation can be constructed independently from the single tube tunnel. These structures can be constructed prior, during or after tunnel boring operations without impact on street level activities. The impacts will be similar to that of any surface building structure affecting only the immediate sidewalk or curbside lane depending on the size of setback(s). Once both the tunnel and vertical circulation structures have been completed, connections between the two elements can be mined.

In all situations, construction of the entrance and emergency exit buildings along with tunneling operations can occur with no impact on day-to-day surface activities. As such, vehicular and pedestrian traffic can flow unimpeded especially at intersections making it ideal for high density areas.

The vertical circulation structure is based upon local building code standards to determine minimum spatial requirements. The depth of a single tube tunnel and associated platforms may not be conducive for passengers' perception in comfort and safety when implementing engineering minimums. Passenger safety and comfort is directly associated with visibility. Visibility from the entrance building, down the vertical circulation structure and through to the platforms can be highly constrained when using engineering minimums typically applied for shallow or deep stations.

The design and functional layout of a single tube station are different from the typical centre platform arrangement found at many of the subway stations. The emphasis on connectivity and accessibility in twin tube stations is on horizontality versus verticality in single tube stations. Passenger accessibility for single tube stations should not be perceived as a drastic departure from that for twin tube, since the vertical circulation structure is more similar to that found in buildings especially office towers or shopping centers with multilevel atria.

The perception of depth is a social and psychological factor that is premised upon the volumetric confines within which a person either feels forced or choosing to enter. The ability to see where one is travelling to is an important criterion affecting people's perception of depth and safety. Designing the vertical circulation shaft in a single tunnel station to be spacious and open is equally as important as managing direct and well lit access from individual entrances to a mezzanine or concourse level in twin tube stations.

The introduction of an atrium-type space within the vertical structure would greatly increase visibility between levels and allow passengers to not only see where they are going and want to go but also assists in reducing the perception of depth and distance. Atria have been highly successful in higher density developments of all types, worldwide. The application of an atrium-type volume would allow the vertical circulation structure to function as the backbone for ancillary underground passages to adjacent properties especially in high density downtown areas.

The single tube tunnel concept can be made compliant with all the requirements of the latest version of NFPA 130 and other local codes. Stations will be designed to be evacuated within four minutes with all passengers reaching full safety within six minutes. This is an easily achievable goal given the excellent location potential of fire separations in the structures immediately beside the platforms of the single tube tunnel. It is reasonable to consider the single tube tunnel performs more safe in this regard as fire separations can be placed anywhere along the platforms in the connections to the vertical transportation elements.

## SUMMARY

In this paper, two TBM tunneling methods for transit projects are compared. The alternatives considered in this evaluation are:

- Twin tube alternative—two conventional size TBMs for excavating two running tunnels and cut-and-cover tunneling method for the construction of underground stations and crossovers;
- Single tube alternative—one large diameter TBM for excavating one running tunnel which includes all the stations and crossovers.

These alternative tunneling methods were studied as part of the Environmental Impact Studies for several North American transit projects to determine the most cost-effective method of construction.

Construction staging and maintenance of traffic is determined to be more favorable for the single tube alternative. Major cut-and-cover operations for the twin tube option will result in difficult construction conditions in congested areas along the street. Environmental impact in general is again more favorable towards the single tube alternative.

Over the past twenty years the TBM technology has continued to evolve around the world where TBMs with diameters greater than 12 meters have become common. Conceptually, the primary feature that would make single tube the preferred configuration is that the running tracks, station platforms and ancillary spaces are all contained within a large diameter tunnel. As such, provisions for station entrances, ventilation shafts, etc. can be taken "off alignment" of the tunnels and greatly reduce the need for roadway and intersection disruptions at the surface as would be required with cut and cover construction.

The time savings in construction of the single tube are realized mainly from the construction of the stations. Both the time savings and minimization of roadway right-of-way disruptions, due to cut and cover construction, translate into significant construction cost savings and much more manageable public relations effort.

Given the significant cost savings, overall shorter project construction time, simpler station design and construction with significant less surface disruption and reduced need for complicated cut and cover and maintenance and protection of vehicular and pedestrian traffic, the single tube tunnel solution can be considered as a serious alternative in the study of transit projects.

## REFERENCES

Della Valle, N. (2003), The New Line 9 of Barcelona Metro. Rapid Excavation and Tunneling Conference 2003, New Orleans, Louisiana, June 16–18, 2003. pp. 1280–1293.

Herrenknecht, M. and Bäppler, K. (2008), Tunnel Boring Machine Development. North American Tunneling Conference 2008, San Francisco, California, June 8–11, 2008, pp.52–57.

# Digging Deep to Save Green While Being Green and Sustainable

S. Nielsen, J. Morgan
Department of Public Works, Indianapolis, Indiana

J. McKelvey, J. Trypus
Black and Veatch Corporation, Indianapolis, Indiana

ABSTRACT: Like many other cities, the city of Indianapolis has been tackling their combined sewer overflow (CSO) and general wastewater services issues. The city's overall long term control plan price tag exceeds $3 Billion (B) with the primary program component being the deep tunnel system estimated at approximately $1.2 B. Faced with steep rate increases over the next 20 years, the city decided to investigate other sustainable alternatives with potential cost savings while meeting the objectives of the Consent Decree. This paper presents considerations and information on the evaluations required to construct a deep tunnel system, which is expected to result in over $140 Million in cost savings over a combined shallow and deep rock tunnel system.

## INTRODUCTION

### Combined Sewer Overflows

Many cities are facing the challenges of controlling combined sewer overflows (CSO), and progressing toward the design and construction of CSO abatement facilities to improve water quality. One of the primary abatement measures typically considered is the storage of CSO until after the wet weather event where it is then sent for treatment at the wastewater treatment plant. Many older cities are served by combined sewers that carry both storm water runoff and sewage. During rain events, combined sewers fill to their capacity and discharge a mixture of storm water runoff and sewage directly into waterways, which adversely affects water quality. To control CSOs, the Environmental Protection Agency (EPA) requires jurisdictions with CSOs to develop a Long Term Control Plan (LTCP) to identify how the untreated discharges will be abated. The overall goal of a LTCP is to identify alternatives to reduce and/or eliminate the volume and frequency of CSOs well into the future. One of the remedial measures to address this issue is to build consolidation sewers and convey the CSOs to a storage and conveyance tunnel.

### Storage Tunnels for CSO Abatement

Tunnels provide a long term solution for abating CSOs by storing the combined flows underground. Following the rain event, the tunnel conveys stored CSO to a deep pump station. Stored CSO in the tunnel is pumped out and treated at the wastewater treatment plant. The design and construction of a storage tunnel for controlling and conveying CSOs is a significant undertaking and a large capital investment. As such, careful consideration during the facility planning and design phase of a tunnel project can help to mitigate risk and potential pitfalls during construction.

### Deep Rock Tunnel Connector and Fall Creek/White River Tunnel System Evaluation

An evaluation study and advanced facility plan of CSO abatement facilities to supplement the city of Indianapolis' (city) LTCP was completed in 2005 and 2009. The pre-design efforts were completed to develop information for the overall best cost-benefit for constructing the Deep Rock Tunnel Connector (DRTC), which was formerly the Interplant Connector project, and the Fall Creek/White River Tunnel System (FCWRTS) in consideration of the city's CSO LTCP consent decree. A total of 44 CSO outfalls (27 along Fall Creek and 17 along White River) will be consolidated and conveyed to the DRTC and FCWRTS.

In March 2008, the city of Indianapolis decided to re-evaluate the proposed shallow soft ground tunnel known as the Interplant Connector project. The Interplant Connector project was under design for construction of a 12-foot finished diameter soft ground tunnel to be constructed utilizing a pressurized-face tunnel boring machine with segmental precast concrete lining. The original project was evaluated to determine program cost savings that would result from constructing the Interplant Connector as a deep rock tunnel (subsequently know as the DRTC) instead of a shallow soft ground tunnel, and reducing the proposed diameter of the upstream FCWRTS rock tunnel that was planned to discharge into the system. The evaluation determined that converting to the DRTC alternative to address the issue for CSO control and abatement along Fall Creek and White River resulted in savings over $140 Million.

Information is included in this paper for the DRTC and FCWRTS projects as it pertains to the considerations that should be completed during preliminary project phases of any CSO storage and conveyance tunnel project. The primary components resulting in the substantial cost savings for the DRTC was the reduction in risk by moving from soft ground conditions to deep bedrock. As part of that assessment, the quality and borability of the bedrock ultimately proved to present less risk and more favorable tunneling conditions. As such, the geology and hydrogeology of the tunnel system is presented as a primary component of this paper.

## EVALUATION STUDY AND PRELIMINARY DESIGN

### Geology and Hydrogeology

Available literature on the regional geology and hydrogeology provided information for the preliminary development of the proposed tunnel system corridor. In addition to known geological data, information on public wells and private wells was evaluated based on available literature, including the review of previously completed well logs and geotechnical borings in the project area.

The soils along the tunnel corridor are thick unconsolidated deposits from three glacial ages. Dating from oldest to youngest, they are: Kansan, Illinoian, and Wisconsinan. The soils consist primarily of unconsolidated glacial and glaciofluvial sediments. Throughout most of the project corridor, the average sediment thickness is 100 feet. However, the deposit thickness ranges from less than 15 feet to more than 300 feet. The unconsolidated deposits are generally differentiated into two formations by age, the pre-Wisconsin (Illinoian and older) Jessup Formation and the late Wisconsin Trafalgar Formation.

The overburden along the proposed tunnel corridor consists primarily of sand and gravel. The thickness generally ranges between 65 and 80 feet in the north-east to about 120 feet in the southwest. The entire column of the deposit is sand and gravel and lies directly on bedrock except for two short sections along the proposed tunnel corridor. These sections have sand and gravel lying on or intercalated with discontinuous late Wisconsin till sheets resting on the bedrock.

The bedrock surface along the proposed tunnel corridor is developed on the Devonian carbonates except for the extreme southern end. It generally rests at an approximate elevation of 650 feet mean sea level (msl) to the north and slopes to about 600 feet msl on the up-per southern end, where the elevation remains relatively constant to the far southern end of the proposed tunnel corridor. Although more prevalent in southwestern Indiana, it is generally accepted that this surface may have been subjected to the solution activity of groundwater. It has resulted in open joints and other karstic features in the upper 100 feet of bedrock in the area along the proposed tunnel corridor.

At the far southern end of the proposed corridor, the Devonian carbonates are overlain by the New Albany shale, but the karst features are known to be developed for some distance under the New Albany. Devonian rock, mainly dolomite and limestone, of the Muscatatuck Group underlie the northeast corner of the county except in the deeper bedrock valleys that are floored by Silurian limestone, dolomite, and argillaceous dolomite of the Wabash and Pleasant Mills Formations. The Silurian is underlain by Ordovician limestone and shale of the Maquoketa Group. The Maquoketa Group is a regional aquiclude that forms the base for the local bedrock groundwater aquifer system in the Devonian and Silurian carbonate sequence.

Devonian and Silurian carbonate rock underlie most of Indianapolis, the valleys of the White River and the lower reaches of Fall Creek, which includes the proposed tunnel system corridor. The formations, with an aggregate thickness of nearly 300 feet, include in descending order the North Vernon Limestone; the Jeffersonville Limestone; the Wabash Formation consisting of dolomitic limestone and shale members; the Pleasant Mills Formation; the Salamonie Dolomite; and the Brassfield Limestone. At the extreme southern end of the project corridor, the Devonian carbonates are overlain by shale of the New Albany Formation.

The city of Indianapolis falls within the Environmental Protection Agency (EPA) defined seismic impact zone due to the thickness of the soils that could result in amplification of earthquake shaking. Several paleoseismic studies have been previously completed to document major prehistoric earthquakes, which have generally occurred in the Wabash Valley and southeastern Indiana. However, a report completed in 1995 on Indiana's seismic risk concluded that additional geotechnical and hydrologic data is needed to better define the future risk of liquefaction in central and southern Indiana. Considering that no past indications of earthquake liquefaction features have been confirmed for Indianapolis, specialized seismic zone building codes are not currently planned for design and construction.

Several aquifer systems produce groundwater in Indianapolis. Water level data from wells, piezometers, and subsurface geology indicate that sand and gravel deposits at and above the pre-Wisconsin surface constitute a shallow aquifer. The shallow aquifer is generally confined in the upland areas and unconfined in the lower reaches of the tunnel corridor. Below this are three generally confined aquifer systems, corresponding to the three relatively continuous horizons of sand and gravel in the pre-Wisconsin deposits, and the bedrock aquifer system. In general, these aquifer systems are directly connected along the tunnel corridor. To a limited extent, they may locally function as independent systems separated by cohesive fine grained till and till-like deposits.

The tunnel corridor is located in a geological environment that is capable of producing large amounts of groundwater. By studying local well yields, it is estimated that the outwash has hydraulic conductivities ranging from 40 to 415 feet per day ($10^{-2}$ to $10^{-1}$ centimeters per second). The potentiometric surface along the proposed corridor is shallow, ranging from an approximate elevation of 700 feet msl in the northeast to about elevation 670 feet msl at the south end. This corresponds to an estimated depth to groundwater of 15 to 25 feet adjacent to the floodplains along the proposed tunnel corridor.

The rock is well jointed, but the productivity of the aquifer system varies greatly depending on the extent of solution activity. The rock currently overlain by drift was subject to solution activity before the drift was deposited. The solution activity opened joints and produced karst features in the upper 100 feet of rock, which increased the aquifer transmissivity. The rock overlain by shale had retarded solution activity. The shale also restricts the downward percolation of recharge. The bedrock aquifer is most productive where it is in contact with the glacial outwash complex in most of the tunnel corridor within the upper 50 feet of bedrock. Recharge is greatly accelerated by the rapid percolation of water through the outwash. In this upper 50 feet of the bedrock, individual wells can commonly produce several hundreds of gallons per minute from the bedrock. Groundwater inflow will be a concern during construction of the tunnel and more importantly the shafts in the overburden or bedrock, and will be evaluated in more detail during the ongoing geotechnical exploration program in the design phase.

**Construction and Project Considerations**

The DRTC and FCWRTS system will consist of a main tunnel, up to three primary working shafts, up to three tunnel boring machine (TBM) retrieval shafts, a deep tunnel pump station, consolidation sewers, drop shafts, and connection tunnels. Figure 1 presents an illustrative representation of the overall tunnel system proposed for the project. The tunnel will be sized based on the required percent capture for CSO abatement along the Fall Creek and White River in the LTCP currently under negotiation with regulatory agencies. Main tunnel sizes that provide 95 percent (up to four overflows per year on average) and 97 percent (up to two overflows per year on average) capture along White River and Fall Creek, respectively, are currently under design. The percent capture volume and quantity of CSO events are based on the annual average CSO flows in the system.

Construction considerations for the tunnels include safety concerns; main tunnel and connection tunnel construction techniques; shaft construction techniques; power availability; handling and disposal of tunnel and shaft spoils; handling, treatment and discharge of water present during tunnel and shaft construction; and protection of existing structures. Several project components including access shafts, drop shafts, force mains, and outfall structures are near environmentally-sensitive areas. During design and prior to construction, environmental surveys will be performed along the tunnel system corridor.

By their very nature, infrastructure projects impact the community in which they are located. Construction of infrastructure projects involves the introduction of additional traffic, noise, vibrations, dust, and heavy machinery. Therefore, the impact of construction on the community and environment was considered. Project considerations include the assessment of community outreach and coordination efforts, especially as related to odor, traffic, noise and lighting concerns.

The process of dropping CSOs from the near-surface collection system to the main tunnel will entrain air. To prevent a reduction in the tunnel's hydraulic capacity and transient releases of high pressure air from the drop shafts, a venting system should be installed. The vented air may require treatment to reduce odors. Activated carbon is generally the best choice for odor control for this application. However, an evaluation is required during the detailed design phase to determine the liquid-phase and vapor-phase odor control potential of the existing CSOs near the proposed drop shaft locations.

A limited Phase I Environmental Site Assessment (ESA) was completed for the project corridor to identify areas that may pose a risk to locating the tunnel alignment, drop shafts, working shafts, or other necessary surface facilities in areas that may contain recognized hazardous, toxic, or radioactive waste (HTRW) conditions. The limited Phase I ESA revealed evidence of recognized environmental conditions (REC) within the project area. During design of the DRTC and FCWRTS projects, a full Phase I and Phase II

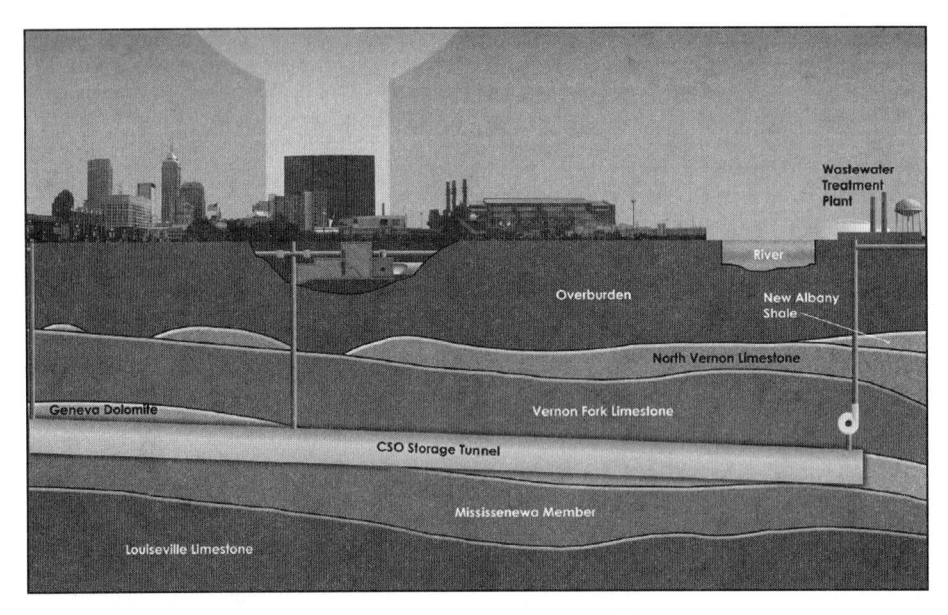

**Figure 1. DRTC and FCWRTS CSO storage and conveyance tunnel illustration**

ESA will be conducted at all proposed shaft locations and along the project corridor. If present and anticipated, issues regarding the handling and disposal of contaminated soils and groundwater resulting from project operations will be evaluated and recommendations will be developed.

Permanent subterranean easements are necessary along the entire main tunnel and connection tunnel alignments. The number and size of the easements will depend on the diameter of the main tunnel. The width of the tunnel subterranean easement is expected to be 50 feet. A 50-foot wide underground easement has been assumed for the 18-foot finished diameter main tunnel for the DRTC project. Each connection tunnel for the project will require a subterranean easement with a maximum width of 40 feet. Property may need to be purchased, if not currently owned by the city. Land should be acquired for each shaft and consolidation sewer location. However, the purchasing and acquisition of property should be evaluated by the city on a site-specific basis during the design phase.

Permits need to be considered during the design for the construction of the tunnel system because they impact the construction schedule and costs. Local, state, and federal regulations need to be reviewed to ensure all permits are obtained in a timely manner, as to not delay the project. A preliminary list of regulatory and private agencies and their required permits for the projects has been developed, and will be drafted and submitted during the design phase of the projects.

**Geotechnical Exploration Program**

Geotechnical site investigations were recommended to be conducted in multiple phases to obtain data for use during the planning and design of the tunnel system. Geotechnical investigations were recommended in two or three phases during the preliminary and design stages. Geotechnical investigations provide the required data for finalizing the tunnel alignment, evaluating construction methods and developing an accurate opinion of probable construction costs. The geotechnical data obtained during the investigations will also provide a basis to identify potential geologic hazards, determine ground conditions, characterize soil and rock mass, and establish baseline conditions along the tunnel alignment.

The initial phase of the geotechnical exploration program was conducted during the planning stage in order to obtain data that provides a better understanding of the existing ground conditions and characteristics. Although multiple data gaps exist to completely design the tunnel, the Phase 1 geotechnical exploration program scope should focus on planning level answers. Phase 1 data is critical for providing an understanding of the subsurface materials in order to determine the appropriate construction methodologies, identify the risks that require mitigation and initiate development of the preliminary opinion of probable construction costs.

The work performed during the Phase 1 geotechnical exploration program should focus on the following aspects: depth to the bedrock surface, soil and rock mass engineering characteristics, presence of obstructions in the soil, extent and properties of clay layers in the overburden, stratigraphic profile along the alignment, extent of karst development in the bedrock, and hydrogeologic properties of the soil and rock.

The hydrogeologic properties of the rock should be tested in distinct intervals to allow the tunnel to be placed in rock with lower permeability, if present. Locating the tunnel in rock with lower permeability would help minimize the amount of groundwater inflows during tunnel construction and exfiltration during tunnel operation, as well as reduce the requirements and cost for pre-excavation grouting.

There were 26 borings drilled along the tunnel alignment for the preliminary phase of the FCWRTS project, and five borings for the preliminary phase of the DRTC project. These borings provided a representation of the subsurface conditions across the entire alignment and at the most critical shaft locations. For the geotechnical program, most of the borings were advanced vertically to allow soil and rock sampling and testing. Piezometers were installed in the completed boreholes to determine the groundwater conditions before construction, and can be used in later design and construction phases. To establish bedrock conditions and identify any irregularities, these borings were drilled an additional 1 to 2 tunnel diameters below the proposed tunnel invert elevation. To vertically locate the tunnel, these boreholes were utilized to identify zones of lower permeability rock. Several borings located along the tunnel corridor were also drilled at a 15 degree angle from vertical in an attempt to intersect the near vertical jointing in the rock where solution activity may be most prevalent.

The geotechnical exploration program also includes overburden sampling using split barrels with standard penetration tests in granular soils and thin-walled tubes in cohesive soils. Rock, soil and groundwater samples were collected for physical and chemical analysis. Groundwater levels are being monitored from the piezometers installed in the boreholes to aid in the pre-construction assessment of groundwater table variation along White River and Fall Creek. Coring of bedrock was completed to perform water pressure tests to determine permeability by using a dual (straddle) packer assembly in bottom-up stages throughout the full depth of rock in each boring.

During the field investigations, an experienced geologist was onsite to keep a descriptive geotechnical log of each boring. Drilling observations and explosive gas monitoring results were included in the drilling records. The following testing was performed on the overburden soil samples: index and strength testing, grain size analysis and Atterberg limits, consolidation of cohesive soils, moisture content, unconfined compressive strength, and unconsolidated-undrained (U-U) triaxial compression.

Laboratory testing on the core samples was completed to determine the physical and mechanical properties of the bedrock. Various tests were performed on multiple samples of each rock type identified along the project alignment to establish a range of values. The laboratory tests on the rock core samples include: bulk density and moisture content, unconfined compressive strength with elastic moduli determination, indirect (Brazilian) tensile strength, petrographic analyses, punch penetration index, cerchar abrasivity index, and slake durability.

Based on the data obtained from the preliminary Phase 1 geotechnical exploration program, the Phase 2 program was developed and is being conducted during detailed design to further define the existing conditions. Phase 3 geotechnical exploration programs are conducted during later stages of detailed design to fill any data gaps identified during the Phase 2 program. To address any data gaps, identify hazards or conduct further hydrogeologic testing along the tunnel alignment, development of the scope for the Phase 2 and 3 geotechnical exploration programs should consider the following: additional borings, additional piezometer installations, soil and rock aquifer pump tests, and geophysical surveys.

The Phase 2 and 3 geotechnical exploration programs are expected to concentrate on data collection for the soft ground and rock connection tunnels. In addition, alternative drilling and sampling techniques should be considered during the Phase 2 and 3 geotechnical exploration programs based on data collected during the Phases 1 program.

Upon completion of the Phase 1 geotechnical exploration program, a preliminary Geotechnical Data Report (GDR) was prepared to document the findings. The GDR included the following: description of exploration program, description of region, site geology and hydrogeology, testing results, piezometric data, boring and well construction logs, and recommendations for the Phase 2 and 3 geotechnical exploration programs.

Upon completion of the Phase 2 and 3 geotechnical exploration programs, the preliminary GDR will be revised to include all additional geotechnical and hydrogeological project data from work performed. In addition to the GDR, a Geotechnical Baseline Report (GBR) should be prepared during the detailed design to be included with the contract documents.

The GBR is a contractual statement quantifying the baseline for the geotechnical conditions. The baseline statements in the GBR are not necessarily geotechnical facts. The contractor can assume soil conditions to be encountered during construction are as described in the GDR. Risks associated with conditions consistent with or less adverse than the baseline are allocated to the contractor while those more adverse than the baseline are typically accepted by the owner.

**Risk Management Strategies**

Planning, design and construction of a tunnel system are subject to a number of technical risks and contractual challenges that are inherent to large underground civil projects. Effective risk management and risk reduction, through continuous assessment, mitigation and contingency planning, are an essential and prudent management strategy. A systematic risk identification, evaluation, and management strategy will lead to early identification of risks and allow deployment of appropriate mitigation of possible onerous situations. A risk registry can be used to establish the basis for management of technical, contractual and socio-economic risks in the planning phase, subsequent preliminary engineering and design phases, and ultimately in the contract documents and construction management process.

A risk registry will be developed for the project and include detailed information on technical, contractual, and socio-economic risks. Technical risks are project-specific and construction-related, including cost increases; property and economic damage; failures; potential loss of life; delays; not attaining design, operational, and quality standards; claims; disputes and differing site conditions. Contractual risks are related to the management of geotechnical reports, design approach, and construction. Socio-economic risks include impacts on the communities, businesses, and profit or non-profit interest groups.

## CONCLUSIONS

An evaluation study and advanced facility plan of CSO abatement facilities to supplement the city of Indianapolis' LTCP was completed in 2005 and 2009. The pre-design efforts were completed to develop information for the overall best cost-benefit for constructing the DRTC, which was formerly the Interplant Connector Sewer project, and the FCWRTS in consideration of the city's CSO LTCP consent decree. A total of 44 CSO outfalls (27 along Fall Creek and 17 along White River) will be consolidated and conveyed to the DRTC and FCWRTS.

In March 2008, the city of Indianapolis decided to re-evaluate the proposed shallow soft ground tunnel known as the Interplant Connector project. The Interplant Connector project was under design for construction of a 12-foot finished diameter soft ground tunnel to be constructed utilizing a pressurized-face tunneling machine with segmental precast concrete lining. The original project was evaluated to determine program cost savings that would result from constructing the Interplant Connector as a deep rock tunnel (subsequently know as the DRTC) instead of a shallow soft ground tunnel, and reducing the proposed diameter of the upstream FCWRTS rock tunnel that was planned to discharge into the system.

Based on considerations of the geology, hydrogeology, construction impacts and overall risk as summarized in this paper, the most favorable cost-benefit for the city of Indianapolis was to convert the Interplant Connector shallow tunnel design into the DRTC project. The project modification included revising the 12-foot diameter of the original Interplant Connector into the 18-foot diameter DRTC, and revising the planned 30-foot diameter FCWRTS main tunnel to a more cost effective and less risky 18-foot diameter bedrock tunnel. In addition to cost savings associated with the smaller diameter FCWRTS main tunnel needs, there were approximately $55 million in cost savings as a result of eliminating improvements at the nearby Belmont Advanced Wastewater Treatment Plant (AWTP) headworks associated with the capture and treatment of one additional combined sewer overflow. The final evaluation determined that converting to the DRTC alternative and extending in into the FCWRTS to address the issue for CSO control and abatement along Fall Creek and White River will result in savings over $140 Million.

The city of Indianapolis is also evaluating and incorporating sustainable alternatives into both the designs of the DRTC and FCWRTS projects. Green Infrastructure (GI) best practices are being included in the design to manage new impacts during construction of structures at the surface such as working shafts, drop shafts and other appurtenant structures. Project sustainability considerations include the beneficial reuse and recycling of spoils during tunnel construction; the use of variable frequency drives for deep tunnel dewatering pumps to lower energy demand; and the use of the tunnel to house force mains for potential flow augmentation and sludge transportation force mains. In addition, operation flexibility is enhanced and the carbon footprint is reduced by having one deep underground pump station at the downstream end of the DRTC to serve both the DRTC and FCWRTS. As originally envisaged a deep pump station was required for FCWRTS, and another, albeit shallower, underground pump station for the Interplant Connector. The city of Indianapolis, in keeping with its goals to be more cost effective and sustainable, is "Digging Deep to Save Green While Being Green and Sustainable."

# Design and Construction Considerations for Shafts at Grand Central Terminal for the MTACC's East Side Access Project

A.J. Thompson
Hatch Mott MacDonald, New York, New York

**Abstract:** To link the 7 miles of running tunnels and station caverns being constructed as part of the East Side Access Project (ESA) that will bring the Long Island Railroad (LIRR) from Queens into a new terminal located beneath the existing Grand Central Terminal (GCT) in Manhattan, to the surface, five vertical and four inclined shafts are being constructed for passengers, utilities and ventilation.

Although the shafts are neither particularly deep, nor large, the complexity of the environment within which they are being built has had a significant effect on the design and construction including; the proximity of columns and footings that support the existing GCT structure, multi-storey buildings and roadways; utilities including gas lines and steam distribution lines: an operational railroad terminal immediately adjacent to and above the shafts.

This paper discusses the issues connected with the vertical and inclined access shafts that have influenced the design and construction process.

## INTRODUCTION

The Metropolitan Transportation Authority (MTA) East Side Access Project will bring the LIRR directly from Long Island into a new station located 120ft beneath the existing Grand Central Terminal (GCT) located in the heart of Manhattan. The new route alignment will connect the LIRR's mainline tracks in Queens via the existing 63rd Street Tunnel under the East River, to a new LIRR station constructed within and beneath the existing Grand Central Terminal. Work on the project commenced in the 1960s with the construction of the bi-level 63rd Street Tunnel across the East River, as a bi-level immersed tube with the upper level used for New York City Transit's F subway line and the lower level reserved for ESA. Approximately 50,000 feet of new tunnels together with a number of caverns, shafts and other structures will be excavated for the project.

ESA is the largest project undertaken by the MTA and at a cost of over $7 billion, one of the largest infrastructure projects currently underway in the United States.

## PROJECT OVERVIEW

### Tunnel Construction

In Manhattan, approximately 38,000 ft of tunnel will be excavated using Tunnel Boring Machines (TBM) from the end of the previously excavated tunnels at 63rd Street/2nd Avenue. The tunnel alignment heads south west passing beneath the Lexington Avenue Subway line at 59th Street before turning south beneath Park Avenue and finishing south of GCT at 37th Street.

### Cavern Construction

To house the various railroad and station facilities, 12 separate enlargements, totaling over 400,000 cubic yards will be excavated around the previously mined TBM tunnels. These include the two 60ft wide, 60 ft high and 1,200ft long East and West Station Caverns, located beneath GCT with cover that varies from 45 to 60ft. Excavation of these caverns is currently underway using both roadheaders and drill and blast. Initial support consists of combinations of tensioned rock bolts, weld mesh and shotcrete. Once cavern excavation is completed, a hybrid pre-cast and cast-in-place fully waterproofed final lining will be installed prior to installing the railroad systems and facilities.

### Concourse to Cavern Connections

The 350,000 sq ft concourse for the new Long Island Rail Road station will be located in what was the Madison Yard Storage Area on the lower level of GCT. The concourse will be connected to the station caverns by four inclined escalator shafts and five vertical shafts. These shafts are being constructed from the top down using drill and blast and mechanical methods and it is the issues that have affected the design and subsequent construction activities that will be discussed further.

## PROJECT ENVIRONMENT
### Planning Phase

Early in the planning phase it was recognized that the design and construction of these shafts would be significantly impacted by the physical environment within which they were to be constructed. As the shafts are being constructed in Manhattan Schist the predominant means of construction would be by drill and blast with mechanical excavation as an alternative, both of which methods introduce issues that had to be addressed.

The major items identified as impacting both design and construction included:

- Protection of Metro North Railroad (MNR) personnel, customers and critical operational infrastructure
- Construction was occurring within a 100 year old historical bi-level train shed
- Extent and accuracy of information available from as–built records
- Proximity of structural columns and footings for highway viaducts and air rights structures to proposed excavation perimeter
- Presence of high pressure steam lines and other utilities within the footprint of the Work Site
- Effect of close proximity blasting and mechanical excavation on the surrounding structures, utilities and infrastructure

### Metro North Railroad

One of the fundamental principles adopted was that construction work associated with ESA should cause no impact to MNR's revenue operations or affect the safety of MNR's customers, personnel and operational infrastructure. As with any operating railroad there are elements of the systems infrastructure such as signals, communications cables, switch machines, and instrument huts that have a degree of sensitivity to construction activities. Older instruments that included relays rather than solid state electronics were identified as specific items of concern that could be impacted by the vibration associated with construction activities.

As the work site from where the shafts would be constructed is located in a former train storage yard on the lower level of the existing GCT this means that in many places construction is happening only 22 ft below and 10 ft away from operational track.

### GCT Train Shed Structure

The existing GCT structure is a bi-level steel framed structure constructed prior 1913 in a large open cut excavated using drill and blast methods. The train shed extends from 42nd St to 52nd St and stretches between Madison and Lexington Avenue. Over time many structures and cross streets have been built over the structure requiring columns to be constructed through the train shed to take the loads to the rock beneath the lower level. Once the structures above were completed the underside of their basement slabs became the roof of the upper level of the train shed. In many locations the condition of these slabs and the fireproofing of the structural steel have deteriorated to the extent that chunks of concrete would fall from them onto the track and platforms below. Although critical repair and maintenance work had been undertaken over the years the general superficial condition of the structure, especially on the upper level, was considered poor considering the work to be undertaken. The train shed itself is separated into a number of discrete sections with expansion joints in between them, so whilst giving the appearance of a massive structure it is in fact comprised of a number of individual sections.

In addition given the age of the structure the availability and accuracy of as-built records was considered questionable. This was compounded as the original design calculations were not available for the train shed structure which meant that the available additional stress that could be carried by the structure as a result of construction impacts was difficult to determine.

### Structural Footings

There are over 250 columns within the footprint of Madison Yard that sit on grillages constructed in pits excavated into the rock which had to be taken into account when determining the location and shape of the shafts. These columns support different structural elements and different categories were identified during the planning and design phase. These include upper level train shed support columns, New York City Department of Transportation (NYCDOT) viaduct and building columns that pass through the upper level and are not connected to the train shed structure.

A related factor that had to be considered was the transmission of vibration form the demolition, mechanical and drill and blast excavation that would be needed to create the space for the concourse and sink the shafts.

### Utilities

Contained within the train shed are a large number of utilities including sewer and water lines suspended from the roof of the upper level, high voltage power feeders, gas lines, communication lines and high pressure steam lines. When GCT was constructed tunnels were excavated beneath the lower level to permit the transmission of high pressure steam throughout the terminal and to the properties above GCT. While the steam lines had been identified as an item of concern this was reinforced in 2007 when a high pressure steam line located just outside the footprint of GCT on Lexington Avenue ruptured leading to significant damage.

## DESIGN CONSIDERATIONS

### Shaft Geometry

Typically, shafts are designed to be as circular as possible, thereby allowing the final lining to act in compression, reducing the structural requirements for the final lining. However, circular shafts may require a larger footprint than is strictly necessary to house the various facilities included in them and the final cross section is usually a compromise.

On ESA it was recognized that circular shafts would not be practical. As noted previously there are a huge number of columns and footings of differing ownership that support various types of structures within the Madison Yard and it was decided that minimal interference with the structures and footings was the preferred option. Given the linear nature of the column lines that flanked the old train storage tracks, the shafts had to be shoehorned into the existing spaces leading to rectangular shaft layouts. While the use of non-circular shafts has resulted in more heavily reinforced final linings the layout of the elevators, stairwells and utilities within the shafts has been optimized and there is little non-used space being excavated.

The same principle of non-interference with the existing structure was adopted for the escalator shafts although in this case structural reframing was required and to minimize impacts on private structures wellways were located beneath cross streets and the reframing required is either to trainshed or roadway columns. Only at well way 1, where there are 5 escalators and hence a wider well way, an inclined shaft is required to a privately owned structure.

### Shaft Initial Support Design

Having determined the shape, size and location of the shafts the initial support could be designed. In some locations, the shaft excavation line is within 5 ft of existing footings and the effect of the loads being transferred through these footings had to be considered when developing the initial support design. During the design analysis it became evident that there could be a potential for wedges to develop beneath the footings which could result in significant structural movement and potential failure of a footing. As a result of this 15 ft long tensioned strand anchors were included as underpinning elements in addition to the regular initial support which consisted of #9 bolts and 4 inches of shotcrete. In addition vertical grouted dowels were also included to prevent any potential movement of ground from beneath the footings while initial excavation was underway and prior to the ground anchors being installed.

### Construction Effects on Existing Structures

It was recognized that the excavation of the shafts close to the existing footings would give rise to a number of concerns including:

- Stability of ground underneath the footings
- Transmission of vibration and noise through the footings and columns causing disturbance to building tenants
- Ability of the existing footings and columns to withstand the additional loads imposed on them as a result of the vibrations
- Accelerated deterioration of existing fireproofing and under basement slabs

### Stability of Ground Underneath Footings

Based on the information concerning method of construction and geological investigations, it was recognized that there was a significant chance of ground sliding from beneath existing footings during excavation and the initial support system was designed to minimize this possibility. In addition it was recognized that limits should be imposed on the amount of ground that could be exposed during excavation and specific restrictions were developed and included in the contract specifications to minimize the potential for wedge movement.

### Good Vibrations

The effect of close in blasting on existing structures is a subject that is less well understood than perhaps it should be. In addition to the various technical issues that need to be resolved one of the major obstacles that had to be addressed and overcome was the innate concern and fear related to the effects of blasting that exists. This issue was exacerbated by the fact that the single biggest entity affected by the blasting operations was MNR who had both operational infrastructure and the safety of personnel and customers to consider.

In order to assess the effects of vibration on the overlying and adjacent structures a series of studies were undertaken to establish baseline vibration levels in the structures resulting primarily from the movement of trains on the bi-level structure. A separate study was undertaken to establish the potential levels of vibration that the structures may experience as a result of blasting operations. Experience from other parts of the project was used to determine the appropriate scaled distance (k) in the calculations to predict peak particle velocity. Vibration levels were calculated for both the upper and lower levels of the train shed and contour maps of potential vibration levels that could arise from shaft and cavern blasting were superimposed onto plans of Grand Central Terminal.

This initial assessment of blasting impacts indicated that the whole-building response to blast events from the caverns and other underground structures would be negligible. The response of individual footings and columns to the close proximity shaft blasting becomes the controlling element in developing blasting criteria. This was predominantly due to the increased distance away from columns and footings, the crown of the caverns being at a minimum 45 ft below the level of the lowest structures at 45th and 48th St Cross Passages. Based on the information available and the parameters assumed for blasting it was recognized that levels of peak particle velocity in excess of 2 in/sec could be experienced in the footings and columns immediately adjacent to the shaft excavations. It was also recognized that the accuracy of seismograph readings taken immediately adjacent to the shaft may be questionable. Following this initial assessment two separate, although linked, strands were further investigated:

- Alternative measurements to establish affect of vibrations on structure
- Need for repairs, remedial and protective measures to existing structures/facilities

**Effects of Vibrations on Structures**

For above ground structures the impact of blasting is usually considered to be primarily related to the free response of the structure, a racking movement, related to the relative movement of parts of the structures in response to an imposed vibration by the ground. The blast wave, and associated vibration, decreases both in amplitude and frequency as it propagates away from the blasting location. For close in blasts such as for the ESA shafts, the dominant frequency at the receiver will therefore be high, in the hundreds to thousands of hertz. Structures have resonant frequencies that are primarily related to the stiffness and dimensions of the structure with large engineered structures typically having resonant frequencies in the neighborhood of 1 Hz, far below the frequencies of vibration generated by blasting.

Damage caused by vibration is due to movement of one part of the structure relative to another which can be characterized by strain, which is dimensionless, and described as the change of length ($\Delta l$) divided by the length ($l$):

$$\varepsilon = \frac{\Delta l}{l}$$

It would seem logical therefore to use strain to determine building response. However strain varies substantially within a structure, concentrating at corners and the middle of beams and walls, and varies for different materials, so it is difficult to determine a representative strain to a complete structure. However strain as a criterion for specific structural elements such as columns, is an appropriate criterion as they can be analyzed in terms of their load capacity and failure criteria. To undertake this analysis an approach that calculated the bending strains of the columns from the displacement of the base of the columns was used. The strain estimates were then calculated as a fraction of the capacity of the columns, as determined from the as-built information obtained.

It was determined that the structural behavior of the columns to a rigid displacement of the base of the columns. As a vibration input to these calculations, the following methodology was used:

- Vibration levels were calculated based upon the 95% confidence level from the regression equation from blasting undertaken previously on the project
- Displacements are calculated from the peak particle velocities using the sinusoidal approximation, and assuming that the frequency is 100 Hz, displacement $D$ is in inches, $PPV$ is in in/s, and frequency f is in Hz.

$$D = \frac{PPV}{2\pi f}$$

These then gave a failure limit based upon excess capacity, which was then used to develop the strain criteria that incorporate safety factors and the excess capacity.

In order to measure the strain resistive foil strain gages connected to data loggers were specified to be installed on columns where maximum strain is anticipated.

**Repairs, Remedial and Protective Measures**

As noted previously the GCT structure is over 100 years old and the roof of the upper level has been closed with a variety of structures and roads and these have become the ceiling of the train shed. Leakage of water lines, salt effects on the roadway viaducts has all contributed to a situation where significant portions of the concrete fireproofing and slabs has delaminated from the steel.

Based on the results of vibration analysis on the structure and visual inspections of the condition of the structures it was determined that it would be prudent to undertake remedial works to all concrete within 100ft of the location of the shaft blasts and install a protective mesh system above platform areas and critical MNR infrastructure to prevent spalling pieces of concrete falling on passengers, personnel or this infrastructure.

Associated with this the MTA undertook a more detailed visual survey of the train shed structure to establish a baseline of the condition of the structure prior to construction works commencing. This survey also identified a number of locations where additional repairs may be necessary and the results of the surveys were forwarded to MNR for their action as many of the identified locations were outside the blasting zones of influence.

**Metro North Railroad**

The effects of having an operational railroad 22 ft above and in some places 10 ft from the edge of excavations cannot be overstated. In addition to minimizing the potential impacts of blast induced vibrations on operational infrastructure the effects of noise, flyrock, fumes and dust also had to be taken into account and measures introduced to minimize or eliminate their effect. Major restrictions identified were periods during the day when blasting could not be permitted due to the intensity of train movements and the identification of windows within which the blast and post blast inspection could be undertaken prior to train movements restarting.

**Design Phase Summary**

Coming out of the design phase the following parameters were identified to minimize the impacts of shaft construction on the overlying and adjacent structures and protect MNR operations:

- Inspection and remediation of defective concrete above upper level platforms and critical infrastructure
- Installation of protective nets on upper level
- Use of both ppv and microstrain to measure effect of blasting
- Restrictions on round lengths and bench heights for both shaft and cavern excavation.
- Mechanical excavation in all shafts for 16ft below the lowest footing level
- Blasting only permitted at times that do not require MNR to suspend railroad operations
- Pre and post blast inspections of facilities and structures
- Extensive instrumentation program

- Shaft covers to be used to minimize potential for flyrock, limit airblast and control blasting fumes
- Use of channel and line drilling in shaft excavations to minimize vibration transmission
- Positive ventilation systems to control fumes and dust

**CONSTRUCTION PHASE**

In addition to the items listed above in the design phase and which were identified in the Specifications other issues that had to be taken into account by the Contractor during the construction phase including:

- *Restricted headroom*: As mentioned previously the work is occurring only 22 ft beneath the upper operational level of the GCT train shed. This means that lifting equipment has to be specially designed and constructed for these works and the Contractor elected to design and procure low height portal cranes for maximum operational flexibility.
- *Staging Area*: The staging area for this work is located some 9 miles away in the Bronx at the BN Yard. All materials and equipment has to be delivered to BN Yard and then transferred to MTA supplied rolling stock for onward shipment into GCT. Only men, explosives and small hand tools can be brought in through dedicated project entrance located on Madison and 48th St.

**Environmental Management**

To minimize the migration of dust and fumes from the work site to the MNR operational areas a number of measures were implemented, including:

- Installation of tarpaulins to act as a physical barrier to contaminant flow
- Installation of a ventilation system within the work site designed to draw air away from MNR areas and evacuate to atmosphere through existing ventilation shafts
- Exhausting fans and scrubbers were used to suck blast fumes and dust from below the shaft collars to capture and control these before they could enter the main body of air

The natural air flow within the GCT train shed is from North to South where the concourse is located so it was important to ensure dust and fumes did not migrate through to the dining concourse and passenger waiting areas. MTA performed monitoring at the closest operational platform and other locations within GCT and to date minimal impacts to the air quality have been recorded.

## Pre-excavation Coordination

Prior to blasting starting at a particular location a detailed score card system was implemented to ensure that the various elements that needed to be in place had been completed. It should be pointed out that the installation of structural instrumentation as well as the mesh installation and concrete repair works were undertaken under separate contracts. It had been determined during the planning phase that construction impacts would be limited to a 100 ft influence zone as measured in plan from the location of the blast. Therefore, prior to excavation commencing influence drawings were produced and all support works required to be undertaken were identified, prioritized and tracked to completion. Only when all items on the score card were identified as complete was excavation permitted to start.

Due to the restrictions on blasting times a detailed notification process was also developed in conjunction with the Contractor and Metro North Railroad. Once the Contractor was ready to blast notification was provided to Metro-North Railroad track controllers who would then provide the time, to the nearest minute, when the blast could be detonated. Typically blasting was permitted when there was a minimum 12 minute window where no train movements were scheduled through the influence zones which would permit the blast to be detonated, initial vibration results obtained, confirmation of no misfires which was achieved through the use of a tattle tale blasting cap attached to each delay sequence and a basic visual survey of the train shed within the influence zone be undertaken. Once the "all clear" had been received to these items train service could be resumed.

## Results of Blasting

To date one shaft has been completed from the lower level to the caverns and excavation of caverns and access tunnels by drill and blast has also progressed and there have been only a limited number of occasions when the pre-determined vibration limits, either ppv or microstrain have been exceeded. When such an event occurs blasting is suspended at that particular location until the cause of the exceedance has been identified. Typically such issues are due to increased burden due to drill hole alignment, incorrect wiring of the round or cut-offs causing later charges to break more rock. In addition, problems have been experienced with electrical interference on the strain gages causing false readings to be generated although enhanced grounding and shielding of the units and cables has generally eliminated this.

It should be noted that the maximum strain measured due to blasting is similar to that imposed on the structure by the movement of trains, 60 to 70 microstrains, and it is considered unlikely that the blasting will significantly affect the integrity of the structure.

## CONCLUSIONS

In general, few or minimal issues arose which was primarily due to the detailed planning and coordination that was undertaken prior to the blasting. The detailed studies undertaken during the planning phase and the early engagement of MNR into the process has certainly contributed to this outcome even though the various restrictions identified have caused modest advance rates during shaft construction.

# The Urban Ring Project: Planning a New Bus Rapid Transit Tunnel for Boston

**D.M. Watson**
Hatch Mott MacDonald, New York, New York

**J.P. Davies**
Hatch Mott MacDonald, Westwood, Massachusetts

**J.A. Doyle**
AECOM, Concord, Massachusetts

**ABSTRACT:** This paper describes the key planning phase issues and tradeoffs associated with a potential tunneled portion of the proposed Urban Ring Project in Boston, Massachusetts. The project is intended to provide new and improved transit service along an orbital corridor in Boston and several surrounding municipalities linking up the existing radial transportation system outside the downtown core with a Bus Rapid Transit system. The tunnel portion of the corridor would be approximately 1.5 mi in length and would be constructed under the densely developed Longwood Medical and Academic area (LMA) and Fenway neighborhood of Boston. The dense urban environment coupled with highly sensitive surface and subsurface conditions imposed many constraints on the planning and conceptual design. Consequently, a wide range of alignments and profiles were considered along with different tunnel construction techniques including: single bore or twin bore pressurized face tunnel boring machines up to 42.0 ft in diameter; sequential excavation methods; and cut and cover techniques.

## PROJECT BACKGROUND

The idea of transportation improvements in the Urban Ring Corridor in Boston has been the subject of considerable public debate and analysis, dating back to the era when a system of circumferential highways was proposed, and later abandoned. The 1970s marked a period of fundamental change in state policy away from highways and in favor of transit based solutions to mobility problems. Starting with the Circumferential Transit Feasibility Study in 1989, potential ridership in the corridor began to be quantified and the cost and feasibility of various alternatives was examined in greater detail. The Urban Ring Major Investment Study (MIS) completed in 2001 presented the approach of three additive phases to transit improvements in the corridor:

- Phase 1: New and improved cross-town bus routes on existing streets.
- Phase 2: Addition of Bus Rapid Transit (BRT) routes with new and improved inter-modal connections.
- Phase 3: Addition of rail rapid transit.

The MIS included an evaluation of various tunneled alternatives for Phase 3 rail transit. As part of the Phase 2 Revised Draft Environmental Impact Report (RDEIR)/Draft Environmental Impact Statement (DEIS) documentation filed in 2008/2009, the possibility of including a portion of the Phase 3 tunnel alignment during Phase 2 was considered. In this scenario, the tunnel alignment would be used by the BRT service during Phase 2, and subsequently converted to rail transit usage in Phase 3.

### The Urban Ring Corridor

The corridor identified within the MIS is shown in Figure 1. The MIS also identified the section of the corridor to be converted to rail, extending from Sullivan Square to Dudley Square in Boston. Subsequent to the MIS, planning and environmental work commenced on the Phase 2 BRT system, which considered a transit tunnel beneath the Fenway/LMA portion of the corridor due to the high levels of traffic congestion and lack of available right-of-way (ROW) for a surface busway. The effectiveness of a surface BRT routing would be constrained by the LMA's dense concentration of jobs and travel

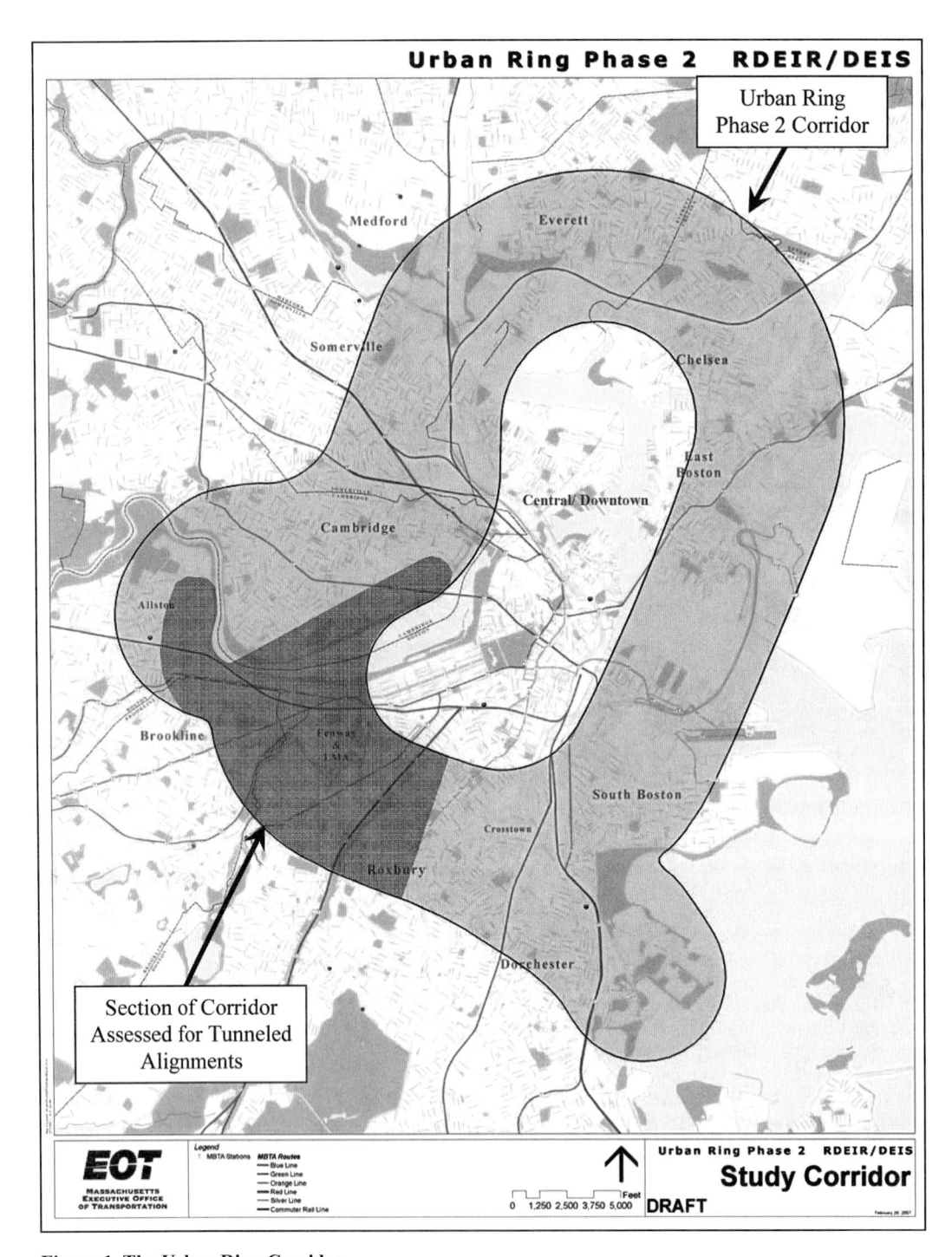

Figure 1. The Urban Ring Corridor

demand, along with its limited and highly congested roadways in close proximity to the sensitive natural and historic resources of the Fenway neighborhood. In response to these constraints, a wide range of potential tunnel alternatives was developed and evaluated during the course of the RDEIR/DEIS. This led to the identification of the most promising tunnel alignment through the Fenway/LMA neighborhoods, as well as an assumption about the construction method (for cost estimating purposes, based on best available information at the conceptual level of planning and engineering).

This paper describes the planning process and conceptual engineering which led to the identification of the tunnel alignment and station locations developed for operation with BRT in Phase 2, and addresses potential conversion of the tunnel from Phase 2 BRT to Phase 3 rail transit. The relationship between surface land use and the choice of tunnel alignment and construction technique is described for both the basic alignment and those areas where alignment options remain to be analyzed in greater detail during subsequent planning and engineering.

**Project Objectives for the Tunnel**

The primary objectives of the Phase 2 BRT tunnel alignments are to:

- Reduce transit trip times.
- Increase quality and reliability of service.
- Minimize impacts of surface transit operations in sensitive locations (e.g., Emerald Necklace parkways).
- Provide subsurface routing of BRT services between Ruggles Station and the Emerald Necklace.
- Improve connectivity with existing transportation services.
- Provide compatibility for future conversion with Phase 3 rail transit.
- Minimize environmental impacts.
- Maximize use of public right-of-way.

During development of alignment alternatives, regular meetings with key stakeholders in the LMA and surrounding neighborhoods were held to ensure that the alternatives developed addressed their concerns. The Citizen's Advisory Committee (CAC) and the Medical Academic and Scientific Community Organization, Inc. (MASCO) have been intimately involved in this process.

## REGIONAL GEOLOGY

At present there is a lack of site-specific geological or geotechnical information along the potential Urban Ring tunnel alignments. The following description of the regional geology has been summarized from secondary sources. Future engineering analysis will require more detailed information on subsurface conditions from a project-specific site investigation program and other primary sources.

In general, the areas through which the Urban Ring tunnels will be constructed comprise the site of an ancient estuary. As a result, the area is typically characterized by marine and glacial deposits and extensive layers of organic silts and clays. Upland areas are generally overlain by glacial till (a typically hard and compact mixture of clay, silt, sand, pebbles, cobbles and boulders); and lowland areas typically have stratified deposits near the surface, which may include both sands and gravels, and fine-grained silts and clays. In the lowest lying areas, near the Charles River and the Muddy River, an extensive, fine-grained deposit known as the Boston Blue Clay was deposited under shallow marine conditions. This is overlain by recent estuarine deposits, which in turn are overlain by artificial fill in some areas. At the west end of Ruggles Street, the upper soils include about 10.0 ft of fill underlain by up to 20.0 ft of organic silts underlain by a thin layer of sand and then Boston Blue Clay. Toward the east the organics taper out and the fill is underlain by sand and sand with gravels which in turn is underlain by clay. This clay deposit is thick in the Ruggles Street area, extending to depths in excess of 150.0 ft.

The bedrock of the lower Charles River watershed comprises a sequence of sedimentary and volcanic rocks that were deposited about 600 million years ago. The rock layers vary from relatively soft siltstones and slates (Cambridge Argillite), to harder conglomerates consisting of pebbles and cobbles in a sand matrix (Roxbury Conglomerate). Uplands in Newton, Brookline, and the southern portion of Boston are underlain by the hard conglomerate and volcanic rocks; lowlands in Cambridge and the northern portion of Boston are underlain by the argillite. The bedrock elevations in the project area vary and are expected to have a high point in the vicinity of Harvard Medical School on Longwood Avenue at around 70.0 to 95.0 ft below ground level.

## CONSTRAINTS

### Right of Way and Land Use

A central consideration in planning for the proposed Urban Ring Phase 2 project is the provision of as much dedicated right-of-way (ROW), in the form of busways and bus lanes, as possible. Dedicated ROW is essential to providing a high-speed and reliable service, especially in areas of heavy traffic congestion. Federal guidelines for BRT projects call for a minimum 50.0% of the project route to be in dedicated ROW. Maximizing the amount of dedicated

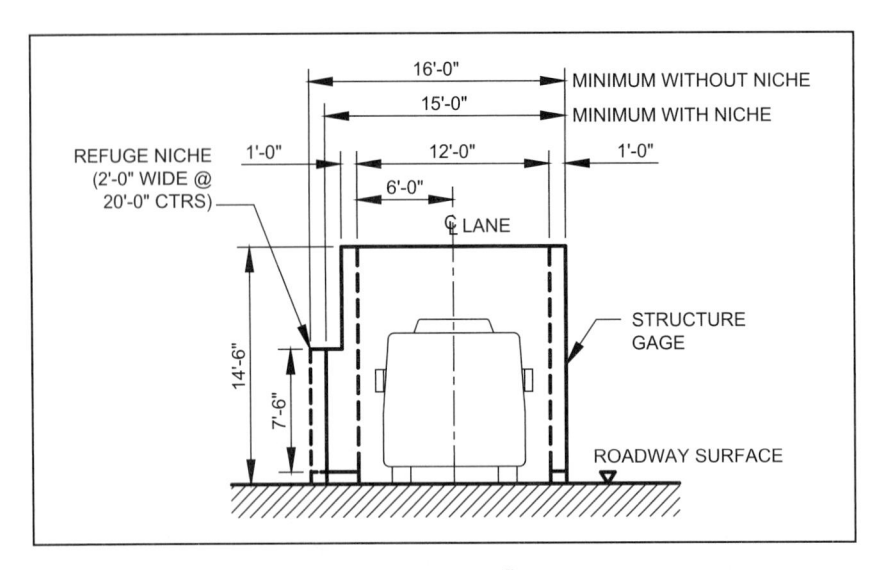

**Figure 2.** BRT clearance envelope (one-way busway tunnel)

ROW, and optimizing its efficiency and effectiveness, have been among the central focuses of the Urban Ring Phase 2 project team and the CAC.

ROW and land use constraints have been one of the major factors in determining viable tunnel alignments. One of the primary objectives has been to follow the alignment of public roads as far as practicable but without compromising other project goals. There are also numerous developments in varying stages of implementation, from master planning level to construction, that have been accounted for in the development of alignment options. In an already well developed corridor of the Urban Ring, this has proven particularly challenging.

## Utilities

Utility diversions are costly and disruptive in their own right and therefore any alternative should seek to minimize the impact on existing utilities where possible. Some major utilities have been identified and, where possible, alignment options have been developed to minimize disruption to these utilities.

## TUNNEL DESIGN CRITERIA

### Tunnel Cross Section

The design vehicle for the BRT system is a 60.0 ft long articulated bus. The minimum clearances required for this vehicle in a one-way busway tunnel are presented in Figure 2. Walkways are provided throughout the tunnel to allow for safe access during routine maintenance operations, without the need to close the tunnel. A minimum walkway width of 3.0 ft has been adopted. Where it is not practicable to provide a 3.0 ft wide walkway, then a 2.0 ft wide walkway is provided with refuge niches, sized at 7.5 ft high by 2.0 ft wide and 1.0 ft deep, spaced at 20.0 ft centers, and protected from errant vehicles.

To ensure compatibility for conversion to rail usage in Phase 3, the requirements for light rail and heavy rail were developed based on the existing Massachusetts Bay Transportation Authority (MBTA) Green Line and Orange Line, respectively. It was observed that the clearance requirements presented in Figure 2 for the BRT system were the controlling factor in developing tunnel cross sections.

### Tunnel Alignment

The alignment requirements vary considerably between BRT and rail. These factors have been critical in the development of the tunnel alignment options. The main criteria are presented in Table 1.

### Underground Stations

Underground stations were developed to meet the following criteria:

- 12.0 ft nominal platform width.
- 8.0 ft minimum platform width at objects/stairs/escalators etc.
- 10.0 ft vertical clearance above platform to any overhead signage, lighting etc.
- 4,000.0 ft minimum horizontal radius for convex platforms.
- 5,000.0 ft minimum horizontal radius for concave platforms.

## Table 1. Tunnel alignment design criteria

| Alignment Component | | Phase 2 BRT | Phase 3 Light Rail | Phase 3 Heavy Rail |
|---|---|---|---|---|
| **Horizontal Alignment** | Minimum horizontal radius (general) | 250.0 ft | 250.0 ft | 1800.0 ft |
| | Minimum horizontal radius (absolute minimum) | 100.0 ft | 150.0 ft | 700.0 ft |
| **Vertical Alignment** | Preferred maximum grade | 5.0% | 5.0% | 3.0% |
| | Absolute maximum grade | 8.0% | 7.0% | 4.0% |
| **Underground Stations** | Platform length | 220.0 ft | 300.0 ft | 410.0 ft |

### Ventilation

The tunnel ventilation system would need to address the implications of vehicle engine choice for the BRT vehicles. The four main options considered are:

- Emission Controlled Diesel (ECD).
- Compressed Natural Gas (CNG).
- Dual Mode (electrified trolley bus in the tunnels).
- Hybrid Electric (battery powered in the tunnels).

Preliminary assessments of the tunnel ventilation system requirements indicate that jet fans would not be required in the running tunnels. The conceptual tunnel ventilation system would however require fan plants to be located at each end of underground stations and at tunnel portals. For longer sections of tunnel, ventilation shafts may be required at intermediate locations. Based on the design criteria and assumptions, none of the alignment alternatives identified for the LMA Tunnel would require ventilation ducting within the running tunnels.

Initial assessment of vehicle technologies, based on the tunnel ventilation requirements and capital and operational and maintenance costs, indicate that the currently preferred technology is hybrid electric. If buses other than hybrid electric vehicles are to use the tunnel then this could impact the tunnel ventilation design.

### Fire Life Safety

The fire life safety and fire protection of the tunnel require assessment of, and planning for, the following features: emergency egress; emergency ventilation; fire protection of structures; fire detection, fire suppression, and fire fighting equipment and systems; communication systems; traffic control; drainage; and emergency response plans.

The key criteria at this stage are the emergency egress requirements for road tunnels (NFPA 502), which states that the spacing between emergency exits should not exceed 1000.0 ft. Where tunnels are divided by a minimum of 2 hour fire-rated construction or where the tunnels are in twin bores, cross passageways can be used instead of emergency exits. Cross passageways should have a maximum spacing of 656.0 ft. Future conversion to light rail will invoke the requirements of NFPA 130 with the less stringent requirement of maximum cross passageway spacing of 800.0 ft.

Additional items considered during the planning stage include: MBTA operations, including recovery of disabled vehicles; tunnel lighting; electrical and safety equipment; drainage; and security.

## TUNNEL ALIGNMENT ALTERNATIVES

The tunnel alternatives that have been analyzed encompass a significant range of lengths, number of underground stations, connections, and costs. All of the options include a minimum tunnel segment beneath the LMA. This is because the LMA is a critical activity center with a combination of characteristics that create the greatest challenges for surface BRT connections; it has a very high density of travel demand, a limited roadway network, significant traffic congestion, and limited opportunities for roadway expansion or new roadway connections. In addition, it is bounded on the north by the Fenway, a parkway that is a component of the Emerald Necklace park system. This results in a minimum tunnel segment extending from the vicinity of Ruggles Station in the southeast to the Sears Rotary in the northwest. Beyond this segment, the tunnel options encompass a range of lengths, alignments, and connections.

The tunnel alignment alternatives were developed based on the project goals and technical constraints, in addition to significant consultation and input from the project CAC and other stakeholders. The alignment options can be broadly classified into two categories—short tunnel options and long tunnel options. The short tunnel alternatives begin immediately west of Ruggles Station, pass beneath the LMA, and extend to either Yawkey Station, the vicinity of the BU Bridge, or Allston Landing, and stay to the south of the Charles River.

The longer tunnel alternatives provide connection from the Melnea Cass Boulevard corridor,

Table 2. Tunnel alignment design criteria

| Alignment Alternative | Length of Running Tunnel (ft) | |
|---|---|---|
| | Constructed in Phase 2 | Abandoned on Conversion to Phase 3 Rail |
| Western | 6,293 | 2,710 |
| Central | 5,710 | 1,630 |
| Eastern | 5,895 | 765 (light rail) |
| | | 2,545 (heavy rail) |

with an underground connection to Ruggles Station, beneath the LMA, to Allston and Cambridge, requiring bifurcations in the tunnel alignment and passing beneath the Charles River.

Preliminary ridership analysis and evaluation of cost and benefit indicated that the increase in ridership for the long tunnel options compared with the short tunnel options was marginal, but came at a greatly increased cost as a result of the considerable additional length of tunnel and increased number of underground stations. This, in combination with feedback from various public meetings, suggested that further development of the short tunnel options was warranted, particularly in the LMA. Following further consultation with key stakeholders and engineering analysis, the most promising alternatives for the busway tunnel include the following features:

- East portal location and configuration at Leon Street within the MBTA ROW.
- Alignment of the tunnel from the east portal at Leon Street to the proposed underground station beneath Longwood Avenue (following the alignment of Huntington Avenue).
- Location of the proposed underground station beneath Longwood Avenue in the vicinity of Avenue Louis Pasteur.
- Location of the west portal adjacent to the Landmark Center and the Green Line "D" Branch within the abandoned CSX ROW.
- Inclusion of an underground BRT station as part of the west portal structure adjacent to the Green Line "D" Branch to provide better connectivity with existing Green Line stations.

There are currently three different options for the section of busway tunnel alignment between Longwood Avenue and the west portal, referred to as eastern, central and western. A summary of the lengths of running tunnel for each alignment alternative and the length of tunnel that would need to be abandoned on conversion to Phase 3 rail transit is presented in Table 2. The LMA busway tunnel including the western, central, and eastern alignment options is presented in Figure 3.

## CONSTRUCTION TECHNIQUES

There are a number of potentially feasible construction methodologies that could be used to construct the Urban Ring tunnels. The methodologies can be grouped into three main types: cut and cover tunnels (including the top-down method); sequential excavation method (SEM) mined tunnels with a sprayed concrete lining; and tunnel boring machine (TBM) bored tunnels.

The three tunneling methods were evaluated to determine which method or methods would be appropriate for the Urban Ring Phase 2 busway tunnel structures, given the requirements and constraints of the project and the corridor. The intent of this evaluation process was to make an initial recommendation of viable construction methods for alignment alternatives analysis to allow a more transparent comparison of the numerous alternatives. Some alternatives require several different construction methods to be employed. The final choice of running tunnel construction method and configuration will depend on: the final busway tunnel alignment chosen, the geology and hydrogeology, the vertical alignment, the anticipated ground movements and building settlement assessments, and noise and vibration impacts on sensitive hospital and research operations. These issues will be addressed during subsequent engineering studies as more information becomes available.

### Tunnel Portal Structures

The tunnel portal structures would comprise an open cut approach ramp ("boat" section) and a section of cut and cover tunnel. The construction would most likely require temporary earth support systems to be installed to enable the construction of a cast-in-place concrete structure that provides permanent ground support. The construction of the tunnel portals would be one of the more challenging aspects of this project owing to the extensive constraints at the portal locations, including: the Landmark Center shopping mall, the Green Line "D" Branch portal structure and

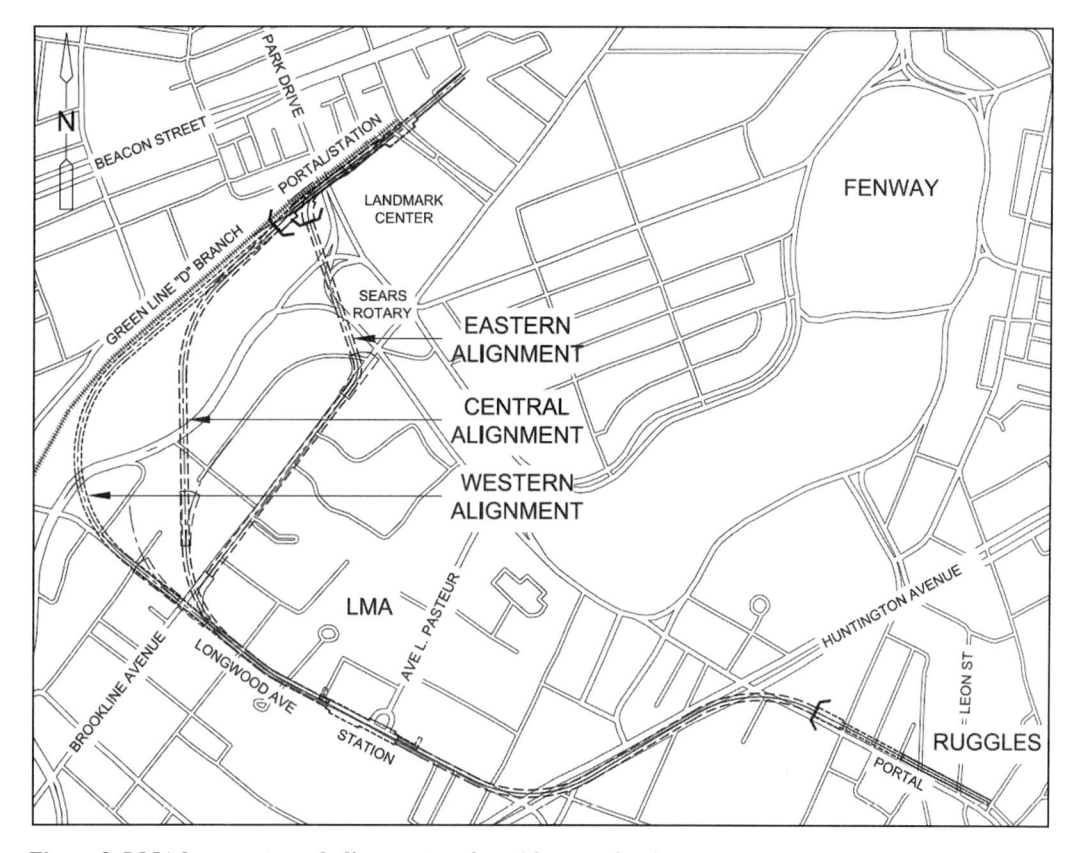

**Figure 3.** LMA busway tunnel alignment options (short options)

Fenway Station, the historic Emerald Necklace park, historic buildings, residential areas, and heavy pedestrian traffic. The portals and their construction sites would need to be extremely compact as a result, with construction techniques that make the best use of the available space. A typical cross section through the Ruggles Street portal at the east end of the alignment is included in Figure 4.

**Running Tunnels**

The three construction methods mentioned above were considered for the construction of the running tunnels. Single bore or twin bore configurations can be achieved with any construction method. These configurations are discussed later. The cut and cover method was not recommended for the following reasons:

- Physical constraints and heavy traffic demand.
- Impacts on sensitive environmental and open space resources (e.g., the Emerald Necklace parkway).
- Lack of available public ROW corridors for key components of the corridor would require significant land takings to allow cut and cover construction, resulting in additional cost and disruption.
- Disturbance over a long period of time due to the slow anticipated advance rates and utility impacts.

Limited geotechnical information is currently available. Consequently the cost of the SEM mined tunnel is at greater risk of significant increases owing to the potential for an extensive amount of ground treatment. Therefore the SEM mined tunnel method was not recommended for planning and alternatives analysis (it should be noted that this decision does not preclude this method from further evaluation in future stages of the project).

The TBM bored tunnel method was recommended for the following reasons:

- Potential to minimize surface disruption and reduce environmental impacts.

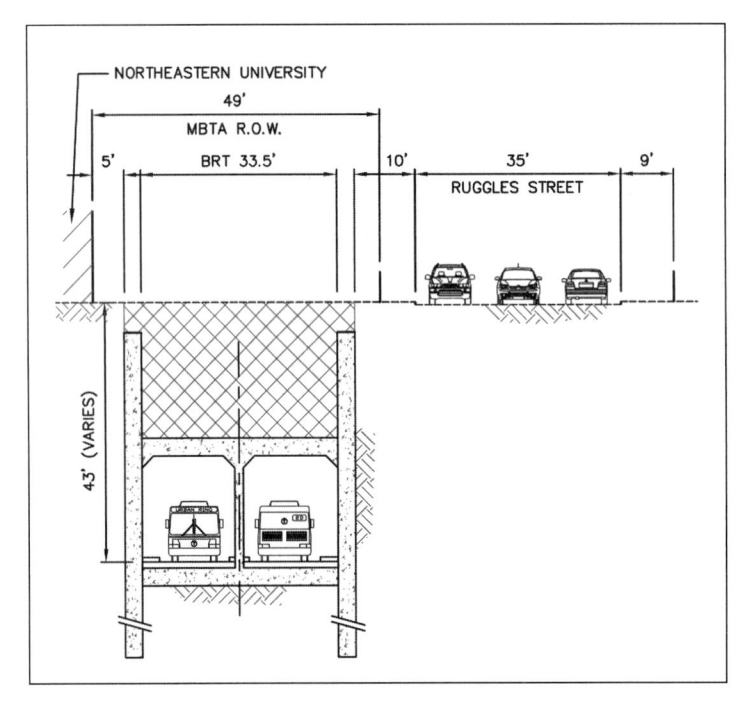

**Figure 4. Conceptual cross section through tunnel portal at Ruggles Street**

- Pressurized face TBMs can safely construct tunnels in soft ground and mixed face conditions, while minimizing impacts on surrounding structures.
- Consideration of environmentally sensitive zones (e.g., Emerald Necklace and the Muddy River) would favor methods that do not require excavation from the surface or ground treatment methods.

As a result of this review, it was determined that TBM construction was an environmentally acceptable solution offering the potential to minimize disruption and provide the most cost-effective approach for the planning of the Urban Ring Phase 2 running tunnels. The cut and cover and SEM methods could be appropriate for discrete lengths of some tunnel options where site constraints, alignment geometry, project requirements or other factors favor these methods of construction. Noise and vibration impacts will require further assessment once project-specific geotechnical information is available.

**Underground Stations**

Alternatives for station construction using cut and cover, the SEM, and TBM methods were investigated. The selected method depends on a number of factors, including the location of the station, the site constraints, and the geology and groundwater conditions.

The use of an over-sized TBM which would accommodate station platforms within the bored tunnel was rejected owing to spatial constraints, right-of-way issues, impacts on portal structures and difficulties converting to Phase 3 rail use. The SEM method is a viable solution, and can reduce surface impacts. However, the SEM method will still require two large shafts at each end of the station to accommodate ventilation equipment and vertical circulation elements. Given the lack of geotechnical information, the desire to keep the stations relatively shallow, and the relatively short length of the stations, it was considered prudent at this stage in the planning process to adopt cut and cover for the full length of the station.

The top-down method of cut-and-cover construction, where the main excavation occurs below a roof deck, has been identified as the preferred method for constructing the LMA station. This method would minimize disruption to the surrounding communities in this densely developed and heavily trafficked location. The increase in cost and construction duration associated with this method needs to be balanced with the potential to minimize disruption.

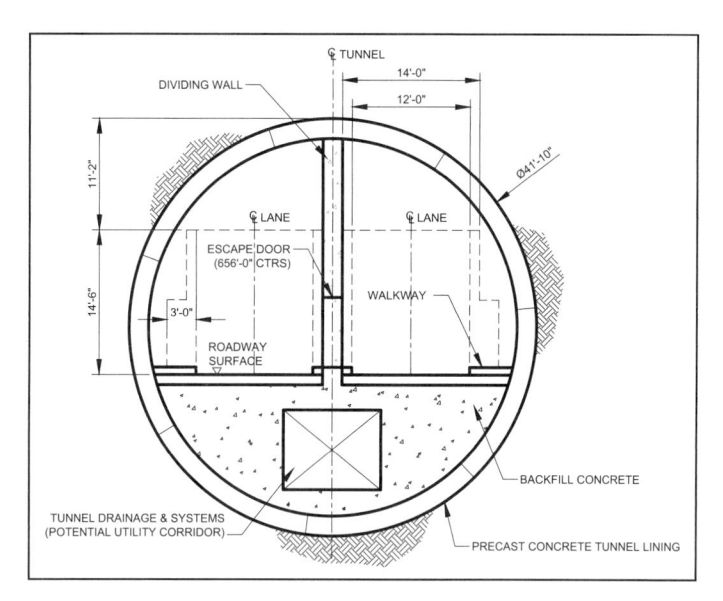

**Figure 5. Conceptual cross section of single bored tunnel**

## SINGLE BORED TUNNEL VERSUS TWIN BORED TUNNELS

The running tunnels can be configured to provide either a single tunnel carrying two lanes separated by an internal dividing wall or two separate tunnels each carrying one lane. These alternatives are evaluated below.

### Twin Bored Tunnels

Some potential benefits of the twin bored tunnel solution compared with the single bored tunnel may include:

- Reduced ground surface settlements.
- Reduced cost for bored tunnels (although the savings may be offset by deeper stations and construction of egress shafts or cross passages or both).
- Reduced volume of excavated material.
- Shorter portal structures.
- Higher utilization of the underground space formed within the tunnel.

A major challenge associated with the twin bored tunnel in relation to the LMA Tunnel alignment is the heavily constrained horizontal corridor width. The available corridor width would require twin bored tunnels to be vertically separated and may require escape shafts to provide emergency escape facilities, since ADA-compliant cross passages could not be constructed between the twin bored tunnels. This stacked configuration would also constrain the tunnel alignments requiring transitions from horizontal to vertical separation configurations. Another challenge related to the twin bored tunnel concept is the potential for greater construction phase impacts both spatially, owing to the wider plan footprint and potential escape shafts, and temporally, because the bores would be made either sequentially using one TBM or concurrently using two TBMs with a lag between the drives.

### Single Bored Tunnel

The potential benefits and challenges of a twin bore approach are outweighed by the benefits of the single bored tunnel and this was therefore selected as the preferred option. The single bored tunnel conceptual cross section is shown schematically in Figure 5. Some of the benefits of the single bored tunnel include:

- Smaller plan footprint than twin bored tunnels—critical where public ROW corridors are very narrow;
- A single bored tunnel with a central dividing wall allows greater flexibility in BRT recovery operations: access doors can be provided in the central dividing wall thereby limiting the distance over which a recovery vehicle must reverse to reach a disable bus;
- Greater flexibility in locating track crossovers on conversion to Phase 3 rail use: a length of the central dividing wall would be removed to install the necessary switches and crossings, potentially reducing the length of station excavations;

- Drainage and cross passage connections are all contained within the bored tunnel, eliminating the need to form breakouts from the running tunnels; and
- Opportunity to include revenue-generating utilities within a service corridor below the road deck.

## CONVERSION TO RAIL FOR PHASE 3

The Downtown Seattle Transit Tunnel is the only BRT tunnel that has undergone conversion from BRT to light rail usage, and now operates with shared use of buses and light rail vehicles. General planning considerations for inclusion of LRT convertability within BRT facilities are discussed by Wood *et al.* (2006). The primary aspects identified in relation to the Urban Ring Project include:

- Cross sectional geometry: determined by the size of the vehicle, kinematic envelope and structure gauge, including provision of walkways and equipment within the tunnel;
- Horizontal and vertical alignment: operational capabilities of the vehicles to be used, and consideration of life-cycle costs associated with gradients;
- Stations and platforms: platform height, lengths, and geometry; and
- In future stages of design, more detailed assessments of the conversion process should be considered, including: structure loading; stray current protection; installation of rail (either during Phase 2 or Phase 3); and accommodation and installation of rail systems.

The design and construction of the BRT tunnel needs to consider the future conversion to rail transit to avoid significantly higher costs of conversion. Where cut and cover structures are required for tight turns in Phase 2, these would be built to incorporate Phase 3 turnouts. In addition, where portals are required to be re-graded during Phase 3 conversion, the portals would be designed and constructed to accommodate these requirements in Phase 2.

Major structural works required for Phase 3 that could be built during Phase 2 may include:

- Dedicated underground turnout structures to suit Phase 3 rail alignments;
- Longitudinal extension of underground stations to allow for Phase 3 platform lengths;
- Vertical extension of underground stations to allow Phase 3 station platforms to be built beneath the Phase 2 station (such that both BRT and rail could operate simultaneously); and
- Construction of a larger diameter tunnel to incorporate two decks—an upper deck for BRT, fitted out during Phase 2, and a lower deck provided during Phase 2 and fitted out for rail during Phase 3.

## PLANNING PROCESS

The Federal Transit Administration (FTA) is serving as the lead federal agency in the preparation of the DEIS for the Urban Ring Phase 2 project to address the federal environmental process under the National Environmental Policy Act (NEPA) and related federal requirements. The DEIS was prepared and filed in November 2008 in combination with a RDEIR to address Massachusetts state environmental requirements under the Massachusetts Environmental Policy Act (MEPA). The RDEIR is a revision to a stand-alone Draft Environmental Impact Report (DEIR) that was filed under MEPA in November 2004; the revision to the original DEIR is intended to address public comments on the DEIR, and to enable a reconnection of the state MEPA and federal NEPA environmental review processes.

The RDEIR project proponent was the EOT (now MassDOT). Throughout the planning and environmental review process, EOT coordinated its actions with the MBTA, the Boston Region Metropolitan Planning Organization, and the project's CAC, which includes representatives from the municipalities in the project corridor (Boston, Brookline, Cambridge, Chelsea, Everett, Medford, and Somerville, MA), neighborhood and citizens groups, and the many educational and medical institutions in the corridor.

The combined RDEIR/DEIS for Phase 2 of the Urban Ring project was the latest step in a decades-long planning process for public transit improvements in the corridor. During that time, new residential, commercial and institutional development in the corridor has increased travel demand and worsened congestion, and transit needs in the corridor have grown.

### Capital Cost and Funding

The capital cost estimate for the LMA Tunnel, based on the central tunnel alignment option, is approximately $1.5 billion in 2007 dollars. This cost estimate has been developed based on conceptual designs and is subject to change following site investigations, selection of a preferred alignment, and further engineering design. It is recognized that the incremental benefits of including the LMA Tunnel within the Urban Ring Project need to be balanced with the significant cost increase that the tunnel would have on the project. The project is not currently included in the financially constrained Boston Region Metropolitan Planning Organization's long-range Regional Transportation Plan (RTP), which includes all of the projects in the Boston region that have demonstrable funding sources for the next 20 years. A project must be included in a region's RTP in order to receive federal funding.

**Phasing and Implementation**

It is recognized that the current state and federal financial environment presents significant constraints to implementation of the LMA Tunnel. The schedule for final environmental analysis, preliminary engineering, design, and construction is dependent on identification of project funding sources.

## CONCLUSIONS

This paper has highlighted some of the key considerations in planning a Bus Rapid Transit tunnel in Boston as part of the Urban Ring Project. Design criteria for tunnel cross sectional geometry and tunnel alignment geometry have been defined for both BRT and rail usage. The cross sectional geometry required for BRT was found to be the controlling factor in determining the tunnel cross sections. A wide range of tunnel alignments has been considered during the RDEIR/DEIS process, including tunnels that cross the Charles River to Cambridge. The tunnel planning, engineering, and ridership analysis identified and addressed a number of key issues relative to portal locations, tunnel horizontal and vertical alignment, station locations, and ventilation. As a result, the costs and benefits of a BRT tunnel in the Fenway/LMA area were documented and subjected to public review and comment.

The existing and future land use on the surface in the Fenway and the LMA established the parameters within which the choice of tunnel alignments and station locations were set for the LMA Tunnel. Construction techniques have been evaluated for the major components of the LMA Tunnel, and the configuration of the tunnel in twin bore or single bore arrangements was assessed, with single bore being recommended at this stage in the planning phase. Some of the major issues associated with conversion of the tunnel to rail usage have also been presented.

The significant cost that the LMA Tunnel would add to the Urban Ring Project has to be balanced with the incremental benefits it provides. Subsequent planning and engineering phases for the LMA Tunnel are needed to develop detailed information on subsurface conditions to verify the optimal tunnel configuration and construction techniques, and identify related construction and operations phase impact avoidance and mitigation measures.

## ACKNOWLEDGMENTS

The authors wish to thank MassDOT (formerly the Massachusetts Executive Office of Transportation) and their Project Manager, Ned Codd, for consenting to the publication of this paper.

## REFERENCES

NFPA 130. 2007. *Standard for Fixed Guideway Transit and Passenger Rail Systems.*
NFPA 502. 2008. *Standard for Road Tunnel, Bridges and Other Limited Access Highways.*
Wood, E., Shelton, D.S., Shelden, M. 2006. *Designing BRT for LRT Convertability: An Introduction for Planners and Decision-Makers.* Transportation Research Record: Journal of the Transportation Research Board, Issue 1955, 2006, pp 47–55.

# TRACK 3: PLANNING

Bradford Townsend, Chair

*Session 4: Project Risk, Budget, and Schedule*

# Blast and Post Blast Behavior of Tunnels

Sunghoon Choi, Ian Chaney, Taehyun Moon
Parsons Brinckerhoff, New York, New York

ABSTRACT: A numerical approach for tunnel blast behavior has been developed and introduced by Choi et al (2003, 2006). The method uses coupled Euler-Lagrange analysis with consideration of strain rate and dynamic material properties. Even though this approach accurately predicts the damaged area of the tunnel structures, proper assessment requires post blast analysis. Since a tunnel is loaded by surrounding ground and water, sudden loss of structural components to carry the loads results in redistribution of the loads and may induce progressive failure of tunnel structures, ground, adjacent other structures. Flood Potential is another part of post blast concern if the surrounding ground is sufficiently permeable. This paper introduces blast and post blast behavior of various types of tunnels. Outlined also are recently developed blast protection measures.

## INTRODUCTION

Terrorism remains a constant, serious threat to the public transportation. A key area of concern that could threaten our national economy is an explosion that cripples one of our nation's major tunnel thoroughfares. This has the potential to cause a substantial loss of life, a major disruption in our national economy, and a significant change in our confidence to use public transportation. Protective design of tunnels, especially passenger tunnels, becomes an important element of nation security due to their vital roles in modern transportation system.

Authors have published a tunnel security design guideline in 2006 (Choi & Munfakh, 2006). Since that time the authors have found the method to be very practical and useful and the guideline has been implemented in many tunnel security designs and assessments. As the blast protective design was being performed, however, up to date approaches and methodologies have been added into the guideline. An updated version of the design guideline was published in Choi (2009). The new proposed guideline is intended to clearly explain how to assess the risk and vulnerability, how to analyze the blast impact on the structures including progressive failure potential identification, and to present the latest developed tunnel blast protection measures.

Figure 1 shows protective design steps proposed in Choi (2009). The new protective design of tunnels consists of three elements: (1) threat, vulnerability and asset criticality evaluation through a risk assessment, (2) blast analysis and post-blast stability analysis including evaluation of progressive failure potential, and finally (3) development of tunnel blast protection measures and system countermeasures. This paper mainly discusses the background and implementation of blast analysis and post blast analysis of a tunnel when the tunnel is subject to a blast loading. It should be noted that the illustrations and analyses presented herein are not related with actual projects or real-world infrastructures but performed for research purpose only.

## ANALYSIS APPROACH

### Computer Program

Three-dimensional coupled Euler-Lagrange nonlinear transient dynamic analysis is conducted using the commercial computer program LS-DYNA developed by Livermore Technology Co. (LSTC). LS-DYNA is a general purpose multiphysics simulation software package for analyzing large deformation dynamic responses of structures, including structures coupled to fluids, and is a fully-integrated analysis tool specifically designed for nonlinear dynamic problems. Its core competency is highly nonlinear dynamic finite element analysis (FEA) using explicit time integration. Solid Lagrange elements are used to model the most structural elements while Eulerian elements are used to model liquid or gaseous elements such as air, water and explosives. In addition, shell Lagrange elements are routinely used to model thinner structural elements such as steel plates and CFRP sheets. The Eulerian elements and the Lagrange structural elements are fully coupled during the entire modeling process.

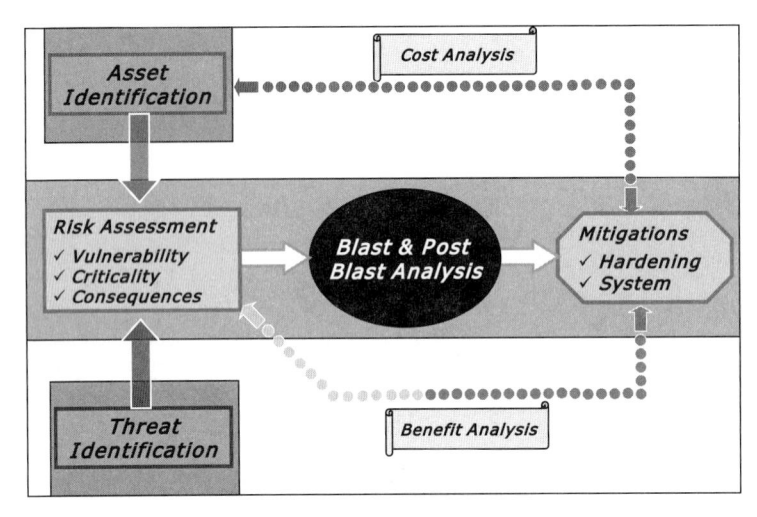

**Figure 1.** Protective design steps for tunnel security (Choi, 2009)

## Material Models and Equations of State

Prediction of material damage under dynamic loadings is complex behavior and is a function of several factors such as loading magnitude, loading rate and loading duration, among others. If the material experiences the same magnitude of loading at a different loading rate, the response is different. For instance, as the loading rate increases, the material can sustain higher loads for the same respective duration. The analyses presented herein fully incorporated dynamic properties of materials as well as strain rate effects so that reliable dynamic structural response could be predicted when the structure was subject to dynamic loadings such as blast loadings.

## Jones-Wilkins-Lee Equation of State for High Explosives

The expansion of high explosive products is modeled using the Jones, Wilkins and Lee (JWL) equation of state. High explosives (HE) are chemical substances which, when subjected to suitable stimuli, react chemically to give a very rapid (order of microseconds) release of energy. In the hydrodynamic theory of detonation, this very rapid time interval is shrunk to zero and a detonation wave is assumed to be a discontinuity which propagates through the unreacted material instantaneously liberating energy and transforming the explosive into detonating products. HE equation of state is expressed as the conservation of mass, momentum and energy across the discontinuity and energy conservation equation. By comparing hydro calculations with experiments on propelling metal plates by normally incident detonation waves, an energy equation is produced as:

$$p = Av^{-\gamma} + Bv^{-(1+\omega)} + Ce^{-rv} \quad (1)$$

where $p$ is pressure, $v$ is specific volume, $e$ is specific internal energy, $\gamma$ is the adiabatic exponent evaluated at the CJ point, $A, C, r, \omega$ are constants and $B$ is a function of entropy, differing for each adiabat. The Wilkins form of equation, as presented in equation (1), is capable of predicting the motion correctly until the pressure in the products falls below approximately 5 kbar but then becomes insufficiently energetic. Lee et al (1968) updated the energy equation by replacing the first term (the power law term) in the Wilkins equation by a second exponential term as:

$$p = C_1\left(1 - \frac{\omega}{r_1 v}\right)e^{-r_1 v} + C_2\left(1 - \frac{\omega}{r_2 v}\right)e^{-r_2 v} + \frac{\omega e}{v} \quad (2)$$

This equation is known as the Jones-Wilkins-Lee (JWL) equation of state and is currently used in standard practice for hydrodynamic calculations of detonation product expansions to pressures down to 1 kbar. The values of the constants $C_1$, $r_1$, $C_2$, $r_2$, $B$ and $\omega$ for many common explosives have been determined from dynamic experiments over the last several decades.

### Reinforced Concrete

The damage level of concrete materials under explosion loads is highly dependent upon the strain rate effect (dynamic response). To incorporate dynamic response of concrete materials, the current analysis utilizes a strain-rate dependent concrete model in LSDYNA material model library, MAT_CONCRETE_DAMAGE concrete model. This material model has been successfully used to model the behavior of standard reinforced concrete subjected to blast loads.

Generally, the behavior of an element in a reinforced concrete structure cannot be satisfactorily modeled using uniaxial stress-strain characteristics and a consideration of triaxial stress conditions is desirable for better understanding of the behavior. The failure limits in concrete can then be represented as surfaces in a three-dimensional principal stress state. Failure surfaces can be combined with plasticity-based constitutive models for the analysis of three-dimensional reinforced concrete structures. The concrete model used in this study is a plasticity-based formulation with three independent failure surfaces known as maximum, yield and residual, which change shape depending on the pressure (Tavarez, 2001). Because the failure surface depends on the pressure, the material model must be used in conjunction with an equation of state, which gives the current pressure as a function of current and previous volumetric strain. Once the pressure is known, the stress tensor can be calculated as being a point of a moveable surface that can be a yield surface, failure surface or residual surface.

Concrete reinforcement is typically modeled using a smeared model in which the reinforcement is assumed to be uniformly distributed over the concrete elements. As a result, the properties of the composite material in the element are constructed from individual properties of concrete and reinforcement using composite theory. This technique is usually applied for large structural model, such as the current study, where reinforcement details are not essential to capture the overall response of the structure (Tavarez, 2001). Furthermore, this is usually sufficient because most of the difficulties in modeling reinforced concrete behavior rely in the development of an effective and realistic concrete material formulation, and not in the modeling of the reinforcement. The concrete model utilized herein requires that a percent of reinforcement as an input and creates a composite reinforced concrete model from the material properties of both the steel and the concrete.

### Steel

In order to model various steel elements used in the blast analyses such as segment bolts, dowels and protective steel plating, the MAT_PLASTIC_KINEMATIC material model is utilized. This model is suited to model isotropic and kinematic hardening plasticity and can account for rate effects using the Cowper & Symonds model which scales the yield stress by a strain rate dependent factor.

### Geomaterial

Mechanical properties for geomaterials are important for accurate predictions of ground shock wave propagation and the behavior of the structure. Proper assessment of dynamic response of geomaterials requires a selection of adequate material models. In the current study, several material models were used.

The MAT_JOINTED-ROCK material model is typically used to model soft soil materials. This material type can incorporate the effects of ubiquitous rock joints existing in a general rock mass but in cases involving soil-like materials, the number of rock joint planes is set to zero. The intact material strength (matrix behavior) follows a modified Drucker-Prager strength criterion, so that the yield surface never infringers the yield surface described by the Mohr-Coulomb criterion. Depending on the shear strength of the soil it is possible that premature failure can be predicted at the early stages of stress initialization in the model, before blasting effects are modeled. A significant feature of the MAT_JOINTED-ROCK model is that it allows for additional shear strength to be used, at the stress initialization before the execution of the blast analysis. During blast analysis, the shear strength is reset to its normal level. This material was used to simulate the behavior of soil around the structures.

The elastic properties for the soils are paramount importance in blast analyses; specifically required for the said analyses were bulk modulus, $K$ and Young's Modulus, $E$. The bulk modulus relates the change in volume of a solid as the pressure varies and is defined as:

$$K = -V\left(\frac{dP}{dV}\right) \quad (3)$$

The Young's modulus is quite simply defined as the ratio of stress divided by strain. In typical geotechnical practice, it is not uncommon to use Young's Modulus and because of such, many published correlations are available for the value. However, there are not many standard published values for bulk modulus of different soils. Therefore, the Bulk modulus values that are typically used are theoretically derived from a relationship with the P-wave velocity. The P-wave velocity is a very commonly used parameter in earthquake engineering and a large amount of research and field tests have been published regarding the subject. The Bulk Modulus, $K$, is related to the P-wave velocity from the following equation:

$$K = \rho V_p^2 - \tfrac{4}{3}\mu \quad (4)$$

where $\rho$ is density, $\mu$ is Shear Modulus, also called rigidity, which is experimentally observed to relate stress to strain according to Hooke's Law.

It is related to the Young's Modulus through the equation:

$$\mu = \frac{E}{2(1+\nu)} \quad (5)$$

where $\nu$ is Poisson's Ratio.

It is worth noting that blast will occur in a very short time frame. In a saturated materials there will be no significant drainage of water nor any significant dissipation of excess pore pressures, which will lead to the analyses being considered undrained, regardless of the cohesive nature of the material. This is in contrast to typical geotechnical practice where cohesionless materials such as sands and gravels are usually considered to be drained irrespective of the loading rate.

*Energy Absorbing Layer*

One of the retrofit measures is the addition of an energy absorbing layer to dissipate the blast energy before it reaches the critical structural elements. In practice, energy absorbing materials, such as honeycomb and foam, are commonly used in blast applications. The energy absorbing layer in the current simulations is modeled as honeycomb using the MAT_HONEYCOMB material model. The parameters representing elastic and shear modulus for uncompacted configuration and stress-volumetric strain response were based on the values reported by Zaouk & Marzougui (2001).

**Analysis Assumptions**

The blast analysis results are referenced to the blast pressure output of bare spherical TNT explosives. The data can be extended to include other potentially mass-detonating materials by relating the explosive energy of the "effective charge weight" of those materials to that of an equivalent weight of TNT. The effective charge weight ratio is defined as the ratio between TNT charge weight and equivalent charge weight of the specific explosive. For example, the effective charge weight ratio for ANFO (Ammonium Nitrate-Fuel Oil), one of the most common explosive types, is 0.82, in other words, 100 kg of ANFO is equivalent to 82 kg of TNT.

In the current blast study, the effect of flying debris or shrapnel was not evaluated and it was assumed that damage to the structures was due primarily to the shockwave propagation and reflection. The primary loading mechanism on a structure resulting from a blast is the shock front that is reflected and diffracted as it encounters the structure.

It is noted that the current study does not consider shape charge (flat, square, round, etc.), the number of explosive items and explosive confinement (casing, containers, etc.), since the impacts of those factors are relatively less important in determining structural response subject to the blast loadings. In the analyses presented herein, all charges were assumed to be non-confined and spherical in shape.

**BLAST ANALYSIS**

Prediction of material damage under a dynamic loading is a complex function of several factors such as loading magnitude, loading rate and loading duration. If the material experiences the same magnitude of loading at a different loading rate, the response will be different. The current paper introduces a blast analysis method, which fully incorporates dynamic properties of materials as well as strain rate effects, so that a dynamic structural response could be reliably predicted when the structure is subject to a blast loading.

The distinct behavior of an explosion in a tunnel was investigated by authors (Choi et al., 2003; Choi et al., 2006, Choi & Munfakh, 2006; Choi, 2009). Those publications introduced a three dimensional coupled Euler-Lagrange nonlinear dynamic analysis for modeling tunnel explosion. Authors proposed conceptual evaluation charts for conceptual and practical explosion impact assessment, expressed in terms of effective strain of the concrete liner for various types of surrounding grounds, charge weights, standoff distances, and sizes of the tunnel (Choi et al., 2006). This approach provides a conceptual evaluation of the blast loading on a tunnel structure.

For the precise and reliable evaluation, engineers need to conduct blast analysis for the given tunnel structures. Proper assessment of the impact of explosions on the underground facilities requires sophisticated analytical simulations and the application of numerical analyses that take into account several factors representing the explosion, the structure, and the ground and water. The following sections introduce blast behavior of various types of tunnel structures.

**Bored/Mined Tunnel**

In cases where a tunnel is located in significant depth or overlying structures exist above the tunnel alignment, bored or mined underground tunnel construction is typically preferred. The bored tunnel structure is usually composed of concrete. Even though the ballast fill concrete and concrete walk benches of a typical tunnel provide some cushion against an interior blast, the bored tunnel is likely to be very vulnerable to an interior blast due to the brittle nature of concrete. In addition, if the tunnel has relatively shallow cover, an external blast set from the water or land above could be detrimental to the structure. When the tunnel is underneath a body of water, cracking or failure of the concrete liner may allow inundation with water and result in high flooding potential in the transportation system if the tunnel is connected to underground transportation network. When the ground stability is not preserved, failure of the tunnel liner would also impact adjacent surface

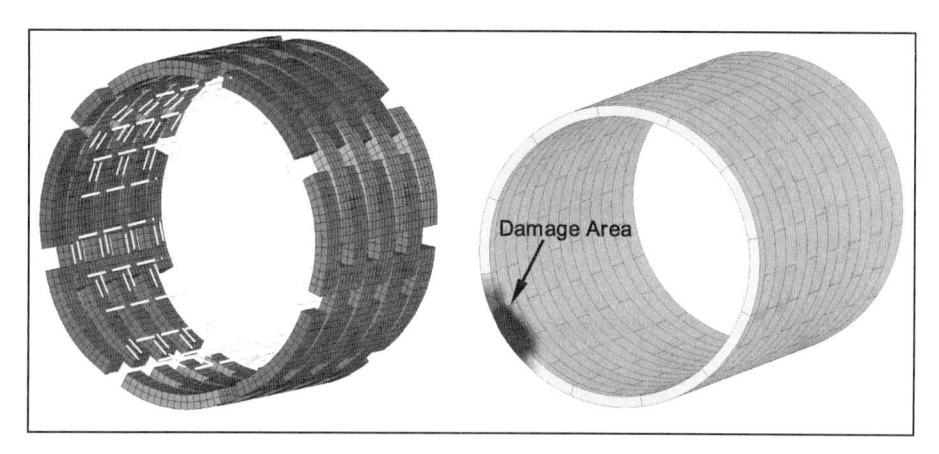

Figure 2. Three-dimensional blast analysis on bolted precast segment lining

and underground structures due to the loss of ground into the tunnel and the resulting instability of the structure and ground.

When segmented precast concrete lining is used, the behavior of the lining is also of concern (in addition to just the area of damage of the concrete). For instance, in this case, the bolts should not be overstressed and the outward displacement of the segments should not be large enough that the connections of gaskets and waterproofing are permanently rendered useless. Three-dimensional blast analysis enables to model behavior of a bolted segment to investigate the performance of the segments and to recommend a proper dimension and number of bolts to be used (Figure 2). In this case, the connecting elements (dowels and bolts) are explicitly modeled and pretension and interface strength can also be modeled.

## Cut-and-Cover Tunnel

Shallow depth tunnels and approaches in land are frequently designed using the cut-and-cover method. This tunneling method involves braced, trench-type excavation and placement of fill materials over the finished structure. The excavation is typically rectangular in cross section and in relatively shallow depth. Intuitively, the cut-and-cover tunnel is extremely vulnerable to an interior blast due to less confinement from surrounding ground. However, it is probable that the cut-and-cover tunnel is less vulnerable to a surface blast due to the open-air nature of the blast and soil cover over the tunnel. For U-tunnel section, where the tunnel structure is open and not covered, the vulnerability is extremely high due to relatively easy delivery method of explosives. Conversely, since the blast is not as confined as in a tunnel structure, some energy may dissipate and less reflection of the blast waves could occur.

A unique example of a cut-and-cover structure subjected to blast loadings is where a water conduit crosses over a cut-and-cover station and the water conduit is connected to the infinite supply of water. In this case, blast induced damage on the station ceiling could result in uncontrollable water inflow and the ultimate flooding of the interior structure. Figure 3 presents a blast detonation at the conduit and damage areas at the conduit invert and station ceiling.

## Immersed Tube

Immersed tube tunnel is employed to traverse a body of water. The tunnel construction method involves (1) construction of tunnel sections in an offsite casting or fabrication facility that are finished with bulkheads and transported to the tunnel site; (2) placement of the sections in a pre-excavated trench, jointing and connecting together and ballasting/anchoring; and (3) removal of temporary bulkheads and backfilling the excavation. The top of the tunnel is usually at least 1.5 meter (5 feet) below the original bottom to allow for an adequate protective backfill. The typical immersed tube consists of concrete liner, steel shell and concrete tube. The concrete liner and concrete tube are load bearing elements, while the steel shell is usually not considered to be a load bearing element but rather acts as a water-proofing membrane. The joints between segments may be the most vulnerable if it is not covered with tremie concrete. Local breach of the main tunnel structures would induce complete inundation with water and cause flooding in the underground transportation system. Flooding may also introduce large quantities of sand, silt or gravel. Significant lengths of tunnel can become filled with debris or backfill in a short period of time. For this reason, the immersed tube structure is considered to be the most vulnerable

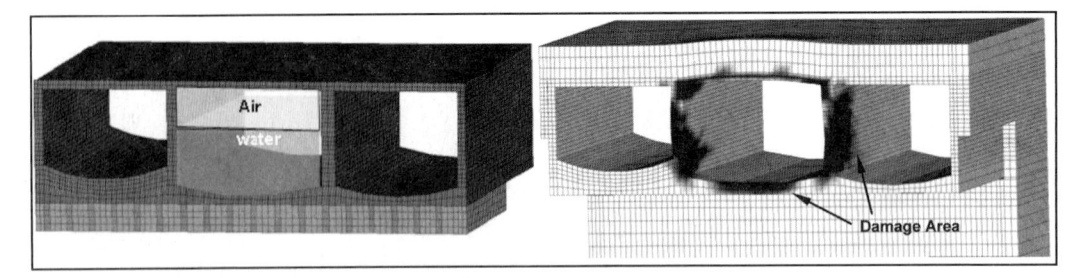

**Figure 3. Three-dimensional blast analysis on cut-and-cover station underneath water conduits**

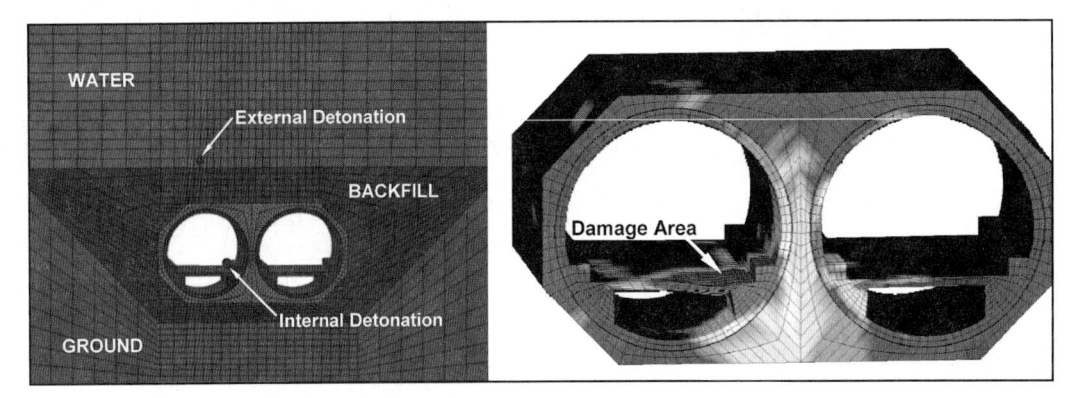

**Figure 4. Three-dimensional blast analysis for immersed tube**

element. Figure 4 presents a typical immersed tube tunnel subjected to a blast loading.

## Underground Station

Underground station is constructed by either cut-and-cover method or using bored/mined technique. Considering the high level of access by pedestrians and riders, the number of exposed persons and consequences, the station structures are likely to be very vulnerable to an interior blast of even a relatively small magnitude charge weights. For the underground stations, hand-carried satchel bombs and suitcase bombs are the predominant mode of explosive attack, while subway storage yards, service facilities, and the shipping channels above the tunnels allow for the potential delivery of much larger explosive/incendiary materials.

## Ventilation Shaft

Ventilation shafts are typically reinforced concrete shafts extended from the land surface to the underground structure. At the interface of the shafts and the bored tunnel, they may be more vulnerable to damage because high stress concentrations may occur at these junctions. However, due to access restrictions, only a small amount of explosive is likely to be brought into the shafts, therefore an interior threat within the emergency exit shafts is not considered critical. A more critical threat would be one introduced to the interface from the tunnel structure where a large amount of explosives can be carried by vehicle.

## Cross Passageway

A cross passageway may be vulnerable to damage because high stress concentrations may occur at the junctions with a main tunnel and given the same amount of explosive charge, the resulting blast peak pressure in a cross passageway tunnel may be greater than that in the main tunnel due to its smaller cross-sectional geometry. However, from an operational standpoint, the cross passageway is not considered to be more critical than the main tunnel because of their greater degree of redundancy due to a number of cross passageways. Furthermore, local failure or collapse of one or more of the cross passageway tunnels may not necessarily affect the stability of the main tunnels or prevent their continuous use if the water inflow is controlled.

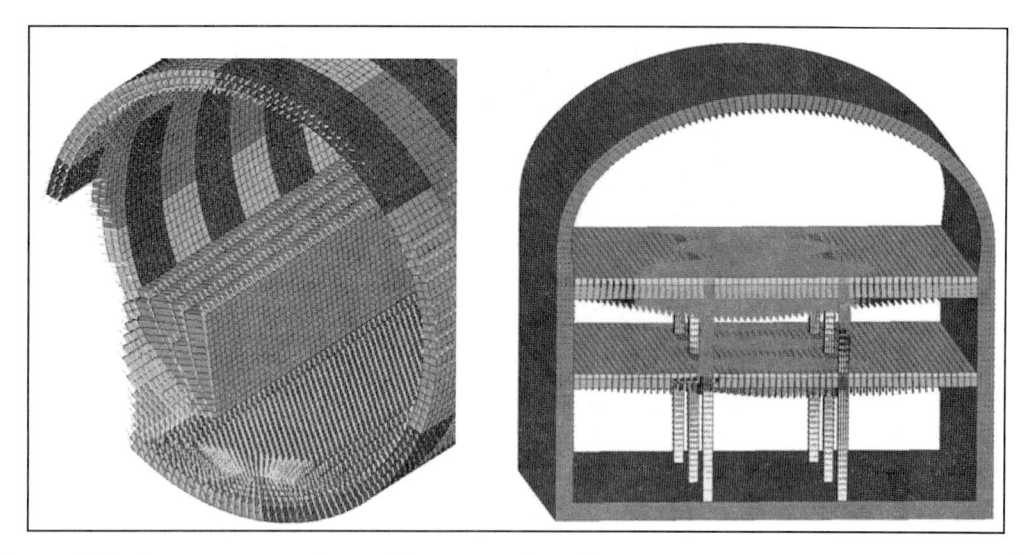

Figure 5. Displacement vectors in post blast progressive collapse analysis

## Portal

From a stability standpoint, the tunnel portal area is generally one of the critical locations due to the inherent slope stability problem and/or retaining structure failure. Tunnel portals are therefore considered to be particularly vulnerable during extreme events. Nevertheless, the consequences of a portal failure are generally considered to be less severe than those of main tunnel element failure because the repair can be done in an open space. The flooding is not an issue when a portal is damaged or collapses, so the repair time and associated costs are relatively low compared to the other parts of the tunnel. Furthermore, at the portal, the blast is less confined and the energy dissipates away rapidly than it will in the confined tunnel environment.

## POST BLAST ANALYSIS

In addition to the evaluation of damage extents caused by the blast-induced loading, post-blast behavior should be analyzed to evaluate progressive type failure/collapse where continued failure may occur due to the structural weakening, load redistribution, excessive displacements, water inflow or running ground into the tunnel.

Progressive collapse of underground structures is of great concern, even if the underground structural elements are not completely damaged during the blast. They may be weakened or softened, at which point the normal loading imparted during operations would cause further damage or failure to the structures. The progressive failure analysis considers the structure in operation, subsequent to the blast loading. The post blast progressive collapse analysis is performed in such a way where the damaged area(s) are removed or weakened, and by applying the applicable load combinations to the structures. Figure 5 presents post blast progressive collapse analysis for a tunnel and complex underground structure. The tunnel is loaded by surrounding ground and water pressure. The complex underground structure is assumed to be fully loaded with trains on each track level, with the mezzanine level and platforms assumed to be fully occupied with the full live load and dead load expected.

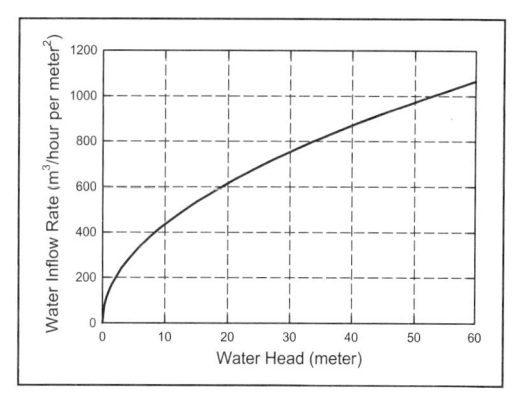

Figure 6. Predicted water inflow rate toward tunnel

The stability of the ground is also of concern with the breach failure of the liner when the stability of surrounding ground is not preserved. For soft ground tunnels under the water, ground failure and subsequent flowing condition into the tunnel associated with the breached concrete liner failure is a

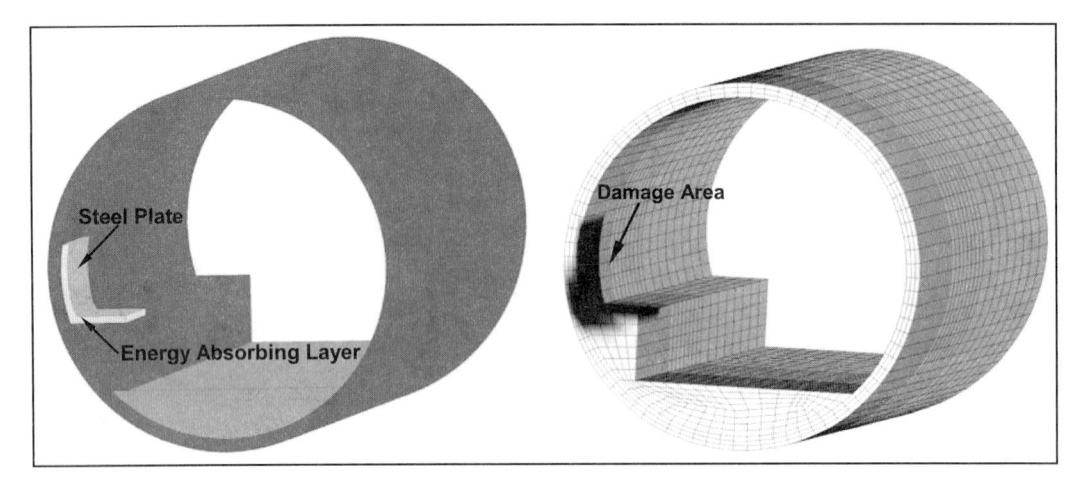

**Figure 7. Thin steel plate with energy-absorbing layer**

serious issue. This ground failure and/or flowing condition may result in large water inflow because of the high water pressure and infinite water supply.

A qualitative evaluation of flood potential for various charge weights of internal explosions at various detonation points of the tunnel should be conducted. A sample water inflow rate is presented in Figure 6 based on the channel flow toward the holes in the liner. Figure 6 indicates that 1 square meter of hole in the liner under the 30-meter water head could induce flooding of the 40-ft diameter tunnel in approximately 25 minutes.

## BLAST PROTECTION MEASURES

The blast protection measures available for a tunnel were introduced in Choi & Munfakh (2006). This paper discusses effectives of typical blast protection measures. The effectiveness was evaluated with three-dimensional blast analyses as described above.

### Thin Steel Plate with Energy Absorbing Layer

This protection measure considers a relatively thin steel plate with an energy absorbing layer (blast mat or crushable foam). The example analysis shown in Figure 7 consists of 6.35 mm (¼ inch) thick steel plate and 150 mm (6 inch) thick energy absorbing layer inside the tunnel liner. The damage contours, due to the internal detonation, indicate that the tunnel liner is not damaged, but slight permanent plastic displacement is observed, while the sacrificial, energy-absorbing layer is completely damaged. The relative cost of this measure is estimated to be "medium to high" while the relative effectiveness is considered to be "very high."

### Steel Plate Outside of the Tunnel Liner

This protection measure considers a confining 50.8 mm (2 inch) thick steel plate installed outside of the tunnel liner. The damage contours indicate that the concrete liner is locally damaged, however, the steel plate is not damaged due to ductile behavior of steel, resulting in a system that will require some concrete lining repair, but will prevent water inflow and a subsequent flowing soil condition. Issues to be considered in this measure may be difficulty of fabrication and installation. The relative cost is considered to be "high" while the relative effectiveness is considered to be "high."

### Increase of Lining Thickness, Strength, or Reinforcing Ratio

This protection measure considers increasing the liner thickness, concrete strength, or the percentage of reinforcing steel. The damage at the tunnel liner is reduced, however, the soil instability and significant water inflow is still expected due to breaching of the concrete liner. The relative effectiveness is "Low to Medium," while the relative cost can be considered "Moderate."

## SYSTEM COUNTERMEASURES

The system countermeasures should be implemented in combination with the tunnel blast protection measures. Choi & Munfakh (2006) proposed three categories of system countermeasures such as (1) basic measures of safety and security, (2) measures deployed for an elevated threat (hazard) level in response to a specific threat condition, and

(3) permanent enhancements to the tunnel structure or systems. Details of system countermeasures are discussed author's previous paper, Choi & Munfakh (2006).

## CONCLUSIONS

Numerical modeling has been driven by a perceived need in recent times. It has led to large, sophisticate and complex numerical models not always because they are an integral part of the design process, but sometimes because it is considered irresponsible to not bolster a design with plots of stress and displacement contours. Advanced numerical modeling is not a subject that leads to a research proposal, higher degree or paper publication. Properly performed numerical model analysis will lead engineers to think about why they are building it—why build one model rather than another—and how the design can be improved and effectively constructed.

A blast and post-blast analysis is an essential to the security design of tunnels. Proper assessment of the impact of explosions on a tunnel requires sophisticated analytical simulations and the application of numerical analyses that take into account several factors representing explosion, structure, and ground and water. Properly selected numerical approach in this case should, therefore, properly incorporate explosive-air-structure-ground interaction. Approximation and simplification are essential in any modeling process, however, the modeler should be aware of the limitations and application of the assumptions in order to produce a reasonable engineering solution. Comparison with analytical/empirical solutions is a good endorsement to check the results. A parameter sensitivity analysis may be a good practice to increase credibility of the modeling results.

## ACKNOWLEDGMENTS

The authors gratefully acknowledge the Career Development Committee of Parsons Brinckerhoff for its support of the William Barclay Parsons Fellowship Research program. The authors would also like to thank George Munfakh, Nasri Munfah and Sanja Zlatanic of Parsons Brinckerhoff for their technical review and advice in preparation of this paper.

## REFERENCES

Choi, S., *Tunnel Stability under Explosion*, William Barclay Parsons Fellowship Monograph, Parsons Brinckerhoff, 2009

Choi, S., and Munfakh, G., "Tunnel Design Under Explosion," keynote paper on *Third International Conference on Protection of Structures against Hazards*, September 2006, Padova, Italy

Choi, S., Wang, J., Munfakh, G., and Dwyer, E., "3D Nonlinear Blast Model Analysis for Underground Structures," *GeoCongress06*, Geo-Institute of American Society of Civil Engineers, February 2006, Atlanta, Georgia U.S.A.

Choi, S., Wang, J., and Munfakh, G., "Tunnel Stability under Explosion—Proposed Blast Wave Parameters for a Practical Design Approach," *First International Conference on Design and Analysis of Protective Structures against Impact/Impulsive/Shock Loads*, 15–18 December 2003, Tokyo, Japan

Lee, E.L., Hornig, H.C. and Kury, J.W., *Adiabatic Expansion of High Explosive Detonation Products*, UCRL-50422, Lawrence Radiation Lab., University of California, Livermore, 1968

Tavarez, F., *Simulation of Behavior of Composite Grid Reinforced Concrete Beams Using Explicit Finite Element Methods*, Master's Thesis, University of Wisconsin—Madison, 2001

Transportation Research Board, *Making Transportation Tunnels Safe and Secure*, TCRP Project J-10G/NCHRP Project 20–67, 2006

Zaouk, A., Marzougui, D., "Development and Validation of a US Side Impact Moveable Deformable Barrier FE Model." *Third European LS-DYNA Conference*, Paris, France, June, 2001

# Decision-Making Case History for Municipal Infrastructure Improvements

**Lizan N. Gilbert, Jennifer Stark**
URS Corporation, Austin, Texas

**Gopal Guthikonda, Joe Hoepken**
Austin Water Utility, Austin, Texas

**ABSTRACT:** The City of Austin ("City") is providing water and wastewater service for future development in the southeast portion of the City's service area, including 12,000 feet of wastewater tunnels. Selecting a tunnel alignment was challenging because an existing homeowner's association ("HOA") had disagreed with previous routing decisions made by the City on another wastewater project. Four alignment alternatives were determined and evaluated to maximize the tangible benefits, e.g., cost, as well as non-tangible benefits, e.g., environmental protection, of the options and adhered to the physical constraints of the area, e.g., connections. A multi-attribute utility theory framework was developed to consider the non-tangible attributes of the alternatives. The evaluation successfully identified one preferred alternative. This alternative was presented at a public meeting and was supported by the HOA. The process presented in this paper reflects the successful application of a multi-attribute utility function that facilitated the selection of an alignment.

## INTRODUCTION

The southeastern area of the City's Extraterritorial Jurisdiction (ETJ) is currently undergoing significant development. To meet the needs and schedule of the proposed development, the City has created the South IH-35 Water and Wastewater Infrastructure Improvements Program. This program, made up of City staff and consultants, is responsible for managing the design and construction of approximately 70,000 linear feet of large-diameter water line and 16,000 linear feet of wastewater interceptor, approximately 12,000 of which will be tunneled. The tunnel portion runs adjacent to two creeks, and under a private golf course and an existing residential neighborhood. Figure 1 presented the project area.

Alignment selection began with a standard process focusing on engineering considerations and logistical constraints, such as system connections. Initially, 35 alignments were considered that included various configurations for manholes, launch and retrieval shaft locations, and alignments under the Onion Creek Golf Course. After analysis, the 35 possible alignments were reduced to four as shown on Figure 2. These four alignments were determined to maximize the tangible benefits, e.g., cost, as well as non-tangible benefits, e.g., environmental protection, and adhered to the physical constraints of the various creeks crossing the area, and existing and proposed infrastructure.

Each alternative alignment consisted of a route to convey wastewater flows from the upstream location at Onion Creek and IH-35 to the downstream location at an existing junction box at the convergence of Slaughter and Onion Creeks. The primary area of focus for this alignment selection study was from the existing Onion Creek Package Wastewater Treatment Plant to the existing junction box since this portion of the alignment cuts through mostly residential single-family homes within the HOA. The south alignment alternatives (shown in black in Figure 2) were evaluated in the same manner, but received far less interest from the HOA because impact to the HOA was fully dependent on which north alignment alternative was selected. The four alignments are as follows:

- Yellow—shortest route between connection points and aligned to allow for the future abandonment of existing WWTP and lift station in poor condition;
- Purple—shortest route between connection points that stays within the public Right of Way (ROW) and aligned to allow for the future abandonment of existing WWTP and lift station in poor condition;

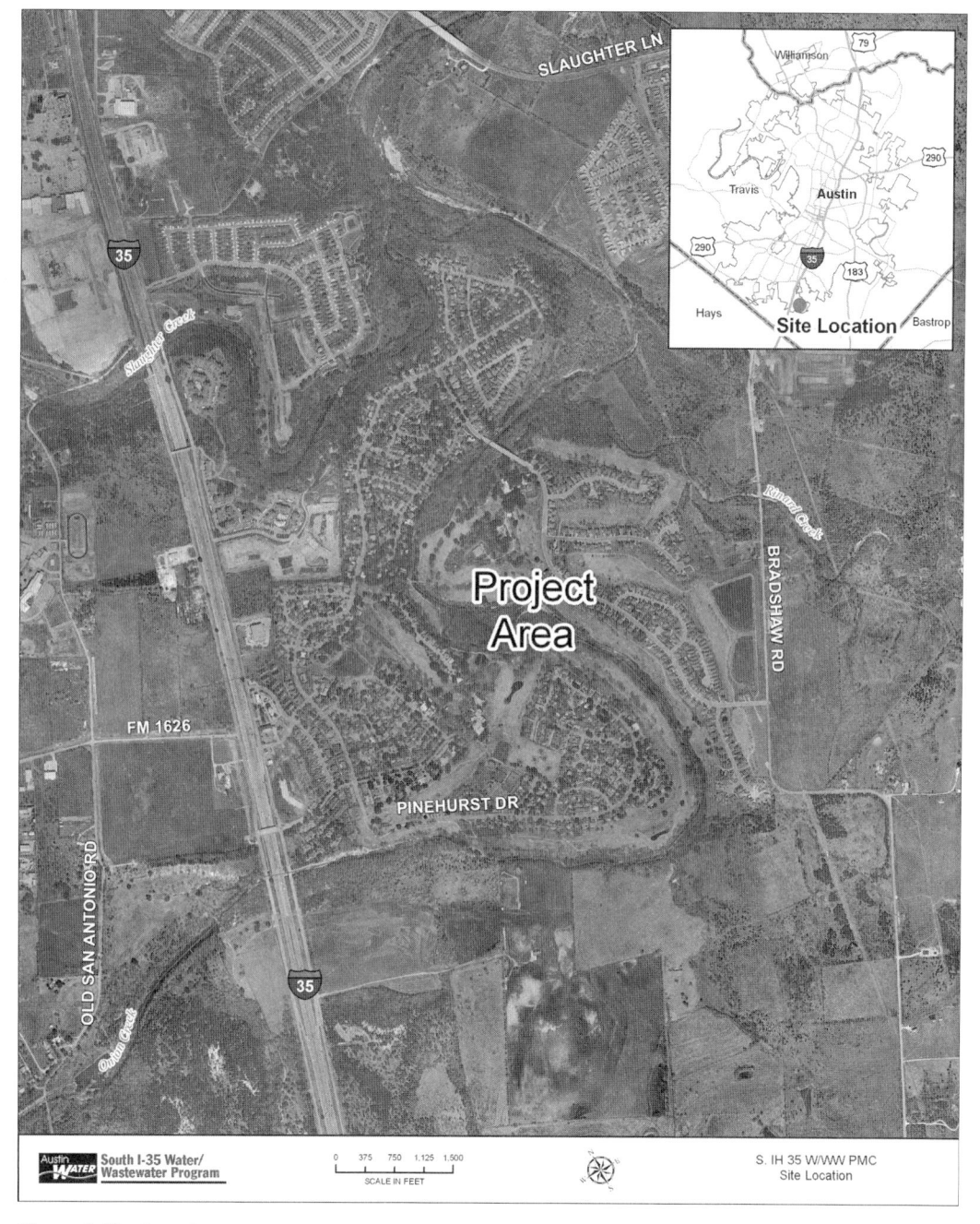

**Figure 1. Site location map**

- Orange—outside of the neighborhood and consistent with historical City alignments, i.e., parallel to creeks and aligned to allow for the future abandonment of existing WWTP;
- Green—outside of the neighborhood and along public ROW, but requires re-design of

proposed Zachary Scott Line to meet regional flow requirements.

These four alignments were presented to the public, who expressed a desire to evaluate a fifth alignment (Figure 2. Residential Proposed Route

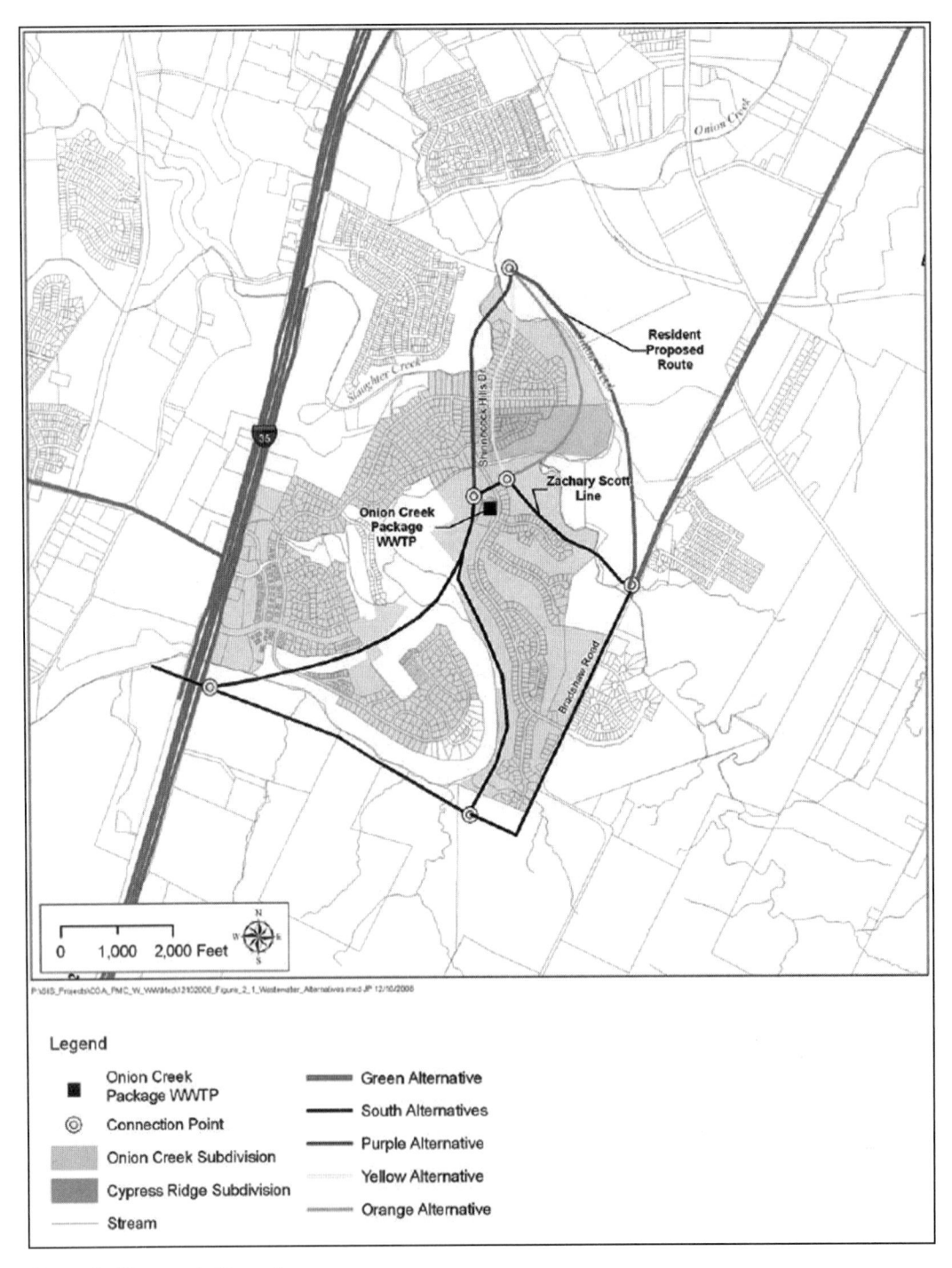

Figure 2. Alignment alternatives

labeled pink) because it connected with the northern connection point and was aligned outside of the neighborhood.

- Pink—outside of the neighborhood and in creek greenbelt area, but requires re-design of proposed Zachary Scott Line to meet regional flow requirements.

While looking at the pros and cons of each alignment, the program team worked to identify the group of factors that would affect the alignment decision. These factors were revealed during the basic alignment alternative identification. They consisted of the following:

- Feedback received at meetings with the stakeholders which revealed a strained relationship between the HOA and the City on various past projects since the City annexed the HOA in 2004.
- Discussions within the program team to clarify the City's focus on environmental protection. Historically the City had aligned wastewater infrastructure within creeks but now has abandoned that practice to minimize negative environmental impacts to those areas. Current requirements include additional permitting whenever crossing or coming within 400 feet of the creek centerline.
- The challenges of crossing under the private golf course including limited surface access, limited hours for access, and a vested (?) neighborhood.
- Affluent and well-educated HOA members primarily consisting of retirees, with the resources available to ensure that their objectives are well represented.

With these diverse and occasionally conflicting factors in play, the program team recognized the need for a transparent decision-making process that considered tangible attributes of each alignment, including cost and schedule, and non-tangible attributes, including engineering feasibility, environmental impacts, neighborhood impacts, and constructability.

The approach used to select the preferred alignment by considering the tangible and non-tangible features of each alternative is presented in this paper. First, the estimated costs will be presented. Next, the multi-attribute utility framework that was used to assess the non-tangible features will be described. Finally, the process used to select the final alignment based on the cost and utility information will be explained.

**Table 1. Unit costs for tunnel estimating**

| Attributes | Estimated Unit Cost |
|---|---|
| Construction Capital Costs | |
| *Length* | |
| Tunnel (LF) | $2,074 |
| *Construction Shaft Types* | |
| Launch (EA) | $124,050 |
| Intermediate (EA) | $87,035 |
| Receiving (Launch/Exit) (EA) | $124,050 |
| *Permanent Shafts* | |
| Access (EA) | $6,997 |
| Non Construction Capital Costs | |
| *Mobilization* | |
| Tunnel Boring Machine Mobilization (LS) | $18,000 |
| Contractor Mobilization (LS) | $397,561 |

## ESTIMATE OF CONSTRUCTION COST

The total construction costs were calculated by adding the estimated cost of tunneling with the estimated cost of construction delays for each alternative. The estimated cost of tunneling was calculated by using historical bid tabulations from tunnel projects constructed locally. Unit prices for each of the "major" items included in the construction costs for each alternative (e.g., tunnel construction, launch/exit shafts), the number of items included for each alternative (e.g., 12,800 lineal feet of tunnel and 2 launch/exit shafts), and the total cost of these items were evaluated. These unit values for the alternatives (i.e., how many lineal feet, shafts, etc.) were estimated considering the field conditions and the items required for each alternative. Table 1 represents the unit costs considered for each alignment.

The estimated costs for the schedule were based on delays. Schedule ties directly to the cost since increased construction duration will result in increased costs, all else being equal. The basis for evaluating the impact of schedule on cost was to first estimate the total construction time, as the sum of the construction time plus the time to acquire easements plus the additional time due to constructability. From this total, baseline construction duration was subtracted to arrive at a total delay in months. First, the baseline construction time of 12 months was used based on a proposed City schedule of completion. The critical factor in this evaluation was to set this baseline so that a relative comparison could

**Table 2. Schedule evaluation for orange alternative**

| Attributes | Orange Schedule Months |
|---|---|
| Construction time | 31 |
| Time to acquire easements due to condemnation | 12 |
| Additional construction time due to curved alignment | 1 |
| Total construction time (months) | 44 |
| Baseline construction time (months) | (12) |
| Delay (months) | 32 |
| 32 × (0.8% × construction cost) | $9,800,000 |

be made. The scheduled construction time was based on a 2-month mobilization period, a 2-month shaft construction period, an advance rate for the tunnel of 40 ft/day in a 10-hour shift, and a pipe installation rate of 80 ft/day.

The two additional components considered in the schedule to estimate construction costs were easement acquisition and constructability. Easement acquisition is defined as the additional time due to acquiring easements, and on the conservative side, having to proceed through condemnation for acquiring private easements. These times are shown as the "Time to Acquire Easements due to Condemnation." The constructability timeframe comes from the additional time required to advance a tunnel along a curve as opposed to straight runs—the additional time to survey points, evaluate alignment, etc. and were simplified into "Additional Construction Time due to Curved Alignment." For each alternative, the portion of the length within a curve was determined and an increased construction time of 10% of the standard 40-ft/day duration to traverse this portion was included. It should be noted that only the Orange alternative was assigned an increased duration due to curved alignment. In the case of the Orange alternative shown, the 32-month delay is estimated from the total construction time of 44 months minus the 12 month baseline schedule as shown in Table 2.

The total additional cost due to schedule is based on the assumption that construction costs generally increase at an average value of inflation of approximately 10% per year, or 0.8% per month. This additional schedule cost is calculated as the product of the estimated delay and the monthly inflation rate. Continuing with the Orange alternative, the estimated construction cost was $38M. This would result in a schedule cost of $9.8M that was added to the total construction cost to arrive at an overall cost of $48M.

When reviewing the final costs, shown in Table 3, it is clear that the relative difference between alternatives is small. Using these metrics alone, no single alternative clearly stands out as preferred. Therefore, consideration of alternative attributes, the non-tangible factors, was necessary to identify the preferred alternative.

## ASSESSMENT OF NON-TANGIBLE FACTORS

A framework was developed to consider the non-tangible attributes of the alternatives. The intent of this framework was to incorporate the value systems of the City and the stakeholders (the HOA) in the selection of a preferred alignment. This framework was based on multi-attribute utility theory, where the multiple factors that were considered to be of importance to the City and stakeholders are the attributes and the potential outcome for that attribute with each alternative reflected by a numerical utility value.

### Multi-attribute Utility Evaluation

The total utility score for each alternative was calculated as follows:

$$(\text{utility}_{total})_{\text{alternative k}} = \sum_{i=1}^{n} \{[(\text{utility}_{\text{attribute i}})_{\text{alternative k}}] w_{\text{attribute i}}\} \quad (1)$$

Where $(\text{utility}_{total})_{\text{alternative k}}$ is the total utility value for alternative k; $(\text{utility}_{\text{attribute i}})_{\text{alternative k}}$ is the utility value for attribute i if alternative k is selected, where a utility value of 1 is the best possibility and a utility value of 0 is the worst possibility; and $w_{\text{attribute i}}$ is the relative importance of that particular attribute to the public. The maximum possible utility score is the sum of the weights, which corresponds to the case where an alternative provides the best possible outcome for every attribute. The minimum possible utility score is zero.

### Selection of Attributes

The non-tangible attributes to be considered were selected through a series of meetings with the City. The program team first established a preliminary list of proposed factors. The intent of this preliminary list was to capture the values of the City and the basic concerns raised by the public in early public meetings, and also to motivate and facilitate an organized discussion about what the most important non-tangible factors were. This list and its rationale were then presented to various groups and parties within the City through a series of five meetings with representatives from the following departments: Contracts

## Table 3. Total costs per alternative

| Measurement | Alignment Alternatives | | | | |
|---|---|---|---|---|---|
| | Purple | Yellow | Orange | Green | Pink |
| Cost ($M) | $42 | $43 | $48 | $52 | $45 |

## Table 4. Decision attributes

| No. | Attribute | Description | Metric |
|---|---|---|---|
| 1. | Relative length of route | The relative length of alignment is directly related to the environmental disturbance occurring throughout the construction and tunneling process. The shorter the alignment, the less potential disturbance is sustained. The potential impacts are related to underground and natural features, such as paleochannels, where a tunnel may intercept such a feature and potentially affect the overall water quality. | Total length of alignment (ft) |
| 2. | Percentage of route within critical water quality zone (CWQZ)/environmentally sensitive areas | The CWQZ is an area defined by either 400 ft from the center line of the creek or the floodplain limits as defined by the City. The length of the alignment within this area is directly proportional to the environmental impacts. | Length within the CWQZ (ft) |
| 3. | Plant reliability/back-up system | This attribute evaluates if a back-up system exists to support the Onion Creek Package Wastewater Treatment Plant should mechanical or electrical failure or a biological process upset occur that renders the plant unable to produce a quality effluent. Rather than send poor quality effluent to the ponds or experience an overflow to the creek, the alignments that provide plant reliability would allow the raw wastewater to be directed downstream to the South Austin Regional Wastewater Treatment Plant (SARWWTP). Alignments with the plant reliability feature also simplify maintenance operations where it is desirable to take a tank off line for scheduled equipment repairs. This criterion reflects the availability of a backup option for the alignment option. | Indicates whether the alignment reaches the plant (yes or no) |
| 4. | Percentage of route within private property | This attribute evaluates the percentage of private properties that each alignment intersects. Private property is considered any land parcels that are not within the City ROW or are not public utility easements, such as the Onion Creek Golf Course. | Length of alignment (by percent) that is aligned across private property (%) |
| 5. | Number of private easements required | Acquisition of easements takes time and costs money. The greater the number of easements, the more effort required, and the greater the likelihood of going through condemnation on at least one property. | Number of private easements each alignment intersects (no.) |
| 6. | Temporary easements required during construction | The need for temporary working space easements for construction activities would be necessary if the surface activities were located in areas not owned or operated by the City. | Indicates whether or not temporary working space easements are required (yes or no) |

(continues)

**Table 4. (continued)**

| No. | Attribute | Description | Metric |
|---|---|---|---|
| 7. | Solids handling elimination | Currently, solids generated at OCPWWTP are trucked from the plant to Walnut Creek WWTP several times per day. Upon completion of the proposed interceptor, these solids will be routed into the Interceptor and routed to SARWWTP for treatment. | Indicates whether the alignment provides an alternative to route solids away from the plant (yes or no) |
| 8. | Pinehurst lift station abandonment | Abandonment of the Pinehurst Lift Station would eliminate the long term capital, operation, and maintenance costs associated with this process component. With abandonment of the lift station, localized wastewater flows will be routed into the proposed Interceptor and treated at SARWWTP instead of at OCPWWTP. | Indicates whether or not the alignment allows for the abandonment of the Pinehurst Lift Station (yes or no) |
| 9. | Onion creek package wastewater treatment plant (OCPWWTP) abandonment potential | Location of the proposed interceptor adjacent to OCPWWTP will ultimately allow for future abandonment of OCPWWTP. As the plant ages the time will come when a determination must be made as to whether it is cost-effective to rebuild the plant or to abandon the plant, sending wastewater flows to SARWWTP, and replacing the golf course irrigation water with reclaimed water piped from an extension of the reclamation system that originates at the SARWWTP. | Indicates whether or not the alignment allows for the future abandonment of the OCPWWTP (yes or no) |
| 10. | Duration of construction impacts | Construction of shafts within the adjacent neighborhood would increase noise pollution due to removal of dirt and rock cuttings, as well as truck traffic. If the alignment is located such that a shaft would have to be constructed within the neighborhood, these impacts are related to the duration of those impacts. | Duration of impacts to residential areas due to construction (months) |
| 11. | Historical sites | If a historical site were identified along the route, the potential for delay or re-design is likely. This criterion indicates the number of historical sites the alignment intersects. | Number of historical sites intersected by the alignment (no.) |
| 12. | Shafts in the critical water quality zone | Shafts within the CWQZ indicate a likely significant impact to the environment due to construction activities as well as future maintenance activities. | Number of shafts located within the CWQZ (no) |
| 13. | Hydraulic/grades, geotechnical, maintenance | These criteria reflect the engineering feasibility of the alignments and whether or not standard construction techniques and locally available equipment could be used. | Indicates engineering feasibility (yes or no) |

and Land Management, Austin Water Utility, and Watershed Protection. A public relations consultant was also involved in these meetings. In this process, various attributes were revised, added, or removed.

An important goal in finalizing the list of attributes was that each be measureable. For example, one attribute was creek crossings, where the metric for this attribute was the actual numbers of creeks each alignment crossed with a smaller number of creeks crossed being preferred over a greater number. Another attribute was treatment plant reliability, with the metric for this being the ability to provide an alternative to sending flows to the treatment plant and the future decommissioning of the plant. The final list of 13 attributes resulting from this process is summarized in Table 4.

## Weighting of Attributes

Assigning weights to the attributes was also accomplished at the referenced series of five meetings.. The starting point of the conversation was a discussion regarding the values outlined in the list of attributes in the context of historical decisions made with the infrastructure projects proposed and/or constructed within the HOA as well as on other City projects. The goal was to identify the innate preferences that typically exist when decisions are made, even when the decision maker is reluctant to assign a preference.

For example, a previous wastewater project in the neighborhood had been designed as an open-cut installation. However, after significant public opposition, the City revised the design to a tunnel installation at an additional cost of approximately 100 percent. With regard to the environmental impacts, the depth of the tunnel in many locations is within the depths that could be accomplished using open-cut installation. However, the impacts to the creeks and within the CWQZ are prioritized such that the additional permitting effort and construction costs for restoration are less appealing than the additional cost of tunneling. Although these anecdotes could potentially be used to quantify a preference, the City preferred the alternative of considering each decision independently.

The final decision was to assign equal weight to each attribute, i.e., $w_{attribute\ i} = 1.0$ for all n attributes. Therefore, the maximum possible value for the total utility score was 13.0.

## Assignment of Utility Values

The utility values were assigned by first ranking the alternatives by the outcomes for a given attribute. For example, the alignments were ordered by the length of the route. The alignment with the longest length (Green alternative with 22,900-foot length) was given a utility value of 0.0, and that with the shortest length (Yellow alternative with 16,750 foot length) was given a utility value of 1.0. The utility values for the other alternatives were then assigned between 0.0 and 1.0 in proportion to their lengths relative to the maximum and minimum. The length of the Orange alternative is 17,800 feet, giving it a utility value of 0.8 (or 80 percent of the way between the longest and shortest lengths possible). Also, the number of private easements ranged from 0 to 5 with the Purple alternative requiring 4 easements, giving it a utility value of 0.2. For those attributes that were evaluated using a 'yes' or 'no', such as the Solids Handling Elimination or Pinehurst Lift Station Abandonment, utility values of 1.0 or 0.0 were assigned.

For three attributes, the utility values were the same for each alternative. These utility values were the same because although the attribute was important in the alignment selection process, it did not differ for any of the alignments. For example, the historical sites attribute did not differ between alignments because there were no historical sites identified on any of the alignments. However, it was important to include these attributes in the discussion so as to convey the message that they were considered in the evaluation. These attributes were not considered further because they did not distinguish the alternatives. Hence, the list of attributes was shortened to 10 and the maximum possible utility score was reduced to 10.0.

## Assessment of Total Utility Value for Each Alternative

The calculated value for the total utility is presented in Table 5. The maximum total utility is 7.3 and the minimum is 1.7. Two values are higher than the rest because of four primary attributes for which these two alternatives scored a 1.0: 1) Plant Reliability, 2) Solids Handling Elimination, 3) Pinehurst Lift Station Abandonment, and 4) OCPWWTP Potential Future Abandonment. One is well below the others because it accumulated scores for only three of the attributes: 1) Relative Length of Alignment, 2) No. of Private Easements Required, and 3) Duration of Construction Impacts, while all others were scored at 0.0.

These results were presented to the City one final time to see if the factors that should be driving the utility scores were being captured properly. Internally, the team evaluated the outcome to ensure that the values were being accurately reflected in the selected alignment by varying the weights of the attributes to determine the sensitivity of the decision. In addition, the elimination of various criteria were evaluated to determine if, for example, an alternative that was relatively short and went through private easements, but had disturbances to the environment would be preferred if all environmental criteria were eliminated. These evaluations supported the utility function and its outcomes.

## SELECTION OF FINAL ALIGNMENT

The estimated cost and total utility values are summarized in Table 5 and on Figure 3 for each alternative. The two alternatives that stand out are the yellow and purple due to their ability to allow for the relief or future abandonment of existing infrastructure, to minimize the private easements required and to avoid environmentally sensitive areas. These two alternatives were also appealing based on costs because they had the lowest cost impacts.

Although the yellow and purple alternatives were very close, the utility score and cost for purple made it the preferred alternative. This difference is

Table 5. Total utility value by alternative

**South I-35 Water / Wastewater Program**
**North-Reach Interceptor Route Selection Evaluation Matrix**
**September 23, 2008**

| Evaluation Criteria | Possible Options | Score | Purple | Yellow | Orange | Green | Pink |
|---|---|---|---|---|---|---|---|
| 1. Relative Length of Route | shortest → longest | = 1.0<br>= 0.9<br>= 0.8<br>= 0.7<br>= 0.0 | 0.9 | 1.0 | 0.8 | 0.0 | 0.7 |
| 2. Percentage of Route within Critical Water Quality Zone (CWQZ)/ Environmentally Sensitive Areas | 70%<br>30% | = 0.0<br>= 1.0 | 1.0 | 1.0 | 0.0 | 1.0 | 0.0 |
| 3. Plant Reliability / Back-up System | no<br>yes | = 0.0<br>= 1.0 | 1.0 | 1.0 | 1.0 | 0.0 | 0.0 |
| 4. Percentage of Route Within Private Property | least → most | = 1.0<br>= 0.3<br>= 0.2<br>= 0.0 | 0.3 | 0.2 | 0.0 | 1.0 | 0.0 |
| 5. Number of Private Easements Required | least → most | = 1.0<br>= 0.5<br>= 0.3<br>= 0.2<br>= 0.0 | 0.2 | 0.0 | 0.5 | 1.0 | 0.3 |
| 6. Temporary Easements Required during Construction | yes<br>no | = 0.0<br>= 1.0 | 0.0 | 0.0 | 0.0 | 1.0 | 0.0 |
| 7. Solids Handling Elimination | no<br>yes | = 0.0<br>= 1.0 | 1.0 | 1.0 | 1.0 | 0.0 | 0.0 |
| 8. Pinehurst Lift Station Abandonment | no<br>yes | = 0.0<br>= 1.0 | 1.0 | 1.0 | 0.0 | 0.0 | 0.0 |
| 9. Onion Creek Package WWTP Abandonment Potential | no<br>yes | = 0.0<br>= 1.0 | 1.0 | 1.0 | 1.0 | 0.0 | 0.0 |
| 10. Duration of Construction Impacts | shortest → longest | = 1.0<br>= 0.9<br>= 0.8<br>= 0.7<br>= 0.0 | 0.9 | 1.0 | 0.8 | 0.0 | 0.7 |
| 11. Historical Sites | | | Consistent for All Routes ||||| 
| 12. Shafts in Critical Water Quality Zone | | | Consistent for All Routes |||||
| 13. Hydraulics/Grades, Geotechnical, Maintenance | | | Consistent for All Routes |||||
| **Total Score** | | | 7.3 | 7.2 | 5.1 | 4.0 | 1.7 |
| **Cost (millions)** | | | $42 | $43 | $48 | $52 | $45 |

Austin Water — South I-35 Water/ Wastewater Program

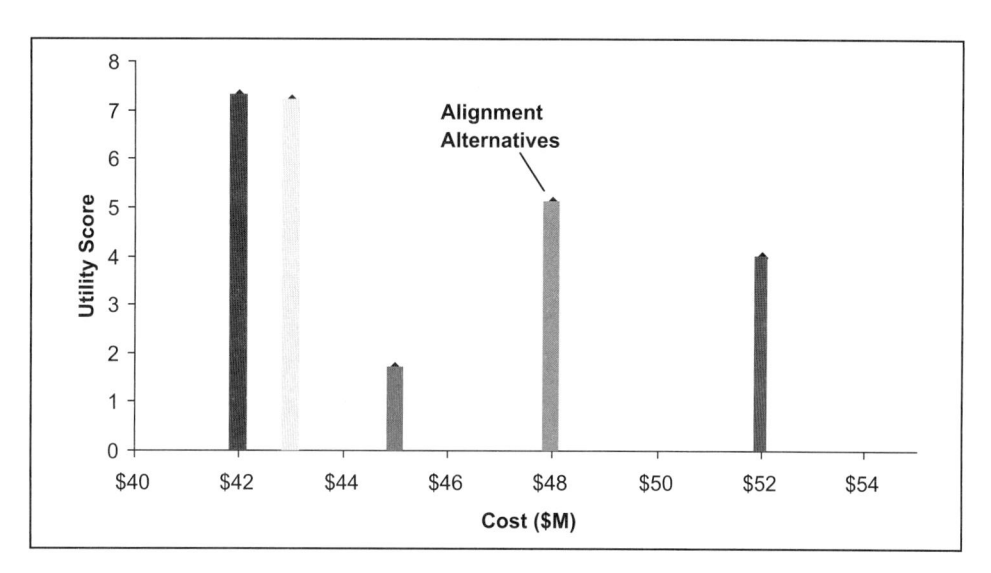

**Figure 3.** Alignment alternatives summary statistics

**Table 6.** Alignment alternatives summary statistic

|  | Alignment Alternatives ||||| 
| Measurement | Purple | Yellow | Orange | Green | Pink |
| --- | --- | --- | --- | --- | --- |
| Total Utility Value | 7.3 | 7.2 | 5.1 | 4.0 | 1.7 |
| Cost ($M) | $42 | $43 | $48 | $52 | $45 |

based on private easements, where the yellow alternative traversed a private neighborhood that the purple alternative did not. This led to a reduced utility score and an increased cost for the yellow. When reviewing these results with the city and working through a brief and informal sensitivity analysis, it was agreed that these differences were significant enough to differentiate between these alternatives. It was also clear that because these alternatives were quite similar (the primary difference being the private easements required), any additional easements required for the Purple alternative would have resulted in a higher utility score for the Yellow alternative and a change in the decision.

An alternative to using utility scores would have been to use cost as the common value for comparison. The benefit of using cost is that engineering (and many other disciplines) decisions can be related to costs. The simplicity of this metric allows for easy comparison where the magnitude of the difference between costs is unimportant and only the relative value is needed to make a decision. For example, if the cost of disturbing a creek can be boiled down to one cost that reflects the value of that creek to the owner, say $10M, then incorporating the environmental challenges of working within a creek are no longer discussed on the basis of emotion or preference, but rather the relative costs of alternatives. The significant challenges with using cost include perception of both the owners and stakeholders. It is unpalatable to discuss a value system in terms of cost.

In the case discussed in this paper, private easement acquisition and environmental impacts were the 'hot topics' in the discussions. As such, the times where these attributes were discussed in terms of costs were extremely challenging and unproductive. In other contexts, such as in the health industry for example, costs (as unpalatable as it sounds) for various illnesses, care services and deaths are assigned to evaluate the best allocation of resources. However, this approach is highly inflammatory to the public. For all of these reasons, utility is a far easier value for comparison even though it is less precise.

The final step was to present this information to the public. Table 5 was distributed to the public with a table of the attributes and descriptions similar to those shown in Table 4. A presentation was given that outlined the broad categories used to make these types of alignment decisions: Environmental Impacts, Public Impacts, Engineering Feasibility, Constructability, Cost, and Schedule, and the 13

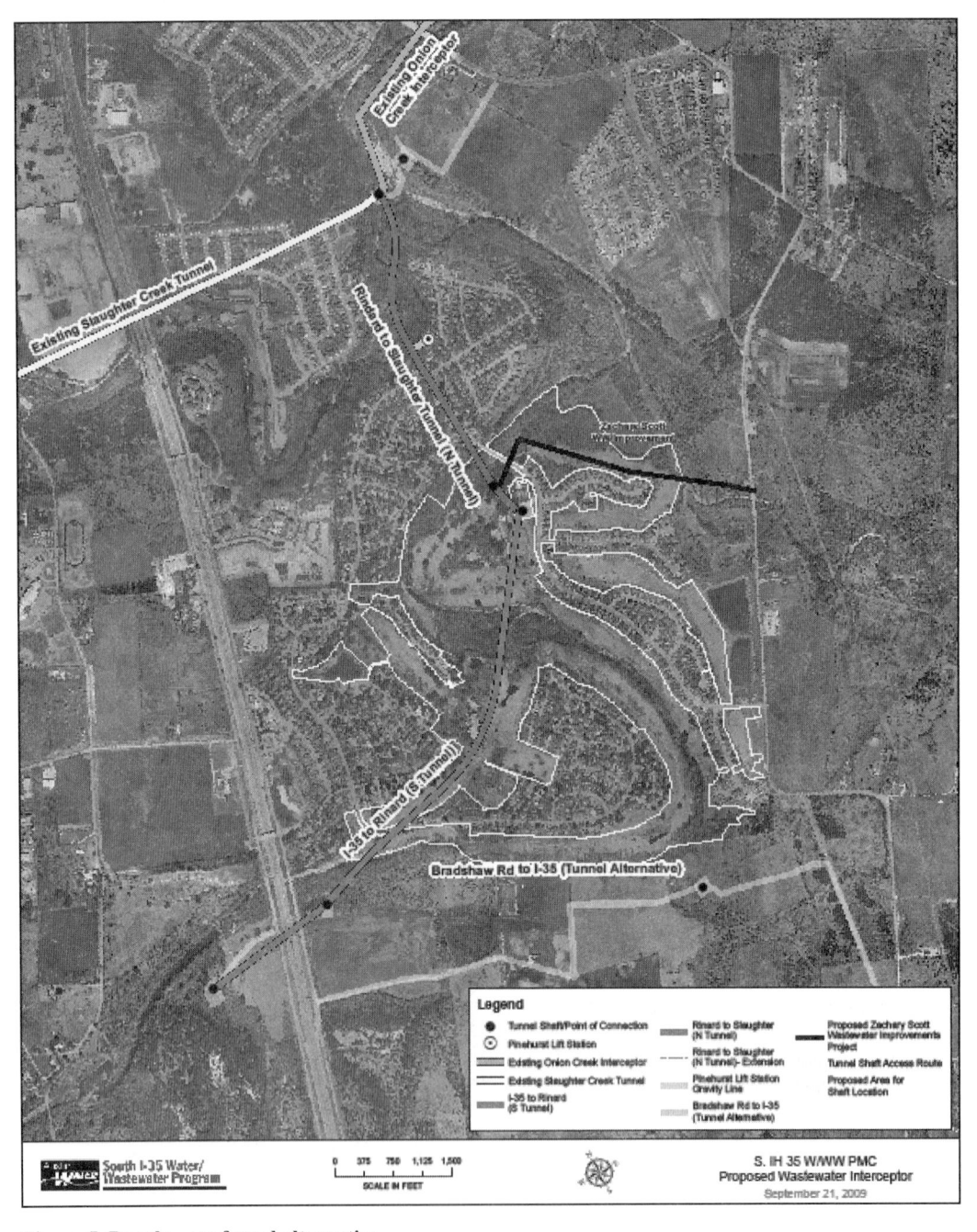

Figure 5. Purple—preferred alternative

site-specific attributes. The process was explained and the selected alignment—the purple alternative, as presented in Figure 5 was shown. On the whole, the public received the alignment selection well. A handful of residents were very vocal about their opposition; however, the majority of the comments received were that the process had represented their concerns and presented a process that was clear and equitable. It is possible that had the opposition group been larger, the team would have needed to refine the process to include their concerns under one of the existing categories. In the end, the HOA supported the decision as it was presented and the project moved forward on the selected alignment.

## CONCLUSIONS

The process presented in this paper reflects the successful application of a multi-attribute utility function that facilitated the selection of an alignment. This process did a good job of balancing cost and intangible factors. The importance of using this particular process was that it allowed a wide group of stakeholders to provide input to the decision in a logical and measurable manner. In addition, it allowed these same stakeholders to trust and support the outcome using this process.

With this decision process, as with most, the importance of the process is what is paramount and particularly the need for finding a tool that motivates and facilitates the discussion, rather than focusing on the numerical values that are used in the analysis. It is challenging to take a large diverse group of stakeholders and find a common ground in which their values can be included in a decision. This process was successful because in a relatively short amount of time, all stakeholders agreed and supported the decision.

Last, the importance of communication, both internally to elicit information from the City and externally to present the information to the public, should not be minimized. Communication is central to successfully working through a decision. Many times engineers are challenged to develop a consensus, and it is important to recognize that the calculations and supporting data are not sufficient justification. Emotions always play a role in decision making and the ability to reduce the influence of emotion in finding common ground is increased if a reliable process is used.

There are a couple of lessons that the program team learned through this process. First, the basis of comparison should be established by all team members. In this process, cost was first discussed as the basis of comparison and led to many challenging discussions before utility value was implemented and supported. Second and most important, open communication is central to arriving at a decision that all stakeholders will support. It is important to create an atmosphere where each stakeholder is able to present their value system so that the decision attributes and outcomes fully reflect their input.

## ACKNOWLEDGMENTS

The URS Program Management Consulting staff would like to acknowledge the following City of Austin departments for their support and work that made this process such a success: Austin Water Utility, Contracts and Land Management, and Watershed Protection.

## REFERENCES

Ang, A. H-S., and Tang, W.H. 1984. Probability Concepts in Engineering Planning and Design. John Wiley & Ang, A.H-S., and Tang, W.H. 1984. Probability Concepts in Engineering Planning and Design. John Wiley & Sons, New York.

Dawes, S.S., Pardo, T.A., Simon, S., Cresswell, A.M., LaVigne, M.F., Andersen, D.F., and Bloniarz, P.A. 2004. *Making Smart IT Choices: Understanding Value and Risk in Government IT Investments.* Center for Technology in Government, University of Albany. 2004.

Keeney, R.L., and Raiffa, H. 1976. Decisions with Multiple Objectives. John Wiley & Sons, New York.

Nelson, P.E. 1999. Multiattribute Utility Models. *Elgar Companion to Consumer Research and Economic Psychology.* 392–400.

# Risk Management to Make Informed, Contingency-Based CIP Decisions

**Paul Gribbon, Christa Overby**
Bureau of Environmental Services, Portland, Oregon

**Gregory Colzani**
Jacobs Associates, Portland, Oregon

**Julius Strid**
EPC Consultants, Portland, Oregon

**ABSTRACT:** The City of Portland, Oregon's Bureau of Environmental Services (BES) initiated the use of risk registers and risk assessment practices early in the development of the $800 million (USD) Willamette River Combined Sewer Overflow (CSO) Program. As the program progressed into final design and construction, BES expanded the risk management process to include transfer of risk registries from design entities to a project team that included owner, construction contractor, construction manager, and design consultants. Qualitative and quantitative re-evaluation of the design phase risk registries by the project team lead to the development of project contingencies as well as advantageous procurement of Owner Controlled Insurance (OCIP) Coverage.

This paper will discuss the evolution of the risk management process utilized by BES throughout the CSO Program and explore the implementation of the bureau-wide systematic project risk management program to develop and manage BES-wide CIP contingencies.

## INTRODUCTION

The City of Portland entered into an Amended Stipulation and Final Order (ASFO) agreement with the Oregon Department of Environmental Quality in August 1994. The agreement requires the City to control its 55 combined sewer overflows (CSOs) by 2011 with interim deadlines imposed to complete specific portions of the work prior to that date. The purpose of the Willamette River CSO Program (WRCSO) is to reduce the frequency and volume of combined sewer overflows from the drainage areas surrounding the Willamette River. The program, managed by the Bureau of Environmental Services (BES), commenced in 1991 with a set of initial cornerstone projects, including installation of stormwater sumps and sedimentation manholes in residential areas, building separate sewers for stormwater in some neighborhoods, downspout disconnection programs, and removal of certain streams from the combined sewer system. The program then continued with the development, design, and construction of two major capital improvement programs:

1. The Columbia Slough CSO Program included a 3.65-m (12-ft) diameter tunneled pipeline and open-cut section in addition to wastewater treatment plant improvements and was completed by its deadline of December 1, 2000.
2. The West Side Willamette River CSO Project (WSCSO) included a 5,633-m (18,481-ft) long, 4.3-m (14-ft) diameter conveyance tunnel and the 220 million gallons per day (MGD) Swan Island Pump Station and was completed ahead of its December 1, 2006 deadline.
3. The East Side Willamette River CSO Project (ESCSO) includes an 8900-m (29,200-ft) long, 6.7-m (22-ft) diameter storage and conveyance tunnel and is scheduled to be completed in 2011.

In addition to these three major projects, several appurtenant pumping and connection facilities are under construction.

In accordance with the ASFO, all facilities associated with the WRCSO Program must be constructed and operational by December 1, 2011. Otherwise, substantial fines may be imposed.

Historically within BES, inaccurate project schedules, cost projections, and unmitigated risk often negatively impacted fiscal year capital expenditure projections. Recognizing that a failure to properly manage risk and allocate realistic contingency budgets could jeopardize the program, BES adopted a risk recognition, assessment, and mitigation strategy in the design phase of the WRCSO program. As the program advanced into final design and construction, BES expanded the risk management process to include transfer of risk registries from design entities to a project team that included owner, construction contractor, construction manager, and design consultants. Subsequently, BES incorporated the risk program into all the WRCSO design and construction phases.

## RISK MANAGEMENT PHILOSOPHY

Construction of any kind exposes the participants (i.e., owner, engineer, contractor, etc.) to some level of risk and liability. When risk is defined and managed ahead of time, projects are more likely to be delivered on schedule and budget.

A risk is an uncertain event or condition that, if it occurs, has a consequence. Risk can be considered a function of both frequency (probability or how often a particular risk may occur) and consequence (the impact or outcome of the risk). For example, a risk management process should identify the following types of risks on a project:

1. Risk to the health and safety of workers
2. Risk to the health, safety, and property of third-party people
3. Risk of schedule delay to completing the project
4. Risk of financial losses and unplanned costs
5. Risk to the environment
6. Risk associated with political and public issues
7. Risk of construction claims

A proper risk assessment should be conducted at defined milestones throughout the project and should generally include the following:

1. **Risk Management Objectives/Planning:** Determine how to implement the risk assessment process to best suit your project.
2. **Risk Identification:** Identify potential risks on your project and describe them. This involves the creation of a Risk Register.
3. **Risk Qualitative Analysis:** Assess the probability and consequence of the risks.
4. **Risk Quantitative Analysis:** Complete a numerical analysis of risk probabilities and their consequence on project objectives.
5. **Risk Mitigation:** Develop, quantify and implement plans to reduce either the probabilities or the consequences of identified risks. The cost-benefit of mitigation measures should be analyzed if they are not considered to be mandatory.
6. **Risk Management and Monitoring:** Track identified risks and mitigate for them. This involves monitoring risk, identifying new risks, requantifying existing risks, and evaluating the effectiveness of actions taken.

In general, a typical risk assessment involves assigning numerical values to both the probability of a risk occurring and to the severity of the consequences of the risk item in order to obtain a risk rating. When analyzing risk on a project, several risks may have a low or high probability of occurring along with a minor consequence, making them an acceptable risk to the project. Unacceptable risks will also be encountered, which may have a high likelihood of occurring with a major consequence to the project. The overall project risk management philosophy is to minimize and ultimately eliminate "unacceptable" risks by reducing the likelihood of occurrence of an event with large consequences.

## RISK ASSESSMENT PROCESS

### Risk Management Objectives/Planning

As the program headed into the design phase of its first major project, the West Side CSO Tunnel Project, BES set forth the objective to recognize and reduce all risks to their lowest possible level while properly allocating risk contingency monies for recognized risks. The cornerstone to achieving this objective was to formalize the risk assessment process into the future projects from early design through construction, including contractor participation.

### Risk Identification

Brainstorming workshops were used to identify project risks. Typical risks include major unforeseen geologic conditions, equipment failure or malfunction, and the occurrence of extreme events (or hazards) not considered or remediated in the design/planning stages of the project. The purpose of the brainstorming is to identify and develop a registry of risk events that have the potential to cause undesirable impacts.

### Risk Qualitative Analysis

Once the working list of identified risks/events was compiled, an analysis was performed to determine the likelihood of each risk occurring and the possible consequence if the risk actually did occur. Values for likelihood and consequence were assigned to each risk in accordance with those summarized in Tables 1 and 2. To the extent possible, historical information and experience with similar projects was relied upon to help in assigning the values.

## Risk Quantitative Analysis

The overall significance of a risk is defined as the likelihood value multiplied by the consequence value. The significance of a risk is defined as substantial if its numerical designation is greater than 12 (shaded areas in Table 3). For significance values of 8 to 12, the risk is considered moderate. Acceptable risks are those for which the numerical designation for significance is less than 8.

In addition to the social analysis discussed above, a quantitative risk model was used to quantify the uncertainties in capital cost and schedule associated with risks with a significance of 8 or higher following the development of the risk registry. The risks were input into a risk model that uses a mathematical approach to assess the likelihood of the occurrence of a certain risk (event), and the cost (in time and money) to the project if such an event occurs. The model was run for each phase of review. The mathematical model (Monte-Carlo simulation) ran 10,000 iterations with events occurring at random according to the assigned probability. The result of the analysis was a probability distribution indicating contingency amount confidence levels based on the input risk probability and impact. The results of the model were used to provide additional calibration of the social model and to develop a confidence level for the appropriation of financial contingency for risks that could not be mitigated to the acceptable level. Examples of the model runs are presented in Figure 1.

## Risk Mitigation

The mitigation process involves identifying measures that can be taken to reduce the probability or consequence of the risk/event. These measures fall into one of four categories:

- **Mitigate:** Implement an action to reduce the probability and severity of the risk. For example, additional data from a geotechnical investigation can reduce uncertainty about subsurface conditions, thus reducing the probability that a hazard will occur.
- **Transfer:** Direct the consequence of risk by allocating the risk contractually either to the contractor or to another party such as an insurance carrier.
- **Avoid:** Change the project plan to eliminate the risk.
- **Accept**: Accept the risk and do not assign a risk control measure, but monitor it in case it changes and a risk control measure becomes necessary.

Several workshops were held to identify appropriate risk mitigation strategies. The risk registry was updated, revised cost and schedule impacts for each of the risk mitigation measures were developed, and the numerical models rerun.

Table 1. Likelihood categories

| Value | Likelihood | Probability (%) |
|---|---|---|
| 5 | Very Likely | >70 |
| 4 | Likely | 50–70 |
| 3 | Possible | 30–50 |
| 2 | Unlikely | 10–30 |
| 1 | Very Unlikely | <10 |

Table 2. Consequence categories

| Value | Consequence | Increase of Cost or Time (% of cost or time) |
|---|---|---|
| 5 | Catastrophic | >20 |
| 4 | Severe | 10–20 |
| 3 | Substantial | 2.5–10 |
| 2 | Moderate | 0.5 –2.5 |
| 1 | Minor Impact | <0.5 |

Table 3. Risk matrix for evaluating significance of a risk

| Likelihood of Risk Event | Consequence (Severity) of a Risk Event ||||| 
|---|---|---|---|---|---|
|  | Catastrophic | Severe | Substantial | Moderate | Minor |
| Very Unlikely | 5 | 4 | 3 | 2 | 1 |
| Unlikely | 10 | 8 | 6 | 4 | 2 |
| Possible | 15 | 12 | 9 | 6 | 3 |
| Likely | 20 | 16 | 12 | 8 | 4 |
| Very Likely | 25 | 20 | 15 | 10 | 5 |

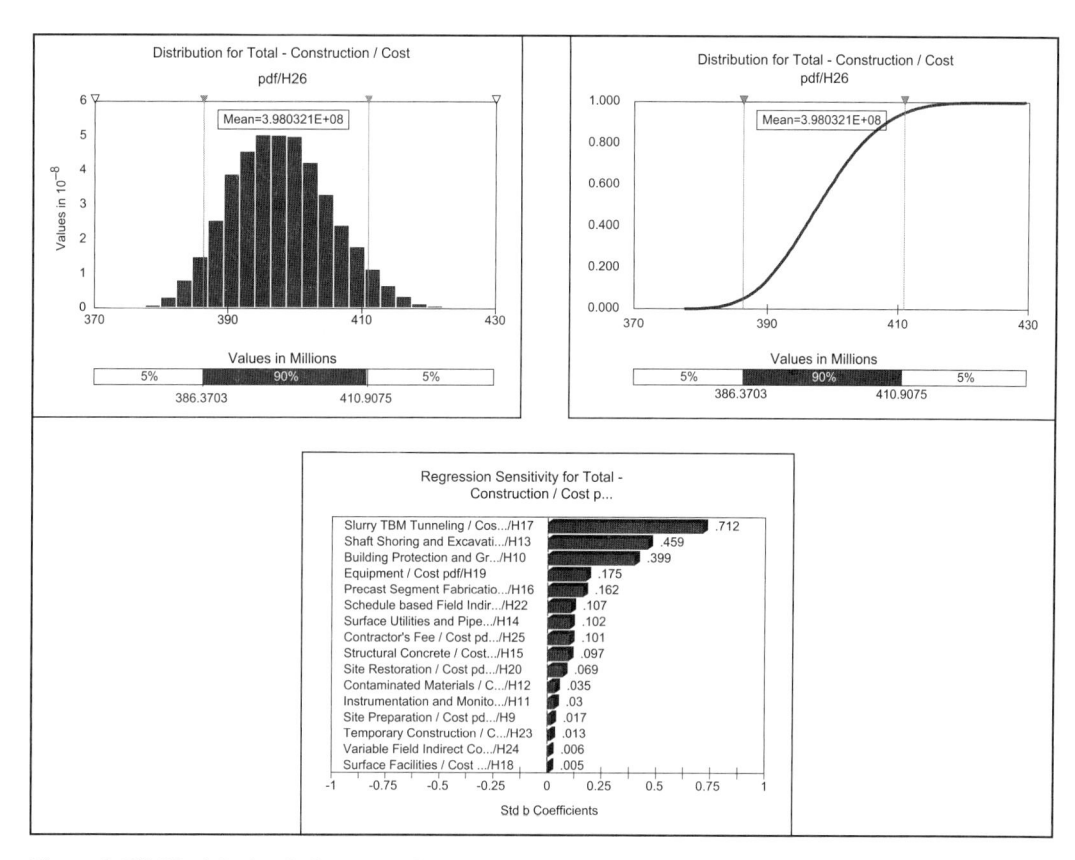

Figure 1. ESCO risk simulation examples

## CASE STUDIES

### West Side Willamette River CSO Project

Early in the design process of the WSCSO project, BES had concerns that the existing contracting practices commonly implemented by the City of Portland may be unsuitable for the WSCSO. The City of Portland standard practice is to use the design-bid-build approach to contracting, as required by Oregon Revised Statutes (ORS) Section 279. The design-bid-build projects constructed under the City's previous CSO milestone were done utilizing partnering, escrowing bid documents, geotechnical baseline reports, differing site condition clauses and, in some cases, dispute review boards. However, unresolved disputes still led to claims for additional compensation as a result of a number of differing site condition claims. Consequently, with the higher level of construction risk and complexity of the West Side CSO project, BES chose to utilize an alternate contracting structure, modeled after but with significant differences from a construction manager/general contractor (CM/GC) approach. The contract approach was to procure a prime contractor using a qualifications-based process, and establish a contract as cost reimbursable with a fixed fee. The significant differences with CM/GC contracts researched by BES is that the prime contractor would not be limited in the amount of work to be self-performed (on the contrary, BES wanted to know who would be doing the tunnel and shaft work), and there would be an estimated reimbursable cost developed rather than a guaranteed maximum price (GMP).

Oregon Revised Statutes (ORS) 279 allows public agencies to utilize alternate procurement methods for public contracts, provided they offer the best value to the public. Subsequently, BES evaluated several alternate delivery methods for applicability to the WSCSO project. The Bureau chose to utilize a cost reimbursable fixed fee (CRFF) contract for the WSCSO project. The selected contractor, Impregilo-S.A. Healy Joint Venture, entered into a two-phase contract that included a preconstruction services phase and construction phase. The preconstruction services phase scope of work included the following to be performed by the contractor: design review, cost saving suggestions, project planning, schedule development, risk assessment, reimbursable cost estimate, safety plans, and subcontracting procurement plans.

The shared risk assessment during the preconstruction services phase turned out to be a significant and critical activity. In the normal course of design development, the design engineer had performed a risk assessment prior to selection of the contractor. Some 41 risk items were identified and quantified. Upon mobilization, the contractor was tasked to facilitate a formal risk assessment workshop for the end of the first month of the preconstruction phase. As the development of the design continued during the contractor procurement period, this month allowed the contractor to get familiar with the current stage of the design and with other members of the project team. A professional facilitator was engaged, and team members from the owner, design engineer, contractor, construction manager, and the City's advisory board were included in the workshop. During the workshop, a total of 251 risks were identified for the various categories of work, as shown in Table 4.

Each of the identified risk items was evaluated following the risk assessment process described above. Upon rating the various identified risks, a series of risk mitigation proposals was developed with the intent of reducing risk to the project during the preconstruction phase. Methods of mitigation included revisions to the design and the contract documents, additional geotechnical investigation or instrumentation, and development of a subcontracting plan that addressed packaging of subcontracts and management of subcontractor claims and disputes. In addition, the access/permit and financial/other risks were to be managed by including the costs thereof in the estimated reimbursable cost (project construction budget).

This risk assessment was then used by the contractor and BES to independently develop proposed contingencies to be applied to the budget. BES elected to assign costs to items that had a risk condition of 8 or more. Each party provided more detailed cost range estimates for each risk. The risk-estimating exercise resulted in a total estimated risk impact amount of $34 million. The total amount was then evaluated. BES assigned a contingency value of $17 million based upon the assumption that 50% of the items could be expected to occur.

Additionally, the risk data were statistically modeled by Jacobs Associates. This was done to provide an independent secondary check of the contingency amount. The estimated construction amounts along with the lowest cost and higher cost percentages were evaluated by the model. The model calculated the ranges and ran through a number of iterations on a Monte Carlo simulation of probability of increased or decreased costs. This resulted in

**Table 4. WSCSO risk categories**

| Risk Category | No. |
|---|---|
| Access/Permit Risks | 35 |
| Tunnel Construction Risks | 47 |
| Ground Improvement Construction Risks | 58 |
| Shaft Construction Risks | 25 |
| Pump Station Construction Risks | 20 |
| Microtunneling Construction Risks | 33 |
| Completion/Startup/Operation/Maintenance Risks | 10 |
| Financial/Other Risks | 23 |
| **Total Identified Risks** | **251** |

a median cost, lowest cost, highest cost, standard deviation, confidence level, and required contingency based upon the confidence level. This check confirmed the assumptions made by the project team, estimating a contingency value of $17 million at the 99 percent confidence level.

Subsequently, the contingency amount of $17 million, representing approximately 6% of the construction contract value, was approved by the City Council along with the construction contract. During construction, risk mitigation was a continuing process through value engineering, regular and detailed schedule management, and aggressive cost management. The risk registers and corresponding contingency budgets were reviewed on a quarterly basis and at each new major construction activity or phase. During construction, a few of the funded risks did occur and remedial actions were successful. Additionally, a couple of identified risks that scored in the acceptable range occurred with a greater consequence than initially anticipated. For example, the subcontractor "low bid" for the slurry walls and ground improvement exceeded the project budget by $20 million and consumed the contingency budget prior to commencement of any construction work. This event, coupled with delays caused by difficulties in achieving groundwater cut-off for the pump station excavation, could have had a devastating impact on the contingency budget and the overall project cost. However, the contract method allowed for continual value engineering, and as a result of several collaborative efforts to implement innovative solutions coupled with good contractor performance, the project team recovered the majority of the contingency budget and lost schedule. At project completion, the majority of the remaining contingency budget was then utilized to refund $14 million to the Bureau's CIP budget for escalation associated with the WSCSO project.

It is clear that the contractor's input into risk assessment and constructability reviews in the midst of the design process provided a fresh outlook. The pre-construction services agreement of the type used here provided a timely opportunity for this activity.

**East Side Willamette River CSO Project**

Similar to its requirements for the WSCSO project, BES required a rigid risk assessment and mitigation program to be carried forth, beginning with the design of the ESCSO project. Once again the CRFF contracting method was utilized to procure the ESCSO contractor. The joint venture of Kiewit Bilfinger-Berger (KBB) was selected, based on qualifications, as the construction contractor for the ESCSO Project. Springboarding off the success of the WSCSO risk assessment program, BES made some refinements and continued the process with KBB. The approach utilized to assess risk and develop contingency budgets for the ESCSO included:

- Design stage risk assessment
- BES budget risk calculation
- KBB budget risk calculation
- Statistical modeling

*Risk Assessment*

The design-stage risk assessment was performed by the design team based upon risks that could be foreseen at the time. Two additional risk calculations were performed during the preconstruction stage of the project. These calculations were based upon a risk assessment developed jointly by BES, designers, construction managers, and the contractor. The list of items to be considered were derived from the West Side CSO project risk register and supplemented by the designer's risk assessment listing and additional items brought up by the contractor. Two hundred and eighty items were identified during this phase. Each item derived from this assessment was then evaluated following the previously discussed process.

*Contingency Development*

This risk assessment was then used by the contractor and BES to independently develop proposed contingencies to be applied to the budget. BES elected to assign costs to items that had a risk condition of 8 or more. The total BES contingency amount developed was $33.5 million. The KBB risk assessment resulted in a total contingency amount of $39.7 million.

Additionally, a statistical method of analyzing the risk was prepared by Jacobs Associates. The information was provided by BES and included a probable, low, and a high estimate of the costs for the major construction portions of the project. The simulation model estimated a value of $31 million of contingency at a 95 percentile level of confidence. An example of the risk simulation is shown in Figure 1.

*Follow-on Review of Risk and Contingency*

The project risk register, project budgets, and contingency budgets are reviewed on a quarterly basis, and risk contingencies are adjusted to release contingency as risk contingency items are completed or are not realized. Contingency monies are often redistributed to other items or new risk items based upon project experience. Every six months a comprehensive cost to complete forecast is conducted by both KBB and BES. To date, the ESCSO project is approximately 80% complete, the contingency budget is relatively untouched, and the project has been able to release a portion of the contingency budget back to the Bureau CIP budget.

*Owner Controlled Insurance (OCIP) Coverage*

Concurrent with the ESCSO Project preconstruction phase, BES was in the process of marketing the OCIP coverage for the remainder of the WRCSO Program. Initial carrier quotations were significantly higher than estimated. BES and its broker agreed that more favorable terms and conditions and cost savings could be realized if BES marketed the program directly to the London Underwriters Market. This action required BES to demonstrate proof of compliance with the ITIG Code of Practice for Risk Management of Tunnels. The strong risk management approach implemented by BES coupled with the methods to select and utilize the contractor for preconstruction input were looked upon favorably by insurers and delivered favorable results. As part of the process, the underwriters required that an independent engineering group evaluate the program on a quarterly basis. Part of this review includes review and updating of the project risk register.

## MOVING FORWARD

Based upon the success of the risk assessment process utilized on the WSCSO and ESCSO projects, BES expanded the process to include additional projects, for example, the Portsmouth Force Main Project Segments 1 and 2 and the Balch Consolidation Conduit contracts. To date, the use of this process has generally been limited to projects with tunnel or microtunnel construction elements. However, the favorable experience has BES exploring the implementation of a formalized risk assessment process for all projects in the Bureau's Capital Improvement Program (CIP). To this extent, BES has developed a risk assessment manual and is preparing to add the process into the CIP Implementation Plan.

## CONCLUSION

By implementing a risk assessment program, BES found it possible to more accurately identify and quantify project risks and their impacts to cost and schedule in advance, and to account for this as the project contingency in schedule and cost projections. In addition, implementation of a risk assessment program brings together design and construction staff to identify and manage potential project issues ahead of time, prior to their occurring during construction.

However, in our experience, risk management processes are also fraught with their own set of risks. For example, as discussed above, the WSCSO encountered risks that were evaluated as "acceptable" individually but, when coupled together, the "consequence" pushed the risk value to the unacceptable range. If this had occurred on a low bid contract, both of these together would have severely hindered the project from a schedule/cost standpoint and would not have been covered by a risk contingency.

Additionally, the ESCSO Project may be impacted by the acceleration of the proposed streetcar and light rail expansion construction running through the middle of the project's main staging area. This event was never considered as a possibility during the risk assessment process. Hopefully, these impacts will be manageable, but the project won't have the ability to have control over the direct costs and impacts, and this certainly was not included in the risk contingency.

The point that it is impossible to predict and analyze the unknown is emphasized in the book *The Black Swan, the Impact of the Highly Improbable* (Taleb 2007). In some instances, this can be a role for insurance, but since there are numerous exclusions in most insurance policies and it is equally as difficult to buy insurance for something that cannot be described or statistically presented, it should be understood that the process remains "risk management" not "risk elimination."

## REFERENCES

Actuarial Profession and Institute of Civil Engineers. 2002. *Risk Analysis and Management for Projects (RAMP)*.

Bureau of Environmental Services, City of Portland, Oregon. 2009. *Risk Assessment Manual—Draft* (Internal Correspondence).

Gribbon, P., Irwin, G., Colzani, G., Boyce, G., and McDonald, J. 2003. Portland, Oregon's alternative contract approach to tackle a complex underground project. *RETC Proceedings*.

Gribbon, P., Colzani, G., McDonald, J., 2005. Portland, Oregon's alternative contract approach—A work in progress. *RETC Proceedings*.

Eskesen, S.D., Tengborg, P., Kampmann, J., Veicherts, T.H. 2004. Guidelines for tunnelling risk management. *International Tunnelling Association, Working Group* No. 2 q, Research, ITA-AITES, c/o EPFL, Bat GC, CH 1015 Lausanne, Switzerland.

Taleb, N.N. 2007. *The Black Swan: The Impact of the Highly Improbable*. Random House.

ITIG, February 2005. *A Code of Practice for Risk Management of Tunnel Works—Final Draft*

State of Oregon, 2001. *Oregon Revised Statutes (ORS) Section 279*.

# Linear Schedules for Tunnel Projects

**Mun Wei Leong**
Jacobs Associates, Seattle, Washington

**Daniel E. Kass**
Jacobs Associates, San Francisco, California

**ABSTRACT:** Tunneling projects lend themselves to the use of linear schedules for planning, executing, and monitoring the progress of tunnel work. However, construction contracts generally require designers, construction managers, and contractors to use critical path scheduling (Critical Path Method) techniques. The techniques used in creating linear schedules and their benefits for planning, investigating alternatives, determining cost/time benefits of multiple heading, and evaluating actual performance will be presented.

## INTRODUCTION

Most construction projects use the Critical Path Method (CPM) to plan and schedule the work. Construction contracts typically require the use of commercially available scheduling programs such as Primavera Project Planner because these programs calculate the critical path(s) for the project. CPM, however, is not the best tool to model linear projects or portions of projects. CPM schedule tasks for linear work usually lack sufficient detail to provide a useful management tool to evaluate progress on the linear tasks and in turn on the project's critical path. This is because linear tasks are logically more spatially related than sequentially related. Moreover, updates and changes to CPM schedules can be complicated and time consuming when dealing with linear work, whose durations are controlled by production rates. CPM schedules also do not provide management with a simple graphic to visualize progress on linear projects or portions of projects. The linear scheduling method is not uncommon in the tunneling industry; however, this method of scheduling is typically not incorporated into project scheduling specifications. This paper discusses the advantages of using the linear scheduling method to plan, monitor, and measure progress during the various stages of a project—design through construction—on an exemplar tunnel project consisting of five shafts, various outfall structures, microtunneling, and an odor control building to be constructed. Examples of how to use the linear scheduling method for delay analysis will also be presented.

## CPM SCHEDULING METHOD

CPM scheduling calculates the critical path through a network model of the work. This path of activities or tasks consists of those activities that control the overall duration of the project; changes to their durations change the overall duration of the project. Noncritical activities have "float," which allows those activities to be delayed or postponed within the float values without delaying the overall project completion. Scheduling programs such as Primavera are commonly used to plan and monitor projects because of this ability to determine the critical path and the float for noncritical activities. Managers are able to use CPM scheduling programs to effectively plan and monitor complex projects, including cost and resource loading the schedule, and producing detailed graphs and reports on the status of the project. The CPM schedule provides information such as project duration, early and late start and finish dates, and float values for the project activities.

Figure 1 shows a simplified CPM schedule for a tunnel project using the Primavera P3 scheduling program. This example plan anticipates using one tunnel boring machine (TBM) to mine the entire reach of the tunnel, starting at Shaft 2, going to Shaft 5, remobilizing the TBM back to Shaft 2 to complete the remaining tunnel, and demobilizing the TBM at Shaft 1. The duration of the project is 67 months, with a Notice to Proceed (NTP) in mid-May 2008 and a Substantial Completion in mid-November 2013.

A CPM schedule for this type of project typically includes thousands of activities. Linear activities are usually modeled in CPM by breaking the operation(s) into a sequential series of tasks of shorter duration to account for variations in production rates for linear activities over significant distances (i.e., for variations in expected ground conditions). This complicates the CPM model, making the plan in the CPM schedule more difficult for project personnel unfamiliar with its structure to fully understand. In addition, if actual production rates vary from the plan, periodic updates to the CPM schedule become a monumental task, and the calculated critical path becomes unreliable.

## LINEAR SCHEDULING METHOD

The linear scheduling method uses a diagram to graphically show the location and time of each work activity. This scheduling method is well suited for linear-type projects such as tunnels, pipelines, and road construction, where repetitive tasks are performed over a distance. In linear schedules, continuous tasks performed over a distance are represented by lines composed of a continuous set of points. From this graphic depiction, the location of work in progress can be determined at any point in time. Nonlinear activities are best planned with a network schedule (CPM) and then graphically shown in the linear schedule as boxes or vertical bars.

A typical linear schedule identifies the length of the linear project on the x-axis and time of performance on the y-axis. The scheduler first places all known constraints (contractual, physical, and environmental) on the linear schedule, followed by key features at their physical locations. After which, the scheduler determines the duration of the major aspects of work such as mobilization, shaft construction, TBM setup and mining time, and other time-related activities. Once the duration for each major aspect of the work is determined, the scheduler can draw several drafts of the linear schedule, varying the starting shaft location, number of TBMs, number of shaft crews, production rates, etc., to determine the most optimum plan for the project. Finally, the controlling critical path in the linear schedule can be determined manually (Harmelink 1998), and the float (rates) can be determined for the noncontrolling linear activities.

Once the plan is completed, the scheduler can use the information from the baseline linear schedule (Figure 2) to create the CPM schedule (Figure 1). Among its many advantages, the linear schedule allows the scheduler to have an overview of the entire project in a single graphic. This is not the case with the CPM schedule. For the nonlinear portions of the project, however, the CPM schedule provides a network for the work and can be used to determine the performance time for those tasks included in the linear schedule.

Some of the advantages of the linear scheduling method are:
- It is easy to understand and present graphically,
- It provides the scheduler with a simple overview of the project by identifying the location for each activity,
- Relationships between different construction activities, such as shaft and microtunnel construction, are easily identified,
- Required resources for the linear tasks can be identified at any time,
- Contractual, weather, environmental, and other constraints can be easily identified,
- Changes to the schedule are easy to make, and
- It is easier to measure progress and identify and evaluate performance-improving opportunities with the schedule.

The disadvantages of the linear scheduling method are:

- It cannot use computer programs to determine the critical path,
- Nonlinear portions of the project are not sufficiently detailed,
- Tasks cannot be cost loaded or the total project costs easily determined,
- Activities may not represent the true complexity of the work, and
- Features included in CPM scheduling program, such as resource leveling or the determining of float values, cannot be used.

Despite these disadvantages, the linear scheduling method is a superior tool for planning, scheduling, and monitoring linear projects such as tunnels. In the following three examples from one project, we will demonstrate how the linear scheduling method can be used during design and construction, and to evaluate the impact of a change on the time of performance.

## EXAMPLE 1: USING THE LINEAR SCHEDULING METHOD FOR THE DESIGN OR BIDDING PHASE

The linear scheduling method can be used effectively during the design phase. A planner can provide the owner with various options such as using multiple TBMs or varying the shaft location. The planner can evaluate the effect of a late permit or other restrictions such as real estate negotiations or environmental restrictions. Once the planner identifies the impact of these potential issues, the owner can make informed decisions to address the various concerns by either revising the specifications and design drawings, or

| Activity ID | Activity Description | Original Duration | Cal ID | Early Start | Early Finish |
|---|---|---|---|---|---|
| **PROJECT SUMMARY** | | | | | |
| **SHAFTS** | | | | | |
| PM0009100 | Shaft 2 Construction to Tunnel Lining | 282 | 1 | 15MAY08 | 15JUL09 |
| PM0009110 | Shaft 3 Construction to Tunnel Lining | 274 | 1 | 01AUG08 | 20SEP09 |
| PM0009150 | Shaft 5 Construction to Tunnel Lining | 484 | 1 | 15OCT08 | 15OCT10 |
| PM0009130 | Shaft 1 Construction to Tunnel Lining | 254 | 1 | 15DEC08 | 07JAN10 |
| PM0009140 | Shaft 4 Construction to Tunnel Lining | 300 | 9 | 15JAN09 | 15JUN10 |
| PM0008020 | Shaft 3 Structural Concrete | 110 | 1 | 15MAY10 | 20OCT10 |
| PM0008040 | Shaft 5 Structural Concrete | 199 | 1 | 20OCT10 | 20AUG11 |
| PM0008030 | Shaft 4 Structural Concrete | 109 | 9 | 28MAY11 | 21JAN12 |
| PM0008010 | Shaft 1 Structural Concrete | 151 | 1 | 20AUG11 | 15APR12 |
| PM0009170 | Odor Control Building Construction | 64 | 1 | 15AUG13 | 15NOV13 |
| **TUNNEL** | | | | | |
| PM0008100 | Setup TBM At Shaft 2 | 170 | 1 | 07DEC08 | 20AUG09 |
| PM0009210 | Tunnel 2 - Mine from Shaft 2 to Shaft 3 | 148 | 5 | 21AUG09 | 02APR10 |
| PM0009220 | Tunnel 3 - Mine from Shaft 3 to Shaft 4 | 205 | 5 | 02MAY10 | 03MAR11 |
| PM0009240 | Tunnel 4 - Mine from Shaft 4 to Shaft 5 | 202 | 5 | 28APR11 | 26FEB12 |
| PM0009250 | Demobilize TBM from Shaft 5 to Shaft 2 | 112 | 5 | 27FEB12 | 02AUG12 |
| PM0009280 | Tunnel 1 - Mine from Shaft 2 to Shaft 1 | 220 | 5 | 03AUG12 | 26JUN13 |
| PM0009290 | Demobilize TBM and Trailing Gear | 38 | 5 | 27JUN13 | 20AUG13 |
| **PIPELINE** | | | | | |
| PM0009360 | Outfall Diversion and Outfall Structures | 311 | 1 | 29APR09 | 08AUG10 |
| PM0008220 | Complete Micro Tunneling Near Shaft 5 | 111 | 1 | 20NOV09 | 15MAY10 |
| **SYSTEM TIE IN** | | | | | |
| PM0009630 | Shaft 3, 4, and 5 Finish Operations | 403 | 1 | 28APR11 | 27DEC12 |
| PM0009610 | Shaft 1 and 2 Finish Operation | 25 | 1 | 15AUG13 | 19SEP13 |
| PM0009730 | System Tie-In | 39 | 1 | 20SEP13 | 15NOV13 |

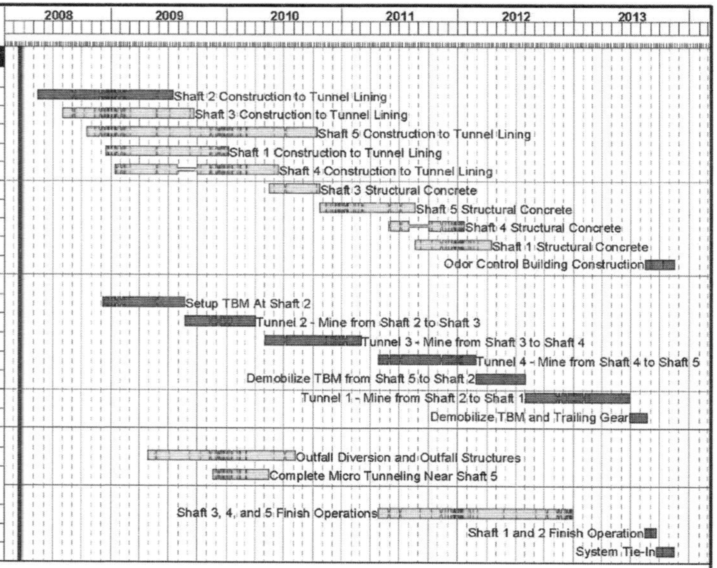


Start Date 20MAR08
Finish Date 08NOV11
Data Date 20MAR08
Run Date 09NOV09 09:12

Early Bar
Progress Bar

NATS
Sheet 1 of 1

TUNNELS AND PIPELINE SCHEDULE

Summarize Schedule



Figure 1. Simplified CPM schedule for a tunnel project using Primavera P3

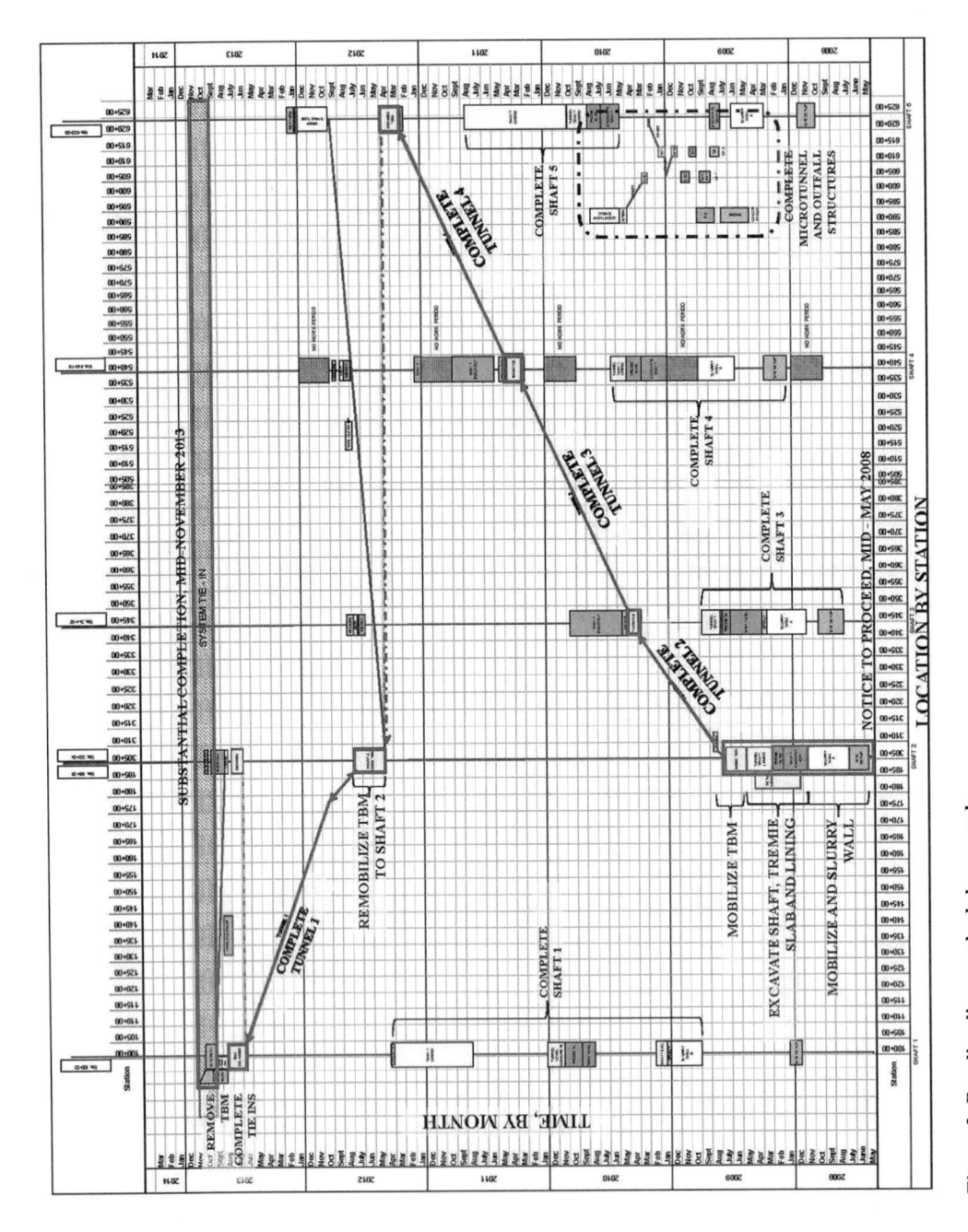

Figure 2. Baseline linear schedule example

Table 1. Comparison of the three construction options

| Item | Scenario 1: 1 TBM from Shaft 2 (Fig. 2) | Scenario 2: 2 TBMs (Fig. 3) | Scenario 3: 1 TBM from Shaft 1 (Fig. 4) |
|---|---|---|---|
| 1. Project Duration | 67 months | 56.5 months | 68 months |
| 2. Number of TBMs | 1 | 2 | 1 |
| 3. Site Prep and Earthwork | 1 crew | 1 crew | 1 crew |
| 4. Shaft Construction | 1 crew | 1 crew | 1 crew |
| 5. Micro Tunnel | 1 crew | 1 crew | 1 crew |
| 6. Finishing | 1 crew | 2 crew | 2 crew |
| 7. Building | 1 crew | 1 crew | 1 crew |
| 8. Other and Risk | Medium | Low | High |

addressing the potential issues as a contingency in the budget. In the following two examples, we will evaluate three scenarios for the project:

- Scenario 1: Plan the project using one TBM, performing the mining from Shaft Number 2
- Scenario 2: Plan the project using two TBMs
- Scenario 3: Plan the project using one TBM, performing the mining from Shaft Number 1

The results of the three schedules are summarized in Table 1.

**Scenario 1: Using One TBM on the Project, Mining from Shaft 2**

The linear schedule in Figure 2 shows the plan for using one TBM on the project, mining from Shaft 2. This plan involves the contractor mobilizing and mining Tunnel 2 from Shaft 2. Once the TBM breaks through at Shaft 5, the contractor would remobilize the TBM at Shaft 2 to mine Tunnel 1. After Tunnel 1 is complete, the contractor would demobilize the TBM and complete the odor control building, remaining shaft lining at Shafts 1 and 2, and the system tie-in.

The advantage of this plan is that the contractor is able to save costs by using one TBM to complete the project. The use of Shaft 2 allows the contractor to minimize the risk by having the mining site close to the middle of the project, which reduces the requirements of the support mining equipment. This method of construction also allows the contractor to perform the cleanup work for Tunnels 2, 3, and 4 after completion of the mining. This step moves a significant amount of work off the critical path. Another advantage of this plan is that the contractor would be able to service the TBM after it is removed from Shaft 5 and before it is reassembled at Shaft 2. This allows for greater success with the TBM in completing Tunnel 1.

The disadvantage of this plan is it would take 67 months to complete the project with one TBM. Additionally, it may be more expensive to set up the TBM from Shaft 2 because of the proximity of the Shaft 2 location to the middle of the project. This plan also requires the contractor to remobilize the TBM from Shaft 5 to Shaft 2, which would take 5 months to complete.

**Scenario 2: Using Two TBMs on the Project**

The linear schedule in Figure 3 shows the advantages of using two TBMs on the project. Essentially, the planner assumes that the contractor would begin mobilizing and mining Tunnel 2 from Shaft 2. Once the TBM breaks through at Shaft 3, the contractor would begin mobilizing the second TBM at Shaft 2 to mine Tunnel 1. The mining of both tunnels would be performed from Shaft 2.

This construction method allows for much quicker project completion, with a total project duration of 56.5 months. Initially, it looks as if the project could save 14 months, with 11 months on mining Tunnel 1 and another 3 months on remobilizing the TBM from Shaft 5 to Shaft 2. However, the critical path of the tunnel also changes from completing removal of the TBM at Shaft 1 and completing the odor control building and tie-ins, to removing the TBM at Shaft 5 and completing the cleanup and tie-ins and drop structure at Shaft 5. Therefore, the net savings using two TBMs is 10.5 months because of the additional time needed to complete the cleanup for Tunnels 2 through 4 and time to complete the drop structure at Shaft 5.

The disadvantages of using two TBMs are cost and space: the additional cost for a second TBM and whether the Shaft 2 location has sufficient space to allow for the mining of two TBMs. These are considerations the owner must take into account. Additionally, the plan must account for the increased capacity to handle the additional spoils, increased use of precast segments, as well as logistics such as

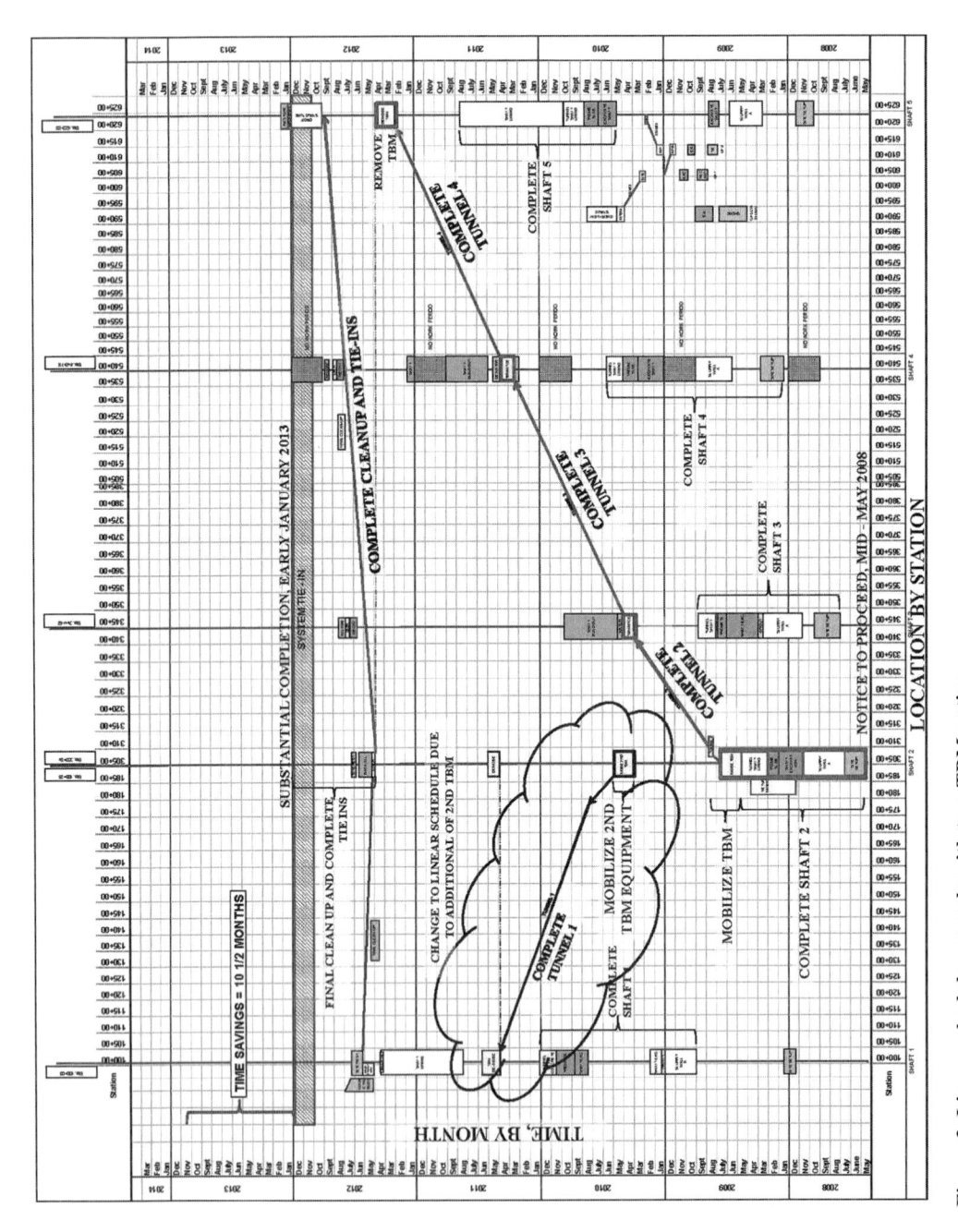

Figure 3. Linear schedule example with two-TBM option

traffic control, handling water treatment, and even the availability of skilled labor to run the two TBMs.

## Scenario 3: Using One TBM on the Project, Mining from Shaft Number 1

In the third scenario, a different mining shaft site is considered because of space or site condition concerns. The linear schedule in Figure 4 shows the projected schedule using Shaft 1 instead of Shaft 2 as the mining shaft. In this scenario, the contractor can continuously mine the TBM and eliminate the need to remobilize the TBM from Shaft 5 to complete Tunnel 1. This step saves three months on the schedule. However, Shaft 1 has a final lining that only can be placed after the TBM is removed. This adds an additional four months to the schedule. The net effect of mining from Shaft 1 is one additional month to the schedule (Table 1).

The advantage of this construction sequence is that it would allow the contractor to complete the work from one mining shaft without the need to remobilize the TBM. The disadvantages could include the technical difficulty of moving material from Tunnel 4 to Shaft 1. Additionally, it could delay the work to complete the lining at Shaft 1, the odor control building, and the start of cleaning up and tie-ins for the tunnel until after the TBM is removed.

The linear schedule allows the project team to discuss the advantages and disadvantages of these construction schemes. The owner can use the linear schedule to evaluate the risk of each option and determine the reasonable completion time for the project. The contractor can use the linear schedule during bid to determine the best means and methods of construction, identify potential risk to the work, and adjust its bid accordingly.

## EXAMPLE 2: USING THE LINEAR SCHEDULING METHOD DURING CONSTRUCTION

The owner decided to go with Scenario 2: constructing the tunnel with one TBM, mining from Shaft 2. This is due to many factors, among them the additional cost of a second TBM, the reduced length of time for maintenance because of mining from Shaft 2, and the reduced risk of completing a majority of the shaft concrete and cleanup work off the critical path. However, even with the best option, as we all know, changes will occur during construction. The ability to adapt to changing conditions greatly reduces the risk and increases the success of a project. The linear scheduling method provides a good tool for the planner to evaluate the effect of a change and appropriately plan the proper response.

Figure 5 shows that the mining rate for the TBM is 30% faster than was anticipated during the bid. As a result, the contractor is concerned that the receiving shaft may not be sufficiently completed to receive the TBM. In this instance, the planner adjusts the rate of production for mining by 30%, using the baseline linear schedule in Figure 2. The linear schedule in Figure 5 shows that the increase in production rate would possibly result in the TBM arriving at Shaft 4 during a nonwork period. The analyses show that Shaft 3 and Shaft 4 would be sufficiently completed prior to the arrival of the TBM. The overall project would be completed 9.5 months ahead of schedule. Using this information, the project team can evaluate the need to increase the capacity of the segment plant to accommodate the increase in production rate. The project team can begin investigating the need to apply for a special permit to work during the nonworking period at Shaft 4 or even consider slowing down the production of the TBM.

In the second instance, the contractor discovered issues concerning late permits at Shaft 3. This delay impacted the start of the construction on this shaft. The planner can insert the delay due to the permit in the linear schedule and determine if the delay would impact the critical path. The linear schedule in Figure 6 shows that the start of Shaft 3 would be delayed by four months. However, there was sufficient float in the schedule to not impact the critical path, although the permit delay did impact the sequence of shaft construction. In an effort to mitigate the delay, the contractor could mobilize to Shaft 5 after completing the slurry walls at Shaft 2. Additionally, the slurry wall work at Shaft 4 could also extend into the three-month nonworking period between the months of October and December. The permit delays also caused the float available between completing Shaft 2 and the arrival of the TBM to be reduced from six months to two months. The last issue regarding float may not be a problem if the production rate of the TBM remains the same and there are no further delays to shaft construction. The planner would be able to present the information to the project team to discuss the impact of the permit delay and any further mitigation efforts to implement on the project.

In both of these examples, the planner could use the CPM schedule to evaluate the impacts. However, making these changes to the CPM schedule can be complex and takes time to complete. The use of the linear schedule allows the project team to make sound and informed decisions in a timely manner.

## EXAMPLE 3: USING THE LINEAR SCHEDULING METHOD FOR CLAIMS AND DISPUTES

Changes and unforeseen delaying events do occur during construction. Events that impact time are

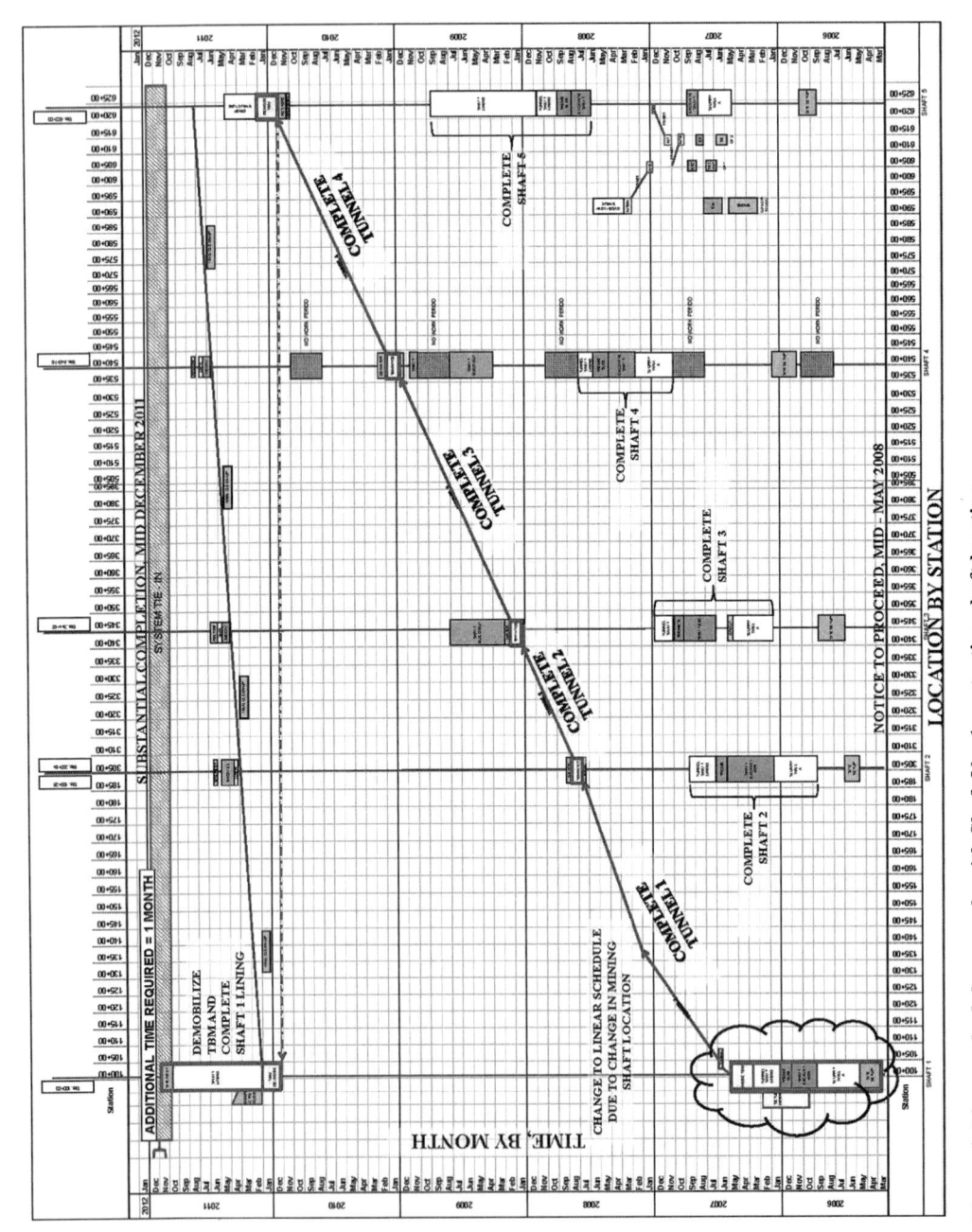

Figure 4. Linear schedule example with Shaft No. 1 as starting shaft location

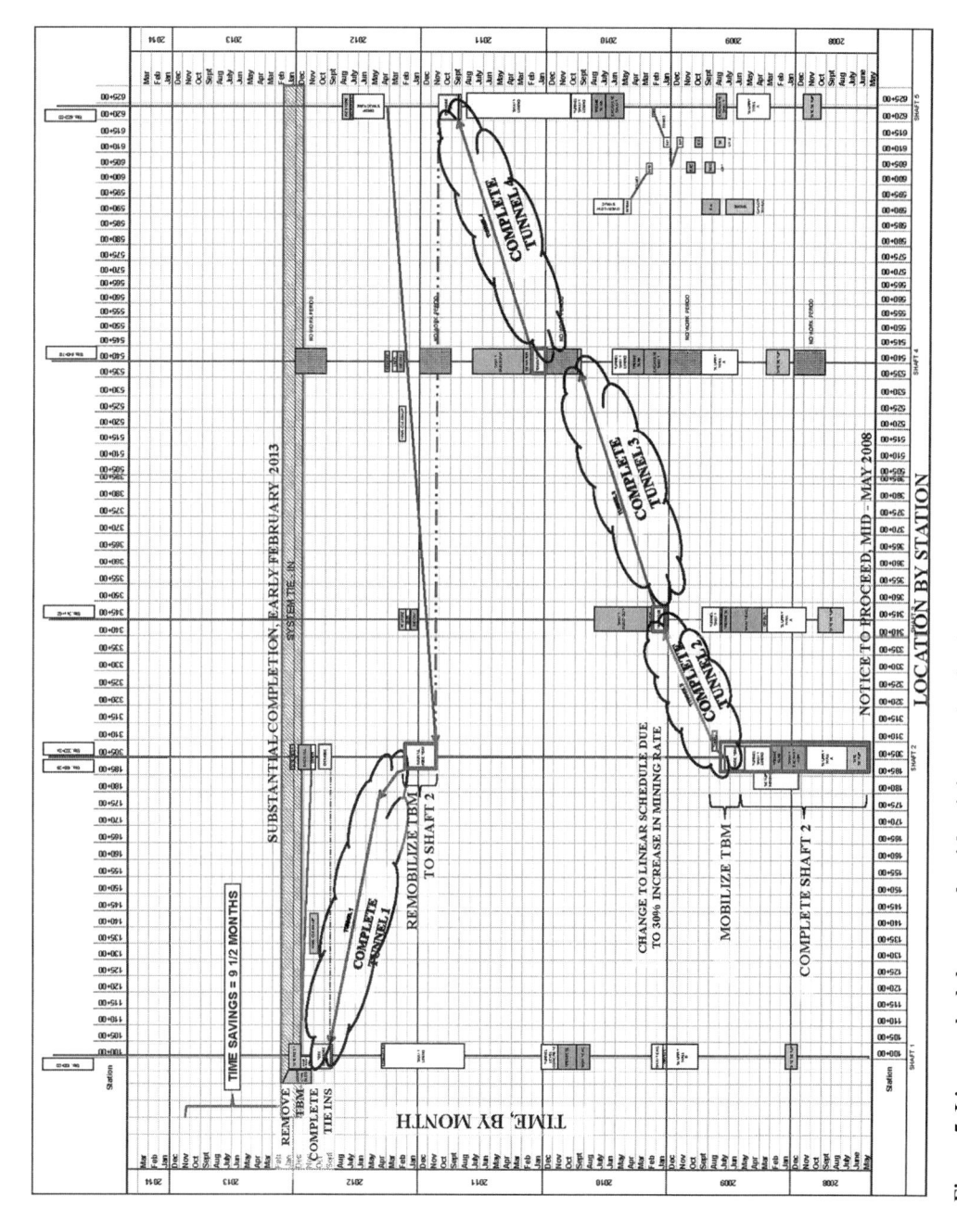

Figure 5. Linear schedule example with mining production increased 30%

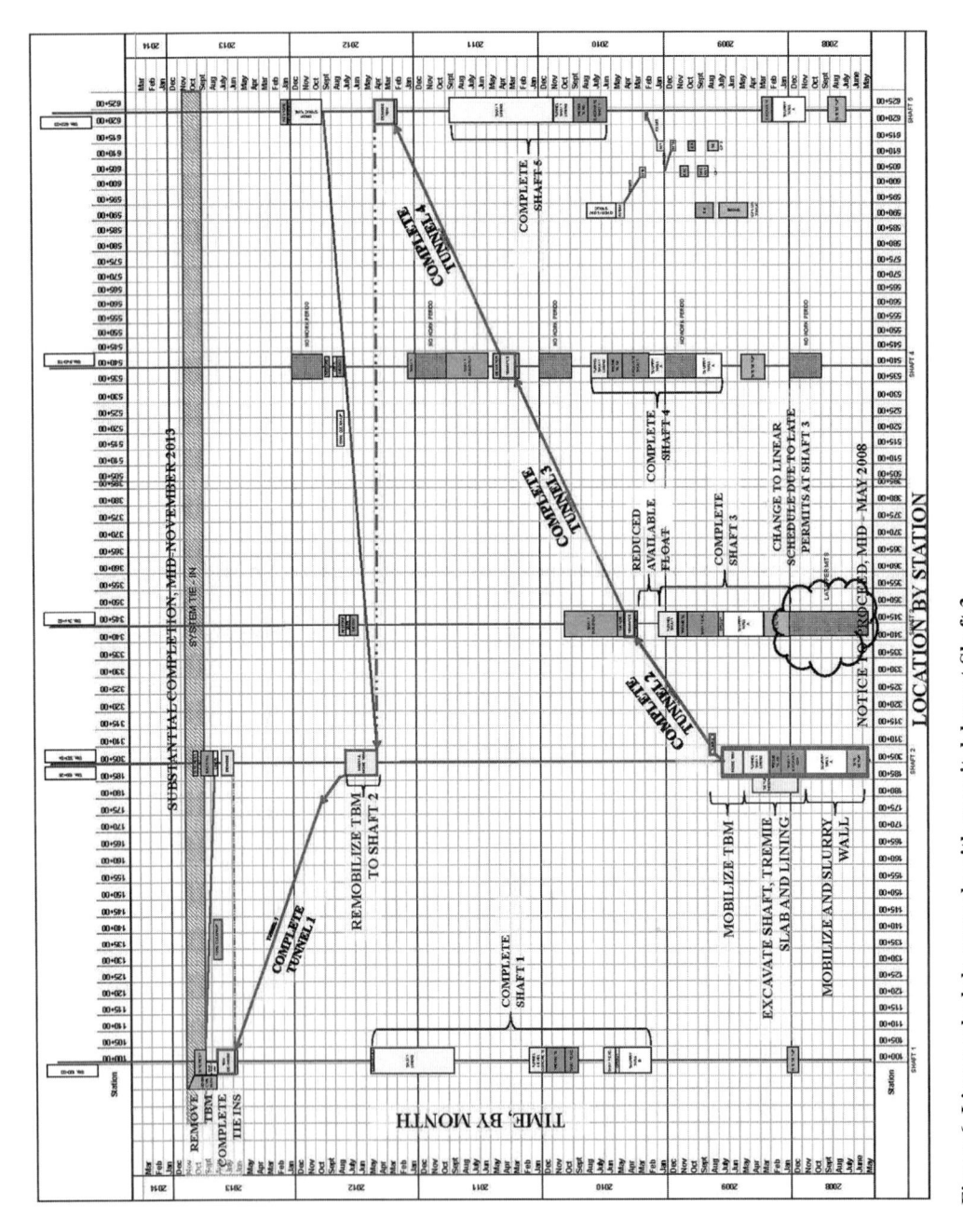

Figure 6. Linear schedule example with permit delays at Shaft 3

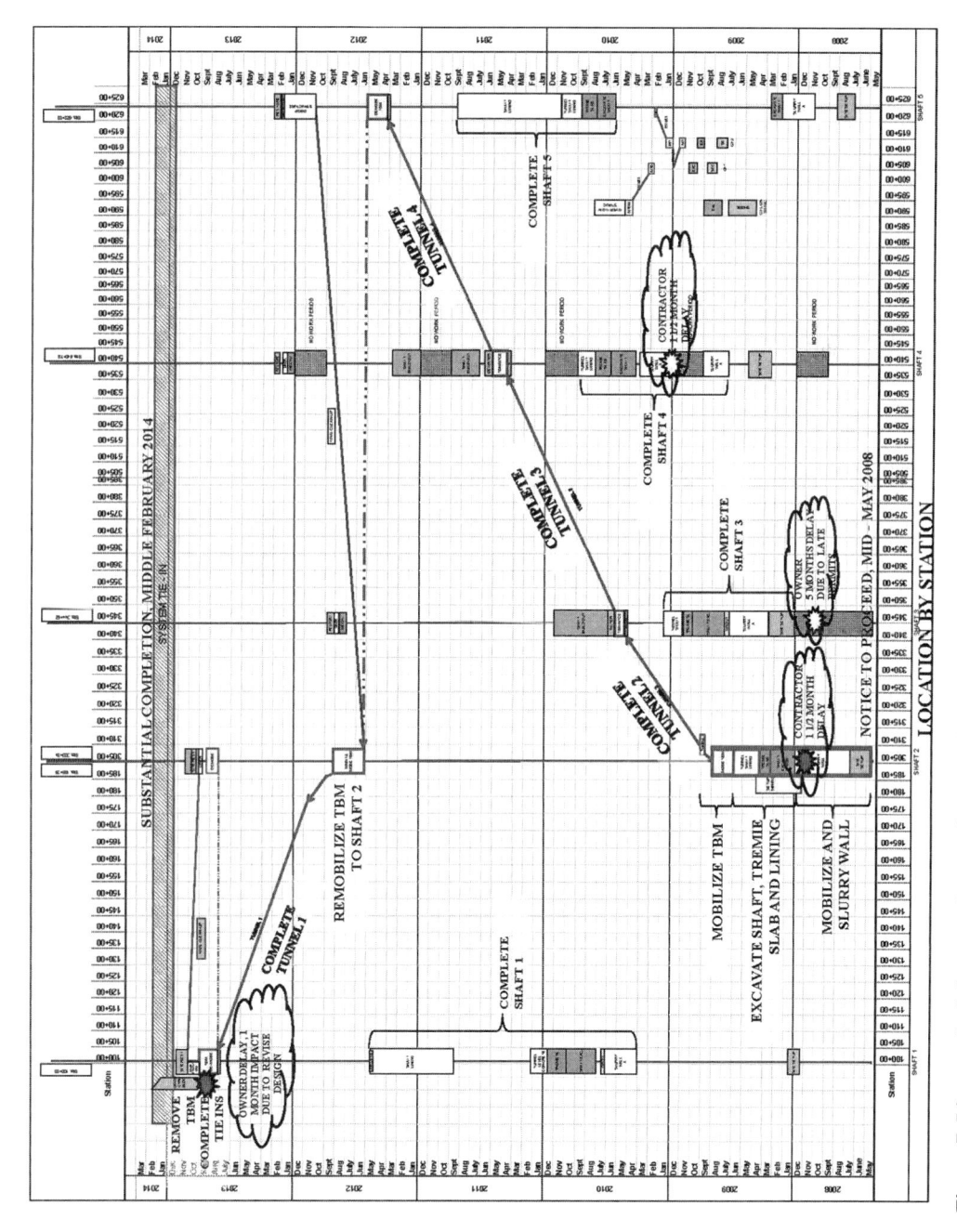

Figure 7. Linear schedule example to evaluate project delays

typically a point of contention between the owner and the contractor. The use of the linear schedule can be used to evaluate the impact of the change to the overall project schedule and help determine who was responsible for the impact to the critical path.

In this example, the contractor on the project encounters several delays related to late permits to Shaft 3 and design changes to the odor control building. These are owner-caused delay. Additionally, the contractor also encountered problems while excavating Shaft 4. This issue is attributable to the contractor and delayed the completion of excavation by 1.5 months. The linear schedule can be used to evaluate the impact of the delays on the overall project schedule and determine if a delay is a critical path delay or a concurrent delay.

Figure 7 shows the linear schedules with the delays mentioned above incorporated into the schedule. The overall impact of all the delays impacted the project by 2.5 months. The linear schedule in Figure 7 shows that the contractor delay while excavating the shaft caused a 1.5-month delay to the critical path. The design changes to the odor control building delayed the critical path by one month. The delay due to the late permits at Shaft 3 and the delay in completing the slurry wall at Shaft 4 did not cause delays to the critical path.

An analysis determines that the owner should compensate the contractor for one month of time extension and delay damages to the overall project. The owner should also compensate the contractor for the actual cost due to the permit at Shaft 3 but should not compensate the contractor for time or time-related cost for the Shaft 3 delays.

## CONCLUSION

The linear scheduling method is a much better tool for planning a tunnel project than the CPM schedule. Both owners and contractors can use the linear schedule to evaluate the project from the design and bid phase, through construction, to evaluating disputes. The linear scheduling method is easy to use and understand. However, it should be used in conjunction with the CPM schedule because of the CPM schedule's ability to record details and provide the many different tools not available with a linear schedule.

## REFERENCE

Harmelink, D.J., and Rowings, J.E. 1998. Linear Scheduling Model: Development of Controlling Activity Path. *Journal of Construction Engineering and Management* July/August 1998, 263–268.

# Building Mined Underground Stations in Soft Ground with NATM Construction Practices

**Gerhard Sauer, Juergen Laubbichler, Sebastian Kumpfmueller**
Dr. G. Sauer Corporation, Herndon, Virginia

**Thomas Schwind**
Dr. G. Sauer Corporation, New York, New York


**ABSTRACT:** Shotcrete was invented over 100 years ago by Carl Akeley, a taxidermist from Chicago. The first shotcrete machine was built in Allentown, PA in 1908. These inventions revolutionized the tunneling industry and became key components of the New Austrian Tunneling Method, which was developed for Alpine tunnels in the 1950s. Due to the versatility and flexibility of its application, shotcrete was also found to be an ideally ground support for mined underground facilities in urban areas. Many major cities were naturally founded close to bodies of water (sea, waterways, river deltas or alluvial plains), hence urban stations are often located in soft ground and below the groundwater table. Special construction techniques and ground support means and methods had to be developed to facilitate their safe and cost effective construction. This paper summarizes some of the main technical elements and principles of those soft ground NATM/SEM construction techniques. It further presents a collection of international station samples (our ongoing survey), and addresses some lessons learned.

## INTRODUCTION

The New Austrian Tunneling Method (NATM) dates back to 1954, when it was first used to construct tunnels through squeezing ground conditions in the Alps. Shotcrete and rock bolts were used systematically as initial ground support to stabilize the excavated tunnel and control deformations. From this initial application, the method quickly spread around the world and was used on road and rail tunnels in rural and mountainous areas. Increasing development, growing density and a rising traffic volume led planners to investigate the application of this technique also in urban areas, in lieu of traditional cut-and-cover and shielded TBM methods. After conducting extensive research, tunnel engineers found ways to make urban NATM tunneling, nowadays sometimes reffered to as Sequential Excavation Method (SEM) not only safe and feasible but also very cost-effective.

In order to handle soft ground conditions, low overburden, existing buildings and utilities, advances in theory and practice were called for, and adaptations to the design and construction process were required. Here are some of its most important rules in soft ground:

- Systematical geotechnical investigation, to gain profound knowledge about the expected ground conditions.
- Ground variability, asks for a certain level of flexibility in the contract documents and the design.
- The excavated cross section should be of ovoid shape! As a result ground and groundwater loads generate normal forces rather than bending moments in the tunnel linings.
- Evaluate time-dependent, three-dimensional stress redistribution and design the excavation support sequences accordingly, particularly where multiple openings are planned.
- Excavation and support sequences need to be designed also to minimize deformations, and to protect existing structures and utilities.
- Immediate shotcrete support is required to maintain stability and minimize initial ground movement.
- Adequate means and methods for groundwater control need to be designed.
- Ring closure of the tunnel lining needs to be achieved as quickly as possible, and within one tunnel diameter from the advancing face to create a stable load bearing structure.

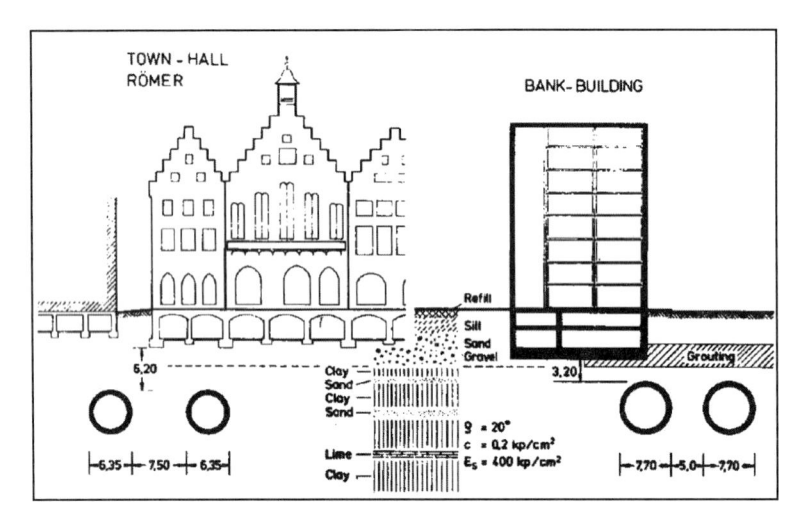

**Figure 1. Frankfurt 1968**

- Monitoring of loads and deformations on the initial lining, the surrounding ground and structures.
- It is essential to guarantee safe and sound construction procedures to verify the design assumptions.

With these requirements properly implemented, underground stations can be safely constructed through virtually all ground conditions. Any failures or collapses arising, can usually be attributed to the violation of one or several of these basic rules.

## NATM IN URBAN AREAS

### Brief History of Urban NATM

In 1968, contractor Beton- und Monierbau, with consultant and tunnel expert Leopold Müller, was the low bidder on an urban metro tunnel project in Frankfurt, Germany (Krimmer, Sauer 1985). They suggested a value engineering alternative using NATM construction practices. The owner required an approx. 262ft (80m) test tunnel and the possibility of reverting back to shield tunneling in case of problems. The test tunnel was unnecessary, and instead, excavation of the first NATM tunnel for an urban metro project proceeded with great success (see Figure 1).

Other German cities followed in quick succession, and the first hand mined underground metro stations were constructed during the 1970s. In 1973 NATM was introduced in Bochum where a 700ft$^2$ (65m$^2$) cross section was excavated in complex ground conditions with low overburden. That same year NATM was applied for the first time at Nuremberg's metro for the Lorenzkirche station with less than 33ft (10m) of soil cover and completion in 1976. Munich's Department of Construction implemented hand mined tunneling techniques in the early 70s and in the following years constructed Poccistrasse, Sendlinger Tor, Theresienwiese and Karlsplatz stations. About 60% of the running tunnels and station tunnels in Munich's challenging soft ground were excavated shortly after by using hand mined tunneling (25 Jahre U-Bahn-Bau 1990). Additionally, dewatering and compressed air schemes were developed which allowed for controlled handling of the occurring groundwater. Mined tunneling techniques advanced quickly during these first years and were introduced in Austria's capital Vienna in the late 70s, followed by other cities around the world.

The introduction of NATM station construction into the U.S. market occurred in 1984 in Washington D.C. The contract to construct two station tunnels, each about 700ft (210m) long, and about 8,200ft (2,500m) of running tunnels for the Wheaton Station was awarded to Contractor Ilbau. The Contractor proposed an NATM alternative for excavation and a then new waterproofing system consisting of a flexible PVC membrane sandwiched between initial and final lining.

While Wheaton Station was constructed mainly in rock, the first U.S. NATM station in soft ground was the Fort Totten Station, constructed in 1988. The 300ft (91m) underground portion of the station was excavated in overconsolidated gravels, sands, silts and plastic clays. As Figure 2 shows, results from evaluations of instrumentation readings and back analysis led to a refinement of the bid design, resulting in a faster, more efficient construction (Donde, Heflin and Wagner 1991). The experiences from both projects were vital for further improvement of the design of Washington Metro projects.

**Reasons for Utilizing NATM to Construct Underground Stations**

When considering the appropriate technique for the excavation of a station at a given location, a number of factors need to be taken into account including:

- Cost,
- Surface constraints (buildings, utilities…),
- Surface disruption and its cost to local businesses
- Construction duration,
- Risk.

The selection of NATM as the method of construction needs to be made on a case-by-case basis. The direct comparison with cut-and-cover methods reveals that utilizing hand mined tunneling for urban underground stations allows for alignment flexibility, minimized surface disruptions, avoidance of noise, pollution and vibration problems, and less utility relocation. Considering these advantages, mined tunneling is the more cost effective alternative. Several studies over the years have shown that cost savings in the order of 10–30% can be realized compared to cut-and-cover construction (Greifeneder 2003).

Two recent examples for reconsideration of the planned concept were the Red Line in Tel Aviv, Israel and the metro expansion in Santiago, Chile. In Tel Aviv, the city changed its strategy for urban underground stations from cut-and-cover to mined tunneling after professional consulting, in order to benefit from above mentioned advantages. The goal was to preserve the city's daily routine with little disturbance. Construction of the first phase of the system began in 2007.

In Santiago, cut-and-cover stations were replaced by mined stations in the late 1990s, and value engineering was introduced as a tool for the contractors to optimize the designs. All station work was performed using mined tunneling techniques, and costs were reduced by up to 40% through further improvements over the last ten years.

**KEY ELEMENTS OF SOFT GROUND TUNNELING**

**Prescriptive Excavation Sequences to Maintain Face Stability**

Maintaining the stability of the exposed excavation face is key for minimizing surface and subsurface

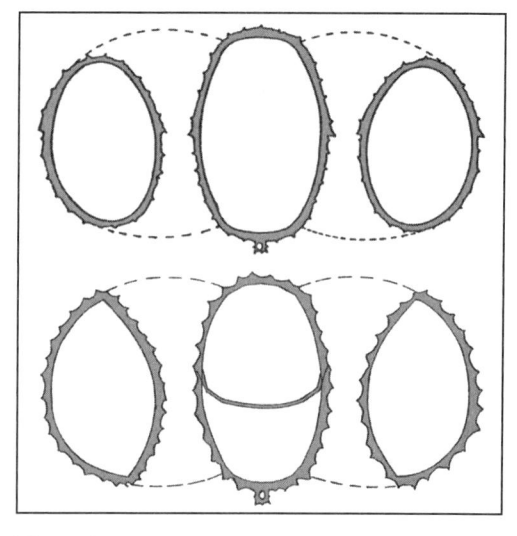

Figure 2. Fort Totten station, initial design (top) and redesign (bottom) (Donde, Heflin and Wagner 1991)

Figure 3. Cut-and-cover in Washington D.C. and downtown Boston

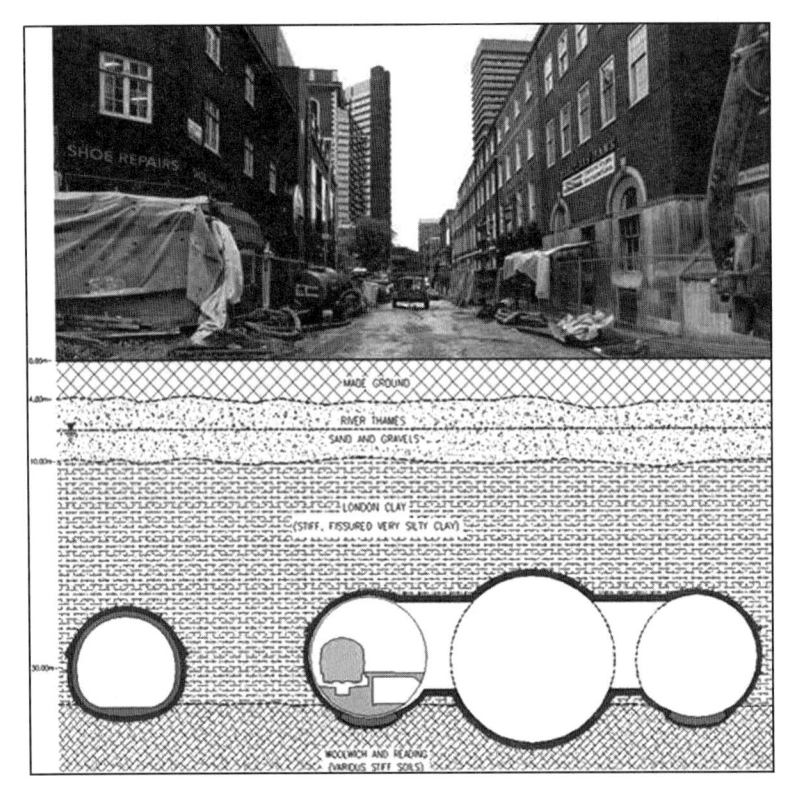

**Figure 4. Mined tunneling within city—London Bridge station and pedestrian access to Northern Line, London, UK 1997**

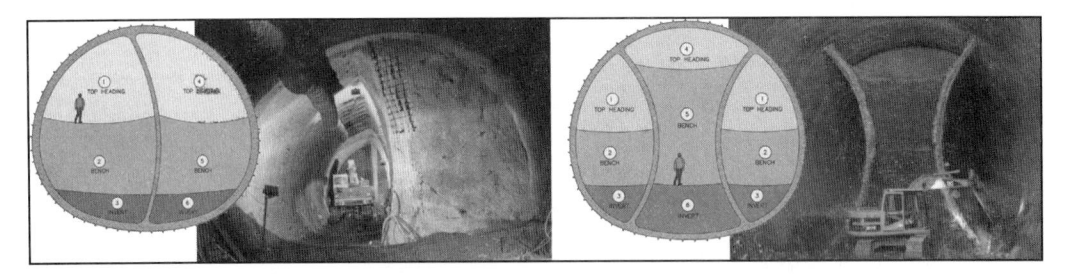

**Figure 5. Typical single and dual sidewall drift**

deformations and ensuring the safety of the tunneling operation. Since the stability of an excavated tunnel face is in inverse proportion to its size, and the size increases with the square of the diameter, a subdivision of the cross section into multiple drifts is required for large cross sections (see Figure 5 and 8). How these multiple-drifts are excavated, needs to be carefully designed, both in time and spatial succession of the individual drifts.

**Ring Closure**

To minimize ground movements, immediate ring closure is essential in soft ground, particularly in urban environments. Due to the ongoing excavation, a re-distribution of stresses occurs in the surrounding ground, as shown in Figure 6. While the stresses are reduced in the zone of active excavation, they are increased ahead of and behind the tunnel face. The

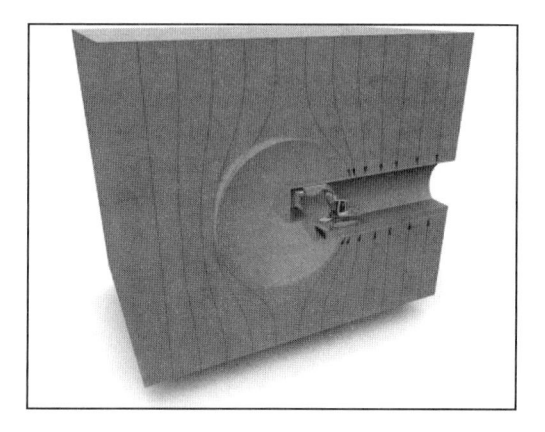

Figure 6. Load distribution

Figure 7. Shotcrete application at face

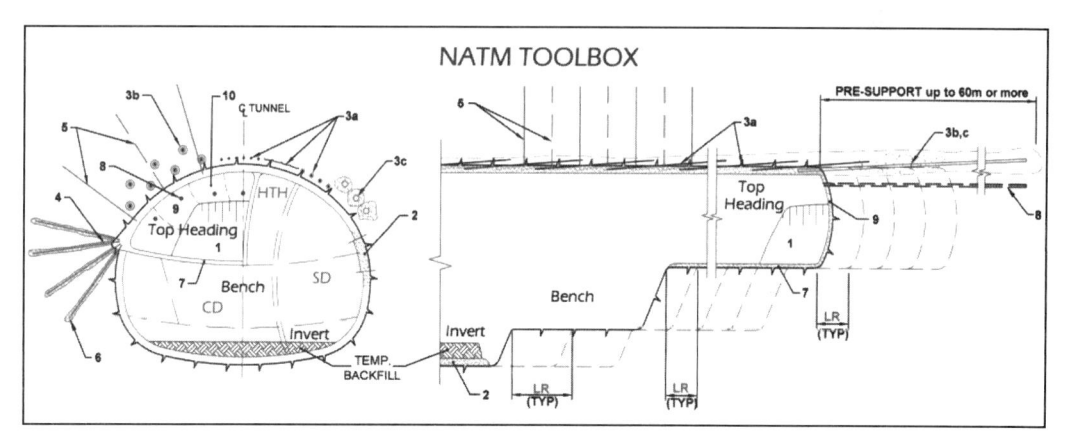

Figure 8. Toolbox items

peak stress occurs at a distance of about one and a half times the tunnel diameter, before and after the tunnel face. At this point the structural tunnel lining must be effective, which is achieved by closing the tunnel lining in the invert within one time the tunnel diameter behind the face. Sometimes, this is compromised to speed up the progress, which is one of the leading causes of major collapses.

**Toolbox Items and Ground Treatment/ Improvement**

In soft ground conditions, the exposed soil can exhibit the following, time dependent behavior:

- Ravelling,
- Running, or
- Flowing.

While pre-treatment methods, such as grouting or dewatering are utilized to address this problem, it is also cost effective to seal the excavation faces immediately after exposure with a layer of shotcrete (see Figure 7). If the standup time is still not sufficient, the excavation face can be further subdivided into small pockets, which are shotcreted in succession with the ongoing excavation. This technique can be used to mine through the most challenging and variable ground conditions.

Today, there are a variety of support, face support and ground improvements (toolbox items) available to address nearly all complex and challenging ground conditions. While a detailed description of each of these items is beyond the scope of this paper, Figure 8 displays the available toolbox items graphically. The paper "Ground Support and its Toolbox" by Dr. G. Sauer, 2003 addresses this topic in more detail.

The most important supports in mined urban soft ground tunneling used today are:

1. Stabilization of the Face with Earth/Face Wedge

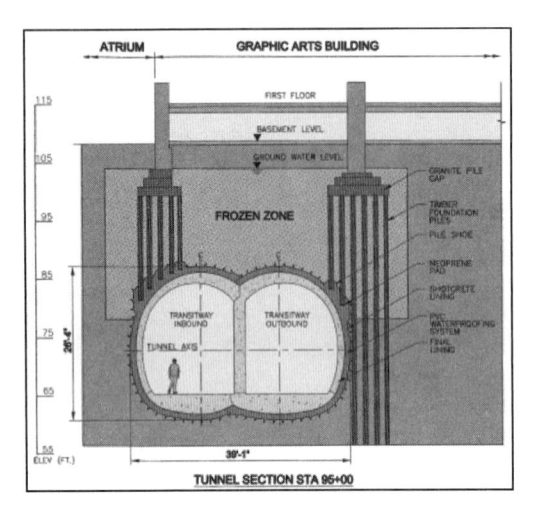

Figure 9. Frozen ground scheme (Russia Wharf—Boston)

Figure 10. Air lock at Orme Street III sewer project (Atlanta, Georgia)

   2. Shotcrete Lining
   3a. Spiling with Rebars or Grouted Pipe Spiling
   3b. Barrel Vault Method (BVM)
   3c. Horizontal Jet Grouting Method
   4. Increasing Width of Shotcrete Foundation
   5. Improving bearing capacity at springline with Grouting or Grouted Pipe Spiling
   6. Utilization of a Temporary Shotcrete Invert
   7. Dewatering of the Excavation Area
   8. Stabilization of the Face with 2" Flashcrete
   9. Pocket Excavation Method
     – Subdivision of the Cross Section by Side Wall Drift (SD), Center Drift (CD) and Multiple Drifts like Top Heading, Bench and Invert.
     – Determining proper Round Length (LR)

**Special Construction Methods**

In situations where conventional mining methods and NATM toolbox items are not sufficient, special construction methods are available.

*Freezing*

Ground stabilization and groundwater cut-off utilizing ground freezing is a well tested method. In 2003, it was applied for the first time in the U.S. in combination with NATM, on Boston's Russia Wharf project. A 100-year-old historic building supported by timber logs had to be mined under (Figure 9).

*Compressed Air*

If kept below about 20psi/1.4bar (economical decompression times still possible) (TUM 2009), compressed air beneficially contributes to the tunneling process and the tunnel stability during construction in three ways:

- It avoids lowering of the groundwater table and its detrimental effect on buildings.
- It balances the external head of water and prevents water inflow into the tunnel.
- It provides a direct face support and thus minimizing ground movement.

Previously, this method was very popular in the U.S. in conjunction with shield tunneling, but has been widely replaced by the development of face pressurized TBMs. In combination with NATM, compressed air was applied in the U.S. for the first time in 2001 on a sewer tunnel project in Atlanta, Georgia (Burke 2001). With a bulkhead situated close to or at the tunnel portal, excess air pressure can be applied inside the tunnel. Men and material locks allow transfer of people, machinery, and material in and out of the tunnel (Figure 10).

*Horizontal and Vertical Jet Grouting*

Jet grouting is an alternative to common pre-support measures. Pipes are drilled down from the surface to the depth of the future tunnel alignment where the grout will be injected into the ground. If the cover above the tunnel is too high, or the surface space is limited (i.e., in dense urban areas) horizontal jet grouting can be utilized. Overlapping jet grout columns are installed in 40–50 ft (12–15 m) advance of the tunnel face, forming a protective grouted umbrella around the excavation perimeter.

## Monitoring

During construction, geotechnical monitoring and instrumentation cross sections are used along the tunnel alignment. Such sections can consist of a variety of instruments such as convergence bolts, concrete pressure cells, and strain gauges recording stresses and strain in the tunnel lining as well as earth pressure cells to record the ground loads on the tunnel lining.

Surface monitoring is necessary to document pre-construction conditions and changes during tunneling, and after the excavation has passed.

Deformations, in and above the tunnel structure, need to be continuously monitored and interpreted. It verifies the design assumptions, to confirm adequacy of the applied excavation sequence and installed support system. Should monitoring indicate that basic design assumptions (mainly soil strength, layering etc.) were incorrect; a reassessment of the current design is required. Constant feedback ensures safe underground construction.

## SOFT GROUND NATM STATIONS AROUND THE WORLD

### Overview

The success of hand mined tunneling in soft ground urban areas with NATM construction practices is evident by the fact that more than 150 mined stations have been constructed around the world to date. Information about these stations is being assembled in a NATM station database. Table 1, which is an excerpt from this database, presents an overview and provides information about some selected cities. The database is a work in progress for research on various stations, approaches to design and construction methodologies, as well as the lessons learned.

For compiling data, the authors have received support from various owners and consultants around the globe. However, the authors also encountered difficulties in gathering information in certain cities due to political or security concerns. Input is welcome from owners and professionals, working in the industry, in making this database more complete, comprehensive, and accurate. Everyone interested in the collected information can access the database under www.dr-sauer.com.

The compiled information in this database allows running statistics on a variety of topics. It allows for general overviews and/or specific evaluations, ranging from geology and station alignment comparison to equipment or special construction methods that were used. Also station geometry, shape and arrangement as well as the comparison of applied waterproofing systems are available. Tunnel spans and range of overburden of the stations can be compared as well.

### Geometry, Shape, Arrangement

Comparison of the design approach of various cities in terms of station geometry, shape, and arrangement provides some interesting results. Influenced by design philosophies or technical possibilities at the time of construction, the decision for a particular station configuration is determined by a number of factors, including access and passenger flow, geologic conditions, space constraints, and existing structures in its vicinity. It can be noticed that certain cities have a preference for particular configurations, for example Vienna/Austria (two platform tunnels), Munich/Germany (binocular shape) or Santiago/Chile (large-span cavern). Figure 11 depicts the most common options of station tunnel designs.

### Waterproofing

Four commonly used waterproofing systems for hand mined stations are:

- PVC or other plastic material membranes
- Impermeable concrete
- Bentonite material
- Bituminous based material

In evaluating more than 150 stations some interesting outcomes were revealed, shown in Figure 12. For almost 60% of the stations, a waterproofing system with flexible membranes was chosen. The statistics show that this system presents a preferred choice for mined tunnels. Impermeable concrete was used on more than 25% of the stations, mostly in the German-speaking countries, where the concrete technology is highly developed. It is also a result of the much longer history of mined tunneling in these countries. Bentonite and bituminous based waterproofing materials were used on the least number of projects, mainly in conjunction with extensions of existing stations that used these materials when they were built originally. Only for those projects where conditions are optimal there is the option of not using a waterproofing system. One example is Santiago, Chile where due to favorable ground conditions stations are designed and constructed without a waterproofing system.

### Tunnel Spans of Soft Ground Stations

Figure 13 shows a very even distribution among tunnel spans ranging from 16.5ft (5.0m) to 65.5ft (20.0m), which usually covers the two platform tunnel and binocular tunnel shapes. On the other hand, a high percentage of tunnel spans in the range of 65.5ft (20.0m) to 82.0ft (25.0m) can be noticed.

These results indicate that there is a tendency towards large span caverns for mined soft ground stations. Today's mined tunneling techniques

Table 1. Mined stations overview

| City | Country | Number of Stations | Tunnel Span [ft/m] | Depth [ft/m] | Equipment | Water-proofing |
|---|---|---|---|---|---|---|
| Algiers | Algeria | 6 | 69.0/21.0 | 41.0/12.5 | Excavator | Plastic membrane |
| Athens | Greece | 12 | 65.5/20.0 | 49.0/15.0 | Excavator | Plastic membrane |
| Bangkok | Thailand | 2 | 39.0/12.0 | 33.0–65.5/ 10.0–20.00 | Excavator | N/A |
| Barcelona | Spain | 4 | 98.0/30.0 | 98.0/30.0 | Excavator | Plastic membrane |
| Budapest | Hungary | 4 | 36.0/11.0 | 98.0/30.0 | Excavator | Plastic membrane |
| Caracas | Venezuela | 17 | 72.0/22.0 | 42.5/13.0 | Excavator | Plastic membrane |
| Frankfurt | Germany | 2 | 36.0/11.0 | 16.5/5.0 | Excavator | impermeable concrete |
| Lisbon | Portugal | 8 | 69.0/21.0 | 52.5/16.0 | Excavator | Plastic membrane |
| London | United Kingdom | 3 | 33.0/10.0 | 33.0/10.0 | Excavator | Plastic membrane |
| Munich | Germany | 9 | 39.0/12.0 | 49.0/15.0 | Roadheader Excavator | impermeable concrete |
| New Delhi | India | 2 | 33.0/10.0 | 82.0/25.0 | Excavator | impermeable concrete |
| Nuremberg | Germany | 4 | 26.0/8.0 | 23.0/7.0 | Roadheader | Impermeable concrete |
| Porto | Portugal | 10 | 65.5–105.0/ 20.0–32.0 | 33.0/10.0 | Excavator | Plastic membrane |
| Santiago | Chile | 8 | 46.0–59.0/ 14.0–18.0 | 26.0/8.0 | Roadheader Excavator | No waterproofing |
| San Francisco | United States | 1 | 56.0/17.0 | 33.0/10.0 | Excavator | Plastic membrane |
| Sao Paulo | Brazil | 15 | 59.0–98.0/ 18.0–30.0 | 56.0/17.0 | Excavator | Plastic membrane |
| Seattle | United States | 1 | 36.0/11.0 | 131.0/40.0 | Excavator | Plastic membrane |
| Tel Aviv | Israel | 9 | 72.0/22.0 | 36.0/11.0 | Excavator | Plastic membrane |
| Vienna | Austria | 15 | 33.0/10.0 | 10.0–65.5/ 3.0–20.0 | Excavator | impermeable concrete |
| Washington | United States | 1 | 69.0/21.0 | 98.0/30.0 | Excavator | Plastic membrane |

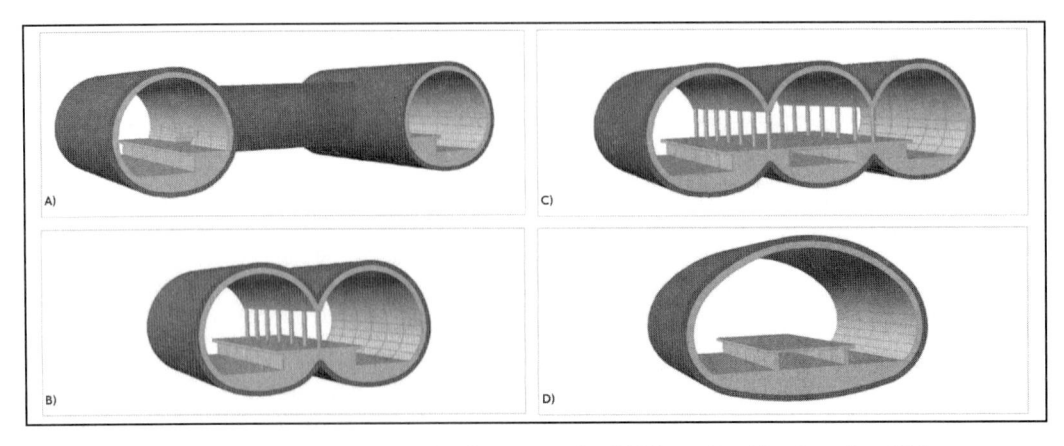

Figure 11. Station tunnel shapes: (a) two platform tunnels, (b) binocular, (c) trinocular, (d) large-span cavern

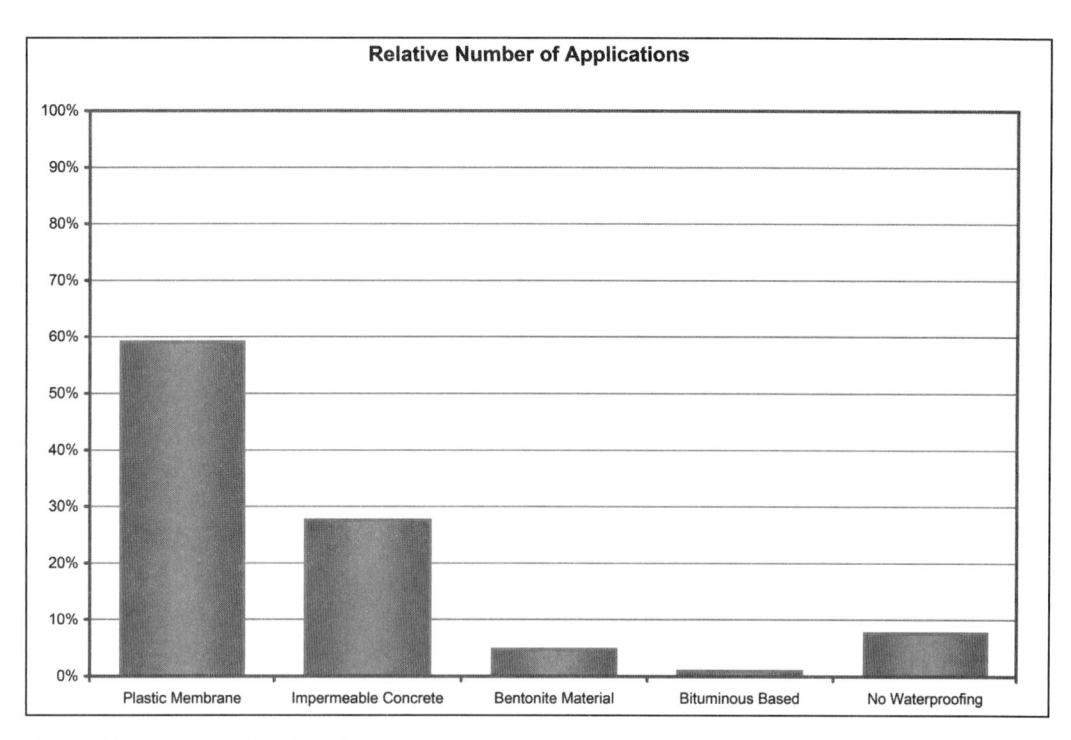

Figure 12. Waterproofing statistics

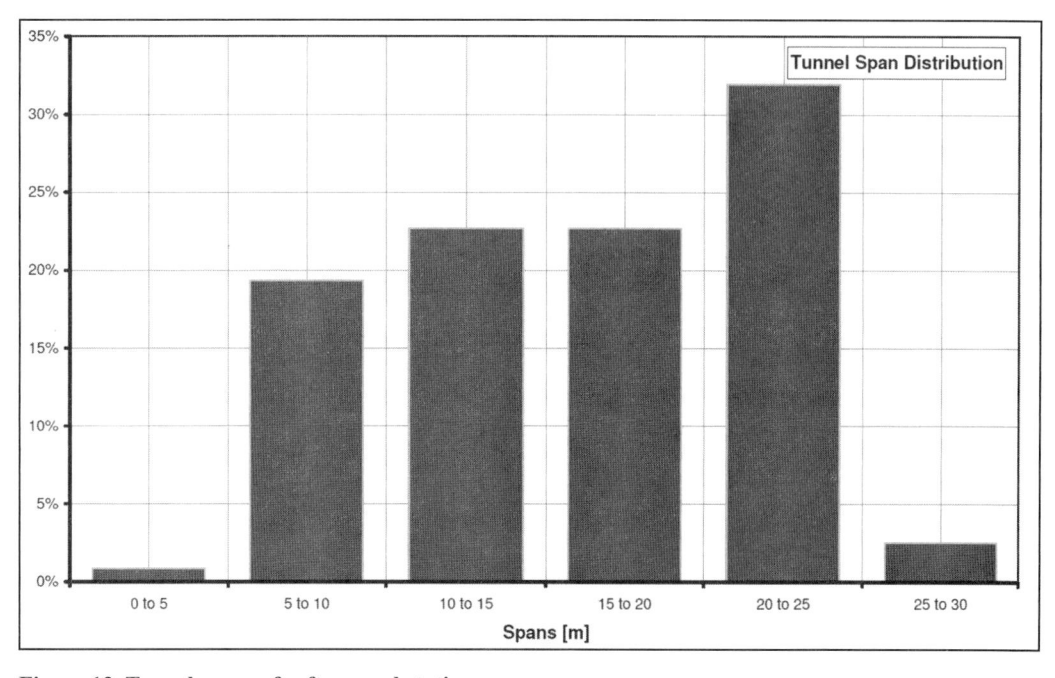

Figure 13. Tunnel spans of soft ground stations

703

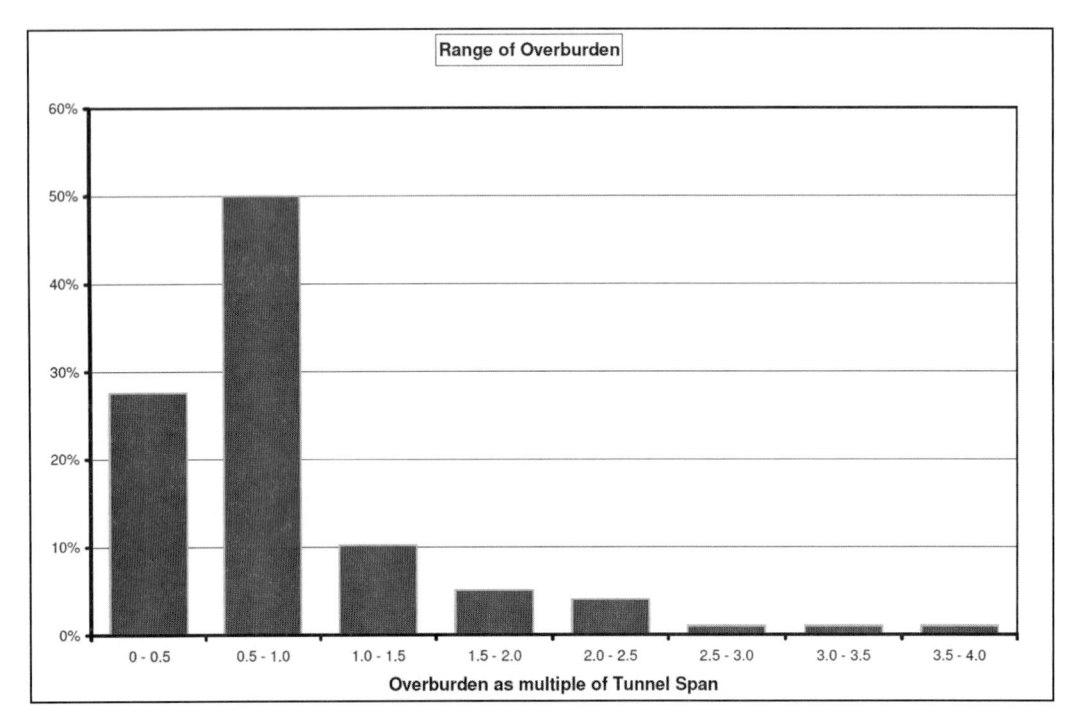

**Figure 14. Overburden as multiple of tunnel span**

provide owners with the opportunity to manage only one tunnel excavation during the project. While the construction of a large cavern with NATM construction practices does not generate substantial differences to a smaller tunnel diameter, the large cavern design can be more advantageous. Some owners prefer to combine everything under one roof, which means housing tracks and associated platforms for both directions, transit levels, control and maintenance facilities, ventilation and utilities within one large cavern.

However, many different factors are involved and have to be considered when deciding on the size of a station, and as the chart presents, there is a wide range of spans possible.

**Range of Overburden**

Usually one and a half times the diameter is assumed overburden to allow for stress re-distribution. In mined soft ground station tunneling and especially in urban areas we are often faced with overburdens ranging from 0.5 to 1.0 and 1.0 to 1.5 times the tunnel diameter as shown in Figure 14. Instead of ground loads being diverted around the tunnel perimeter properly, they will act as dead load on the tunnel lining. To minimize settlements it is necessary to utilize key elements and support tools such as the ones described above.

The combination of depth and tunnel span from more than 150 evaluated stations is presented in Figure 15. Keep in mind that there are stations with equal diameter, which are located at the same depth and therefore show up as one point on the chart.

**SIGNIFICANT PROJECTS**

**Beacon Hill Station, Seattle, Washington**

The Beacon Hill Station is part of the 14 mile initial segment of the Sound Transit Central Link Light Rail Line that establishes a high capacity commuter connection from downtown Seattle to SeaTac International Airport, and to Tacoma going south. The deep mined station was constructed using slurry walls for the shafts and NATM construction practices for the tunnel excavations (Figure 16).

The station scheme includes a 181ft (55m) deep × 46ft (14m) inner diameter main shaft that today houses four high speed elevators, emergency staircases, ventilation shafts, and mechanical and electrical equipment. A 26ft (8m) inner diameter ancillary shaft was lowered to accommodate another set of emergency staircases and ventilation shafts. From the main shaft, the 41ft (12.5m) wide concourse cross adit was excavated to the north and south to provide passenger and emergency access to the platform tunnels. These are 380ft (116m) long by 32ft (10m) wide and were designed to accommodate the

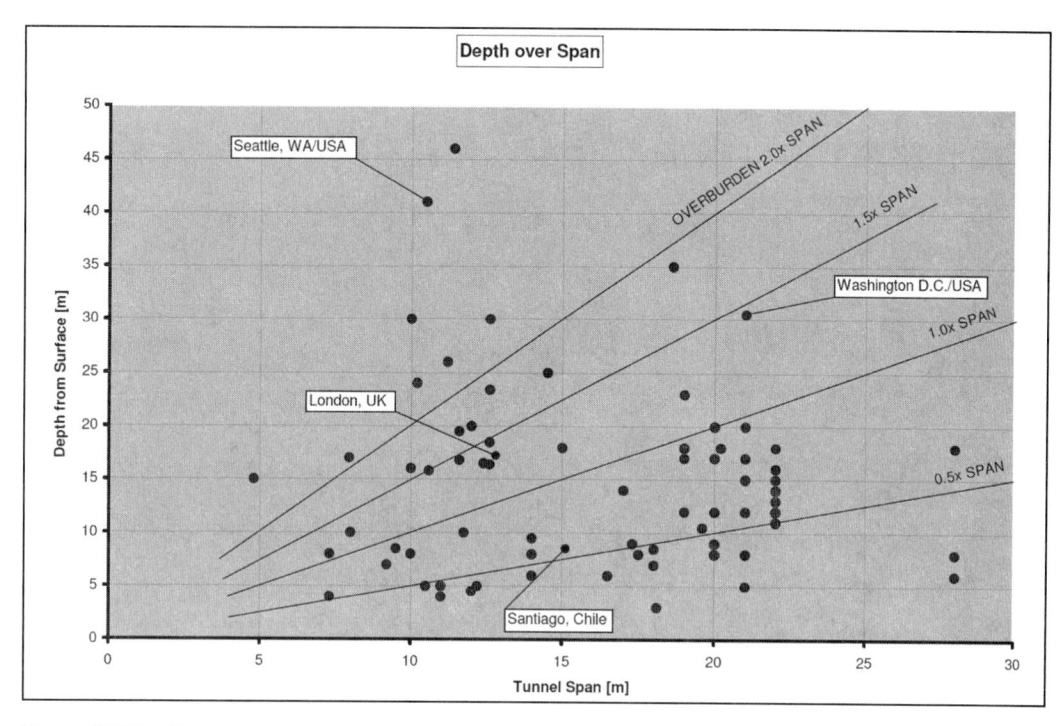

Figure 15. Depth over span

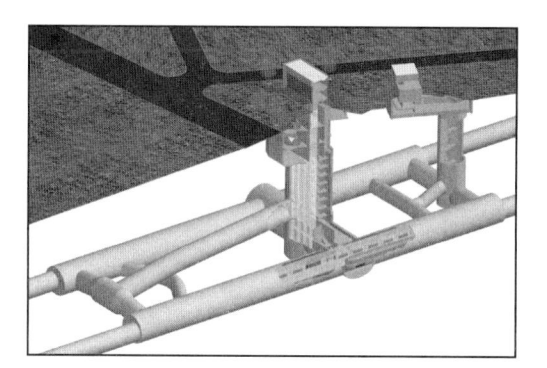

Figure 16. Final station layout of Beacon Hill Station

platforms, artwork, architectural finishes, and the light rail tracks. Two additional cross adits connect the platform tunnels, and ventilation tunnels provide air flow in normal operation and for emergencies.

Due to the large diameters of the tunnels, multiple drift excavation sequences were utilized. Grouted pipe spiling, barrel vault umbrellas, and jet grouting were used as special construction methods in the varying and challenging ground conditions.

Most of the Beacon Hill Station was excavated within glacial, overconsolidated, partly fractured or slickensided clays and tills. Intermittent sand and silt layers were present with multiple perched groundwater horizons. The high variability of the geology in the area of the station posed the main challenge for design and construction.

### Santiago, Chile

By the 1980s Santiago faced increasing public objection to surface disruption. Until then all work on the city's metro was carried out by cut-and-cover construction. Authorities reacted promptly on the public's demand by investigating less intrusive alternatives and subsequently moved from cut-and-cover construction to mined tunneling. In 1993 mined tunneling techniques were applied to a running tunnel portion for the first time. In the last ten years two lines were extended and two new lines constructed. With the constant improvement of mined tunneling techniques, NATM became the exclusively used method for excavating running tunnels and stations in the city. Good working relationships of Joint Ventures between local firms and experienced international NATM design companies allowed for innovative new approaches in urban underground station design. One of them was the introduction of the dual-sidewall excavation sequence with top heading, bench and invert excavation for the large caverns which are up to 56ft (17m) wide and 46ft (14m) high (Figure 17). This design became a preferred shape for Santiago's underground stations. It allowed shallow urban tunneling, with overburdens less than half the tunnel diameter, without the need of forepoling or spiling.

Santiago's ground proved to be well situated to mined tunneling, as the low water table and the Ripio de Santiago, a quaternary conglomerate, deposited by fluvial and glaciofluvial processes provides a stable matrix. The decision to use mined techniques exclusively helped reduce construction costs significantly. This could be achieved by avoiding surface disruption and the use of a lining concept comprising a combi-shell system of shotcrete initially, and a finial lining without a waterproofing system. So far, Santiago's strategy has been extremely successful and those responsible, are continuing to improve it with the goal of achieving even better results in the future.

**Fort Totten Station, Washington, D.C.**

In 1988, NATM construction practices were used to excavate the Fort Totten Station, which was the first application of hand mined tunneling in soft ground in the United States. The successful completion of the station is exemplary for mined tunneling in the U.S. A typical top heading, bench and invert excavation sequence was chosen for the binocular shaped, 69ft (21m) wide and 29ft (9m) high cross section.

The initial support measures included a shotcrete layer of eight inch, at minimum, welded wire fabric coupled with lattice girders on three foot centers, and twelve foot long soil anchors, which were required above and below the tunnel springline. Additionally, shotcrete was used at the face because of the short stand-up time of the encountered sands. Steel sheet forepoling with grout was installed locally to stabilize the crown and to increase safety during tunneling (Figure 18).

**London Bridge Station, London**

London's Underground operates one of the oldest underground networks in the world. The first line opened in 1863. Construction of the London Bridge Station in the 1990s was part of the Jubilee Line Extension (JLE), London's greatest expansion to its underground lines since the 1960s.

Extensive tunneling was required for London Bridge Station with its concourses, platform tunnels, interchange passages, emergency stairs, ventilation shafts, and about 18 escalators. Diameters ranged from 25ft (7.5m) to 39ft (11.8m). A required reduction of maximum allowable operating air velocities in the station tunnels resulted in significantly larger tunnels and junctions than used before in London's long history of underground design (Figure 19).

Utilizing hand mined tunneling techniques was needed to accommodate the required criteria and was beneficial for the flexible adaptability of tunnel shapes and sizes.

Figure 17. Typical excavation sequence for Santiago's La Cisterna station

Figure 18. Scheme of Fort Totten station

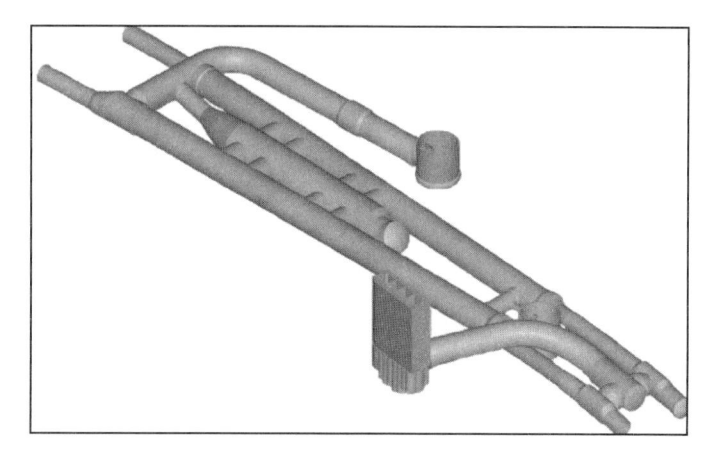

**Figure 19. London Bridge 3D station model**

Due to the stations location beneath the London Bridge Railway Station and other historic listed and sensitive surface structures, compensation grouting was utilized to treat the ground and limit deformations to contractually established threshold values.

## COLLAPSES AND LESSONS LEARNED

### Introduction

There have been collapses and downfalls during minded tunneling projects in the past, the most recent ones in Sao Paulo, Brazil (2007), Barcelona, Spain (2005) and Lausanne, Switzerland (2005). Evidence shows that the collapses were not the result of a fundamental flaw in the method, but rather a collapse in the application (Figures 20 and 21).

Collapses and downfalls happen mainly on night shifts and/or weekends, and are in most cases a management problem. Countless successful projects demonstrate that if managed and handled professionally, accidents like collapses and downfalls can be avoided. Specifying the unit price system in the bid documents has proven to be an effective tool in preventing such accidents, and their negative impact on the contract.

Cities like Sao Paulo, Munich and London, where major collapses have occurred, have also completed many successfully projects in similar environments and under comparable conditions.

Some findings from the investigation of collapses and the lessons learned are summarized in following:

**Figure 20. Collapsed Capri shaft in Sao Paulo, Brazil**

### *During Design*

- Geomechanical models and structural concepts should not be over-simplified.
- The adequacy of the excavation and support sequence design (i.e., missing or too late ring closure and a lack of adequate support are critical design mistakes in soft ground conditions) should always be validated.
- Risk assessment and management, as well as proper quality control should be addressed through special contractual arrangements.

**Figure 21. Sinkhole at Placa St-Laurent above Saint Laurent tunnel in Lausanne, Switzerland**

- Monitoring threshold values for warning and emergency levels should be defined.
- Emergency communication between authorities and project members should be established and a contingency plan designed to ensure quick reaction in case of an accident.
- Emergency plans for labor, but also for third parties (residents, transportation, etc.) should be established.
- Owners should contain a certain level of control throughout design and construction.

*During Construction*

- Design changes without proper reports or calculations to support them should not be allowed.
- Design during construction by observation, monitoring, and mapping of the geology should be validated.
- Measures to limit the probability of negative events should be applied.
- Contingency measures to reduce the consequences if a negative event occurs should be utilized.
- Quality control as specified in the design, plus a level of self-certification should be performed and maintained. Utilization of the established remedial actions in case of negative results. This is very important, especially for support elements that are used during the tunneling operation.
- Sufficient communication among the parties should be maintained.

## CONCLUSION

Hundreds of underground stations have been built to date in soft ground, often ground water bearing, around the world using hand mined tunneling techniques such as NATM. To date, two (Fort Totten Station, Washington D.C. and Beacon Hill Station, Seattle) were built in the fast expanding urban areas and public transportation systems of the United States with a third one currently under design (Chinatown Station, San Francisco). Today, the tunneling industry has a viable and proven alternative for the excavation of soft ground urban underground stations. Hand mined tunneling techniques, such as NATM, provide reliable, fast, economical, flexible, and interesting solutions for increasingly complex projects. Owners and contractors together with skilled and experienced designers can benefit from hundreds of past projects and the many lessons learned around the world from the construction of them. The U.S. market in hand mined soft ground station tunneling provides the potential to improve processes, techniques, and strategies today more than ever.

## REFERENCES

Barros, J.M., Eisenstein, Z., and Assis, A.P. 2008. Lessons from Brazil: Pinheiros examined. Tunnels & Tunnelling International. Nov. 2008: 16–21.

Burke, J. 2001: NATM with compressed air support for Orme street III. World Tunnelling. Dec. 2001: 480–483

Chamorro, G.S., Egger, K., Mercado, C.H. 2004. Santiago's Metro Expands. Proceedings of the North American Tunneling Conference 2004.: 195–200

Donde, P., Heflin, L.H., and Wagner, H. 1991. U.S. Approach to Soft Ground NATM. Rapid Excavation Tunneling Conference Proceedings. June 1991: 141–155.

Edgerton, W., ed. 2008. Recommended Contract Practices for Underground Construction. Colorado: Society for Mining, Metallurgy, and Exploration, Inc.

Federal Transit Administration (FTA). 1998. Lessons Learned Program—Waterproofing and Its Effect on Operation and Maintenance of Underground Facilities. www.fta.dot.gov/funding/oversight/about_FTA_9648.html. Accessed Nov. 2009.

Gaj, F., and Badoux, M. 2008. Building Lausanne's new m2 metro. World Tunnelling. May 2008: 14–16.

Gildner, J., and Urschitz, G. 2004. SEM/NATM Design and Contracting Strategies. NAT 2004 Conference. June 2004.

Greifeneder, G. 2003: Comparison of Cut-and-Cover Tunneling Method vs. New Austrian Tunneling Method (NATM) for Urban Tunnels with Shallow Overburden. Master's Thesis. Technical University Vienna.

Krimmer, H., and Sauer, G. 1985. Die Neue Oesterreichische Tunnelbauweise im U-Bahnbau—Rückblick und Ausblick. Felsbau 3. Nr. 3. 129–135.

Moss, J., and Murphy, P. 1993. Jubilee Line at London Bridge. World Tunnelling. Dec. 1993. 467–470.

Müller, L. 1978. Der Felsbau-Dritter Band: Tunnelbau. Stuttgart: Enke Verlag.

Sauer, G. 1998. Shallow Tunneling in urban areas. ASCE MET Section Seminar, New York.

Sauer, G. 1998. Urban Tunnelling Consequences. World Tunnelling. March 1998. 92–98.

Sauer, G. 2003. Ground Support and its Toolbox. Earth Retention Systems 2003: A Joint Conference. May 2003.

Sauer, G. 2008. Tunneling and Beyond. William Barclay Parsons Lecture. April 2008, New York.

TUM Zentrum Geotechnik 2009. Tunnelbau Spritzbetonbauweise. Lehrstuhl fuer Grundbau, Bodenmechanik, Felsmechanik und Tunnelbau. http://www.lrz-muenchen.de/~t5412cs/webserver/webdata/download/tb/spritzbeton.pdf. Accessed Dec. 2009

Wiener Linien 2008. Die Linie U2, Geschichte, Technik, Zukunft.

# Cost and Schedule Contingency for Large Underground Projects: What the Owner Needs to Know

**Christopher Laughton**
Fermi Research Alliance, Batavia, Illinois

**ABSTRACT:** Contingency planning is a key element of any large construction project. When a significant amount of the construction work takes place underground, the contingency planning process is a critical early duty of the project management team. A contingency plan must lay out a defensible rationale for the establishment of levels of cash reserve and schedule float that are adequate to support a project through construction completion.

Before contingency values can be set in terms of days and dollars, the planner will first need to establish a robust schedule and estimate that provide a sound foundation for the compilation of a risk registry and the performance of a risk analysis.

The paper will note some major insurance losses reported on underground projects and suggest that despite subsequent improvements in risk management practices, underground projects continue to all too frequently exceed budgets and incur delay. The findings of studies on a number of past construction projects will be reviewed relative to performance against cost and schedule. These studies identify an underlying trend or bias towards the underestimation of capital cost and duration. To reduce such a bias, improved methods for establishing project duration, budget, and allocating contingencies are required.

To best protect an underground project from cost overruns and delays, an owner needs to actively promote the development of objective estimates, and understand and take early control of the risk management and contingency setting processes. A contingency setting and tracking strategy for an underground project will be outlined within the context of establishing a pre-project plan for a major new underground physics experiment sponsored by the US Department of Energy and National Science Foundation.

## INTRODUCTION

Underground construction can offer public and private owners attractive options for the construction of new infrastructure networks and space creation in both urban and rural settings. Once in service, the host excavations can deliver excellent long-term, low-maintenance solutions that both contribute to an improved quality of life, and greater cost effectiveness relative to other construction options. However, although planners may find underground options attractive, owners, sponsors, loan officers, and insurance underwriters may wish to remove such options from consideration claiming them to be prohibitively expensive, risky, and difficult to manage. To support these latter claims, project sponsors can point to a good number of high profile construction cases where underground problems during design and construction led to significant cost overruns and delivery delays. Overruns, delays, protracted disputes, and litigation have turned otherwise technical successes into financial failures for one or more of the construction partners. A case in point is the Channel Tunnel Project, both a magnificent engineering achievement and a "Financial Black Hole" (Hout, 1995).

At one extreme, failure can be attributed to a single, unexpected event, such as a collapse that causes years of delay and results in millions of extra dollars spent; at the other extreme, failure might be simply the result of overly optimistic projections of cost, schedule, and contingency. To reduce the frequency of disappointing project performance, more objective and transparent methods of estimating costs and schedules and allocating contingencies are needed. These methods would also serve to restore confidence in planning practices and increase the probability of success on future underground ventures. However, developing more reliable methods for estimating and contingency setting for underground work is no simple matter. A number of poorly quantified, high-impact variables enter into the equation. Adequate reference to objective data and sound engineering judgment are not always achieved prior to the setting of a project budget, and despite recent improvement in management practices construction issues and overruns continue.

In this paper, construction problems associated with tunneling and mining projects are noted and briefly discussed. Specific reference is made to a number of studies undertaken in the mining industry that report on the prevalence and severity of cost overruns and delays on large capital construction projects. Data from these studies provides prospective owners and sponsors with an objective framework against which to benchmark deterministic and probabilistic predictions of project performance, expressed in hard dollars and calendar days.

Contingency for underground projects is reviewed within the context of a typical US Department of Energy project, and guidance is provided on early steps that the owner and sponsor can take to validate contingency plans.

## PERFORMANCE ON RECENT CIVIL TUNNELING AND MINING PROJECTS

### Major Losses on Recent Tunnel Projects

In the past decade there has been a concerted effort made within the underground industry to improve underground construction performance. This move was driven by the insurance industry in the wake of a spate of major losses. A partial list noting some more notable losses incurred in the 1990s and early 2000s is provided below:

- Heathrow Express Link, London, UK, 1994
- Pinheiros, Station, Subway 4, Sao Paulo, Brazil, 2007
- Taegu Underground, South Korea, 2000
- Tseung Kwan O Underground Line, Hong Kong, 2001
- Socatop Tunnel, Paris, France, 2002
- Shanghai Underground's Pearl Line, Peoples' Republic of China, 2003
- Circle Line, Singapore, 2004
- Orange Line, Kaohsiung, Taiwan, 2005

Underlining the singularly poor performance in the underground construction industry, Wannick (2007) notes, "Since the early 1990s, no other area of the construction industry has been as adversely affected by major losses as tunneling." Wannick further noted that tunnel failures experienced in this period involved the loss of life, property, and third party damages with resulting payments in excess of $600M.

In response to these losses, more rigorous risk management strategies have been promoted. In 2003, the British Tunnelling Society (BTS) and Association of British Insurers published a Joint Code of Practice for the Risk Management of Tunnel Works. The UK initiative was followed by the release of a Code of Practice by the International Tunnelling Insurance Group in 2006. These texts are good references for owners at the pre-design phase of a project providing timely guidance on the development of a project execution plan, allocation of risks and the definition of roles and responsibilities for design and construction contractors.

Despite the heightened emphasis on risk management, problems continue. Recent newspaper headlines and paper titles serve to document the range of potential challenges that an underground planner may face:

- "Law Suit Against Caldecott Fourth Bore Settled" (Cuff, 2009)
- "National Road Authority Prepares to Sue Port Tunnel builders," (Melia, 2008)
- "Gotthard Tunnel Faces Year Delay After Rival Complains," (Anon., 2006)
- "Light-rail Transit Tunnel Project Jolted by Construction Bid," (Grata, 2005)
- "CERN's LHC Project: Large Caverns in Soft Rock, at the Edge of Feasibility (Kurzweil & Mussger, 2003)

Problems such as these, involving third party approvals, deficiencies in estimating, design and contracting practices, and adverse events during construction can all rapidly lead to blown budgets. To convince sponsors that these risks can be accommodated within larger construction programs, more focus must be placed on understanding the limitations of the cost estimate and scheduling process and the development of a strong rationale that supports the establishment of realistic contingencies of time and money. The baseline plan that incorporates such contingencies must not only demonstrate that risks are properly characterized and addressed, but also show that there are adequate on-project contingencies to cover a degree of technical shortfall in design and uncertainties during construction.

### Estimating Project Duration and Cost

Although some underground construction risks will be underwritten, others will not. Risk retained "on-project" must be managed to retirement by the project team, either through the implementation of mitigation measures or the set-aside of contingency. Contingency needs to be explicitly included in the budget to accommodate known but unmitigated risks and a project-specific level of uncertainty associated with the types of adverse events that occur during construction. Contingency will not guarantee that a project will be delivered in a time and for a cost certain, but rather it should identify a percent probability that the sponsor-approved baselines can be met.

Project managers of capital projects in the mining and civil engineering arenas should expect

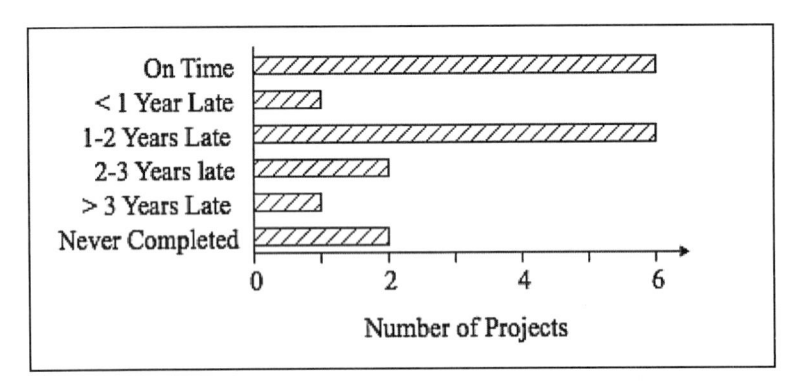

Figure 1. Project performance versus estimated duration for a set of mining projects (after Castle, 1985)

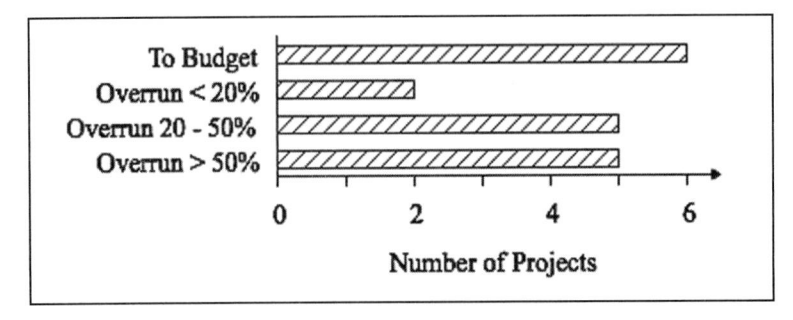

Figure 2. Project performance versus estimated cost for a set of mining projects (after Castle, 1985)

to add significant contingencies to run-of-industry estimates of duration and cost. Reference to performance against the estimate for mining projects indicates shows that underestimate of cost and duration are significant and commonplace. Figures 1 and 2 summarize the schedule and cost outcome for a set of 18 mine projects, reported by Castle (1985). Castle compared actual duration and cost against pre-construction estimates. With respect to duration, roughly one third of the projects were completed on time. The majority of projects were late, over a year behind schedule. Two of the projects in the sample set were never completed.

Figure 2 summarizes cost performance for the same set of projects. Again, approximately one third of the projects were completed on budget. A cost overrun of over twenty percent was observed on the majority of projects.

Perhaps the most notable observation made from a cursory review of the data set is the absence of projects completed substantially early or under-budget. For the set of projects reported by Castle, the pre-construction estimates that were used to establish cost and schedule prior to construction represented near minimums of cost and duration. Owners setting budgets based on run-of-industry estimates of cost and schedule would be unlikely to receive an early facility or cash-back and would have just a two-to-one chance of achieving the cost and schedule performance goals established at the moment of project approval.

**Over-Optimism in Project Baseline Estimates**

The data presented in the histogram shown in Figure 3 summarizes cost performance for a sample set of 63 mining case histories (Bertisen and Davis, 2008). The data underlines the marginal opportunities for project cost reduction relative to pre-construction estimates. The histogram shows actual cost normalized as a percentage of the estimated cost. The best performance reported in this data set was a cost at completion some seven percent below the estimated cost. The worst case reported an increase over the pre-construction estimate of 114%. The average cost overrun for this data set was reported at 25%. This latter number is consistent with a similar study undertaken by Gypton (2002) who calculated an average cost overrun of 22% for a set of 60 case histories.

Based on the findings of their 2008 study, the authors discussed the underlying causes behind the trend towards underestimation. They suggested that over-optimism within the basis of estimate gave rise

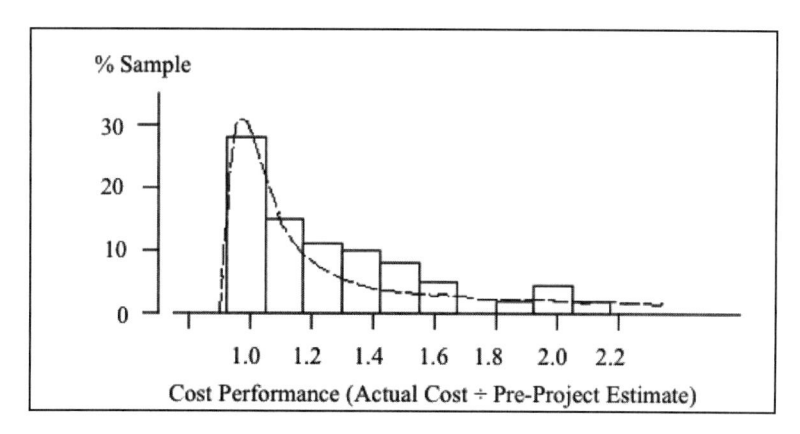

**Figure 3. Cost performance for a set of mining projects (after Bertisen and Davis)**

to significant, systematic underestimation of capital construction cost and duration.

Underestimates undermine project viability. A case in point is the Galore Creek Gold Project, British Columbia Canada. The mining company started construction of the project based on an estimated cost of some two billion Canadian Dollars. Within a year the estimated cost had jumped to nearly five billion Canadian dollars. An engineer explained the jump by noting that "When you're doing an early stage study, if you're in doubt about something you try to put an optimistic spin on it. You don't want to kill a project at the study stage" (Vanderklippe, 2007). Such spin is not confined to the mining industry, Flyvberg et al have reported on under performance in the arena of publics works with articles such as "Underestimating Costs in Public Works, Error or Lie" (2002) and "Megaprojects and Risk: an Anatomy of Ambition (2006)." Regardless of the source of error, owners need to be aware that estimates are inherently inaccurate. In developing estimates, the owner needs to ensure that basis-of-estimate is well documented, weaknesses well understood, and companion contingencies set to compensate for inherent shortcomings in the estimating process.

## ESTIMATING AND CONTINGENCY SETTING ON UNDERGROUND PHYSICS PROJECTS

### Underground Physics Projects

The physics community has constructed a number of large underground projects. With the notable exception of the Superconducting Super Collider, these projects have been successfully completed. However, the construction of these facilities has not been problem-free. Several projects have cost substantially more to build than estimated and were completed late. Most recently, the cost of a cavern at the European Particle Physics Laboratory increased from roughly 112 to 480 million Swiss Francs (Osborne, 2000, Peoples et al., 2003). The overrun was largely attributed to changed ground conditions. Such overruns are problematic for any project, but are particularly onerous when funds are secured on an annual basis. Proponents of new underground research ventures are striving to develop better budgeting practices more consistent with the constraints of year-on-year funding profiles.

### Long Baseline Neutrino Experiment (LBNE)

A new flagship experiment, the Long Baseline Neutrino Experiment (LBNE), is under development in the US to study neutrino particles. The experiment calls for the construction of new underground facilities at Fermilab and at the Deep Underground Science and Engineering Laboratory (DUSEL). DUSEL is located in Lead, South Dakota, within the footprint of the Homestake Mine. The project is estimated to cost between 500 and 900 million dollars. For the elements of the project sponsored by the US Department of Energy (DOE), the Project will be managed using procedures laid-out in Order 413.3A "Program and Project Management for the Acquisition of Capital Assets." The DOE management process involves the use of standard instruments used to monitor and control the design and construction progress. Perhaps most importantly, over the life of the project DOE will make five critical decisions (CD's) that allow the project to advance through the various design and construction phases of the Project to completion. These Critical Decisions are noted in Table 1.

The CD points for the LBNE project are placed within the context of a simplified underground design and construction process in Figure 4. Before moving from one phase of design and construction to the next, the project is subject to peer review at which

Table 1. Critical decision steps in the design and construction of capital projects

| CD | Description of the Critical Decision Approvals |
|---|---|
| 0 | Mission need statement |
| 1 | Alternative selection and cost range |
| 2 | Performance baseline |
| 3 | Start of construction |
| 4 | Start of operations or project completion |

time the readiness of the project to advance to the next phase is assessed relative to CD-specific performance metrics and documentation requirements, as described in Manual 413.3A. At these reviews key engineering components and contract issues are discussed in detail, and cost, schedule, and contingency projections updated. A fundamental element of passing a CD review is a demonstration that the estimates and contingency allocations are appropriate for the level of design and construction completeness and consistent with the findings of contemporary risk assessments. During construction, unexpected or excessively rapid draws on contingency are good indicators of problems developing on-site.

The early development of a reliable budget and duration that incorporate objectively assessed contingencies is important to the long-term success of any project. For research projects the ability to stay on time and within budget is particularly important. The value of experimental work is likely be time- and cost sensitive and scientific merit may be undermined by delay or overrun. Precedent exists for the cancellation of projects. These cancellations have been made even if conventional construction is well advanced.

**Development of the LBNE Cost and Schedule for Planning Purposes**

Good budgets and schedules underpin successful projects. Without a clear understanding of cost drivers and the critical path, project management's ability to make good decisions is compromised. The project must place a high priority on obtaining reliable estimates of cost and schedule. Estimates should be based on the use of field-proven methods and means and incorporate site-specific data for labor, equipment, materials, crew sizes and productivity rates. To maximize transparency, the basis of estimate should be well documented. For greater objectivity owners may consider using an independent team of cost engineers with recent experience bidding similar projects.

Schedule development is integral to the estimating process. More mature schedules should be resource-loaded and show critical path activities. Owners should be alert to the fact that for "underground-centric" projects the critical path is likely to pass through a series of underground excavation and lining activities. There are limited opportunities to accelerate such work. Delays to mining and lining typically result in delays to completion. Critical path delays drive up capital costs, as Salvucci (2003) noted in reporting on lessons learned from the Central Artery/Tunnel Project, Boston "Delay is the most significant driver of cost increases and reduced project benefit."

Given the necessary continuity in the budgeting and scheduling process from CD-0 through CD-4, all bases of estimates must be well documented, noting the key assumptions that drive cost, duration and risk. Engineering changes and their impacts on cost and schedule should be traceable over the life of the project. Ideally, even the earliest estimate should allow for the identification of major cost items and risks, as it is early in the project cycle when there is the greatest opportunity for achieving project optimization for a minimal effort.

Estimates should always be benchmarked against subsets of similar projects. Although each underground project is somewhat unique, there is always value in identifying other projects with similar design and construction criteria. An early demonstration that cost and/or duration are in the right ballpark can improve confidence in the overall plan and help broaden the pool of ideas for improved value management. Contingencies should reference the basis of estimate and risk assessment findings. Resulting contingencies, whether expressed as single numbers or probability density functions, should be benchmarked against the like-project sample sets, such as the one shown in Figure 4.

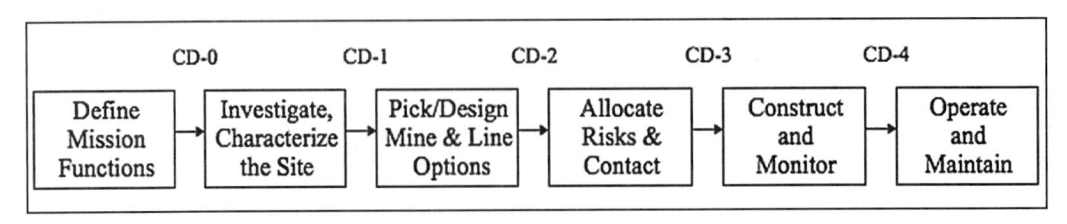

Figure 4. Design and construction of an underground project in a critical decision framework

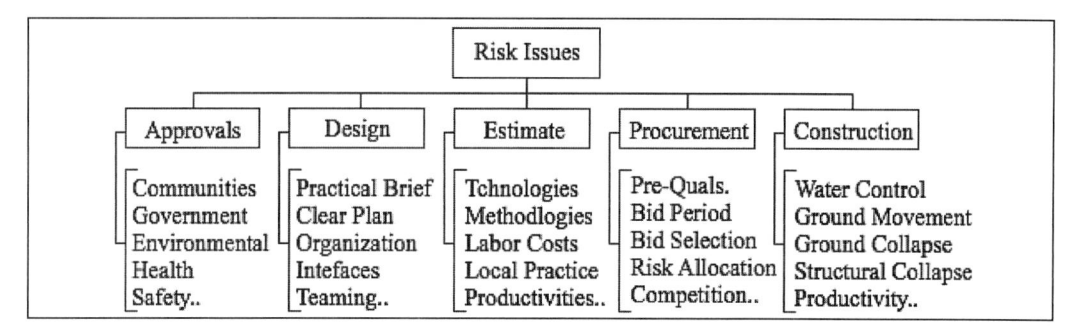

| | | Risk Issues | | |
|---|---|---|---|---|
| Approvals | Design | Estimate | Procurement | Construction |
| Communities | Practical Brief | Tchnologies | Pre-Quals. | Water Control |
| Government | Clear Plan | Methodlogies | Bid Period | Ground Movement |
| Environmental | Organization | Labor Costs | Bid Selection | Ground Collapse |
| Health | Intefaces | Local Practice | Risk Allocation | Structural Collapse |
| Safety.. | Teaming.. | Productivities.. | Competition.. | Productivity.. |

Figure 5. Sources of risk checklist for use in developing a project risk registry

### Risk Management of the LBNE Project

The key to an effective risk management program is the early identification of risk. At the start of the project cycle, focus should be placed on developing a comprehensive risk inventory. Figure 5 labels some of the risks to which an underground project is subject. In the graphic risks are placed in five categories. One category is associated with off-project approvals, including interfacing with stakeholders and third parties; the other four categories are associated with on-project work related to design, estimating, procurement, and construction work. More comprehensive listings of "threats and opportunities" are found in the appendices of the BTS Tunnel Lining Guide Handbook (2004) and in Edgerton et al (2008). The categories noted are largely consistent with those noted by Sperry (1985) in a discussion of contingency allocation for tunnel construction.

Although not specifically noted as categories in the figure, poor organizational and communication structures are often cited as underlying causes of problem performance. Forming and maintaining technically strong teams and establishing robust communications structures are key factors that undoubtedly contribute to the success of an underground project (Laughton, 2004). A chain is only as strong as its weakest link. It is important to convince the owner early in the project cycle that changes in their normal staffing, management and procurement practices may be worthwhile in order to better guarantee satisfactory performance on underground projects. Underground, staff with project-specific qualification and experience should fill key technical positions within the organization.

When a significant amount of construction is to take place underground, the inexperienced owner should specifically be prepared to import the engineering talent necessary to manage the specialist design and construction components of the project. These skills need to be present project-wide within both management and engineering structures. Access to these skills is particularly critical at the start of the project when first estimates of cost, duration and contingency are developed. As noted by Loofborow (1979) "The first stage should be directed and at least in part performed by those with the broadest understanding of the objectives, the conditions, the likely construction methods as well as engineering geology."

### LBNE: Tracking Contingency Allocations Over the Life of the Project

Figure 6 shows the projected evolution of contingency during the design and construction of a hard rock physics cavern under a DOE management framework. The contingency allocations at each CD-milestone and at bid time are shown in stacked histogram form with contingency values expressed deterministically as a percentage of the estimated construction cost. The contingency categories noted in the figure are aligned with the risks labeled in Figure 5.

Contingency values are reduced as the design matures and construction proceeds. A minor amount of contingency is left at CD-4 to represent a residual uncertainty and additional operating expenses that may be incurred during the early operating years of the facility.

The graphic shows a range of positive and negative values compared against a baseline construction estimate. In the graphic negative construction contingency is attributed to the presence of opportunities to reduce the baseline cost by relocating facilities to areas of better ground, and reducing design overconservatism embedded in the basis-of-estimate. The value of potential cost-savings associated with opportunities to reduce construction cost reduces rapidly as early siting decisions are made. Geology has a dominant impact on the setting of construction contingency.

There may also be opportunities to improve constructability and reduce risk by changing design and procurement provisions, for example if original specifications or risk allocations prove to have been

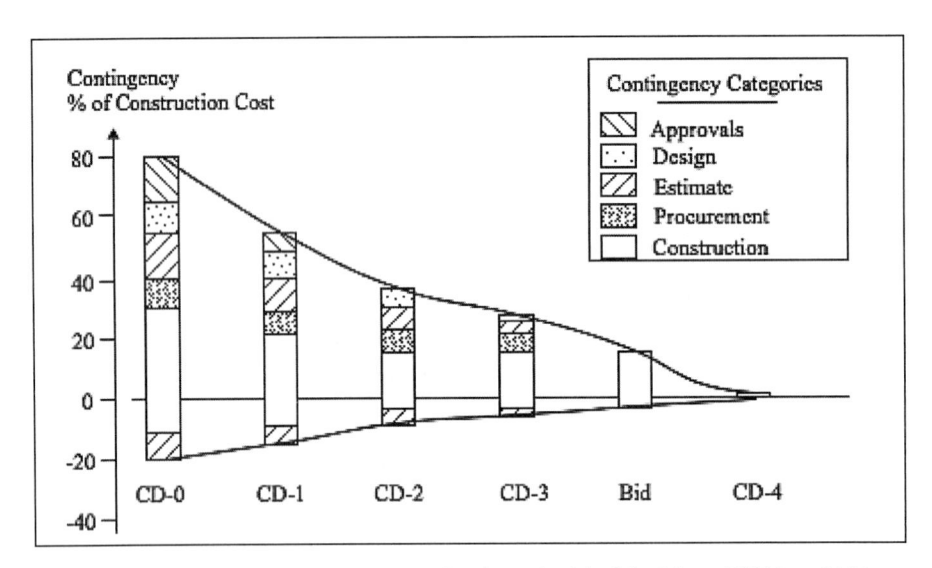

Figure 6. Stacked histograms for contingency at the time of critical decisions (CD's) and bid

too conservative. However, such savings may be discounted in even the earliest estimates.

The graphic provides owners and sponsors with a longer-term perspective of contingency needs, from pre-conceptual study through construction. Contingency allocations at each stage of the project and the contingency-time profile are both highly project-specific. However, the graphical representation does allow management greater insight into the overall process and encourages a deeper understanding of the individual design and construction steps. Developing a long-range expectation of contingency evolution can help better identify threat and opportunities in each contingency category and encourage the timely initiation of measures that can optimize the project at an earlier point in the life cycle.

## SUMMARY

Despite the adoption of more rigorous risk management processes, projects continue to be delivered late and incur cost overruns. A substantial amount of this underperformance may be attributed to the adoption of overly optimistic estimates of cost, schedule and contingency.

Improved estimating and contingency setting practices are needed to protect the project from the vagaries of underground construction. To this end, greater focus needs to be placed on establishing objective estimates of cost and schedule. These values should be routinely benchmarked against like projects. Such actions better inform contingency setting, a key element of large underground physics projects, where budget and contingency must be managed to conform to rigid year-on-year funding profiles and meet time-sensitive research deadlines.

To improve the chance of success underground, the owner does not simply need to ensure that best practices are followed; the owner also needs to understand, scrutinize, and coordinate estimating and contingency setting processes. Good funding decisions require good estimates of cost and schedule and realistic contingencies that reflect the true level of uncertainty that the execution of underground work entails.

## REFERENCES

Anon. 2006. Gotthard Tunnel Faces Year Delay After Rival Complains. European Foundations. 2:4.

Bertisen, J. and Davis, G. 2008. Bias and Error in Mine Capital Cost Estimation. SME Preprint No. 07-082, Littleton, CO: SME.

British Tunnelling Society and Association of British Insurers. 2003. Joint Code of Practice for Risk Management of Tunnel Works in the UK. London:British Tunneling Society.

Castle, G.R. 1985. Feasibility Studies and Other Pre-Project Estimates, How Reliable are They? Finance for the Mineral Industry.

Cuff, D. 2009. Law Suit Against Caldecott Fourth Bore Settled. Oakland Tribune.

Edgerton, W.W. (ed). 2008. Recommended Contract Practices for Underground Construction. Littleton, CO: SME.

Flyvberg, B, Holm, M.S. and Buhl, S. 2002. Underestimating Costs in Public Works, Error or Lie. *American Planning Association Journal* 68(3):279.

Flyvbjerg, B. Bruzelius, N., and Rothengatter, W. 2003. Megaprojects and Risk: An Anatomy of Ambition. Cambridge University Press.

Grata, J. 2005. Light-Rail Transit Tunnel Project Jolted by Construction Bid. Pittsburg Post-Gazette.

Gypton, C. 2002. How Have We Done?—Feasibility Performance Since 1980. *Engineering and Mining Journal* 203(1):40.

International Tunnel Insurance Group. 2006. A Code of Practice for Risk Management of Tunnel Works. Munich:Munich-Re.

Kurzweil, H-C and K. Mussger. 2003. CERN's LHC Project—Large Caverns in Soft Rock, at the Edge of Feasibility. RETC.

Loofburrow, R.L. 1979. Drilling and testing sites for underground construction. Conference on the Economical Subsurface Space Using Modern Mining Techniques.

Melia, P. 2008. National Road Authority Prepares to Sue Port Tunnel Builders. Irish Independent.

Osborne, J. 2000. Shaft Excavation in Frozen Ground at Point 5. Report ST-2000-051, CERN, Geneva, Switzerland.

Peoples, J., Bacher, R., Cox, M., Harrison, M., and J-O Joranli. 2003. LHC Cost and Schedule Review Committee Report. CERN.

Sperry, J. 1988. Costing Contingencies. Civil Engineering. 4:68.

Vanderklippe, N. 2007. NovaGold takes big hit on Galore Creek stoppage. Toronto Financial Post.

Wannick, H.P. 2006. Risk management for tunneling projects. Munich:Munich-Re.

# Management of Cost and Risk to Meet Budget and Schedule

**John Reilly**
John Reilly Associates International, Framingham, Massachusetts

**Dwight Sangrey**
Consultant, Portland, Oregon

**Steve Warhoe**
Consultant, Seattle, Washington

**ABSTRACT:** Costs of complex and/or underground projects have been consistently underestimated and this has attracted significant concern by agencies and the profession. Use of probabilistic cost-risk estimating procedures, such as the Washington State Department of Transportation's CEVP® process (developed by the authors and others) can produced more reliable estimates. The CEVP® process is currently being upgraded to including Value Engineering (VE) and enhanced Risk/Opportunity management to meet aggressive cost and schedule targets for Seattle's Alaskan Way SR99 Deep-Bore tunnel. This paper updates previous presentations related to CEVP® (NAT, ITA) and reports on the current process improvements and actions to control cost and schedule.

## INTRODUCTION

Even when a major infrastructure project is well planned and managed, conditions change and problems can arise. Technical issues may be a common reason for change but, in a significant number of cases, political changes seem to have the most significant impact (Salvucci 2003, Flyvbjerg 2002). These changes and impacts have resulted in significant and undesirable consequences which include cost and schedule over-runs, resource competition between projects, negative media attention and, consequently, public mistrust.

Thus we find that the public is skeptical of our ability, as a profession, to accurately develop initial budget estimates for the final costs of large, complex public projects. They are also skeptical of our ability to manage these projects to established budgets. Questions the public has asked include:

"Why do costs seem to always go up?"
"Why can't the public be told exactly what a project will cost?"
"Why can't projects be delivered at the cost you told us in the beginning?"

Our inability to answer these questions consistently and clearly is a consequence of many factors, including the large uncertainties associated with long project time-frames and, up to now, our inability to identify and correct inadequate estimating practices. Additionally, the effects of poor project management and poor communication with the public has further added to the problem—resulting in unfortunate results, including rejection of funding for proposed transportation projects.

Many government agencies have recognized this problem and in response are now requiring risk-based probabilistic cost and schedule estimating, as well as formal risk management plans (FTA 2004, FHWA 2006). In many cases, the development of budget estimates and political or legislative action now requires enhanced cost-estimating including use of probalistic, risk-based processes.

A variety of approaches using probalistic, risk-based methods have been developed in an attempt to provide better cost and schedule estimates. Most of these methods incorporate one or more of the following changes from traditional estimating.

- Replace traditional contingency-based deterministic (single value) approaches with a risk-based analysis that presents estimates as ranges with probabilistic weighting.
- Consider the uncertainties that would potentially impact a project by an developing an explicit listing of risk factors (or risk events) that may be candidates for risk management.

- Recognize the importance of schedule uncertainty and cost uncertainty. This may be addressed using integrated cost and schedule models, or by equivalent methods.

The specifics of these various risk-based approaches vary widely in the level of detail and in the techniques used to gather data for input to the risk process. The authors, clients and colleagues (WSDOT 2007, Roberts & McGrath 2005, Grasso et al. 2002, Reilly et al. 2004) believe that a flexible (depth and breadth of detail and degree of approximation), probabilistic, risk-based approach using an integrated cost and schedule model is the most appropriate way to quantify uncertainties for complex projects and to guide risk management in order to better define and control costs and schedules.

## THE CEVP® PROCESS, HISTORY AND DEVELOPMENT

In January 2002, the Washington State Department of Transportation (WSDOT) Secretary was challenged by the state legislature regarding the poor reliability of cost estimating and a history of increases to the cost estimate for a large highway project. WSDOT managers and key consultants were asked to develop a better cost estimation process. As part of defining the problem, a review of relevant data led to the following findings:

- There was a general failure to adequately recognize that an estimate of future cost or schedule involves substantial uncertainty (risk).
- Uncertainty must be included in cost estimating.
- Cost estimates must be validated by qualified professionals including experienced construction personnel who understand real-world bidding and construction.
- Large projects often experience large scope and schedule changes creep which affect the final out-turn cost. Provision for this must be made in the cost estimates and management must deal competently with managing potential changes.
- Inadequate communication with the public compounds the problem of poor estimating.

WSDOT decided to act on these findings by developing an improved cost estimating methodology that would incorporate a higher level of cost validation with a comprehensive assessment of those risks that could impact a project and an analysis approach that would quantify these risk impacts for risk management.

WSDOT's strategy also included policy changes that would deal openly with the process of public infrastructure cost estimating so that the public would better understand, and would be better informed, as project managers and elected officials make critical project funding decisions. WSDOT decided to open the "black box" of estimating and present a candid assessment of the range of potential project costs, including acknowledgment of the uncertainty of eventual project scope, the inevitable consequence of cost escalation fluctuations, and other major risks. Key concepts that were identified as principles for the new approach included:

- Avoid single number estimates. Recognize that at any point in the development of a project, from initial conceptualization through the end of construction, an estimate will require selecting a representative value, considering many factors that are inherently variable.
- Use a collaborative and consistent assessment process that combines high levels of critical external peer review expertise, particularly in construction cost estimating in a competitive environment, with appropriate roles and responsibilities for the Project Team and the independent experts.
- Acknowledge that both cost uncertainty and schedule uncertainty are major contributors to problems with project estimating, and incorporate both in the evaluation methodology. WSDOT foresaw the clear advantage, in fact the necessity, to integrate the effects of cost and schedule uncertainty.
- Use a high level of rigor identifying and quantifying probabilities and consequences of risks.
- Be practical and use common sense notions of risk descriptions and quantification. The new WSDOT method was to be completely rigorous and treat uncertainty in ways that acknowledged correlation, independence and other probability principles. However, the sources of information and definition of uncertainty were likely to encompass a range which might extend from highly quantified issues to those where subjective opinions of the contributors were all that would be available. This range of uncertainty needed to be captured objectively.
- Produce data that could be understood by the ultimate audience, the public.

The resulting Cost Estimate Validation Process or CEVP® (Reilly et al. 2004) develops a probabilistic cost and schedule model to define the probable ranges of cost and schedule required to complete

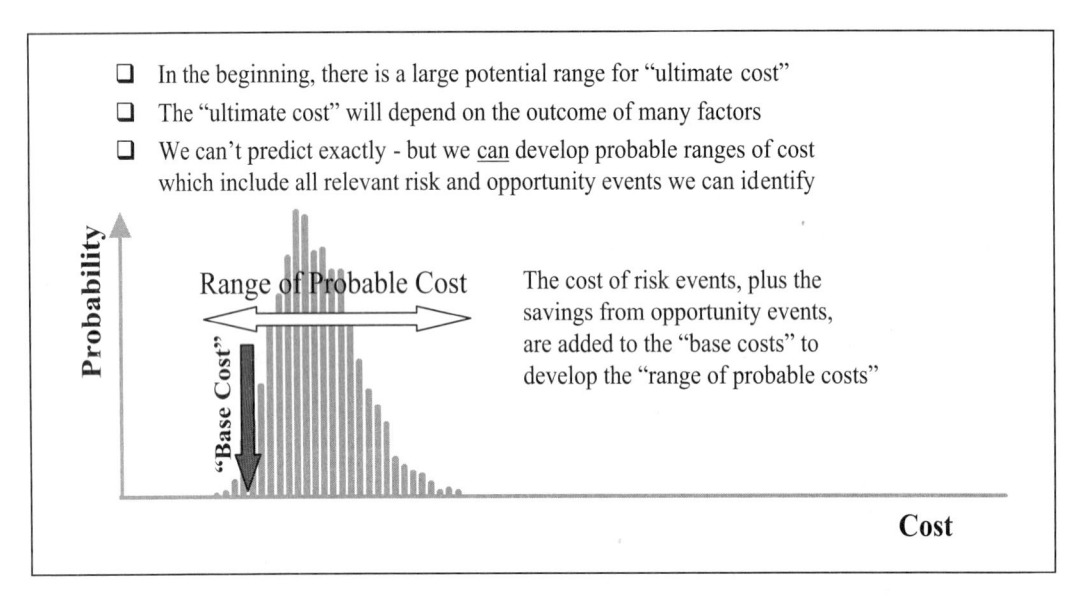

**Figure 1. Future costs are a range of probable cost**

each project. There are three principal and integrated components to the CEVP® process: 1) Cost validation, 2) Risk identification, and 3) Modeling, as described in Figure 1.

### Cost Validation

The cost validation process includes a critical examination of the details of the cost estimate presented by the Project Team. These details include assumptions, unit rates, prices and quantities. The basis of each element is critically examined and either accepted or modified.

The scope of the project, as reflected in the cost estimate is examined and the estimate adjusted if significant changes are found. The completeness of the estimate is compared to the scope and any cost elements that may have been excluded or neglected are included and quantified.

The cost validation part of the process is led by a cost validation facilitator with extensive estimating and program delivery experience, supplemented by team members with both design and real-world construction experience. The use of personnel with experience in contractor's methods is necessary to bring that perspective into the cost review for a well-shaped determination of base cost—the cost without contingency—that is, the cost if all goes as planned and assumed.

The usual contingency that is included in each unit price and quantity—or which has been applied to the entire estimate—is identified and removed from the cost estimate to define the base cost. The uncertainties that are typically the basis for a contingency in an estimate will eventually be addressed (added) through the risk assessment and model development. The schedule for the project is reviewed and assumptions, constraints and logic are critically examined—so that a base schedule can be defined.

During the discussions, and upon completion of the validation reviews, items of work that may have been missed, and the over- or under-estimated quantities and unit prices are identified and recorded. Estimates for missing items are developed and recommendations for adjustments are made. Finally, an agreed base cost is determined—this becomes the base to which the cost of potential risk and opportunity events are added by the cost/schedule uncertainty model.

### Risk Identification and Quantification

In the CEVP® process, risk identification and quantification is led by an experienced risk elicitator/analyst who is familiar with uncertainty theory, debiasing techniques and the structure of a subsequent cost and risk model. Other workshop participants include representatives from the project team who have familiarity with the plans, strategies, assumptions and constraints on the project, plus the Subject Matter Experts (SME's) who bring an independent perspective on important areas of project uncertainty.

The risk and opportunity events that are the output of the risk workshop are defined and evaluated with respect the validated base cost and schedule. Other factors such as correlation or dependencies among events must be defined and accounted for. In addition, each risk or opportunity event must be allocated to the project activities that are affected by it or, if a given event affects multiple project activities, significant correlations among occurrences need to be addressed. Significant uncertainties and correlations among event impacts also need to be defined.

Risk elicitation in the workshop is an iterative process that combines subjective and objective information. Uncertainty characterizations and probabilities are defined simultaneously to provide reasonable, practical descriptions of uncertainty.

## Modeling

The base cost and schedule, plus the quantified risk information, is analyzed with respect to a modeling framework that describes the planned project, its strategy and the schedule of activities required to deliver the final project. Several analytical methods have been used but most CEVP® analysis is done using simulation techniques (WSDOT 2007, Roberts & McGrath 2005, Grasso et al. 2002, Reilly et al. 2004, Einstein & Vick 1974, Isaksson 2002). The output of analysis is in the form of a "range of probable cost and schedule" and other characteristics of the project of interest to management, such as cashflow and risk ranking.

Following the successful development of CEVP® and its implementation for WSDOT "mega projects" in 2002, WSDOT applied the risk-based cost and schedule evaluation techniques broadly within the organization. The range of project sizes and types where CEVP® was used ranged widely from projects with budgets as low as $20 million or less up to the $multi-billion level mega projects. The detailed application of CEVP® over this range of projects has varied—in some cases with a number of simplifications and compromises made to the original principles of CEVP®. These simplifications were formally acknowledged and, in practice, different levels of risk-based analysis were given different names, such as Cost Risk Assessment (CRA).

The CEVP® approach has also been used to quantify uncertainty in programmatic measures such as program expenditure and cash flow and for programs consisting of a large number of individual projects, each of which have specific uncertainties but are often related to some degree. Specifics of the development of this approach have been reported (Reilly et al. 2004, Reilly 2009). The approach has been used or adapted by numerous Agencies including the U.S. Federal Highway Administration and numerous State Departments of Transportation.

## IMPROVING THE CEVP® APPROACH FOR THE ALASKAN WAY VIADUCT PROJECT

In early 2009, the Alaskan Way Viaduct Replacement Project (AWV), a multi-billion dollar highway program in Seattle, was Legislatively authorized by the State of Washington in cooperation with the City of Seattle and King County. An early management decision was to use CEVP® systematically in the evaluation of cost and schedule estimates and use of risk management. In this case, since there was a hard limit on available budget, WSDOT management and its advisors were clear in wanting to use a risk-based approach that was based on the highest level of principle and rigor—not a more simplified approach that had been used in some cases previously. Over the past seven years, WSDOT has performed hundreds of CEVP® workshops, however, there were concerns that some parts of the process, over time, had been overly simplified. In particular, these concerns led to identification of the following six elements where an improvement was considered necessary:

1. Improve the accuracy and validation of the base cost and schedule estimates.
2. Involve a sufficient number of appropriately qualified independent Subject Matter Experts to support review of all key areas of the project to be constructed.
3. Conduct risk assessments and risk management workshops in a well managed and professional manner.
4. Do not bias the analysis and reporting of a cost and schedule assessment by introducing constraining assumptions. While acknowledging that the results of analysis using certain assumptions may be helpful when making comparative decisions, all reports for such a major project should include a "reference assessment" that is based on a complete description of all uncertainties to which the project is exposed, as may be possible with the data which is available.
5. It will frequently be advantageous to consider and report higher level risk factors that have not been included in the analysis, recognizing that such higher level risk factors may be outside the control of the project team.
6. Apply consistency and rigor when managing to budget by adjusting scope, methods, requirements and/or schedule as part of the CEVP® process.

A discussion of why these six topics were considered as priorities for improving the implementation of CEVP® and how these improvements were to be implemented is discussed in the following sections. The CEVP® process including the application of these principles was referred to as "CEVP+."

## PRIORITIES FOR CEVP+

### Improve the accuracy and validation of the cost and schedule estimates.

From the initial applications of CEVP® in 2002 it has been recognized that a high quality, comprehensive estimate of cost and schedule should be prepared as a starting point for defining the base cost and base schedule. Depending on the level of design, different levels of detail in the base estimates would be appropriate with general conceptual and parametric estimates being used in early stages of design and more specific line-item estimates being the standard as the design matured. Large percentages in allowances and other systematic adjustments were appropriate in early stages but should be minimized as final design progresses. Finally, as indicated by the V in CEVP®, the base estimates must be sufficiently validated.

In subsequent applications of the CEVP® process the principles of base preparation and validation were simplified and streamlined, especially when working with smaller and less complex projects. This simplification was justified by the smaller size of the projects and the lower level of project complexity. However, a much higher standard of base cost definition was required for the large, complex and high priority AWV project, particularly considering the constrained budget and tight schedule as required by the governor and legislature.

The improved CEVP+ process returned to basic principles with a focus on developing the most complete and accurate base cost possible through the following requirements:

- The base cost estimate must include a comprehensive basis of estimate document that includes a clear statement of the scope of work the estimator is addressing, calculations made, sources for all data, and assumption made.
- The base cost estimate must be consistent with the base schedule in terms of productivity assumptions and activity durations.
- Validation of base cost and schedule estimates should be done by a team of at least two independent experts who have sufficient time to examine the cost and schedule estimates in detail.
- The proposed cost and schedule estimates (including all supporting materials) should be provided to the validation team at least one week prior to the validation review or the use of these estimates in a workshop to define a base cost or schedule.

### Involve a sufficient number of independent subject matter experts for all key project areas.

WSDOT has used Subject Matter Experts (SMEs) to assist in the CEVP® assessments since 2002. Many of these SMEs have come from outside organizations but the Agency has also been successful in using WSDOT staff that were not involved directly in a subject project and could be judged as "independent." Because of budget constraints on smaller projects, the number of independent SME's was sometimes limited to as few as two or three.

For the CEVP+ level of assessment used in the AWV Project, the need for independent SMEs with a high level of relevant experience dictated that almost all of these experts should be brought in from outside. WSDOT has limited experience with many of the key elements of the AWV project including tunnels, underground construction, design/build contracting and projects of the size and complexity of AWV. For the first two CEVP® assessments conducted in July and October 2009, 20 or more outside SMEs, including international experts, were engaged. These individuals were selected because of their specific high-level experience and expertise and their prior experience as consultants, public agency officials, constructors, and academics. In a few cases the SMEs came from WSDOT staff. For a few of the more salient subject areas, SMEs with overlapping expertise areas were brought in to help eliminate the possibility of subjectivity in the risk assessment process.

### Conduct risk assessments and workshops in a well-managed and professional manner.

To assure that workshops and other data gathering sessions were conducted in an efficient and professional manner, the CEVP+ process used for the AWV Project adopted the highest standards and practices that had evolved and been used by WSDOT for risk assessments since 2002. Among the most important aspects of this process were:

- A risk elicitation professional should lead and control the process of information gathering. This responsibility acknowledges that the primary objective of a risk assessment or risk management process is to gather and balance information from two perspectives:

  1. The professionals on the project team who best understand the designs and estimates.
  2. Un-biased independent subject matter experts.

- Achieving this balance the responsibility of the experienced risk elicitors who have developed techniques for gathering and balancing information from such perspectives.
- The appropriate individuals to contribute ideas or opinions on a specific risk topic are those with specific expertise on this topic. In most cases the appropriate contributors should be therefore limited to the relevant design team members and the SMEs selected for their expertise and experience on this specific topic.
- The ideal size for risk elicitation workshops varies. It is often effective and most efficient to collect information by working with small groups of project team members together with independent experts in a small workshop format. In other cases it may be most effective to conduct individual or small group interviews.
- Large group meetings with design team members and independent experts representing a variety of issues may be appropriate for training or to start off a risk assessment workshop with a general project overview. However, this format is not appropriate for gathering specific information about the details of project uncertainties.

**Do not bias the analysis and reporting of cost and schedule by introducing constraining assumptions.**

The integrity and accuracy of a cost and schedule estimate, or risk-based assessment, depends on considering the full range of potential outcomes and uncertainties (risks). No element of the risk assessment process is more important than recognition and acceptance of the fact that there is always uncertainty in basic constraints such as scope, financing scenarios, schedule for environmental processes or escalation rates. This principle has been recognized since the beginning of CEVP® in 2002. However, in the application of CEVP® this principle is frequently compromised by applying constraining assumptions in the interest of focusing attention on those design and construction risks which the project team can best control, leaving the higher-level risks to management or political representatives. Often, the higher-level risks have not been sufficiently evaluated.

The policy adopted in CEVP+ was to always conduct an unconstrained analysis as a reference for the consideration of potential project outcomes. It may be appropriate, and valuable, to do subsequent analysis that introduces one or more of the constraining assumptions. Such an analysis may provide information that is very useful in comparisons among projects and other considerations. However, the unbiased, unconstrained analysis is the most fundamental assessment of the actual project outcome and should always be the primary statement of the expected outcome, which may then be modified for the constraints that are to be applied.

When reasonable constraining assumptions about some elements of the project are used in analysis, these should be reported in perspective. Following the reporting of the "unconstrained analysis" it is appropriate to present:

1. Clear and unambiguous communication about the assumptions that have been made for any constrained analysis.
2. A quantitative evaluation of what the potential impacts of these constraining assumptions may have on the final project outcome models.

**Consider and report higher-level risk factors that have not been included in the analysis.**

An analysis that is unconstrained from the perspective of the project team or other body responsible for the project success is, nevertheless, subject to a set of uncertainties that may impact project outcomes but which are beyond the knowledge and control of the project leaders. Such factors as major change in the political environment and support for a project (election of a different mayor or governor with a different vision of the project, for example) may be recognized but are often inappropriate to try to incorporate in analysis.

Senior managers for the project team are usually aware of these higher level risks even if they are not included in analysis. In the interest of providing the most complete reporting of potential cost and schedule, it is recommended that higher level risk factors be identified and reported, along with the clear statement that uncertainty about these high level risks are not incorporated in the quantitative analysis.

## CONCLUSIONS

The CEVP® process was initially developed as a tool for Agency managers to use in evaluating cost and schedule estimates for projects with a defined scope and delivery strategy. In application by WSDOT, however, the results of the CEVP® assessments soon became the basis for risk management and managing to fixed or highly constrained budgets. WSDOT had traditionally managed projects tightly and this practice was reinforced when, in 2003, the State Legislature passed regulations that required all projects to be delivered at or near a line-item budget set by the legislature.

The practice that evolved within WSDOT to meet these requirements was to use the results of a CEVP® analysis to 1) first identify the likelihood that a project would be delivered within budget and schedule and then, 2) if the project is projected to come in above budget, to cut scope or take other necessary actions using the CEVP® model to guide these changes. This approach has proved to be successful for WSDOT in managing most of its larger construction projects.

A similar approach is being used on the AWV project. There is a clearly defined limit to the cost for this project set by the state legislature considering funding from the City of Seattle and other agencies. Successful design and construction of the project must fall within the budget limits that have been authorized. CEVP+ is being used to rigorously evaluate the probable range of cost for the project considering risk, to manage those risks defined by the CEVP® process, to conduct value-engineering to reduce cost and identify opportunities (better alternatives), to adjust the scope and modify the contractual environment (risk-sharing, collaborative process)—all to increase the likelihood that the final project will be delivered within the budget.

As of this writing, the CEVP+ process has resulted in WSDOT's ability to aggressively advance the design of the project keeping the probable cost and schedule within the authorized budget and schedule.

**REFERENCES**

Einstein, H.H. & Vick, S.G. 1974, 'Geological model for a tunnel cost model' Proc Rapid Excavation and Tunneling Conf, 2nd, II:1701–1720.

FHWA 2006, "Guide to Risk Assessment and Allocation for Highway Construction Management," Report FHWA-PL-06-032.

Flyvbjerg, B., Holm, M.S., and Buhl, S. 2002, "Underestimating Costs in Public Works Projects: Error or Lie?" Journal of the American Planning Association, Summer, 2002. Vol. 68, Issue 3; pg. 279–296.

FTA, 2004 "Risk Assessment Methodologies and Procedures," Project No. DC-03-5649, Work Order No. 6, May.

Grasso, P., Mahtab, M., Kalamaras, G. & Einstein, H. 2002, 'On the Development of a Risk Management Plan for Tunnelling', Proc. AITES-ITA Downunder 2002, World Tunnel Congress, Sydney, March.

Isaksson, T., 2002, 'Model for estimation of time and cost, based on risk evaluation applied to tunnel projects', Doctoral Thesis, Division of Soil and Rock Mechanics, Royal Institute of Technology, Stockholm.

Reilly, J.J 2009 "Probable Cost Estimating and Risk Management Part 2" Proc. International Tunneling Association, World Tunnel Conference, Budapest, May.

Reilly, J.J., McBride, M., Sangrey, D., MacDonald, D. & Brown, J. 2004 "The development of CEVP—WSDOT's Cost-Risk Estimating Process" Proceedings, Boston Society of Civil Engineers, Fall/Winter ed.

Roberds, W. & McGrath, T., 2005 "Quantitative Cost and Schedule Risk Assessment and Risk Management for Large Infrastructure Projects," Project Management Institute October.

Salvucci, F.P. 2003 "The 'Big Dig' of Boston, Massachusetts: Lessons to Learn," Tunnels & Tunnelling, North America, May and Proc. International Tunnelling Association Conference, Amsterdam, April.

WSDOT 2007, Guidelines for CEVP® and Cost Risk Assessment, see http://www.wsdot.wa.gov/Projects/ProjectMgmt/RiskAssessment/default.htm

# Overhead and Uncertainty in Cost Estimates: A Guide to Their Review

**John M. Stolz**
Jacobs Associates, San Francisco, California

**ABSTRACT:** There are few industry guidelines that help owners better understand the cost estimates upon which project budgets are based. This paper begins by briefly amplifying the industry consensus that cost estimates for heavy civil and tunnel projects must be based on a "bottom-up" approach. The main focus is on explaining the categories of indirect costs typically used by contractors and their perhaps surprising contribution to overall cost. It then moves on to emphasize the need for an integrated project schedule that quantifies the duration over which these indirect costs are incurred. The paper concludes with a brief examination of and recommendation on the issue of estimating accuracy. These are some of the issues often overlooked when owners review a cost estimate.

## INTRODUCTION

In the past few years, the tunnel industry has experienced an unprecedented surge in material and equipment price volatility, skilled labor shortages, and a shortage of bidders themselves. This confluence of volatility and shortages took the expected toll on owners' construction budgets with fewer project bids that were not only higher than the engineer's estimate, but also more dispersed. Arguably, nobody could agree on what it cost to build tunnels!

The industry responded with calls for improving the quality of the cost estimating process, and for larger projects, integrating that process into a risk assessment program to help determine the amount of additional contingency that should be carried in the project budget without limiting owners' program budgets. The most comprehensive opus, *Recommended Contract Practices for Underground Construction,* published by SME in 2008, provides owners with valuable information that integrates the cost estimating process into a comprehensive suite of practices and disciplines that are the basis for crafting an effective contract.

In support of the industry's general recommendation for preparing "bottom-up" estimates, the following discussion provides information for owners to use in evaluating the categories of so-called indirect costs. These costs, while contributing a significant percentage of the total bottom-up estimated construction costs, are similarly organized on any large underground construction project.

## INDIRECT COSTS

To review briefly, construction estimates consist of so-called direct costs, indirect costs, and profit. Direct costs are those costs that can be directly ascribed to the performance of a specific construction task—excavating a tunnel or shaft—and are grouped into direct cost items. Indirect costs, on the other hand, are costs expended in support of the construction project as a whole and are often referred to as overhead costs. As can be expected, sometimes the division between direct and indirect costs can get blurry, but indirect costs usually are not estimated until such time as a draft estimate is made of the direct costs, including a preliminary construction schedule. The following categories of indirect costs are typically always used when preparing a tunnel cost estimate.

### Equipment Ownership

This indirect cost category captures the capital costs of providing the equipment needed to perform the work. In contrast, the equipments' operating and maintenance costs are carried with the direct costs. If the equipment is purchased or is part of the contractor's fleet, there is an associated acquisition cost based on either a purchase price or book value, some or all of which must be written off against the project. Rental equipment is usually a straight charge to the job. Then there are the applicable taxes, freight, and erection and dismantling costs in arriving at the cost charged to the project. Perhaps a piece of equipment needs some modification for use in the work.

As can be expected, a detailed project schedule is needed to determine how many pieces of equipment are required—a piece of equipment probably would not be able to support two or more concurrent tasks, especially if they are some distance apart. Also, the total number of operating hours for a particular piece of equipment must be known so that the appropriate write-off can be estimated based on how hard the equipment is used.

Some plant and equipment do not have operating costs, nor do they require maintenance in the sense that they consume fuel, oil, and grease. For tunnels, linear plant items (track, utility lines, ventilation fans and ducts, and so on) are examples of such plant and equipment. Nevertheless, they have an oftentimes significant capital cost and must be included in the equipment ownership item, as are other general plant items such as maintenance shop equipment and survey instruments.

Sometimes equipment ownership costs may be allocated to direct cost items in proportion to individual operating hours. This is especially true for heavy civil applications involving a significant amount of earth moving or concrete placement, such as for highway or dams. Contractors engaging in such work usually already have a fleet of equipment to perform this work. As such, the corporate equipment department usually charges the projects much like an equipment rental company. Tunnel projects, on the other hand, require a fleet of specialized equipment that usually cannot be used on a subsequent project without heavy overhaul or modification. Furthermore, a contractor's equipment department has no historical costs upon which to base an ownership rate. It is for this reason that equipment ownership for tunnel projects tends to be carried as an indirect cost.

**General Mobilization and Demobilization**

These two indirect cost items are always estimated and scheduled separately. For tunnel projects, these activities can be quite involved because a significant amount of plant and equipment is needed to support underground construction activities, including maintenance shops; warehouse areas; worker changing and shower facilities; fuel, oil, and grease areas; power drops, electrical substations, and power distribution systems; compressed air and distribution systems; and water supply and distribution systems. There could be a batch plant on site, room must be established for muck handling and loading for off-site disposal (and there may even be a muck processing facility such as a slurry separation plant), and the work area must be fenced. Often the contractor is required to provide a functionally complete project office for the owner in addition to one for the contractor's own use. Access roads may be blazed or temporary bridges built. Also, the contractor typically establishes the project erosion control facilities during general mobilization. These are all estimated as separate line items.

There may be cases where a particular item of work such as a drop shaft requires mobilization of certain equipment. In such cases, the mobilization and demobilization are usually priced in the direct cost items associated with that item of work, i.e., the drop shaft construction.

Equipment freight in and out and its erection and dismantling are usually carried in the equipment ownership item. However, their costs may be transferred to this item, especially when mobilization costs must be justified.

For general demobilization, punch list items are usually performed while select elements of the above plant and facilities are being dismantled and shipped off site.

These costs can be substantially higher than costs for building or industrial construction. Since many owners limit mobilization payments to some fraction of the contract and even meter payments based on some schedule of earned value for contract work, the cost estimate serves as a useful tool to help owners understand when it may be beneficial to make changes to the standard contract language.

**General Plant Operation and Maintenance**

This item identifies the cost for operating and maintaining the contractor's general plant, described in the General Mobilization and Demobilization items, and serves the project as a whole. This item could therefore include everything from providing drinking water supplies to monthly estimated power costs for facilities that have no operating costs, such as office trailers. Costs for office supplies and connectivity are also estimated here.

Costs are usually estimated by the month, starting after mobilization is complete and the contractor is ready to start contract work, and ending at the beginning of demobilization, when contract work is completed.

**Weekend Maintenance**

This captures the cost of manning pumps or performing other maintenance activities on weekends. Weekends are used for performing site safety inspections and performing noncritical but essential preventive maintenance or overhauls on equipment. For a tunnel boring machine (TBM) job, cutters might be changed on weekends, and the surveyors are almost always on site. Without weekends, this kind of nonproduction work would need to be scheduled during the week, invariably at the expense of performing production work. In addition, since tunnel work is usually performed on two long shifts or three regular shifts per day/five days per week, weekends are a valuable resource for another reason: they allow the contractor to accelerate its work if needed to mitigate the contractor's own delays since tunnel work does not lend itself well to accelerating the pace of work simply by adding crews or starting work in another area.

Clearly, estimating these costs requires knowing the number of project weekends. Since the level of weekend efforts will vary over the life of the project depending on what types of contract work is being performed, this information can only be reasonably estimated using a detailed project schedule.

**Field Supervision**

The cost of project field supervision—i.e., personnel above foreman classification—is summarized here. Various personnel are carried for the amount of time that their expertise is needed throughout the project, so their involvement is always estimated based on the project schedule. Some classifications, such as the project manager, project engineer, and business manager, are chargeable to the project from the date of award; other staff such as equipment superintendents, staff level engineers, and the purchasing agent start ramping up their involvement after the Notice To Proceed is given. Until the field offices are established, these personnel usually work from the corporate office or apartments near the site.

By the time general mobilization is complete, almost all field personnel are on site, including the safety manager and his/her on-site emergency medical technicians, the field and office engineers, the superintendents, QC manager, and clerical help. By the time tunnel excavation starts, the walkers and field engineers for each shift will be on site.

Field supervisory personnel may be salaried or hourly, exempt or nonexempt. They may be local hires, meaning they will be terminated after the project is completed, or they may be permanent employees relocated from the corporate office or another project. Regardless, each carries a separate set of benefits. Field supervision is therefore estimated by the month.

**Overhead Maintenance and Service**

This is a broad cost category that captures everything associated with the contractor's operation as a corporation. Typical corporate charges include those from the accounting, IT, HR, design, and corporate departments. Sometimes allowances are made for legal reviews and audits. These costs may be estimated monthly, but are usually applied as a percentage of expected contract billings since contractors usually absorb home office overhead costs for each construction project as a function of total yearly revenue from all construction projects undertaken.

Since the level of effort for some tasks varies from project to project, separate charges may be customary for assisting in the initial high volume of submittals, or for engaging the corporate design department in custom plant and equipment design.

To this, noncorporate project-specific requirements must be added for services such as maintaining project outreach, screening for drug tests, providing for the contractor's share of partnering and DRB expenses, and travel and inspection of off-site material and equipment fabrications by contractor and owner personnel. Site conditions may require engaging the services of a noise or blasting consultant. Usually the corporation will rent a number of apartments for key site personnel. Any costs for preconstruction surveys might be carried here. Any warranties required under the contract are usually priced in this item. Collectively, these kinds of overhead and maintenance costs are usually estimated by the month over the life of the project.

**Bond, Insurance, and Taxes**

Conventions for what to include under this item vary from contractor to contractor. Some of these types of costs—most often employee general liability and workers' compensation insurances—are usually carried in the unit rates for labor resources, and therefore are spread throughout the direct costs since they are a function of the base plus vacation portion of the wage rate. Those costs that are calculated based on the contract value or some other basis are carried here, including:

- A bid bond, which is submitted with the bid as a guarantee that the bidder will undertake the terms of the contract. If the bidder is found nonresponsive, the bid bond assures the owner that it will receive the difference between the nonresponsive bid and the next lowest responsive bid.
- A performance bond, which assures that the owner receives payment for the cost to complete the project in the event of the contractor's default. The bond is usually written for the contract amount and replaces the bid bond on award of the contract.
- A payment bond, which assures that the contractor will pay subcontractors, laborers, and suppliers on the project. This protects owners against mechanics liens—or claims to title—on the project.

Insurance costs are more project specific. Certainly the item will include premiums for builder's risk (insurance against damage to the project while under construction), the usual contractor's automobile and equipment insurance, and perhaps an allowance for insurance deductibles that may be paid out during the course of a project. Other project-specific insurance may also be needed, such as railroad protective insurance.

Taxes include such items as state or local taxes, including property taxes and permit and license fees.

### Financing Charges

This is the contractor's cost for complying with contract requirements for submitting a balanced bid. Since the contractor's cost in such matters is also a cost to the owner, these financing costs must be estimated, usually by generating a cash-flow curve for the project that takes into account the expected timing of contractor expenditures relative to the contract provisions that dictate when and how revenue is earned.

The financing cost is the sum total of the monthly interest expenses or revenues, calculated on the net difference between the revenue and expenditure cash streams for each month. When the net difference is negative—meaning the contractor is spending more than it earns for that month—the contractor must borrow funds at its borrowing rate. Conversely, when the net difference is positive, it can be invested at the investment rate.

The most significant contributors to financing charges that are within the control of the owner include mobilization payment caps or schedules that drag out the payments, the absence of a specialized equipment mobilization item, and timing of progress payments after an acceptable payment application has been made. However, owners should also review the specifications regarding retained earnings. Even though many states require owners to substitute securities in lieu of cash retention, there will still be a net (albeit smaller) cost since the rate for borrowing is almost always greater than the rate of return on investments.

### Contractor Contingency

The amount of contingency that a contractor may carry for whatever risk elements it must bear responsibility for is quantified in this item. As an example of contractor contingency, many contracts require that a "normal" amount of inclement weather days based on a specified actuarial publication be built into the schedule. Since inclement weather day delays are usually considered excusable and non-compensable—meaning that the contractor can get a time extension but no reimbursement of monetary damage—bidders would be wise to examine the project schedule and have some way of determining the number of man-days for activities that experience normal inclement weather so that an estimate can be made of their impacts—for example, in call-in pay. Similarly, when subcontractors quote prices for performing work, they may not be given details of the season in which they are working, so they may exclude inclement weather days from their quote unless otherwise directed. Again, a bidder may wish to know the value of subcontract work performed during a period when a potential inclement weather day might be encountered so that an assessment of the average daily standby cost to a subcontractor is known.

### Contribution of Indirect Costs to the Bid Amount

When indirect costs are compiled as described above, they represent a sizable amount of the total bid—typically on the order of 70% of the direct costs when profit is included. This often surprises owners who may expect this ratio to be more like the language in the Changes Clause of their Standard Specifications; 15% or 20% for Overhead and Profit. However, neither the contractor nor the owner is wrong, since the Changes Clause is intended to apply to the pricing of change orders that are relatively small compared to the total contract value.

## PROJECT SCHEDULE

The foregoing discussion underscores the need for a comprehensive project schedule since so many indirect costs are time-dependent. The project schedule must also be cost-loaded, since not only does this assist in generating a cash flow schedule from which financing costs can be determined, but it also assists owners with their budgets in the timing and amounts of disbursements made via progress payments over the life of the contract.

## BASE ESTIMATE UNCERTAINTY v. ACCURACY

The lead project estimator wields a large amount of influence over cost, and the project costs estimated are influenced largely by estimating experience and judgment. The spread in bids from responsible and responsive bidders on bid day is not only a result of market factor adjustments, it is also generated by the experience and judgment of the bidders as to the level of effort required and the production rates that can be achieved in performing the work. Many refer to this as estimating accuracy, with such accuracy increasing with project definition. However, it may be better to think of this as estimating *uncertainty*, since the term *accuracy* imparts a tendency to consider low accuracy estimates of inferior quality. This should be discouraged, because the accuracy *must always* be high. Nothing should be left out except math errors, quotes should be firm and supportable, labor rates must be researched, and so forth. And yet, one can still be unsure of an accurate estimate, especially when project definition is low.

Note that this discussion is *not* about risk. The base estimate should not include risk, but only a best assessment of the cost and time required for construction of the project. Risk is more properly handled separately, since there are many more kinds of risk than construction or bid market risks.

The issue is really about *base uncertainty*: how much can things vary and still be considered reasonable? For example, an estimator may generate a cycle time analysis for tunnel excavation and support resulting in some production rate—say 40 feet per day. However, that production rate is founded on a number of conditions or parameters, such as TBM instantaneous penetration rates, that must necessarily contain some measure of uncertainty itself. When pressed to justify the 40 feet per day figure, an estimator is likely to concede that the production rate could just as easily be as high as, say, 45 feet per day (bump up the TBM instantaneous penetration rate 5% and cut the 30 minute segment build times by 5 minutes) or as low as 36 feet per day, just by varying some of the basic assumptions of the cycle time analysis by small amounts. And yet, such base uncertainties on individual direct costs can have a significant impact on total project estimated cost and duration. It would seem prudent to examine these kinds of base uncertainties for the major project cost elements to determine their overall impact to schedule and cost. This gives owners another tool to reduce subjectivity when establishing the overall project budget as a measure of confidence that the project cost will not be exceeded.

For things like labor rates, manning provisions and to a lesser extent crew sizes and subcontractor or vendor quotes are more certain because for labor rates, these are usually published. However, it should be expected that vendors will tend to not give as much thought to providing competitive quotes for engineers' estimates as they do for those of bidding contractors who will then purchase their product.

On the other hand, construction productivity is usually the biggest uncertainty in an estimate. For underground construction where subsurface conditions can have a profound influence on means and methods, uncertainty should be expected, not only in production rates, but also in whether some work, like pre-excavation grouting, is needed or not.

In general, an uncertainty analysis should be made in these major categories. For tunnel construction, these categories include production rates for tunnel or shaft excavation and support, and tunnel or shaft final lining; separate variables affecting tunnel production that are occasioned by the uncertainty in groundwater control or ground improvement, interventions, and other more geotechnical-related unknowns; and other variables that represent a more pure cost, such as a range of the anticipated amount of muck that might be classified as hazardous and therefore cost more to handle and dispose.

## SUMMARY

Having a thorough understanding of these three areas—indirect cost categories, project schedules, and estimating uncertainty—not only helps owners establish criteria for preparing cost estimates, but also guides them to the cost elements that may suggest when it is beneficial to make changes to the standard contract language. Understanding the categories of indirect costs helps owners better comprehend how contractors price their work for these costs and why these costs are so much higher than the standard allowances for overhead an markup in the Standard Specifications. It also explains why having a detailed project schedule is key because so many of these indirect costs are time dependent. Finally, understanding base uncertainty in an estimate gives owners a better idea about how confident they can be in the numbers when establishing a project budget. Supplied with these tools, owners can review a cost estimate with more confidence and greater effectiveness.

# TRACK 3: PLANNING

Bradford Townsend, Chair

*Session 5: Project Planning and Implementation II*

# Sedimentary Rock Tunnel for CSO Storage and Conveyance in Cincinnati, Ohio

## Michael Deutscher, Samer Sadek
HNTB Corporation, Boston, Massachusetts

## Roger Ward
HNTB Corporation, Cincinnati, Ohio

**ABSTRACT:** The Metropolitan Sewer District of Greater Cincinnati (MSDGC) is in the study and planning stage for the West Branch Muddy Creek Project Bundle, part of the EPA-mandated program to reduce combined sewer overflows and eliminate sanitary sewer overflows throughout the District's wastewater collection system. MSDGC is considering an alternative to construct a deep storage/conveyance tunnel with a finished diameter of up to 9.1 m (30 ft) to provide a storage capacity of up to 150 million liters (39 million gallons). The tunnel would be located at approximately 46 m (150 ft) below grade in Ordovician sedimentary rocks consisting of inter-bedded shale, siltstone and limestone. This paper addresses the design and construction issues for a possible deep storage tunnel with consideration to local geology and tunneling experience in sedimentary rock.

## INTRODUCTION

The Metropolitan Sewer District of Greater Cincinnati (MSDGC) is considering the use of a deep storage/conveyance tunnel to help achieve the performance goals of their Wet Weather Improvement Plan (WWIP) to reduce combined sewer overflows and eliminate sanitary sewer overflows as mandated by the EPA. The tunnel would receive wet weather flows from the existing Muddy Creek Pump Station (MCPS), the existing Muddy Creek Interceptor and excess flows to the Muddy Creek Waterwater Treatment Plant (MCWWTP) and store these flows until such time that capacity at the MCWWTP was available for treatment. The locations of the existing MCPS, Muddy Creek Interceptor and MCWWTP are given in Figure 1 along with the proposed alignment of the storage/conveyance tunnel. The tunnel under consideration is approximately 2,300 m (7,500 ft) long with a finished diameter of up to 9.1 m (30 ft) to provide a storage capacity of up to 150 million liters (39 million gallons).

## INITIAL SITE INVESTIGATION

An initial phase site investigation, consisting of a desk study as well as a drilling and laboratory testing program, was performed in the spring of 2009. The drilling consisted of six (6) borings: four (4) of the borings were extended to the top of rock only, while the remaining two (2) borings, located near the ends of the proposed alignment, were each advanced approximately 49 m (160 ft) into the rock. Monitoring was performed during drilling for naturally occurring combustible or toxic gas (e.g., methane ($CH_4$), hydrogen sulfide ($H_2S$), carbon monoxide (CO) and oxygen ($O_2$)). Packer testing was performed at 6 m (20 ft) intervals for the entire depth of rock core to obtain an estimate of in situ hydraulic conductivity. Soil samples and rock cores were collected from the two deeper borings for laboratory testing. Testing for soils consisted of moisture content and classification tests (Atterberg limits and grain size analysis). Testing for rocks consisted of unconfined compressive strength testing, slake durability testing, and swell testing (both free swell and null pressure swell tests).

## PROJECT GEOLOGY

The project site is situated along the demarcation between a regional lowland (elevations circa 137 m (450 ft) to 168 m (550 ft)) and upland (elevations circa 168 m (550 ft) to 259 m (850 ft)) in west Hamilton County just north and northeast of the Ohio River (HCN 2008). The proposed tunnel alignment follows a north-south route beneath Hillside Avenue (refer to

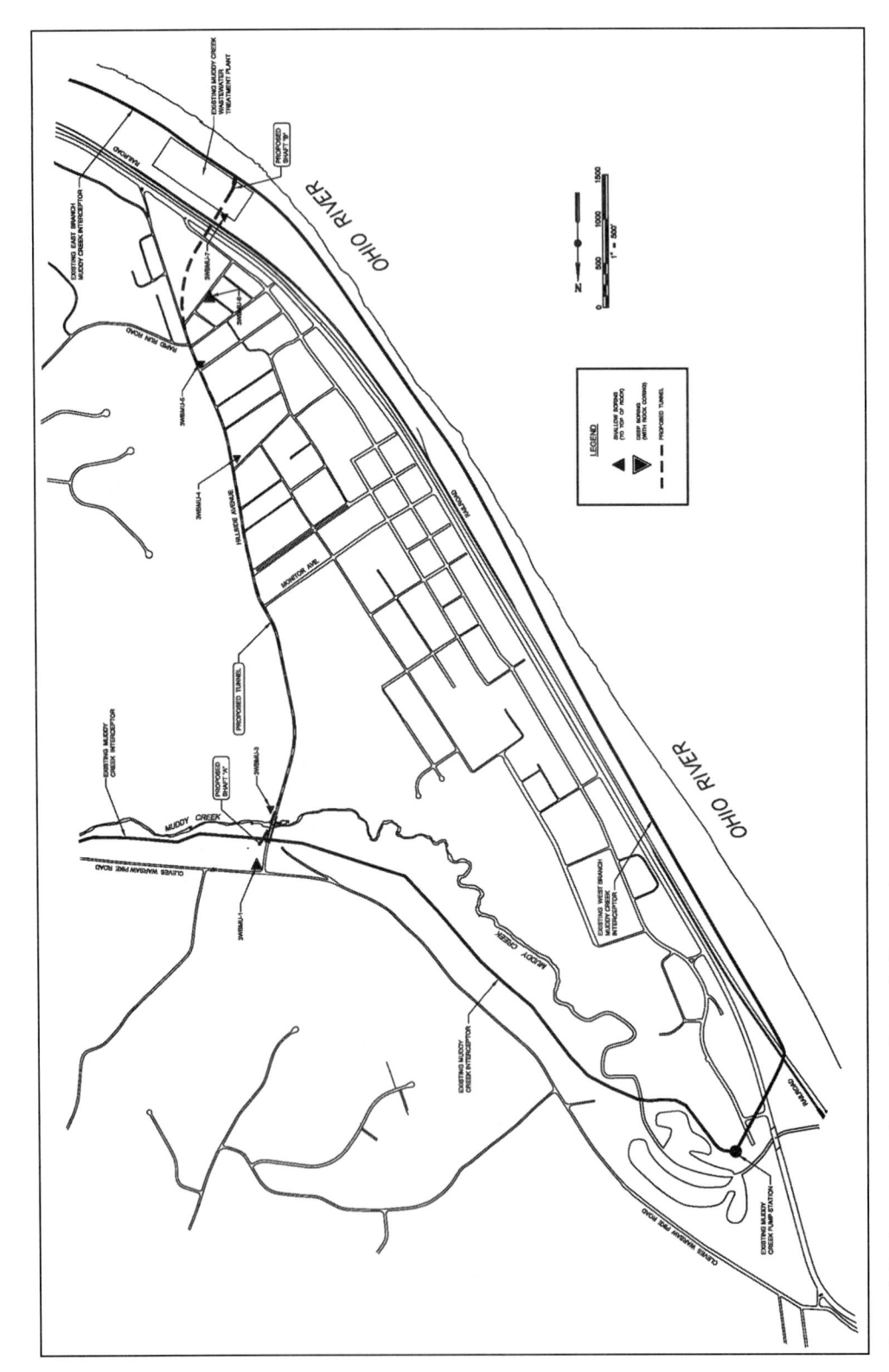

Figure 1. Project area and proposed tunnel alignment

Figure 1), which generally divides the lowland from the upland. A geological cross-section along the proposed tunnel alignment is shown in Figure 2.

## Overburden Soils

The overburden soils in the project area consist of:

- Glacial outwash (sands and gravels),
- Glaciolacustrine soils (lake bed silts and clayey silts)
- Alluvial soils
- Colluvial soils
- Fill

The soil conditions can vary over short spatial distances as result of the complex geological history of the area. The overburden soils are relatively thick (i.e., 15 m (50 ft) to 30 m (100 ft)) in the lowland area of the project site (south and west of Hillside Avenue). Conversely, the overburden soils become much thinner (only a few feet) in the upland area (north and east of Hillside Avenue).

## Groundwater

At the lower elevations of the project area, within 1.6 km (one mile) of the Ohio River, perched groundwater can be found in the glacial outwash and alluvial sands. The elevation of the water table is close to, and influenced by, the pool elevation of the river (typical elevation 137 m (455 ft), but higher during rainy periods), which in turn is regulated by the Markland dam. Farther back from the river, where the outwash sands thin and pinch out, and where the colluvial mantle increases in thickness, groundwater within the overburden soils occurs locally as perched water within alluvial silt lenses, and generally as a water table overlying the shale-dominated bedrock. In the bedrock uplands, where the overburden soil mantle is thin, groundwater occurs as a water table overlying the shale-dominated bedrock under wet conditions, and may not be encountered under drought conditions (HCN 2008).

## Bedrock

The Point Pleasant Formation bedrock, which underlies the overburden soils at the site, is an inter-bedded shale, calcareous siltstone and limestone. The Point Pleasant Formation generally consists of three zones:

- Based on published geologic mapping, the uppermost zone (the River Quarry beds) contains about 45% to 70% limestone overall, in beds ranging from 5 cm (2 inches) to 1.2 m (4 ft) thick; the shale and siltstone beds are up to 38 cm (15 inches) thick. The River Quarry beds were not encountered in the site investigation.
- The middle zone is the Bromley Shale Member, which contains up to 30% limestone in beds ranging from 1.3 cm (0.5 inches) to 25 cm (10 inches) thick; the shale beds are 5 cm (2 inches) to 41 cm (16 inches) thick. It is believed the bottom of the Bromely Shale Member was encountered during the initial phase site investigation as evidenced by the low limestone contents of the first several core runs in each of the two deeper borings.
- The lower zone contains up to 70% limestone, but typically up to 50%, in beds up to 38 cm (15 inches) thick; the shale and siltstone beds are up to 20 cm (8 inches) thick.

The Lexington Limestone Formation is located below the Point Pleasant Formation and is also subdivided into three members: the Undifferentiated Strata, the Logana Member and the Curdsville Member. Only the Undifferentiated Strata were encountered in the Initial Phase Borings (the borings were terminated above the Logana Member). The Undifferentiated Strata typically consist of approximately 70% thin to medium bedded gray crystalline limestone and 30% shale. The unit is up to 30 m (100 ft) thick and contains a 15 m (50 ft) to 18 m (60 ft) thick zone of argillaceous, fine to medium bedded, crystalline, fossiliferous limestone interbedded with thin beds of calcareous, medium strong shale. A zone noted as the Westboro K-bentonite zone is said to occur in the top 3 m (10 ft) to 12 m (40 ft) of the Member. The Westboro K-bentonite zone consists of four impure bentonite beds consisting of a mixture of altered volcanic ash and marine sediments. The beds range from 2.5 cm (1 inch) to 15 cm (6 inches) in thickness. Bentonite beds were not obvious in the Initial Phase borings, which were not gamma logged.

## TUNNEL ALIGNMENT AND PROFILE

The pre-conceptual tunnel alignment and profile are given in Figures 1 and 2, respectively, and are discussed in the sections that follow.

### Tunnel Alignment

The alignment begins at a starter shaft within a decommissioned ash-lagoon at the existing Muddy Creek Waste Water Treatment Plant (MCWWTP). The ash currently stored in the ash lagoon is not considered to be a hazardous material and has been disposed at a local landfill as recently as the late 1990s. The lagoon was constructed in the early 1970s and is not known to have a liner. The starter shaft will house intake facilities and part or all of a permanent pump station. Alternatively, the pump station could

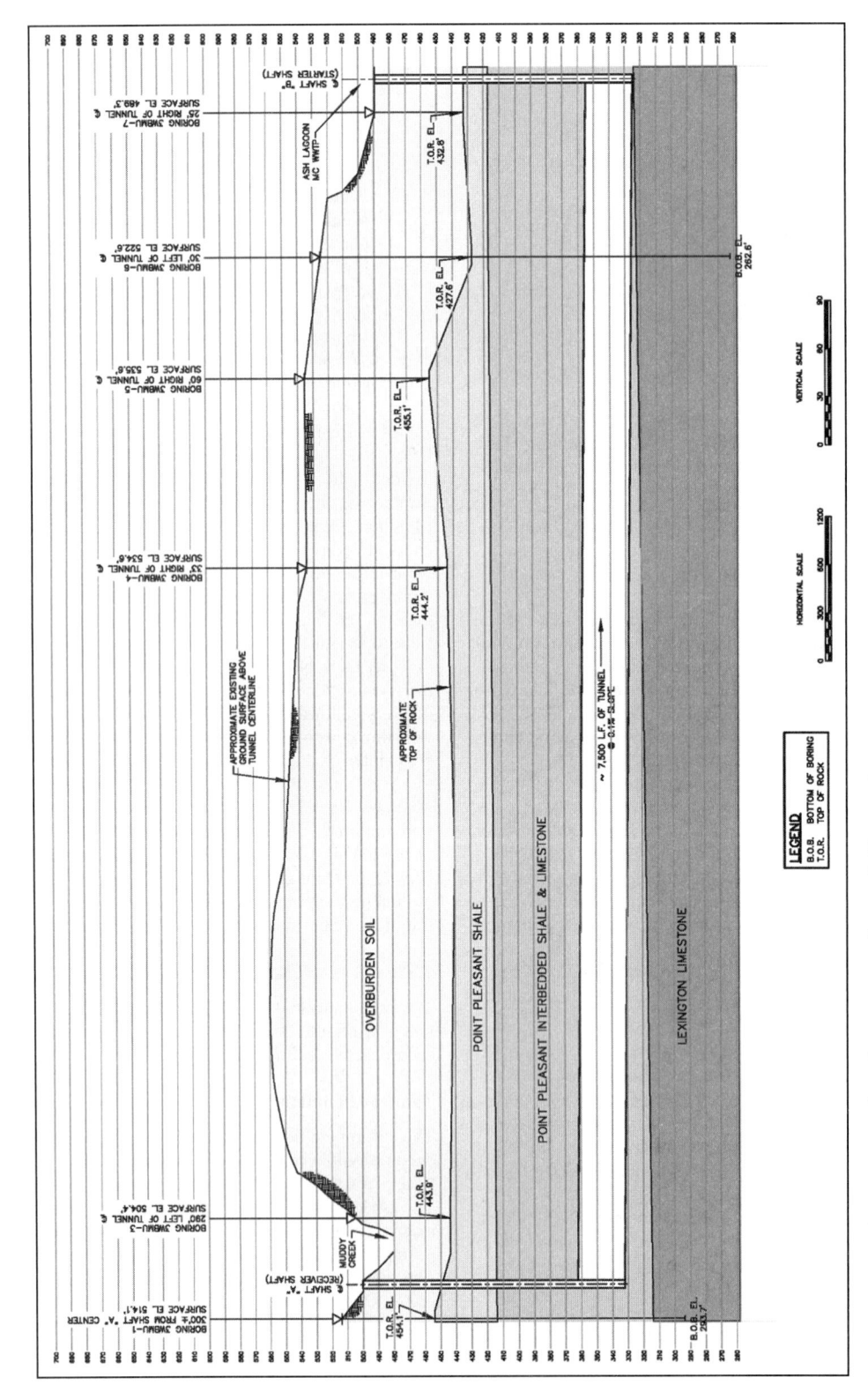

Figure 2. Geological cross-section along proposed tunnel alignment

be located in a rock cavern, which would be excavated adjacent to the shaft. The pump station could be configured to increase firm pumping capacity at the plant, if so desired.

From the starter shaft, the planned alignment of the tunnel heads in a north-northeasterly direction, proceeding in a straight line under a set of railroad tracks and US-50 before turning north-northwest to meet Hillside Avenue on a tangent near the intersection with Whipple Street. The alignment continues north-northwest, following the corridor of Hillside Avenue, to the southern bank of Muddy Creek at which point the tunnel alignment diverges from Hillside Avenue, turning eastward to a receiver shaft at the southeast corner of the intersection of Hillside Avenue and Cleves Warsaw Pike Road.

The receiver shaft would house a permanent drop structure, which would be used to drop diverted gravity flows from the Muddy Creek Interceptor as well as pressurized flows from the Muddy Creek Pump Station (via a forcemain) to the proposed tunnel. The receiver shaft site was selected because of its accessibility to main roads (haul routes), its relative isolation from local residences and its close proximity to the existing Muddy Creek Interceptor.

**Tunnel Profile**

Two conceptual tunnel profiles are under consideration:

- Point Pleasant Formation profile
- Lexington Limestone profile

Both of the above profiles are located in bedrock (i.e., rock tunnels). Shallower tunnel profiles located in overburden soils (i.e., soft ground tunnels) are not considered to be economically competitive with rock tunnels for this project. Additionally, soft ground tunnels are generally considered to carry a greater level of risk than hard rock tunnels, particularly given the highly variable geology of the overburden soils and the fact the tunnel is located in a residential area.

The Point Pleasant Formation profile is the preferred tunnel profile at the conceptual level of design. The main advantages for the Point Pleasant Formation profile, as compared to a Lexington Limestone profile, are as follows:

- Since the Point Pleasant Formation is the top-most rock formation, the depth of shafts can be minimized, which in turn minimizes the cost of shaft construction and the cost of future pumping (operational) requirements.
- The results of the initial phase site investigation indicate the permeability of the Point Pleasant Formation is lower than that of the Lexington Limestone (this finding will be confirmed during subsequent geotechnical investigation). This is a benefit both to tunnel construction (less groundwater inflow to manage) and future operations (less potential for infiltration into the finished tunnel).
- The results of the desk study, and to a lesser extent the borings drilled to date, indicate the potential for encountering hazardous gas is lower in the Point Pleasant Formation than in the underlying Lexington Limestone Formation, which is a historic producer of natural gas.
- Reduced risk of encountering bentonite beds, which are more prevalent in the Lexington Limestone than in the Point Pleasant Formation.

The main disadvantages with tunnel construction in the Point Pleasant Formation as compared to the Lexington Limestone are as follows:

- The Point Pleasant Formation has a higher percentage of weak shale layers as compared to the Lexington Limestone. As such, a tunnel excavated through the Point Pleasant Formation could be anticipated to have greater initial support requirements than a similar tunnel excavated through rock with a lower percentage of shale. Furthermore, the weakness of the rock could impact the ability of a tunnel boring machine to propel itself forward using the rock for a reaction.
- The alternation of weak shale layers with strong limestone layers, as confirmed by the Initial Phase borings and laboratory testing, could negatively impact the performance of a Tunnel Boring Machine (TBM), including reduced penetration rates, increased cutter tool wear and damage and increased difficulty in maintaining the alignment and profile of the tunnel excavation.
- Concerns of swelling and slaking behavior in the shale layers, which are more prevalent in the Point Pleasant Formation, although the results of the laboratory testing to date have reduced these concerns somewhat; six slake durability tests performed on shale specimens were found to have a have a Slake Durability Index ranging from 84.6% to 96.3% with an average of 91.0%, which corresponds to a "medium high" durability.
- Increased risk of encountering a buried glacial valley as compared to a deeper tunnel profile.

The depth of the tunnel in the Point Pleasant Formation has been selected to provide a minimum of two tunnel diameters of rock cover over the crown of the tunnel. It is possible the required rock cover may be reduced once more information is gathered regarding the variability of the top of rock elevation.

## DESIGN AND CONSTRUCTION CONSIDERATIONS

### Tunnels

A tunnel boring machine (TBM) is likely to be the most economical choice to excavate the tunnel given the significant length of the tunnel. However, other methods (drill and blast, road-headers, etc.) may be employed for short reaches to provide a starter or tail tunnel or to excavate adits or chambers for hydraulic structures such as drop structure connections or de-aeration chambers.

Construction of the tunnel is currently anticipated to use a two-pass system of excavation and support (i.e., initial (temporary) support of the tunnel concurrent with the excavation followed by installation of a final (permanent) lining upon completion of the excavation). However, this decision is subject to change based on future ground investigation and the final diameter of the tunnel.

*Initial Support Design*

The initial support requirements are dependent upon the size (diameter) and orientation of the tunnel coupled with the geological conditions. Considerations for initial support design in sedimentary rocks include:

- Construction loads
- Strength of the rock relative to in situ stress conditions
- Orientation and frequency of rock mass discontinuities (joints, faults, shears, seams, etc.)
- Permeability of the rock
- Presence of hazard gas (in the vapor phase or entrained in the groundwater)
- Squeezing and swelling potential, and
- Slaking potential

Initial support requirements for the project could range from a minimum pattern rock dowel reinforcement with C-channels and welded wire mesh to a more robust system of full-circle expanded steel ribs (with hardwood lagging) or even expandable pre-cast segmental linings (i.e., junk segments).

*Final Lining Design*

The final lining design must consider the following loads:

- Construction loads
- Ground (earth, rock) loads
- Water pressures (both internal and external),
- Squeezing pressures
- Swelling pressures

The final lining design must also give proper consideration to long-term durability against corrosion and sulfate attack. Sulfate attack of concrete is perhaps the more common problem facing sewer tunnels. Sulfate-resistant cement and pozzolans are commonly used in concrete mix designs for concrete-lined CSO storage tunnels to protect the concrete from sulfate attack. Secondary liners (PVC or HDPE) are commonly used in sewer tunnels that convey dry weather flows. However, secondary liners are not commonly used for sewer tunnels that store and/or convey wet weather flows only.

### Shafts

The methods of shaft construction must account for the significant depth of overburden soils, which include unstable soils under a significant head of water as well as the presence of over-sized material (boulders). Shaft construction will likely require the installation of vertical support in advance of the excavation (e.g., slurry wall, secant pile wall or perhaps ground freezing). Excavation of the rock is likely to be performed by drill and blast with initial support consisting of rock dowels, welded-wire mesh and shotcrete. Grouting may be required in advance of the rock excavation to reduce the anticipated ground water inflows. Final lining for shafts, where required, will consist of reinforced concrete. Water infiltration will most likely be mitigated utilizing pre-excavation and post-excavation grouting systems that will be designed to meet the operational requirements of the client, which will be established during subsequent phases of design.

### Geohazards

The list of possible geo-hazards includes:

- In-filled valleys
- Excessive faulting or broken zones of rock
- Excessive rock permeability
- Rock mass instability
- Hazardous gas
- Hydrocarbons
- Clay seams and/or bentonite beds

- Compressible soil layers above the tunnel alignment (i.e., sensitive to potential under-drainage by the tunnel)

The above geo-hazards, as well as any other geo-hazard identified by subsequent geotechnical investigation, will be investigated to the extent required for planning and diligent design. Any geo-hazard that cannot be completely mitigated by planning or design will be elevated to a risk, and documented in the project risk register (along with other risks of a non-geotechnical nature) in due course.

## FUTURE ALTERNATIVES ANALYSIS

MSDGC and their consultants are evaluating several alternatives to achieve the EPA-mandated requirements for reduction/elimination of combined/sanitary overflows. As an alternative to a large diameter, deep storage tunnel in rock, MSDGC is also considering the use of a smaller diameter conveyance tunnel, possibly with some storage capacity, in conjunction with surface storage and/or high-rate treatment at the MCWWTP. The smaller diameter conveyance tunnel would also be located in the Point Pleasant Formation and would follow a similar alignment to the deep storage/conveyance tunnel. An alternatives analysis was underway at the time of writing. All of the alternatives currently under consideration include a rock tunnel of some size.

## FUTURE GROUND INVESTIGATION

Additional drilling and laboratory testing will be performed once the alternatives selection process is advanced and the size and depth of the tunnel and the layout of tunnel components at the end points of the alignment are better established. The future investigation will be performed in two additional stages and will include up to 15 additional borings. The future drilling program will include continuous soil sampling and rock coring, in situ permeability testing and gas detection. Piezometers will be installed in selected completed borings to obtain information about groundwater levels for subsequent design. The future laboratory program will include additional testing to characterize the unconfined compressive strength, slake durability and swell potential of the relevant rock formations. Direct shear tests are planned for shale samples to establish strength parameters along bedding planes. Future laboratory testing will also include testing to establish the borability of the rock (Cerchar Abrasivity Index, Brazilian Tensile Strength, Punch Penetration, etc.), soil and pore-water chemistry and rock mineralogy.

Down-the-hole geophysical exploration (scanning) of rock bores may be performed along the tunnel alignment and at the locations of the shafts and the (possible) pump station cavern. Down-the-hole geophysical exploration is anticipated to consist of both acoustic televiewer (ATV) and gamma-ray scanning to determine the orientation/spacing/aperture of fractures and to detect the presence of clay seams/bentonite beds, respectively.

In addition to down-the-hole geophysical exploration, geophysical exploration may also be performed from the ground surface to locate buried valleys in the rock, if such valleys are suspected from the outcome of the additional drilling program. The surface geophysical exploration would consist of seismic refraction techniques.

## CONCLUSIONS

MSDGC is considering the use of a deep storage/conveyance rock tunnel to help meet the goals of their Wet-Weather Improvement Plan (WWIP) to reduce CSOs and eliminate SSOs as mandated by the EPA. Alternatives to a deep storage/conveyance rock tunnel include the use of a smaller conveyance rock tunnel in conjunction with surface storage and high-rate treatment. All of the alternatives currently under consideration include a rock tunnel of some shape or form. Key design and construction issues for a rock tunnel were identified with consideration to the local geology. Future investigation and design effort required to identify, quantify and mitigate the risks moving forward were identified.

## ACKNOWLEDGMENTS

The authors wish to acknowledge Tom Roe and Roger McClellan of HNTB Corporation, who assisted in the production of the figures in this paper. The authors also wish to acknowledge the contributions of John Nealon of Thelen Associates who executed the drilling and laboratory testing program. Finally, the authors would like to thank the Metropolitan Sewer District of Greater Cincinnati for granting their permission to publish this paper and for their valued on-going input into the design process. The authors thank Tom Crawford for his contribution to this paper.

## REFERENCES

H.C. Nutting Company, 2008. Geological and Geotechnical Overview for the Proposed Muddy Creek Basin Storage Program, Cincinnati, Ohio. Unpublished report prepared for the Metropolitan Sewer District of Greater Cincinnati.

# Tunneling to Preserve Tollgate Creek

**Michael Gilbert, Jacqueline Wesley**
CDM, Cambridge, Massachusetts

**Swirvine Nyirenda**
City of Aurora, Aurora, Colorado

## INTRODUCTION

The design for the Tollgate Creek Parallel Sewer project initially was thought to be a straight forward open cut sewer project. As the design details developed it became clear that risk management of the design issues would dictate both the design and means and methods of construction. The design development ultimately became a case study in a comparison of open cut versus microtunneling. How the key risk issues were evaluated with consideration of what the issue was, how it could be mitigated and who was in the best position to manage the mitigation are the items discussed in this paper.

### Background and Project Planning

The city of Aurora is one of the fastest growing communities in Colorado, and currently the state's third-largest city with a 2007 estimated population of 312,000. As the community continues to grow, Aurora Water continues to maintain and expand its wastewater collection system. The city has developed a prioritized capital investment program to provide a continued high level of wastewater service to its customers. Part of this commitment to service includes increasing capacity in key segments of the wastewater collection system to accommodate flows associated with growth in upstream reaches of the collection system.

With this growing population, the capacity of several key segments of the wastewater collection system must be increased. The Tollgate Interceptor conveys wastewater collected from the Tollgate Creek drainage basin to the Sand Creek Interceptor, which then conveys the wastewater to treatment at the city's Sand Creek Water Reuse Facility and the Metro Wastewater Reclamation District's Robert W. Hite (formerly "Central") Wastewater Treatment Plant (Figure 1).

A portion of the Tollgate Interceptor between East 2nd and East 11th Avenues was paralleled in the early 1990s. More recently, interceptor capacity needs have been addressed via paralleling portions of the interceptor upstream (south) of East Alameda Parkway. The remaining portions of the interceptor (identified herein as the south and north projects) require parallel pipelines.

In 2007, a collection system hydraulics model was developed to determine system needs. Based upon the results and predicted flows, the two parallel pipes were sized at 36-inch diameter and 42-inch diameter. These new pipelines are required to be completed before flow increases above the carrying capacity of the new parallel. The city reports the condition of the "old" interceptor as poor and in need of rehabilitation. Therefore, the city has requested that the new interceptor be designed to transport all flow for ten years after completion of construction, to allow the older segments to be taken completely offline for rehabilitation during that time period.

The city's main objective for this project is to develop a long-term solution for meeting the future flow capacity needs of the Tollgate Creek Interceptor from about East Alameda Parkway to the Sand Creek Interceptor tie-in. The interceptor improvements are designed to meet the flow capacity requirements through the year 2030.

The project was initially considered as a single bid package. However, the project is split by an area that was previously paralleled and routing issues differed between the south and north portions of the project. The south project recommended route was largely on city-owned property (historic Delaney Farm) and was thought to be more quickly able to bid due to less property acquisition requirements. The north project recommended route is much more complex with extensive utilities, major roadway crossings, and narrow construction corridor. In an effort to move quickly forward with a portion of the

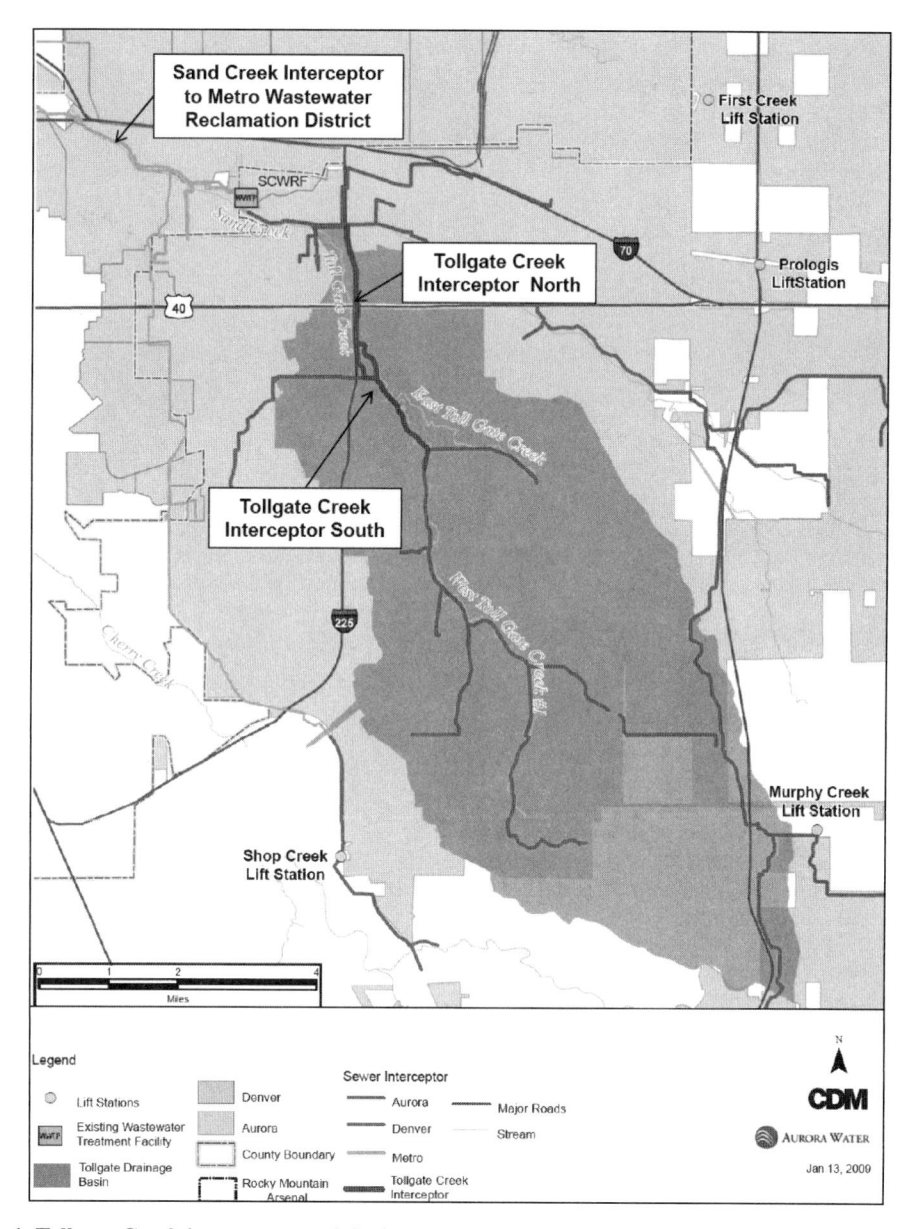

**Figure 1. Tollgate Creek interceptor and drainage basin**

parallel and spend a bond which will expire at the end of 2020, the city elected to separate the projects.

## PROJECT ROUTE DESCRIPTIONS

### South Route

The south project route passes through historic Delaney Farm, which is managed by Aurora Parks and Open Space (Figure 2). This farm is located in the center of Aurora and contains old farm buildings, an organic farm, and is the future site for a school house (to be relocated to the farm site in 2010). The Parks and Open Space department was very concerned about maintaining the character of the farm during and after construction and avoiding impact to the historic portions of the facility.

Tollgate Creek passes through the center of the Farm and must be crossed by the parallel interceptor. The invert of the new interceptor place the top of pipe close to the creek bottom and the city was

**Figure 2. Delaney Farm historic site**

concerned about protecting the pipe from scour in the creek. The creek banks on Delaney Farm also serve as native bird habitat and the city desired to protect as much habitat from construction impacts as possible.

## North Route

Habitat considerations for the north route were limited to a small area of wetlands located near the upstream end of the project. The areas surrounding the two crossings of Tollgate Creek were not determined to be sensitive habitat and presented little environmental concern. However, there are extensive conflicts for the north Project with other on-going infrastructure projects that have also been deemed by the city to be in the interest of the citizens of Aurora. The Colorado Department of Transportation (CDOT) has two active projects in the same area as the north project—expanding the I-225 and Colfax Boulevard interchange and improving access to the Fitzsimon's area and 17th Street through a new elevated bridge and road. The 17th Avenue project includes two water quality ponds which will be located above or adjacent to the proposed parallel. One of the discharge pipes from the water quality pond is in direct conflict with the new interceptor parallel. The 17th Avenue project has been released for bidding and will be in construction concurrently with the north project. Both projects pass along the same corridor on the east side of Tollgate Creek.

In addition, the Regional Transportation District (RTD) has developed plans to extend light rail service along the same corridor as the north interceptor project. A new station is planned for the area south of Colfax, which is the most congested area of the project with extensive utilities and limited access. The date for construction of the light rail extension has not been determined and is contingent upon funding; however, the city of Aurora considers the project important and requires the design of the interceptor to consider the future light rail extension by placing of manhole access and pipe construction.

In addition to conflicts with these transportation projects, most of the north route is also parallel to or crosses Xcel Energy's main transmission power lines. Large towers are located along a portion of the route. Xcel limits construction around the power poles and lines thereby further restricting the available corridor for pipeline construction activities.

## RISK ISSUES

The design evolved as a risk management issue. As part of the design process, CDM and the city of Aurora reviewed the recommended routes to consider impacts of construction.

### Project Funding

The first risk issue that was obviously an owner controlled and managed risk was funding for the project. The original basis of design was the typical open cut method. The project was initially considered as a single bid package. However, the project is split by an area that was previously paralleled and routing issues differed between the south and north portions of the project. The south project recommended route was largely on city-owned property (historic Delaney Farm) and was thought to be more quickly able to bid due to less property acquisition requirements. The north project recommended route is much more complex with extensive utilities, major roadway crossings, and narrow construction corridor. In an effort to move quickly forward with a portion of the parallel and spend a bond which will expire at the end of 2020, the city elected to separate the projects.

### Easement and Construction Work Space Risks

In addition, the Parks and Open Space Department requested the construction corridors be limited in areas of environmental or historic significance (i.e., Delaney Farm) and that the creek banks remain in their "natural" state to preserve the nesting areas along the banks of Tollgate Creek on Delaney Farm. The construction corridor was limited to no more than 75 feet in width. Access to some areas was limited and the minimum recommended construction corridor was not attainable due to existing structures, utilities, and the location of Tollgate Creek. In addition, depth-of-cut for some portions of the interceptor exceed 35 feet, impacting construction rate and methods. These considerations and limitations formed the basis from which all risk issues were judged and mitigation methods developed.

**Technical Risk Issues**

CDM and the city reviewed the plan and profile for each area of the interceptor and considered: contractor experience with the different tunneling techniques required to maintain control of the tunnel heading for the given ground conditions; access to the pipe length; removal of excavated material; and, limitations affecting construction methods. It quickly became apparent that in some areas, alternative methods of construction (over traditional open-cut) would be required to provide access to the contractor and specific methods of ground support or modification would best be utilized to successfully complete the tunneling. An evaluation of several segments also suggested that the overall construction cost would decrease with trenchless installation of the pipe. The entire south and north routes were evaluated for potential construction corridor width versus depth of cover, access restrictions, and ability to maintain the Tollgate Creek bank.

The subsurface soil and groundwater conditions and ground behavior of unsupported ground were based on the geology as briefly described below. Aurora lies near the western boundary of the Colorado Piedmont portion of the Great Plains Physiographic Province. Relief in the project area is generally low with gentle slopes. However, moderately steep to steep slopes are present along the Tollgate Creek Drainage, the Highline Canal and some roadway embankments. Locally slopes of up to 65%formed by erosional cutting of streams feeding the creek and the creek itself. Several of these steep embankments will either be crossed by the pipeline or the run parallel to them at very close distances. Much of the ground surface within the project area has been modified by human activity. Elevations range from approximately 5332 feet above sea level at Sand Creek, which is just north of the north end of the northern contract, to 5458 feet above sea level at Alameda Avenue near the south end of the southern contract.

The geology within the project area generally consists of unconsolidated deposits of Quaternary (less than 1.8 million years before present age) unconformably overlying Late Cretaceous (66 to 100 million years before present) age bedrock of the Denver Formation. The unconsolidated deposits consist of fine sand, silt and clay deposited by wind and alluvium containing clay, silt, sand and gravel that has been deposited by streams. The Denver Formation is comprised of claystone, siltstone, sandstone and conglomerate with varying degrees of cementation. Some parts of the Denver Formation contain montmorillonite clay that swells upon wetting.

No mapped faults have been reported within the project area. Nor were other potential geologic hazards, such as rockfall, landsliding, significant erosion, or abandoned coal mines observed during the various site reconnaissance trips.

The topography along the entire alignment is a result of the geologic deposition and the erosional working of Tollgate Creek. The terrain is relatively low relief except where the creek has cut into the ground leaving deep aroyoys. This erosional process is sharply defined because of the composition of soil and underlying rock. The soils primarily consist of Alluvium overlying Residuum and claystone. For tunnel evaluation and behavior characteristics the Alluvium is subdivided into fine-grained, primarily clay and silt with undrained shear strength values ranging from 500 to 3000 psf; and coarse-grained soil consisting of poorly grained silty sand to well graded sand with $\phi'$ ranging from 30° to 38° and averaging 33°. The Residuum is the result of in-place weathering and decomposition of bedrock and represents the transitional phase between rock and soil. The stratum has the appearance of rock but the engineering property values of soil. As such Residuum is generally characterized as stiff to hard, gray-brown to black or orange mottled, silty clay or fine sandy, silty clay with undrained shear strength in the range of 1500 to 4000 psf. When described as a rock, Residuum was classified as very low strength, highly to completely weathered claystone, with thin beds of fine- to medium-grained sandstone, comprising the Denver Formation. The contact between Residuum and the claystone is gradational and not usually distinctive except that the Residuum tended to exhibit some degree of yellow or orange staining/ mottling. of and the rock in most places is a soft claystone with compressive strength of 3 to 8 ksi.

Upon review of the project construction corridors and depths, the following areas were recommended for trenchless installation.

- The segment on Delaney Farm that passes under Tollgate Creek. The creek banks in

Figure 3. Tollate Creek bank near crossing on Delaney Farm

this area are steep and marginally stable (Figure 3).
- The crossings of I-225, Colfax Avenue, Chambers Avenue, and Alameda Avenue due to the heavy traffic on both these roads and the impacts that would result from closing them for open cut construction.
- The segment from 13th Street to north of the Xcel Energy substation due to construction access and the associated cost if the interceptor was installed by open cut.
- The segments from Potomac Street to the camp ground and from north of Sable Ditch to the future city park due to the proximity to Tollgate Creek and existing structures. The construction corridor in this area is very limited and access is restricted.

There are segments where becasue of the dictates of system hydraulics cover over a tunnel at stream crossings is less than half a tunnel diameter. There are also both creek and roadway crossings that have to be tunneled. Because of the generally relatively low strength of the claystone, high groundwater levels, proximity of the steep creek embankment to alignment and deep excavations a viable construction method alternative to the open cut is to microtunnel in most of the of alignment. The various issues that were considered in making this overall evaluation of open cut vs. tunneling are discussed in the following text. Because of requirement of manhole locations for inspection and maintenance reasons as well as the alignment dictates tunneling was considered to be most appropriately done by pipe-jacking. With consideration of the casing diameter microtunneling using either a slurry or earth pressure balance microtunnel machine was considered. However, use of these types of machines were not required at all crosssings and required use of them by experieced contractors would restrict the limit of qualified bidders.

**Easement Widths Required for Construction Method**

Oringially the construction easement was set at a maximum width of 75 feet and in many area were limited to 30 feet because of structures and the embankment of Tollgate Creek. This reduced easement width existed forever 2,700 feetf of the alignment by Tollgate Creek. As a result of this limited easement width all the construction equipment would be required to work on one side of the pipeline trench. This restriction would affect stockpiling of materials such as stone bedding, pipe going into trecnh excavation as well as materials to support the trench and removal of trench spoil material. A further reducion is space would be required to install a dewatering system in several segments of the alignment. This easement width limitation would dramatically affect the rate of production of open cut work and subsequently the cost.

The microtunnel alternative would mitigate this potential issue to only the immediate areas of the jacking and recieving shafts. Restrictions regarding working or driving construction equipment too close to the top of the Tollgate embankment could be imposed without a significant cost implication.

This space restriction was consider an unacceptable risk that could be mitigated by increasing the easement in several areas. It was also considered to be an issue that was in the control of the owner. As a result of discussions with the city of Aurora Parks and Open Space Department agreed to allow a wider construction corridor on Delaney Farm, thereby making open cut installation more competitive. These areas are being bid as either open cut or micro-tunnel to allow best pricing from contractors. All other segments except two crossings of Tollgate Creek can be either open cut or microtunnelled, depending upon the contractors bid and method.

**Alignment Grade**

Because the middle segment between the south and north contract limits is already in place there is very little vertical play in either contract and therefore the methods of tunneling had to consider viable methods of maintaining grade and also providing a stable tunnel heading in various ground conditions. These restrictions could have a significant impact on the ability of the local tunneling firms to show experience with the tunneling methods deemed the most appropriate when considering all of the potential crossings in the north contract.

**Creek Crossings, Highway and Major Roadway Crossings**

As described above there are a total of seven creek and roadway crossings that are mandatory tunnel segments and an additional two creek crossings that are open cut segments (due to shallow depth of cover). In addition there are areas of historical importance, limited surface construction area and utility obstruction with open cut that account for over 5,800 linear feet of pipeline. In most of these locations the method of construction was dictated by non-technical issues and therefore the method was stated as mandatory on the drawings. The reasoning behind the required method of construction—either open cut or trenchless—is addressed in the following text.

In the southern contract there are two creek crossings. In the first crossing there is only 3.0 feet of cover over the 36-inch carrier pipe and that would be further reduced if microtunneling were required and a 66-inch casing was used. Although this crossing would be in the Residuum stratum the risk of blowing the face was deemed to high and therefore open cut is required in this segment. The second water crossing has 3.6 feet of cover between stream bottom and top of steel casing. The tunnel heading is located in the claystone near the interface with the Residuum. Allowing for the potential that a mixed face of stiff clay overlying soft rock was considered a calculated risk that is worth the savings in environmental impact, time and cost. Finally there is one roadway crossing that is under a very busy stree. The tunnel depth results in a Z/D ratio of 4 in the coarse-grained Alluvium and under 0.33 bar of hydrostatic head. At this depth open cut would require sheeting and dewatering and would result in major economical impact to commuters, and the city.

As indicated in Figure 4, the remaining segments both have very solid reasons and advantages for construction by either open cut or microtunneling. There was no obvious preferred method. The design process and considerations of several risk issues were evalauted. The major items are addressed here.

**Embankments: Restoration and Erosion Over Installed Pipe**

There was concern of embankment failure. The open cut excavation would be 25 to 30 feet along most of the alignment parallel to the creek. The composition of the steep embankment slopes (1 H to 0.65 V) is composed of medium stiff low plastic silt overlying the stiff to very stiff clay. Shear strength of the Alluvium ranges from 500 to 3000 psf. The lower end of the range is the more silty material and the clay at the higher end of the range. Because of the easement width limitations weight of construction equipment would have to be restricted to a minimum distance to the top of the embankment. This restriction would then require the alignment to be moved closer to the top of the embankment to maximize surface working space. An excavation at a limited distance behind the embankment was considered a high risk of failure of the embankment. Damage to the enviroment, possibly to construction equipment and personnel and schedule impacts plus the cost of ensuing restoration efforts were considered to be significant.

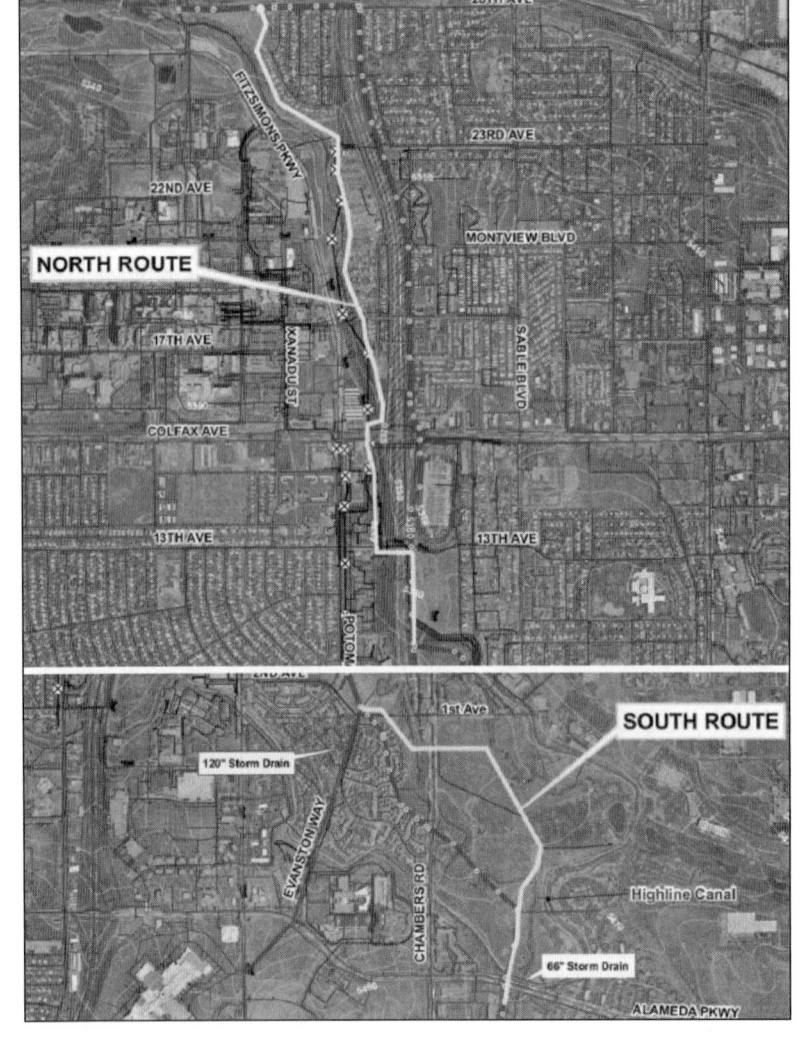

**Figure 4. Tollate Interceptor north and south**

Microtunneling work was considered to be less likely to cause slope failures provided a stable heading was maintained. Even with this method there is still one location where the microtunnel alignment will cross a streamlet to the creek with very shallow overburden, less than half a tunnel diameter. At this locations control of the stream flow could be controlled to reduce the risk of losing the face and flooding the tunnel. This remedial work would involve installing a temporary CMP with sandbags to direct flow channel flow through the pipes while the tunnel work proceeded under the CMP. Once completed these diversion works would be removed. This work would be accomplished using a crane with sufficient reach so that the natural steep slopes of the streams would be preserved.

### Dewatering and Grouting

Correlations of grain size to permeability and soil density resulted in establishing strata permeability that ranged from $5 \times 10^{-2}$ cm/sec to $5 \times 10^{-4}$ cm/sec in the upper portion of the Alluvium. In those areas of the aligment deep wells with spacing of 25 to 50 feet could be considered. In alignment segments where the pipeline is founded in the finer-grained Alluvium with $10^{-4}$ t o$10^{-6}$ would require well points. In either case there is a restriction on working space for construction equipment because of the dewatering. In addition to construction rate of progress impact there was the consideration of the effective drawdown radius of a dewatering system. This radius of influence of drawdown would have an affect on structures but was generally considered to be minor for this project.

With the microtunneling system the issue of dewatering is addressed with the closed-face support and therefore not required except at the shafts.

Crossing a roadway limits the options of ground modification to either face support with a closed-face tunnel machine or to grout the soil permeation grouting if the soil is capable of being grouted for the intended ground behavior modification.

**Affects to Structures**

The load increase to the underlying soil due to dewatering would result in some minor consolidation of the clay stratum. However of more concern for open cut work would be the lost ground due to either trench support using stacked trench boxes or the additional cost to mitigate that potential issue by requiring driven steel piles. A trench box is used for worker safety there is excess soil excavated along the trench. Because of the swelling nature of this material resulting ground movement could be excessive.

There were several segments of pipeline where the nearby structures could be affected. There are also areas where the existing utilities will require extensive relocation and shoring. With both of these issues the proximity of the construction to the structure as much as the method of construction will affect the mitigation approach. However, there is an added risk with deep excavations requiring lateral support. The city requires all sheeting to be removed. Since most of the open cut excavation is through the fine-grained Alluvium, excessive ground loss due to the soil sticking to pulled sheets could be expected. The alternative would be to use stacked trench boxes. This latter alternative is designed for the safety of personnel in the trench and not to mitigate ground loss. In either case when in the near vicinity of structures there is a higher potential of damage to the structures.

The risk of damage to nearby structures may be manageable provided the contractor using the proper equipment and good workmanship. Management of this risk is also dependent upon the owner to accept that certain methods of construction such as tunneling without a means of maintaining a stable heading and controlling the behavior of the various soil conditions or open cut with only sump pumps and double stacked trenc boxes will increase the probability of a problem occurring. Therefore the risk mitigation was considered in two parts. One, a specification that included a Geotechnical Baseline Report and limited the methods of construction in certin areas; and, two, allowing the contractor the option on the general method of construction while providing for a level playing field during the bid process.

Because either construction method was deemed viable the risk and entity in the best positon to manage the risk was considered to be the contractor and therefore the construction method was established as optional. With regards to trenchless methods microtunneling was selected over conventional tunneling because of alignment route that follows Tollgate Creek with its many turns and the need to tie into the sewer at junctions. However, there are also segments where the consequences of excessive ground movement did not impose the same economical risk and in those areas tunneling methods using pipe jacking with an open face and required ground modification primarily by dewatering was deemed an acceptable risk. Also the city wanted MH access for maintenance at no more than 600 ft spacing. The following text addresses the design and construction issues that were considered in the devlopment of the design documents.

## CONCLUSIONS

As a result of those disucssion and considerations of the various risk issues and who—owner or contractor—was in the best position to control the risk the following conclusions and resulting decisions were made.

**Funding**

The time expirtion for city bond was an immediate issue that affected both the design effort and future construction. The entity best situtated to control this risk was the the city. The method of mitigation was to separte the work into two contracts which was also done. Technically this worked nicely because of the physical layout of the project and the size of the carrier pipes.

**Easements**

Open cut too close to the Creek embankment was deemed an unacceptable risk of damage to the environment. It was also considered a risk that the Owner had control of. As a result the width of the easement was widened to allow for a more reasonable construction approach with regards to placement of construction equipment. This addressed some but not all of the concerns with embankment concerns.

**Open Cut or Microtunneling**

The alignments of south and north contracts both follow the creek. To a large extent the result is that when evaluated in terms of segments from MH to MH the optimal method of construction varies by segment. For projects like the Tollgate north interceptor, where subsurface conditions vary widely along the project alignment, there are two general approaches for development of a contractor bid package: evaluate the overall project and specify a single method of ground support, or evaluate each tunnel segment as a stand-alone construction item and allow for differing tunneling methods and ground support.

The risk was mitigated by the city by use of a GBR and specifications limiting methods of construction and requiring certain ground modification controls.

At that point, it becomes the contractor's decision on how to manage the risk of excessive ground movement by either including cost to cover mitigating measures with open cut or to change the method of construction to pipe jacking using a closed-face microtunneling system where required or in other tunneled segments construct the tunnel with an open-face pipe jacking operation with required ground modifications and manage the risk via that means and methods.

We evaluated the north project as a single contract and determined that the best tunneling method for the project in its entirety would be microtunneling. This conclusion was basically controlled by critical segments in terms of economic impact in terms of a failure due subsurface conditions that the contractor will encounter: loose soils and high groundwater that would result in a flowing behavior.

The second bid approach (allowing multiple tunneling methods and ground support) presents a higher risk to the city for claims. There is a probability that a bidder would submit a low bid that does not adequately address the need for ground modifications. For example, one mandatory tunnel segment is about 530 lf and less than 8 percent of the alignment is shown to encounter medium dense silty sand below the groundwater. This material is very susceptible to flowing, difficult to grout, and because of the underlying clay it is very difficult to lower the groundwater to the interface level when encountered along the tunnel. Because of the high cost to mobilize grouting equipment, stop the tunnel work to grout and use relatively little quantity of grout there is probable willingness by a bidder to risk not including a cost for ground modification here. This approach is in large part due to the higher unit price for grouting brought about by cost of the mobilization and limited grout quantities at this crossing.

It is our experience that because of these higher unit prices, some bidders will have a tendency to take a risk of not having to use any ground modification in segments where the ground support may be less than very obvious. When a ground condition requiring some type of soil modification is not adequately addressed, the result is usually excessive ground loss and a change condition claim is made.

## Construction Method Recommendations

The construction method recommendation is really addressing the risk of face instability along various segments of Tollgate north. CDM made recommends that the city accepted that included requiring use of a closed-face microtunneling in five segments that total 1,691 linear feet. Soil in these segments has a high potential to flow and groundwater is high.

Three tunnel or open cut segments, approximately 1,041 linear feet, are in fine-grained alluvium and have less potential for flowing ground and face instability. However, if a coarser grained alluvium is encountered, the material at the heading will behave as a fast-raveling material immediately followed by a flowing material. Required tunneling between three other segments, 1,225 linear feet is also in the alluvium, but the groundwater data indicate that the tunnel is above the groundwater level. If the tunnel heading does encounter a higher groundwater level, this material would change behavior from a slow raveling soil to a flowing soil. Conventional open faced pipe jacking can be used in all of these areas but modification of the ground (such as grouting and/or dewatering) to maintain a stable heading and tunnel face is required and design and costs for this work is included in the bid documents.

For the remaining segments designated for trenchless installation, CDM recommended and again the city accepted the method of trenchless installation be left to the contractor with the requirement of meeting the specifications (including dewatering and ground stabilization). This may result in a reduced cost for the overall bids and allows the contractors to assume the risk for method of construction. These segments are either above the groundwater level, or below the groundwater level in the residuum or claystone and are expected to exhibit slow raveling or stable conditions.

There are also five segments of pipeline totaling 3,517. lf that allow the contractor the option of either open cut or tunnel. Four of these segments will require ground modification or support for tunneling to be acceptable.

## Bidding Recommendations

To open the bidding up to as many local tunnel contractors as possible while managing the risk of bidders including the cost of ground modifications, the contract documents were set up to allow the bidders the option of method of tunneling. A set price for a designed ground modification at each crossing was included in the bid form (fixed cost for ground modification set by the city). This set cost per segment would be added to the bid price for the tunneling within each segment.

The method of ground support consists of permeation grouting or dewatering. These ground modification methods were specified based on constructability constraints. Dewater ing with well points could not be used along segments that have limited clearance for close well spacing and operation due crossing under a roadway. Where success of a dewatering system was considered a very high risk because of anticipated face conditions—mixed-face of permeable soils over lying impermeable soils grouting was required.

The option of using a closed-face microtunnel machine without ground stabilization for all of tunneled segments.

This approach opens the bidding up to local tunneling contractors. The approach also provides the city with a method to manage the risk that the contract will be awarded to a bidder with insufficient funds to perform the necessary ground modifications. And hopefully it provides some assurances to all the bidders that a risk of any one bidder low-balling a bid that could win the project and result in a future claim is mitigated and the "playing field is level."

# Tunneling Under Downtown Los Angeles

**Mohammad Jafari**
CDM, Providence, Rhode Island

**Jeffrey Woon, Osman Pekin**
CDM, Irvine, California

**Amanda Elioff**
Parsons Brinckerhoff, Los Angeles, California

**Girish Roy**
Los Angeles County Metropolitan Transportation Authority, Los Angeles, California

ABSTRACT: The Regional Connector Transit Corridor project will connect "the missing link" between Metro's existing Gold Line and Blue Line light rail transit systems through downtown Los Angeles. Three alternative alignments, approximately 1.8 miles each, are under study for the Regional Connector LRT extension. For the underground options, the construction methods being considered include cut-and-cover, tunnel boring machine, and sequential excavation method. This paper highlights various elements of the project and some challenges associated with planning and conceptual engineering for underground construction close to significant buildings, utilities and bridge foundations along Flower and Second streets in downtown Los Angeles.

## INTRODUCTION

Prompted by the recent success of the Metro Gold Line Eastside Extension (MGLEE) project and anticipated ridership needs, the Los Angeles County Metropolitan Transportation Authority (Metro) is planning new rail lines with underground alternatives, including the Regional Connector Transit Corridor project. The regional connector project will add approximately 1.8 miles (2.9km) of light rail through downtown Los Angeles and connect existing light rail lines, the existing 7th Street/Metro Center Blue Line station, and the Little Tokyo/Arts District Gold Line station. Stations along the new downtown segment will serve the city's core not easily accessible to Metro's heavy rail Red and Purple lines. This new link would enable continuous travel between Pasadena and Long Beach (37 miles/59 km) and from East Los Angeles to Culver City (15.5 miles/25 km) without transferring, and relieve congestion at the existing transfer stations—7th/Metro and Union Station. The trip between Pasadena and Long Beach currently requires two transfers, and takes approximately 115 minutes. When the regional connector is complete, this trip will be shortened to about 95 minutes (Elioff et al., 2009).

## PROJECT REVIEW AND ALTERNATIVES

The Regional Connector Transit Corridor project is located primarily within the central business district of Los Angeles. The location of the existing Metro's Red and Purple lines and the proposed regional connector alignments under planning in the downtown area is shown in Figure 1. During the alternative analysis (AA) studies and the current environmental and advanced conceptual engineering (ACE) phases, two alternatives were identified for further consideration. The alternatives are presented in Figure 1, and are named the at-grade emphasis alternative and the underground emphasis alternative. Both alternatives will extend from the existing underground Seventh Street/Metro Center Station, head northeast under Flower Street, enter the Bunker Hill area, and turn southeast along Second Street. Recently, a third alternative, the fully underground alternative was added. This variation is essentially the same as the underground emphasis alternative for the majority of the alignment west of the intersection of Second Street and Central Avenue, but it would add an underground station and extend portal structures to the northeast of Second Street. Trains would continue to travel underground northeast from under the intersection of Second Street and Central Avenue in a

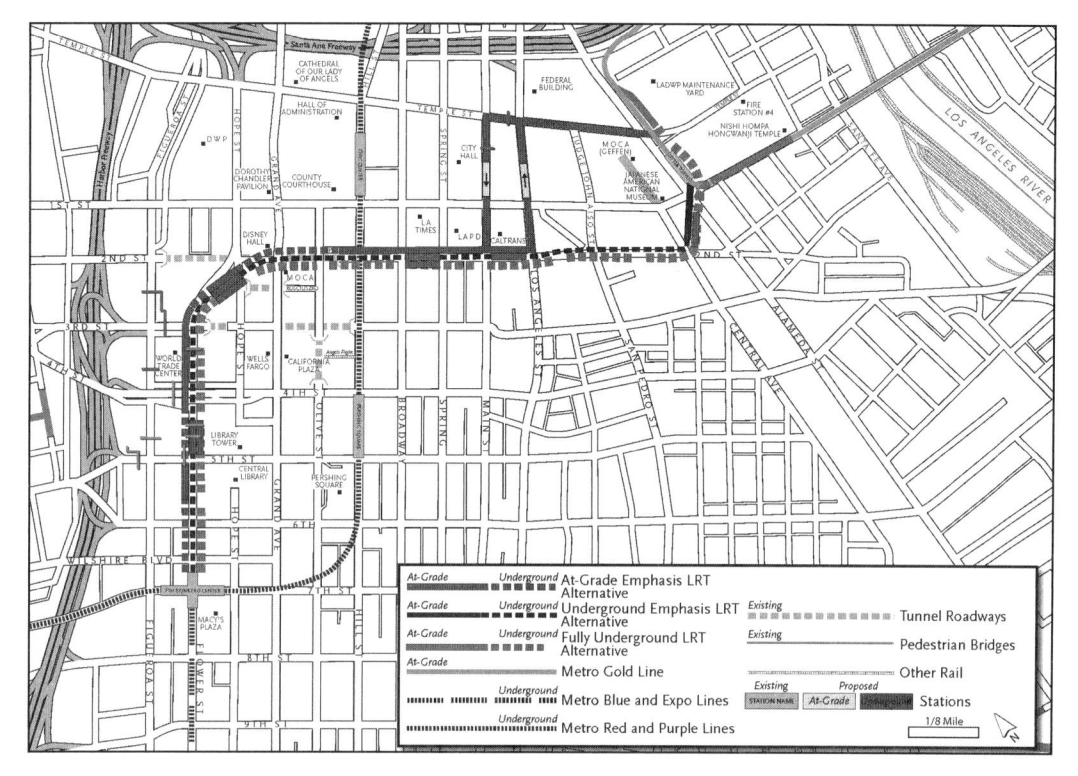

**Figure 1. Alignments of existing metro lines and alternatives under consideration**

cut-and-cover tunnel to just east of the intersection of Alameda and First streets, making First and Alameda a fully grade separated intersection. Alameda Street would remain at-grade.

Connection with the existing MGLEE would be south of the existing Little Tokyo station. The alignments of all alternatives are adjacent to major highrise buildings with underground basements that used temporary shoring and tieback systems during their original construction. The tiebacks were typically left in place and decommissioned after basement construction, in accordance with local practice in southern California. The abandoned tiebacks could be encountered along many parts of Flower Street. Tieback cables could pose a problem for a tunnel boring machine (TBM) construction. Because of the presence of tiebacks, along with a side platform station (and side-by-side track configuration) at Fifth and Flower streets, the use of cut-and-cover method is preferred over TBM use for the segment along Flower Street.

In addition to the abandoned tiebacks, the downtown area has numerous utilities, documented and undocumented, due to the age of the developed area. Major storm drains exist along Flower, Second, and Alameda streets that would require relocation or support-in-place during tunnel and station construction. A large box culvert storm drain runs under most of Second Street and it is above the tunnels or in direct conflict with the tunnels close to the portal. The existing Red Line tunnels which cross the alignment at the intersection of Hill and Second streets will likely require special consideration.

## At-Grade Emphasis Alternative

The major elements of the At-Grade Emphasis alternative consist of approximately 1,600 feet of cut-and-cover tunnel along Flower Street between the existing Seventh Street/Metro Center Station and a portal just south of the proposed Second Street/Hope Station. Construction at the Bunker Hill area would require a connection into the existing Second Street tunnel from the proposed Second Street/Hope Station area, supporting and breaking into an existing vehicular tunnel, and construction of new supports for the openings into the tunnel. At-grade stations are planned on Main and Los Angeles Streets, while a cut-and-cover underground station is planned along Flower Street, and a retained cut station is planned at Second and Hope streets. This alternative would involve the construction of an underpass and a pedestrian bridge at the intersection of Alameda and Temple Streets, and require the modification

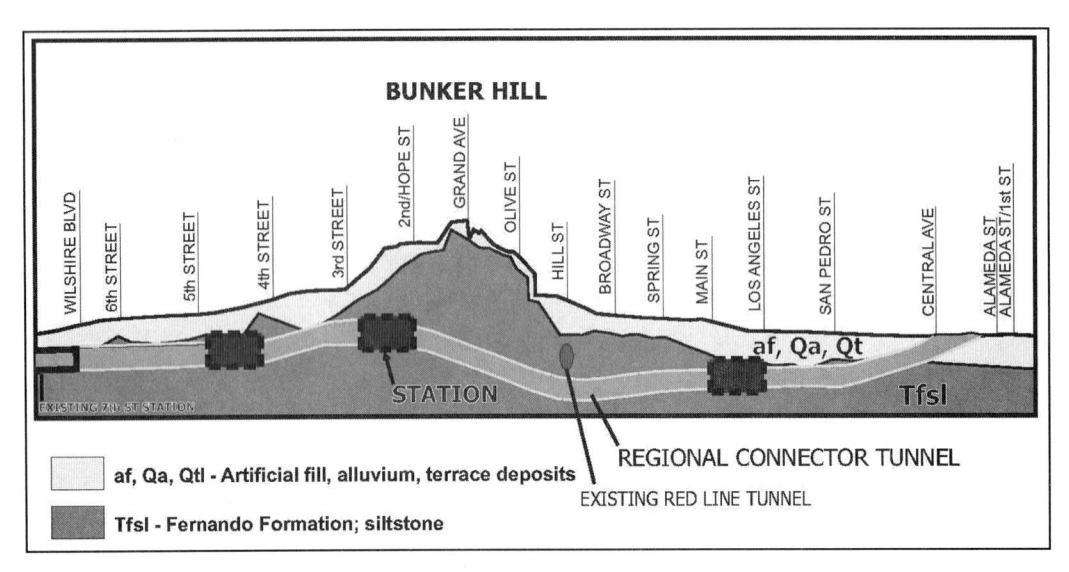

**Figure 2.** Geologic profile (underground alternatives)

of an existing mechanically stabilized earth (MSE) embankment north of the Little Tokyo/Arts District Station, on the eastside of Alameda Street.

## Underground Emphasis Alternative and Fully Underground Alternative

The underground emphasis alternative would be in a tunnel for almost the entire alignment with three underground stations and a portal just west of Alameda Street. The tunnel would be cut-and-cover construction along Flower Street and twin TBM bored tunnels along Second Street. The stations are generally planned to be constructed with the cut-and-cover method, except for the deeper station at Second /Hope streets where consideration is being given to the sequential excavation method (SEM).

A third alternative, the fully underground alternative was recently added to study an underground section under Alameda Street, with portals to meet the existing at-grade tracks along First Street north of the existing Little Tokyo/Arts District station and east of Alameda Street on First Street. This alternative would include a fourth station located within the lot bounded by First and Second streets, Central Avenue, and Alameda Street.

The Bunker Hill area is naturally elevated ground. In order to pass beneath the hill and potential underground parking structure, as well as to pass under the existing Red Line tunnels further east, the portion of the tunnels and station under Bunker Hill are being planned as deep as 120 feet (36.6m). This is more than typically practical depths for cut-and-cover construction. In addition, this section of the alignment will require a short turning radius of about 400 feet (122m) which cannot be easily accommodated using a TBM. Therefore, mining using SEM is being considered for this area. Near the eastern toe of Bunker Hill, as the alignment dips under the existing Red Line Tunnel, the proposed Second/Hope Street station will be about 110 feet (33.5m) below grade.

## GEOLOGY AND SUBSURFACE CONDITIONS

The proposed tunnel alignment along Flower to Second streets will encounter several geologic units ranging in age from Miocene to Holocene. The geologic units anticipated within the alignment, from the oldest to the youngest in geologic ages are: the Miocene-age sedimentary rock of the Puente Formation, Pliocene-age sedimentary rock of the Fernando Formation, Pleistocene-age alluvium, and Holocene-age alluvium. The Pliocene and Miocene formations are exposed in the Bunker Hill area in the northern part of downtown Los Angeles. Beyond the Bunker Hill area, these formations are overlain by Pleistocene and Holocene age alluvial sediments. Artificial fill had been placed at various locations along the alignment from previous construction and grading including areas overlying the existing tunnels. Portions of the alignment are also mapped within the city's methane risk zone. Buildings in this zone must comply with special codes to exclude and monitor gas.

A preliminary geologic profile was prepared by Mactec (the geotechnical consultant for the conceptual engineering study) and the interpretation for the underground alternative is presented in Figure 2. The interpretation of the subsurface contacts between various geologic units was obtained from new borings, and Mactec's archives that included 75 borings, drilled over a period of many years for various projects along or adjacent to the alignment.

## CONSTRUCTION METHODS

The tunnel and stations would be constructed with a number of tunneling techniques, depending on the geological and environmental conditions, cost, schedule, alignment, and other factors.

**Cut-and-Cover Construction**

This construction method entails excavating from the ground surface after a temporary excavation support system is provided to stabilize the ground and the adjacent properties. A temporary concrete decking is then placed over the cut immediately following the first lift of excavation (about 12 to 15 feet/3.7 to 4.6m), to allow traffic to pass above. Once the deck is in place, excavation and internal bracing is continued below, to the required depth. When the construction is completed within the excavated area, backfill is placed around the structural elements and the surface is restored permanently. The tunnels under Flower Street for both alternatives, as well as all underground stations, with the exception of Bunker Hill area, are planned to be constructed with the cut-and-cover method.

Depending on the depth of excavation and ground conditions, the excavation support could consist of reinforced concrete drilled-in-place piles (tangent pile wall), secant pile wall, soldier piles and lagging, or slurry walls. In some cases, sheet pile walls can also be used, but installation vibrations would probably preclude this method from consideration. These excavation support systems would be braced with internal struts or supported by tiebacks as the excavation progresses. However, right-of-way restrictions may preclude the use of tiebacks in many areas.

**Tunnel by Tunnel Boring Machine**

The basic tunneling considerations were based on ground conditions and recent success and tunneling experience of the MGLEE project. On the MGLEE project, earth pressure balance (EPB) TBMs were successfully used to bore about 1.4 miles (2.2 km) of tunnel, a similar length to the Regional Connector underground alternative. Tunneling conditions for the regional connector will need to consider both soft grounds (alluvial soils) where the profile is relatively shallow and "bedrock" conditions for deeper segments. Bedrock in the regional connector area is expected to consist predominantly of the Fernando Formation and the Puente Formation in the Bunker Hill area. Both the Fernando and Puente formations are expected to be weathered to highly weathered siltstone and clay stones. Groundwater may be perched on the Fernando Formation and in other clay layers. Gassy tunneling conditions are also anticipated based on previous studies and construction in the area. EPB TBMs are generally well suited for soft ground (sand and clays) and may also be adapted for harder materials.

The portion of the underground alternative to be bored with TBM along Second Street would likely consist of twin tunnels with outside diameters of up to about 22 feet (6.7m). The launch shaft for the TBM is planned near the east end of the project, on Second Street between Central Avenue and Alameda Street. From there, the machine would bore westward along Second Street towards the Second/Hope Street Station site, passing through the proposed Second Street station, either between Broadway and Spring streets or Main and Los Angeles streets. The TBM would be dismantled and retrieved through a vertical shaft created by cut-and-cover method adjacent to the Second/Hope Street Station. It would then be transported back to the launching shaft, reassembled, and repeat its journey for the second twin tunnel. Alternatively, two TBMs could be launched at the same time or driven from the Bunker Hill area towards the southeast.

**Sequential Excavation Method (SEM)**

Due to the depth of the Second/Hope Street Station for the underground alternative, the use of SEM is being evaluated as an alternative to the cut-and-cover method. The cut-and-cover technique may be less cost-effective due to the depth of the station in this area and the curve restriction would not permit the use of a TBM. Application of the SEM would have less surface disruption since the excavation would be performed mostly underground and accessed via vertical shafts.

If selected, this method would be the first application of the SEM in the Los Angeles area for a subway station. Generally, SEM is applied for large non-circular tunnels or short tunnels where TBMs are not economical or feasible. The SEM calls for the ground to be excavated incrementally in small areas and supported with shotcrete and steel supports advanced beyond the opening. After the crown (roof) area is excavated and supported, the larger area of the station or the tunnel can be completed. Access to the excavated opening would be required to remove excavated materials and bring in supplies. This construction technique is being considered as it may be more cost effective than cut-and-cover in the Bunker Hill area.

The sequence of excavation for the SEM method would be determined during design and, controlled and modified as needed during construction based on actual conditions encountered. In addition to the

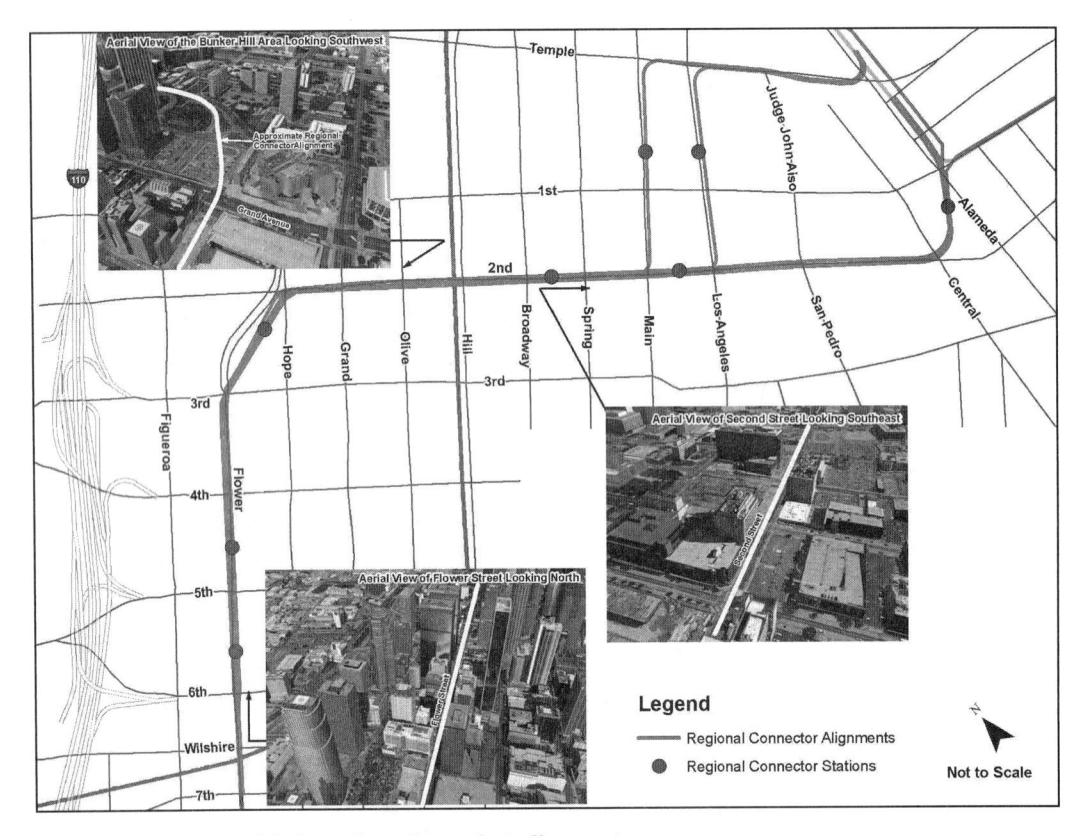

Figure 3. Plan and aerial views along the project alignment

Second/Hope Street station, SEM is also under consideration for approximately 350 feet (107m) of the curved portion of the underground alternative alignment west of the same station.

## CONSTRUCTION SCENARIOS AND ASSOCIATED CHALLENGES

### Overview of Existing Structures Along Alignment

The alignment of both alternatives had been planned to avoid existing structures. However, a majority of the existing structures on either side of the alignment on Flower and Second streets would be close to the cut-and-cover excavation sites or the bored tunnel alignment. In many areas along Flower and Second streets, there may be approximately 10 to 30 feet (3 to 9m) minimum separation between the proposed excavation and the existing building foundations.

Many of the existing structures along Flower Street have underground basements that utilized temporary shoring and tieback systems during their original construction. The abandoned tiebacks could be encountered in many parts of Flower Street and pose problem to TBM.

Along Second Street, the narrow street right-of-way and conflict with existing building foundations and utilities are major considerations for the proposed station or bored tunnels. Some of the buildings are historic, including the oldest (former) cathedral in the city and could be less tolerant to construction-induced movements.

In general, most buildings and structures are supported on foundations bearing above and to the side of the proposed tunnel and stations excavations. When buildings are within the potential influence zone of the excavation, special protective measures will be required. Figure 3 shows a plan and aerial views (from Google Earth Pro) along the alignment on Flower and Second streets. The aerial views can provide a general sense of the downtown setting and the underground construction challenges that one might expect in such an area.

### Anticipated Ground Deformation from Tunneling and Underground Excavation

One of the major concerns for underground construction is ground deformation caused by volume loss and stress relief due to excavation and tunneling.

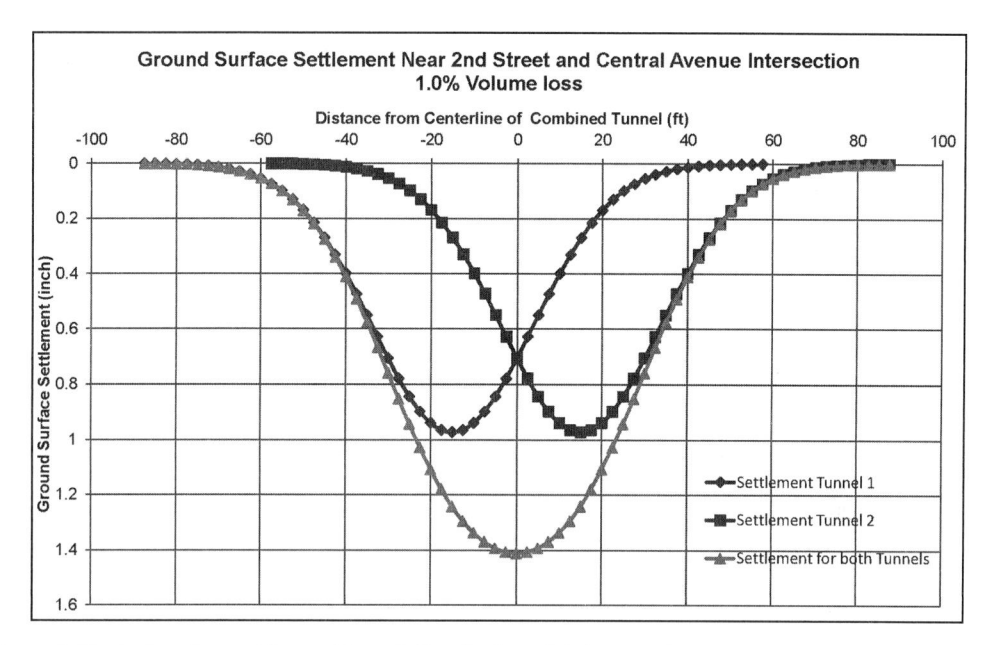

Figure 4. Typical surface settlement trough for single and twin tunnels

Ground movements are of particular significance for the regional connector project because of the urban environment, proximity of adjacent buildings, other structures (bridges, subway tunnels, etc.), and numerous utilities.

During TBM tunneling, some ground loss would likely occur in the alluvial soils, which could produce surface settlement. The amount of settlement would be a function of the sequence of excavation, amount of ground support, precast segment installation, and thickness of shotcrete support, each of which are adjusted during mining to control ground loss. Settlement prediction is typically based on a percentage of ground loss in proportion to the area of the excavated tunnel. Assuming good construction quality, the ground loss volumes when utilizing pressurized face mining techniques, such as those planned for the proposed project are generally less than 1 to 2 percent. For the MGLEE project, an estimated ground loss of 1 percent was used for estimates during design; however much less ground loss (less than measurable) was realized (Choueiry et. al., 2007).

Preliminary settlement calculations were performed at several locations along the underground alternative in accordance with Peck's (1969) method. Settlements were estimated based on assumed ground loss values of 1 to 1.5 percent. To account for twin tunnels, settlements above each tunnel were estimated separately and then superimposed as illustrated in Figure 4. Note this figure is based on an assumption of 1 percent volume loss with corresponding settlement estimates of about 1 inch (25mm) above the center of a single tunnel due to its excavation only, and about 1.4 inches (36mm) above the center of both tunnels due to the excavation of both tunnels.

On this project, the footprints of all buildings are located at some distance from the tunnel centerline and thus the ground surface settlement at the building line would be smaller as can be seen in Figure 4. For example, at a lateral distance of 30 feet (9m), the estimated settlements are about half the corresponding values above the centerline.

Underground excavation for stations and tunnels using the cut-and-cover technique would also result in ground relaxation and deformation of the retained earth. The magnitude of ground movement for that case will depend on the strength of the earth materials, groundwater conditions, building surcharge, and the rigidity of the shoring system. For cut-and-cover excavation, the zone susceptible to ground movement generally extends a lateral distance of approximately one to one-and-the-half times the depth of the excavation. Accordingly, structures located within this settlement/deformation zone would be further evaluated for potential impact.

## PROTECTION OF STRUCTURES AND POTENTIAL MITIGATION METHODS

Maintaining the integrity and protection of existing facilities are of utmost importance due to the proximity of structures to the proposed alignment. The need

Table 1. Typical values for maximum building/ground slope or settlement for damage risk assessment

| Risk Category | Deformation and/or Tilt Criteria | | Description of Risk |
| --- | --- | --- | --- |
| | Rankin (1988) | Boscardin and Cording (1989) | |
| 1 | <1/500<br><0.4" | <1/600 | Caution. Continue monitoring.<br>(Negligible to very slight: damage unlikely) |
| 2 | 1/500–1/200<br>0.4"–2.0" | 1/600–1/300 | Consider and apply mitigative measures (grouting etc.) as appropriate. Continue monitoring.<br>(Slight: possible superficial damage; unlikely to have structural significance) |
| 3 | 1/200–1/50<br>2.0"–3.0" | 1/300 to 1/150 | Plan for mitigative measures in advance and have in place prior to tunneling. Continue monitoring.<br>(Moderate to severe: expected superficial damage; possible structural damage to buildings; possible damage to relatively rigid pipelines) |
| 4 | >1/50<br>>3" | >1/150 | Plan for mitigative measures in advance and have in place prior to tunneling. Continue monitoring.<br>(High to very severe: expected structural damage to buildings; expected damage to rigid pipelines; possible damage to other pipelines) |

for protection of the existing structures is a function of the anticipated ground movement due to the proposed excavation. Depending on the magnitude of ground deformation and the deformation tolerance of these structures, mitigation such as underpinning or ground improvement could be required.

Risk levels for structural deformation ranges have been developed by various researchers, including Rankin (1988) and Boscardin and Cording (1989). Boscardin and Cording (1989) criterion was used during construction of the MGLEE project (Choueiry, et. al., 2007). Table 1 provides a comparison of these two criteria.

A specific criterion has not been adopted at this study phase, but risk categories 3 and 4 listed above are considered unacceptable. The criteria for tolerable movements depend on various factors including the type, age, and significance of the buildings in question. Evaluations during future phases will help determine the appropriate levels of monitoring, protection, and mitigation measures required during construction.

## Geotechnical Instrumentation and Monitoring

As part of the construction mitigation program, survey of structures within the anticipated zone of construction influence will be done prior to construction, to establish baseline conditions. Geotechnical instrumentation program would be designed to provide essential information to monitor vertical and where needed, lateral ground movements, and groundwater levels. Data gathered during construction would then be compared to baseline conditions and the agreed upon trigger levels, so that appropriate actions could be taken.

## Potential Mitigation Methods

To reduce surface settlement and the potential for ground loss and soil instability (i.e., sloughing, caving) at the tunnel face due to tunneling, pressure-face TBMs and pre-cast, bolted, gasketed lining systems would be employed. In combination with the face pressure, grout is injected immediately behind the TBM between the installed precast concrete liners (tunnel rings) and the ground. The pressure-face TBM can tunnel below the groundwater table without requiring dewatering or lowering of the groundwater table.

Following is a brief summary of the protective methods against TBM-induced settlement that have been considered during conceptual development.

### Grout Stabilization

Ground stabilization can be achieved by permeation grouting, compaction grouting, or compensation grouting. Grouting operations can be conducted either from the ground surface or from the tunnel face. The specific details of each grouting method will be further evaluated during the design phase of the project. To grout from inside the tunnel, the TBM would need to be stopped for grouting operations. Some disruption, traffic control, and lane closure would be involved for grouting from the street prior to the TBM reaches the area in question.

Permeation grouting uses a sodium silicate or cement injection method. This has been used successfully for the Metro Red Line in instances where the tunnel passed under potentially sensitive or important structures, such as the US 101 freeway (three locations: downtown, Hollywood, and at Universal City).

Compaction grouting uses a stiff mix, typically sand with small amounts of cement above the tunnel crown as the tunnel advances. This method was used in several instances for the Metro Red Line project, including the downtown Los Angeles area and along portions of Hollywood Boulevard.

Compensation grouting involves injection of grout between the intended tunnel position and the structures requiring protection, in advance of tunneling. Monitoring of both structure and ground movements would be used to optimize the grouting operations. Grout pipes are generally reused to inject grout before, during, and after tunneling as needed.

*Protective Soldier Piles*

In areas where the estimated settlement trough could extend into the adjacent building foundations, a line of soldier beams can be installed on one or both sides of the tunnel prior to tunnel advancing to the subject area. These would be somewhat a closely spaced row of cast-in-drilled-hole (CIDH) piles that would serve to truncate the settlement trough. Along Second Street, this could provide an attractive alternative to the grouting methods because it can be done without temporarily taking up the center of the street or entering the buildings or the basements. Also, soldier piles and lagging type shoring systems may likely be used for the removal of the existing box culvert under Second Street. Accordingly, it may be possible to leave the same soldier beams in place to control and truncate the TBM-induced settlement trough at a lower cost.

*Underpinning*

Underpinning involves supporting the foundations of an existing building by carrying its load bearing element to deeper levels than its previous configuration. This helps protect the building from settlement that may be caused by nearby excavation. This can be accomplished by providing deeper piles adjacent to or directly under the existing foundation and transferring the building foundation loads onto the new system. There are a few cases where existing buildings are very close to the alignment and underpinning may be needed.

## CONCLUSIONS

The Regional Connector Transit Corridor project will face design and construction challenges typically associated with a highly developed downtown urban environment. The critical issues to be dealt with include limited right-of-way, protection of existing buildings, potential obstructions, utility relocations, and considerations of construction methods to minimize community impact. Additional engineering analysis, survey, and geotechnical exploration will be required to further characterize the subsurface conditions for final design. New technology to the local area, including SEM, could be used. The recent success and experience of the MGLEE project will be utilized in the future design and construction of this project.

## ACKNOWLEDGMENT

Particular thanks are due to Pierre Romo (MACTEC) for providing the geologic profile used in the paper, and John Newby (CDM) and Kimberly Scheller (CDM) for review and assistance in preparing this paper.

The consultant team for the Regional Connector Transit Corridor project was led by CDM, with Parsons Brinckerhoff and MACTEC as the engineering and geotechnical subconsultants, respectively.

## REFERENCES

Boscardin, M.D. and Cording, E.J. *"Building Response to Excavation-Induced Settlement,"* ASCE Journal of Geotechnical Engineering, Vol. 115, No. 1, January 1989.

Choueiry, E., Elioff, A., Richards, J. and Robinson, B. (2007), *"Planning and Construction of the Metro Goldline Eastside Extension Tunnels, Los Angeles, California,"* 2007 Rapid Excavation and Tunneling Conference Proceedings.

Elioff, A., Perry, D., Roy, G., and Romo, P., (2009), *"Planning New Metro Subways—Los Angeles, California,"* 2009 Rapid Excavation and Tunneling Conference Proceedings.

Lee, C., Wu, B. and Chiou, S. (1999), *"Soil Movements Around a Tunnel in Soft Soils,"* Proc. Nat'l. Sci. Council, ROC(A) Vol. 23, No. 2, pp. 235–247

Peck, R.B. (1969), *"Deep Excavations and Tunneling In Soft Ground,"* Proceedings of the 7th International Conference on Soil Mechanics and Foundation Engineering, Mexico City, pp. 225–290

Rankin, W.J. (1988), *"Ground Movements Resulting From Urban Tunnelling: Predictions and Effects,"* from Engineering Geology of Underground Movements (Bell et al. eds), Geological Society Engineering geology Special Publication No. 5, pp. 79–92.

Robinson, B. and Bragard, C. (2007), *"Los Angeles Metro Gold Line Eastside Extension—Tunnel Construction Case History,"* 2007 Rapid Excavation and Tunneling Conference Proceedings.

# Selecting an Alignment for the Blacklick Creek Sanitary Interceptor Sewer Tunnel—Columbus, Ohio

**Valerie R. Rebar, Heather M. Ivory**
URS Corporation, Columbus, Ohio

**ABSTRACT:** Several alignment alternatives were developed during preliminary design of the Blacklick Creek Sanitary Interceptor Sewer. The local geology varies greatly throughout each alternative considered, with paleovalleys of soft ground containing sands, gravels, cobbles, and boulders with transition zones to hardpan clay, weathered and non-weathered shales, siltstones and sandstones. The range of subsurface conditions within any single alignment presents more than ten transition zones requiring different construction methods, materials, final linings, ancillary items, and other considerations. This paper examines the process used to compare alternatives leading ultimately to the selection of an optimal tunnel alignment in this challenging ground.

## PROJECT OBJECTIVES AND BACKGROUND

Upon the development of the City of Columbus, Ohio Master Plan, it was determined that the Blacklick Creek Sanitary Interceptor Sewer would be necessary to support the anticipated development of the area. The nominal sewer diameter would be 66 inches; however, the excavated diameter would depend on the selected constructed method and materials. The primary sewershed to be serviced by the project is the Blacklick Creek Service Area; although, considerations were made to accept additional flow volumes (either full or partial flows) from the Rocky Fork Service Area during the development of alignment alternatives. When the Big Walnut Trunk Sewer was constructed (see Figure 1), it had not been designed to serve most of the Rocky Fork Service Area; however, sewage from this area is currently being directed to the Big Walnut Trunk Sewer. Since this flow is using the existing sewer's capacity development in the area is restricted. Consequently, in the future, the City of Columbus plans to construct the Rocky Fork Pump Station that will redirect flow from the Rocky Fork Service Area to the Blacklick Creek Sanitary Interceptor Sewer.

From 2000 to 2003, 3,550 linear feet of the Blacklick Creek Sanitary Interceptor Sewer was fully designed for construction by microtunneling methods. Two contractors bid the project, both of which significantly exceeded the project budget (Budgetary Estimate—$9.0 million, Bid Estimates—$15.4 million and $18.8 million). Consequently, the City canceled the project indefinitely.

Since additional sanitary sewer service remained necessary for future development of the Blacklick Creek area, in 2007 the City of Columbus hired another design team to reinvestigate the previous design and reassess the available options. This preliminary redesign of the Blacklick Creek Sanitary Interceptor Sewer would extend an existing 66-inch gravity sewer at its southernmost terminus 24,000 LF (4.5 miles) to some undetermined location on Morse Road.

Numerous constraints, including crossing the Blacklick Creek, heavily trafficked suburban area, and deep inverts indicated that tunneling construction methods would prove to be the best option. However, the challenging glacial geology made selecting the alignment (both horizontal and vertical) as well as construction method an arduous task. The following process was utilized to compare various alternatives and select an optimal tunnel alignment to be used for final design of this sanitary sewer.

The local geology varies greatly throughout the project area. The soil in the project area consists of glacial till and outwash, with cobbles and boulders. Below the soil, the bedrock is composed of layers of Cuyahoga, Sunbury, Bedford, and Ohio Shales as well as the Berea Sandstone. The elevation of the top of bedrock varies significantly within the projected area with numerous bedrock highs and buried valleys. This bedrock topography suggests that the final tunnel alignment will pass through areas of rock, soil and transition zones between rock and soil as shown in Figure 2. In addition to these bedrock valleys along a north-south trend, there is also a bedrock valley to the east of that appears to be a major carrier of

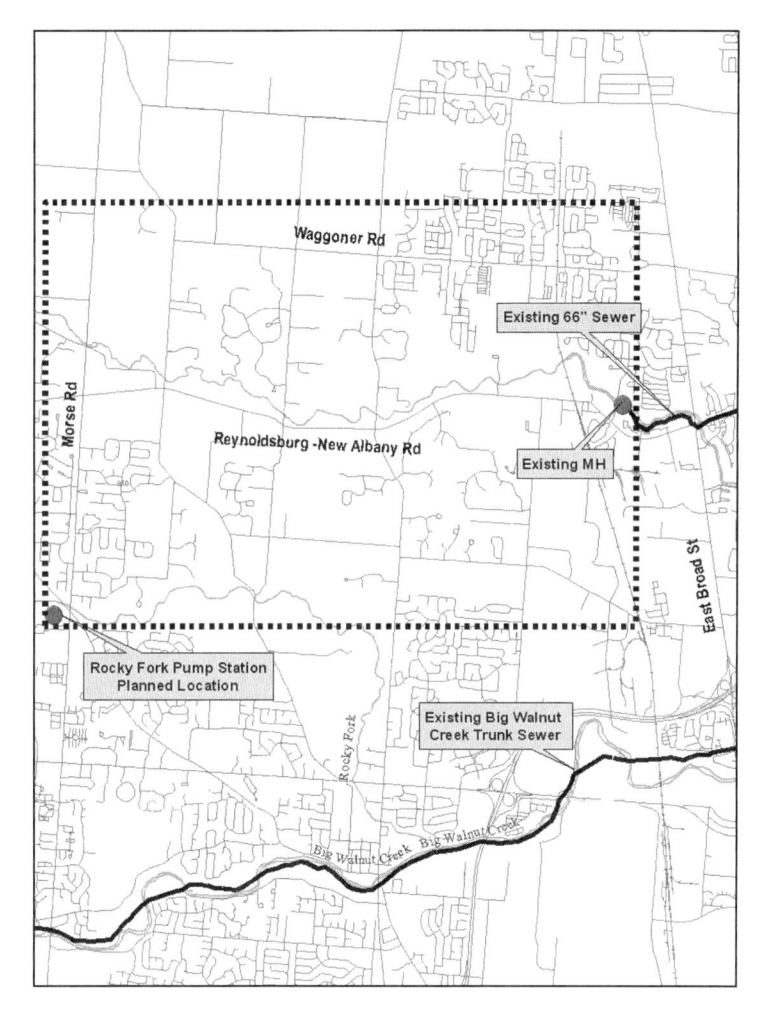

**Figure 1. Overall project area**

groundwater. This valley is peculiar in that the bedrock valley wall may be a sharp declined or a moderate one. The valley wall may also be extremely close to the alignment or far enough away to minimize impacts to construction. The first step in the process of developing a successful project would be to gain a thorough geotechnical understanding and anticipated ground conditions throughout the project area.

## HORIZONTAL ALIGNMENT CORRIDORS

Determining the subsurface conditions and extent of the mixed face and soil encountered is imperative to design and construction as well as the cost implications that would arise if the extents are much longer or wider than anticipated through the geotechnical investigation. First, the overall project area (Figure 1) was narrowed down into horizontal corridor alignment areas where soil borings would be drilled. These horizontal corridors are shown in Figure 3. Horizontal alignment corridors were developed in order to try to minimize anticipated construction issues that could occur because of the changing geology along a possible alignment. Variations within the corridors were possible as the first step in the process was to focus the alignment alternatives into generalized areas. Once a horizontal alignment corridor could be selected, then various vertical alignments within that corridor, each having individual feasible construction methods dependant on the anticipated geology, could be investigated.

The design team looked outside of the scoped terminus location in an attempt to create the best overall design for the sewer tunnel construction as well as operation and maintenance of the sewer system. Extending the alignment corridors an additional 10,000 feet beyond Morse Road to the site of the future Rocky Fork Pump Station (Rocky Fork Extension) was considered. The City plans to connect the Blacklick Creek Sanitary Interceptor Sewer to the Rocky Fork Pump Station within the next 10 years, and the design team knew that constructing one longer tunnel all at once is much more cost effective than mobilizing twice to construct two shorter tunnels. Continuing the sewer to this location would also allow immediate relief to the Big Walnut Trunk Sewer of flows from the Rocky Fork Service Area. Consequently, each corridor was extended the original project length in order to investigate the feasibility of connecting the Blacklick Creek Sanitary Interceptor Sewer to the proposed Rocky Fork Pump Station.

The Reynoldsburg-New Albany Road (RNA) Alignment Corridor follows the alignment recommended in the previous design along Reynoldsburg-New Albany Road but was anticipated to encounter multiple bedrock highs and valleys. The Waggoner Road Alignment Corridor is almost twice as long as the RNA Alignment Corridor. However the Waggoner Road Alignment was assumed to be located in two distinct sections, one in soil and one in rock; consequently reducing the amount of mixed-face conditions and ultimately significantly reducing the construction cost per foot. The modified Waggoner Road Alignment Corridor follows the same geology as the Waggoner Road Alignment Corridor but avoids municipal water well fields near the southern portion of the Waggoner Road Alignment.

## DYNAMIC DRILLING PROGRAM

The geotechnical investigation for the proposed sewer was dynamic in nature in that a formal and predefined drilling plan was abandoned in lieu of one that was accommodating to change, allowing flexibility in boring locations, depths, and data obtained. The preliminary geologic investigation for this project concentrated on determining the location

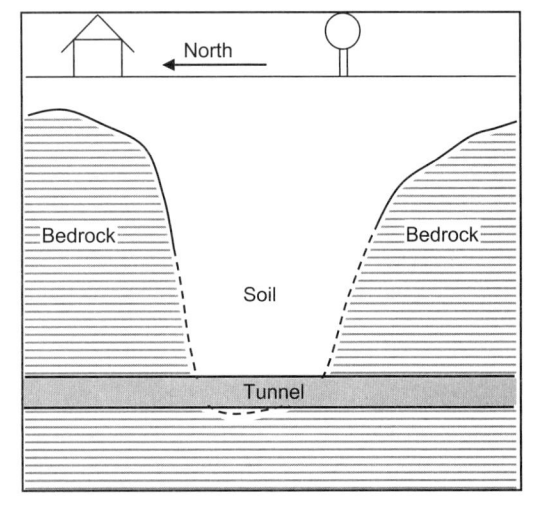

**Figure 2. Top of bedrock elevation varies significantly**

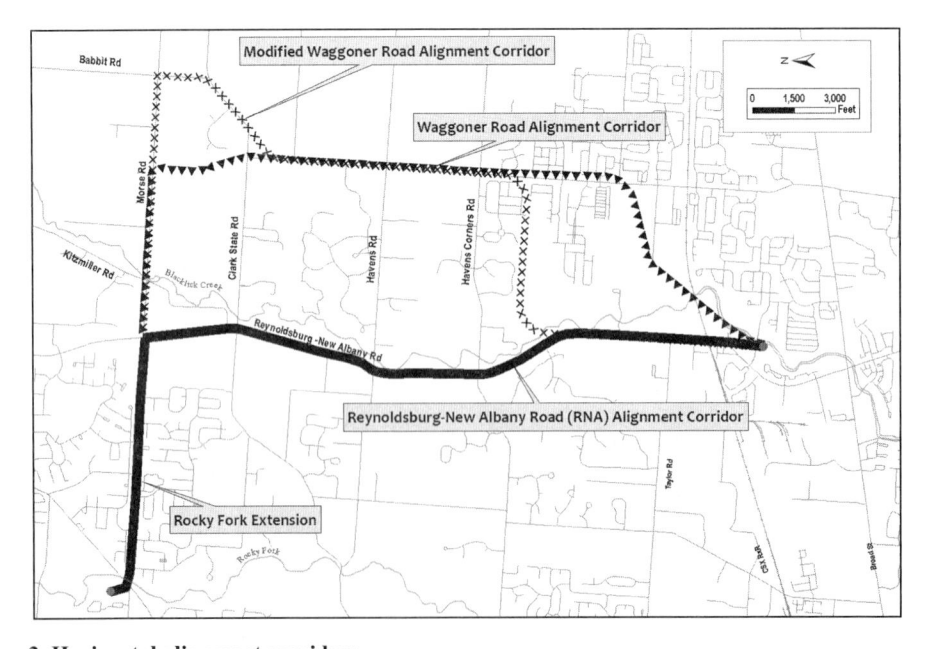

**Figure 3. Horizontal alignment corridors**

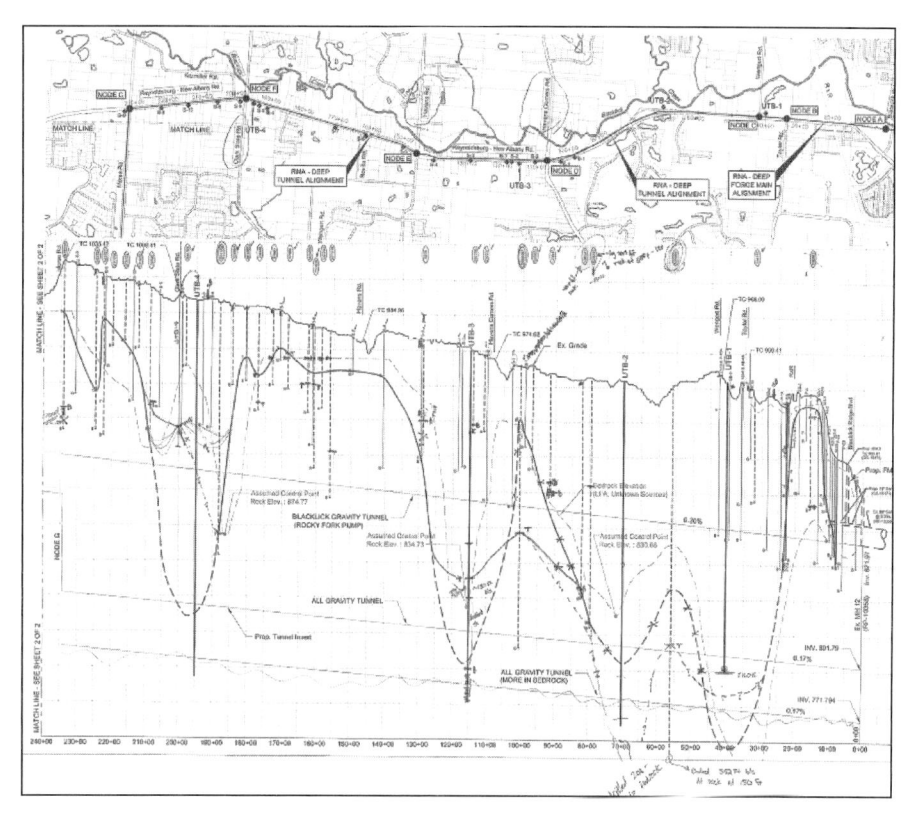

Figure 4. Initial RNA horizontal alignment profile and revisions from first phase of geotechnical borings

and extent of the buried valleys and soil and rock formations and characteristics thereof. This would provide detailed information for more accurate horizontal and vertical alignment selection, construction method evaluation, material selection and cost estimation. The preliminary borings were focused in the portions of the alignment corridors as defined by the project scope. During the boring process, real time data from the field was utilized for revising the subsurface investigation plan, which included relocating or eliminating unnecessary borings and adjusting planned boring depths. After each set of geotechnical borings was completed, the bedrock profiles were adjusted accordingly. The following figures illustrate the extent that the bedrock profile was revised as more data was acquired throughout the drilling program.

Figure 4 shows the initial RNA Alignment Corridor profile based on existing data. During the planning of the first phase of geotechnical borings, large bedrock valleys were assumed to be present in areas where no bedrock had been previously identified (shown as dashed lines). In order to verify or disprove this assumption, borings were drilled in these valleys to determine the precise bedrock surface profile as well as to identify the depth and general size of these deep buried valleys for tunnel design. As was anticipated, several bedrock highs and valleys were present along the alignment. Figure 4 shows how the bedrock profile was adjusted as the first phase of borings were drilled. These borings were the first step to identifying the actual size and depth of the bedrock valleys. They provided a solid foundation to locate the next phase of borings.

Figure 5 depicts the geological conditions initially anticipated for Waggoner Road Horizontal Alignment and the Modified Waggoner Road Alignment based on existing data. Since these alignments are slight variations of each other, initial borings for each were in the same location; from this point forward, these alignments are referred to collectively as the Waggoner Road Corridor. The reasoning behind this corridor alternative was to have a soil tunnel in the large valley that spanned the alignment and to have a rock tunnel in the northern section as was anticipated based on existing data as shown in the Figure. Even though this alignment corridor is much longer than the RNA corridor, it was

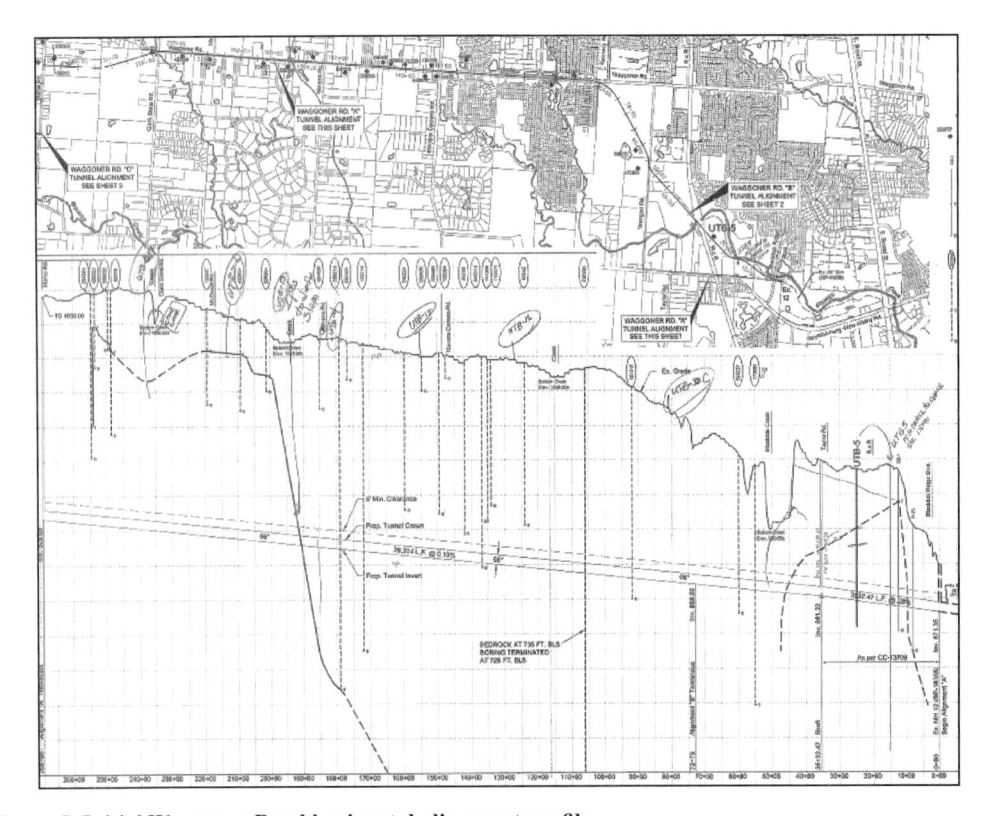

**Figure 5. Initial Waggoner Road horizontal alignment profile**

initially believed to be more cost effective because it would have far less soil/rock transitions than the RNA Alignment.

When Figure 5 is compared with Figure 6, it becomes obvious that the actual bedrock profile in this corridor is similar to that in the RNA Corridor with each having many bedrock highs and valleys. This alignment corridor was deemed undesirable because of its length being around twice as long as RNA Alignment with the same geological challenges and rock/soil transitions as the RNA alignment. These borings indicated that the two alignment alternatives would have similar costs per foot. Since the Waggoner Road Corridor is almost twice the length as the RNA Alignment Corridor, it was concluded that the construction cost in the Waggoner Road Corridor would be around double that in the RNA Alignment. Therefore, further geotechnical investigations for the Waggoner Road Corridor were canceled when it became evident that this alternative would not be cost effective when compared with the RNA Alignment.

Since the second phase of borings in the Waggoner Road Corridor was canceled once the alternative became impractical, the nature of the dynamic boring plan allowed the borings that would have been drilled within that corridor to be moved to the RNA corridor. This allowed the design team to focus on the horizontal alignment that was more favorable, including the area that would be included if the alignment were extended an additional 10,000 feet to the proposed Rocky Fork Pump Station (Rocky Fork Extension).

Although the process for adjusting the investigation plan initially increased the schedule duration of the drilling phase, it ensured better knowledge of the subsurface conditions and the efficiency for ranking proposed tunnel alignments. As a result, lessons learned through using this dynamic process will help to plan the next phase of borings during detailed design more effectively.

## VERTICAL ALIGNMENT ALTERNATIVES WITHIN SELECTED HORIZONTAL CORRIDOR

The dynamic nature of the geotechnical drilling program allowed all but one horizontal alignment to be eliminated fairly quickly. The RNA horizontal alignment was carried forward beyond an initial screening due to the impractical nature of the majority of the alternatives as well as the geologic conditions encountered during drilling. Due to the bedrock

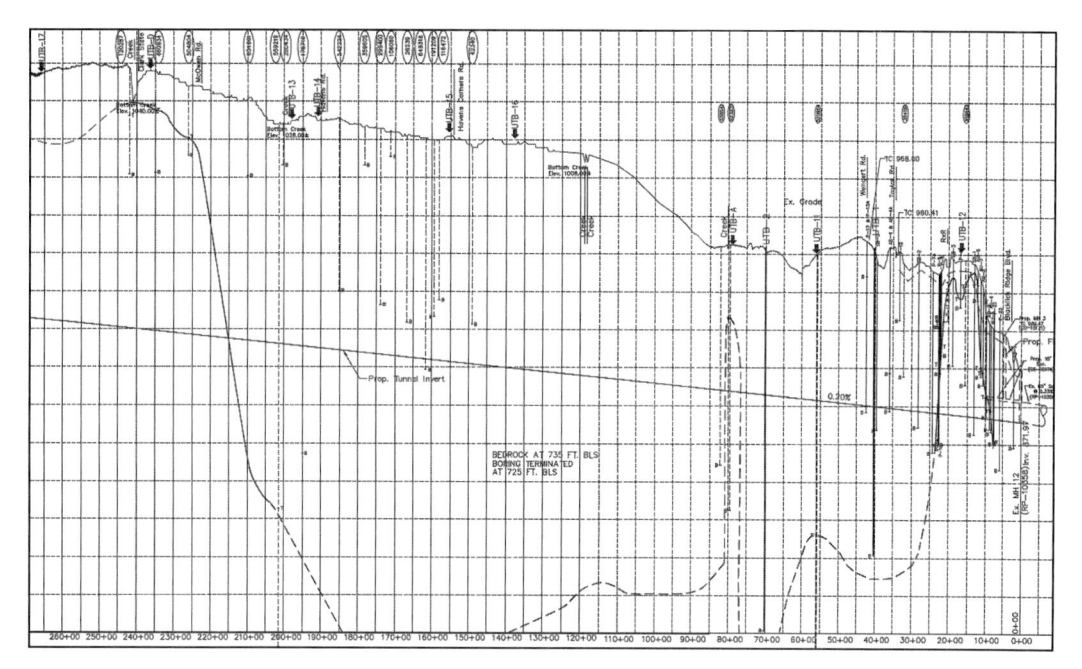

Figure 6. Waggoner Road Alignment Corridor revised profile following first phase of geotechnical borings

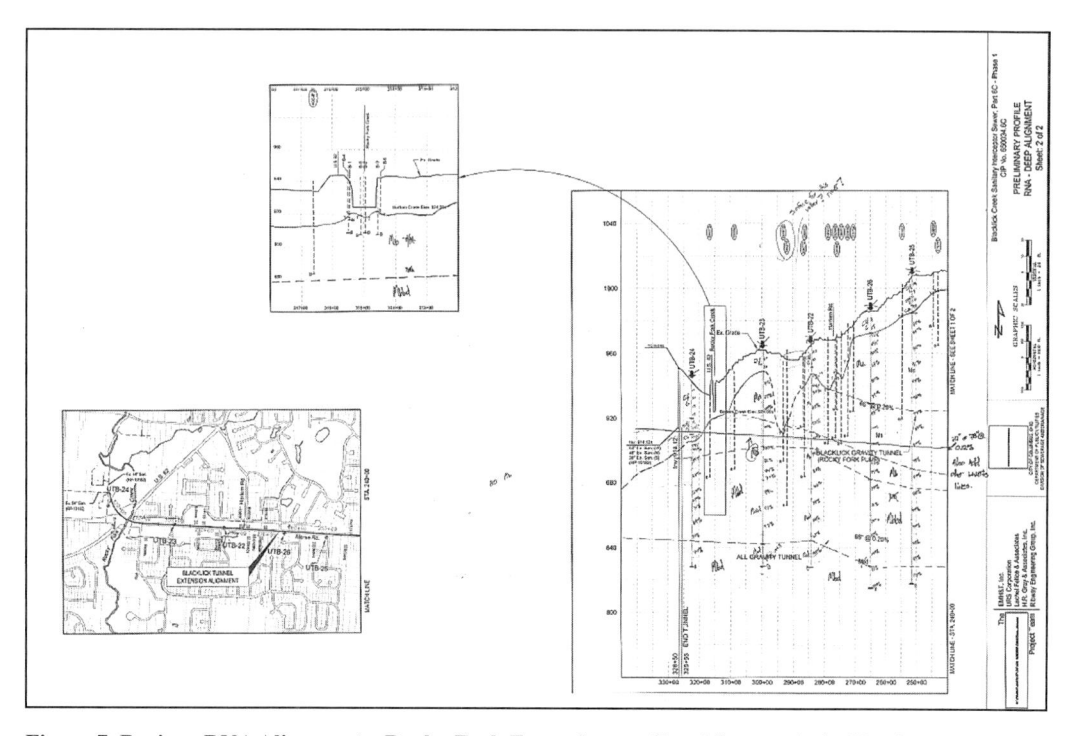

Figure 7. Borings RNA Alignment—Rocky Fork Extension profile with geotechnical borings

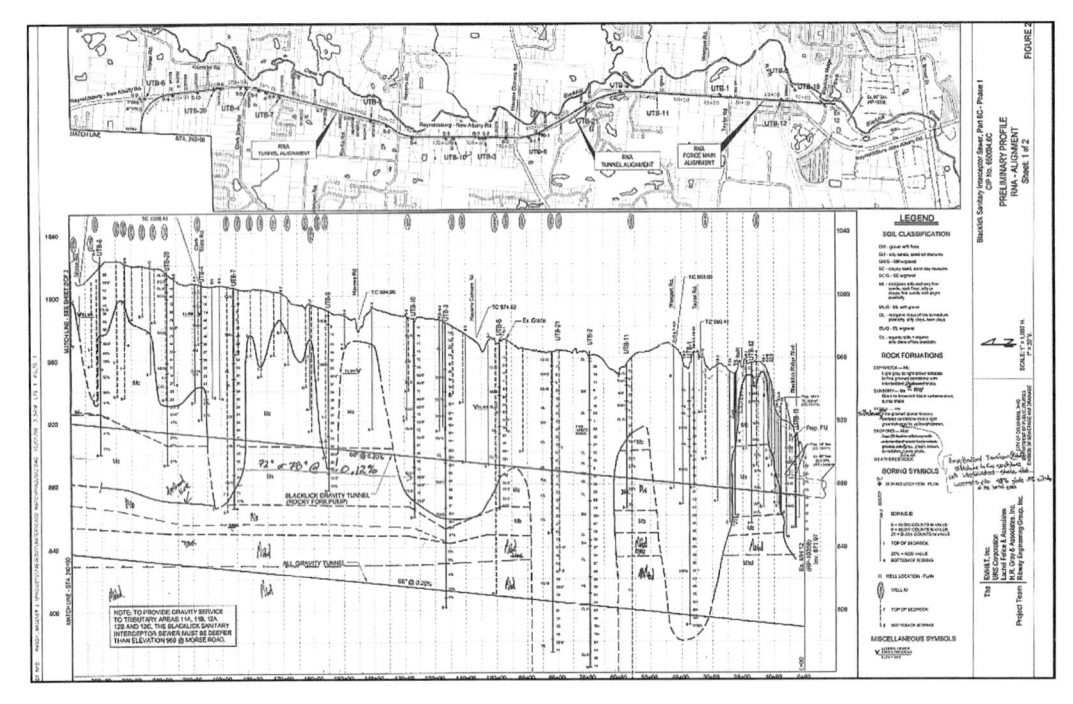

**Figure 8. RNA alignment profile after Phase 1-Step 2 geotechnical**

highs and valleys, ground conditions could be quite different depending on the depth. Consequently, eight different vertical alternatives were then developed along this horizontal alignment with various geology, risk, cost, and sewer service considered to determine the optimal tunnel alignment. These alignments including the Rocky Fork Extension are shown in Figure 9.

The various vertical alignments evaluated all have different possible construction methods associated with them and varying geological conditions. The geological conditions and the excavation size also dictate the cost-effective methods available for each alignment. The bedrock highs and valleys affect the vertical alignments as well as machine selection.

### Alignment Alternative 1

The vertical alignment for this option is expected to occur in mixed-face conditions of soil and rock. This option incorporates a combination of open cut construction at the southern terminus and access shaft locations, and the use of a Tunnel Boring Machine (TBM). Average invert depths range from approximately 28 feet to nearly 100 feet. The use of a TBM is advantageous because it can be used mixed face conditions. Additionally, TBM construction requires fewer access shafts than microtunneling, thus reducing the surface disturbance during construction. This alternative would require a pump station near the original Rocky Fork Pump Station planned location.

### Alignment Alternative 2

This alternative would follow the same alignment as Alignment Alternative 1. However, since the soil borings indicated soft rock such as sandstone and shale, it may be beneficial to substitute open face tunneling with an earth pressure balance machine (EPBM) or a slurry pressure balance machine (SPBM). Both of these technologies resist earth and hydrostatic pressure at the cutting face of the machine. This alternative also includes a pump station near the original Rocky Fork Pump Station location.

### Alignment Alternative 3

This alternative would follow the same alignment as Alignment Alternatives 1 and 2, but would utilize a microtunnel boring machine (MTBM) for construction. The disadvantage of this method is that the length of the runs is limited by the jacking capacity of the MTBM. It is anticipated that at least 23 access shafts would be required for this scenario, depending on the equipment used, surface characteristics and shaft site availability. While the tunneling cost may be less than TBM methods, the costs of environmental clearance, easements, surface restoration, and access shaft construction is expected to be much

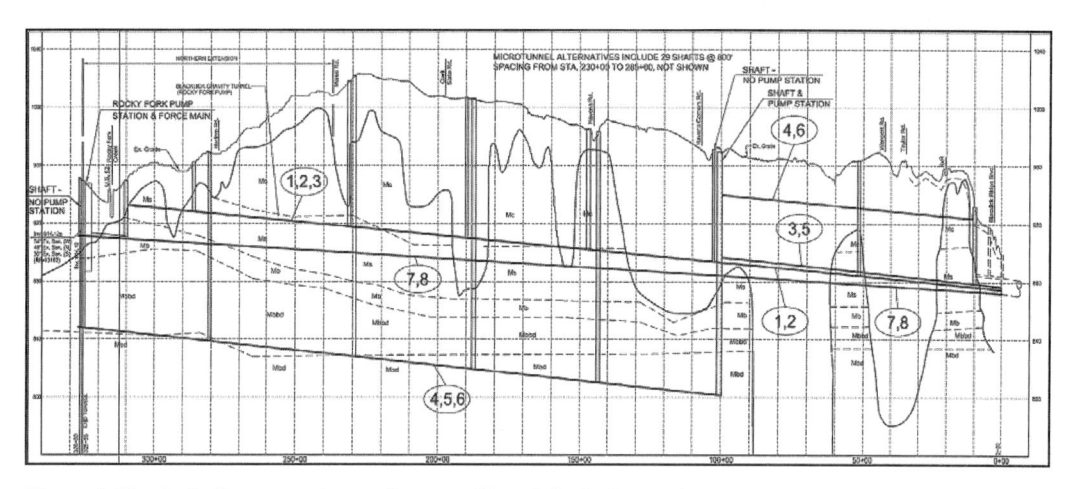

Figure 9. Vertical alignment alternatives considered (includes Rocky Fork Extension)

greater. A pump station would be required at the Rocky Fork Pump Station location.

### Alignment Alternative 4

This option consists of 18,000 feet of deep tunnel from the Rocky Fork Pump Station location (no pump station required at that site) that would flow by gravity at a depth around 70 feet, allowing the TBM tunnel to be constructed completely in rock. It would then flow to a pump station located approximately 9,000 feet from the southern terminus of the project. The flow would then be pumped up to a shallow (20-ft to 40-ft deep) gravity sewer segment (approximately 9,000 feet) of the interceptor between the pump station and the tie-in location that would be built using open cut construction methods.

### Alignment Alternative 5

This option consists of 18,000 feet of deep tunnel from the Rocky Fork Pump Station location (no pump station required at that site) that would flow by gravity at a depth around 70 feet, allowing the tunnel to be constructed completely in rock, just as in Alignment Alternative 4. It would then flow to a pump station located approximately 9,000 feet from the southern terminus of the project. The flow would then be pumped up to a shallow (20–40 ft deep) gravity sewer (approximately 9,000 feet) between the pump station and the tie-in location built using microtunneling methods.

### Alignment Alternative 6

This option consists of 18,000 feet of deep tunnel from the Rocky Fork Pump Station location (no pump station required at that site) that would flow by gravity at a depth around 70 feet, allowing the tunnel to be constructed completely in rock, just as in Alignment Alternatives 4 and 5. It would then flow to a pump station located approximately 9,000 feet from the southern terminus of the project. The flow would then be pumped up to a shallow (10-ft deep) forcemain (approximately 9,000 feet) between the pump station and the tie-in location built using horizontal directional drilling (HDD) methods.

### Alignment Alternatives 7 and 8

The vertical alignment for this option is located at depths between those previously discussed for the entire length between the southern terminus and the Rocky Fork Extension (approximately 32,000 feet). If further geotechnical investigations indicate dry mixed-face conditions of soil and rock, then the entire sewer would be constructed by TBM (Alternative 7). If it is determined that wet mixed-face conditions are anticipated, then the entire sewer would be constructed by EPBM or SPBM (Alternative 8). This alternative would not require a pump station anywhere between the Rocky Fork Service Area and the entire Blacklick Sanitary Interceptor Sewer.

As shown in Figure 9 and Table 1, Alignment Alternatives 1, 2 and 3 have the same vertical alignment but would use different construction methods in the saturated soil, rock and mixed face conditions. Each of those alternatives would require a pump station at the northern end of the project in order to provide sewer service to the Rocky Fork Service Area. Alignment Alternatives 4, 5 and 6 would be a deep rock tunnel for the majority of the alignment with a pump station to lift the sewage to shallower depths at the southern end of the project. These alignments would eliminate the need for a pump station at the northern end of the project. Alignment Alternatives 7 and 8 are located at an elevation between the

Table 1. Summary of vertical alignment alternatives

| Anticipated Item/Feature | | Alignment Alternative | | | | | | | |
|---|---|---|---|---|---|---|---|---|---|
| | | 1 | 2 | 3 | 4 | 5 | 6 | 7 | 8 |
| | Blacklick Service Area | Gravity | Gravity | Gravity | Gravity | Deep Tunnel & Pump Station | Deep Tunnel & Pump Station | Deep Tunnel & Pump Station | Gravity |
| | Rocky Fork Service Area | Pump Station | Pump Station | Pump Station | Pump Station | Gravity | Gravity | Gravity | Gravity |
| | Geology | Saturated Soil & Rock | Saturated Soil & Rock | Rock & Saturated Soil/ Mixed Face | Rock & "Dry" soil | Rock & Saturated Soil/ Mixed Face | Rock & "Dry" soil | Rock & "Dry" soil | Rock & Saturated Soil/ Mixed Face |
| | Construction Method(s) | TBM | Mixed Face Machine EPBM/ SPBM | MTBM | TBM & Open Cut | TBM & MTBM | TBM & HDD | TBM | EPBM/ SPBM |
| | Excavation Support Materials | Ribs & Lagging | Segmental Tunnel | Segmental Tunnel | Ribs & Lagging | Ribs & Lagging, MTBM Jacked Pipe | Ribs & Lagging, HDD pipe | Ribs & Lagging | Ribs & Lagging, Segmental Tunnel |
| | Final Liner/ Pipe | Corrosion protection liner or Pipe-in-Tunnel | Corrosion protection liner or Pipe-in-Tunnel | Corrosion protection liner or Pipe-in-Tunnel | Pipe-in-Tunnel | Pipe-in-Tunnel | Pipe-in-Tunnel | Pipe-in-Tunnel | Corrosion protection liner or Pipe-in-Tunnel |

previous alternatives, and would not require a pump station anywhere in order to serve both the Blacklick and Rocky Fork Service Areas. By eliminating a pump station in the sanitary sewer system, the lifecycle costs of Alternatives 7 and 8 are drastically less than all other alternatives, even though they require the sewer to be longer than originally planned. After preliminary cost estimations were completed, Alternatives 7 and 8 were deemed to provide the best solution for the Blacklick Creek Sanitary Interceptor Sewer. Further geotechnical investigation that will occur during detailed design will determine which of these two alternatives is the best option.

## CONCLUSIONS

The process used to compare alternatives leading to the selection of an optimal tunnel alignment in this challenging ground proved to be a success. The original project design was revisited so that the design team could figure out the plan's strengths and weaknesses so that the City's needs could be met in the most cost effective way possible. Before developing a geotechnical drilling plan, two different general horizontal alignment corridors in the project area were defined. The corridors incorporated the original design alignment as well as alternatives that were anticipated to minimize the soil/rock interfaces and other associated issues that would be encountered based on geotechnical information. The original project length was also extended in order to investigate the feasibility of connecting the Blacklick Creek Sanitary Interceptor Sewer to an existing sanitary sewer and eliminate a future pump station to the north of the original project extents.

The dynamic boring plan allowed for real time data from the field to be utilized in revising the geologic profile and investigation as it progressed, which included relocating or eliminating unnecessary borings and adjusting drilling depths. Because of this dynamic drilling process, the Waggoner Road Corridor was eliminated early in the preliminary design due to the geologic conditions encountered during drilling. The remaining borings that would have been drilled in that corridor were then moved to the corridor that revealed itself as being the most cost effective through this process, the RNA alignment. This allowed the design team to focus their time and drilling budget on the area where the tunnel would actually be constructed.

Once the project area could be focused to a single horizontal alignment corridor, the RNA alignment was carried forward beyond an initial screening. Eight different vertical alternatives were then developed along the RNA alignment with alternate construction methods, materials and geology considered during development.

By utilizing the process discussed in this paper for selecting a tunnel alignment, the design team was able to develop the optimum alignment for the Blacklick Creek Sanitary Interceptor Sewer. Through this analysis, the design team reduced the budgeted construction cost by $100 million, while also extending the tunneled sewer with the added benefit of relieving existing sewers immediately after it is placed in operation. Additionally, the longer alignment eliminates the need for a pump station that had been planned for a future connection of the Rocky Fork Service Area to the Blacklick Creek Interceptor Sewer, greatly reducing the lifecycle cost of the sewer system in the area.

# TRACK 4: CASE HISTORIES

Larry Eckert, Chair

*Session 1: Small Diameter*

# Microtunneling Challenges: Crossing Under Major Railroad and Highways in Very Soft Glacial Soils—The Evolution of a Ground Treatment Assessment Process

**Philip W. Lloyd, Zhenqi Cai**
Hatch Mott MacDonald, New York, New York

**Glenn Duyvestyn**
Hatch Mott MacDonald, Cleveland, Ohio

**ABSTRACT:** Bergen County Utilities Authority (BCUA), New Jersey has initiated an ambitious program to invest US$65 million to carryout a significant CSO improvement program to meet a State DEP consent order by 2010. The New Overpeck Valley Parallel Relief Sewer Project is part of the overall scheme to provide additional capacity and redundancy for sewage conveyance from the BCUA service areas to the BCUA Wastewater Treatment Plant (WTTP). A major portion of this CSO improvement program utilized microtunneling techniques to complete 1.2 miles (1.8 km) of new sewer. A Herrenknecht AVND1800AB microtunnel boring machine (MTBM) was used to install 72 in (1800 mm) diameter reinforced concrete pipe (RCP). This machine was equipped with two articulation joints and was selected by the Contractor due to its increased ability to steer to maintain line and grade. The microtunnel alignment included drives through 2,788 ft (850 m) of extremely soft glaciolacustrine varved clay, transitioning to Deltaic deposits of loose to medium dense mixed soils containing silty sands, gravels, and boulders. Included within these soft ground drives were crossings beneath the New Jersey Turn Pike and the 12 track CSX Railroad Intermodal Yard and a branch mainline. During construction, the Contractor proposed a significant jet grouting program within the sensitive varved clays to increase the bearing capacity of the soft material required to support the weight of their MTBM and trailing equipment. A thorough bearing capacity analysis and advanced geotechnical investigation program was developed and used to eliminate the need for extensive jet grouting in the soft soils along the entire alignment. Recommendations were also provided during construction to change the Contractor's microtunnel cutting head to reduce the weight of the MTBM at the front of the machine and improve the overall factor of safety for bearing capacity for the proposed equipment. Challenging conditions were also encountered when microtunneling under US Highway Route 46. These challenges included excavation within soft varved clays with also the potential to encounter boulder(s) within rock fill. This paper discusses the extremely challenging ground conditions encountered along the alignment and presents findings of the detailed ground treatment assessment process used on this project which can be applied to other similar soft soil installations. This paper also discusses as a case history the significant engineering challenges related to all drives that include ground characterization, key MTBM features and performance, and ground monitoring results.

## INTRODUCTION

In order to eliminate wet weather overflows into surrounding waterways from the existing 50-year old Overpeck Valley Trunk Sewer System, a 60 in Reinforced Concrete Pipe gravity sewer by 2010, Bergen County Utilities Authority, (BCUA), commissioned the design and construction of a new relief sewer installed in a parallel alignment to the existing system. The New Overpeck Valley relief sewer line is an important part of a broader program of works initiated to provide additional sewerage capacity and extra conveyance redundancy to BCUA's wastewater treatment plant (WWTP). The overall program consists of installation of 5.3 miles (8.45 km) of new interceptor pipelines that vary in diameter from 42 in (1050 mm) to 96 in (2400 mm) through major highways, railroads and densely industrialized areas in Bergen County, NJ.

The overall sewer improvement project encountered a wide range of soil conditions. Microtunneling, conventional open-cut and sub aqueous methods were all used to construct various portions of the relief sewer. The project was separated into three separate contracts—one contract with primarily

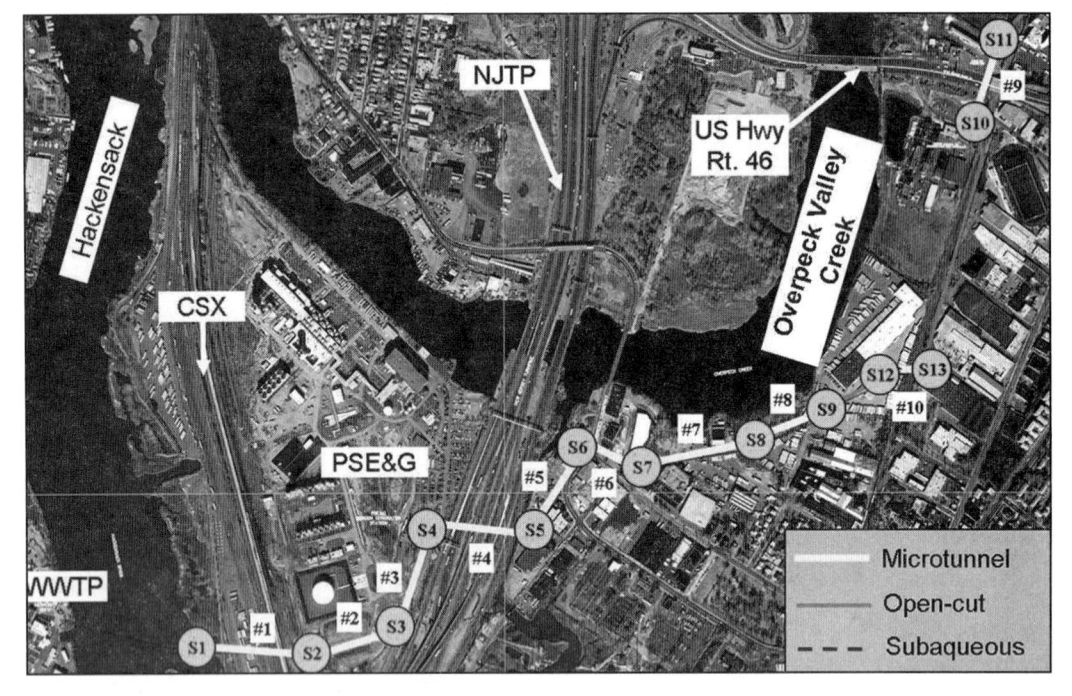

**Figure 1. New Overpeck relief sewer microtunnel alignment and crossings**

microtunneling and open-cut, two fully open-cut contracts and, a sub aqueous pipe installation across the Hackensack River. Hatch Mott MacDonald was appointed as the Engineer for the planning, design and construction management of the whole project by BCUA. Northeast Remsco Construction was awarded the microtunnel construction contract. This paper discusses the microtunneling portion of the project including crossing of major highways, railroads within difficult ground conditions, ground treatment evaluation and construction performance compared to instrumentation results.

## MICROTUNNEL ALIGNMENT AND MAJOR CROSSINGS

The New Overpeck Valley Relief Sewer was excavated using microtunnel along ten (10) microtunnel drives with varying lengths between 200 ft and 1000 ft. In total, approximately 6,000 ft (1.8 km) was installed with microtunneling. Figure 1 illustrates the microtunnel drives and shaft locations and provides a clear perspective of the alignment area showing the major structures and crossings encountered along the alignment. Table 1 summarizes each microtunnel drive.

A one pass microtunnel installation strategy, where the jacking pipe also served as the carrier pipe to convey flows, was used to install the new sewer. The jacking pipe consisted of 10 ft (3.0 m) long, 89-in (1830 mm) outer diameter reinforced concrete pipe. Microtunneling was selected as the optimum method to install the new sewer pipe based on its ability to install a new sewer beneath critical features (such as the New Jersey Turnpike and major railroad arteries) and utilities and the decreased construction footprint.

The BCUA WWTP is situated to the west of the Hackensack River. The new relief sewer was constructed across this River using double piles with concrete caps. This new sewer extended BCUA's service areas from the east side of the river (Cai et al 2009). This method of installation was selected because of the very low cover at the WWTP intake elevation and extremely weak adverse soil conditions in the riverbed, producing factors that were considered to be unsuitable for microtunneling.

### Drive #1—CSX Railroad Intermodal Yard

Drive #1 was at a depth of 6 m (20 ft) and crossed beneath the CSX Railroad Intermodal Yard consisting of 12 railroad tracks and a rail freight storage area. This drive was also completed within very soft varved clays. Beneath the railroad tracks the microtunnel alignment parallels the existing Overpeck Valley trunk sewer and crosses under an existing

## Table 1. Summary of microtunnel drives and major crossings in order of construction

| Drives | Direction of Drive | Drive Distance m (feet) | Microtunnel Major Crossings (In order of completion) |
|---|---|---|---|
| #1 | S2 to S1 | 256 (839) | Microtunnel Under-crossing of CSX Intermodal Rail 12 tracks; In close proximity to existing 760 mm (30 in) force main; In parallel to existing 1520 mm (60 in) trunk sewer. |
| #2 | S2 to S3 | 198 (650) | Microtunneling within close proximity to existing 1520 mm (60 in) trunk sewer; Numerous buried utilities, above ground electrical equipment and overhead HV power lines, beneath electricity and gas utility (PSE & G) property. |
| #3 | S4 to S3 | 242 (794) | |
| #4 | S4 to S5 | 187 (613) | Microtunnel crossing of New Jersey Turnpike (NJTP); At-grade and elevated sections of highway with drainage channels, in vicinity of viaduct piles. |
| #5 | S6 to S5 | 205 (674) | Driven in major industrialized zone; very close to existing 1520 mm (60 in) trunk sewer and underground chambers/pipelines. |
| #6 | S6 to S7 | 69 (227) | This drive is also in very close proximity to existing sewer and underground utilities. |
| #7 | S8 to S7 | 308 (1010) | Beneath Bell Drive, driven to the west direction in very close proximity to existing sewer and underground chambers and pipelines; drive within a busy industrialized zone. |
| #8 | S8 to S9 | 157 (515) | Low cover and in close proximity to the existing Overpeck Valley Trunk Sewer as well as number of underground chambers and pipelines; drive within a busy industrialized zone. |
| #9 | S10 to S11 | 148 (487) | Microtunnel crossing beneath US Highway Route 46. |
| #10 | S12 to S13 | 52 (170) | Microtunnel crossing under CSX Railroad branch mainline. |

water main situated less than one tunnel diameter away from the MTBM, (Figure 2).

### Drives #2 and #3—Public Service Electric and Gas Facility, (PSE&G)

These two drives were advanced through difficult ground conditions consisting of very soft varved clays. Drive #2 was driven from shaft 2 to shaft 3. Drive #3 was driven from shaft 4 to reception shaft #3. The Overpeck Valley existing 1520 mm (60 in) RCP trunk sewer, various other potential obstructions that included electrical transmission towers, overhead power lines, numerous gas and water mains, and buried communication cables, had to be identified and negotiated by the microtunnel drives #3 and #2.

### Drive #4—New Jersey Turnpike (NJTP)

Drive #4 was beneath the New Jersey Turnpike (NJTP) (Figure 3) and presented one of the biggest geotechnical challenges to constructing this project. The soils along this alignment included very soft varved clays and dense sands and gravels with cobbles and boulders. The drive depth was 6 to 7 m (20–23 ft) below grade in a west to east alignment from shaft 4 to shaft 5. Above the microtunnel crossing, the NJTP consists of lanes at-grade, an elevated section supported on battered piles, and on and off ramps as well as drainage channels. The alignment was selected to provide adequate clearance from the piles supporting the abutments of the elevated section. A thorough study reviewing the existing data and information was conducted to gain a better understanding of the construction history of the NJTP at this location to assess, identify and ultimately avoid any potential obstructions to the crossing.

### Drives #5 and 6—Industrial Zone

Drives #5 and #6 located to the east of New Jersey Turnpike were driven next. These alignments were located within densely industrialized areas occupied by a variety of businesses ranging from trucking to storage facilities. Drive #5 was driven from Shaft 6 to Shaft 5 beneath Hendricks Causeway. Drive #6 was driven beneath Edgewater Road from a dual purpose jacking shaft at Shaft 6 to a reception shaft at Shaft 7. These drives were of particular concern as

Figure 2. Drive #1—microtunnel crossing beneath CSX railroad yard

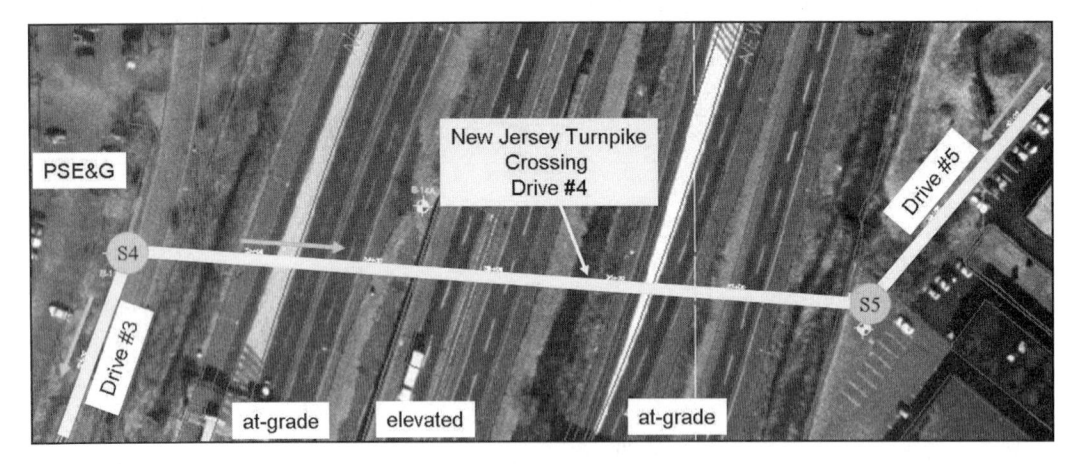

Figure 3. Drive #4—microtunnel crossing under New Jersey Turnpike (NJTP)

the existing BCUA trunk sewer and its service chambers were located within the vicinity of these drives. The existing trunk sewer was constructed using traditional open cut or trenching methods with timber lagging and a gravel bedding.

**Drives #7—Industrial Zone**

The second microtunnel drive also crossed beneath Bell Drive. This drive was constructed from Shaft 8 to the reception shaft located at Shaft 9. This drive was in particularly close proximity to the existing 1520 mm (60 in) trunk sewer and a number of underground chambers and pipelines; clearance between Drive #7 and the existing trunk sewer was less than one tunnel diameter for a stretch more than 10 m (33 ft) and the clearance reduced to 1 m (3 ft) near the reception shaft.

**Drive #8—Crossing of Bell Drive**

This microtunnel drive crossed beneath Bell Drive with a shallow depth of cover of 4.5m (15 ft). Shaft 8 served as the dual purpose jacking shaft and Shaft 9 served as the reception shaft. This drive was in close proximity to the existing 1520 mm (60 in) trunk sewer and a number of underground chambers and pipelines. In addition, it was located within a busy industrialized zone that included electrical facilities, trucking transportation, storage buildings, and other industrial businesses.

### Drive #9—US Highway Rt. 46

Drive #9 crossed beneath US Highway Rt. 46, a major regional Highway linking Northern New Jersey with New York City. Drive #9 passed under the highway at a location where a 9 m (30 ft) high elevated embankment section of US HWY Rt 46 was constructed in the 1930s. The tunnel crown at this crossing point was approximately 4.5 m (15 ft) below the toe of the elevated embankment.

### Drive #10—CSX Railroad Mainline

Drive #10 was a relatively short drive under a branch mainline railroad owned by CSX. This drive of the pipeline was originally planned to be driven by a conventional jack and bore installation. However, the decision was taken to use the MTBM due to the presence of saturated silty sand and weak clay found in site investigations, it was concluded that these ground conditions were adverse and challenging for a mainline railroad crossing and it would be too difficult to control the face of the excavation face using conventional jack and bore.

## GEOTECHNICAL CHARACTERIZATION

This geotechnical conditions were investigated by a site investigation program that broadly covered the microtunnel alignment prior to the design phase. An additional investigation program was performed during the construction phase of the project in the vicinity of microtunnel drives 1 through 4 to further define the soil conditions within very soft varved clays.

### General Geological Conditions

The soil conditions encountered along the entire microtunnel alignment generally consist of two separate and distinct geological glacial depositional environments, divided, coincidentally, by the New Jersey Turn Pike, (NJTP), see Figure 1. To the west of the NJTP, the soil consisted of deep beds of glaciolacustrine deposits (Qhkl) underlain by a dense till (Qt). The glaciolacustrine deposits were overlain by estuarine deposits (Qm) of organic silt and clay and salt marsh peat forming the natural ground surface now covered by variable amounts of artificial fill (af). Alluvium deposits (Qal) consisting of silts and sands with minor amounts of clay and gravels were found directly beneath the estuarine deposits.

The glaciolacustrine soils deposit consists of alternating thin layers of silt and clay, reaching an overall thickness of about 18 m (60 ft). The microtunnel alignments immediately to the west of NJTP (Drives 1 through 4) are located within this "varved" soil deposit. These clays and silts were deposited in a calm glacial lake environment and are fine grained and very soft.

Where the microtunnel alignment crosses beneath the New Jersey Turnpike in an easterly direction, the varved clays and silts deposit transition into deltaic deposits (Qhk) of medium dense to dense granular sands, gravels and some cobbles. This deposit is the more prevalent material within the microtunnel alignment to the east of the NJTP (Drives 5 through 8). The deltaic deposits were transported and deposited in more turbulent depositional events within glacial lake environments. Typical of Qhk deposits, the gradation of this stratum becomes finer with depth. The upper portion of this deposit was observed as being loose to medium dense fine grained silty sand giving way to a sequence of interbedded stiff silt and clay. Published data maps estimate this deposit to up to 15 m (50 ft) thick.

### Site Specific Geotechnical Investigation and Testing Program, Phase 1

The difficult crossings of the major highways and railroads required the design and implementation of a comprehensive geotechnical exploration program that was undertaken and outlined in a contractual Geotechnical Data Report. The exploration program consisted of the following elements:

- A total of 31 standard soil borings were drilled along the microtunnel alignment to recover targeted soil samples for laboratory testing, with field tests (SPT's) conducted within each boring to establish subsurface stratigraphy;
- Seven (7) directional geoprobes to detect potential obstructions at locations where timber piles and gravel beddings were used for construction of the existing trunk sewer;
- Environmental sampling conducted at 12 locations along the alignment;
- Laboratory tests on soils conducted for classification, estimation of material properties and assessment of design properties, to assess and evaluate the feasibility of microtunneling through encountered soils.

### Site Specific Geotechnical Investigation and Testing Program, Phase 2 (Completed During Construction)

- 15 cone penetration tests conducted during construction stage to supplement original soil borings and tests for the purpose of assessing,

Table 2. Summary of soil characterization for microtunnel alignment (phase 1)

| Drives | Direction of Drive | Drive Distance m (ft) | Soil Characterization |
|---|---|---|---|
| #1 | S2 to S1 | 256 (839) | Glaciolacustrine deposits (Qhkl) Soft varved clay Low plasticity CL Baseline undrained Cu = 20 kPa (400 psf) Sticky and squeezing behavior Wetting and remolding lowers strength "Weight of rod" & "weight of hammer" SPTs. (tunnel face) |
| #2 | S2 to S3 | 198 (650) | |
| #3 | S4 to S3 | 242 (794) | |
| #4 (half) | S4 to S5 | 187 (613) | |
| #4 (half) | | | Deltaic deposits (Qhk) Mixed soil conditions Low to medium dense fine silty sand Baseline strength φ=32° Flowing behavior when not supported Gravel bedding and timber piles Defined boulders 750 mm (30 in) in diameter with UCS=175 MPa (25 ksi) |
| #5 | S6 to S5 | 205 (674) | |
| #6 | S6 to S7 | 69 (227) | |
| #7 | S8 to S7 | 308 (1010) | |
| #8 | S8 to S9 | 157 (515) | |
| #9 | S10 to S11 | 148 (487) | Estuarine deposit (Qm) and Deltaic Deposit (Qhk) Soft marsh Mixed soil conditions Soft to medium stiff silty clay Highway embankment containing rock fills Defined rock fill sizes 750 mm (30 in) in diameter with UCS=175 MPa (25 ksi) |
| #10 | S12 to S13 | 52 (170) | Deltaic (Qhk) Mixed soil conditions Loose to medium dense silty sand Soft to medium stiff silty clay |

evaluating strength and identifying weak soil locations for ground treatment.
- Eight (8) additional site investigation geotechnical borings were drilled and further additional undisturbed soil samples recovered for laboratory testing to correlate with CPT results and original test data.

## Soil Characterization

The original site investigation borings (Phase 1), laboratory test data and analyses help with the assessment of the general stratigraphic sequence along the alignment. A geotechnical and geological model of the encountered soils was developed that identified the two distinct soil environments—the glaciolacustrine (varved clay/silts) and more granular saturated mixed ground conditions within the proposed alignment to the east of the NJTP, across. The Geotechnical Baseline Report described the characterized the ground conditions, and presented the interpreted subsurface stratigraphy, the expected soil property ranges were baselined, and the range of anticipated characteristic in situ soil behavior was discussed along with any potential obstructions. A summary of the generalized ground characterizations for all the drives is given in Table 2.

### Varved Clay/Silts

The microtunnel alignment Drives #1, #2 and #3 and approximately half of Drive #4 (see Figures 1 thru 3) are excavated in the glaciolacustrine varved clay/silts deposit. The laboratory test results for soil characterization and selected strength are presented in Figure 4. The results suggested that the varved soil possesses a low plasticity to semi-liquid consistency at the majority of locations where undisturbed samples were recovered. Laboratory soil samples subjected to undrained triaxial shear strength tests were exhibited relatively low undrained peak shear strengths ranging from 15 kPa (300 psf) to 35 kPa (700 psf).

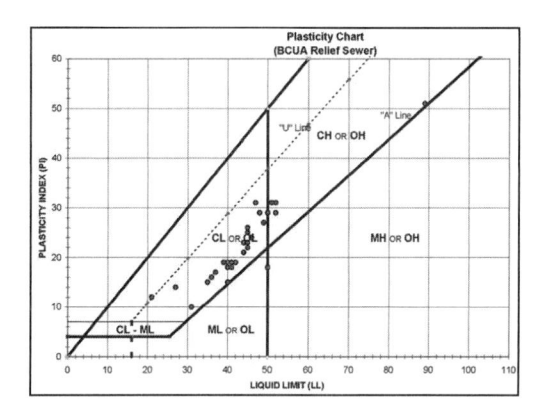

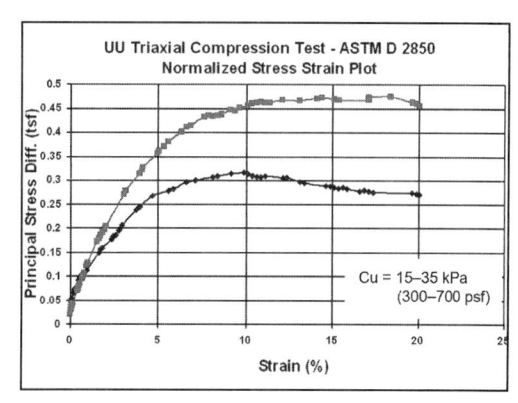

Figure 4. Varved clay/silt (Qhkl Deposit) plasticity chart and example triaxial test results

The Phase 1 field information and laboratory data indicated that this varved clay/silt deposit was extremely soft with zero to very low SPT blow counts and may be susceptible to significant settlement and possibly liquefaction during microtunneling operations. The MTBM weight, steering loading and vibration may result in excessive settlement beneath the MTBM with the risks of the machine "sinking," with loss of line and grade, and significant distortion of pipe joints. The GBR outlined these potential risks and as a result, two definitive measures were adopted:

1. Specifications set out the required specific MTBM design components prior to manufacture, in order to facilitate excavation by microtunneling operation within the anticipated adverse soil conditions—this is further elaborated in Section 4; and
2. Bids included a substantial ground treatment allowance was incorporated in the contract to provide the requisite ground treatment and soil stabilization where required during the construction stage—this is further elaborated in Section 6.

*Deltaic Mixed Soils*

The microtunnel drives to the east of NJTP are within mixed ground conditions. From the microtunnel Drive #4 under the NJTP through to Drive #10 (see Figure 1) the alignment runs primarily in saturated mixed granular soils composed of sands, silty sands, gravels, and scattered artificial fills containing debris, blocks and stones from previous constructions. Drives #9 and #10 were excavated in soft to medium stiff silty clay. Figure 5 shows a collection of gradation curves of samples from all deposits encountered along the alignment.

*Boulders*

Boulders were encountered and excavated successfully during Drive #7, taking approximately four hours for the MTBM to excavate. The microtunnel crossing of Drive #9 under US Highway Rt.46 also encountered boulders potentially from the rock fill used to construct the elevated embankment section.

## MICROTUNNEL BORING MACHINE SPECIFICATION

Project specifications were tailored for the anticipated geotechnical materials and included the following main fundamental requirements:

- A MTBM with a pressurized closed-face was especially necessary for the soil conditions beneath highways and railroad crossings and drives in the near vicinity of buried chambers and pipelines, existing utilities and surface structures;
- A MTBM with a cutterhead with the ability to excavate through mixed soils including artificial fills, boulders, cobbles, and other buried obstructions;
- A MTBM with the capability to maintain line and grade in the soft soil conditions by compensation through steering.

**MTBM Cutterhead**

The Contractor chose to use an MTBM equipped with mixed soil cutterhead that was fitted with disc cutters and soft soil cutting tools. The disc cutters were provided in case any boulders were encountered during the drives through the Deltaic Deposits. The soft cutting tools were provided to excavate soils containing cobbles, gravels, sands, silts and clays. While this cutterhead is well equipped to handle the coarse soils associated with the Deltaic Deposits, it is

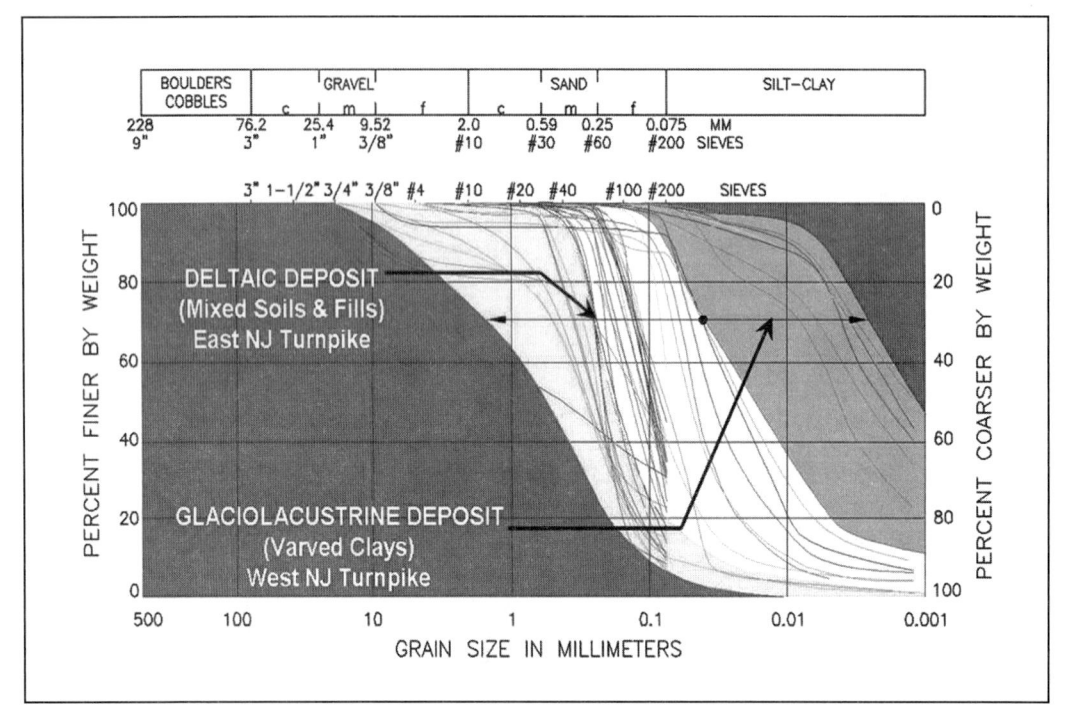

**Figure 5. Gradation curves of soil samples**

not well suited to the excavation of the microtunnels through the very soft varved clays.

The Contractor initially elected to purchase a second articulated section to provide additional steering capability in an attempt to excavate the very soft varved clays while maintaining line and grade. The location of this second steering joint is shown in Figure 6.

During construction of the drives through the Deltaic Deposits, the Contractor expressed concerns that the MTBM equipment and their approach may not be sufficient to enable construction of the microtunnel drives through the varved clays. The Contractor elected to engage an independent engineering consultant to assess the ability to maintain line and grade through the soft soils. This review expressed serious concerns about the bearing capacity of the site soils and the potential sinking of the MTBM during microtunneling. Their initial review questioned the undrained shear strength value of 400 psf provided in the Geotechnical Baseline Report developed specifically for this project. To overcome their concerns, the Contractor proposed consideration of ground improvements in the form of jet grouting along the entire alignment where the varved clays would be encountered.

Realizing that these drives included crossings of CSX rail yard, New Jersey Turnpike, and several critical utilities, HMM conducted additional site investigations using Cone Penetration Tests to verify the conditions established in the Geotechnical Baseline Report. The assessment included using the full geotechnical data set to assess, sensitivity of the soils, the potential for liquefaction and bearing capacity of the soil during microtunneling. The results of the additional undrained shear strength tests are summarized in Table 3.

*Phase 2 Site Investigation Results*

Based on the available additional geotechnical information, the baselined undrained shear strength of 400 psf was considered an accurate representation of the anticipated geotechnical materials.

The Contractor was also concerned about the sensitivity of the soils through the varved clays. The results of HMM's assessment of soil sensitivity for Drives 1, 2 and 3 is summarized in Table 4 and Table 5.

The Rosenqvist (1953) classification of clays was used to characterize the sensitivity of the varved clays using the following relationship:

$$\text{Sensitivity, } S_t = \frac{q_u(undisturbed)}{q_u(remolded)}$$

Soil sensitivity was highest in the varved clays associated with Drive 1 in the vicinity of the CSX

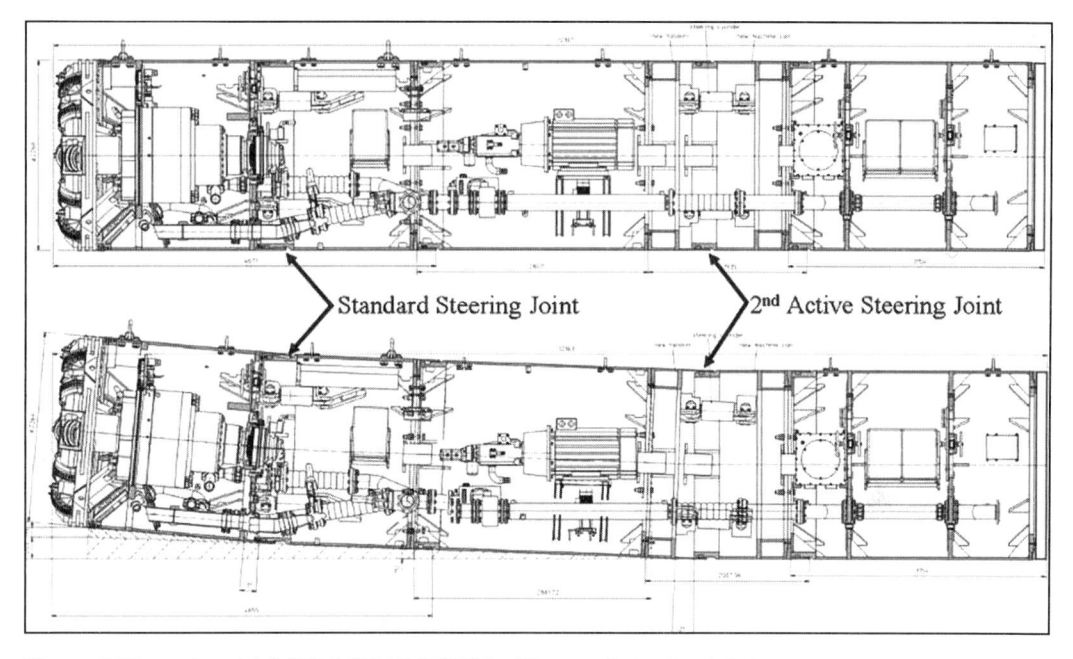

Figure 6. Herrenknecht AVND 1800AB MTBM with second steering joint

rail yard. The Contractor raised issues with the rotation of the MTBM cutterhead causing an annular remolded zone of material that will quickly regain strength following initial disturbance. It was HMM's opinion that the extent of such a disturbed remolded zone would not be sufficient to cause substantial settlement of the machine.

Liquefaction concerns were also raised by the Contractor. Their concern was that the vibrations associated with operation of the MTBM would result in sinking of the MTBM and an inability to maintain line and grade. HMM revisited their liquefaction analysis completed during design and updated it with the additional geotechnical information collected during the second geotechnical investigation. Assessing the liquefaction potential for the varved clays involved determining the Atterberg limits and other soil properties and comparing the results to the guidelines shown in Table 6. The results of these comparisons are shown in Table 7.

As can be seen from the Tables 6 and 7, the liquefaction analysis does not provide a definitive answer with respect to the behavior of the varved clays amongst the various approaches. The Wang criteria appear to be too general, since it classifies all of the soil samples as susceptible to liquefaction. While this may be true, it is unclear what site conditions would impose the liquefaction. The Andrews and Martin criteria are more detailed, and identify 4 soils samples that are likely to liquefy. They also rule out some samples as unlikely to cause problems, but it is uncertain how reliable this identification will prove, since one sample that was classified as "not susceptible" by Andrews and Martin appears to be susceptible according to Seed's criteria, while others that are not susceptible are recommended for more detailed testing by other constraints. The Bray criteria identify 5 samples as susceptible, with all other requiring more laboratory testing. Seed, et al further refine these samples by identifying 7 samples as susceptible, 24 as not likely to cause a problem, 3 samples with an unknown behavior, and further testing for the balance of samples. The final analysis, according to Boulanger and Idriss, identifies all samples as being clay-like in nature, and thus susceptible to clay-like cyclic softening. While they are not identified as samples that behave like fine-grained soils, which may liquefy, there is not enough information soil information to use this as the sole method for analysis.

Based on the analysis, the following soils were deemed to be susceptible to liquefaction included:

- B5A:S-19
- BH103A:U-2
- BH103A:U-3
- BH105:S-4
- B13A:S-4B
- B7A:S-13

Although the additional geotechnical investigations confirmed the baseline conditions presented in the Geotechncial Baseline Report, the Contractor

Table 3. Summary of soil CPT and laboratory tests (phase 2)

| Station | Borehole | Lab Test (psf) Peak | Lab Test (psf) Post | CPT (psf) | SPT | Elevation Range (ft) | Position |
|---|---|---|---|---|---|---|---|
| 11+76 | BH101 | 550 | 530 | 500 | WOH | −26.6 to −28.6 | Face |
|  |  | 550 | 470 | 500 | 5 | −32.6 to −34.6 | 0.5 diameters below invert |
|  |  | 630 |  | 500 | WOR | −38.6 to −40.6 | 1 diameter below invert |
| 11+81 | CPT101 |  |  |  |  |  | Below invert |
| 12+35 | CPT109 |  |  |  |  |  | Below invert |
| 13+00 | BH108 | 670 | 650 | 630 | WOH | −30.2 to −32.2 | Directly below invert |
|  |  | 680 | 680 | 600 | WOH | −36.2 to −38.2 | Directly below invert |
|  |  | 230 |  | 650 | WOH | −37.2 to −39.2 | 1 diameter below invert |
|  |  | 650 |  | 800 | 2 | −24.2 to −26.2 | Face |
| 13+05 | CPT108 |  |  |  |  |  | Below invert |
| 14+20 | CPT110 |  |  |  |  |  | Below invert |
| 16+20 | BH102 | 520 | 520 | 650 | WOH | −32.1 to −34.1 | Directly below invert |
|  |  | 360 |  | 600 | WOH | −26.1 to −28.1 | Face |
|  |  | 470 |  | 700 | WOH | −40.1 to −42.1 | 1 diameter below invert |
|  |  | 550 |  | 700 | WOH | −34.1 to −336.1 | 1.5 diameters below invert |
|  |  | 550 |  | 600 | WOH | −32.1 to −34.1 | 0.5 diameters below invert |
| 16+10 | CPT102 |  |  |  |  |  | Below invert |
| 20+05 | BH103A | 690 | 630 |  | WOH | −30.2 to −32.2 | Directly below invert |
|  |  | 490 |  |  |  | −30.2 to −32.2 | Directly below invert |
|  |  | 690 |  |  |  | −25.2 to −27.2 | 1.5 diameters below invert |
|  |  | 670 | 630 | 700 | WOH | −25.2 to −27.2 | Face |
|  |  | 630 | 480 | 700 | WOH | −25.2 to −27.2 | Face |
|  |  | 580 | 500 | 690 | WOH | −25.2 to −27.2 | Face |
| 20+10 | CPT103 |  |  |  |  |  | Below invert |
| 27+20 | BH104 | 530 | 320 | 670 | WOH | −30.7 to −32.7 | 0.5 diameters below invert |
|  |  | 520 | 360 | 550 | WOR | −24.7 to −26.7 | Face |
|  |  | 230 | 220* | 600 | 1 | −24.7 to −26.7 | Face* |
| 27+15 | CPT104 |  |  |  |  |  | Below invert |
| 29+85 | BH105 | 550 | 490 | 600 | 1 | −23.4 to −25.4 | 0.5 diameters below invert |
|  |  | 460 | 380 | 400 | WOH | −23.4 to −25.4 | Face |
| 30+05 | CPT105 |  |  |  |  |  | Below invert |
| 31+95 | BH106 | 446 |  |  |  | −33 to −35 | 0.5 diameters below invert |
|  |  | 370 | 340 | 450 | 1 | −27 to −29 | Invert* |
|  |  | 310 | 290 | 500 | 1 | −27 to −29 | Invert* |
| 31+95 | CPT106 |  |  | 620 | WOR |  | Below invert |
| 34+20 | CPT107 |  |  | 1600 | 22 |  | Below invert |

* Indicates specimen disturbed.

Table 4. Soil sensitivity

| Station | Sample ID | Depth (feet) | $(s_u)_{lv}$ (ksf) | $(s_{ur})_{lv}$ (ksf) | Sensitivity, $S_t$ | General Classification |
|---|---|---|---|---|---|---|
| 11+76 | BH101:U1 | 27.7 | 0.69 | 0.10 | 6.90 | Very Sensitive |
| 11+76 | BH101:U4 | 46.7 | 0.63 | 0.11 | 5.73 | Very Sensitive |
| 11+76 | BH101:U5 | 56.7 | 0.52 | 0.07 | 7.43 | Very Sensitive |
| 13+00 | BH108:U1 | TBD | TBD | TBD | TBD | N/A |
| 13+00 | BH108:U2 | 32.5 | 0.65 | 0.11 | 5.91 | Very Sensitive |
| 13+00 | BH108:U3 | TBD | TBD | TBD | TBD | N/A |
| 13+00 | BH108:U4 | 46.5 | 0.23 | 0.04 | 5.75 | Very Sensitive |
| 16+20 | BH102:U1 | 33.7 | 0.36 | 0.05 | 7.20 | Very Sensitive |
| 16+20 | BH102:U2 | 39.6 | 0.55 | 0.07 | 7.86 | Very Sensitive |
| 16+20 | BH102:U3 | 41.7 | 0.50 | 0.07 | 7.14 | Very Sensitive |
| 16+20 | BH102:U4 | 47.7 | 0.47 | 0.04 | 11.75 | Slightly Quick |
| 16+20 | BH102:U5 | 56.7 | 0.55 | 0.04 | 13.75 | Slightly Quick |
| 20+05 | BH103A:U2 | 40.7 | 0.49 | 0.07 | 7.00 | Very Sensitive |
| 20+05 | BH103A:U3 | 46.8 | 0.69 | 0.07 | 9.86 | Slightly Quick |
| 20+05 | BH103A:U4 | 56.8 | 0.71 | 0.09 | 7.89 | Very Sensitive |
| 27+20 | BH104:U1 | 21.7 | 0.56 | 0.06 | 9.33 | Slightly Quick |
| 27+20 | BH104:U2 | TBD | TBD | TBD | TBD | N/A |
| 27+20 | BH104:U3 | TBD | TBD | TBD | TBD | N/A |
| 29+85 | BH105:U2 | TBD | TBD | TBD | TBD | N/A |
| 29+85 | BH105:U3 | 51.7 | 0.25 | 0.04 | 6.25 | Very Sensitive |
| 31+95 | BH106:U2 | TBD | TBD | TBD | TBD | N/A |
| 34+20 | BH107:U2 | 28.7 | 0.43 | 0.07 | 6.14 | Very Sensitive |

was still concerned about their MTBM performance within the varved clays and requested an extensive ground improvement program to increase the bearing capacity of the varved clays. Their proposed solution was to construct jet grout columns extending down to bearing soils at depth along the entire alignment within the varved clays. The proposed jet grouting solution would increase the contract price by 8 to 10 million dollars.

While HMM firmly believes that the anticipated geotechnical conditions were properly characterized in the Geotechnical Baseline Report, HMM evaluated the Contractor's microtunnel equipment and MTBM cutterhead in terms of the appropriateness for excavating the varved clays. This analysis involved determining the bearing pressure of the disproportioned MTBM due to its weight and the increased bearing pressure required to induce a steering correction. This required bearing pressure was then compared to the bearing capacity of the soil.

Based on the Contractor provided information, the mixed face cutterhead had a weight of

Table 5. Sensitivity classification

| Sensitivity | Descriptive term |
|---|---|
| <2 | Insensitive |
| 2–4 | Moderately sensative |
| 4–8 | Sensative |
| 8–16 | Very sensative |
| 16–32 | Slightly quick |
| 32–64 | Medium quick |
| >64 | Quick |

approximately 16,940 lbs. This cutter wheel is located at the front of the machine and produces a disproportioned bearing pressure on the soils beneath the front few feet of the MTBM. The first section of the MTBM (including the first articulation joint) had a weight of approximately 45,560 lbs. The combined the total weight was 62,480 lbs. For the MTBM to maintain line and grade the MTBM needs

Table 6. Liquifaction criteria

| | |
|---|---|
| Wang | Clay soils having<br>• Less than 15% finer than 0.005 mm<br>• Liquid limit, LL < 35<br>• Water content $w_c$ > 0.9LL<br>May be vulnerable to severe strength loss as a result of earthquake shaking. |
| Andrews & Martin | • Soils susceptible to liquefaction if:<br>– <10% finer than 0.002 mm and,<br>– LL < 32<br>• Soils not susceptible if:<br>– >10% finer than 0.002 mm and,<br>– LL > 32<br>• Study is needed for soils that meet only one criteria. |
| Bray et al. | • PI < 12<br>• Susceptible to liquefaction,<br>• $w_c$ > 0.85LL<br>while soils with<br>• 12 < PI < 20<br>• Are more resistant to liquefaction; still susceptible to cyclic mobility<br>• $w_c$ > 0.8LL |
| Seed et al. | • PI < 12 and LL < 37<br>  * considered potentially susceptible if $w_c$ > 0.8LL<br>• PI < 20 and LL < 47<br>  * considered potentially liquefiable; lab testing needed if $w_c$ > 0.85LL<br>• Soils with PI > 20 or LL > 47<br>  * generally not susceptible to liquefaction |
| Boulanger & Idriss | • Clay-like for PI > 7<br>– Includes all CL soils<br>– For CL-ML, PI > 5<br>• Fine-grained soils not meeting above: liquefiable<br>– True unless testing shows otherwise<br>• Intermediate behavior: PI b/w 3–6<br>– Need more testing |

to be properly supported by the site soils. An additional bearing capacity is required to induce a steering correction to maintain line and grade.

The bearing capacity of the varved clays can be assessed using Terzaghi, Hansen, Vessic criteria (Das 2002), and compared to the loads in the form of a safety factor. Pressure distribution diagrams were developed for the microtunnel machine to determine the load distributions attributed to the machine weight, soil loads, and steering corrections exerted onto the soft soils. The resulting pressure distributions were then compared to bearing capacity of the varved clays to determine whether ground improvements were required.

The same analysis was completed assuming a soft soil cutting wheel would be used to complete the drives in the very soft varved clays. This analysis showed that, by replacing the 16,940 lbs mixed cutter wheel with a 10,000 lb (representing a weight reduction of 41 percent) soft cutter wheel, the bearing pressure is significantly reduced. The lighter soil cutter wheel is equivalent to an increase in the undrained shear strength of 86 psf, in comparison to the heavier mixed soil cutter wheel. See Figure 7.

The results of the bearing capacity analysis for the mixed face and soft ground cutterheads are provided in Table 8.

Based on the results of the analysis, it was determined that the heavier mixed face cutterhead did not provide a sufficient factor of safety to maintain line and grade in the site soils. Hence, if the Contractor were to proceed with the proposed mixed face cutterhead, jet grouting would need to be completed along the entire microtunnel alignment within the varved clays.

Replacing the mixed face cutter wheel to a soft soil cutterhead decreases the required bearing capacity of the MTBM thereby increasing the factor of safety associated with bearing capacity. Aside from a reduction in the required bearing pressure, HMM's analysis demonstrated that the microtunnel drives could be completed as originally designed without ground improvements along the entire alignment. It should be noted that it was HMM's original design

## Table 7. Liquefaction assessment

| Station | Sample | Depth | Wang | Andrews & Martin | Bray et al. | Seed et al. | Boulanger & Idriss |
|---|---|---|---|---|---|---|---|
| 10+86 | B-5A:S-4 | 12–14' | Susceptible | Further study | Further study | Potential-need testing | Clay-like |
| 10+86 | B-5A:UD-1 | 30–32' | Susceptible | Not susceptible | Further study | Not likely | Clay-like |
| 10+86 | B-5A:S-19 | 56–58' | Susceptible | Susceptible | Susceptible | Susceptible | Clay-like |
| 11+76 | BH101:S-5 | 20–22' | Susceptible | Further study | Further study | Potential-need testing | Clay-like |
| 11+76 | BH101:U-1 | 26–28' | Susceptible | Further study | Further study | Potential-need testing | Clay-like |
| 11+76 | BH101:U-2 | 33–35' | Susceptible | Not susceptible | Further study | Potential-need testing | Clay-like |
| 11+76 | BH101:U-4 | 45–47' | Susceptible | Further study | Further study | Not likely | Clay-like |
| 11+76 | BH101:U-5 | 55–57' | Susceptible | Further study | Further study | Not likely | Clay-like |
| 12+80 | B-6A:S-9 | 18–20' | Susceptible | Further study | Further study | Potential-need testing | Clay-like |
| 12+80 | B-6A:U-1 | 30–32' | Susceptible | Further study | Further study | Potential-need testing | Clay-like |
| 12+80 | B-6A:S-18 | 38–40' | Susceptible | Further study | Further study | Potential-need testing | Clay-like |
| 13+00 | BH108:S-4 | 20–22' | Susceptible | Further study | Further study | Susceptible | Clay-like |
| 13+00 | BH108:U-1 | 26–28' | Susceptible | Not susceptible | Further study | Potential-need testing | Clay-like |
| 13+00 | BH108:U-2 | 32–34' | Susceptible | Further study | Further study | Not likely | Clay-like |
| 13+00 | BH108:U-3 | 38–40' | Susceptible | Further study | Further study | Potential-need testing | Clay-like |
| 13+00 | BH108:U-4 | 45–47' | Susceptible | Further study | Further study | Not likely | Clay-like |
| 13+00 | BH108:U-5 | 55–57' | Susceptible | Not susceptible | Further study | Not likely | Clay-like |
| 16+20 | BH102:S-5 | 20–22' | Susceptible | Further study | Further study | Potential-need testing | Clay-like |
| 16+20 | BH102:U-1 | 32–34' | Susceptible | Further study | Further study | Potential-need testing | Clay-like |
| 16+20 | BH102:U-2 | 38–40' | Susceptible | Further study | Further study | Potential-need testing | Clay-like |
| 16+20 | BH102:U-3 | 40–42' | Susceptible | Further study | Further study | Potential-need testing | Clay-like |
| 16+20 | BH102:U-4 | 46–48' | Susceptible | Further study | Further study | Not likely | Clay-like |
| 16+20 | BH102:U-5 | 55–57' | Susceptible | Further study | Further study | Not likely | Clay-like |
| 16+47 | B-7A:S-5 | 14–16' | Susceptible | Further study | Further study | Unknown | Clay-like |
| 16+47 | B-7A:S-10 | 24–26' | Susceptible | Further study | Further study | Potential-need testing | Clay-like |
| 16+47 | B-7A:S-12 | 28–30' | Susceptible | Further study | Further study | Potential-need testing | Clay-like |
| 16+47 | B-7A:S-13 | 30–32' | Susceptible | Further study | Susceptible | Susceptible | Clay-like |
| 16+47 | B-7A:S-15 | 39–41' | Susceptible | Further study | Further study | Potential-need testing | Clay-like |
| 16+47 | B-7A:U-1A | 34–36' | Susceptible | Further study | Further study | Not likely | Clay-like |
| 18+22 | B8A:S-7 | 24–26' | Susceptible | Further study | Further study | Unknown | Clay-like |
| 18+22 | B8A:S-10 | 30–32' | Susceptible | Further study | Further study | Not likely | Clay-like |
| 18+22 | B8A:U-1 | 34–46' | Susceptible | Not susceptible | Further study | Potential-need testing | Clay-like |
| 20+05 | BH103A:U-1 | 34–36' | Susceptible | Not susceptible | Further study | Susceptible | Clay-like |
| 20+05 | BH103A:U-2 | 39–41' | Susceptible | Further study | Susceptible | Susceptible | Clay-like |
| 20+05 | BH103A:U-3 | 45–47' | Susceptible | Susceptible | Susceptible | Susceptible | Clay-like |
| 20+05 | BH103A:U-4 | 55–57' | Susceptible | Not susceptible | Further study | Potential-need testing | Clay-like |
| 21+65 | B9A:S-9 | 27–29' | Susceptible | Further study | Further study | Potential-need testing | Clay-like |
| 21+65 | B9A:S-11 | 31–33' | Susceptible | Further study | Further study | Not likely | Clay-like |
| 21+65 | B9A:S-13 | 35–37' | Susceptible | Further study | Further study | Potential-need testing | Clay-like |
| 24+40 | B10A:S-8 | 25–27' | Susceptible | Further study | Further study | Not likely | Clay-like |
| 24+40 | B10A:U-1 | 31–33' | Not enough info | Not susceptible | Further study | Not likely | Clay-like |
| 24+40 | B10A:S-15 | 41–43' | Susceptible | Further study | Further study | Not likely | Clay-like |
| 27+02 | B11A:S-10 | 29–31' | Susceptible | Further study | Further study | Potential-need testing | Clay-like |
| 27+20 | BH104:S-5 | 15–17' | Susceptible | Further study | Further study | Potential-need testing | Clay-like |
| 27+20 | BH104:U-1 | 20–22' | Susceptible | Further study | Further study | Potential-need testing | Clay-like |
| 27+20 | BH104:U-2 | 32–34' | Susceptible | Not susceptible | Further study | Not likely | Clay-like |
| 27+20 | BH104:U-3 | 38–40' | Susceptible | Further study | Further study | Not likely | Clay-like |
| 29+73 | B12A:S-8 | 25–27' | Susceptible | Further study | Further study | Potential-need testing | Clay-like |
| 29+73 | B12A:S-10 | 29–31' | Susceptible | Further study | Further study | Not likely | Clay-like |

(a) Mixed ground cutterhead
Weight=7700 kg (17 kips)

(b) Soft ground cutterhead
Weight=4550 kg (10 kips)

**Figure 7. MTBM cutterheads**

**Table 8. Bearing capacity analyses**

| Bearing Capacity Factors of Safety ||||||||
| :---: | :---: | :---: | :---: | :---: | :---: | :---: | :---: |
| Heavier Mixed Head Cutter Wheel |||| Lighter Soil Cutter Wheel ||||
| Borehole ID | Shear Strength psf | Factor of Safety | FS <1.5 | Borehole ID | Shear Strength psf | Factor of Safety | FS <1.5 |
| BH101 | 530 | 1.53 | no | BH101 | 530 | 1.66 | no |
| BH101 | 550 | 1.59 | no | BH101 | 550 | 1.72 | no |
| BH108 | 650 | 1.88 | no | BH108 | 650 | 2.03 | no |
| BH108 | 680 | 1.97 | no | BH108 | 680 | 2.13 | no |
| BH102 | 520 | 1.50 | no | BH102 | 520 | 1.62 | no |
| BH102 | 480 | 1.39 | yes | BH102 | 480 | 1.50 | yes |
| BH102 | 500 | 1.44 | yes | BH102 | 500 | 1.56 | no |
| BH102 | 630 | 1.82 | no | BH102 | 630 | 1.96 | no |
| BH104 | 320 | 0.92 | yes | BH104 | 320 | 1.01 | yes |
| BH105 | 490 | 1.42 | yes | BH105 | 490 | 1.53 | no |
| Baseline | 400 | 1.14 |  | Baseline | 400 | 1.23 |  |

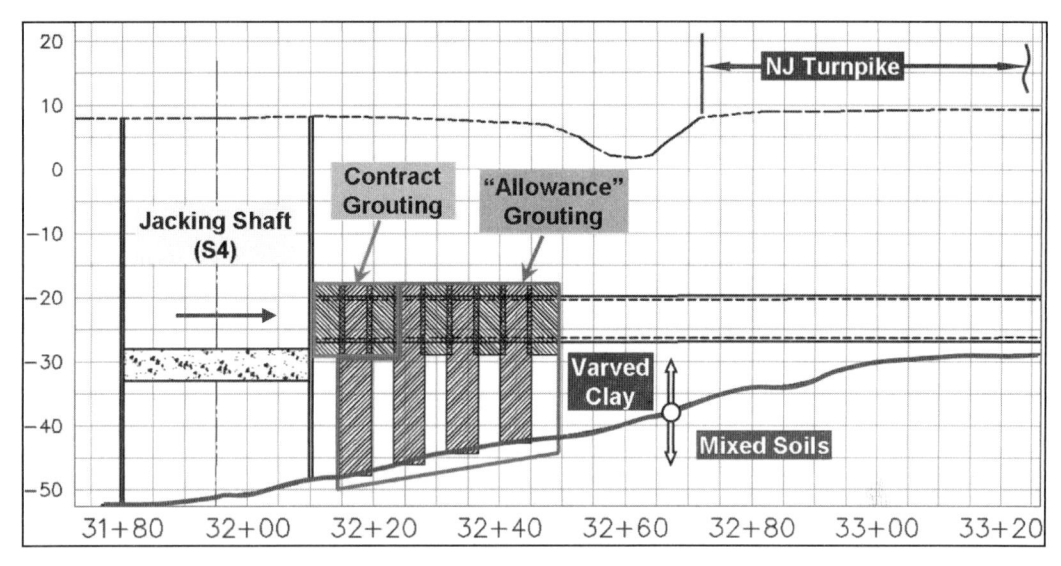

**Figure 8. Reach 4 ground improvements**

for the Contractor to provide a MTBM with multiple cutterheads that would be capable of excavating both the coarse grained soils and the soft varved clays. However, the Contractor did not have a soft cutterhead for their MTBM.

Based on the bearing capacity analysis, HMM recommended the use of a soft ground cutterhead without extensive ground improvements for Drives 1, 2, and 3. However, ground improvements were recommended for Drive 4 where the MTBM would need to excavate soft varved clays and coarse grained Deltaic Deposits beneath the New Jersey Turnpike, as the mixed face cutterhead would be required to excavate potential boulders within the coarse grained soils, (Figure 8). The costs for the ground improvements would be taken out of the ground conditioning improvement allowance included in the Contractor's bid.

The cost for a replacement cutterhead was also taken out of the ground conditioning improvement allowance. The decision to reimburse the Contractor for the replacement cutterhead was negotiated between the Contractor, HMM and Owner. The cost of the soft ground cutterhead was significantly less than the overall costs for ground improvements over the 2,200 feet of microtunneling for Drives 1, 2 and 3.

## MICROTUNNELING

All drives were been completed within schedule. The microtunneling operations were staged from six jacking shafts. It generally took two 10-hour days to launch all the MTBM sections and a third day to install the first few pipes that contain the cooling pipes, the tunnel slurry pump, and an Intermediate Jacking Station. On day four in general efficient productivity began. Summaries of construction of the sewer in each reach are discussed in more detail in the following sections.

### Drive #1—CSX Railroad Intermodal Yard

The rail crossing from jacking shaft 2 to reception shaft 1 was driven within varved clay/silts, with control of the face better achieved by reducing the open area of the soft soil cutter head, This was identified as necessary from the experience gained within drive #3 where less efficient control of the face and soil volume through the cutterhead may have contributed to relatively high surface settlements of ~76 mm (3 inches). Although conditions within the soil was markedly similar, Drive #1 under the CSX rail yard experienced 0.48 inches of surface settlement within one section of the drive. The maximum jacking force was 340 tonnes after a delay, reducing significantly during construction operations. Twenty pipes were driven over 24 hours the total crossing time was 9 days.

### Drives #2 and #3—Public Service Electric and Gas Facility, (PSE&G)

Drives #2 and #3 within the PSE&G property were within weak varved clay. Drive #2 was 200.5m (665ft) in length, with Drive #3 nearly 250m (820 ft) long. These drives installed the sewer pipe operating on a 1 × shift per day basis. The maximum number of pipes driven in a single shift during Drive #2 was 10, with an average of about 7 pipes per day driven over the 9-day duration of the drive. The maximum jacking force was approximately 269 tonnes (296 tons). A maximum of 8 pipes were driven during one day during Drive #3, with an average of about 5 pipes per day driven over the 14-day duration of the drive. The maximum jacking force was 366 tonnes (403 tons), which occurred at the start of the day, after the ground had converged around the MTBM following downtime of over 24 hours.

### Drive #4—New Jersey Turnpike (NJTP)

The New Jersey Turn Pike crossing from jacking shaft 4 to reception shaft 5 was driven from within varved clay/silts that transitioned to deltaic deposits of saturated granular soils, The mixed face cutter head, was used for this drive, with jet grouting of the initial varved soil area near the jacking shaft. The drive length was 189m (621ft) and took 9 days to complete. The maximum jacking force was 540 tonnes with the best productivity of 19 pipes installed over a 24 hour period. The cutterhead settled 31mm (1.25 inches) on leaving the ground improvement above which was the maximum surface settlement of 15mm.( 0.6 inches) The drive was completed successfully without any significant issues.

### Drives #5 and 6—Industrial Zone

Drives #5 and #6, approximately following Hendricks Causeway and Edgewater Avenue, were within deltaic deposits. Drive #5 was nearly 203m (667ft) in length, with Drive #6 approximately 69m (227 ft) long. These drives installed the sewer pipe operating on a 1 × shift per day basis. The maximum number of pipes driven in a single shift during Drive #5 was 8, with an average of 5 pipes per day driven over the 13-day duration of the drive. The maximum jacking force of 357 tonnes (393 tons) occurred at the start of the day. A maximum of 6 pipes were driven during one day during Drive #6, with an average of about 4 pipes per day driven over the 5-day duration of the drive. The maximum jacking force was 188 tonnes (207 tons).

### Drives #7—Industrial Zone

Drive #7 approximately following Bell Drive, was within deltaic deposits. For approximately 762m (250 ft) of the alignment, fill falls within the tunnel alignment. The drive was approximately 300m (985ft) in length. This drive installed the sewer pipe operating on a 1 × shift per day basis. The maximum number of pipes driven in a single shift during the drive was 9, with an average of 4 pipes per day driven over the 21-day duration of the drive. The maximum jacking force was 322 tonnes (355 tons), which occurred at the start of the day, after the ground had converged around the MTBM after being left overnight.

### Drive #8—Industrial Zone Crossing of Bell Drive

Drive #8 approximately following Bell Drive, was within deltaic deposits. The drive was approximately 181m (593ft) in length. This drive installed the sewer pipe operating on a 1 × shift per day basis. The maximum number of pipes driven in a single day during the drive was 9, with an average of 5 pipes per day driven over the 9-day duration of the drive.

### Drive #9—US Highway Rt. 46

Drive #9 crossing below US Highway Route 46, was within glaciolacustrine deposits. For approximately 45.7m(150ft) of the alignment, till encroaches into the invert of the tunnel alignment. In addition, there is about 25 feet of fill over the alignment, where it passes below Rt. 46. The drive was approximately 154m (505ft) in length. This drive installed the sewer pipe operating on a 1 × shift per day basis. The maximum number of pipes driven in a single shift during the drive was 7, with an average of 5 pipes per day driven over the 10-day duration of the drive. The maximum jacking force was 443 tonnes (488 tons), which occurred at the start of the day, after the ground had converged around the MTBM after a two-day pause in advancing the drive.

### Drive #10—CSX Railroad Mainline

Drive #10 crossing below CSX Railroad tracks, through clays, a section of artificial fill in the crown and till rising from the invert in the section half of the drive. The drive was approximately 50m (164ft) in length. This drive installed the sewer pipe operating on a 1 × shift per day basis. The maximum number of pipes driven in a single shift during the drive was 3, with an average of 2 pipes per day driven over the 4-day duration of the drive. The maximum jacking force was 324 tonnes (357 tons).

## CONCLUSIONS

Microtunneling was used to successfully construct the New Overpeak Valley Relief Sewer. The mixed face cutterhead performed well in the coarse soils. The soft ground cutterhead performed very well in the very soft varved clay. Use of the soft ground cutterhead significantly increased the factor of safety associated with bearing capacity while significantly decreasing the costs associated with jet grouting.

## ACKNOWLEDGMENTS

The authors would like to acknowledge Bergen County Utilities Authority for permission to publish this paper. They also wish to acknowledge the work of. Monica J. Paciorek, Tunnel Engineer (HMM) New York office, for her help with the preparation of this paper.

**REFERENCES**

Andrews, D.C.A. and Martin, G.R. (2000). "Criteria for Liquefaction of Silty Soils." 12th World Conference on Earthquake Engineering, Proceedings, Auckland, New Zealand.

Boulanger, R.W. and Idriss, I.M. (2006). "Liquefaction Susceptibility Criteria for Silts and Clays." Journal of Geotechnical and Geoenvironmental Engineering, 132(11), 1413–1426.

Bray, J.D. and Sancio, R.B. (2006). "Assessment of the Liquefaction Susceptibility of Fine-Grained Soils. Journal of Geotechnical and Geoenvironmental Engineering, 132(9), 1165–1177.

Cai, Z, Solano A.G. O'Connor N., and Lloyd P.W. 2009, "Microtunneling 1.2-Mile, 72-IN RCP with Crossings of New Jesey Turnpike and CSX rail roads," RETC Conference, Las Vegas, Nevada.

Das B., M. 2002, Principles of Geotechnical Engineering," 5th Edition, pub. Brooks/Cole, USA.

Rosenqvist 1953 "Considerations of the Sensitivity of Norwegian Quick Clays," Goetechnique, vol. 3, No. 5, 195–200.

Seed, R.B., et al. (2003). "Recent Advances in Soil Liquefaction Engineering: A Unified and Consistent Framework." EERC-2003-06, Earthquake Engineering Research Institute, Berkeley, California.

Terzaghi, K., Peck, R.B., and Mesri, G. "Liquefaction of Saturated Loose Sands." Soil Mechanics in Engineering Practice, 3rd Ed. 193–208.

Wang, W. (1979). "Some Findings in Soil Liquefaction." Research Report, Water Conservancy and Hydroelectric Power Scientific Research Institute, Beijing.

# Marysville Trunk Interceptor Project: A Case History

**Paul de Verteiul**
DLZ Ohio, Inc., Columbus, Ohio

**ABSTRACT:** The Marysville Trunk Interceptor was designed and constructed to transport existing and future flows from the existing Marysville Waste Water Treatment Plant to the new Water Reclamation Facility and is part of the Wastewater Treatment Expansion Project for the City of Marysville, Ohio. The Wastewater Treatment Expansion also includes the Crosses Run Pump Station, the Crosses Run Pump Station Force main and the new Water Reclamation Facility. The trunk interceptor is a 60 inch gravity sewer approximately 20,000 feet long constructed entirely in soil. Design constraints, such as an urban setting, potentially contaminated near surface soils, and restrictive easements resulted in approximately 14,000 lineal feet of the sewer being designed and constructed utilizing trenchless (Microtunneling) techniques and the remaining 6,000 feet being installed using the open trench (cut and cover) method of construction. This paper presents the design and construction methods used to successfully complete the project.

## INTRODUCTION

### Project Location

The Trunk Interceptor is located in Union County Ohio, within the City of Marysville, beginning at the existing Waste Water Treatment Plant at the north of the City, proceeding south across Mill Creek and then along Industrial Parkway to the new Crosses Run Pump Station at Scottslawn Road. The project setting is urban within the City and mostly rural farmland along Industrial Parkway. The route of the sewer crosses Mill Creek, and also crosses the railroad at two locations.

### Project Description

The Trunk Interceptor sewer was constructed entirely in soil and at a depth below the normal ground water table. The type of sewer pipe selected for the project was Centrifugally Cast Fiberglass Reinforced Polymer Mortar (CCFRPM) Pipe; with the exception of the PVC lined reinforced concrete pipe (RCP) used at the two railroad crossings. Reinforced concrete pipe was the pipe specified for use by the Railroad Company. Table 1 summarizes the design features of the project.

**Table 1. Design features**

| | |
|---|---|
| Average daily flow | 12.94 MGD |
| Project construction | Microtunneling/open trench |
| Length | 20,000 LF (total) |
| | 14,000 LF (microtunneling) |
| | 6,000 LF (open trench) |
| Diameter | 60 inch |
| Pipe | CCFRPM |
| | RCP with PVC T-Lock liner at RR crossings |
| Depth | 35'-45' |
| Ancillary structures | Shafts: 23 |
| | Manholes: 34 |
| | Drop structures: 5 |
| Notice to proceed | 6/11/07 |
| Completion date | 5/01/09 |
| Construction cost | $35.5 million |

### Project Team

Owner: City of Marysville
Lead Design Engineer: DLZ Ohio, Inc.
Tunnel Design Engineer: Jenny Engineering Corporation
Hydrogeologist: Herb Eagon & Associates, Inc.
Construction Administration: DLZ Ohio, Inc.
Contractor: Super Excavators, Inc. (SEI)

## SITE GEOTECHNICAL CONDITIONS

### Geologic Setting

Most of the native soils in the area of Union County are of glacial origin (glacial drift) having been deposited either directly by glacial ice (till), by glacial melt water streams (glaciofluvial), or by glacial lakes (lacustrine deposits). The earliest glaciations of

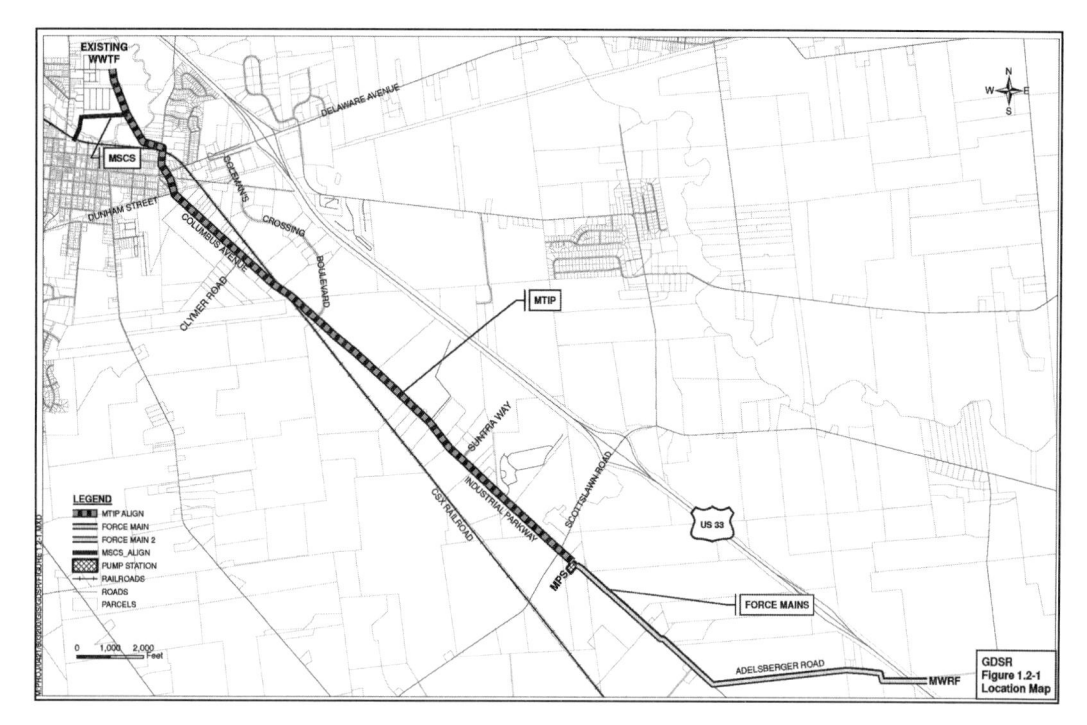

**Figure 1. Location map**

note occurred during the Illinoian stage, and the second during the Wisconsin stage. Glaciation occurred at several intervals and the repeated advance and retreat of the glaciers has resulted in complex subsurface conditions, in which soil types can change dramatically and radically over short distances.

The glacial till is not homogeneous. It varies in texture and is composed of a varying mixture of all sizes of soil particles, including cobbles and boulders and contains seams, lenses or layers of sand and gravel interbedded in the glacial till mass. Recessional moraines transect the County from northeast to southwest and are characterized by broad belts of sloping topography. The glaciers deposited stratified sand and gravel outwash, mostly along a few of the principal streams in the County. Lacustrine material was deposited in relatively small areas on the bottoms of temporary glacial lakes over glacial till. Recent alluvial material was deposited on the flood plains of recent streams, such as Mill Creek.

No bedrock was encountered at the level of the trunk interceptor.

**Ground Water Conditions**

Throughout much of Union County the glacial drift is relatively thin and not considered to be an important water source. Sand and gravel lenses, containing perched water were however, encountered interbedded in the more clayey glacial till. However, in the vicinity of the project the glacial drift varies widely and is as much as 250 feet thick in a buried pre-glacial valley. A sand and gravel aquifer that is potentially hydraulically connected to the underlying bedrock aquifer was encountered in this area. Portions of the project were constructed within water bearing sand and gravel layers. Other parts of the project encountered minimal groundwater and other areas of the project fell somewhere in between these two scenarios.

**PROJECT CHALLENGES**

The design and construction teams faced a variety of challenges on this project, including: variable soil conditions, saturated ground conditions, urban setting, and a river crossing.

**Variable Soil Conditions**

By nature glacial deposits are variable. As discussed in the previous section, the soils encountered on the project consisted of: glacial till containing cobbles and boulders, with interbedded granular layers of sand and gravel; alluvial deposits along Mill Creek; and manmade fill.

The tunneling machine selected by Super Excavators, Inc. (SEI) to handle the anticipated soil

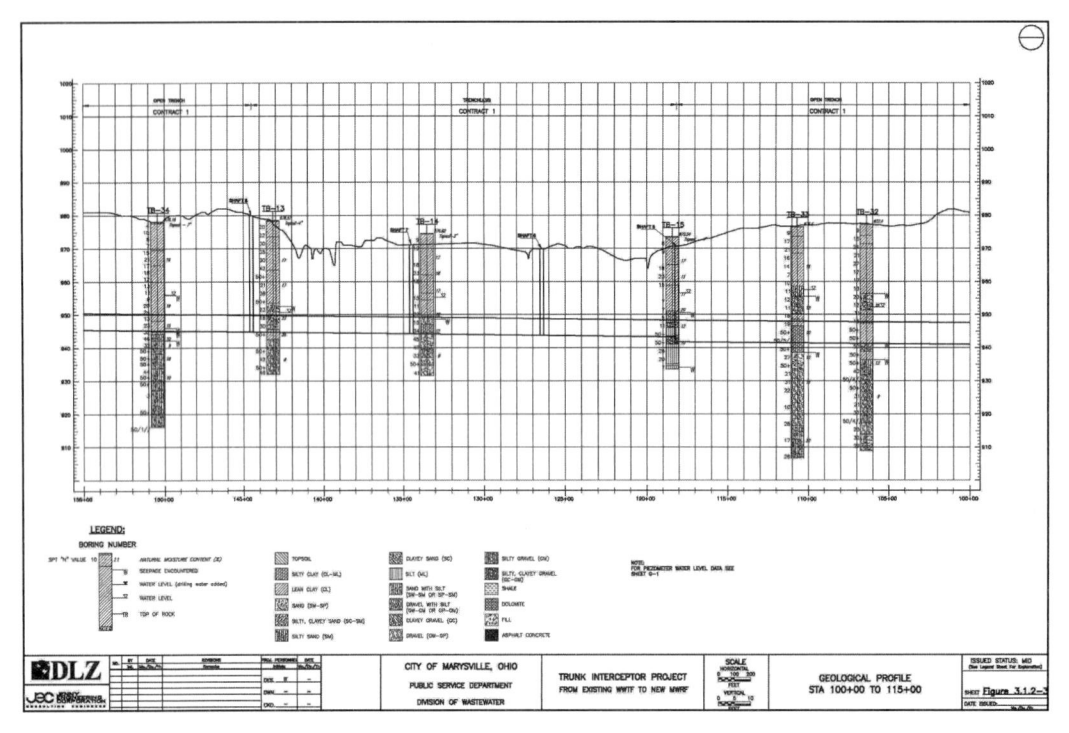

**Figure 2.** Typical geologic profile of the subsurface conditions encountered along the trunk interceptor alignment

conditions was an Ackerman SL60 Microtunnel Boring Machine (MTBM). Microtunneling is the process that uses a remotely controlled boring machine combined with pipe jacking techniques to directly install product pipelines in a single pass. The MTBM has an earth pressure balanced cutting chamber and uses a closed loop slurry system to remove the excavated tunnel spoil, a slurry cleaning system to remove the spoil from the slurry water, a lubrication system (bentonite) to lubricate the exterior of the pipeline during installation, and a guidance system to provide installation accuracy. The main drawback of the MTBM is that it can be stopped by large quantities of cobbles and/or large boulders. For this project the MTBM was equipped with a cutter head capable of mining through soft ground containing cobbles and boulders. The cutter head was fitted with drag (chisel) teeth for excavating through soft ground, and bullet teeth and roller cutters for mining through cobbles and boulders.

**Saturated Ground Conditions**

The majority of the sewer was located below the normal ground water table and in water bearing sand and gravel layers causing concern for inflow into excavations. Ground water was handled differently at the shaft locations, open cut and tunneled sections.

**Figure 3.** Ackerman SL60 Microtunneling Boring Machine (MTBM)

Two shaft designs were provided giving the contractor the option to either dewater the shaft locations or construct the shafts to handle the anticipated hydrostatic pressures. SEI opted to dewater each of the shaft locations prior to excavating the shafts and keeping them dewatered until the shafts were backfilled.

For open cut construction, the specifications called for ground water levels to be maintained

**Figure 4. MTBM cutting head, fitted with drag teeth, bullet teeth, and roller cutters**

**Figure 5. Cutter head after first run showing wear on teeth**

at a minimum of six feet below the bottom of the trench excavation until backfilling was completed. The open cut sections were dewatered, for the most part, to six feet below proposed bottom of the excavation prior to excavating and the ground water did not present any problems during the installation and backfilling of the sewer pipe.

The tunneled sections did not require dewatering as the MTBM was equipped with an earth pressure balanced cutting chamber and was capable of mining both cohesive and non-cohesive soils in a dry or water bearing condition.

**Urban Setting**

Approximately one third of the sewer alignment was located in urban areas. In order to minimize disruptions to the public, the entire length of the sewer in these areas was tunneled. The alignment and shaft locations were chosen so as to minimize the effect of the construction on businesses and homes in the area. Structures that were adjacent to the sewer were also monitored during the tunneling operations for possible settlement. The measured settlement was less than the 0.25 inches and within design limits.

**River Crossing**

The sewer crossed Mill Creek south of the existing Wastewater Treatment Plant between station 303+40 and station 304+90. Because the depth to the top of the sewer below the creek bed was as little as two feet, the amount of soil cover over the sewer pipe was not sufficient to allow for tunneling to be used.

The Contract Documents called for the river crossing to be open trench with either sheet pile cells or cofferdams constructed to dam the flows during the installation of the sewer pipe. SEI proposed the use of water inflated portable dams (Aqua Barriers)

**Figure 6. Tunneling plant layout at shaft**

as an alternative method to dam the flow during construction. This alternative method was accepted by the City's Management Team (CMT) and by performing the work in September, when the flows in the creek were at a minimum, the river crossing was successfully completed.

## CONSTRUCTION

### Open Cut

The open trench portion of the project was accomplished using a Liebherr R 984 trackhoe with a 7.2 cubic yard bucket and a maximum digging depth of 26 feet. The depth of the open cut excavation was generally in the order of 40 to 45 feet. In order for the trackhoe to reach down to this depth SEI first had to pre-cut or bench down to 15 to 20 feet and use stacked trench boxes to support the lower 20+ feet of the excavation. A second smaller trackhoe (Kabelco—SK480) with a combination hoe pack and bucket was used to backfill and compact the

Figure 7. Placing portable dam in Mill Creek before inflating with water

Figure 9. Installing pipe at Mill Creek crossing

Figure 8. Portable dam inflated with water at Mill Creek crossing

Figure 10. Pipe installation in open trench section

lower part of the excavation. A Caterpillar 815 self propelled sheepsfoot roller was used to compact the upper benched area. The Contractor had little problem meeting the specified density in the backfill.

The installation rates achieved by the Contractor for the open cut ranged from 60 feet to 150 feet per shift, with an average of 90 feet per shift. A shift is considered to be 10 hours per day.

**Shaft Excavation and Support**

The shafts were constructed using steel ribs or ring beams with vertical wood lagging for the upper approximately 30 feet and with steel liner plates for the bottom portion of the shaft. The top 16 feet of the shaft support, referred to as the "Pickle Barrel" by the Contractor, was assembled above ground. The upper approximately 12 feet of the shaft was excavated using a trackhoe and the "Pickle Barrel" was lowered into the excavation and grouted in place. The remainder of the excavation was completed using a Clamshell bucket and hand digging with pneumatic spades. The openings in the steel liner plates, where the tunnel pipe entered or exited the shaft, were reinforced using steel H-Beams. Pressure grouting was also performed to fill any voids behind the liner plates, with special emphasis in the area where the pipe entered and exited the shaft.

The construction of a shaft was generally accomplished within seven 10-hour work days.

**Tunneling**

The tunneling was performed using an Ackerman SL60 Microtunneling Boring Machine (MTBM) as previously described.

The length of the tunnel runs between the shafts ranged from 350 feet to 1,260 feet. The Contractor had concerns that the longer runs( greater than 1000') would produce jacking pressures that would exceed design limits and also that the MTBM would not be able to hold line and grade because of inherent inaccuracies in the Laser equipment at these distances. Because of these concerns SEI opted, at their

Figure 11. Compacting backfill with bucket plate compactor

Figure 14. Shaft excavation by hand with pneumatic spades

Figure 12. Pickle barrel

Figure 15. MTBM control panel

Figure 13. Shaft excavation with clamshell bucket

Figure 16. Launching the MTBM from the first shaft

**Figure 17. Intermediate jacking station**

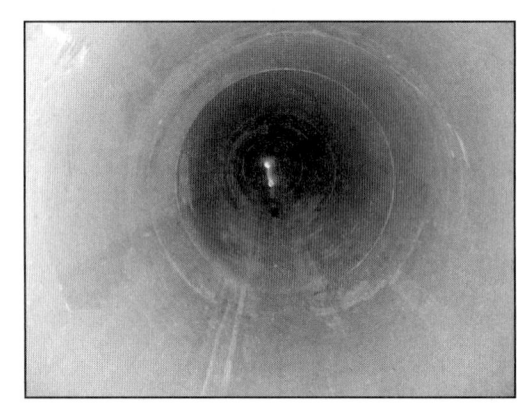

**Figure 19. CCFRPM sewer pipe installed**

**Figure 18. Jacking RCP pipe and advancing the MTBM**

expense, to construct two additional shafts halfway along the two longest runs. The job Specifications also called for the use of intermediate jacking stations to be used at the Contractor's discretion.

Problems were encountered during the second tunnel run of approximately 1000 feet. The MTBM became stuck at approximately 840 feet out from the shaft and a recovery shaft had to be excavated to retrieve the boring machine. The remaining 160 feet of tunnel was completed by hand mining. Initially SEI had opted not to use the intermediate jacking stations. However, after the MTBM became stuck intermediate jacking stations were introduced into the pipe string on the longer runs, as a precaution. If jacking pressures became too high the intermediate jacking stations would be used. However, the jacking pressures were such that the intermediate jacking stations did not need to be activated on any of the runs.

After each tunneling run, the cutter head on the MTBM was rebuilt because of wear. This involved replacing cutting head and hand facing the outside of the cutting head. This repair work generally took two to three days to complete.

The tunneling rates achieved by the Contractor for a 12-hour shift ranged from 45 feet per shift to 105 feet per shift with an average of 70 feet per shift. This rate does not include the hand mined section.

The condition of the in-place product was visually checked by walking the entire length of the completed Trunk Interceptor Sewer. The condition of the in-place sewer was found to be good with the exception of some slight infiltration at some of the grout ports and leakage at one of the pipe joints in the open cut section. The grout ports were repaired by replacing the bungs at the leaking ports and the leakage at the pipe joint was sealed by chemically grouting around the joint.

**Laser Guidance**

A surveying laser mounted independently of the thrust block was used to maintain the alignment of the MTBM. Although we live in the era of star wars, where lasers are used as part of our missile defense, there is a limit to the distance that a commercial laser can be considered accurate. Since 9/11 the US government has banned the commercial sale of lasers that are accurate over long distances, for obvious reasons. The accuracy of the laser, therefore, has to be taken into account in determining the length of the tunnel run.

The laser beam is also affected by temperature differential. The temperature at the head of the MTBM can exceed 100° F and the temperature in the shaft may be as low as 60° F. This difference in temperature causes the laser beam to bend downwards. SEI used an air duct system to continuously pump air from the shafts to the back of the MTBM during tunneling and was successful in minimizing the downward bend in the laser beam.

**Figure 20. Checking laser alignment during tunneling**

The tunnel pipe was installed essentially within the specified tolerances of not more than 1 inch from the design grade in the vertical direction and 3 inches from the design alignment in the horizontal direction.

**Settlement and Displacement Monitoring**

Monitoring instruments were specified to determine ground behavior for comparison with design assumptions and to provide timely warning for the implementation of remedial measures to prevent damage to structures, CSX railroad tracks, equipment and utilities.

Inclinometers were installed at all of the shaft locations to measure the lateral ground displacements during construction of the shafts. The lateral displacement measured at the shaft locations did not exceed the specified 0.5 inches and was generally less than 0.25 inch.

Settlement monitoring devices were installed at strategic locations along the tunnel route to measure heave/settlement. The maximum settlement measured was 0.25 inches.

## CONCLUSIONS

Although the Marysville Trunk Interceptor Project could be considered a relatively straight forward sewer line job there are always challenges and uncertainties with underground construction and if not handled properly may result in significant problems and cost overruns. The fact that the construction was completed within budget, four months ahead of schedule, with change orders on the project amounting to less than 0.5 percent and with no lost time due to injuries (SEI received the Ohio Contractors Association 2008 Safety Award in Division III) would suggest a very successful project and speaks well for the cooperation that existed between the project team members in resolving any issues that arose so that there were minimal disruptions to the project.

The success of the project can be attributed to several factors, namely: 1. A knowledgeable and informed owner, who understood the process and remained closely involved in the decision making from the start, all the way to completion of the project, 2. A well prepared set of plans and specifications, 3. An experienced contractor whose work force was not only efficient but took pride in the work they performed, and 4. The construction management staff who interfaced smoothly with both the City of Marysville and the Contractor, to resolve problems, and respond to Requests for Information, Change Orders, etc. in a timely manner.

# Case History: Innovative CSO Pipe Installation in a Congested Urban Setting

**Emad Farouz**
CH2M Hill, Chantilly, Virginia

**ABSTRACT:** Contract 8 is a part of the Narragansett Bay Commission Combined Sewer Overflow (NBC CSO) project located in Providence, Rhode Island. Contract 8 included the installation of 550 feet long 60-inch Interior Diameter (ID) consolidation conduit, at depth of more than 25-ft below ground surface; which is located in a very congested area with various utilities and existing structures. The subsurface conditions were challenging, including very soft clays with shear strength of less than 250 pounds per square foot and average blow counts of 2 blows-per-foot. Moreover, groundwater was encountered at about 10 feet below ground surface. Open-cut construction was not feasible, given the site's extensive utility lines, ground water, and existing structures. A further complication was the presence of a retaining wall that supported a traffic ramp for I-95. The retaining wall was constructed on battered piles that were as close as 2 feet from the outside pipe alignment. The project team contemplated the use of microtunneling but the soft soils presented a challenge for the alignment of the microtunneling. To overcome the challenging soft ground, ground improvement using jet grouting was utilized ahead of the microtunneling to provide adequate bearing capacity for the microtunnel boring machine. The jet grouting consisted of 3-ft diameter columns, and improved a 13-foot by 13-foot cross-sectional area, centered on the pipe axis. The pipe invert was approximately 25 feet below ground surface. Despite the limited space, number of existing structures, and soil conditions, the project was successfully executed due to the innovative practices of jet-grouting and microtunneling. The daily production rates averaged about 60-foot of installed pipe per 10-hour shift per day.

## INTRODUCTION

### Project Overview

The Narragansett Bay Commission's (NBC's) Main Spine Combined Sewer Overflow (CSO) Tunnel went on line Saturday, November 1, 2008. The NBC owns and operates the interceptor and wastewater treatment facilities serving 10 communities with a total population of 360,000. The project includes five CSOs that discharge to the Woonasquatucket, West, Moshassuck, Seekonk, and Providence Rivers. These rivers are tributaries to Rhode Island's Narragansett Bay, an "estuary of national significance."

NBC developed a comprehensive program in order to abate CSO pollution in the Upper Narragansett Bay. Phase 1 of the program includes a 7.9-m (26-ft) diameter, 70.1-m (230-ft) deep storage tunnel, drop shafts, a CSO pump station, and several near-surface interceptors. The project also included detailed design and services during construction of near-surface CSO control facilities, consisting of diversion structures, screening structures, consolidation conduits, approach and vortex structures, and inflow control gates. Of these, the installation of Consolidation Conduit, with insider diameter of 60-inch, between Diversion Structure 04 and the Gate and screening 04/61 facility is the focus of this paper.

The NBC CSO Project site for the Construction Package 8 (CP8) is located in Providence, RI, near the I-95, as shown in Figure 1. Site plan 1. A 152-cm (60-in.) diameter Consolidation Conduit, approximately 165.7 m (550 ft) in length, was to connect the two structures, as shown in Figure 1. Site plan 2. The excavations for the Diversion Structure and the Gates and Screening Structure would also be used as work shafts for the microtunnel excavation to install the Consolidation Conduit.

### Subsurface Conditions

The subsurface soils consisted of the following:

- Medium Dense Granular Fill to a depth of approximately 10–12 feet.
- Alluvial-Estuarine Deposits in the form of loose to medium dense silty-fine sand to depths of approximately 13.5–16 feet.
- At the northern end of the alignment, 12-foot thick layer of soft to medium dense organic silt and fibrous peat was identified.

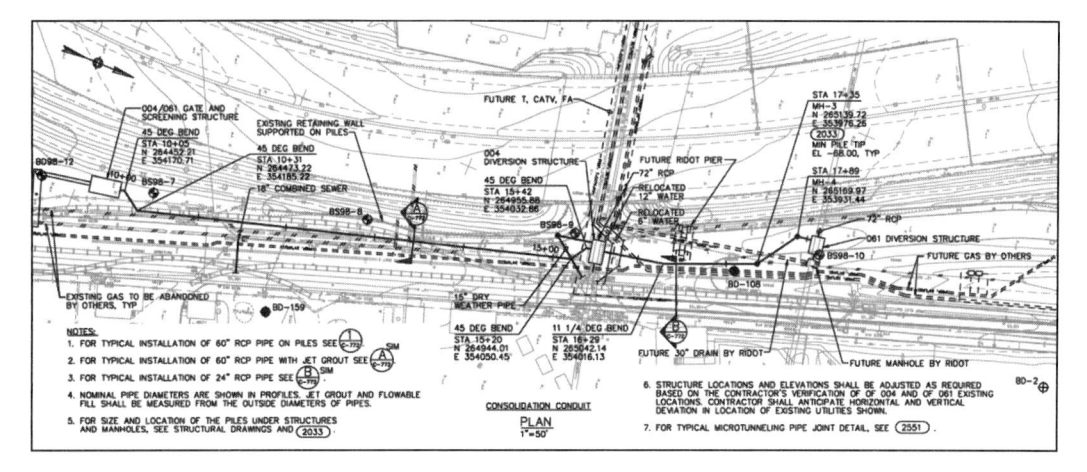

**Figure 1. Site plan**

- Glaciolacustrine deposits underlay the entire site to the bottom of the investigated depth. They consist of stiff to medium stiff silt with clay.

The invert of the proposed conduit is in the Alluvial-Estuarine Deposits, with the exception of one area, where the alignment is within the organic deposits. The ground water table is at approximately elevation 2.0 feet, or approximately 3.3 m (10-ft) below ground surface. The invert of the proposed consolidation conduit is at elevation—9.35 feet.

## Site Constraints

There are numerous existing utilities present at the project site, especially along consolidation conduit alignment. Those that posed challenges for the installation of the consolidation conduit included the following:

- Existing reinforced concrete retaining wall No. 2 that is carrying ramp for Interstate 95. The as built drawings for the wall indicated that the wall is supported on two rows of piles. The piles are 12-inch Cast-In-Place (CIP) Concrete piles with spacing ranging between a minimum of 0.914 m (3-ft) to a maximum of 1.83 m (6-ft) c-c. In particular, at the southern end of the wall, where the existing piles are closer to the proposed pipe, the battered piles appear to be at their maximum spacing.
- Electrical conduits above the consolidation conduit.
- Overhead Electric Poles.
- Gas lines to be abandoned 76-cm (30-in.) and 46.6-cm (16-in.) above the consolidation conduit.
- A 38 cm (15 in.) sewer line above the consolidation conduit.

Figure 2 shows a plan that illustrates the existing utilities along the conduit.

Additionally, the microtunnel excavation for the Consolidation Conduit crosses directly under and within approximately one tunnel diameter of an existing manhole.

## EVALUATION OF INSTALLATION TECHNIQUES

Two installation techniques were considered for the consolidation conduit installation:

1. Open-cut construction, with a three meter wide (ten foot) wide excavation supported by jet grouting walls, and jet grouted bottom plug to provide both for temporary support and permanent pipe support.
2. Microtunneling installation through improved ground by jet grouting.

Several issues have been identified that could potentially impact the design and construction of the jet grouting, as described below. All of these have evident economic impact on the construction. Listed in no particular order:

- The cut-and-cover design for this relatively deep cut, which is approximately over 6 m (20-ft) below the water table, will most likely require a very thick jet grouted wall. The soilcrete thickness will also have to account for the

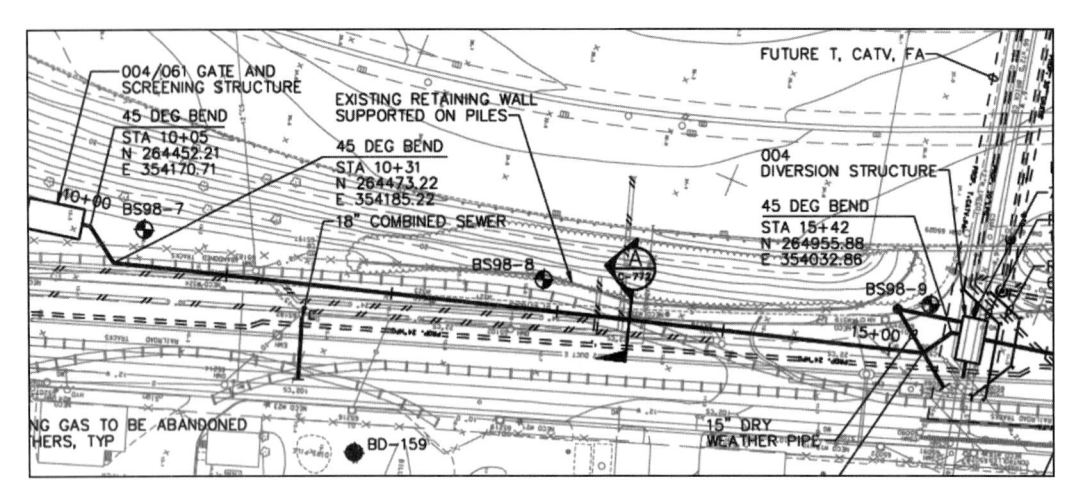

Figure 2. Consolidation conduit and existing utilities plan

variability in the soil profile, i.e., the jet grouting treatment will have to be designed for the worst case scenario. As a result, the jet grouting treatment will have to be extended a considerable distance outside of the anticipated ten foot wide trench. This will increase the number of utilities (underground and aerial) and structures impacted by the jet grouting work on both sides of the trench. In particular, along the side closer to I-95, the jet grouting may impact the retaining wall and its foundations. Also, along the other side, the jet grouting may encroach on the abandoned sheeting.

- The top elevation of the jet grouting for the bottom plug will have to be extended to a considerable elevation below the microtunnel invert 1 meter (3-ft), to ensure proper bearing against the pipe foundation bedding. In the area where the soft organic silts/peat is present, the deepening of the jet grouting plug to the top of the glaciolacustrine deposits will have to be implemented to avoid differential settlements and damages to the pipe.

In both alternatives the jet grouting may be installed in the vicinity of the battered piles. Depending on the condition of the existing retaining wall and on the pile tip elevation, which was uncertain at the time of evaluation, it was decided that it will be necessary to spread-out the installation sequence of the jet grout columns in order to minimize potential instabilities to the existing wall or its foundation. The wall has been constructed with joints every 5.5 to 6.4 m (18 to 21-ft) of wall length. It was planned that the number of jet grout columns installed at the same shift should be limited to 2 columns per wall length section, to limit the undermining of the piles supporting the existing retaining walls.

Based on the above and evaluating the risks associated with both alternatives, the microtunneling scheme would present the following advantages:

1. Minimize the jet grouting volume, therefore the total cost of ground treatment and also the impact on the surrounding neighborhood is considerably less than the open cut alternative. This means less spoil to be trucked off site, less cement to be delivered, shorter construction duration, and overall less risk.
2. Reduce the soil types that need to be treated, thus providing for a more efficient and potentially more cost effective ground treatment.
3. Reduce impact on existing utilities by minimizing footprint of ground treatment. Also, the ground treatment would be at a greater depth, thus minimizing the risk of undermining existing utilities and structures above the proposed pipe alignment.
4. Attenuate the interference with the existing piles by maintaining the treatment at a greater distance from the piles and by allowing for a more flexible layout (i.e., battered holes where necessary).

Based on the above evaluation and cost considerations; the option including pre-treatment of the ground using jet-grouting then microtunneling through the soilcrete mass was selected.

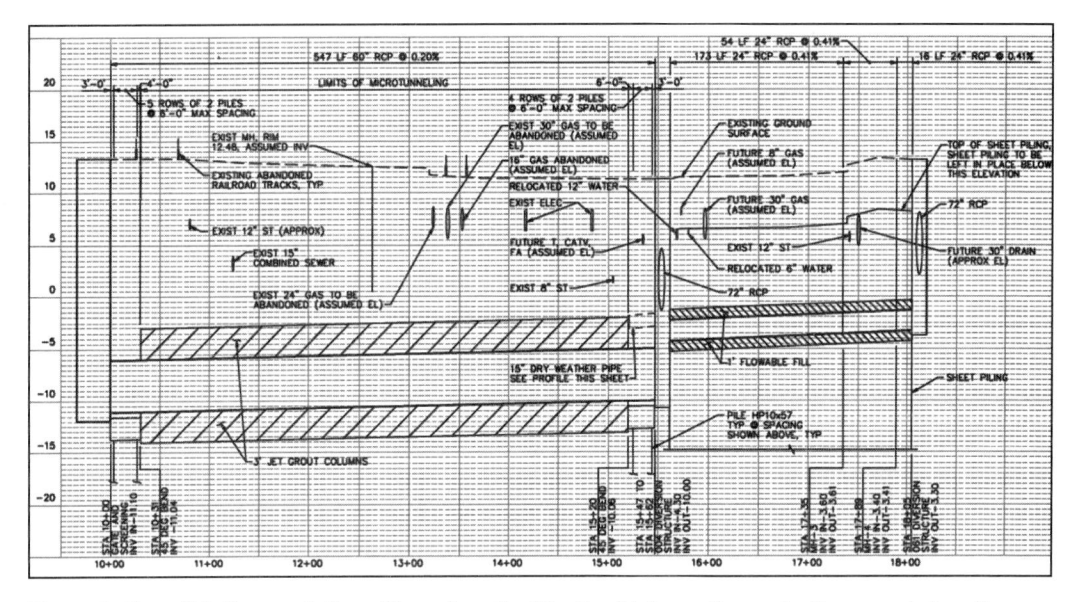

Figure 3. Consolidation conduit profile and vertical limits of jet grouting and pile supported sections

## JET-GROUTING AND MICROTUNNELING DESIGN

The Consolidation Conduit is 152 -cm (60-in.) ID and approximately 167.7-m (550-ft) long that connects the Gates and Screening Structure to the Diversion Structure. During design, it was anticipated that one of the most likely tunneling methods for installing the conduit would involve the use microtunneling through the jet grouted mass. The use of tunnel shield without applying face pressure was considered, however, the risk of imperfect jet grout mass and risks associated with a blow in and settlement lead the team to select microtunneling with face pressure. The tunnel excavation was anticipated to be primarily in alluvial deposits and Peat below the groundwater table, and ground improvement would be required in order to maintain the microtunneling alignment. Additionally, the alignment of the conduit was as close as 0.67 m (2-ft) from the battered piles supported the existing retaining wall. It was anticipated that the piles may not have been installed exactly as planned on the design drawings; therefore accurate tunnel alignment was considered very important. Additionally, there were concerns that settlements caused by tunneling could damage surface facilities located approximately 4 m (13 ft) above the tunnel zone.

### Ground Improvement along the Consolidation Conduit Tunnel Alignment

The limits of the jet grout zone along the Consolidation Conduit tunnel alignment are shown in Figures 3 and 4. Plaxis was used to model the size of the jet grout zone and the required strength. The types of jet grout zones along the tunnel alignment are described below:

- Based on the site constraints and discussion with spatiality contractor, it was estimated that the soil-crete mass will achieve on average 500 psi in 28 days. The area where peat was encountered may only have strength of 100 was specified.
- The Jet Grout mass will have coefficient of permeability not to exceed $1 \times 10^{-5}$ cm/sec.
- Figure 5 shows the Plaxis output showing maximum deformation that predicted within the jet grout mass and soil above it. The predicted movement was considered acceptable.

As a result the size of the jet grout zone specified was 3.4 m by 3.4 m (13 f by 13) as shown in Figure 4.

As shown in Figure 3, the alignment of the consolidation conduit, was not conducive to have the entire 167.7 (550ft) to be pipe jacked. Therefore, the first and the last 30-ft of the alignment were open cut and supported on piles. At these locations no existing utilities were in conflict with the consolidation conduit. Since the gate and screening and Diversion structure were also supported on piles, it was decided that the launching shaft and receiving shafts, will be incorporated within the same excavation for the gate and screening and Diversion structure, respectively.

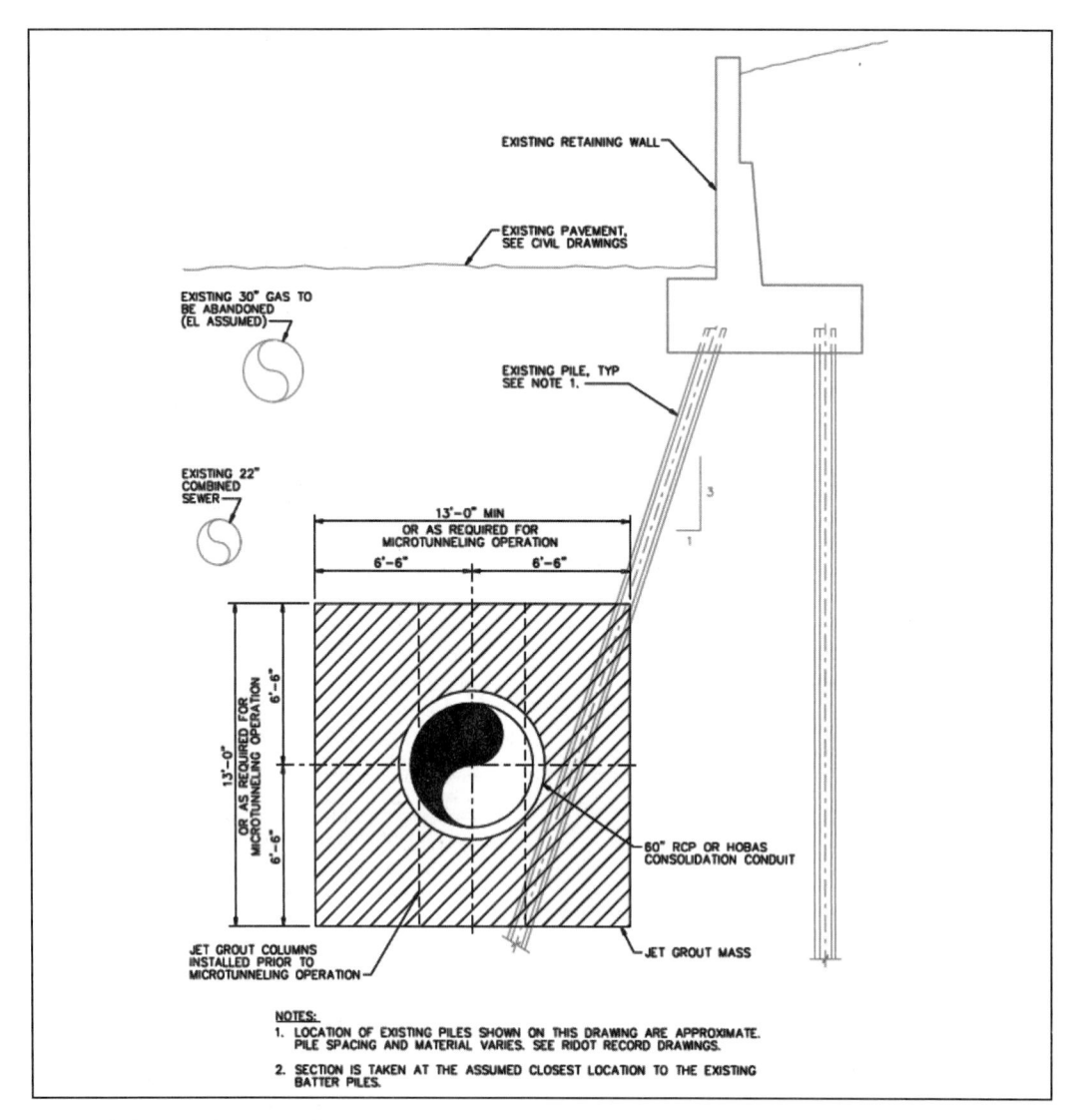

Figure 4. Jet grouting and microtunnel near the existing wall piling

The pile supported sections of the consolidation conduit were support one two 10×57 H-piles, which are spaced approximately at 6-ft along the consolidation conduit. Figure 6 shows the pile supported section of the conduit.

The excavation and support sequence for the construction of the consolidation conduit is outlined below. These steps have been generalized for discussion purposes.

1. Relocate overhead lines and certain underground utilities in the vicinity of the proposed construction that will interfere with the equipment.
2. Install excavation support for the Gate and screening structure and launching shaft for the microtunneling within the same excavation support system. Tight sheeting braced internally, with entrance and exit eyes using jet grouting were specified.
3. Perform test pits to locate the as built location of the battered piles supporting the existing retaining wall.
4. Once the battered piles are located, plan the locations of the jet grout columns centers, which will be located in between the existing piles.

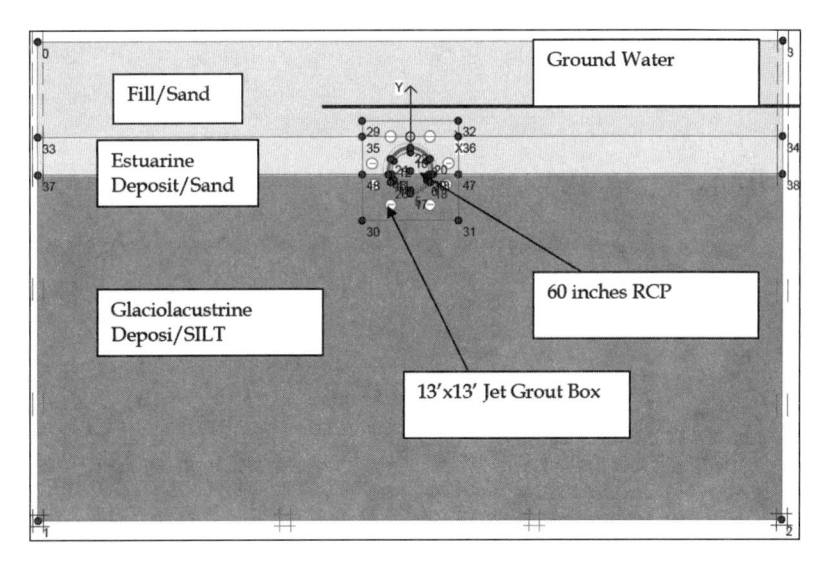

Figure 5. Jet grouting and microtunnel near the existing wall piling

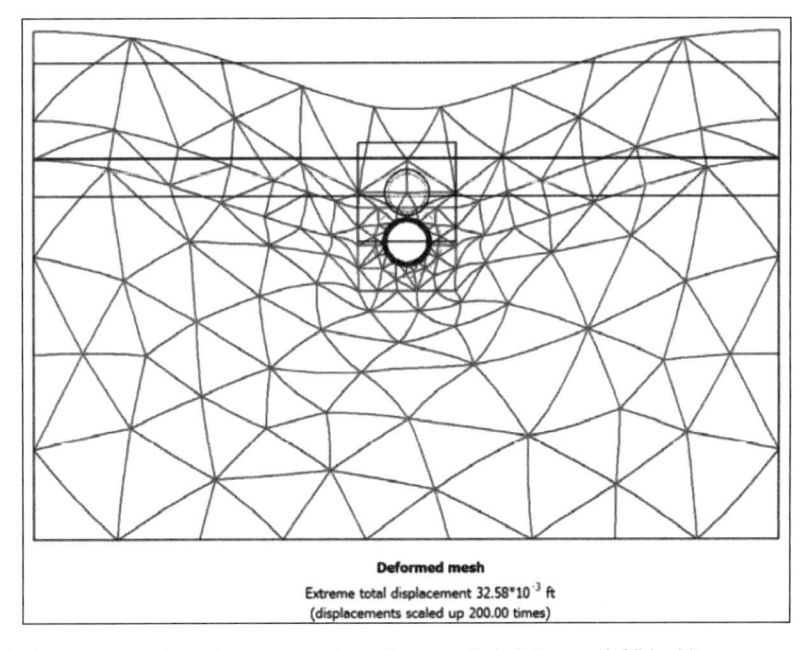

Figure 6. Maximum predicted displacement based on plaxis is 7.5 mm (0.35 inch)

5. Install at most two jet grout columns at the same shift within one existing wall panel. The existing wall panel is 5.5 to 6.4 m (18 to 21 ft) long. This is intended such as the piles will not be undermined while the soilcrete is gaining strength.
6. Perform strength and Permeability tests to confirm the jet grouting meets the specified parameters.
7. Lower the MTBM into the Launching Shaft.
8. Start Tunneling once the Jet grout has attained the specified strength.
9. Perform excavation support for the Diversion structure and receiving launching shaft for the microtunneling within the same excavation support system.
10. Remove the second-level excavation support frame. Finish the final walls of the Diversion

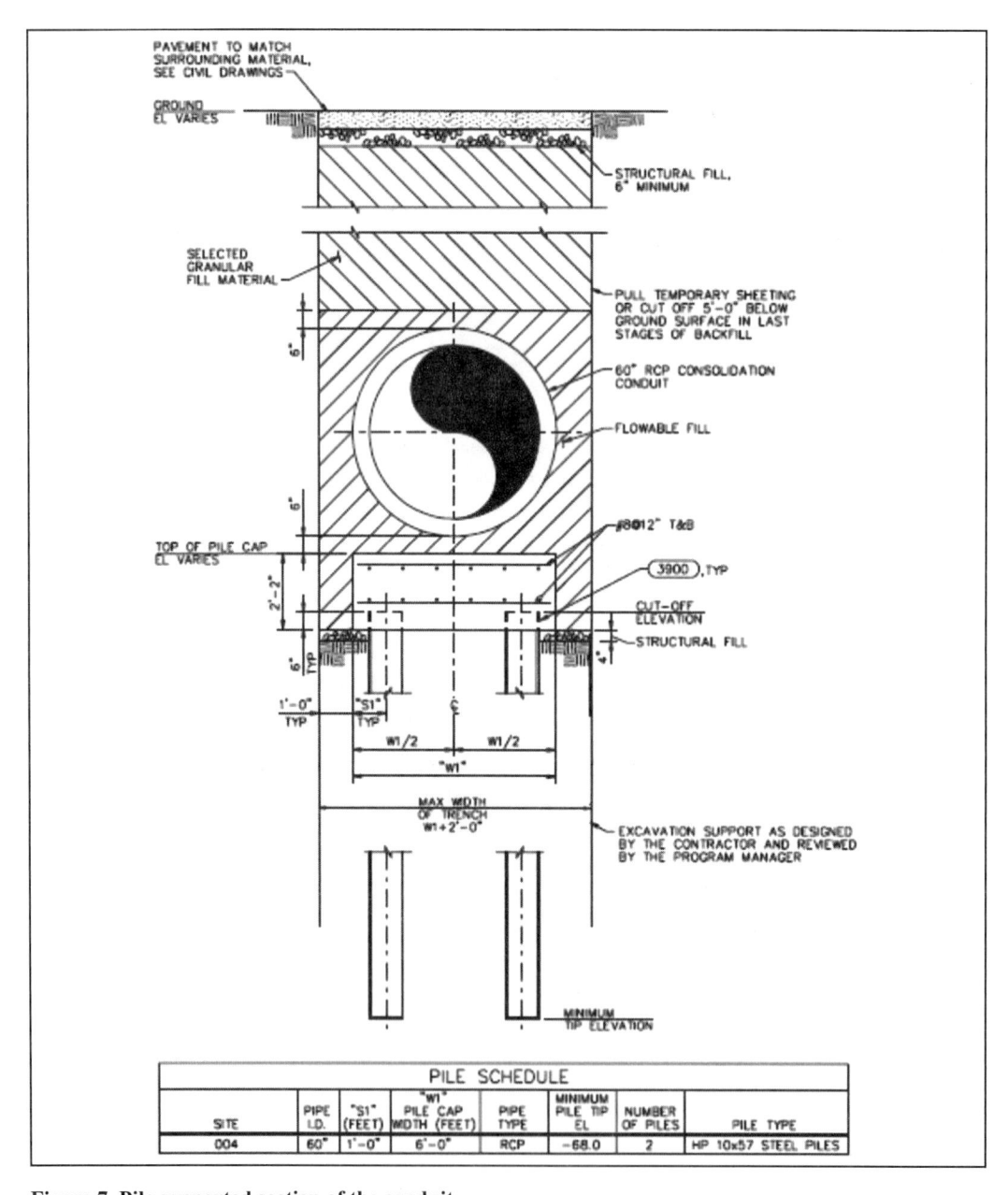

**Figure 7. Pile supported section of the conduit**

Structure by completing the walls at the previously blocked out areas.

## CONSTRUCTION

The construction generally followed the construction sequence listed above. Figure 8 shows a photo of the launch of the MTBM through the launching shaft located near the excavation support for gate and screening structure. Figure 9 shows a photo of tunnel shield breaking through the jet grouting at the Diversion Structure. Figure 10 shows the MTBM machine breaking through the Diversion Structure. Hobas pipe was jacked behind the MTBM. The installation progressed very successfully, without any significant issues. The entire microtunnel was completed within two weeks with peak daily production of 100-ft per day.

**Figure 8.** Photo of the diversion structure after completion of the secant pile walls and jet grouting

**Figure 9.** Tunnel shield breaking through the jet grouting at the diversion structure

## CONCLUSIONS

The project was completed successfully in spite of the many challenges during the design and construction. The following conclusions are made by the authors from the design and construction:

- Jet Grouting provided a versatile ground treatment that enables this challenging and complex underground project to be completely successfully. The use of jet grouting on this project was quite successful in lowering the risk of damaging the existing utilities and other facilities from ground movements due to tunnel and shaft excavations.
- It is essential to have adequate cover of Jet grout that is at least 50% of the excavated tunnel diameter to avoid excavating in and out of soilcrete zones. This could have posed significant challenge to steering the microtunnel and maintaining the planned alignment.
- Imperfection of the jet grouting process should be recognized and planned for, especially where grouting is planned adjacent to extensive existing utilities and piling and challenging subsurface conditions such as organic silt and Peat.
- Developing a realistic jet grout specification that can be achieved is an important step during the design.
- Specifying conditions such as limiting the installation of jet grout columns to two jet grout columns at the same time within an existing wall panel is essential to the existing wall stability and to provide the contractor

**Figure 10.** Photo showing MTBM breaking through the receiving shaft

bidding on the project with adequate information to budget money and schedule time appropriately. In the author's opinion, despite the fact that this can be considered contractor's "means and methods," the engineer should layout these conditions during design, if the "means and methods" will impact the stability of existing structures.
- The use of qualified contractor is essential in completing this challenging project successfully.
- The successful design and construction of the NBC consolidation conduit provides a clear example of the level of complexity that can be achieved in large urban construction even under severe site and operational constraints.

# Pipe Jacking Through Hardpan: A Case History—North Gratiot Interceptor Drain Phase I

**Joseph B. Alberts, Jason R. Edberg**
NTH Consultants, Detroit, Michigan

**Keith Graboske**
Office of the Macomb County Public Works Commissioner, Macomb County, Michigan

**Gordon Wilson**
Anderson, Eckstein, and Westrick, Shelby Township, Michigan

**Steven Mancini**
Ric-Man Construction, Inc., Sterling Heights, Michigan

**ABSTRACT:** The North Gratiot Interceptor Drain Phase I project is being constructed in Macomb County, Michigan using Pipe Jacking. The project is unique for tunneling considerations because it is being built using open face tunnel boring equipment through hardpan materials. The hardpan has limited clay content and contains boulders. The available cover between the hardpan and overlying wet granular materials is very limited. The contractor has been successful in constructing the project with long jacking runs and relatively low jacking pressures, while overcoming boulders.

## PROJECT BACKGROUND

The North Gratiot Interceptor Drain project is being implemented by the Office of the Macomb County Public Works Commissioner in Macomb County, Michigan in order to provide additional service capacity to northeastern Macomb County. Phase I of the project is the southernmost end of the proposed project and consists of constructing approximately 13,000 lineal feet of 66-inch inside diameter finished sewer. The Phase I sewer will receive flow at the northern end from both existing sewers and future project phases and will outlet at the south end to an existing 11-foot diameter sewer constructed in tunnel in the 1970s.

The Phase I sewer is being constructed beneath and immediately adjacent to an existing 42-inch sewer constructed in open excavation in the late 1980s. Due to the 60-foot wide sewer easement, proximity to the existing 42-inch sewer, large billboards, open storm drains, an interstate freeway, and an existing 42-inch diameter Detroit Water and Sewerage Department water main, trenchless methods were selected for construction.

To carry the required flow, the sewer is designed as a 66-inch inside diameter conduit at a depth of generally 45 to 50 feet below the ground surface.

Concrete pipe with Type IP cement was selected as the conduit material to resist deterioration due to hydrogen sulfide attack.

## SUBSURFACE CONDITIONS

As part of the project design, a geotechnical investigation was conducted incorporating previous information from the open cut sewer project as well as performing new test borings resulting in an effective test boring spacing of approximately 500 feet. The investigation revealed a design soil profile generally consisting of an upper layer varying between soft clay and granular materials and a lower layer of very compact clayey sand and silt locally termed *hardpan* materials. Groundwater was generally identified at a depth of approximately 10 feet below the ground surface which roughly correlates to the water surface of Lake Saint Clair approximately one mile east of the project. Based on hydraulic design considerations and site constraints, the proposed sewer vertical alignment was located within the hardpan materials.

Several tunnels have been constructed through the hardpan materials in the general vicinity of the project, namely the 11-foot Lakeshore Interceptor, to which the North Gratiot Interceptor

Figure 1. Map of Michigan, Macomb County highlighted

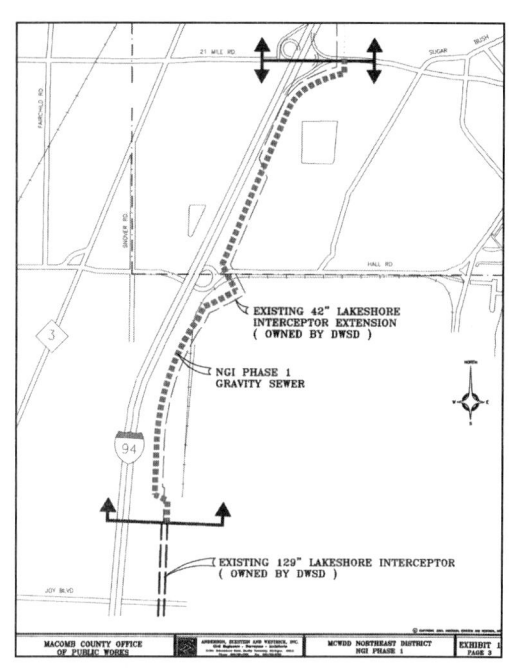

Figure 2. Phase 1 project area

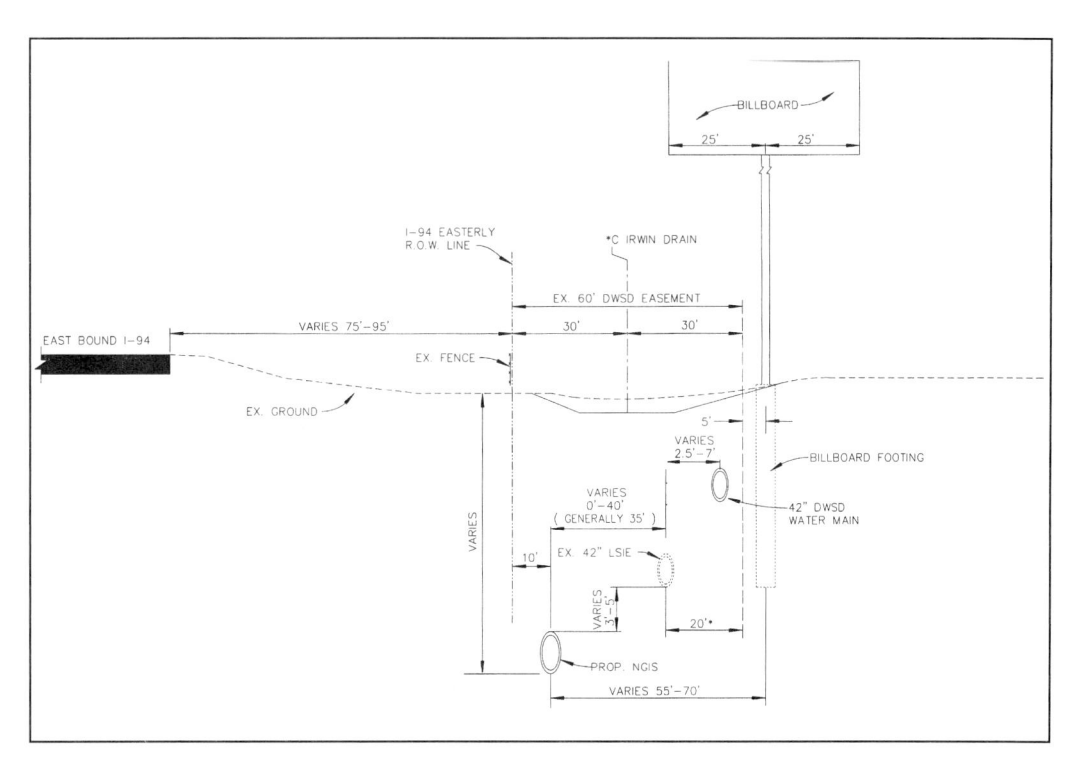

Figure 3. Cross-section showing surrounding infrastructure

805

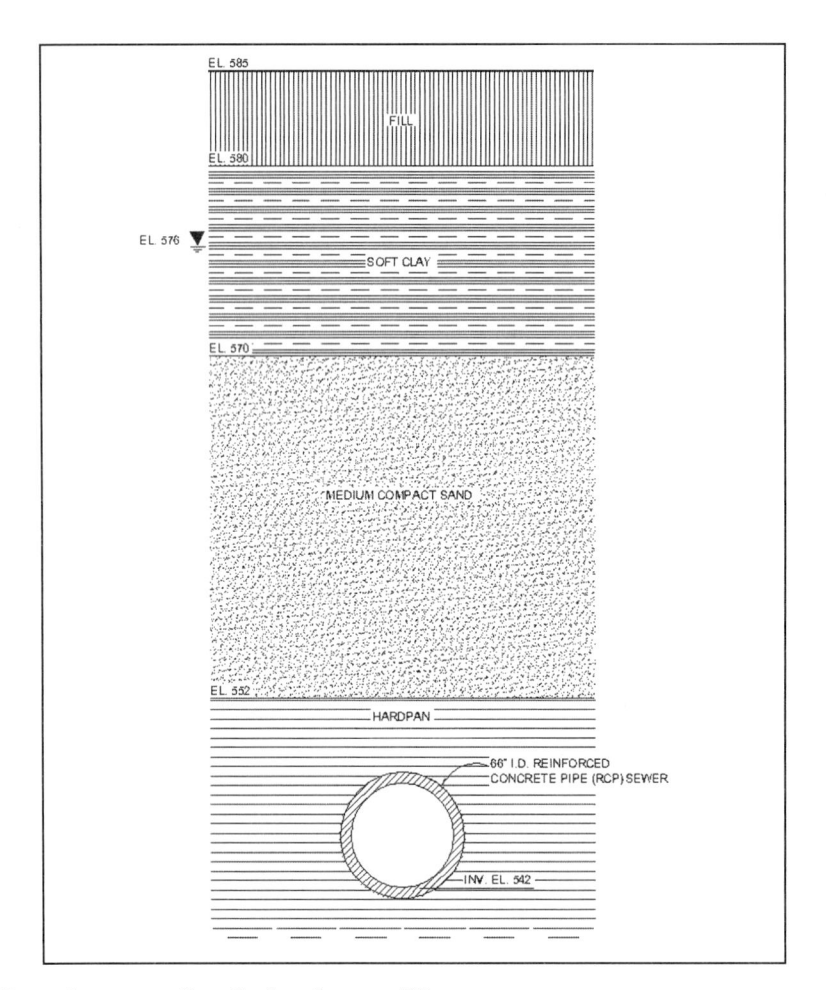

**Figure 4. Example cross-section of subsurface conditions**

Drain project connects at the downstream end, and a recent shorter tunnel constructed in nearby Mt. Clemens, Michigan. The Lakeshore Interceptor was constructed in the early 1970s with a primary liner and cast-in-place secondary liner under air pressure. During construction of the Lakeshore Interceptor, an explosion occurred in the tunnel when high concentrations of methane gas were encountered. Information on boulders encountered during construction of the Lakeshore Interceptor was unavailable. The Mt. Clemens project was constructed using a primary liner with pre-cast concrete pipe secondary liner. Approximately 10 boulders of approximately 24 inch size were encountered in the 1,200 lineal feet of 54-inch finished diameter tunnel. Methane was also encountered on the Mt. Clemens project within granular seams in the hardpan.

## TUNNEL DESIGN CONSIDERATIONS

The design team chose to use trenchless methods to install the sewer based on the collected information and surrounding infrastructure. Major design and construction considerations included: identification of anticipated hardpan behavior during tunneling, boulders within the hardpan, required competent hardpan cover between the tunnel and overlying wet granular materials, dewatering requirements, and the resulting selection of tunneling methods. Open cut excavation was deemed infeasible due to the depth of the proposed sewer and proximity of the existing infrastructure.

Based on laboratory testing, the hardpan materials contained on the order of 13 to 17 percent clay, with the remaining material consisting of approximately equal amounts of silt and sand. Liquid and plastic limits generally ranged from 16 to 11 percent

**Table 1. Average composition of hardpan soils**

| Colloids/Clay | Silt | Fine Sand | Medium Sand | Coarse Sand | Gravel |
|---|---|---|---|---|---|
| 16% | 34% | 31% | 8% | 5% | 6% |

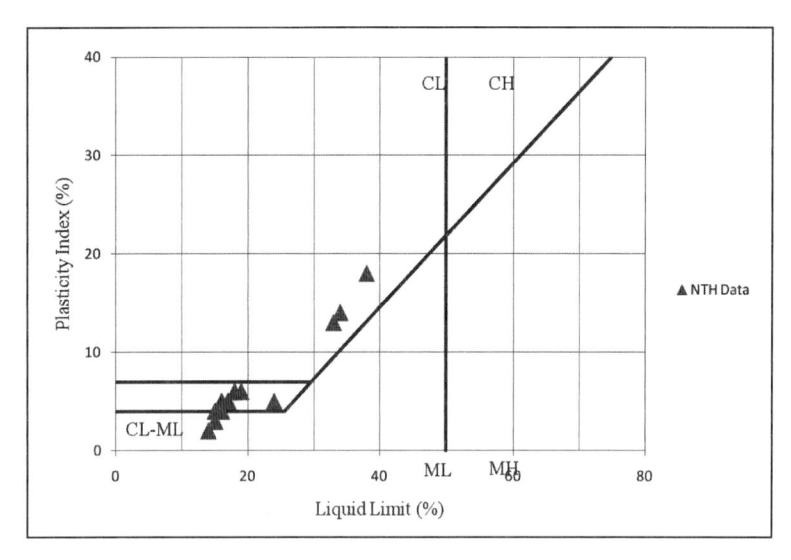

**Figure 5. Plasticity of hardpan soils**

with a corresponding plasticity index ranging from 3 to 5. Moisture contents were approximately 7 to 10 percent. Based on plasticity, the material would be classified as ML bordering on CL-ML material. Unconfined compressive strength testing measured strengths ranging from 3,000 to 24,000 psf. Evaluation of hydrometer and limit data raised significant concerns regarding the ability of the hardpan to present stable face conditions for open air tunneling with conventional mining and development of possible friction due to collapsing of overcuts for jack pipe tunneling. After reviewing historical data, laboratory testing, and prospective tunneling methods, the design team concluded that the hardpan materials were capable of limited stand-up time and would have a relatively low permeability; however some areas may present raveling and/or running face conditions.

Boulders within the hardpan were not encountered during the investigation; however boulders were expected based on records from previous tunnels constructed in the area and discussions with persons involved in the construction of those tunnels. The design team estimated two boulders having an average diameter greater than 24 inches would be encountered within each 100 feet of tunnel, based primarily on the adjacent Mt. Clemens project. It was also expected that many small-diameter boulders (less than 24 inches) would be encountered, making the presence of boulders a major factor in determining the tunneling methods specified for the project.

Hardpan cover between the top of the proposed tunnel bore and overlying granular material generally ranged from 4 to 10 feet, which roughly corresponds to ½ to 1½ tunnel diameters. Generally two tunnel diameters are recommended for design. However, the high compressive strength, low permeability, and load-carrying capabilities of the hardpan materials due to soil arching led the design team to determine that the available cover was acceptable and that an appropriate tunneling method could be selected to address the condition. Simple-shear methods for overhead arching, as well as methods presented in *Earth Tunneling with Steel Supports* by Proctor and White, were used for this analysis. To provide some reserve capacity and to compensate for potential lack of cover, dewatering was recommended in low hardpan cover areas to reduce hydrostatic pressure and minimize risk of breaching the cover into an open-atmosphere tunnel.

Several tunneling methods were evaluated for use on the project. Due to the expected presence of boulders and availability of other feasible options, micro-tunneling methods did not appear to be the best option for this project. Local practice and approved methods of governmental agencies leaned towards a final product of concrete lining. A primary liner of steel rib and wood lagging while technically feasible in the ground conditions and advantageous due to the length of possible tunnel runs, was eliminated due to the larger tunnel bore diameter required. This would have further reduced the hardpan cover by one-foot resulting in a 25% decrease in low cover areas. Use of concrete jack pipe methods combined with an open atmosphere tunnel boring machine was finally selected as the preferred method. Manholes were spaced at approximately 500 to 800 feet along the alignment and corresponding horizontal alignment changes were made at the manholes to provide straight runs between the manholes and provide an optimum location within the available easement.

The final project design consisted of ASTM C-76 Class V concrete pipe installed using jack pipe methods with a conventional tunnel boring machine. Removal of boulders less than 24 inches in average diameter was classified as incidental to the project. Removal of two boulders per 100 feet of tunnel between 24 inches and 48 inches in diameter were included as pay items. Removal of boulders greater than 48 in diameter were to be compensated for on a time and material basis. Dewatering of water bearing granular materials was also required as a baseline condition in low hardpan cover areas to reduce risk of inundating the tunnel with flowing materials and breaching of the hardpan.

## CONSTRUCTION

### Hand Mining

The construction phase of the project was awarded to Ric-Man Construction, Inc. through the low bid process. The first construction element performed was a short connector between the proposed sewer and the existing 11-foot sewer. This connector was performed via hand-mining techniques from within the concrete pipe. The hand mining was the first test on the behavior of the hardpan materials to tunneling operations. The hardpan material exhibited almost unlimited stand-up time and very little raveling was experienced. Little groundwater and no significant boulders were encountered in this first 50-foot long run. This experience gave both the contractor and designers confidence that the hardpan would perform as anticipated in the project design and the possibility of increased jacking distances beyond the project design.

A second hand mine operation was performed where the proposed sewer crossed approximately 3 feet beneath the existing DWSD sewer. This mining was also performed from within the concrete pipe and the tunnel face again performed well with unlimited stand-up time. However, just past the invert of the existing sewer, a large boulder was encountered in the upper portion of the bore. After some discussion, this boulder was blasted in a few hours and the mining operation continued.

**Figure 6. Photo of hand-mined tunnel face**

### Main Tunneling

Mainline pipe jacking operations began in the spring of 2008 with an open-face tunnel boring machine. At first, the TBM face was relatively closed with a face consisting of button rollers in combination with spade or picker teeth. The TBM was soon modified to increase the opening at the tunnel face, and replace the button rollers with disk cutters. The cutter head configuration was modified several times throughout the first tunnel run, and it appeared that utilizing bullet teeth provided the best production while mining through the hardpan soils.

The RCP was calculated to have a joint capacity on the order of 1,200 tons using the ASCE Standard Practice for Direct Design of Precast Concrete Pipe for Jacking in Trenchless Construction (ASCE27-00) and concentric loading conditions. The joint contact area used for design was for the spigot shoulder lead to trailing bell and this was where wood packers were used. Contingency packers for the spigot face were also on site in case jacking stresses became large and the overall pipe area would be needed to distribute jacking forces more evenly. Based on the capacity of the pipe and adhesion from the hardpan soils, tunneling runs on the order of 600 to 800 feet initially appeared feasible during the beginning phases of construction. Ric-Man proposed tunnel runs of up to approximately 1,600 feet in length. This was thought to be optimistic by the design team.

Jacking operations were launched in August, 2008 from within a mining shaft using a thrust block at the back of the shaft and a jacking frame within the shaft. On subsequent tunnel runs, the jacking frame was recessed into a tail tunnel to decrease the required shaft size. The jacking frame had a capacity of approximately 1,600 tons. Medium viscosity bentonite was continuously pumped in a sequential pattern approximately 30 feet behind the heading to reduce skin friction. Jacking pressures and resulting thrust were monitored during tunneling by hand recording on a regular basis during mining operations. In general, the jacking forces required to advance the pipe were relatively low compared with theoretical estimated values. This was attributed to the tunneled hardpan materials remaining in place and not caving into the tunnel overcut resulting in an open tunnel annulus. Based on this information longer tunneling runs were then attempted. This behavior allowed the relatively long tunnel runs to be accomplished without overloading the concrete jack pipe. Overall, the required jacking force to advance the pipe was on the order of 200 pounds (0.1 tons) per linear foot of pipe. The jacking forces vs. distance for one of the longer runs are presented in Figure 8.

**Figure 7. Photo of reinforced concrete pipe showing wood packers**

These forces vs. distance remained relatively consistent during tunneling of other runs including steep grade changes during jacking due to normal tunneling variations, sand and silt zones encountered within the bore, and shut downs during weekends and holidays.

## Boulders

As on any tunnel project, boulders were an issue during construction. Identification of boulders was somewhat difficult because boulders greater than approximately 12 inches were generally cut or broken by the TBM into unidentifiable pieces. Larger boulders generally greater than 48 inches were only occasionally encountered (approximately one per one thousand feet) and usually required drilling and blasting to remove. Ric-Man was compensated for these boulders on a time and material basis, and was able to remove these boulders within one working day. Actual sizes of smaller boulders were often difficult to establish because they generally did not require additional effort to advance through and Ric-Man determined that the pay item amount did not justify stopping tunneling operations measure boulders.

During a portion of the tunnel, the Ric-Man raised concerns about boulder frequency. An engineer was placed in the tunnel heading to obtain boulder data on a continuous basis. The engineer documented boulders observed in the tunnel face and where hard grinding was felt at the cutting face. The engineer then recorded the advancement of the TBM until the grinding stopped. The length

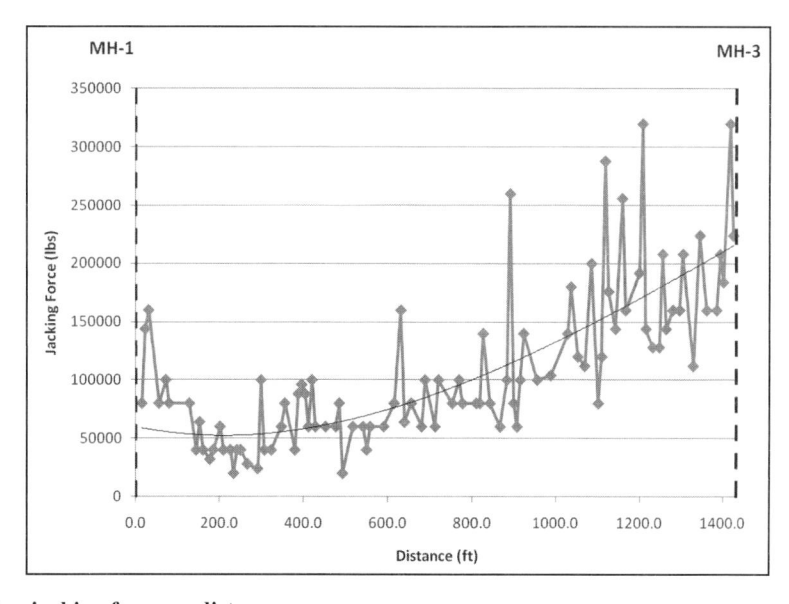

**Figure 8. Pipe jacking force vs. distance**

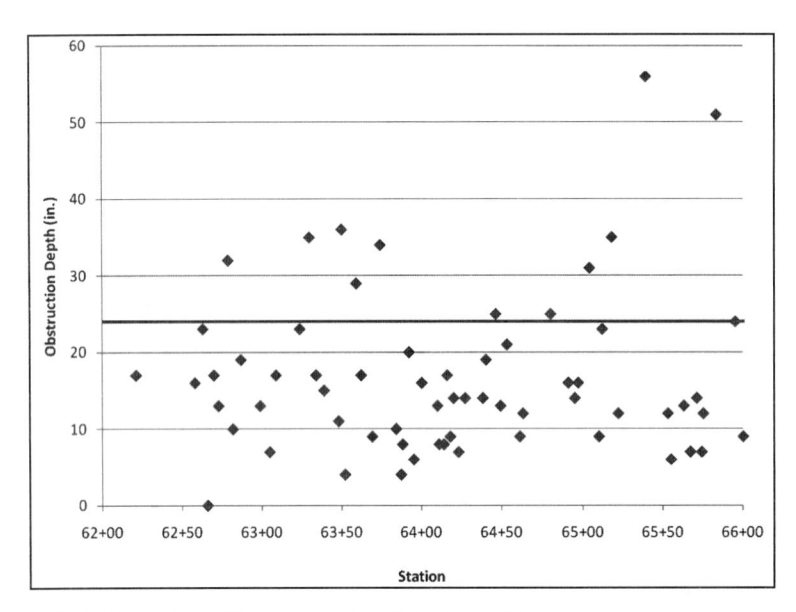

**Figure 9. Sizes of cobbles and boulders encountered during mining**

of advancement through an individual boulder was assumed to approximate at least one dimension of the boulder. A graphical representation of the approximate size and frequency of boulders observed during this period is presented below. In general, it appears one small-diameter boulder was being encountered approximately every 10 feet within this area.

**Differing Subsurface Condition**

During the fourth tunnel run, the TBM encountered a wet granular sand layer in the upper portion of the tunnel face. The project team was concerned about loss of ground into the tunnel and the possibility of compromising the limited hardpan cover. As the tunnel was advanced, the sand layer dipped within the face and remained wet. Groundwater infiltration into the tunnel increased to approximately 15 to 20 gallons per minute and carried sand and silt through the tunnel face, thereby inundating the TBM and creating a void in the tunnel face that extended above the TBM. To stabilize the hardpan, the tunnel was stopped, a bulkhead was built in the TBM face, and the void was filled with cement grout.

The tunnel was advanced approximately 20 feet with continuing ground loss and water infiltration. On multiple occasions, water, sand, and silt infiltration inundated the motor. The TBM was then stopped, the motor cleaned, the tunnel face grouted before advancement could continue. The TBM then encountered large-diameter, nested boulders which stopped the advancement of the TBM. Ric-Man attempted to enter the TBM face to remove the boulders but was unsuccessful, due to the continued water, sand, and silt infiltration and newly-encountered methane gas concentrations on the order of 65% of the lower explosive limit.

Under the Engineer's direction, dewatering wells where installed immediately adjacent to the tunnel bore. However, the wells were unsuccessful in diverting the water flow from the tunnel. Groundwater infiltration rates remained steady, and still carried sand and silt through the tunnel face. Potential loss of ground adjacent to and above the TBM was becoming a major concern to the project team. The Engineer performed a series of test borings adjacent to and ahead of the TBM to determine what ground conditions would be encountered over the next 25 feet of mining. The borings identified that hardpan cover remained intact above the tunnel zone, but a thick seam of sand and silt within the hardpan continued ahead of the TBM. The project team then conceded the ground to Mother Nature and excavated a recovery shaft to remove the sand, silt, and boulders ahead of the TBM. The shaft was constructed using steel sheet piling for earth support and a trench shield for internal bracing.

At the far side of the shaft during excavation, cobbles and boulders were encountered in a wet sand layer within the tunnel zone at a frequency on the order of 8 to 10 per foot of alignment. The project team determined that attempting to re-launch the TBM into this ground was not a desirable course of action. Extensive evaluations were performed and discussions were held to determine the best method of advancing the tunnel. Options discussed included open-cut excavation, soil stabilization grouting, and

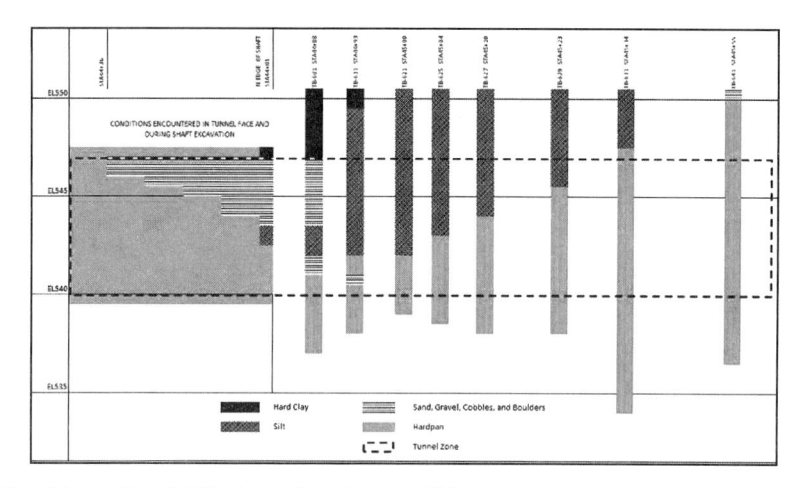

**Figure 10. Graphic profile of differing subsurface conditions**

dewatering. Another series of test borings were performed. These borings encountered sand and silt within the tunnel zone, but no cobbles or boulders. Additional cobbles and boulders were expected to exist within the granular soils encountered, but could not be confirmed in the test borings. Evaluation of the test boring data led to the conclusion that a full face of hardpan would be encountered approximately 50 feet beyond the shaft wall. Given the close proximity to a full hardpan face and the lack of cobbles and boulders encountered in the test boring, the Engineer directed Ric-Man to attempt advancement of the TBM through the remaining wet granular soil zone without special measures. Through exhaustive effort, Ric-Man was able to advance the TBM and again return to a full face of hardpan. The remaining tunnel run was completed to the next shaft without experiencing an excessive increase in jacking pressure.

The differing condition was eventually identified as an alluvial deposit incised into the hardpan. The total length of the deposit was on the order of 100 feet within the tunnel zone. In order to have identified this feature during a geotechnical investigation, test borings would need to have been performed on an extremely tight spacing. Overall, this feature cost the project approximately $2.5M and three months of schedule. The differing subsurface condition was recognized by the owner and entitlement granted to the Contractor shortly after the feature was encountered. Recognizing the feature as a differing subsurface condition at the outset, performing additional test borings, and working with Ric-Man to determine the most cost-effective and time-saving course of action allowed the feature to be overcome, and the cost and schedule impacts to the work, as substantial as they were, to be mitigated.

**Construction Shafts**

Two types of construction shafts were used on the project; mining shafts and intermediate manhole shafts. Both types of shafts were designed by the Ric-Man. Ric-Man used a rectangular steel sheet pile shaft for the initial mining shaft, and circular shaft geometry with corrugated metal pipe (CMP) supports for the remaining mining shafts. The CMP-supported mining shafts were excavated in multiple stages using augers. The upper stage utilized a combination of 20-foot diameter horizontal steel ribs and vertical timber lagging for temporary earth support. The middle stage utilized an 18-foot diameter CMP, and the lower stage utilized a 16-foot diameter CMP for earth support. Depending on the soil conditions encountered, some stages of the shafts were excavated under flooded conditions, or "in the wet," and required placement of the concrete base slabs and annular grout by tremie methods. Others were able to be excavated to their design depths in open, dry excavations. The CMP supports were abandoned in-place one mining operations were complete and the manholes were constructed.

Intermediate shafts were constructed using 8-foot diameter temporary steel casing supports, and were excavated in multiple stages using augers. Similar to the mining shafts, depending on the soil conditions encountered, some of the intermediate shafts were excavated under flooded conditions, or "in the wet," while others were able to be excavated to their design depths in open, dry excavations. Penetrations into the newly-constructed sewer liner were accomplished using cast-in-place concrete "collars" designed by Ric-Man. These collars were connected to the RCP sewer using reinforcing steel dowels installed in epoxy-grouted holes in the RCP. The dowels were then cast within reinforced concrete that, upon achieving the required design strength, supported the proposed opening in the sewer crown as well as the proposed manhole risers. Ric-Man utilized an interior form when casting the collar, which allowed them to avoid having to core through the reinforced concrete collar, and line-drilled and saw-cut the crown of the RCP to create the required manhole penetration into the newly-constructed sewer.

## CONCLUSIONS

This project advanced the field of knowledge for tunneling and concrete jack pipe operations in several ways:

1. The particular hardpan materials for this project, while low in clay content and plasticity, were capable of stable face conditions to allow open face tunneling;
2. Hardpan cover on the order of 3 feet appears capable of supporting the overlying wet partially dewatered soils in these conditions to allow open face jack pipe methods;
3. Long pipe jacking runs are feasible in the hardpan materials and relatively low jacking pressures were developed using the construction methods employed for this project;
4. In the project area, test boring spacings on the order of 200 feet appear warranted based on the DSC encountered;
5. Large boulders appear to be present in the hardpan at the rate of approximately two per 100 feet. Smaller boulders are present on a closer spacing.

## REFERENCES

Proctor and White, 1977, "Earth Tunneling With Steel Supports," Commercial Shearing, Inc.

# Construction Challenges for Small Diameter Soft Ground Tunnels

**William Bergeson, Verya Nasri**
AECOM, New York, New York

**Alan Pelletier**
Hartford Metropolitan District Commission, Hartford, Connecticut

**Leo Martin**
AECOM, Hartford, Connecticut

**Richard Palmer**
Northeast Remsco Construction, Farmingdale, New Jersey

**ABSTRACT:** The Hartford Metropolitan District Commission (District) of Connecticut serves 8 municipalities to provide water and sewer services for approximately 400,000 residents. In response to two issued consent decrees, the District initiated a clean water program with total estimated costs of $1.6 billion to eliminate sanitary sewer overflows and to reduce combined sewer overflows that pollute the Connecticut River. The Homestead Avenue Interceptor Extension (HAIE) is the first major project of the Clean Water Program; the HAIE extends the current interceptor approximately 3,600 linear feet (1,100 m) to serve as the planned downstream conduit for several planned upstream separation projects.

This paper presents the construction challenges associated with tunnel installation for the HAIE. Challenges for this project included low ground cover, crossing major transportation routes, and limiting settlement within a highly urbanized downtown environment.

A slurry machine with an adjustable compressed air cushion was used for installing the 72-inch (1,830 mm) PVC lined, class V, reinforced concrete pipe within five drives located in soft to very soft, varved silt and clay with occasional mixed face conditions being encountered. The muck processing system consisted of coarse screening conveyors, primary and secondary shakers, vortex de-silting cones, vertical clarifier, three centrifuges, two 20,000 gallon Baker slurry storage tanks, and polymer flocculent. Jacking loads were controlled using an automatic bentonite lubrication pumping system.

## INTRODUCTION

The Hartford Metropolitan District Commission (District) is a municipal corporation that was chartered by the State of Connecticut in 1929, which includes the municipalities of Bloomfield, East Hartford, Hartford, Newington, Rocky Hill, West Hartford, Wethersfield and Windsor. The District is governed by a Board of twenty-nine Commissioners, seventeen of whom are appointed by the legislative bodies from the eight member municipalities, eight by the Governor of Connecticut, and four by the leadership within the Connecticut General Assembly. A Chief Executive Officer manages the more than 600 full-time employees of the District, which is composed of four functional divisions to include an Administrative Division, Finance Division, Operations Division, and a Program Management Division.

Water and sewer services are provided by the District to approximately 400,000 residents. The water distribution system consists of upland impoundments in the Farmington River watershed, two filtration plants, and approximately 1,500 miles of distribution mains. Flows in the system are by gravity except for some pumping of treated water to higher elevations. Average treated water use is about 55 million gallons per day. The sewage collection system serves the member municipalities and consists of almost 1,200 miles of sanitary and combined sewers. The combined sewer system serves Harford and portions of West Hartford with 38 active Combined Sewer Overflows (CSOs); separate sanitary sewer collection systems serve the remaining communities with 8 active structural Sanitary Sewer Overflows (SSOs) within the system. The District maintains four wastewater treatment plants with a combined average of about 79 million gallons per day (mgd) for 2008 to include Hartford at 62 mgd, East Hartford at 8 mgd, Rocky Hill at 6.5 mgd, and Poquonock at 2.5 mgd.

The District also operates hydroelectric facilities at the Goodwin and Colebrook River Dams on the West Branch of the Farmington River. Furthermore, the District is under contract with the Connecticut Resources Recovery Authority to take a major part in the Mid-Connecticut Project operating a 2000 ton per day resource recovery plant for municipal solid water and related waste transfer subsystems. The Mid-Connecticut Project serves over 70 municipalities.

In 2006, the District entered into a Consent Order with the State of Connecticut Department of Environmental Protection (DEP) to reduce Combined Sewer Overflows (CSO) to a one (1) year level of control, within fifteen (15) years. The District also entered into a Consent Decree that same year with the Environmental Protection Agency (EPA) and the U.S. Department of Justice to implement a Sanitary Sewer Overflow (SSO) Abatement Program to eliminate structural SSO's over a seven (7) year period for the communities of Rocky Hill, Wethersfield and Windsor, and to eliminate SSO's over a twelve (12) year period for the communities of West Hartford and Newington. In reaction to the two mentioned consent decrees, the District formed a Program Management Division to oversee, design, manage, and implement a Clean Water Program (CWP), which was tasked to ensure compliance with both regulatory orders.

The work under the CWP includes three major elements: (1) construction of new sanitary sewers, interceptors and tunnels that reduce CSOs within the District's collection system; (2) rehabilitation of existing sanitary sewers and construction of new interceptors that eliminate structural and non-structural SSOs from East Hartford, Bloomfield, Wethersfield, West Hartford, Windsor, Rocky Hill and Newington; and (3) improvements at the Water Pollution Control Facilities (WPCF) to include Hartford, Rocky Hill and the East Hartford plants to increase treatment flow capacity and to reduce nitrogen discharge levels.

Funding for Phase I of the CWP has been approved through a referendum vote on November 7, 2006. Phase I funding is $800,000,000. Additional authorizations will be required over the fifteen (15) year life of the CWP. At this time, the total program cost is estimated to be about $1.6 billion based on 2006 dollars, Bergeson, et al. (2009).

**Homestead Avenue Interceptor Extension Project**

The Homestead Avenue Interceptor Extension Project (HAIE) is the first major project within the CWP; the purpose of the project is to extend the current Homestead Avenue Interceptor from its current discharge point at the Gully Brook Conduit to the Park River Interceptor where the flow, up to a 1 year storm event, will be directed to the Harford WPCF. The HAIE will serve as the downstream conduit for several future upstream separation projects that are planned for eliminating combined sewer overflow points and for reducing combined sewer overflows discharges that eventually drain into the Connecticut River. This project received funding from the ARRA program.

This paper discusses the construction challenges associated with the project. As of the time of writing for this paper, three of the five drives have been completed in a highly urbanized section of downtown Hartford as shown in Figure 1. This new PVC lined 72-inch (1,830 mm) reinforced concrete sewer pipeline crossed major transportation routes, encroached upon critical utilities, and passed in close proximity to historically significant buildings. Four building footprints were located within 20 feet (6 m) of the alignment. Important transportation routes crossed by the alignment include interstate I-84, which contains multiple lanes of traffic and a rail corridor containing 4 sets of tracks with one of the tracks operated by Amtrak. Three locations along the alignment were identified as being in close proximity to critical utilities. AECOM provided final design and construction management services for this project; in December 2008, Northeast Remsco Construction (NRC) of Farmingdale, NJ was awarded the construction contract.

Three jacking shafts and three receiving shafts were used to launch the five drives, which are being completed at an average rate of about 60 ft of pipe per day during driving. The daily progress rate was inextricably linked to the muck processing system since the slurry became over-laden with fines and had to be heavily processed. As such, selection of both the microtunneling boring machine and the slurry processing plant were critical elements of the tunneling operation performance.

**Subsurface Conditions**

Subsurface Conditions at the Project Site: included geologic units such as the bedrock, glacial till, varved silt and clay, and emplaced miscellaneous fill. The bedrock of the Connecticut Valley generally consists of conglomerates, feldspar-rich sandstone (arkose), or red and black shale formations; the overlying glacial till has been characterized as very stiff to hard reddish brown material with sandy and gravelly

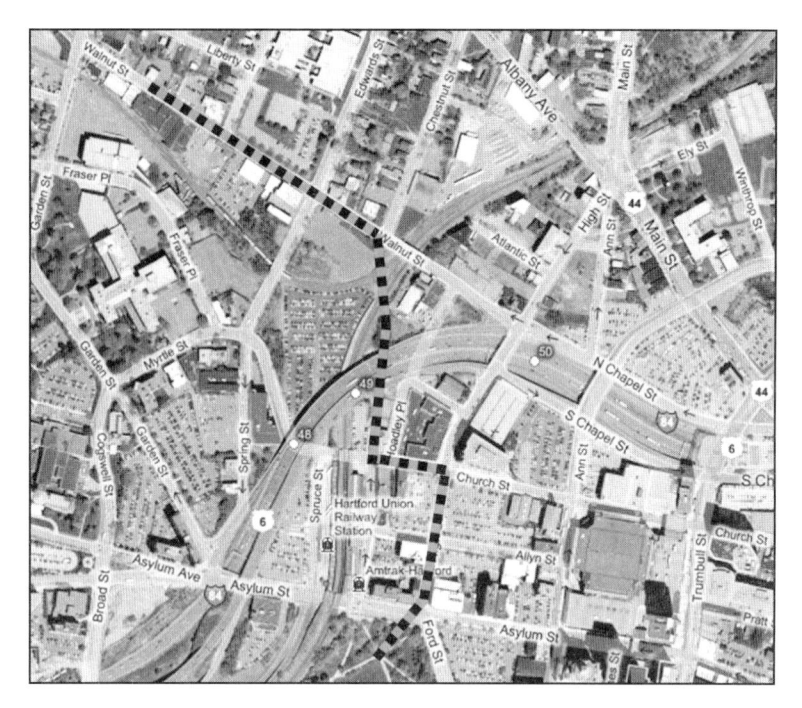

Figure 1. HAIE sewer alignment in downtown Hartford, Connecticut (www.bing.com/maps)

silts and clays occasionally containing boulders; the varved silt and clay deposit that was some 10 ft to 30 ft (3 to 9 m) thick at the project site has been described as soft to very soft material having been deposited within a glacier lake environment over many alternating cycles of freezing and thawing; the miscellaneous fill has been characterized as highly variable mixtures of sand, clay, gravel, boulders, and even construction debris that was roughly 10 ft thick. The groundwater table was generally located near the interface of the fill and the varved silt and clay nearly 10 feet (3 m) below the ground surface.

In the preliminary design stage, 12 exploratory boreholes were drilled. At the 30-percent design stage, 10 additional test borings were drilled; and falling-head tests were performed in 7 of the 10 test borings; and monitoring wells were installed in 8 of these test borings. At the 60-percent design stage, 4 additional test borings were drilled at select locations to confirm the interface depths between the softer varved silt and clay and the stiffer materials located at the site to include the bedrock, till, and fill.

Laboratory testing was performed on 37 samples to provide information on moisture content, grain-size distribution, Atterberg limits, oedometer, unconfined-compression, unconsolidated-undrained triaxial, consolidated-undrained triaxial, electrical resistivity, chlorides, and sulfates.

In addition to the test borings and the laboratory testing program, geophysical methods were used to map the surface of the glacial till and the bedrock along the alignment. The geophysical methods included seismic refraction, Multi-channel Analysis of Surface Waves (MASW), and low frequency Ground Penetrating Radar (GPR). Typical profiles are shown in Figure 2a for MASW and Figure 2b for GPR. Seismic refraction did not produce useful results because of interference between the pavement layer and the equipment, Hager GeoScience, Inc. (2008).

The logs from the mentioned 26 test borings; and the results from the mentioned geophysical surveying methods were all used to develop a working geologic profile of the alignment. The saturated, soft to very-soft, varved silt and clay was selected to locate the tunnel alignment since the extent of other materials at the site was typically intermittent; or else these materials were generally located at depths that were inconsistent with other alignment selection criteria such as adequate ground cover and settlement considerations or concerns for utility and obstruction avoidance.

Mixed face conditions were occasionally encountered at the interface between the upper soft to very soft varved silt and clay and the lower very stiff glacial till; however, there were no excessive settlements observed from ground loss due to these

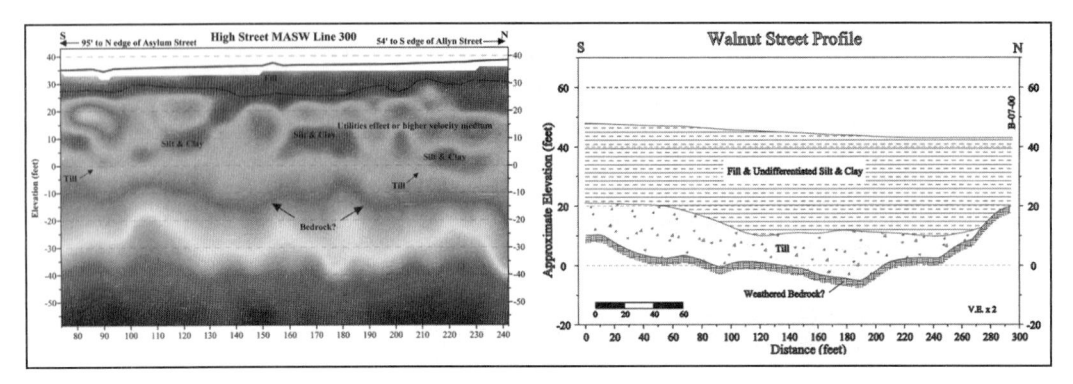

**Figure 2.** (a) Multi-channel analysis of surface waves (MASW); (b) Ground penetrating radar (GPR)

mixed ground conditions even though these mixed conditions were encountered for nearly 180 ft on the first drive.

## SHAFT PREPARATION AND EQUIPMENT LAYOUT AREA

Generally, the jacking and receiving shafts were set-up in advance before completing the previous drive. The construction manager and contractor worked with the various utility providers to identify any existing utilities, which were later exposed as part of the pre-trenching activities prior to the installation of any sheeting. The sheeting was vibrated through the varved clay unit into the glacial till until refusal. The shafts were built at depths ranging from 20 feet to 47 feet with roughly 25 ft diameters except for one of the shafts, which was constructed rectangular in cross-section to accommodate planned structural work. The excavation and installation of the support system followed the installation of the sheets. All of the shafts had a 6 to 12 inch concrete working slab installed.

Following shaft construction, the areas around the entry and exit locations of the shaft were improved using grout columns to support the MTBM during the launch and retrieval process. Generally a pattern of 21 columns was installed at the shafts, although the number was reduced at certain locations due to utility conflicts. Analysis of settlement markers adjacent to the entry and exit locations of the shafts indicated that the grout columns were moderately effective in supporting the machine during periods of break-in and break-out but allowed for some ground loss to take place.

After the shafts and the ground support were completed, the preparation for tunneling continued; the entry/exit portals were prepared, and once the previous drive had been completed, the necessary equipment used with the slurry plant, pipe jacking operation, control cabin, and pipe lubrication was then transported and set up for the next drive.

Figure 3 shows the set-up for a typical driving operation. Since the drive shaft locations did not provide sufficient space to stockpile more than a few sections of the 72-inch pipe, the pipe was stored at other easement locations, delivered as needed to the shaft locations where it was prepared by the tunneling crew for installation.

## SLURRY PROCESSING SYSTEM

The daily progress rate was inextricably linked to the muck processing plant since only about eight to nine 10-ft pipe sections generally could be advanced before the slurry became over-laden with fines; as a result, the contractor used the 12-hour night shift to clean the slurry, which was stored in two 20,000 gallon Baker after accumulating during the 12-hour tunneling shift during the day. The slurry processing plant consisted of coarse screening conveyors and primary shale-shaker (mid-screens) working in parallel. These were attached to vortex de-silting cones and secondary shale-shakers (fine screen). A vertical clarifier was used in conjunction with three centrifuges that were injected with polymer flocculent, which generally worked effectively as a coagulant aid or sludge conditioning agents as part of the liquid-solid separation processes.

The slurry plant was set up to process the fine materials through the use of scalpers, desanding and desilting equipment. The slurry was directed into the slurry storage tanks located below the processing units; and the sand and silt was directed in front of the plant, which was then loaded onto trucks and transported to the soil stockpile areas. From the storage tanks, the slurry was pumped through a vertical clarifying tank and three centrifuges to force the fines out of the mix. With the silty-clay materials encountered along the tunnel, the centrifuges worked the hardest and separated the majority of the slurry. The scalpers and coarser screens provided mainly for the removal of balled and clumps of clay that worked their way through the pumping system to the separation unit.

**Figure 3. Shaft with equipment layout for tunneling**

For reasons pertaining to economics, the mucking and separating systems should always keep pace with the excavation process; however, the slurry processing was a challenge for this project because of the high percentage of ultra-fine material in suspension. The generated muck was loose and wet. Flocculants were used to improve the separation of the fines, even still, overnight processing of the slurry using the vertical clarifying tank and centrifuges was needed. This process was repeated daily until each of the drives were completed.

Figure 4 shows the slurry processing equipment to include the desander and desilter in the foreground, and the vertical cone clarifying tank in the background. There are also three centrifuges with the two smaller centrifuges behind the desander and desilter units and the larger centrifuge shown in the upper left hand photograph. Also shown are small piles depicting the wet condition of the processed muck as well as the slurry over-laden with fines.

## MACHINE SELECTION

Since settlement was critical along the entire alignment, the contract specifications required the contractor to use a pressurized face microtunneling boring machine to install the pipe sections. The contractor selected a Herrenknecht AVND 1800AB microtunneling machine, which is a slurry machine that has an adjustable air cushion D-Mode as shown in Figure 5a lowered into the first jacking pit; the schematic for this machine is shown in Figure 5b. This machine was able to precisely balance the face pressure especially in areas with low ground cover. Similarly equipped machines have been used successfully within comparable ground conditions worldwide; however, selection of an EPB machine would have been acceptable according to contract documents, especially because of the heavy fines in the ground.

The machine was fitted with a second articulated steering joint near the aft of the can assemblies to provide increased steering performance in the very soft ground conditions. Steering of this particular machine was enhanced in the weak soils through the use of a secondary steering joint that was positioned approximately 25 feet back from the cutting wheel. The second steering joint allowed the MTBM to engage three times more surface area than the conventional articulated steering joint located at the front of the machine. The rear steering mechanism provided an immediate reaction when engaged even in the very soft ground conditions; however, tempered use of this steering joint was warranted in order to maintain a smooth alignment.

The machine components were arranged within the various machine cans to achieve nearly neutrally buoyant conditions, which limited any sinking tendencies of the machine; however since the balance point of any tunnel machine will be significantly influenced by the heavy weight of the cutting head and the main drive, compensation for the heavy front end had to be achieved by coupling the machine cans together.

**Figure 4. Slurry processing system**

Slurry machines perform well in ground with hydraulic conductivity values ranging from 1E–6 to 1 cm/s, as well as under variable groundwater pressures. Using mixshield principles that incorporate a cushion of air behind the bulkhead of the machine, the face could be carefully balanced by the operator; however, very good operating techniques had to be incorporated especially in areas of low ground cover where inadvertent slurry returns to the surface had to be controlled while simultaneously maintaining sufficient pressure at the face to minimizing any ground loss and the resulting settlement.

## JACKING FORCES AND ADVANCE RATES

The contractor installed at least one intermediate jacking station (IJS) in each of the tunnel drives and lubrication was automatically injected. Each of these IJS was recovered at the end of the drive; and afterwards, the steel casings were pushed tight to form a water tight seal. The hydraulic rams, as shown in Figure 6a, did not have to be engaged to control the jacking loads in any of the drives thus performed.

The jacking loads were primarily controlled using an automatic lubrication system. Figure 6b shows some of the automatic lubrication equipment, which consisted of an Ackerman mixer unit feeding the automated injection system built into the Herrenknecht MTBM. This mixer contained two 250-gallon tubs with sufficient bentonite slurry for two 10-ft. pipe sections. The pumping unit was set-up to pump at a rate of about 11 gallons per foot.

Soil freeze was observed at the start of the shift; typically, it was about 180% of the normal operating loads of the previous day's drive. The jacking forces ranged from 90 to 480 tons on the longest drive. For the two shorter drives, the jacking forces ranged from 61 to 125 tons on the second drive and from 58 to 211 tons on the third drive. The jacking forces were estimated in the design phase to be less 900 tons on the long drive as long as good practice installation guidelines were maintained to include the continuous lubrication of the pipe string.

To date three of the tunnel drives have been completed. The first drive was started on September 1, 2009 and was completed on September 24, 2009. The 1,180 foot drive began in mixed face conditions that persisted for approximately 180 feet. The lower section of the tunnel was located in glacial till whereas the upper portion of the face was located in the varved silt and clay. Once past this point, the tunnel was entirely within the varved silt and clay unit. The second drive was initiated on October 21st; and this 317 foot long drive was completed on October

Figure 5. a. Slurry machine being lowered into jacking shaft; b. schematic (Herrenknecht)

26th. The third drive began on November 6th and was completed on November 17th. It was 657 feet long. The fourth drive is anticipated to begin in mid-January 2010, which is the time of writing for this paper.

## INSTRUMENTATION

Since the tunneling took place adjacent to some large structures, beneath active rail lines and between abutments and support piers for Interstate Highway I-84, as well as the overlying utilities, a detailed geotechnical instrumentation program was developed. Monitoring devices were placed on adjoining buildings, on utilities and structures that had the potential to be impacted by the tunneling operation or the shaft installation. The instrument locations were depicted on the contract drawings. All of the data that was recorded had to be entered into a website monitoring program (Argus) such that all parties involved in the contract could review the data in a timely manner. To aid in determining which points were of any concern, the monitoring points turned yellow when viewing on the computer screen once the instrument deflection approached 75% of the allowable value; and then turned red when the deflection limit was reached. For the railroad crossing the points were automatically downloaded to "Argus" while for the

Figure 6. (a) Intermediate jacking station (IJS); (b) automatic lubrication equipment

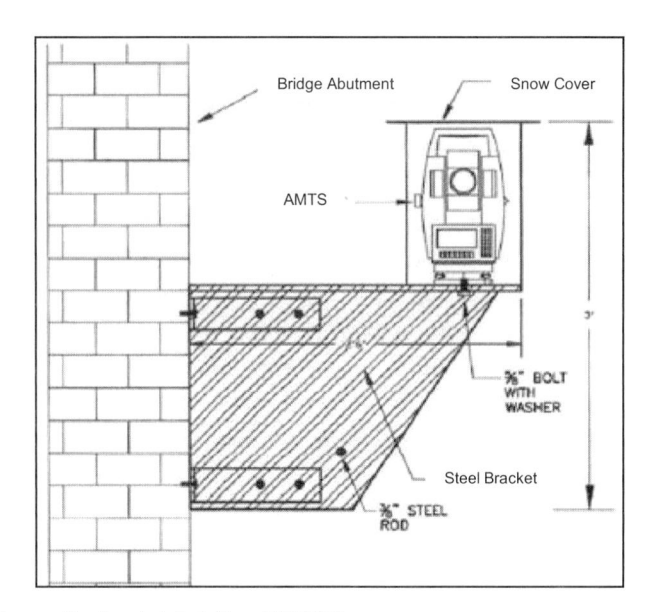

Figure 7. Automatic monitoring total station (AMTS)

rest of the project points, the data had to be manually input.

The second drive passed below four sets of railroad tracks. Three of them were area spur lines used primarily for storing railcars serving nearby businesses; and the fourth rail line belonged to Amtrak having a mainline through Hartford. The crown of the tunnel was just 15 feet below the top of rail for the main track; and Amtrak required the monitoring of five points on both sides of the centerline for each of the 8 rails. To monitor all 88 points at the rail crossing, the contract required the contractor to have installed an automated monitoring system; therefore, the contractor mounted an automatic monitoring total station (AMTS) theodolite on a nearby bridge abutment that allowed for line of sight to each of the reflectors at the prescribed locations along the rails. The automated system was set up to sweep the points twice daily to meet the requirements set forth by Amtrak. Since theodolites use electromagnetic energy to determine distances and angles, and because they have small built-in computers, their accuracy is generally much greater than that achieved using classical optical surveying methods.

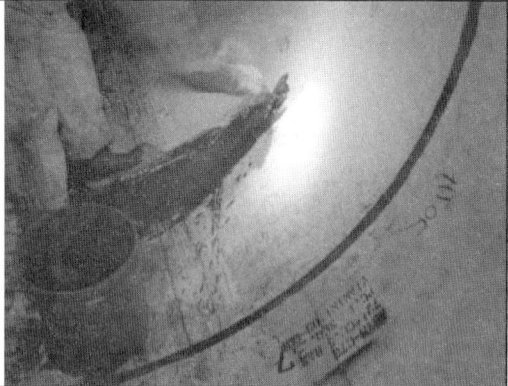

Figure 8. Contact grouting of the pipe: (a) batching; (b) verification

The AMTS unit was set-up one month prior to the start of the drive, which allowed for temperature variation of the rails to be taken into account. The target prisms, which are 2 to 3 inches in diameter, can generally be read as far as 300 feet away (see Figure 7).

The system showed some settlement along the Amtrak rails with approximately ¾-inch after the TBM passed under the mainline. The automated monitoring system as well as traditional surveys completed on all the rails shows that the overall settlement to date has been somewhere between ¼ to 1½ inches. The readings indicate that the settlement trough extended about 7 to 15 feet either side of the tunnel centerline. A section of approximately 15 feet to either side of the tunnel centerline had to be immediately re-ballasted. The monitoring of the rails is scheduled to continue until the backfilling of the shaft adjacent to the railroad tracks.

## POST TUNNELING ACTIVITIES

Upon completion of each drive, the tunnels were stripped of the support piping and control wires. Once removed, contact grouting operations were initiated. Then low strength grout was injected under low pressure into the annulus using calibrated equipment as shown in Figure 8.

The MTBM laser system was checked using traditional survey methods at least twice each drive to verify compliance with the specifications; as-built surveys were performed at the completion of each tunnel drive to identified locations where the specified tolerances have been exceeded. The contract allowed for vertical deviations from the proposed alignment of 1.2-inches (30.5 mm) when a constant grade had been maintained. When these values were exceeded, the potential causes were evaluated amongst those involved in the work to determine if any modifications were needed to the tunneling operations.

## SUMMARY AND CONCLUSIONS

The new sewer pipeline projects consists of about 3,600 linear feet (1,100 m) of 72-inch (1,830 mm) reinforced concrete pipe installed using pressurized face microtunneling in downtown Hartford, CT with the potential to impact several historical buildings, critical utilities, passenger rail tracks, city streets, and interstate I-84; however, some of these impacts were mitigated to the extent possible by alignment considerations in the design. The alignment is located within varved silt and clay with areas of low ground cover. The response of the machine was improved in the soft ground by using a long stable base, formed by connecting the machine, machine can, and trailing pipe as well as by using secondary steering articulation at the machine aft. The use of slurry machine in the clayey ground resulted in wet muck that was very hard to separate to the point of impeding progress; however, the pressure at the face was carefully controlled using the available air cushion, which limited critical settlements within this highly-urbanized, downtown project environment where transportation corridors, historic buildings, and critical utilities were located. The automatic bentonite lubrication system helped to keep the jacking loads well within the design limits of the pipes.

## REFERENCES

Bergeson, W., Nasri, V., Sullivan, J., Pelletier, A. "*Microtunneling Challenges in Soft Ground of Downtown Hartford, CT.*" RETC-2009, Las Vegas, NV.

Hager GeoScience, Inc. "*Geotechnical Report Homestead Avenue Interceptor Extension—Hartford, Connecticut.*" January 2008.

# TRACK 4: CASE HISTORIES

Larry Eckert, Chair

*Session 2: NATM/SEM*

# Past and Present Soft Ground NATM for Tunnel and Shaft Construction for the Washington, D.C. Metro

**John Rudolf**
Bechtel Infrastructure Corporation, Vienna, Virginia

**Vojtech Gall, Timothy O'Brien**
Gall Zeidler Consultants, LLC, Ashburn, Virginia

**ABSTRACT:** The first application of soft ground NATM in the US was on Washington, D.C. Metro Section E5 Fort Totten Station and running tunnels. Due to its success it was subsequently applied for tunneling under Rock Creek Cemetery and Section F6b, south of Congress heights station on the Outer Branch Route. Tunneling for Section E4b also called for the construction of a large NATM oval shaft that was used to stage tunnel construction. In its final build-out it is configured as an emergency egress structure. Application of the NATM and its successes laid the ground work for the method's increased use on urban soft ground tunnels, recently the Beacon Hill station in Seattle. NATM tunneling is currently being employed in a design-build framework for the extension of Washington's Metro to Dulles International Airport. This paper provides a concise case history summary and focuses on the developments and refinements of the NATM in Washington ranging from technical, to contractual and newly risk management aspects and its use on the Dulles Corridor Metrorail project at Tysons Corner, Virginia. This project involves two shallow 520 m long single track tunnels that feature a systematic implementation of the pipe arch canopy method for pre-support for the entire length of the tunnels where settlement is critical to control. The latter is one of the examples of the many innovations that led to the method's refinements and were pioneered on the Metro system in Washington, D.C.

## INTRODUCTION

Soft ground NATM's first application in the US was to Section E5 Fort Totten Station and running tunnels of the Washington, D.C. Metro system in 1988. In addition, this method was applied to the Section F6b tunnels under Rock Creek Cemetery and for the tunnels and shaft of Section E4b. All three projects are significant in their own right, being the testing ground of what were innovative and novel concepts at the time. These refinements of the NATM in Washington, D.C. range from technical, to contractual and new risk management concepts. With the start of construction of the latest soft ground NATM tunnels in Tysons Corner for the metrorail extension, the concepts and the construction method itself that were once so novel have been refined further, and are now being successfully applied to the Dulles Corridor Metrorail Project.

## FORT TOTTEN STATION AND RUNNING TUNNELS, SECTION E5

### Project Description

Section E5 of the Washington Metro at Fort Totten Station in Northeast Washington, D.C. on the Green Line represents the first Washington Metro station and running tunnels to be constructed using the soft ground NATM. The underground structures of the project consisted of a station chamber, a vent shaft, twin tunnels, and a fan shaft. The station chamber measures approximately 20 m in width by 10 m in height by 91 m in length, while the twin tunnels are approximately 305 m in length. Overburden ranges from approximately 6 m at the least over the station to as much as 30.5 m over the tunnels. The construction contract was awarded in August 1988. The contractors had the option to submit bids based upon conventional construction methods, NATM, or a combination of both. The contractor, a joint venture between Mergentime and HT Construction, chose the NATM option with Hochtief providing specialized equipment as well as management personnel with prior NATM experience (Darmody 1991).

### Geologic and Hydrologic Conditions

Washington D.C. is located on the boundary between Virginia's Piedmont and Coastal Plain physiographic provinces. This boundary is commonly referred to as the Fall Line. Section E5 is located east of the Fall Line within the Coastal Plain sediments identified

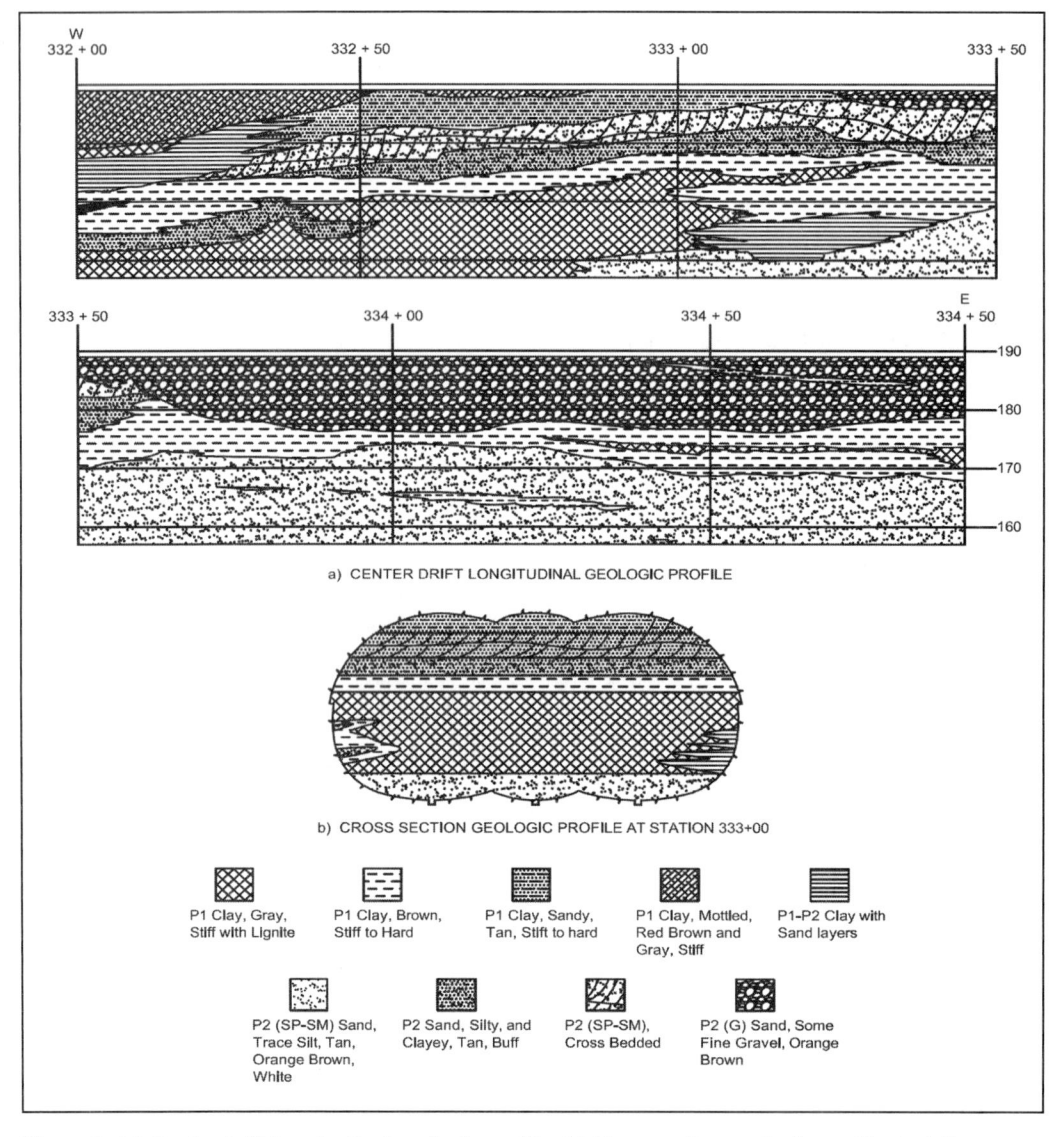

**Figure 1.** (a) Center drift longitudinal geologic profile, (b) Cross section geologic profile at station 333+00 (after Darmody 1991)

collectively as the Potomac Group. More specifically, construction was within the Patuxent Formation, the lowest member of the Potomac Group, which is predominately unconsolidated sand and gravel with lesser amounts of clay. The sediments are overconsolidated and contain some irregular beds of iron-oxide cemented sand or gravel. The Potomac Group sediments at Section E5 were categorized into two primary groups: P1 clays and P2 sands following the general and system wide nomenclature established by WMATA's General Soils Consultant.

The ventilation and fan shafts were excavated within interlayered sands and clays while the two single track tunnels were excavated primarily through cross-bedded, silty or clayey, fine to medium sands. The station was excavated within loose to compact interlayered sand and gravels, stiff clay, and clean to silty sand in the eastern half and interlayered clay and sandy clay in the western half. Figure 1 shows a geological profile longitudinal section and cross section from the center drift of the station as obtained from borings and field mapping.

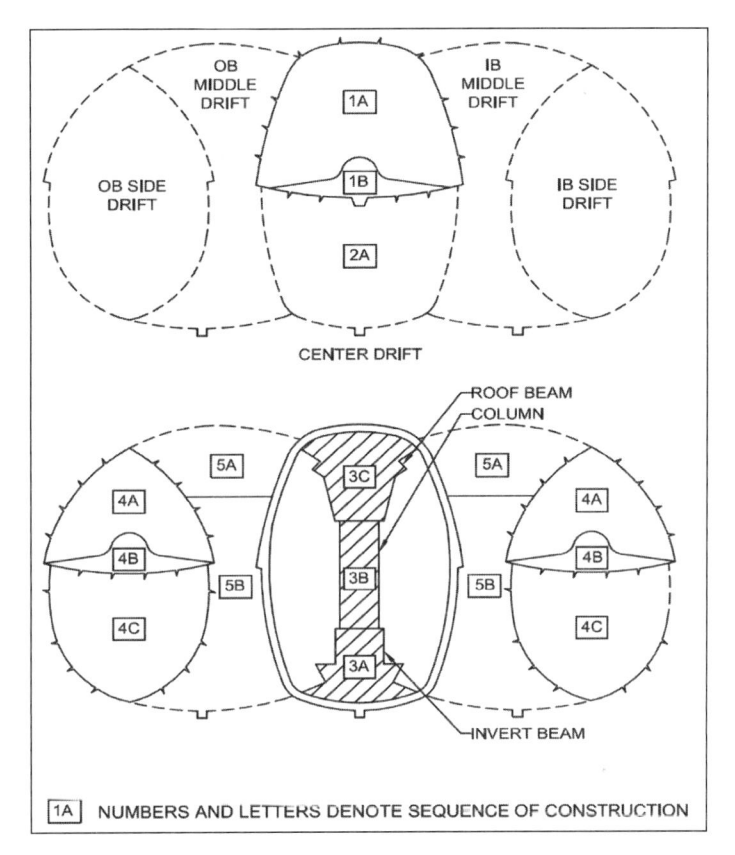

Figure 2. Excavation and support sequence of NATM station

Piezometer information obtained prior to construction indicated the static groundwater table to generally be below invert elevation of the NATM structures. Due to the lenticular nature of the Patuxent Formation sediments, subsurface water inflows and perched groundwater were more of a concern during construction. Therefore prior to the start of excavation, the contractor drilled two 24 m long, 152 mm horizontal drains at the portal wall on either side of the center drift for drainage. Initial high flow rates were attributed to the drainage of a large perched groundwater basin. The subsequent flow rate of 2.8 liters per minute from one of the two holes for the remainder of the station excavation was attributed to groundwater recharge of the basin (Darmody 1991).

**NATM Station Tunneling**

The station excavation sequence was broken down into multiple drifts. The center drift was the first to be excavated, with excavation proceeding in 0.9 m increments westward from the portal wall. The top heading was excavated in its entirety, followed by the bench. Upon completion of the center drift, station excavation activities paused for approximately 4 months while a cast-in-place concrete center drift support frame consisting of an invert beam, nine columns, and a roof beam was constructed. See Figure 2 for the excavation sequence.

Once the center drift and its support frame were completed, the contractor proceeded with simultaneous excavation of the IB and OB side drifts in 0.9 m increments, alternating between the top heading and bench. Excavation of the twin tunnels proceeded in 0.9 m increments as well, with excavation alternating between top heading, bench, and invert. The initial support for each excavation round consisted of the application of a sealing layer of shotcrete to the excavated ground, followed by installation of a lattice girder, installation of the first layer of welded wire fabric (WWF), shotcrete to the outer edge of the lattice girder, excavation and shotcrete of a temporary invert, installation of a second layer of WWF, and finally the application of a second layer of shotcrete.

The occasional running or raveling ground encountered during excavation was successfully controlled using pre-support means; however, only with one significant instance of ground loss. In this instance, some 38 m³ of sand fell from a small hole in the IB tunnel crown. The resulting surface settlement was less than 0.5 mm and the ground loss area was stabilized with additional shotcrete and positively drained by installing drain pipes. Once excavation had progressed 4.5 m beyond the area of settlement, the contractor grouted the area of ground loss with water-cement grout to further stabilize the ground. Overlapping forepoling sheets were effectively used to control raveling ground during excavation.

Upon completion of the initial lining for all structures, the station, twin tunnels, and shafts were waterproofed using a full round or "tanked" PVC waterproofing membrane. Subsequently, the cast-in-place concrete final lining was installed.

## ROCK CREEK CEMETERY, SECTION E4B

### Project Description

Section E4b was constructed as part of WMATA's $642 million Mid City E-Route construction that provided connection between the U Street/Shaw station and the Fort Totten Station in northeast Washington, D.C. Construction began in the mid 90s and included two cut-and-cover stations, 4.7 kilometers of single-track tunnels constructed mainly by TBMs and a series of ventilation and emergency access and egress shafts. The Farragut shaft, part of Section E4b, was the only shaft constructed by NATM. Section E4b adjoined the running tunnels of Section E5a that were constructed several years earlier by NATM. From the Farragut Shaft, which adjoined Section E5a, to the Buchanan shaft, located further south, the tunnels had to pass under the historic Rock Creek cemetery, the oldest in Washington, D.C. The tunnel alignment passes beneath the cemetery compound at a depth of approximately 15–25 meters. Prior to the start of the project, the cemetery stated that any tunneling approach under the cemetery would have to be without any "discernible settlements." WMATA's Board of Engineering Consultants (BOEC, a select group of US tunneling experts) and Design Engineers responded with an investigation into a series of tunneling and ground conditioning options. With initial hesitation mainly due to the fact that soft ground NATM was still a relatively untested technique in the US at that time, an engineering recommendation was made to apply NATM with the aid of ground improvement and pre-support measures to suit local conditions. In the end, it was concluded that only the NATM with such additional measures would be able to meet the requirement for close to negligible surface settlements, thus satisfying the restrictions imposed by the cemetery. Tunneling surface settlements were estimated to be less than 1 cm and were projected to spread over a wide surface settlement trough.

Figure 3. Section E4b—NATM tunneling under Rock Creek Cemetery (Courtesy Paul Madsen, Kiewit Construction)

### NATM Tunneling

The two single-track tunnels, separated by approximately one tunnel diameter pillar width were about 940 meters in total length, each. The contractor, a joint venture between Kiewit and Kenny Construction, used the fully designed classical soft ground NATM excavation and support sequence with top heading and bench excavation sequencing and the top heading advance limited to a maximum of 1 meter per round. The bench excavation followed at a distance of 2 meters (Figure 3). The initial shotcrete lining was 18 cm thick and reinforced with welded wire fabric. The tunnel was waterproofed using a PVC continuous membrane and unreinforced cast-in-place 30 cm thick concrete was installed as final lining (Irshad 1995). Depending on geologic conditions, pre-support in the crown involved pipe spiling, and forepoling sheets were used to control potentially running soils such as soft silts and silty sands. Pipe spiling was used as needed in stiffer silts. Chemical-grout was specified in portions of the alignment where silty sands with occasional pockets or zones of clean sand and perched water were anticipated within the P2 materials. Installation of a chemically improved soil arch above the tunnel crown was carried out using the horizontal directional drilling technique, a novelty for tunnel pre-support installation at that time.

### Grouted Pre-Support Arch—Horizontal Directional Drilling and Chemical Grouting

Chemical grouting was required above both the inbound and outbound NATM tunnel crowns for about 100 meters starting from the Farragut shafts towards Ft. Totten Station, from Farragut shaft for about 250 meters towards the Buchanan shaft and from the Buchanan shaft towards the Farragut shaft for about 160 meters, for a total length of about 2 × 510 meters. The grouting subcontractor, Hayward Baker, Inc. proposed a Horizontal Directional Drilling (HDD) program for the installation of steel TAM pipes to allow unimpeded access to the heading for continuous mining, better control of the grout injections by minimizing the number of faces exposed to the grouting process. After weighing schedule and cost advantages/disadvantages of HDD vs. conventional drilling per design the joint venture of Kiewit/Kenny proposed and WMATA accepted the HDD installation of TAMs for chemical grouting.

A total of more than 8,000 m in TAM pipes was installed either from within the shafts or from grouting chambers created near the shafts. HDD use enabled NATM tunneling in silty and clean sands on Section E4b without the need for stoppages during tunneling.

In summary, use of chemical grouting and standard pre-support measure including spiling and sheeting in combination with NATM sequencing and early support installation achieved project objectives and guaranteed the tunneling performance with "no discernible" surface settlements within the cemetery.

## SECTION F6B—PRE-SUPPORT OPTIMIZATION

### Project Description

Section F6b included the construction of Congress Heights Station using cut-and-cover techniques and two 460 m long, single track tunnels. This section is part of the Branch Avenue Route and is situated in southeastern Washington, D.C. NATM was selected for tunnel construction because of (1) the short length of tunnels, (2) short mobilization duration, and (3) the ability to mine from multiple headings. It was anticipated that with rapid mobilization and the use of multiple headings to advance through a relatively short length and comparatively homogeneous geology, the NATM tunnels could be completed during the early part of the overall project construction. The Contract for Section F6b was awarded to a joint venture of Clark/Shea with J.F. Shea bseing responsible for NATM tunneling which started in October 1996 and the "hole through" in the IB tunnel completing both tunnel drives was made in September 1997.

A number of pre-support measures were used to stabilize the tunnel crown in areas where saturated sand layers were known to occur and in adjacent sections where the thickness, extent and stability of firm clays over the tunnel were uncertain. These measures included pre-support grouting using directional drilling, grouted pipe spiling, and rebar spiling, in combination with dewatering in advance of the face. These measures were required to increase the stand up time and cut off groundwater. Pre-support chemical grouting of the northernmost 70 m was required to consolidate a saturated sand layer in the crown to enable successful NATM tunneling.

Per WMATA design criteria the tunnel cross section was designed as a dual lining structure comprising a shotcrete lining and cast-in-place unreinforced concrete lining. The shotcrete lining was reinforced with welded wire fabric, lattice girders, splice bars, and splice clips. The tunnel is waterproofed by a flexible plastic membrane around its entire circumference.

The tunneling of the outbound tunnel proceeded as specified in the contract documents. During mining of the inbound tunnel, however, interception of 4 unmapped sand lenses in the crown resulted in unexpected inflow of groundwater and work stoppages. With chemical grout injection primarily from inside the tunnel and installation of drain pipes ahead of the face, the contractor was able to resume safe mining below a hydrostatic pressure of up to 8 m (Gall, Zeidler, Bohlke, and Alfredson 1998).

### Geologic Conditions

Based on exploratory borings along the NATM tunnels, the first approximately 70 meters of the tunnel beginning at the station in the north were mapped as a thick sequence of firm overconsolidated (P1) clays with up to 1.5 m of saturated sands (P2) in and above the tunnel crown. The presence of the P2 sand layer dictated the use of pre-support methods to stabilize the crown and prevent the inflow of flowing sands and uncontrolled loss of ground. Directional drilling, pipe spiling, and grout injection were specified to assist in stabilizing the crown of the tunnel through this particular reach. Beyond the first 70 m south of Congress Heights Station, the predominant soil type at the face and above the tunnel was anticipated as a thick sequence of over-consolidated Cretaceous P1 clays with occasional sand lenses.

### Pre-Support Methods

Based on ground conditions the main types of pre-support foreseen F6b included rebar spiling, grouted pipe spiling, and installation of a chemical grouting canopy.

Rebar spiling consists of steel rebar rammed into the ground in soils where a bridging effect can be achieved between closely spaced bars. Rebar spiling was also specified in the area of the chemical grout canopy to limit over break in the tunnel arch. Experience during construction of past projects has shown that even when the P2 material is improved by chemical grouting, over break in the crown and tunnel shoulders may occur because of inhomogeneities in the ground. To limit such over break No. 8 (25 mm) rebar spiling installed at 0.3 m distances was specified on an as required basis.

Grouted pipe spiling consisted of perforated steel pipes installed above the tunnel crown at a centerline spacing of approximately 0.3 m in pre-drilled holes at a look-out angle of 5 degrees and through which grout was injected to stabilize the ground. These were installed systematically between the ground improved by chemical grouting and to a location where the P1 Clays would extend a minimum of 1.5 m above the tunnel crown.

Installation of the grout canopy was specified to ensure that ground water is prevented from draining into the tunnel section, the ground properties are improved, and the steel grouting pipes provide longitudinal reinforcement by forming a supporting arch extending into the P1 clays above the tunnel crown. The chemical grouting canopy was to consist of steel pipes installed into the ground ahead of the tunnel face for a distance of up to 100 m using directional drilling. Following placement of the pipes chemical grout was to be injected to stabilize the ground in sandy P2 materials along the perimeter of the tunnel crown.

However, shortly after award the contractor proposed to install roughly 45 m long drills from within the Congress Heights Station excavation into the arch (and saturated sand layer) above the future tunnel using conventional methods. The remaining length of this canopy was then grouted from within the tunnel during advancing northward from the south using shorter, overlapping 21 m long TAMs.

### Dewatering/Groundwater Control

To dewater the saturated P2 sands in advance of the tunnel heading after installation of the chemical grout arch the contract foresaw drilling of horizontal dewatering pipes to lower the groundwater table for the tunnel construction in the P2 material. The groundwater table was to be lowered to the interface between the P2 and P1 material. To achieve a continuous dewatering system drilling for the horizontal dewatering pipes was also to be carried out using directional drilling.

### Tunneling

The tunnel excavation and support sequence followed basic considerations for soft ground NATM tunneling. The excavation face was subdivided in top heading and bench with excavation rounds of up to

**Figure 4. Section F6b—top heading in P1 clays**

1 m length in the top heading. The typical excavation sequence for tunneling in P1 clays called for a ring closure distance of maximum 5 m. The actual support installation in the top heading is shown in Figure 4. For tunneling beneath the chemically grouted sand canopy the excavation procedure was modified, calling for an earlier shotcrete ring closure, allowing a maximum distance between advancing top heading and closed shotcrete ring in the invert of only 4 m. Tunnel excavation began in October 1996 from a southern shaft at Mississippi Avenue with two headings being driven simultaneously.

Geologic conditions observed in the outbound tunnel (OB) were in overall agreement with, or better than those established in the geotechnical documents and tunneling proceeded with pre-support as specified or less. In contrast to the outbound tunnel, the inbound tunnel excavation was affected by occasional, large inflow of water at the face when the heading intercepted unmapped saturated sand layers. Tunnel advance was stopped four different times with groundwater inflow rates of 3.2 to 5.0 liters per second. In those instances tunnel excavation was halted until the water bearing P2 material affecting the tunnel was sufficiently grouted to enable tunneling beneath and through a soil improved ground with additional dewatering through drainage pipes as required. Due to the required use of probe drilling to check for presence of water saturated sand lenses known to be present in the P1 clay material, the sand lenses were detected early on. This exploratory measure provided knowledge of ground conditions ahead of the face, and repeatedly, proved essential to assessing tunnel face stability and tunneling safety.

In one instance inflows were of such magnitude resulting in ground loss estimated at 50–100 cubic meters which required the use of cement grouting to fill the void created by the loss of ground ahead of the tunnel face. This was achieved by a surface grouting program which led to an interruption of tunneling of about 2 months (Gall, Zeidler, Bohlke, and Alfredson 1998). Four work stoppages occurred

**Figure 5. View of NATM tunnel east portal in relation to high-rise buildings near the alignment**

during inbound tunneling and in each case, tunneling was able to resume once the crown was adequately stabilized with the use of pre-support chemical grouting in advance of the face. In summary NATM tunneling was carried out successfully for the first time under a hydrostatic head of up to about 8 m.

## TUNNELING AT TYSONS CORNER FOR DULLES METRORAIL

### Project Description

The Metropolitan Washington Airports Authority (MWAA) is currently undertaking the first major expansion of the WMATA system that reached a total rail length of 172 kilometers with the completion of the Largo Line in 2004 (Rudolf and Gall 2007). This current expansion is known as the Dulles Corridor Metrorail Extension Project (DCMP) and will significantly improve the service of WMATA's metrorail system in the Capitol Region in Northern Virginia and will connect the Washington Dulles International Airport (IAD) with Washington D.C. The expansion will add a total length of 37 kilometers of rail. Its first phase is currently being implemented in a design-build contract by Dulles Transit Partners (DTP) a joint Venture of Bechtel, Inc. and URS. This Phase 1 involves NATM tunneling as the most feasible tunneling option at Tysons Corner.

The NATM tunnel segment includes twin single-track tunnels at a length of approximately 520 m each. A short cut-and-cover section adjoins the NATM tunnels at the east portal and a longer cut-and-cover section exists at the west portal. These tunnels are being constructed in soft ground and are located adjacent to existing structures and utilities that are sensitive to ground movements (Figure 5). The alignment and elements of the short tunnels at Tysons Corner are shown in Figure 6. For this very shallow overburden alignment, a busy 4-lane thoroughfare at International Drive is located about only 4.6 m above the tunnel crown. The deepest overburden cover exists at about mid-point of the alignment with nearly 11.6 m. At the west portal and the transition to the cut-and-cover box the overburden is about 6 m.

### Geologic Conditions

The soils along the tunnel alignment include mainly residual soils and soil-like completely decomposed rock. The residual soils are the result of in-place weathering of the underlying bedrock and are typically fine sandy silts, clays and silty fine sands. According to the project classification, the residual soils are identified as Stratum S, which can be divided into two substrata (S1 and S2) based on the consistency and degree of weathering. Only to a limited extent where the tunnel is deepest will tunneling encounter decomposed rock referred to as "D1" in bench and invert. The decomposed rock is a soil-like material but has higher strength. Ground water at portal locations is generally at invert elevation, at the mid-point of the tunnel alignment it rises up to the tunnel spring line.

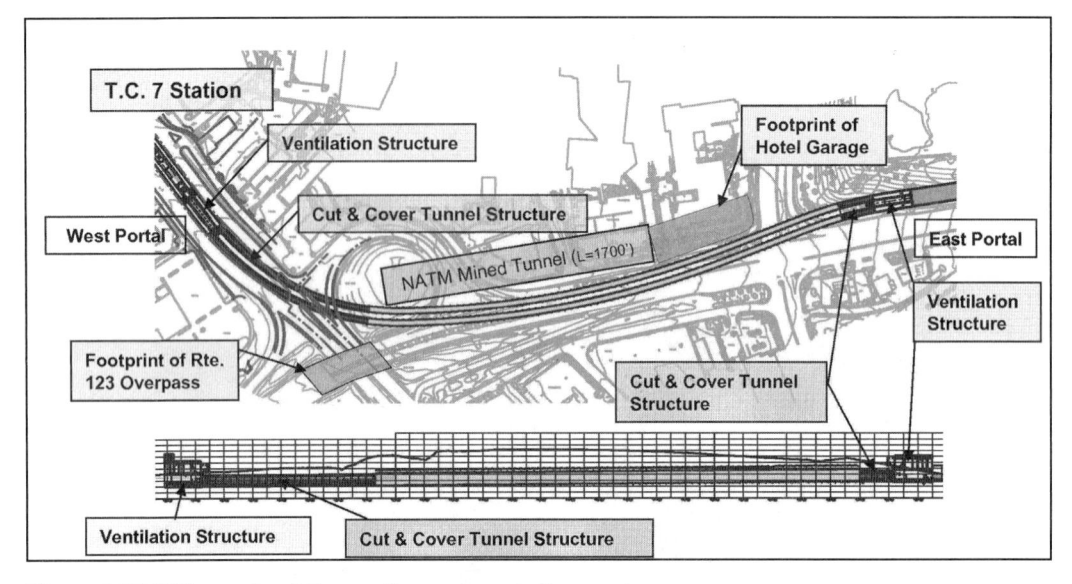

Figure 6. NATM tunnel and Tysons Corner tunnel alignment

## Tunneling

Due to WMATA's long history of tunneling the agency has developed a comprehensive and detailed set of design criteria that often impose "mandatory" design requirements on the tunnel engineers. The NATM design at Tysons Corner, however, is using a newly developed NATM tunnel cross section that is wider than the classical and previously mandatory WMATA NATM regular cross section. Just as the previous section for soft ground NATM, this section is of dual lining character and features a waterproofing wrap around system but is wider to accommodate fire-life-safety considerations called for by the National Fire Protection Association's (NFPA) 130 code requirements for walkway width (Figure 7).

Because of the shallow depth, the prevailing soft ground conditions, the need to control settlements, and risk mitigation issues the NATM is supplemented by a grouted pipe arch canopy for the entire length of the tunnels (Figures 8, 9 and 10a). This will be sufficient for pre-support where the overburden is greater and surface structures are less sensitive. An additional row of pipe arch umbrellas, using closely spaced, approximately 114 mm diameter grouted steel pipes will be used on the first 90 m length at the east portal where tunneling is shallow with 4.6 m overburden. The pipes will be installed at 30 cm center-to-center distances around the tunnel crown. Construction of the NATM tunnels at Tysons Corner is underway, and thus far the Pipe Arch Canopy pre-support is working as it should with no significant ground settlements observed.

Presently, the excavation of outbound tunnel is proceeding through a combination of residual Piedmont soil, found in the bench/invert, and ancient Coastal Plain material found in the top heading (Figure 10b). The Coastal Plain material contains distinct bands of clay between layers of silty sand and gravels/cobbles. The sand and gravel/cobbles are rounded and smooth due by mechanical action of water when the layers were originally deposited. As excavation progresses, the tunnel will eventually move entirely within the residual Piedmont soils.

## Design-Build Contract

The project is being realized under a design-build contract. The design-builder, Dulles Transit Partners, was required to develop preliminary engineering for the rail project. The preliminary engineering then formed the basis to develop a fixed firm price by the design-builder. The need to maintain previously established budget resulted in design challenges and the need to optimize design and construction methods. Value Planning (VP) and Value Engineering (VE) exercises were a central activity of the design development in pursuit of the most economical approach that would impart the least impact on the surroundings. For Phase I of the project, these exercises led to a series of changes in the underground segment at Tysons Corner. The alignment was originally envisioned as deep, 1.6 kilometer long twin TBM tunnels an approximately 24 meter deep underground station constructed by cut-and-cover methods within Route 7, a major traffic artery. However,

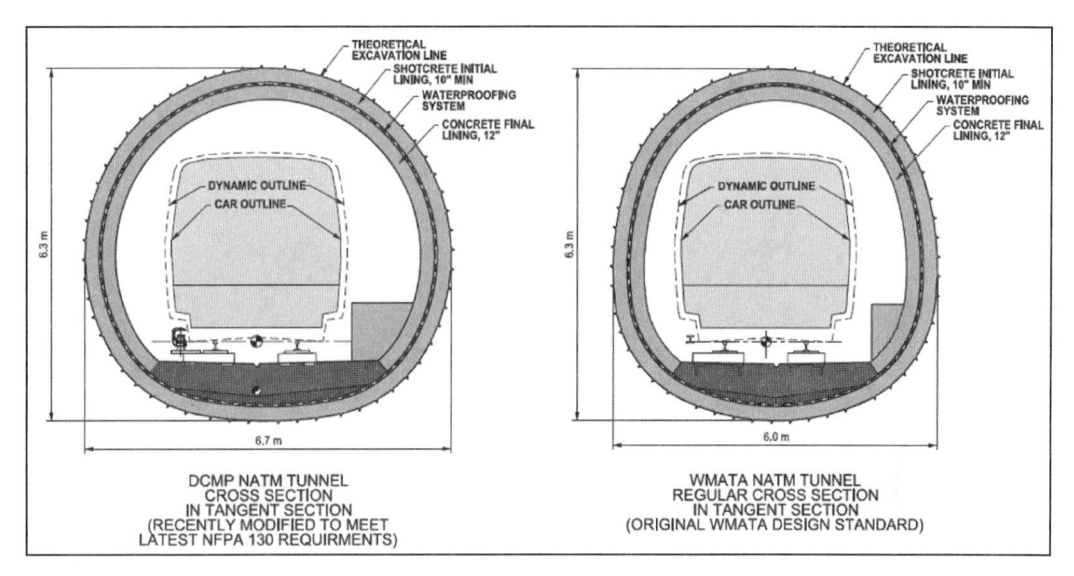

**Figure 7. DCMP new cross section vs. WMATA NATM regular cross section**

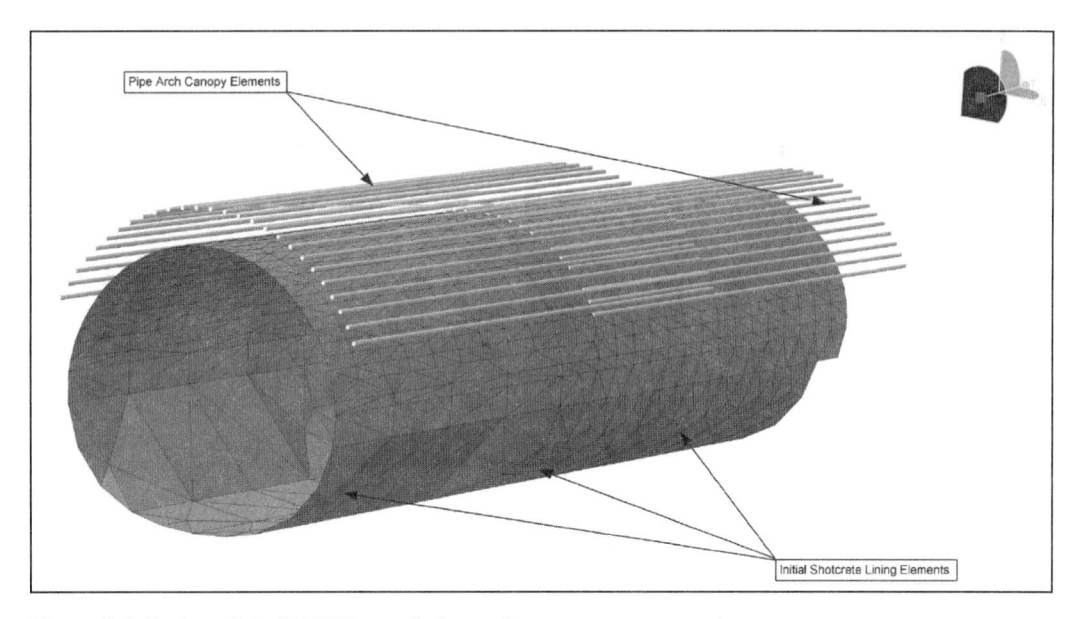

**Figure 8. 3-D view of the NATM tunnel pipe arch canopy pre-support**

analysis of construction cost favored implementation of the current alignment consisting of the short NATM tunnels with a quasi at-grade station within the median of Route 7 at a cost savings of roughly $200 million.

## SUMMARY AND CONCLUSION

With the commencement of construction on the Dulles Corridor Metrorail Extension, the Washington Metro system is once again undertaking soft ground NATM tunneling through a sensitive urban area. All four of the tunnel projects discussed presented their own unique challenges, while adapting state-of-the-art pre-support techniques to meet those challenges. Table 1 summarizes the main characteristics of NATM tunneling for each of the four discussed projects.

The use of the NATM within the Washington D.C. Metrorail system has made a significant contribution

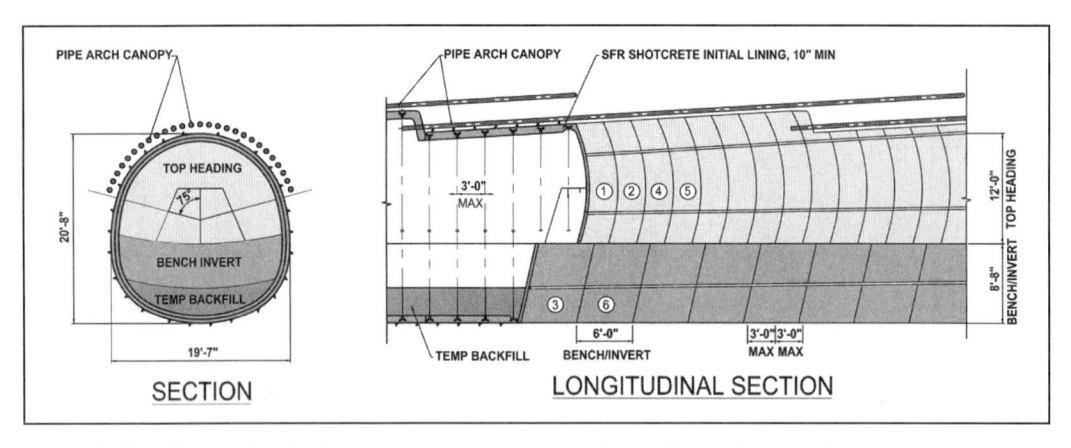

Figure 9. NATM tunnel with single pipe arch pre-support for shallow soft ground tunneling

Figure 10. (a) View of the NATM tunnel pipe arch canopy pre-support installation, (b) View of coastal plain material in outbound NATM tunnel top heading

to soft ground NATM in the U.S. Construction of the Section E5, F6b, and E4b essentially used continuously updated, state-of-the-art pre-support techniques in soft ground NATM tunneling, which have now become accepted and widely used. These experiences and technologies are now being applied in the challenging task of constructing the twin soft ground tunnels through the urban Tysons Corner.

**BIBLIOGRAPHY**

Blakita, P.M. 1995. Directional Soil Stabilization in D.C. Subway Construction. *Trenchless Technology.* 3(10).

Darmody, J. 1991. *Geology and Construction of Fort Totten Station Washington D.C.—A Case History of NATM in Soft Ground Mining.* Department of Design and Construction—Geotechnical Branch Report. Washington, D.C.: Washington Metropolitan Area Transit Authority.

Gall, V., Zeidler, K., Bohlke, B.M., and Alfredson, L.E. 1998. Optimization of Tunnel Pre-Support—Soft Ground NATM at Washington, D.C. Metro's Section F6b. In Proceedings, *North American Tunneling '98.* Edited by L. Ozdemir. Rotterdam, Netherlands: A.A. Balkema Publishers

Table 1. Summary of characteristics for Washington Metro NATM tunnels

| Project/Elements | E5 Fort Totten | E4b New Hampshire Ave. | F6b Congress Heights | DCMP Tysons |
|---|---|---|---|---|
| Ground conditions | Overconsolidated clays and sands | Overconsolidated clays with saturated sands | Overconsolidated clays with saturated sands | Coastal plain silty sands & gravels; piedmont residual soil |
| Groundwater elevation | Below invert | Invert | Above crown | Springline |
| Cross section | Trinocular and single track standard | Single track standard | Single track wmata standard | Single track widened per nfpa 2003 |
| Overburden | 20 to 100 feet | 50 to 147 feet | 35 to 60 feet | 8 to 35 feet |
| Surface features | Park and residential | Rock creek cemetery | Green field, apartment building | International drive, rte. 123 |
| Excavation and support sequence | Top heading -bench/invert 3–4" top heading max. | Top heading -bench/invert 3–4" top heading max. | Top heading -bench/invert 3–4" top heading max. | Top heading -bench/invert 3–0" top heading max. |
| Pre-support | Spiles, sheets | Spiles, sheets, pipe spiling | Spiles, pipe spiling, chemical grouting | Pipe arch canopies: double and single rows |
| Ground improvement | N/a | Chemical grouting with directional drilling/dewatering by vacuum lances | Permeation grouting for differing site conditions | Ground water pressure relief |
| Formal risk assessment | No | No | No | Yes |
| Contract | Design-bid-build | Design-bid-build | Design-bid-build | Design-build |

Irshad, M. 1995. NATM Developments 1994–1995. NATM Subcommittee Report presented to TRB Committee A2C04, January 24, 1995. Washington, D.C. Unpublished Report.

Jenny, R.J., Donde, P.M., and Wagner, H. 1987. New Austrian Tunneling Method Used for Design of Soft-Ground Tunnels for Washington Metro. *Transportation Research Record.* 1150: 11–14.

Myers Böhlke, B., Rosenbaum, A. and Boucher, D. 1996. Complex soil Conditions and Artesian Groundwater Pressures Dictate Different Tunneling Methods and Liner Systems for Sections F6a, F6b and F6c of the Washington D.C. Metro. In Proceedings, *North American Tunneling '96*. Edited by L. Ozdemir. Rotterdam, Netherlands: A.A. Balkema Publishers.

Rudolf, J. and Gall, V. 2007. The Dulles Corridor Metrorail Project—Extension to Dulles International Airport and its Tunneling Aspects. In *Rapid Excavation and Tunneling Conference Proceedings*. Edited by M.T. Traylor and J.W. Townsend. Littleton, CO: SME

Rudolf, J., V. Gall and Wagner H. 2009. History and recent developments in soft ground NATM tunneling for the Washington, DC Metro. In *Proceedings, ITA-AITES World Tunnel Congress 2009—Safe Tunnelling for the City and for the Environment*. Edited by P. Kocsonya. Budapest, Hungary: Hungarian Tunnelling Association.

# The Lincoln Square Tunnel: Tunneling Between Two Parking Garages Using Sequential Excavation Mining

**Chris D. Breeds, Larry Leone**
SubTerra, Inc., North Bend, Washington

**Don Gonzales**
Northwest Boring, Inc., Woodinville, Washington

ABSTRACT: This paper presents the design and construction of the Lincoln Square Tunnel, a 35-ft wide, 14-ft high traffic and pedestrian tunnel constructed 25-ft deep under NE 8th in Bellevue, WA. This design-build project was built from the third level of the parking garage in Bellevue Place, the busiest hotel complex on Seattle's eastside, to the third floor of the parking garage beneath Lincoln Square. Restrictions were imposed by weight limits on the access floors, the presence of soldier piles, tie backs, and soil nails installed during parking garage construction, mining near the garage walls, zero allowable settlement, and concurrent operation of the facilities. The project was successfully competed in less than a year from starting the design.

## INTRODUCTION

The Lincoln Square Tunnel is a privately funded project that connects two of the Owner's underground parking garages in a downtown urban area. With the addition of the tunnel, office workers, shoppers, hotel guests, and condominium residents are given another option to reach their destination, which in turn, as the owner predicted, has alleviated traffic congestion on the surface streets.

The Tunnel connects the P3 levels of Bellevue Place and Lincoln Square crossing under NE 8th Street immediately East of its intersection with Bellevue Way as shown in Figure 1. The alignment runs approximately 78° to the buildings with an elevation drop of 6.2 feet from Bellevue Place to Lincoln Square. The completed tunnel featured two automobile lanes and a sidewalk on either side for pedestrians. The tunnel cross-section incorporated a 35 feet wide 35-ft radius, shallow arch 14 feet high at the center and 9 feet high on either side.

Ground conditions included very dense glacial till consisting of clayey soil with gravel and cobbles up to 6 inches in size. A few sand lenses were encountered as well as three glacial erratic boulders with dimensions up to 33 inches. The stand-up time for this material was on the order of weeks which contributed to safe working conditions near the face.

Numerous tiebacks and soil nails that were originally installed to support the excavation of the two, multi-level subgrade parking garages were also located in the mining horizon. Tiebacks (three ½-inch wire ropes with grouted anchors) and soil nails (#10, 150 ksi rebar grouted over their entire length) were incrementally cut away as the tunnel advanced.

## GARAGE WALL PRE-SUPPORT

Early in the design process, the Owner's Structural Engineer determined that the existing garage walls could not react the redistributed ground pressure concentrated at each concrete footing. An alternative footing design was prepared using a #14 bar pin pile to transfer the arch loads below the base of each parking garage (see Figure 1). This alternative was later dropped and the original 4-ft by 4-ft contract footing designed with a subgrade bearing capacity of 12,000 psf was installed.

The P4 and P5 walls immediately below the tunnel portal were reinforced using 10,000 psi rebar reinforced shotcrete that was coupled to the existing wall using epoxied dowels. Floor P2, immediately above the portal area, was supported using 12-inch channel spanning beyond each of the four portal entrances. All pre-support was in place prior to cutting the concrete in the Bellevue Place wall.

## TUNNEL PRE-SUPPORT

Before tunneling began, a pre-support system consisting of 32 spiles was drilled from the P2 level of the Bellevue Place garage to the Lincoln Square garage along the upper profile of the tunnel. The spiling, each a 6⅝-inch diameter, ⅜-inch-thick wall drill pipe, was drilled on one-foot centers and

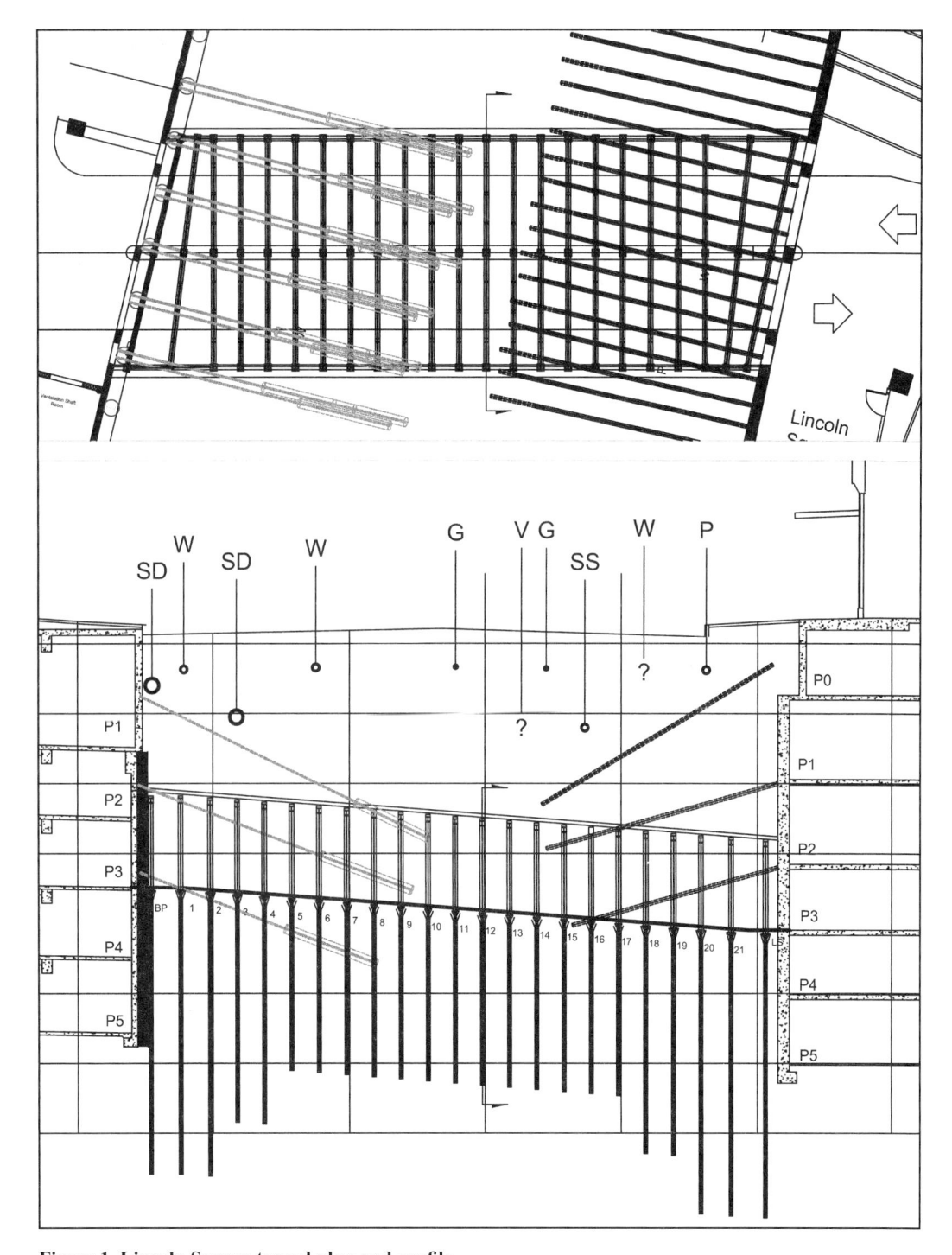

Figure 1. Lincoln Square tunnel plan and profile

Figure 2. Pre-coring spiling holes from P2

Figure 3. Exposed spiling inside tunnel portal

injected with grout. As the tunnel face was advanced, the spiling was exposed and supported between the last rib set and the face. This was the ground support system used to protect workers in otherwise unsupported ground.

**Spile Design**

Spiling was designed to carry a spanning load of 12-ft as well as cantilevered loads at each end of the tunnel to accommodate the SE's requirement that the tunnel ground support not contact the parking garage walls. This allowed the Contractor to mine for two sets before installing the shotcrete lagged steel arch ground support.

Design reviews illustrated potential problems associated with hitting boulders, tie-backs and soil nails and this eventuality was taken into account during the design. Holes were started 4-inches above the planned top arch profile with provisions for adding spacing plates at the base of each arch to optimize shotcrete quantities.

**Survey Control and Installation**

A detailed radar survey of the parking garage wall was completed prior to drilling the spiling and the location of the existing wall reinforcement was marked on the inside of the garage wall. This process was conducted to ensure that spile drilling did not sever any of the existing reinforcement. Precise surveying was used to accurately locate the existing reinforcement so that an optimum spiling layout could be developed. Precise surveying was also used to locate the center of each spile hole and a back site that could be used by the Contractor for alignment and declination control. Spile drilling tolerances were nominally set at ±1-inch on line and ±0.5% on grade.

Coring was used to form a starter hole (see Figure 2) ensuring preservation of the wall reinforcement and avoiding the need for spile drilling to contact the wood lagging or steel soldier piles. Each spile was installed and incrementally surveyed using the Flexit Gyrosmart system, a specialized and very accurate borehole survey tool.

As anticipated, at least 30% of the spiles encountered both de-tensioned tiebacks near the portal wall and soil nails approaching Lincoln Square. Only one of the spiles was significantly deflected downward while three of the holes were stopped short of the Lincoln Square wall. Overall accuracy was excellent completing almost 90% of the planned "umbrella" without impacting the clearance envelope for steel arch and shotcrete installation.

**Underground Conditions**

The spiling pre-support provided excellent tunneling conditions enabling complete control of the roof and ensuring zero subsidence at the road surface just 25-ft above. Figure 3 shows this condition near the portal wall where use of the boom mounted, NDCO cutterhead was essential to mining to 15-ft, some 7-ft above the portal entrance.

The Flexit Gyrosmart survey data was used to map out the spiling to an accuracy of ±0.5-inches enabling confirmation of the arch canopy clearance and vertical positioning requirements for each arch footing.

**GARAGE FLOOR REINFORCEMENT**

The relatively low bearing capacity of the Bellevue Place parking garage floor (40 to 50 psf) was a major issue faced during equipment mobilization. This issue was resolved using a walking roof jack system to locally support the under floors while spile drilling equipment was moved to the P2 portal. Mining equipment weighing up to 13 tons was sequentially moved to the P3 tunnel portal area approximately one month later. The floors below the access tunnel portal area in Bellevue Place were reinforced using a closely spaced pattern of floor jacks that remained in place throughout the construction period (see Figure 4). This allowed floor loads up to 1,900 psf at P3 and 1,300 psf on P2.

## MOBILIZATION AND SITE SET UP

Two of the Owner's primary criteria included preventing impacts to property users and ensuring that there were no impacts to parking outside of the dedicated portal work areas on P2 and P3. This led the Contractor to create a completely sealed portal area venting air from the portal area to the parking garage through HEPA filters. This precaution was rewarded by zero complaints and zero requests for car cleaning.

**Figure 4. Floor jacks beneath the P3 portal area**

## TUNNEL GROUND SUPPORT

### Design

The structural members of the tunnel consisted of steel rib sets founded on concrete footings and fully lagged into contact with the ground with steel fiber reinforced shotcrete. A total of twenty-three steel rib sets were erected on 4-foot centers each consisting of two W6×25 floor struts, three W8×48 columns, and arches built up from W8×48 and WT12×52 beams (see Figure 5).

W8×48 beams were rolled to a radius of 35-ft and originally designed to transfer overburden loads to the tunnel ribs. This design was subsequently modified to provide full bending resistance assuming that the set would need to carry the full depth of overburden as well as loads from two very large mobile cranes that were used to install pedestrian bridges. The composite beam including the reinforced shotcrete was analyzed but the additional strength provided by the shotcrete was not relied on for ground support.

### Installation

Tolerances were established for positioning the floor struts and outer ribs requiring that each center rib be perfectly centered. Concrete arch footings were installed with a tolerance of +0, and –1-inch. These careful preparations facilitated installation and bolting the footing blocks and assembly of each set.

### Shotcrete

Shotcrete specifications for the project included:

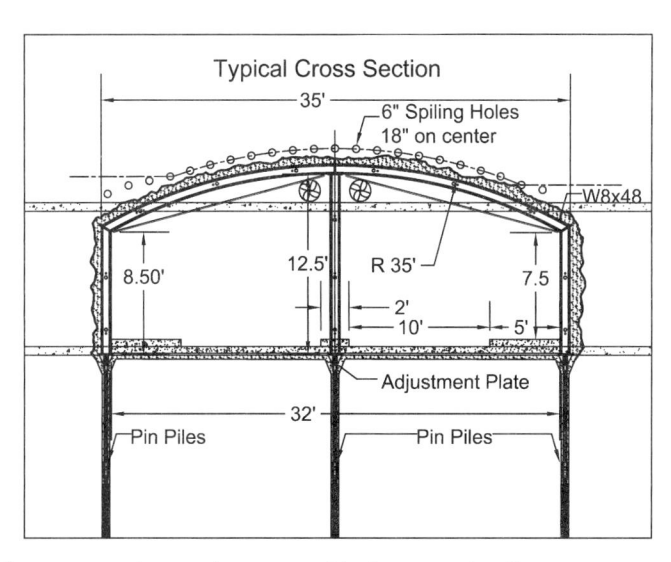

**Figure 5. Lincoln Square tunnel ground support with alternate pin piles**

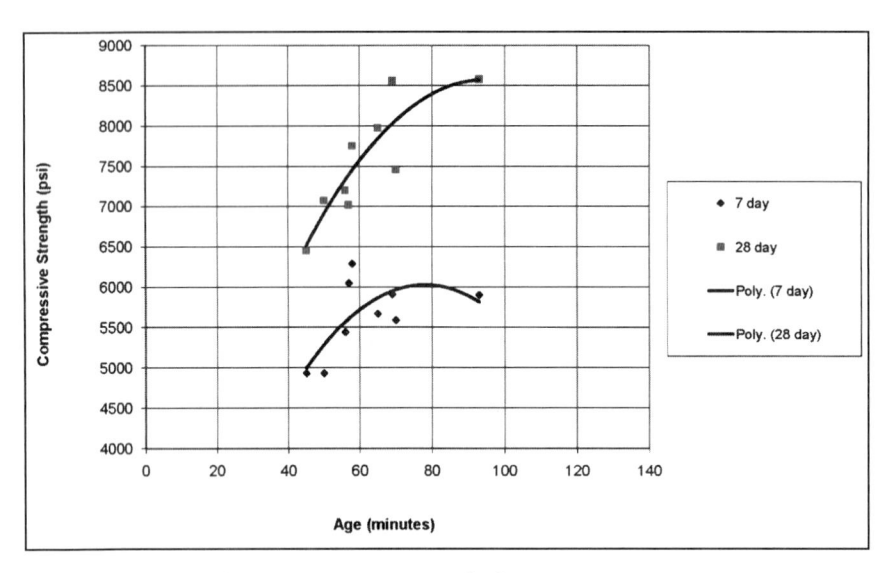

Figure 6. Shotcrete compressive strength versus age of mix

- Minimum steel fiber content of 60 lbs. per cubic yard
- Discharge from the mixer within 90 minutes from batch time
- Minimum thickness of 6 inches at any measured point
- Average 8 inches of thickness
- Minimum compressive strength of 3500 psi at 7 days and 5000 psi at 28 days

The Contractor elected to initially use pre-bagged, Target Superstick shotcrete pre-blended with 30mm Dramix fibers and locally dosed with a superplasticizer to enable wet mixing. This product was successfully used to line the first ten feet of the tunnel. For future applications the Contractor elected to use a ready-mixed product in order to increase the production rate. A Warrior model 3050 shotcrete pump was used to pump the mix from the street, down the ventilation shaft of the garage, and into the tunnel. Glacier NW mix 0876 was batched with five-cubic-yards in each 10 cubic yard truck using Novocon 730 steel fibers. Because of heavy traffic in the Seattle area, trucks typically arrived on site between 60 and 80 minutes old. Shotcrete placement frequently exceeded the specified 90 minutes, and occasionally was being placed beyond 100 minutes old. As noted below, testing revealed that this was not detrimental to the strength of the final product.

An independent laboratory performed compressive strength tests on cores taken from a 12-by-12-by-6-inch test panel. A single core was tested at 7 days and three more were tested at 28 days. Typically the Inspector or the Engineer would choose which truck the sample would be collected from with a goal of capturing the "worst case" by selecting a truck with an older mix. Once shot, the test panel was left to cure for at least 24 hours in the tunnel before being transported to the testing lab where it was cured under controlled conditions.

For the ready-mixed product, the average 7-day strength was 5520 psi, which exceeded the 28-day strength requirement. The 28-day average strength was 7380 psi and the average mix age was 62 minutes as it began to discharge into the shotcrete pump. These averages include one sample that tested at 4530 psi at 7 days, and 5733 psi at 28 days. Although this sample meets the specification, it is considered an outlier for the purposes of this analysis and was removed from the data set used to produce Figure 6.

An analysis of compressive strength results versus the mix age was performed and an interesting trend emerged. Using second order polynomial regression, the maximum 7-day compressive strength was achieved at 78 minutes while the oldest mix tested at 93 minutes resulted in the highest 28-day strength. These results are presented in the graph in Figure 6. It should be noted that the age of the mix in this analysis represents the time elapsed from batching to discharge into the shotcrete pump. A more accurate analysis would have used the time the test panel was shot, however, these times were not consistently available. A separate analysis of the site-mixed products was not practical due to a very small data set.

Figure 7. Boom mounted NDCO used during portal development

Figure 8. West side of completed tunnel

## SEQUENTIAL EXCAVATION MINING

### Breakout

The typical tunneling cycle began with two days of excavation and muck haulage. This was accomplished with an Alpine F16A roadheader and an NDCO Construction EC25 transverse cutting head operated by a John Deere JD50D excavator. The outer limits of the excavation were guided by the spiling canopy above and by measuring out from a vertical laser plane offset fifteen feet either side of the centerline. The tunnel advanced by eight feet during each cycle requiring the excavation and transport of 130 cubic yards of the native till. Excavated material was mucked out by a Caterpillar 257 skid steer loader and loaded into Akkerman dirt buckets normally used with microtunnel boring machines. To meet the maximum allowable floor loading, the 1.25-cubic-yard flat-bottomed dirt buckets were hauled out of the garage on tandem axle flat bed trailers towed behind pickup trucks. A Grove RT635C crane was used to dump the dirt buckets into a stockpile that was later hauled off site.

The next step in the cycle was to build wooden forms for the footings for two steel rib sets. A vertical laser plane projected by a Spectra HV301 rotary laser was used to position the forms and a conventional surveyor's level was used to set them at the design elevation. The East and West footings were 3 feet wide and 4 feet long and the center footings were 4-foot square. The forms were built 8 feet long with a step in the middle to account for the elevation change of the second of the two rib sets. Ready-mixed concrete was transferred into the garage in the bucket of the skid steer loader and dumped directly into each of the forms. A #6 rebar mat was set 6 inches from the bottom of each of the forms and concrete was mechanically consolidated.

The day after the concrete was poured two rib sets were erected and anchored into the footings. One inch spacers were used at the base of rib set closest to Bellevue Square to ensure the arches were positioned within 8-inches of the spiling canopy. Data available from the Flexit Gyrosmart survey was invaluable in planning these activities. Welded wire mesh and rebar was attached to the outer perimeter for shotcrete reinforcement. On the final day of the cycle the ribs and back received an application of shotcrete that filled the gap from the steel rib sets to the native ground and the spiling.

## FINISHES AND FIREPROOFING

A final mesh reinforced shotcrete lining was installed to form the walls and a scored concrete floor was placed and finished. Code requirements dictated that all steel components be fire-proofed so the roof and central pillars were wrapped and finished as shown in Figure 8.

## CONCLUSIONS AND LESSONS LEARNED

A very successful project was realized due, in part, to the flexibility built into the Design Build process that facilitated the interaction of the Engineer and Contractor when planning and executing the work. Precision was dictated at every stage of construction starting with the need to avoid reinforcing steel in the portal wall, continuing through spile drilling and steel set placement to installation of the fire-proofing and final lining. Several lessons were learned during this process and are noted below.

Spiling Pre-Support: Meticulous attention to detail, very careful drilling, and intermittent hole surveys allowed the spiling to be installed a distance of 96-ft with just under 90% coverage.

Site Enclosures: A completely sealed portal enclosure ensured compliance with the Owner's requirements not to affect the building users and not to impact parked cars.

Shotcrete: On-site batching of wet mix shotcrete had both positive and negative attributes. It was certainly a labor-intensive process to break open the bags of pre-blended products and a dusty atmosphere resulted. A better use of curtains could have been used to control airflow through filtered exhaust fans installed in the work area. Conversely, on-site batching offered the ability to control the age of the mix and the concern to place the product within 90 minutes was eliminated. Another advantage of the site-mixed product was the quality of the product itself. The Target Superstick shotcrete was observed to hang overhead better than the ready-mixed product. The ready-mixed product offered the advantages of higher production rates and fewer required personnel, which was ultimately preferred by the Contractor.

Overall the project was a great success. The owner was very pleased with the rate at which the tunnel was completed and is considering building two more in the area for the same purpose.

## ACKNOWLEDGMENTS

The authors wish to acknowledge the following organizations that contributed to the success of the project: General Contractor, GLY; Design-Build Subcontractor, Northwest Boring; Design subcontractor, SubTerra, Inc.; Spiling subcontractor Northwest Cascade; Shotcrete subcontractor, Johnson Western Gunite; the Owner, Kemper Development Company; the Owner's Structural Engineer CKC and the Owner's Architect, Sclater Partners.

# Case History: Complex Design and Construction of Tunnel and SOE to Accommodate Challenging Site Conditions

**Emad Farouz**
CH2M Hill, Chantilly, Virginia

**John I. Yao**[*]
Jacobs Associates, Pasadena, California

**ABSTRACT:** The Narragansett Bay Commission Combined Sewer Overflow (CSO) Project includes the deep-storage Main Spine CSO Tunnel, a pump station, shafts, and near-surface facilities. This paper presents a case history of the complex design and construction of a tunnel for the installation of a Consolidation Conduit, as well as the support of excavation (SOE) for a 12.2-m (40-ft) deep Diversion Structure that also served as a tunnel receiving shaft during construction. This shaft consists of an innovative combination of secant piles, soilcrete walls, and bracings, and also supported a 259-cm (102-in.) pipe that maintained the existing sewer flow during construction.

## INTRODUCTION

### Project Overview

The Narragansett Bay Commission's (NBC's) Main Spine Combined Sewer Overflow (CSO) Tunnel went on line Saturday, November 1, 2008. The NBC owns and operates the interceptor and wastewater treatment facilities serving 10 communities with a total population of 360,000. The project includes five CSOs that discharge to the Woonasquatucket, West, Moshassuck, Seekonk, and Providence Rivers. These rivers are tributaries to Rhode Island's Narragansett Bay, an "estuary of national significance."

NBC developed a comprehensive program in order to abate CSO pollution in the Upper Narragansett Bay. Phase 1 of the program includes a 7.9-m (26-ft) diameter, 70.1-m (230-ft) deep storage tunnel, drop shafts, a CSO pump station, and several near-surface interceptors. The project also included detailed design and services during construction of near-surface CSO control facilities, consisting of diversion structures, screening structures, consolidation conduits, approach and vortex structures, and inflow control gates. Of these, the Diversion Structure and the tunnel for the Consolidation Conduit associated with the contract for OF 067. Facilities are the focus of this paper.

The NBC CSO Project site for the OF 067 Facilities is located in Providence, RI, near the Ernest Street Sewage Pumping Station, which is bounded by Ellis Street on the east and Ernest Street on the south, as shown in Figure 1. The Diversion Structure and the Gates and Screening Structure had been planned for construction at Ernest Street and Ellis Street, respectively. A 274-cm (108-in.) diameter Consolidation Conduit, approximately 37.8 m (124 ft) in length, was to connect the two structures, as shown in Figure 1. During preliminary design, it was decided that the Gates and Screening Structure would be designed by the contractor, while the Diversion Structure and the Consolidation Conduit would be designed by the engineer as part of the contract for OF 067 Facilities. The excavations for the Diversion Structure and the Gates and Screening Structure would also be used as work shafts for the tunnel excavation to install the Consolidation Conduit.

### Subsurface Conditions

The subsurface soils consist of sands with various amounts of silt (SM and SP-SM) and nonplastic silts (ML). The soils are typically medium dense, but range from loose to very dense. The groundwater levels measured from boreholes are typically 4.6 m (15 ft) or less below the ground surface. Ground behavior of these soils in unsupported excavations can range from rapid raveling to running condition above the groundwater table, or flowing condition below the groundwater table.

---

[*] The author's involvement in the project discussed in this paper was during his previous employment with CH2M Hill.

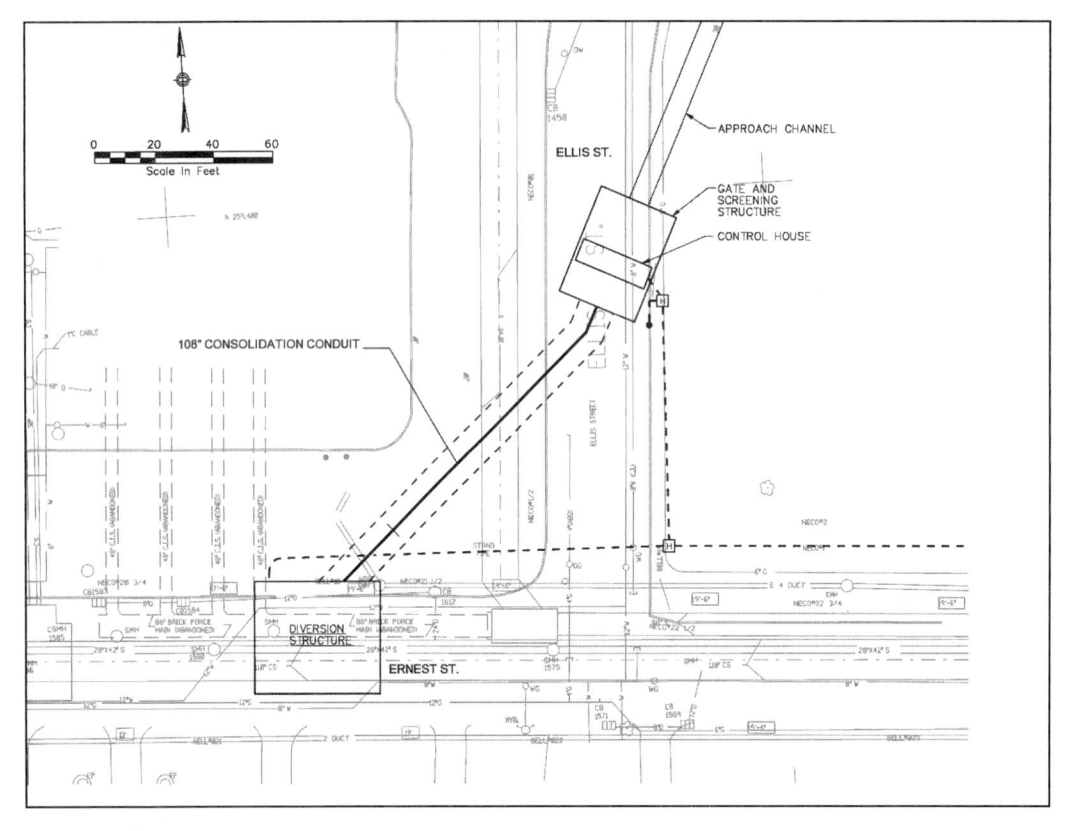

**Figure 1. Site plan**

## Site Constraints

There are numerous existing utilities present at the project site, especially along Ernest Street. Those that posed challenges for the construction of the Diversion Structure included the following:

- Overhead electrical lines along the north and south sides of Ernest Street, and overhead telecommunication line along the north side of Ernest Street.
- Electrical poles in the immediate vicinity of the Diversion Structure footprint.
- An abandoned 224-cm (88-in.) brick force main and associated abandoned 122-cm (48-in.) lateral lines and manhole.
- A 71 by 107 cm (28 by 42 in.) sewer line and a 279-cm (110-in.) brick sewer line in the middle of Ernest Street. The 279-cm (110-in.) sewer line must maintain active flow during construction.
- A 30.5-cm (12-in.) drain line, a 30.5-cm (12-in.) gas main, and a 20.3-cm (8-in.) water line.

Figure 2 shows a schematic cross section of Ernest Street that illustrates the existing utilities at the site of the Diversion Structure.

Additionally, the tunnel excavation for the Consolidation Conduit crosses directly under and within approximately one tunnel diameter of an existing box conduit running west of and parallel to Ellis Street. There are also overhead electrical lines running along the west side of Ellis Street.

## DIVERSION STRUCTURE

Two concepts were initially considered for the excavation support system of the Diversion Structure:

1. Temporary excavation support using reinforced soilcrete walls consisting of soldier beams encased within soilcrete (i.e., jet grout), with multiple levels of bracing.
2. Permanent secant pile walls, with multiple levels of temporary bracing.

For both concepts, plain soilcrete walls were to be used to support the excavation where penetration through the shaft wall was required for either the

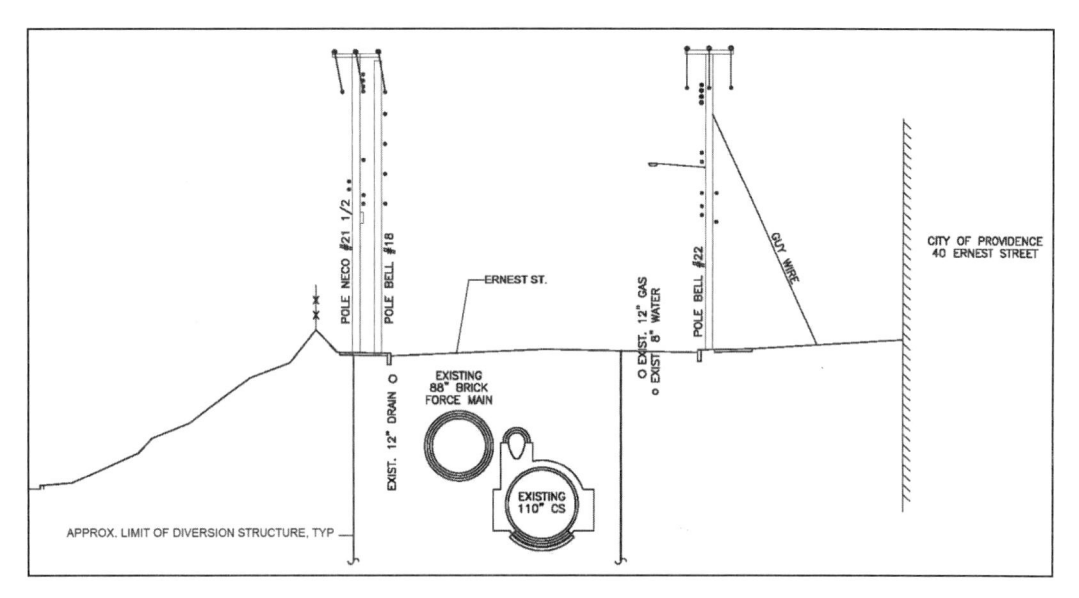

**Figure 2. Schematic cross section of Ernest Street looking east showing a number of existing utilities**

tunnel excavation for the Consolidation Conduit or allowing existing utility to pass through the excavation. Additionally, a soilcrete plug was to be installed below the bottom of the excavation to maintain the bottom stability under the high groundwater head and to laterally brace the excavation support system.

The second concept involving the use of secant pile walls was eventually selected as the excavation support system for the Diversion Structure. There were several reasons for its selection. First, the secant pile walls can be part of the walls for the final structure, whereas the soilcrete walls can only be considered as temporary support of the excavation. Second, installation of soilcrete by jet grouting was expected to be very challenging at the project site. The presence of numerous existing utilities above and below the area of the Diversion Structure meant that some of the jet grouting would have to be done at an inclined angle, and there would be increased risk of ungrouted pockets or windows of soil. Therefore, it was decided that soilcrete would be used only where its use was necessary, such as for the bottom plug and the shaft wall penetrations. Third, the use of secant pile walls would require only two levels of bracing, but the reinforced soilcrete walls would require three levels of bracing because the soilcrete has lower material strength and stiffness than the secant piles.

**Design of the Excavation Support System**

The excavation and support sequence for the Diversion Structure is complex. The loads on the structural support elements vary depending on the stage of the excavation as well as the previously installed support elements. Once the Consolidation Conduit is completed, the temporary support elements need to be removed in a staged process as the final structure is being constructed. Each support element has to be designed for the maximum loading condition anticipated during the entire construction process of the Diversion Structure.

A two-dimensional finite element model (FEM) was set up using the computer software PLAXIS to model each excavation and support sequence. Major elements of the excavation support system—including secant piles, two levels of bracing, and the soilcrete bottom plug—were incorporated into the model. The model was used to evaluate the overall performance of the support system and to optimize each major support element—including the design of the secant piles, the size and vertical location of the bracings, and the bottom plug. The maximum loads from the FEM analysis were then used for the detail design of the various structural components such as the braces, walers, and connections.

In addition, FB-Pier, a computer software by Bridge Software Institute, was used as an alternative method to check the performance of the secant pile wall at the stage when the excavation reached the bottom. The software is most commonly used for the design of bridge pier structures, and couples nonlinear structural finite element analysis with nonlinear static soil models.

**Final Design**

The final design of the excavation support system for the Diversion Structure is shown in Figure 3 through Figure 5. The excavation, approximately 10.1 m

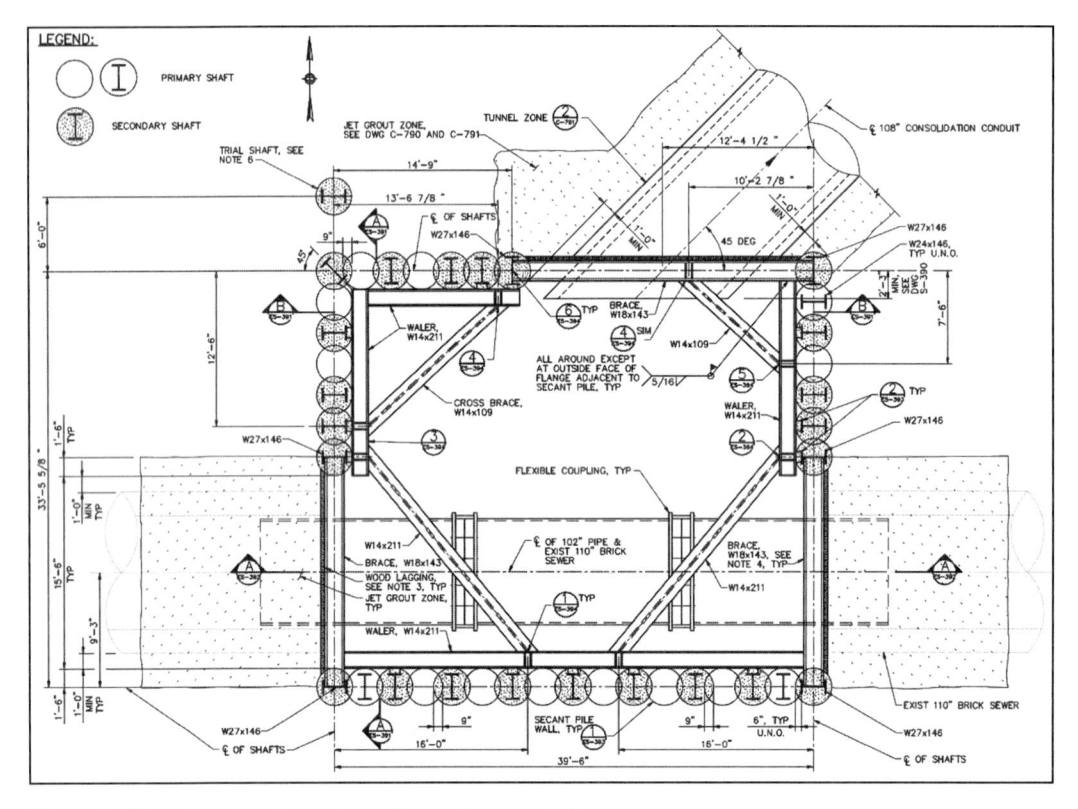

**Figure 3. Diversion structure excavation and support plan**

wide by 12.2 m long by 12.2 m deep (33-ft wide by 40-ft long by 40-ft deep), has the following main support elements:

- Secant pile walls consisting of 91.4-cm (36-in.) diameter drilled shafts with generally 15.2-cm (6-in.) overlaps. As shown in Figure 3, typically every other shaft is reinforced with a W27×146 soldier beam, except adjacent to each wall opening, where three adjacent drilled shafts on each side of the opening are reinforced. Although the soils behind the wall openings are jet grouted so that there is little or no lateral earth pressure from the soilcrete, the reinforced drilled shafts adjacent to each penetration are designed to take any remaining lateral earth pressure transferred to them through the wood lagging and the braces across the wall openings.
- The first two levels of the excavation support frames consisting of walers and cross braces to provide lateral support to the walls of the excavation. The locations of the cross braces were designed to provide as much clear space as possible in the excavation for the retrieval of the tunnel shield and other construction equipment. In addition to providing lateral support to the shaft walls, two of the second-level cross braces, as shown in Figure 4, were also designed to support a temporary 259-cm (102-in.) pipe running through the excavation to maintain flow from the existing 279-cm (110-in.) brick sewer during construction.
- A soilcrete plug is installed below the bottom of the excavation. In addition to providing bottom stability, the plug reduces the embedment length of the secant piles to 3.4 m (11 ft) by providing lateral support to the shaft walls.

Since the design of the excavation support system was based on an assumed excavation and support sequence, the contractor was required to follow the detailed excavation and support sequence indicated on the contract plans. Otherwise, it would be possible for the anticipated maximum loading that the support system was designed for to be exceeded. The excavation and support sequence for the Diversion Structure is outlined below. These steps have been generalized for discussion purposes.

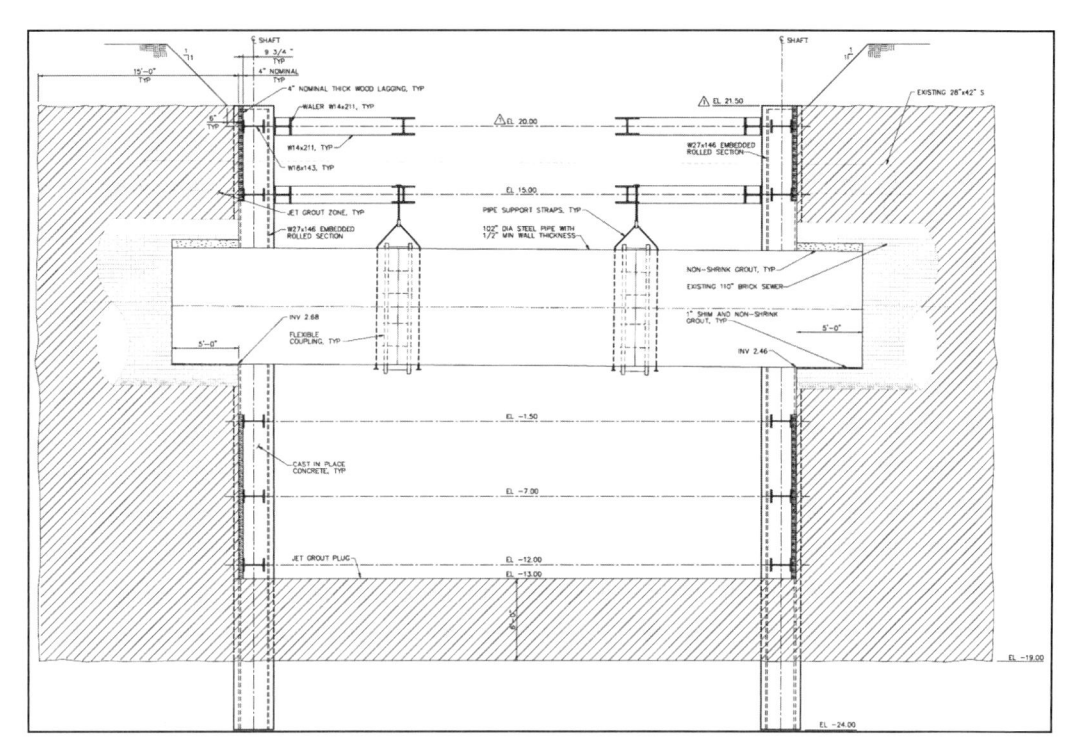

Figure 4. Diversion structure section A-A

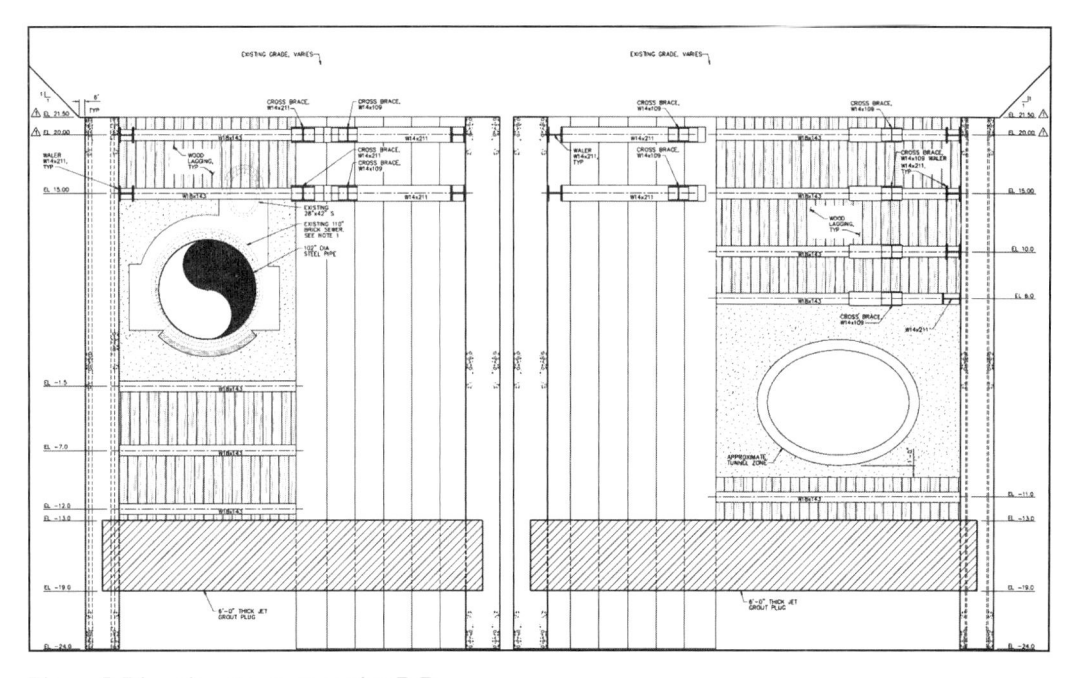

Figure 5. Diversion structure section B-B

1. Relocate overhead lines and certain underground utilities in the vicinity of the Diversion Structure site.
2. Perform jet grouting to form zones of soilcrete that will be between the shaft wall openings and to create the shaft bottom plug.
3. Construct the secant pile walls.
4. Excavate to 0.6 m (2 ft) below the first level of bracing and install the walers, cross braces, and other support elements.
5. Excavate to 0.6 m (2 ft) below the second level of bracing and install the support elements per previous step. Remove and dispose the abandoned portion of the existing 71×107 cm (28 × 42 in.) sewer.
6. Continue to excavate and install excavation support elements, including braces and wood lagging across the shaft wall openings. Do not excavate more than 0.6 m (2 ft) below any braces.
7. Once the excavation reaches the existing 279-cm (110-in.) brick sewer, install the temporary 259-cm (102-in.) steel pipe by slip lining. Cut the top half of the existing brick sewer and install the two end sections of the steel pipe. Grout the ends of the steel pipe to the brick sewer. Cut the bottom half of the brick sewer and remove any flow into the excavation as needed to install the middle section of the steel pipe. Connect the flexible coupling to hang the steel pipe from the second-level cross braces (Figure 4) before continuing to excavate and install excavation support elements per the previous step.
8. After the excavation reaches the bottom and the installation of the Consolidation Conduit is complete, construct the bottom slab of the Diversion Structure. Braces supporting the openings between the secant pile walls are to remain in place and embedded within the slab and walls of the final structure.
9. The corner braces on the third and fourth levels can be removed after the bottom slab has been constructed.
10. Construct the final exterior walls of the Diversion Structure without removing the excavation support frames on the second level by blocking out the support elements and constructing the walls around the support elements. The first-level excavation support frame may be removed when the top of the final walls is 0.6 m (2 ft) below the second-level frame.
11. Continue the construction of the final exterior walls to the top. Construct the interior walls.
12. Construct the final roof structure.
13. Remove the second-level excavation support frame. Finish the final walls of the Diversion Structure by completing the walls at the previously blocked out areas.

## CONSOLIDATION CONDUIT

The Consolidation Conduit is a 274-cm (108-in.) ID, approximately 37.8-m (124-ft) long, reinforced concrete pipe (RCP) that connects the Gates and Screening Structure to the Diversion Structure. During design, it was anticipated that one of the most likely tunneling methods for installing the RCP would involve the use of an advancing shield that allows a primary lining such as liner plates to be erected behind the shield. Upon completion of tunneling and the removal of the tunnel shield, the RCP would be installed inside the primary lining and the annular space between the RCP and primary lining would be filled with grout. The tunnel excavation was anticipated to be primarily in sands and silts below the groundwater table, and ground improvement would be required if an open-face tunneling method was to be used. There were also concerns that settlements caused by tunneling could damage surface facilities located approximately 6.1 m (20 ft) above the tunnel zone, and an existing box conduit located approximately 3.7 m (12 ft) above the tunnel zone. A pressurized face tunneling method such as a slurry tunnel boring machine was technically suitable, but was determined to be uneconomical because of the short length of the tunnel drive. Since jet grouting equipment would be mobilized on site for the construction of the Diversion Structure, it was decided that the ground along the tunnel alignment would be improved so that an open-face tunneling method could be used.

### Ground Improvement Along the Consolidation Conduit Tunnel Alignment

Figure 6 shows the limits of the various types of jet grout zones along the Consolidation Conduit tunnel alignment. The minimum cross-sectional area required for jet grouting for each type of jet grout zone is shown in Figure 7. More than one type of jet grout zone was specified to optimize the use of soilcrete such that jet grouting would be performed only in areas necessary to achieve the intended design function(s) along that particular section of the alignment. The types of jet grout zones along the tunnel alignment are described below:

- Jet grout zone Type I was to be installed adjacent to the Gates and Screening Structure and Type III, was to be installed adjacent to the Diversion Structure (Figure 7). The entire tunnel zone was to be fully grouted because the tunnel shield would be launched through

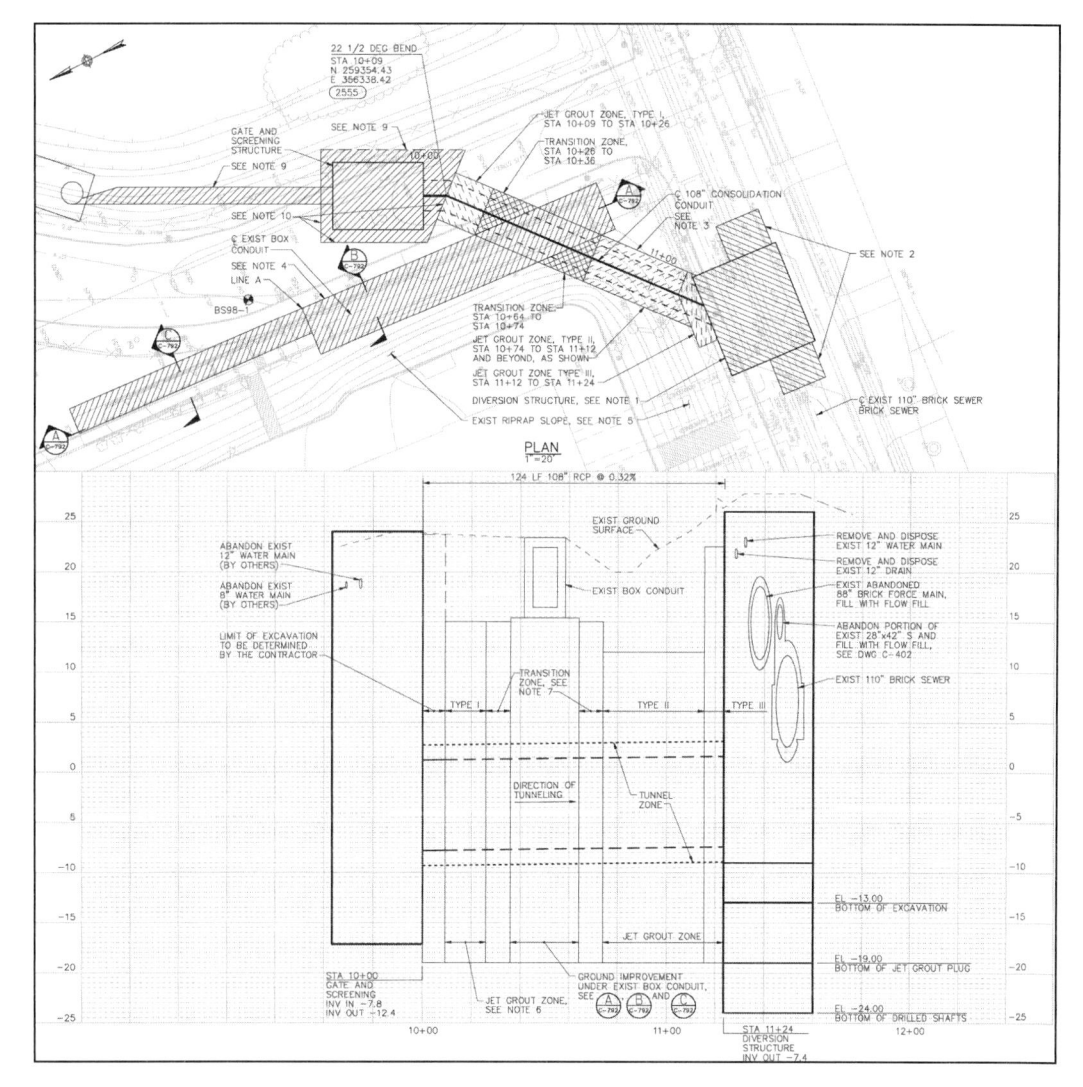

**Figure 6. Plan showing limits of various types of ground improvement zones**

the Type I zone and received from the Type III zone. The Type III zone extended to near the ground surface because the soilcrete formed the shaft wall at the opening between the secant pile walls.

- The purposes of jet grout zone Type II, located between the "transition zone" and Type III zone, were to allow open face tunneling and minimize settlement caused by tunneling. The square area shown in the Type II cross section was left unimproved because: (1) the soilcrete surrounding the unimproved area should be sufficient to prevent the settlement trough due to tunneling from forming or reaching the ground surface; (2) there were no critical facilities directly above the limits of the Type II zone; and (3) the tunnel advance rate may be higher because of the relative ease of excavating unimproved ground.

- As shown Figure 6, a jet grout zone was specified below the existing box conduit. In order to minimize the risk for any amount of settlement, the area within the vertical limits of the jet grout zone, including the tunnel zone, was to be fully grouted. The transition zones under both sides of the existing box conduit were also to be fully grouted.

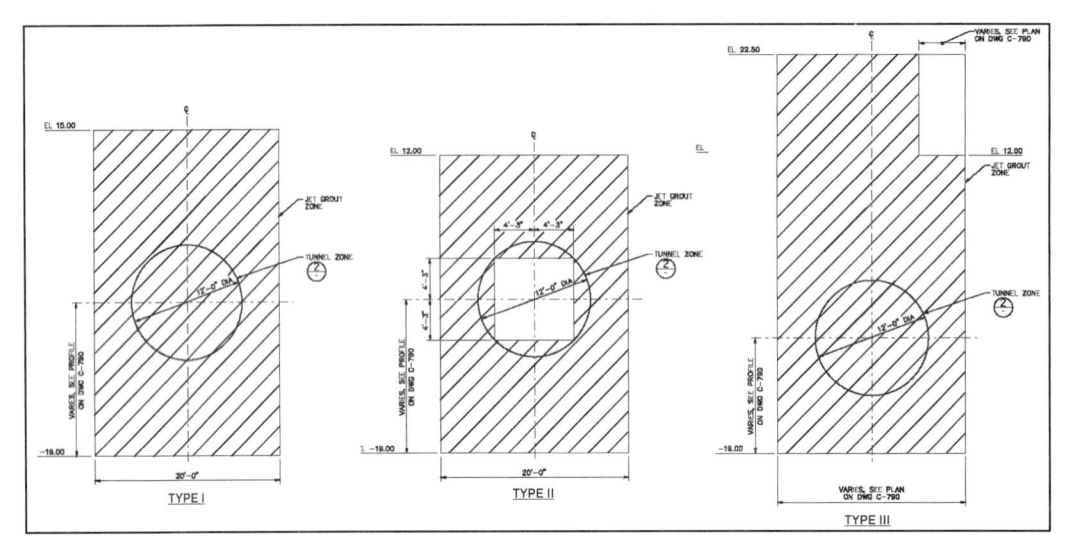

Figure 7. Cross sections showing three types of jet grout geometries along the tunnel alignment

### Ground Improvement Along the Existing Box Conduit

The existing box conduit is a reinforced concrete structure, approximately 3.0 m (10 ft) wide by nearly 1.5 m (5 ft) high. Ellis Street is on the east side of the box conduit. On the west side of the box conduit, a slope with a grade of approximately 2:1 (H:V) slopes down away from the box conduit. As discussed above, the soil below the box conduit where the tunnel crosses under the box conduit would be jet grouted. However, there were also concerns that ground movements associated with the excavations for the Gates and Screening Structure could cause damage to the box conduit. Therefore, it was decided that soilcrete would be used to underpin an approximately 73.2-m (240-ft) long section of the box conduit prior to the start of excavation.

As shown on the plan in Figure 6, only the southern section of the box conduit, approximately 39.6 m (130 ft) in length, would be fully supported on soilcrete. It was envisioned during design that the jet grouting would have to be performed at inclined angles from possibly both the east and west sides of the box conduit (i.e., from Ellis Street and the slope). The existing 0.6-m (2-ft) thick sand and gravel fill immediately below the box conduit would be grouted using chemical grout. For the remaining length of the box conduit further north, only the east side of the box conduit would be supported on soilcrete. Figure 8 shows cross sections of the box conduit at the fully supported section and the partially supported section. The main reason for underpinning only the east side of the box conduit was that several large existing pipes extended laterally from the northern section of the box conduit. Since the partially supported section of the box conduit was further away from the excavation for the Gates and Screening Structure, it was determined that the lower risk of damaging ground movements was not worth further complicating the jet grouting operation by requiring the underpinning of the west side of the box conduit.

### CONSTRUCTION SEQUENCE

One of the project requirements was that either Ernest Street or Ellis Street must be open to traffic. Both streets cannot be closed to traffic at the same time. Therefore, a general construction sequence for site 067 was indicated on the contract plans. These steps of the construction sequence have been generalized for discussion purposes.

1. Close Ernest Street. Perform the required utility relocations and abandonment. Underground utilities to be abandoned (e.g., existing 224-cm [88-in.] brick force main) are to be filled with flow fill.
2. Perform ground improvement within the footprint of the support of excavation for the Diversion Structure and along the tunnel alignment without interrupting traffic on Ellis Street.
3. Construct the secant pile walls.
4. Provide temporary cover over the Diversion Structure excavation to allow traffic on Ernest Street. Open Ernest Street and close Ellis Street.

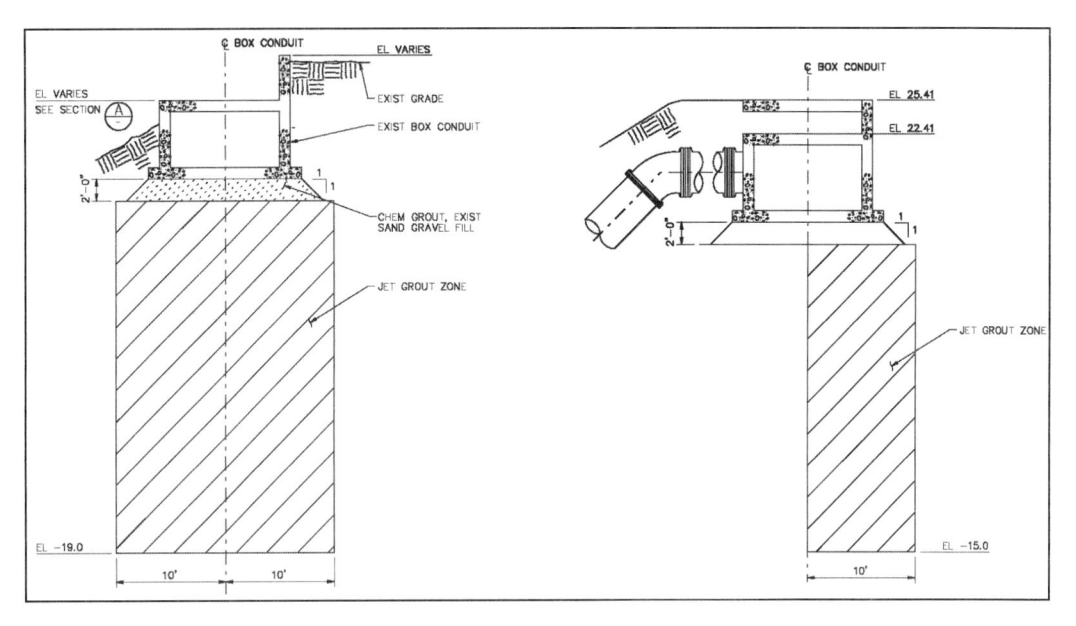

**Figure 8. Cross sections showing the underpinning of the existing box conduit**

5. Underpin the existing box conduit using jet grout and chemical grout prior to installing the support of excavation for the Gate and Screening Structure.
6. The support of excavation for the Gate and Screening Structure is to be designed by the contractor. Complete any ground improvement work prior to excavating the work shaft for the Gate and Screening Structure.
7. Once the work shaft for the Gate and Screening Structure is completed, begin tunneling from the work shaft toward the work shaft at the Diversion Structure. When tunneling reaches the Diversion Structure, install the 274-cm (108-in.) RCP to complete the Consolidation Conduit.
8. Construct the final structure and complete any remaining work associated with the Gate and Screening Structure. Restore Ellis Street.
9. Open Ellis Street and close Ernest Street.
10. Construct the final structure and complete any remaining work associated with the Diversion Structure. Restore Ernest Street.
11. Open Ernest Street and close Ellis Street. Complete construction of the above-grade control house of the Gate and Screening Structure.

The construction generally followed the construction sequence listed above. Figure 9 shows a photo of the Diversion Structure after completion of the secant pile and jet grouting and after supporting the temporary 259-cm (102-in.) pipe to maintain flow of the existing 279-cm (110-in.) sewer line. The Diversion Structure jet-grouted bottom plug thickness was a subject of debate during construction. The debate was about the adequacy of the thickness of the plug. The design approach assumed that achieving perfection in jet grout is not possible given the site constraint and presence of existing structures and utilities that are in the way during the installation of the bottom plug. Ultimately, the design jet grout plug was implemented successfully. Figure 10 shows a photo of tunnel shield breaking through the jet grouting at the Diversion Structure. Figure 11 shows the placement of the final RCP jacked in place after the tunnel excavation was completed.

## CONCLUSIONS

The project was completed successfully in spite of the many challenges during the design and construction. The following conclusions are made by the authors from the design and construction:

- In the authors' opinion, designing the excavation support systems and underpinning systems by a qualified design team with adequate time and resources instead of leaving it to the contractor is recommended for underground structures that are located near major existing facilities that cannot be disrupted during construction. The designers should strive to present one constructable method for the various facilities, with adequate details to convey the intent of design to the contractor.

Figure 9. Photo of the diversion structure after completion of the secant pile walls and jet grouting

Figure 10. Tunnel shield breaking through the jet grouting at the diversion structure

Figure 11. Pushing of the RCP into the completed tunnel

- In the authors' opinion, leaving the excavation support system design to the contractor for similarly complex projects could potentially unfairly burden the contractor and increase the overall project risk in the traditional low-bid environment. The contractor typically has a short period of time to develop a bid, which may be insufficient to learn the details, develop the design, and determine the construction cost for complex excavation support systems. If the design of the excavation support system for the Diversion Structure was to be performed by the contractor, costly change orders and claims would have likely occurred.
- In the authors' opinion, it is appropriate to leave excavation support system design up to the contractor if the construction will not impact other existing facilities that must be operational during all times and the site does not have severe site constraints. The construction of the Consolidation Conduit and the Diversion Structure was limited by these constraints.
- The use of jet grouting on this project was quite successful throughout in lowering the risk of damaging the existing box conduits and other facilities from ground movements due to tunnel and shaft excavations.
- The successful design and construction of the NBC CSO Project provides a clear example of the level of complexity that can be achieved in large urban construction even under severe site and operational restraints.

# TRACK 4: CASE HISTORIES

Larry Eckert, Chair

*Session 3: Challenging Conditions*

# Tunneling on Brightwater West

## Glen Frank, Mina M. Shinouda, Greg Hauser
Jay Dee Contractors Inc., Seattle, Washington

**ABSTRACT:** Brightwater West (BT4) represents the state of the art in utility tunneling, namely a long relatively small diameter, soft ground tunnel, with no intermediate shafts, under significant active earth and groundwater pressures, requiring very precise survey control in order to hit a small exit window. The main tunnel is over 6.4 km (4 miles) in length, and has encountered active earth pressures of over 5 bars in glacial geology, and a planned hole through into a shaft eye constructed at 45.7 meter (150 feet) below the water table. Despite the fact that all of these challenges have been previously overcome in larger diameter tunnels, the technical solutions are much more challenging in a smaller diameter due to the lack of space available for implementing the equipment and techniques required. In addition to the technical challenges, the project is faced with many of the constraints designed to minimize the impact of these types of projects on the neighboring community and environment.. The paper is a case history of the Brightwater West project. and briefly addresses the work performed during the preparation for tunneling stage of this project, but is primarily focused on the tunneling phase of the work which is nearly 90% complete at the time this paper is being prepared.

## INTRODUCTION

The Brightwater Conveyance Project is being constructed as a regional wastewater treatment facility, to cope with the growth of the greater Seattle region. It consists of a series of tunnels and associated structures; which include a new treatment plant, 21.7 kilometers (13.5) of conveyance lines (influent and effluent) and five portals. Construction was divided into three main contracts: East, Central, and West.

The Brightwater East tunnel was completed in November 2008, by the joint venture group of Kenny/Shea/Traylor. This tunnel was 14,050 ft (4.2 km) long and was lined with 16'-8" internal diameter concrete segments.

The Brightwater Central tunnel is currently ongoing, and the work is being performed by the joint venture group of Vinci/Parsons RCI/Frontier-Kemper. This tunnel consists of two drives, in opposite directions, from a central access shaft at North Kenmore. The two tunnels consist of 11,600 ft (3.5 km) (BT-2) east drive and a 20,100 ft (6.2 km) (BT-3) west drive. The Ballinger way receiving portal forms part of the central contract; however it also serves as the receiving portal for the West Contract (BT-4) Tunnel Boring Machine (TBM).

The Brightwater West Tunnel Contract was awarded to the joint venture group of Jay Dee Contractors/Frank Coluccio Construction/Taisei Corporation (JCT). This tunnel (BT-4) is 21,000 ft (6.4 km) long and is lined with 13ft (4m) internal diameter concrete segments. The machine chosen for the project is a LOVAT RME184SE Series 23600 Earth Pressure Balance (EPB) Tunnel Boring Machine (TBM). The tunnel was driven up-gradient in an easterly direction, extending from the Point Wells Portal Structure to the Ballinger Way Portal, which is the terminus for the BT-4 TBM.

The west tunnel is the longest single heading on the Brightwater project, traversing glacially deposited outwash and tills, under hydrostatic heads exceeding 5 bar pressure, with no intermediate access points from the surface, making cutting head inspections and maintenance a challenge.

The focus of this paper is on the excavation of the main tunnel on the Brightwater West Contract, which is approximately 90% complete at the time this paper is being prepared.

## PREPARATION FOR TUNNELING

Notice to Proceed was issued to JCT on February 20, 2007 and a significant amount of planning was required prior to occupying the site and beginning the process of preparing the site for the main tunnel excavation. JCT mobilized to the site in late spring of 2007 and began site preparation work which is summarized here, as it was the subject of a paper presented at RETC 09 (Shinouda, Frank, and Hauser, 2009).

### Railroad Crossing within 100 feet of the Launch Shaft

The Project had a very challenging start where the launch face station was about 30.5 m (100 ft.) away from the Burlington Northern Santa Fe (BNSF) railroad right-of-way (ROW), also, the portion of tunnel under the railroad was on a 304.8 m (1,000 ft.) radius curve and was crossed with less than a tunnel diameter of cover to the bottom of the ballast. Moreover, it was not possible to assemble the entire TBM before launch due to shaft length restrictions, thus the assembly of the TBM had to be staged as the mining progressed under the BNSF ROW.

### No Possibility for an Intermediate Shaft

The tunnel alignment starts at Point Wells portal shaft and goes eastward passing under BNSF railroad, residential area, local highway (US 99), commercial area, Interstate highway (I-5), and ending at Ballinger Way shaft. There are no intermediate shafts allowed in the contract, and in actual fact, there is no real possibility in finding a suitable location for any shafts along the whole alignment. This fact introduced several challenges to the Project, including survey accuracy, tunnel ventilation, and cutterhead maintenance.

### High Probability of Hyperbaric Work at Above 4 Bar

JCT anticipated needing to access the cutterhead chamber to change teeth and do required maintenance at least every 1,524 lineal meter (5,000 feet) of tunnel advance. Since over 2,740 lineal meters (9,000 feet) of the tunnel is more than 30.5 meters (100 feet) below the water table at least one of these "interventions" will be done with very high static water pressures. It was anticipated that this (these) interventions will require hyperbaric pressures of above 50 psig, which is the highest pressure allowed for hyperbaric work in the state of Washington.

### Challenging Exit Shaft Scenario

The exit shaft for the TBM was constructed by the Central Contractor at the Ballinger Way Portal. The shaft is 60.9 meters (200 feet) deep with static groundwater levels of more than 45.7 meters (150 feet) above the tunnel invert and was constructed using ground freezing as the primary means of earth support. The unknown behavior of the ground that has been frozen then thawed is a significant concern for JCT in planning the final few feet of the BT4 tunnel.

### No Trucking of Tunnel Muck

With few exceptions, excavated spoils from the jobsite are not permitted to be trucked out. The contract mandates the haul out of spoils by barges via the adjacent Puget Sound. Since there was an extensive amount of work needed to be done to get the barge hauling system in place, a limited number of trucks was allowed at the early stage of the project, but after a specific cutoff date, only contaminated spoils were to be trucked out.

An existing pier and wharf, owned by Paramount Petroleum, was the only way available to load the barges. An conveyor belt system was developed to transfer the muck from the jobsite to the barge docking location at the wharf. Some restrictions were imposed by Paramount Petroleum for the use of their pier and wharf (see Figure 1).

### Geologic Conditions

The project is located within the Puget Trough, which is a structural basin located between the Olympic and Cascade Mountains, formed by the Juan de Fuca oceanic plate being thrust beneath the North American Continental plate. The bedrock contact is over 305 meter (1,000 feet) below the surface, and is overlain by glacial and non-glacial sediment through which the tunnel will be constructed.

The geologic history of the project site is dominated by at least six different episodes of advance-retreat cycles of continental glaciers during the Pleistocene era. Each of these glacial advances partially eroded the pre-existing stratigraphy, and deposited a fresh sequence of sediment.

The stratigraphy along BT4 is complex due to the multiple erosion/deposition cycles that have occurred during the time that these materials were at the surface, and the orientation between the depositional glaciers, rivers, streams, and lakes, and the tunnel alignment, which is generally perpendicular to the advance direction of the glacial and glaciofluvial flows.

During the last glacial period, which ended about 10,000 years ago in the project area, large quantities of sediments ranging from clay-sized to large boulder-sized were deposited in the Puget Trough. Each depositional event was followed by one of erosion during which large and small channels, ravines and valleys were incised into the previously deposited sediments. Subsequent deposition either filled or partly filled those channels, ravines and valleys, then the process was repeated again and again. Consequently, many if not all of the formations are only remnants, and refilled channels are common.

The depositional environment for the soils expected to be encountered during the excavation of BT4 is of two basic types. The soils expected in the tunnel envelope west of Station 636+00 will consist of alluvial sands, gravels, silts and clays, and lacustrine clays deposited during the interglacial period between glacial advances. The soils expected to be encountered east of Station 636+00 are primarily glacial and glaciofluvial silts, clays, and sands. The entire project is located under the water table in glacial and interglacial sediments, containing boulders, and significant amounts of flowing ground. The primary geologic concerns for this project has been the groundwater pressure felt at the TBM and the abrasiveness of the soil.

## RESULTS

In general the preparations for tunneling on site was hampered by the uncertainty associated with a delayed delivery of the TBM. The TBM manufacture was hampered by unanticipated difficulties in meeting all of the requirements for the hyperbaric work being integrated into a TBM of this diameter. The delivery of the TBM was 5 months later than was originally programmed and tunneling didn't begin until early September of 2008.

### Tunneling Under the BNSF Railroad

The first 200 feet of tunnel included the crossing of 2 mainline railroad tracks with less than a diameter of cover to the bottom of the ballast. This line carries an average of 42 passenger and freight trains per day, and is the mainline between Seattle WA, and Vancouver, BC.

The launch and tunneling under the railroad went very well despite the fact that the full muck train could not be utilized. Each 5 ft advance for a ring build was accomplished through a cycle of 3 pushes. The only unusual complication was associated with ensuring that any overexcavation was immediately compensated with extra backfill grout. The belt conveyor on the TBM was not installed (the muck cars were loaded directly from the end of the second screw conveyor) so the belt scale was not in use. JCT utilized a system consisting of the crane operator calling out the weight of each car as it was hoisted and dumped, and then the weight of the empty car. This data was recorded by the engineering staff and compared to the anticipated weight of the muck from the advance of the TBM, resulting in the theoretical amount of over excavation. Based on these calculations, the amount of grout that should be injected from the tail shield (40 feet behind the cutting head) was prescribed and communicated to the heading.

The result of this work was that several advances were overweight, and several rings took more grout than would have been attributable to normal overcut, but there was no measurable settlement at the surface in the railroad right of way.

Figure 1. Photograph showing low angle belt conveyor and muck barge modified for fluid muck

### Cutting Head Inspection and Maintenance

Once the Railroad had been successfully negotiated with no settlement, the primary technical focus was on minimizing cutting head wear. This was a major concern in the TBM design as is mentioned above as well as in the plan for tunneling operations.

In addition to the TBM design, all of the muck handling aspects of the project were set up around handling a very fluid muck, since a fluid muck resulted in less wear on the consumable components of the TBM. This included the use of low angle belt conveyors, and specially modified muck barges capable of transporting muck that was nearly a fluid.

Thus far on the project 31 cutting head inspection/maintenance stops have been performed. Of these 1 has been under hyperbaric conditions, 7 have been under free-air conditions and the remaining 23 have been remote camera (periscope) inspections.

#### Remote Cutting Head Inspections

Of the 23 remote camera inspections 4 showed a distinct image of a ripper and clearly indicated the condition of the cutting tools on the head. The remaining 19 attempts failed to yield video evidence of the condition of the cutters, but valuable information was typically obtained. Since the crew that performed the remote inspection always included the TBM operator and the TBM mechanic and usually the Tunnel Foreman, the activity of pushing the periscope out through the cutting head chamber and flood door served a dual purpose as both a periscope and a probe.

The periscope had a camera on the end of it, which was used to record video and was successful

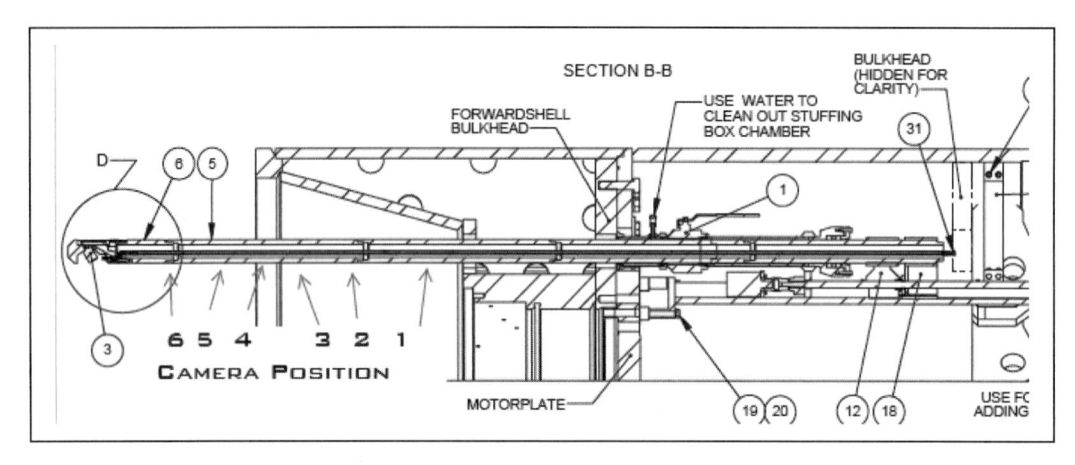

**Figure 2. Figure showing the remote camera probe (periscope) within the cutting head**

in capturing images of the rippers about 20% of the time, but it also acted as a probe and provided information to the personnel who were operating it. As the probe was extended, the amount of force required to push it out, coupled with the length it was extended allowed the personnel familiar with the TBM to determine whether the cutters were at full length or not (see Figure 2).

Figure 3, illustrates the information that the remote camera could provide given the right conditions at the face.

*Man Entry Cutting Head Maintenance*

As is mentioned above 8 man-entry cutting head maintenance stops have been performed 7 of these were under atmospheric conditions and 1 was under hyperbaric conditions. The hyperbaric work was carried out at just under the 50 psi threshold that would have required Tri-Mix Gas.

On average the amount of maintenance has been essentially as anticipated, with the exception that the majority of the wear on the cutting head has occurred over the latter half of the drive rather than the first half.

**Muck Barging**

The transport of the muck by conveyor and barging has been problematic due to several factors, primarily due to the variability in the ground coming out of the heading, but the implemented system has had enough flexibility to keep up with the faster than anticipated advance rate of the TBM. The muck coming from the heading was often too sticky to pass through the hopper that fed the conveyor from the shaft to the muck holding bin and would have to be dumped on the ground and trammed to the holding bin with the front end loader.

**Figure 3. Screen shot showing the video image of a ripper at ambient pressure at the face**

The only time that the project was muck bound and mining had to be halted was when the weather conditions prevented the loading of barges. This occurred on a couple of occasions but was not a serious impact to the overall progress of the project.

Another problem associated with the muck barging was damage to the pier, which was owned by a third party. Several days of production were impacted by repair work that had to be done to the pier due to damage caused by the heavy day to day use of loading barges. This was not completely unexpected due to the preexisting condition of the pier and the fact that the loading of muck onto barges was outside of what the pier was originally designed to handle.

The photos presented as Figure 4 illustrates the overall layout of the muck conveyance and barge loading system utilized by JCT on this project.

**Figure 4. Photograph showing the overall layout of the main shaft location and muck handling system**

## Subsurface Conditions

The abrasiveness of the ground, and the groundwater head and the behavior of the ground in reaction to that head, have been the primary geologic concerns during the tunneling portion of this project. While both of these parameters have proven to be somewhat different than what was originally anticipated, the tunneling systems have proven flexible enough to maintain production. This flexibility was critical as the geologic conditions constantly changed, sometimes within an individual push of 5 feet and often from push to push. Soil conditioning and cutter head lubrication were constantly adjusted by the TBM operator to compensate for these varying conditions.

### *Abrasive Conditions and Soil Conditioning*

Currently there remains no standard test method for measuring how abrasive a given soil will be in regards to pressure balance tunneling. However, the Geotechnical Baseline Report did provide some guidance concerning the abrasive nature of the soil using x-ray diffraction methods to estimate the mineralogy composition of the soils, as well as slurry abrasivity (Miller Number) tests, and abrasivity results from the Norwegian University of Science and Technology performance prediction model (AVS). The two relevant statements in the Brightwater West GBR are as follows:

1. "Based on the AVS values, the quartz content and soil gradation along the alignment the abrasiveness of the soil will be higher at the western end of the project and be less abrasive toward the east."
2. "Due to the abrasive nature of the soils, attempting to excavate through any portion of BT4 alignment using a pressurized-face TBM without the use of the proper soil conditioners *may* (JCT assumed "will") result in severe wear of the TBM and associated cutting tools."

The first statement has not proven to be true with the abrasion mitigation methods chosen by JCT. The cutting head wear experienced during the first half of the alignment was very minimal. JCT did change out most of the cutters in free-air maintenance stops during this section of the drive, but based on the condition of the tools replaced, it is probable the TBM could have done the entire first half without changing any of the cutting tools at all. This would have amounted to an almost incredible 10,000 lineal feet of tunneling without changing cutters.

The wear encountered on the second half of the drive on the other hand has been significantly higher than was anticipated. The TBM was completely stopped at around 13,000 feet due to the cutting tools being worn down to the structure of the cutting head itself (though the cutting head structure showed little

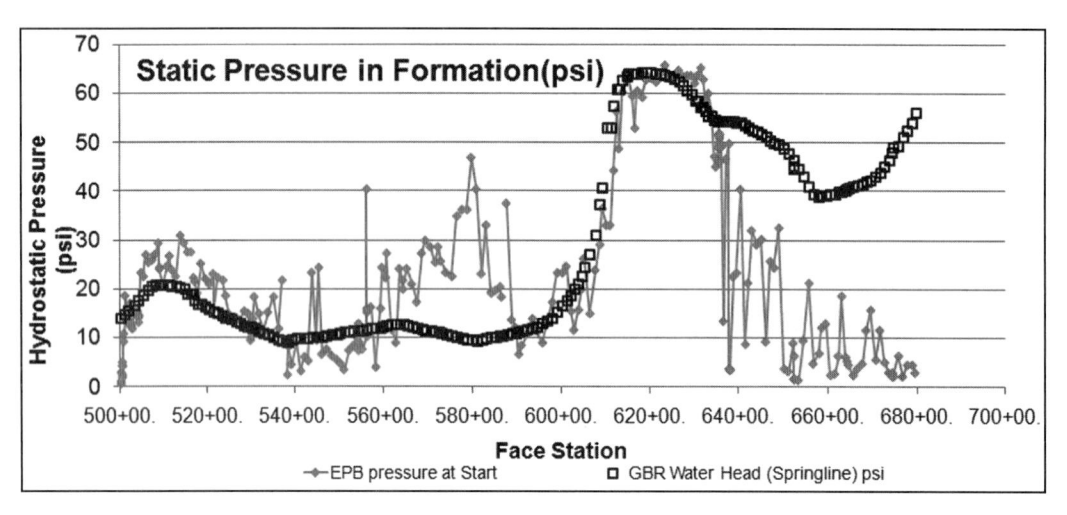

**Figure 5. Illustrates the anticipated groundwater pressure (black squares no line) vs. the pressure in the cutting head chamber at the beginning of each day shift**

wear being protected by the scrapers). The tools have shown considerable more wear during the maintenance stops conducted since the halfway point of the alignment.

JCT believes that the second statement in the GBR is absolutely true and has not mined a foot of this project without the proper use of soil conditioning, which consisted mostly of foam for lubrication and to decrease the stiffness of the muck (ground conditions did occasionally require the use of polymer, however this was to reduce the permeability of the muck and had little wear reduction effect).

*Groundwater Head and Soil Behavior*

Figure 5 illustrates the anticipated static groundwater head and the encountered groundwater head for the project thus far.

Analysis of this figure results in the following observations:

1. The location and magnitude of the highest groundwater pressure was accurately predicted in the GBR.
2. The actual groundwater pressure was typically higher than anticipated during the first half of the tunnel drive.
3. The actual groundwater pressure was typically lower than anticipated during the second half of the tunnel drive.
4. The accuracy and location of the lowest groundwater pressure was not accurately predicted in the GBR.
   a. The predicted location (580+00) of the lowest groundwater pressure had an actual groundwater pressures 4 times that of the predicted value
   b. The actual lowest was essentially zero.

Item 1 above ensured that JCT was prepared for the worst case scenario (hyperbaric work at above 4 bar). Item 4 above led to some difficulty in the management of the backfill grouting, as well as an increased opportunity to perform maintenance under atmospheric conditions.

The higher than anticipated static groundwater pressure encountered during the first half of the drive, was due to the fact that the ground at the tunnel envelope consisted of a vertically fractured impermeable clay, which was overlaid by an aquifer made up of permeable glacial outwash and underlain by a permeable layer that carried the water to Puget Sound. The water in the vertical fractures was actually moving downward and the pressure of this water was dependent on how far above the draining layer the tunnel was located.

It is hypothesized that the groundwater monitoring wells that we installed as part of the geologic investigation were not affected by the hydrostatic pressure in these fractures due to the fact that both the wells and the fractures were vertical structures and they did not intersect. The wells only registered the pressure in the underlying drainage layer. The tunnel on the other hand is a horizontal structure and it intersected the vertical fractures and therefore was subject to the hydrostatic pressure in the fractures.

Figure 6 illustrates how the mining pressure related to the hydrostatic pressure during the first half of the alignment, much of which was in an aquitard with the water moving through vertical fractures. In such formations a small amount of water is removed

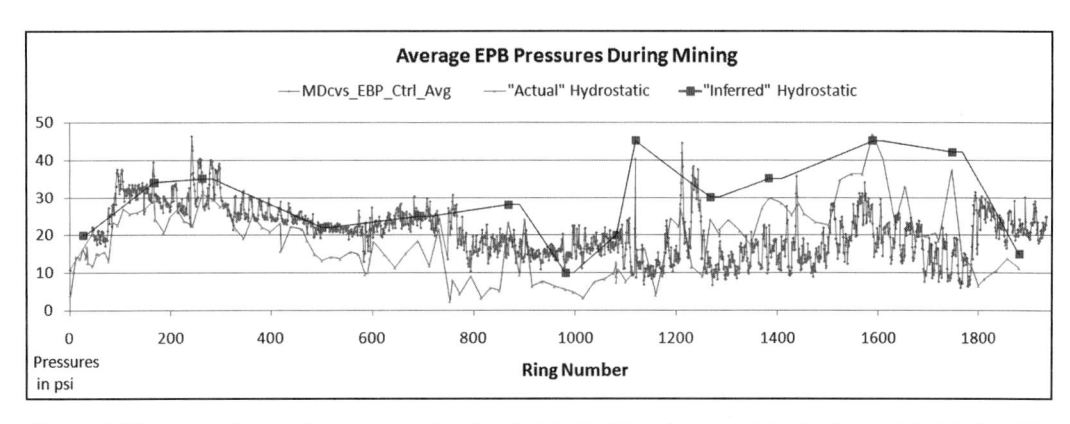

Figure 6. Illustrates the mining pressure for the first half of the alignment (each ring is 5 feet in length)

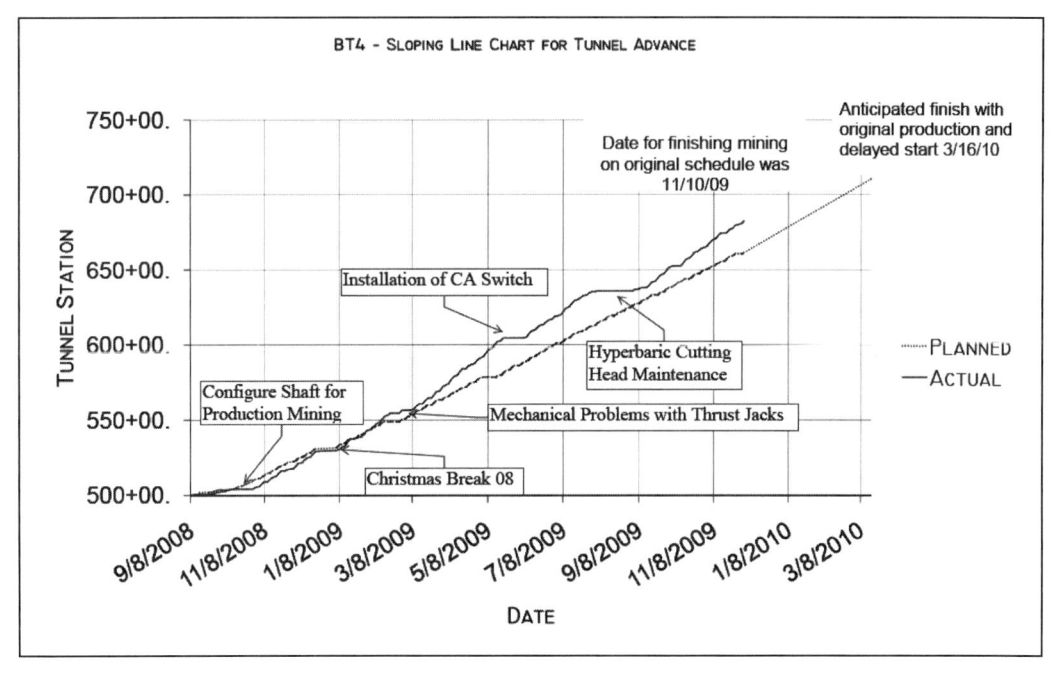

Figure 7. Comparing the actual production to the planned production for the tunneling portion of the project

during mining which significantly drops the groundwater pressure without loss of ground.

In Figure 6:

- The blue line shows the average pressure in the cutting head chamber during each advance (one advance per Ring). This line is spiky since there is a data point for every ring, and it also reflects the tendencies of the day shift TBM operator versus the night shift TBM operator (see Ring # 350 to 450).

- The red line ("Actual" Hydrostatic) shows the pressure in the cutting head chamber prior to the first push each day. This represents the static state after a typical 8 hours (longer on Mondays) of no impact from the mining.
- The line with the boxes ("Inferred" Hydrostatic) represents what the pressure in the formation might have been if the TBM and tunnel was not there. Due to the low permeability of the majority of the soils in the tunnel alignment even a very small amount of water leaking through the screw conveyor

**Figure 8. Preparatory work for ground freezing at the exit shaft**

could have a significant effect on the pressure being measured in the mining chamber.

**Production**

Through 18,226 lineal feet of tunnel or about 87% of this project the production has averaged 285 lineal feet a week including shutdowns for maintenance, holidays, and mechanical problems.

Figure 7 is a sloping line schedule comparing the planned production versus the actual production. The average production on days when only routine maintenance was performed is 80 feet.

The best day of production was April 7, 2009 with 130 lf, working two 10 hour shifts and only tunneling for about 6 hours on the second shift. The best week was 580 lf working the same two 10 hour shifts, and April 2009 the best month with 1,960 lf of lined tunnel completed.

**EXIT STRATEGY**

The Ballinger Way Portal is the reception shaft for the West Tunnel on the Brightwater Conveyance Project, and was constructed by the Central Contractor as is described above.

At the time of this writing, JCT is installing a ground freezing system to freeze a ring of soil around the tunnel envelope at the shaft eye location. The ring of frozen soil will be 50 feet long, have an inside diameter of approximately 13 feet and an outside diameter of approximately 20 feet. Installation of the freezing system requires installation of 19 freeze pipes and 4 temperature probes in 23 horizontally drilled boreholes around the entrance eye for the TBM.

Figure 8 is a photo showing the installation of a 60 foot section of HDPE freeze pipe into one of the horizontal holes at the bottom of the Ballinger Way Portal shaft.

**SUMMARY**

At the time of writing this article, JCT has completed nearly 90% of the 21,000 lineal feet of main tunnel and is currently preparing for a hole-through in early February 2010.

After completion of tunneling, approximately 2,500 lineal feet of the main tunnel will be lined with steel pipe, the remainder of the tunnel will be left unlined and used to convey the effluent to the sampling facility and metering vault. The sampling facility and metering vault will be constructed inside the main shaft on the at the Point Wells site during the remainder of 2010.

The final work of connecting the main tunnel to the already completed outfall is currently scheduled for May 2011.

**REFERENCES**

Shinouda, M. Frank, G., and Hauser, G. 2009., Planning and Preparation for Tunneling at Brightwater, Proceedings Rapid Excavation and tunneling Conference, Almeraris, G., Mariucci, B. (eds) Littleton, CO. pp.1154–1170, 2009.

# New York City Transit No. 7 Subway Extension Underpinning and Construction Under the 8th Ave. Subway

**Aram Grigoryan, Raymond J. Castelli, Fuat Topcubasi**
Parsons Brinckerhoff, New York, New York

**Sankar Chakraborty**
MTA Capital Construction, New York, New York

**ABSTRACT:** This paper describes a combination of underpinning and other construction methods used for the No. 7 Subway Line Extension under the existing 8th Avenue subway. Underpinning methods were developed in collaboration with the designer to address actual structure and geotechnical field conditions. Combinations of mini-piles, steel framing and concrete supports were used to ensure structural integrity, safe and efficient construction, and uninterrupted subway service. The paper also describes the mini-pile field test procedures performed to verify load capacity in the variable quality rock at the site, and summarizes the load test results.

## INTRODUCTION

This paper describes the underpinning and construction of the No. 7 Subway Line Extension beneath the active and heavily used 8th Avenue subway in New York City. An earlier paper (Grigoryan, 2009) focused on the challenges of developing a feasible and constructible structure for the No. 7 Line Extension, and emphasized the correlation of the final structure with the construction method and underpinning, and the challenges of designing and integrating the underpinning with the new structure.

The earlier papers (Grigoryan, 2006 and 2009) described:

- History of the existing station complex.
- Existing structure condition and deteriorations.
- Some of the repairs that were made during the recent station rehabilitation.
- Structural design challenges associated with the design of No. 7 subway extension under the 42nd Street subway station structure.

The present paper, intended as a continuation of this project case history, describes the development and evolution of the construction methods and underpinning, and final structural configuration from design to actual construction.

At the beginning of construction three major requirements and challenges had to be met:

1. Public safety.
2. Structural integrity of the existing structures.
3. Subway service and operations continuity.

## PROJECT INFORMATION

The design team appreciated the challenges of the underpinning construction and shared their prior experience of the existing subway structure, as well as potential pitfalls associated with various methods and procedures, during the development of the underpinning scheme by the contractor. Even small details of the existing structure, if not considered properly during the underpinning design, might have significant impact in terms of cost and schedule. For example, one of the proposed temporary works for gaining access from the tail track tunnel to the abandoned lower level tunnel required temporary removal of seemingly insignificant horizontal struts that were spanning over the Independent Subway (IND) abandoned lower platform. From prior experience with the structure the design team was aware that these 6" H-beams had buckled horizontally probably due to interaction between two adjacent subway structures that were built some 25 years apart by two then competing entities—the Interborough Rapid Transit Subway (IRT) and the Independent Subway (IND). These members were reinforced by the designer during the last station rehabilitation work in 2004. Grigoryan (2009) described this condition in more detail. This information assisted the team in

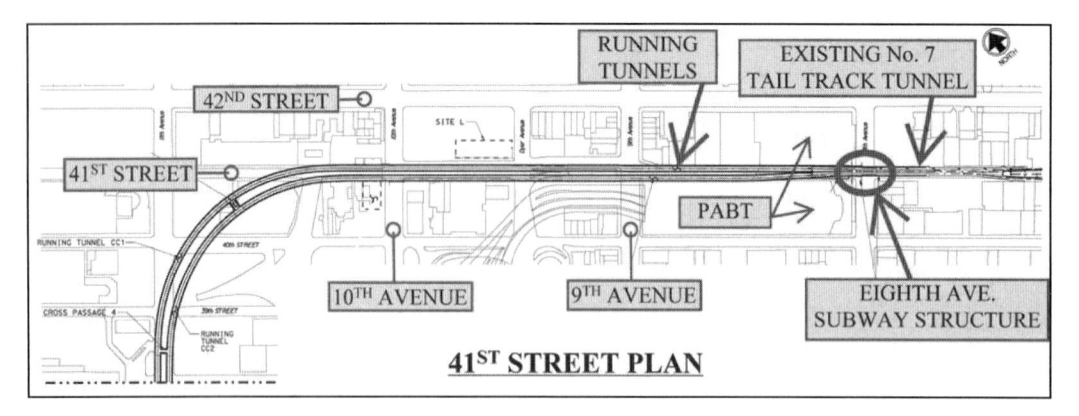

**Figure 1. Project location plan**

developing a construction method that would keep these structural members undisturbed.

The existing No. 7 Line tunnel west of Times Square Station was constructed circa 1915 beneath West 41st Street. The tunnel consists of two tracks, and was originally 624 feet long, which accommodated storage of one train on each track. At the western end of the Times Square Station, each track continued in a single-track tunnel for 190 feet. These tunnels converged into a single two-track tunnel that continued for 434 feet to the middle of 8th Avenue (above). The tunnels were constructed by conventional mining methods, while the Times Square Station was constructed by open-cut methods.

In the 1960s, the stations along the No. 7 Line were lengthened to accommodate an eleven-car train. At the Times Square Station, the platform was extended 85 feet to the west by enlarging the two existing tunnels. The tracks west of the station now measured only 539 feet, which was insufficient to store a full length train. The tracks west of the station were then used primarily for over-run protection for trains entering the terminal station, allowing for higher entrance speeds. The tracks were also used for the occasional storage of work trains.

**Infrastructure Projects—Post 1915**

The original design of the No. 7 Line allowed for a future westward extension. However, infrastructure projects over the years added complexities to such an extension. Two main projects added structures directly in the path of a future extension.

In 1932, the 8th Avenue IND Subway was constructed. The 42nd Street Station along the line is a three-level structure located beneath 8th Avenue. At the intersection of West 41st Street and 8th Avenue, the 8th Avenue Subway consists of a mezzanine level, an upper track level with four tracks and platforms for downtown local and express tracks, and a lower track level with a single track and platform. The lower track level was constructed at approximately the same elevation as the existing No. 7 Line tracks and immediately adjacent to the existing tunnel bulkhead. The upper level tracks were constructed above the No. 7 Line tunnel and required reconstructing portions of the existing tunnel.

In 1975, the Port Authority of NY & NJ expanded the Port Authority Bus Terminal (PABT). The expansion included constructing a lower bus level and bus tunnel beneath West 41st Street between 8th and 9th Avenues. The elevation of the lower level and bus tunnel is approximately at the elevation of the 8th Avenue Subway upper track level, which is slightly above the existing No. 7 Line tunnels.

**Proposed Extension to 34th Street**

In 2002, engineering design began on extending the No. 7 Line to a new terminal station at 11th Avenue and West 34th Street (Figure 1). Both the 8th Avenue Subway lower track level and the PABT lower level and bus tunnel presented obstacles for extending the No. 7 Line westward.

To avoid the 8th Avenue Subway and the PABT and bus tunnel, the design required a profile for the No. 7 Line tracks to get below those two facilities. However, all plans for the extension had to maintain current operations at the Times Square Station. In order to not affect normal peak-hour train service, 200 feet of the existing tracks west of the station had to be maintained to provide minimum over-run protection for trains entering the terminal station. Similarly, appropriate track grades had to be utilized for the proposed extension.

Profiles could not be developed which achieved all of the goals stated above. For example, a profile required to avoid the 8th Avenue lower track level and utilize appropriate grades would require lowering of the Times Square Station. MTA New York

City Transit recognized that the 8th Avenue Subway lower track level was used only sporadically since 1932 and had not been used for revenue service since the late 1970s. Also, the track connections to the lower level track had been removed in the late 1990s when the lower level was permanently abandoned. Using the abandoned space for the No.7 Line Extension allowed for a suitable profile to be developed for the No. 7 Line that maintained Times Square service, avoided the 8th Avenue Subway upper level tracks (A, C, and E Lines), and avoided the PABT and bus tunnel.

The profile developed requires 270 feet of modifications to the existing No. 7 Line tunnel to the 8th Avenue Subway to allow for the deepening of the No. 7 Line in this area. Also, this requires underpinning and reconfiguration of the existing structure. All of the underpinning and reconfiguration works have been developed to minimize impacts to the revenue service of the A, C, and E lines along 8th Avenue, and the No. 7 Line.

## SELECTED PROJECT CRITERIA FOR UNDERPINNING AND TEMPORARY SUPPORTS

In order to ensure that all project requirements were met, project criteria were developed for the design of underpinning and temporary works. A specific requirement was that the existing structures not be affected during any phase of construction. To achieve this challenging task the designers presented a suggested construction method. Many valuable comments and observations were incorporated or addressed in refining the underpinning procedures and construction sequences. However, it was clear that the final structure configuration in many ways would depend on the actual construction method. The contract documents required the contractor to provide detailed design and procedures for all temporary works for supporting the existing structure, transferring loads from the existing structure to temporary supports, and transferring loads from temporary supports to the permanent structure. Preloading the support structures was critical for preventing undesirable and uncontrollable deformations of the existing structure. The project criteria required the entire load to be removed from an existing structural support (column) prior to column removal. This requirement ensures that the existing structure and deformations remain within the range of the underpinning criteria.

## DESIGN AND CONSTRUCTION UNDER PABT AND NYCT INTERFACE

Based on the history of construction and details of the existing NYCT and PABT structures, it was

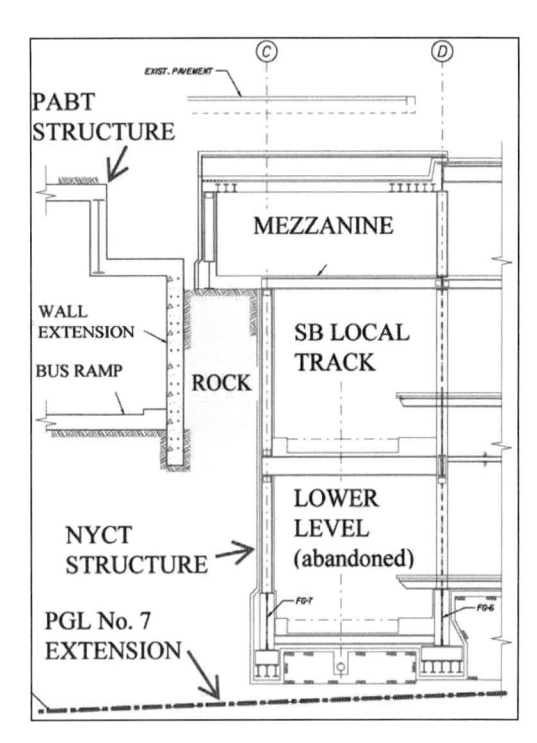

**Figure 2. Typical section showing interface of PABT and NYCT prior to construction**

realized that special design solutions and construction methods would be required to ensure the structural integrity of both structures during the underpinning and construction.

As shown on the typical section (Figure 2), the overhang part of the existing NYCT 8th Avenue mezzanine structure rests on rock that is sandwiched between the PABT underground retaining wall and by the west wall of the NYCT subway structure. The width of the rock pillar is approximately 10 to 13 feet. Historical construction sequence indicates that the rock has been excavated on the east side approximately 50 feet deep for the construction of the 8th Avenue subway, and approximately 35 feet deep on the west side for the construction of the PABT. Therefore, it was concluded that the rock might not be capable of safely supporting the imposed loads during construction, and should therefore be considered as a dead load in developing and designing the underpinning method and the final structure.

The designer developed a roof support shielding system that was to be installed by horizontal drilling from an excavation under the bus ramps (supported by decking), or from within the 8th Avenue subway's lower abandoned level, depending on the contractor's preference. The suggested roof support

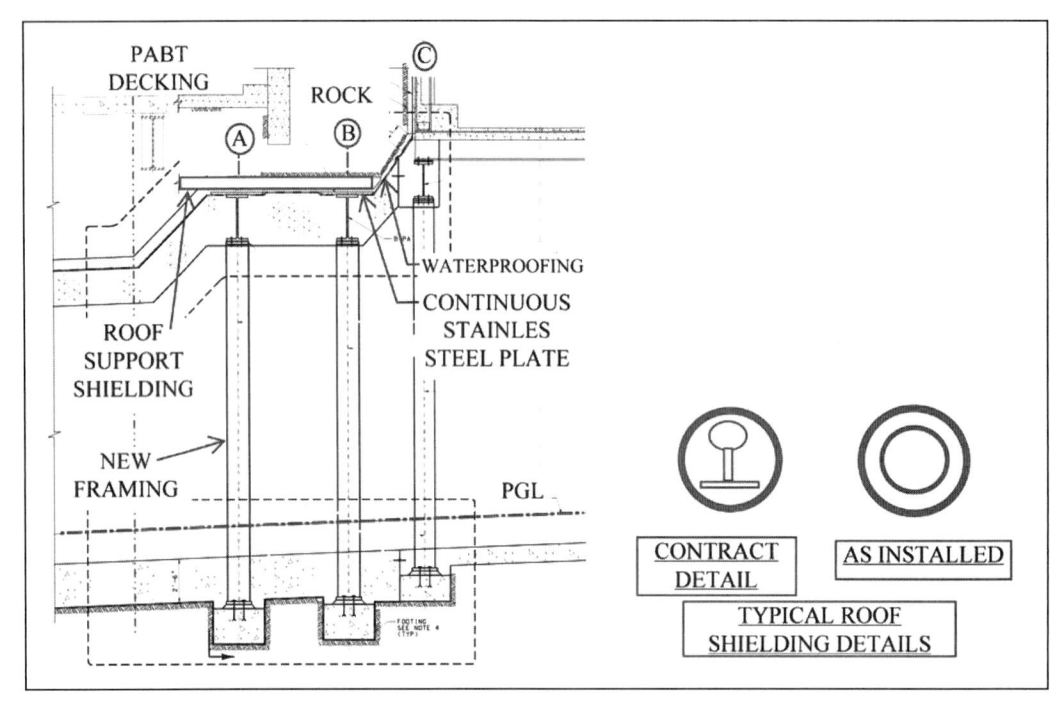

**Figure 3.** Contract drawings showing completed structure with roof shielding and detail

shielding consisted of 8-inch diameter steel pipes, reinforced with rails and filled with grout (Figure 3).

The contractor's initial proposed construction method did not fully address this issue. The proposed design apparently assumed the arching effect of the rock and focused on supporting only the PABT wall and bus ramps (Figure 4).

During the submittal review process, the contract intent was discussed, and the contractor proceeded in developing and customizing the roof support shielding method as was intended in the contract. The contractor system used 7⅝" diameter steel pipes reinforced with 5½ steel pipes and filled with grout (Figure 3) as an alternative to the concept shown on the contract drawings. The proposed substitution was deemed adequate, was approved, and the contractor successfully installed the roof support shielding from an excavation under the PABT bus ramps using a hydraulic drill rig.

Several critical revisions were made to the contractor's initially proposed construction procedures. After installation of roof support shielding, the contractor suggested to progress the excavation under the protection of shielding, from west to column line A, and install the permanent columns and framing. However, the proposed method would leave a cantilevered end of the shielding. During the submittal review, the designer recommended providing either an additional portal frame at the west end of the roof support shielding, or support the PABT ramps independently so that the cantilevered ends of shielding are not loaded. It was recommended to limit the excavation stages, to limit the exposed unsupported span of roof support shielding to approximately 6 feet. Although the strength of the shielding was adequate, it was also considered important to maintain the deformations of the roof support shielding within the permissible limit by minimizing the unsupported span length. The designer's recommendations were implemented by the contractor.

Another critical element of the underpinning and construction staging was sequencing the work at the interface. The existing west exterior wall of the 8th Avenue subway structure was supported by transfer girders at the lower track level. The transfer girder, in turn, is supported on steel grillages founded on rock. The construction at the interface would require excavation in the influence zone of the grillages. Column line C is structurally critical since its columns support the operating IND local southbound track. To secure column line C the designer recommended the contractor to complete the underpinning of column line C and transfer of existing loads from columns on line C to the underpinning frame prior to excavating within the grillage foundation influence zone. This recommendation was implemented by the contractor. Figures 5 and 6 show completed new framing at the interface of the PABT and NYCT structures.

## DESIGN AND CONSTRUCTION UNDER THREE IND OPERATING TRACKS

One of the most challenging areas was construction under the three operating IND tracks. In order to accommodate the No. 7 Line extension the existing profile had to be lowered up to 7 feet. The existing tail track tunnel invert slab had to be demolished, the existing rock supporting the invert and the sidewalls had to be excavated, the sidewalls extended down to the new elevation, and a new invert slab constructed (see Figure 7).

The existing center columns and grillage foundations support up to 1250 Kips load. The IND structure is supported by transfer girders that are

Figure 4. Sections showing contractor's initially proposed support method

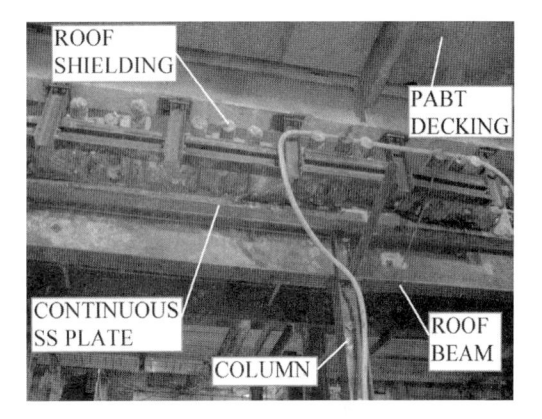

Figure 6. Construction under PABT showing roof shielding and stainless steel (SS) plates

Figure 5. Completed steel framing under PABT showing roof shielding pipes

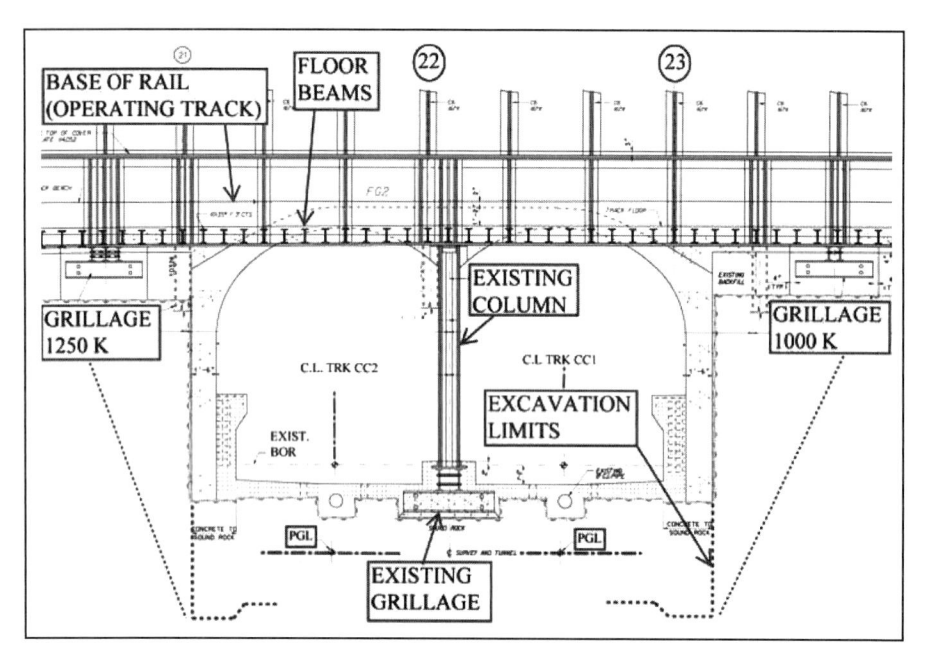

Figure 7. Typical section—tail track under IND subway

supported on grillages on either side of the tail track tunnel and by the center columns.

There were two very challenging issues that needed to be addressed:

1. Underpinning of side walls
2. Underpinning of center columns

**Underpinning of Side Walls**

Considering high grillage loads (up to 1250 Kips) adjacent to the existing tail track tunnel walls, the designer developed an underpinning method and sequence to ensure that the effects of the new construction on the existing IND structure are minimized as much as possible. One of the main requirements was to stagger the work for excavating under and lowering the sidewalls, preventing the entire cross section from being undermined at the same time (Figure 8).

The method ensures structural integrity of the supported structure by exposing a limited length of the wall, excavating and extending the wall down before excavating under the adjacent part of the wall. However, this method limits the contractor's work area, requires careful planning and coordination by the contractor, and limits contractor's ability to use heavy equipment. The contractor proposed to excavate under both walls at the same time for the entire length of this area, and to support the rock using SWELLEX rock bolts. During submittal review and discussions with the contractor, the designer noted the concerns in respect to the stability of the sidewalls and overall structural integrity. After several revisions, approval was given for a method that provided support during the excavation using Dywidag rock bolts to allow immediate engagement of the bolts to stabilize the rock wedge under the grillages and to prevent sidewall deformations. The basic steps of the method are illustrated in Figure 9 and listed below:

- Vertical line drilling at the face of the existing wall
- Excavating rock, leaving a rock bench under the existing walls
- Installing rock bolts to stabilize the rock bench
- Excavating rock under the wall and installing steel supports
- Extending the wall down to the new level

The modified method allowed the contractor to streamline the construction process and efficiently use his equipment.

**Underpinning of Center Columns**

Some of the factors that complicated the underpinning of center columns were:

- Three operating subway tracks above
- No access to transfer girders for strengthening at intended pickup points

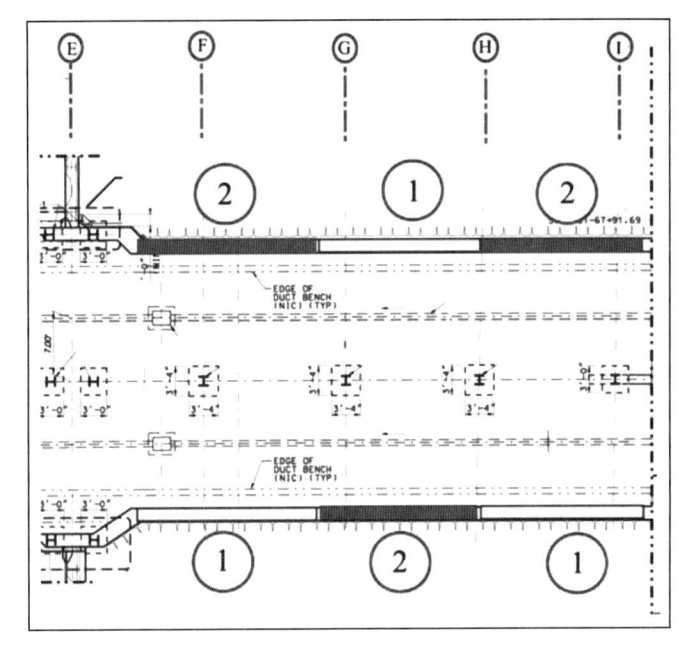

Figure 8. Plan showing intended excavation sequence for the construction of sidewalls

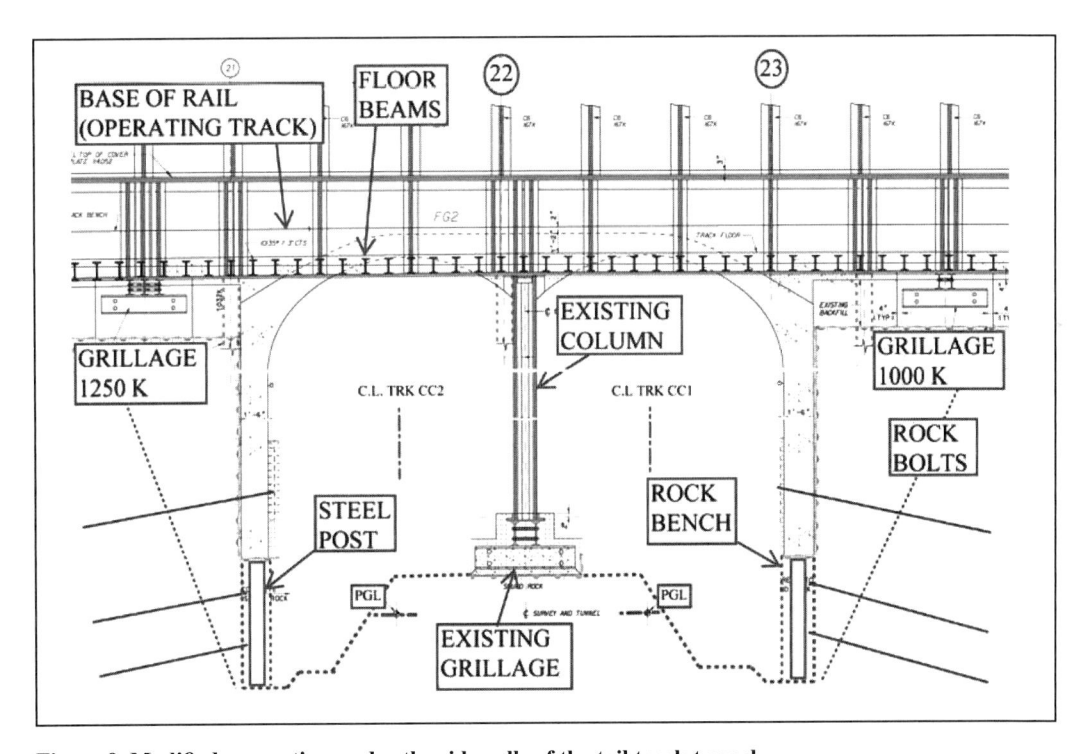

Figure 9. Modified excavation under the sidewalls of the tail track tunnel

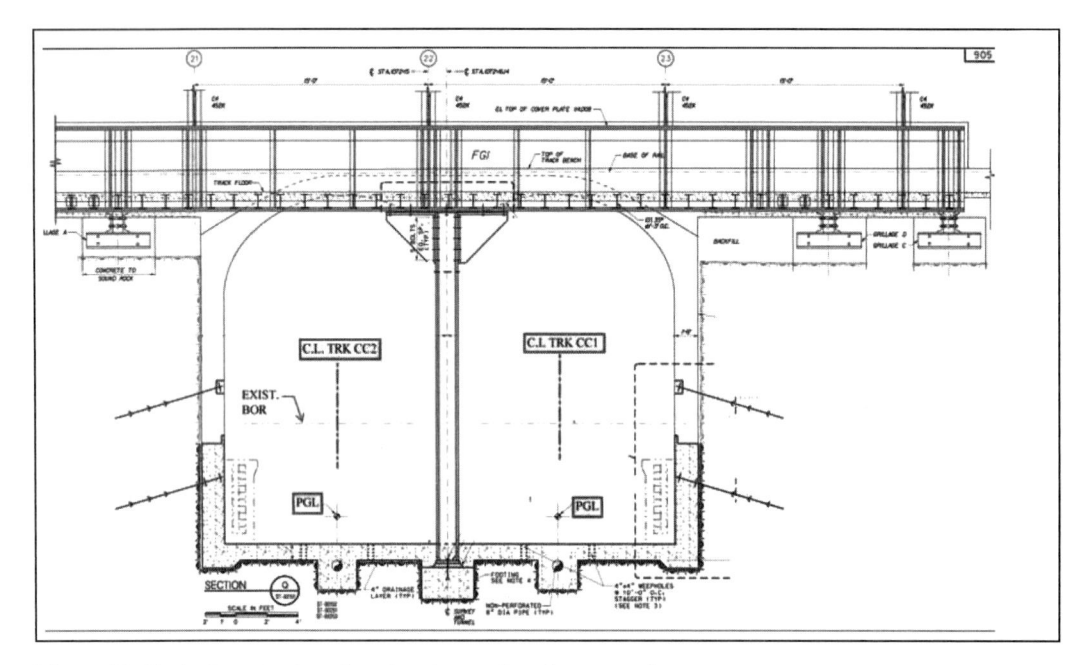

**Figure 10. Typical contract section showing replaced center column**

- Confined space within the tail track tunnel
- Significant column loads—up to 1200 Kips
- Built-up riveted structural members
- Unknown conditions of structural elements

Based on the designer's previous experience with the existing structure it was deemed prudent to assume that the actual conditions of the center columns, especially at the bearing base, might require extensive repairs. Therefore, the design called for the replacement of the critical center columns (Figure 10). An underpinning and construction method was developed that was both constructable and feasible, and addressed all the concerns that were expressed by the client and by the reviewers. One of the main concepts was based on using four mini-piles to temporarily support each of the columns.

During construction the contractor was concerned with the limited space, and explored alternative options that were based on using two mini-piles for temporarily supporting each column. Several alternative options were presented for review; however, the designer had certain concerns with the presented options, including:

- Overloading the mini-piles
- Load transfer procedures that did not satisfy project criteria
- Overall safety and redundancy of the temporary support system

- High concentrated loads at pick-up points of the existing transfer girders that cannot be reinforced
- Deformation of the existing structure that did not meet design criteria

**Underpinning Piles**

The underpinning piles selected by the contractor consisted of 12-inch diameter mini-piles with a design (allowable) capacity of 600 kips and a factor of safety of 2.0. These piles were composed of a single #28, Grade 75 threadbar, centered within a 12-inch diameter borehole, using concrete with a minimum 28-day compressive strength of 6,000 psi. The original design required a minimum 9-ft long socket into rock of Class 2-65 ("medium hard rock" with a "basic allowable bearing value" of 40 tsf) or better. The piles were designed to achieve their full capacity from side friction (bond) in the rock socket, while ignoring any contribution from end bearing.

Three core borings were performed within the lower, abandoned tunnel by the contractor in advance of underpinning operations to supplement the design stage borings. These core borings encountered about a 2-foot zone of broken rock directly beneath the base of the existing structure, indicative of damage probably caused by previous construction operations. This fractured rock zone was generally followed by competent pegmatite to the maximum 14-foot depth of the borings.

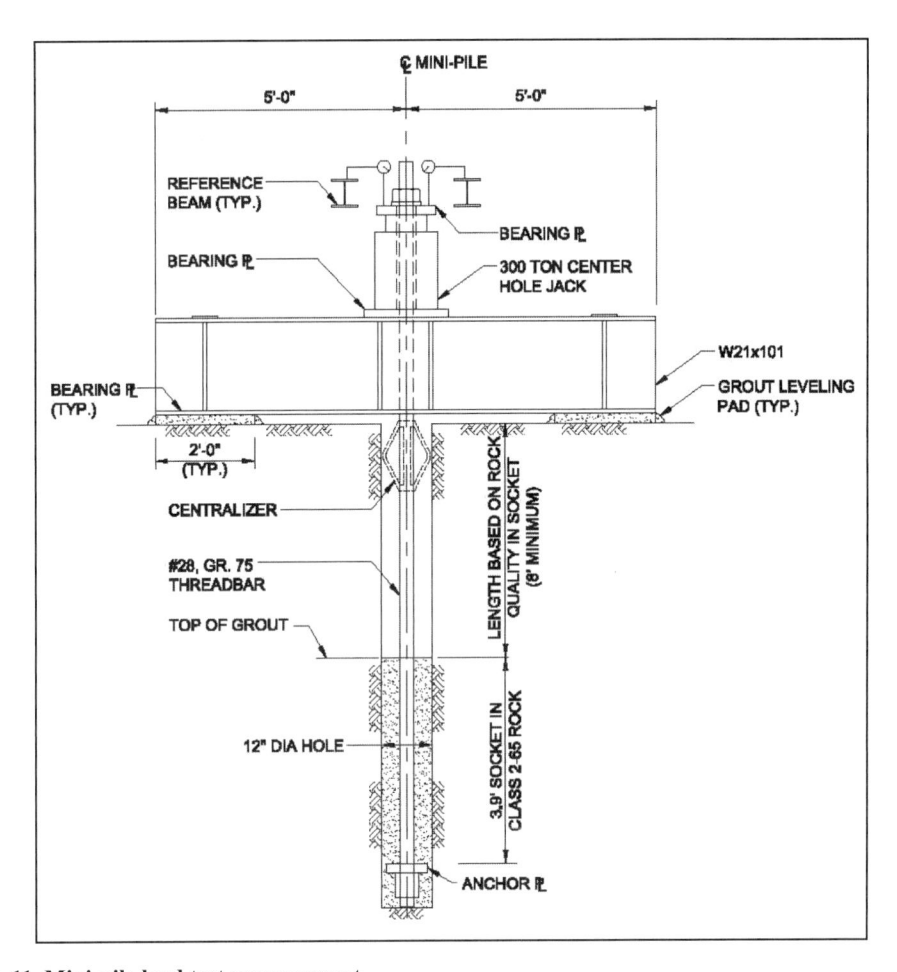

Figure 11. Mini-pile load test arrangement

**Pile Test Program.** In accordance with contract requirements, a load test program was implemented to confirm the design unit bond resistance in rock. The program included a single test pile, constructed to the same cross section as the planned production piles. To facilitate testing in the confined space of the existing abandoned subway tunnel, the contractor was permitted to perform a pull-out test in lieu of the specified compression load test. In addition, the required bond length of the pile was reduced to 4 feet to reduce the maximum test load and the bearing requirements for the load test reaction system. The test set-up, illustrated in Figure 11, consisted of a hollow core jack that transmitted the test load to a reaction beam supported by grout leveling pads directly on the rock surface. An outer isolation sleeve, not shown, was placed around the reinforcing bar to the top of the bonded length of the pile to prevent contact between the grout and the rock above the test length. The actual bond length of the test pile was 3.9 feet.

**Test Results.** Load was applied in three load-unload cycles. The first load cycle was performed to assess the performance of the test set-up; and the second and third cycles were performed to the maximum test load of 450 kips. For the second load cycle, the test load was applied in increments of about 56 kips, or 12.5% of the maximum applied test load of 450-kips, and each load increment was maintained for a period of about 60 minutes, except the 450 kip load which held for about 15 ½ hours. A plot of applied test load versus displacement of the rock socket is presented in Figure 12. The socket displacement was determined using the measured displacement at the top of the reinforcing bar, and subtracting the elastic elongation of the bar. The elastic elongation was computed based on a bar length extending from the connection nut at the top of the bar to the middle of the bond length, or about 15 ft.

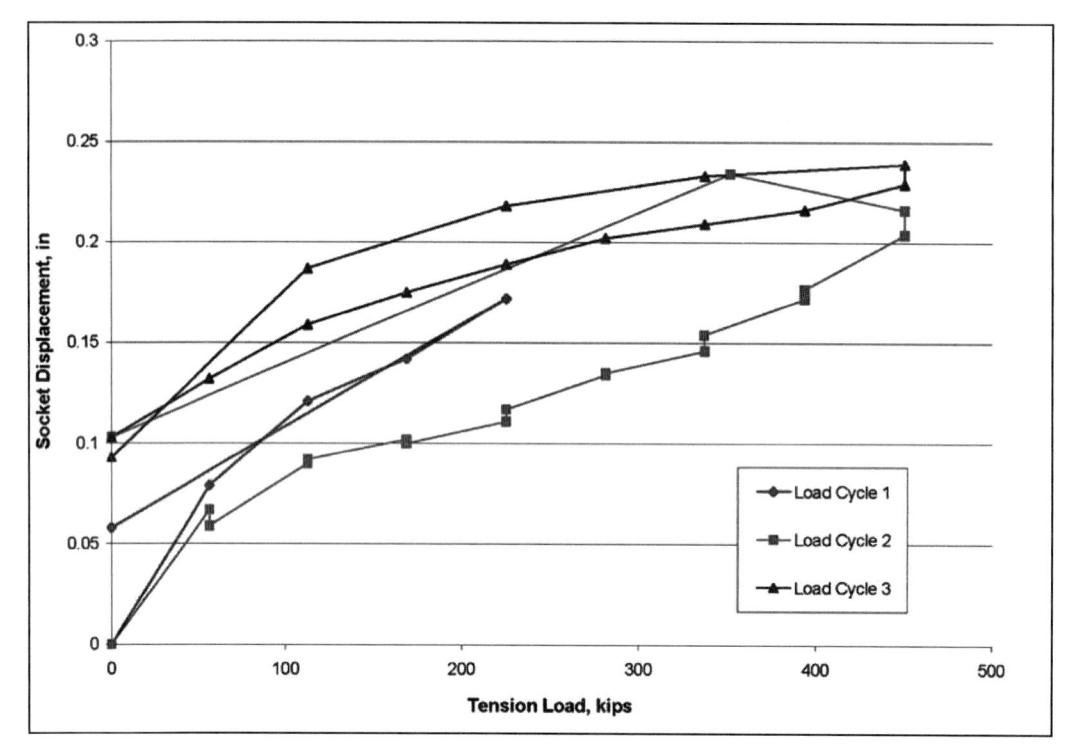

Figure 12. Results of load test on mini-pile

As shown in Figure 12, the socket displacement at the maximum test load of 450 kips was only 0.274 inches, equal to the sum of the net displacement of 0.058 inches from the first load cycle plus 0.216 inches from the second load cycle. At the design load of 225 kips, socket displacement was 0.172 inch, from the first loading stage, well within the specified acceptance criterion of 0.25 inch. Creep at the maximum test load was 0.0035 inch between the 1 and 10 minute readings, well below the criterion of 0.04 inch. The average unit bond resistance at the design load was calculated to be 127.5 psi.

Based on the load test results, the design criteria for the production piles established an allowable bond stress of 128 psi, with a factor of safety of 2.0; and required the socket to be within the competent rock. To confirm the required rock quality, a borehole video camera inspection was required for all production piles, and the socket limits were confirmed by a geotechnical engineer. Based on the above criteria, the minimum rock socket length for the mini-piles was 11 feet.

**Column Pedestal**

A pedestal support option was used for the design and construction of the new tale track tunnel east of the column line I. However, at column lines F, G, H, and I, where the existing columns are spaced at 15 feet, the structure is different from the eastern part of the tail track where the center columns are spaced at 5 feet, carry lighter loads, and pedestals are shorter. The condition of the existing built-up columns, and base bearing conditions were not known. The design decision was based on the assumption that it is likely that the existing columns would need to be replaced.

During construction, after all the encased elements of these columns were exposed, the contractor was able to inspect the existing columns according to the contract requirements. The inspection indicated that only one of these columns—on column line I—was in poor condition and required reinforcing. The finding prompted the contractor to consider saving these columns. Following submittal review meetings and several revisions, the design shown in Figures 13 and 14 were approved for construction.

Following are the main elements of the design that requested special attention:

- Base of the columns: Since the columns are built up sections, special reinforcing details were required to transfer the significant column load to new bearing plates and pedestals. The reinforcing plates are connected to the existing built up column web plate and

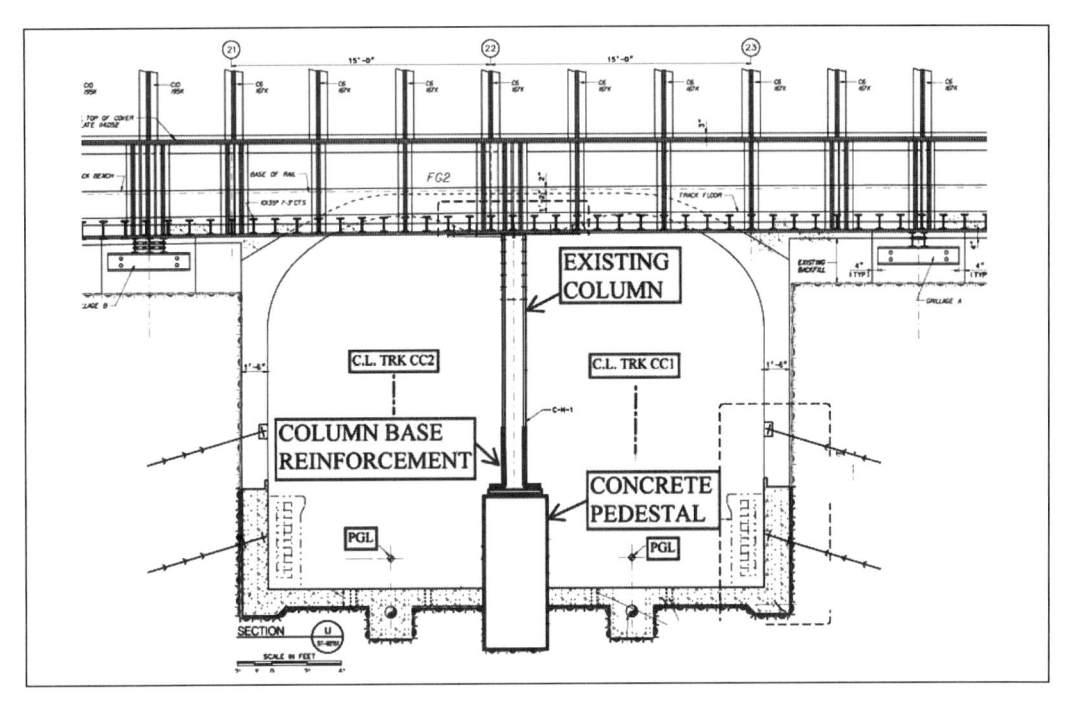

Figure 13. Modified section showing existing center column supported by a concrete pedestal

Figure 14. Existing column underpinned on mini-piles and new reinforced concrete pedestal

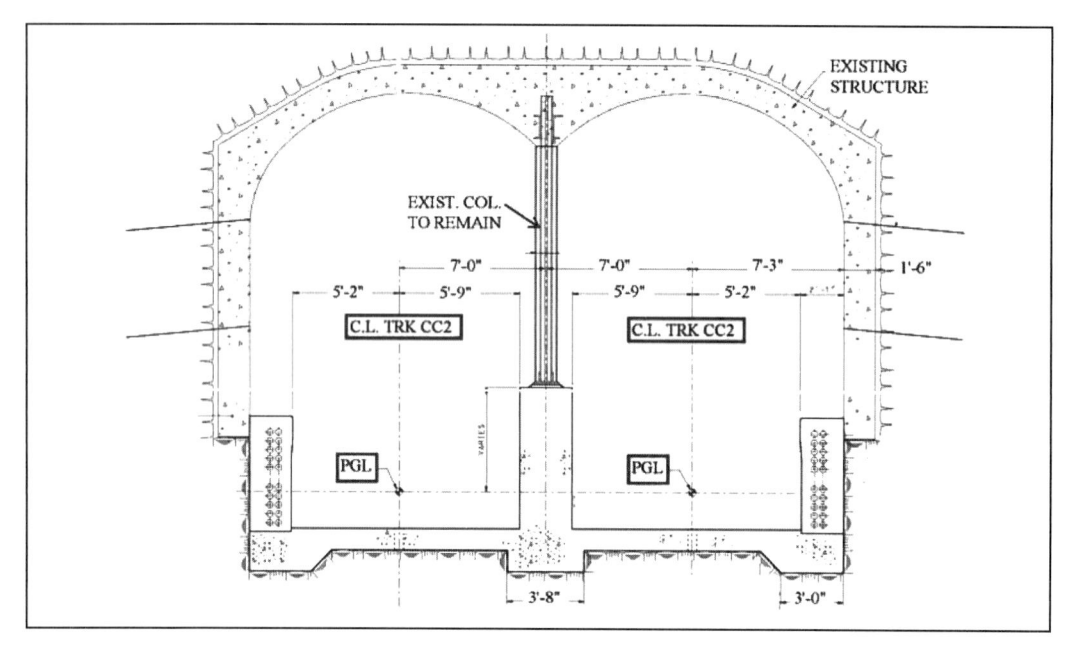

**Figure 15.** Typical section showing the completed structure

flanges that are comprised of angles and cover plates.
- Pedestals: One of the constraints imposed on the pedestals was the transverse dimension that was limited by the structural clearance line. It was a challenge for the contractor to provide sufficient support area for 1200 Kips column load. The contractor's initial submission utilized the entire cross sectional area of the pedestal to meet the bearing criteria. However, in response to submittal review comments, the column bearing area was limited to the area within the lateral reinforcement.

## RECONSTRUCTION OF THE TAIL TRACK TUNNEL AT TIMES SQUARE STATION

The existing tail track tunnel consists of an unreinforced double arch tunnel with structural steel center columns spaced at 5 foot centers. To construct the guide-ways for the No. 7 Line extension to the west, the invert slab of the tunnel had to be lowered gradually to a maximum of 7 feet. The existing tunnel was constructed in rock and was designed as a drained structure by providing the existing invert slab with weep holes to relieve the hydrostatic pressure. Along both exterior walls there are concrete benches housing the cable ducts which were cast together with the invert slab.

### Design Approach

To replace the tunnel with a cut and cover box was not feasible due to adjacent building entrances, underground pedestrian walkway connecting Times Square Station to the 8th Avenue subway, and all the utilities and ventilation louvers located along 41st Street. Demolishing and rebuilding the tunnel was also considered to be costly and a high-risk operation.

The selected construction method required only lowering the invert slab and the two side duct benches. The exterior walls and rock underneath the exterior walls were left intact and stabilized using two rows of rock bolts. The new invert slab and the side benches were designed as a drained structure by providing weep holes at 10-foot centers in the new invert slab. The invert slab is designed for 10% of the full hydrostatic pressure and the exterior walls are designed for 25% of the full hydrostatic pressure in addition to expected rock pressures and lateral reactions from the existing side walls. Due to expected lateral forces, the limited cross sections of the side benches were heavily reinforced with reinforcing bars and, where required, with structural sections.

The existing steel center columns were underpinned in groups of four by using needle beams and mini-piles. After completion of the excavation, a continuous concrete plinth was cast together with the invert slab, and the columns were supported on the plinth. During excavation, a lateral bracing system was installed to provide lateral stability. The existing concrete walls between the columns were demolished and re-built after completion of the plinth. The completed structure is shown on Figure 15.

## CONCLUSION

This paper addresses a rarely explored stage of the project—transition from final design to construction; the part that is very dynamic, challenging and creative, that requires close coordination between all parties—the owner, the engineer, and the contractor. The project benefits when the team functions in a way that capitalizes on the strengths of the team members. The contractor brings to the table valuable experience, creativity, and practical approach, familiarity with latest advances in construction materials, equipment, and his "know-how." The engineer's role is critical for fully communicating the design intent, for understanding the contractor's proposed methods, and for envisioning and analyzing all the consequences and implications of various construction methods and sequences. In dealing with such an important infrastructure as an operating subway, it is the engineer's responsibility to protect the client and the general public by reviewing contractor's proposed method of construction, and ensuring that the existing structures remain safe and sound. Such approach creates a positive effect on the project by clearly establishing the expected level of effort and various constraints, saving valuable time for the project, while not restricting in any way the contractor's creativity, and allowing the contractor to focus on developing the construction method that would address all the engineer's and client's concerns.

## DISCLAIMER

The opinions expressed in this paper are solely the authors' point of view and may not necessarily be considered the opinions of their respective employers.

## REFERENCES

Grigoryan, A. 2009. *No. 7 subway extension crossing under an existing subway station: Challenges and integration of underpinning into the design of new tunnels* (Part 1). In 2009 RETC proceedings, page 802. Edited by Gary Almeraris and Bill Mariucci, SME

Grigoryan, A. 2006. *Case studies: Structural investigation and rehabilitation of the 42nd Street & 8th Avenue subway station, New York City.* In 2006 NAT proceedings, page 401. Edited by Levent Ozdemir.

# Consolidation Grouting of the Riverbank Filtration Tunnel

**Adam L. Bedell, Steve Holtermann**
Jordan, Jones, and Goulding, Norcross, Georgia

**Kay Ball**
Louisville Water Company, Louisville, Kentucky

**ABSTRACT:** During construction of Louisville Water Company's Riverbank Filtration Tunnel, a fault zone was encountered which yielded a continuous inflow of approximately 300 gallons per minute over 60 linear feet. Through the owner's desire for the lowest practical inflow criteria and restriction of potential undesirable groundwater chemistry from the bedrock, a consolidation grouting program was designed and implemented. The consolidation grouting program had minimal impact to construction and schedule. Consolidation grouting was performed after concrete forms had passed the zone but before they were removed. This negated the need to remobilize equipment and go through additional cleanup. Consolidation grouting resulted in greater than 99 percent reduction in groundwater inflow.

## BACKGROUND

Construction of the Riverbank Filtration (RBF) Tunnel began in 2007. The tunnel is approximately 150 feet deep, 10 feet in finished diameter and 7,750 feet long. The tunnel was excavated at zero-grade within interlayered limestone and shale bedrock using a 12 foot diameter Robbins tunnel boring machine. The purpose of the tunnel is to collect bank-filtered raw water from the alluvial sand and gravel aquifer along the Ohio River and convey it to a new pump station at the B.E. Payne Water Treatment Plant as shown in Figure 1.

At approximately Sta. 15+00, a fault zone was encountered which continued for 100 feet to Sta. 16+00. Inflows occurring along the fault zone averaged 300 gallons per minute (gpm). Geochemical analysis of groundwater samples from the bedrock indicated high chloride and sulfide concentrations which were undesirable. The construction management team and design engineer devised a method to decrease the unwanted groundwater infiltration.

Excavation of the tunnel was completed on October 9, 2008. Production placement of the cast-in-place (CIP) concrete liner began on January 22, 2009. The construction management team met with the contractor, Mole Constructors, Inc., to discuss grouting options and impacts to the construction schedule. The contractor expressed a desire to proceed with grouting while all tunnel concrete forms and concrete equipment were in the tunnel. This would eliminate remobilization of additional equipment into a lined tunnel. The construction manager and owner agreed with the contractor. It was determined to stop concrete operations and consolidation grout the fault zone while concrete equipment was in the tunnel. It was anticipated that two weeks would be required from the schedule to complete all activities associated with consolidation grouting of the fault zone.

## CONSOLIDATION GROUTING WORK PLAN

The designed consolidation grouting work plan was based on methods developed on past and current tunnel projects in Atlanta, Georgia. Prior to placement of panning to control groundwater flows during concrete placement, a panning map was developed using geotechnical maps made during excavation. Drill holes were then laid out to ensure adequate pressure relief would be provided for groundwater and grout during grouting. These holes also allowed grout injection to occur as close as possible to the points of high groundwater inflow (Figure 2).

A method to capture and divert groundwater was installed first. After the concrete lining was placed to approximate station 16+30, an eight-inch deep trench was saw cut into the invert of the tunnel extending from Sta. 16+00 to Sta. 11+85. Two 6-inch invert drains were installed in the trench. The invert drains consisted perforated pvc pipes covered by a clean gravel within the 100 foot long fault zone. From Sta. 15+00 to Sta. 11+85, the invert drains consisted of blank pipe with no gravel backfill. Figure 3 shows the trench, invert drains, and gravel backfill.

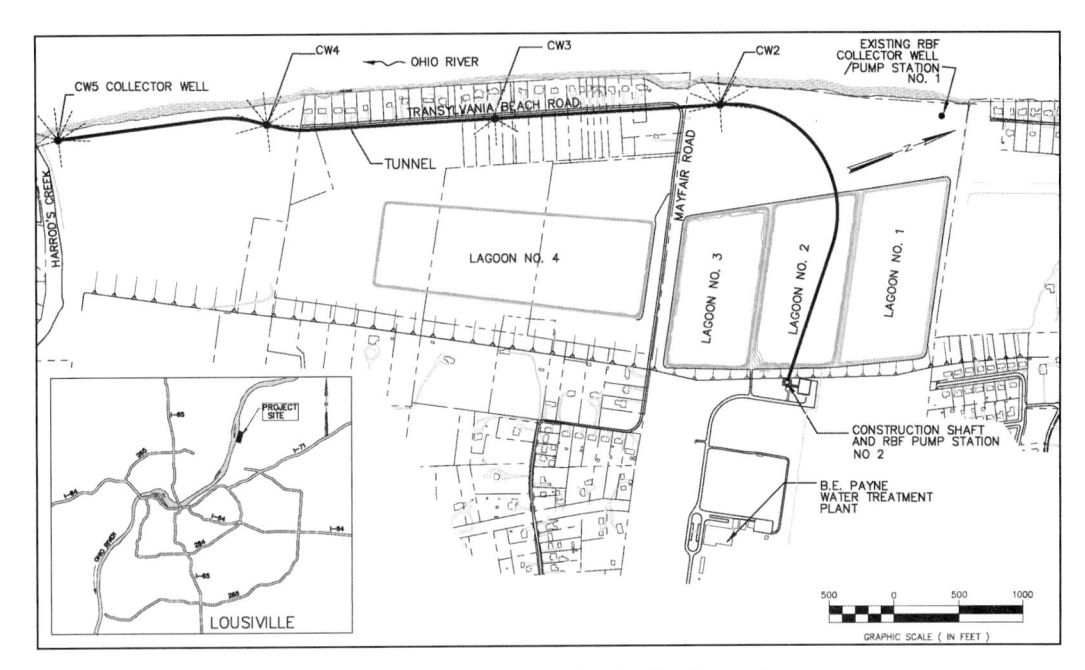

Figure 1. General location map for the RBF Tunnel in Louisville, Kentucky

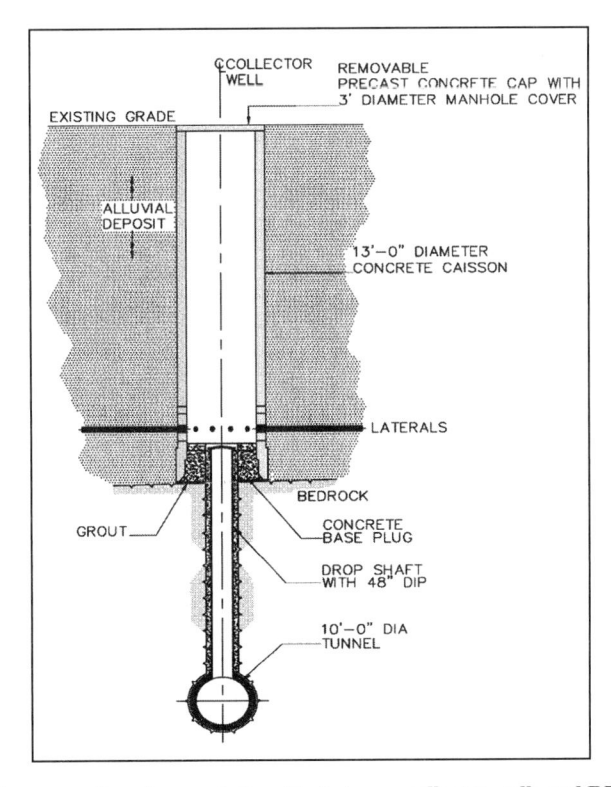

Figure 2. Generalized cross section shows relationship between collector wells and RBF tunnel. Groundwater is collected from 4 radial collector wells which transmit water from the aquifer to the tunnel.

**Figure 3. Clean gravel at the bottom of the photo and the saw-cut trench with invert drains at the top. In the fault zone, panning was installed on the tunnel walls running over the invert drain trench.**

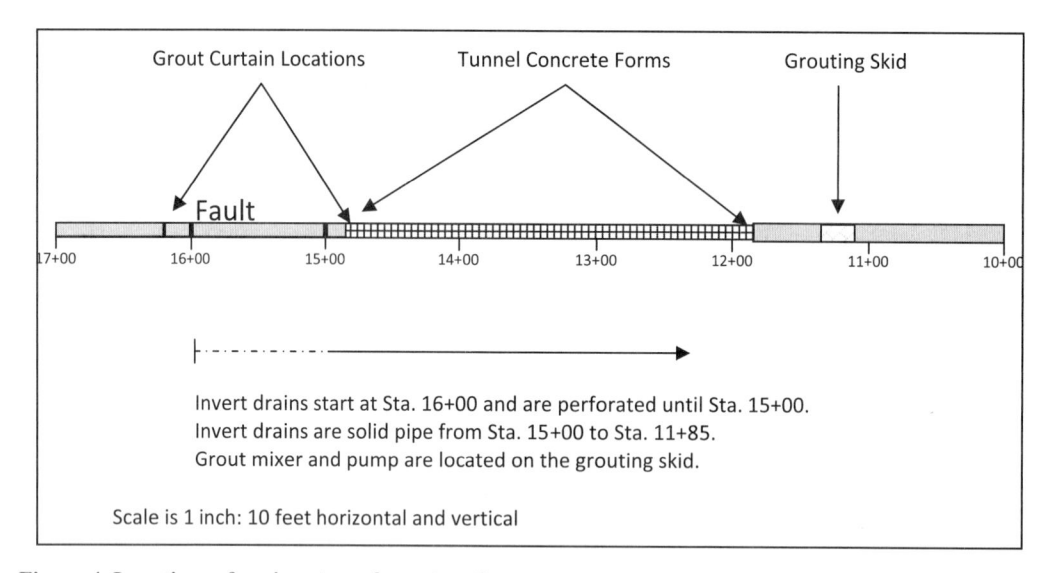

**Figure 4. Locations of various tunnel construction apparatus**

After groundwater inflows were controlled by the panning and invert drain system, concrete operations resumed and continued to approximate Sta. 11+85. Figure 4 illustrates the locations of equipment and features.

At approximate Sta. 11+85, the concrete lining operation was temporarily halted to allow for consolidation grouting to be performed. Grouting work was subcontracted out to Nicholson Construction Company. Grout curtains were installed to ensure that grout did not travel away from the fault zone along the crown or elsewhere. The grout curtain consisted of 8 holes drilled radially around the tunnel 20 feet up-station (Sta. 16+20) and down-station (Sta. 14+80) of the fault zone. A thicker grout mix was used (0.8:1 water: cement by volume) under moderate pressure (50 psi) to seal the interval.

Consolidation grouting began after concrete in the tunnel liner had reached compressive strengths of 5,000 psi. Test cylinders were broken to verify the strength of the tunnel concrete prior to consolidation grouting. The maximum allowable grouting pressure determined by the design engineer was 110 psi. A single Y connection was fabricated to attach to the end of both invert drains. This allowed simultaneous grouting of both invert drains. Grouting began and continued until the refusal pressure was reached. Communication was established to all of the grout holes drilled in the panning.

Establishing communication to the panning holes from the invert drain verifies an open pathway from the invert drain, up the panning and to the fractures bearing water. This communication helps delineate where the grout is traveling. After communication to the packers in the panning holes was established, the thin, diluted grout was allowed to flow out of the packer and into the invert until a thick grout representative of the mix appeared. Once observed, then the valve on the packer is closed.

After grouting reached refusal on the invert drains, panning drill holes were then grouted. Placement of grout would then start up-station and progress down-station, connecting to every hole drilled, including ones that had previously exhibited communication while the invert drain was grouted, and injection would continue until refusal pressure was reached.

## CONSOLIDATION GROUTING RESULTS

Concrete was placed to approximate St. 118+85 on May 20, 2009. Nicholson arrived on site May 20, 2009. On May 21, 2009, the grout curtains were drilled and grouted. A total of 6.4 gallons of grout was placed in both grout curtains.

Grouting began on May 26, 2009. The grout mix was 1.5:1 water: cement by volume. Grouting continued until refusal. A total of 5,638 gallons of grout were pumped into the invert drain on May 26, 2009. On May 27, 2009, 2,507 gallons were pumped through 12 panning holes for a total of 8,426 gallons. Nicholson demobilized and left the tunnel on Thursday, May 28, 2009.

Groundwater inflow estimates prior to grouting activities were approximately 300 gpm. Groundwater inflow estimates from the tunnel following consolidation grouting of the fault zone were estimated at approximately 3 gpm.

On Monday, May 29, 2009 concrete placement for the cast in place concrete liner resumed following a two week period for consolidation grouting of the fault zone. The contractor was able to resume concrete operations without additional clean up or additional remobilization of equipment into the tunnel.

## CONCLUSION

During excavation of the RBF tunnel, a 100 foot long fault zone was encountered which provided a large, sustained inflow during the remaining excavation and subsequent concrete lining operations. Development and implementation of a consolidation grouting program to deal with groundwater inflow from the fault zone resulted in a 99 percent reduction in inflow while minimizing the impact to the overall construction schedule.

# Gotthard Base Tunnel: Micro Tremors and Rock Bursts Encountered During Construction

**Michael Rehbock-Sander, Rolf Stadelmann**
Amberg Engineering Ltd., Regensdorf, Switzerland

ABSTRACT: The Gotthard Base Tunnel is a 57 km long railway tunnel through the Swiss Alps. During construction of the multifunction station (MFS) at the construction section Faido a until than unknown fault zone system was encountered. Additionally rock bursts have occurred since March 2004 and the Swiss Seismological Service recorded an accumulation of seismic activity in the area of the MFS Faido. This paper deals with the coping of the challenges of the construction works at the MFS. The results from seismic measuring and numerical simulations are shown. The reasons which give rise for rock bursts and the so called micro tremors are explained. The taken measures to ensure the workers safety during construction works are described and the risks of a micro tremor for the tunnel under operation are estimated.

## INTRODUCTION

The Gotthard Base Tunnel (GBT) is the core of the NEAT (New Alpine Transversal) through the Swiss Alps. The entire 57 km long tunnel is divided into five construction sections in order to attain a reasonable construction time and for ventilation purposes. Excavations started from the portals at Erstfeld and Bodio as well as from three intermediate attacks located in Amsteg, Sedrun and Faido (Figure 1).

The tunnel consists of two parallel single track tubes which are linked by cross-passages every 300 m. Multifunction stations are located at two locations one-third and two-thirds along the length of the tunnel. These will be utilized for the diversion of trains to the other tube via crossover tunnels, to house technical infrastructure and equipment, and as an emergency station for the evacuation of passengers. More information can be found under www.alptransit.ch.

From north to south, the GBT passes through mostly crystalline rock, the massifs which are interrupted by narrow sedimentary tectonic zones. The three crystalline rock sections include the Aare massif to the north, the Gotthard massif and the Pennine gneiss zone to the south. These massifs consist mainly of high strength igneous and metamorphic rock. More than 90% of the total tunnel length consists of these types of rock. The maximum overburden is about 2350 meters (Figure 1).

During construction of the Multifunction Station (MFS) Faido frequent and often massive rock bursts have occurred since March 2004. Additionally the Swiss Seismological Service (SED) recorded an accumulation of seismic activity in the area of the MFS Faido. In July 2005 the tunnel's owner ATG (AlpTransit Gotthard Base Tunnel AG) formed a working group, Micro Tremors, to investigate all aspects related to the seismic activity, especially the impact of a seismic event on the tunnel under operation.

This paper deals with the reasons which give rise for rock bursts and micro tremors. The results from seismic measuring and numerical simulations are shown. The taken measures to ensure the worker's safety during the construction works are described and the risks of a micro tremor for the tunnel under operation are estimated.

## MULTIFUNCTION STATION FAIDO

The intermediate points of attack at Sedrun and Faido shall serve as multifunction stations (MFS) during operation of the tunnel. These MFS enable the trains to change the tunnel tube in case of maintenance works and the specially ventilated emergency sections serve the rescue of people in emergency cases. Drill & Blast was used for the construction of both MFS. The MFS Faido had to be accessed by a 2.7 km long access tunnel declining at 12.7% from the portal. The overburden in the MFS Faido is between 1500 and 1800 meters.

### Predicted Geology and Hazards of the MFS Faido

The location of the MFS Faido originally was predicted in Leventina Gneiss of good quality. The outcrops from quarries in the area of the MFS Faido, the

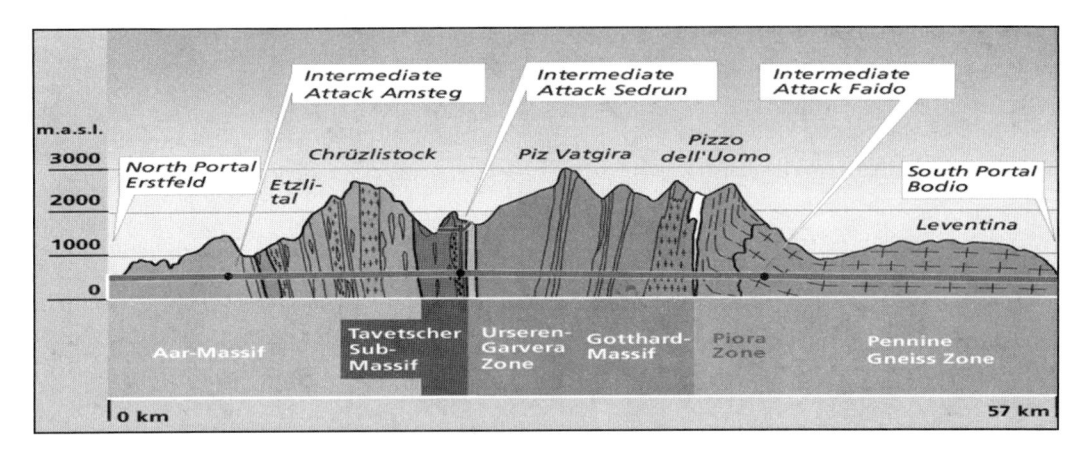

**Figure 1. Geological longitudinal profile of the GBT**

experiences made during construction of the investigation system for the Triassic Piora basin as well as vertical exploration drilling confirmed a favorable geological section.

The most relevant hazards for the Faido MFS were:

- Detaching of wedges (brittle failure): the danger of detaching of wedges depends of frequency, distance and quality of joints and discontinuities in the rock mass.
- Loosening: in highly jointed rock mass loosening can occur in the crown area. The loosened rock mass is additionally loading the support and the lining.
- Squeezing (plastic deformations): squeezing properties in deep tunnels in hard rock normally are to be expected in the mylonitic and cataclastic zones of fault zones.
- Rock burst: (brittle failure).

### Difficulties During Construction

During construction of the cross cavern a break down of fine grained quartz occurred in the cavern's roof forming a cavity of 8 meters in height. Therefore extensive exploration drillings and seismic reflection measurements were effectuated during further construction. The results of these investigations revealed a, until then unknown, large fault system in the area of the MFS Faido. The main kernel of this fault strikes at an average angle of about 20° to 15° to the tunnel axis and dips at about 80° to the East (Figure 2). In the fault's kernel, layers of partially completely decomposed rock (kakirite) are embedded. Adjacent to the east of the fault hard and brittle Leventina gneiss is located. To the west of the fault the rock mass consists of hard but less brittle Lucomagno gneiss.

As a result of the aforementioned investigations, it was decided to adapt the layout of the MFS Faido with the aim of placing the large caverns in good rock conditions. Different alternative layouts were investigated. The brunch-off structures were shifted in the southern part of the MFS finally (Figure 2).

Besides the layout of the MFS the geology encountered also made it necessary to carry out a critical review of excavation support means to be applied in the relevant cross-sections. With the initially designed support consisting of pattern bolting steel meshes and shotcrete no stability could be achieved in the single-track tunnel west/north (EWN). So the section of the EWN tunnel was rebuilt with a support consisting of steel arches HEM 200 backfilled with 40 cm concrete. The support was installed immediately after each excavation step of 1 m. This rigid support was intended to cater for the heavy pressure and especially to protect workforce from break in the working area. However, the displacements developed immediately after excavation. In the rebuilt section, on a length of 250 m from the cross cavern to the north the support's loading gave rise to displacements of up to 1 m. Strain measurements revealed yielding of the steel arches already four days after backfilling. Severe damage of the support developed (Figure 3) and the critical section had again to be rebuilt with an enlarged excavation radius of 1.5 m to allow additional displacements. A flexible support had been installed successfully. TH profiles with sliding connections were used. Horizontal shotcrete slots at the level of the arches' clutches were left open for unhindered sliding of the arches and to avoid damage of the shotcrete lining in case of increasing displacements (Figure 4).

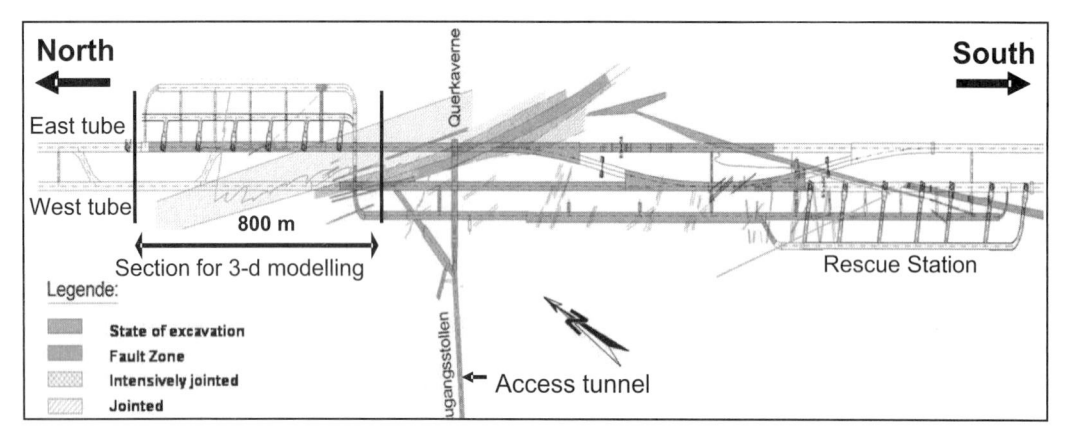

Figure 2. MFS Faido, actual geology recognized in January 2003

Figure 3. Sheared off steel arch in the invert

Figure 4. Rebuilding of the critical section

## COMPUTATIONAL MODEL USED FOR NUMERICAL SIMULATIONS

For simulating the encountered phenomena such as the large deformations and the seismic impacts which occurred in a generally hard rock formation a discontinuum model was preferred. The 2D and 3D distinct element codes UDEC (Universal Distinct Element Code) and 3DEC respectively were used.

The discontinuum model enables to realistically simulate shear movements and opening/closing of joints. The rock mass is assembled by blocks formed by joint systems. The joints behave as interfaces and the blocks of rock behave as continua. The deformations in such a block system are formed by rigid body movements (translations and rotations) along joints and elastic or elastic plastic deformations of the blocks.

Movements and displacements are caused by disturbances extending in the block system. The disturbances are caused by body forces and external forces acting on the block system. The speed of the disturbance's distribution depends on the physical properties of the block system i.e., the blocks and the joints. A dynamic approach following the second law of Newton (force = mass × acceleration) to determine the block's movements. With a force—displacement relation the contact forces are determined from known displacements.

A time stepping approach with a Finite Difference solution is used for the dynamic process. Velocities and accelerations are constant within a time step (Figure 5).

## NUMERICAL SIMULATION OF THE STATIC LOAD CASE

In order to better understand the failure mechanisms in the area around the tunnels of the MFS numerical

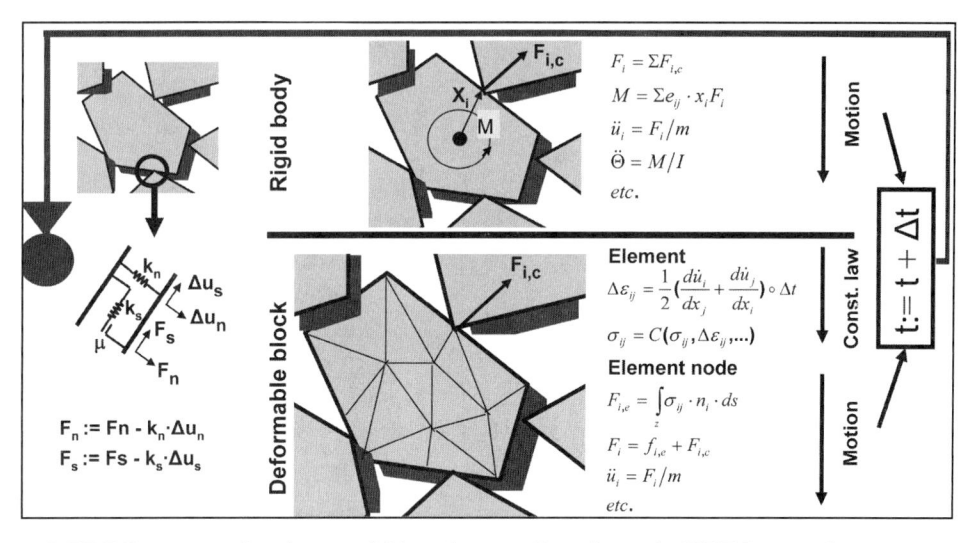

Figure 5. UDEC: computational steps within a time step Dt, schematic (UDEC manual)

modeling was carried out (static load case). The model section is indicated in Figure 2. The investigation was based on a parametric study comprising the variation of rock and joint properties. The 2D UDEC model is shown in Figure 6. The different fault regions in the model have been selected according to the geologist's findings based on interpretation of borehole results.

The results clearly show a considerable extension of the stress redistribution due to the excavation of the tunnels. The stress concentration at tunnel level amounts to σyy = 60 to 80 MPa (Figure 6). The area of this stress is in accordance with the hypocenters of the micro tremors determined by the SED (Figure 9).

## ROCK BURSTS AND MICRO TREMORS

In this paper, micro tremors are defined as a seismic event generally occurring at a larger distance from the tunnel (< hundred meters) whereas a rock burst occurs in the direct vicinity of the tunnel. Richter scale M is used for magnitudes.

### Impacts of Rock Bursts on the Tunnel

During the excavation of the north eastern section of the MFS a large number of rock bursts occurred. At that time 75 % of all events take place at the face during the first three hours after a drill and blast round and are perceived in the form of vibrations and loud cracking or bangs. In May 2004 a rock burst occurred for the first time in the side wall of the single-track tunnel east/north (EON) that had already been secured for several months. Rock suddenly loosened and the vault was deformed over a distance

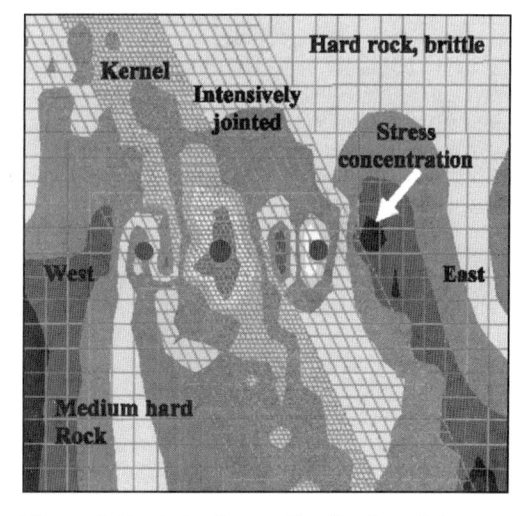

Figure 6. Static load case: distribution of the vertical stresses σyy

of about 30 m. Some days later a major rock burst occurred in the single-track tube east/south (EOS), resulting in rock loosening in the left side wall. This also destroyed the shotcrete lining over a length of 30 m.

Rock burst's damage potential is illustrated in Figure 7. The left picture shows the damage of the support with a shotcrete plate ejected into the EON. This rock burst occurred together with the M1.9 micro tremor of July 2005. The EON's invert heave presented in the right picture was caused by the M2.4 micro tremor of March 2006. The invert heave due

Figure 7. Single-track tube east/north. Left: damage of the support's shotcrete. Right: invert heaves.

to the seismic impact solely was smaller than it is shown in the picture taken two days later.

## Measurements During Construction Works

Rock bursts can neither be predicted exactly nor prevented by excavation support measures. Nevertheless, action must be taken to ensure safety of workforce and equipment. Constructional adaptations of the support and excluding critical tunnel section for access were required. A prognosis of the rock burst risks was undertaken along with the envisioned measures.

A rock burst information sheet and lists of progressive measures (Table 1) had been prepared. Specific dangers and the necessary actions were being determined in advance and are continually adapted as new knowledge was obtained during tunneling operations. Various preventive measures were prescribed:

- Sealing of the face with steel-fibre shotcrete (to prevent loosening of small particles from the face)
- Face anchors with large anchor plates
- Leaving a pile of material in front of the face (to prevent access close to the face)
- Switching to top heading excavation method
- Arched face formation (to anticipate the excavation form usually resulting from stresses)
- Prohibiting of manual work around the face for the first three hours after a drill and blast round

In addition, excavation support measures are altered to meet the different levels of rock burst risk. Special yielding support elements are used to absorb dynamic loading, such as rock bolts (Swellex or Yielding Swellex bolts) and flexible steel arches.

During the excavations the potential of heavy rock bursts in the different parts of he MFS have been predicted, the necessary support measures and additional means have been fixed. This finally lead to the closure of some sections for all traffic with major impact on the logistics of the site.

During the construction works at the MFS Faido several hundred rock bursts occurred. With the aforementioned countermeasures no injuries or accidents due to a rock burst were reported.

## Development of the Seismic Activity

Between March 2004 and June 2005 the Swiss Seismological Service (SED) recorded an accumulation of seismic activity in the area of the MFS Faido, a region, normally with a very low seismicity. During the above mentioned period the permanently installed Swiss Digital Seismic Network (SDSNet) registered 10 seismic events with local magnitudes M between 0.9 and 1.9. With the SDSNet located at the surface, the epicenters could be associated with the area of the MFS Faido within an accuracy of one kilometer. Together with the M1.9 tremor of 1.6.2005 a strong rock burst could be associated. The same holds for two additional tremors of similar magnitude. On the other hand no relations to rock bursts could be identified for tremors during the period March to April 2004. On March 25, 2006 the strongest micro tremor of M 2.4 was registered. This tremor was felt by the inhabitants of the village Faido close to the jobsite.

## Additional Seismic Measuring Stations

The working group, Micro Tremors, decided to install additional seismic stations at the surface and in the MFS Faido. For precise monitoring and location of the seismic activity's sources a special local seismic network consisting of nine stations at the surface, including one station from the SDSNet, were installed in a circular arrangement 10 to 15 km around the MFS Faido. In addition, two stations were installed at different locations in the tunnels of

Table 1. Rock burst classes, perceived phenomena, and measures

|  | Term (event) | Perceived Phenomena | Measures |
|---|---|---|---|
| E1 | Release | • Cracking sound<br>• Rumbling | • Document observations in shift and daily report of contractor<br>• Intensify observations<br>• Partial excavation (top heading)<br>• Overhead protective mesh over entire crown<br>• Shotcrete on face, vaulted face<br>• Face anchor bolts |
| BS1 | Light rock bursts | • Vibrations<br>• Dust dispersion<br>• Heavy face spalling during loosening in the form of plates up to about 5 m³ and about 1 m deep | Additional to E1:<br>• strengthen face bolting<br>• leave wedge of material in front of face<br>• intensify and strengthen system bolts<br>• strengthen overhead mesh |
| BS2 | Medium rock bursts | • Extreme vibrations<br>• Vibrations 3–6 h after-excavation<br>• Dust clouds from crown<br>• Face spalling less than 5 m³ prior to loosening<br>• End anchor plates into L2 | Additional to BS1:<br>• denser, yielding rock burst anchoring |
| BS3 | Extreme rock bursts | • Extreme vibrations<br>• Several successive vibrations<br>• Concussions after more than 3 h<br>• Shotcrete spalling in L2<br>• Cracking of shotcrete on face<br>• Face spalling greater than 5 m³ before loosening<br>• Overbreak formation<br>• Anchor heads torn off near abutment | Additional to BS2:<br>• Denser bolt pattern if damage pattern<br>• Re-pattern bolting/strengthening<br>• Face bolting, reduction<br>• Bolt spacing, reduction<br>• Pressure relief blasting<br>• Submit event report (short report) |

the MFS Faido. The circular position of the seismic measuring equipment allows a precise determination of the epicenters whereas the measuring stations directly above and inside the MFS Faido serve the evaluation of the depths of the micro tremors' sources. Accurate seismic wave velocities required for the hypocenter's determinations were derived from two calibration shots carried out in the MFS Faido. An average P-wave velocity of 5.33 km/s was calculated. The readings of the measuring stations were integrated in the SED's data acquisition system. The real time transmission of the measuring data guaranteed a continuous survey of the seismic activity allowing for an immediate alert of the responsible organizations such as ATG, supervision and authorities in case of a strong tremor. This was of particular importance in case of the M2.4 tremor occurring March 2006.

### Chronology of the Seismic Events

Figure 8 shows the development of the recorded micro tremors' number and magnitudes as functions of time. The highest seismic activity took place during December 2005, March 2006 and May 2006. The highest magnitude of 2.4 occurred on 25th March 2006. From October 2005 to February 2008 112 micro tremors were recorded.

The magnitudes of most of the tremors were below 1.0. With termination of excavation in the MFS Faido the micro tremors' number and magnitudes decreased continuously. Since September 2007 no more micro tremors have been recorded above the measuring threshold of $M = -1.0$ in the area of the MFS Faido.

### Epicenters

The epicenters of all registered micro tremors during October 2005 to February 2008 are depicted in Figure 9. The micro tremors are concentrated in the rock mass to the north of the MFS Faido close to the eastern part of the tunnel system. The accuracy of the epicenters' localization is less than 100 m and less than 250 m in focal depth as determined by relocation of the calibration shots. Within the error ellipsoid the tremors' sources are at tunnel level.

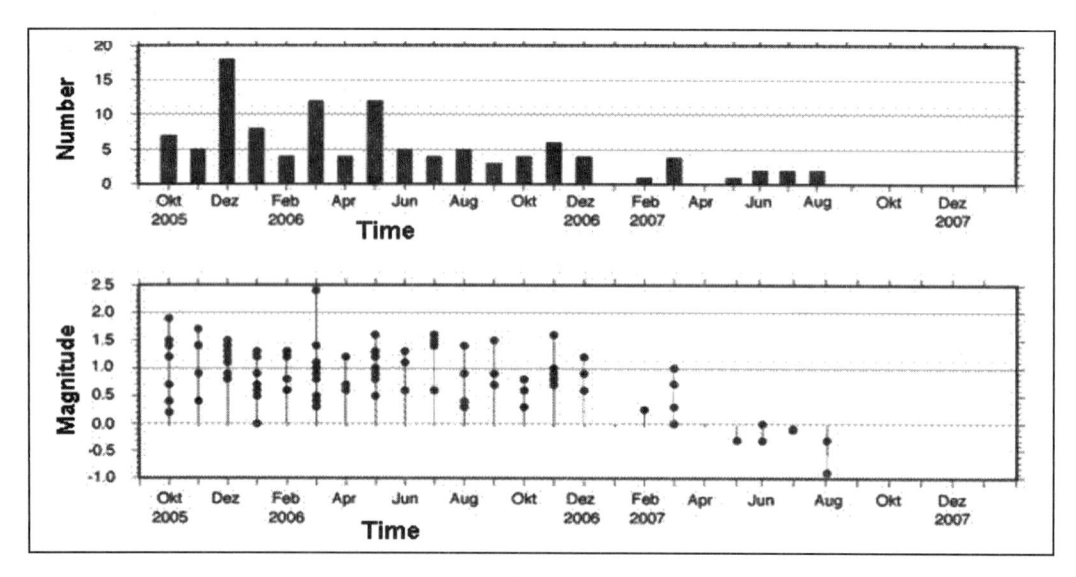

Figure 8. Chronological development of number and magnitudes of the micro tremors (courtesy of SED)

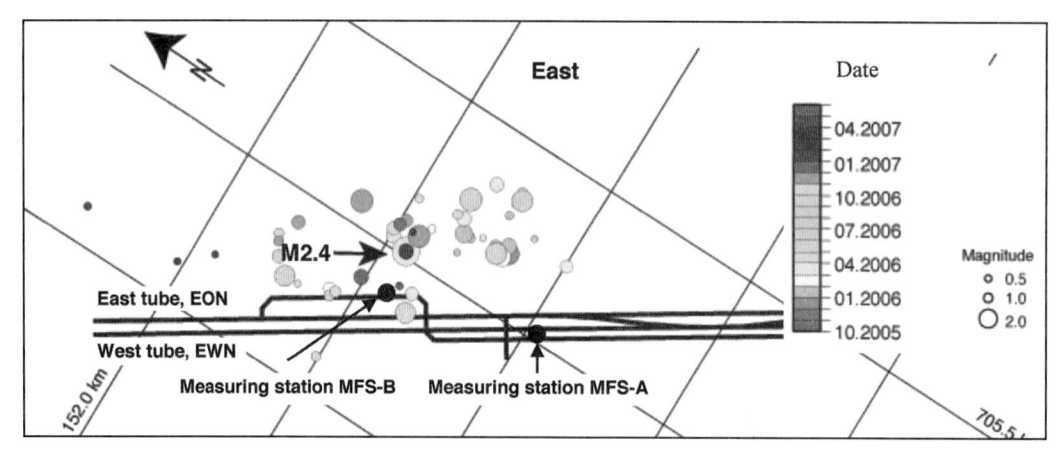

Figure 9. Epicenters of the micro tremors from October 2005 to February 2008 (courtesy of SED)

### Findings from the Seismic Measurements

The micro tremors in the northern part of the MFS tend to form clusters i.e., the sources of several tremors are located within the same area. Considering the predominant steep west–east dipping joint system striking sub parallel to the tunnel axis (Figure 2 and 6) shear failures along joints are most likely. The micro tremor's source locations in the hard Leventina gneiss to the east of the fault corresponds to the location of the vertical stress concentrations resulting from the computations of the static load case in Figure 6. There is a general tendency of the micro tremors to move together with the excavation of the tunnels form the cross cavern's area to the north. Very few micro tremors occurred in the southern part of the MFS Faido. Many of the rock bursts causing support's damages in the tunnels were triggered by micro tremors.

## DYNAMIC MODELING

### Aims of Modeling

The aim of the numerical modeling was to provide basic appraisals with respect to the structural safety

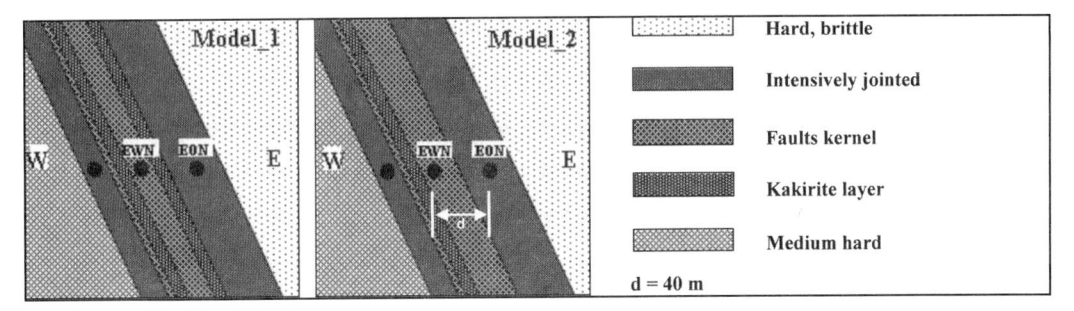

Figure 10. 2D-Model_1 and Model_2 with 2 and 1 layers of kakirite in the fault's kernel

and usability of the tunnels during and after a seismic event. The investigation's result should furthermore disclose the residual risks to be accepted and its impact potential on the tunnels concrete linings. The structural design has been completed prior to the occurrence of rock bursts. Therefore, it had to be controlled whether the lining designed for the static load case still fulfills the tunnels' safety and usability requirements during and after a micro tremor's impact.

### Computational Methods and Models

The results presented refer solely to the dynamic load case. In a first step the static equilibrium of the supported tunnel system was computed. In a second step the final lining was inserted in the model and the design micro tremor's load superimposed to the static case. Weak rock properties were assigned to the layers containing kakirite. For modeling support and final lining in the 2D models block elements and structural bar elements respectively, were used. Special investigations have shown the deviation of seismic waves along weak rock formations. Therefore two different 2D models with one and with two kakirite layers, respectively, were investigated (Figure 10).

### Specification of the Model Design Loading Wave

The work group, Micro Tremors, decided to consider the M2.4 micro tremor of 25 March 2006 as the decisive design tremor.

First, the dynamic model load (input wave) corresponding to the design tremor had to be determined. The emitted wave at the tremor's source first had to pass the tunnel system and the fault zone prior to arrive at measuring station MFS-A. Between source and measuring station the wave was attenuated. This means that the measured signal at MFS-A corresponds to a damped wave. This design load covers the strongest micro tremor identified since starting the extended seismic monitoring in the MFS Faido (Figure 8) and it is conservative.

## RESULTS FOR OPERATIONAL PERIOD OF THE TUNNEL

The probability of the occurrence of the design micro tremor and the corresponding triggering of a stress drop has been assessed by the experts as very low. This assessment is based on the fact that after termination of the excavation in the MFS no more micro tremors occurred. This confirms the correlation between excavation activity and micro tremors. Furthermore, the general seismic activity in the area of the MFS Faido is very low. On 21.01.2008 a M4.0 earthquake occurred at a distance of approximately 50 km from the MFS Faido, in an area with a little higher (but still low) seismic activity compared to the Faido-area. This earthquake was recognized by the measuring stations in the MFS. No triggering of a micro tremor could be identified. Based on above mentioned aspects the case of a stress drop near a tunnel in the MFS Faido has been accepted as a residual risk. However, a residual risk cannot be completely excluded. Therefore the owner of the tunnel, ATG, decided to install seismic measuring equipments and vibration sensors in the linings of the tunnels in the MFS Faido for a permanent seismic monitoring during operation.

## CONCLUSIONS

- During tunneling at great depth in geological conditions as encountered in the MFS Faido micro tremors triggering rock bursts are likely to occur.
- The stress redistribution due to tunneling and the stress concentration in hard rock in combination with an existing zone of weakness (fault) favors the occurrence of micro tremors.
- The micro tremors did clearly correlate with the excavation activities. After terminating the excavation no more micro tremors have been identified.
- Micro tremors cannot be avoided. During construction precaution measures such as

closing of critical sections and flexible support consisting of flexible rock bolts and steel arches are to be applied.
- A seismic wave is deviated by a weak zone. With the orientation of the weak zones in the MFS Faido a seismic wave is deviated towards the tunnels.
- The dynamic impact on a lining of a tunnel in front of a weak zone and exposed to a more or less unhindered micro tremor's wave is considerably higher compared to the impact on a liner of a tunnel in the 'shelter' of a weak zone.
- The specification of the design micro tremor for the MFS Faido is conservative.
- Excluding the additional loading due to a spontaneous 'stress drop' in the direct vicinity of a tunnel there has been no need for improving (thickness, additional reinforcing) the linings designed for the static load case in the MFS Faido.

## ACKNOWLEDGMENT

The authors thank the AlpTransit Gotthard AG for the permission to publish this lecture.

## REFERENCES

"Gotthard Base Tunnel, section Faido, handling of risks and opportunities in the most challenging part of the Gotthard Base Tunnel." S. Flury, Swiss Tunnel Congress 2009, Luzern, Switzerland

"Gotthard Base Tunnel, section Sedrun, squeezing rock conditions in the northern part of the Tavetsch intermediate massif—project design and implementation." H. Ehrbar, K. Kovári, Swiss Tunnel Congress 2008, Lucerne, Switzerland

"Gotthard-Basistunnel—Bergschläge und Mikrobeben in der MFS Faido." E. Kissling, M. Rehbock-Sander, Swiss Tunnel Congress 2008, Lucerne, Switzerland

"Experience Gained in Mechanical and Conventional Excavations in Long Alpine Tunnels in Switzerland," Y. Boissonnas, M. Rehbock-Sander, RETC 2009 Las Vegas, USA

# Optimization in Blasting Production and Vibration Mitigation for Shaft and Tunnel Construction at Lake Mead

**Jordan Hoover**
Barnard Construction Company, Inc., Bozeman, Montana

**Dan Brown, Caroline Boerner**
Barnard of Nevada, Las Vegas, Nevada

**Shimi Tzobery**
Parsons Corporation Inc., Las Vegas, Nevada

**Erika Moonin**
Southern Nevada Water Authority, Las Vegas, Nevada

**ABSTRACT:** Near the city known for taking risks, the close proximity of the two main Southern Nevada Water Authority (SNWA) pumping stations left no room to gamble when it came to blasting operations for new shaft and tunnel construction near Las Vegas, Nevada. The pumping stations deliver water from Lake Mead to Las Vegas that is vital to the city's way of life. Drill and blast excavation as close as 55 feet to these facilities posed exceptional constraints and engineering challenges. This, coupled with a looming limited shutdown period, made this project uniquely challenging. This paper describes these challenges, how creative design approaches were implemented, how monitoring efforts were managed, and ultimately, how the team successfully protected sensitive multi-million dollar pumping equipment during the construction of the Intake No. 2 Connection and Modification Project.

## INTRODUCTION

A prolonged drought and increased demand for water have caused water levels at Lake Mead to drop significantly in recent years. Located 20 miles southeast of Las Vegas, this man-made reservoir provides water for users in Nevada, Southern California, Arizona, and Mexico. The decline in water level has pushed the SNWA to construct a third intake, Intake No. 3, to withdraw water from a deeper area in the lake.

The Lake Mead Intake No. 2 Connection and Modifications Project was a vital part for the creation of the new Intake No. 3 as it connects existing Intake No. 2 (IPS-2) to the prospective Intake No. 3 (IPS-3). Barnard of Nevada, Inc., a subsidiary of Barnard Construction Company, Inc., was awarded this contract on May 15, 2008 for $30.09 million following a completive bid process. The project consists of a new 22-ft final diameter Isolation Gate shaft, and approximately 570-ft long of 14-ft wide × 16-ft high horseshoe-shaped tunnel. This shaft extends 380 feet in depth where two tunnel headings have been designed to tee, one "tying in" to the existing IPS-2 Tunnel and the other terminating for future connection to IPS-3 Tunnel. At the base of the shaft a 90-ft long × 26-ft high × 22-ft wide concrete transition structure was constructed. The transition structure houses the embedded gate sealing frame for the Isolation Gate. The Isolation Gate is an electrically-driven hoisted slide gate constructed to isolate water in either the IPS-2 side or IPS-3 side upon requirement. A gate guide system was installed throughout the shaft for raising and lowering the gate from closed position to upper inspection position.

To facilitate the tie in to the existing IPS-2 Tunnel a tight six-week shutdown period was allowed following the shaft and gate construction to plug and dewater the existing tunnel, and excavate a 50-foot-rock plug left in place to prevent water from entering the working area. Major maintenance and cleaning within IPS-2 Tunnel was also conducted, including installation of new water sampling and chemical feed lines. This six-week shutdown period was contractually limited to occur only during the months of December 2009 through February 2010 resulting from predicted low water demand. This critical constraint left little room for delays.

Table 1. Test blast design data

| Test # | Hole Depth (ft) | Stemming (ft) | Powder Column (ft) | Exp. Per Hole (lbs) | Holes per Delay | Pounds per Delay | Expected Vibration (in/sec) | Number of Test Holes | Total lbs. Used |
|---|---|---|---|---|---|---|---|---|---|
| 1 | 2 | 1 | 1 | 0.88 | 1 | 0.88 | 0.394 | 2 | 1.76 |
| 2 | 3 | 1.5 | 1.5 | 1.32 | 1 | 1.32 | 0.546 | 4 | 5.28 |
| 3 | 4 | 2 | 2 | 1.76 | 1 | 1.76 | 0.687 | 6 | 10.56 |
| 4 | 5 | 2 | 3 | 2.64 | 1 | 2.64 | 0.95 | 1 | 2.64 |
| 5 | 5 | 2 | 3 | 2.64 | 1 | 2.64 | 0.95 | 4 | 10.56 |
| 6 | 4 | 2 | 2 | 1.76 | 1 | 1.76 | 0.687 | 25 | 44 |
| 7 | 5 | 2 | 3 | 2.64 | 1 | 2.64 | 0.95 | 27 | 71.28 |
| 8 | 6 | 2 | 4 | 3.52 | 1 | 3.52 | 1.17 | 8 | 28.16 |
| 9 | 8 | 2 | 6 | 5.28 | 1 | 5.28 | 1.65 | 8 | 42.24 |
| 10 | 8 | 2 | 6 | 5.28 | 1 | 5.28 | 1.65 | 18 | 95.04 |

## SHAFT BLASTING

As the shaft, transition structure, and tunnels were all excavated in amphibolite and gneiss using drill-and-blast method, a unique challenge was introduced as a result of the close proximity to the two existing pumping stations and pump wells. The pump wells and housed pumps extend to depths of 250 feet parallel to the shaft. A significant risk was identified as minor vibrations at the pump wells may have been able to damage the pumps causing a tremendous impact and a potential of major disruptions of water supply to users.

## Optimizing Shaft Blasting Design

To prevent damage to these pumps, a comprehensive monitoring program was implemented for measuring and analyzing the resulted blasting vibrations. This consisted of multiple seismograph monitoring points and a thorough test blasting program, through which a vibration classification of the rock was dynamically developed. The test blast program consisted of 16 initial blasts ranging in size from less than 1 lb/delay to 9 lbs/delay. Table 1 shows the first 10 blasts in the test blast program with varying hole lengths, powder columns, and pounds per delay for each blast. Based on the results, the optimal powder column length of 5 feet was selected for a total of 4.5 pounds per 8-foot drilled hole. Seismographs were closely monitored during the test blast program to determine the maximum powder column without exceeding the allowable maximum 0.5 peak particle velocity (PPV) at the surrounding structures.

Each of the blasts in the test blast program was plotted on a regression curve. The regression analysis plotted the scaled distance versus PPV which was read from each seismograph. Ideally, seismographs would be installed in a cross pattern propagating from the shaft in four perpendicular directions. Due to the confined work area and close proximity of the blasting to Lake Mead, the ideal layout for the seismographs to perform a regression analysis could not be utilized because many seismographs would have required placement in inaccessible locations. Alternatively, seismographs were installed within and outside the existing pumping stations near the closest most critical structures and pumps.

In practice, readings from the installed seismographs were routinely plotted on the same logarithmic graph showing scaled distance versus PPV. The graph was consistently used to derive the anticipated PPVs per determined distance. From this graph, two constants were determined: (n) the slope of the graph, and (k) the Y-intercept of the line where scaled distance equals one. From this information, the maximum charge was derived and adjusted depending on the distance from the source of the blast to the surrounding structures. The following established equation was used (Dyno Nobel, 2009):

$$PPV = \frac{\sqrt{\text{pounds per delay}}^n}{\text{feet to seismograph}} * k$$

Regression graphs were only used to calculate predicted vibrations when the $R^2$ value was close to the value of one. The closer the $R^2$ value was to the value of one, the less deviation existed between seismograph readings. Using a regression analysis it could be determined that the pumping stations and pump wells would be subjected to a maximum PPV of 0.5 in/sec with a charge of 9 pounds per delay. Therefore, a maximum of two holes per delay was used when timing the round (Figure 1).

Close analysis was also focused on the blast pattern. During the blast design, it was recognized that an optimum amount of relief would grant a production efficiency and reduced vibration. Hence, a half

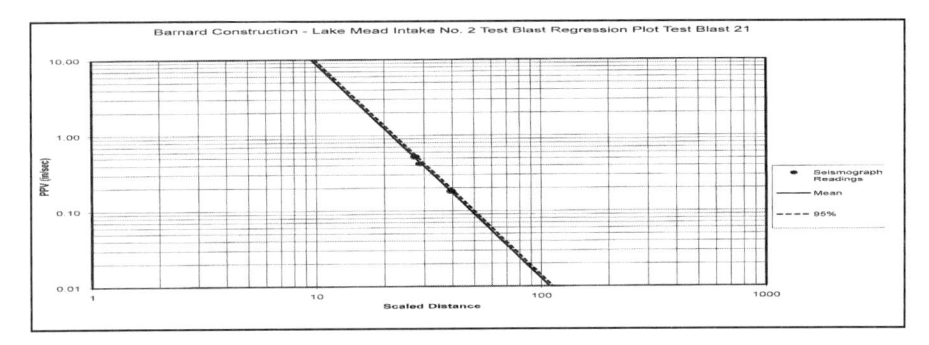

**Figure 1. Regression analysis**

shaft benched method of excavation was selected, as illustrated in Figure 2. The typical blast pattern is shown in Figure 3. The designed blasting pattern proposed an optimal angle of relief for the blasted holes. Using this blast pattern, a PPV of well under 0.5 in/sec was anticipated at the pumping stations and pump wells.

### Execution of Shaft Blasting and Construction

Following the design, blast holes were 1.75 inches in diameter and 8 feet long each. In the shaft, blast holes were drilled using sinking hammers. The production holes had a powder column of 5 feet with 3 feet of stemming, loaded with Dyno Xtra. Perimeter holes were angled out 2 to 4 degrees. They were loaded with DynoSplit D. Perimeter holes were spaced at 1.5 feet rather than the 3-foot spacing of the production holes. This spacing reduced the overbreak in the shaft.

A non-electric initiation system was used with 18 grain detonation cord and long period delays. Each blast required two to three loops of detonation cord delayed with a 42 ms nonel TD. This allowed up to 45 delays, rather than a maximum of 15 delays with one loop. Each blast was covered with blasting mats to prevent flyrock until an adequate depth was achieved. After the smoke had cleared, crews performed scaling, installed any needed ground support, and mucked out the shaft using mini excavators and a muck bucket. Two blasts were routinely achieved in the shaft per day when working a 24-hr work day. This fit well within a specified 12-hr daily blasting window but left little room for delays. Following shaft excavation, a final reinforced cast-in-place lining was placed in the shaft.

## TRANSITION STRUCTURE AND TUNNEL BLASTING

The transition structure at the invert of the shaft was 26 feet high at maximum height, 22 feet wide, and 90 feet long. It was excavated in three lifts due to the height of the cavern and size of equipment used to drill and muck. The blast holes at the transition structure were drilled using sinking hammers and jacklegs. Blast rounds ranged from 4 feet to 10 feet in length. It was not until the entire transition structure was excavated that a drill jumbo could be lowered into the shaft to be assembled and complete the remaining portions of the tunnels.

### Optimizing Tunnel Blasting Design

Blasting within the transition structure and tunnels became more critical than the shaft since the tunnel blasting face was closer to the pumps at several points in the tunnel than at any point in the shaft. The closest distance between the blasting face to the pumping equipment was 120 feet. The blasting was designed to achieve a high production while mitigating the risk of damaging the pumps. Compared to the explosive load during shaft excavation, a higher amount of explosive per delay of 25 was used within the tunnels and transition structure. To monitor the potential differences in vibrations propagated by horizontal blast holes versus vertical blast holes, two geophones were installed within the shaft. As both were located in the shaft, one was located 6 ft above the transition structure; the other was located at the same elevation of the well pumps. A small-scale test blasting program was also conducted in the transition structure.

To develop the optimal blasting pattern, quantity of explosives, and loading, a plan similar to the Test Face Program outlined in "Construction Vibrations" by PhD Charles Dowding, was adopted. This advocated blast monitoring method optimizing the implemented blasting technique by analyzing the results of minor adjustment in blast pattern and loading schemes. Each blast design scheme was monitored to determine which blast had the least amount of overbreak beyond the neat line excavation, the best-sized muck for removal with the available equipment onsite, and the least resulting vibration as monitored at the seismographs located both on the surface and in the shaft. Note that the Test Face monitoring program was slightly modified as full-sized rounds were drilled rather than shorter rounds as suggested by the program.

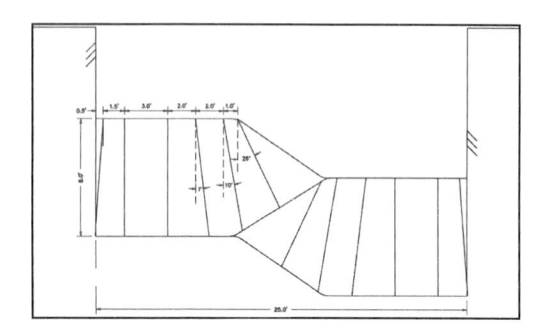

**Figure 2.** Side view of shaft blasting pattern

A close awareness for reduction in vibrations was accounted for in the design, and successfully accomplished in the field. Opposed to the shaft, the tunnel cross section has a limited area for blast relieving; therefore lower production and higher vibrations were a concern. To compensate for the lack of relief area, burn holes were intensively used. Initially, 4 holes 1.75-inch-diameter each were drilled for the burn. This burn often left bootlegs of about 2 feet in length. The burn holes were consequently switched to 3 holes of 3-inch-diameter. As experienced, the increase in relief by doubling the area tremendously improves the blasting performance from the production and vibration prospective. The typical blast pattern had a hole spacing of 3 feet with a powder factor of 4 pounds per CY. This pattern was changed to a tighter spacing with less powder in each hole with no change in the powder factor. Using a tighter hole spacing produced smaller-sized muck, and the reduced amount of powder in each hole mitigated the vibrations. Figure 4 shows the typical tunnel blast pattern used in the tunnel.

It was found that the largest vibrations recorded by the seismographs often occurred on the first delay of the blast, as shown in Figure 5. The first delay was designed to have the least charge per delay of the entire blast due to the limited relief for the blasted material. Upon switching to 3 each 3 inch diameter relief holes the high vibrations due to the limited relief of the first delay were reduced as shown in Figure 6. The burn holes were enlarged not only for better round pull, but also for preventing damage to any surrounding pumps and underground structures resulted by the reduced vibration.

Initial analyses of blasted round indicated that excessive vibrations extended beyond the perimeter holes. This could be recognized by the increased overbreak and lack of half-cast signatures on the exposed rock surface. This was not only a vibration issue, but also a safety issue due to increased fractures in the rock at the tunnel crown and walls. Subsequent to a thorough evaluation, the blast pattern design was modified for resolving the resulted

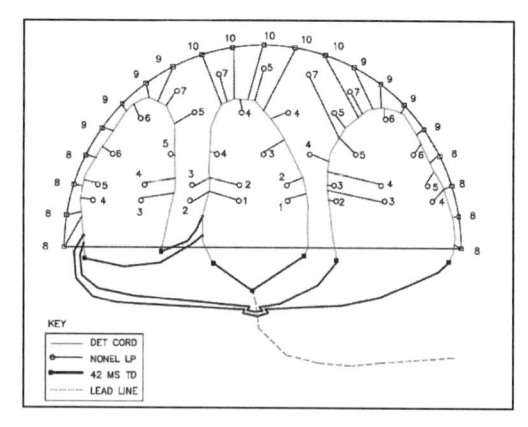

**Figure 3.** Plan view of shaft blasting pattern

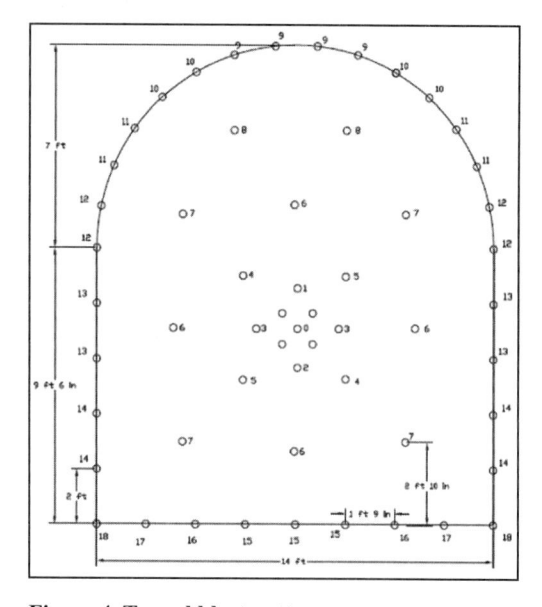

**Figure 4.** Tunnel blast pattern

overbreak. The row of holes adjacent to the perimeter holes were moved further from the perimeter holes to prevent fractures from propagating beyond the perimeter holes. The perimeter holes were also spaced tighter, loaded with less Dynosplit D, and timed so more holes were initiated on the same delay. When timed, loaded and laid out in this array, the cracks propagated from one perimeter hole to the next, limiting fractures and overbreak beyond the perimeter of the tunnel and reducing the amount of ground support needed.

### Execution of Tunnel Excavation

Blasting schedule was a crucial issue in the tunnel operation due to the 12-hrs limited blasting window

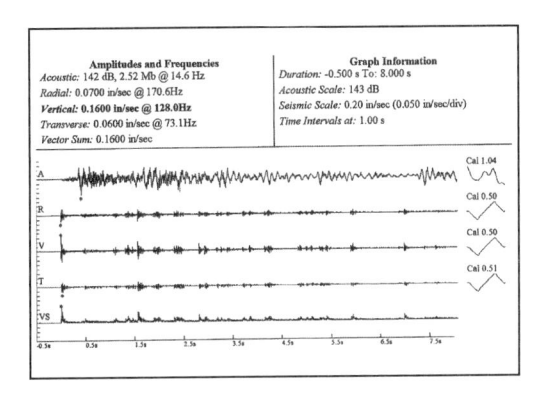

**Figure 5.** Vibration readings from a blast with limited relief

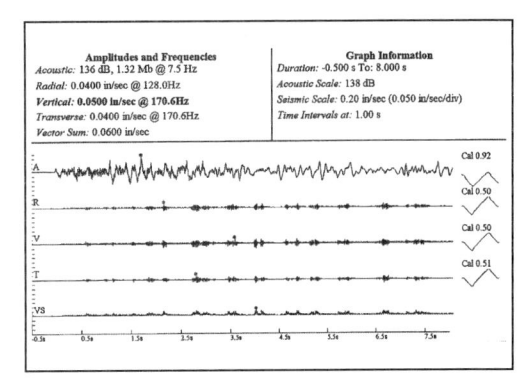

**Figure 6.** Vibration readings from a blast with adequate relief

as permitted in the specified work. Since the blast cycle could be up to 18 hours, the blasting window frequently forces the blast to be postponed until the following day causing delays and reduced production. Barnard worked closely with SNWA to analyze the effects of tunnel blasting on the pumping operation for improving the resulted reduced progress. After compiling and analyzing all seismograph data and deriving the "best fitted" regression, it was recognized that a maximum charge of 25 pounds per delay would eliminate the potential of pump damage. Barnard and SNWA provided expert consultation for determining the impact of the proposed blast design on the surrounding equipment and structures. An engineering evaluation and cooperation between the owner and contactor allowed for temporarily suspension of the blasting window limitation with no shutoff of pumps, thus improving the excavation progress on the project.

Similar to the shaft rounds, the tunnel rounds used Dyno Xtra in the production holes and Dynosplit D in the perimeter holes. The rounds were initiated non-electrically with long period delays and 18-grain detonation cord. Many of the shots in the transition structure were so large, delay wise, that three loops of detonation cord had to be used, allowing for 57 delays. Each loop was delayed with a 42 ms non-electric TD to keep the maximum pounds per delay beneath the 25 pounds per delay established limit. While using three delayed loops in a tunnel round, close attention was paid to tying the round in to ensure that no more than three holes initiated simultaneously.

A two-boom jumbo drilled the remaining tunnel rounds once enough room had been made to assemble and maneuver the machine underground. The jumbo was soon followed by a small scooptram.

The perimeter holes in the tunnel looked-out 2 to 3 degrees to create enough room to drill the following round. Two operating headings provided the ability to alternate between drilling in one heading and ground support installation and mucking in the other. On average, two blasts per were achieved per day with a 24 hour work day.

## CONCLUSION

Although many people flock to Las Vegas to gamble, Barnard of Nevada, Inc. came to Las Vegas with the intention of taking no risks. The Lake Mead Intake No. 2 Connection and Modifications Project was successful in protecting surrounding critical pumping equipment and structures during close-proximity blasting. With a proven cooperation between the SNWA and Barnard, each round was analyzed and adjusted to reduce vibration, decrease overbreak, and increase excavation production. Careful tracking and analysis of each blast provided Barnard the ability to successfully predict the propagation of vibration through the surrounding rock. Detailed analysis of blast patterns, relief, loading, and timing of each round showed which method of blasting created the least amount of vibration. Using these results, Barnard was able to reduce the amount of vibration from each blast, allowing for larger round length and greater production.

## REFERENCES

Dowding, Charles, *Construction Vibrations*, 2000.
Dyno Nobel, *Explosives Engineer's Guide*, 2009.
Bickel, John and Kuesel, T.R., *Tunnel Engineering Handbook*, Krieger Publishing Company, Malabar, Florida, 1992.

# TRACK 4: CASE HISTORIES

Larry Eckert, Chair

*Session 4: Conventional Tunneling*

# A Conventionally Tunneled River Undercrossing

**Adrian A.J. Holmes**
Shannon and Wilson, Inc., Portland, Oregon

**Sangyoon Min**
Parsons Corporation, San Diego, California

**Klaus G. Winkler**
Shannon and Wilson, Inc., Seattle, Washington

**Jim Brunkhorst**
Kiewit Pacific Co., Omaha, Nebraska

**ABSTRACT:** The Portland Water Bureau opted to relocate two of Portland's primary water supply conduits from an aging and vulnerable 113 year old bridge into a tunnel beneath the Sandy River. Two replacement 6-foot diameter gravity pipelines drop through 80 to 100-foot deep shafts on either side of the river channel, and pass beneath the river in a 16 ft wide by 10 ft high by 400-foot long tunnel. The tunnel was advanced conventionally by road header, 40 ft below the lowest known paleo river channel. The open tunnel approach was chosen over microtunnel options, and measures were taken to reduce the risk of an open excavation. Site geology consisted of low strength sedimentary rock overlain by variable thickness alluvial deposits. A comparison of expected conditions versus encountered ground, on-site condition assessment, and approach to overcome design and construction challenges are presented.

## PROJECT BACKGROUND

The primary source of water for the City of Portland, Oregon is the Bull Run watershed, which is located about 25 miles east of the city near Mt. Hood. Within the watershed, surface water and runoff is stored in 2 reservoirs, each created by a dam. Water from the Bull Run reservoirs is treated in the watershed and transported to distribution reservoirs in Portland through 3 large-diameter gravity pipelines. At the lowest elevation of the conduit alignment between the Bull Run watershed and Portland, the conduits cross over the Sandy River on two aging pin-truss steel bridges. One of the bridges, adjacent to a public park (Dodge Park), parallels a local roadway bridge and carries 2 of the 3 conduits. The third conduit lies on a smaller pipeline bridge that crosses the Sandy River nearly 0.5 miles downstream. The two conduits paralleling the local roadway bridge carry about two thirds of the Bull Run system's capacity. Their location adjacent to a local roadway on a 113 year old bridge make them vulnerable to both malicious anthropogenic activity and natural catastrophes such as torrential floods, landslides, volcanic lahars and earthquakes.

In March 2007, the City of Portland Water Bureau solicited design-build teams to relocate the two water conduits sharing the local roadway bridge into a tunnel underneath the Sandy River. Their decision to relocate the conduits underground was based on numerous feasibility and preliminary engineering studies dating back to March of 1998. While the City's primary goal was to reduce the conduits' vulnerability, they also wanted to complete the project using a quality, cost-effective, and safe approach with minimal impacts to the environment, local community, travelling public, and adjacent park.

### The Decision Process

The local geology consists of low strength sedimentary rock overlain by variable thickness alluvial deposits, as presented in a Geotechnical Baseline Report from 2007. Three design-build teams submitted cost proposals to the City of Portland to build the project. Two teams planned to construct the new conduit undercrossing using microtunneling and pipe jacking techniques. The third bid, submitted by Kiewit Pacific Co. (KPC), offered a conventional alternative, proposing to excavate the tunnel using

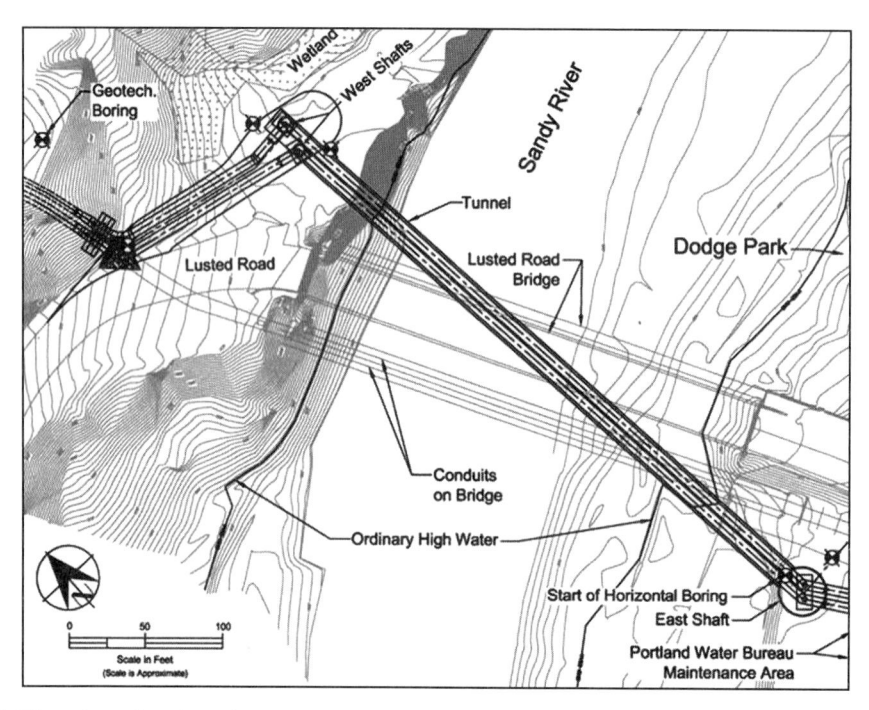

**Figure 1. Plan view of project elements**

a road header machine. While the conventional proposal was the most expensive bid, it provided clear advantages that ultimately led the City of Portland to choose it over the microtunneling approaches. The first advantage was security. The proposed depths combined with the annular concrete surrounding the steel pipes provide additional security against scour due to 500-year storm events and volcanic lahar events. The second advantage was the flexibility and ability to reduce construction risk that is provided by conventional excavation operations. The geotechnical documents and preliminary engineering provided by the City indicated the possibility of encountering boulders. Conventional tunneling provides the flexibility to better handle boulders without the risk intervention should the microtunneling machine encounter obstructions, particularly when tunneling below Sandy River. The third advantage was a reduced footprint on the west bank of Sandy River where space is limited between the river's ordinary high water and canyon walls. Another advantage was the elimination of slurry treatment facilities with substantial footprint and risk of river contamination. Figure 1 shows the project area, being constrained by a winding and limited local roadway shoulder, a public park, and a Portland Water Bureau maintenance facility. Since equipment could be moved both directions in a conventional tunnel at any point during excavation, the same shaft could be used for equipment entry and exit, eliminating the need for a large-diameter retrieval shaft on the west side of the river.

## Conventional Tunneling and Shaft Construction Approach

The conventional tunneling approach was started by constructing a 30-foot diameter, 80-foot deep shaft on the east side of the river. The east shaft was excavated with a track-hoe and supported by pre-assembled steel liner plate and ring beams, forming a steel can in the alluvial section. Shotcrete support was utilized in the low strength mudstone. All equipment and personnel required for tunneling utilized the east shaft for entry and exit. The 16 foot wide, 10 foot tall, and 405 feet long tunnel was excavated using an Alpine AM50 conventional roadheader machine and was supported using lattice girders and shotcrete. On the west side of the river, near its terminus, the tunnel opened into an 18-foot wide, 13-foot tall, 30 ft long cavern intersected by two 100-foot deep drilled shafts; one 9-foot diameter and one 11-foot diameter. Figure 2 shows a profile view of the excavations in relation to the general topography and geology of the site. The twin 6 foot diameter welded steel pipe conduits were then installed through the shafts and tunnel, which were later backfilled with concrete.

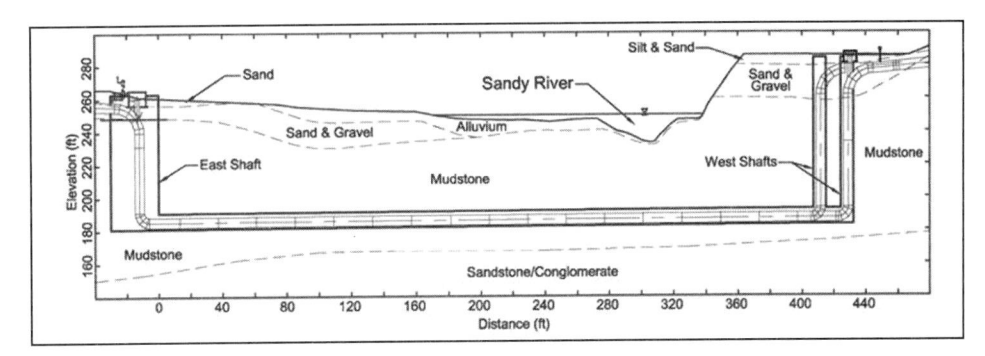

Figure 2. Profile view of project elements shown with respect to generalized geology

## GEOTECHNICAL STUDIES

### Explorations

Shannon & Wilson, Inc. (S&W) performed final design geotechnical explorations, instrument installation and in-situ testing near each of the proposed shaft locations and directly above the proposed tunnel. The final design explorations supplemented data provided in the geotechnical baseline report with additional data obtained directly adjacent to the proposed structures. To support final design, six geotechnical borings were drilled at the locations shown in Figure 1. One boring was drilled adjacent to the east shaft. Two boring were drilled near the west shafts. One boring was drilled in the slope above the west shafts, and one was drilled west of the shafts in the vicinity of a proposed thrust block. The borings on the west side of the river were performed both for information at the proposed structure locations and for overall slope stability analysis. In the material overlying the mudstone, the borings were advanced using mud rotary techniques and standard penetration tests were conducted. After reaching the mudstone contact, HQ coring was typically used to obtain continuous core specimens. During construction, one boring was drilled horizontally from within the east shaft approximately ten feet above tunnel alignment. In each of the borings, S&W installed geotechnical instrumentation for monitoring groundwater levels, slope movements, and/or settlement prior to and during construction.

The mudstone encountered in the borings, known as Sandy River Mudstone, is a generally uniform, very low to low strength, gray, near-horizontally bedded, fresh to slightly weathered, and micaceous mudstone. Structurally, it contains widely spaced, discontinuous, generally clean, tight, slightly rough to slickensided, planar joints. Core samples parted easily along wide spaced bedding planes. Composition of the mudstone is predominantly silt with clay and a lesser percentage of fine sand. The sand content was variable, and occasionally dominated in 1 to 4 foot thick poorly indurated sand lenses.

### Testing

#### In-Situ Testing

Groundwater was of particular concern in planning the tunnel excavation, particularly the risk of an undetected river connection. The hydraulic conductivity of the mudstone was tested initially during preliminary design studies and presented in the Geotechnical Baseline Report. Preliminary design testing yielded hydraulic conductivity values less than $7 \times 10^{-5}$ centimeters per second. The design-build team's concern focused on the possibility that the tunnel might encounter one or more joints or fractures that might be hydraulically connected to the Sandy River. To address this issue, S&W performed a series of large-interval packer tests in the horizontal boring, located about ten feet above the proposed tunnel crown. Results indicated mass hydraulic conductivities of $1.5 \times 10^{-5}$ to $3.7 \times 10^{-5}$ centimeters per second, further supporting the assumption that groundwater inflow would be minimal and that it could be managed by a sump pump at the bottom of the east shaft.

#### Laboratory Testing

A suite of laboratory tests were performed on core samples obtained from the geotechnical borings to provide parameters for engineering and design. Tests included moisture content and Atterberg Limit determinations, grain size analyses, swell tests, and unconfined compressive strength testing. Moisture contents in the mudstone ranged from about 24 percent to 45 percent. Atterberg Limit tests indicated plasticity indices from about 21 percent to 29 percent. Grain size analyses indicated the typical sand content of the mudstone to be about 10 to 15 percent. Swell tests in the mudstone indicated swell pressures ranging from 100 to 1,700 pounds per square foot. Unconfined compressive strengths of tested mudstone cores ranged from about 50 to 495 pounds per square inch.

## Instrumentation

### Vibrating Wire Piezometers

Five vibrating wire piezometers were installed: one at the gravel-mudstone contact in the boring on the slope above the west shafts; two in a boring adjacent to the west shafts (one at the gravel-mudstone contact and one deep in the mudstone); one in the boring adjacent to the thrust block; and one in the boring adjacent to the east shaft, set deep in the mudstone. Having vibrating wires installed at various depths across the site allowed us to monitor pore water pressures in the alluvium and mudstone prior to and during construction. We observed groundwater both deep within the mudstone and perched on top of it in the alluvium.

### Standpipe Piezometer

One shallow open-tube piezometer was installed in the boring near the east shaft, screened in the gravels directly above the Sandy River Mudstone. This provided a way to manually monitor the elevation of water perched on top of the mudstone, and also allowed us to conduct falling-head permeability tests on the gravel unit to estimate dewatering requirements for the upper portion of the east shaft excavation. Based on the falling head test, a hydraulic conductivity of $4 \times 10^{-4}$ centimeters per second was estimated for the alluvial gravel.

### Inclinometers

Slope inclinometers were installed in the boring on the slope above the west shafts and in the boring between the west shafts and Sandy River bluff. These instruments allowed us to monitor slope movements above and below the west shafts prior to and during construction.

### Vibrating Wire Settlement Profiler

Components for a vibrating wire settlement profiler were installed in the horizontal core boring, approximately 10 feet above the tunnel crown. The in-hole components consisted of one 2-inch schedule 80 polyvinyl chloride (PVC) pipe for the settlement profiler probe, one ⅜-inch schedule 80 PVC pipe for a wire pull cable, and one dead-end pulley assembly at the end of the hole to transfer the wire pull cable from the 2-inch probe pipe to the ⅜-inch return pipe. The instrument consisted of a vibrating wire pressure sensor connected to a vented signal cable and a liquid–filled tube mounted on a portable reel. The reel contained a reservoir with sight tube and leads for a vibrating wire readout box. The reel was mounted on a pedestal located on a platform secured to the wall of the east shaft, above the tunnel. To take readings, the sensor was positioned at repeatable locations along the PVC using the pull cable. The instrument would be allowed to stabilize and a vibrating wire measurement of head pressure would be recorded. By keeping the fluid in the sight tube at a consistent elevation, changes in the elevation of the probe at various positions along the casing could be measured at regular time intervals during tunneling. The best repeatability of readings we were able to achieve prior to tunnel excavation was approximately ±0.7 inches. Unfortunately, due to friction and temperature affecting the density of the fluid in the long fluid cable, the entire hole needed to be read at the same interval each time to obtain repeatable results. The purpose of the instrument was to determine the extent to which settlements observed by convergence measurements in the tunnel extended above the crown and, as such the instrument was well suited to its purpose. However, it was time-consuming to read and sensitive to the operators' practices and background activity. No measureable deflection of the settlement sensor was observed during tunnel construction.

## DESIGN

A preliminary Geotechnical Baseline Report was provided by the Portland Water Bureau with the bidding documents. Details of the Geotechnical Baseline Report are described in detail in these proceedings by Collins et al. (2010). After the supplementary geotechnical investigation described above, a final baseline report was prepared with the expected geotechnical and tunneling conditions.

The new geotechnical borings were evaluated and the baseline conditions and weak rock/firm ground designations were refined in terms of tunneling. The Rock Types I (very good) through V (very poor), based on RMR and Q ratings, were provided and the various loading conditions were implemented into the design. Two approaches were used to design the initial support, which consisted of shotcrete and lattice girders. The first method (Harrison, 1993) was a conventional static analysis using gravity load factoring based on Terzaghi rock loading for projects in similar weakly cemented siltstone and sandstone. With this simplified method, the maximum moment, thrust and shear capacities of shotcrete and lattice girders were compared to the induced loads in the liner for the worst rock loading case, Rock Types V.

The second method used a finite element method (Plaxis) to determine the load reduction factor considering stand-up time (time lag for support), analysis of stress/strain profiles around the tunnel opening, and the structural forces around the tunnel support. The two critical loading conditions were for

RMR Class IV, poor rock, and swelling ground conditions, which were expected based on the geotechnical borings. The analysis of the swelling ground conditions are discussed in Min and Kaneshiro (2010). The design analysis incorporated standup time (lag time) and the relaxation parameter (i.e., load reduction factor) to determine the loading and stress concentrations in the ground and on the liner based on the excavation and support cycle.

## Numerical Model

To consider three dimensional arching effects for surrounding rock, the tunnel construction process was modeled in two different stages (Phase I: tunnel excavation, Phase II: installation of lining) using the load reduction factor method, or the so-called Beta ($\beta$) method (Schikora and Fink, 1982). This is so because tunneling causes a transfer of the ground load by arching to the sides of the opening, and at the heading the arching effect is three-dimensional, locally creating a ground dome in which the load is arched not only to the sides but also forward and back. The load reduction factor is directly related to time-lag of the tunnel support (i.e., stand-up time). The initial geostatic stresses ($p_k$) acting around the tunnel location prior to the mining process was divided into a part (1-$\beta$) $p_k$ which was applied to the unsupported excavated tunnel face and a portion of the initial stress $\beta$ $p_k$ was gradually transmitted to the tunnel lining according to the construction staging. The two phases of the load reduction method, being related to the so-called ground-response curve or Fenner-Pacher curve are shown in Figure 3. Inside the tunnel a support pressure of the amount $\beta$ $p_k$ (with $0 < \beta < 1$) is left to account for the missing three dimensional arching at the tunnel heading, and for the effect of the time lag of the support. The amount of support pressure $\beta$, which determines the moment of placement of tunnel lining installation, directly influences the magnitude of both settlements and structural forces.

A larger $\beta$ factor corresponds to an early installation (i.e., $\beta$=0.999 means immediate support) of lining with larger structural forces acting on the support but less displacements. On the other hand, a smaller factor indicating a delayed installation of lining (before a tunnel collapses) leads to lower structural forces in the lining and larger displacements correspondingly. This is so because the surrounding rock takes greater portion of the initial stresses around the tunnel, but smaller portion of the initial stresses is acting on the lining. Theoretically, the optimum time for installation of support ("best support") is determined based on the ground reaction curve and stiffness of support when the load reduction factor was at the boundary between elastic and plastic behaviors (see Figure 4). Reaching this best support case, however, is not always realistic in the field. Nevertheless, lag time factors were based on case histories including displacement rate of the rock or the unsupported length of tunnel, and the excavation support cycle.

Immediate supports for the springline and the side walls were recommended for Rock Type IV since the simulation results showed that the stress concentration factors (compressive strength/maximum stress, $q_u/\sigma_{max}$) for "$\beta$ for the best support" at these locations were lower than 1.5~3 (see Figure 5 for stress profiles around tunnel opening) where possible crack growth or loosening zones at the springlines and slabbing and slaking at the wall could occur. The simulation results also showed that the structural forces including the maximum moment, thrust and shear friction on the tunnel support were within the capacities of the shotcrete and lattice girders for Rock Types I ~ III while a closer spacing of lattice girders and 1-inch thicker shotcrete was required for Rock Types IV and V.

Based on the Plaxis results, the vertical displacements increase as the Rock Types become worse (i.e., III → V). For the same rock type, the vertical displacements increase as the load reduction factor, $\beta$ decreases with delayed support. The Plaxis analysis shows that very little to no settlement would be expected for the "best support" both at the crown and at the location of the settlement profiler (0.13~0.25-inches). Even for "delayed support (before collapse)," the displacements (0.3~0.85-inches) were less than 2-inches of threshold value which was set by using the ultimate strength concrete theory. While the best achievable accuracy of the settlement profiler was approximately 0.7 inches, the actual readings taken prior to and during tunneling showed no measurable movement, supporting the results of the FEM analysis.

## CONSTRUCTION

### Expected vs. Encountered Conditions

Conditions encountered in the shaft and tunnel excavations were, in general, consistent with those disclosed by the geotechnical explorations. The core samples of mudstone from the geotechnical holes well represented the character and quality of the mudstone encountered in excavation. The ground conditions encountered during construction were mapped daily, and disclosed generally one RMR rating higher than predicted based on the geotechnical borings. Consequently, additional as required support measures were not implemented for Type IV and V ground.

Figure 6 shows a comparison of the Rock Mass Rating (RMRs) range estimated for the tunnel based on the horizontal boring to RMRs observed during

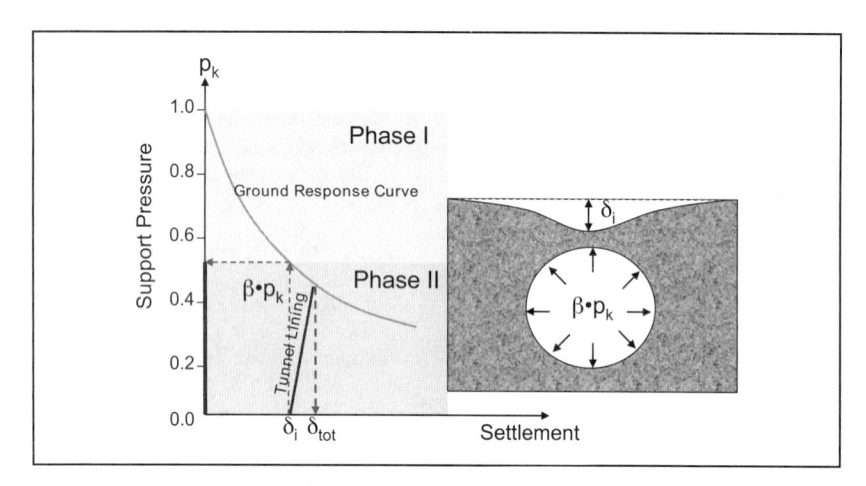

Figure 3. β (load reduction factor) method adopting ground response curve

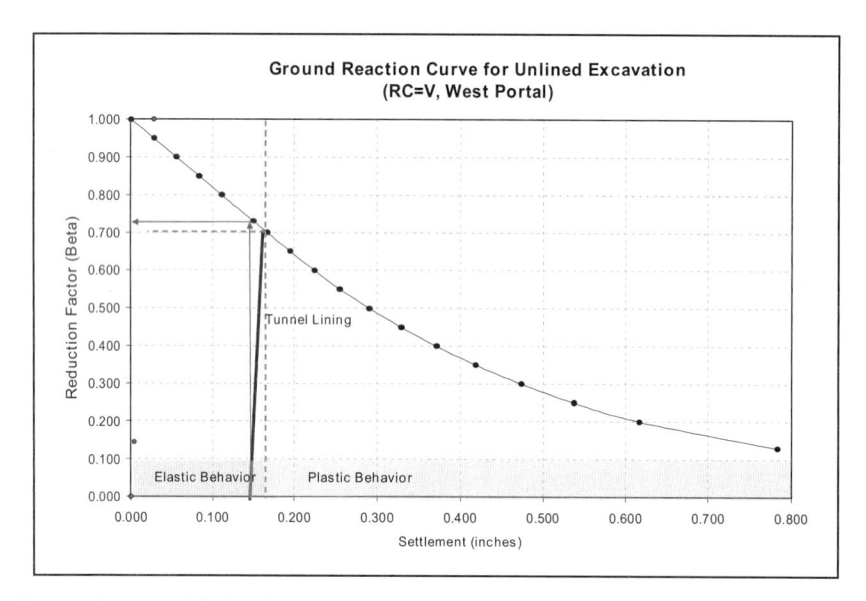

Figure 4. Determination of the load reduction factor for the best support

tunneling. The upper range of the estimated RMRs corresponded well with what was encountered. In retrospect, RMRs were conservatively estimated from the core. This is attributable to a desire to be conservative, as well as a slight degradation of the core quality resulting from the drilling process.

During the explorations, it was observed that the core samples parted easily along bedding. This observation proved particularly relevant to the tunnel excavation in that most overbreak occurred near the tunnel crown where 2 to 6 inch slabs of mubstone fell away, separating from the crown along bedding planes. Small wedges of similar thickness often fell out where weak bedding planes intersected joints in the quarter arches. The weak bedding planes, along which overbreak occurred, typically contained very thin beds of very fine cohesionless sand and silt or organic material. Often, the bedding plane partings displayed organic debris from leaves and twigs buried by subsequent beds. The only condition encountered in the tunnel that was unforeseen in the geotechnical borings was the presence of one erratic andesitic boulder, about 1.5 feet in diameter (which had no impact on the construction of the tunnel). Figure 7 is a photograph showing typical conditions at the face of the tunnel excavation.

Geotechnical studies had identified 1 to 4 foot thick sand lenses in the mudstone as a potential

| Horizontal Stresses Around Tunnel Opening | Vertical Stresses Around Tunnel Opening |

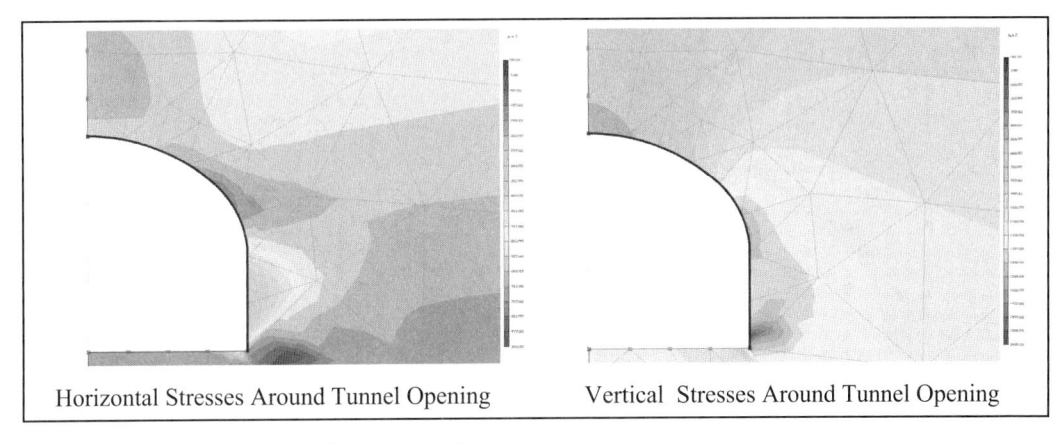

**Figure 5. Stress profiles around tunnel opening**

problem for shaft and tunnel excavation, either because they might produce more water than expected for the mudstone or because they might be a source of running ground in excavations. In open excavation, the sand lenses were only encountered in the shafts and actually posed few problems. While they appeared to lack cementation or induration in the borings, the sand materials did not run into the shaft excavations and only produced manageable groundwater inflow.

Groundwater inflow to the shafts and tunnel was close to that anticipated by the later explorations. The majority of the mudstone produced little to no water. Water that was encountered typically flowed through exposed joints, fractures, and occasionally through bedding planes. Where joints were encountered, inflows of 1 gallon per minute or much less were typical, even directly under the river channel. During the entire tunnel excavation, very few joints produced as much as 2 to 3 gallons per minute. Commonly, inflows would slow or stop during excavation, suggesting that the joint represented a finite source of water with no direct connection to the river or perched water.

After the completion of tunnel excavation, an as-built profile was created to summarize encountered conditions, including RMR, material contacts within the Sandy River Mudstone, and some of the data collected from prominent discontinuities observed during construction. A simplified version of the as-built profile is provided in Figure 8.

### Constructability

#### East Shaft

Excavation through the 14 feet of the overburden was accomplished with a Cat 320 excavator leaving a glory hole that the 10ft section of assembled liner plate can was set on. The planned liner plate support to −18ft was obstructed by an array of 3–4ft boulders nested above the mudstone which caused us to transition to the shotcrete support. Daily excavation and support progress through the mudstone averaged 4ft per day. Water infiltrations were encountered at the sand lenses which were treated by use of panning and cementatious grouting.

#### Tunnel

Mining of the mudstone with the AM50 proved to be a good selection as advancement of the excavation did not pose any problems with cutting rates. Forward probe drilling was performed at 100 foot reaches with 3–2inch holes located in the crown of the face. Water was encountered in 2 of the 3 probe holes at Sta. 2+50 that measured >5gpm. These holes were grouted to seal of the source.

Rutting of the invert was controlled by pouring an invert slab during the night shift at approximately 60 ft increments. This proved to be a benefit to the maintenance of the constructed tunnel for foot traffic and small consumables storage.

### LESSONS LEARNED

- The City of Portland wanted a conservative approach to the project and conventional tunnel excavation provided that. The conventional method easily managed the conditions encountered.
- The conditions encountered were generally better than those anticipated based on the Geotechnical Baseline Report and subsequent geotechnical borings, both because of a desire to be conservative as well as a slight degradation of the core quality resulting from the drilling process.
- The vibrating wire settlement profiler was installed for added security and conservatism in the event that substantial settlement

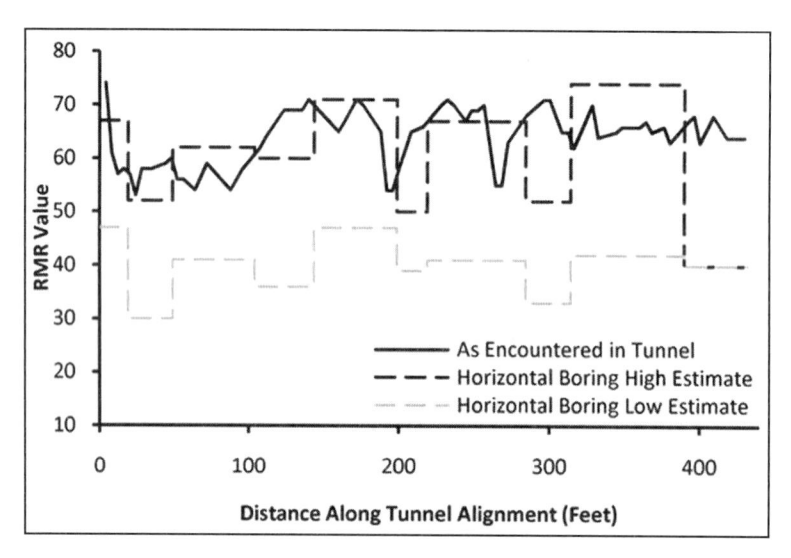

Figure 6. Graph comparing the RMRs anticipated in the tunnel based on the horizontal boring and RMRs actually observed during excavation

occurred in the tunnel crown. The minute settlements predicted by modeling were generally less than the instruments margin of error as used in the field.
- Design of the various portions of this project were packaged, which allowed for an early start of the east shaft.
- Successful execution of the design-build process requires a good design build coordinator and adherence to the processes set up in the design build documents.
- Early coaching of and by the owner to understand the design-build process to capitalize on the efficiencies of the design-build process is mandatory in order have a smooth flow of information.

## ACKNOWLEDGMENTS

The authors would like to thank their respective employers and the City of Portland for permission to publish this paper. Thanks is also given to G. Peterson, T. Nguyen, and D. Higgins of Shannon & Wilson; J. Kaneshiro, D. Duprey, L. Weinbrenner, M. Sinha, N. Ta, and P. Creegan of Parsons; J. Carlson, B. Mariucci, M. Hanley, B. Bridges, W. Robertson, C. Whitesel, J. Devers of Kiewit; and J. Horne and T. Willoughby of PB Americas.

## REFERENCES

Harrison, W. (1993), "Lattice girders for shotcrete support," in Wood, D.F. and Morgan, D.R. eds., Shotcrete for Underground Support VI, Proceedings of the Engineering Foundation Conference, ASCE Geotechnical Engineering Division, Niagara-on-the-Lake, Canada, May 2–6.

Collins, T. and Horne, J. (2010) "Managing risk in the procurement and design/build delivery of the Sandy River conduit relocation project," 2010 North American Tunneling Conference, Portland, Oregon.

Min, S.Y. and Kaneshiro, Y.K. (2010), "Rational design and construction approach for evaluating swelling ground and time effects on tunnel liners," International Tunneling Association (ITA) World Tunnel Congress 2010.

Figure 7. Photograph of typical tunnel excavation showing freshly exposed massive sedimentary bedrock, widely spaced joints, a lattice girder, and a flash-coat of shotcrete

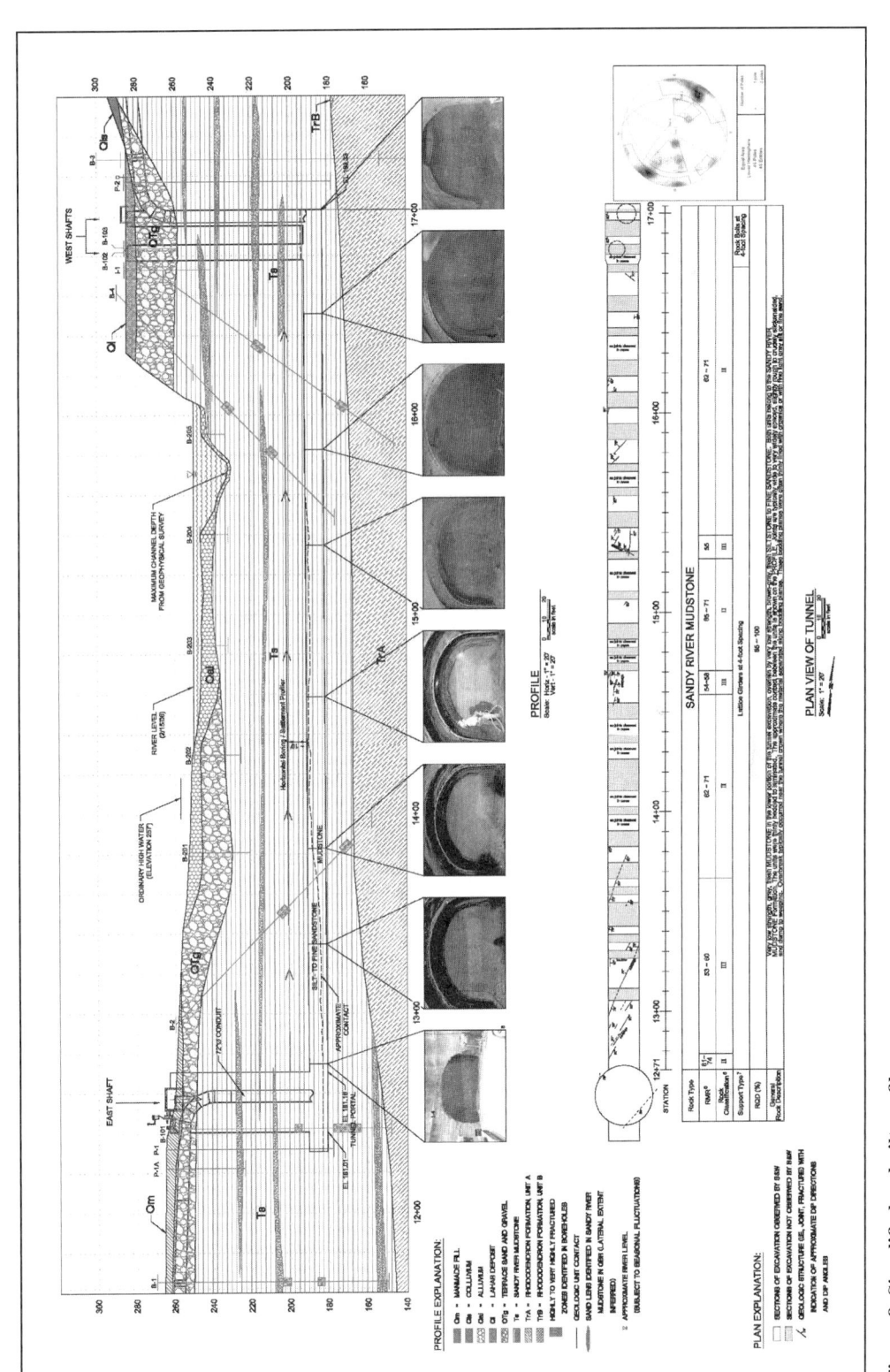

Figure 8. Simplified as-built profile

# Tunneling Ground Reinforcement by TAM Grouting: A Case History

## Ahmad Samadi, Gary Seifert
Philadelphia Water Department, Philadelphia, Pennsylvania

ABSTRACT: The objectives of this project were to reinforce a section of the decomposed rock above a 16.5-foot diameter tunnel and to improve the permeability of the zone above the tunnel to reduce seepage into the tunnel during boring.
The work was performed in two phases. Phase one consisted of installing seventy five (75), fifty foot long, 1-in. diameter fiberglass rods, spaced 3.5-feet apart in a fifty-foot section. Phase two consisted of injecting grout from the ground level through twenty seven (27) TAM pipes into the rock, over a section of about 225 feet over the tunnel.
To install the fiberglass soil nails, 6-in diameter cased holes were advanced to about 50 feet below grade. The soil nails were 1-in. diameter fiberglass bars and 50 feet long. The soil nails were placed in the holes and the boreholes were filled with 4000-psi grout. The installation of seventy five (75) soil nails took about 10 weeks.
The grouting operation involved injecting grout into pre-determined pattern locations horizontally and vertically into the rock formation using tube-a-manchette (TAM) methods, also known as permeation grouting technique. The grout was injected into the rock, below ground water table, starting from the spring line of the tunnel to about 30' above the spring line of the tunnel using cement grout material.
The TAM pipes were constructed of 1.5-inch diameter PVC material and were about 50 feet long. The TAM grout pipes were installed by advancing 6-in diameter holes, then setting the pipes in the boreholes. The annular space of each of the borehole was grouted with a brittle sheathing grout.
To determine the approximate in-situ permeability of the rock, a Falling Head permeability test was performed in each borehole prior to the TAM pipe installation.
The grouting operation typically began with a 3:1, water cement ratio. Grout was injected and thickened to 2:1, 1:1, and 0.5:1 until refusal was reached. A total of 250 cubic yards of grout was injected into the rock. The grouting operation took about 12 weeks.
A detailed quality control procedure was accomplished through the use of an automatic data recording system (HANY) connected to the grouting equipment. This equipment monitored grout intake and pressure over time, and provided a detailed record of the grouting operation.
Three months after the completion of the grouting operation a series of exploration was performed in the grouted area to determine and measure the effectiveness of the grouting operation. That exploration testing reveled presence of grout in the overburden soil and in some accessible joints within the rock and the permeability of the grouted zone were reduced by up to 500 times. The grouting operation achieved its objectives.

## BACKGROUND

Approximately 3600 linear feet of 12'-6" diameter cast-in-place concrete storm water conduit in rock tunnel was to be built under this Contract. The storm sewer conduit was to be constructed of a 16.5-ft diameter tunnel bored beneath Allegheny Avenue, excavated through rock using a TBM (Tunnel Boring Machine).

## INTRODUCTION

This project consisted of reinforcement of a section of the rock above the 16.5 foot diameter Dobson Run tunnel. For most of the tunnel alignment, the tunnel would be constructed with sufficient rock cover. However, during the preliminary soil investigation it was determined that some of the weathered rock above the tunnel was not as competent as expected and required reinforcement and improvement.
The improvement consisted of ground reinforcement by soil nails and grouting. The area to be improved was divided into two sections. Section one was 50 feet long and section two was 175 feet long. Section one required both Soil Nails and grout injection whereas section two required only grout injection. The work was performed in two phases. Phase-I

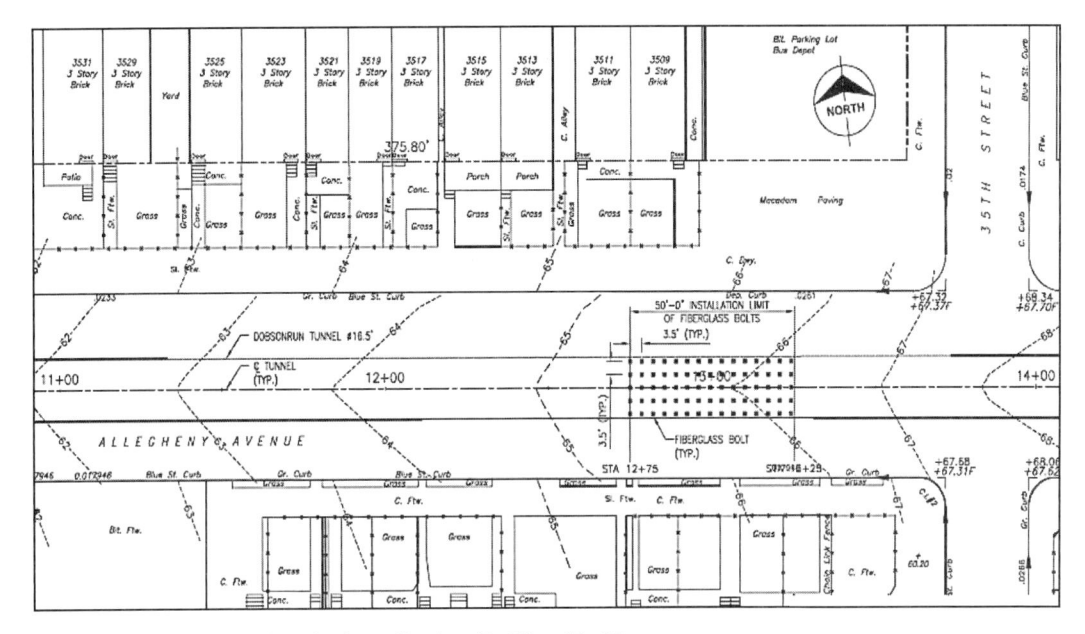

Figure 1. Location of soil nails from Station 12+75 to 13+25

consisted of installing seventy-five, fifty foot long, 1-in. diameter fiberglass soil nails spaced 3.5-foot apart in the fifty foot section (as shown in Figure 1).

Phase-II consisted of injecting grout into the rock from the ground level through 27 TAM pipes over a section of about 225 feet long over the tunnel, (as shown in Figure 2).

## PERMEATION (TAM) GROUTING

Permeation grouting is the injection of a fluid grout into granular, fissured or fractured ground to produce a solidified mass to carry increased load and/or fill voids and fissures to reduce water flow. It can be utilized in sands, gravels and coarser open materials, fissured, jointed and fractured rock. It was used in this project to improve the condition and stability of the weathered rocks with fissures, joints and cracks.

Permeation grouting is done mostly by utilizing "tube-a-manchette" (TAM) methods. Permeation grouting of the ground or rock is achieved by high-pressure injection through discreet ports at specified designed intervals, rates and pressures to fully treat the targeted areas. Permeation grouting also reduces the soil permeability by filling up its voids.

## GROUT INJECTION

After the drilling is completed and the slough removed, the hole is fitted with a TAM pipe. The sleeved-port grout pipe, also known as a tube-a-manchette, is a system consisting of a 1.5–2.0 inch diameter PVC pipe with rubber jackets covering small grout holes, which are equally spaced along the pipe. These jackets act as check-valves that permit the grout to flow only out of the pipe and preventing the grout getting back into the grout pipe.

A device called an internal packer (see Figure 3) isolates the immediate vicinity of the grout hole on the inside of the TAM pipe. When the rubber gaskets at both ends of the packer are inflated, the packer creates a tight seal with the inside wall of the TAM pipe.

After grout pipe installation the annular space between the TAM pipe and the borehole wall is sealed with a brittle cement-bentonite mortar, sometimes referred to as annular grout (see Figure 4).

During grouting, the injection pressure expands the rubber jacket that covers the grout holes and moves it away from the grout hole. The grout then fractures the mortar seal allowing grout to flow through into the soil or rock. After the grout is pumped from that confined pipe section, the packer is deflated and pulled up to the next sleeved-port for the subsequent injection. (See Figure 3)

## PHASE I: SOIL NAIL INSTALLATION

Seventy five (75) fiberglass reinforcing bars of 1-in diameter and 50 feet in length, referred to as Soil Nails (shown in Figure 5) were installed in a 3.5 ft × 3.5 ft grid between stations 12+75 and 13+25 as shown in Figure 1.

To install each soil nail a 6-in diameter cased hole was advanced using duplex drilling method to about 50 feet below grade as shown on the drawings below. The soil nails were placed in the boreholes

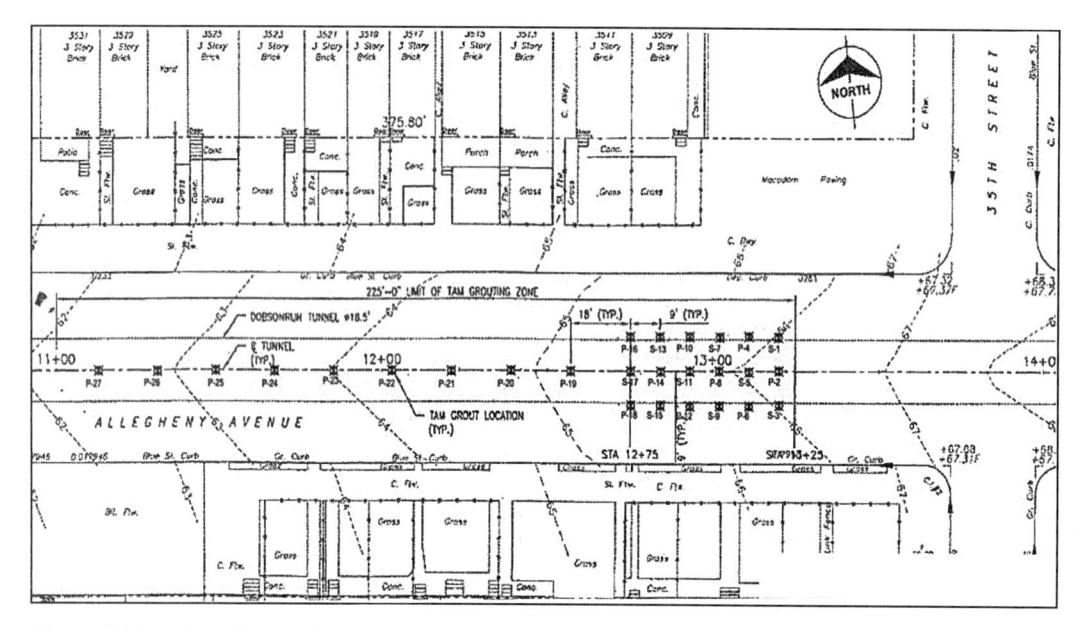

Figure 2. Location of TAM pipes from Station 11+00 to 13+25

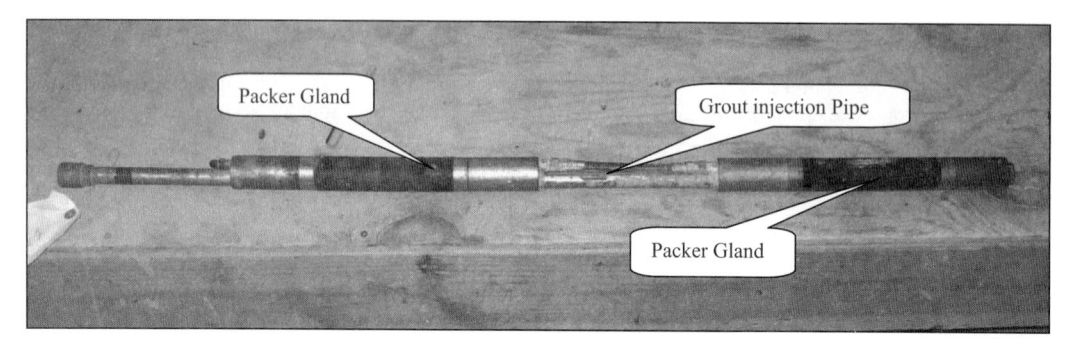

Figure 3. Section of a packer pipe

and the holes were filled with 4000 psi grout (see Figure 6).

## PHASE II: GROUTING

The grouting operation involved injecting grout into concentrated locations in the rock formation, starting from the spring line of the tunnel to about 30' above the spring line of the tunnel. To perform the TAM grouting eighteen (18) TAM pipes were installed in three rows spaced nine (9) foot apart in section 1 and in one row spaced eighteen (18) foot apart in section 2.

The TAM locations were designated as primary and secondary holes, with one hole being a primary and the next being a secondary hole, so that no two primary or secondary holes would be next to each other. The injection ports were about 3 feet apart and were designated as primary and secondary ports, with the first port being a primary port and the next a secondary, and vice versa, as shown in the schematic in Figure 7.

The grout was injected starting from the bottom primary port of primary TAM pipe moving upward to the next primary port; refer to the grouting sequence below for more information.

The TAM grout pipes were constructed of 1.5 inch diameter PVC material and were about 50 feet long. The TAM pipes were installed by advancing a 6-in diameter hole using duplex drilling methods, and then setting the pipe in the borehole. The annular space between the each TAM pipe and borehole wall was grouted with a brittle sheathing grout.

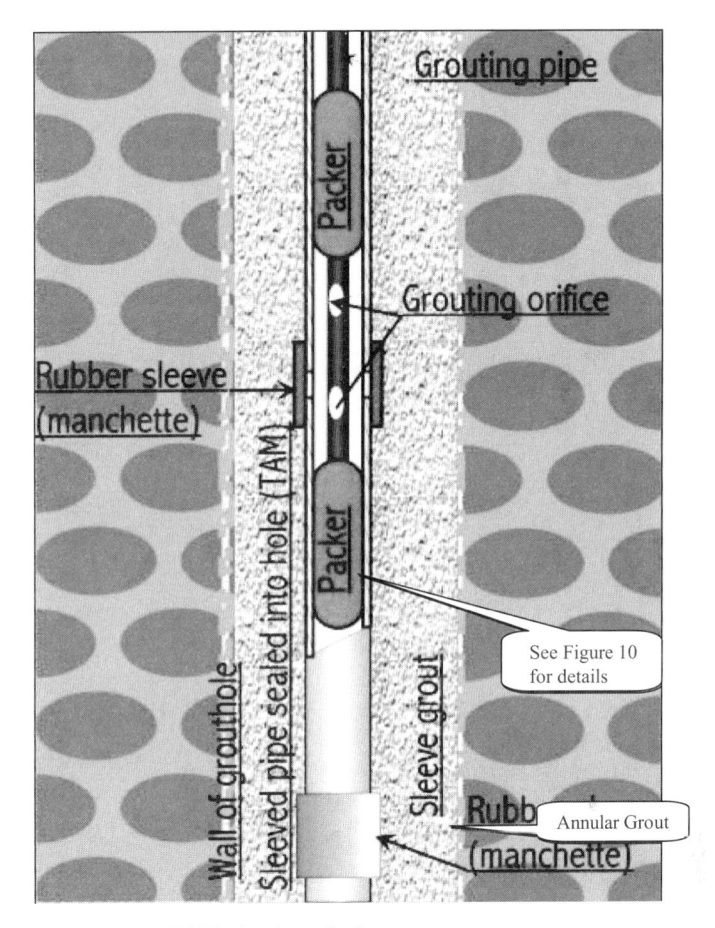

Figure 4. Schematic of a typical TAM pipe installation

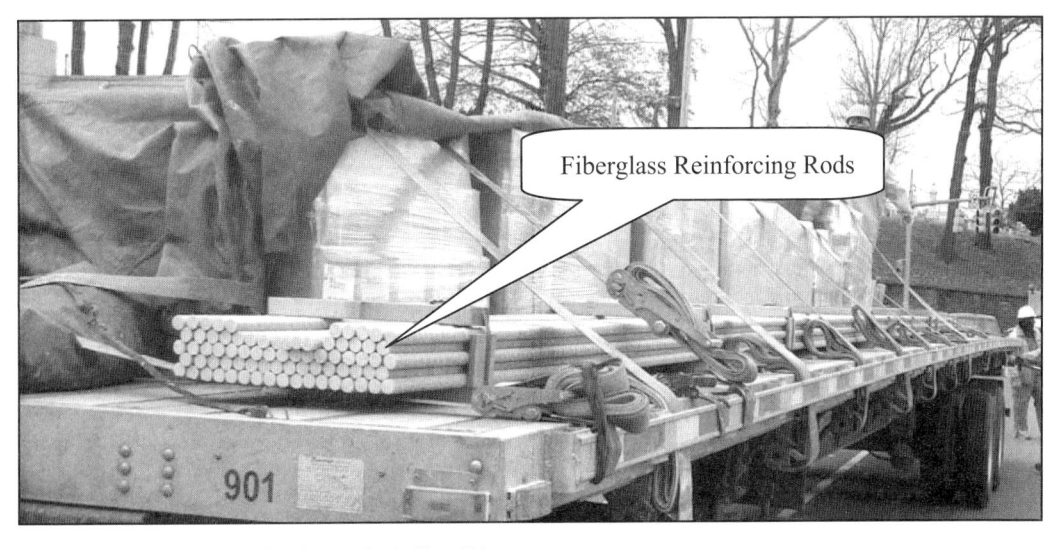

Figure 5. Fiberglass reinforcing rods (soil nails)

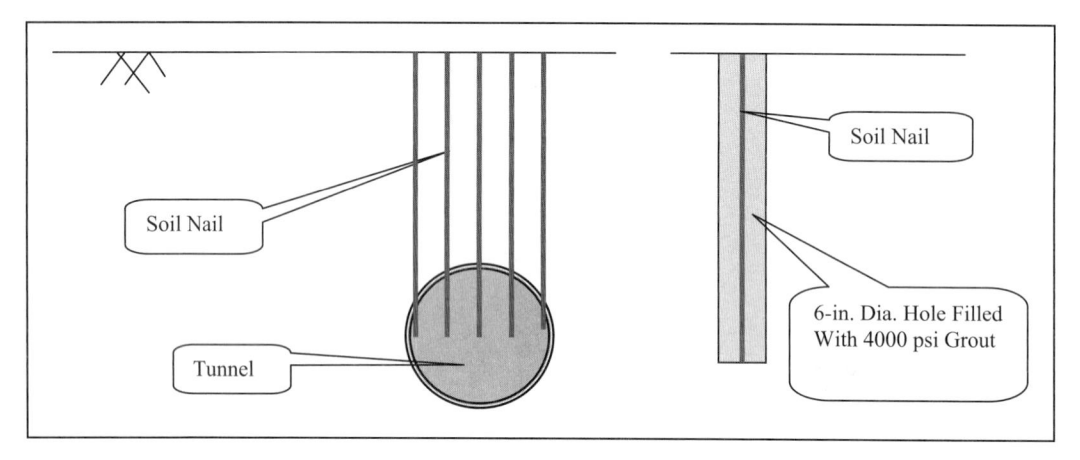

Figure 6. Typical installation of soil nails with respect to the tunnel

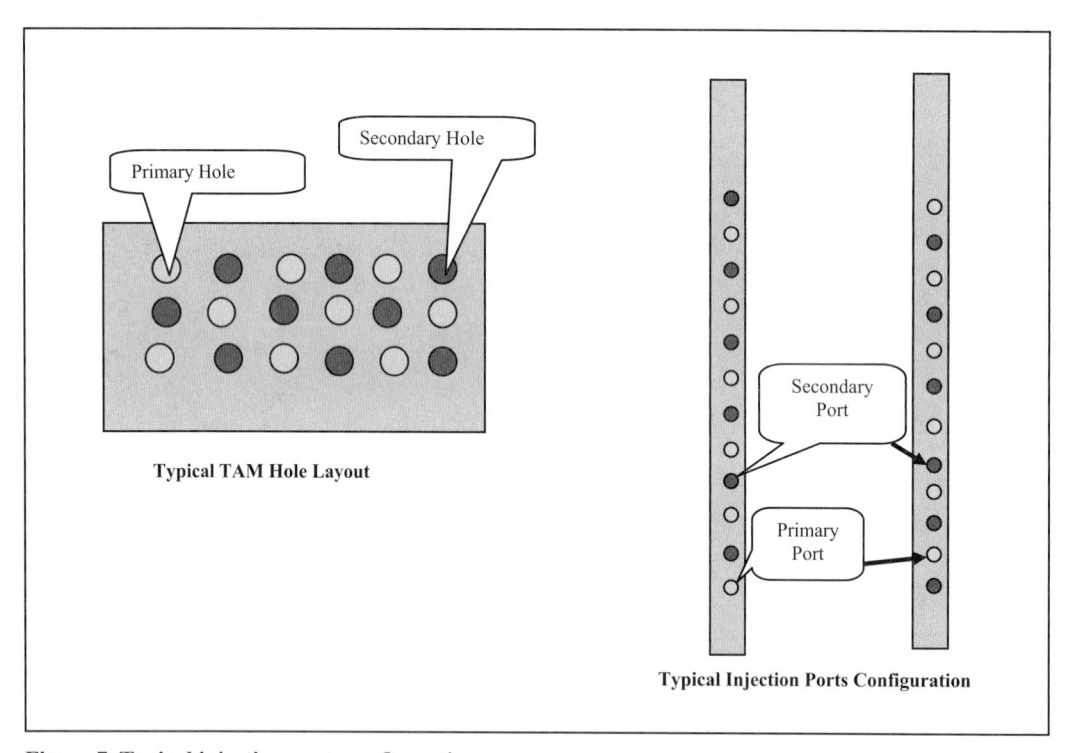

Figure 7. Typical injection ports configuration

After the grout pipes were installed cement grout material was injected into the rock using a HANY two pump system shown below (Figure 8). The overburden was not grouted.

The grouting operation was preformed using Type-III Portland cement. The grouting typically began with a 3:1, water cement ratio mix (by weight). Grout was injected and thickened to 2:1, 1:1, and 0.5:1 as described by the work plan until refusal was reached.

A detailed quality control procedure was accomplished through the use of an automatic data recording system (shown in Figure 9) connected to the grouting equipment. This equipment monitored grout take and pressure over time, and provided a detailed record of the grouting operation. A total of

**Figure 8. Hany pumps**

**Figure 9. Automatic data recording system (Hany Computer)**

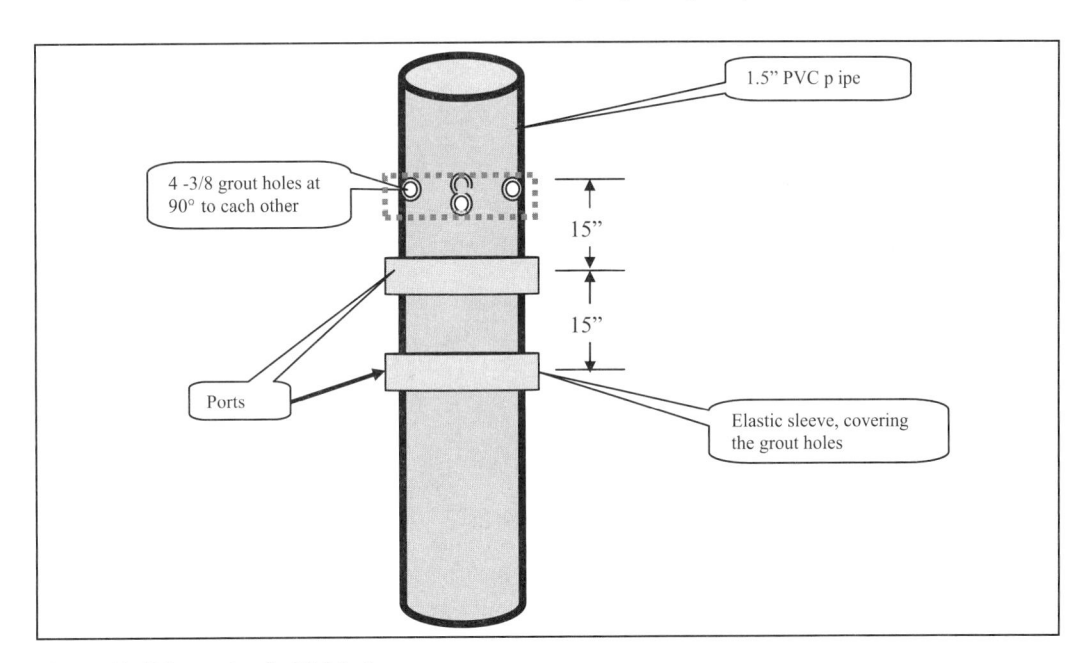

**Figure 10. Schematic of a TAM pipe**

250 cubic yards of grout was injected into the rock. The grouting operation started on March 3, 2009 and was completed on May 4, 2009. One hundred and fifty cubic yards of grout was injected into the section one where one hundred cubic yards was injected into the section two.

## TAM GROUT PIPE

To inject grout we used 1.5 inch diameter PVC pipes with rows of 4 holes of ⅜-in. diameter, at 90-degree from each other, spaced at 15 inches interval along the length of the pipe as shown below (Figures 10, 11, 12). The collection of 4 holes in each row, are referred to as ports. Each port was covered by an

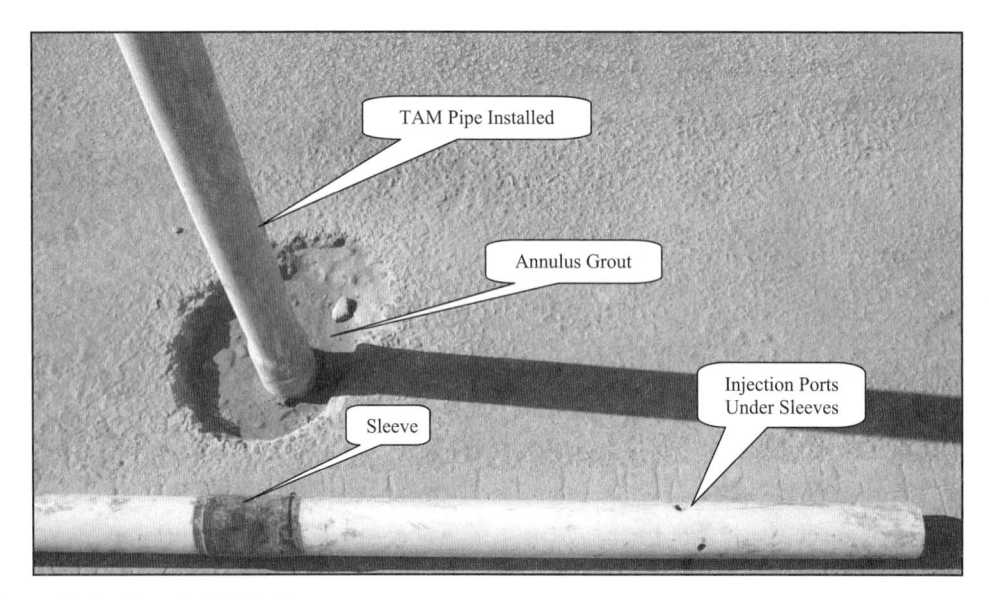

Figure 11. Section of a TAM pipe

Figure 12. An installed TAM pipe

elastic sleeve to allow the grout to go out but prevent it from getting back into the pipe.

## GROUTING SEQUENCE

Grouting was performed in five stages as follow. Prior to each stage of grouting the ports to be used for grouting were fractured (opened up) using water to make sure that the ports could receive grout. The grouting operation in each stage started from the lowest port moving upward. After completing the grouting operation at Primary Ports, grouting was performed through the ports in-between Primary Ports at about 3-feet from each Primary port. These ports were referred to as Secondary Ports.

**Stage 1:** Grouting of primary ports in primary holes
**Stage 2:** Grouting of primary ports in secondary holes
**Stage 3:** Grouting of secondary ports in primary holes
**Stage 4:** Grouting of secondary ports in secondary holes
**Stage 5**: After completing the grouting operation at primary and secondary ports, confirmation grouting was performed through all the ports with a 3:1 grout mix

At the completion of the grouting operation all the TAM pipes were filled with a 4000 psi grout.

## REFUSAL CRITERIA

In order to achieve the optimum grout injection a set of criteria was set to define the maximum rate, volume, and pressure as described below:

- Maximum amount of 3:1 grout allowed to be injected was about 15 cf. (112 gal)
- Maximum amount of 2:1 grout allowed to be injected was about 15 cf. (112 gal)
- Maximum amount of 1:1 grout allowed to be injected was about 20 cf. (150 gal)
- Grouting would stop if less than ½ cubic foot (3.75 gal.) of grout was injected in 10 minutes (less than 0.5 gpm)
- Grouting would stop after 40 minutes of grout take
- Grouting would stop after 50 cubic foot. (375 gal) of grout take
- Grouting would stop if there were ground heave in excess of 0.5 in.
- Most grouts were injected at 2 to 5 gallon per minute
- Grout pressure did not exceed 1.5 psi per foot of depth below the ground plus the required pressure to open the sleeve

- The pressure required to open the sleeves based on the four grout mixes were as follow:
  a. For 3:1 grout: 25 psi
  b. For 2:1 grout: 45 psi
  c. For 2:1 grout: 65 psi
  d. For 0.5:1 grout: 65 psi
- The net grouting pressure ranged from 65 to 145 psi as measured at the grout manifold

## PERMEABILITY TEST

During drilling operation of the TAM pipes, a falling-head permeability test was performed in all of the boreholes to determine the approximate in-situ permeability of the rock. The permeability was performed in the uncased section of the borehole. The results are presented in Figure 13.

The permeability test results were used as guide to determine the capacity of each location for grouting. As it can be seen from the test results provided below, the permeability varied between locations and ranged from 3.42 E-01 to 4.6 E-04 cm/s. However, since these tests were performed over a length of 15 to 20 feet of borehole, they represent very general values of permeability at each location.

## GROUTING OPERATION FINDINGS

During the grouting operation, the flow rate, pressure, and total volume of each type of grout mix at each port were measured. To determine which elevation took the most grout the total volume of the grout intake at each TAM pipe, and at each elevation for all the TAM pipes were compared. Also the total grout intake in all the Primary and Secondary Ports in all of the TAM pipes was compared. The results are presented below in Figures 14 and 15.

Figure 14 illustrates the variation of grout intake at same port (elevation) of TAM pipes at different locations, indicating the variability of the rock formation from one TAM location to the next.

Figure 15 illustrates the Primary Ports took about 10 times more grout than the Secondary Ports. Hardly any grout could be injected during the third stage.

## DETERMINATION OF EFFECTIVENESS OF THE GROUTING OPERATION

Three months after the completion of the grouting operation a series of exploratory boring was performed in the grouted area to determine and measure the effectiveness of the grouting operation. To achieve this goal three borings were performed within the grouted area.

The borings were advanced to about fifty feet deep. A three inch diameter split spoon was derived through the overburden into the decomposed rock to refusal (roughly 30 feet below surface) for retrieving

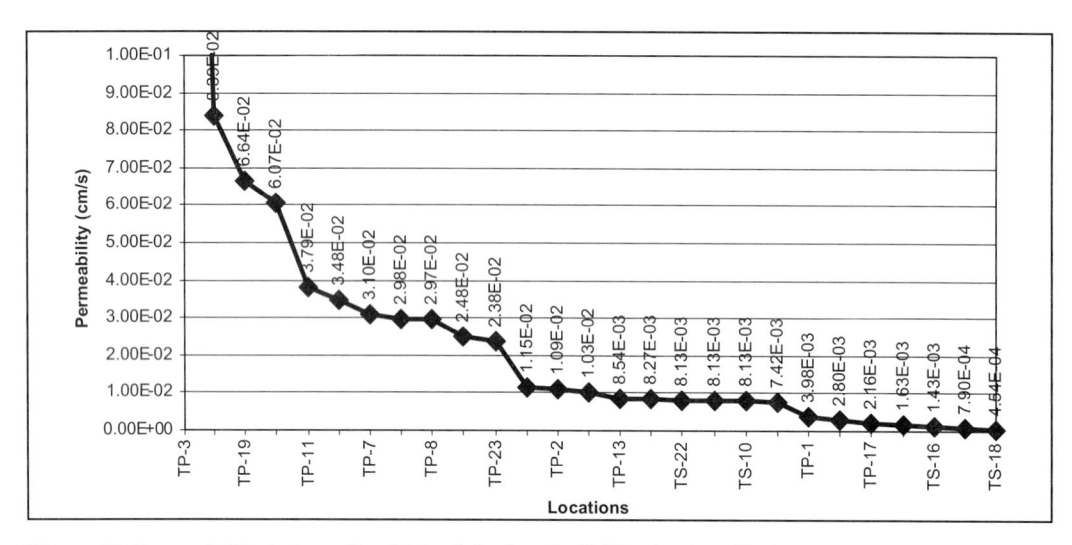

Figure 13. Permeability test results obtained during the TAM pipe installation

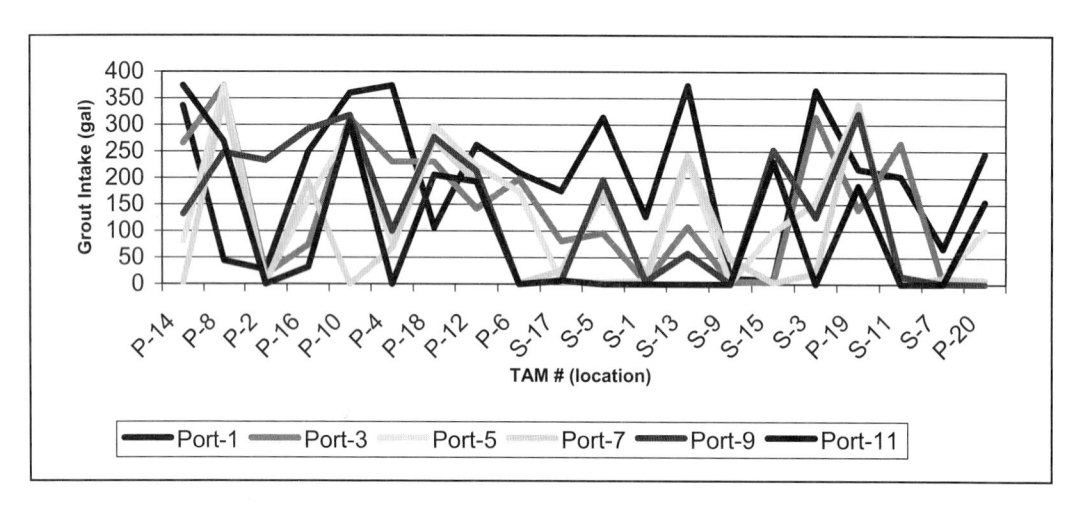

Figure 14. Comparison of total grout intake in each primary port in all TAM pipes

undisturbed samples for evaluation. From 30 feet to 50 feet below surface (spring line of the tunnel) the rock was cored using 2.5-in. diameter core barrel. Subsequently the permeability of the 15-feet cored section of the rock was tested by using a double packer permeability test.

The samples from the split-spoon and cores were physically examined and tested for presence of grout using Phenolphthalein solution. Even though some of the grout in the cores samples seemed to have been washed away during the coring operation, the testing reveled presence of grout in the overburden soil and in some joints within the rock and the permeability of the grouted zone was reduced by up to 500 times at various locations.

## CONCLUSIONS

- The ground improvement achieved its objectives to reinforce and improve the relatively weaker rock formation above the tunnel at an isolated area. The improvement incorporated the installation of seventy-five fiberglass reinforcing bars (soil nails) and injection of cement grout at 27 locations at various elevations.
- The entire grouting operation (phase-I and phase-II) took approximately eighty (80) working days and was successfully implemented in accordance with the pre-established protocol and specification.

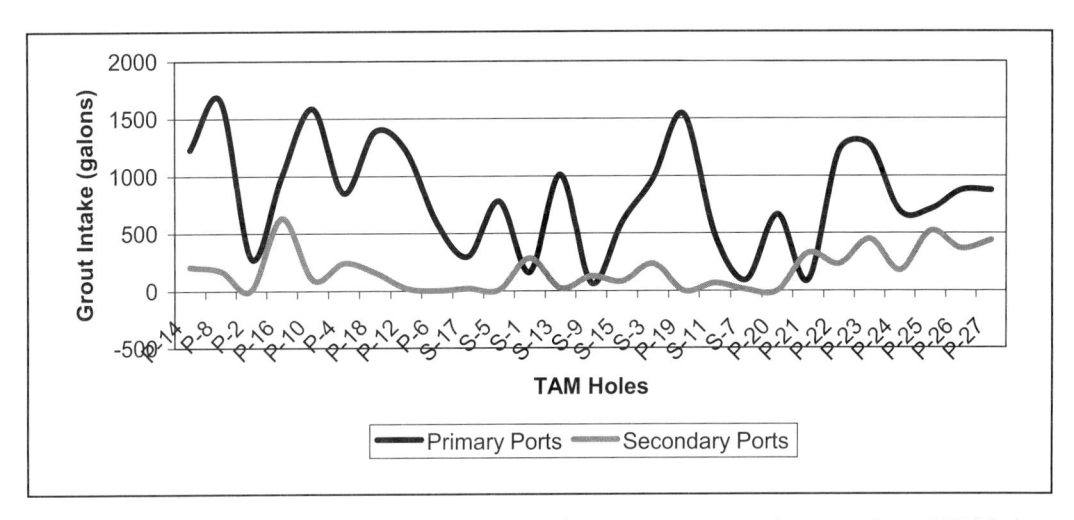

Figure 15. Comparison of total grout intake in all the primary ports vs. secondary ports in each TAM pipe

- Monitoring the day-to day operation and performing quality control/quality assurance, and the implementation of strict procedures by PWD engineers, yielded great savings to the Philadelphia Water Department.
- Even though there was not much grout evidence of presence of grout in the core samples the permeability of the grouted zone was reduced by up to 500 times at various locations.
- The reduction of the permeability indicated the presence of the grout within that region. Thus concluding that the grouting operation had achieved its objectives of reducing the permeability of the treated zone.
- The reduction of the permeability resulted in reduction of seepage of water into the tunnel during tunneling. Seepage of water can result in loss of fines within the joints and fissures within the rock and possible collapse of the tunnel.
- It was determined that the best way to determine the effectiveness of the grouting is the comparison of the permeability before and after grouting operation.
- About six months after the grouting operation the TBM passed through the reinforced rock. The rocks held up and the TBM successfully proceeded with completing the tunneling.

# The Construction of the Tunnels and Shafts for the Project XFEL (X-Ray Free Electron Laser)

Paul Erdmann
Amberg Engineering Ltd., 8105 Regensdorf-Watt, Switzerland

ABSTRACT: With the European XFEL a unique free-electron laser research centre in the area of Hamburg, Germany, will be erected. The facility consists of a two kilometer long accelerator tunnel, five finger tunnels starting from the end of the accelerator tunnel and leading to the underground experimental hall, a number of underground halls and shafts as well as additional infrastructural installations on the ground surface. The overall construction length of the facility is about 3.4 km. The tunnels are situated about 6 to 30 m under the surface underneath the ground water level. The construction pits are up to 40 m deep. The demand of straightness and therefore the demands on the tolerances of the construction is extremely high.

The construction started 2009, initial operation will be in 2014. The tunnels are driven with hydroshield TBM's with segmental lining. The paper will describe the experience gained in the first 18 month of work.

## INTRODUCTION

A new physical high-performance facility for research with light is being produced at the German Electron Synchroton DESY in Hamburg in the form of the European X-Ray Laser XFEL. With its extremely short and intensive x-ray flashes nuclear processes can be observed in real time.

The DESY in Hamburg, Germany is a worldwide accepted research center for physics concerning to particle, acceleration and laser technology.

The European X-Ray free electron laser project (XFEL) is located on the border between the states of Hamburg and Schleswig-Holstein in Germany. The bulk of the plant will be built underground in tunnels and shafts. Access to these structures is via the three sites Bahrenfeld, Osdorfer Born and Schenefeld, on which special facilities and additional infrastructures will be built. 13 countries are shareholders of the European XFEL GmbH.

The XFEL structure will be constructed in two stages, with the 1st stage being the accelerator tunnel, the tunnel branch, the first experimental hall, a part of the facility buildings and the necessary outbuildings to operate the plant. The 2nd stage includes the second half of the tunnel branch, the second experimental hall as well as the infrastructure buildings. These will be built at a later date. The design fort he necessary tunnels and shafts has not already started.

The Linac tunnel XTL follows the direction of the electron beam of the injector XTIN. Herein all essential components necessary for the acceleration of the electron packages in the facility XFEL have been installed with the exception of the source to generate the electron beam.

To supply the experiments in the experimental hall with radiation, the electro beam which is generated in the linear accelerator will be directed on to various kinds of X-rays undulators. This serves the branched structure, which consists of branching ducts (XS1-4), which includes associated halls (XHE1-4), absorber shafts (XSDU1 and XSDU 2), and undulator and photon tunnels (XTD1-10).

Already in the XTL tunnel the electron beams are separated into three distinct beams. One of them runs in the XTD20 tunnel, which is the connecting tunnel for the prospective expansion stage 2. The other two electron beams pass through two separate undulators and each end up in an absorber (XSDU1-2). In order to divide the rays a shaft structure is needed at the mouth of the tunnel which then separates into two new tunnels.

The design of the construction was carried out by Amberg Engineering AG in an engineering JV association. Amberg Engineering AG is responsible for the design and site/construction supervision. An overview showing the facilities is given in Figures 1 and 2.

## PHYSICAL BACKGROUND

At the European XFEL ultra short X-ray flashes are produced. To generate the X-ray flashes, firstly the electron packets are accelerated to high velocities within the XTL tunnel and finally with specifically positioned magnets are steered into the tunnels XTD 1–XTD 5.

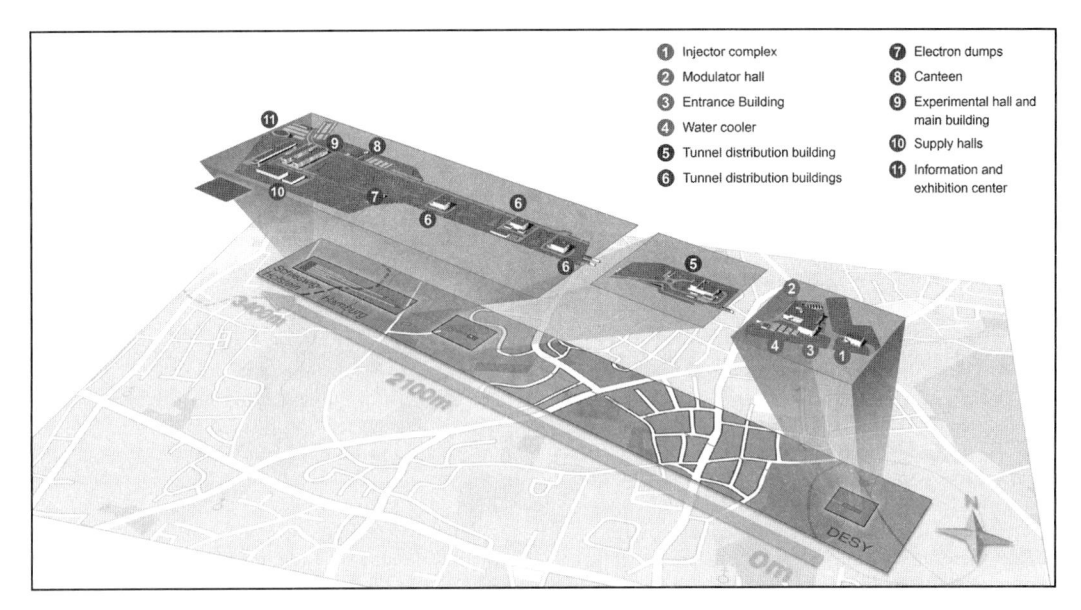

Figure 1. XFEL overview with visualisation of the buildings on completion

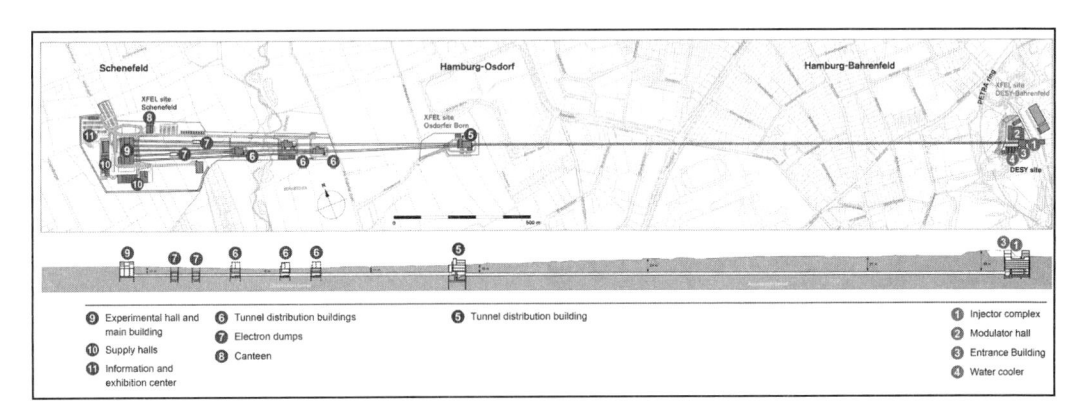

Figure 2. XFEL cross section with visualisation of the buildings on completion

Thereby small particles of light are emitted, which are increasingly strengthened, until finally they create an extremely short and intense X-ray flash. Using the X-ray flashes, molecular structures become visible and thus can be explored.

With these X-ray flashes, for example; the atomic details of viruses can be recognized, the molecular structures of cells deciphered, the making of three-dimensional images from the nanocosmos can be made, as well as the filming of chemical reactions and the investigation of the inner workings of the planet.

## UNDERGROUND

### Geological Conditions

The entire XFEL section lies in glacial deposits of sand, pebbles and cohesive soils, which are partly in a ring formation in the Tertiary substrate. The tunnel alignment is thus mainly in marl carrying groundwater of Pleistocene sands and gravels, and lies partly in tertiary mica fine sands.

In addition, one expects localized and limited disturbances to occur of homogeneous layers through sand, peat and stones of all sizes. The glacial marl is stored with water-bearing sand belts/sandbeds, and stones up to the size of blocks (boulders) are to be expected.

Since the tunneling will to a large extent be carried out in glacial deposits, an overall high abrasiveness is to be expected.

The covering of the tunnel is about 6 m to 38 m, but mainly a covering of 10m, 12m will be executed. The low coverage of 6 m is at the lowest point of the route, which is conditional on the crossing of the Düpenau. The catchment area of the Düpenau is characterized by large sealed areas. As a rule it carries a few liters per second and the water level averages a few centimeters. In heavy rain or during prolonged periods of rain the Düpenau can burst its banks, which means that the river flow increases to 2.5 m/s.

## Groundwater

In the area between the DESY site in Hamburg Bahrenfeld and the border to the municipality of Schenefeld there are free groundwater conditions, ie, the existing water table and the piezometric surface are identical. Minor water-permeable layers at the earth`s surface cause pressurized groundwater conditions to exist in the surrounding areas of the Schenefeld building site.

## Pollution

### *Contamination*

The tunnel alignment passes through old deposits, which were caused by the refills of a former sand mining pit. Additional information on suspected site contamination does not exist.

### *Groundwater Analysis*

The groundwater in the area of the tunnel XTL, XTD1 and XDT2 in terms of hydrologic analysis has been shown to be moderately corrosive to concrete to DIN 4030-1. In the area of the tunnel XTD 3–XTD 10, the groundwater is mostly not considered to be corrosive to concrete. Only in the area of the Düpenau can it be considered to be weak to moderately corrosive.

### *Ordnance*

In the Hamburg metropolitan area no suspicion exists for explosives. In the Schleswig-Holstein part, soundings were made in the construction area by the Ordnance Service for discarded munitions which have been secured.

## CONSTRUCTION

### Injector

The injector consists of the Injector shaft XTIN, Injector XTIN tunnel and the entry shaft XSE and defines the beginning of the XFEL installation. During the construction phase, it serves as the end shaft for driving the XTL. The injector complex will be installed in an open pit as a massive reinforced concrete structure with a total of seven stories below ground.

### Shafts XS1–XS4

All excavations are done in the so-called slurry wall trenching method, that is, without sinking large areas of ground water, and without the use of rams or sheet piling.

At all beam branches shafts need to be arranged. In addition to beam branching, the shafts generally serve as access to the tunnel for persons and material as well as for the services needed in the tunnel (electricity, water, ventilation, etc.). During construction, they serve as necessary start and end shafts for preparatory work for the tunnel.

The lowest floor of each shaft with the beam level is separated from the upper floor by a radiation shield cover. This cover will be made of ordinary concrete with a thickness of 2.0 meters, the necessary opening for installation will have removable stones shaped, such that no continuous vertical gaps are formed.

The requirement is that the tunnel downstream of the beam to the shaft can be traversed individually at any one time with ongoing operations in the neighboring tunnel and shaft. This requires a separate beam protected anteroom within the shaft to the tunnel. This will be separated by an angled access through massive concrete from the main shaft region. The angled access will be made up of movable concrete shielding pieces so as to facilitate a straight passage to the shaft for the transport of bulky items. In the area of the anteroom installation holes will be arranged.

The shafts each have a separate security stairway and a fire brigade lift. This means that an additional room with positive pressurized ventilation at the exits of each level has been arranged. The lift's switch gear includes a safety interlock so that passage is excluded through the Interlock area during operation.

Outside the shafts between the tunnels is an additional shield of heavy concrete, which will be placed in the feeder areas to ensure the accessibility of the photon tunnel with simultaneous operation of the adjacent tunnel.

The retaining wall of the shaft is planned as a waterproof, internally stiffened diaphragm wall. The sealing of the pit from below shall be effected by an underwater concrete base. For the entry of the tunnel boring machine in each end wall an appropriately prepared window reinforced with fiberglass (glass reinforced plastic) will be provided in the trench wall.

The final shaft structure will be made of reinforced concrete. The outer building walls are concreted directly against the trench wall enclosure and will be formed with waterproof concrete.

The shaft XS1 will be larger than the other shafts, because it has 3 tunnel branches. The third branch, the tunnel XTD 20 and the associated second branch construction will be installed during the second stage of construction. It also includes a beam absorber shafts like the dump-pit. The shaft will have a seal in the form of a deep set base.

The beam absorber shafts will not be accessible from the upper surface in their final state. After passing through the undulator stretch the electron beam is no longer needed. At this time it has a power of 300 kW and therefore must be guided into a beam absorber. This dump pit consists of a graphite core, copper jacket and cooling coils and can absorb the resulting radiation.

## Experimental Hall

The experimental hall will be built as a reinforced concrete structure in a water tight excavated pit, which consists of an anchored underwater concrete base enclosed by diaphragm walls.

## Tunnels

The tunnels will be produced with two shield tunnel boring machines working underground and in parallel. The tunneling will be done with a tunnel boring machine with active tunnel face support.

The tunnels will be constructed as single skin lining segments in water-impermeable concrete and provided with circumferential sealing profiles.

When the excavations are ready for tunneling to start the two tunnel boring machines—the larger one with an outer diameter of 6.17 meters, moves toward the Osdorfer Born site, the smaller one with an outer diameter of 5.48 meters working under the Schenefelder site. After completion of a tunnel, the tunnel boring machines will be transported back into the respective starting positions and then go to the next tunnel stretch. Once an excavation is no longer needed for the tunnel as the start or destination of the tunnel boring machine, it will be developed as an appropriate underground structure.

The tunnel lining will be assembled with tunnel segment produced out of reinforced concrete. The thickness of the lining is 300 mm and the width of one ring is 1500 mm. The ring division is constructed with 6 normal and 1 little keystone. It is shown in Figure 3.

Lengths of the tunnel sections are:

- XTL 2011 m
- XTD1 480 m
- XTD2 594 m with an outer diameter of 6.17 m.
- XTD3 263 m
- XTD4 301
- XTD5 204 m
- XTD6 661 m
- XTD7 137 m
- XTD8 365 m
- XTD9 545 m
- XTD10 221 m with an outer diameter of 5.48 m

For reasons of radiation protection, without additional measures, it is necessary to provide a minimum cover above the tunnel crown of approximately 6 meters to the surface. The depth of the tunnel will be fixed by the lowest point along the tunnel routes. This local low point is in the area of the Osdorfer Feldmark Düpenau.

The course of individual tunnel sections is absolutely straight i.e., the tunnels do not follow the curvature of the earth. All tunnel axes are, apart from any vertical Tunnel deviations, parallel to the beam plane which is oriented exactly tangential to an equi potential surface of the earth's gravitational field.

A detailed analysis of the use of space in the tunnel XTL showed that the suspension of the main linear accelerator under the tunnel ceiling, particularly during maintenance periods has considerable advantages. For this purpose, steel bands will be included in part of the lining segments. They will be installed during the tunneling in the ridge area and allow the subsequent installation of the interior and make possible the exact location for the suspension of the main linear accelerator.

The concrete and reinforced concrete work will include the production of the flooring in the tunnels. In the course of concrete and reinforced concrete work, the tunnel drainage system will be made. In tunnel XTL condensation or water leaks will be caught at the low point in the tunnel section using barriers, which are arranged every 48 m. The accumulated water can be extracted with mobile pumps. In the other tunnels, any condensation and water leakages will be collected at the lowest point of the tunnel section in a basin. It can be removed with a suction line. Cleaning shafts regularly spaced will be provided. In Figure 4 the structural works and technical equipment are shown.

### *Pipe Line Excavation*

Because of its depth, the tunnels lie mainly well under the pipe line zones. For the tunneling XTD1 and XDT2 a dirty water sluice DN 1000 will be excavated with a small clearance of about 1.23 meters. The sluice will be taken out of service during tunneling. The waters collected in the sluice will be

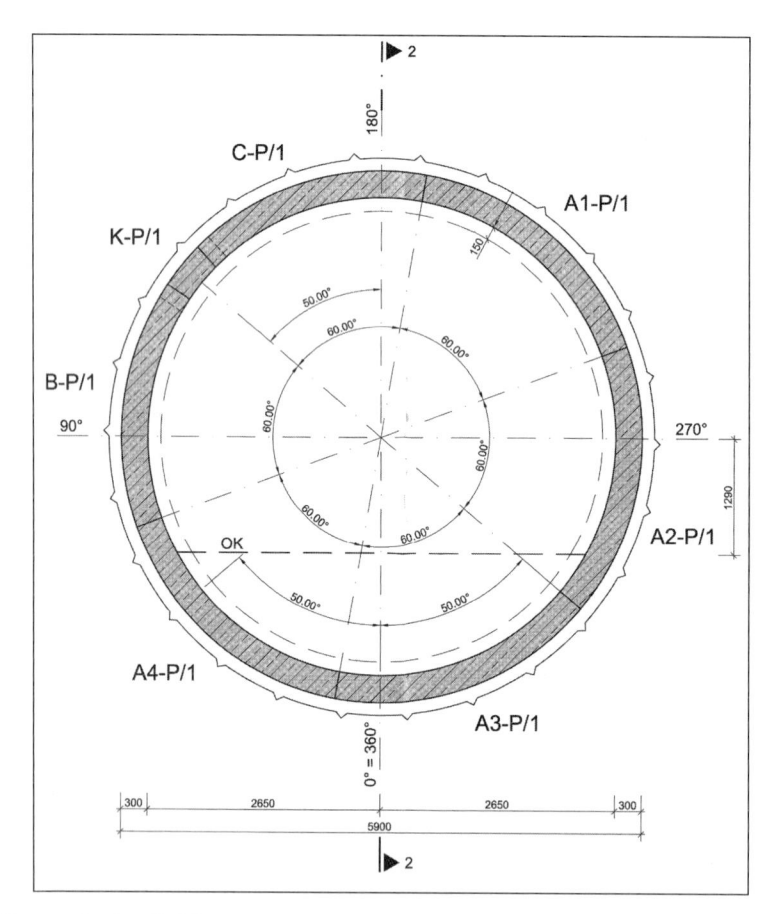

Figure 3. XTL tunnel, ring division

Figure 4. Drawing of the technical equipped acceleration tunnel

temporarily maintained. Before beginning the tunneling of XTD the sluice will be filled so as to be ready for the tunneling work. The permissible settlement of the buildings located above the tunneling line is 2–4 cm.

## THE STATUS OF THE WORK AND OUTLOOK

The work is performed in parallel at three sites.

Construction work on the site DESY-Bahrenfeld began in January 2009 starting with the removal of soil in the building area of the Lise Meitner-park, as it has 10 meters difference in height to the future site. Parallel to this the 100 meter long and 40 meter deep trench for the injector complex is currently being constructed. In the finished excavation after MAY 2010 the construction of the below ground injector complex will be put in place.

The tunnels are to be driven underground using 2 hydroshield-machines with different diameters. The start of the first tunnel constructions with the bigger diameter of the shield has taken place in April/May 2010 Foreseeably in March 2011 the shield tunneling machine will come to the west side of the site, after it has bored the 2,1 kilometer main tunnel between the sites Osdorfer Born and DESY-Bahrenfeld. There it will be dismantled and removed in pieces.

In June 2011 building construction will be started on these sites. The entrance halls to the two shafts of the underground injector complex will be built. The construction of the modulator hall for the power supply of the plant, the erection of additional infrastructure facilities and the landscaping of the site will complete the construction work at the European XFEL site DESY—Bahrenfeld in the autumn of 2012.

After the completion of under-and aboveground structures, the technical equipment of buildings (electricity, water, ventilation, plumbing etc.) will be done. Then the installation work starts: All components for the first under ground section of the 3,4 kilometer long European-XFEL will be brought over the DESY site to both entry shafts on the construction site, and there lowered down, with the help of special vehicles transported to their location in the tunnel and installed into position.

On the Osdorfer Born site the XS1 will be erected by September 2012. In the first months after the start of the construction site it was prepared and technically equipped. Then the preliminary excavation was carried down to the working level. The actual work on the excavation began in August 2009.

Foreseeably in June 2010, the tunnel boring machine will for the first time reach the west side of the excavation. There it will be dismantled and transported back to the main site in Schenefeld, from where it will start a second time in direction Hamburg. It comes to the Osdorfer Born excavation again in September, and in October starts at the east end of the excavation in the direction of DESY—Bahrenfeld. The boring of the 2.1 kilometers long tunnel section will last until March 2011. During this time the machine will be supplied by Schenefeld. Also a rail track in the finished tunnel leads to the current location of the machine.

The largest of the three sites is located in the south of the city Schenefeld (Pinneberg). Here, by August 2010 the excavation of the six pits will have been completed. In Schenefeld on the edge of the site, excavated soil will be temporarily stored, which will later be used for site landscaping. Piles of earth will be used during the construction period for noise and dust protection.

Work on the Schenefeld site began with site clearing, site preparation and the earthworks for the production of the elevated Lise Meixner park area in the region of the experimental hall. In May 2009, the production of the slurry wall for this pit was

**Figure 5. Drawings of the experimental hall aboveground (left) and underground (right)**

started, in which the small shield boring machine will be started five times from September 2010. A visualisation of the finished buildings aboveground and underground is given in Figure 5.

Staggered and sometimes parallel to this, the other five underground shafts and halls will be constructed.

The success of the project essential depends on the possibility to comply with the demands of straightness. The tolerances of the final construction are very small. So it will depends on the experiences and responsibility of the site supervision to finish this project in a satisfied way and for the satisfaction of our client.

**LITERATURE**

[1] Additional information is provided at the homepage of the project XFEL: www.xfel.eu

# Canadian Fast-Track Drill and Blast: Excavating the Rupert Transfer Tunnel at James Bay, Québec, Canada

**C.H. Murdock**
Independent consultant, Vancouver, British Columbia, Canada

**R.W. Glowe**
Independent consultant, Longueuil, Quebec, Canada

The Rupert Transfer Tunnel allows the transfer of a substantial portion of the Rupert River flow into the drainage basin of the Eastmain River, increasing the discharge of the Eastmain by nearly 100%. The increased flow will be turbined at five downstream hydroelectric powerhouses- Eastmain-1-A, then at Sarcelle, both new installations, followed by LG-2, LG-2A and LG-1, all existing powerhouses, all built in the 1970s and 1980s, before reaching the sea at James Bay.

The Rupert Transfer Tunnel is 2,908 meters long, with a height of 18.6 meters and a width of 12.7 meters. It was excavated in 2007 and 2008, using the top heading and bench method. The heading was drilled and blasted from both ends, alternating rounds, while the bench was excavated from one end only. Modern computer-controlled data jumbos were used for the heading. The bench was excavated at 10 meter height, using crawler hydraulic drills. Large side-dump loaders and 50 tonne trucks were employed for mucking both the heading and bench. The excavation of 651,300 cubic meters of Canadian Shield granite took 13 months, working through a tough sub-Arctic winter, with temperatures reaching –40 Celsius.

To put this into a TBM-equivalent context, the excavated drill and blast volume equals 23 kilometers of 6 meter diameter TBM tunnel. The production rate required of that hypothetical 6 meter TBM would be an average of 1,772 meters per month, to excavate the same rock volume.

## BACKGROUND

The owner of the Rupert Tunnel is Hydro-Québec. Hydro-Québec is a Provincial Crown Corporation, owned by the people of Québec. It serves 3.6 million customers in the Province of Québec. H-Q exports energy to New England, New York State and the Province of Ontario, through 18 high voltage interconnections. It also imports energy from those same markets, buying somewhat more than it sells, using a concept it pioneered, that of energy-banking, which follows the old stock-market maxim of buy low-sell high. Low-cost night-time power purchased from nuclear and coal-fired must-run plants allows Hydro-Québec to conserve water in its reservoirs, then sell peak power back later in the day.

Hydro-Québec is the largest hydroelectric generator in the world. H-Q currently operates 59 hydroelectric powerhouses, equipped with 336 turbines, and producing 33,680MW. Three projects involving seven new powerhouses are under construction at this time, planned to come into service between 2011 and 2020, adding a further 2,468MW. They are: Eastmain-1-A, 768MW, in service by late 2011; La Sarcelle, 150MW, in service by late 2011; La Romaine, 1,550MW, in four power plants, the first to be in service by 2014, the second in 2016, the third by 2017 and the fourth plant in 2020.

Within the last decade, H-Q has brought on line seven new plants, totalling to 2,643MW, with 25 turbines. They are: Sainte-Marguerite-3, 884MW; Toulnustouc, 526MW; Eastmain-1, 480MW; Peribonca, 385MW; Grand-Mère, 230MW; Rapide-des-Coeurs, 76MW and Chute-Allard, 62MW.

Hydro-Québec owns and operates the largest underground powerhouse complex in the world, at the Robert Bourassa (LG-2) and LG-2A site. The 22 units installed there have a total capacity of 7,722MW. A dozen H-Q powerhouses are each of over 1,000MW in their installed capacities. In addition to these large hydroelectric resources, Hydro-Québec operates the Gentilly-2 CANDU nuclear reactor- 675MW and the Tracy oil-fired thermal plant of 660MW, infrequently run, in addition to 3 gas turbine plants- Becancour, of 439MW; La Citière, 280MW and Cadillac,162MW, or 217,200 horsepower, by far the most powerful Cadillac ever built. All of these generating facilities- hydroelectric, both those built and those now under construction; nuclear; oil-fired and gas-fired total to over 38,000MW. Over 90% of Hydro-Québec's energy is produced from falling water. H-Q also operates 23 diesel plants, serving off-grid island and northern communities.

During the 1970s and 1980s, Hydro-Québec, through its subsidiary, Société d'énergie de la Baie James, built seven hydroelectric generating stations in cascade on the La Grande River. These stations, with a total of 65 units, have an installed capacity of 16,020MW, with a replacement value in excess of $50 billion. Hydro-Québec has had, since 1971, a power purchase agreement for 5,429MW of power from Newfoundland and Labrador's Churchill Falls plant, with the agreement running until 2041.

Hydro-Québec operates the largest electricity transmission network in North America. It was the first utility in the world to operate transmission lines at over 700,000 volts. The first 735KV line came into service in 1965. The H-Q high voltage (765–735KV) network now extends to 11,422 kilometres. The medium and low voltage distribution network now totals to 110,127 kilometres.

Residential customers of H-Q enjoy the lowest rates in North America- 6.87 cents per KWh. This compares to 7.13 cents in Vancouver, 11.01 cents in Portland, 15.05 cents in Chicago and 25.32 cents in New York City. About 95% of Hydro-Quebec's current production is hydroelectric, but 3,000MW of wind power is either online now or in the supply pipeline, with online dates to 2015.

## TWO NEW GENERATING STATIONS

The Rupert Transfer Tunnel will divert up to 800 cubic meters per second of water from the Rupert River watershed into the Eastmain River watershed, to provide additional hydroelectric generating capacity. This capacity will be realized at two new generating stations now under construction- the 768MW Eastmain-1-A powerhouse and the 150MW Sarcelle powerhouse. The total cost of the project is $5 billion. Further energy production from the existing facilities downstream will add substantially to the energy benefits.

In summary, and this by Terawatthours:

Eastmain-1-A
768MW, in service by 2011            2.3TWh
Sarcelle
150MW, in service by 2011            1.1TWh
Total from new facilities            3.4TWh   = 39%
From the three existing
downstream power plants—
Robert-Bourassa (LG-2), La Grande
2-A and La Grande-1                  5.3TWh   = 61%
Total energy attributable to the
Eastmain-1-A Project                 8.7TWh   = 100%

For comparison purposes, the 4 powerhouse La Romaine complex, now starting construction, will produce 8.0TWh, with a total installed capacity of 1,550MW, and a projected cost of $6.5 billion. In another comparison, the 935MW of new capacity added with the Rupert Transfer Tunnel will produce 9% more energy than the 1,550MW La Romaine project, and at lesser cost, due to greater utilization of the existing downstream plants.

1TWh = 1,000GWh = 1,000,000MWh
     = 1 billion KWh

To put all these electrical terms into an urban context, the annual energy requirements of the Chicago metropolitan area could be met with 8.7TWh + 8.0TWh. (= Eastmain-1-A + La Romaine). An average thermal plant, to generate 8.7TWh, would require over four million tons of coal, represented by a loaded coal train of over 1,000 kilometres in length- Vancouver to Calgary.

## PRINCIPAL ELEMENTS

The Eastmain-1-A- La Sarcelle- Rupert Project is comprised of these three principal elements:

- The construction of the 768MW Eastmain-1-A Powerhouse.
- The construction of the 150MW Sarcelle Powerhouse.
- The partial diversion of the Rupert River towards these two powerhouses, and onward, towards the existing Robert-Bourassa (LG-2), La Grande 2-A and La Grande-1 powerhouses.

## PROJECT SCOPE

The partial diversion of the Rupert River includes these works:

- Four dams.
- A spillway on the Rupert River, regulating the downstream flow into that river.
- 74 dikes around the perimeter of the reservoirs.
- 2 forebay lakes, totalling 346 square kilometres in area.
- A transfer tunnel of 2.9 kilometres in length.
- A network of canals totalling 12 kilometres, to facilitate the flow of water.
- Construction of a series of control weirs on the Rupert River, downstream of the spillway.

## PROJECT DESCRIPTION

Description of the Rupert Transfer Tunnel Project.

- The tunnel is 2,908 meters long.
- Open cut rock excavation for the upstream and downstream portals-467,000m³.
- Rock cuts to 60m depth, done with 10m benches.
- Underground rock excavation:
  Heading      282,000m³
                    8.6m high, 12.7m wide
  Face area    97m²
- Underground rock excavation:
  Benching     369,300m³
                    10.0m high, 12.7m wide
  Face area    127m²
- The heading was excavated with horizontal drilling, using 3 boom Sandvik (Tamrock) T11A data jumbos, with man-baskets. Drill length was 5.8m, while pull averaged 5.4m. For 15m at each end, the pilot & slash technique was used, with the pilot rounds limited to 2.5m pull, leaving at least 2m of rock to the 'A' line for the subsequent slash round.

  Number of pilot rounds required:
  30m/2.5m = 12 rounds
  Number of slash rounds required:
  30m/2.5m = 12 rounds
  Full face rounds required:
  2,878m/5.4m = 533 rounds

  A typical 5.4m heading round broke 524Bm³ of rock, or about 1,400 tonnes of muck.
  For 300m, under a surface lake, probe drilling was required for 10m+ beyond the face.

- Benching was done using 4 to 5 hydraulic crawler drills working at one end only.
- Drill depth was generally 10.6m, with a section of 10–12m in length blasted on each shift.
- The 12m round blasted 1,524Bm³ of rock, or about 4,100 tonnes of muck.
- Mucking for both the heading and the bench rounds was with Cat 988 side-dump loaders, into 4–6 Cat 773 trucks, of 50 tonne capacity. Power scaling was done with Cat 365 hoes.
- Each round was bolted and meshed, to the face, before drilling the next round. The mesh remains in place.

## SCHEDULE

- The job was bid on January 30th and was awarded in late March, 2007.
- Four new drill jumbos were conditionally ordered from Sandvik (Tamrock) in February of 2007- two data T11s, one non-data T11 and a 2 boom T8. The T8 was used for bolting.
- Mobilization, clearing of the portals and construction of temporary power lines and access roads started in April, 2007.
- Overburden excavation in the upstream and downstream portal areas started in May.
- Rock excavation in the portals started in July of 2007.
- In August, 2 of the 4 jumbos were delivered; the other 2 followed shortly thereafter.
- Pilot and slash work at the downstream portal started in September, 2007.
- Due to unforeseen rock excavation difficulties, the upstream portal work was finished later than planned, concluding in October, 2007.
- Tunnelling, on a sustained basis, working at both ends started in mid-October, 2007.
- The Christmas 2007/New Year 2008 holiday shutdown was of 2 weeks duration.
- Top heading work was finished in early June, 2008, for a heading duration of 8 months.
- Benching was completed by early November, 2008, for a benching duration of 5 months.
- Overall heading and benching duration was 13 months.
- The job was substantially de-mobilized by December, 2008, avoiding a second winter.
- The tunnel was finished 12 months ahead of the owner's diversion schedule requirement.
- All the owner's time-related objectives were met.
- Filling of the Rupert River forebay started on November 7th, 2009, a year after the tunnelling was completed.

## GENERAL OBSERVATIONS

- Cost: The low bid submitted was $57 million. Some unforeseeable rock conditions were encountered, resulting in temporary losses in production, tunnelling delays and extra costs. These matters are currently under discussion.
- Camp: The camp is owned and operated by Hydro-Québec, without cost to the contractor or the employees- Yes, there IS a free lunch!
- The four Sandvik (Tamrock) jumbos were maintained by the contractor's mechanics, working under the daily supervision of Tamrock's onsite technician, supported by an adequate parts supply. There was never a round missed due to jumbo non-availability.

# Re-Design of Water Tunnels for Croton Water Treatment Plant, New York City

## Jozef F. Zurawski
Dawn Underground Engineering, Inc., Hazlet, New Jersey

## Paul J. Scagnelli
Schiavone Construction Co., LLC, Secausus New Jersey

**ABSTRACT:** The paper discusses design changes made during construction to the water tunnels for the Croton Water Treatment Plant in New York City. The original contract design consisted of two 9 ft service pipes in a single drill and blast horseshoe tunnel to connect the Jerome Park Reservoir with the new water treatment plant. Twin TBM driven tunnels were proposed by the Contractor as a substitute for the drill and blast excavation. Each TBM drive required that the machine is backed to the starting chamber. The paper discusses details of design changes that were required for the concept to be accepted by the Owner, City of New York DEP.

## INTRODUCTION

The New York City's Croton System is the oldest of three systems (Croton, Catskill and Delaware) that provide drinking water to the City and upstate communities. Croton was once the only reservoir system supplying water from outside the City, but currently the system is the smallest of the three. The Croton watershed is a series of interconnected reservoirs and lakes in northern Westchester and Putnam Counties. The Jerome Park Reservoir which is located at the downstream end of the Croton System is a distribution reservoir from which Croton System water enters City's water distribution system. On an average The Croton System provides about 10 percent of the New York City's daily demand. During droughts, the Croton System provides up to 30 percent of City's consumption. Croton water is primarily used in low-lying areas of the Bronx and Manhattan, where the water is conveyed by gravity. Two existing pump stations, the Jerome Avenue Pump Station and the Mosholu Pump Station, can supply additional Croton water to the Intermediate and High Level service areas that are normally served by the Catskill and Delaware Systems.

The Croton Water Treatment Plant (WTP) at Mosholu Golf Course and the associated water tunnels projects are being constructed to meet the public water supply and public health needs of the New York City, and to comply with State and federal drinking water standards and regulations. In recent history, water quality problems (mostly violations of the aesthetic standard for color) have resulted in the Croton System being removed from service on few occasions, typically during the summer and fall months (in four of the last several years—1992, 1993, 1994 and 1998). The entire system was shut down for most of 2000–2001 because of contaminants that leaked into the New Croton Aqueduct. The construction of the new Croton Water Treatment Plant (WTP) and the associated tunnels are part of the City's goal to provide high quality water to all its users while minimizing the risks associated with the use of chemicals.

Bids for the Contract CRO-313 were taken by the City on February 23, 2006 and the successful bidder was the Joint Venture of Schiavone Construction Co. LLC and John P. Picone Inc. In general the scope of work included in the contract consisted of the following:

- Construction of drill and blast tunnel with two 9 ft in diameter carrier (service) pipes to connect the Jerome Park Reservoir with the new Croton Water Treatment Plant. A steel service pipe was designated for high pressure service (HS) and a Reinforced Concrete Cylinder Pipe (RCCP) for the low pressure service (LS).
- Rehabilitate portions of the New Croton Aqueduct, to allow the treatment plant to be connected to the existing city water supply.
- Construct tunnel to supply raw water from the New Croton Aqueduct (NCA) to the new Croton Water Treatment Plant.

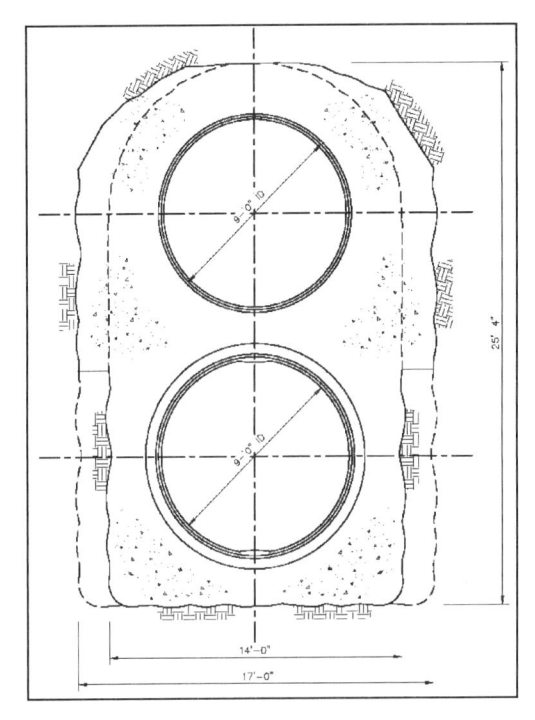

**Figure 1. High and low service treated water pipes in a tall horseshoe tunnel**

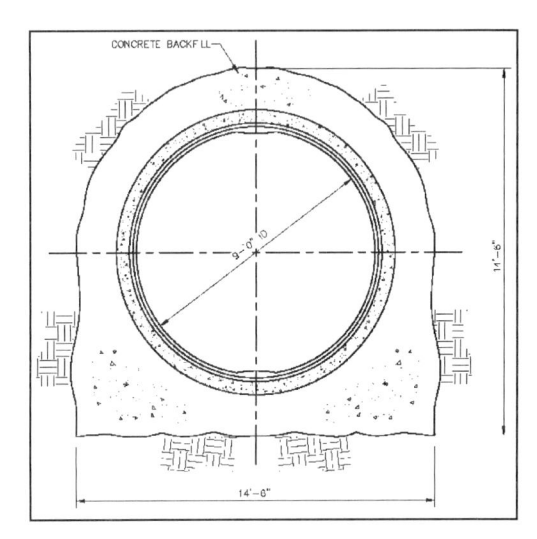

**Figure 2. Low service treated water pipe in a horseshoe tunnel**

## TREATED WATER TUNNELS

The original contract design called for excavation of a horseshoe shaped tunnel to carry two treated water service pipes: steel pipe for the high pressure service (HS) and RCCP pipe for the low pressure service (LS). The tunnel cross section, between WTP and Jerome Park Reservoir shafts (JPR), was a tall horseshoe shape 25'-4" high by 14'-0" or 17'-0" wide, depending on the support type required. The total length of the tall horseshoe tunnel was 3,658 feet. The pipes in the tunnel were stacked vertically one over the other with HS pipe on top (Figure 1). Both pipes are 108 inch internal diameters.

The LS RCCP pipe continued from JPR shafts to connect with the existing New Croton Aqueduct. This LS connection run was 14'-6" by 14'-6" horseshoe drill and blast tunnel about 615 feet long (Figure 2). The installation of the pipes in the tunnels included backfilling of the pipes with 4,000 psi concrete.

The anticipated ground conditions along the treated water tunnel and the corresponding support types, including support details, were shown on the contract drawings. Plan and profile drawings also showed anticipated limits for each support type, Rock Mass Quality Index (Q) values for crown and sidewalls; top of rock and groundwater elevations and the tunnel slope. The Q values indicated that the tunneling will be in very poor to good rock conditions. Suggested tunnel grouting details were also shown on the drawings along low rock cover reaches, fault zone and along crossing of the existing City Tunnel No 1.

The contract Geotechnical Baseline Report identified the location of the Mosholu Fault zone between Stations 20+00 to 30+00. North of the Mosholu Fault zone, anticipated Q values were in the range 2 to 6 (poor to fair). Within the Mosholu Fault zone the tunnel was expected to be in very poor to poor quality rock with Q values identified to vary from 0.3 to 4. Past construction records from City Tunnel No.1 and No.3 through the Mosholu Fault reported similar rock conditions when crossing the fault. South of the fault zone the rock quality was anticipated to be mostly fair.

In general, higher ground water infiltration was anticipated in the treated water tunnel excavation along the fault zone and along sections with low rock cover for total of about 1,400 feet. Poor rock quality was anticipated along the same zone for 400 feet requiring Type TD II support (Figure 3) and very poor conditions were anticipated for 1,000 feet requiring Type TD III support (Figure 4).

## RAW WATER TUNNEL

The Raw Water tunnel (RW) design consisted of 14'-6" by 14'-6" horseshoe excavation with 12 ft ID cast in place unreinforced concrete lining (Figure 5). Short sections of the lining were reinforced at the connection to the new WTP and at the crossing of the existing City Tunnel No. 1. Anticipated limits for

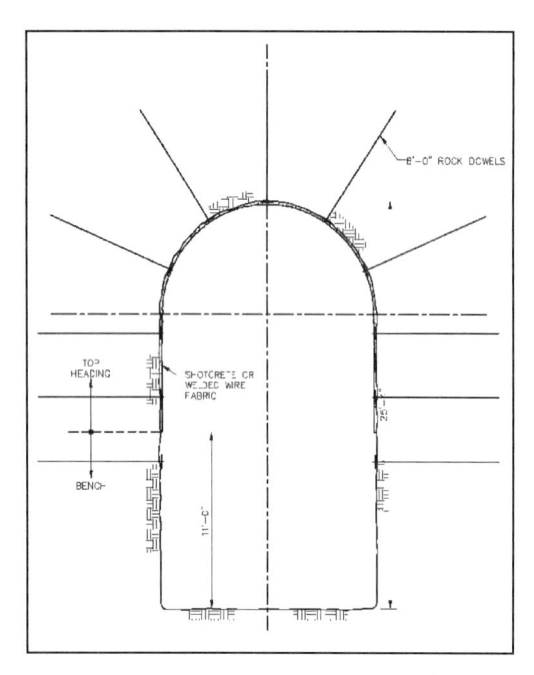

**Figure 3.** Support type TD II

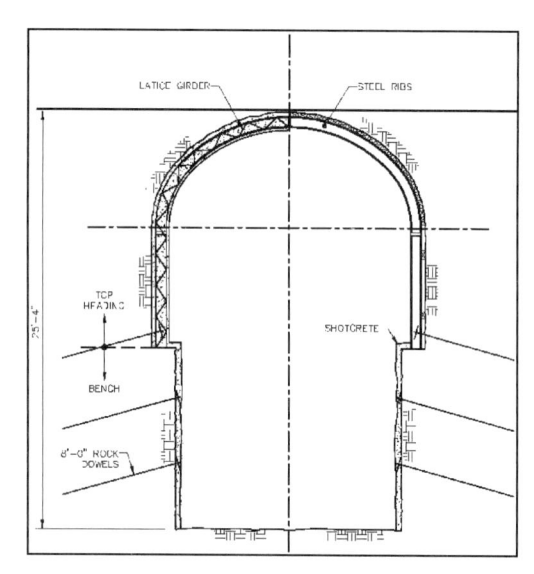

**Figure 4.** Support type TD III

different tunnel support types, ground conditions, and profile and alignment data were also identified on the contract drawings. Total length of the tunnel was 853 feet. Rock conditions along the Raw Water Tunnel were anticipated to be from poor to fair based on Rock Mass Quality (Q) requiring patterned rock bolts and shotcrete as initial supports.

Groundwater inflows were expected at isolated locations in the RW tunnel excavation, and were anticipated to intensify within the Mosholu Fault zone and along the crossing with City Tunnel No. 1. The predicted steady state inflow rate was in the range of 50 gpm (ungrouted tunnel) for the entire length of the excavation.

## SHAFTS

One construction/access shaft (Treated Water Shaft—TWS) was identified on the contract drawings for the Treated Water tunnels (TW) and another construction /access shaft (Raw Water Shaft—RWS) for the Raw Water (RW) tunnel (Figure 6). These shafts are located near the WTP site and are separated by approximately 30 feet. Stub tunnels extend from the TW and RW tunnels shafts for about 120 feet toward WTP for connection to the plant. The shafts by the WTP are the only construction/access shafts for tunnel excavations, pipes and final lining placements.

This construction access and connections to NCA and JPR chamber required that all tunnels were to be excavated down grade to the existing facilities.

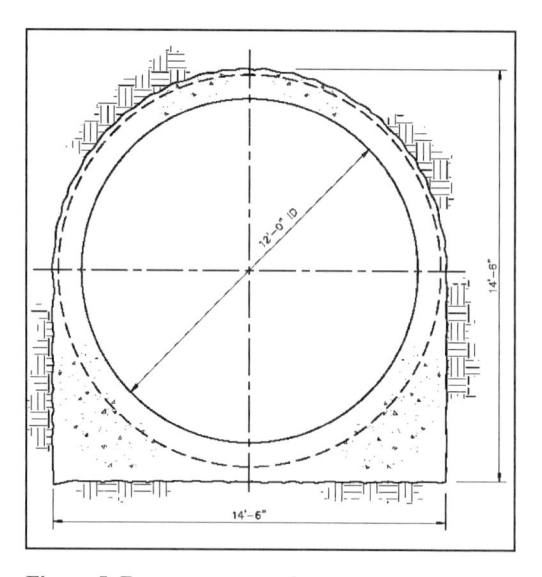

**Figure 5.** Raw water tunnel

The Low Service and Raw Water tunnels terminate at connection points with the existing NCA tunnel. The HS treated water pipe terminates at the JPR shaft chamber (Figure 7) with a riser which is located about 3,630 feet down stream from Treated Water Shaft. The LS treated water tunnel also has a riser at the JPR shaft chamber from which point it continues for another 613 feet to a blind connection at NCA tunnel.

The Raw Water Shaft, as originally designed, was approx. 160 feet deep. The connection of the RW

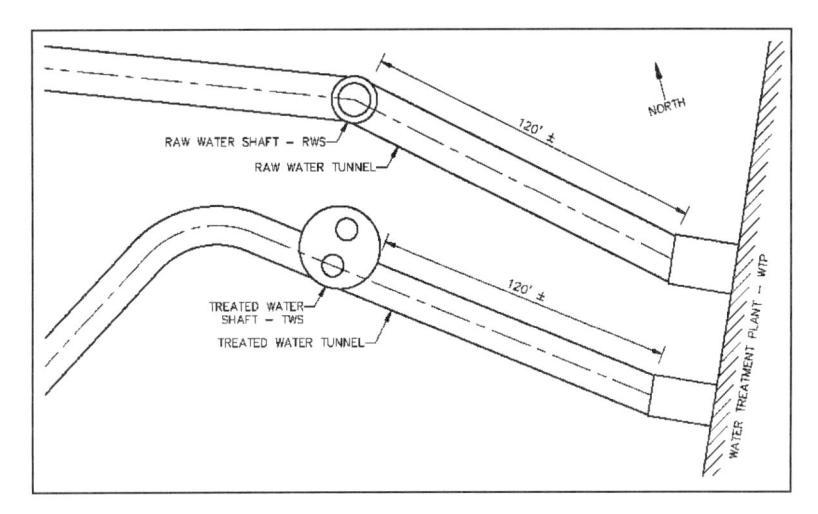

**Figure 6. Original shaft location layout at Croton Water Treatment Plant**

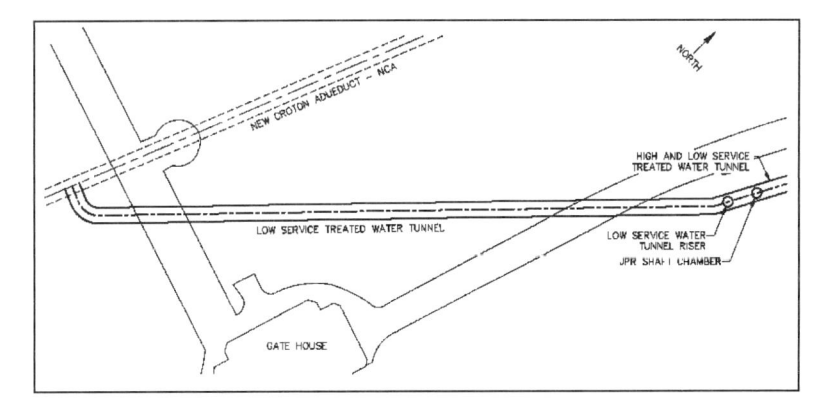

**Figure 7. Treated water tunnel risers at Jerome Park Reservoir**

tunnel to the WTP was made with a 55 foot vertical riser in the shaft. The riser continued to the ground surface where a cover was provided for inspections and maintenance access to the tunnel (Figure 8).

The Treated Water Shaft at the WTP was designed for two risers with access covers at the ground surface. Since the two treated water pipes were designed for installation one over the other in the tunnel, the access to the treated water pipe required a TEE and 90° elbow fitting to offset the riser (Figure 9).

The shafts for the risers at the JPR chamber were designated for excavation by raise-bore method since blasting was prohibited at this location. In the plan the two raise bore shafts were separated by 28 ft from center to center. The JPR raise-bored shafts terminate at the rock surface with temporary covers.

## PROPOSED REVISIONS

The Contractor proposed major design changes to the RW and TW tunnels and shafts for the Croton Water Treatment Plant:

- **Combined Raw and Treated Water Shafts**
  The Contractor proposed to realign the tunnels at the WTP to combine RW and TW shafts into a single larger shaft. The shaft geometry was changed to a single elliptical shaft allowing excavation of the RW and TW tunnels from a single access point. Since the TW tunnels were at higher elevation than RW tunnel, the foot print of the new proposed shaft was within acceptable size of 49 feet by 31 feet (Figures 10 and 11). This proposed revision required that the TW and RW tunnels be realigned to meet at the common shaft.

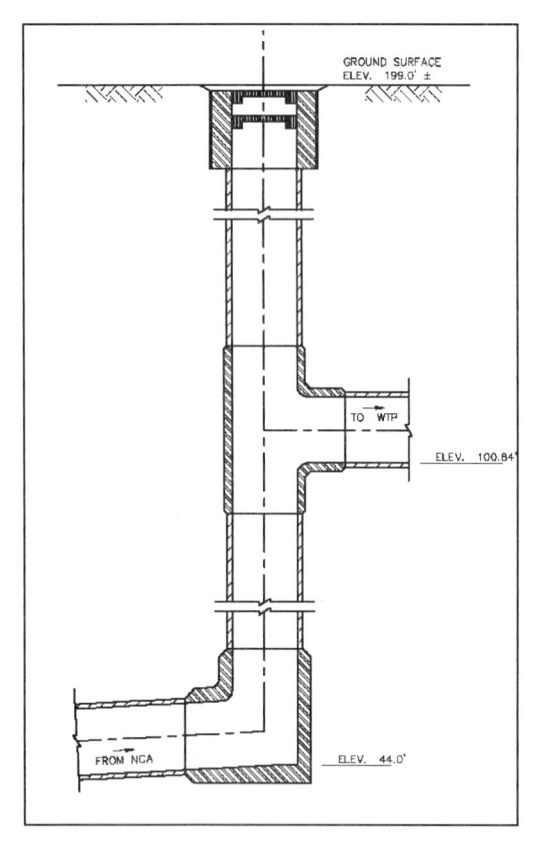

**Figure 8. Raw water tunnel risers at water treatment plant**

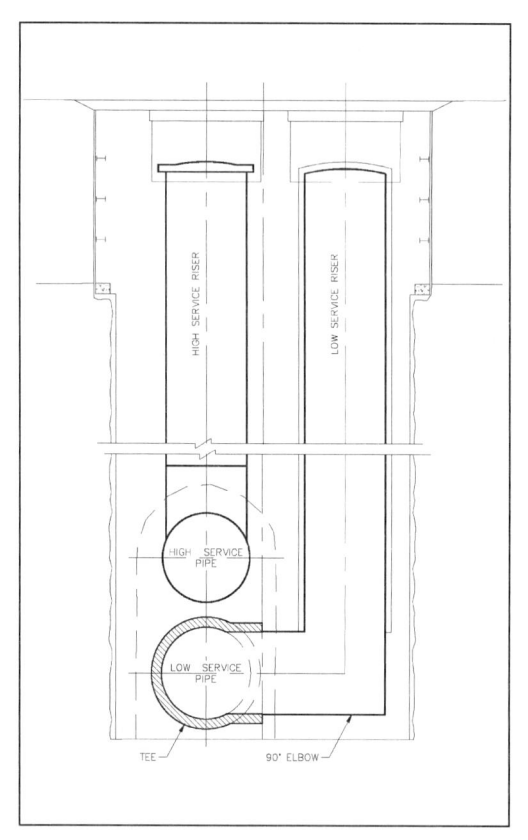

**Figure 9. Treated water tunnel risers at water treatment plant**

- **Revised Raw Water Tunnel Slope**
  The RW tunnel slope was revised to a single run from NCA connection to WTP without a riser. This proposal more than doubled the original 5.08% tunnel slope to 11.15% but it eliminated the need for a deeper shaft and a vertical riser between the tunnel and the connection point with the WTP (Figure 12).
- **Excavate TW Tunnels with TBM**
  This revision proposed to change drill and blast excavation of the double-high horseshoe tunnel for treated water pipes (LS and HS) to excavation with a TBM of two parallel tunnels—one for LS and the other for HS pipe. The tunnel excavation diameters would be the same for each service line and at the end of the each tunnel run the machine would be backed out into the Water Treatment Plant shaft (WTS). This change also required the establishment of a new alignment for the two parallel TBM tunnels. The tunnels would be started from a single heading tunnel at the WTS (Figure 13).

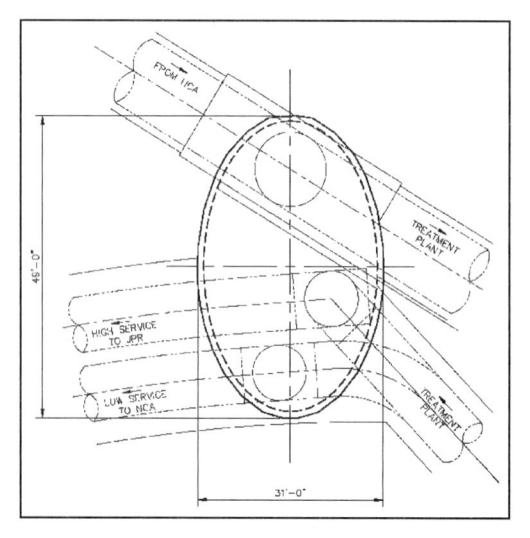

**Figure 10. Outline of the combined treated water shaft**

- **Permalok Pipe Joints for HS and LS TW Pipes.**
  Two types of pipes were required by the original contract—RCCP for the LS line and a steel pipe for the HS line. RCCP joints are steel with neoprene gasket and the steel pipe required welded joints. The Contractor proposed to install steel pipe with machined Permalok joints in place of the specified pipes. The use of steel pipes for each service line would result in a smaller excavation diameter and it would eliminate welding steel joints in the tunnel. Each Permalok joint was machined steel bell and spigot with two neoprene gaskets (Figure 14).

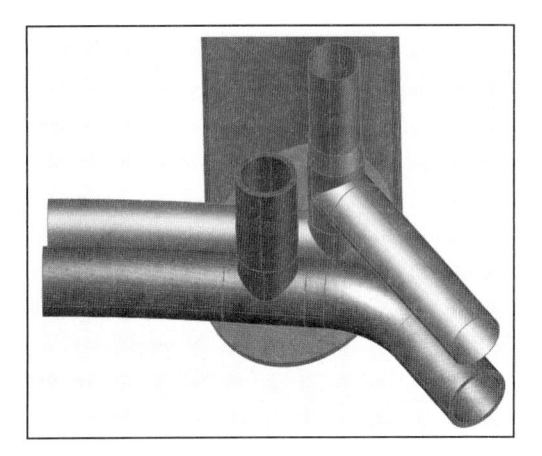

Figure 11. Isometric view of the combined treated water pipes in the shaft

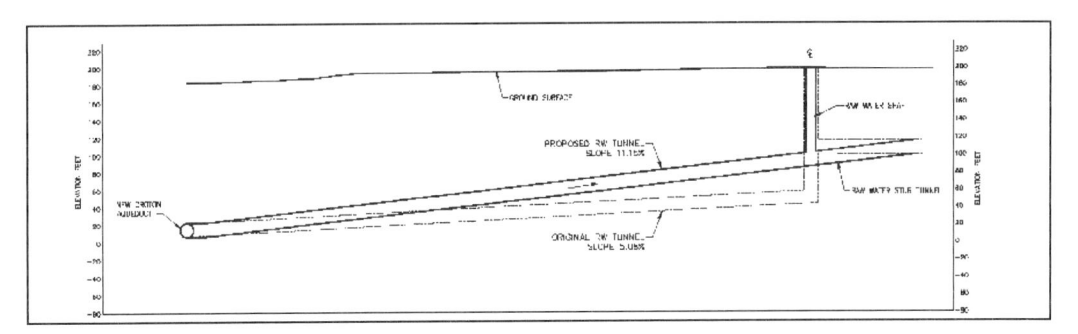

Figure 12. Revised raw water tunnel profile

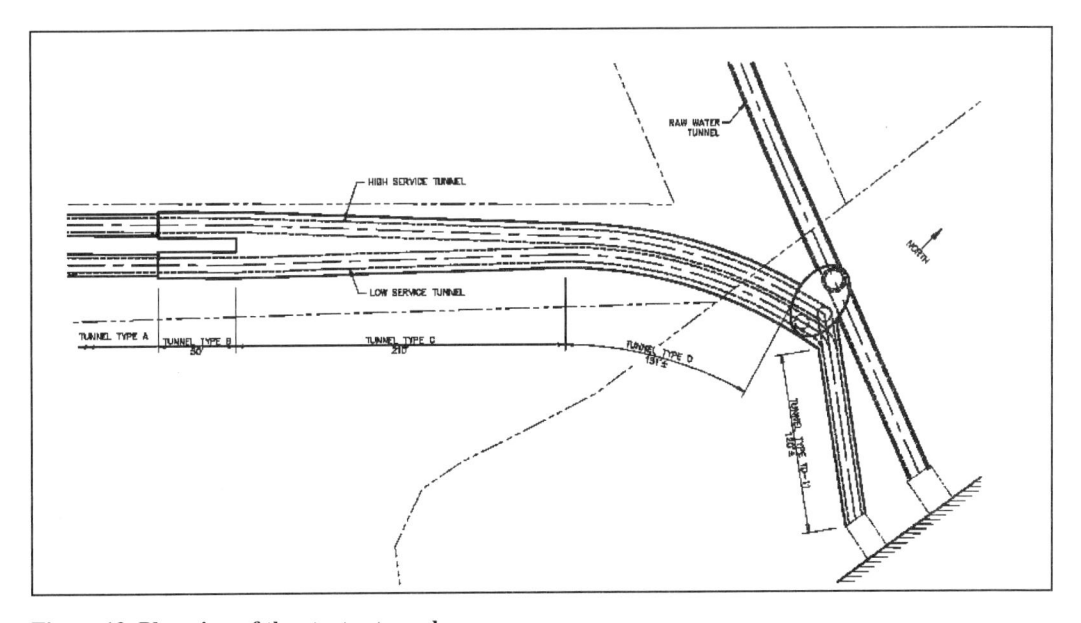

Figure 13. Plan view of the starter tunnel

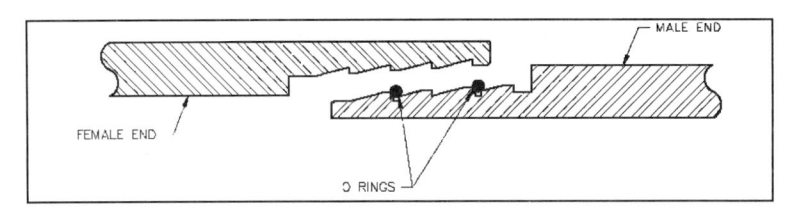

**Figure 14. T7 Permalok joint**

## BENEFITS AND PURPOSE OF THE REVISIONS

Combining the TW and RW shafts into a single shaft would result in one shaft overhead arrangement and a single mobilization. The excavation by drill and blast of the shaft, the Raw Water tunnel and starter chamber for both Treated Water tunnels could be accomplished more effectively with the same crew.

Revising the Raw Water tunnel slope would result in shallower shaft excavation with savings to the Owner. Higher tunnel profile provided 11 feet of additional vertical separation between the existing City Water Tunnel No 1 and the RW tunnel. In addition a 90° elbow required by the contract design at the bottom of the riser would be eliminated. The revision in the tunnel profile resulting in a shallower shaft combined in a single shaft excavation reduced the drill and blast volume in comparison to original design concept of two shafts.

One of the major reasons to propose a twin tunnel TBM excavation was that Schiavone had an access to a 12'-7" diameter TBM. For this reason, mobilizing the machine to drive only two 4000 foot tunnels was economically feasible. The particular TBM (Robbins) proposed for the Croton Tunnels was already proven performer in New York City on Con Edison's 1st Ave. Steam Tunnel and on NYC DEP's City Water Tunnel #3. The Croton Tunnels contract allowed for drill and blast tunnel excavation to proceed on three shifts. This presented certain concerns that the community may object and organize to prohibit blasting on the third (night) shift. Such actions by communities were successful in the past on other projects in New York City. Such a possibility in restricting the tunnel mining on the third shift could result in a delay to the project completion. That possibility would also jeopardize meeting the DEP's Consent Order milestone date.

In addition, excavating two tunnels by TBM instead of a drill-and-blast double-high horseshoe tunnel would address the following concerns:

- Major reduction in noise, dust and vibration
- Reduce water infiltration into excavations from rock fractures caused by blasting
- Reduce or eliminate impact on existing Water Tunnels #1 and #3
- Reduce safety risks—TBM excavations are safer
- Reduce transportation, storage and use of explosives
- Reduce security issues
- TBM excavation reduces disturbance of ground—an important factor for mining across Mosholu fault and sections with low rock cover
- With less disturbance of the ground, TBM excavation minimizes ground support requirements and ground water control
- Minimizes disturbance to all other above and below ground facilities
- Excavation of the TW tunnels presented 12 to 15 months potential early project completion

Proposal for substitution of welded steel pipe joints with Permalok Type 7 joint would eliminate majority of field welding, all in tunnel. With 40 feet steel pipe sections and machined Permalok joint, fewer joints were proposed thus increasing the installation quality on the project. Additional savings would be realized for the Owner if the RCCP was also substituted with ⅝ inch steel pipe with Permalok joints. Using steel pipe instead of the RCCP would result in smaller tunnel excavations since the placement of concrete pipe required one foot large excavation diameter by the TBM. The Permalok Type 7 (T7) joint is a machined joint with higher tolerances for roundness than AWWA standard. The T7 joint is a snap-on joint consisting of female and male ends (bell and spigot) with two neoprene gaskets. The T7 joint with neoprene O-ring gaskets mimics the RCCP joint which is also of steel with neoprene gaskets. The T7 joint was tested with potable water to pressure of up to 300 psi by the Permalok and the pipe was used for potable water high pressure service already on a project in California.

## INITIAL REVIEW AND ACCEPTANCE

The initial presentation of the proposed revisions met with favorable review results. The initial assessment presented to the NYC DEP indicated that a potential cost savings to the owner would be in the range of $11,300,000. Some of the sources of the anticipated savings were as follows:

- Reduction in tunnel excavation support items—Less ground disturbance is experienced with smaller size TBM excavations than drill-and-blast large openings. Potential for savings from reduced usage of the tunnel support items was anticipated at $2,200,000.
- Since early project completion was projected with use of TBM an earlier access to the JPR shaft and NCA tunnel could potentially result in additional $1,700,000 to $2,300,000 savings. Coordination with NYC DEP was required to attain an earlier access to the tunnel and the shaft site.
- Combining the two shafts by the TWP, RW and TW shafts, into a single shaft would produce another $900,000 of savings.
- Substituting welded joint on the HS steel pipe with T7 Permalok joint and reducing the pipe wall thickness from 1 inch to ⅞ inch would result in $2,900,000 savings.
- Additional saving $900,000 was offered for substituting RCCP in the LS tunnel with ⅝ inch thick steel pipe with T7 Permalok joints. In an alternate, a cast in place concrete tunnel lining with waterproofing membrane instead of RCCP would provide $3,000,000 savings to the owner.

However, the proposed substitution offered additional cost benefits to the owner in a form of reductions in city's potential liability costs associated with the project. $10,000,000 to $15,000,000 in liability cost avoidance could be realized by adopting the TBM excavation alone since the possibility of blasting restrictions on third shift could be avoided. As with every project in congested city setting, claims for damages resulting from blasting were anticipated.

With anticipated delays to the project completion resulting from the possibility of the elimination of third shift blasting potential costs associated with the consent decrees penalties could be avoided if TBM excavations were accepted. These penalties were anticipated to reach $55,000,000. Also, with shorter project duration possible, additional savings in a range of $5,000,000 to $10,000,000 would be realized from reduced project supervision and inspection costs.

In principal, the proposal met with favorable acceptance. However, before implementation, detail data, information, design documentation with revised set of construction drawings was needed as part of the review process and change order processing. Since the TBM excavation diameter depended on the pipe external diameter, the first item requiring resolution was the pipe and joint type. After, presenting additional technical documentation on the Permalok pipe joint, the owner ruled out the reduced steel pipe thickness for HS pipe as well as replacing RCCP with steel pipe and providing Permalok joint 7T instead a welded joint.

**RE-DESIGN**

Once the decision on the pipe types was made by the Owner (RCCP for LS and 1 inch thick steel pipe with welded joints for HS), the TBM tunnel excavation diameter was set at 13'-6". This excavation diameter would allow a 40 foot steel pipe sections to be transported for placement within curved alignment of the mined tunnel.

Plan and profile selection for the two TBM tunnels was influenced by two major factors:

- One—the tunnels must connect to the same end point locations and maintain the same surface access locations at JPR shaft.
- Second—the two TBM tunnels alignments must be separated by a safe distance based by rock quality and safe level of stress distribution in the rock pillar during TBM driving and from internal pressure service loads.

Phase 2D program was utilized to assess the stress distribution in the rock next to tunnels during TBM operation and under the service loads. Stress from The TBM griper pressure was modeled by conventional hand calculations to determine stress distribution in the rock pillar next to the already excavated tunnel. Internal working and surge service loads were analyzed for conditions when two tunnels are in service and when one is taken out of service and is emptied for inspection (Figure 15 and 16). A 25 feet center-to-center separation between the two TBM drives proved to be an acceptable distance resulting in safe level of stress changes in the rock.

Geotechnical profiles included in the bid documents summarized ground and water conditions and identified limits of tunnel support types that may be required along the drives. Because of the proposed changes, the original contract geotechnical profiles required revisions. The Raw Water tunnel required the revision because of the profile change and the Treated Water tunnels because of the change in excavation method (D&B to TBM) and the size of the excavation. To annotate the limits of the support types anticipated and to develop the corresponding detail of installation along each excavation reach, UNWEDGE software was utilized to predict sizes of rock wedge loosening and to select rock bolt lengths and spacing for the tunnels. Rock joints set used by the original designers were made available to the Contractor for the re-design purposes. Figure 17 shows an example of UNWEDGE analysis output.

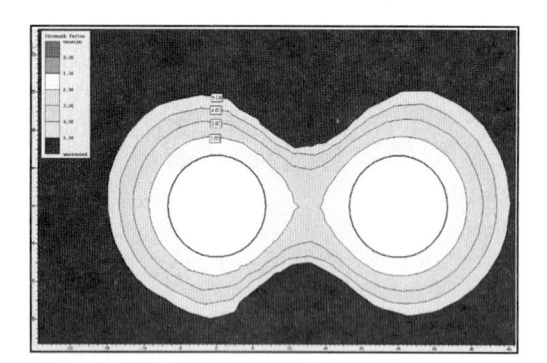

**Figure 15. Stress distributions—excavated tunnels**

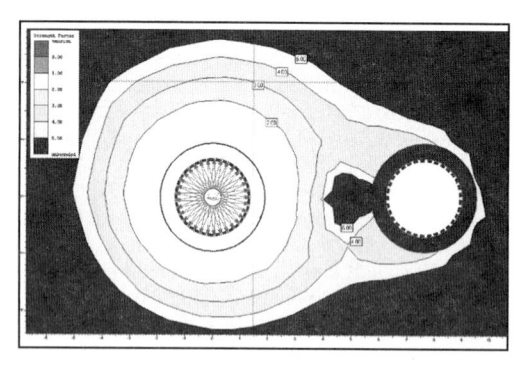

**Figure 16. Stress distributions—service condition**

However, to drive two parallel TBM tunnels from a single shaft at WTP required wide underground starting chamber. The chamber starts with a single tunnel drive at the shaft and wideness to a shallow horseshoe cross section to provide two portals for TBM starter stubs. The geometry of the starting chamber (starter tunnel) is shown on Figure 18. The chamber was situated in good quality rock, resulting in excavation supports consisting only of rock bolts. The starting chamber needed to be large enough to assemble the TBM and its trailing gear, to re-assemble it again when it would be backed up from the first drive and to provide sufficient width to accommodate a rail switch for muck cars and pipe carriers.

Backfilling of steel cement lined pipes with concrete was one of the major design challenges. It was desirable to place the pipe in forty foot sections, already lined with ½ inch cement lining at the manufacturing plant. Forty foot pipe sections minimize number of joints and welding within the tunnel, lower the number of trips by pipe carrier from the shaft to the heading and ultimately result in shorter placement time. The pipe selected for the project was SpiralWeld pipe, cement lined at the manufacturing plant. The cement lining was omitted from both ends at the joint area. The steel pipe joints were bell and spigot; with a ⅜ inch weld to be completed after the pipes were backfilled with concrete (Figure 19). The cement lining at the joint area was applied in the field after the pipe was backfilled and the joints were welded.

The original contact documents called for backfilling of pipes in three foot lifts with 4500 psi concrete and for a distance of not more than 140 feet. This was an acceptable requirement since the pipes were being placed one over the other within a large horseshoe cross section, with sufficient room for personnel access for slick line placement, removal and relocation. With TBM driven tunnels clearance between the steel pipe and the excavated diameter

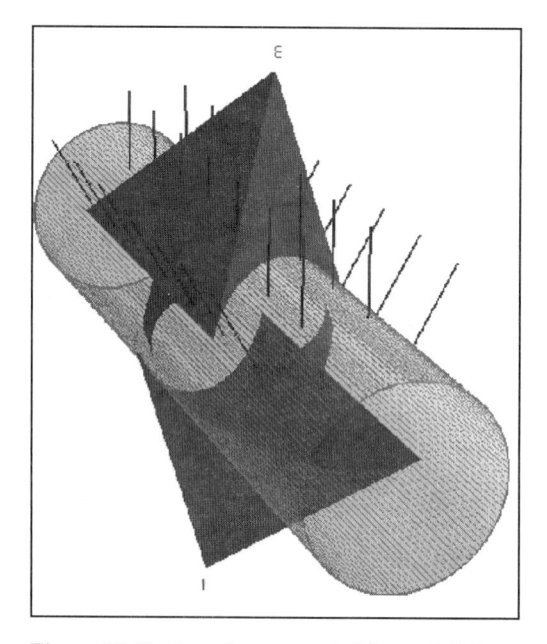

**Figure 17. Rock wedge supported by rock bolts**

was approximately 24 inches all around and with RCCP the clearance was only 1'-5". With ribs in place another 6 inches of clearance would be lost. In either case, this was not sufficient space for personnel access to handle slick line or perform any other work on the pipe exterior once the pipe was in the tunnel. As a result, it was obvious that the pipe sections must be supported at ends only: by the joint of previously placed section and at the free end by blocking against the rock before the next pipe section arrives.

In planning backfilling of pipe with grout or concrete, the backfill lift height is selected to prevent the pipe from floatation, excessive deflection and/or buckling collapse. It is a common problem

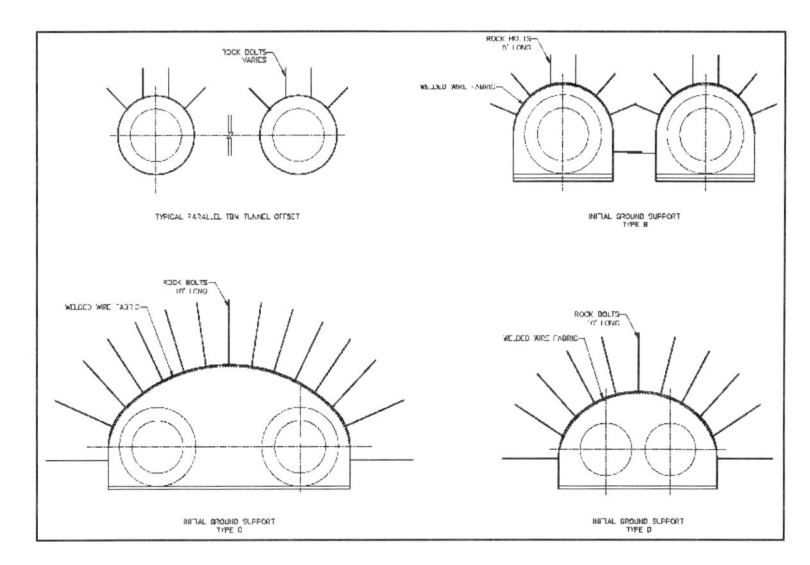

**Figure 18. Cross sections of the starter tunnel**

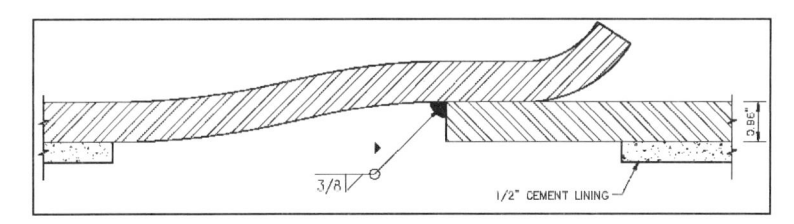

**Figure 19. Steel pipe joint details**

that contractors are faced with while backfilling relatively thin wall flexible pipes (steel, plastic or fiber reinforced resin) on slip lining projects or when placing large diameter steel pipes in tunnels. Most of the time on slip line projects to overcome floatation problems, the pipes are placed for a long distance, flooded internally with water and backfilled in lifts with low density flowable foam grouts. On the Croton Tunnels project, placement of 4,500 psi concrete on the outside of the steel pipe results in very high uplift force. The backfill operation resembled more concrete lining placement behind steel pipes than simple pipe backfilling. Flooding of short pipe reaches with water was not practical and would not sufficiently offset uplift (buoyant) forces.

In order to analyze the stresses in the steel pipe during backfilling, the steel pipe sections with cement lining as a composite structure were modeled in 3D structural program RISA. The purpose of the modeling was to test different blocking arrangements at the pipe section ends that would result in acceptable stress level in the cement lining (without developing cracks) and maintain deflections in the open end joint within tolerance to allow for subsequent pipe section placement when the concrete sets. Three pipe sections were modeled to mimic installation method: first pipe section joint was fully supported all around since it was pushed "home" into pipe joint fully supported by the concrete backfill. Second joint was modeled to allow movement (sliding) along longitudinal axis of the pipe joint, but was coupled as slave joint in $x$ and $y$ axis, mimicking movement and force transfer along the pipe joint that is not welded.

After many trial blocking arrangements, including placement of spuds through the pipe for support against tunnel walls, the acceptable solution was to block the pipe with timber blocking on the exterior and to install spider enragement on the interior at each pipe joint. Trials of pipe blocking with spuds along the pipe, to minimize exterior blocking efforts, indicated that unacceptably high stress levels were developing at the spud locations, resulting in tension stresses which would cause cracking and spalling in the cement lining.

The arrangement of exterior blocking and internal bracing at each joint proved to be the simplest combination of pipe supports to allow full depth concrete backfill placement while maintaining

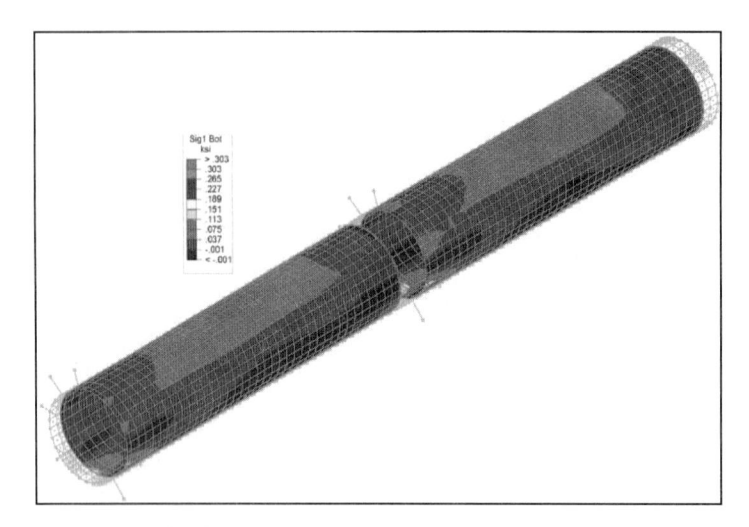

Figure 20. Stress in cement lining Sig1 Bot

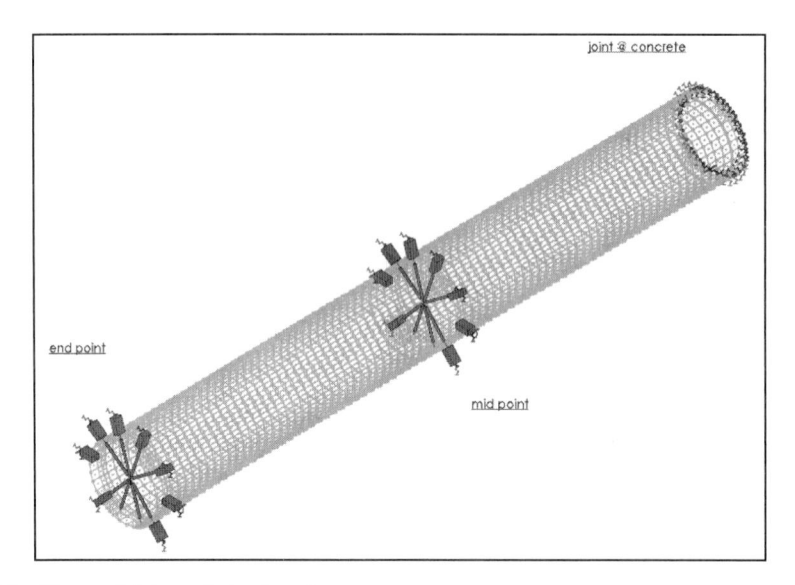

Figure 21. Blocking and internal pipe bracing

acceptable stress levels within the cement lining (Figure 20). Structural modeling of pipe sections as a composite member of steel and cement allowed for a more realistic prediction of stress levels in the steel and cement lining and allowed for predicting more accurately deflections of the pipe during concrete backfill placement. Introduction of the cement lining into the structural model allowed the pipe "member" to benefit from the added stiffening effect of otherwise flexible thin wall steel pipe. Figure 21 shows typical blocking and internal bracing (spider) arrangement. This support combination resulted in end pipe joint deflections that were within the desired tolerance of pipe joint to allow for the relative ease in placement of the subsequent pipe section.

Once the pipe support arrangements for concrete backfilling were selected, the structural pipe section model was tested for other handling and placement loads such as lifting, placement on the pipe carrier and supporting the pipe during final positioning for grade and alignment.

The first steel pipe sections placed in the tunnel were instrumented to observe deflections and stress in the steel during concrete placement. The initial recordings indicated that all stresses and deflections were within acceptable limits; however the final assessment of the instrumentation results was not completed at the time of this paper writing.

The analysis of RCCP for support and blocking arrangement for full depth concrete placement proved to be unexciting. The shorter pipe sections selected for the placement resulted in more frequent blocking resulting in lesser reactions and stress in the concrete pipe walls. Also, the heavier and stiffer concrete wall provided better counter-action to the uplift force than the cement lined steel pipe.

## PROJECT DESIGN DATA

The Raw Water Tunnel internal pressures design parameters were as follows:

Maximum tunnel flow:           290 MGD
Maximum short-term internal pressure:   58 psi

The Treated Water Tunnels internal pressure design parameters were as follows:

### Treated Water Tunnels:

*Low Level Service*

Maximum tunnel flow:           190 MGD
Maximum short-term internal pressure:   59 psi
Pipe (tunnel final lining):    9'-0" dia. RCCP

*High Level Service*

Maximum tunnel flow:           290 MGD
Maximum short-term internal pressure:   250 psi (revised)
Pipe (tunnel final lining):    9'-0" dia. cement lined steel pipe

## PROJECT STATISTICS

- Bid date: February 23, 2006
- Bid amount: $212,227,000.00
- Projected completion date: December, 2010
- Seven supplemental borings were taken by the joint venture to gain additional information in the fault zone and to learn more about rock permeability and water infiltration into tunnels during excavation. This additional information prompted grouting program conducted from the surface prior to treated water tunnels excavation.
- Water encountered in the tunnel: 150–200 gal/minute.
- After the fist tunnel drive (Low Level Service) the High Service Treated Water profile was lowered to avoid poor rock quality encountered along the profile with low ground cover.
- During the construction the surge pressure was revised by the owner in the High Level Service Treated Water tunnel to 250 psi. This change resulted in design revisions to access riser cover at water treatment plant shaft.

## PROJECT STAKEHOLDERS

Owner: New York City Department of Environmental Protection

Contractor: Schiavone Construction Co., LLC and John P. Picone Inc., a joint venture.

Revision designer: Dawn Underground Engineering, Inc.

# Keys to Success in Managing a Complicated Tunnel Project: City of Columbus—Big Walnut Sanitary Trunk Sewer Extension

**Michael J. Hall**
H.R. Gray, Columbus, Ohio

**John G. Newsome**
City of Columbus, Division of Sewerage and Drainage, Columbus, Ohio

**ABSTRACT:** The Big Walnut Sanitary Trunk Sewer Extension project, completed in the fall of 2009, offers many lessons learned in terms of a cost-effective approach to managing a tunnel project. Key lessons relate to the use of new technology, claims avoidance and communication with area residents, including 24/7 responsiveness. The project consisted of the installation of 9,900 lineal feet of 72-inch sanitary sewer, installation of five access shaft/manhole structures, installation of a tangential inlet drop structure and associated deaeration chamber/appurtenances, as well as the installation of 36-inch sanitary sewer and manholes. The construction management team encountered many challenges that could have resulted in potential claims situations.

## OVERVIEW OF THE PROJECT

H.R. Gray, under contract to DLZ Ohio, Inc., the design professional on this project, provided engineering and construction management services on the Big Walnut Sanitary Trunk Sewer Extension project. The project team, otherwise known as the City's Construction Management Team (CMT), consisted of field engineers and managers from H.R. Gray and designer engineers from DLZ. The CMT worked directly for and with John Newsome, project manager for the City of Columbus Division of Sewerage and Drainage (DOSD).

The $25-million contract was awarded to KMM&J Joint Venture. The contract Notice to Proceed was issued in January 2007 and Final Completion and project closeout was issued in November 2009.

This project utilized a 105-inch diameter tunnel boring machine (TBM) to tunnel through 9,300 linear feet of Ohio shale and 600 linear feet of soft-ground/mixed-face conditions. The TBM was a LOVAT RM 99/105 consisting of gage cutters, center cutters, shale rockers, and front-loading "ripper" teeth. The TBM cutterhead configuration was modified for the last 600 feet of soft ground by eliminating the shale rockers and installing front-loading "tiger" teeth and flood doors. Upon completion of the mining, the contractor utilized a pipe carrier for installation of 9,900 lineal feet of 72-inch sanitary sewer using 8-foot sections of Reinforced Concrete Pipe (RCP) with PVC T-lock liner for the tunnel lining. This project also included the installation of five access shaft/manhole structures and the installation of a tangential inlet drop structure and associated deaeration chamber and appurtenances as well as the installation of 15 lineal feet of 16-inch and 18-inch DIP, 20 lineal feet of 36-inch sanitary PVC sewer pipe and two 48-inch pre-cast manholes.

## PROJECT CHALLENGES AND CLAIM MITIGATION

The City and H. R. Gray encountered two major challenges on this project which led to claims that could have been significant. One challenge was mining through 600 feet of soft-ground/mixed-face conditions containing several nests of boulders and cobbles underneath and within the groundwater table. The contractor originally filed a claim for additional time and money to remove the multiple boulders and cobbles they encountered through this zone. The contract documents contained a pay item for boulder obstruction removal time, and Geotechnical Design Summary Report requirements that the TBM must be able to excavate through ground with boulder obstructions with a maximum dimension of two feet. The H.R. Gray team was on hand to take measurements of boulders immediately after they were excavated in order to get a factual detailing of the situation and document the associated downtime. Previous experience has shown that if this documentation

**Figure 1. Tunnel boring machine—hard-rock configuration**

**Figure 2. Tunnel boring machine—soft-ground configuration**

**Figure 3. Boulder and cobbles excavated in soft-ground zone**

**Figure 4. City's construction management team measures obstruction**

doesn't take place immediately, messages tend to get distorted from field foremen to contractor project managers and then back to the CMT field office. It is a best practice to provide complete documentation as soon as possible on any situation that may result in a potential claim. H.R. Gray was on site to take measurements and pictures of the boulders and cobbles immediately after excavation.

In the end, H.R. Gray provided extensive documentation about all aspects of the claim as well as documents about ambiguities in the dewatering system. With the assistance of this documentation, the claim was minimized.

Another major challenge was mining through a 200-foot soft-ground buried valley with a hard rock TBM. The contract documents required pregrouting of this area in order to stabilize the excavation and permit the rock TBM to excavate through the area. Permeation grouting, utilizing sodium silicate grout was conducted between station 11+00 and 13+00, where soil dips into the tunnel excavation. The purpose was to solidify the soft ground material and retard groundwater infiltration to allow mining through this zone using the TBM. Because the borings were further apart—the zone was different than what was predicted — the contractor filed a claim for $1 million for additional time and materials. The CMT was able to settle the claim for $250,000 with extensive documentation and calculations about what work should have taken place and materials that should have been used versus what actually took place. The claim was resolved concurrently with the field work enabling the completion of the permeation grouting in time for the arrival of the TBM through the zone, thus no additional claims or delays were encountered.

Often in claims reporting, there is a trend in which contractors provide excessive amounts of information to the CMT to cloud the waters with both pertinent and not pertinent information. In this case, H.R. Gray conducted an independent study and calculated what additional work was determined to be fair and justified. Because the results were from an independent source, it was easy to compare the data and the contractor had a better understanding of the situation. They agreed to the results of the study and reduced their claim by 75% and less than 1% of the entire project budget.

Both of these challenges resulted in claim situations for additional time and compensation. H.R. Gray's on-site management, extensive documentation, rescheduling and negotiations with the contractor were able to keep the project on time and budget. Several other challenges were overcome by the contractor and the CMT maintaining open communications and addressing every issue immediately and in good faith. In the end, the project finished on time and UNDER budget by $2.2 million or 9% of the total budget.

In the end, mitigating claims relies on two items: communication and documentation. Communication keeps all parties up-to-date on the status of issues and problems arising on the project. Successful communication must be timely, clear and effective. Likewise, complete and accurate documentation of all aspects of the project is critical to mitigating claims. Contemporaneous documentation is the key.

## PROACTIVELY DEALING WITH CLAIMS AND CLAIMS AVOIDANCE

When dealing with claims, it is important to take proactive measures. By doing so, an organization will save both time and money. To save time in the claims process, an organization must have continuous review of the schedule. Knowledge of the day-to-day changes and events will help a firm identify potential claims. Also, the firm will be able to identify the activities or events that have caused a delay or acceleration. Identifying claims early will help an organization collect and create the proper documentation to win the claim. Early identification also allows timely and proper notice pursuant to the contract requirements.

Saving time means saving money. To ensure money is saved in the claims process, an organization must identify and document a claim in order to effectively recover from, or defend against, the claim. Researching the claim and recreating the

**Figure 5. Permeation grouting setup**

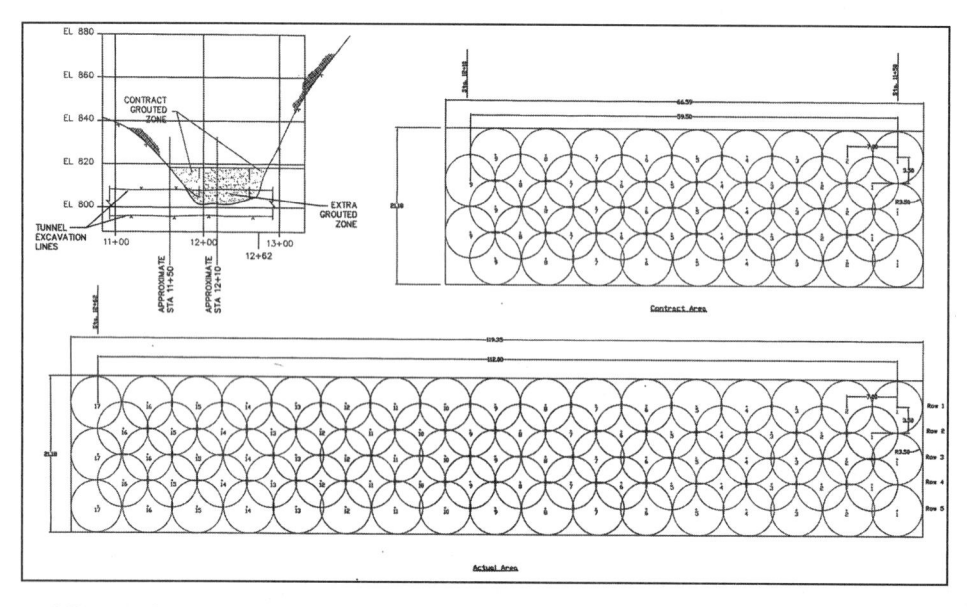

**Figure 6. Permeation grouting setup with tunnel profile**

documentation after the fact is costly. In addition, numerous factors related to time can wreak havoc on a project's bottom line. These issues may include overhead costs, equipment rental, price escalation, labor costs, lost profit, lost productivity, impacts incurred by subcontractors and other third parties, lost profits to businesses, and fines from governmental agencies.

To mitigate or minimize impacts on a project, it is necessary to identify ways the project can make up for lost time and money. Often, rescheduling or resequencing work will help recover lost time.

Claims avoidance begins with knowing the contract requirements. The contract helps avoid problems. It is imperative to know your organization's responsibilities, other parties' responsibilities and the ramifications of any party's failure to fulfill its responsibilities. Understanding the contract allows one to prepare a strategy to deal with problems before they actually arise. Identifying the vague areas and developing a plan to deal with potential problems not addressed by the contract are important. Such knowledge and understanding of the contract will help address the everyday problems that occur. It is critical to follow the contract.

If problems or issues occur, they need to be documented as they arise. Ongoing documentation has numerous benefits. First, it allows your organization to remain aware of continuing problems in order to follow up on a regular basis. Problems will be identified in the early stages before they have a significant impact on the schedule and budget. Finally, authorization will be obtained (or provided) before performing any work outside the scope of the contract.

In addition to documentation, communication is key in avoiding claims. Effective written communication must include both formal and informal methods. Communication keeps all parties aware of the status of important activities and issues.

## ON-SITE FIELD ENGINEERS AND CONSTRUCTION MANAGERS

As was mentioned earlier, one way to avoid and mitigate claims is accurate, contemporaneous project documentation. For this project, the CMT provided engineers and construction managers on site every day work was performed, seven days per week and sometimes two shifts per day. This extensive oversight ensured project success for all parties involved.

The CMT also had a good working relationship with the contractor, which was another key to success. When questions arose during construction, the CMT was able to translate exactly what the designer had intended immediately. If the contractor had any field issues, the team responded with an answer and potential solutions. Whenever questions surfaced, the CMT was there to provide direction immediately—thus saving costly downtime and/or contractor error. The constant on-site management also provided for an improved quality of work as well as ensured that materials were used per specification.

This contract had both lump-sum and unit-cost activities. Because all the quantities going into the work were documented and reviewed by the CMT, the team was able to save the City approximately $500,000 in unit-cost pay items.

## PROACTIVE COMMUNICATION KEY IN AVOIDING PROBLEMS

Before the construction of shafts #3 and #4, it was determined that due to the dewatering of the area, the potential of disturbing the water table would affect some domestic water wells. The homeowners that may have been affected were notified in advance. Homeowners who were notified of potential problems on this tunnel project responded overwhelmingly to this proactive notification. The CMT also notified the township officials of potential problems.

The township received numerous inquiries about the CMT homeowner notification. The township Board requested a CMT presence at their meeting. Prior to the Board meeting, the CMT met with the contractor to formulate a response plan for any water-well problems that might arise.

The prime contractor secured a subcontractor to provide a temporary water supply if needed. The CMT procured a bottled-water supplier to provide drinking water. The CMT also procured cases of bottled water that would be delivered by the CMT when they were notified of disruption of water supply to a homeowner. The CMT also provided telephone numbers for both the contractor and the CMT that would respond 24 hours a day. The CMT was successful in accommodating all homeowners who experienced a well problem.

The CMT attended a township Board meeting to explain the plan to accommodate homeowners if needed. Several homeowners attended the meeting and expressed their concerns to the Board members. They stated that due to the proactive approach the CMT was taking, they were confident that the homeowners would be well taken care of. The CMT not only attended this meeting, but as dewatering took place, the CMT attended two other meetings to keep everyone informed of the success of the plan for accommodating the homeowners.

The CMT did inform the township Board every two weeks (day before their Board meeting) of any problems and the response to resolve them.

**Figure 7. Vertical boring machine (VBM) designed by The Kassouf Company, utilized for shaft excavation through shale**

## 24/7 RESPONSIVENESS TO HOMEOWNERS IN CONSTRUCTION AREA

The CMT developed an extensive plan to communicate with homeowners affected by the blasting and construction. The blasting took place in the first stages of the shaft construction while encountering limestone. Thereafter, the vertical boring machine (VBM) was used for the remaining excavation through Ohio shale. The VBM excavated upwards of 20–25 feet per shift. It began with a door-to-door handout to residents within a 300-foot radius of the proposed work area. The handout explained the process that takes place when blasting is performed. For claims documentation purposes, the handout included a form that could be completed if the homeowner wanted a survey performed of their residence. The survey consisted of both video and still photos of the inside and outside of the home. It was explained that the construction crews would follow the City of Columbus ordinances pertaining to both noise and working hours. Each homeowner was asked to respond within five days to schedule an appointment to have a survey performed or to decline the offer. The handout included a list of telephone numbers to contact the CMT if they had any questions.

With a 90% response rate to the handout, the team received telephone calls from more than 100 homeowners requesting a survey of their home. The remaining 10% of the homeowners that did not respond were sent an additional copy of the original handout via certified mail and again a response was requested within five days. After the five day period, only two residents did not respond. A registered letter was sent to these homeowners and in the end, the goal of receiving responses from 100% of the residents was reached.

The CMT also attended the Blendon Township Board of Trustees meetings to explain what affect the tunnel construction would have on their community. In addition, the Board was updated monthly with information about any inquiries received from the residents affected by the construction. The homeowners were assured that the CMT would respond to any inquires or complaints 24/7 and any construction updates would be distributed immediately. Understanding that it may be more convenient for residents to contact the CMT after traditional work hours, homeowners were told that their inquiries were welcome at any time of the day. The CMT contacted the president of the homeowners' association and conveyed that she or any homeowner was welcome at the CMT office to review the construction drawings and ask any questions they had about the project. Finally, school bus pick-up and drop-off locations were analyzed with the Westerville City School District and, upon further review, two locations were temporarily relocated for safety reasons.

In another instance, the CMT assisted the City during the construction of the outfall shaft, which was designed and constructed in the center of the Remington Station Apartment complex, less than 100 feet away from single family homes. The City acquired a temporary construction easement for this shaft construction and, in return, had to meet time and noise restrictions. The CMT assisted the City by coordinating with the apartment management and residents to appease everyone and minimize complaints that the City would have to handle.

## TUNNEL RESCUE TEAM

Before tunnel construction could begin, it was mandated under federal law that a rescue team be formed. This team is usually assembled using both contractor employees and the local fire department. On this project, the tunnel rescue team was comprised of the contractor's employees and a member of the CMT staff. The local fire department was a backup to the contractor's team. The team trained every other Saturday and participation was on a volunteer basis for the CMT. The participation of the CMT showed the contractor and the City that the CMT was involved with every facet of the project.

The CMT's position on the team was that of the communication officer. This position not only dealt with the City, press and the public, but allowed the CMT frontline information not secondhand information.

The tunnel rescue team was prepared to handle any emergencies that would arise, but was never needed.

Figure 8. Close proximity of outfall shaft to residents

## ACTIVE OWNER PARTICIPATION IN THE PROJECT

By maintaining the City's project manager as an active member of the CMT structure, many Requests for Information (RFI), Requests for Proposal (RFP), and Change Orders (CO) were turned around quickly and mitigated in a responsive fashion leading to potential cost and time savings to the contractor. With the Owner being active in the decision-making process and a vital part of the interaction with the contractor, an open line of communication was maintained allowing for a quick resolution of potential risks, both to the contractor and the Owner, as they arose.

Having active owner participation in the management of the process also allowed the City to auction off materials (property) that were no longer required upon construction completion. This allowed the City to recover some costs for items which in the past would have been stored and unlikely to have ever been used. By coordinating the efforts of the CMT and the contractor, auction items were able to be quickly identified, sold, and these additional monies were returned to the City.

## POSITIVE CONTRACTOR CONTRIBUTIONS

The Kassouf Company's willingness and dedication to use a proactive project approach, communication and a knowledgeable construction superintendent contributed to a foundation of cooperation that permitted a successful project completion. This resulted in open communication between all parties and issues were dealt with on an immediate basis minimizing claims and delays.

## DISPUTES REVIEW BOARD (DRB)

Within the scope of the contract, the City made an allowance of $250,000 for a Disputes Review Board (DRB), an entity that was established to assist in the resolution of disputes and claims arising out of work performed under this contract. The partnering between the City, the contractor and CMT enabled claims to be dealt with on a project level and negated the need to utilize the DRB which in turn saved the City money on the project.

## CONCLUSION

Through this project, the City of Columbus, Division of Sewerage and Drainage, gained further understanding of the value of an on-site construction management team as an integral part of public works construction. The CMT's proactive communication addressed many potential issues before they became problematic and reassurances that the area residents had a system in place to communicate their concerns placated the area elected township trustees. Quick response to issues aided in community acceptance of the project. CMT construction managers were able to document and monitor the construction activities and materials, providing accurate on-site project information from which to make decisions. Active participation of the CMT and the owner, created the ability to address potential claims at the construction site which led to fewer claims and claims at a reduced cost. Because of these activities, the City received a refund in the amount of $2.25 million by written change order.

## ACKNOWLEDGMENTS

The authors would like to thank the following for their assistance in the preparation of the paper and those team members involved in this successful project:

- City of Columbus, DOSD
- H.R. Gray
- DLZ, Inc.
- Jenny Engineering Corp.
- The Kassouf Company
- Constructive Communication, Inc.

# Drop Structures and Diversion Structures for the East Side Combined Sewer Overflow Project, Portland, Oregon

**Roy F. Cook**
Parsons Brinckerhoff, Los Angeles, California

**Tammy R. Cleys**
Bureau of Environmental Services, City of Portland, Oregon

**Tony O'Donnell**
Kiewit Underground District, Omaha, Nebraska

**Tom Corry**
Kiewit Bilfinger Berger JV, Portland, Oregon

**ABSTRACT:** The East Side Combined Sewer Overflow Project is being built to control combined sewer overflows (CSOs) from existing outfalls located along the east bank of the Willamette River in Portland Oregon. This paper describes the design and construction of the system of shafts and diversions that will divert flows to the deep storage and transport tunnel. These drop and diversion structures, and access shafts were excavated to depths of up to 180 feet in challenging alluvial soil conditions below the groundwater table.

The drop structures and access shafts were constructed within large diameter slurry wall supported excavations that allowed build-out to occur while tunneling activities continued below. Diversion structures were constructed in fill built-up over a century of industrial development along the bank of the river as well as the underlying alluvial soils. A major consideration in the design and construction of diversions is the need to maintain discharge at each outfall during overflow events and until the system becomes operational.

## INTRODUCTION

The East Side CSO (ESCSO) Tunnel Project is the last major component of the City of Portland's CSO Program, the purpose of which is to reduce overflows into the City's Willamette River and Columbia Slough from its combined sewer system. The program was started in 1991 and is scheduled for completion in 2011.

The ESCSO Project as well as a deep tunnel, includes the diversion and transfer of combined sewer overflows from thirteen outfalls. This paper describes the design and construction of the system of shafts and diversions that will divert flows to the deep storage and transport tunnel. These drop and diversion structures, and access shafts were excavated to depths of up to 180 feet in challenging alluvial soil conditions below the groundwater table.

### Project

Along the east bank of the river, overflows from existing outfalls will be rerouted by diversions to the deep tunnel. The tunnel is approximately 5½ miles long and has an internal diameter of 22 feet. At its downstream end, it connects with the West Side CSO (WSCSO) Tunnel. The storage capacity of the combined ESCSO and WSCSO tunnels is 101 million gallons. Figure 1 schematically shows the layout of the system on the east side of the river identifying the outfalls being diverted and the locations of the shafts where drop structures transfer flows to the tunnel. Table 1 indicates the outfalls to be diverted and the flows used to estimate the sizes of the diversions and drop shafts. It also identifies the locations where overflows to the river will be allowed during rain events that exceed the system's design storms.

## DROP STRUCTURES

Drop structures transfer flows from near surface systems to deep collector tunnels. Major projects, most notably Chicago's Tunnel and Reservoir Plan (TARP) in the 1970s through 2000s, and Milwaukee's In-Line Storage System Project in the 1980s and 1990s led the development of dropshaft

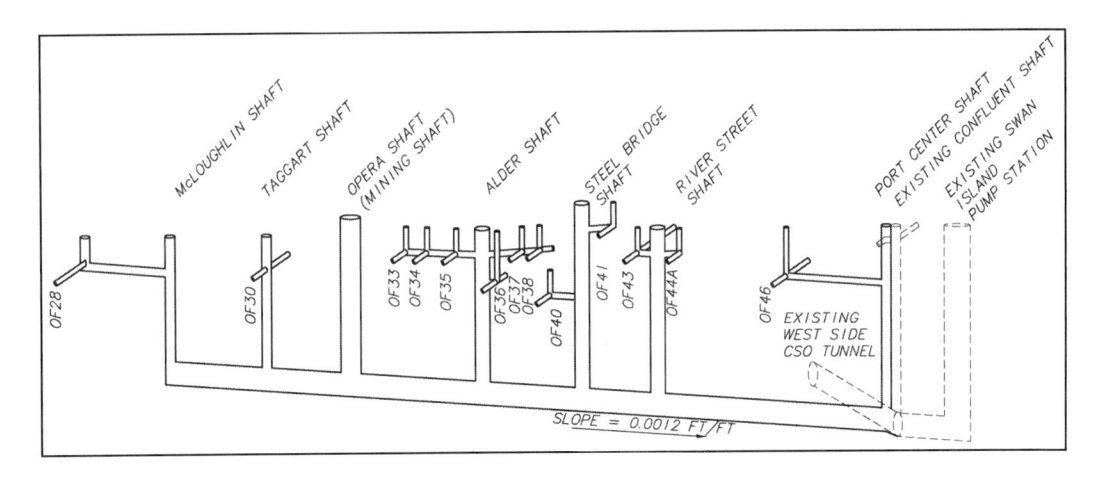

Figure 1. Schematic of east side CSO diversions

Table 1. Proposed diversions

| Shaft | Outfall | Final Configuration | 25-year/6-hour (cfs) | Summer 6 4th Storm (cfs) |
|---|---|---|---|---|
| McLoughlin | Insley (OF 28) | Closed | 380 | 66 |
| Taggart | Taggart (OF 30) | Overflow to river | 965 | 318 |
| Alder | Alder (OF 33) | Closed | 6 | 3 |
|  | Alder (OF 35) | Closed | 8 | 5 |
|  | Alder (OF 34) | Closed | 4 | 1 |
|  | Alder (OF 36) | Overflow to river | 297 | 121 |
|  | Stark (OF 37) | Closed | 521 | 129 |
|  | Oak (OF 38) | Closed | 114 | 26 |
| Steel Bridge | Sullivan (OF 40) | Closed | 507 | 205 |
|  | Holladay (OF 41) | Closed | 199 | 60 |
| River Street | Beech/Essex (OF44/44A) | Closed | 50 | 26 |
|  | Wheeler (OF 43) | Overflow to river | 151 | 46 |
| Port Center | Beech/Essex (OF 46) | Overflow to river | 258 | 89 |

design in the U.S as the need for more efficient and economic structures was recognized. Drop structure design is important because:

- The turbulence generated by the free-falling waste water flows can release hydrogen sulfide (H2S) and other malodorous gases that can lead to public complaints regarding smell.
- Turbulence can introduce air into the tunnel that reduces storage capacity and causes sewer pressurization and leading to violent releases of air and wastewater at access locations.
- Access to carry out drop structure maintenance once in operation is likely to be difficult.

As part of the design for the CSO projects, a study was undertaken to select the appropriate type of drop structure for the conditions. This study looked at how to:

- Transfer flow down with minimum turbulence and dissipate energy from the falling flow.

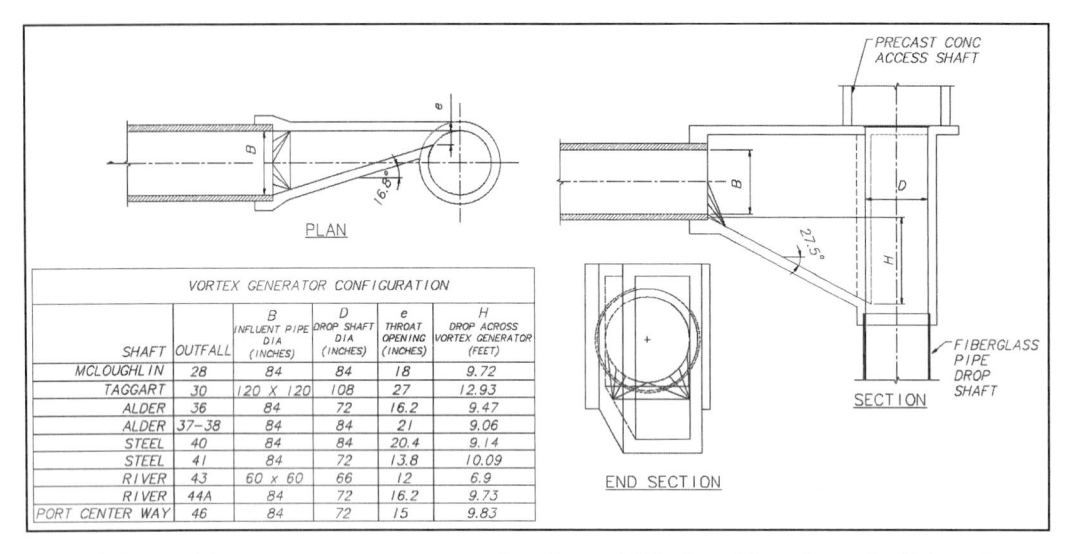

**Figure 2** Tangential vortex generator geometry (based on published model study results, Jain and Kennedy 1983)

| SHAFT | OUTFALL | B INFLUENT PIPE DIA (INCHES) | D DROP SHAFT DIA (INCHES) | e THROAT OPENING (INCHES) | H DROP ACROSS VORTEX GENERATOR (FEET) |
|---|---|---|---|---|---|
| MCLOUGHLIN | 28 | 84 | 84 | 18 | 9.72 |
| TAGGART | 30 | 120 X 120 | 108 | 27 | 12.93 |
| ALDER | 36 | 84 | 72 | 16.2 | 9.47 |
| ALDER | 37–38 | 84 | 84 | 21 | 9.06 |
| STEEL | 40 | 84 | 84 | 20.4 | 9.14 |
| STEEL | 41 | 84 | 72 | 13.8 | 10.09 |
| RIVER | 43 | 60 x 60 | 66 | 12 | 6.9 |
| RIVER | 44A | 84 | 72 | 16.2 | 9.73 |
| PORT CENTER WAY | 46 | 84 | 72 | 15 | 9.83 |

- Minimize the quantity of air drawn down into the tunnel potentially reducing storage volume.
- Ensure operational reliability given the difficulties with access to the drops during operations.
- Minimize structure degradation from corrosion and abrasive wear.

A drop shaft design in which the flow spirals around the shaft as a vortex was selected for the projects. To generate the vortex flow, the inlet structure to the drop shaft was designed as a tangential vortex flow generator. This type of inlet structure causes the approach flow to enter the dropshaft tangentially such that the flow clings to the wall and as it drops, maintains a central air-core along the length of the shaft. The geometric configuration for the tangential vortex generators used on the project is shown in Figure 2. The geometry was originally developed from the hydraulics model work carried out (Jain and Kennedy, 1983) for the Milwaukee drop shafts.

As well as requiring effective transfer of the flows, the drop structures must be simple and readily constructible. The construction of drop structure's ancillary components—underground chambers for inlet structures, de-aeration chambers, air relief shafts and dropshafts—in the saturated ground conditions found along the banks of the Willamette River was recognized as challenging. As a result, an alternative approach to that historically used to construct these structures within separate underground excavations was needed. For the project, it was determined that these structures would be most safely and most effectively constructed within large diameter shafts that serve multi-purpose uses. These uses include locations for:

- Dropshafts and associated tangential vortex generators.
- Access shafts used for:
- Tunnel maintenance and inspection during operations
  – Removing air from the tunnel introduced by the flows down the dropshafts
  – Providing additional system storage capacity and dampen surges
  – Inspection and maintenance of the TBM during construction.

With this approach, the drop structure including all its components including tangential vortex inlet, drop shaft and de-aeration facility can be constructed as part of the shaft build-out. And where necessary, several drop structures and an access shaft can be fitted within the shaft footprint. The shaft configurations used for the ESCSO Project are shown in Table 2.

**Shafts**

For the ESCSO Project, six shaft excavations, with inside diameters ranging from 49 to 56 feet and with depths ranging from 100 to 180 feet were constructed. These shafts were built out with drop structures transferring flows to the tunnel from thirteen outfalls, with access shafts that would be used both for monitoring, inspection and maintenance during operations and as relief shafts for air introduced into

Table 2. ESCSO shafts

| Shaft Name | Existing Outfall | Final Shaft Type | Dropshaft Depth | Dropshaft Diameters | Internal Slurry Wall Diameter | Purpose of Shaft During Construction |
|---|---|---|---|---|---|---|
| McLoughlin | OF 28 | Open | 65 ft | 7 ft | 49 ft | Retrieval shaft |
| Taggart | OF 30 | Open | 68 ft | 9 ft | 56 ft | TBM maintenance |
| Opera | | Backfilled | | None | | Mining shaft |
| Alder | OF 33/35 | 22 ft. dia. access shaft | 84 ft | 4 ft | 49 ft | TBM maintenance |
| | OF 36 | | 79 ft | 6 ft | | |
| | OF 37/8 | | 63 ft | 7 ft | | |
| Steel | OF 40 | 22 ft. dia. access shaft | 74 ft | 7 ft | 49 ft | TBM maintenance |
| | OF 41 | | 111 ft | 6 ft | | |
| River | OF 43 | 22 ft. dia. access shaft | 86 ft | 5.5 ft | 49 ft | TBM maintenance |
| | OF 44A | | 81 ft | 6 ft | | |
| Port Center Way | OF 46 | Open | 72 ft | 6 ft | 49 ft | Retrieval shaft |

the tunnel from the dropshafts. Four of the shafts (Alder, Steel, River and Taggart) also serve for inspection and maintenance of the TBM during construction. The shafts at Port Center and McLoughlin are used for retrieval of the TBM at the end of the north and south drives. A seventh shaft built at the Opera Site serves as a construction shaft only and will be abandoned prior to start of operations.

The design for these shafts takes into account the geometric configurations of the inlet structures (vortex generators) and dropshafts within the shaft footprint, hydraulic and air relief considerations, property needs, site constraints, proximity to outfalls and proximity of the lowest sanitary connection.

*Shaft Excavation*

Excavations for the shafts are primarily within the Troutdale Formation—a dense gravel and cobble material in a sandy/silty matrix—but also encounter Artificial Fill, Sand/Silt Alluvium and Gravel Alluvium. For each shaft:

- A diaphragm wall is constructed using slurry wall construction techniques. The wall—nominally 42-inches thick—is installed prior to shaft excavation. It provides a continuous concrete lining around the perimeter of the shaft and is taken down below the proposed depth of the excavation. The wall which acts as a cylinder in ring compression supporting the ground and cutting off lateral groundwater inflows.
- Excavation within the diaphragm wall is performed in dry and wet soils using a clamshell in combination with a chisel for harder material.
- Once the excavation is complete, the bottom is sealed by means of a tremie plug with a keyway and steel dowels tying the plug into the diaphragm wall.
- The shaft is then dewatered and built-out.

As the ground conditions are heterogeneous, slurry wall excavation had to deal with such challenges as: removal of boulders in the Troutdale and Gravel Alluvium; loss of bentonite slurry in permeable open zones above the groundwater table, particularly within the Gravel Alluvium and open zones of the Troutdale Formation; and caving and sidewall over-excavation.

*Shaft Bottom Build-Out*

With the shaft excavated and dewatered, the shaft bottom is built out to let the break-in to the shaft be accomplished under flooded conditions. Reinforced concrete side walls form a pathway into which the TBM can tunnel. Within these walls, the bottom sections of the drop shafts are constructed and connected into the tunnel flow channel by an adit. Once the construction of these sidewalls is completed, the pathway between them is filled with a combination of granular fill and compacted density fill (CDF). After this, water is used to fill the remainder of the shaft to equalize the external groundwater pressure during the tunnel break-in.

*Tunnel Break-In and Break-Out of Shaft*

With the shaft flooded, the TBM mines through the slurry wall and into the granular/CDF-filled pathway. Once the break-in is sealed by grouting from the tunnel, the shaft is dewatered and the CDF excavated from around the TBM, allowing its inspection and maintenance or for the shafts at each end of the drives for TBM retrieval.

- Grouting through the tunnel lining is carried out to ensure that the penetration through the slurry wall was sealed.
- The shaft is dewatered and the gravel removed from the shaft bottom and the CDF removed to expose the TBM.
- Maintenance of the TBM including the cutter head is then carried out in the safety of the shaft.
- Once the TBM is ready to proceed, the granular fill is replaced and the shaft once again flooded. Then with the hydrostatic conditions equalized between the inside and outside of the shaft, the TBM breaks through the slurry wall and out of the shaft.
- Once the TBM has exited the shaft, the tunnel segments are installed throughout the breakout and the penetration sealed by grouting from the tunnel.
- The shaft is dewatered again and the granular fill is removed again and the tunnel lining exposed.

*Completion of Shaft Build-Out*

Two basic shaft build-out configurations were developed to satisfy hydraulic design and operational considerations. Both configurations provided an equipment and personnel access to the tunnel for operational maintenance purposes, drop shafts, and surge volume and air relief (see Table 1). They are termed—Open Shafts and Closed Shafts (see Figure 3):

Open Shafts are used at the termination of tunnel drives (Port Center Way and McLoughlin) where the start of shaft build-out must wait until the tunnel drive is completed and where flow surge and air relief volume are needed by the design (see Figure 3—Open Shaft Concept). Each open shaft consists of:

- Flow channel through the shaft bottom that channels the tunnel flow through the shaft.
- Concrete side walls to the flow channel containing adit connections and vertical shaft components of the drop shafts. Side walls are constructed to the top elevation of the vortex generator. The side walls are the foundation for the base of the vortex generator.
- Drop shaft(s) formed using fiberglass pipe stacked vertically and embedded within the concrete side walls.
- Vortex generators built within the concrete side walls, where feasible, or built as a structure bridging across the open area within the shaft.

Closed Shafts are configured to allow for their build-out while tunneling continues. Each closed or mine-through shaft consists of:

- Flow channel (using the segmental tunnel lining) through the shaft bottom that channels the tunnel flow through the shaft. Openings were cut in the tunnel lining for the adit connections and the connection between the access shaft and the tunnel.
- Concrete side walls outside the flow channel containing adit connections and vertical shaft components of the drop shafts. The side walls were constructed above the tunnel elevation and support a beam and slab structure that supports the circular access shaft.
- Concrete beam and slab structure that frame a 14 foot by 14-foot square opening made in the crown of the tunnel and supports the lining of the access shaft and transfers the loads from the backfill to the shaft base.
- Access Shaft forming a 22-foot diameter vertical opening (using tunnel segmental rings as a lining) from the 14-foot by 14-foot penetration for the tunnel crown to the ground surface.
- Backfill formed from CDF and placed between the slurry wall excavation support and the access shaft lining.
- Drop shaft formed using fiberglass pipe stacked vertically within the CDF backfill.
- Vortex generators built within the CDF backfill that forms the foundation for the base of the vortex generator

Figure 4 shows the pipelines stacked above one another in a closed shaft. The vortex generators are yet to be constructed from these pipeline terminations and connect to the dropshafts. Figure 5 provides a view of shaft construction during build-out with the central access shaft and the formwork for one of the vortex inlet structures that will generate the vortex flow in the dropshaft.

## DIVERSION STRUCTURES

The diversion structures intercept and divert flows from existing outfalls and reroute them to the drop structures (within the large shafts) that then transfer them to the storage and conveyance tunnel. The diversions must be built out during construction to ensure that the existing outfalls remain operational and discharging to the river until the system is ready in 2011 to be connected to the tunnel and made active. During the pre-construction phase of the Contractor's contract, diversions were modified to plan for this and to allow effective implementation

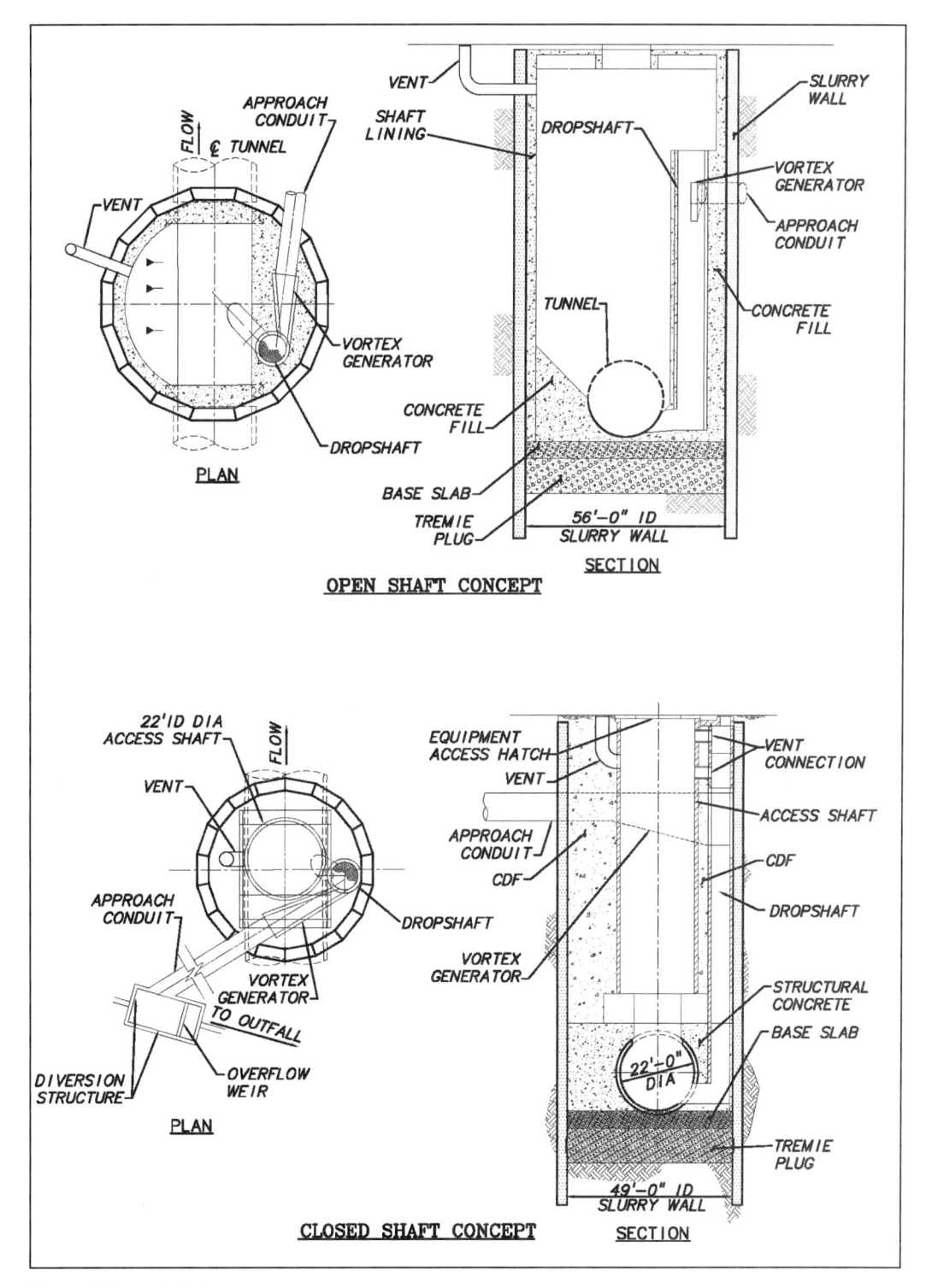

Figure 3. Dropshaft layouts

**Figure 4. Stacking of OF36 and OF37 pipelines in Alder shaft**

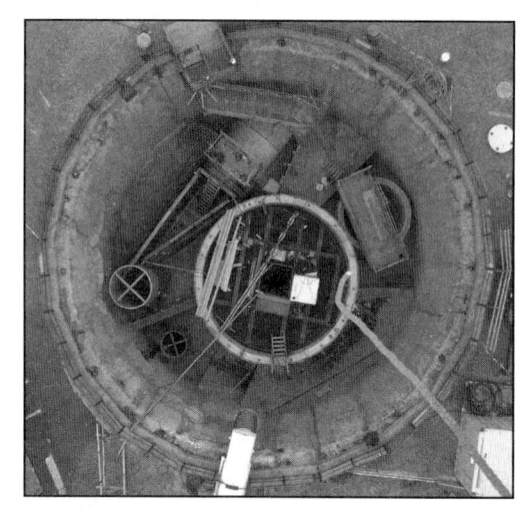

**Figure 5. Alder shaft build-out**

of construction to ensure that the system would be operational on schedule.

Primary means for ensuring that the outfalls remain operational during construction but can be readily diverted at the end of the construction period was the incorporation of Y-shaped structures at the diversions that provided flow-through pipes and diversion pipes (temporarily left plugged during construction). To activate the system, the flow-through pipes are permanently sealed and the temporary plugs in the diversion pipes are removed.

Four of the thirteen outfall pipelines will remain with a relief to the Willamette River. The tunnel system cannot accept CSOs beyond the design limits and must overflow to the river to prevent back-ups within the system upstream. The overflow elevation within the system is set to El. +18. The outfalls allowing relief to the river have backflow gates to prevent the river flowing back into the diversion structures and then into the tunnel system. The other outfalls are designed for the complete diversion of flows to the tunnel system with the outfall pipes plugged downstream of the diversion structure and with storm flows from pipes downstream of the plug allowed to continue to discharge to the river.

Detailed hydraulic modeling was used to analyze the complex hydraulics for each diversion structure individually and to evaluate the performance of alternative diversion structure configurations. Since the outfalls are generally perpendicular to the river, each diversion generally included a ninety degree bend to connect to a diversion pipeline or drop structure as part of its configuration. Available site areas near the river were a major consideration in locating the diversions, particularly at sites with overflow weirs that required additional space.

Diversions at OFs 28, 36, 37, 38 40, 41, 43 and 46 were connected by 84-inch microtunneled pipelines to drop structures in the shafts. The microtunneling is described in Overby et al. (2009) and has included a 3,055 foot long single drive microtunnel. OFs 30 and 43 were sufficiently close to shafts that they were connected by concrete structures built in open excavations and OFs 33, 34 and 35 were diverted by means of approximately 1,100 feet of open cut, 18-inch and 30-inch diameter reinforced concrete pipe at depths of up to 25 feet.

These near surface structures had to deal with areas of fill built up over a century of industrial development along the banks of the Willamette River as well as poor, soft underlying soft alluvial deposits. Various methods were used to control groundwater, depending upon ground conditions including dewatering with wells, sumping, and jet grouting. Excavation support generally used soldier piles and lagging support and secant piled shafts.

**OF28 Diversion**

OF28 is diverted within City right-of-way at a street intersection located within a quiet residential neighborhood. At this location, the existing 72-inch diameter concrete outfall pipe is approximately 54 feet below the ground surface and built within the Troutdale Formation. Since the diversion structure is

**Figure 6. OF28 "Y" diversion structure**

**Figure 7. OF38 precast concrete diversion structure**

approximately 900 feet from the McLoughlin Shaft, a diversion pipeline connects the structure to the drop structure within the McLoughlin Shaft.

To minimize disturbance to the neighborhood, the construction took place in two phases. The first phase was the construction of a 22-foot inside diameter secant piled circular shaft used as the retrieval shaft for the MTBM constructing the diversion pipeline. The second phase was the construction of an L-shaped excavation, approximately 40 feet long by 25 feet wide at its maximum and supported with soldier piles and lagging and internally braced. Since the excavation bottom is close to the groundwater table, groundwater inflows to the excavation is controlled by sumping. Within this excavation, a reinforced concrete Y structure for the diversion and the temporary flow-through pipe was built (see Figure 6). To activate the system in 2011, the diversion is built out by penetrating through the secant wall and connecting to the diversion pipeline. At this stage, the temporary flow-through pipe is permanently plugged.

## OF30 Diversion

OF30 is diverted close to the Taggart Shaft and does not require a pipeline to connect the diversion to the drop structure in the shaft. This site is located on a bluff overlooking the river. At this location, the existing nominal 120-inch diameter concrete outfall pipe is approximately 70 feet below ground surface and is built within the Troutdale Formation.

The diversion is built within an irregular shaped excavation supported with soldier piles and lagging and internally braced that abuts the Taggart Shaft. Within the approximately 60-foot long by 50-foot wide (at its maximum) excavation, a reinforced concrete Y structure is built. It consists of the diversion connecting to the vortex generator built within the shaft, a temporary flow-through pipe that allows flows to the river to continue until the system becomes operational, and an overflow weir wall (top elevation set at El +18) as this outfall will remain open. This site will be used as a swale for storm runoff from a nearby Oregon State highway project.

## OF33, 34 and 35 Diversions

OFs33, 34 and 35 are 18-inch concrete, 15-inch concrete and 16-inch sewer pipes respectively. They are diverted and consolidated by a pipeline along Second Avenue in the Central Eastside Industrial District. The diversion pipeline was constructed by open cut and pipe ramming methods with the diversions constructed as manholes. The diversion pipeline for these consolidated flows terminates in a plunge type drop structure at the Alder Shaft—the only plunge used on the project.

The diversion manholes and diversion pipelines are constructed at depths up to about 25 feet primarily within artificial fill. The fill was placed at the end of the nineteenth century to elevate ground levels and provide land for industrial development. Within this area, the streets had originally been plank roads supported on timber piles. These and trestle-supported railroad tracks were abandoned in place when the ground level was raised. As a result, the open cut encountered difficult excavation conditions. As well as coping with the ground, the construction had to ensure that business activities along Second Avenue were not disrupted.

## OF36 Diversion

OF36 is diverted within City right-of-way at a location a couple of hundred feet from the Alder Shaft. At this location, it is an 84-inch brick sewer. The irregular shaped excavation for the diversion is approximately 50 feet long by 25 feet wide at its maximum. It is supported with soldier piles and lagging and internally braced. The excavation depth is approximately 31 feet through artificial fill and very soft low to medium plasticity silt. A permeable saturated gravel layer is found beneath the silt. Sumping within the excavation was required to maintain bottom stability. Nearby structures were monitored to establish that the dewatering did not result in building settlement or damage.

The diversion is built as a reinforced concrete structure within the excavation. The structure includes the connection to the OF 36 diversion pipeline, the temporary flow-through pipe that allows flows to the river to continue until the system becomes operational, an overflow weir wall (set at El +18), and a flap gate to prevent back flows from the river entering the system during operations.

**OF37 and 38 Diversions**

OF37, a 72-inch horse-shoe shaped pipeline with a stone block invert and OF38, two pipes (a 42-inch and a 24-inch pipe) are located in the central Eastside Industrial District on Third Avenue. A series of structures and a microtunneled pipeline divert the flows from the outfalls, consolidate them and transfer them to a drop structure at the Alder Shaft. These structures are built to depths of about 40 feet in artificial fill and the underlying Sand/Silt Alluvium.

OF37 diversion is built within City right-of-way in a rectangular excavation supported by a slide rail support system with sheeting used around the existing outfall pipe. Within the 16-foot deep excavation, the diversion and the temporary flow-through pipe that allows flows to the river to continue until the system becomes operational. Since the excavation is located at a busy intersection, traffic control and the need to minimize impacts to the local businesses were major considerations at this site.

OF38 diversion is constructed within City right-of-way as a secant piled circular shaft. Since the shaft bottom is beneath the groundwater table, it was excavated in the wet and the shaft bottom placed as a four-foot thick reinforced concrete tremie slab. To hold the slab down against uplift pressure, vertical steel posts were embedded in the tremie and anchored to the secant piled wall. The diversion structure was built out within the excavated shaft with a connection to the diversion pipeline (see Figure 7).

**OF40 Diversion**

OF40 is diverted at a location beneath the I-5 overpass that is routed along the east river bank. At this location, the existing 72-inch diameter concrete outfall pipe is approximately 25 feet below the ground surface and excavated through artificial fill and Sand/Silt Alluvium.

The construction took place in two phases. The first phase was the construction of a circular excavation supported with steel ribs and shotcrete. Ribs and lagging were primarily selected because of the limited headroom under the overpass. This excavation was used as the retrieval shaft for the MTBM constructing the diversion pipeline and for the build-out of part of the diversion and a maintenance hole. The second phase was the construction of an irregular shaped excavation supported with soldier piles and lagging and internally braced. Within this excavation, a reinforced concrete Y structure for the diversion and the temporary flow-through pipe was built. The diversion connects to the OF40 diversion pipeline that terminates at a drop structure located at the Steel Bridge Shaft. The temporary flow-through pipe allows flows to the river to continue. To activate the system in 2011, the temporary flow-through pipe will be permanently plugged and the flow routed 90 degrees through the diversion.

**OF41 Diversion**

OF41 is diverted at a location in front of the Rose Garden, the professional sports and entertainment facility in Portland. This is a heavily visited area and is served by the MAX, Portland's light rail system. At this location, the existing 60-inch diameter outfall pipe is approximately 30 feet below the ground surface.

The construction took place in two phases. The first phase was the construction of a rectangular shaped excavation supported with soldier piles and lagging and internally braced. The excavation is through artificial fill and Sand/Silt Alluvium. This excavation was used as the retrieval shaft for the MTBM constructing the diversion pipeline and for the build out of the connection between the diversion and the pipeline. The second phase was the construction of another excavation, in which a reinforced concrete Y structure for the diversion and the temporary flow-through pipe was built. The diversion connects to the OF41 diversion pipeline that terminates at a drop structure located at the Steel Bridge Shaft. The diversion incorporated a drop from the outfall pipe to the diversion pipeline so that it would have sufficiently cover over the MAX light rail tracks. A spiral drop was used to minimize odors. The temporary flow-through pipe allows flows to the river to continue. To activate the system in 2011, the temporary flow-through pipe will be permanently plugged and the flow routed through the diversion structure.

Since the construction takes place in a public area where crowds congregate, construction activities were scheduled to avoid disruption to the public during events at the Rose Garden.

**OF43 Diversion**

OF43 is located close to the River Street Shaft and does not require a pipeline to connect the diversion

**Figure 8. OF46 diversion pipe base slab layout**

to the drop structure in the shaft. At this location, the existing 62-inch diameter outfall pipe is approximately 35 feet below ground surface and is built within the artificial fill and Sand/Silt Alluvium on the bank of the river

The construction takes place in two phases. The first phase was the construction of a rectangular shaped excavation, approximately 43 feet long by 23 feet wide and supported with soldier piles and lagging and internally braced. Because of groundwater considerations, the bottom of the excavation was jet grouted to prevent groundwater inflows. Within this excavation, a rectangular reinforced concrete box diversion structure is built containing the temporary flow-through pipe and an overflow weir wall. The second phase is the construction of a short length of open cut reinforced concrete box connecting the diversion structure to the vortex generator in the River Street Shaft. To activate the system in 2011, the temporary flow-through pipe in the diversion will be permanently plugged and the flows routed through the diversion.

**OF44A Diversion**

OF44A is diverted at a location within City right-of-way near the river. At this location, the existing 72-inch diameter outfall pipe is approximately 28 ft below ground surface and is built within artificial fill and Sand/Silt Alluvium.

The diversion is built within an irregular shaped excavation approximately 30 feet long by almost 30 feet maximum width. Because of groundwater considerations, the bottom of the excavation was jet grouted to prevent groundwater inflows. Within this excavation, a reinforced concrete Y structure for the diversion and the temporary flow-through pipe are. To activate the system in 2011, the temporary flow-through pipe will be permanently plugged and the flow diverted through the diversion.

**Figure 9. Flow-through pipe formwork**

**OF46 Diversion**

OF46 is a 78-inch pipe that runs beneath the Albina Railroad Yard and flows into the river close by an active cement storage silo facility. The only available location for the diversion structure was at the river bank.

The diversion is built within a rectangular shaped excavation, approximately 38 feet long and 26 feet wide (see Figure 8) and supported with soldier piles and lagging, walers and tiebacks. The excavation is approximately 31 feet deep through artificial fill and Sand/Silt Alluvium. Within the excavation, a reinforced concrete Y structure was built. It consists of the diversion connecting to a pipeline that connects to the vortex generator built within the Port Center Way Shaft, a temporary flow-through pipe (see Figure 9) that allows flows to the river to continue until the system becomes operational, and a weir wall as this outfall will remain open.

The excavation is on the river bank with its west side close to the steep riprap protected river bank. In the past, this site had been a dock supported on timber piles. The piles were still in place and had to be removed from within the excavation as construction proceeded. The existing OF 46 concrete pipe was also supported on timber piles and the pipe had to be temporarily supported in place as excavation was carried out beneath the pipe and the timber piles removed.

## SUMMARY

The ESCSO Project is a large and complex construction project. It includes seven major shafts, nine vortex type drop structures and thirteen diversions. By using similar designs and construction techniques for each of these similar project elements, efficiencies were achieved in terms of the use of resources and the application of lessons learned. This was particularly successful with regard to the shafts and the drop structures. However, the variable ground conditions, the work sites and their associated constraints, especially with regard to community impacts and the different functions associated with their operation resulted in a wide range of different solutions being applied to the construction of the diversion structures.

The City of Portland's contracting method for the ESCSO allowed for pre-construction involvement of Contractor staff on the constructability issues of the design. The flexibility of this contracting approach also adds value during construction with the Contractor able to submit value-added changes that can be acted upon in a timely manner. Such changes have saved the City money, the Contractor time, and the public inconvenience.

## ACKNOWLEDGMENTS

The authors wish to acknowledge the work of Mr. David Dailer (of CH2MHILL) with regard to design for the dropshafts and the Kiewit/Bilfinger Berger JV and Portland's Bureau of Environmental Services Construction Management staff.

## REFERENCES

Jain, S.C. and J.F. Kennedy, 1983, "Vortex-Flow Drop Structures for the Milwaukee Metropolitan District Inline Storage System." Prepared for the Milwaukee Metropolitan Sewerage District, Iowa Institute of Hydraulics Research Report No. 264, University of Iowa.

St. Anthony Falls Hydraulics Laboratory, University of Minnesota, 1975, "Model Studies of Dropshafts for the Tunnel and Reservoir Plan." Prepared for the Department of Public Works, Chicago, Illinois.

Iowa Institute of Hydraulic Research, University of Iowa, February 1998, "Hydraulic Design of Vertical Drop Structures" for Massachusetts Water Resource Authority Subcontract No. 104274, February 1998.

Overby C., M. Roberts, and C. Kolell, 2009, "The Longest Drive—Portland's CSO Microtunnels" North American Society for Trenchless Technology (NASTT) and International Society for Trenchless Technology (ISTT), International No-Dig Show 2009.

*Special Session: Operational Criteria
and Functionality for Highway Tunnels*

# National Tunnel Inspection Standards (NTIS)

## Jesus M. Rohena y Correa
Federal Highway Administration, Washington, DC

**ABSTRACT:** The safety and security of the highway bridges and tunnels in the country is a priority for the FHWA. Recent events such as the I-35 Bridge collapse in MN, and the Suspended Ceiling collapse in the Central Artery Tunnel in Boston, MA, have highlighted the need for inspection of bridges and tunnels. There is no requirement to inspect highway tunnels and to report the findings to FHWA, as is the case for bridges.

## INTRODUCTION

FHWA estimates that there are more than 300 highway tunnels in the United States. FHWA bridge inspection regulations were developed as a result of the Federal Aid Highway Act of 1968 (23 U.S.C. 151). But tunnels were not included in these inspection regulations. Therefore, there is no existing mandate to inspect tunnels and to report those inspections to FHWA. However, there are some States that have voluntarily furnished this information to FHWA.

In 2003 FHWA contracted Gannett Fleming (GF) to develop a Tunnel Management System. GF conducted a survey among all the States as part of that contract. The findings related to the inspection of tunnels are listed below.

1. Currently, there is no uniformity on how frequently tunnels are inspected.
2. The frequency of tunnel inspections varies from 1 month to 10 years.
3. Some owners in cold climates walk through air ducts on a daily basis to identify potential icing problems due to water leakage.
4. Some owners inspect mechanical and electrical equipment on a daily basis and many others perform such inspections on a monthly basis.

The average age of tunnels is 44 years. The average age of the tunnels in the Interstate system is 39 years. Since a tunnel can be a very complex transportation facility, some tunnels have components that should also be inspected such as electrical, mechanical, ventilation, fire safety, and security systems.

## Why an NTIS?

There are many reasons why I believe an NTIS should be developed. The issue of tunnel inspection has gained more public attention recently after a fatal accident in the Central Artery Tunnel in July 2007. A section of the suspended ceiling fell onto a car traveling to the airport. In the accident, the driver was injured and the passenger died (Figures 1 and 2). The National Transportation Safety Board (NTSB), FHWA, FBI, police, and local authorities moved quickly and started the investigation. The possibility of a terrorist act was discarded promptly, and the investigation focused on the failure of structural elements of the suspended ceiling. In the final report, the NTSB recommended that FHWA, in collaboration with AASHTO, develop an NTIS to prevent this failure from happening again.

As can be seen in the Table 1, highway tunnels are getting older. The majority of tunnels in the United States were built before the 1970s. The condition of these tunnels should be available to FHWA and others that have a valid need to access this information.

Table 1

| Tunnel Age | Year Constructed | Number of Tunnels |
|---|---|---|
| < 10 Years | 1991 or Later | 31 |
| 11 to 20 | 1981 to 1991 | 24 |
| 21 to 30 | 1971 to 1980 | 23 |
| 31 to 40 | 1961 to 1970 | 59 |
| 41 to 50 | 1951 to 1960 | 48 |
| 51 to 100 | 1901 to 1950 | 139 |
| >100 | 1900 or prior | 13 |

**Figure 1.** Suspended ceiling collapse in the Central Artery Tunnel, Boston, MA

**Figure 2.** Condition of an anchor found after inspection. Central Artery Tunnel, Boston, MA

The NTIS will be used to develop a National Tunnel Inventory, which will help determine how many tunnels we have in the United States more accurately. A database system is envisioned that would be modeled after the current NBIS, but preferably separate from the NBIS database.

The NTIS will also establish requirements for inspectors. Proper training is important to make sure that the tunnels are inspected by qualified individuals. The NTIS will identify and recommend any new training that needs to be developed to assure that inspectors are well trained.

The NTIS would require the proper safety inspection and evaluation of all Federal-aid highway tunnels on public roads. National Tunnel Inspection standards are needed to ensure that all structural, mechanical, electrical, hydraulic, and ventilation systems, and other major elements of our Nation's tunnels, are inspected and tested on a regular basis. The NTIS would also ensure safety for the surface transportation users of our Nation's highway tunnels, and would make tunnel inspection standards consistent across the Nation. Additionally, tunnel inspections would help protect Federal investment in such key infrastructure. Timely tunnel inspection is vital to uncovering safety problems and preventing failures. When corrosion or leakage occurs, electrical or mechanical systems malfunction, or concrete cracking and spalling signs appear, they may be symptomatic of dire problems (see Figure 3). The importance of tunnel inspection was demonstrated once again in the summer of 2007 in the I-70 Hanging Lake tunnel in Colorado. After the Central Artery ceiling collapse in Boston, the Colorado Department of Transportation moved promptly to inspect the ceiling and roof of the I-70 Hanging Lake tunnel and uncovered a crack in the roof that was compromising the structural integrity of the tunnel (see Figure 4). This discovery prompted the closure of the tunnel for several months for needed repairs. The repairs included

**Figure 3.** In some tunnels it is necessary to remove precast panels to inspect the tunnel wall

removal of more than 30 feet of soil fill material from the top of the tunnel roof, temporary support of the roof from the inside of the tunnel, removal of the suspended ceiling, and the design and construction of a new slab cast on top of the existing roof to reinforce and add extra structural capacity. To accomplish the repair, the eastbound tube under the cracked roof was closed to traffic, and the adjacent westbound tube was converted to a tube with bi-directional traffic. The eastbound tunnel was closed for 7 months, and the repairs cost approximately $6 million, but the repairs helped prevent a potential safety incident.

Some tunnels have other components that need to be inspected. These components include lighting, mechanical, fire, and traffic management systems. The systems are needed to ensure the safety and security of the facility and the traveling public.

The frequency of tunnel inspection should be in accord with the complexity, importance, and age of the tunnel. It could be possible to have varying

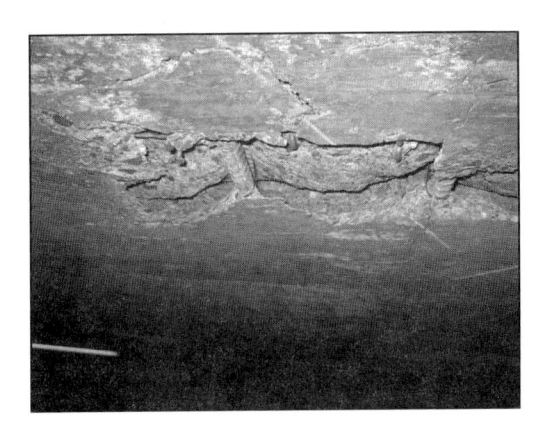

**Figure 4. Crack on tunnel roof. Hanging Lake Tunnel, I-70, Colorado**

frequencies for inspection of structural elements and safety and security systems.

A qualified bridge inspector should be able to inspect all structural components of the tunnels. However, for other systems such as tunnel ventilation, the owner might need to find a private firm that specializes in ventilation systems for its testing and inspection.

## CONCLUSIONS

1. Tunnels should be inspected and the inspection findings should be reported to the FHWA.
2. A national tunnel inventory should be developed. The system should be modeled after the current National Bridge Inventory, but since tunnels are more complex than bridges, the new NTI should be a separate system.
3. Inspection training courses should be developed for those responsible to inspect highway tunnels.
4. The NTIS should establish inspection frequency, inspector's qualification, training requirements, and a national inventory database system.

# U.S. Domestic Scan Program—Best Practices for Roadway Tunnel Design, Construction, Operation, and Maintenance

## Jesus Rohena
Federal Highway Administration, Washington D.C.

## OVERVIEW

Most highway facility components in the United States are governed by design, construction, operation, and maintenance codes and regulations of the American Association of State Highway and Transportation Officials (AASHTO) and the Federal Highway Administration (FHWA). However, to date highway tunnels in the U.S. do not have comparable national codes and regulations. Recent events such as the July 2006 ceiling collapse of the I-90 Central Artery Tunnel in Boston, Massachusetts, have called attention to the need for such national standards.

The *Best Practices for Roadway Tunnel Design, Construction, Operation, and Maintenance* domestic scan, conducted August-September 2009, is one of the activities initiated to assist in addressing the need for national tunnel standards. The nine-member team consisted of two representatives from FHWA, five representatives from State Departments of Transportation (DOT), an academic member representing the Transportation Research Board (TRB) Tunnels and Underground Structures Committee (AFF60), and the report facilitator. The scan was sponsored by the FHWA, AASHTO, and the National Cooperative Highway Research Program (NCHRP).

Scan hosts were agencies that have significant tunnels in their inventories. Hosts along the east coast were the Chesapeake Bay Bridge and Tunnel District, the Massachusetts Turnpike Authority, the Port Authority of New York and New Jersey, and the Virginia Department of Transportation (DOT). Hosts in the western U.S. were Caltrans, the Colorado DOT, and the Washington State DOT. In addition to site visits with scan hosts, the team held web conferences with representatives from the Alaska DOT, the District of Columbia DOT, and the Pennsylvania DOT.

The scan team investigated tunnels on the state, regional, and local highway systems. The focus was on tunnel maintenance and inspection practices, safety as related to emergency response capability, and design and construction standards practiced by state DOTs and other tunnel owners. Included were consideration of fire suppression, traffic management, incident detection and management, and security features. Also included were forensic inspection, analysis, design, and construction repairs of existing tunnels.

General topics of interest to the scan team were:

1. Specialized technologies currently used for existing and new U.S. roadway tunnel design, construction, maintenance, inspection, and operations.
2. Standards, guidance, and "best practices" for existing and new roadway tunnels in the United States.
3. Current criteria used by owners and states to identify tunnels in their inventory.

## SUMMARY OF INITIAL FINDINGS AND PRELIMINARY RECOMMENDATIONS

The scan team identified highway tunnel initiatives or practices of interest for nationwide implementation or for further evaluation for potential nationwide implementation. The team recommended that eight of these initiatives or practices, briefly described below, be implemented first.

### 1. Develop standards, guidance, and best practices for roadway tunnels

Design criteria for new roadway tunnels should consider:

- Performance-based construction specifications.
- Design recommendations for extreme events (manmade and natural) and tunnel security and blast, lifeline, etc.
- Design criteria for vertical, horizontal clearances, and sight distance.
- Criteria for tunnel design life and future maintenance for structural, mechanical, electrical, and electronic systems.
- Criteria for new tunnel load rating.
- Seismic design criteria for one-level versus two-level design events.
- Americans with Disabilities Act (ADA) requirements for emergency egress.
- Placement and layout of the tunnel operations center.

Rehabilitation of existing tunnels should consider obsolescence, tunnel design life, high-performance materials, and existing geometry to maximize safety and system operation.

Tunnel systems are generally complex and expensive in terms of capital costs. The use of peer review teams and technical advisory panels with subject matter expertise should be considered in developing site-specific criteria. Risk management of complex systems is important. Redundancy of systems (SCADA, etc.) is important.

Develop contract procurement guidelines for roadway tunnels to include design-bid-build, design-build, design-build-operate-finance, etc., considering to the extent applicable the Underground Construction Association's "Recommended Contract Practices for Underground Construction."

Develop design and construction standards and guidelines for tunnel construction methods such as the use of Tunnel Boring Machines versus conventional tunneling, design criteria including seismic design, and lifeline requirements. Conventional tunneling methods include the Sequential Excavation Method (SEM), the New Austrian Tunneling Method (NATM), the Analysis of Controlled Deformations (ADECO), and cut-and-cover.

The above topics will be addressed in a proposed research project to develop LRFD design specifications and guidance for new and existing tunnels that was submitted to NCHRP in 2009 by the AASHTO Subcommittee on Bridges and Structures.

## 2. Develop an emergency response system plan unique to each facility which takes into account human behavior, facility ventilation, and fire mitigation

The design of a tunnel to adequately address emergencies should take into account human behavior, the realistic spread of fire and smoke in the tunnel including toxic gases and heat, and the effect of different types of ventilation systems on the fire, including fire suppression and deluge systems if so equipped. The fire plan should be consistent with users' instinctive response to a fire, and the operation of all tunnel fire response systems should be consistent with this behavior. Every facility should do a study of their system and adopt procedures based on that study.

In general, the scan team finds that facilities should improve their procedures to direct the public to safety. Further study and research is needed on how fire and smoke spread in a tunnel and how people react in emergencies. Consider better signage, intelligible public address systems, etc., including recommendations for these from the 2005 international tunnels scan.

## 3. Develop and share inspection practices among tunnel owners

The scan team found the best tunnel inspection programs have been developed under bridge inspection programs, and the team recommends that tunnel inspection programs be as similar as possible to bridge inspection programs. In many cases, bridge inspectors also perform the structural inspection of tunnels.

Those components of the tunnel that carry or affect traffic should be load rated in accordance with the *AASHTO Manual for Bridge Evaluation* to the extent possible, e.g., roadway slabs and floor systems that carry traffic. In the analyses, consider different operational conditions. Structural analyses should be performed on non-traffic-carrying components such as plenums, plenum walls, and hangers as their physical conditions change, as they are modified or as the loads that they are to be subjected to change, such as air forces if fans are upgraded.

Develop recommended practices for inspection frequencies, minimum coding requirements, and a federal coding manual. Current practice is one-to-five years for structural inspections, and daily to yearly for mechanical and electrical inspections depending on the level of inspection. Maximum frequencies should be set, and owners should be encouraged to develop actual frequencies based on a risk-based analysis of hazards due to condition, deterioration, and performance history.

Develop a baseline data inventory for tunnels for submission to the FHWA in conjunction with NCHRP 20-07, Task 261, Task 4.

Inspection practices need to be shared among tunnel owners in four areas. First, the scan team identified a best practice for the inspection of submerged tunnels using multi-beam sonar scans. Second, tunnel inspection training needs to be developed taking into consideration all aspects of the tunnel structure and systems. Third, tools need to be developed to find voids behind tunnel linings. Fourth, coordinated closing of the tunnel overnight to do as much maintenance and inspection as possible.

## 4. Consider inspection and maintenance operations during the design stage

The scan team found that during the design phase, inviting all disciplines into the design results in a better product. The design of a tunnel should address future inspection and maintenance of all tunnel systems and equipment by providing for adequate, safe and unimpeded access to all components by bringing together all engineering disciplines that will have to be accommodated in the tunnel. While the scan team understands that tradeoffs must be made between access and a practical design, these tradeoffs could have cost and safety impacts for maintenance and inspection over the life of the tunnel. There is a need to be able to reach all components.

## 5. Develop site-specific plans for the safe and efficient operation of roadway tunnels

Develop a concise site-specific operations manual to include tunnel emergency response procedures; safe ventilation procedures; safe traffic control guidelines; and general maintenance procedures such as tunnel washing guidelines, fan and bearing maintenance, etc. The manual should include training guidelines and training schedules for all personnel.

Tunnel owners should implement state-of-the-art video surveillance and communication systems. These systems provide numerous benefits, e.g., incident response, traffic management, and increased security. The scan team found a best practice of lane closure or changing traffic direction, e.g., pneumatically-activated lane delineators and zipper barriers that provide for reversible lanes and barriers through tunnels and tunnel approaches. The owners should have an operating procedure that considers safety for the public and owner personnel.

A separate incident response manual should be developed to outline procedures that will require various community, police, fire, and emergency services response in the event of catastrophic incidents. Perform periodic drills including table-top exercises with appropriate agencies.

The scan team findings support restricting hazardous cargo through tunnels. In the event of no alternate route, a well-defined emergency response and fire ventilation plan should be in place. Restricted hours of tunnel operation for hazardous cargo are an option, e.g., hours from 3 a.m. to 5 a.m. under controlled conditions. In any case, a fire ventilation study and fire ventilation plan should be performed for each facility.

## 6. Develop a proactive plan for tunnel design and funding that considers life-cycle costs for preventive maintenance, upgrading of systems, and training and retention of operators

The decision to build a tunnel is a long-term commitment on the part of the owner. The tunnels which include functional systems such as ventilation, fire suppression, and electrical/mechanical components are complex structures with more intensive needs for maintenance and operation than traditional transportation facilities. A proactive plan, considering life-cycle costs, must be developed to address needs for preventive maintenance, upgrading of systems, and training and retention of operators. A target level of condition, system reliability, and performance should be established for the facility to guide operators and owners for current and future decisions which will require manpower or funding.

System components become obsolete and replacement parts will be difficult to find as equipment ages. In particular, electronic equipment such as computers, SCADA systems, and sensors become obsolete or are no longer supported by their original manufacturers sooner than mechanical equipment. Periodic upgrades are vital to keep all systems functioning reliably.

Owner agencies should develop tunnel preservation guidelines for funding purposes, e.g., for concrete repair and washing of walls.

A separate fund should be dedicated for tunnels. Agencies should work with local planning organizations to accomplish this task. The financial management plan for tunnels should not only include first costs for construction, but should also address future preservation and upgrading needs. The scan team found that without this dedicated fund, the funding for tunnel upgrades does not compete well with system-wide needs for traffic signals, pavement preservation, etc.

Training, retention, and a succession plan should be developed for tunnel operators. The scan team found best practices that fostered pride of ownership, a "home away from home" culture and can-do-anything attitude.

## 7. Share existing technical knowledge within the industry to design a tunnel

Technical knowledge that exists within the industry should be shared with tunnel owners in order to not reinvent the wheel. Owners would benefit because they would be provided with a range of practical options in for the design of their tunnels than if they relied only on one tunnel design engineer. This would include using domestic and international tunnel scan information, past project designs, construction practices, emergency response best practices, and subject matter experts. Value engineering can improve technology transfer with limited owner experience in tunnel systems, e.g., Value Engineering/Accelerated Construction Technology Transfer (VE/ACTT).

Design documents including calculations and as-built documents should be filed and easily retrieved by the controlling owner. Recognizing security concerns of tunnel owners, the scan team believes that actual details and best practices used in tunnels should be shared with prospective and existing tunnel owners without identifying the specific facilities where these details and practices are used.

*8) Provide education and training in tunnel design and construction*

The scan team findings support training and development for owner agencies. Currently, there are few Civil Engineering programs in the U.S. that offer a graduate course in tunneling. It is very likely (99% or more) that civil engineers are not exposed to tunneling. Many DOTs do not have tunnels in their transportation systems, others built their last tunnel 20–30 years ago and, therefore, the in-house expertise is either non-existent or out of date. The number and magnitude of tunneling projects is projected to increase dramatically in the next few years. The current offering of short courses allows engineers to acquire the nomenclature in tunneling, but not the working knowledge necessary to design, manage, review, and specify tunnel projects.

Highway tunnel owners and FHWA should provide their engineers with access to education and training on tunnels available through academia and industry. This involvement would also help direct academic research on tunneling. On-line courses and certificates on tunneling of international reputation would allow one to acquire up-to-date information and working knowledge in design and construction of tunnels.

**Planned Implementation Actions**

The implementation of scan team recommendations will be a step in the process of developing national standards and guidance. Scan findings will also provide data for consideration in the development of a national tunnel inventory. These activities will assist the AASHTO Highway Subcommittee on Bridges and Structures' Technical Committee for Tunnels (T-20) and FHWA in developing best practices for roadway tunnel design, construction, operation, and maintenance of existing and new tunnels.

The lead group for implementation of scan recommendations will be T-20 in conjunction with FHWA and the TRB Tunnels and Underground Structures Committee, and working with the National Fire Protection Association (NFPA) and other tunnel organizations. Initial scan team efforts include distribution of the FHWA Tunnel Safety brochure that was developed following the 2005 international tunnels scan and providing additional information on the FHWA tunnels website. Other planned activities include coordination and development of research statements related to tunnel needs. The scan team also plans technical presentations, webinars, and written papers at national meetings and conferences sponsored by FHWA, AASHTO, and other organizations to disseminate information from the scan.

# Index

## A

Abu Dhabi, United Arab Emirates
    instrumentation in preliminary geotechnical investigation (STEP), 307–315
    quantitative risk assessment of fire danger in choice of design approach for Yas Island Southern Crossing Tunnel, 375–386
    Strategic Tunnel Enhancement Programme (STEP), 304–307
Acid resistant concrete additives, 236–237
Acid resistant gunned cementitious lining, 239
Alaskan Way Viaduct and Seawall Replacement (Seattle, Washington), and size considerations, 534
Alliance (risk sharing) contracts, 570–575
Anacostia River CSO Control Project (District of Columbia), 397–399
    geotechnical investigations, 399–406
Analysis of Controlled Deformations (ADECO), as alternative to NATM for full-face, 22-m-wide tunnel through clay, 96–109
Anchored thermoplastic lining (PVC, HDPE), 233–235
Arizona, and sophisticated drilling and downhole survey technology for deep inclined water intake shafts (Navajo Generating Station), 3–9
Aurora, Colorado
    analyzing where to use open-cut or microtunneling (Tollgate Interceptor), 741–749
    Tollgate Interceptor, 740–741
Austin, Texas
    ground conditions and design (Waller Creek Tunnel), 576–587
    Jollyville Tunnel environmental preservation and community impact mitigation criteria, 609–612
    multi-attribute utility theory in decision-making process involving stakeholders (South IH-35 Water and Wastewater Infrastructure Improvements Program), 667–675
    South IH-35 Water and Wastewater Infrastructure Improvements Program, 664–667
    Waller Creek Tunnel, 576
Australia. *See* Brisbane North-South Bypass Tunnel
Austria. *See* Brenner base tunnel rail access

## B

BAM Combinatie Eemstunnel, 164
Barnard of Nevada, Inc., 889–893
Beacon Hill Station and Tunnels (Seattle, Washington)
    and NATM construction in soft ground, 704–705
    and size considerations, 533–534
Belden Tunnel 2 (California), 249–251
    Carpi PVC membrane in remediation of tunnel leakage and slope stability, 251–256
Bellevue, Washington
    Lincoln Square Tunnel, 836
    SEM with spile presupport through difficult ground, connecting two underground garages (Lincoln Square Tunnel), 836–842
Bergen County, New Jersey
    microtunneling through challenging soft ground and mixed soils (New Overpeck Valley Parallel Relief Sewer Project), 772–787
    New Overpeck Valley Parallel Relief Sewer Project, 771–772
Best Practices for Roadway Tunnel Design, Construction, Operation, and Maintenance domestic scan (US), 960–963
Big Walnut Sanitary Trunk Sewer Extension (Columbus, Ohio), 938
    claims mitigation and public communication in construction management, 938–943
Black and Veatch, 502, 516, 609
Blacklick Creek Sanitary Interceptor Sewer Tunnel (Columbus, Ohio), 758–759
    alignment selection involving highly varied ground conditions, 759–767
Blast and post-blast behavior of tunnels, 655–663
Block 37 project (Chicago, IL), 269–272
    instrumentation in analysis of Chicago freight tunnel cracking and planning for future instrumentation improvements, 269–282
Bosphorus Strait. *See* Railway Bosphorus Tube Crossing
Boston, Massachusetts
    alternative approaches considered for Urban Ring Project, 645–651
    binocular vs. stacked alignment and multiple design factors for Silver Line Phase III Project, 461–474
    Central Artery Tunnel Project (Boston, MA), 533
    Silver Line Phase III Bus Rapid Transit Tunnel Project, 461–463
    Urban Ring Project, 641–645
Brenner base tunnel rail access (Austria), and large-diameter TBMs, 91–92

Brightwater Conveyance East Contract (King County, WA), 298
    monitoring of settlement and comparison with modeling results, 298–303
Brightwater Conveyance West Contract (Seattle, Washington), 855
    EPB TBM in construction of 6.4-km, small-diameter tunnel in soft ground with small exit window, 855–862
Brisbane North-South Bypass Tunnel (Australia), and large-diameter TBMs, 92–94

## C

California
    Carpi PVC membrane in remediation of tunnel leakage and slope stability (Belden Tunnel 2), 249–256
    concrete final lining for SEM tunnel (Devil's Slide project), 35–42
    design and planning of Transbay-Downtown Rail Extension (DTX) Project (San Francisco), 475–482
    design challenges and tunneling methods considered for Regional Connector Transit Corridor project (Los Angeles), 750–757
    EPB TBMs with continuous conveyor systems (Sacramento Interceptor Sewers), 136–139
    ground characterization and design challenges (New Irvington Tunnel), 326–336
    Grout Intensity Number (GIN) methodology in leakage remediation (Helms Pumped Storage Project, Fresno County), 215–223
    Irvine-Corona Expressway tunnels geologic challenges and TBM considerations, 147–157
    plastic (PVC) lined concrete segments for Upper Northwest Interceptor Sewer (Sacramento), 191–199
    precast PVC-lined concrete segments for Upper Northwest Interceptor Sewer (Sacramento), 242–248
    seismic design criteria for State Route 75/282 Transportation Corridor Project, 365–372
    sophisticated ventilation system for long-drive, small-diameter tunneling project (San Diego), 55–61
California High Speed Rail (CAHSR) system, 532
Carpi PVC membrane in remediation of tunnel leakage and slope stability (Belden Tunnel 2, California), 249–256
Cassia twin tunnels (Rome, Italy), 96
    Analysis of Controlled Deformations (ADECO), as alternative to NATM for full-face, 22-m-wide tunnel excavation through clay, 96–109

Central Artery Tunnel Project (Boston, MA), 533
CEVP. *See* Cost Estimate Validation Process
CH2M HILL, 304
Charleston, South Carolina
    flood abatement program (US 17 Septima Clark Parkway Project), 613–620
    US 17 Septima Clark Parkway Transportation Infrastructure Reinvestment Project, 614–615
Chicago, Illinois
    Block 37 project, 269–272
    design and planning of McCook Reservoir Main Tunnel System, 502–516
    instrumentation in analysis of freight tunnel cracking and planning for future instrumentation improvements (Block 37 project), 272–282
Chile, and NATM construction in soft ground (Santiago), 705–706
China. *See* Shanghai Changxing Under River Tunnel
Cincinnati, Ohio
    deep storage CSO tunnel in sedimentary rock (West Branch Muddy Creek Project), 733–739
    West Branch Muddy Creek Project Bundle, 733
Colorado
    analyzing where to use open-cut or microtunneling (Tollgate Interceptor, Aurora), 740–749
    inspections of alpine highway tunnels, 224–232
Columbus, Ohio
    alignment selection involving highly varied ground conditions (Blacklick Creek project), 759–767
    Big Walnut Sanitary Trunk Sewer Extension, 938
    Blacklick Creek Sanitary Interceptor Sewer Tunnel, 758–759
    claims mitigation and public communication in construction management (Big Walnut project), 938–943
Computational fluid dynamics, in quantitative risk assessment of fire danger in choice of design approach for Yas Island tunnel (United Arab Emirates), 381–382
Concrete faced with glass reinforced plastic, 235–236
Connecticut
    Heroes Highway Tunnel, 257–259
    Homestead Avenue Interceptor Extension (Hartford), 813–814
    options for rehabilitation of Heroes Highway Tunnel, 259–266

slurry TBM in construction of small-diameter tunnel through soft ground in congested urban environment (Homestead Avenue project), 814–821
Construction
    combined design-bid and design-bid-build approach (District of Columbia Water and Sewer Authority CSO Program), 593–600
    design and construction considerations for station caverns at Grand Central Terminal in complex urban environment (East Side Access Project, New York City), 635–640
    of megaprojects, 566–569
    owner procurement of materials and equipment (York, Ontario), 601–606
    scheduling, 530
    *See also* Cost estimates; Critical Path Method; Public private partnerships; Risk assessment and management
Contingency planning, 710–717
Continuous conveyor systems, in EPB TBM applications, 136–139
Contracts, 544–547
    alliance (risk sharing), 570–575
    and cash flow, 543, 546, 547
    and cumulative expenditure, 546
    and cumulative revenue, 546
    and equipment mobilization, 544
    and interest rate on borrowed or invested capital, 546
    and mobilization cap and payments, 544
    and modified internal rate of return (MIRR), 546–547
    and net project investment cost/income, 547
    typical, 541–543
Conventional tunneling, and Sandy River conduit crossing (Portland, OR), 897–905
Coronado, California
    seismic design criteria for State Route 75/282 project, 366–372
    State Route 75/282 Transportation Corridor Project, 365–366
Corrosion resistance, 233, 240–241
    acid resistant concrete additives, 236–237
    acid resistant gunned cementitious lining, 239
    anchored thermoplastic lining (PVC, HDPE), 233–235
    concrete faced with glass reinforced plastic, 235–236
    deformed pipe lining, 239
    fiber reinforced pipe, 237
    high density polyethylene studliners, 238–239
    liquid applied polymer based protective lining, 238
    polymer concrete pipe, 237
    post-installed polyvinyl chloride, 238
    precast polymer concrete segments, 236
    reinforced precast concrete pipe, with liner, 237–238
    slip lining, 239–240
Cost Estimate Validation Process (CEVP), 718–724
Cost estimates, 529, 548–550, 555–556, 725
    components, 550–555
    emphasizing benefits, 538–540
    ground support, 552–555
    indirect costs, 725–728
    realistic public works budgets, 536–538, 539–540
    and scheduling, 728
    shaft excavation, 550
    and tunnel advance, 555
    tunnel excavation, 550
    and uncertainty vs. accuracy, 728–729
Critical Path Method (CPM), 530
    linear scheduling as preferred approach, 683–694
Croton Water Treatment Plant (New York City), 926
    TBMs in place of drill and blast for twin water tunnels, 927–937
CSO tunnels. *See* Sewer, CSO, and wastewater tunnels
Cut-and-cover tunnels
    assessed for Tollgate Interceptor project (Aurora, CO), 740–749
    considered for Regional Connector Transit Corridor (Los Angeles, CA), 753
    in construction of Marysville Trunk Interceptor (Marysville, OH), 788–795
    in construction of Sunnydale Auxiliary Sewer Project (San Francisco, CA), 337–345
    in construction of Transbay-Downtown Rail Extension (DTX) Project (San Francisco, CA), 475–482
    *Technical Manual for Design and Construction of Road Tunnels—Civil Element* (FHWA) on, 394

## D

Deep Rock Tunnel Connector (Indianapolis, Indiana), 629
    chosen in place of shallow soft ground connector for CSO abatement, 629–634
Deep Underground Science and Engineering Laboratory (DUSEL; South Dakota), 409–411
    contingency planning for Long Baseline Neutrino Experiment, 713–716
    design issues for rock cavern storage of cryogenic fluids, 411–414
    size considerations of Large Water Neutrino Detector Project, 534
Deformed pipe lining, 239

Design and planning
   accounting for geotechnical variability and uncertainty in GBRs, 316–322
   alternative approaches considered for Urban Ring Project (Boston, MA), 641–651
   analyzing where to use open-cut or microtunneling (Tollgate Interceptor, Aurora, CO), 740–749
   Best Practices for Roadway Tunnel Design, Construction, Operation, and Maintenance domestic scan (US), 960–963
   and contingency planning, 710–717
   continuum and discontinuum modeling methods in design of Second Avenue Subway Project rock caverns (New York City), 423–432
   of deep storage CSO tunnel in sedimentary rock (West Branch Muddy Creek Project Bundle, Cincinnati, OH), 733–739
   design and construction considerations for station caverns at Grand Central Terminal in complex urban environment (East Side Access Project, New York City), 635–640
   design challenges and tunneling methods considered for Regional Connector Transit Corridor project (Los Angeles, California), 750–757
   instrumentation in preliminary geotechnical investigation (Strategic Tunnel Enhancement Programme; Abu Dhabi, UAE), 307–315
   McCook Reservoir Main Tunnel System (Illinois), 502–516
   multi-attribute utility theory in decision-making process involving stakeholders (Austin, TX), 664–675
   multiple factors, including nongeotechnical, in design of mass transit tunnels (Boston, MA), 461–474
   multiple tunneling methods required in urban environment (Sunnydale Auxiliary Sewer Project, San Francisco, CA), 337–345
   New Irvington Tunnel (California), 325–336
   for pressurized face tunneling in construction of New York (City) Harbor Siphon, 346–354
   quantitative risk assessment of fire danger in choice of design approach for Yas Island tunnel (United Arab Emirates), 375–386
   for rehabilitation, with cross-section widening, of German rail tunnels, 355–364
   seismic design criteria for State Route 75/282 Transportation Corridor Project (California), 365–372
   structural analysis methodology for large-span rock caverns beneath dense urban environments, 441–458
   *Technical Manual for Design and Construction of Road Tunnels—Civil Element* (FHWA), 387–396
   of Transbay-Downtown Rail Extension (DTX) Project (San Francisco, CA), 475–482
   tunnel alignment selection involving highly varied ground conditions (Blacklick Creek Sanitary Interceptor Sewer Tunnel, Columbus, OH), 758–767
   of University Link tunnel connection to Pine Street Stub Tunnel (Link Light Rail Project, Seattle, WA), 517–524
   *See also* Construction; Contracts; Cost estimates; Public private partnerships; Risk assessment and management; Scheduling
Devil's Slide Tunnel Project (California), 35
   concrete final lining for SEM tunnel, 35–42
Difficult ground
   geologic challenges and TBM considerations (Irvine-Corona Expressway tunnels, California), 147–157
   ground freezing for horizontal connection between shafts in difficult geologic and hydrostatic conditions (Oregon), 26–34
   hybrid (slurry shield & open mode) TBM and geotechnical aspects of operation in poor ground, 125–135
   microtunneling through challenging soft ground and mixed soils (New Overpeck Valley Parallel Relief Sewer Project, Bergen County, NJ), 771–787
   and microtunneling with jet grouting in congested urban setting (Narragansett Bay Commission Combined Sewer Overflow), 796–803
   New Irvington Tunnel (California), 325–336
   pipe jacking in hardpan with boulders (North Gratiot Interceptor Drain, Macomb County, MI), 804–812
   SEM with spile presupport through dense glacial till, connecting two underground garages (Lincoln Square Tunnel), 836–842
   *Technical Manual for Design and Construction of Road Tunnels—Civil Element* (FHWA) on, 391
District of Columbia Water and Sewer Authority
   Anacostia River CSO Control Project, 397–399
   CSO Program project delivery approach (combined design-bid and design-bid-build), 593–600
   CSO tunnels, 483, 486
   geotechnical investigations (Anacostia River CSO Control Project), 399–406
   SHAFT model in analysis of surge and pneumatic forces, 483–495, 499–501

Drill and blast, in fast-track construction of Rupert Transfer Tunnel (James Bay, Québec), 923–925
DUSEL. *See* Deep Underground Science and Engineering Laboratory

E

Earth pressure balance TBMs
    and continuous conveyor systems, 136–139
    for construction of Emisor Oriente Wastewater Tunnel Project in soft ground (Mexico City), 158–163
    in construction of 6.4-km, small-diameter tunnel in soft ground with small exit window (Brightwater Conveyance West Contract, Seattle, WA), 855–862
    in construction of Sunnydale Auxiliary Sewer Project (San Francisco, CA), 337–345
East Side Access Project (New York City), 635
    design and construction considerations for station caverns at Grand Central Terminal in complex urban environment, 635–640
East Side CSO Tunnel (Portland, Oregon), 26, 52–53
    drop and diversion structures, 944–954
    ground freezing for horizontal connection between shafts in difficult geologic and hydrostatic conditions (Oregon), 26–34
Eisenhower-Johnson Memorial Tunnel (Colorado), 224–225
    inspection of 225–229, 231–232
Emisor Oriente Wastewater Tunnel Project (Mexico City), 158–159
    EPB TBMs for construction in soft ground, 159–163
Ems-Dollard estuary crossing (Germany-Netherlands), 164
    slurry TBM in construction of small-diameter gas pipeline tunnel, 165–169
EPB TBMs. *See* Earth pressure balance TBMs
Euler-Lagrange analysis, 655–663
European Global Navigation Satellite System (GNSS). *See* Galileo
European X-ray free electron laser (XFEL) project, 916
    hydroshield TBM with segmental lining in construction of, 916–922

F

Federal Highway Administration (FHWA), *Technical Manual for Design and Construction of Road Tunnels—Civil Element*, 387–396
Fiber reinforced polymer pipe, 237

Finite element analysis
    in modeling design of segmental linings for large-diameter, shallow, circular underground structures, 87–88
    in settlement modeling, and comparison with monitoring results, 298–303
    in structural analysis methodology for large-span rock caverns beneath dense urban environments, 441–458
    *See also* LS-DYNA software
Fire safety, and quantitative risk assessment of fire danger in choice of design approach for Yas Island tunnel (United Arab Emirates), 375–386
Fort Totten Station (Washington, DC), and NATM construction in soft ground, 706, 825–828

G

Galileo (European Global Navigation Satellite System), 116–117
Geotechnical baseline reports (GBRs)
    accounting for geotechnical variability and uncertainty in, 316–322
    *Technical Manual for Design and Construction of Road Tunnels—Civil Element* (FHWA) on, 390
Geotechnical investigations
    Anacostia River CSO Control Project (District of Columbia), 397–406
    Strategic Tunnel Enhancement Programme (STEP; United Arab Emirates), 307–315
    *Technical Manual for Design and Construction of Road Tunnels—Civil Element* (FHWA), 387–396
Germany
    Ems-Dollard estuary crossing, 164
    hydroshield TBM with segmental lining in construction of European X-ray free electron laser (XFEL) project (Germany), 916–922
    rehabilitation, with cross-section widening, of German rail tunnels, 355–364
    slurry TBM in construction of small-diameter gas pipeline tunnel (Ems-Dollard estuary crossing), 165–169
Global Navigation Satellite System (GLONASS), 116
Global Positioning System (GPS), 116. *See also* Positioning systems
Gotthard Base Tunnel (Switzerland)
    and large-diameter TBMs, 94–95
    responses to micro tremors and rock bursts during construction of Multifunction Station Faido, 880–888
GPR. *See* Ground penetrating radar

Green Building Council (US), 183
Green Globes system, 200
Ground conditions
    and design of Waller Creek Tunnel (Austin, TX), 576–587
    history of Portland (Oregon) tunneling projects in diverse geological conditions, 43–54
    New York (City) Harbor Siphon project, 346–354
    rock assessment systems (RQD, RMR, Q), 577–587
Ground freezing, for horizontal connection between shafts in difficult geologic and hydrostatic conditions (Oregon), 26–34
Ground penetrating radar (GPR), 118, 119
Groundwater control, for New Irvington Tunnel (California), 331–332
Grout Intensity Number (GIN) methodology, 215–223
Grouting
    consolidation grouting for fault zone (Riverbank Filtration Tunnel, Louisville, KY), 876–879
    jet grouting with microtunneling in difficult ground and congested urban setting (Narragansett Bay Commission Combined Sewer Overflow, Providence, RI), 796–803
    options for rehabilitation of Heroes Highway Tunnel (Connecticut), 257–266
    of segmental linings for large-diameter, shallow, circular underground structures, 81–82, 83–87
    tube-a-manchette (TAM) permeation grouting (Allegheny Avenue storm water tunnel, Philadelphia), 906–915
Gyroscope probes, 119
Gyroscope stations, 119–120

## H

Hamburg, Germany
    European X-ray free electron laser (XFEL) project, 916
    hydroshield TBM with segmental lining in construction of XFEL, 916–922
Hanging Lake Tunnel (Colorado), 224–225
    inspection of, 230–232
Hartford, Connecticut
    Homestead Avenue Interceptor Extension, 813–814
    slurry TBM in construction of small-diameter tunnel through soft ground in congested urban environment (Homestead Avenue project), 814–821
Hatch Mott MacDonald, and seismic design criteria for Coronado (California) highway tunnel, 365–372

Helms Pumped Storage Project (Fresno County, CA), 215
    Grout Intensity Number (GIN) methodology in leakage remediation, 215–223
Heroes Highway Tunnel (Connecticut), 257–259
    options for rehabilitation of, 259–266
High density polyethylene studliners, 238–239
Highway tunnels. *See* Road tunnels
Homestake Gold Mine (South Dakota). *See* Deep Underground Science and Engineering Laboratory (DUSEL; South Dakota)
Homestead Avenue Interceptor Extension (Hartford, Connecticut), 813–814
    slurry TBM in construction of small-diameter tunnel through soft ground in congested urban environment, 814–821
Hungary
    M6 Motorway tunnel collapse, reconstruction, and risk allocation, 589–592
    M6 Motorway tunnels, 589
Hydro-Québec, 923–925

## I

Illinois
    Block 37 project (Chicago), 269–272
    design and planning of McCook Reservoir Main Tunnel System, 502–516
    instrumentation in analysis of freight tunnel cracking and planning for future instrumentation improvements (Block 37 project, Chicago), 272–282
    McCook Reservoir Main Tunnel System (Illinois), 502
Immersed tunnels
    in Railway Bosphorus Tube Crossing (Turkey), 10, 13–18
    *Technical Manual for Design and Construction of Road Tunnels—Civil Element* (FHWA) on, 393–394
Indianapolis, Indiana
    choice of Deep Rock Tunnel Connector in place of shallow soft ground connector for CSO abatement, 629–634
    CSO abatement considerations, 629
Instrumentation
    in analysis of Chicago freight tunnel cracking and planning for future instrumentation improvements, 269–282
    in preliminary geotechnical investigation (Strategic Tunnel Enhancement Programme; Abu Dhabi, UAE), 307–315
Intake shafts, and sophisticated drilling and downhole survey technology (Navajo Generating Station, Page, AZ), 3–9
International Tunneling Insurance Group, 559, 563
Irvine-Corona Expressway tunnels (California), 147
    geological studies, 148–153

TBM considerations, 153–155
tunneling concepts considered, 148
Istanbul, Turkey
multiple tunneling methods for Bosphorus railway crossing, 13–25
Railway Bosphorus Tube Crossing, 10–13
Italy
Analysis of Controlled Deformations (ADECO), as alternative to NATM for full-face, 22-m-wide tunnel excavation through clay, 96–109
Cassia twin tunnels (Rome), 96

## J

James Bay, Québec
fast-track drill and blast in construction of Rupert Transfer Tunnel, 923–925
Rupert Transfer Tunnel, 923
Jet grouting, in construction of Consolidation Conduit (Narragansett Bay Commission Combined Sewer Overflow, Providence, RI), 843, 848–852
Jollyville Tunnel (Austin, Texas), 609
environmental preservation and community impact mitigation criteria, 609–612

## K

Kentucky
consolidation grouting for fault zone (Riverbank Filtration Tunnel), 876–879
Riverbank Filtration Tunnel (Louisville, Kentucky), 876
King County, Washington
Brightwater Conveyance East Contract, 298
monitoring of settlement and comparison with modeling results (Brightwater Conveyance East Contract), 298–303
See also Seattle, Washington; Seattle (Washington) Public Utilities

## L

Lake Mead Intake No. 3 Tunnel (Nevada), 125, 283
geologic mapping and annotated photo documentation, 283–297
hybrid (slurry shield & open mode) TBM and geotechnical aspects of operation in poor ground, 125–135
joint risk management approach between owner and contractor, 559–565
optimization of blasting production and vibration mitigation in construction, 889–893
temporary support and permanent lining solutions for shaft, cavern, and starter tunnel, 433–440

Large Water Neutrino Detector Project. See under Deep Underground Science and Engineering Laboratory
LaserShell linings, 174–175, 181
LBNE. See Long Baseline Neutrino Experiment
Leadership in Energy and Environmental Design (LEED) program, 183–184, 189–190, 200
LIDAR. See 3D laser scanner-LIDAR
Lincoln Square Tunnel (Bellevue, Washington), 836
SEM with spile presupport through difficult ground, connecting two underground garages, 836–842
Linear scheduling, 683–694
Linings
acid resistant concrete additives, 236–237
acid resistant gunned cementitious lining, 239
anchored thermoplastic lining (PVC, HDPE), 233–235
Carpi PVC membrane in remediation of tunnel leakage and slope stability (Belden Tunnel 2, California), 249–256
concrete faced with glass reinforced plastic, 235–236
concrete final lining for SEM tunnel (Devil's Slide project), 35–42
concrete segmental lining reinforced with steel fibers for small-diameter gas pipeline tunnel (Ems-Dollard estuary crossing), 164–169
connectors for one-pass precast segmental linings, 140–146
deformed pipe lining, 239
design issues for rock cavern storage of cryogenic fluids (DUSEL), 409–414
fiberglass bolts and shotcrete for support and lining of cavern (Lake Mead Intake No. 3 Tunnel, Nevada), 433–440
high density polyethylene studliners, 238–239
high performance fibre reinforced cement composites (HPFRCC), 179–182
LaserShell process, 174–175, 181
liquid applied polymer based protective lining, 238
material and structural properties and structural design of reinforced concrete liners, 446–447, 455–457
options for rehabilitation of Heroes Highway Tunnel (Connecticut), 257–266
plastic (PVC) lined concrete segments, 191–199
post-installed polyvinyl chloride, 238
precast polymer concrete segments, 236
precast PVC-lined concrete segments for large-diameter sewer tunnel, 242–248
in rehabilitation, with cross-section widening, of German rail tunnels, 355–364

segmental lining in construction of European X-ray free electron laser (XFEL) project (Germany), 916–922
segmental linings for large-diameter, shallow, circular underground structures, 75–88
slip lining, 239–240
sprayable waterproof membranes, 175–179, 181
and sustainability, 173–182
traditional sprayed concrete, 173
two-pass lining system for New Irvington Tunnel, 325, 333–334
UltraShell, 180–182
Link Light Rail Project (Seattle, Washington)
design and planning of University Link–Pine Street Stub Tunnel connection, 517–524
University Link tunnel connection to Pine Street Stub Tunnel, 517
Liquid applied polymer based protective lining, 238
London, United Kingdom
London Bridge Station and NATM construction in soft ground, 706–707
SHAFT model in analysis of surge and pneumatic forces (Thames Water Utilities CSO tunnels), 483–485, 495–501
Thames Water Utilities CSO tunnels (London), 483, 495
Long Baseline Neutrino Experiment (LBNE), 409
design issues for rock cavern storage of cryogenic fluids (DUSEL), 409–414
Los Angeles, California
design challenges and tunneling methods considered for Regional Connector Transit Corridor project, 750–757
size considerations for Metro Red Line North Hollywood Extension, 532
Louisville (Kentucky) Water Company
consolidation grouting for fault zone (Riverbank Filtration Tunnel), 876–879
Riverbank Filtration Tunnel, 876
Lower Northwest Interceptor Sewer (Sacramento, CA), 136
EPB TBM with continuous conveyor system, 136
LS-DYNA software, 655–663

# M

Macomb County, Michigan
North Gratiot Interceptor Drain, 804
pipe jacking in hardpan with boulders (North Gratiot Interceptor Drain), 804–812
Mapping
geologic mapping and annotated photo documentation (Lake Mead Intake No. 3 Tunnel, Nevada), 283–297
published guidance and methods, 283–284

Marmaray project (Turkey). See Railway Bosphorus Tube Crossing
Marysville Trunk Interceptor (Marysville, Ohio), 788
cut and cover and microtunneling in construction of, 788–795
Massachusetts
binocular vs. stacked alignment and multiple design factors for Silver Line Phase III Project, 461–474
Silver Line Phase III Bus Rapid Transit Tunnel Project (Boston, MA), 461–463
McCook Reservoir Main Tunnel System (Illinois), 502
design and planning, 502–516
Megaprojects, 566–569
Mexico City
Emisor Oriente Wastewater Tunnel Project, 158–159
EPB TBMs for construction of Emisor Oriente Wastewater Tunnel Project in soft ground, 159–163
Michigan
North Gratiot Interceptor Drain (Macomb County), 804
pipe jacking in hardpan with boulders (North Gratiot Interceptor Drain), 804–812
Microtunneling
assessed for Tollgate Interceptor project (Aurora, CO), 740–749
in challenging soft ground and mixed soils (New Overpeck Valley Parallel Relief Sewer Project, Bergen County, NJ), 771–787
in construction of Marysville Trunk Interceptor (Marysville, OH), 788–795
in construction of Sunnydale Auxiliary Sewer Project (San Francisco, CA), 337–345
in construction of underground pump plant in pristine watershed (Washington), 206–212
with jet grouting in difficult ground and congested urban setting (Narragansett Bay Commission Combined Sewer Overflow, Providence, RI), 796–803
Mission Valley East Light Rail Extension (San Diego, CA), 533
Modeling
accounting for geotechnical variability and uncertainty in GBRs, 316–322
continuum and discontinuum methods in design of Second Avenue Subway Project rock caverns (New York City), 423–432
finite element modeling in design of segmental linings for large-diameter, shallow, circular underground structures, 87–88
finite element modeling in structural analysis methodology for large-span rock caverns beneath dense urban environments, 441–458

settlement modeling by finite element analysis and other modeling approaches, and comparison with monitoring results, 298–303
SHAFT model in analysis of surge and pneumatic forces (London and Washington, DC, CSO tunnels), 483–501
Monitoring of settlement and comparison with modeling results (Brightwater Conveyance East Contract, King County, WA), 298–303
Morse Lake Pump Plant and Intake (North Bend, WA), 206
microtunneling in construction of underground pump plant in pristine watershed (Washington), 206–212
M6 Motorway tunnels (Hungary), 589
tunnel collapse, reconstruction, and risk allocation, 589–592
Multi-attribute utility theory, 664–675

## N

Narragansett Bay Commission Combined Sewer Overflow (Providence, Rhode Island)
jet grouting and construction of Consolidation Conduit, 843, 848–852
Main Spine Tunnel, 843–844, 851–852
microtunneling with jet grouting in difficult ground and congested urban setting (Contract 8), 796–803
support of excavation for Diversion Structure, 843, 844–848, 851–852
NATM. *See* New Austrian Tunneling Method
Navajo Generating Station (Page, AZ), 3
sophisticated drilling and downhole survey technology for deep inclined water intake shafts, 3–9
Netherlands
Ems-Dollard estuary crossing, 164
slurry TBM in construction of small-diameter gas pipeline tunnel (Ems-Dollard estuary crossing), 165–169
Nevada
geologic mapping and annotated photo documentation (Lake Mead Intake No. 3 Tunnel), 283–297
hybrid (slurry shield & open mode) TBM and geotechnical aspects of operation in poor ground (Lake Mead Intake No. 3 Tunnel), 125–135
joint risk management approach between owner and contractor (Lake Mead Intake No. 3 Tunnel), 559–565
Lake Mead Intake No. 3 Tunnel, 125
optimization of blasting production and vibration mitigation in construction (Lake Mead Intake No. 3 Tunnel), 889–893

temporary support and permanent lining solutions for shaft, cavern, and starter tunnel (Lake Mead Intake No. 3 Tunnel), 433–440
New Alpine Transversal (NEAT). *See* Gotthard Base Tunnel
New Austrian Tunneling Method (NATM)
Analysis of Controlled Deformations (ADECO) as alternative to NATM for full-face, 22-m-wide tunnel through clay, 96–109
in construction of mined underground stations in soft ground, 695–709
in construction of Transbay-Downtown Rail Extension (DTX) Project (San Francisco, CA), 475–482
history of, 695–696
principles applied to SEM tunnel (California), 35
in Railway Bosphorus Tube Crossing (Turkey), 10, 20–25
in soft-ground tunnel and shaft construction (Washington, DC), 825–835
*See also* Sequential Excavation Method
New Irvington Tunnel (California), 325–326
ground characterization and design challenges, 326–336
New Jersey
microtunneling through challenging soft ground and mixed soils (New Overpeck Valley Parallel Relief Sewer Project, Bergen County), 772–787
New Overpeck Valley Parallel Relief Sewer Project, 771–772
New Overpeck Valley Parallel Relief Sewer Project (Bergen County, New Jersey), 771–772
microtunneling through challenging soft ground and mixed soils (Bergen County, New Jersey), 772–787
New York City
continuum and discontinuum modeling methods in design of Second Avenue Subway Project rock caverns, 423–432
design and construction considerations for station caverns at Grand Central Terminal in complex urban environment (East Side Access Project), 635–640
pressurized face tunneling in construction of Harbor Siphon, 346–354
TBMs in place of drill and blast for twin water tunnels (Croton Water Treatment Plant), 926–937
underpinning and construction of No. 7 Subway Line Extension under 8th Ave. Subway, 863–875

No. 7 Subway Line Extension (New York City), 863
  underpinning and construction under 8th Ave. Subway, 863–875
North Gratiot Interceptor Drain (Macomb County, Michigan), 804
  pipe jacking in hardpan with boulders, 804–812

## O

OFTA. *See* Onsite first time assembly (OFTA) of TBMs
Ohio
  claims mitigation and public communication in construction management (Big Walnut Sanitary Trunk Sewer Extension, Columbus), 938–943
  cut and cover and microtunneling in construction of Marysville Trunk Interceptor, 788–795
  deep storage CSO tunnel in sedimentary rock (West Branch Muddy Creek Project Bundle, Cincinnati), 733–739
  tunnel alignment selection involving highly varied ground conditions (Blacklick Creek Sanitary Interceptor Sewer Tunnel, Columbus), 759–767
Onsite first time assembly (OFTA) of TBMs, 65–66
  projects utilizing, 66–74
Open-cut tunnels. *See* Cut-and-cover tunnels
Oregon
  East Side CSO Project (Portland), 26–34, 52–53, 944–954
  ground freezing for horizontal connection between shafts in difficult geologic and hydrostatic conditions (Portland), 26–34
  history of tunneling projects in diverse geological conditions (Portland), 43–54
  road header conventional open tunneling for Sandy River conduit crossing, 897–905

## P

Parsons Brinckerhoff, 387
Peachtree Center station (Metropolitan Atlanta Rapid Transit Authority), 532
Pennsylvania
  Allegheny Avenue storm water tunnel (Philadelphia Water Department), 906–908
  tube-a-manchette (TAM) permeation grouting (Allegheny Avenue storm water tunnel, Philadelphia), 906, 908–915
Philadelphia (Pennsylvania) Water Department
  Allegheny Avenue storm water tunnel, 906–908
  tube-a-manchette (TAM) permeation grouting (Allegheny Avenue storm water tunnel), 906, 908–915
Pipe jacking
  in construction of Sunnydale Auxiliary Sewer Project (San Francisco, CA), 337–345
  in hardpan with boulders (North Gratiot Interceptor Drain, Macomb County, MI), 804–812
Planning. *See* Design and planning
Polymer concrete pipe, 237
Portland, Oregon
  Balch Consolidation Conduit, 53–54
  Columbia Slough Consolidation Conduit, 50
  drop and diversion structures (East Side CSO Tunnel), 944–954
  early highway tunnels, 45
  East Side CSO Project, 26–34, 52–53, 944–954
  history of tunneling projects in diverse geological conditions, 43–54
  Peninsular Tunnel, 45–48
  Portsmouth Force Main, 53
  risk management approach for insurance decisions in CSO programs, 677–682
  Southeast Relieving Interceptor, 50
  Tanner Creek Stream Diversion Project, 50–51
  Vista Ridge Tunnel, 48
  West Side CSO Project, 51–52
  Westside Light Rail Transit Tunnels, 50
  Willamette River CSO Program, 676–677
Portland (Oregon) Water Bureau
  road header conventional open tunneling for Sandy River conduit crossing, 897–905
  Sandy River conduit crossing, 897
Positioning systems, 116
  above ground, 116–117
  underground, 117–121
  underwater, 117
Post-installed polyvinyl chloride, 238
Precast polymer concrete segments, 236
Pressurized face TBMs, in construction of New York (City) Harbor Siphon, 346–354
Providence, Rhode Island
  jet grouting and construction of Consolidation Conduit (Narragansett Bay Commission Combined Sewer Overflow), 843, 848–852
  Main Spine Tunnel (Narragansett Bay Commission Combined Sewer Overflow), 843–844, 851–852
  microtunneling with jet grouting in difficult ground and congested urban setting (Narragansett Bay Commission Combined Sewer Overflow), 796–803
  Narragansett Bay Commission Combined Sewer Overflow, 796

support of excavation for Diversion Structure (Narragansett Bay Commission Combined Sewer Overflow), 843, 844–848, 851–852
Public private partnerships, 588
　and risk management, 588–592
　*See also* Construction; Contracts; Cost estimates; Design and planning

## Q

Q system, 577–587
Québec
　fast-track drill and blast in construction of Rupert Transfer Tunnel (James Bay), 923–925
　Rupert Transfer Tunnel (James Bay), 923

## R

Railroad tunnels
　design and planning of University Link–Pine Street Stub Tunnel connection (Link Light Rail Project, Seattle, WA), 517–524
　linings in rehabilitation, with cross-section widening, of German rail tunnels, 355–364
　Westside Light Rail Transit Tunnels (Portland, OR), 50
　*See also* Brenner base tunnel rail access (Austria); Railway Bosphorus Tube Crossing (Istanbul, Turkey); Subway tunnels; Transbay-Downtown Rail Extension (DTX) Project (San Francisco, California)
Railway Bosphorus Tube Crossing (Istanbul, Turkey), 10–13
　immersed tunnels in, 10, 13–18
　NATM in, 10, 20–25
　shield tunneling in, 10, 19–20
Regional Connector Transit Corridor (Los Angeles, California), 750
　design challenges and tunneling methods considered, 750–757
Reinforced precast concrete pipe, with liner, 237–238
Remediation and rehabilitation
　Carpi PVC membrane in remediation of tunnel leakage and slope stability (Belden Tunnel 2, California), 249–256
　with cross-section widening, of German rail tunnels, 355–364
　Grout Intensity Number (GIN) methodology in leakage remediation (Helms Pumped Storage Project, California), 215–223
　inspections of Colorado alpine highway tunnels, 224–232
　options for Heroes Highway Tunnel (Connecticut), 257–266

*Technical Manual for Design and Construction of Road Tunnels—Civil Element* (FHWA) on, 395–396
　*See also* Corrosion resistance
Reverse Curve Tunnel (Colorado), 224–225
　inspection of, 230–232
Rhode Island
　Main Spine Tunnel, Diversion Structure, and Consolidation Conduit (Narragansett Bay Commission Combined Sewer Overflow), 843–844, 851–852
　microtunneling with jet grouting in difficult ground and congested urban setting (Narragansett Bay Commission Combined Sewer Overflow, Providence), 796–803
Risk assessment and management
　approach for insurance decisions in CSO programs (Portland, OR), 676–682
　and Cost Estimate Validation Process (CEVP), 718–724
　joint approach between owner and contractor for Lake Mead Intake No. 3 (Nevada), 559–565
　on megaprojects, 566–569
　quantitative risk assessment of fire danger in choice of design approach for Yas Island tunnel (United Arab Emirates), 375–386
　and public private partnerships, 588–592
　risk sharing, 570–575
　in selection of open-cut, microtunneling, or both, 740–749
　in tunnel construction, 530–531
Riverbank Filtration Tunnel (Louisville, Kentucky), 876
　consolidation grouting for fault zone, 876–879
Road header tunneling, in Sandy River conduit crossing (Portland, OR), 897–905
Road tunnels
　binocular vs. stacked alignment and multiple design factors for Silver Line Phase III Bus Rapid Transit Tunnel Project (Boston, MA), 461–474
　early Portland (OR) highway tunnels, 45
　inspections of Colorado alpine highway tunnels, 224–232
　need for national tunnel inspection standards (US), 957–959
　*Technical Manual for Design and Construction of Road Tunnels—Civil Element* (FHWA), 387–396
　*See also* Devil's Slide Tunnel Project (California); Heroes Highway Tunnel (Connecticut); Irvine-Corona Expressway tunnels; M6 Motorway tunnels (Hungary); State Route 75/282 Transportation Corridor Project (Coronado, CA); Urban Ring Project (Boston, Massachusetts); Yas

Island Southern Crossing Tunnel (United Arab Emirates); *and under* Colorado
The Robbins Company, and cutter instrumentation system for TBMs, 110–115
Rock caverns
    continuum and discontinuum modeling methods in design of Second Avenue Subway Project caverns (New York City), 423–432
    for cryogenic fluid storage and particle physics research, 409, 411–412
    design issues for storage of cryogenic fluids (DUSEL), 409–414
    fiberglass bolts and shotcrete for support and lining of cavern (Lake Mead Intake No. 3 Tunnel, Nevada), 433–440
    load analysis, 447–455
    structural analysis methodology for large-span caverns beneath dense urban environments, 441–458
Rock Creek Cemetery (Washington, DC), and NATM construction in soft ground, 828–829
Rock mass rating (RMR), 577–587
Rock quality designation (RQD), 577–587
Rock tunneling
    Deep Rock Tunnel Connector (Indianapolis, IN), 629–634
    deep storage CSO tunnel in sedimentary rock (West Branch Muddy Creek Project Bundle, Cincinnati, OH), 733–739
    *Technical Manual for Design and Construction of Road Tunnels—Civil Element* (FHWA) on, 390–391
Rome, Italy
    Analysis of Controlled Deformations (ADECO), as alternative to NATM for full-face, 22-m-wide tunnel excavation through clay (Cassia twin tunnels), 96–109
    Cassia twin tunnels, 96
Rupert Transfer Tunnel (James Bay, Québec), 923
    fast-track drill and blast in construction of, 923–925

**S**

Sacramento, California
    EPB TBM with continuous conveyor system (Lower Northwest Interceptor Sewer), 136
    EPB TBM with continuous conveyor system (Upper Northwest Interceptor Sewer), 138–139
    plastic (PVC) lined concrete segments for Upper Northwest Interceptor Sewer, 191–199
    precast PVC-lined concrete segments (Upper Northwest Interceptor Sewer), 242–248
San Diego, California
    San Vicente Pipeline Tunnel, 55
    sophisticated ventilation system for long-drive, small-diameter tunneling project (San Vicente Pipeline), 55–61
San Francisco, California
    design and planning, including tunneling methods, for DTX Project, 477–482
    Transbay-Downtown Rail Extension (DTX) Project, 475–477
San Francisco (California) Public Utilities Commission
    ground characterization and design challenges (New Irvington Tunnel), 325–336
    multiple tunneling methods required in urban environment (Sunnydale Auxiliary Sewer Project), 337–345
San Vicente Pipeline Tunnel (San Diego, CA), 55
    sophisticated ventilation system for long-drive, small-diameter tunneling project, 55–61
Santa Cruz (California) Landfill Water Bypass Tunnel, 531–532
Santiago, Chile, and NATM construction in soft ground, 705–706
Scheduling
    construction, 530
    and contingency planning, 710–717
    and Cost Estimate Validation Process (CEVP), 718–724
    and cost estimates, 728
    Critical Path Method (CPM), 530
    Critical Path Method compared with linear approach, 683–684
    linear, 683–694
Seattle, Washington
    Alaskan Way Viaduct and Seawall Replacement, and size considerations, 534
    Beacon Hill Station and Tunnels, and size considerations, 533–534
    Brightwater Conveyance West Contract, 855
    design and planning of University Link–Pine Street Stub Tunnel connection, 517–524
    EPB TBM in construction of 6.4-km, small-diameter tunnel in soft ground with small exit window (Brightwater Conveyance West Contract), 855–862
    University Link tunnel connection to Pine Street Stub Tunnel (Link Light Rail Project), 517
    *See also* King County, Washington
Seattle (Washington) Public Utilities
    microtunneling in construction of underground pump plant in pristine watershed (Washington), 206–212
    Morse Lake Pump Plant and Intake, 206

Second Avenue Subway Project (New York City), 423–424
    continuum and discontinuum modeling methods in design of Second Avenue Subway Project rock caverns (New York City), 423–432

Seismic design
    criteria for State Route 75/282 Transportation Corridor Project (California), 365–372
    New Irvington Tunnel (California), 332–334
    *Technical Manual for Design and Construction of Road Tunnels—Civil Element* (FHWA) on, 394

Sequential Excavation Method (SEM)
    concrete final lining for SEM tunnel (California), 35–42
    considered for Regional Connector Transit Corridor (Los Angeles, CA), 753–754
    with spile presupport through difficult ground, connecting two underground garages, 836–842
    *Technical Manual for Design and Construction of Road Tunnels—Civil Element* (FHWA) on, 391–393, 415–422
    *See also* New Austrian Tunneling Method

Sewer, CSO, and wastewater tunnels. *See* Anacostia River CSO Control Project (District of Columbia); Big Walnut Sanitary Trunk Sewer Extension (Columbus, Ohio); Blacklick Creek Sanitary Interceptor Sewer Tunnel (Columbus, Ohio); Brightwater Conveyance East Contract; Brightwater Conveyance West Contract; Deep Rock Tunnel Connector (Indianapolis, Indiana); District of Columbia Water and Sewer Authority; East Side CSO Tunnel (Portland, Oregon); Emisor Oriente Wastewater Tunnel Project (Mexico City); Homestead Avenue Interceptor Extension (Hartford, Connecticut); Lower Northwest Interceptor Sewer (Sacramento, California); Marysville Trunk Interceptor (Marysville, Ohio); McCook Reservoir Main Tunnel System (Chicago, Illinois); Narragansett Bay Commission Combined Sewer Overflow; New Overpeck Valley Parallel Relief Sewer Project (Bergen County, New Jersey); North Gratiot Interceptor Drain (Macomb County, Michigan); South IH-35 Water and Wastewater Infrastructure Improvements Program; Southeast Collector trunk sewer (York, Ontario); Sunnydale Auxiliary Sewer Project (San Francisco, California); Thames Water Utilities (London, UK); Tollgate Interceptor (Aurora, Colorado); Upper Northwest Interceptor Sewer (Sacramento, CA); West Branch Muddy Creek Project Bundle (Cincinnati, Ohio); West Side CSO Project (Portland, Oregon); Willamette River CSO Program

SHAFT model, 483–501

Shanghai Changxing Under River Tunnel (China), and large-diameter TBMs, 90–91

Shield tunneling
    hybrid (slurry shield & open mode) TBM and geotechnical aspects of operation in poor ground (Lake Mead Intake No. 3 Tunnel, Nevada), 125–135
    hydroshield TBM with segmental lining in construction of European X-ray free electron laser (XFEL) project (Germany), 916–922
    in Railway Bosphorus Tube Crossing (Turkey), 10, 19–20

Silver Line Phase III Bus Rapid Transit Tunnel Project (Boston, MA), 461–463
    binocular vs. stacked alignment and multiple design factors, 461–474

Slip lining, 239–240

Slurry TBMs
    considered for Irvine-Corona Expressway tunnels (California), 147–157
    in construction of small-diameter gas pipeline tunnel (Ems-Dollard estuary crossing, Germany-Netherlands), 164–169
    in construction of small-diameter tunnel through soft ground in congested urban environment (Homestead Avenue Interceptor Extension, Hartford, CT), 813–821

Soft ground
    EPB TBM in construction of 6.4-km, small-diameter tunnel in soft ground with small exit window (Brightwater Conveyance West Contract, Seattle, WA), 855–862
    EPB TBMs for construction in soft ground (Emisor Oriente Wastewater Tunnel Project, Mexico City), 159–163
    and NATM in construction of mined underground stations, 695–709
    NATM in tunnel and shaft construction (Washington, DC), 825–835
    and slurry TBM in construction of small-diameter tunnel in congested urban environment (Homestead Avenue Interceptor Extension, Hartford, CT), 813–821
    *Technical Manual for Design and Construction of Road Tunnels—Civil Element* (FHWA) on, 391

South Carolina
    US 17 Septima Clark Parkway project flood abatement program (Charleston), 613–620
    US 17 Septima Clark Parkway Transportation Infrastructure Reinvestment Project (Charleston), 614–165
South IH-35 Water and Wastewater Infrastructure Improvements Program (Austin, Texas), 664–667
    multi-attribute utility theory in decision-making process involving stakeholders, 667–675
Southeast Collector trunk sewer (York, Ontario), 601
    owner procurement of materials and equipment, 601–606
Stability and strength assessment
    continuum and discontinuum modeling methods in design of Second Avenue Subway Project rock caverns (New York City), 423–432
    and Deep Underground Science and Engineering Laboratory (South Dakota), 409–414
    for shaft, cavern, and starter tunnel (Lake Mead Intake No. 3 Tunnel, Nevada), 433–440
    structural analysis methodology for large-span rock caverns beneath dense urban environments, 441–458
State Route 75/282 Transportation Corridor Project (Coronado, CA), 365–366
    seismic design criteria, 366–372
Strategic Tunnel Enhancement Programme (STEP; Abu Dhabi, United Arab Emirates), 304–307
    instrumentation in preliminary geotechnical investigation, 307–315
Subway tunnels
    large-bore TBMs for single-tube subway tunnels, 623–624
    and single-tube, large-bore tunnels with stacked tracks, 621–628
    *See also* No. 7 Subway Line Extension (New York City); Second Avenue Subway Project (New York City)
Sunnydale Auxiliary Sewer Project (San Francisco, CA), 337
    multiple tunneling methods required in urban environment, 337–345
Sustainability
    defined, 183, 200
    environmental preservation and community impact mitigation criteria (Jollyville Tunnel, Austin, TX), 609–612
    guideline systems, 201
    and infrastructure, 200–201
    and linings, 173–182
    microtunneling in construction of underground pump plant in pristine watershed (Washington), 206–212
    and plastic (PVC) lined concrete segments, 191–199
    rating systems, 201
    and tunneling, 183–190
    and underground construction, 201–205
    and US 17 Septima Clark Parkway Transportation Infrastructure Reinvestment Project flood abatement program (Charleston, SC), 613–620
Switzerland
    and large-diameter TBMs (Gotthard Base Tunnel), 94–95
    responses to micro tremors and rock bursts during construction of Multifunction Station Faido (Gotthard Base Tunnel), 880–888

## T

TBMs. *See* Tunnel boring machines
*Technical Manual for Design and Construction of Road Tunnels—Civil Element* (FHWA), 387–396, 415–422
Terrorism. *See* Blast and post-blast behavior of tunnels
Texas
    environmental preservation and community impact mitigation criteria (Jollyville Tunnel, Austin), 609–612
    ground conditions and design (Waller Creek Tunnel, Austin), 576–587
    multi-attribute utility theory in decision-making process involving stakeholders (South IH-35 Water and Wastewater Infrastructure Improvements Program), 664–675
Thames Water Utilities (London, UK)
    CSO tunnels, 483, 495
    SHAFT model in analysis of surge and pneumatic forces, 483–485, 495–501
3D laser scanner-LIDAR, 119
Tollgate Interceptor (Aurora, Colorado), 740–741
    analyzing where to use open-cut or microtunneling (Tollgate Interceptor, Aurora, Colorado), 741–749
Transbay-Downtown Rail Extension (DTX) Project (San Francisco, CA), 475–477
    design and planning, including tunneling methods, 477–482
Transportation tunnels. *See* Railroad tunnels; Road tunnels; Subway tunnels
Traylor-Shea Joint Venture, 191

Tunnel boring machines (TBMs)
    considered for Regional Connector Transit Corridor (Los Angeles, CA), 753
    cutter instrumentation system for, 110–115
    hybrid (slurry shield & open mode) TBM and geotechnical aspects of operation in poor ground, 125–135
    hydroshield TBM with segmental lining in construction of European X-ray free electron laser (XFEL) project (Germany), 916–922
    large-bore TBMs for single-tube subway tunnels, 623–624
    large-diameter machine development for traffic tunnels, 89–95
    onsite first time assembly (OFTA), 65–66
    projects utilizing OFTA, 66–74
    replacing drill and blast for twin water tunnels (Croton Water Treatment Plant, New York City), 926–937
    *See also* Earth pressure balance TBMs; Microtunneling; Shield tunneling; Slurry TBMs

Tunneling. *See* Conventional tunneling; Cut-and-cover tunnels; Drill and blast; Microtunneling; New Austrian Tunneling Method; Rock tunneling; Sequential Excavation Method; Shield tunneling; Tunnel boring machines

Tunnels
    blast and post-blast behavior of, 655–663
    construction scheduling, 530
    size considerations, 527–528

Turkey
    Marmaray project, 10
    multiple tunneling methods for Railway Bosphorus Tube Crossing, 10–25

Tysons Corner (Washington, DC), and NATM construction in soft ground, 831–833

## U

UltraShell linings, 180–182
Underground positioning systems, 117
    ferrous objects, 117
    ground penetrating radar (GPR), 118, 119
    gyroscope probes, 119
    gyroscope stations, 119–120
    non-ferrous objects, 117–118
    3D laser scanner-LIDAR, 119
Underwater positioning systems, 117
United Arab Emirates
    instrumentation in preliminary geotechnical investigation for STEP (Abu Dhabi), 307–315
    quantitative risk assessment of fire danger in choice of design approach for Yas Island tunnel (Abu Dhabi), 375–386
    Strategic Tunnel Enhancement Programme (STEP; Abu Dhabi), 304–307

United Kingdom
    SHAFT model in analysis of surge and pneumatic forces (Thames Water Utilities CSO tunnels), 483–485, 495–501
    Thames Water Utilities CSO tunnels (London), 483, 495

United States
    Best Practices for Roadway Tunnel Design, Construction, Operation, and Maintenance domestic scan, 960–963
    Green Building Council, 183
    need for national tunnel inspection standards, 957–959

Upper Northwest Interceptor Sewer (Sacramento, California), 138, 191
    EPB TBM with continuous conveyor system, 138–139
    plastic (PVC) lined concrete segments, 191–199
    precast PVC-lined concrete segments, 242–248

Urban Ring Project (Boston, Massachusetts), 641–645, 650–651
    alternative tunnel construction techniques, 646–648
    single bored tunnel vs. twin bored tunnels, 649–650
    tunnel alignment alternatives, 645–646

US 17 Septima Clark Parkway Transportation Infrastructure Reinvestment Project (Charleston, South Carolina), 614–615
    flood abatement program, 613–620

US Federal Highway Administration (FHWA), *Technical Manual for Design and Construction of Road Tunnels—Civil Element*, 387–396

## V

Ventilation, sophisticated system for long-drive, small-diameter tunneling project (San Diego), 55–61

## W

Waller Creek Tunnel (Austin, Texas), 576
    ground conditions and design, 576–587

Washington, DC
    Anacostia River CSO Control Project (District of Columbia), 397–399
    District of Columbia Water and Sewer Authority CSO Program project delivery approach (combined design-bid and design-bid-build), 593–600
    District of Columbia Water and Sewer Authority CSO tunnels, 483, 486

geotechnical investigations (Anacostia River CSO Control Project), 399–406

NATM in soft-ground tunnel and shaft construction, 706, 825–835

SHAFT model in analysis of surge and pneumatic forces (CSO tunnels), 483–495, 499–501

Washington State

Alaskan Way Viaduct and Seawall Replacement, and size considerations (Seattle), 534

Department of Transportation and CEVP development, 719

EPB TBM in construction of 6.4-km, small-diameter tunnel in soft ground with small exit window (Brightwater Conveyance West Contract, Seattle), 855–862

microtunneling in construction of underground pump plant in pristine watershed (Seattle), 206–212

monitoring of settlement and comparison with modeling results (Brightwater Conveyance East Contract, King County), 298–303

NATM construction in soft ground (Beacon Hill Station and Tunnels, Seattle), 704–705

SEM with spile presupport through difficult ground, connecting two underground garages (Lincoln Square Tunnel, Bellevue), 836–842

size considerations (Beacon Hill Station and Tunnels, Seattle), 533–534

University Link tunnel connection to Pine Street Stub Tunnel (Link Light Rail Project, Seattle), 517

Wastewater tunnels. *See* Sewer, CSO, and wastewater tunnels

Water tunnels. *See* Croton Water Treatment Plant (New York City); Jollyville Tunnel (Austin, Texas); Lake Mead Intake No. 3 Tunnel (Nevada); Morse Lake Pump Plant and Intake (North Bend, Washington); New Irvington Tunnel (California); Portland (Oregon) Water Bureau; Rupert Transfer Tunnel (James Bay, Québec); San Vicente Pipeline Tunnel (San Diego, California); US 17 Septima Clark Parkway Transportation Infrastructure Reinvestment Project (Charleston, South Carolina); Waller Creek Tunnel (Austin, Texas)

West Branch Muddy Creek Project Bundle (Cincinnati, Ohio), 733

deep storage CSO tunnel in sedimentary rock, 733–739

West Side CSO Project (Portland, Oregon), 51–52

Willamette River CSO Program (Portland, Oregon), 676–677

risk management approach for insurance decisions in CSO programs, 677–682

## Y

Yas Island Southern Crossing Tunnel (United Arab Emirates), 375–376

quantitative risk assessment of fire danger in choice of design approach, 375–386

York, Ontario

owner procurement of materials and equipment (Southeast Collector trunk sewer), 601–606

Southeast Collector trunk sewer, 601